DIGITAL
AND
STATE VARIABLE METHODS

CONVENTIONAL AND NEURAL-FUZZY
CONTROL SYSTEMS

Second Edition

About the Author

M. Gopal is presently Professor, Department of Electrical Engineering, Indian Institute of Technology (IIT), Delhi. His teaching and research stints span over three decades at institutes such as IIT Delhi; IIT Bombay; BITS Pilani; REC Jaipur; City University London; and Universiti Teknologi Malaysia. He has won international recognition as a distinguished individual in the field of education.

Dr. Gopal is the author/co-author of six books, a video course in Control Engineering, and a large number of research publications.

His current research interests are in the areas of Robust Control, Neural and Fuzzy Technologies, Process Control and Robotics.

DIGITAL CONTROL AND STATE VARIABLE METHODS

CONVENTIONAL AND NEURAL-FUZZY CONTROL SYSTEMS

Second Edition

M GOPAL
Professor
Department of Electrical Engineering
Indian Institute of Technology
Delhi

Boston   Burr Ridge, IL   Dubuque, IA   Madison, WI   New York   San Francisco   St. Louis
Bangkok   Bogotá   Caracas   Kuala Lumpur   Lisbon   London   Madrid   Mexico City
Milan   Montreal   New Delhi   Santiago   Seoul   Singapore   Sydney   Taipei   Toronto

Digital Control and State Variable Methods
Conventional and Neural-Fuzzy Control Systems, 2nd edition
International Edition 2004

Exclusive rights by McGraw-Hill Education (Asia), for manufacture and export. This book cannot be re-exported from the country to which it is sold by McGraw-Hill. The International Edition is not available in North America.

Copyright © 2003, 1997, Tata McGraw-Hill Publishing Company Limited. All rights reserved. No part of this publication may be reproduced or distributed in any form or by any means, or stored in a database or retrieval system, without the prior written consent of The McGraw-Hill Companies, Inc., including, but not limited to, in any network or other electronic storage or transmission, or broadcast for distance learning.
Some ancillaries, including electronic and print components, may not be available to customers outside the United States.

Information contained in this work has been obtained by Tata McGraw-Hill, from sources believed to be reliable. However, neither Tata McGraw-Hill nor its authors guarantees the accuracy or completeness of any information published herein, and neither Tata McGraw-Hill nor its authors shall be responsible for any errors, omissions, or damages arising out of use of this information. This work is published with the understanding that Tata McGraw-Hill and its authors are supplying information but are not attempting to render engineering or other professional services. If such services are required, the assistance of an appropriate professional should be sought.

---

10 09 08 07 06 05 04 03 02
20 09 08 07
FC   BJE

ISBN 0-07-048302-7

When ordering this title, use ISBN 007-123725-9

Printed in Singapore

*Dedicated
with all my love to my
son Ashwani,
and
daughter Anshu*

Preface to Second Edition

To develop this edition of *Digital Control and State Variable Methods*, the goal was to improve the textbook without disturbing basic foundation. Highlights of major additions to our treatment of *model-based control* are as follows:
- Introduction to CIPS (Computer Integrated Process Systems) and CIMS (Computer Integrated Manufacturing Systems), with detailed account of programmable logic controllers, and tunable PID controllers.
- Introduction to System Identification and Adaptive Control.
- A chapter on Nonlinear Control Systems.
- An appendix on MATLAB Aided Control System Design.

In the process of understanding and emulating salient features of biological control functions, three fields namely fuzzy logic, artificial neural networks, and genetic algorithms have grown into three distinct disciplines with the aim of designing "intelligent" systems for scientific and engineering applications. The theory of fuzzy logic provides a mathematical morphology to emulate certain perceptual and linguistic attributes associated with human cognition. It aims at modelling the inexact modes of reasoning and thought processes that play an essential role in the remarkable human ability to make rational decisions in an environment of uncertainty and imprecision. Artificial neural networks attempt to emulate the architecture and information representation schemes of human brain. Genetic algorithms provide an adaptive, robust, parallel and randomized searching technique where a population of solutions evolves over a sequence of generations to a globally optimal solution.

While the development of individual tools—fuzzy logic, artificial neural networks, genetic algorithms; was in progress, a group of researchers felt the need of integrating them in order to enjoy the merits of different biologically inspired techniques into one system. The result is the development of several hybrid paradigms, like neuro-fuzzy, neuro-genetic, fuzzy-genetic, and neuro-fuzzy-genetic. These hybrid paradigms are suitable for solving complex real-world problems for which only one tool may not be adequate.

We are passing through a phase of rosy outlook for control technology: there has been a tendency to portray neural network and/or fuzzy logic based controller as magical; a sort of black box that does amazing things. Enthusiastic researchers have dreamed big dreams and made sweeping predictions on the potential of *intelligent control systems* with neural processor/fuzzy logic chips in control loop. The growth of the new field during the past few years has been so explosive that the traditional *model-based* control theory and the traditional op amp/computer-based implementations of controllers are under threat.

New design methods are required to prove themselves in actual practice before they can displace well-accepted techniques. Intelligent control technology is slowly gaining wider acceptance among academics and industry. Even if there has sometimes been some over enthusiastic description of the field, the scientific community and industry are converging to the fact that there is something fundamentally significant about this technology. Also preparations are underway to accord a warm welcome to integrated control technology–integration of intelligent control theory with traditional model-based theory, and integration of the VLSI microprocessors with neuro-fuzzy processors to implement the controller.

This edition includes a tutorial introduction to the knowledge-based tools for control system design. Rigorous characterization of theoretical properties of intelligent control methodology will not be our aim in our tutorial presentation; rather we focus on the development of systematic engineering procedures which will guide the design of controller for a specific problem.

The arrangement of topics in the text is a little different from the conventional one. Typically, a book on digital control systems starts with transform-domain design and then carries over to state space. These books give a detailed account of state variable methods for discrete-time systems. Since the state variable methods for discrete-time systems run quite parallel to those for continuous-time systems, full-blow repetition is not appreciated by the readers conversant with state variable methods for continuous-time systems. And for readers with no background of this type, a natural way of introducing state variable methods is to give the treatment for continuous-time systems, followed by a brief parallel presentation for discrete-time systems.

For the purpose of organizing different courses for students with different background, the sequencing of chapters, and dependence of each chapter on previous chapters has been properly designed in this text. For a course on digital control systems that would follow a course in continuous-time linear systems which includes state variable methods, the following sequence of various chapters is suggested.

 Chapters 1 to 4
 Chapter 6 (skip Chapter 5)
 Chapter 7 (Sections 7.9 and 7.10)
 Chapter 8 (Section 8.8)
 Chapter 9 (Section 9.6)

For a course on state-space analysis and design, with a pre-requisite of only a basic course on feedback control systems, the following sequence is suggested.

    Chapters 5 to 9 (skip Chapters 1 to 4)

The typical electrical engineering curriculum at the first degree level includes a basic course on feedback control systems with just one more course on the subject. Different modules under the title "advanced control systems" are being offered as a second course. The breadth and depth of the coverage and sequencing of the chapters in this book will enable the instructors to design their "second course". Two representative modules are given below.

- Chapters 1 to 3
  Chapters 5 to 7 (skip Chapter 4)
  Chapter 10 (skip Chapters 8 and 9)
- Chapters 1 to 3
  Chapters 5 and 6 (skip Chapter 4)
  Chapters 11 and 12 (skip Chapters 7–10)

Over the last three years, it has come to my attention (mostly through direct e-mail contact with users) that certain technical points needed clarification. I have attempted to address each concern and have corrected every known error. I appreciate the efforts of students, faculty, and other users who have kept in touch with me. An instructor using this text may obtain the *Solution Manual* from the accompanying website. The URL of the website is www.tatamcgraw-hill.com/digital_solutions/gopal/controls.

M GOPAL
mgopal@ee.iitd.ac.in

Preface to First Edition

The dramatic development of computer technology has radically changed the boundaries of practical control system design options. It is now possible to employ very complicated high-order digital controllers, and to carry out the extensive calculations required for their design. These advances in implementation and design capability can be achieved at low cost due to the widespread availability of inexpensive, powerful digital computers and related devices. There is every indication that a high rate of growth in the capability and application of digital computers will continue far into the future.

Fortunately, control theory has also developed substantially over the past 35 years. The *classical design* methods have been greatly enhanced by the availability of low cost computers for system analysis and simulation. The graphical tools of classical design like root locus plots, Nyquist plots, Bode plots, and Nichols chart can now be more easily used with computer graphics. Coupled with hardware developments such as microprocessors and electro-optic measurement schemes, the classical control theory today provides useful design tools to practising control engineers.

The *modern control* theory (which can't be termed modern any longer) refers to the state-space based methods. Modern control methods initially enjoyed a great deal of success in academic circles, but did not perform very well in many areas of application. Modern control provided a lot of insight into system structure and properties, nevertheless it masked other important feedback properties that could be studied and manipulated using classical control. During the past two decades, a series of methods, which are a combination of modern state-space methods and classical frequency-domain methods, have emerged. These techniques are commonly known as *robust control*.

The rapid development in digital computers and microelectronics has brought about drastic changes in the approach to analysis, design, and implementation of control systems. The flourishing of digital control has just begun for most industries and there is much to be gained by exploiting the full potential of new technology.

The purpose of this book is to present control theory that is relevant to the analysis and design of computer-control systems. The prerequisite on the part of the reader is that he/she has had an introductory course in control engineering concentrating on the basic principles of feedback control and covering various classical analog methods of control system design. The material presented in this book is closely related to the material already familiar, but towards the end a direction to wider horizons is indicated.

There are nine chapters in the book. Chapter 1 gives an overview of control system terminology. A brief account of major historical landmarks in the development of the fascinating area of control systems engineering is also given. The chapter concludes with an outline of the scope and organization of the book.

Chapters 2–4 are devoted to the classical sampled-data theory. The digital methods of design are developed, paralleling and extending considerably the similar topics in analog control.

Chapters 5–9 deal with the state variable analysis and design methods. It is assumed that the reader is not exposed to these techniques of the so called modern control theory. Our approach is to first discuss the state variable techniques for continuous-time systems and then proceed to a compact presentation of the techniques for discrete-time systems, using the analogy with the continuous-time case.

The familiarity with Fourier transforms, z-transforms, and elementary matrix algebra is helpful but not necessary; these mathematical techniques are reviewed at appropriate places in adequate depth in the book.

A modern development that is significant in practical design is the wide availability and low cost of Computer Aided Design (CAD) software. Although a practical designer is handicapped without access to such tools, and students definitely benefit from their hands-on use, the text has been designed in such a way that personal access to CAD software facility is not essential. It is advisable to first create a firm knowledge of the fundamental results of feedback control theory without any diversion to other domains of learning; this knowledge can then be supplemented by the CAD facility to solve practical design problems wherein it is desirable to vary several parameters during the design stage to investigate their effect on the system performance. Some problems for computer solution may be assigned to familiarise the students with CAD tools. This book offers computer aided learning environment with any commercially available CAD software.

The material in this book has been class tested. The sequencing and internal organization of various chapters is such that it permits flexibility of adaptation to variations in students' prior training and curricula. Greater stress is placed on the interdisciplinary nature of the subject, making the book useful for all engineering disciplines that have control courses as part of their curricula. Practical problems with careful and complete explanations makes the book appealing for self-study to practising engineers. The book should also serve as a guide for applying design techniques to industrial control problems.

Generous participation of instructors, students, and professionals to eliminate errors in the text and to refine the presentation, will be gratefully acknowledged.

M GOPAL

Contents

Preface to Second Edition vii
Preface to First Edition xi

1. Introduction 1

1.1 Control System Terminology *1*
1.2 Computer-Based Control: History and Trends *8*
1.3 Control Theory: History and Trends *13*
1.4 An Overview of the Classical Approach to Analog Controller Design *16*
1.5 Scope and Organization of the Book *22*

2. Signal Processing in Digital Control 26

2.1 Why Use Digital Control *26*
2.2 Configuration of the Basic Digital Control Scheme *29*
2.3 Principles of Signal Conversion *30*
2.4 Basic Discrete-Time Signals *38*
2.5 Time-Domain Models for Discrete-Time Systems *41*
2.6 Transfer Function Models *51*
2.7 Stability on the z-Plane and the Jury Stability Criterion *59*
2.8 Sampling as Impulse Modulation *71*
2.9 Sampled Spectra and Aliasing *77*
2.10 Filtering *83*
2.11 Practical Aspects of the Choice of Sampling Rate *86*
2.12 Principles of Discretization *89*
2.13 The Routh Stability Criterion on the r-Plane *111*
2.14 Review Examples *112*
Problems *120*

3. Models of Digital Control Devices and Systems 127

3.1 Introduction *127*
3.2 z-Domain Description of Sampled Continuous-Time Plants *129*
3.3 z-Domain Description of Systems with Dead-Time *139*
3.4 Implementation of Digital Controllers *143*

3.5 Tunable PID Controllers *152*
3.6 Digital Temperature Control System *171*
3.7 Digital Position Control System *176*
3.8 Stepping Motors and Their Control *184*
3.9 Programmable Logic Controllers *192*
3.10 Review Examples *214*
Problems 219

4. Design of Digital Control Algorithms   231

4.1 Introduction *231*
4.2 z-Plane Specifications of Control System Design *233*
4.3 Digital Compensator Design using Frequency Response Plots *251*
4.4 Digital Compensator Design using Root Locus Plots *267*
4.5 z-Plane Synthesis *284*
4.6 Review Examples *296*
Problems 304

5. Control System Analysis Using State Variable Methods   314

5.1 Introduction *314*
5.2 Vectors and Matrices *315*
5.3 State Variable Representation *328*
5.4 Conversion of State Variable Models to Transfer Functions *341*
5.5 Conversion of Transfer Functions to Canonical State Variable Models *347*
5.6 Eigenvalues and Eigenvectors *360*
5.7 Solution of State Equations *372*
5.8 Concepts of Controllability and Observability *384*
5.9 Equivalence between Transfer Function and State Variable Representations *398*
5.10 Multivariable Systems *403*
5.11 Review Examples *409*
Problems 417

6. State Variable Analysis of Digital Control Systems   429

6.1 Introduction *429*
6.2 State Descriptions of Digital Processors *430*
6.3 State Description of Sampled Continuous-Time Plants *438*
6.4 State Description of Systems with Dead-Time *445*
6.5 Solution of State Difference Equations *449*
6.6 Controllability and Observability *454*
6.7 Multivariable Systems *461*
6.8 Review Examples *464*
Problems 472

7. Pole-Placement Design and State Observers   480

7.1 Introduction *480*

7.2 Stability Improvement by State Feedback *482*
7.3 Necessary and Sufficient Conditions for Arbitrary Pole-Placement *486*
7.4 State Regulator Design *490*
7.5 Design of State Observers *494*
7.6 Compensator Design by the Separation Principle *505*
7.7 Servo Design: Introduction of the Reference Input by Feedforward Control *512*
7.8 State Feedback with Integral Control *515*
7.9 Digital Control Systems with State Feedback *518*
7.10 Deadbeat Control by State Feedback and Deadbeat Observers *530*
7.11 Introduction to System Identification and Adaptive Control *533*
7.12 Review Examples *547*
Problems 554

8. Lyapunov Stability Analysis 567

8.1 Introduction *567*
8.2 Basic Concepts *568*
8.3 Stability Definitions *571*
8.4 Stability Theorems *573*
8.5 Lyapunov Functions for Nonlinear Systems *581*
8.6 Lyapunov Functions for Linear Systems *586*
8.7 A Model Reference Adaptive System *589*
8.8 Discrete-Time Systems *593*
8.9 Review Examples *595*
Problems 599

9. Linear Quadratic Optimal Control 601

9.1 Parameter Optimization and Optimal Control Problems *601*
9.2 Quadratic Performance Index *605*
9.3 Control Configurations *612*
9.4 State Regulator Design through the Lyapunov Equation *616*
9.5 Optimal State Regulator through the Matrix Riccati Equation *625*
9.6 Optimal Digital Control Systems *633*
9.7 Critique of Linear Quadratic Control *640*
9.8 Review Examples *642*
Problems 651

10. Nonlinear Control Systems 658

10.1 Introduction *658*
10.2 A Class of Nonlinear Systems: Separable Nonlinearities *662*
10.3 Filtered Nonlinear System: The Describing Function Analysis *666*
10.4 Describing Functions of Common Nonlinearities *668*
10.5 Stability Analysis by the Describing Function Method *675*
10.6 Nonlinear Sampled-Data Systems *685*

xvi   Contents

    10.7 Second-Order Nonlinear System on the Phase Plane *685*
    10.8 Fundamental Types of Phase Portraits *688*
    10.9 System Analysis on the Phase Plane *695*
    10.10 Optimal Switching in Bang-Bang Control Systems *700*
    10.11 Review Examples *703*
    Problems 711

11. Neural Networks for Control   **718**

    11.1 Introduction *718*
    11.2 Neuron Models *724*
    11.3 Network Architectures *735*
    11.4 Learning in Neural Networks *745*
    11.5 Training the Multilayer Neural Network–Backpropagation Tuning *749*
    11.6 Function Approximation with Neural Networks *757*
    11.7 System Identification with Neural Networks *758*
    11.8 Control with Neural Networks *765*
    11.9 Review Examples *772*
    Problems 779

12. Fuzzy Control   **782**

    12.1 Introduction *782*
    12.2 Fuzzy Quantification of Knowledge *790*
    12.3 Fuzzy Inference *807*
    12.4 Designing a Fuzzy Logic Controller *812*
    12.5 Genetic Algorithms *825*
    12.6 Review Examples *839*
    Problems 844

Appendix A: Mathematical Background   **852**

    A.1 Introduction *852*
    A.2 Fourier Series and Fourier Transforms *852*
    A.3 Laplace Transforms *856*
    A.4 z-Transforms *863*
    A.5 Table of Transforms *872*

Appendix B: MATLAB Aided Control System Design: Conventional   **875**
    Problems 893

Appendix C: MATLAB Aided Control System Design: Neural–Fuzzy   **911**
    Problems 921

References   *940*
Companion Book   *949*
Answers to Problems   *951*
Index   *984*

chapter 1
Introduction

1.1 CONTROL SYSTEM TERMINOLOGY

A *Control System* is an interconnection of components to provide a desired function. The portion of the system to be controlled is given various names: *process*, *plant*, and *controlled system* being perhaps the most common. The portion of the system that does the controlling is the *controller*. Often, a control system designer has little or no design freedom with the plant; it is usually fixed. The designer's task is, therefore, to develop a controller that will control the given plant acceptably. When measurements of the plant response are available to the controller (which in turn generates signals affecting the plant), the configuration is a *feedback control system*.

A *digital control system* uses digital hardware, usually in the form of a programmed digital computer, as the heart of the controller. In contrast, the controller in an *analog control system* is composed of analog hardware; an electronic controller made of resistors, capacitors, and operational amplifiers is a typical example. Digital controllers normally have analog devices at their periphery to interface with the plant; it is the internal working of the controller that distinguishes digital from analog control.

The signals used in the description of control systems are classified as continuous-time and discrete-time. *Continuous-time signals* are defined for all time, whereas discrete-time signals are defined only at discrete instants of time, usually evenly spaced steps. The signals for which both time and amplitude are discrete are called *digital signals*. Because of the complexity of dealing with quantized (discrete-amplitude) signals, digital control system design proceeds as if computer-generated signals were not of discrete amplitude. If necessary, further analysis is then done to determine if a proposed level of quantization is acceptable.

Systems and system components are termed continuous-time or discrete-time according to the type of signals they involve. They are classified as being *linear* if signal components in them can be superimposed—any linear combination of signal components applied to a system produces the same linear combination of corresponding output components; otherwise a system is *nonlinear*. A system or component is *time-invariant* if its properties do not change with time—any time shift of the inputs produces an equal time shift of every corresponding signal. If a system is not time-invariant, then it is *time-varying*.

A typical topology of a computer-controlled system is sketched schematically in Fig. 1.1. In most cases, the measuring transducer (sensor) and the actuator (final control element) are analog devices, requiring, respectively, *analog-to-digital* (A/D) and *digital-to-analog* (D/A) conversion at the computer input and output. There are, of course, exceptions; sensors which combine the functions of the transducer and the A/D converter, and actuators which combine the functions of the D/A converter and the final control element are available. In most cases, however, our sensors will provide an analog voltage output and our final control elements will accept an analog voltage input.

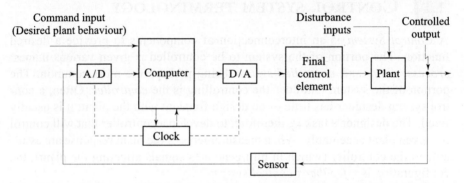

Fig. 1.1 Basic structure of a computer-controlled system

In the control scheme of Fig. 1.1, the A/D converter performs the sampling of the sensor signal (analog feedback signal) and produces its binary representation. The digital computer (*control algorithm*) generates a digital control signal using the information on desired and actual plant behaviour. The digital control signal is then converted to analog control signal via the D/A converter. A real-time clock synchronizes the actions of the A/D and D/A converters, and the *shift registers*. The analog control signal is applied to the plant actuator to control the plant's behaviour.

The overall system in Fig. 1.1 is *hybrid* in nature; the signals are in the *sampled* form (discrete-time signals) in the computer and in a continuous form in the plant. Such systems have traditionally been called *sampled-data systems*; we will use this term as a synonym for *computer control systems/digital control systems*.

The word '*servomechanism*' (or *servosystem*) is used for a *command-following system* wherein the controlled output of the system is required to follow a given command. When the desired value of the controlled outputs is more or less fixed and the main problem is to reject disturbance effects, the control system is sometimes called a *regulator*. The command input for a regulator becomes a constant and is called *set-point*, which corresponds to the desired value of the controlled output. The set-point may however be changed in time from one constant value to another. In a *tracking system*, the controlled output is required to follow or track a time-varying command input.

To make these definitions more concrete, let us consider some familiar examples of control systems.

Example 1.1

Servomechanism for Steering of Antenna

One of the earliest applications of radar tracking was for anti-aircraft fire control, first with guns and later with missiles. Today many civilian applications exist as well, such as satellite-tracking radars, navigation-aiding radars, etc.

The radar scene includes the radar itself, a target, and the transmitted waveform that travels to the target and back. Information about the target's spatial position is first obtained by measuring the changes in the back-scattered waveform relative to the transmitted waveform. The time shift provides information about the target's range, the frequency shift provides information about the target's radial velocity and the received voltage magnitude and phase provide information about the target's angle[1].

In a typical radar application, it is necessary to point the radar antenna toward the target and follow its movements. The radar sensor detects the error between the antenna axis and the target, and directs the antenna to follow the target. The servomechanism for steering the antenna in response to commands from radar sensor is considered here. The antenna is designed for two independent angular motions, one about the vertical axis in which the azimuth angle is varied, and the other about the horizontal axis in which the elevation angle is varied (Fig. 1.2). The servomechanism for steering the antenna is described by

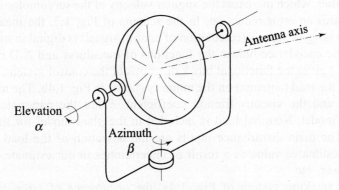

Fig. 1.2 Antenna Configuration

---

1. The bracketed numbers coincide with the list of references given at the end of the book.

4 Digital Control and State Variable Methods

two controlled variables—azimuth angle β and elevation angle α. The desired values or commands are the azimuth angle β_r and the elevation angle α_r of the target. The feedback control problem involves error self-nulling under conditions of disturbances beyond our control (such as wind power).

The control system for steering antenna can be treated as two independent systems—the azimuth-angle servomechanism, and the elevation-angle servomechanism. This is because the interaction effects are usually small. The operational diagram of the azimuth-angle servomechanism is shown in Fig. 1.3.

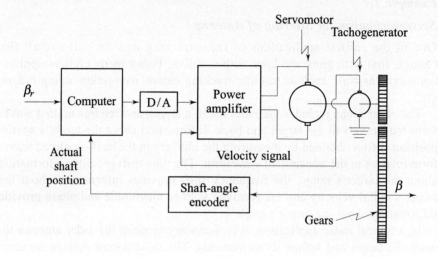

Fig. 1.3 Azimuthal servomechanism for steering of antenna

The steering command from the radar sensor which corresponds to target azimuth angle is compared with the azimuth angle of the antenna axis. The occurrence of the azimuth-angle error causes an error signal to pass through the amplifier, which increases the angular velocity of the servomotor in a direction towards an error reduction. In the scheme of Fig. 1.3, the measurement and processing of signals (calculation of control signal) is digital in nature. The shaft-angle encoder combines the functions of transducer and A/D converter. Figure 1.4 gives the functional block diagrams of the control system. A simple model of the load (antenna) on the motor is shown in Fig. 1.4b. The moment of inertia J and the viscous friction coefficient B are the parameters of the assumed model. Nominal load is included in the plant model for the control design. The main disturbance inputs are the deviation of the load from the nominal estimated value as a result of uncertainties in our estimate, effect of wind power, etc.

In the tracking system of Fig. 1.4a, the occurrence of error causes the motor to rotate in a direction favouring the dissolution of error. The processing of the error signal (calculation of the control signal) is based on the proportional control logic. Note that the components of our system cannot respond

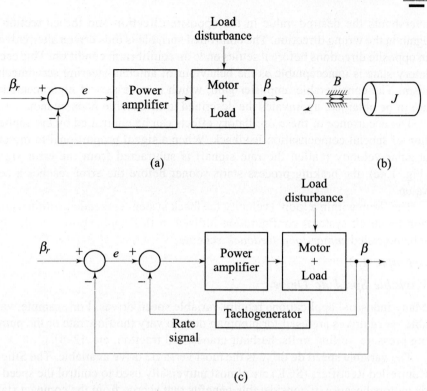

Fig. 1.4 Functional block diagrams of azimuthal servomechanism

instantaneously since any real-world system cannot go from one energy level to another in zero time. Thus, in any real-world system there is some kind of dynamic lagging behaviour between input and output. In the servosystem of Fig. 1.4a, the control action on occurrence of the deviation of the controlled output from the desired value (the occurrence of error) will be delayed by the cumulative dynamic lags of the shaft-angle encoder, digital computer and digital-to-analog converter, power amplifier, and the servomotor with load. Eventually, however, the trend of the controlled variable deviation from the desired value will be reversed by the action of the amplifier output on the rotation of the motor, returning the controlled variable towards the desired value. Now, if a strong correction (high amplifier gain) is applied (which is desirable from the point of view of control system performance, e.g., strong correction improves the speed of response), the controlled variable overshoots the desired value (the 'run-out' of the motor towards an error with the opposite rotation), causing a reversal in the algebraic sign of the system error. Unfortunately, because of system dynamic lags, a reversal of correction does not occur immediately and the amplifier output (acting on 'old' information) is now actually driving the controlled variable in the direction it was already heading, rather than opposing its excursions, leading to a larger deviation. Eventually, the reversed error does cause a reversed correction but the controlled variable overshoots the desired value in the opposite direction and the correction is again in the wrong direction. The controlled variable is thus driven alternatively in opposite directions before it settles on to an equilibrium condition. This oscillatory state is unacceptable as the behaviour of antenna-steering servomechanism. The considerable amplifier gain, which is necessary if high accuracies are to be obtained, aggravates the described unfavourable phenomenon.

The occurrence of these oscillatory effects can be controlled by the application of special compensation feedback. When a signal proportional to motor's angular velocity (called the rate signal) is subtracted from the error signal (Fig. 1.4c), the braking process starts sooner before the error reaches a zero value.

The 'loop within a loop' (velocity feedback system embedded within a position feedback system) configuration utilized in this application is a classical scheme called *minor-loop feedback* scheme.

Example 1.2

Variable Speed dc Drive

Many industrial applications require variable speed drives. For example, variable speed drives are used for pumping duty to vary the flow rate or the pumping pressure, rolling mills, harbour cranes, rail traction, etc. [2–4].

The variable speed dc drive is the most versatile drive available. The Silicon Controlled Rectifiers (SCR) are almost universally used to control the speed of dc motors because of considerable benefits that accrue from the compact static controllers supplied directly from the ac mains.

Basically all the dc systems involving the SCR controllers are similar but with different configurations of the devices, different characteristics may be obtained from the controller. Figure 1.5 shows a dc motor driven by a full-wave rectified supply. Armature current of the dc motor is controlled by an SCR which is in turn controlled by the pulses applied by the SCR trigger control circuit. The SCR controller thus combines the functions of a D/A converter and a final control element.

Firing angle of the SCR controls the average armature current which in turn controls the speed of the dc motor. The average armature current (speed) increases as the trigger circuit reduces the delay angle of firing of the SCR, and the average armature current (speed) reduces as the delay angle of firing of the SCR is increased.

In the regulator system of Fig. 1.5, the reference voltage which corresponds to the desired speed of the dc motor, is compared with the output voltage of tachogenerator corresponding to the actual speed of the motor. The occurrence of the error in speed causes an error signal to pass through the trigger circuit which controls the firing angle of the SCR in a direction towards an error reduction. When the processing of the error signal (calculation of the control signal) is based on the proportional control logic, a steady-state error between the actual speed and the desired speed exists. The occurrence of steady-state error can be eliminated by generating the control signal with two components: one component proportional to the error signal, and the other proportional to the integral of the error signal.

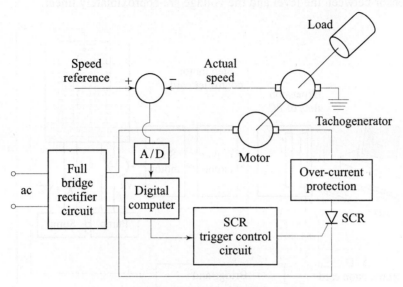

Fig. 1.5 Variable speed dc drive

Example 1.3

Liquid-level Control System

This example describes the hardware features of the design of a PC-based liquid-level control system. The plant of our control system is a cylindrical tank; liquid is pumped into the tank from the sump (Fig. 1.6). The inflow to the tank can be controlled by adjusting valve V1. The outflow from the tank goes back into the sump.

Valve V1 of our plant is a rotary valve, a stepping motor has been used to control the valve. The stepping motor controller card, interfaced to the PC, converts the digital control signals into a series of pulses which are fed to the stepping motor using a driver circuit. Three signals are generated from the digital control signal at each sampling instant, namely, number of steps, speed of rotation, and direction of rotation. The stepping-motor driver circuit converts this information into a single pulse train which is fed to the stepping motor. The valve characteristics between the number of steps of the stepping motor and the outflow from the valve are nonlinear.

The probe used for measurement of liquid level consists of two concentric cylinders connected to a bridge circuit to provide an analog voltage. The liquid partially occupies the space between the cylinders, with air in the remaining part. This device acts like two capacitors in parallel, one with dielectric constant of air ($\simeq 1$) and the other with that of the liquid. Thus the variation of the liquid level causes variation of the electrical capacity measured between the cylinders. The change in the capacitance causes a change in the bridge output voltage which is fed to the PC through an amplifier circuit. The characteristics of the sensor between the level and the voltage are approximately linear.

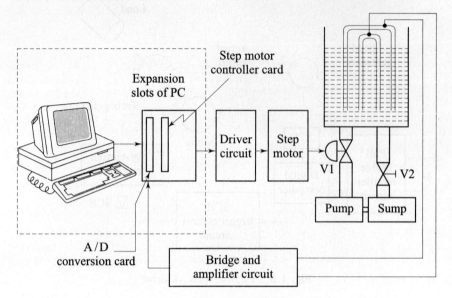

Fig. 1.6 Liquid-level Control System

In the liquid-level control system of Fig. 1.6, the command signal (which corresponds to the desired level of the liquid in the cylinder) is fed through the keyboard; the actual level signal is received through the A/D conversion card. The digital computer compares the two signals at each sampling instant, and generates a control signal which is the sum of two components: one proportional to the error signal, and the other proportional to the integral of the error signal.

1.2 COMPUTER-BASED CONTROL: HISTORY AND TRENDS

Digital computers were first applied to the industrial process control in the late 1950s. The machines were generally large-scale 'main frames' and were used in a so-called *supervisory control mode*; the individual temperature, pressure, flow and the like, feedback loops were locally controlled by electronic or pneumatic analog controllers. The main function of the computer was to gather information on how the overall process was operating, feed this into a technical-economic model of the process (programmed into computer memory), and then periodically send signals to the set-points of all the analog controllers so that each individual loop operated in such a way as to optimize the overall operation.

In 1962, a drastic departure from this approach was made by Imperial Chemical Industries in England—a digital computer was installed which measured 224 variables and manipulated 129 valves directly. The name *Direct Digital Control* (DDC) was coined to emphasize that the computer controlled the process directly. In DDC systems, analog controllers were no longer used. The central computer served as a single time-shared controller for all the individual feedback loops. Conventional control laws were still used for each loop, but the digital versions of control laws for each loop resided in the software in the central computer. Though digital computers were very expensive, one expected DDC systems to have economic advantage for processes with many (50 or more) loops. Unfortunately, this did not often materialize. As failures in the central computer of a DDC system shut down the entire system, it was necessary to provide a fail-safe back-up system, which usually turned out to be a complete system of individual loop analog controllers, negating the expected hardware savings.

There was a substantial development of digital computer technology in the 1960s. By the early 1970s, smaller, faster, more reliable, and cheaper computers became available. The term *minicomputers* was coined for the new computers that emerged. DEC PDP11 is by far the best-known example. There were, however, many related machines from other vendors.

The minicomputer was still a fairly large system. Even as performance continued to increase and prices to decrease, the price of a minicomputer main frame in 1975 was still about $10,000. Computer control was still out of reach for a large number of control problems. But with the development of *microcomputer*, the price of a card computer, with the performance of a 1975 minicomputer, dropped to $500 in 1980. Another consequence was that digital computing power in 1980 came *in quanta* as small as $50. This meant that computer control could now be considered as an alternative, no matter how small the application.

Microcomputers have already made a great impact on the process control field. They are replacing analog hardware even as single-loop controllers. Small DDC systems have been made using microcomputers. Operator communication has vastly improved with the introduction of colour video-graphics displays.

The variety of commercially available industrial controllers ranges from single-loop controllers through multiloop single computer systems to multiloop distributed computers. Although the range of equipment available is large, there are a number of identifiable trends which are apparent.

Single-loop microprocessor-based controllers, though descendant of single-loop analog controllers, have greater degree of flexibility. Control actions which are permitted include on/off control, proportional action, integral action, derivative action and the lag effect. Many controllers have self-tuning option. During the self-tune sequence, the controller introduces a number of step commands, within the tolerances allowed by the operator, in order to characterize the system response. From this response, values for proportional gain, reset time, and rate time are developed. This feature of online tuning in industrial controllers is interesting and permits the concept of the computer automatically adjusting to changing process conditions.

Multiloop single computer systems have variability in available interface and software design. Both the single-loop and multiloop controllers may be used in stand-alone mode or may be interfaced to a host computer for distributed operation. The reducing costs and increasing power of computing systems has tended to make distributed computing systems for larger installations far more cost effective than those built around one large computer. However, the smaller installation may be best catered for by a single multiloop controller or even a few single-loop devices.

Control of large and complex processes using *distributed computer control systems* (DCCS), is facilitated by adopting a multilevel or hierarchical view point of control strategy. The multilevel approach subdivides the system into a hierarchy of simpler control design problems. *On the lowest level of control (direct process control level)*, the following tasks are handled: acquisition of process data, i.e., collection of instantaneous values of individual process variables and status messages of plant control facilities (valves, pumps, motors, etc.) needed for efficient direct digital control; processing of collected data; plant hardware monitoring, system check and diagnosis; closed-loop control and logic control functions based on directives from the next 'higher' level.

Supervisory level copes with the problems of determination of optimal plant work conditions and generation of relevant instructions to be transferred to the next 'lower' level. Adaptive control, optimal control, plant performance monitoring, plant coordination and failure detections are the functions performed at this level.

Production scheduling and control level is responsible for production dispatching, inventory control, production supervision, production re-scheduling, production reporting, etc.

Plant(s) management level, the 'highest' hierarchical level of the plant automation system, is in charge of wide spectrum of engineering, economic, commercial, personnel, and other functions.

It is, of course, not to be expected that in all available distributed computer control systems all four hierarchical levels are already implemented. For automation of small-scale plants, any DCCS having at least two hierarchical levels, can be used. One system level can be used as a direct process control level, and the second one as a combined plant supervisory, and production scheduling and control level. Production planning and other enterprise-level activities can be managed by the separate mainframe computer or the computer centre. For instance, in a LAN (Local Area Network)-based system structure shown in Fig. 1.7a, the 'higher' automation levels are implemented by simply attaching the additional 'higher' level computers to the LAN of the system.

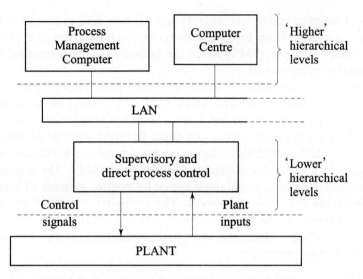

Fig. 1.7a Hierarchical levels in Computer Integrated Process Systems (CIPS)

For complex process plant monitoring, SCADA (*Supervisory Control And Data Acquisition*) systems are available. The basic functions carried out by a SCADA system are:
- Data acquisition and communication
- Data processing
- Events and alarms reporting
- Partial process control

The full process control functions are delegated to the special control units, connected to SCADA system, and are capable to handle the emergency shut down situations.

The separation of SCADA and DCCS is slowly vanishing and the SCADA systems are being brought within the field of DCCS; hierarchical, distributed, flexible and extremely powerful Computer Integrated Process Systems (CIPS) is now a technical reality.

The other main and early application area of digital methods was *machine tool numerical control*, which developed at about the same time as computer control in process industries. Earlier numerically controlled (NC) machines used 'hard-wired' digital techniques. As the microcomputer price and performance improved, it became feasible to replace the hard-wired functions with their software-implemented equivalents, using a microcomputer as a built-in component of the machine tool. This approach has been called *Computerized Numerical Control* (CNC) [29–30]. Industrial *robots* were developed simultaneously with CNC systems.

A quiet revolution is going on in the manufacturing world that is changing the look of the factory. Computers are controlling and monitoring the manufacturing processes. The high degree of automation that until recently was reserved for mass production only, is applied now also to small batches. This requires a change from hard automation in the production line to a *flexible manufacturing system* (FMS) which can be more readily rearranged to handle new market requirements.

Flexible manufacturing systems combined with automatic assembly and product inspection on one hand, and CAD/CAM systems on the other, are the basic components of the modern *Computer Integrated Manufacturing System* (CIMS). In a CIMS, the production flow, from the conceptual design to the finished product, is entirely under computer control and management.

Figure 1.7b illustrates hierarchical structure of CIMS. The lowest level of this structure contains stand-alone computer control systems of manufacturing processes and industrial robots. The computer control of processes includes all types of CNC machine tools, welding, electrochemical machining, electrical discharge machining, and a high-power laser, as well as the adaptive control of these processes.

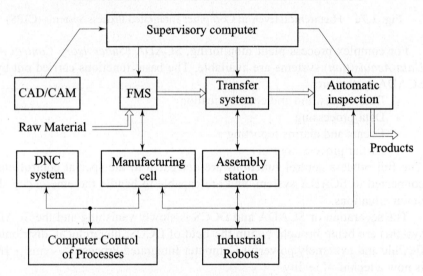

Fig. 1.7b Hierarchical levels in Computer Integrated Manufacturing Systems (CIMS)

When a battery of NC or CNC machine tools is placed under the control of a single computer, the result is a system known as *direct numerical control* (DNC).

The operation of several CNC machines and industrial robots can be co-ordinated by systems called *manufacturing cells*. The computer of the cell is interfaced with the computer of the robot and CNC machines. It receives "completion of job" signals from the machines and issues instructions to the robot to load and unload the machines and change their tools. The software includes strategies permitting the handling of machine breakdown, tool breakage, and other special situations.

The operation of many manufacturing cells can be coordinated by flexible manufacturing system (FMS). The FMS accepts incoming workpieces and processes them under computer control, into finished parts.

The parts produced by the FMS must be assembled into the final product. The parts are routed on a transfer system to assembly stations. In each station, a robot will assemble parts either into a subassembly or (for simple units) into the final product. The subassemblies will be further assembled by robots located in other stations. The final product will be tested by an automatic inspection system.

The FMS uses CAD/CAM systems to integrate the design and manufacturing of parts. At the highest hierarchical level, there will be a supervisory computer which coordinates participation of computers in all phases of a manufacturing enterprise: the design of the product, the planning of its manufacture, the automatic production of parts, automatic assembly, automatic testing, and of course computer-controlled flow of materials and parts through the plant.

In a LAN-based system, the 'higher' automation levels (production planning and other enterprise-level activities) can be implemented by simply attaching the additional 'higher' level computers to the LAN of the system.

One of the most ingenious devices ever devised to advance the field of industrial automation is the *Programmable Logic Controller* (PLC). The PLC, a microprocessor-based general purpose device, provides a 'menu' of basic operations that can be configured by programming to create logic control system for any application [25–28]. So versatile are these devices that they are employed in the automation of almost every type of industry. CIPS and CIMS provide interfaces to PLCs for handling high-speed logic (and other) control functions. Thousands of these devices go unrecognized in process plants and factory environments—quietly monitoring security, manipulating valves, and controlling machines and automatic production lines.

We thus see that the recent appearance of powerful and inexpensive microcomputers has made digital control practical for a wide variety of applications. In fact, now every process is a candidate for digital control. The flourishing of digital control is just beginning for most industries and there is much to be gained by exploiting the full potential of new technology. There is every indication that a high rate of growth in the capability and application of digital computers will continue far into the future.

1.3 CONTROL THEORY: HISTORY AND TRENDS

The development of control system analysis and design can be divided into three eras. In the first era, we have the *classical control theory*, which deals with techniques developed during 1940s and 1950s. Classical control methods—Routh-Hurwitz, Root Locus, Nyquist, Bode, Nichols—have in common the use of transfer functions in the complex frequency(s) domain, and the emphasis on the graphical techniques. Since computers were not available at that time, a great deal of emphasis was placed on developing methods that were amenable to manual computation and graphics. A major limitation of the classical control methods was the use of single-input, single-output (SISO) control configurations. Also, the use of the transfer function and frequency domain limited one to linear time-invariant systems.

In the second era, we have *modern control* (which is not so modern any longer), which refers to state-space based methods developed in the late 1950s and early 1960s. In modern control, system models are directly written in the time domain. Analysis and design are also carried out in the time domain. It should be noted that before Laplace transforms and transfer functions became popular in the 1920s, engineers were studying systems in the time domain. Therefore, the resurgence of time-domain analysis was not unusual, but it was triggered by the development of computers and advances in numerical analysis. As computers were available, it was no longer necessary to develop analysis and design methods that were strictly manual. Multivariable (multi-input, multi-output (MIMO)) control configurations could be analysed and designed. An engineer could use computers to numerically solve or simulate large systems that were nonlinear and/or time-varying. Important results of this era—Lyapunov stability criterion, pole-placement by state feedback, state observers, optimal control—will be discussed in this book.

Modern control methods initially enjoyed a great deal of success in academic circles, but they did not perform very well in many areas of application. Modern control provided a lot of insight into system structure and properties, but it masked other important feedback properties that could be studied and manipulated using the classical control theory. A basic requirement in control engineering is to design control systems that will work properly when the plant model is uncertain. This issue is tackled in the classical control theory using gain and phase margins. Most modern control design methods, however, inherently require a precise model of the plant. In the years since these methods were developed, there have been few significant implementations and most of them have been in a single application area—the aerospace industry. The classical control theory, on the other hand, is going strong. It provides an efficient framework for the design of feedback controls in all areas of application. The classical design methods have been greatly enhanced by the availability of low-cost computers for system analysis and simulation. The graphical tools of classical design can now be more easily used with computer graphics for SISO as well as MIMO systems.

During the past two decades, the control theory has experienced a rapid expansion as a result of the challenges of the stringent requirements posed by modern systems, such as flight vehicles, weapon control systems, robots, and chemical processes; and the availability of low-cost computing power. A body of methods emerged during this third era of control-theory development, which tried to provide answers to the problems of plant uncertainty. These techniques, commonly known as *robust control*, are a combination of modern state-space and classical frequency-domain techniques. For a thorough understanding of these new methods, we need to have adequate knowledge of state-space methods, in addition to the frequency-domain methods. This has guided the preparation of this text.

The modern era of control-theory development has also given an alternative to model-based design methods: the *knowledge-based control*. In this approach, we look for a control solution that exhibits *intelligent behaviour*, rather than using purely mathematical methods to keep the system under control.

Model-based control techniques have many advantages. When the underlying assumptions are satisfied, many of these methods provide good stability, robustness to model uncertainties and disturbances, and speed of response. However, there are many practical deficiencies of these 'crisp' ('hard' or 'inflexible') control algorithms. It is generally difficult to accurately represent a complex process by a mathematical model. If the process model has parameters whose values are partially known, ambiguous, or vague, crisp control algorithms that are based on such *incomplete information*, will not usually give satisfactory results. The environment with which the process interacts may not be completely predictable and it is normally not possible for a crisp algorithm to accurately respond to a condition that it did not anticipate and that it could not 'understand'.

Intelligent control is the name introduced to describe control systems in which control strategies are based on AI techniques. In this control approach, which is an alternative to *model-based control* approach, a behavioural (and not mathematical) description of the process is used, which is based on qualitative expressions and experience of people working with the process. Actions can be performed either as a result of evaluating rules (reasoning) or as unconscious actions based on presented process behaviour after a learning phase. Intelligence comes in as the capability to reason about facts and rules and to learn about presented behaviour. It opens up the possibility of applying the experience gathered by operators and process engineers. Uncertainty about the knowledge can be handled, as well as ignorance about the structure of the system.

Fuzzy logic, and neural networks are very good methods to model real processes which cannot be described mathematically. Fuzzy logic deals with linguistic and imprecise rules based on expert's knowledge. Neural networks are applied in the case where we do not have any rules but several data.

The main feature of *fuzzy logic control* is that a control engineering base (typically in terms of a set of rules) created using expert's knowledge of process behaviour, is available within the controller and the control actions are generated by applying existing process conditions to the knowledge base, making use of an *inference mechanism*. The knowledge base and the inference mechanism can handle noncrisp and incomplete information, and the knowledge itself will improve and evolve through learning and past experience.

In the *neural network based control*, the goal of *artificial neural network* is to emulate the mechanism of human brain function and reasoning and to achieve the same intelligence level as the human brain in learning, abstraction, generalization and making decisions under uncertainty.

In the conventional design exercises, the system is modelled analytically by a set of differential equations, and their solution tells the controller how to adjust the system's control activities for each type of behaviour required. In a typical intelligent control scheme, these adjustments are handled by an intelligent controller, a logical model of thinking processes that a person might go through in the course of manipulating the system. This shift in focus from the process to the person involved, changes the entire approach to automatic control problems. It provides a new design paradigm such that a controller can be designed for complex, ill-defined processes without knowing quantitative input-output relations which are otherwise required by conventional method.

The ever-increasing demands of the complex control systems being built today and planned for the future dictate the use of novel and more powerful methods in control. The potential for intelligent control techniques in solving many of the problems involved is great, and this research area is evolving rapidly. The viewpoint is taken that model-based control techniques should be augmented with intelligent control techniques in order to enhance the performance of the control systems. The developments in intelligent control methods should be based on firm theoretical foundations (as is the case with model-based control methods), but this is still at its early stages. Strong theoretical results guaranteeing control system properties such as stability are still to come, although promising results reporting progress in special cases have been reported recently. The potential of intelligent control systems clearly needs to be further explored and both theory and applications need to be further developed.

1.4 AN OVERVIEW OF THE CLASSICAL APPROACH TO ANALOG CONTROLLER DESIGN

The tools of classical linear control system design are the Laplace transform, stability testing, root locus, and frequency response. Laplace transformation is used to convert system descriptions in terms of integro-differential equations to equivalent algebraic relations involving rational functions. These are conveniently manipulated in the form of transfer functions with block diagrams and signal flow graphs.

The block diagram of Fig. 1.8 represents the basic structure of feedback control systems. Not all systems can be forced into this format, but it serves as a reference for discussion.

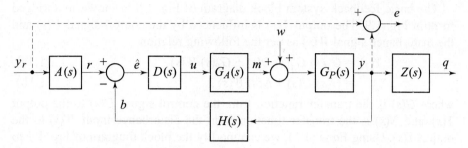

Fig. 1.8 Generalised operational block diagram of a feedback system

In Fig. 1.8, the variable $y(t)$ is the *controlled variable* of the system. The desired value of the controlled variable is $y_r(t)$, the command input. $y_r(t)$ and $y(t)$ have the same units. The *feedback elements* with transfer function $H(s)$ are system components that act on the controlled variable $y(t)$ to produce the *feedback signal* $b(t)$. $H(s)$ typical represents the sensor action to convert the controlled variable $y(t)$ to an electrical sensor output signal $b(t)$.

The *reference input elements* with transfer function $A(s)$ convert the command signal $y_r(t)$ into a form compatible with the feedback signal $b(t)$. The transformed command signal is the actual physical input to the system. This actual signal input is defined as the *reference input*.

The comparison device (*error detector*) of the system compares the reference input $r(t)$ with the feedback signal $b(t)$ and generates the *actuating error signal* $\hat{e}(t)$. The signals $r(t)$, $b(t)$, and $\hat{e}(t)$ have the same units. The *controller* with transfer function $D(s)$ acts on the actuating error signal to produce the *control signal* $u(t)$.

The control signal $u(t)$ has the knowledge about the desired control action. The power level of this signal is relatively low. The *actuator elements* with transfer function $G_A(s)$ are the system components that act on the control signal $u(t)$ and develop enough torque, pressure, heat, etc. (*manipulated variable* $m(t)$), to influence the *controlled system*. $G_P(s)$ is the transfer function of the controlled system.

The *disturbance* $w(t)$ represents the undesired signals that tend to affect the controlled system. The disturbance may be introduced into the system at more than one location.

There are situations wherein the variable $q(t)$ which we really wish to control is not used to generate feedback signal. The transfer function $Z(s)$ in Fig. 1.8 is the transfer function of the *indirectly controlled system elements*. It relates the *indirectly controlled variable* $q(t)$ to the controlled variable $y(t)$. Note that $Z(s)$ is outside the feedback loop.

The dashed-line portion of Fig. 1.8 shows the *system error* $e(t) = y_r - y(t)$. Note that the actuating error signal $\hat{e}(t)$ and the system error $e(t)$ are two different variables.

18 Digital Control and State Variable Methods

The basic feedback system block diagram of Fig. 1.8 is shown in abridged form in Fig. 1.9. The output $Y(s)$ is influenced by the control signal $U(s)$ and the disturbance signal $W(s)$ as per the following relation:

$$Y(s) = G_P(s)\, G_A(s)\, U(s) + G_P(s)\, W(s) \tag{1.1a}$$
$$= G(s)\, U(s) + N(s)\, W(s) \tag{1.1b}$$

where $G(s)$ is the transfer function from the control signal $U(s)$ to the output $Y(s)$, and $N(s)$ is the transfer function from the disturbance input $W(s)$ to the output $Y(s)$. Using Eqns (1.1), we can modify the block diagram of Fig. 1.9 to the form show in Fig. 1.10. Note that in the block diagram model of Fig. 1.10, the plant includes the actuator elements.

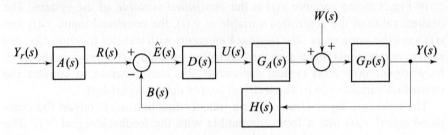

Fig. 1.9 A general linear feedback system

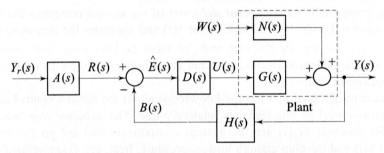

Fig. 1.10 Equivalent representation of the block diagram of Fig. 1.9

The actuating error signal
$$\hat{E}(s) = R(s) - B(s)$$
$$= A(s)\, Y_r(s) - H(s)\, Y(s)$$

The control signal
$$U(s) = D(s)\, A(s)\, Y_r(s) - D(s)\, H(s)\, Y(s) \tag{1.2a}$$
$$= D(s)\, H(s) \left[\frac{A(s)}{H(s)} Y_r(s) - Y(s) \right] \tag{1.2b}$$

Using Eqns (1.2a) and (1.2b), we can simplify Fig. 1.10 to obtain the structure shown in Fig. 1.11.

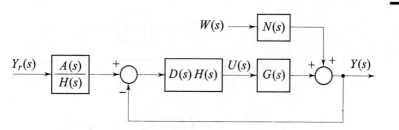

Fig. 1.11 Simplification of the block diagram of Fig. 1.10

A further simplification of Fig. 1.11 is possible if $H = A$. In this case, which is quite common, we can model the system as the *unity feedback system* shown in Fig. 1.12, and take advantage of the fact that now the actuating signal is the system error $e(t)$.

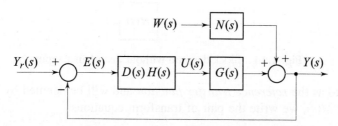

Fig. 1.12 Unity feedback system

The block diagrams in Figs 1.10–1.12 are very useful for the purpose of system design. However, it should be clear that these block diagrams have lost the physical significance. For example, the block in Fig. 1.11 with transfer function $A(s)/H(s)$ does not refer to any physical portion of the original system. Rather, it represents the result of manipulating Eqn. (1.2a) into the form given by Eqn. (1.2b).

Thus the reader is advised to think in terms of the equations that the block diagrams represent, rather than to attach any special significance to the block diagrams themselves. The only role played by a block diagram is that it is a convenient means of representing the various system equations, rather than writing them out explicitly. *Block diagram manipulation* is nothing more than the manipulation of a set of algebraic transform equations.

▲▲

For the analysis of a feedback system, we require the transfer function between the input—either reference or disturbance—and the output. We can use block diagram manipulations to eliminate all the signals except the input and the output. The reduced block diagram leads to the desired result.

Consider the block diagram of Fig. 1.13. The feedback system has two inputs. We shall use superposition to treat each input separately.

When disturbance input is set to zero, the single-input system of Fig. 1.14 results. The transfer function between the input $R(s)$ and the output $Y(s)$ is

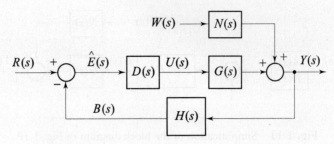

Fig. 1.13 A typical feedback system with two inputs

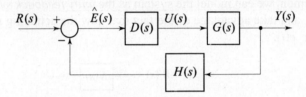

Fig. 1.14 Block diagram without disturbance input

referred to as the *reference transfer function* and will be denoted by $M(s)$. To solve for $M(s)$, we write the pair of transform equations

$$\hat{E}(s) = R(s) - H(s)\,Y(s)$$
$$Y(s) = G(s)\,U(s) = G(s)\,D(s)\,\hat{E}(s)$$

and then eliminate $\hat{E}(s)$ to obtain

$$[1 + D(s)\,G(s)\,H(s)]\,Y(s) = D(s)\,G(s)\,R(s)$$

which leads to the desired result:

$$M(s) = \left.\frac{Y(s)}{R(s)}\right|_{W(s)=0} = \frac{D(s)G(s)}{1+D(s)G(s)H(s)} \quad (1.3)$$

Similarly, we obtain the *disturbance transfer function* $M_w(s)$ by setting the reference input to zero in Fig. 1.13 yielding Fig. 1.15, and then solving for $Y(s)/(W(s))$. From the revised block diagram,

$$\hat{E}(s) = -H(s)\,Y(s)$$
$$Y(s) = G(s)\,D(s)\,\hat{E}(s) + N(s)\,W(s)$$

from which $\hat{E}(s)$ can be eliminated to give

$$M_w(s) = \left.\frac{Y(s)}{W(s)}\right|_{R(s)=0} = \frac{N(s)}{1+D(s)G(s)H(s)} \quad (1.4)$$

The response to the simultaneous application of $R(s)$ and $W(s)$ is given by
$$Y(s) = M(s) R(s) + M_w(s) W(s) \tag{1.5}$$
Figure 1.16 shows the reduced block diagram model of the given feedback system.

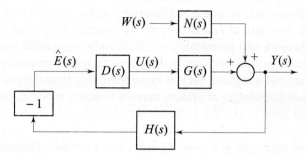

Fig. 1.15 Block diagram without reference input

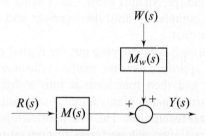

Fig. 1.16 Reduced block diagram model for system of Fig. 1.13

The transfer functions given by Eqns (1.3) and (1.4) are referred to as *closed-loop transfer functions*. The denominator of these transfer functions has the term $D(s)G(s)H(s)$ which is the multiplication of all the transfer functions in the feedback loop. It may be viewed as the transfer function between the variables $R(s)$ and $B(s)$ if the loop is broken at the summing point. $D(s)G(s)H(s)$ may therefore be given the name *open-loop transfer function*. The roots of denominator polynomial of $D(s)G(s)H(s)$ are the *open-loop poles*, and the roots of numerator polynomial of $D(s)G(s)H(s)$ are the *open-loop zeros*.

The roots of the characteristic equation
$$1 + D(s)G(s)H(s) = 0 \tag{1.6}$$
are the *closed-loop poles* of the system. These poles indicate whether or not the system is bounded-input bounded-output (BIBO) stable, according to whether or not all the poles are in the left half of the complex plane. Stability may be tested by the Routh stability criterion.

A root locus plot consists of a pole-zero plot of the open-loop transfer function of a feedback system, upon which is superimposed the locus of the poles of the closed-loop transfer function as some parameter is varied. Design of the *controller (compensator) $D(s)$* can be carried out using the root locus plot. One begins with simple compensators, increasing their complexity until the performance requirements can be met. Principal measures of transient performance are peak overshoot, settling time, and rise time. The compensator poles, zeros, and multiplying constants are selected to give feedback system pole locations that result in acceptable transient response to step inputs. At the same time, the parameters are constrained so that the resulting system has acceptable steady-state response to important inputs, such as steps and ramps.

Frequency response characterizations of systems have long been popular because of the ease and practicality of steady-state sinusoidal response measurements. These methods also apply to systems in which rational transfer function models are not adequate, such as those involving time delays. They do not require explicit knowledge of system transfer function models; experimentally obtained open-loop sinusoidal response data can directly be used for stability analysis and compensator design. A stability test, the Nyquist criterion, is available. Principal measures of transient performance are gain margin, phase margin, and bandwidth. The design of the compensator is conveniently carried out using the Bode plot and the Nichols chart. One begins with simple compensators, increasing their complexity until the transient and steady-state performance requirements are met.

There are two approaches to carrying out the digital controller (compensator) design. The first approach uses the methods discussed above to design an analog compensator and then transform it into a digital one. The second approach first transforms analog plants into digital plants and then carries out the design using digital techniques. The first approach performs discretization after design; the second approach performs discretization before design. The classical approach to designing a digital compensator directly using an equivalent digital plant for a given analog plant parallels the classical approach to analog compensator design. The concepts and tools of the classical digital design procedures will be given in Chapters 2–4. This background will be useful in understanding and applying the state-variable methods to follow.

1.5 | SCOPE AND ORGANIZATION OF THE BOOK

This text is concerned with digital control and state variable methods for a special class of control systems, namely time-invariant, lumped, and deterministic systems. We do not intend to solve all the problems that can be posed under the defined category. Coverage of digital control theory and practice is modest. Various concepts from interdisciplinary fields of computer science and computer engineering which relate to digital control system development—number representation, logical descriptions of algorithmic processes, computer arithmetic operations, computer system hardware and software—are beyond the scope of this book. In fact, a course on control engineering need not include these topics because specialized courses on computer-system architecture are normally offered in undergraduate curricula of all engineering disciplines.

It is assumed that the reader has had an introductory course in control engineering concentrating on the basic principles of feedback control and covering various classical analog methods of control system design. The classical digital methods of design are developed in this book, paralleling and extending considerably the similar topics in analog control.

State variable analysis and design methods are usually not covered in an introductory course. It is assumed that the reader is not exposed to the so-called modern control theory. Our approach is to first discuss state variable methods for continuous-time systems, and then give a compact presentation of the methods for discrete-time systems, using the analogy with the continuous-time case.

This text also prepares a student for the study of advanced control methods developed during the past two decades. However, detailed coverage of these methods is beyond the scope of the book.

There are eleven chapters in the book in addition to this introductory chapter. In each chapter, analysis/design examples are interspersed to illustrate the concepts involved. At the end of each chapter, there are a number of review examples that take the reader to a higher level of application; some of these examples also serve the purpose of extending the text material. The same approach is followed in unsolved problems. Answers to problems have been given to inspire confidence in the reader.

The examples we have considered in this book are generally low-order systems. Such a selection of examples helps in conveying the fundamental concepts of feedback control without the distraction of large amounts of computations inherent in high-order systems. Many of the real-life design problems are more complex than the ones discussed in this book. High-order systems are common and, in addition, several parameters are to be varied during design stage to investigate their effect on the system performance. CAD tools are extremely useful for complex control problems. Several software packages with computer graphics are available commercially for CAD of control systems [169–179].

Let us now go through the organization of the chapters. Chapters 2–4 deal with digital control theory and practice. The philosophy of presentation is that the new material should be closely related to material already familiar, and yet, by the end, a direction to wider horizons should be indicated. The approach leads us, for example, to relate the z-transform to the Laplace transform and to describe the implications of poles and zeros in the z-plane to those known meanings attached to poles and zeros in the s-plane. Also, in developing the design methods we relate the digital control design methods to those of continuous-time systems.

Chapter 2 introduces the sampling theorem and the phenomenon of aliasing. Methods to generate discrete-time models which approximate continuous-time dynamics, and stability analysis of these models are also included in this chapter.

Chapter 3 considers the hardware (analog and digital) of the control loop with emphasis on modelling. Models of some of the widely used digital control systems are also included.

Chapter 4 establishes a toolkit of design-oriented techniques. It puts forward alternative design methods based on root-locus and Bode plots. Design of digital controllers using z-plane synthesis is also included in this chapter. References for the material in Chapters 2–4 are [51], [55–56], [78–101].

Chapters 5–9 deal with state variable methods in automatic control. Chapter 5 is on state variable analysis. It exposes the problems of state variable representation, diagonalization, solution, controllability, and observability. The relationship between the transfer function and state variable models is also given. Although it is assumed that the reader has the necessary background on vector-matrix analysis, a reasonably detailed account of vector-matrix analysis is provided in this chapter for convenient reference.

State variable analysis concepts, developed in continuous-time format in Chapter 5, are extended to digital control systems in Chapter 6.

The techniques of achieving desired system characteristics by pole-placement using complete state variable feedback are developed in Chapter 7. Also included is the method of using the system output to form estimates of the states for use in state feedback. Results are given for both continuous-time and discrete-time systems. A introduction to system identification and self-tuning control is also included.

Direct method of Lyapunov, the most general method for stability analysis of nonlinear and/or time-varying systems, is covered in Chapter 8. In addition to stability analysis, Lyapunov functions are useful in solving some optimization problems; examples are included in this and the next chapter. Concepts are first developed in continuous-time format, and then extended to digital control systems. An introduction to model reference adaptive control is also included.

In Chapter 9, we learn how to optimize the gain matrix of the state-feedback control law by choosing it to minimize a quadratic integral performance index. Results are given for both continuous-time and discrete-time systems. References for the material in Chapters 5–9 are [32–33], [102–115], [123–127].

Describing function approach which has demonstrated great utility in analysis of nonlinear systems, has been paid considerable attention in Chapter 10. Also included is analysis of second-order nonlinear systems on phase plane. The concepts of feedback linearization, and variable structure control are briefly introduced. For detailed study, refer [116–122].

A tutorial introduction to knowledge-based tools (neural networks, fuzzy logic, genetic algorithms) for control system design is given in Chapters 11 and 12. Rigorous characterization of theoretical properties of intelligent control methodology will not be our aim in our tutorial presentation; rather we focus on the development of systematic engineering procedures which will guide the design of controller for a specific problem. For detailed study, refer [158–168].

The text presumes on the part of the reader some knowledge of basic tools like the Fourier transform, the Laplace transform, and the z-transform. However Appendix A on mathematical background has been included to make the book reasonably self-contained. Appendices B and C provide an introduction to the MATLAB environment for computer-aided control system design.

chapter 2
Signal Processing in Digital Control

2.1 WHY USE DIGITAL CONTROL

Digital control systems offer many advantages over their analog counterparts. Of course, there are possible disadvantages also. Let us first look at the advantages of digital control over the corresponding analog control before we talk of the price one has to pay for the digital option.

Advantages Offered by Digital Control

Flexibility

An important advantage offered by digital control is in the flexibility of its modifying controller characteristics, or of adapting the controller if plant dynamics change with operating conditions. The ability to 'redesign' the controller by changing software (rather than hardware) is an important feature of digital control as against analog control.

Wide selection of control algorithms

Implementation of advanced control techniques was earlier constrained by the limitations of analog controllers and the high costs of digital computers. However, with the advent of inexpensive digital computers with virtually limitless computing power, the techniques of modern control theory may now be put to practice. For example, in multivariable control systems with more than one input and one output, modern techniques for optimizing system performance or reducing interactions between feedback loops can now be implemented.

Integrated control of industrial systems

Feedback control is only one of the functions of the computer. In fact, most of the information transfer between the process and the computer is of an on-off nature and exploits the logical decision-making capability of the computer. Real-time applications of information processing and decision-making, e.g., production planning, scheduling, optimization, operations control, etc., may now be integrated with the traditional process control functions.

To enable the computer to meet the variety of demands imposed on it, its tasks are time-shared.

Future generation control systems

The study of emerging applications shows that artificial-intelligence (AI) will affect design and application of control systems as profoundly as the impact of the microprocessor has been in the last two decades. It is clear that future generation control systems will have a significant AI component; the list of applications of computer-based control will continue to expand.

Implementation Problems in Digital Control

The main problems associated with implementation of digital control are related to the effects of sampling and quantization.

Most processes that we are called upon to control operate in continuous time. This implies that we are dealing largely, with an analog environment. To this environment, we need to interface the digital computer through which we seek to influence the process.

The interface is accomplished by a system of the form shown in Fig. 2.1. It is a cascade of analog-to-digital (A/D) conversion system followed by a computer which is in turn followed by a digital-to-analog (D/A) conversion system. The A/D conversion process involves deriving samples of the analog signal at discrete instants of time separated by sampling period T sec. The D/A conversion process involves reconstructing continuous-time signals from the samples given out by the digital computer.

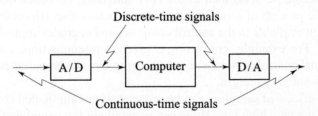

Fig. 2.1   Discrete-time processing of continuous-time signals

Quantization effects

The conversion of signals from analog into digital form and *vice versa* is performed by electronic devices (A/D and D/A converters) of finite resolution. A device of n-bit resolution has 2^n quantization levels. Here the analog signal gets tied to these finite number of quantization levels in the process of conversion to digital form. Therefore, by the sheer act of conversion, a valuable part of information about the signal is lost.

Furthermore, any computer employed as a real-time controller must perform all the necessary calculations with limited precision, thus introducing a truncation error after each arithmetic operation has been performed. As the computational accuracy is normally much higher than the resolution of real converters, a further truncation must take place before the computed data are converted into the analog form. The repeated process of approximate conversion-computation-conversion may be costly, if not disastrous, in terms of control system performance.

The process of quantization in signal conversion systems will be discussed shortly.

Sampling effects

The selection of a sampling period is a fundamental problem in digital control systems. Later in this chapter we will discuss the *sampling theorem* which states that the sampling period T should be chosen such that

$$T < \pi/\omega_m$$

where ω_m is the strict bandwidth of the signal being sampled. This condition ensures that there is no loss of information due to sampling and the continuous-time signal can be completely recovered from its samples using an ideal low-pass filter.

There are, however, two problems associated with the use of this theorem in practical control systems:
  (i) Real signals are not band-limited and hence strict bandwidth limits are not defined.
  (ii) The ideal low-pass filter, needed for the distortionless reconstruction of continuous-time signals from its samples, is not physically realizable. Practical devices, such as the D/A converter, introduce distortions.

Thus the process of sampling and reconstruction also affects the amount of information available to the control computer and degrades control system performance. For example, converting a given continuous-time control system into a digital control system without changing the system parameters degrades the system stability margin.

The ill-effects of sampling can be reduced, if not eliminated completely, by sampling at a very high rate. However, excessively fast sampling ($T \to 0$) may result in numerical ill-conditioning in the implementation of recursive control algorithms (described later in this chapter).

▲▲

With the availability of low-cost, high-performance digital computers and interfacing hardware, the implementation problems in digital control do not pose a serious threat to its usefulness. The advantages of digital control outweigh its disadvantages for most of the applications.

Signal Processing in Digital Control 29

This book attempts to provide a modest coverage of digital control theory and practice. In the present chapter, we focus our attention on the digital computer and its interface to signal conversion systems (Fig. 2.1). The goal is to formulate tools of analysis necessary to understand and guide the design of programs for a computer acting as a control logic component. Needless to say, digital computers can do many things other than control dynamic systems; our purpose is to examine their characteristics when doing the elementary control task.

2.2 CONFIGURATION OF THE BASIC DIGITAL CONTROL SCHEME

Figure 2.2 depicts a block diagram of a digital control system showing a configuration of the basic control scheme. The basic elements of the system are shown by the blocks.

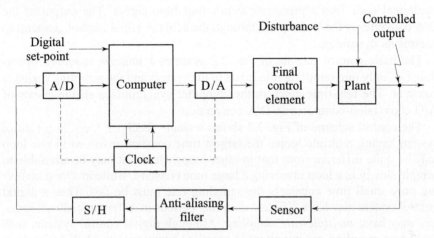

Fig. 2.2 Configuration of the basic digital control scheme

The analog feedback signal coming from the sensor is usually of low frequency. It may often include high frequency 'noise.' Such noise signals are too fast for control system to correct; low-pass filtering is often needed to allow good control performance. The anti-aliasing filter shown in Fig. 2.2 serves this purpose. In digital systems, the phenomenon called *aliasing* introduces some new aspects to the noise problems. We will study this phenomenon later in this chapter.

The analog signal, after anti-aliasing processing, is converted into digital form by the A/D conversion system. The conversion system usually consists of an A/D converter preceded by a sample-and-hold (S/H) device. The A/D converter converts a voltage (or current) amplitude at its input into a binary code representing a quantized amplitude value closest to the amplitude of the input. However, the conversion is not instantaneous. Input signal variation during the conversion time of the A/D converter can lead to erroneous results. For this reason, high performance A/D conversion systems include a S/H device which keeps the input to the A/D converter constant during its conversion time.

The digital computer processes the sequence of numbers by means of an algorithm and produces a new sequence of numbers. Since data conversions and computations take time, there will always be a delay when a control law is implemented using a digital computer. The delay, which is called the *computational delay*, degrades the control system performance. It should be minimized by the proper choice of hardware and by the proper design of software for the control algorithm. Floating-point operations take a considerably longer time to perform (even when carried out by arithmetic co-processor) than the fixed-point ones. We therefore try to execute fixed-point operations whenever possible. Alternative realization schemes for a control algorithm will be given in the next chapter.

The D/A conversion system in Fig. 2.2 converts the sequence of numbers in numerical code into a piecewise continuous-time signal. The output of the D/A converter is fed to the plant through the actuator (final control element) to control its dynamics.

The basic control scheme of Fig. 2.2 assumes a *uniform sampling* operation, i.e., only one sampling rate exists in the system and the sampling period is constant. The real-time clock in the computer synchronizes all the events of A/D conversion-computation-D/A conversion.

The control scheme of Fig. 2.2 shows a single feedback loop. In a control system having multiple loops, the largest time constant involved in one loop may be quite different from that in other loops. Hence, it may be advisable to sample slowly in a loop involving a large time constant, while in a loop involving only small time constants the sampling rate must be fast. Thus a digital control system may have different sampling periods in different feedback paths, i.e., may have *multiple-rate sampling*. Although digital control systems with multi-rate sampling are important in practical situations, we shall concentrate on single-rate sampling. (The reader interested in multi-rate digital control systems may refer Kuo [80]).

The overall system in Fig. 2.2 is hybrid in nature; the signals are in a sampled form (*discrete-time signals/digital signals*) in the computer and in continuous-time form in the plant. Such systems have traditionally been called *sampled-data control systems*. We will use this term as a synonym to *computer control systems/digital control systems*.

▲▲

In the present chapter, we focus our attention on digital computer and its analog interfacing. For the time being, we delink the digital computer from the plant. The link will be re-established in the next chapter.

2.3 PRINCIPLES OF SIGNAL CONVERSION

Figure 2.3a shows an analog signal $y(t)$—it is defined at the continuum of times, and its amplitudes assume a continuous range of values. Such a signal cannot be stored in digital computers. The signal therefore must be converted to a form that will be accepted by digital computers. One very common method to do this is to record sample values of this signal at equally spaced instants. If we sample the signal every 10 msec, for example, we obtain the *discrete-time signal* sketched in Fig. 2.3b. The *sampling interval* corresponds to a *sampling rate* of 100 samples/sec. The choice of the sampling rate is an important one, since it determines how accurately the discrete-time signal can represent the original signal.

In a practical situation, the sampling rate is determined by the range of frequencies present in the original signal. Detailed analysis of uniform sampling process, and the related problem of *aliasing* will appear later in this chapter.

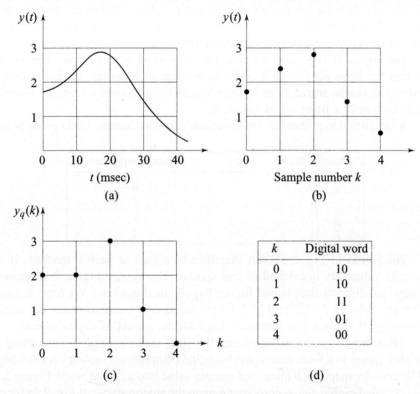

Fig. 2.3 Sampling, quantization and coding of an analog signal

Notice that the time axis of the discrete-time signal in Fig. 2.3b is labelled simply 'sample number' and index k has been used to denote this number ($k = 0, 1, 2, ...$). Corresponding to different values of sample number k, the discrete-time signal assumes the same continuous range of values assumed by the analog signal $y(t)$. We can represent the sample values by a sequence of numbers y_s (refer Fig. 2.3b):

$$y_s = \{1.7, 2.4, 2.8, 1.4, 0.4, ...\}$$

In general,
$$y_s = \{y(k)\}, \; 0 \leq k < \infty$$
where $y(k)$ denotes the kth number in the sequence.

The sequence defined above is *one-sided sequence*; $y_s = 0$ for $k < 0$. In digital control applications, we normally encounter one-sided sequences.

Although, strictly speaking, $y(k)$ denotes the kth number in the sequence, the notation given above is often unnecessarily cumbersome, and it is convenient and unambiguous to refer to $y(k)$ itself as a sequence.

Throughout our discussion on digital control, we will assume *uniform sampling*, i.e., sample values of the analog signal are extracted at equally spaced sampling instants. If the physical time corresponding to the sampling interval is T seconds, then the kth sample $y(k)$ gives the value of the discrete-time signal at $t = kT$ seconds. We may, therefore, use $y(kT)$ to denote a sequence wherein the independent variable is the physical time.

The signal of Fig. 2.3b is defined at discrete instants of time. The sample values are however tied to a continuous range of numbers. Such a signal, in principle, can be stored in an infinite-bit machine because a finite-bit machine can store only a finite set of numbers.

A simplified hypothetical 2-bit machine can store four numbers given below:

Binary number	Decimal equivalent
00	0
01	1
10	2
11	3

The signal of Fig. 2.3b can therefore be stored in such a machine if the sample values are quantified to four *quantization levels*. Figure 2.3c shows a quantized discrete-time signal for our hypothetical machine. We have assumed that any value in the interval [0.5, 1.5) is rounded to 1, and so forth. The signals for which both time and amplitude are discrete, are called *digital signals*.

After sampling and quantization, the final step required in converting an analog signal to a form acceptable to digital computers is *coding* (or *encoding*). The encoder maps each quantized sample value into a digital word. Figure 2.3d gives coded digital signal corresponding to the analog signal of Fig. 2.3a for our hypothetical 2-bit machine.

The device that performs the sampling, quantization, and coding is an A/D *converter*. Figure 2.4 is a block diagram representation of the operations performed by an A/D converter.

It may be noted that the quantized discrete-time signal of Fig. 2.3c and the coded signal of Fig. 2.3d carry exactly the same information. For the purpose of analytical study of digital systems, we will use the quantized discrete-time form for digital signals.

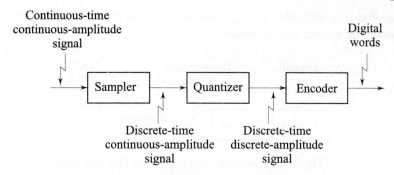

Fig. 2.4 Operations performed by an A/D converter

The number of binary digits carried by a device is its *word length*, and this is obviously an important characteristic related to the *resolution* of the device—the smallest change in the input signal that will produce a change in the output signal. The A/D converter that generates signals of Fig. 2.3 has two binary digits and thus four quantization levels. Any change, therefore, in the input over the interval [0.5, 1.5) produces no change in the output. With three binary digits, 2^3 quantization levels can be obtained, and the resolution of the converter would improve.

The A/D converters in common use have word lengths of 8 to 16 bits. For an A/D converter with a word length of eight bits, an input signal can be resolved to one part in 2^8, or 1 in 256. If the input signal has a range of 10 V, the resolution is 10/256, or approximately 0.04 V. Thus the input signal must change by at least 0.04 V in order to produce a change in the output.

With the availability of converters with resolution ranging from 8 to 16 bits, the quantization errors do not pose a serious threat in the computer control of industrial processes. In our treatment of the subject, we assume quantization errors to be zero. This is equivalent to assuming infinite-bit digital devices. Thus we treat digital signals as if they are discrete-time signals with amplitudes assuming a continuous range of values. In other words, we make no distinction between the words 'discrete-time' and 'digital.'

A typical topology of a single-loop digital control system is shown in Fig. 2.2. It has been assumed that the measuring transducer and the actuator (final control element) are analog devices, requiring respectively A/D and D/A conversion at the computer input and output. The D/A conversion is a process of producing an analog signal from a digital signal and is, in some sense, the reverse of the sampling process discussed above.

The D/A converter performs two functions: first, generation of output samples from the binary-form digital signals produced by the machine, and second, conversion of these samples to analog form. Figure 2.5 is a block diagram representation of the operations performed by a D/A converter. The *decoder* maps each digital word into a sample value of the signal in discrete-time form. It is usually not possible to drive a load, such as a motor, with these samples. In order to deliver sufficient energy, the sample amplitude might have to be so large that it is infeasible to be generated. Also large-amplitude signals might saturate the system being driven.

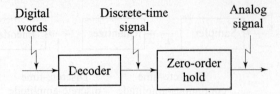

Fig. 2.5 Operations performed by a D/A converter

The solution to this problem is to smooth the output samples to produce a signal in analog form. The simplest way of converting a sample sequence into a continuous-time signal is to hold the value of the sample until the next one arrives. The net effect is to convert a sample to a pulse of duration T—the sample period. This function of a D/A converter is referred to as a *zero-order hold* (ZOH) operation. The term *zero-order* refers to the zero-order polynomial used to extrapolate between the sampling times (detailed discussion will appear later in this chapter). Figure 2.6 shows a typical sample sequence produced by the decoder, and the analog signal[1] resulting from the zero-order hold operation.

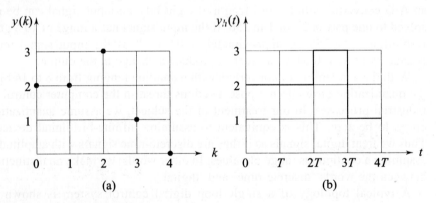

Fig. 2.6 (a) Sampled sequence (b) Analog output from ZOH

D/A Converter Circuits

Most D/A converters use the principle shown in the three-bit form in Fig. 2.7 to convert the HI/LO digital signals at the computer output to a single analog

---

1. An analog signal is defined over a continuous range of time whose amplitude can assume a continuous range of values. Figure 2.3a shows an analog signal, and Fig. 2.6b shows a continuous-time quantized signal. In the literature, including this book, the terms 'continuous-time signal' and 'analog signal' are frequently interchanged, although strictly speaking the analog signal is a special case of the continuous-time signal.

Signal Processing in Digital Control 35

voltage. The circuit of Fig. 2.7 is an 'R–2R' ladder; the value of R typically ranges from 2.5 to 10K ohms.

Suppose a binary number $b_2 b_1 b_0$ is given. The switch (actually electronic gates) positions in Fig. 2.7 correspond to the digital word 100, i.e., $b_2 = 1$ and $b_1 = b_0 = 0$. The circuit can be simplified to the equivalent form shown in Fig. 2.8a. The currents in the resistor branches are easily calculated and are indicated in the circuit (for the high gain amplifier, the voltage at point A is practically zero [180]). The output voltage is

$$V_0 = 3R \frac{i_2}{2} = \frac{1}{2} V_{ref}$$

If $b_1 = 1$ and $b_2 = b_0 = 0$, then the equivalent circuit is as shown in Fig. 2.8b. The output voltage is

$$V_0 = 3R \frac{i_1}{4} = \frac{1}{4} V_{ref}$$

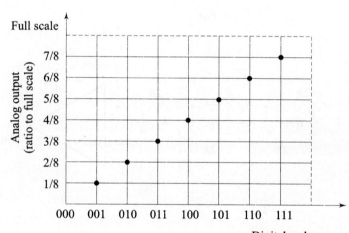

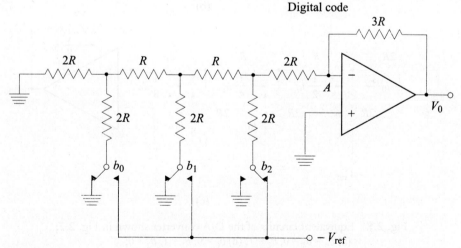

Fig. 2.7 Three-bit D/A converter

Similarly, if $b_0 = 1$ and $b_2 = b_1 = 0$, then the equivalent circuit is as shown in Fig. 2.8c. The output voltage is

$$V_0 = 3R \frac{i_0}{8} = \frac{1}{8} V_{ref}$$

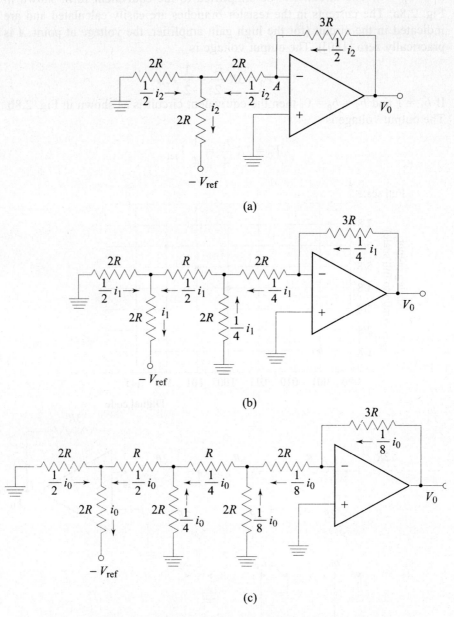

Fig. 2.8 Equivalent circuits of the D/A converter shown in Fig. 2.7; (a) $b_0 = b_1 = 0, b_2 = 1$; (b) $b_0 = 0, b_1 = 1, b_2 = 0$; (c) $b_0 = 1, b_1 = b_2 = 0$.

In this way, we find that when the input data is $b_2 b_1 b_0$ (where the b_i's are either 0 or 1) then the output voltage is
$$V_0 = (b_2 2^{-1} + b_1 2^{-2} + b_0 2^{-3}) V_{FS} \qquad (2.1)$$
where $V_{FS} = V_{ref}$ = full scale output voltage.

The circuit and the defining equation for an n-bit D/A converter easily follow from Fig. 2.7 and Eqn. (2.1) respectively.

A/D Converter Circuits

Most A/D converters use the principle of successive approximation. Figure 2.9 shows the organization of an A/D converter that uses this method. Its principal components are a D/A converter, a comparator, a successive approximation register (SAR), a clock, and control and status logic.

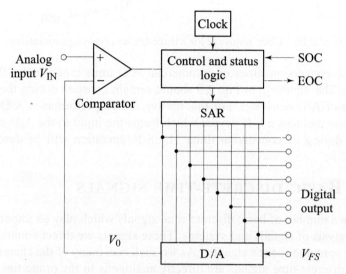

Fig. 2.9 Organization of a successive approximation A/D converter

On receiving the (start-of-conversion) SOC command, the SAR is cleared to 0's and its most significant bit is set to 1. This results in a V_0 value that is one-half of the full scale (refer Eqn. (2.1)). The output of the comparator is then tested to see whether V_{IN} is greater than or less than V_0. If V_{IN} is greater, the most significant bit is left on; otherwise it is turned off (complemented).

In the next step, the next most significant bit of the SAR is turned on. At this stage, V_0 will become either three-quarters or one-quarter of the full scale depending on whether V_{IN} was, respectively, greater than or less than V_0 in the first step. Again the comparator is tested and if V_{IN} is greater than the new V_0, the next most significant bit is left on. Otherwise it is turned off.

The process is repeated for each remaining SAR bit. When the process has been carried out for each bit, the SAR contains the binary number that is proportional to V_{IN}, and the (end-of-conversion) EOC line indicates that comparison has been completed and digital output is available for transmission. Figure 2.10 gives the code sequence for a three bit successive approximation.

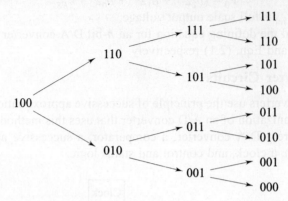

Fig. 2.10 Code sequence for a three-bit successive approximation

Typical conversion times of commercial A/D units range from 10 nsec to 200 μsec. The input V_{IN} in Fig. 2.9 should remain constant during the conversion time of A/D converter. For this reason, a high performance A/D conversion system includes a S/H device which keeps the input to the A/D converter constant during its conversion time. The S/H operation will be described in Section 2.8.

2.4 BASIC DISCRETE-TIME SIGNALS

There are a number of basic discrete-time signals which play an important role in the analysis of signals and systems. These signals are direct counterparts of the basic continuous-time signals.[2] As we shall see, many of the characteristics of basic discrete-time signals are directly analogous to the properties of basic continuous-time signals. There are, however, several important differences in discrete time, and we will point these out as we examine the properties of these signals.

Unit Sample Sequence

The unit sample sequence contains only one nonzero element and is defined by (Fig. 2.11a)

$$\delta(k) = \begin{cases} 1 & \text{for } k = 0 \\ 0 & \text{otherwise} \end{cases} \quad (2.2a)$$

The *delayed* unit sample sequence, denoted by $\delta(k - n)$, has its nonzero element at sample time n (Fig. 2.11b):

---

2. Section 2.5 of reference [180].

$$\delta(k-n) = \begin{cases} 1 & \text{for } k = n \\ 0 & \text{otherwise} \end{cases} \quad (2.2b)$$

One of the important aspects of the unit sample sequence is that an arbitrary sequence can be represented as a sum of scaled, delayed unit samples. For example, the sequence $r(k)$ in Fig. 2.11c can be expressed as

$$r(k) = r(0)\delta(k) + r(1)\delta(k-1) + r(2)\delta(k-2) + \cdots$$

$$= \sum_{n=0}^{\infty} r(n)\delta(k-n) \quad (2.3)$$

$r(0), r(1), \ldots$, are the sample values of the sequence $r(k)$. This representation of a discrete-time signal is found useful in the analysis of linear systems through the principle of superposition.

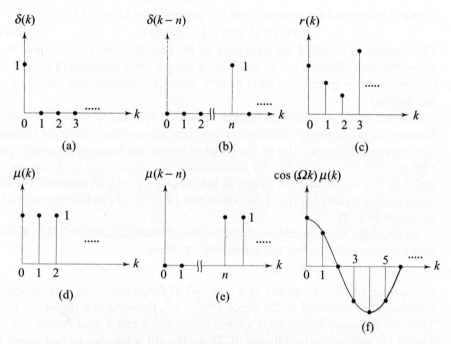

Fig. 2.11 Basic discrete-time signals

As we will see, the unit sample sequence plays the same role for discrete-time signals and systems that the unit impulse function does for continuous-time signals and systems. For this reason, the unit sample sequence is often referred to as the *discrete-time impulse*. It is important to note that a discrete-time impulse does not suffer from the mathematical complexity that a continuous-time impulse suffers from. Its definition is simple and precise.

Unit Step Sequence

The unit step sequence is defined as[3] (Fig. 2.11d)

$$\mu(k) = \begin{cases} 1 & \text{for } k \geq 0 \\ 0 & \text{otherwise} \end{cases} \quad (2.4)$$

The delayed unit step sequence, denoted by $\mu(k - n)$, has its first nonzero element at sample time n (Fig. 2.11e):

$$\mu(k - n) = \begin{cases} 1 & \text{for } k \geq n \\ 0 & \text{otherwise} \end{cases} \quad (2.5)$$

An arbitrary discrete-time signal $r(k)$ switched on to a system at $k = 0$ is represented as $r(k)\mu(k)$.

Sinusoidal Sequence

A one-sided sinusoidal sequence has the general form (Fig. 2.11f)

$$r(k) = A \cos(\Omega k + \phi) \mu(k) \quad (2.6)$$

The quantity Ω is called the *frequency* of the discrete-time sinusoid and ϕ is called the *phase*. Since k is a dimensionless integer, the dimension of Ω must be radians (we may specify the units of Ω to be radians/sample, and units of k to be samples).

The fact that k is always an integer in Eqn. (2.6) leads to some differences between the properties of discrete-time and continuous-time sinusoidal signals. An important difference lies in the range of values the frequency variable can take on. We know that for the continuous-time signal $r(t) = A \cos \omega t = $ real $\{Ae^{j\omega t}\}$, ω can take on values in the range $(-\infty, \infty)$. In contrast, for the discrete-time sinusoid $r(k) = A \cos \Omega k = $ real $\{Ae^{j\Omega k}\}$, Ω can take on values in the range $[-\pi, \pi]$.

To illustrate the property of discrete-time sinusoids, consider $\Omega = \pi + x$, where x is a small number compared with π. Since

$$e^{j\Omega k} = e^{j(\pi + x)k} = e^{j(2\pi - \pi + x)k} = e^{j(-\pi + x)k}$$

a frequency of $(\pi + x)$ results in a sinusoid of frequency $(-\pi + x)$. Suppose now that Ω is increased to 2π. Since $e^{j2\pi k} = e^{j0}$, the observed frequency is 0. Thus, the observed frequency is always between $-\pi$ and π, and is obtained by adding (or subtracting) multiples of 2π to Ω until a number in that range is obtained.

The highest frequency that can be represented by a digital signal is therefore π radians/sample interval. The implications of this property for sequences obtained by sampling sinusoids and other signals will be discussed in Section 2.9.

---

3. In discrete-time system theory, the unit step sequence is generally denoted by $u(k)$. In control theory, $u(k)$ is used to represent the control signal. In this book, $\mu(k)$ has been used to represent the unit step sequence while $u(k)$ denotes the control signal.

2.5 TIME-DOMAIN MODELS FOR DISCRETE-TIME SYSTEMS

A discrete-time system is defined mathematically as a transformation or an operator that maps an input sequence $r(k)$ into an output sequence $y(k)$. Classes of discrete-time systems are defined by placing constraints on the transformation. As they are relatively easy to characterize mathematically, and as they can be designed to perform useful signal processing functions, the class of *linear time-invariant* systems will be studied in this book. In the control structure of Fig. 2.2, the digital computer transforms an input sequence into a form which is in some sense more desirable. Therefore the discrete-time systems we consider here are, in fact, *computer programs*.

As we shall see, there is a similarity in the structure of models of continuous-time and discrete-time systems. This has resulted in the development of similar methods of analysis. For example, the simulation diagrams of discrete-time systems are similar to those for continuous-time systems, with only the dynamic element changed from an integrator to a delayer. The convolution summation is similar to convolution integral, and the z-transform method, tailored especially for linear discrete-time systems, bears many similarities to the Laplace transform. There are differences also between the properties of discrete-time and continuous-time systems. In this chapter, we are concerned with the analysis tools for discrete-time systems. Similarities with the tools for continuous-time systems will be obvious. The differences will be pointed out specifically.

For linear time-invariant discrete-time systems, four different ways of mathematical representation will be discussed. Time-domain models are described in this section, and a transform-domain model will be given in the next section.

State Variable Models

Consider a simple computer program expressed in MATLAB:

$$y(1) = 0$$
$$\text{for} \quad i = 2, N \quad\quad\quad (2.7)$$
$$y(i) = r(i-1) + 0.1 * y(i-1)$$
$$\text{end}$$

where $r(i)$ is the ith sample of the input sequence, $y(i)$ is the ith sample of the output sequence, and N is the total length of the signal record. We must define the value $y(1)$ in order to start signal processing. This value is the initial condition of the signal processor. For the signal processor represented by the computer program (2.7), the initial condition has been taken as zero.

In the computer program (2.7), the initial condition is represented by $y(1)$ and not by $y(0)$ because MATLAB does not allow arrays to be indexed starting with zero. For the analytical study of discrete-time systems, starting a sequence $y(k)$ with $y(0)$ is more convenient.

42 Digital Control and State Variable Methods

Simulation diagrams

It is obvious that the computer program (2.7) is characterized by the three basic operations:
(i) multiplication of a machine variable by a constant;
(ii) addition of several machine variables; and
(iii) storage of past values of machine variables.

These basic operations are diagrammatically represented in Fig. 2.12. The *unit delayer* represents a means for storing previous values of a sequence. If the signal $x_1(k); k \geq 0$ is the input to the unit delayer, its output sequence $x_2(k)$ has the sample values:

$$x_2(0) = \text{specified initial condition}$$
$$x_2(1) = x_1(0)$$
$$x_2(2) = x_1(1)$$
$$\vdots$$

A specified initial condition is stored before the commencement of the algorithm, in the appropriate register (of the digital computer) containing $x_2(\cdot)$. This can be diagrammatically represented by adding a signal $x_2(0)\delta(k)$ to the output of the delayer, where $\delta(k)$ is the unit sample sequence defined by Eqn. (2.2a).

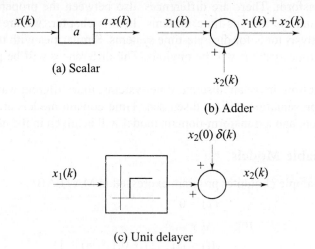

(a) Scalar

(b) Adder

(c) Unit delayer

Fig. 2.12 Basic building blocks of linear time-invariant discrete-time systems

The signal processing function performed by the computer program (2.7) can be represented by a block diagram shown in Fig. 2.13. Various blocks in this figure represent the basic computing operations of a digital computer. The unit delayer is the only *dynamic element* involved. The signal processing configuration of Fig. 2.13, thus, represents a first-order discrete-time system. The output $x(k)$ of the dynamic element gives the *state* of the system at any k. If the signal $r(k)$ is switched on to the system at $k = 0$ ($r(k) = 0$ for $k < 0$), the sample value $x(0)$ of the output sequence $x(k)$ represents the initial state of the system.

Since the initial state in the computer program (2.7) is zero, a signal of the form $x(0)\delta(k)$ does not appear in Fig. 2.13.

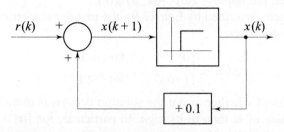

Fig. 2.13 A first-order linear discrete-time system

The defining equation for the computer program (2.7), obtained by forming an equation of the summing junction in Fig. 2.13, is

$$x(k + 1) = 0.1\, x(k) + r(k); \quad x(0) = 0 \qquad (2.8)$$

The solution of this first-order linear difference equation for given input $r(k)$ applied at $k = 0$, and given initial state $x(0)$, yields the state $x(k)$; $k > 0$. Equation (2.8) is thus the *state equation* of the discrete-time system of Fig. 2.13. Conversely, Fig. 2.13 is the *simulation diagram* for the mathematical model (2.8).

To solve an equation of the form (2.8) is an elementary matter. If k is incremented to take on values $k = 0, 1, 2, ...$, etc., the state $x(k)$; $k = 1, 2, ...$, can easily be generated by an iterative procedure. The iterative method however generates only a sequence of numbers and not a closed-form solution.

Example 2.1

In order to establish a feeling for the discrete-time systems, we study the signal processing algorithm given by the difference equation

$$x(k + 1) = -\alpha\, x(k) + r(k); \quad x(0) = 0 \qquad (2.9)$$

where α is a real constant.

We shall obtain a closed-form solution of this equation by what amounts to a brute force method (z-transform method of solving linear difference equations will appear in the next section). When solved repetitively, Eqn. (2.9) yields

$$\begin{aligned}
x(0) &= 0 \\
x(1) &= r(0) \\
x(2) &= -\alpha\, r(0) + r(1) \\
x(3) &= (-\alpha)^2 r(0) - \alpha\, r(1) + r(2) \\
&\vdots
\end{aligned}$$

The general term becomes ($r(k) = 0$ for $k < 0$),

$$x(k) = (-\alpha)^{k-1} r(0) + (-\alpha)^{k-2} r(1) + \cdots + r(k-1) \qquad (2.10)$$

Examining this equation, we note that the response $x(k)$ is a linear combination of the input samples $r(0), r(1), ..., r(k-1)$, and there appears to be a definite structure of the various weights.

The response of linear discrete-time systems to an impulse input $\delta(k)$ (defined in Eqn. (2.2a)) will be of special interest to us. Let us denote this response, called the *impulse response*, by $g(k)$.

For the system described by Eqn. (2.9), the impulse response obtained from Eqn. (2.10) is given by

$$g(k) = \begin{cases} 0 & \text{for } k = 0 \\ (-\alpha)^{k-1} & \text{for } k \geq 1 \end{cases} \quad (2.11)$$

The question of whether or not the solution decays, is more closely related to the *magnitude* of α than to its *sign*. In particular, for $|\alpha| > 1$, $g(k)$ grows with increasing k while it decays when $|\alpha| < 1$. The nature of time functions of the form (2.11) for different values of α will be examined in Section 2.7.

▲▲

A discrete-time system is completely characterized by the output variables of independent dynamic elements of the system. The outputs of independent dynamic elements thus constitute a set of characterizing variables of the system. The values of the characterizing variables at instant k describe the *state* of the system at that instant. These variables are therefore the *state variables* of the system.

The discrete-time system shown in Fig. 2.14 has two dynamic elements; the outputs $x_1(k)$ and $x_2(k)$ of these elements are therefore the state variables of the system. The following dynamical equations for the state variables easily follow from Fig. 2.14:

$$x_1(k + 1) = x_2(k); \quad x_1(0) = x_1^0 \quad (2.12a)$$

$$x_2(k + 1) = \alpha_1 x_1(k) + \alpha_2 x_2(k) + r(k); \quad x_2(0) = x_2^0$$

The solution of these equations for a given input $r(k)$ applied at $k = 0$, and given initial state $\{x_1^0, x_2^0\}$, yields the state $\{x_1(k), x_2(k)\}$, $k > 0$.

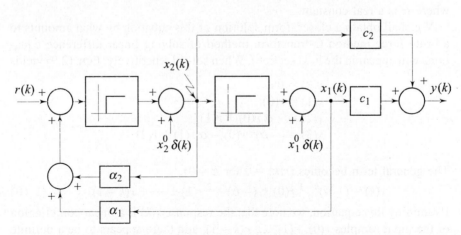

Fig. 2.14 A second-order discrete-time system

If $y(k)$ shown in Fig. 2.14 is the desired output information, we have the following algebraic relation to obtain $y(k)$:

$$y(k) = c_1 x_1(k) + c_2 x_2(k) \qquad (2.12b)$$

Equations (2.12a) are the *state equations*, and Eqn. (2.12b) is the *output equation* of the discrete-time system of Fig. 2.14.

In general, the state variable formulation may be visualized in block diagram form as shown in Fig. 2.15. We have depicted a multi-input, multi-output (MIMO) system which has p inputs, q outputs, and n state variables; the different variables are represented by the *input vector* $\mathbf{r}(k)$, the *output vector* $\mathbf{y}(k)$ and the state vector $\mathbf{x}(k)$, where

$$\mathbf{r}(k) \triangleq \begin{bmatrix} r_1(k) \\ r_2(k) \\ \vdots \\ r_p(k) \end{bmatrix} ; \mathbf{y}(k) \triangleq \begin{bmatrix} y_1(k) \\ y_2(k) \\ \vdots \\ y_q(k) \end{bmatrix} ; \mathbf{x}(k) \triangleq \begin{bmatrix} x_1(k) \\ x_2(k) \\ \vdots \\ x_n(k) \end{bmatrix}$$

Assuming that the input is switched on to the system at $k = 0$ ($\mathbf{r}(k) = \mathbf{0}$ for $k < 0$), the initial state is given by

$$\mathbf{x}(0) \triangleq \mathbf{x}^0, \text{ a specified } n \times 1 \text{ vector}$$

The dimension of the state vector defines the *order* of the system. The dynamics of an nth-order linear time-invariant system are described by equations of the form

$$\begin{aligned}
x_1(k+1) &= f_{11} x_1(k) + f_{12} x_2(k) + \cdots + f_{1n} x_n(k) + g_{11} r_1(k) \\
&\quad + g_{12} r_2(k) + \cdots + g_{1p} r_p(k) \\
x_2(k+1) &= f_{21} x_1(k) + f_{22} x_2(k) + \cdots + f_{2n} x_n(k) + g_{21} r_1(k) \\
&\quad + g_{22} r_2(k) + \cdots + g_{2p} r_p(k) \\
&\vdots \\
x_n(k+1) &= f_{n1} x_1(k) + f_{n2} x_2(k) + \cdots + f_{nn} x_n(k) + g_{n1} r_1(k) \\
&\quad + g_{n2} r_2(k) + \cdots + g_{np} r_p(k)
\end{aligned} \qquad (2.13)$$

where the coefficients f_{ij} and g_{ij} are constants.

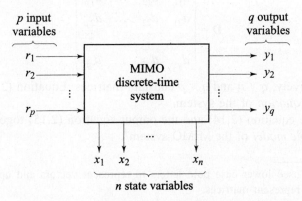

Fig. 2.15 Structure of a general discrete-time system

In the vector-matrix form, Eqns (2.13) may be written as

$$\mathbf{x}(k+1) = \mathbf{F}\mathbf{x}(k) + \mathbf{G}\mathbf{r}(k); \; \mathbf{x}(0) \triangleq \mathbf{x}^0 \qquad (2.14)$$

where

$$\mathbf{F} = \begin{bmatrix} f_{11} & f_{12} & \cdots & f_{1n} \\ f_{21} & f_{22} & \cdots & f_{2n} \\ \vdots & \vdots & & \vdots \\ f_{n1} & f_{n2} & \cdots & f_{nn} \end{bmatrix}$$

and

$$\mathbf{G} = \begin{bmatrix} g_{11} & g_{12} & \cdots & g_{1p} \\ g_{21} & g_{22} & \cdots & g_{2p} \\ \vdots & \vdots & & \vdots \\ g_{n1} & g_{n2} & \cdots & g_{np} \end{bmatrix}$$

are, respectively, $n \times n$ and $n \times p$ constant matrices. Equation (2.14) is called the *state equation* of the system.

The output variables at $t = kT$ are linear combinations of the values of the state variables and input variables at that time, i.e.,

$$\mathbf{y}(k) = \mathbf{C}\mathbf{x}(k) + \mathbf{D}\mathbf{r}(k) \qquad (2.15)$$

where

$$\mathbf{C} = \begin{bmatrix} c_{11} & c_{12} & \cdots & c_{1n} \\ c_{21} & c_{22} & \cdots & c_{2n} \\ \vdots & \vdots & & \vdots \\ c_{q1} & c_{q2} & \cdots & c_{qn} \end{bmatrix}$$

and

$$\mathbf{D} = \begin{bmatrix} d_{11} & d_{12} & \cdots & d_{1p} \\ d_{21} & d_{22} & \cdots & d_{2p} \\ \vdots & \vdots & & \vdots \\ d_{q1} & d_{q2} & \cdots & d_{qp} \end{bmatrix}$$

are, respectively, $q \times n$ and $q \times p$ constant matrices. Equation (2.15) is called the *output equation* of the system.

The state equation (2.14) and the output equation (2.15) together give the *state variable model* of the MIMO system[4]:

---

4. We have used lower case bold letters to represent vectors and upper case bold letters to represent matrices.

$$\mathbf{x}(k+1) = \mathbf{F}\mathbf{x}(k) + \mathbf{G}\mathbf{r}(k); \ \mathbf{x}(0) \triangleq \mathbf{x}^0 \qquad (2.16a)$$
$$\mathbf{y}(k) = \mathbf{C}\mathbf{x}(k) + \mathbf{D}\mathbf{r}(k) \qquad (2.16b)$$

For single-input ($p = 1$) and single-output ($q = 1$) system, the state variable model takes the form

$$\mathbf{x}(k+1) = \mathbf{F}\mathbf{x}(k) + \mathbf{g}\, r(k); \ \mathbf{x}(0) \triangleq \mathbf{x}^0 \qquad (2.17a)$$
$$y(k) = \mathbf{c}\mathbf{x}(k) + d\, r(k) \qquad (2.17b)$$

where $\mathbf{g}$ is $n \times 1$ column vector, $\mathbf{c}$ is $1 \times n$ row vector and d is a scalar:

$$\mathbf{g} = \begin{bmatrix} g_1 \\ g_2 \\ \vdots \\ g_n \end{bmatrix}; \ \mathbf{c} = [c_1 \ c_2 \ \cdots \ c_n]$$

Example 2.2

The discrete-time system of Fig. 2.16 has one dynamic element (unit delayer); it is therefore a first-order system. The state of the system at any k is described by $x(k)$—the output of the dynamic element.

The equation of the input summing junction is

$$x(k+1) = 0.95\, x(k) + r(k); \ x(0) = 0 \qquad (2.18a)$$

This is the state equation of the first-order system.

The output $y(k)$ is given by the following output equation:

$$y(k) = 0.0475\, x(k) + 0.05\, r(k) \qquad (2.18b)$$

Equations (2.18a) and (2.18b) together constitute the state variable model of the first-order system.

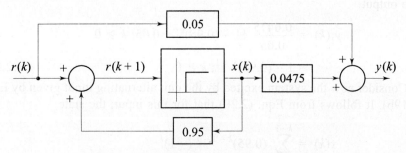

Fig. 2.16 A first-order linear discrete-time system

Let us study the response of the system of Fig. 2.16 to the unit step sequence

$$\mu(k) = \begin{cases} 1 & \text{for } k \geq 0 \\ 0 & \text{for } k < 0 \end{cases} \qquad (2.19a)$$

48 Digital Control and State Variable Methods

and the unit alternating sequence

$$r(k) = \begin{cases} (-1)^k & \text{for } k \geq 0 \\ 0 & \text{for } k < 0 \end{cases} \quad (2.19b)$$

We will first solve Eqn. (2.18a) for $x(k)$ and then use Eqn. (2.18b) to obtain $y(k)$.

The solution of Eqn. (2.18a) directly follows from Eqn. (2.10):

$$x(k) = (0.95)^{k-1} r(0) + (0.95)^{k-2} r(1) + \cdots + r(k-1)$$

$$= \sum_{i=0}^{k-1} (0.95)^{k-1-i} r(i) \quad (2.20)$$

For the unit step input given by Eqns (2.19a), we have[5]

$$x(k) = \sum_{i=0}^{k-1} (0.95)^{k-1-i}$$

$$= (0.95)^{k-1} \sum_{i=0}^{k-1} \left(\frac{1}{0.95}\right)^i$$

$$= (0.95)^{k-1} \left[\frac{1 - \left(\frac{1}{0.95}\right)^k}{1 - \frac{1}{0.95}} \right]$$

$$= \frac{1}{0.05} [1 - (0.95)^k]$$

The output

$$y_1(k) = \frac{0.0475}{0.05} [1 - (0.95)^k] + 0.05; \; k \geq 0$$

$$= 1 - (0.95)^{k+1}; \; k \geq 0 \quad (2.21)$$

Consider now the system excited by the unit alternating input given by Eqn. (2.19b). It follows from Eqn. (2.20) that for this input, the state

$$x(k) = \sum_{i=0}^{k-1} (0.95)^{k-1-i} (-1)^i$$

$$= \frac{1}{1.95} [(0.95)^k - (-1)^k]$$

---

5. $\sum_{j=0}^{k} a^j = \dfrac{1 - a^{k+1}}{1 - a}; a \neq 1$

The output
$$y_2(k) = 0.0475\, x(k) + 0.05\, (-1)^k$$
$$= \frac{0.05}{1.95}[(-1)^k + (0.95)^{k+1}];\; k \geq 0 \qquad (2.22)$$

From Eqns (2.21) and (2.22), we observe that the steady-state values of $y_1(k)$ and $y_2(k)$ are
$$y_1(k) = 1 \text{ for large } k$$
$$y_2(k) = \frac{1}{39}(-1)^k \text{ for large } k$$

Thus, the discrete-time system of Fig. 2.16 readily transmits a unit step and rejects a unit alternating input (reduces its magnitude by a factor of 39). Since the unit alternating signal is a rapidly fluctuating sequence of numbers while the unit step can be viewed as a slowly fluctuating signal, the discrete-time system of Fig. 2.16 represents a low-pass digital filter. In the next example, we will study the frequency-domain characteristics of this filter.

Difference Equation Models

Consider the single-input, single-output (SISO) system represented by the state model (2.17). The system has two types of inputs; the external input $r(k)$, and the initial state $\mathbf{x}(0)$ representing initial storage in the appropriate registers (of the digital computer) containing $x_i(.)$.

If the dynamic evolution of the state $\mathbf{x}(k)$ is not required, i.e., we are interested only in the input-output relation for $k \geq 0$, a linear time-invariant discrete-time system composed of n dynamic elements can be analysed using a single nth-order difference equation as its model. General form of nth-order linear difference equation relating output $y(k)$ to input $r(k)$ is given below:
$$y(k+n) + a_1 y(k+n-1) + \cdots + a_n y(k)$$
$$= b_0 r(k+m) + b_1 r(k+m-1) + \cdots + b_m r(k)$$

The coefficients a_i and b_j are real contants; k, m, and n are integers with $m \leq n$.

We will derive our results for the case of $m = n$, i.e., we consider the general linear difference equation in the following form:
$$y(k+n) + a_1 y(k+n-1) + \cdots + a_n y(k)$$
$$= b_0 r(k+n) + b_1 r(k+n-1) + \cdots + b_n r(k) \qquad (2.23)$$

There is no loss of generality in this assumption; the results for $m = n$ can be used for the case of $m < n$ by setting appropriate b_j coefficients to zero.

To solve an equation of the form (2.23) is an elementary matter. Substituting $k = 0$, we observe that the output at instant n is expressed in terms of n values of the past outputs: $y(0), y(1), ..., y(n-1)$, and in terms of inputs: $r(0), r(1), ..., r(n)$. If k is incremented to take on values $k = 0, 1, 2, ...,$ etc., the $y(k)$;

$k = n, n + 1, \ldots$, can easily be generated by the iterative procedure. The iterative method however generates only a sequence of numbers and not a closed-form solution.

Since the difference equation model (2.23) represents a time-invariant system, the choice of the initial point on the time scale is simply a matter of convenience in analysis. Shifting the origin from $k = n$ to $k = 0$, we get the equivalent difference equation model:

$$y(k) + a_1 y(k-1) + \cdots + a_n y(k-n)$$
$$= b_0 r(k) + b_1 r(k-1) + \cdots + b_n r(k-n) \qquad (2.24)$$

If the input is assumed to be switched on at $k = 0$ ($r(k) = 0$ for $k < 0$), then the difference equation model (2.24) gives the output at instant '0' in terms of the past values of the output; $y(-1), y(-2), \ldots, y(-n)$ and the present input $r(0)$. Thus the initial conditions of the model (2.24) are $\{y(-1), y(-2), \ldots, y(-n)\}$. These initial conditions uniquely determine the initial state $\mathbf{x}(0)$ of the corresponding state variable model (refer Review Example 2.2).

Given $\{y(-1), \ldots, y(-n)\}$, the initial conditions $\{y(0), \ldots, y(n-1)\}$ of the model (2.23) can be determined by successively substituting $k = -n, -n+1, \ldots, -2, -1$, in Eqn. (2.23) (refer Review Examples 2.1 and 2.3).

In this book, we have not accommodated the classical methods of solution of linear difference equations of the form (2.24) for given initial conditions and/or external inputs. Our approach is to transform the model (2.24) to other forms which are more convenient for analysis and design of digital control systems. Our emphasis is on the state variable models and transfer functions.

In Chapter 6, we will present methods of conversion of difference equation models of the form (2.24) to state variable models. We will use state variable models to obtain the system response to given initial conditions and external inputs; to construct digital computer simulation diagrams; and to design digital control algorithms using modern methods of design.

In the next section, we will present the z-transform technique for transforming difference equation model (2.24) to transfer function form. We will use transfer function models to study input-output behaviour of discrete-time systems; and to design digital control algorithms using classical methods of design.

Impulse Response Models

Consider the SISO system represented by the state model (2.17) or the difference equation model (2.24). The system has two types of inputs: the external input $r(k); k \geq 0$, and initial state $\mathbf{x}(0)$.

A system is said to be *relaxed* at $k = 0$ if the initial state $\mathbf{x}(0) = \mathbf{0}$. In terms of the representation (2.24), a system is relaxed if $y(k) = 0$ for $k < 0$.

We have earlier seen in Eqn. (2.3) that an arbitrary sequence $r(k)$ can be represented as a sum of scaled, delayed impulse sequences. It follows from this result that a linear time-invariant initially relaxed system can be completely characterized by its impulse response. This can be easily established.

Let $g(k)$ be the response of initially relaxed liner time-invariant discrete-time system to an impulse $\delta(k)$. Due to time-invariance property, the response to $\delta(k-n)$ will be $g(k-n)$. By linearity property, the response to an input signal $r(k)$ given by Eqn. (2.3) will be

$$y(k) = r(0)\,g(k) + r(1)\,g(k-1) + r(2)\,g(k-2) + \cdots$$

$$= \sum_{j=0}^{\infty} r(j)\,g(k-j); \quad k \geq 0 \qquad (2.25)$$

As a consequence of Eqn. (2.25), a linear time-invariant system is completely characterized by its impulse response $g(k)$ in the sense that given $g(k)$, it is possible to use Eqn. (2.25) to compute the output to any input $r(k)$. Equation (2.25) is commonly called the *convolution sum*.

It should be pointed out that the summation in Eqn. (2.25) is not really infinite in a practical situation, since for causal systems, $g(k-j)$ resulting from the input $\delta(k-j)$ is zero for $k < j$. This is because a causal system cannot respond until the input is applied. Thus for a causal system with input $r(k)\mu(k)$, Eqn. (2.25) modifies to

$$y(k) = \sum_{j=0}^{k} r(j)\,g(k-j); \quad k \geq 0 \qquad (2.26)$$

Another important observation concerns the symmetry of the situation. If we let $k - j = m$ in Eqn. (2.26), we get

$$y(k) = \sum_{m=k}^{0} r(k-m)\,g(m)$$

Reversing the order of summation,

$$y(k) = \sum_{m=0}^{k} g(m)\,r(k-m) \qquad (2.27)$$

The symmetry shows that we may reverse the roles of $r(\cdot)$ and $g(\cdot)$ in the convolution formula.

We may remind the reader here that whenever impulse response models are used to describe a system, the system is always implicitly assumed to be linear, time-invariant, and initially relaxed.

2.6 TRANSFER FUNCTION MODELS

The application of z-transform to Eqn. (2.25) gives an extremely useful mathematical description of a linear time-invariant discrete-time system. The fundamental definition for the z-transform[6] $F(z)$ of a discrete-time sequence $f(k)$; $k \geq 0$, is (refer Eqn. (A.36) in Appendix A),

$$F(z) \triangleq \mathscr{Z}[f(k)] \triangleq \sum_{k=0}^{\infty} f(k) z^{-k}; \qquad (2.28)$$

$z = a$ complex variable.

The application of z-transform to Eqn. (2.25) yields

$$Y(z) = \mathscr{Z}[y(k)]$$

$$= \sum_{k=0}^{\infty} y(k) z^{-k}$$

$$= \sum_{k=0}^{\infty} \left[\sum_{j=0}^{\infty} r(j) g(k-j) \right] z^{-k}$$

Changing the order of summations, gives

$$Y(z) = \sum_{j=0}^{\infty} r(j) z^{-j} \sum_{k=0}^{\infty} g(k-j) z^{-(k-j)}$$

Since $g(k-j) = 0$ for $k < j$, we can start the second summation at $k = j$. Then, defining the index $m = k - j$, we can write

$$Y(z) = \sum_{j=0}^{\infty} r(j) z^{-j} \sum_{m=0}^{\infty} g(m) z^{-m}$$

$$= R(z) G(z) \qquad (2.29)$$

where $\qquad R(z) \triangleq \mathscr{Z}[r(k)]$

and $\qquad G(z) \triangleq \mathscr{Z}[g(k)]$

We see that by applying the z-transform, a convolution sum is transformed into an algebraic equation. The function $G(z)$ is called the *transfer function* of the discrete-time system.

The transfer function of a linear time-invariant discrete-time system is, by definition, the z-transform of the impulse response of the system.

An alternative definition of transfer function follows from Eqn. (2.29).

$$G(z) = \left. \frac{\mathscr{Z}[y(k)]}{\mathscr{Z}[r(k)]} \right|_{\substack{\text{System} \\ \text{initially relaxed}}} = \left. \frac{Y(z)}{R(z)} \right|_{\substack{\text{System} \\ \text{initially relaxed}}} \qquad (2.30)$$

Thus, *the transfer function of a linear time-invariant discrete-time system is the ratio of the z-transforms of its output and input sequences, assuming that the system is initially relaxed.*

---

6. Use of z-transform technique for the analysis of discrete-time systems runs parallel to that of Laplace transform technique for continuous-time systems. The brief introduction to the theory of z-transforms, given in Appendix A, provides the working tools adequate for the purposes of this text.

Figure 2.17 gives the block diagram of a discrete-time system in transform domain.

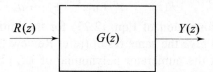

Fig. 2.17 Block diagram of a discrete-time system

Let us use the definition given by Eqn. (2.30) to obtain transfer function model of a discrete-time system represented by a difference equation of the form (2.24), relating its output $y(k)$ to the input $r(k)$. We assume that the discrete-time system is initially relaxed:

$$y(k) = 0 \text{ for } k < 0$$

and is excited by an input

$$r(k); k \geq 0$$

Therefore

and
$$y(-1) = y(-2) = \cdots = y(-n) = 0$$
$$r(-1) = r(-2) = \cdots = r(-n) = 0$$

z-transformation of delay operation requires the following result (Eqn. (A.39) in Appendix A):
Given

$$x(-1) = x(-2) = \cdots = x(-n) = 0,$$
$$\mathscr{Z}[x(k-n)] = z^{-n} X(z); n \geq 0$$

Taking z-transform of all the terms of Eqn. (2.24), under the assumption of zero initial conditions, we obtain

$$Y(z) + a_1 z^{-1} Y(z) + \cdots + a_n z^{-n} Y(z)$$
$$= b_0 R(z) + b_1 z^{-1} R(z) + \cdots + b_n z^{-n} R(z)$$

where

$$Y(z) \triangleq \mathscr{Z}[y(k)] \text{ and } R(z) \triangleq \mathscr{Z}[r(k)]$$

Solving for $Y(z)$:

$$Y(z) = \frac{(b_0 + b_1 z^{-1} + \cdots + b_n z^{-n}) R(z)}{1 + a_1 z^{-1} + \cdots + a_n z^{-n}}$$

Therefore, the transfer function $G(z)$ of the discrete-time system represented by difference Eqn. (2.24) is

$$G(z) = \frac{Y(z)}{R(z)} = \frac{b_0 + b_1 z^{-1} + \cdots + b_n z^{-n}}{1 + a_1 z^{-1} + \cdots + a_n z^{-n}} \qquad (2.31)$$

Equivalently, we can express the transfer function model as

$$G(z) = \frac{Y(z)}{R(z)} = \frac{b_0 z^n + b_1 z^{n-1} + \cdots + b_n}{z^n + a_1 z^{n-1} + \cdots + a_n} \qquad (2.32a)$$

Note that z-transformation of Eqn. (2.23) for the initially relaxed system ($y(k) = 0$, $k < 0$) will give the same result (refer Review Example 2.1).

We will represent the numerator polynomial of $G(z)$ by $N(z)$, and the denominator polynomial by $\Delta(z)$:

$$G(z) = \frac{N(z)}{\Delta(z)} \qquad (2.32b)$$

where

$$N(z) = b_0 z^n + b_1 z^{n-1} + \cdots + b_n$$
$$\Delta(z) = z^n + a_1 z^{n-1} + \cdots + a_n$$

The terminology used in connection with $G(s)$—the transfer function of continuous-time systems[7]—is directly applicable in the case of $G(z)$.

The highest power of the complex variable z in the denominator polynomial $\Delta(z)$ of the transfer function $G(z)$ determines the *order of the transfer function model*. The denominator polynomial $\Delta(z)$ is called the *characteristic polynomial*.

The roots of the equation

$$\Delta(z) = 0 \qquad (2.33a)$$

are called the *poles* of the transfer function $G(z)$, and roots of the equation

$$N(z) = 0 \qquad (2.33b)$$

are called the *zeros*.

Equation (2.33a) is called the *characteristic equation*; the poles are the *characteristic roots*.

By definition (refer Eqn. (A.35a) in Appendix A), the complex variable

$$z = r e^{j\Omega}$$

For a fixed value of r, the locus of z is a circle of radiuis r in the complex z-plane. Circle of radius unity in the complex z-plane is of specific interest to us. This circle is called a *unit circle* (Fig. 2.18).

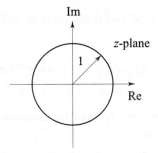

Fig. 2.18 Unit circle in z-plane

---

7. Chapter 2 of reference [180].

Signal Processing and Digital Control

In the next section, we will show that if the poles of the transfer function $G(z)$ of a discrete-time system lie inside the unit circle in the complex z-plane, the discrete-time system is stable.

Transfer Function of Unit Delayer

We now give a simple example of transfer function description of discrete-time systems. Figure 2.12 describes the basic operations characterizing a computer program. The unit delayer shown in this figure is a dynamic system with input $x_1(k)$ and output $x_2(k)$; $x_2(0)$ represents the initial storage in the shift register.

We assume that the discrete-time system (unit delayer) is initially relaxed:

$$x_2(0) = 0 \qquad (2.34a)$$

and is excited by an input sequence

$$x_1(k); \; k \geq 0 \qquad (2.34b)$$

The following state variable model gives the output of the unit delayer at $k = 0, 1, 2, \ldots$

$$x_2(k+1) = x_1(k) \qquad (2.35)$$

The z-transformation of Eqn. (2.35) yields

$$X_2(z) = z^{-1} X_1(z)$$

where

$$X_2(z) \triangleq \mathscr{Z}[x_2(k)]; \; X_1(z) \triangleq \mathscr{Z}[x_1(k)]$$

Therefore, the transfer function of the unit delayer represented by Eqn. (2.35) is

$$\frac{X_2(z)}{X_1(z)} = z^{-1} \qquad (2.36)$$

The discrete-time system of Fig. 2.16 may be equivalently represented by Fig. 2.19a using the transfer function description of the unit delayer. Use of the block-diagram analysis results in Fig. 2.19b, which gives

$$Y(z) = \left[\frac{0.0475}{z - 0.95} + 0.05 \right] R(z)$$

Therefore, the transfer function $G(z)$ of the discrete-time system of Fig. 2.19 is

$$G(z) = \frac{Y(z)}{R(z)} = \frac{0.05z}{z - 0.95} \qquad (2.37)$$

Example 2.3

The discrete-time system of Fig. 2.19 is described by the transfer function (refer Eqn. (2.37))

$$G(z) = \frac{Y(z)}{R(z)} = \frac{0.05z}{z - 0.95}$$

56 Digital Control and State Variable Methods

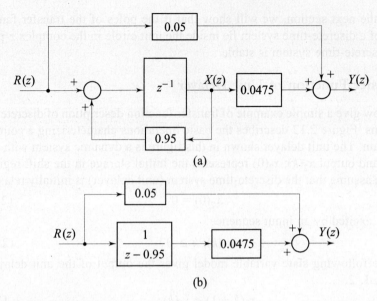

Fig. 2.19 Equivalent representations of the discrete-time system of Fig. 2.16

Find the response $y(k)$ to the input (i) $r(k) = \delta(k)$, (ii) $r(k) = \mu(k)$, (iii) $r(k) = R_0 \cos \Omega k$.

Solution

(i) Equation (2.2a) defines the discrete-time impulse $\delta(k)$. The z-transform of this elementary signal is (refer Table of transforms in Appendix A):

$$\mathscr{Z}[\delta(k)] = 1 \tag{2.38}$$

Letting $R(z) = 1$, we obtain

$$Y(z) = G(z) = \frac{0.05z}{z - 0.95}$$

The impulse response $g(k)$ is therefore (refer Table of transforms in Appendix A),

$$g(k) = \mathscr{Z}^{-1}\left[\frac{0.05z}{z - 0.95}\right]$$

$$= 0.05\,(0.95)^k; \ k \geq 0$$

(ii) Equation (2.4) defines the unit step $\mu(k)$. The z-transform of this elementary signal is (refer Table of transforms in Appendix A):

$$\mathscr{Z}[\mu(k)] = \frac{z}{z-1} \tag{2.39}$$

Letting $R(z) = \dfrac{z}{z-1}$, we obtain

$$Y(z) = \frac{0.05z^2}{(z-0.95)(z-1)}$$

Partial fraction expansion of $Y(z)/z$ is as follows (refer Eqns (A.14)–(A.16) in Appendix A).

$$\frac{Y(z)}{z} = \frac{1}{z-1} - \frac{0.95}{z-0.95}$$

Therefore,

$$Y(z) = \frac{z}{z-1} - \frac{0.95z}{z-0.95}$$

The inverse z-transform operation gives (refer Table of transforms in Appendix A)

$$y(k) = \mu(k) - 0.95(0.95)^k; \ k \geq 0$$

$$= \underbrace{1}_{\text{Steady-state component}} - \underbrace{(0.95)^{k+1}}_{\text{Transient component}}; \ k \geq 0$$

In this case, the transient component vanishes as $k \to \infty$ leaving behind the *steady-state* response.

The steady-state response can be quickly obtained without doing the complete inverse z-transform by use of the *final value theorem* (Eqn. (A.45) in Appendix A):

$$\lim_{k \to \infty} y(k) = \lim_{z \to 1} (z-1) \ Y(z) \tag{2.40}$$

if $(z-1) \ Y(z)$ has no poles on the boundary and outside of the unit circle in the complex z-plane.

(iii) The z-transform of the elementary signal a^k; $k \geq 0$, is (refer Table of transforms in Appendix A)

$$\mathscr{Z}[a^k] = \frac{z}{z-a} \tag{2.41}$$

For $a = e^{j\Omega}$,

$$\mathscr{Z}[e^{j\Omega k}] = \frac{z}{z - e^{j\Omega}} \tag{2.42}$$

The given input sequence is

$$r(k) = R_0 \cos(\Omega k)$$
$$= \text{Re } \{R_0 e^{j\Omega k}\}$$

The output is then given by

$$y(k) = \text{Re } \left\{ \mathscr{Z}^{-1}\left[G(z) \frac{R_0 \ z}{z - e^{j\Omega}} \right] \right\}$$

$$= \text{Re}\left\{\mathcal{Z}^{-1}\left[\frac{0.05 R_0 z^2}{(z-0.95)(z-e^{j\Omega})}\right]\right\}$$

$$= \text{Re}\left\{\mathcal{Z}^{-1}\left[\frac{0.0475 R_0}{0.95 - e^{j\Omega}}\left(\frac{z}{z-0.95}\right) + G(e^{j\Omega})\frac{R_0 z}{(z-e^{j\Omega})}\right]\right\}$$

where

$$G(e^{j\Omega}) = G(z)\bigg|_{z=e^{j\Omega}} = \frac{0.05 e^{j\Omega}}{e^{j\Omega} - 0.95}$$

The inverse z-transform operation yields

$$y(k) = \text{Re}\left[\underbrace{\frac{0.0475 R_0}{0.95 - e^{j\Omega}}(0.95)^k}_{\text{Transient component}} + \underbrace{G(e^{j\Omega}) R_0 e^{j\Omega k}}_{\text{Steady-state component}}\right]$$

This equation shows that as k increases, the transient component dies out. When this happens, the output expression becomes

$$y(k)\bigg|_{k \text{ very large}} = y_{ss}(k) = \text{Re}\{G(e^{j\Omega}) R_0 e^{j\Omega k}\}$$

Let
$$G(e^{j\Omega}) = |G(e^{j\Omega})| e^{j\phi}; \phi = \angle G(e^{j\Omega})$$
Then
$$y_{ss}(k) = \text{Re}\{R_0 |G(e^{j\Omega})| e^{j(\Omega k + \phi)}\}$$
$$= R_0 |G(e^{j\Omega})| \cos(\Omega k + \phi)$$

The steady-state response has the same form as the input (discrete sinusoidal), but is modified in amplitude by $|G(e^{j\Omega})|$ and in phase by $\angle G(e^{j\Omega})$.

The graphs of $|G(e^{j\Omega})|$ and $\angle G(e^{j\Omega})$ as the frequency Ω is varied, are the *frequency-response curves* of the given discrete-time system. The graphs are shown in Fig. 2.20. It is obvious from these curves that the given system is a low-pass digital filter.

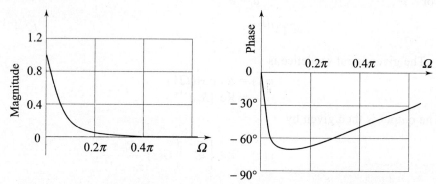

Fig. 2.20 Frequency-response curves of the discrete-time system of Fig. 2.19

2.7 STABILITY ON THE z-PLANE AND THE JURY STABILITY CRITERION

Stability is concerned with the qualitative analysis of the dynamic response of a system. This section is devoted to the stability analysis of linear time-invariant discrete-time systems. Stability concepts and definitions used in connection with continuous-time systems[8] are directly applicable here.

A linear time-invariant discrete-time system described by the state variable model (refer Eqns (2.17)),

$$\mathbf{x}(k+1) = \mathbf{F}\mathbf{x}(k) + \mathbf{g}r(k); \quad \mathbf{x}(0) \triangleq \mathbf{x}^0$$
$$y(k) = \mathbf{c}\mathbf{x}(k) + d\,r(k)$$

has two sources of excitation:
  (i) the initial state $\mathbf{x}^0$ representing initial internal energy storage; and
  (ii) the external input $r(k)$.

The system is said to be in *equilibrium state* $\mathbf{x}^e = \mathbf{0}$, when both the initial internal energy storage and the external input are zero.

In the stability study, we are generally concerned with the questions listed below.
  (i) If the system with zero input ($r(k) = 0;\ k \geq 0$) is perturbed from its equilibrium state $\mathbf{x}^e = \mathbf{0}$ at $k = 0$, will the state $\mathbf{x}(k)$ return to $\mathbf{x}^e$, remain 'close' to $\mathbf{x}^e$, or diverge from $\mathbf{x}^e$?
  (ii) If the system is relaxed, will a bounded input $r(k);\ k \geq 0$, produce a bounded output $y(k)$ for all k?

The first notion of stability is concerned with the 'boundedness' of the state of an unforced system in response to arbitrary initial state, and is called *zero-input stability*. The second notion is concerned with the boundedness of the output of a relaxed system in response to bounded input, and is called *bounded-input, bounded-output* (BIBO) *stability*.

BIBO Stability

A relaxed system (zero initial conditions) is said to be BIBO stable if for every bounded input $r(k);\ k \geq 0$, the output $y(k)$ is bounded for all k.

For a linear time-invariant system to satisfy this condition, it is necessary and sufficient that

$$\sum_{k=0}^{\infty} |g(k)| < \infty \qquad (2.43)$$

where $g(k)$ is the impulse response of the system.

To prove that condition (2.43) guarantees BIBO stability—i.e., sufficiency—we first establish an upper bound on $|y(k)|$.

---

8. Chapter 5 of reference [180].

From Eqn. (2.25), we can express the output as a convolution sum:

$$y(k) = \sum_{j=0}^{\infty} g(j)\, r(k-j)$$

If we consider the magnitude of the response $y(k)$, it is easy to see that

$$|y(k)| = \left|\sum_{j=0}^{\infty} g(j) r(k-j)\right|$$

which is surely less than the sum of the magnitudes as given by

$$|y(k)| \leq \sum_{j=0}^{\infty} |g(j)||r(k-j)|$$

Now take a bounded input, i.e.,

$$|r(k)| < M;\ 0 \leq k < \infty$$

where M is an arbitrary but finite positive constant. With this input,

$$|y(k)| \leq M \sum_{j=0}^{\infty} |g(j)|$$

and if condition (2.43) holds true, $|y(k)|$ is finite—hence the output is bounded and the system is BIBO stable.

The condition (2.43) is also necessary, for if we consider the bounded input

$$r(k-j) = \begin{cases} +1 & \text{if } g(j) > 0 \\ 0 & \text{if } g(j) = 0 \\ -1 & \text{if } g(j) < 0 \end{cases}$$

then the output at any fixed value of k is given by

$$|y(k)| = \left|\sum_{j=0}^{\infty} g(j) r(k-j)\right|$$

$$= \sum_{j=0}^{\infty} |g(j)|$$

Thus, unless the condition given by (2.43) is true, the system is not BIBO stable. ▲▲

The condition (2.43) for BIBO stability can be translated into a set of restrictions on the location of poles of the transfer function $G(z)$ in the z-plane. Consider the discrete-time system shown in Fig. 2.21. The block-diagram analysis gives the following input-output relation for this system.

$$\frac{Y(z)}{R(z)} = G(z) = \frac{z}{z^2 + a_1 z + a_2}$$

or

$$Y(z) = \frac{z}{z^2 + a_1 z + a_2} R(z)$$

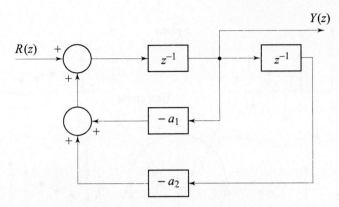

Fig. 2.21 A simple discrete-time system

For the impulse input, $R(z) = 1$. Therefore, response transform

$$Y(z) = \frac{z}{z^2 + a_1 z + a_2}$$

The impulse response of the system is given by

$$y(k) = g(k) = \mathscr{Z}^{-1}\left[\frac{z}{z^2 + a_1 z + a_2}\right]$$

Assume that the poles of the response transform $Y(z)$ are real and distinct:

$$z^2 + a_1 z + a_2 = (z - \alpha_1)(z - \alpha_2)$$

Partial fraction expansion of $Y(z)/z$ is then of the form

$$\frac{Y(z)}{z} = \frac{A_1}{z - \alpha_1} + \frac{A_2}{z - \alpha_2}$$

where A_1 and A_2 are real constants.
This gives

$$Y(z) = \frac{A_1 z}{z - \alpha_1} + \frac{A_2 z}{z - \alpha_2}$$

$$y(k) = A_1(\alpha_1)^k + A_2(\alpha_2)^k$$

The time functions $(\alpha_1)^k$ and $(\alpha_2)^k$ are the response functions contributed by the system poles at $z = \alpha_1$ and $z = \alpha_2$, respectively. These time functions dictate the qualitative nature of the impulse response of the system.

A time function $(\alpha)^k$ either grows or decays depending on $|\alpha| > 1$ or $|\alpha| < 1$ respectively. The growth or decay is monotonic when α is positive and alternates in sign when α is negative. $(\alpha)^k$ remains constant for $\alpha = 1$ and alternates in sign with constant amplitude for $\alpha = -1$ (Fig. 2.22).

Consider now the situation wherein the poles of response transform $Y(z)$ are complex.

For the complex-conjugate pole pair at $z = p = re^{j\Omega}$ and $z = p^* = re^{-j\Omega}$ of $Y(z)$, the response $y(k)$ is obtained as follows.

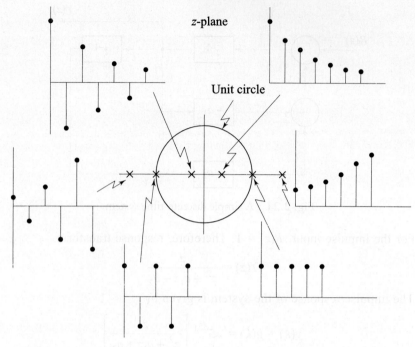

Fig. 2.22 The nature of response function $(\alpha)^k$ for different values of α

$$\frac{Y(z)}{z} = \frac{A}{z-p} + \frac{A^*}{z-p^*}$$

where $A = |A| \angle \phi$ and A^* is complex-conjugate of A.

The impulse response

$$y(k) = A(p)^k + A^*(p^*)^k = A(p)^k + [A(p)^k]^*$$
$$= 2\text{Re}\,[A(p)^k] = 2\text{Re}\,[|A|\,e^{j\phi}\,r^k\,e^{j\Omega k}]$$
$$= 2|A|\,r^k\,\text{Re}[e^{j(\Omega k + \phi)}] = 2|A|\,r^k\,\cos(\Omega k + \phi)$$

Therefore, the complex-conjugate pair of poles of the response transform $Y(z)$ gives rise to a sinusoidal or oscillatory response function $r^k \cos(\Omega k + \phi)$, whose envelope r^k can be constant, growing or decaying depending on whether $r = 1$, $r > 1$, or $r < 1$ respectively (Fig. 2.23).

For an nth order linear discrete-time system, the response transform $Y(z)$ has an nth-order characteristic polynomial. Assume that $Y(z)$ has a real pole at $z = \alpha$ of multiplicity m, and partial fraction expansion of $Y(z)$ is of the form

$$Y(z) = \frac{A_{1(m)}z}{(z-\alpha)^m} + \frac{A_{1(m-1)}z}{(z-\alpha)^{m-1}} + \cdots + \frac{A_{12}z}{(z-\alpha)^2} + \cdots + \frac{A_{11}z}{(z-\alpha)} + \cdots \quad (2.44)$$

where $A_{1(m)}, \ldots, A_{12}, A_{11}$ are real constants.

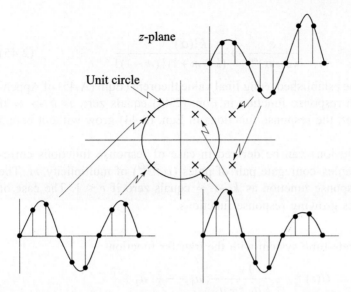

Fig. 2.23 The nature of response function $(r)^k \cos(\Omega k + \phi)$ for different values of r

Response functions contributed by the real pole of multiplicity m can be evaluated as follows.
Consider the transform pair

$$\frac{z}{z-\alpha} \leftrightarrow (\alpha)^k$$

Application of the delay theorem (Eqn. (A.39) of Appendix A) gives

$$\frac{1}{z-\alpha} \leftrightarrow \alpha^{k-1}$$

Using Eqn. (A.44) of Appendix A,

$$-z\frac{d}{dz}\left(\frac{1}{z-\alpha}\right) \leftrightarrow k\alpha^{k-1}$$

or

$$\frac{z}{(z-\alpha)^2} \leftrightarrow k\alpha^{k-1}$$

From this pair, we may write (Eqn. (A.39) in Appendix A),

$$\frac{1}{(z-\alpha)^2} \leftrightarrow (k-1)\alpha^{k-2}$$

Performing differentiation operation once again, we have

$$-z\frac{d}{dz}\left[\frac{1}{(z-\alpha)^2}\right] \leftrightarrow k(k-1)\alpha^{k-2}$$

or

$$\frac{z}{(z-\alpha)^3} \leftrightarrow \frac{1}{2!}k(k-1)\alpha^{k-2}$$

64 Digital Control and State Variable Methods

In general

$$\frac{z}{(z-\alpha)^m} \leftrightarrow \frac{k!(\alpha)^{k-m+1}}{(k-m+1)!(m-1)!} \qquad (2.45)$$

It can easily be established using final value theorem (Eqn. (A.45) of Appendix A) that each response function in Eqn. (2.44) equals zero as $k \to \infty$ if $|\alpha| < 1$. However, the response functions in Eqn. (2.44) grow without bound for $|\alpha| \geq 1$.

Similar conclusions can be derived in case of response functions corresponding to complex-conjugate pair of poles $(re^{\pm j\Omega})$ of multiplicity m. The limit of each response function as $k \to \infty$ equals zero if $r < 1$. The case of $r \geq 1$ contributes growing response functions.

Example 2.4

Consider a discrete-time system with the transfer function

$$G(z) = \frac{1}{z^2 + a_1 z + a_2}; \; a_1 = -\tfrac{3}{4}, \; a_2 = \tfrac{1}{8}$$

With the given values of the system parameters, the transfer function of the system becomes

$$\frac{Y(z)}{R(z)} = G(z) = \frac{1}{z^2 + a_1 z + a_2}$$

$$= \frac{1}{z^2 - \tfrac{3}{4} z + \tfrac{1}{8}} = \frac{1}{\left(z - \tfrac{1}{4}\right)\left(z - \tfrac{1}{2}\right)}$$

For impulse input, $R(z) = 1$ and therefore,

$$Y(z) = \frac{1}{\left(z - \tfrac{1}{4}\right)\left(z - \tfrac{1}{2}\right)} = \frac{4}{z - \tfrac{1}{2}} - \frac{4}{z - \tfrac{1}{4}}$$

$$y(k) = g(k) = 4(1/2)^{k-1} - 4(1/4)^{k-1}; \; k \geq 1$$

The impulse input thus excites the system poles without creating any additional response term.

Let us now study the response of the system to unit step input $r(k) = \mu(k)$. For this input

$$R(z) = \frac{z}{z-1}$$

Therefore, response transform

$$Y(z) = \frac{1}{\left(z - \tfrac{1}{4}\right)\left(z - \tfrac{1}{2}\right)} \left[\frac{z}{z-1}\right] = \frac{z}{\left(z - \tfrac{1}{4}\right)\left(z - \tfrac{1}{2}\right)(z-1)}$$

$$= \underbrace{\frac{\tfrac{16}{3} z}{z - \tfrac{1}{4}} + \frac{-8z}{z - \tfrac{1}{2}}}_{\text{System poles}} + \underbrace{\frac{\tfrac{8}{3} z}{z - 1}}_{\text{Excitation pole}}$$

The inverse transform operation gives

$$y(k) = \left[\underbrace{\tfrac{16}{3}\left(\tfrac{1}{4}\right)^k - 8\left(\tfrac{1}{2}\right)^k}_{\text{Transient response}} + \underbrace{\tfrac{8}{3}(1)^k}_{\text{Steady-state response}}\right]$$

The *transient response* terms correspond to system poles excited by the input $\mu(k)$. These terms vanish as $k \to \infty$.

The second response term arises due to the excitation pole, and has the same nature as the input itself except for a modification in magnitude caused by the system's behaviour to the specified input. Since the input exists as $k \to \infty$, the second response term does not vanish and is called the *steady-state response* of the system.

From the foregoing discussion it follows that the nature of the response terms contributed by the system poles (i.e., the poles of the transfer function $G(z)$) gives the nature of the impulse response $g(k)$ (= $\mathscr{Z}^{-1}[G(z)]$) of the system. This therefore answers the question of BIBO stability through condition (2.43), which says that for a system with transfer function $G(z)$ to be BIBO stable, it is necessary and sufficient that

$$\sum_{k=0}^{\infty} |g(k)| < \infty$$

The nature of response terms contributed by various types of poles of $G(z)$ = $\dfrac{N(z)}{\Delta(z)}$, i.e., the roots of the characteristic equation $\Delta(z) = 0$, has already been investigated. Table 2.1 summarizes the results.

Certain observations are easily made from the study of Table 2.1. All the roots for which $|z| \neq 1$ [cases (i), (ii), (iii), and (iv)], contribute response terms with a multiplying factor of either α^k or r^k. If $|\alpha| < 1$ and /or $r < 1$ (i.e., the roots lie inside the unit circle), the response terms vanish as $k \to \infty$, and if $|\alpha| > 1$ and/or $r > 1$ (i.e., the roots lie outside the unit circle), the response terms increase without bound. Roots on the unit circle with multiplicity two or higher [cases (vi) and (viii)] also contribute terms which increase without bound as $k \to \infty$. Single root on the unit circle [case (v)] or non-multiple root pairs on the unit circle [case (vii)] contribute terms which are constant amplitude or constant amplitude oscillation. These observations lead us to the following general conclusions on BIBO stability.

(i) If all the roots of the characteristic equation lie inside the unit circle in the z-plane, then the impulse response is bounded and eventually decays to zero. Therefore $\sum_{k=0}^{\infty} |g(k)|$ is finite and the system is BIBO stable.

(ii) If any root of the characteristic equation lies outside the unit circle in the

Table 2.1 Response terms contributed by various types of roots.

Type of roots	Nature of response terms
(i) Single root at $z = \alpha$	$A\alpha^k$
(ii) Roots of multiplicity m at $z = \alpha$	$\left[A_1 + A_2 k + A_3 k(k-1) + \cdots + A_m \dfrac{k!}{(k-m+1)!(m-1)!}\right]\alpha^k$
(iii) Complex-conjugate root pair at $z = re^{\pm j\Omega}$	$Ar^k \cos(\Omega k + \phi)$
(iv) Complex-conjugate root pairs of multiplicity m at $z = re^{\pm j\Omega}$	$\left[A_1 \cos(\Omega k + \phi_1) + A_2 k \cos(\Omega k + \phi_2) + \cdots + A_m \dfrac{k!\cos(\Omega k + \phi_m)}{(k-m+1)!(m-1)!}\right]r^k$
(v) Single root on the unit circle (i.e., $\|z\|=1$)	A
(vi) Roots of multiplicity m on the unit circle	$\left[A_1 + A_2 k + A_3 k(k-1) + \cdots + \dfrac{A_m k!}{(k-m+1)!(m-1)!}\right]$
(vii) Single complex-conjugate root pair on the unit circle (i.e., $z = e^{\pm j\Omega}$)	$A \cos(\Omega k + \phi)$
(viii) Complex-conjugate root pairs of multiplicity m on the unit circle	$\left[A_1 \cos(\Omega k + \phi_1) + A_2 k \cos(\Omega k + \phi_2) + \cdots + \dfrac{A_m k!}{(k-m+1)!(m-1)!}\cos(\Omega k + \phi_m)\right]$

z-plane, $g(k)$ grows without bound and $\sum_{k=0}^{\infty} |g(k)|$ is infinite. The system is therefore unstable.

(iii) If the characteristic equation has repeated roots on the unit circle in the z-plane, $g(k)$ grows without bound and $\sum_{k=0}^{\infty} |g(k)|$ is infinite. The system is therefore unstable.

(iv) If one or more non-repeated roots of the characteristic equation are on the unit circle in the z-plane, then $g(k)$ is bounded but $\sum_{k=0}^{\infty} |g(k)|$ is infinite. The system is therefore unstable.

An exception to the definition of BIBO stability is brought out by the following observations.

Consider a system with transfer function

$$G(z) = \frac{N(z)}{(z-1)(z-e^{j\Omega})(z-e^{-j\Omega})}$$

The system has non-repeated poles on the unit circle in the z-plane. The response functions contributed by the system poles at $z = 1$ and $z = e^{\pm j\Omega}$ are respectively $(1)^k$ and $\cos(\Omega k + \phi)$. The terms $(1)^k$ and $\cos(\Omega k + \phi)$ are bounded, $\sum_{k=0}^{\infty} |g(k)|$ is infinite and the system is unstable in the sense of our definition of BIBO stability.

Careful examination of the input-output relation

$$Y(z) = G(z)R(z) = \frac{N(z)}{(z-1)(z-e^{j\Omega})(z-e^{-j\Omega})} R(z)$$

shows that $y(k)$ is bounded for all bounded $r(k)$ unless the input has a pole matching one of the system poles on the unit circle. For example, for a unit step input $r(k) = \mu(k)$,

$$R(z) = \frac{z}{z-1} \text{ and } Y(z) = \frac{zN(z)}{(z-1)^2 (z-e^{j\Omega})(z-e^{-j\Omega})}$$

The response $y(k)$ is a linear combination of the terms $\cos(\Omega k + \phi)$, $(1)^k$, and $k(1)^k$, and therefore $y(k) \to \infty$ as $k \to \infty$. Such a system which has bounded output for all bounded inputs, except for the inputs having poles matching the system poles, may be treated as acceptable or non-acceptable. We will bring the situations where the system has non-repeated poles on the unit circle under the class of *marginally stable systems*.

Zero-input Stability

This concept of stability is based on the dynamic evolution of the system state in response to arbitrary initial state representing initial internal energy storage. State variable model (refer Eqn. (2.17))

$$\mathbf{x}(k+1) = \mathbf{F}\mathbf{x}(k) \tag{2.46}$$

is the most appropriate for the study of dynamic evolution of the state $\mathbf{x}(k)$ in response to the initial state $\mathbf{x}(0)$.

We may classify stability as follows:

(i) *Unstable:* There is at least one finite initial state $\mathbf{x}(0)$ such that $\mathbf{x}(k)$ grows thereafter without bound as $k \to \infty$.
(ii) *Asymptotically stable:* For all possible initial states $\mathbf{x}(0)$, $\mathbf{x}(k)$ eventually decays to zero as $k \to \infty$.

(iii) *Marginally stable:* For all initial states $\mathbf{x}(0)$, $\mathbf{x}(k)$ remains thereafter within finite bounds for $k > 0$.

Taking z-transform on both the sides of Eqn. (2.46) yields
$$z\,\mathbf{X}(z) - z\,\mathbf{x}(0) = \mathbf{F}\mathbf{X}(z)$$
where
$$\mathbf{X}(z) \triangleq \mathscr{Z}[\mathbf{x}(k)]$$
Solving for $\mathbf{X}(z)$, we get
$$\mathbf{X}(z) = (z\mathbf{I} - \mathbf{F})^{-1} z\mathbf{x}(0) = \mathbf{\Phi}(z)\,\mathbf{x}(0)$$
where
$$\mathbf{\Phi}(z) = (z\mathbf{I} - \mathbf{F})^{-1} z = \frac{(z\mathbf{I} - \mathbf{F})^{+} z}{|z\mathbf{I} - \mathbf{F}|} \qquad (2.47a)$$

The state vector $\mathbf{x}(k)$ can be obtained by inverse transforming $\mathbf{X}(z)$:
$$\mathbf{x}(k) = \mathscr{Z}^{-1}[\mathbf{\Phi}(z)]\,\mathbf{x}(0) \qquad (2.47b)$$

Note that for an $n \times n$ matrix $\mathbf{F}$, $|z\mathbf{I} - \mathbf{F}|$ is an nth-order polynomial in z. Also, each element of the adjoint matrix $(z\mathbf{I} - \mathbf{F})^{+}$ is a polynomial in z of order less than or equal to $(n-1)$. Therefore, each element of $\mathbf{\Phi}(z)/z$ is strictly proper rational function, and can be expanded in a partial fraction expansion. Using the time-response analysis given earlier in this section, it is easy to establish that
$$\lim_{k \to \infty} \mathbf{x}(k) \to \mathbf{0}$$
if all the roots of the characteristic polynomial $|z\mathbf{I} - \mathbf{F}|$ lie strictly inside the unit circle of the complex plane. In Chapter 6 we will see that under mildly restrictive conditions (namely the system must be both controllable and observable), the roots of the characteristic polynomial $|z\mathbf{I} - \mathbf{F}|$ are same as the poles of the corresponding transfer function, and asymptotic stability ensures BIBO stability and *vice versa*. This implies that stability analysis can be carried out using the BIBO stability test (or only the asymptotic stability test).

We will use the following terminology and tests for stability analysis of linear time-invariant systems described by the transfer function $G(z) = N(z)/\Delta(z)$, with the characteristic equation $\Delta(z) = 0$.
  (i) If all the roots of the characteristic equation lie inside the unit circle in the z-plane, the system is *stable*.
  (ii) If any root of the characteristic equation lies outside the unit circle in the z-plane, or if there is a repeated root on the unit circle, the system is *unstable*.
  (iii) If condition (i) is satisfied except for the presence of one or more non-repeated roots on the unit circle in the z-plane, the system is *marginally stable*.

It follows from the above discussion that stability can be established by determining the roots of the characteristic equations. All the commercially available CAD packages ([176–179]) include root-solving routines. However, there exist tests for determining the stability of a discrete-time system without finding the actual numerical values of the roots of the characteristic equation. We shall briefly discuss some of these tests from the point of view of their usefulness for the design of digital control systems.

A well known criterion to test the location of zeros of the polynomial

$$\Delta(z) = a_0 z^n + a_1 z^{n-1} + \cdots + a_{n-1} z + a_n$$

where a's are real coefficients, is the Jury stability criterion. The proof of this criterion is quite involved and is given in the literature (Jury and Blanchard [101]). The criterion gives the necessary and sufficient conditions for the roots to lie inside the unit circle. In the following, we present the Jury stability criterion without proof. An alternative stability test based on the bilinear transformation coupled with the Routh stability criterion, will be outlined in a later section.

The Jury Stability Criterion

In applying the Jury stability criterion to a given characteristic equation $\Delta(z) = 0$, we construct a table whose elements are based on the coefficients of $\Delta(z)$.

Consider the general form of the characteristic polynomial $\Delta(z)$ (refer Eqn. (2.32)):

$$\Delta(z) = a_0 z^n + a_1 z^{n-1} + \cdots + a_k z^{n-k} + \cdots + a_{n-1} z + a_n; \quad a_0 > 0 \quad (2.48)$$

The criterion uses the Jury stability table given in Table 2.2.

Table 2.2 General form of the Jury stability table

Row	z^0	z^1	z^2	z^3	...	z^k	...	z^{n-2}	z^{n-1}	z^n
1	a_n	a_{n-1}	a_{n-2}	a_{n-3}	...	a_{n-k}	...	a_2	a_1	a_0
2	a_0	a_1	a_2	a_3	...	a_k	...	a_{n-2}	a_{n-1}	a_n
3	b_{n-1}	b_{n-2}	b_{n-3}	b_{n-4}	...	b_{n-k-1}	...	b_1	b_0	
4	b_0	b_1	b_2	b_3	...	b_k	...	b_{n-2}	b_{n-1}	
5	c_{n-2}	c_{n-3}	c_{n-4}	c_{n-5}	...	c_{n-k-2}	...	c_0		
6	c_0	c_1	c_2	c_3	...	c_k	...	c_{n-2}		
⋮										
$2n-5$	p_3	p_2	p_1	p_0						
$2n-4$	p_0	p_1	p_2	p_3						
$2n-3$	q_2	q_1	q_0							

The Jury stability table is formed using the following rules.

(i) The first two rows of the table consist of the coefficients of $\Delta(z)$, arranged in ascending order of power of z in row 1 and in reverse order in row 2.

(ii) All even-numbered rows are simply the reverse of the immediately preceding odd-numbered rows.

(iii) The elements for rows 3 through $(2n - 3)$ are given by the following determinants:

$$b_k = \begin{vmatrix} a_n & a_{n-1-k} \\ a_0 & a_{k+1} \end{vmatrix} ; k = 0, 1, 2, ..., n - 1$$

$$c_k = \begin{vmatrix} b_{n-1} & b_{n-2-k} \\ b_0 & b_{k+1} \end{vmatrix} ; k = 0, 1, 2, ..., n - 2 \qquad (2.49)$$

$$\vdots$$

$$q_k = \begin{vmatrix} p_3 & p_{2-k} \\ p_0 & p_{k+1} \end{vmatrix} ; k = 0, 1, 2$$

The procedure is continued until the $(2n - 3)$rd row is reached which will contain exactly three elements.

The necessary and sufficient conditions for polynomial $\Delta(z)$ to have no roots on and outside the unit circle in the z-plane are:

$$\Delta(1) > 0$$

$$\Delta(-1) \begin{cases} > 0 \text{ for } n \text{ even} \\ < 0 \text{ for } n \text{ odd} \end{cases} \qquad (2.50a)$$

$$|a_n| < |a_0|$$

$$\left. \begin{array}{c} |b_{n-1}| > |b_0| \\ |c_{n-2}| > |c_0| \\ \vdots \\ |q_2| > |q_0| \end{array} \right\} (n-2) \text{ constraints} \qquad (2.50b)$$

The conditions on $\Delta(1)$, $\Delta(-1)$, and between a_0 and a_n in (2.50a) form *necessary conditions* of stability that are very simple to check without carrying out the Jury tabulation.

It should be noted that the test of stability given in (2.50) is valid only if the inequality conditions provide conclusive results. Jury tabulation ends prematurely if either the first and the last elements of a row are zero, or a complete row is zero. These cases are referred to as *singular cases*. These problems can be resolved by expanding and contracting the unit circle infinitesimally, which is equivalent to moving the roots off the unit circle. The transformation for this purpose is

$$\hat{z} = (1 + \varepsilon)z$$

where ε is a very small real number.

This transformation can easily be applied since

$$(1 + \varepsilon)^n z^n \cong (1 + n\varepsilon)z^n$$

When ε is a positive number, the radius of the unit circle is expanded to $(1 + \varepsilon)$, and when ε is negative, the radius of the unit circle is reduced to $(1 - |\varepsilon|)$. This is equivalent to moving the roots slightly. The difference between the number of roots found inside (or outside) the unit circle when the circle is expanded and contracted by ε is the number of roots on the circle [100].

Example 2.5

Consider the characteristic polynomial
$$\Delta(z) = 2z^4 + 7z^3 + 10z^2 + 4z + 1$$
Employing stability constraints (2.50a), we get
  (i) $\Delta(1) = 2 + 7 + 10 + 4 + 1 = 24 > 0$; satisfied
  (ii) $\Delta(-1) = 2 - 7 + 10 - 4 + 1 = 2 > 0$; satisfied
  (iii) $|1| < |2|$; satisfied

Next we construct the Jury table:

Row	z^0	z^1	z^2	z^3	z^4
1	1	4	10	7	2
2	2	7	10	4	1
3	-3	-10	-10	-1	
4	-1	-10	-10	-3	
5	8	20	20		

Employing stability constraints (2.50b), we get
  (i) $|-3| > |-1|$; satisfied
  (ii) $|8| > |20|$; not satisfied

The system is therefore unstable.

▲▲

Usefulness of the Jury stability test for the design of a digital control system from the stability point of view, will be demonstrated in the next chapter.

2.8 SAMPLING AS IMPULSE MODULATION

In previous sections of this chapter, we have refrained from relating discrete-time and continuous-time signals and systems except for pointing out some of the similarities between important theoretical concepts. Often, however, discrete-time signals are derived from continuous-time signals by periodic sampling. Hence, it is important to understand how the sequence so derived is related to the original signal.

Consider an analog signal $x_a(t)$. The sequence $x(k)$ with values $x(k) = x_a(kT)$ is said to be derived from $x_a(t)$ by periodic sampling and T is called the *sampling period*. The reciprocal of T is called the *sampling frequency* or the *sampling rate*. In a typical digital control scheme shown in Fig. 2.2, the operation of deriving a sequence from a continuous-time signal is performed by an A/D converter. A simple symbolic representation of the sampling operation is shown in Fig. 2.24.

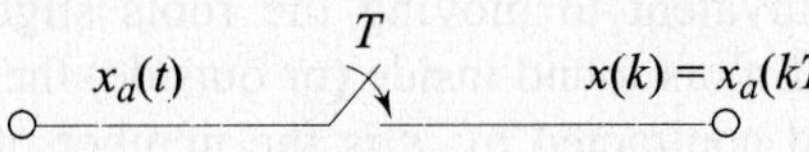

Fig. 2.24 Symbolic representation of the sampling operation

To establish a relationship of the sequence $x(k)$ to the continuous-time function $x_a(t)$ from which this sequence is derived, we take the following approach. We treat each sample of the sequence $x(k)$ as an impulse function of strength equal to the value of the sample (Impulse function $A\delta(t - t_0)$ is an impulse of strength A occurring at $t = t_0$). The idea is to give a mathematical description to periodic samples of a continuous-time function in such a way that we can analyse the samples and the function simultaneously using the same tool (Laplace transform). The sequence $x(k)$ can be viewed as a train of impulses represented by continuous-time function $x^*(t)$:

$$x^*(t) = x(0)\delta(t) + x(1)\delta(t - T) + x(2)\delta(t - 2T) + \cdots$$

$$= \sum_{k=0}^{\infty} x(k)\delta(t - kT) \qquad (2.51)$$

Typical signals $x_a(t)$, $x(k)$ and $x^*(t)$ are shown in Fig. 2.25. The sampler of Fig. 2.24 can thus be viewed as an 'impulse modulator' with the carrier signal

$$\delta_T(t) = \sum_{k=0}^{\infty} \delta(t - kT)$$

and modulating signal $x_a(t)$. The modulation process is schematically represented in Fig. 2.26a and the *impulse train* $\delta_T(t)$ in Fig. 2.26b.

The Laplace transform of $x^*(t)$ is

$$\mathscr{L}[x^*(t)] = X^*(s) \triangleq \int_0^{\infty} x^*(t)\,e^{-st}\,dt$$

$$= \int_0^{\infty} \sum_{k=0}^{\infty} x(k)\delta(t - kT)\,e^{-st}\,dt = \sum_{k=0}^{\infty} x(k)\,e^{-skT} \qquad (2.52)$$

The notation $X^*(s)$ is used to symbolize the (Laplace) transform of $x^*(t)$ — the impulse modulated $x_a(t)$.

It is important to emphasize here that the impulse modulation model is a mathematical representation of sampling; not a representation of any physical system designed to implement the sampling operation. We have introduced this representation of the sampling operation because it leads to a simple derivation of a key result on sampling (given in the next section) and because this approach allows us to obtain a transfer function model of the hold operation.

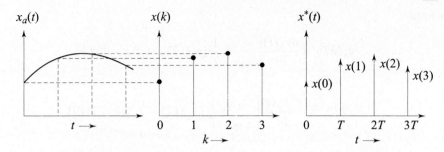

Fig. 2.25 Conversion of a continuous-time signal to a sequence

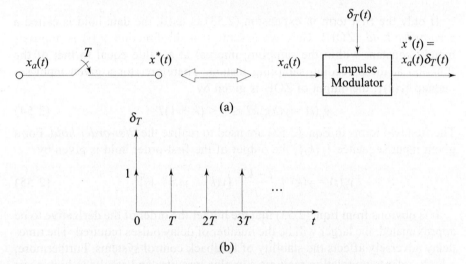

Fig. 2.26 (a) Representation of sampling as impulse modulator
(b) Representation of impulse train

The Hold Operation

It is the inverse of the sampling operation—conversion of a sequence to a continuous-time function. In computer-controlled systems, it is necessary to convert the control actions calculated by the computer as a sequence of numbers, to a continuous-time signal that can be applied to the process.

The problem of hold operation may be posed as follows:
Given a sequence $\{y(0), y(1), ..., y(k), ...\}$, we have to construct $y_a(t)$, $t \geq 0$.

A commonly used solution to the problem of hold operation is polynomial extrapolation. Using a Taylor's series expansion about $t = kT$, we can express $y_a(t)$ as

$$y_a(t) = y_a(kT) + \dot{y}_a(kT)(t - kT) + \frac{\ddot{y}_a(kT)}{2!}(t - kT)^2 + \cdots;$$
$$kT \leq t < (k+1)T \qquad (2.53)$$

where

$$\dot{y}_a(kT) \triangleq \frac{dy_a(t)}{dt}\bigg|_{t=kT} \cong \frac{1}{T}\{y_a(kT) - y_a[(k-1)T]\}$$

$$\ddot{y}_a(kT) \triangleq \frac{d^2 y_a(t)}{dt^2}\bigg|_{t=kT} \cong \frac{1}{T}\{\dot{y}_a(kT) - \dot{y}_a[(k-1)T]\}$$

$$= \frac{1}{T^2}\{y_a(kT) - 2y_a[(k-1)T] + y_a[(k-2)T]\}$$

If only the first term in expansion (2.53) is used, the data hold is called a *zero-order hold* (ZOH). Here we assume that the function $y_a(t)$ is approximately constant within the sampling interval at a value equal to that of the function at the preceding sampling instant. Therefore, for a given input sequence $\{y(k)\}$, the output of ZOH is given by

$$y_a(t) = y(k); \ kT \le t < (k+1)T \tag{2.54}$$

The first two terms in Eqn. (2.53) are used to realize the *first-order hold*. For a given input sequence $\{y(k)\}$, the output of the first-order hold is given by

$$y_a(t) = y(k) + \frac{t-kT}{T}[y(k) - y(k-1)] \tag{2.55}$$

It is obvious from Eqn. (2.53) that the higher the order of the derivative to be approximated, the larger will be the number of delay pulses required. The time-delay adversely affects the stability of feedback control systems. Furthermore, a high-order extrapolation requires complex circuitry and results in high costs of construction. The ZOH is the simplest and most commonly used data hold device. The standard D/A converters are often designed in such a way that the old value is held constant until a new conversion is ordered.

A Model of Sample-and-Hold Operation

In the digital control structure of Fig. 2.2, discrete-time processing of continuous-time signals is accomplished by the system depicted in Fig. 2.27. The system is a cascade of an A/D converter followed by a discrete-time system (computer program) followed by a D/A converter. Note that the overall system is equivalent to a continuous-time system since it transforms the continuous time input signal $x_a(t)$ into the continuous-time signal $\bar{y}_a(t)$. However, the properties of the system are dependent on the choice of the discrete-time system and the sampling rate.

In the special case of discrete-time signal processing with a unit-gain algorithm and negligible time delay (i.e., $y(k) = x(k)$), the combined action of the A/D converter, the computer, and the D/A converter can be described as a system that samples the analog signal and produces another analog signal that is constant over the sampling periods. Such a system is called a *sample-and-hold* (S/H). Input-output behaviour of a S/H system is described diagrammatically in Fig. 2.28. In the following, we develop an idealized model for S/H systems.

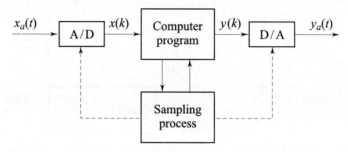

Fig. 2.27 Discrete-time processing of continuous-time signals

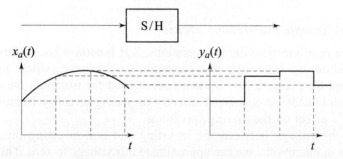

Fig. 2.28 Input-output behaviour of a sample-and-hold system

S/H operations require modelling of two processes:
(i) extracting the samples, and
(ii) holding the result fixed for one period.

The impulse modulator effectively extracts the samples in the form of $x(k)\delta(t - kT)$. The remaining problem is to construct a linear time-invariant system which will convert this impulse into a pulse of height $x(k)$ and width T. The S/H may therefore be modelled by Fig. 2.29a, wherein the ZOH is a system whose response to a unit-impulse $\delta(t)$ is a unit-pulse $g_{h0}(t)$ of width T. The Laplace transform of the impulse response $g_{h0}(t)$ is the transfer function of the hold operation, namely

$$G_{h0}(s) = \mathscr{L}[g_{h0}(t)] = \int_0^\infty g_{h0}(t) e^{-st} dt = \int_0^T e^{-st} dt = \frac{1-e^{-sT}}{s} \qquad (2.56)$$

Figure 2.29b is a block diagram representation of the transfer function model of the S/H operation.

76 Digital Control and State Variable Methods

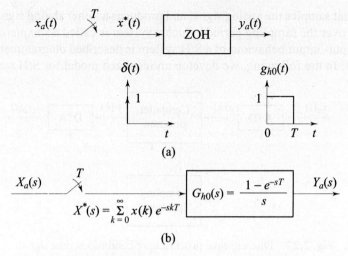

(a)

(b)

Fig. 2.29 A model of a sample-and-hold operation

Practical Sample-and-Hold Circuit

In a majority of practical digital operations, S/H functions are performed by a single S/H device. It consists of a capacitor, an electronic switch, and operational amplifiers (Fig. 2.30). Op amps are needed for isolation; the capacitor and switch cannot be connected directly to analog circuitry because of the capacitor's effect on the driving waveform.

Since the voltage between the inverting and non-inverting inputs of an op amp is in microvolts, we can approximate this voltage to zero. This implies that the voltage from the inverting input (–input) to ground in Fig. 2.30 is approximately V_{IN}; therefore the output of first op amp is approximately V_{IN}.

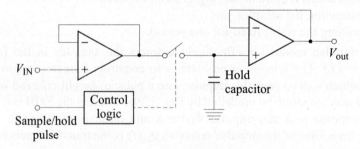

Fig. 2.30 A sample-and-hold device

When the switch is closed, the capacitor rapidly charges to V_{IN}, and V_{OUT} is equal to V_{IN} approximately. When the switch opens, the capacitor retains its charge; the output holds at a value of V_{IN}.

If the input voltage changes rapidly while the switch is closed, the capacitor can follow this voltage because the charging time-constant is very short. If the switch is suddenly opened, the capacitor voltage represents a sample of the input voltage at the instant the switch was opened. The capacitor then holds this sample until the switch is again closed and a new sample taken.

As an illustration of the application of a sampler/ZOH circuit, consider the A/D conversion system of Fig. 2.31. The two subsystems in this figure correspond to systems that are available as physical devices. The A/D converter converts a voltage (or current) amplitude at its input into a binary code representing a quantized amplitude value closest to the amplitude of the input. However, the conversion is not instantaneous. Input signal variation during the conversion time of A/D converter (typical conversion times of commercial A/D units range from 100 nsec to 200 μsec) can lead to erroneous results. For this reason, a high performance A/D conversion system includes a S/H device as in Fig. 2.31.

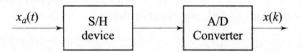

Fig. 2.31 Physical configuration for A/D conversion

Although a S/H is available commercially as one unit, it is advantageous to treat the sampling and holding operations separately for analytical purposes, as has been done in the S/H model of Fig. 2.29b. This model gives the defining equation of the sampling process and the transfer function of the ZOH. It may be emphasized here that $X^*(s)$ is not present in the physical system but appears in the mathematical model; the sampler in Fig. 2.29 does not model a physical sampler and the block does not model a physical data hold. However, the combination does accurately model a sampler/ZOH device.

2.9 SAMPLED SPECTRA AND ALIASING

We can get further insight into the process of sampling by relating the spectrum of the continuous-time signal to that of discrete-time sequence that is obtained by sampling.

Let us define the continuous-time signal by $x_a(t)$. Its spectrum is then given by $X_a(j\omega)$, where ω is the frequency in radians per second. The sequence $x(k)$ with value $x(k) = x_a(kT)$ is derived from $x_a(t)$ by periodic sampling. Spectrum of $x(k)$ is given by $X(e^{j\Omega})$ where the frequency Ω has units of radians per sample interval.

Using Eqns (A.8b) and (A.34a) of Appendix A, we obtain

$$X_a(j\omega) = \int_{-\infty}^{\infty} x_a(t)\, e^{-j\omega t}\, dt \qquad (2.57)$$

$$X(e^{j\Omega}) = \sum_{k=-\infty}^{\infty} x(k)\, e^{-j\Omega k} \qquad (2.58)$$

We use the intermediate function $x^*(t)$—the impulse modulated $x_a(t)$—to establish a relation between $X_a(j\omega)$ and $X(e^{j\Omega})$.

The Fourier transform of $x^*(t)$, denoted by $X^*(j\omega)$ is (refer Eqn. (2.51)) given by

$$X^*(j\omega) = \int_{-\infty}^{\infty} x^*(t) e^{-j\omega t} dt = \int_{-\infty}^{\infty} \left[\sum_{k=0}^{\infty} x(k) \delta(t-kT) \right] e^{-j\omega t} dt$$

$$= \sum_{k=-\infty}^{\infty} \int_{-\infty}^{\infty} \delta(t-kT) x(k) e^{-j\omega t} dt = \sum_{k=-\infty}^{\infty} x(k) e^{-jk\omega T} \qquad (2.59)$$

(The summation over the interval $-\infty$ to ∞ is allowed, since $x(k) = 0$ for $k < 0$). We have arrived at our first intermediate result. By comparing Eqn. (2.59) with Eqn. (2.58), we observe that

$$X(e^{j\Omega}) = X^*(j\omega) \Big|_{\omega = \frac{\Omega}{T}} \qquad (2.60a)$$

$X(e^{j\Omega})$ is thus a frequency-scaled version of $X^*(j\omega)$ with the frequency scaling specified by

$$\Omega = \omega T \qquad (2.60b)$$

We now determine $X^*(j\omega)$ in terms of the continuous-time spectrum $X_a(j\omega)$. From Eqn. (2.51), we have

$$x^*(t) = \sum_{k=0}^{\infty} x(k) \delta(t-kT) = \sum_{k=0}^{\infty} x_a(t) \delta(t-kT) = x_a(t) \sum_{k=0}^{\infty} \delta(t-kT)$$

The summation over the interval $-\infty$ to ∞ is allowed since $x_a(t) = 0$ for $t < 0$. Therefore,

$$x^*(t) = x_a(t) \sum_{k=-\infty}^{\infty} \delta(t-kT)$$

Since $\sum_{k=-\infty}^{\infty} \delta(t-kT)$ is a periodic function, it can be expressed in terms of the following Fourier series expansion (refer Eqns (A.2) in Appendix A):

$$\sum_{k=-\infty}^{\infty} \delta(t-kT) = \sum_{n=-\infty}^{\infty} c_n e^{j\frac{2\pi nt}{T}}$$

where

$$c_n = \frac{1}{T} \int_{-T/2}^{T/2} \left[\sum_{k=-\infty}^{\infty} \delta(t-kT) \right] e^{-j\frac{2\pi nt}{T}} dt$$

$$= \frac{1}{T} \int_{-T/2}^{T/2} \delta(t) e^{-j\frac{2\pi nt}{T}} dt = \frac{1}{T} e^{-j0} = \frac{1}{T} \text{ for all } n$$

Substituting this Fourier series expansion into the impulse modulation process, we get

$$x^*(t) = x_a(t) \sum_{k=-\infty}^{\infty} \delta(t - kT)$$

$$= x_a(t) \frac{1}{T} \sum_{n=-\infty}^{\infty} e^{j\frac{2\pi nt}{T}} = \frac{1}{T} \sum_{n=-\infty}^{\infty} x_a(t) e^{j\frac{2\pi nt}{T}}$$

The continuous-time spectrum of $x^*(t)$ is then equal to

$$X^*(j\omega) = \int_{-\infty}^{\infty} x^*(t) e^{-j\omega t} dt = \frac{1}{T} \int_{-\infty}^{\infty} \left[\sum_{n=-\infty}^{\infty} x_a(t) e^{j\frac{2\pi nt}{T}} \right] e^{-j\omega t} dt$$

Interchanging the order of summation and integration, we obtain

$$X^*(j\omega) = \frac{1}{T} \sum_{n=-\infty}^{\infty} \left[\int_{-\infty}^{\infty} x_a(t) e^{-j\left(\omega - \frac{2\pi n}{T}\right)t} dt \right]$$

$$= \frac{1}{T} \sum_{n=-\infty}^{\infty} X_a\left(j\omega - j\frac{2\pi n}{T}\right) \quad (2.61a)$$

where $X_a(j\omega)$ is the Fourier transform of $x_a(t)$.

We see from this equation that $X^*(j\omega)$ consists of periodically repeated copies of $X_a(j\omega)$, scaled by $1/T$. The scaled copies of $X_a(j\omega)$ are shifted by integer multiples of the sampling frequency

$$\omega_s = \frac{2\pi}{T} \quad (2.61b)$$

and then superimposed to produce $X^*(j\omega)$.

Equation (2.61a) is our second intermediate result. Combining this result with that given by Eqn. (2.60a), we obtain the following relations:

$$X^*(j\omega) = \frac{1}{T} \sum_{k=-\infty}^{\infty} X_a\left(j\omega - j\frac{2\pi k}{T}\right) \quad (2.62a)$$

$$X(e^{j\Omega}) = X^*\left(j\frac{\Omega}{T}\right) = \frac{1}{T} \sum_{k=-\infty}^{\infty} X_a\left(j\frac{\Omega}{T} - j\frac{2\pi k}{T}\right) \quad (2.62b)$$

Aliasing

While sampling a continuous-time signal $x_a(t)$ to produce the sequence $x(k)$ with values $x(k) = x_a(kT)$, we want to ensure that all the information in the original signal is retained in the samples. There will be no information loss if we can exactly recover the continuous-time signal from the samples. To determine the condition under which there is no information loss, let us consider $x_a(t)$ to be a bandlimited signal with maximum frequency ω_m, i.e.,

$$X_a(j\omega) = 0 \text{ for } |\omega| > \omega_m \tag{2.63}$$

as shown in Fig. 2.32a. Figure 2.32b shows a plot of $X^*(j\omega)$ under the condition

$$\frac{\omega_s}{2} = \frac{\pi}{T} > \omega_m \tag{2.64a}$$

Figure 2.32c shows the plot of $X(e^{j\Omega})$, which is derived from Fig. 2.32b by simply scaling the frequency axis.

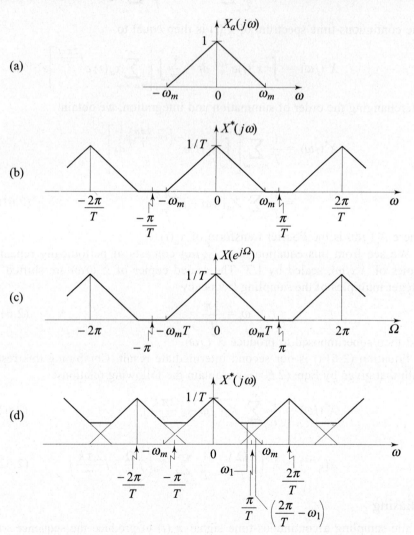

Fig. 2.32 Frequency-domain considerations in sampling and reconstruction

$X^*(j\omega)$ is seen to be a periodic function with period $2\pi/T$ ($X(e^{j\Omega})$ is a periodic function with period 2π). The spectrum $X^*(j\omega)$ for $|\omega| \leq \pi/T$ is identical to the continuous-time spectrum $X_a(j\omega)$ except for linear scaling in amplitude (the spectrum $X(e^{j\Omega})$ for $|\Omega| \leq \pi$ is identical to the continuous-time spectrum $X_a(j\omega)$ except for linear scaling in amplitude and frequency). The continuous-time signal $x_a(t)$ can be recovered from its samples $x(k)$ without any distortion by employing an ideal lowpass filter (Fig. 2.33).

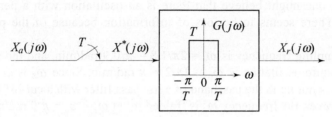

Fig. 2.33 Ideal lowpass filter

Figure 2.32d shows a plot of $X^*(j\omega)$ under the condition

$$\frac{\omega_s}{2} = \frac{\pi}{T} < \omega_m \qquad (2.64b)$$

The plot of $X(e^{j\Omega})$ can easily be derived from Fig. 2.32d by scaling the frequency axis.

It is obvious from Fig. 2.32d that the shifted versions of $X_a(j\omega)$ overlap; $X^*(j\omega)$ in the range $|\omega| \leq \pi/T$ can be viewed as being found by superimposing onto this frequency range the behaviour of the shifted versions of $X_a(j\omega)$.

Consider an arbitrary frequency point ω_1 in Fig. 2.32d which falls in the region of the overlap of shifted versions of $X_a(j\omega)$. The frequency spectrum at $\omega = \omega_1$ is the sum of two components. One of these, the larger one in the figure, has a value equal to $X_a(j\omega_1)$. The other component comes from the spectrum centred at $2\pi/T$, and has a value equal to $X_a\left(j\left(\frac{2\pi}{T} - \omega_1\right)\right)$. Note that the high frequency $\left(\frac{2\pi}{T} - \omega_1\right)$ is 'folded in' about the *folding frequency* π/T; and appears as low frequency at ω_1. The frequency $\left(\frac{2\pi}{T} - \omega_1\right)$ which shows up at ω_1 after sampling is called in the trade as the 'alias' of ω_1. The superimposition of the high-frequency behaviour onto the low frequency is known as *frequency folding* or *aliasing*. Under the condition given by (2.64b), the form of $X^*(j\omega)$ in the frequency range $|\omega| \leq \pi/T$ is no longer similar to $X_a(j\omega)$; therefore the true spectral shape $X_a(j\omega)$ is no longer recoverable by lowpass filtering (refer Fig. 2.33). In this case, the reconstructed signal $x_r(t)$ is related to the original signal $x_a(t)$ through a distortion introduced by aliasing and therefore there is loss of information due to sampling.

Example 2.6

We consider a simple example to illustrate the effects of aliasing

Figure 2.34a shows a recording of the temperature in a thermal process. From this recording we observe that there is an oscillation in temperature with a period of two minutes.

The sampled recording of the temperature obtained by measurement of temperature after every 1.8 minutes is shown in Fig. 2.34b. From the sampled recording, one might believe that there is an oscillation with a period of 18 minutes. There seems to be loss of information because of the process of sampling.

The sampling frequency is $\omega_s = 2\pi/1.8 = \pi/0.9$ rad/min, and the frequency of temperature oscillation is $\omega_0 = 2\pi/2 = \pi$ rad/min. Since ω_0 is greater than $\omega_s/2$, it does not lie in the passband of a lowpass filter with a cut-off frequency $\omega_s/2$. However, the frequency ω_0 is 'folded in' at $\omega_s - \omega_0 = \pi/9$ rad/min which lies in the passband of the lowpass filter. The reconstructed signal has therefore a period of 18 minutes, which is the period of the sampled recording.

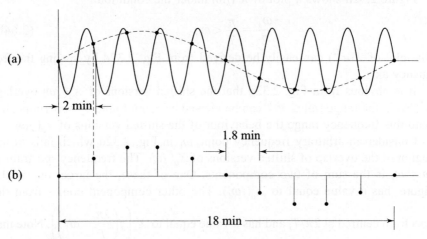

Fig. 2.34 Continuous and sampled recording of temperature

Sampling Theorem

A corollary to the aliasing problem is the sampling theorem stated below.

Let $x_a(t)$ be a band-limited signal with $X_a(j\omega) = 0$ for $|\omega| > \omega_m$. Then $x_a(t)$ is uniquely determined from its samples $x(k) = x_a(kT)$ if the sampling frequency

$$\omega_s \left(= \frac{2\pi}{T} \right) > 2\omega_m,$$ i.e., the *sampling frequency must be at least twice the highest frequency present in the signal.*

We will discuss the practical aspects of the choice of sampling frequency in Section 2.11.

2.10 FILTERING

In this section we discuss filtering requirements in digital control applications.

Reconstruction of Analog Signals

Digital control systems usually require the transformation of discrete-time sequences into analog signals. In such cases, we are faced with the converse problem from that of sampling $x_a(t)$ to obtain $x(k)$. The relevant question now becomes—how can $x_a(t)$ be recovered from its samples.

We begin by considering the unaliased spectrum of $X^*(j\omega)$ shown in Fig. 2.32b. $X_a(j\omega)$ has the same form as $X^*(j\omega)$ over $\frac{-\pi}{T} \leq \omega \leq \frac{\pi}{T}$. $X_a(j\omega)$ can be recovered from $X^*(j\omega)$ by a lowpass filter.

Ideal lowpass filter

Consider the ideal lowpass filter shown in Fig. 2.33. It is characterized by $G(j\omega)$ defined below:

$$G(j\omega) = \begin{bmatrix} T & \text{for } \frac{-\pi}{T} \leq \omega \leq \frac{\pi}{T} \\ 0 & \text{otherwise} \end{bmatrix} \quad (2.65)$$

Note that the ideal filter given by Eqn. (2.65) has a zero phase characteristic. This phase characteristic stems from our requirement that any signal whose frequency components are totally within the passband of the filter be passed undistorted.

The impulse response of the ideal lowpass filter is given by inverse Fourier transformation (refer Eqn. (A.9b) in Appendix A):

$$g(t) = \frac{1}{2\pi} \int_{-\infty}^{\infty} G(j\omega) e^{j\omega t} d\omega = \frac{1}{2\pi} \int_{-\pi/T}^{\pi/T} T e^{j\omega t} d\omega$$

$$= \frac{T}{j2\pi t}(e^{j\pi t/T} - e^{-j\pi t/T}) = \frac{\sin \pi t/T}{\pi t/T} \quad (2.66)$$

Figure 2.35 shows a plot of $g(t)$ versus t. Notice that the response extends

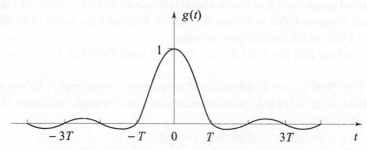

Fig. 2.35 Impulse response of an ideal lowpass filter

from $t = -\infty$ to $t = \infty$. This implies that there is a response for $t < 0$ to a unit impulse applied at $t = 0$ (i.e., the time response that begins before an input is applied). This cannot be true in the physical world. Hence, such an ideal filter is physically unrealizable.

Zero-order hold

We consider polynomial holds as an approximation to the ideal lowpass filter. The ZOH was considered in Section 2.8, and its transfer function was derived to be (Eqn. (2.56))

$$G_{h0}(s) = \frac{1 - e^{-sT}}{s}$$

Its frequency response is consequently given by

$$G_{h0}(j\omega) = \frac{1 - e^{-j\omega T}}{j\omega} = \frac{e^{-j\omega T/2}(e^{j\omega T/2} - e^{-j\omega T/2})}{j\omega}$$

$$= T \frac{\sin(\omega T/2)}{\omega T/2} e^{-j\omega T/2} \qquad (2.67)$$

Plot of $\dfrac{\sin(\omega T/2)}{\omega T/2}$ versus ω will be of the form shown in Fig. 2.35 with sign reversals at $\omega = \dfrac{2\pi}{T}, \dfrac{4\pi}{T}, \cdots$ (i.e., $\omega = \omega_s, 2\omega_s, \cdots$). The sign reversals amount to a phase shift of $-180°$ (it can be taken as $+180°$ as well) at $\omega = k\omega_s$; $k = 1, 2, \ldots$

Equation (2.67) can therefore be expressed as

$$G_{h0}(j\omega) = |G_{h0}(j\omega)| \angle G_{h0}(j\omega)$$

where

$$|G_{h0}(j\omega)| = T \left| \frac{\sin(\omega T/2)}{\omega T/2} \right| \qquad (2.68a)$$

and

$$\angle G_{h0}(j\omega) = \left(\frac{-\omega T}{2} - 180° \right) \text{ at } \omega = \frac{2\pi k}{T}; k = 1, 2, \ldots \qquad (2.68b)$$

A plot of magnitude and phase characteristics of ZOH is shown in Fig. 2.36. The ideal lowpass filter is shown by dashed lines in Fig. 2.36a. The phase of the ideal filter, at all frequencies, is zero.

It is obvious that the hold device does not have the ideal filter characteristics.

  (i) The ZOH begins to attenuate at frequencies considerably below $\omega_s/2$
  (ii) The ZOH allows high frequencies to pass through, although they are attenuated.
  (iii) As a consequence of shifting property:

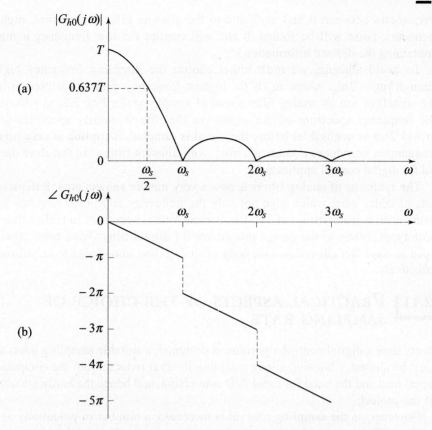

Fig. 2.36 ZOH device (a) Gain, and (b) Phase characteristics

$$\mathcal{F}\left[x\left(t-\frac{T}{2}\right)\right] = e^{-j\omega T/2}\, X(j\omega),$$

the linear phase characteristic introduces a time delay of $T/2$. When ZOH is used in a feedback system, the lag characteristic of the device degrades the degree of system stability.

The higher-order holds, which are more sophisticated and which better approximate the ideal filter, are more complex and have more time delay than the ZOH. As the additional time delay in feedback control systems decreases the stability margin or even causes instability, higher-order holds are rarely justified in terms of improved performance, and therefore the zero-order hold is widely used in practice.

Anti-aliasing filter

In practice, signals in control systems have frequency spectra consisting of low-frequency components as well as high-frequency noise components. Recall that all signals with frequency higher than $\omega_s/2$ appear as signals of frequencies between 0 and $\omega_s/2$ due to the aliasing effect. Therefore, high-frequency noise will be folded in and will corrupt the low frequency signal containing the desired information.

To avoid aliasing, we must either choose the sampling frequency high enough ($\omega_s > 2\omega_m$, where ω_m is the highest-frequency component present in the signal) or use an analog filter ahead of sampler (refer Fig. 2.2) to reshape the frequency spectrum of the signal (so that the frequency spectrum for $\omega > (1/2)\omega_s$ is negligible) before the signal is sampled. Sampling at very high frequencies introduces numerical errors. Anti-aliasing filters are therefore useful for digital control applications.

The synthesis of analog filters is now a very mature subject area. Extensive sets of tables exist which give not only the frequency and phase response of many analog prototypes, but also the element values necessary to realize those prototypes. Many of the design procedures for digital filters have been developed in ways that allow this wide body of analog filter knowledge to be utilized effectively.

2.11 PRACTICAL ASPECTS OF THE CHOICE OF SAMPLING RATE

Every time a digital control algorithm is designed, a suitable sampling interval must be chosen. Choosing a long sampling interval reduces both the computational load and the need for rapid A/D conversion, and hence the hardware cost of the project.

However, as the sampling interval is increased, a number of potentially degrading effects start to become significant. For a particular application, one or more of these degrading effects set the upper limit for the sampling interval. The process dynamics, the type of algorithm, the control requirement and the characteristics of input and noise signals, all interact to set the maximum usable value for T.

There is also a lower limit for the sampling interval. Digital hardware dictates the minimum usable value for T.

We discuss some of the factors which limit the choice of sampling interval. Some empirical rules for the selection of sampling interval are also reported.

Information Loss due to Sampling

The sampling theorem states that a continuous-time signal whose frequency spectrum is bounded by upper limit ω_m, can be completely reconstructed from its samples when the sampling frequency is $\omega_s > 2\omega_m$. There are two problems associated with the use of the sampling theorem in practical control systems.

  1. The frequency spectra of real signals do not possess strictly defined ω_m. There are almost always frequency components outside the system bandwidth. Therefore, the selection of the sampling frequency ω_s using the sampling theorem on the basis of system bandwidth ($\omega_b = \omega_m$) is risky, as frequency components outside ω_b will appear as low-frequency

signals of frequencies between 0 and $\omega_s/2$ due to the aliasing effect, and lead to loss of information.
2. The ideal lowpass filter needed for perfect reconstruction of a continuous-time signal from its samples is not physically realizable. Practical filters, such as the ZOH, introduce reconstruction errors because of the limitations of their operation.

Figure 2.28 clearly indicates that the accuracy of the zero-order hold as an extrapolating device depends greatly on the sampling frequency ω_s. The accuracy improves with increase in sampling frequency.

Information Loss due to Disturbances
In practice, signals in control systems include low-frequency components carrying useful information, as well as high-frequency noise components. The high-frequency components appear as low-frequency signals (of frequencies between 0 and $\omega_s/2$) due to the aliasing effect, causing a loss of information.

To avoid aliasing, we use the analog filter ahead of sampler (refer Fig. 2.2) to reshape the frequency spectrum of the signal so that the frequency spectrum for $\omega > (1/2)\omega_s$ is negligible. The cut-off frequency $\omega_s/2$ of the anti-aliasing filter must be much higher than the system bandwidth, since otherwise the anti-aliasing filter becomes as significant as the system itself in determining the sampled response.

Destabilizing Effects
Due to the conversion times and the computation times, a digital algorithm contains a dead-time that is absent from its analog counterpart. Dead-time has a marked destabilizing effect on a closed-loop system due to the phase shift caused.

A practical approach of selecting the sampling interval is to determine the stability limit of the closed-loop control system as sampling interval T is increased. For control system applications, this approach is more useful than the use of the sampling theorem for the selection of sampling interval. In the later chapters of this book, we will use stability tests, root-locus techniques, and frequency-response plots to study the effect of the sampling interval on closed-loop stability.

Algorithm-accuracy Effects
A number of digital control algorithms are derived from analog algorithms by a process of discretization. As we shall see in the next section, in the transformation of an algorithm, from continuous-time to discrete-time form, errors arise and the character of the digital algorithm differs from that of its analog counterpart. In general, these errors occurring during the discretization process become larger as the sampling interval increases.

This effect should rarely be allowed to dictate a shorter sampling interval than would otherwise have been needed. We will see in Chapter 4 that the direct digital design approach allows a longer sampling interval without the introduction of unacceptable errors.

Word-length Effects

As the sampling interval T becomes very short, a digital system does not tend to the continuous-time case because of the finite word-length. To visualize this effect, we can imagine that as a signal is sampled more frequently, adjacent samples have more similar magnitudes. In order to realize the beneficial effects of shorter sampling, longer word-lengths are needed to resolve the differences between adjacent samples.

Excessively fast sampling ($T \to 0$) may also result in numerical ill-conditioning in implementation of recursive control algorithms (such as the PID control algorithm—discussed in the next section).

Empirical Rules for the Selection of Sampling Rate

Practical experience and simulation results have produced a number of useful approximate rules for the specification of minimum sampling rates.

1. The following recommendations for the most common process variables follow from the experience of process industries:

Type of variable	Sampling time (seconds)
Flow	1—3
Level	5—10
Pressure	1—5
Temperature	10—20

2. Fast-acting electromechanical systems require much shorter sampling intervals, perhaps down to a few milliseconds.
3. A rule of thumb says that a sampling period needs to be selected that is much shorter than any of the time constants in the continuous-time plant to be controlled digitally. The sampling interval equal to one-tenth of the smallest time-constant or the inverse of the largest real pole (or real part of complex pole) has been recommended.
4. For complex poles with the imaginary part ω_d, the frequency of transient oscillations corresponding to the poles is ω_d. A convenient rule suggests sampling at the rate of 6 to 10 times per cycle. Thus, if the largest imaginary part in the poles of the continuous-time plant is 1 rad/sec, which corresponds to transient oscillations with a frequency of 1/6.28 cycles per second, $T = 1$ sec may be satisfactory.
5. Rules of thumb based on the open-loop plant model are risky under conditions where the high closed-loop performance is forced from a plant with a low open-loop performance. The rational choice of the sampling rate should be based on an understanding of its influence on the closed-loop performance of the control system. It seems reasonable that the highest frequency of interest should be closely related to the 3dB-bandwidth of the closed-loop system. The selection of sampling rates can then be based on the bandwidth of the closed-loop system. Reasonable sampling rates are 10 to 30 times the bandwidth.

6. Another rule of thumb based on the closed-loop performance is to select sampling interval T equal to or less than one-tenth of the desired settling time.

2.12 PRINCIPLES OF DISCRETIZATION

Most of the industrial processes that we are called upon to control are continuous-time processes. Mathematical models of continuous-time processes are usually based around differential equations or, equivalently, around transfer functions in the operator s. A very extensive range of well-tried methods for control system analysis and design are in the continuous-time form.

To move from the continuous-time form to the discrete-time form requires some mechanism for time discretization (we shall refer to this mechanism simply as *discretization*). In this section, principles and various methods of discretization will be presented. An understanding of various possible approaches helps the formation of a good theoretical foundation for the analysis and design of digital control systems.

The main point is to be aware of the significant features of discretization and to have a rough quantitative understanding of the errors that are likely to be introduced by various methods. We will shortly see that none of the discretization methods preserves the characteristics of the continuous-time system exactly.

The specific problem of this section is: given a transfer function $G(s)$, what discrete-time transfer function will have approximately the same characteristics?

We present four methods for solution of this problem.
  (i) Impulse-invariant discretization
  (ii) Step-invariant discretization
  (iii) Discretization based on finite-difference approximation of derivatives
  (iv) Discretization based on bilinear transformation

Impulse Invariance

If we are given a continuous-time impulse response $g_a(t)$, we can consider transforming it to a discrete-time system with impulse response $g(k)$ consisting of equally spaced samples of $g_a(t)$ so that

$$g(k) = g_a(t)|_{t=kT} = g_a(kT)$$

where T is a (positive) number to be chosen as part of the discretization procedure.

The transformation of $g_a(t)$ to $g(k)$ can be viewed as impulse modulation (refer Fig. 2.26) giving impulse-train representation $g^*(t)$ to the samples $g(k)$:

$$g^*(t) = \sum_{k=0}^{\infty} g(k)\delta(t - kT) \qquad (2.69)$$

From the discussion in Section 2.9, and specifically Eqns (2.62), it follows that

$$G^*(j\omega) = \frac{1}{T} \sum_{k=-\infty}^{\infty} G_a\left(j\omega - j\frac{2\pi k}{T}\right) \quad (2.70a)$$

$$G(e^{j\Omega}) = \frac{1}{T} \sum_{k=-\infty}^{\infty} G_a\left(j\frac{\Omega}{T} - j\frac{2\pi k}{T}\right) \quad (2.70b)$$

ω, in radians/second, is the physical frequency of the continuous-time function and $\Omega = \omega T$, in radians, is the observed frequency in its samples.

Thus, for a discrete-time system obtained from a continuous-time system through impulse invariance, the discrete-time frequency response $G(e^{j\Omega})$ is related to the continuous-time frequency response $G_a(j\omega)$ through replication of the continuous-time frequency response and linear scaling in amplitude and frequency. If $G_a(j\omega)$ is bandlimited and T is chosen so that aliasing is avoided, the discrete-time frequency response is then identical to continuous-time frequency response except for linear scaling in amplitude and frequency.

Let us explore further the properties of impulse invariance. Applying the Laplace transform to Eqn. (2.69), we obtain (refer Eqn. (2.52))

$$G^*(s) = \sum_{k=0}^{\infty} g(k) \, e^{-skT} \quad (2.71a)$$

On the other hand, the z-transform of $g(k)$ is, by definition,

$$G(z) = \sum_{k=0}^{\infty} g(k) \, z^{-k} \quad (2.71b)$$

Comparing Eqns (2.71a) and (2.71b), it follows that

$$G(z)\bigg|_{z=e^{sT}} = G^*(s) \quad (2.72a)$$

Rewriting Eqn. (2.70a) in terms of the general transform variable s, gives a relationship between $G^*(s)$ and $G_a(s)$:

$$G^*(s) = \frac{1}{T} \sum_{k=-\infty}^{\infty} G_a\left(s - \frac{2\pi k}{T}\right) \quad (2.72b)$$

Therefore, $\quad G(z)\bigg|_{z=e^{sT}} = \frac{1}{T} \sum_{k=-\infty}^{\infty} G_a\left(s - \frac{2\pi k}{T}\right) \quad (2.72c)$

We note that impulse invariance corresponds to a transformation between $G^*(s)$ and $G(z)$ represented by the mapping

$$z = e^{sT} = e^{(\sigma \pm j\omega)T} = e^{\sigma T} \angle \pm \omega T \quad (2.73)$$

between the s-plane and the z-plane. The mapping is illustrated in Fig. 2.37.

Consider the mapping of the *s*-plane path described by *abcdea* traversed in the counter-clockwise direction. The length of the interval *bc* of path is assumed to be infinite.

s-plane	*z*-plane
Interval [*a*, *b*]	Magnitude 1 Angle 0 to $+\pi$
Interval [*b*, *c*]	Magnitude $e^{\sigma T}$ Angle $+\pi$
Interval [*c*, *d*]	Magnitude 0 Angel $(+\pi)$ to 0 to $(-\pi)$
Interval [*d*, *e*]	Magnitude $e^{\sigma T}$ Angle $-\pi$
Interval [*e*, *a*]	Magnitude 1 Angle $-\pi$ to 0

As per the mapping given by Eqn. (2.73), the section $\left[\left(-j\frac{\pi}{T}\right) - 0 - \left(+j\frac{\pi}{T}\right)\right]$ of the $j\omega$-axis maps into the unit circle in the counter-clockwise direction.

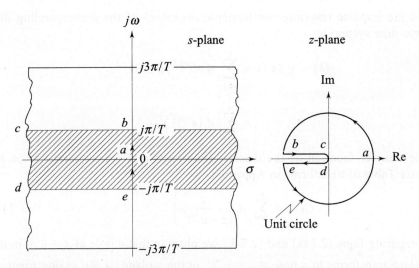

Fig. 2.37 Mapping of the *s*-plane to the *z*-plane using the impulse-invariant transformation

It can easily be demonstrated that the shaded strip in Fig. 2.37 maps into the entire *z*-plane. The points in the shaded strip belonging to the left-half of the *s*-plane map into the interior of the unit circle; the points in the strip belonging to $j\omega$-axis of the *s*-plane map onto the boundary of the unit circle; and the points in the strip belonging to the right-half of the *s*-plane map into the exterior of the unit circle in the complex *z*-plane. In fact, the *s*-plane can be divided into strips of width $2\pi/T$, each of which maps into the entire *z*-plane. The *primary strip* extends from $\omega = -\pi/T$ to $+\pi/T$, and the *complementary strips* extend from $-\pi/T$ to $-3\pi/T$, $-3\pi/T$ to $-5\pi/T$, ..., for negative frequencies and from π/T to $3\pi/T$, $3\pi/T$ to $5\pi/T$, ..., for positive frequencies.

$G(z)$ can be viewed as being found by first superimposing or aliasing into the shaded strip the behaviour of $G_a(s)$ in complementary strips of width $2\pi/T$, and then mapping the shaded strip to the entire z-plane.

When the Laplace transform of $g_a(t)$ is rational, the relationship between $G(z)$ and $G_a(s)$ can be expressed in a more convenient form than Eqn. (2.72c). Let us consider the transfer function of a continuous-time system expressed in terms of a partial fraction expansion. For convenience, we will assume only the first-order poles; the argument can easily be extended to multiple-order poles.

With $G_a(s)$ expressed in the form

$$G_a(s) = \sum_{i=1}^{n} \frac{A_i}{s+a_i} \qquad (2.74a)$$

the corresponding impulse response is

$$g_a(t) = \sum_{i=1}^{n} A_i e^{-a_i t}; t \geq 0$$

and the impulse response (unit-sample response) of the corresponding discrete-time system is

$$g(k) = g_a(kT) = \sum_{i=1}^{n} A_i e^{-a_i kT}; k \geq 0$$

$$= \sum_{i=1}^{n} A_i (e^{-a_i T})^k$$

The transfer function of the discrete-time system is consequently given by (refer Table of transforms in Appendix A)

$$G(z) = \sum_{i=1}^{n} A_i \left\{ \frac{z}{z - e^{-a_i T}} \right\} \qquad (2.74b)$$

Comparing Eqns (2.74a) and (2.74b) we observe that a pole at $s = -a_i$ in the s-plane transforms to a pole at $z = e^{-a_i T}$ in the z-plane. If the analog function $G_a(s)$ is stable, the real part of $-a_i$ will be less than zero and the magnitude of $e^{-a_i T}$ will be less than unity, so that the corresponding pole in the z-plane is inside the unit circle, and consequently the discrete-time function $G(z)$ is also stable.

While poles in the s-plane map to the poles in the z-plane according to the relationship $z = e^{sT}$, the zeros of $G(z)$ are a function of poles and residues in Eqn. (2.74b), and their locations do not in general correspond to a direct mapping of the zeros of $G_a(s)$ through the transformation by which poles are mapped.

In summary, the use of impulse invariance corresponds to converting the continuous-time impulse response to a discrete-time impulse response through sampling. To avoid aliasing, the procedure is restricted to transforming band-limited frequency responses. Except for aliasing, the discrete-time frequency response is a replication of the continuous-time frequency response linearly scaled in amplitude and frequency.

Although useful for discretizing band-limited analog systems, the impulse-invariance method is unsuccessful for discretizing transfer functions $G_a(s)$ for which $|G_a(j\omega)|$ does not approach zero for large ω. In these cases, an appropriate sampling rate cannot be found to prevent aliasing.

To overcome the problem of aliasing, we need a method in which the entire $j\omega$-axis in the s-plane maps uniquely on to the unit circle in the z-plane. This is accomplished by the *bilinear transformation* method described later in this section.

▲▲

For a given analog system $G_a(s)$, the impluse-invariant discrete-time system is obtained following the procedure given below:

(i) Obtain the impulse response
$$g_a(t) = \mathcal{L}^{-1}[G_a(s)]$$

(ii) Select a suitable sampling interval and derive samples $g(k)$ from $g_a(t)$,
$$g(k) = g_a(t)\Big|_{t=kT}$$

(iii) Obtain z-transform of the sequence $g(k)$,
$$G(z) = \mathcal{Z}[g(k)]$$

The three steps given above can be represented by the following relationship:

$$G(z) = \mathcal{Z}\left[\left[\mathcal{L}^{-1}\{G_a(s)\}\right]\Big|_{t=kT}\right] \qquad (2.75a)$$

This z-transform operation is commonly indicated as

$$G(z) = \mathcal{Z}[G_a(s)] \qquad (2.75b)$$

Single factor building blocks of the Laplace and z-transform pairs are given in Table of transforms in Appendix A. Expanding any $G_a(s)$ into partial fractions, $G(z)$ can be found by use of this table.

Example 2.7

With the background on analog design methods, the reader will appreciate the value of being able to correlate particular patterns in the s-plane with particular features of system behaviour. Some of the useful s-plane patterns, which have been used in analog design, are the loci of points in the s-plane with (i) constant damping ratio ζ, and (ii) constant undamped natural frequency ω_n. In this example, we translate these patterns in the primary strip of the s-plane onto z-plane using the basic relation $z = e^{sT}$, where T is some chosen sampling period.

Consider a second-order system with transfer function

$$G_a(s) = \frac{K}{s^2 + 2\zeta\omega_n s + \omega_n^2}$$

where ζ = damping ratio, and ω_n = undamped natural frequency.

The characteristic root locations in the s-plane are

$$s_1, s_2 = -\zeta\omega_n \pm j\omega_n\sqrt{1-\zeta^2} = -\zeta\omega_n \pm j\omega_d$$

Figure 2.38a shows a locus of the characteristic roots with ζ held constant and ω_n varying. Figure 2.38b shows a locus with ω_n held constant and ζ varying. The loci in Figs 2.38a and 2.38b correspond to an underdamped second-order system.

Define the sampling frequency by

$$\omega_s = 2\pi/T$$

Corresponding to each point $s = \sigma + j\omega$ in the primary strip of the s-plane, is a point

$$z = \exp[(\sigma + j\omega)2\pi/\omega_s]$$

in the z-plane.

(i) *Mapping of constant damping ratio loci*

A point on the constant ζ line in the second quadrant (Fig. 2.38a) can be expressed as

$$s = \sigma + j\omega_d = -\zeta\omega_n + j\omega_n\sqrt{1-\zeta^2}$$

Since

$$\cot\theta = \frac{\zeta\omega_n}{\omega_d} = \frac{\zeta\omega_n}{\omega_n\sqrt{1-\zeta^2}} = \frac{\zeta}{\sqrt{1-\zeta^2}},$$

the s-plane point may be described by the relation

$$s = -\omega_d \cot\theta + j\omega_d$$

The z-plane relation becomes

$$z = \exp[(-\omega_d\cot\theta + j\omega_d)\,2\pi/\omega_s]$$
$$= \exp[((-2\pi\cot\theta)/\omega_s)\,\omega_d]\exp[j(2\pi/\omega_s)\,\omega_d]$$

where ω_d varies from 0 to $\omega_s/2$.

As per this equation, for a constant value of ζ (and hence $\cot\theta$), z is a function of ω_d only. The constant damping ratio line in the s-plane maps into the z-plane as a logarithmic spiral (except for $\theta = 0°$ and $\theta = 90°$). The portion of ζ-line between $\omega_d = 0$ and $\omega_d = \omega_s/2$ corresponds to one-half revolution of the logarithmic spiral in the z-plane. Mapping of one representative constant ζ line is shown in Fig. 2.38c.

(ii) *Mapping of constant undamped natural frequency loci*

A point on the constant ω_n locus in the second quadrant (Fig. 2.38b) can be expressed as

$$s = \sigma + j\omega_d = -\zeta\omega_n + j\omega_d$$

It lies on the circle given by

$$\sigma^2 + \omega_d^2 = \omega_n^2$$

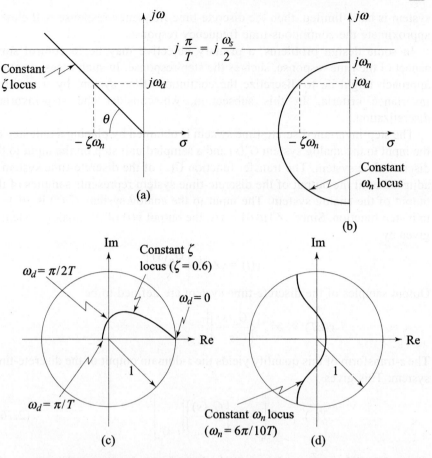

Fig. 2.38 Mapping of constant ζ, and constant ω_n loci from the s-plane to the z-plane

For points in the second quadrant,

$$\sigma = -\sqrt{\omega_n^2 - \omega_d^2}$$

The locus of constant ω_n in the z-plane is given by the relation

$$z = e^{\left(-\sqrt{\omega_n^2 - \omega_d^2} + j\omega_d\right)T}$$

where ω_d varies from 0 to $\omega_s/2$.

Mapping of one representative constant ω_n locus is shown in Fig. 2.38d.

Step Invariance

The basis for impulse invariance is to choose an impulse response for the discrete-time system that is similar to the impulse response of the analog system. The use of this procedure is often motivated not so much by a desire to preserve the impulse-response shape, as by the knowledge that if the analog system is band-limited, then the discrete-time frequency response will closely approximate the continuous-time frequency response.

In some design problems, a primary objective may be to control some aspect of the time response, such as the step response. In such cases, a natural approach might be to discretize the continuous-time system by waveform-invariance criteria. In this subsection, we consider the step-invariant discretization.

The step-invariant discrete-time system is obtained by placing a unit step on the input to the analog system $G_a(s)$ and a sampled unit step on the input to the discrete-time system. The transfer function $G(z)$ of the discrete-time system is adjusted until the output of the discrete-time system represents samples of the output of the analog system. The input to the analog system $G_a(s)$ is $\mu(t)$—a unit step function. Since $\mathscr{L}[\mu(t)] = 1/s$, the output $y(t)$ of the analog system is given by

$$y(t) = \mathscr{L}^{-1}\left\{\frac{G_a(s)}{s}\right\}$$

Output samples of the discrete-time system are defined to be

$$y(kT) = \mathscr{L}^{-1}\left\{\frac{G_a(s)}{s}\right\}\bigg|_{t=kT}$$

The z-transform of this quantity yields the z-domain output of the discrete-time system. This gives

$$Y(z) = \mathscr{Z}\left[\mathscr{L}^{-1}\left\{\frac{G_a(s)}{s}\right\}\bigg|_{t=kT}\right] \qquad (2.76a)$$

Since $\mathscr{Z}[\mu(k)] = \dfrac{z}{z-1}$, where $\mu(k)$ is the unit-step sequence, the output $y(k)$ of the discrete-time system $G(z)$ is given by

$$Y(z) = G(z)\left\{\frac{z}{z-1}\right\} \qquad (2.76b)$$

Comparing Eqn. (2.76b) with Eqn. (2.76a), we obtain

$$G(z) = (1 - z^{-1})\left[\mathscr{Z}\left\{\mathscr{L}^{-1}\left(\frac{G_a(s)}{s}\right)\bigg|_{t=kT}\right\}\right] \qquad (2.77a)$$

or

$$G(z) = (1 - z^{-1})\left[\mathscr{Z}\left(\frac{G_a(s)}{s}\right)\right] \qquad (2.77b)$$

Notice that Eqn. (2.77b) can be rewritten as follows:

$$G(z) = \mathscr{Z}\left[\frac{1-e^{-sT}}{s}G_a(s)\right] \qquad (2.77c)$$

This can easily be established.

Let
$$\mathscr{L}^{-1}\left[\frac{G_a(s)}{s}\right] = g_1(t), \text{ and } \mathscr{Z}[g_1(kT)] = G_1(z)$$

Then
$$\mathscr{L}^{-1}\left[e^{-sT}\frac{G_a(s)}{s}\right] = g_1(t-T)$$
$$\mathscr{Z}\{g_1(kT-T)\} = z^{-1} G_1(z)$$

Therefore,
$$\mathscr{Z}\left[\frac{G_a(s)}{s} - e^{-sT}\frac{G_a(s)}{s}\right] = (1-z^{-1})\mathscr{Z}\left[\frac{G_a(s)}{s}\right]$$

This establishes the equivalence of Eqns (2.77b) and (2.77c).

The right hand side of Eqn. (2.77c) can be viewed as the z-transform of the analog system $G_a(s)$ preceded by zero-order hold (ZOH). Introducing a *fictitious* sampler and ZOH for analytical purposes, we can use the model of Fig. 2.39 to derive a step-invariant equivalent of analog systems. For obvious reasons, step-invariant equivalence is also referred to as ZOH *equivalence*. In the next chapter we will use the ZOH equivalence to obtain discrete-time equivalents of the plants of feedback control systems.

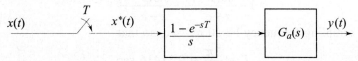

Fig. 2.39 $G_a(s)$ preceded by a fictitious sample-and-hold device

Equivalent discrete-time systems obtained by the step-invariance method may exhibit the frequency folding phenomena and may therefore present the same kind of aliasing errors as found in impulse-invariance method. Notice, however, that the presence of $1/s$ term in $G_a(s)/s$ causes high-frequency attenuation. Consequently, the equivalent discrete-time system obtained by the step-invariance method will exhibit smaller aliasing errors than that obtained by the impulse-invariance method.

As for stability, the equivalent discrete-time system obtained by the step-invariance method is stable if the original continuous-time system is a stable one (refer Review Example 6.2).

Example 2.8

Figure 2.40 shows the model of a plant driven by a D/A converter. In the following, we derive the transfer function model relating $y(kT)$ to $r(kT)$.

The standard D/A converters are designed in such a way that the old value of the input sample is held constant until a new sample arrives. The system of Fig. 2.40 can therefore be viewed as an analog system $G_a(s)$ preceded by zero-order hold, and we can use ZOH equivalence to obtain the transfer function model relating $y(kT)$ to $r(kT)$.

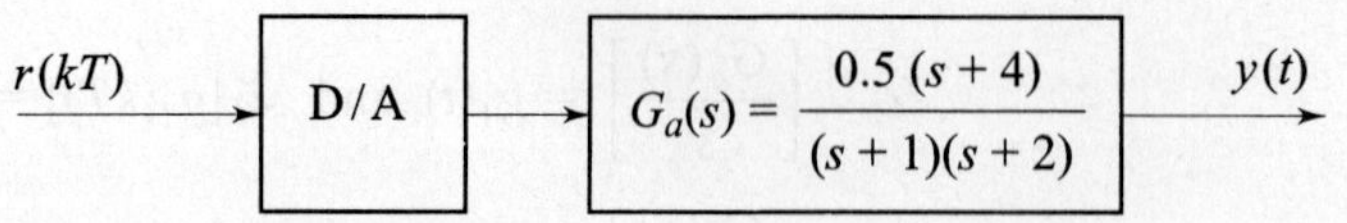

$$r(kT) \longrightarrow \boxed{D/A} \longrightarrow \boxed{G_a(s) = \frac{0.5(s+4)}{(s+1)(s+2)}} \longrightarrow y(t)$$

Fig. 2.40 A plant driven by a D/A converter

ZOH equivalent (step-invariant equivalent) of $G_a(s)$ can be determined as follows. Since

$$\frac{1}{s}G_a(s) = \frac{0.5(s+4)}{s(s+1)(s+2)} = \frac{1}{s} - \frac{1.5}{s+1} + \frac{0.5}{s+2}$$

we have (refer Table of transforms in Appendix A)

$$\mathscr{Z}\left[\frac{1}{s}G_a(s)\right] = \frac{z}{z-1} - \frac{1.5z}{z-e^{-T}} + \frac{0.5z}{z-e^{-2T}}$$

From Eqn. (2.77b),

$$G(z) = \frac{z-1}{z}\left[\frac{z}{z-1} - \frac{1.5z}{z-e^{-T}} + \frac{0.5z}{z-e^{-2T}}\right]$$

$$= 1 - \frac{1.5(z-1)}{z-e^{-T}} + \frac{0.5(z-1)}{z-e^{-2T}}$$

Let the sampling frequency be 20 rad/sec, so that

$$T = \frac{2\pi}{20} = 0.31416 \text{ sec}$$

$$e^{-T} = 0.7304; \; e^{-2T} = 0.5335$$

With these values, we get the following step-invariant equivalent of the given analog system:

$$G(z) = \frac{0.17115z - 0.04535}{z^2 - 1.2639z + 0.3897}$$

Finite-Difference Approximation of Derivatives

Another approach to transforming a continuous-time system into a discrete-time one is to approximate derivatives in a differential equation representation of the continuous-time system by finite differences. This is a common procedure in digital simulations of analog systems and is motivated by the intuitive notion that the derivative of a continuous-time function can be approximated by the difference between consecutive samples of the signal to be differentiated. To illustrate the procedure, consider the first-order differential equation

$$\frac{dy(t)}{dt} + ay(t) = r(t) \tag{2.78}$$

The *backward-difference method* consists of replacing $r(t)$ by $r(k)$, $y(t)$ by $y(k)$; and the first derivative $dy(t)/dt$ by the first backward difference:

$$\left.\frac{dy(t)}{dt}\right|_{t=kT} = \frac{y(k) - y(k-1)}{T} \tag{2.79}$$

This yields the difference equation
$$\frac{y(k) - y(k-1)}{T} + ay(k) = r(k) \qquad (2.80)$$

If T is sufficiently small, we would expect the solution $y(k)$ to yield a good approximation to the samples of $y(t)$.

To interpret the procedure in terms of a mapping of continuous-time function $G_a(s)$ to a discrete-time function $G(z)$, we apply the Laplace transform to Eqn. (2.78) and z-transform to Eqn. (2.80), to obtain
$$sY(s) + aY(s) = R(s)$$

so that
$$G_a(s) = \frac{Y(s)}{R(s)} = \frac{1}{s+a}$$

and
$$\left(\frac{1-z^{-1}}{T}\right) Y(z) + a\, Y(z) = R(z)$$

so that
$$G(z) = \frac{Y(z)}{R(z)} = \frac{1}{\left(\dfrac{1-z^{-1}}{T}\right) + a}$$

Comparing $G_a(s)$ with $G(z)$, we see that
$$G(z) = G_a(s)\Big|_{s=(1-z^{-1})/T}$$

Therefore,
$$s = \frac{1-z^{-1}}{T}; \quad z = \frac{1}{1-sT} \qquad (2.81)$$

is a mapping from the s-plane to the z-plane when the backward-differnece method is used to discretize Eqn. (2.78).

The stability region in the s-plane can be mapped by Eqn. (2.81) into the z-plane as follows. Noting that the stable region in the s-plane is given by $\mathrm{Re}(s) < 0$, the stability region in the z-plane under the mapping (2.81) becomes
$$\mathrm{Re}\left(\frac{1-z^{-1}}{T}\right) = \mathrm{Re}\left(\frac{z-1}{Tz}\right) < 0$$

Writing the complex variable z as $\alpha + j\beta$, we may write the last inequality as
$$\mathrm{Re}\left(\frac{\alpha + j\beta - 1}{\alpha + j\beta}\right) < 0$$

or
$$\mathrm{Re}\left[\frac{(\alpha + j\beta - 1)(\alpha - j\beta)}{\alpha^2 + \beta^2}\right]$$
$$= \mathrm{Re}\left[\frac{\alpha^2 - \alpha + \beta^2 + j\beta}{\alpha^2 + \beta^2}\right] = \frac{\alpha^2 - \alpha + \beta^2}{\alpha^2 + \beta^2} < 0$$

which can be written as
$$(\alpha - 1/2)^2 + \beta^2 < (1/2)^2$$
The stable region in the s-plane can thus be mapped into a circle with centre at $\alpha = 1/2$, $\beta = 0$ and radius equal to 1/2 as shown in Fig. 2.41a.

The backward-difference method is simple and will produce a stable discrete-time system for a stable continuous-time system. Also, the entire s-plane imaginary axis is mapped once and only once onto the small z-plane circle; the folding or aliasing problems do not occur. The penalty is a 'warping' of the equivalent s-plane poles as shown in Fig. 2.41b. This situation is reflected in the relationship between the exact z-transformation and the backward-difference approximation.

Consider the s-plane imaginary axis: $j\omega$. Assuming $\omega < \omega_s/2$, where ω_s is the sampling frequency, the transformation of $j\omega$ without any approximation is given by (refer Eqn. (2.73))
$$z = e^{j\Omega} = e^{j\omega T}$$
The discrete-time frequency response is a replication of the continuous-time frequency response linearly scaled in amplitude and frequency (refer Fig. 2.32).

By backward-difference approximation, $z = e^{j\Omega} = e^{j\omega T}$ is mapped into $j\hat{\omega}$ by the relationship (refer Eqn. (2.81))
$$z = e^{j\omega t} = \frac{1}{1 - j\hat{\omega}T}$$
which can be written as
$$e^{j\omega T} = \frac{1}{2}\left(1 + \frac{1 + j\hat{\omega}T}{1 - j\hat{\omega}T}\right) = \frac{1}{2}[1 + \exp(j2 \tan^{-1} \hat{\omega}T)] \quad (2.82)$$
where the exponent is found from
$$\frac{1 + ja}{1 - ja} = \exp[\ln(1 + ja) - \ln(1 - ja)] = \exp[2 \tanh^{-1} ja] = \exp[j2 \tan^{-1} a]$$
Thus a nonlinear relationship or 'warping' exists between the two frequencies ω and $\hat{\omega}$.

Note that for small ωT, using the first two terms of the exponentials of Eqn. (2.82), yields
$$1 + j\omega T \cong \frac{1}{2}(1 + 1 + j2 \tan^{-1} \hat{\omega}T)$$
or $\tan \omega T \cong \hat{\omega}T$, and $\omega \cong \hat{\omega}$

The 'warping' effect on frequency response is thus negligible for relatively small ωT (about 17° or less).

Let us now investigate the behaviour of the equivalent discrete-time system when the derivative $dy(t)/dt$ in Eqn. (2.78) is replaced by *forward difference*:
$$\left.\frac{dy(t)}{dt}\right|_{t=kT} = \frac{y(k+1) - y(k)}{T}$$

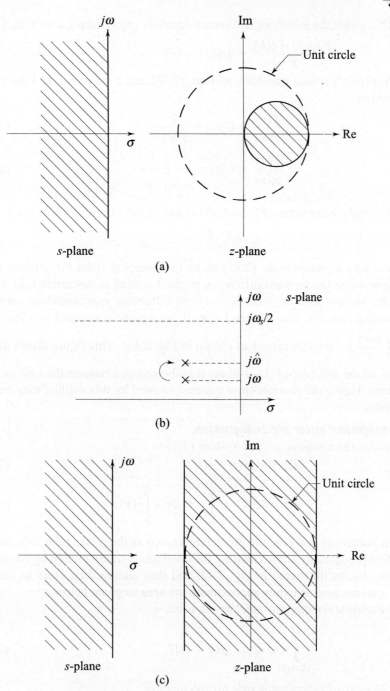

Fig. 2.41 (a) Mapping of the s-plane to the z-plane using backward difference approximation of the derivative (b) Warping effect (c) Mapping of the s-plane to the z-plane using forward difference approximation of the derivative

This yields the following difference equation approximation for Eqn. (2.78):

$$\frac{y(k+1)-y(k)}{T} + ay(k) = r(k) \tag{2.83}$$

Applying Laplace transform to Eqn. (2.78) and z-transform to Eqn. (2.83), we obtain

$$\frac{Y(s)}{R(s)} = G_a(s) = \frac{1}{s+a} \tag{2.84a}$$

and

$$\frac{Y(z)}{R(z)} = G(z) = \frac{1}{\frac{z-1}{T}+a} \tag{2.84b}$$

The right-hand sides of Eqns (2.84a) and (2.84b) become identical if we let

$$s = \frac{z-1}{T} \tag{2.85}$$

We may consider Eqn. (2.85) to be the mapping from the s-plane to the z-plane when the forward-difference method is used to discretize Eqn. (2.78).

One serious problem with the forward-difference approximation method is regarding stability. The left hand side of the s-plane is mapped into the region $\text{Re}\left(\frac{z-1}{T}\right) < 0$ or $\text{Re}(z) < 1$ as shown in Fig. 2.41c. This figure shows that the poles of the left half of the s-plane may be mapped outside the unit circle in z-plane. Hence the discrete-time system obtained by this method may become unstable.

Rectangular rules for integration
Consider the continuous-time system (2.78):

$$\dot{y}(t) = -ay(t) + r(t) \tag{2.86a}$$

or

$$y(t) = y(0) - a\int_0^t y(\tau)\,d\tau + \int_0^t r(\tau)\,d\tau \tag{2.86b}$$

In numerical analysis, the procedure known as the rectangular rule for integration proceeds by approximating the continuous-time function by continuous rectangles, as illustrated in Fig. 2.42, and then adding their areas to compute the total integral. We thus approximate the area as given below:

(i) *Forward rectangular rule for integration*

$$\int_{(k-1)T}^{kT} y(t)\,dt \cong [y(k-1)]T \tag{2.87a}$$

(ii) *Backward rectangular rule for integration*

$$\int_{(k-1)T}^{kT} y(t)\,dt \cong [y(k)]T \tag{2.87b}$$

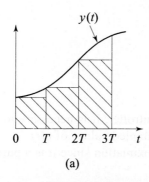 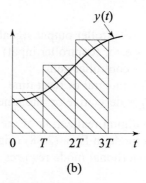

Fig. 2.42 (a) Forward rectangular rule, and (b) backward rectangular rule for integral approximation

With the forward rule for integration, the continuous-time system (2.86) is converted to the following recursive algorithm:

$$y(k) = y(k-1) - aTy(k-1) + Tr(k-1)$$

The z-transformation of this equation gives

$$Y(z) = z^{-1} Y(z) - aT z^{-1} Y(z) + T z^{-1} R(z)$$

or

$$\frac{Y(z)}{R(z)} = \frac{1}{\dfrac{z-1}{T} + a}$$

Laplace transformation of Eqn. (2.86a) gives the transfer function of the continuous-time system:

$$\frac{Y(s)}{R(s)} = \frac{1}{s+a}$$

The forward rectangular rule for integration thus results in the s-plane to z-plane mapping:

$$s = \frac{z-1}{T}$$

which is same as the one obtained by forward-difference approximation of derivatives (Eqn. (2.85)).

Similarly, it can easily be established that the backward rectangular rule for integration results in s-plane to z-plane mapping which is same as the one obtained by backward-difference approximation of derivatives (Eqn. (2.81)).

Example 2.9

The simplest formula for the PID or three-mode controller is the addition of the proportional, integral, and derivative modes:

$$u(t) = K_c \left[e(t) + \frac{1}{T_I} \int_0^t e(t)dt + T_D \frac{de(t)}{dt} \right] \qquad (2.88)$$

where
- u = controller output signal
- e = error (controller input) signal
- K = controller gain
- T_I = integral or reset time
- T_D = derivative or rate time

For the digital realization of the PID controller, it is necessary to approximate each mode in Eqn. (2.88) using the sampled values of $e(t)$.

The proportional mode requires no approximation since it is a purely static part:

$$u_P(k) = K_c\, e(k)$$

The integral mode may be approximated by the backward rectangular rule for integration. If $S(k-1)$ approximates the area under the $e(t)$ curve up to $t = (k-1)T$, then the approximation to the area under the $e(t)$ curve up to $t = kT$ is given by (refer Eqn. (2.87b))

$$S(k) = S(k-1) + Te(k)$$

A digital realization of the integral mode of control is as follows:

$$u_I(k) = \frac{K_c}{T_I}\, S(k)$$

where
$S(k)$ = sum of the areas under the error curve
$= S(k-1) + Te(k)$

The derivative mode may be approximated by the backward-difference approximation:

$$\left.\frac{de(t)}{dt}\right|_{t=kT} = \frac{e(k) - e(k-1)}{T}$$

Therefore,
$$u_D(k) = \frac{K_c T_D}{T}\,[e(k) - e(k-1)]$$

Bringing all the three modes together results in the following PID algorithm:

$$u(k) = u_P(k) + u_I(k) + u_D(k)$$

$$= K_c\left[e(k) + \frac{1}{T_I} S(k) + \frac{T_D}{T}[e(k) - e(k-1)]\right] \quad (2.89a)$$

where $\quad S(k) = S(k-1) + Te(k) \quad (2.89b)$

We can directly use the s-plane to z-plane mapping given by Eqn. (2.81) to obtain the discrete equivalent (2.89) of the PID controller (2.88).

The PID controller (2.88), expressed in terms of operator s, is given by the input-output relation

$$U(s) = K_c\left[1 + \frac{1}{T_I s} + T_D s\right] E(s) \quad (2.90a)$$

The mapping (2.81):

$$s = \frac{1-z^{-1}}{T}$$

corresponds to backward-difference approximation of derivatives. This mapping transforms Eqn. (2.90a) to the following system.

$$U(z) = K_c \left[1 + \frac{T}{T_I}\left(\frac{1}{1-z^{-1}}\right) + \frac{T_D}{T}(1-z^{-1}) \right] E(z) \qquad (2.90b)$$

This is the input-output relation of the PID controller in terms of operator z. By the inverse transform operation, we can express individual control modes by difference equations:

(i) $\qquad u_P(k) = K_c\, e(k)$

(ii) $\qquad u_I(k) - u_I(k-1) = \dfrac{K_c T}{T_I} e(k)$

or $\qquad u_I(k) = u_I(k-1) + \dfrac{K_c T}{T_I} e(k) = \dfrac{K_c}{T_I} S(k)$

where $\qquad S(k) = S(k-1) + T\, e(k)$

(iii) $\qquad u_D(k) = \dfrac{K_c T_D}{T} [e(k) - e(k-1)]$

Bringing all the three modes together results in the PID algorithm given by Eqn. (2.89).

Bilinear Transformation

The technique based on finite-difference approximation to differential equations for deriving a discrete-time system from an analog system, has the advantage that z-transform of the discrete-time system is trivially derived from the Laplace transform of the analog system by an algebraic substitution. The disadvantages of these mappings are that $j\omega$-axis in the s-plane generally does not map into the unit circle in the z-plane and (for the case of forward-difference method) stable analog systems may not always map into stable discrete-time systems.

A nonlinear one-to-one mapping from the s-plane to the z-plane which eliminates the disadvantages mentioned above and which preserves the desired algebraic form is the *bilinear*[9] *transformation* defined by

$$s = \frac{2}{T}\left(\frac{z-1}{z+1}\right) \qquad (2.91)$$

---

9. The transformation is called *bilinear* from consideration of its mathematical form.

This transformation is invertible with the inverse mapping given by

$$z = \frac{1+(T/2)s}{1-(T/2)s} \quad (2.92)$$

The bilinear transformation also arises from a particular approximation method—the *trapezoidal rule* for numerically integrating differential equations.

Let us consider a continuous-time system for which the describing equation is (Eqn. (2.78))

$$\dot{y}(t) = -ay(t) + r(t) \quad (2.93a)$$

or

$$y(t) = y(0) - a\int_0^t y(\tau)\,d\tau + \int_0^t r(\tau)\,d\tau \quad (2.93b)$$

Laplace transformation of Eqn. (2.93a) gives the transfer function of the continuous-time system.

$$\frac{Y(s)}{R(s)} = G_a(s) = \frac{1}{s+a}$$

Applying bilinear transformation (Eqn. (2.91)) to this transfer function, we obtain

$$G(z) = \frac{1}{\dfrac{2}{T}\left(\dfrac{z-1}{z+1}\right)+a}$$

In numerical analysis, the procedure known as the trapezoidal rule for integration proceeds by approximating the continuous-time function by continuous trapezoids, as illustrated in Fig. 2.43, and then adding their areas to compute the total integral. We thus approximate the area

$$\int_{(k-1)T}^{kT} y(t)\,dt \quad \text{by} \quad \frac{1}{2}[y(k) + y(k-1)]T$$

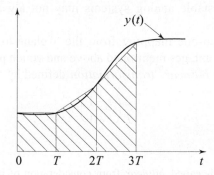

Fig. 2.43 Trapezoidal rule for integral approximation

With this approximation, Eqn. (2.93b) can be converted to the following recursive algorithm:

$$y(k) = y(k-1) - \frac{aT}{2}[y(k) + y(k-1)] + \frac{T}{2}[r(k) + r(k-1)]$$

The z-transformation of this equation gives

$$Y(z) = z^{-1} Y(z) - \frac{aT}{2}[Y(z) + z^{-1} Y(z)] + \frac{T}{2}[R(z) + z^{-1} R(z)]$$

or

$$\frac{Y(z)}{R(z)} = \frac{\frac{T}{2}(1+z^{-1})}{(1-z^{-1}) + \frac{aT}{2}(1+z^{-1})} = \frac{1}{\frac{2}{T}\left(\frac{z-1}{z+1}\right) + a}$$

This result is identical to the one obtained from the transfer function of the continuous-time system by bilinear transformation.

The nature of bilinear transformation is best understood from Fig. 2.44, which shows how the s-plane is mapped onto the z-plane. As seen in the figure, the entire $j\omega$-axis is in the s-plane is mapped onto the unit circle in the z-plane. The left half of s-plane is mapped inside the unit circle in the z-plane, and the right half of s-plane is mapped outside the z-plane unit circle. These properties can easily be established. Consider, for example, the left half of s-plane defined by $\text{Re}(s) < 0$. By means of Eqn. (2.91), this region of s-plane is mapped into the z-plane region defined by

$$\text{Re}\left(\frac{2}{T}\frac{z-1}{z+1}\right) < 0 \text{ or } \text{Re}\left(\frac{z-1}{z+1}\right) < 0$$

By taking the complex variable $z = \alpha + j\beta$, this inequality becomes

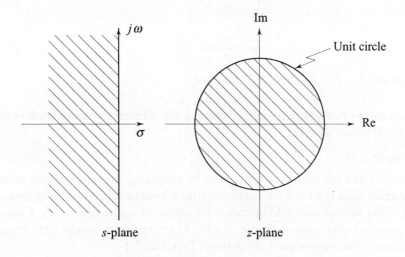

Fig. 2.44 Mapping of the s-plane to the z-plane using bilinear transformation

$$\text{Re}\left(\frac{z-1}{z+1}\right) = \text{Re}\left(\frac{\alpha+j\beta-1}{\alpha+j\beta+1}\right) = \text{Re}\left[\frac{(\alpha-1+j\beta)(\alpha+1-j\beta)}{(\alpha+1+j\beta)(\alpha+1-j\beta)}\right]$$

$$= \text{Re}\left[\frac{\alpha^2-1+\beta^2+j2\beta}{(\alpha+1)^2+\beta^2}\right] < 0$$

which is equivalent to
$$\alpha^2 - 1 + \beta^2 < 0$$
or
$$\alpha^2 + \beta^2 < 1^2$$

which corresponds to the inside of the unit circle in z-plane. The bilinear transformation thus produces a stable discrete-time system for a stable continuous-time system.

Since the entire $j\omega$-axis of the s-plane is mapped once and only once onto the unit circle in the z-plane, the aliasing errors inherent with impulse-invariant transformations are eliminated. However, there is again a warping penalty.

Let $s = j\hat{\omega}$; the bilinear transformation yields (refer Eqn. (2.92))

$$z = \frac{1+j\hat{\omega}T/2}{1-j\hat{\omega}T/2}$$

The exact z-transform yields

$$z = e^{j\omega T}$$

where ω is an equivalent s-plane frequency. Using the approach employed in deriving Eqn. (2.82), we obtain the following relationship between ω and $\hat{\omega}$.

$$e^{j\omega T} = \exp\left(j2\tan^{-1}\frac{\hat{\omega}T}{2}\right)$$

Thus
$$\frac{\omega T}{2} = \tan^{-1}\frac{\hat{\omega}T}{2}$$

or
$$\hat{\omega} = \frac{2}{T}\tan\left(\frac{\omega T}{2}\right) \tag{2.94}$$

When $\omega T/2 < 17°$, or about 0.3 rad, then
$$\hat{\omega} \cong \omega$$
which means that in the frequency domain the bilinear transformation is good for small values of $\omega T/2$.

Example 2.10

A method that has been frequently used by practising engineers to approximate a sampled-data system by a continuous-time system relies on the approximation of the sample-and-hold operation by means of a pure time delay. Consider the sampled-data system of Fig. 2.45a. The sinusoidal steady-state transfer function of the zero-order hold is (refer Eqn. (2.67))

$$G_{h0}(j\omega) = T\frac{\sin(\omega T/2)}{\omega T/2}e^{-j\omega T/2}$$

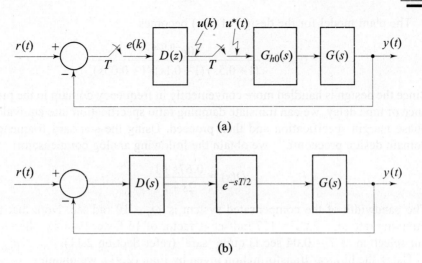

Fig. 2.45 Discretization of analog design

The sinusoidal steady-state transfer function of the impulse-modulation model of the open-loop system is given by (refer Eqn. (2.61a))

$$G_{h0}G^*(j\omega) = \frac{1}{T}\sum_{n=-\infty}^{\infty} G_{h0}\left(j\omega - j\frac{2\pi n}{T}\right)G\left(j\omega - j\frac{2\pi n}{T}\right)$$

Since in most control systems, $G(j\omega)$ has low-pass filter characteristics, we can approximate the right hand side of the equation given above just by the $n = 0$ term. At low frequencies, the magnitude of $\sin(\omega T/2)/(\omega T/2)$ is approximately unity; therefore (refer Eqn. (2.67))

$$G_{h0}G^*(j\omega) = G(j\omega)\, e^{-j\omega T/2}$$

This means that the sample-and-hold operation can be approximated by a pure time delay of one-half the sampling period T. Figure 2.45b shows the continuous-time approximation of the sampled-data system.

The approximating continuous-time system can be used for the design of the discrete-time system as illustrated by the following example.

Consider that the transfer function of the controlled process of the system shown in Fig. 2.45a is

$$G(s) = \frac{10}{(1+0.5s)(1+0.1s)(1+0.05s)}$$

We wish to design a digital controller for the process so that the closed-loop system acquires a damping ratio of 0.4 without loss of steady-state accuracy. We select sampling time $T = 0.04$ sec for the proposed digital controller. Our approach will be to first design an analog controller $D(s)$ for the approximating continuous-time system in Fig. 2.45b (with $G(s) = 10/[(1+0.5s)(1+0.1s)(1+0.05s)]$), that meets the given design specifications and then discretizing $D(s)$ to obtain the digital controller $D(z)$.

The plant model for the design of $D(s)$ becomes

$$G_P(s) = \frac{10e^{-0.02s}}{(1+0.5s)(1+0.1s)(1+0.05s)}$$

Since the design is handled more conveniently in frequency domain in the presence of time delay, we can translate damping ratio specification into equivalent phase margin specification and then proceed. Using the standard frequency-domain design procedure[10], we obtain the following analog compensator:

$$D(s) = \frac{0.67s+1}{2s+1}$$

The bandwidth of the compensated system is $\omega_b = 10$ rad/sec. Note that the sampling rate $\omega_s = 2\pi/T = 157$ rad/sec is factor of 16 faster than ω_b; therefore our selection of $T = 0.04$ sec is quite 'safe' (refer Section 2.11).

Using the bilinear transformation given by Eqn. (2.91), we obtain

$$D(z) = \frac{0.67\left[\frac{50(z-1)}{z+1}\right]+1}{2\left[\frac{50(z-1)}{z+1}\right]+1} = \frac{34.5z-32.5}{101z-99} = \frac{U(z)}{E(z)}$$

which leads to

$$u(k) = 0.9802\ u(k-1) + 0.3416\ e(k) - 0.3218\ e(k-1)$$

where $e(k) = r(k) - y(k)$

This is the proposed algorithm for the digital control system of Fig. 2.45a. To make sure that with the proposed design, the system will behave as expected, we must analyze the system response. Methods for analysis of digital control systems are covered in the next chapter.

▲▲

In this section, we have presented several methods for obtaining discrete-time equivalents for continuous-time systems. The response between sampling points is different for each discretization method used. Furthermore, none of the equivalent discrete-time systems can have complete fidelity. The actual (continuous-time) response between any two consecutive sampling points is always different from the response between the same two consecutive sampling points that is taking place in each equivalent discrete-time system, no matter what method of discretization is used.

It is not possible to say which equivalent discrete-time system is best for any given analog system, since the degree of distortions in transient response and frequency response characteristics depends on the sampling frequency, the cut-off frequency, the highest frequency component involved in the system, transportation lag present in the system, etc. It may be advisable for the designer to try a few alternate forms of the equivalent discrete-time systems for the given analog system.

---

10. Chapters 8–10 of reference [180].

2.13 THE ROUTH STABILITY CRITERION ON THE r-PLANE

The Routh stability test[11] can be applied to the stability analysis of discrete-time systems if a transformation of characteristic equation $\Delta(z) = 0$ can be found which fulfils the following requirements.
   (i) The outside of the unit circle in the z-plane is transformed to the right half of the new complex plane.
   (ii) The boundary of the unit circle in the z-plane is transformed into the imaginary axis of the new complex plane.
   (iii) The inside of the unit circle in the z-plane is transformed into the left half of the new complex plane.

Let us name the new complex plane as w-plane where w is a complex variable. The following bilinear transformation (refer Eqn. (2.92)) satisfies the three requirements stated above:

$$z = \frac{1 + wT/2}{1 - wT/2} \qquad (2.95)$$

The three requirements are also satisfied by the bilinear transformation

$$z = \frac{1+r}{1-r} \qquad (2.96)$$

where r is a complex variable of the r-plane. Transformation (2.96) is in simpler form compared to that given by Eqn. (2.95).

In the stability analysis using the bilinear transformation coupled with the Routh stability criterion, we first substitute $(1 + r)/(1 - r)$ for z in the characteristic equation $\Delta(z) = 0$ to obtain the transformed equation $\Delta(r) = 0$.

Consider a discrete-time system with the characteristic equation

$$\Delta(z) = 4z^3 - 4z^2 - 7z - 2 = 0$$

Applying the bilinear transformation (2.96), we obtain

$$\Delta(r) = 4\left(\frac{1+r}{1-r}\right)^3 - 4\left(\frac{1+r}{1-r}\right)^2 - 7\left(\frac{1+r}{1-r}\right) - 2 = \frac{3r^3 + 17r^2 + 21r - 9}{(1-r)^3} = 0$$

By inspection we see that there is a negative coefficient in the numerator of $\Delta(r)$. Since the necessary conditions for the numerator polynomial of $\Delta(r)$ to have no root with positive real part are that all its coefficients must be of the same sign and none of the coefficients may vanish, we know that $\Delta(r) = 0$ must have at least one root in the right-half of the r-plane. This fact means that $\Delta(z) = 0$ must have at least one root outside the unit circle in the z-plane and that the discrete-time system is unstable.

---

11. Chapter 5 of reference [180].

2.14 REVIEW EXAMPLES

Review Example 2.1

Consider the difference equation
$$y(k+2) + a_1 y(k+1) + a_2 y(k) = b_0 r(k+2) + b_1 r(k+1) + b_2 r(k) \quad (2.97)$$

Assuming that the system is initially at rest, and $r(k) = 0$ for $k < 0$, obtain the transfer function $G(z) = Y(z)/R(z)$.

Solution

The z-transform of Eqn. (2.97) gives (refer Eqn. (A.38) in Appendix A),

$$[z^2 Y(z) - z^2 y(0) - z y(1)] + a_1 [z Y(z) - z y(0)] + a_2 Y(z)$$
$$= b_0 [z^2 R(z) - z^2 r(0) - z r(1)] + b_1 [z R(z) - zr(0)] + b_2 R(z)$$

or $(z^2 + a_1 z + a_2) Y(z) = (b_0 z^2 + b_1 z + b_2) R(z) + z^2[y(0) - b_0 r(0)]$
$$+ z[y(1) + a_1 y(0) - b_0 r(1) - b_1 r(0)] \quad (2.98)$$

Since the system is initially at rest, $y(k) = 0$ for $k < 0$. To determine the initial conditions $y(0)$ and $y(1)$, we substitute $k = -2$ and $k = -1$, respectively, into Eqn. (2.97):

$$y(0) + a_1 y(-1) + a_2 y(-2) = b_0 r(0) + b_1 r(-1) + b_2 r(-2)$$

which simplifies to
$$y(0) = b_0 r(0) \quad (2.99a)$$

and $y(1) + a_1 y(0) + a_2 y(-1) = b_0 r(1) + b_1 r(0) + b_2 r(-1)$

or $y(1) = -a_1 y(0) + b_0 r(1) + b_1 r(0) \quad (2.99b)$

By substituting Eqn. (2.99) into Eqn. (2.98), we get

$$(z^2 + a_1 z + a_2) Y(z) = (b_0 z^2 + b_1 z + b_2) R(z)$$

Therefore $G(z) = \dfrac{Y(z)}{R(z)} = \dfrac{b_0 z^2 + b_1 z + b_2}{z^2 + a_1 z + a_2} = \dfrac{b_0 + b_1 z^{-1} + b_2 z^{-2}}{1 + a_1 z^{-1} + a_2 z^{-2}}$

The same result can be obtained by taking the z-transformation of the shifted difference equation
$$y(k) + a_1 y(k-1) + a_2 y(k-2) = b_0 r(k) + b_1 r(k-1) + b_2 r(k-2)$$

Review Example 2.2

Consider a first-order discrete-time system described by the difference equation
$$y(k+1) + a_1 y(k) = b_0 r(k+1) + b_1 r(k) \quad (2.100)$$

The input is switched to the system at $k = 0$ ($r(k) = 0$ for $k < 0$); the initial state $y(-1)$ of the system is specified. Obtain a simulation diagram for the system.

Solution

State variable models of discrete-time systems can easily be translated into digital computer simulation diagrams. Methods of conversion of difference equation models to state variable models will be presented in Chapter 6.

It can easily be verified that the following state variable model represents the given difference Eqn. (2.100):

$$x(k+1) = -a_1 x(k) + r(k) \qquad (2.101)$$

$$y(k) = (b_1 - a_1 b_0) x(k) + b_0 r(k)$$

In terms of the specified initial condition $y(-1)$ of the difference equation model (2.100), the initial state $x(0)$ of the state variable model (2.101) is given by

$$x(0) = \frac{-a_1}{b_1 - a_1 b_0} y(-1)$$

Note that if the first-order discrete-time system is relaxed before switching on the input $r(k)$ at $k = 0$, the initial condition $y(-1) = 0$ for the model (2.100), and equivalently the initial condition $x(0) = 0$ for the model (2.101).

Figure 2.46 shows a simulation diagram for the given discrete-time system.

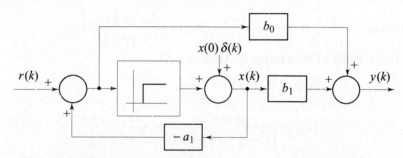

Fig. 2.46 Simulation diagram for the state model (2.101)

Review Example 2.3

Consider a discrete-time system

$$y(k+2) + \frac{1}{4} y(k+1) - \frac{1}{8} y(k) = 3r(k+1) - r(k) \qquad (2.102)$$

with input

$$r(k) = (-1)^k \mu(k)$$

and initial conditions

$$y(-1) = 5, \; y(-2) = -6$$

Find the output $y(k); \; k \geq 0$

Solution

The difference Eqn. (2.102) is first converted to the equivalent form

$$y(k) + \frac{1}{4} y(k-1) - \frac{1}{8} y(k-2) = 3r(k-1) - r(k-2) \qquad (2.103)$$

z-transformation of the linear difference Eqn. (2.103) requires the following results:

114 Digital Control and State Variable Methods

$$\mathscr{Z}[y(k-1)] = \sum_{k=0}^{\infty} y(k-1)z^{-k}$$

$$= y(-1) + z^{-1}\sum_{k=0}^{\infty} y(k)z^{-k} = y(-1) + z^{-1}Y(z) \quad (2.104a)$$

$$\mathscr{Z}[y(k-2)] = y(-2) + z^{-1}[y(-1) + z^{-1}Y(z)]$$

$$= y(-2) + z^{-1}y(-1) + z^{-2}Y(z) \quad (2.104b)$$

z-transformation of each term in Eqn. (2.103) yields

$$Y(z) + \tfrac{1}{4}[z^{-1}Y(z) + y(-1)] - \tfrac{1}{8}[z^{-2}Y(z) + z^{-1}y(-1) + y(-2)]$$
$$= 3[z^{-1}R(z) + r(-1)] - [z^{-2}R(z) + z^{-1}r(-1) + r(-2)]$$

Since $\quad r(-1) = r(-2) = 0$, we have

$$\left(1 + \tfrac{1}{4}z^{-1} - \tfrac{1}{8}z^{-2}\right)Y(z) = (3z^{-1} - z^{-2})R(z) + \tfrac{5}{8}z^{-1} - 2$$

or $\quad \left(z^2 + \tfrac{1}{4}z - \tfrac{1}{8}\right)Y(z) = (3z - 1)R(z) + \tfrac{5}{8}z - 2z^2$

Therefore $\quad Y(z) = \dfrac{3z-1}{z^2 + \tfrac{1}{4}z - \tfrac{1}{8}}R(z) + \dfrac{-2z^2 + \tfrac{5}{8}z}{z^2 + \tfrac{1}{4}z - \tfrac{1}{8}}$

For (refer Table of transforms in Appendix A)

$$R(z) = \mathscr{Z}[(-1)^k] = \dfrac{z}{z+1},$$

$$Y(z) = \dfrac{z(3z-1)}{\left(z^2 + \tfrac{1}{4}z - \tfrac{1}{8}\right)(z+1)} + \dfrac{-2z^2 + \tfrac{5}{8}z}{z^2 + \tfrac{1}{4}z - \tfrac{1}{8}}$$

$$= \dfrac{-2z^3 + \tfrac{13}{8}z^2 - \tfrac{3}{8}z}{\left(z + \tfrac{1}{2}\right)\left(z - \tfrac{1}{4}\right)(z+1)}$$

Expanding $Y(z)/z$ into partial fractions,

$$\dfrac{Y(z)}{z} = \dfrac{-2z^2 + \tfrac{13}{8}z - \tfrac{3}{8}}{\left(z + \tfrac{1}{2}\right)\left(z - \tfrac{1}{4}\right)(z+1)} = \dfrac{\tfrac{9}{2}}{z + \tfrac{1}{2}} + \dfrac{-\tfrac{1}{10}}{z - \tfrac{1}{4}} + \dfrac{-\tfrac{32}{5}}{z+1}$$

Then (refer Table of transforms in Appendix A)

$$y(k) = \left[\tfrac{9}{2}\left(-\tfrac{1}{2}\right)^k - \tfrac{1}{10}\left(\tfrac{1}{4}\right)^k - \tfrac{32}{5}(-1)^k\right]\mu(k)$$

Review Example 2.4

Consider a second-order discrete-time system described by the difference equation

$$y(k+2) - \tfrac{3}{2}y(k+1) + \tfrac{1}{2}y(k) = r(k+1) + \tfrac{1}{2}r(k)$$

The system is initially relaxed ($y(k) = 0$ for $k < 0$) and is excited by the input

Signal Processing and Digital Control

$$r(k) = \begin{cases} 0; & k = 0 \\ 1; & k > 0 \end{cases}$$

Shifting the difference equation by two sampling intervals, we obtain
$$y(k) - \tfrac{3}{2} y(k-1) + \tfrac{1}{2} y(k-2) = r(k-1) + \tfrac{1}{2} r(k-2)$$
z-transformation of this equation gives
$$Y(z) = \frac{z^{-1} + \tfrac{1}{2} z^{-2}}{1 - \tfrac{3}{2} z^{-1} + \tfrac{1}{2} z^{-2}} R(z) = \frac{z + \tfrac{1}{2}}{(z - \tfrac{1}{2})(z - 1)} R(z)$$

The system modes are $\left(\tfrac{1}{2}\right)^k$ and $(1)^k$. The mode $\left(\tfrac{1}{2}\right)^k$ decays as $k \to \infty$, and the mode $(1)^k$ is constant (i.e., it remains within finite bounds for all k).

The input $\qquad r(k) = \mu(k-1)$

Therefore $\qquad R(z) = z^{-1} \left[\dfrac{z}{z-1}\right] = \dfrac{1}{z-1}$

For this input, $\qquad Y(z) = \dfrac{z + \tfrac{1}{2}}{(z - \tfrac{1}{2})(z - 1)^2}$

It is observed that excitation pole matches one of the system poles. Though the system modes as well as the input do not grow with increasing k, the effect of the pole-matching is to give rise to a time function in forced response of the system that grows idenfinitely as $k \to \infty$. This is evident from the inverse transform of $Y(z)$ (refer Eqn. (A.39) and the Table of transforms in Appendix A):

$$Y(z) = \frac{4}{z - \tfrac{1}{2}} + \frac{3}{(z-1)^2} + \frac{-4}{z-1}$$
$$y(k) = 4\left(\tfrac{1}{2}\right)^{k-1} + 3(k-1)(1)^{k-1} - 4(1)^{k-1}$$
$$\qquad = 4\left(\tfrac{1}{2}\right)^{k-1} + 3(k-1) - 4; k \geq 1$$
$$\qquad = 4\left(\tfrac{1}{2}\right)^{k-1} + 3k - 7; k \geq 1$$

Review Example 2.5

Solve for $y(k)$ the equation
$$y(k) = r(k) - r(k-1) - y(k-1), k \geq 0$$
where $r(k) = \begin{cases} 1; & k \text{ even} \\ 0; & k \text{ odd} \end{cases}$; $y(-1) = r(-1) = 0$

Solution

z-transformation of the given equation yields
$$Y(z) = \frac{1 - z^{-1}}{1 + z^{-1}} R(z) = \frac{z - 1}{z + 1} R(z)$$

For the given input,

$$R(z) = 1 + z^{-2} + z^{-4} + \cdots = \frac{1}{1-x}\bigg|_{x=z^{-2}} = \frac{1}{1-z^{-2}} = \frac{z^2}{z^2-1}$$

Thus $Y(z) = \left(\dfrac{z-1}{z+1}\right)\dfrac{z^2}{z^2-1} = \dfrac{z^2}{z^2+2z+1}$

We can expand $Y(z)$ into a power series by dividing the numerator of $Y(z)$ by its denominator:

$$\begin{array}{r}
1 - 2z^{-1} + 3z^{-2} - 4z^{-3} + \cdots \\
z^2+2z+1 \overline{) z^2 } \\
\underline{z^2 + 2z + 1} \\
-2z - 1 \\
\underline{-2z - 4 - 2z^{-1}} \\
3 + 2z^{-1} \\
\underline{3 + 6z^{-1} + 3z^{-2}} \\
-4z^{-1} - 3z^{-2} \\
\vdots
\end{array}$$

Therefore $Y(z) = 1 - 2z^{-1} + 3z^{-2} - 4z^{-3} + \cdots$
and the values of $y(k)$ are $\{1, -2, 3, -4, \cdots\}$

Review Example 2.6

Find the response of the system shown in Fig. 2.47 to a unit impulse input.

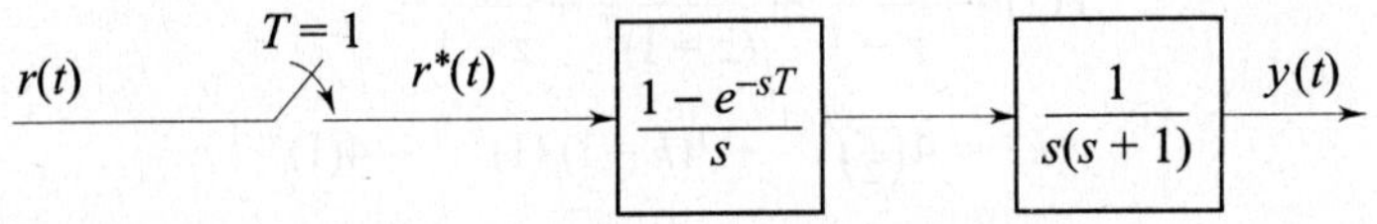

Fig. 2.47

Solution

The discrete-time transfer function of the given system is obtained as follows (refer Eqns (2.77)):

$$G_{h0}(s) = \frac{1-e^{-sT}}{s}, \quad G_a(s) = \frac{1}{s(s+1)}$$

$$\frac{Y(z)}{R(z)} = \mathscr{Z}[G_{h0}(s)G_a(s)] = (1-z^{-1})\,\mathscr{Z}\!\left(\frac{G_a(s)}{s}\right)$$

$$= (1-z^{-1})\,\mathscr{Z}\!\left[\frac{1}{s^2(s+1)}\right] = (1-z^{-1})\,\mathscr{Z}\!\left[\frac{1}{s^2} - \frac{1}{s} + \frac{1}{s+1}\right]$$

Using the Table of transforms given in Appendix A, we obtain

$$\frac{Y(z)}{R(z)} = (1-z^{-1})\left[\frac{Tz}{(z-1)^2} - \frac{z}{z-1} + \frac{z}{z-e^{-T}}\right]$$

$$= \left[\frac{(ze^{-T} - z + Tz) + (1 - e^{-T} - Te^{-T})}{(z-1)(z-e^{-T})}\right]$$

For $T = 1$, we have

$$\frac{Y(z)}{R(z)} = \frac{ze^{-1} + 1 - 2e^{-1}}{(z-1)(z-e^{-1})}$$

$$= \frac{0.3678z + 0.2642}{(z-1)(z-0.3679)} = \frac{0.3678z + 0.2642}{z^2 - 1.3678z + 0.3679}$$

For unit-impulse input, $R(z) = 1$.

Therefore $$Y(z) = \frac{0.3678z + 0.2642}{z^2 - 1.3678z + 0.3679}$$

We can expand $Y(z)$ into a power series by dividing the numerator of $Y(z)$ by its denominator:

$$\begin{array}{r} 0.3678z^{-1} + 0.7675z^{-2} + 0.9145z^{-3} + \cdots \\ z^2 - 1.3678z + 0.3678 \overline{\smash{\big)}\, 0.3678z + 0.2644} \\ \underline{0.3678z - 0.5031 + 0.1353z^{-1}} \\ +0.7675 - 0.1353z^{-1} \\ \underline{+0.7675 - 1.0497z^{-1} + 0.2823z^{-2}} \\ 0.9145z^{-1} - 0.2823z^{-2} \\ \vdots \end{array}$$

This calculation yields the response at the sampling instants and can be carried on as far as needed. In this case we have obtained $y(kT)$ as follows: $y(0) = 0$, $y(T) = 0.3678$, $y(2T) = 0.7675$, and $y(3T) = 0.9145$.

Review Example 2.7

A PID controller is described by the following relation between input $e(t)$ and output $u(t)$:

$$u(t) = K_c\left[e(t) + \frac{1}{T_I}\int_0^t e(t)\,dt + T_D\frac{de(t)}{dt}\right] \quad (2.105)$$

Using the trapezoidal rule for integration and backward-difference approximation for the derivatives, obtain the difference-equation model of the PID algorithm. Also obtain the transfer function $U(z)/E(z)$.

Solution

By the trapezoidal rule for integration, we obtain

$$\int_0^{kT} e(t)\,dt = T\left[\frac{e(0) + e(T)}{2} + \frac{e(T) + e(2T)}{2} + \cdots + \frac{e((k-1)T) + e(kT)}{2}\right]$$

$$= T\left[\sum_{i=1}^{k} \frac{e((i-1)T)+e(iT)}{2}\right]$$

By backward-difference approximation for the derivatives (refer Eqn. (2.79)), we get

$$\left.\frac{de(t)}{dt}\right|_{t=kT} = \frac{e(kT)-e((k-1)T)}{T}$$

A difference-equation model of the PID controller is, therefore, given by

$$u(k) = K_c \left\{ e(k) + \frac{T}{T_I} \sum_{i=1}^{k} \frac{e(i-1)+e(i)}{2} + \frac{T_D}{T}[e(k)-e(k-1)] \right\} \quad (2.106)$$

Let us now obtain the transfer function model of the PID control algorithm given by Eqn. (2.106).

Define (refer Fig. 2.48)

$$\frac{e(i-1)+e(i)}{2} = f(i); \; f(0) = 0$$

Then
$$\sum_{i=1}^{k} \frac{e(i-1)+e(i)}{2} = \sum_{i=1}^{k} f(i)$$

Taking the z-transform of this equation (refer Eqn. (A.42) in Appendix A), we obtain

$$\mathcal{Z}\left[\sum_{i=1}^{k} \frac{e(i-1)+e(i)}{2}\right] = \mathcal{Z}\left[\sum_{i=1}^{k} f(i)\right] = \frac{z}{z-1} F(z)$$

Notice that
$$F(z) = \mathcal{Z}\left[\frac{e(i-1)+e(i)}{2}\right] = \frac{1+z^{-1}}{2} E(z)$$

Hence
$$\mathcal{Z}\left[\sum_{i=1}^{k} \frac{e(i-1)+e(i)}{2}\right] = \frac{1+z^{-1}}{2(1-z^{-1})} E(z) = \frac{z+1}{2(z-1)} E(z)$$

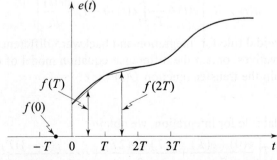

Fig. 2.48 Diagram depicting function $f(kT)$

The z-transform of Eqn. (2.106) becomes

$$U(z) = K_c\left[1 + \frac{T}{2T_1}\frac{1+z^{-1}}{1-z^{-1}} + \frac{T_D}{T}(1-z^{-1})\right]E(z)$$

$$= K_c\left[1 + \frac{T}{2T_1}\left(\frac{z+1}{z-1}\right) + \frac{T_D}{T}\left(\frac{z-1}{z}\right)\right]E(z) \qquad (2.107)$$

This equation gives the transfer function model of the PID control algorithm.

Note that we can obtain the discrete-time transfer function model (2.107) by expressing the PID controller (2.105) in terms of operator s and then using the mapping (2.81) for the derivative term and the mapping (2.91) for the integral term of the controller.

Review Example 2.8

Derive the difference equation model for the numerical solution of the differential equation

$$\frac{d^2y(t)}{dt^2} + a_1\frac{dy(t)}{dt} + a_2y(t) = r(t); \; y(0) = y_1^0, \; \frac{dy(0)}{dt} = y_2^0, \; 0 \le t \le t_f$$

$$(2.108)$$

Use backward difference approximation for the derivatives.

Solution

We divide the interval $0 \le t \le t_f$ into N equal intervals of width equal to *step-length* T:

$$\frac{t_f}{N} = T; \; t = kT, \; k = 0, 1, 2, ..., N$$

By backward difference approximation,

$$\left.\frac{dy(t)}{dt}\right|_{t=kT} \triangleq \dot{y}(k) = \frac{y(k) - y(k-1)}{T} \qquad (2.109)$$

$$\left.\frac{d^2y(t)}{dt^2}\right|_{t=kT} \triangleq \ddot{y}(k) = \frac{\dot{y}(k) - \dot{y}(k-1)}{T} = \frac{1}{T^2}[y(k) - 2y(k-1) + y(k-2)]$$

$$(2.110)$$

From Eqn. (2.109), we have

$$\dot{y}(0) = y_2^0 = \frac{y_1^0 - y(-1)}{T}$$

Substituting Eqns (2.109) and (2.110) into (2.108) at $t = kT$, we obtain

$$\frac{1}{T^2}[y(k) - 2y(k-1) + y(k-2)] + \frac{a_1}{T}[y(k) - y(k-1)] + a_2 y(k) = r(k)$$

or $\left(a_2 + \dfrac{a_1}{T} + \dfrac{1}{T^2}\right) y(k) - \left(\dfrac{a_1}{T} + \dfrac{2}{T^2}\right) y(k-1) + \dfrac{1}{T^2} y(k-2) = r(k);$

$$y(0) = y_1^0, \; y(-1) = y_1^0 - Ty_2^0 \qquad (2.111)$$

Incrementing k to take on values $k = 1, 2, \ldots, N$, we can easily obtain $y(1), \ldots, y(N)$ from Eqn. (2.111) by the iterative procedure.

Review Example 2.9

It is desired to obtain the numerical solution of the nth order state equation

$$\dot{\mathbf{x}}(t) = \mathbf{f}(\mathbf{x}(t), \mathbf{r}(t)) \qquad (2.112)$$

in the time interval $0 \le t \le t_f$; given the input vector $\mathbf{r}(t)$, and the initial state vector $\mathbf{x}(0) \triangleq \mathbf{x}^0$.

We divide the interval $0 \le t \le t_f$ into N equal intervals of step length $T = t_f/N$, and set $t = kT;\; k = 0, 1, 2, \ldots N$.

Integrating both sides of Eqn. (2.112) over the interval $[kT, (k+1)T)$, we obtain

$$\mathbf{x}(k+1) = \mathbf{x}(k) + \int_{kT}^{(k+1)T} \mathbf{f}(\mathbf{x}(t), \mathbf{r}(t))\, dt$$

The forward rectangular rule for integration (refer Eqn. (2.87b)) yields

$$\mathbf{x}(k+1) = \mathbf{x}(k) + T\mathbf{f}(\mathbf{x}(k), \mathbf{r}(k));\; \mathbf{x}(0) = \mathbf{x}^0,\; k = 0, 1, \ldots, N \qquad (2.113)$$

The differential Eqn. (2.112) is thus converted into the difference Eqn. (2.113) which can be solved in an iterative manner using a computer. The smaller T is made (or the larger N is made), the more accurate is our approximation.

This technique of replacing a differential equation with a difference equation is known as the *Euler method*.

PROBLEMS

2.1 Consider the signal processing algorithm shown in Fig. P2.1.
  (a) Assign the state variables and obtain a state variable model for the system.
  (b) Represent the algorithm of Fig. P2.1 by a signal flow graph and from there obtain the transfer function model of the system using Mason's gain formula.

Signal Processing and Digital Control

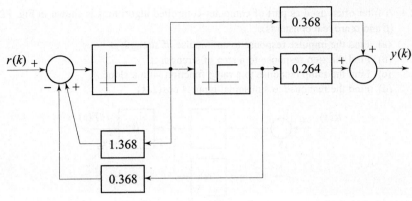

Fig. P2.1

2.2 Consider the signal processing algorithm shown in Fig. P2.2. Represent the algorithm by (a) difference equation model, (b) a state variable model, and (c) a transfer function model.

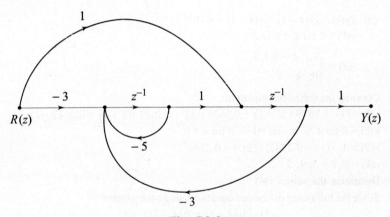

Fig. P2.2

2.3 Consider the discrete-time system shown in Fig. P2.3
   (a) Obtain the difference equation model and therefrom the transfer function model of the system.
   (b) Find the impulse response of the system.
   (c) Find the response of the system to unit step input $\mu(k)$.

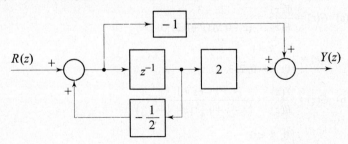

Fig. P2.3

2.4 A filter often used as part of computer-controlled algorithms is shown in Fig. P2.4 (β and α are real constants).
 (a) Find the impulse response to an impulse of strength A.
 (b) Find the step response to a step of strength A.
 (c) Find the ramp response to a ramp function with a slope A.
 (d) Find the response to sinusoidal input $A \cos(\Omega k)$.

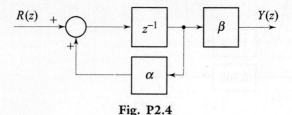

Fig. P2.4

2.5 Solve the following difference equations:
 (a) $y(k+2) + 3y(k+1) + 2y(k) = 0$; $y(-1) = -\frac{1}{2}$, $y(-2) = \frac{3}{4}$
 (b) $2y(k) - 2y(k-1) + y(k-2) = r(k)$
 $y(k) = 0$ for $k < 0$ and
 $$r(k) = \begin{cases} 1; & k = 0, 1, 2, \ldots \\ 0; & k < 0 \end{cases}$$

2.6 Consider the difference equation:
 $y(k+2) - 1.3679\, y(k+1) + 0.3679\, y(k) = 0.3679\, r(k+1) + 0.2642\, r(k)$
 $y(k) = 0$ for $k \leq 0$, and $r(k) = 0$ for $k < 0$,
 $r(0) = 1$, $r(1) = 0.2142$, $r(2) = -0.2142$
 $r(k) = 0$, $k = 3, 4, 5, \ldots$
 Determine the output $y(k)$.

2.7 Solve the following difference equation using z-transforms:
 $$y(k) - 3y(k-1) + 2y(k-2) = r(k)$$
 where $r(k) = \begin{cases} 1 \text{ for } k = 0, 1 \\ 0 \text{ for } k \geq 2 \end{cases}$; $y(-2) = y(-1) = 0$

 Will the final value theorem give the correct value of $y(k)$ as $k \to \infty$? Why?

2.8 For the transfer function models and inputs given below, find the response $y(k)$ as a function of k:
 (a) $G(z) = \dfrac{Y(z)}{R(z)} = \dfrac{2z - 3}{(z - 0.5)(z + 0.3)}$
 $r(k) = \begin{cases} 1; & k = 1 \\ 0; & k = 0, 2, 3, 4, \ldots \end{cases}$
 (b) $G(z) = \dfrac{Y(z)}{R(z)} = \dfrac{-6z + 1}{\left(z - \frac{1}{2} + j\frac{1}{4}\right)\left(z - \frac{1}{2} - j\frac{1}{4}\right)}$
 $r(k) = \begin{cases} 0; & k < 0 \\ 1; & k = 0, 1, 2, 3, \ldots \end{cases}$

2.9 For the transfer function models and inputs given below, find the response $y(k)$ as a function of k:

(a) $G(z) = \dfrac{Y(z)}{R(z)} = \dfrac{1}{(z-0.5)(z+0.3)}$

$r(k) = \begin{cases} 1; & k \text{ even} \\ 0; & k \text{ odd} \end{cases}$

(b) $G(z) = \dfrac{Y(z)}{R(z)} = \dfrac{1}{(z-0.5)^2(z-0.1)}$

$r(k) = \begin{cases} 0; & k < 0 \\ 1; & k = 0, 1, 2, 3, \ldots \end{cases}$

2.10 Determine $y(\infty)$ for the following $Y(z)$ function (a is a real constant):

$$Y(z) = \dfrac{K[z^3 - 2az^2 + (a^3 - a^2 + a)z]}{(z-1)(z-a)(z-a^2)}$$

Assuming stable response, determine what the value of K must be for $y(\infty) = 1$.

2.11 Given: $y(k) = \dfrac{k!(\alpha)^{k-m+1}}{(k-m+1)!(m-1)!}$

Prove that $y(k)$ decays to zero as $k \to \infty$ if $|\alpha| < 1$.

2.12 A system has the transfer function

$$G(z) = \dfrac{Y(z)}{R(z)} = \dfrac{1}{z^2+1}.$$

Show that when the input $r(k)$ is a unit step function, the output $y(k)$ is bounded; and when the input

$$r(k) = \{1, 0, -1, 0, 1, 0, -1, \ldots\},$$

the output $y(k)$ is unbounded.

Explain why a bounded input produces a bounded output in the first case but an unbounded output in the second case.

2.13 Check if all the roots of the following characteristic equations lie within the unit circle:
(a) $z^3 - 1.3z^2 - 0.08z + 0.24 = 0$
(b) $z^4 - 1.368z^3 + 0.4z^2 + 0.08z + 0.002 = 0$

2.14 Using r-transformation followed by the Routh stability criterion, find the number of poles of the following transfer function that lie inside the unit circle on the z-plane.

$$G(z) = \dfrac{3z^4 + 2z^3 - z^2 + 4z + 5}{z^4 + 0.5z^3 - 0.2z^2 + z + 0.4}$$

2.15 Figure P2.15 shows the input-output description of a D/A converter. The converter is designed in such a way that the old value of the input sample is held constant until a new sample arrives. Treating each sample of the sequence $r(kT)$ as an impulse function of strength equal to the value of the sample, the system of Fig. P2.15 becomes a continuous-time system. Determine the transfer function model of the system.

Fig. P2.15

2.16 (a) State and prove the sampling theorem.

(b) Given: $E(s) = \dfrac{10(s+2)}{s^2(s^2+2s+2)}$

Based upon the sampling theorem, determine the maximum value of the sampling interval T that can be used to enable us to reconstruct $e(t)$ from its samples.

(c) Consider a system with sampling frequency 50 rad/sec. A noise signal $\cos 50t$ enters into the system. Show that it can cause a dc component in the system output.

2.17 Draw the magnitude and phase curves of the zero-order hold, and compare these curves with those of the ideal low-pass filter.

2.18 Consider a signal $f(t)$ which has discrete values $f(kT)$ at the sampling rate $1/T$. If the signal $f(t)$ is imagined to be impulse sampled at the same rate, it becomes

$$f^*(t) = \sum_{k=0}^{N} f(kT)\, \delta(t - kT)$$

(a) Prove that $F(z)\big|_{z=e^{sT}} = F^*(s)$

(b) Determine $F(z)\big|_{z=e^{sT}}$ in terms of $F(s)$. Using this result, explain the relationship between the z-plane and the s-plane.

2.19 Figure P2.19 shows two root paths in the s-plane:

(i) roots with the same time-constant $\tau = 1/a$;

(ii) roots with the same oscillation frequency ω_0.

Derive and sketch the corresponding root paths in the z-plane under the impulse-invariant transformation. Sampling frequency is ω_s.

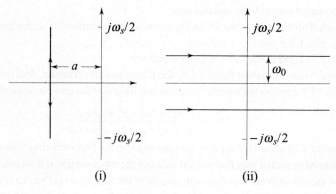

Fig. P2.19

2.20 Figure P2.20 shows a discrete-time system. Determine the transfer function $G(z)$ of this system assuming that the samplers operate synchronously at intervals of T sec. Also find the unit-step response of the system.

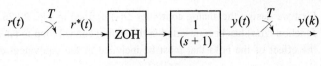

Fig. P2.20

2.21 Figure P2.21 shows the model of a plant driven by a D/A converter. Derive the transfer function model relating $r(kT)$ and $y(kT)$; $T = 0.4$ sec.

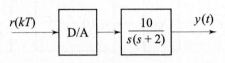

Fig. P2.21

2.22 Show that if y is the integral of a function r, then
   (i) by the backward rectangular rule for integration,
$$Y(z) = \frac{Tz}{z-1} R(z);$$
   (ii) by the trapezoidal rule for integration
$$Y(z) = \frac{T}{2} \frac{z+1}{z-1} R(z)$$

2.23 Consider the discretization method based on the backward-difference approximation of derivatives as a mapping from the s-plane to the z-plane. Show that the mapping transforms the left-half of s-plane into a circle in the z-plane. Is the size of the circle dependent on the choice of the sampling interval?

2.24 Prove that the bilinear transformation maps the left-half of the s-plane into the unit circle in the z-plane.

The transformation $z = e^{sT}$ also maps the left-half of the s-plane into the unit circle in the z-plane. What is the difference between the two maps?

2.25 A PID controller is described by the following relation between input $e(t)$ and output $u(t)$:
$$u(t) = K_c \left[e(t) + \frac{1}{T_I} \int_0^t e(t)dt + T_D \frac{de(t)}{dt} \right]$$

Obtain the PID control algorithm by the discretization of the equation
$$\dot{u}(t) = K_c \left[\dot{e}(t) + \frac{1}{T_I} e(t) + T_D \ddot{e}(t) \right]$$

using the backward-difference approximation of the derivatives. Also find the transfer function $U(z)/E(z)$.

2.26 For a plant $1.57/[s(s + 1)]$, we are required to design a digital controller so that the closed-loop system acquires a damping ratio of 0.45 without loss of steady-state accuracy. The sampling period $T = 1.57$ sec. The following design procedure may be followed.

(i) First we design the analog controller $D(s)$ defined in Fig. P2.26. The transfer function $G_h(s)$ has been inserted in the analog control loop to take into account the effect of the hold that must be included in the equivalent digital control system. Verify that $D(s) = \dfrac{25s+1}{62.5s+1}$ meets the design requirements.

(ii) Discretize $D(s)$ using bilinear transformation.

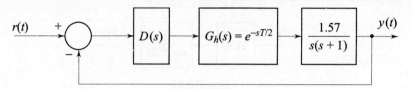

Fig. P2.26

2.27 A PID controller is described by the following relation between input $e(t)$ and output $u(t)$:

$$U(s) = K_c\left[1 + \dfrac{1}{T_I s} + T_D s\right] E(s)$$

(a) Derive the PID algorithm using the s-plane to z-plane maps—bilinear transformation for integration and backward-difference approximation for the derivatives.

(b) Convert the transfer function model of the PID controller obtained in step (a) into a difference equation model.

2.28 Derive difference equation models for the numerical solution of the following differential equation using (a) the backward rectangular rule for integration, and (b) the forward rectangular rule for integration:

$$\dot{y}(t) + ay(t) = r(t); \ y(0) = y^0$$

2.29 Consider the second-order system

$$\ddot{y} + a\dot{y} + by = 0; \ y(0) = \alpha, \ \dot{y}(0) = \beta$$

(a) Approximate this equation with a second-order difference equation for computer solution. Use backward-difference approximation for the derivatives.

(b) Take the state variables $x_1 = y$, $x_2 = \dot{y}$. Derive a state variable model for the system. Discretize the state model by Euler method.

chapter 3
Models of Digital Control Devices and Systems

3.1 INTRODUCTION

Now that we have developed the prerequisite signal processing techniques in Chapter 2, we can use them to study closed-loop digital control systems. A typical topology of the type of systems to be considered in this chapter is shown in Fig. 3.1.

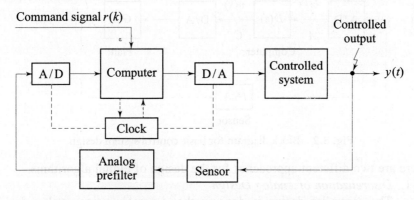

Fig. 3.1 Configuration of the basic digital control scheme

Digital control systems with analog sensors include an analog prefilter between the sensor and the sampler (A/D converter) as an anti-aliasing device. The prefilters are low-pass, and the simplest transfer function is

$$H_{pf}(s) = \frac{a}{s+a}$$

so that the noise above the prefilter breakpoint a is attenuated. the design goal is to provide enough attenuation at half the sample rate ($\omega_s/2$) so that the noise above $\omega_s/2$, when aliased into lower frequencies by the sampler, will not be detrimental to the control-system performance.

Since the phase lag from the prefilter can significantly affect system stability, it is required that the control design be carried out with the analog prefilter included in the loop transfer function. An alternative design procedure is to select the breakpoint and ω_s sufficiently higher than the system bandwidth so that the phase lag from the prefilter does not significantly alter the system stability, and thus the prefilter design problem can be divorced from the control-law design problem. Our treatment of the subject is based on this alternative design procedure. We therefore ignore the prefilter design and focus on the basic control-system design. The basic configuration for this design problem is shown in Fig. 3.2, where

- $G(s)$ = transfer function of the controlled plant (continuous-time system);
- $H(s)$ = transfer function of the analog sensor; and
- $D(z)$ = transfer function of the digital control algorithm.

The analog and digital parts of the system are connected through D/A and A/D converters. The computer with its internal clock drives the D/A and A/D converters. It compares the command signal $r(k)$ with the feedback signal $b(k)$ and generates the control signal $u(k)$ to be sent to the final control elements of the controlled plant. These signals are computed from the digital control algorithm $D(z)$ stored in the memory of the computer.

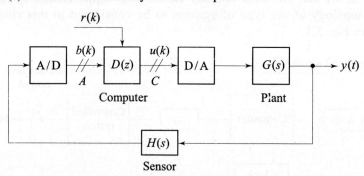

Fig. 3.2 Block diagram for basic control system design

There are two different approaches for the design of digital algorithms.
1. *Discretization of Analog Design*
   The controller design is done in the s-domain using analog design methods.[1] The resulting analog control law is then converted to discrete-time form using one of the approximation techniques given in Section 2.12.
2. *Direct Digital Design*
   In this approach, we first develop the discrete-time model of the analog part of the loop—from C to A in Fig. 3.2—that includes the controlled plant. The controller design is then performed using discrete-time analysis.

---

1. Chapters 7–10 of reference [180].

An actual design process is often a combination of the two methods. First iteration to a digital design can be obtained using discretization of an analog design. Then the result is tuned up using direct digital analysis and design.

The intent of this chapter is to provide basic tools for the analysis and design of a control system that is to be implemented using a computer. Mathematical models of commonly used digital control devices and systems are developed. Different ways to implement digital controllers (obtained by the discretization of analog design (Section 2.12) or by direct digital design (Chapter 4) are also given in this chapter.

3.2 z-DOMAIN DESCRIPTION OF SAMPLED CONTINUOUS-TIME PLANTS

Whenever a digital computer is used to control a continuous-time plant, there must be some type of *interface system* that takes care of the communication between the discrete-time and the continuous-time systems. In the system of Fig. 3.2, the interface function is performed by A/D and D/A converters.

Simple models of the interface actions of A/D and D/A converters have been developed in Chapter 2. A brief review is in order here.

Model of an A/D Converter

A simple model of an A/D converter is shown in Fig. 3.3a. A continuous-time function $f(t)$, $t \geq 0$, is the input and the sequence of real numbers $f(k)$, $k = 0, 1, 2, \ldots$, is the output. The following relation holds between input and output:

$$f(k) = f(t = kT); \quad T \text{ is the time interval between samples} \tag{3.1}$$

The sequence $f(k)$ can be treated as a train of impulses represented by continuous-time function $f^*(t)$:

$$f^*(t) = f(0)\,\delta(t) + f(1)\,\delta(t-T) + f(2)\,\delta(t-2T) + \cdots$$

$$= \sum_{k=0}^{\infty} f(k)\,\delta(t-kT) \tag{3.2}$$

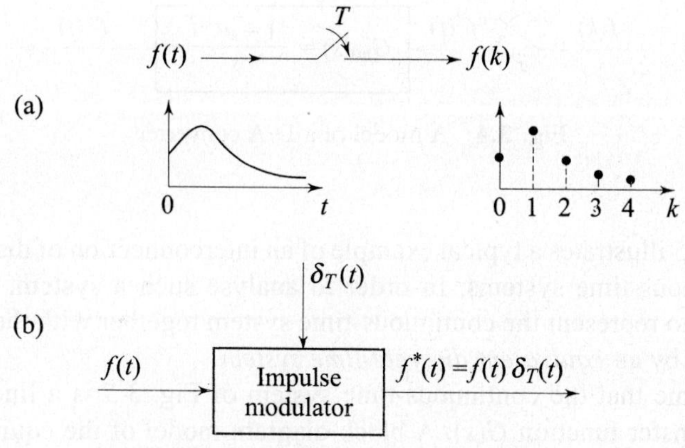

Fig. 3.3 A model of an A/D converter

The sampler of Fig. 3.3a can thus be viewed as an 'impulse modulator' with the carrier signal

$$\delta_T(t) = \sum_{k=0}^{\infty} \delta(t - kT) \qquad (3.3)$$

and modulating signal $f(t)$. A schematic representation of the modulation process is shown in Fig. 3.3b.

Model of a D/A Converter

A simple model of a D/A converter is shown in Fig. 3.4a. A sequence of numbers $f(k)$, $k = 0, 1, 2, ...$, is the input, and the continuous-time function $f^+(t)$, $t \geq 0$ is the output. The following relation holds between input and output:

$$f^+(t) = f(k); \quad kT \leq t < (k+1)T \qquad (3.4)$$

Each sample of the sequence $f(k)$ may be treated as an impulse function of the form $f(k)\delta(t - kT)$. The zero-order hold (ZOH) of Fig. 3.4a can thus be viewed as a linear time-invariant system that converts the impulse $f(k)\delta(t - kT)$ into a pulse of height $f(k)$ and width T. The D/A converter may therefore be modelled by Fig. 3.4b, where the ZOH is a system whose response to a unit impulse $\delta(t)$ is a unit pulse $g_{h0}(t)$ of width T. The Laplace transform of $g_{h0}(t)$ is the transfer function of hold operation, namely (refer Eqn. (2.56))

$$G_{h0}(s) = \mathscr{L}[g_{h0}(t)] = \frac{1-e^{-sT}}{s} \qquad (3.5)$$

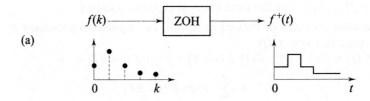

Fig. 3.4 A model of a D/A converter

Figure 3.5 illustrates a typical example of an interconnection of discrete-time and continuous-time systems. In order to analyse such a system, it is often convenient to represent the continuous-time system together with the ZOH and the sampler by an *equivalent discrete-time system*.

We assume that the continuous-time system of Fig. 3.5 is a linear system with the transfer function $G(s)$. A block diagram model of the equivalent discrete-time system is shown in Fig. 3.6a. As seen from this figure, the impulse modulated signal $u^*(t)$ is applied to two s-domain transfer functions in tandem. Since the two blocks with transfer functions $G_{h0}(s)$ and $G(s)$ are not separated by an impulse modulator, we can consider them as a single block with transfer function $[G_{h0}(s)G(s)]$ as shown in Fig. 3.6b. The continuous-time system with transfer function $[G_{h0}(s)G(s)]$ has input $u^*(t)$ and output $y(t)$. The output signal $y(t)$ is read off at discrete synchronous sampling instants kT; $k = 0, 1, ...,$ by means of a mathematical sampler $T(M)$.

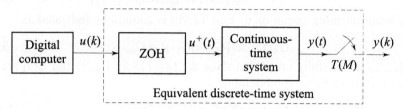

Fig. 3.5 Interconnection of discrete-time and continuous-time systems

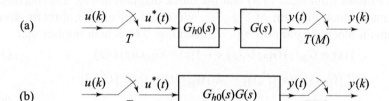

Fig. 3.6 Block diagrams for the equivalent discrete-time system

We assume that $\hat{g}(t)$ is the impulse response of the continuous-time system $G_{h0}(s)G(s)$:

$$\hat{g}(t) = \mathscr{L}^{-1}[G_{h0}(s)G(s)] \tag{3.6}$$

The input signal to the system is given by (refer Eqn. (3.2)),

$$u^*(t) = \sum_{k=0}^{\infty} u(kT)\delta(t - kT) \tag{3.7}$$

This is a sequence of impulses with intensities given by $u(kT)$. Since $\hat{g}(t)$ is the impulse response of the system (response to the input $\delta(t)$), by superposition from Eqn. (3.7),

$$y(t) = \sum_{j=0}^{\infty} u(jT)\hat{g}(t - jT)$$

At the sampling times $t = kT$, $y(t)$ is given by

$$y(kT) = \sum_{j=0}^{\infty} u(jT)\hat{g}(kT - jT) \tag{3.8}$$

We can recognize it at once as discrete-time convolution (refer Eqn. (2.25)). Taking the z-transform of both sides of Eqn. (3.8), we obtain (refer Eqn. (2.29)),

$$Y(z) = U(z)\, \hat{G}(z) \qquad (3.9a)$$

where
$$\hat{G}(z) = \mathscr{Z}\,[\hat{g}(kT)]$$
$$= \mathscr{Z}\,[\mathscr{L}^{-1}\{G_{h0}(s)G(s)\}|_{t\,=\,kT}] \qquad (3.9b)$$

The z-transforming operation of Eqn. (3.9b) is commonly indicated as

$$\hat{G}(z) = \mathscr{Z}\,[G_{h0}(s)G(s)] = G_{h0}G(z) \qquad (3.10)$$

It may be carefully noted that since the two blocks $G_{h0}(s)$ and $G(s)$ are not separated by an impulse modulator,

$$\hat{G}(z) \neq \mathscr{Z}\,[G_{h0}(s)]\, \mathscr{Z}\,[G(s)] \qquad (3.11a)$$
$$\neq G_{h0}(z)\, G(z) \qquad (3.11b)$$

It follows from Eqns (3.9) that the block diagram of Fig. 3.6b becomes the z-domain block diagram of Fig. 3.7. We could, of course, directly draw the z-domain block diagram of Fig. 3.7 from Fig. 3.6b which implies that

$$Y(s) = G_{h0}(s)G(s)U^*(s) \leftrightarrow Y(z) = G_{h0}G(z)U(z) \qquad (3.12a)$$

or
$$\mathscr{Z}\,[G_{h0}(s)G(s)U^*(s)] = \mathscr{Z}\,[G_{h0}(s)G(s)]\, \mathscr{Z}\,[U^*(s)]$$
$$= G_{h0}G(z)\, U(z) \qquad (3.12b)$$

$$U(z) \longrightarrow \boxed{G_{h0}G(z) = \mathscr{Z}\,[G_{h0}(s)G(s)]} \longrightarrow Y(z)$$

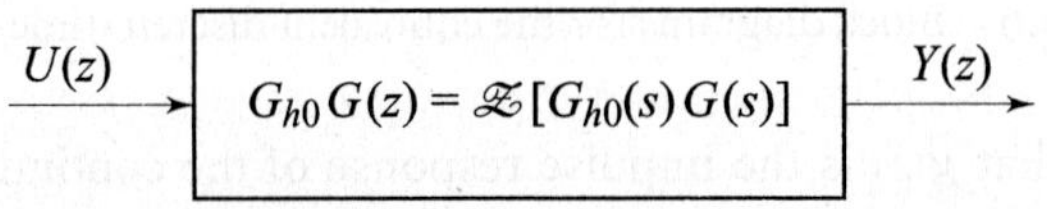

Fig. 3.7 The z-domain equivalent of Fig. 3.6

We can use the following relation (refer Eqns (2.77b) and (2.77c)) for evaluation of $G_{h0}G(z)$:

$$G_{h0}G(z) = \mathscr{Z}\,[G_{h0}(s)G(s)] = \mathscr{Z}\left[(1 - e^{-sT})\frac{G(s)}{s}\right]$$
$$= (1 - z^{-1})\, \mathscr{Z}\left[\frac{G(s)}{s}\right] \qquad (3.13)$$

Single factor building blocks of the Laplace and z-transform pairs are given in Section A.5 of Appendix A. Expanding $G(s)/s$ into partial fractions. $\mathscr{Z}\,[G(s)/s]$ can be found by using the table in this appendix.

▲▲

Consider now the basic sampled-data feedback system whose block diagram is depicted in Fig. 3.8a. In terms of impulse modulation, this block diagram can be redrawn as in Fig. 3.8b.

Using the relations previously established in this section,

we have
$$U(z) = D(z)E(z) \quad (3.14a)$$
$$Y(z) = G_{h0}G(z)U(z) \quad (3.14b)$$

Since
$$e(kT) = r(kT) - y(kT)$$

we have
$$E(z) = R(z) - Y(z) \quad (3.14c)$$

Combining Eqns (3.14a), (3.14b), and (3.14c) gives
$$\frac{Y(z)}{R(z)} = \frac{D(z)G_{h0}G(z)}{1+D(z)G_{h0}G(z)} \quad (3.15)$$

Figure 3.9 gives the z-domain equivalent of Fig. 3.8.

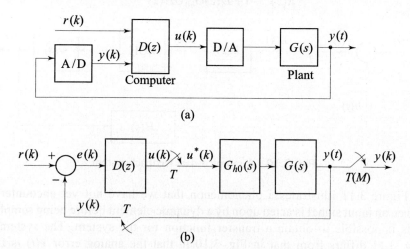

Fig. 3.8 A sampled-data feedback system

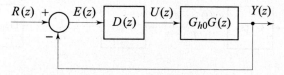

Fig. 3.9 The z-domain equivalent of Fig. 3.8

▲▲

Having become familiar with the technique, from now onwards we may directly write z-domain relationships without introducing impulse modulators in block diagrams of sampled-data systems.

Consider the sampled-data feedback system of Fig. 3.10 where the sensor dynamics is represented by transfer function $H(s)$. The following equations easily follow:

$$E(z) = R(z) - B(z) \quad (3.16a)$$

$$U(z) = D(z)\,E(z) \tag{3.16b}$$

$$Y(z) = G_{h0}G(z)\,U(z) = \mathscr{Z}[G_{h0}(s)G(s)]\,U(z) \tag{3.16c}$$

$$B(z) = G_{h0}GH(z)\,U(z) = \mathscr{Z}[G_{h0}(s)G(s)H(s)]U(z) \tag{3.16d}$$

Equations (3.16a), (3.16b) and (3.16d) give

$$\frac{E(z)}{R(z)} = \frac{1}{1 + D(z)\,G_{h0}GH(z)} \tag{3.17}$$

Combining Eqns (3.16b), (3.16c) and (3.17), we get

$$\frac{Y(z)}{R(z)} = \frac{D(z)G_{h0}G(z)}{1 + D(z)G_{h0}GH(z)} \tag{3.18}$$

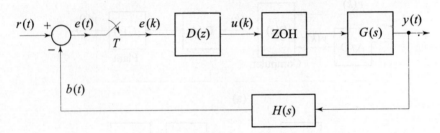

Fig. 3.10 A sampled-data feedback system

Figure 3.11 illustrates a phenomenon that we have not yet encountered. When an input signal is acted upon by a dynamic element before being sampled, it is impossible to obtain a transfer function for the system. The system in Fig. 3.11 differs from that in Fig. 3.10 in that the analog error $e(t)$ is first amplified before being converted to digital form for the control computer. The amplifier's dynamics are given by $G_1(s)$.

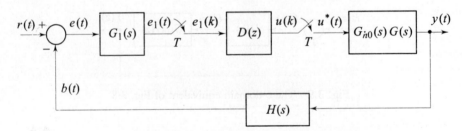

Fig. 3.11 A sampled-data system

Consider first the subsystem shown in Fig. 3.12a. We can equivalently represent it as a block $[G_1(s)E(s)]$ with input $\delta(t)$ as in Fig. 3.12b. Now the input, and therefore the output, does not change by imagining a fictitious impulse modulator through which $\delta(t)$ is applied to $[G_1(s)E(s)]$ as in Fig. 3.12c.

On application of Eqn. (3.12), we can write

$$E_1(z) = \mathscr{Z}[G_1(s)E(s)]\,\mathscr{Z}[\delta(k)] = \mathscr{Z}[G_1(s)E(s)] \tag{3.19}$$

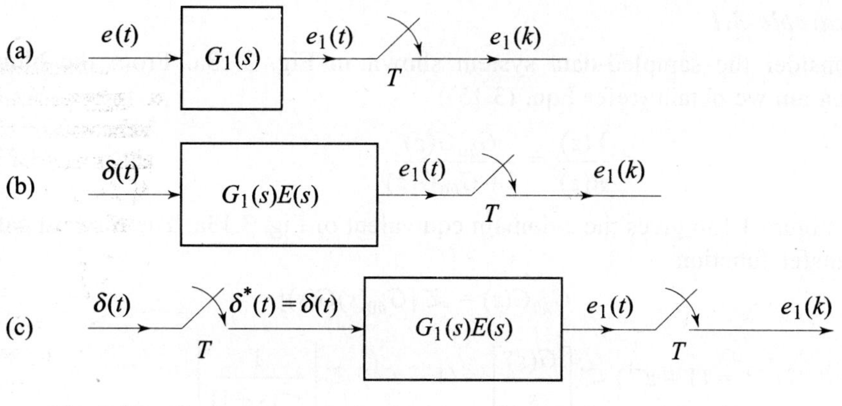

Fig. 3.12 Equivalent s-transfer function with impulse modulated input

Now, for the system of Fig. 3.11,
$$E(s) = R(s) - B(s) = R(s) - H(s)\,Y(s)$$
$$= R(s) - H(s)\,G_{h0}(s)\,G(s)\,U^*(s) \qquad (3.20)$$

Therefore from Eqns (3.19) and (3.20), we obtain
$$E_1(z) = \mathscr{Z}[G_1(s)R(s)] - \mathscr{Z}[G_1(s)H(s)G_{h0}(s)G(s)]U(z)$$
$$= G_1R(z) - G_1G_{h0}GH(z)\,U(z) \qquad (3.21)$$

Since $\qquad U(z) = D(z)\,E_1(z),$

we can write from Eqn. (3.21),
$$E_1(z) = \frac{G_1R(z)}{1 + D(z)G_1G_{h0}H(z)}$$

The output $Y(z) = G_{h0}G(z)\,U(z) = D(z)\,G_{h0}G(z)\,E_1(z)$
$$= \frac{D(z)G_{h0}G(z)G_1R(z)}{1 + D(z)G_1G_{h0}GH(z)} \qquad (3.22)$$

Since it is impossible to form the ratio $Y(z)/R(z)$ from Eqn. (3.22), we can not obtain a transfer function for the system of Fig. 3.11, and we cannot analyse it further without specifying a functional form for the input $r(t)$.

It is not permissible to create a transfer function for the system in Fig. 3.11 by inserting a fictitious sampler at the input, because this would change the physics of the situation represented by the diagram (the analog input would be replaced by the train of impulses). Fictitious samplers are permissible at only the output, because they are simply a means of selecting the values of the output at the times of interest to us, namely, the sample times.

Example 3.1

Consider the sampled-data system shown in Fig. 3.13a. From the block diagram we obtain (refer Eqn. (3.15))

$$\frac{Y(z)}{R(z)} = \frac{G_{h0}G(z)}{1+G_{h0}G(z)} \tag{3.23}$$

Figure 3.13b gives the z-domain equivalent of Fig. 3.13a. The forward path transfer function

$$G_{h0}G(z) = \mathscr{Z}[G_{h0}(s)G(s)]$$

$$= (1-z^{-1})\mathscr{Z}\left[\frac{G(s)}{s}\right] = (1-z^{-1})\mathscr{Z}\left[\frac{1}{s^2(s+1)}\right]$$

$$= (1-z^{-1})\mathscr{Z}\left[\frac{1}{s^2}-\frac{1}{s}+\frac{1}{s+1}\right] = (1-z^{-1})\left[\frac{Tz}{(z-1)^2}-\frac{z}{z-1}+\frac{z}{z-e^{-T}}\right]$$

$$= \frac{z(T-1+e^{-T})+(1-e^{-T}-Te^{-T})}{(z-1)(z-e^{-T})}$$

When $T = 1$, we have

$$G_{h0}G(z) = \frac{ze^{-1}+1-2e^{-1}}{(z-1)(z-e^{-1})} = \frac{0.3679z+0.2642}{z^2-1.3679z+0.3679}$$

Substituting in Eqn. (3.23), we obtain

$$\frac{Y(z)}{R(z)} = \frac{0.3679z+0.2642}{z^2-z+0.6321}$$

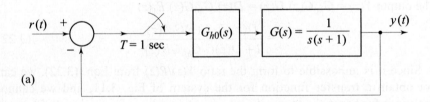

(a)

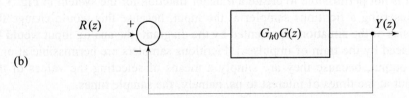

(b)

Fig. 3.13 A closed-loop sampled-data system

For a unit-step input,

$$R(z) = \frac{z}{z-1}$$

and therefore

$$Y(z) = \frac{z(0.3679z + 0.2642)}{(z-1)(z^2 - z + 0.6321)} = \frac{0.3679z^2 + 0.2642z}{z^3 - 2z^2 + 1.6321z - 0.6321}$$

By long-division process, we get

$$Y(z) = 0.3679\,z^{-1} + z^{-2} + 1.3996\,z^{-3} + 1.3996\,z^{-4} + 1.1469\,z^{-5}$$
$$+ 0.8944\,z^{-6} + 0.8015\,z^{-7} + \cdots$$

Therefore, the sequence ($k = 1, 2, \ldots$)

$$y(kT) = \{0.3679, 1, 1.3996, 1.3996, 1.1469, 0.8944, 0.8015, \ldots\}$$

Note that the final value of $y(kT)$ is (refer Eqn. (A.45) in Appendix A),

$$\lim_{k \to \infty} y(kT) = \lim_{z \to 1} (z-1)Y(z) = \frac{0.3679 + 0.2642}{0.6321} = 1$$

The unit-step response is shown in Fig. 3.14. Also shown in this figure is the unit-step response of the continuous-time system (i.e., when $T = 0$). The overshoot of the sampled system is 45%, in contrast to 17% for the continuous-time system.

The performance of the digital system is thus dependent on the sampling period T. Larger sampling periods usually give rise to higher overshoots in the step response, and may eventually cause instability if the sampling period is too large.

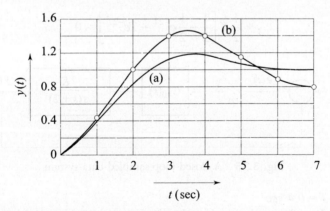

Fig. 3.14 The response of a second-order system: (a) analog; (b) sampled

Example 3.2

Let us compare the stability properties of the system shown in Fig. 3.15, with and without a sample-and-hold on the error signal.

Without sample-and-hold, the system in Fig. 3.15 has the transfer function
$$\frac{Y(s)}{R(s)} = \frac{K}{s^2 + 2s + K}$$
This system is stable for all values of $K > 0$.

For the system with sample-and-hold, the forward-path transfer function is given by

$$G_{h0}G(z) = (1 - z^{-1}) \mathscr{Z}\left[\frac{K}{s^2(s+2)}\right]$$

$$= (1 - z^{-1}) \mathscr{Z}\left[\frac{K}{2}\left(\frac{1}{s^2} - \frac{1/2}{s} + \frac{1/2}{s+2}\right)\right]$$

$$= \frac{K}{2}(1 - z^{-1})\left[\frac{Tz}{(z-1)^2} - \frac{(1/2)z}{z-1} + \frac{(1/2)z}{z-e^{-2T}}\right]$$

$$= \frac{K}{2}\left[\frac{2T(z - e^{-2T}) - (z-1)(1 - e^{-2T})}{2(z-1)(z - e^{-2T})}\right]$$

The characteristic equation of the sampled-data system is
$$1 + G_{h0}G(z) = 0$$

or $\quad 4(z-1)(z - e^{-2T}) + 2KT(z - e^{-2T}) - K(z-1)(1 - e^{-2T}) = 0$

or
$$z^2 + z\left[\frac{1}{2}K\left(T - \frac{1}{2} + \frac{1}{2}e^{-2T}\right) - 1 - e^{-2T}\right] + e^{-2T}$$
$$+ \frac{1}{2}K\left[\frac{1}{2} - \frac{1}{2}e^{-2T} - Te^{-2T}\right] = 0$$

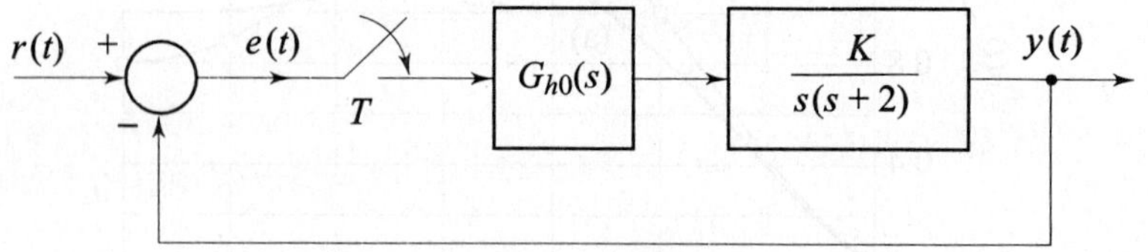

Fig. 3.15 A closed-loop sampled-data system

Case 1: $T = 0.4$ sec

For this value of sampling period, the characteristic polynomial becomes
$$\Delta(z) = z^2 + (0.062K - 1.449)z + 0.449 + 0.048K$$

Applying the Jury stability test (refer Eqns (2.48)–(2.50)), we find that the system is stable if the following conditions are satisfied:
$$\Delta(1) = 1 + 0.062K - 1.449 + 0.449 + 0.048K > 0$$

$$\Delta(-1) = 1 - 0.062K + 1.449 + 0.449 + 0.048K > 0$$

$$|\,0.449 + 0.048K\,| < 1$$

These conditions are satisfied for $0 < K < 11.479$.

Case II: $T = 3$ sec

For this value of sampling period, the characteristic polynomial becomes

$$\Delta(z) = z^2 + (1.2506K - 1.0025)\,z + 0.0025 + 0.2457K$$

The system is found to be stable for $0 < K < 1.995$.

Thus the system which is stable for all $K > 0$ when $T = 0$ (continuous-time system) becomes unstable for $K > 11.479$ when $T = 0.4$ sec. When T is further increased to 3 sec, it becomes unstable for $K > 1.995$. It means that increasing the sampling period (or decreasing the sampling rate) reduces the margin of stability.

3.3 z-DOMAIN DESCRIPTION OF SYSTEMS WITH DEAD-TIME

Figure 3.16 is the block diagram of a computer-controlled continuous-time system with dead-time. We assume that the continuous-time system is described by transfer function of the form

$$G_p(s) = G(s)\,e^{-\tau_D s} \tag{3.24}$$

where τ_D is the dead-time, and $G(s)$ contains no dead-time.

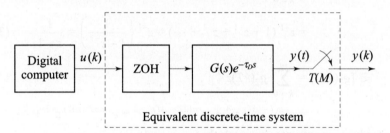

Fig. 3.16 A sampled continuous-time system with dead-time

The equivalent discrete-time system shown by dotted lines in Fig. 3.16 is described by the model

$$\frac{Y(z)}{U(z)} = \mathscr{Z}[G_{h0}(s)G_p(s)] = G_{h0}G_p(z) \tag{3.25a}$$

$$= (1 - z^{-1})\,\mathscr{Z}\!\left[\frac{1}{s}e^{-\tau_D s}G(s)\right] \tag{3.25b}$$

If N is the largest integer number of sampling periods in τ_D, we can write

$$\tau_D = NT + \Delta T;\ 0 \le \Delta < 1 \tag{3.26}$$

140 Digital Control and State Variable Methods

Therefore $\quad G_{h0}G_p(z) = (1 - z^{-1})z^{-N} \mathscr{Z}\left[\dfrac{1}{s}e^{-\Delta Ts}G(s)\right]$

Let us take an example where

$$G(s) = \dfrac{1}{s+a} \qquad (3.27)$$

For this example, $\quad \dfrac{Y(z)}{U(z)} = G_{h0}G_p(z) = (1 - z^{-1})z^{-N} \mathscr{Z}\left[\dfrac{e^{-\Delta Ts}}{s(s+a)}\right]$

$$= \dfrac{1}{a}(1 - z^{-1})z^{-N} \mathscr{Z}\left[\dfrac{e^{-\Delta Ts}}{s} - \dfrac{e^{-\Delta Ts}}{s+a}\right] \qquad (3.28)$$

Now $\quad \mathscr{L}^{-1}\left[\dfrac{e^{-\Delta Ts}}{s}\right] = g_1(t) = \mu(t - \Delta T)$

$\quad \mathscr{L}^{-1}\left[\dfrac{e^{-\Delta Ts}}{s+a}\right] = g_2(t) = e^{-a(t - \Delta T)}\mu(t - \Delta T)$

where $\mu(t)$ is a unit-step function.

Therefore $\quad g_1(kT) = \mu(kT - \Delta T)$

$\qquad g_2(kT) = e^{-a(kT - \Delta T)}\mu(kT - \Delta T)$

$$\mathscr{Z}[g_1(kT)] = \sum_{k=0}^{\infty} g_1(kT)z^{-k} = z^{-1} + z^{-2} + z^{-3} + \cdots$$

$$= z^{-1}(1 + z^{-1} + z^{-2} + \cdots) = z^{-1}\left(\dfrac{1}{1 - z^{-1}}\right) = \dfrac{1}{z - 1} \qquad (3.29)$$

$$\mathscr{Z}[g_2(kT)] = \sum_{k=0}^{\infty} g_2(kT)z^{-k}$$

$$= e^{-a(T - \Delta T)}z^{-1} + e^{-a(2T - \Delta T)}z^{-2} + e^{-a(3T - \Delta T)}z^{-3} + \cdots$$

We introduce a parameter m, such that

$$m = 1 - \Delta$$

Then $\quad \mathscr{Z}[g_2(kT)] = e^{-amT}z^{-1} + e^{-amT}e^{-aT}z^{-2} + e^{-amT}e^{-2aT}z^{-3} + \cdots$

$$= e^{-amT}z^{-1}[1 + e^{-aT}z^{-1} + e^{-2aT}z^{-2} + \cdots]$$

$$= e^{-amT}z^{-1}\left[\dfrac{1}{1 - e^{-aT}z^{-1}}\right]$$

$$= \dfrac{e^{-amT}}{z - e^{-aT}} \qquad (3.30)$$

Substituting the z-transform results given by Eqns (3.29) and (3.30) in Eqn. (3.28), we get

Models of Digital Control Devices and Systems

$$\frac{Y(z)}{U(z)} = G_{h0}G_p(z) = \frac{1}{a}(1-z^{-1})\,z^{-N}\left[\frac{1}{z-1} - \frac{e^{-amT}}{z-e^{-aT}}\right]$$

$$= \frac{\frac{1}{a}\left[(1-e^{-amT})z + e^{-amT} - e^{-aT}\right]}{z^{N+1}(z-e^{-aT})} \qquad (3.31)$$

Table 3.1 has been generated by applying the procedure outlined above to commonly occurring functions.

Table 3.1 Table of transform pairs for systems with dead-time

Laplace transform $F(s)e^{-\Delta Ts};\ 0 \le \Delta < 1$	z-transform $\mathscr{Z}[F(s)e^{-\Delta Ts}];\ m = 1-\Delta$
$\dfrac{e^{-\Delta Ts}}{s}$	$\dfrac{1}{z-1}$
$\dfrac{e^{-\Delta Ts}}{s^2}$	$\dfrac{mT}{z-1} + \dfrac{T}{(z-1)^2}$
$\dfrac{2e^{-\Delta Ts}}{s^3}$	$T^2\left[\dfrac{m^2z^2 + (2m - 2m^2 + 1)z + (m-1)^2}{(z-1)^3}\right]$
$\dfrac{e^{-\Delta Ts}}{s+a}$	$\dfrac{e^{-amT}}{z-e^{-aT}}$
$\dfrac{e^{-\Delta Ts}}{(s+a)(s+b)}$	$\dfrac{1}{(b-a)}\left[\dfrac{e^{-amT}}{z-e^{-aT}} - \dfrac{e^{-bmT}}{z-e^{-bT}}\right]$
$\dfrac{ae^{-\Delta Ts}}{s(s+a)}$	$\dfrac{(1-e^{-amT})z + (e^{-amT} - e^{-aT})}{(z-1)(z-e^{-aT})}$
$\dfrac{ae^{-\Delta Ts}}{s^2(s+a)}$	$\dfrac{T}{(z-1)^2} + \dfrac{amT-1}{a(z-1)} + \dfrac{e^{-amT}}{a(z-e^{-aT})}$

Example 3.3

The scheme of Fig. 3.17 produces a steady stream flow of fluid with controlled temperature θ. A stream of hot fluid is continuously mixed with a stream of cold fluid in a mixing valve. The valve characteristic is such that the total flow rate Q (m³/sec) through it is maintained constant but the inflow q_i (m³/sec) of hot fluid may be linearly varied by controlling valve stem position x. The valve stem position x thus controls the temperature θ_i (°C) of the outflow from the mixing valve. Due to the distance between the valve and the point of discharge into the tank, there is a time delay between the change in θ_i and the discharge of the flow with the changed temperature into the tank.

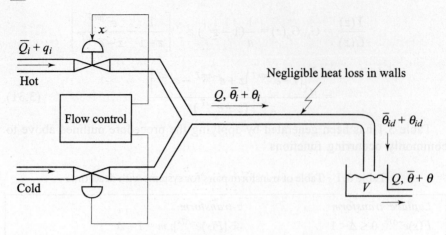

Fig. 3.17 Tank fluid temperature control

The differential equation governing the tank temperature is (assuming an initial equilibrium and taking all variables as perturbations)

$$V\rho c \frac{d\theta}{dt} = Q\rho c (\theta_{id} - \theta) \qquad (3.32)$$

where

θ = tank fluid temperature, °C
   = temperature of the outflowing fluid from the tank;
c = specific heat of the fluid, Joules/(kg)(°C);
V = volume of the fluid in the tank, m^3;
ρ = fluid density, kg/m^3;
Q = fluid flow rate, m^3/sec; and
θ_{id} = temperature of the fluid entering the tank, °C.

The temperature θ_{id} at the input to the tank at time t, however, is the mixing valve output temperature τ_D seconds in the past, which may be expressed as

$$\theta_{id}(t) = \theta_i(t - \tau_D) \qquad (3.33)$$

From Eqns (3.32)–(3.33), we obtain

$$\dot{\theta}(t) + a\theta(t) = a\theta_i(t - \tau_D)$$

where $\qquad a = Q/V$

Therefore $\qquad G_p(s) = \dfrac{\theta(s)}{\theta_i(s)} = \dfrac{ae^{-\tau_D s}}{s+a} \qquad (3.34)$

To form the discrete-time transfer function of $G_p(s)$ preceded by a zero-order hold, we must compute

$$G_{h0}G_p(z) = \mathscr{Z}\left[\left(\frac{1-e^{-sT}}{s}\right)\frac{ae^{-\tau_D s}}{s+a}\right]$$

$$= (1 - z^{-1}) \mathscr{Z}\left[\frac{a}{s(s+a)}e^{-\tau_D s}\right] \quad (3.35)$$

For the specific values of $\tau_D = 1.5$, $T = 1$, $a = 1$, Eqn. (3.35) reduces to

$$G_{h0}G_p(z) = (1 - z^{-1}) \mathscr{Z}\left[\frac{1}{s(s+1)}e^{-s}e^{-0.5s}\right]$$

$$= (1 - z^{-1}) z^{-1} \mathscr{Z}\left[\frac{1}{s(s+1)}e^{-0.5s}\right]$$

Using transform pairs of Table 3.1, we obtain

$$G_{h0}G_p(z) = (1 - z^{-1}) z^{-1} \left[\frac{(1-e^{-0.5})z + (e^{-0.5} - e^{-1})}{(z-1)(z-e^{-1})}\right]$$

$$= \frac{0.3935(z + 0.6066)}{z^2(z - 0.3679)} = \frac{\theta(z)}{\theta_i(z)} \quad (3.36)$$

The relationship between x and θ_i is linear, as is seen below.

$$(\overline{Q}_i + q_i)\rho c \theta_H + [Q - (\overline{Q}_i + q_i)] \rho c \theta_C = Q\rho c (\overline{\theta}_i + \theta_i)$$

where θ_H and θ_C are constant temperatures of hot and cold streams respectively.

$$q_i = K_v x$$

where K_v is the valve gain.

The perturbation equation is obtained as (neglecting second-order terms in perturbation variables),

$$K_v(\theta_H - \theta_C) x(t) = Q\theta_i(t)$$

or
$$x(t) = K\theta_i(t); \; K = Q/[K_v(\theta_H - \theta_C)]$$

Therefore

$$\frac{\theta(z)}{X(z)} = \frac{(0.3935/K)(z + 0.6066)}{z^2(z - 0.3679)}$$

3.4 IMPLEMENTATION OF DIGITAL CONTROLLERS

The application of conventional 8- and 16-bit microprocessors to control systems is now well established. Such processors have general purpose architectures which make them applicable for a wide range of tasks, though none are remarkably efficient. In control applications, such devices may pose problems such as inadequate speed, difficulties with numerical manipulation, and relatively high cost for the completed system; the latter being due to both the programming effort and the cost of the peripheral hardware (memories, I/O ports, timers/counters, A/D converters, D/A converters, PWM circuit, etc.).

In applications requiring small amounts of program ROM, data RAM, and I/O ports, single-chip microcontrollers are ideally suited. In these chips, the capabilities in terms of speed of computation, on-chip resources, and software facilities are optimized for control applications. Should the on-chip features be insufficient to meet control requirements, the microcontroller chips allow for easy expansion.

The Intel microcontroller family (MCS-48 group, MCS-51 group, MCS-96 group) includes 8- and 16-bit processors with the following on-chip resources—ROM, RAM, I/O lines, timer/counter, A/D converter, and PWM output. The Motorola microcontroller family (HC 05 group, HC 11 group, HC 16 group) also provides microcontroller chips with similar features.

In many application areas, processing requirements for digital control systems, such as execution time and algorithm complexity, have increased dramatically. For example, in motor control, short sampling time constraints can place exacting requirements on algorithm execution time. New airframe designs and extended aircraft performance envelopes increase the complexity of flight control laws. Controller complexity also increases with number of interacting loops (e.g., in robotics) or the number of sensors (e.g., in vision systems). For a growing number of real-time control applications, conventicnal single-processor systems are unable to satisfy the new demands for increased speed and greater complexity and flexibility.

The dramatic advances in VLSI technology leading to high transistor packing densities have enabled computer architects to develop parallel-processing architectures consisting of multiple processors; thus realizing high-performance computing engines at relatively low cost. The control engineer can exploit a range of architectures for a variery of functions.

Parallel-processing speeds up the execution time for a task. This is achieved by dividing the problem into several sub-tasks and allocating multiple processors to execute multiple sub-tasks simultaneously. Parallel architectures differ from one another in respect of nature of interconnectivity between the processing elements and the processing power of each individual processing element.

The transputer is a family of single-chip computers which incorporates features to support parallel processing. It is possible to use a network of transputers to reduce the execution time of a real-time control law.

Digital signal processors (DSPs) offer an alternative strategy for implementation of digital controllers. DSPs use architectures and dedicated arithmetic circuits that provide high resolution and high speed arithmetic, making them ideally suited for use as controllers.

Many DSP chips are available commercially that can be applied to a wide range of control problems. The Texas Instruments TMS 320 family provides several beneficial features through its architecture, speed, and instruction set.

TMS 320 is designed to support both numeric-intensive operations, such as required in signal processing, and also general purpose computation, as would be required in high speed control. It uses a modified architecture, which gives it speed and flexibility—the program and data memory are allotted separate sections on the chip permitting a full overlap of the instruction fetch and execution cycle. The processor also uses hardware to implement functions which previously had been achieved using software. As a result, a multiplication takes only 200 nsec, i.e., one instruction cycle, to execute. Extra hardware has also been included to implement shifting and some other functions. This gives the design engineer the type of power previously unavailable on a single chip.

Implementation of a control algorithm on a computer consists of the following two steps:
 (i) Block diagram realization of the transfer function (obtained by the discretization of analog controller (Section 2.12) or by the direct digital design (Chapter 4) that represents the control algorithm.
 (ii) Software design based on the block diagram realization.

In the following, we present several different structures of block diagram realizations of digital controllers using delay elements, adders, and multipliers. Different realizations are equivalent from the input-output point of view if we assume that the calculations are done with infinite precision. With finite precision in the calculations, the choice of the realization is very important. A bad choice of the realization may give a controller that is very sensitive to errors in the computations.

Assume that we want to realize the controller

$$D(z) = \frac{U(z)}{E(z)} = \frac{\beta_0 z^n + \beta_1 z^{n-1} + \cdots + \beta_{n-1} z + \beta_n}{z^n + \alpha_1 z^{n-1} + \cdots + \alpha_{n-1} z + \alpha_n} \quad (3.37a)$$

where the α_i's and β_i's are real coefficients (some of them may be zero).

Transfer functions of all digital controllers can be rearranged in this form. For example, the transfer function of PID controller, given by Eqn. (2.90b), can be rearranged as follows:

$$D(z) = \frac{U(z)}{E(z)} = K_c \left[1 + \frac{T}{T_I}\left(\frac{1}{1-z^{-1}}\right) + \frac{T_D}{T}(1-z^{-1}) \right]$$

$$= K_c + \frac{K_c T}{T_I}\left(\frac{1}{1-z^{-1}}\right) + \frac{K_c T_D}{T}(1-z^{-1})$$

$$= \frac{\beta_0 z^2 + \beta_1 z + \beta_2}{z^2 + \alpha_1 z + \alpha_2}$$

where

$$\alpha_1 = -1$$

$$\alpha_2 = 0$$

$$\beta_0 = K_c\left(1 + \frac{T}{T_I} + \frac{T_D}{T}\right)$$

$$\beta_1 = -K_c\left(1 + \frac{2T_D}{T}\right)$$

$$\beta_2 = \frac{K_c T_D}{T}$$

We shall now discuss different ways of realizing the transfer function (3.37a), or equivalently the transfer function

$$D(z) = \frac{U(z)}{E(z)} = \frac{\beta_0 + \beta_1 z^{-1} + \beta_2 z^{-2} + \cdots + \beta_{n-1} z^{-(n-1)} + \beta_n z^{-n}}{1 + \alpha_1 z^{-1} + \alpha_2 z^{-2} + \cdots + \alpha_{n-1} z^{-(n-1)} + \alpha_n z^{-n}} \quad (3.37b)$$

The methods for realizing digital systems of the form (3.37) can be divided into two classes—*recursive* and *nonrecursive*. The functional relation between the input sequence $e(k)$ and the output sequence $u(k)$ for a *recursive realization* has the form

$$u(k) = f(u(k-1), u(k-2), \ldots, e(k), e(k-1), \ldots) \quad (3.38)$$

For the linear time-invariant system of Eqn. (3.37b), the recursive realization has the form

$$u(k) = -\alpha_1 u(k-1) - \alpha_2 u(k-2) - \cdots - \alpha_n u(k-n)$$
$$+ \beta_0 e(k) + \beta_1 e(k-1) + \cdots + \beta_n e(k-n) \quad (3.39)$$

The current output sample $u(k)$ is a function of past outputs and present and past input samples. Due to the recursive nature, the errors in previous outputs may accumulate.

The impulse response of the digital system defined by Eqn. (3.39), where we assume not all α_i's are zero, has an infinite number of non-zero samples although their magnitudes may become negligibly small as k increases. This type of digital system is called an *infinite impulse response* (IIR) system.

The input-output relation for a *nonrecursive realization* is of the form

$$u(k) = f(e(k), e(k-1), \ldots) \quad (3.40a)$$

For a linear time-invariant system, this relation takes the form

$$u(k) = b_0 e(k) + b_1 e(k-1) + b_2 e(k-2) + \cdots + b_N e(k-N) \quad (3.40b)$$

The current output sample $u(k)$ is a function only of the present and past values of the input.

The impulse response of the digital system defined by Eqn. (3.40b) is limited to a finite number of samples defined over a finite range of time intervals, i.e., the impulse response sequence is finite. This type of digital system is called a *finite impulse response* (FIR) system.

The digital controller given by Eqn. (3.37b) is obviously an FIR digital system when the coefficients α_i are all zero. When not all α_i's are zero, we can obtain FIR approximation of the digital system by dividing its numerator by the denominator and truncating the series at z^{-N}; $N \geq n$:

$$\frac{U(z)}{E(z)} = D(z) \cong a_0 + a_1 z^{-1} + a_2 z^{-2} + \cdots + a_N z^{-N}; \ N \geq n \qquad (3.41)$$

Notice that we may require a large value of N to obtain a good level of accuracy.

In the following, we discuss the most common types of recursive and non-recursive realizations of digital controllers of the form (3.37).

Recursive Realizations

The transfer function (3.37) represents an nth order system. Recursive realization of this transfer function will require at least n unit delayers. Each unit delayer will represent a first-order dynamic system. Each of the three recursive realization structures given below uses the minimum number (n) of delay elements in realizing the transfer function (3.37).

Direct realization

Let us multiply the numerator and denominator of the right-hand side of Eqn. (3.37b) by a variable $X(z)$. This operation gives

$$\frac{U(z)}{E(z)} = \frac{(\beta_0 + \beta_1 z^{-1} + \beta_2 z^{-2} + \cdots + \beta_{n-1} z^{-(n-1)} + \beta_n z^{-n}) X(z)}{(1 + \alpha_1 z^{-1} + \alpha_2 z^{-2} + \cdots + \alpha_{n-1} z^{-(n-1)} + \alpha_n z^{-n}) X(z)} \qquad (3.42)$$

Equating the numerators on both sides of this equation gives

$$U(z) = (\beta_0 + \beta_1 z^{-1} + \cdots + \beta_n z^{-n}) X(z) \qquad (3.43a)$$

The same operation on the denominator brings

$$E(z) = (1 + \alpha_1 z^{-1} + \cdots + \alpha_n z^{-n}) X(z) \qquad (3.43b)$$

In order to construct a block diagram for realization. Eqn. (3.43b) must first be written in a cause-and-effect relation. Solving for $X(z)$ in Eqn. (3.43b) gives

$$X(z) = E(z) - \alpha_1 z^{-1} X(z) - \cdots - \alpha_n z^{-n} X(z) \qquad (3.43c)$$

A block diagram portraying Eqns (3.43a) and (3.43c) is now drawn in Fig. 3.18 for $n = 3$. Notice that we use only three delay elements. The coefficients α_i and β_i (which are real quantities) appear as multipliers. The block diagram schemes where the coefficients α_i and β_i appear directly as multipliers are called *direct structures*.

Basically, there are three sources of error that affect the accuracy of a realization (Section 2.1):
  (i) the error due to the quantization of the input signal into a finite number of discrete levels;
  (ii) the error due to accumulation of round-off errors in the arithmetic operations in the digital system; and
  (iii) the error due to quantization of the coefficients α_i and β_i of the transfer function. This error may become large for higher-order transfer functions. That is, in a higher-order digital controller in direct structure,

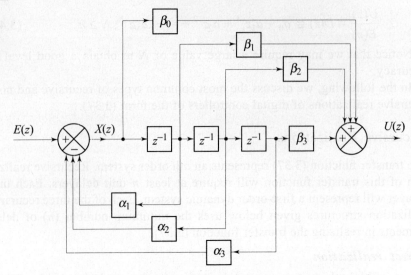

Fig. 3.18 Block diagram realization of the transfer function of Eqn. (3.37) with $n = 3$

small errors in the coefficients α_i and β_i cause large errors in the locations of the poles and zeros of the controller (refer Review Example 3.3).

These three errors arise because of the practical limitations of the number of bits that represent various signal samples and coefficients. The third type of error listed above may be reduced by mathematically decomposing a higher-order transfer function into a combination of lower-order transfer functions. In this way, the system may be made less sensitive to coefficient inaccuracies.

For decomposing higher-order transfer functions in order to reduce the coefficient sensitivity problem, the following two approaches are commonly used. It is desirable to analyse each of these structures for a given transfer function to see which one is better with respect to the number of arithmetic operations required, the range of coefficients, and so forth.

Cascade realization

The sensitivity problem may be reduced by implementing the transfer function $D(z)$ as a cascade connection of first-order and/or second-order transfer functions. If $D(z)$ can be written as a product of transfer functions $D_1(z), \ldots, D_m(z)$, or
$$D(z) = D_1(z) D_2(z) \cdots D_m(z),$$
then a digital realization for $D(z)$ may be obtained by a cascade connection of m component realizations for $D_1(z), D_2(z), \ldots,$ and $D_m(z)$, as shown in Fig. 3.19.

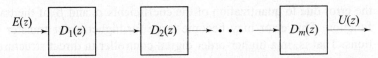

Fig. 3.19 Cascade decomposition of $D(z)$

In most cases, the $D_i(z)$; $i = 1, 2, ..., m$, are chosen to be either first-order or second-order functions. If the poles and zeros of $D(z)$ are known, then $D_i(z)$ can be obtained by grouping real poles and real zeros to produce first-order functions, or by grouping a pair of complex-conjugate poles and a pair of complex-conjugate zeros to produce a second-order function. It is, of course, possible to group two real poles with a pair of complex-conjugate zeros and *vice versa*. The grouping is, in a sense, arbitrary. It is desirable to group several different ways to see which one is best with respect to the number of arithmetic operations required, the range of coefficients, and so forth.

In general, $D(z)$ may be decomposed as follows:

$$D(z) = \prod_{i=1}^{p} \frac{1+b_i z^{-1}}{1+a_i z^{-1}} \prod_{j=p+1}^{m} \frac{1+e_j z^{-1}+f_j z^{-2}}{1+c_j z^{-1}+d_j z^{-2}}$$

The block diagram for

$$D_i(z) = \frac{1+b_i z^{-1}}{1+a_i z^{-1}} = \frac{U_i(z)}{E_i(z)}$$

and that for

$$D_j(z) = \frac{1+e_j z^{-1}+f_j z^{-2}}{1+c_j z^{-1}+d_j z^{-2}} = \frac{U_j(z)}{E_j(z)}$$

are shown in Figs 3.20a and 3.20b respectively. The realization for the digital controller $D(z)$ is a cascade connection of p first-order systems of the type shown in Fig. 3.20a, and $(m - p)$ second-order systems of the type shown in Fig. 3.20b.

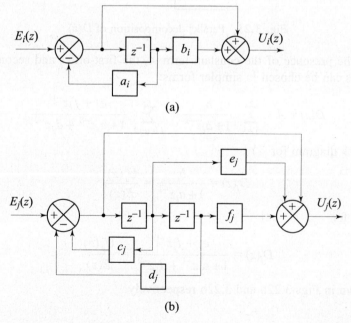

Fig. 3.20 Realizations of (a) first-order and (b) second-order functions

Parallel realization

Another approach to reduce the coefficient sensitivity problem is to expand the transfer function $D(z)$ into partial fractions. If $D(z)$ is expanded so that

$$D(z) = A + D_1(z) + D_2(z) + \cdots + D_r(z),$$

where A is simply a constant, then a digital realization for $D(z)$ may be obtained by a parallel connection of $(r + 1)$ component realizations for A, $D_1(z)$, ..., $D_r(z)$, as shown in Fig. 3.21.

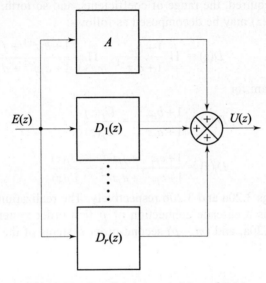

Fig. 3.21 Parallel decomposition of $D(z)$

Due to the presence of the constant term A, the first-order and second-order functions can be chosen in simpler forms:

$$D(z) = A + \sum_{i=1}^{q} \frac{b_i}{1+a_i z^{-1}} + \sum_{j=q+1}^{r} \frac{e_j + f_j z^{-1}}{1+c_j z^{-1} + d_j z^{-2}}$$

The block diagram for

$$D_j(z) = \frac{b_i}{1+a_i z^{-1}} = \frac{U_i(z)}{E(z)}$$

and that for

$$D_j(z) = \frac{e_j + f_j z^{-1}}{1+c_j z^{-1} + d_j z^{-2}} = \frac{U_j(z)}{E(z)}$$

are shown in Figs 3.22a and 3.22b respectively.

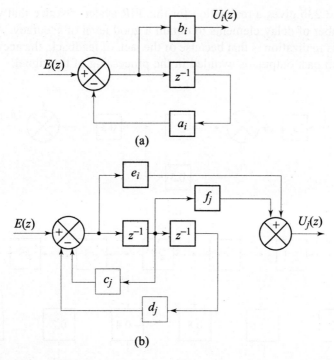

Fig. 3.22 Realization of (a) first-order and (b) second-order functions

Nonrecursive Realizations

Nonrecursive structures for $D(z)$ are similar to the recursive structures presented earlier in this section. In the nonrecursive form, the direct and cascade structures are commonly used; the parallel structure is not used since it requires more elements.

Example 3.4

Consider the digital controller with transfer function model

$$D(z) = \frac{U(z)}{E(z)} = \frac{2 - 0.6z^{-1}}{1 + 0.5z^{-1}}$$

Recursive realization of $D(z)$ yields the block diagram shown in Fig. 3.23a. By dividing the numerator of $D(z)$ by the denominator, we obtain

$D(z) = 2 - 1.6z^{-1} + 0.8z^{-2} - 0.4z^{-3} + 0.2z^{-4} - 0.1z^{-5} + 0.05z^{-6} - 0.025z^{-7} + \cdots$

Truncating this series at z^{-5}, we obtain the following FIR digital system:

$$\frac{U(z)}{E(z)} = 2 - 1.6\,z^{-1} + 0.8\,z^{-2} - 0.4\,z^{-3} + 0.2\,z^{-4} - 0.1\,z^{-5}$$

Figure 3.23b gives a realization for this FIR system. Notice that we need a large number of delay elements to obtain a good level of accuracy. An advantage of this realization is that because of the lack of feedback, the accumulation of errors in past outputs is avoided in the processing of the signal.

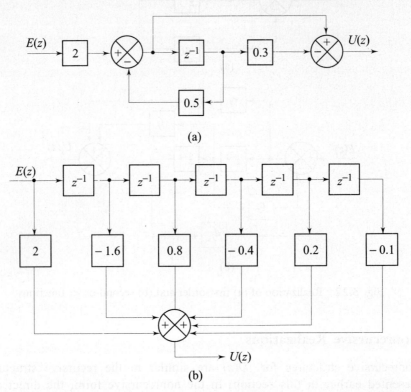

Fig. 3.23 (a) Recursive realization (b) nonrecursive realization of a first-order function

3.5 TUNABLE PID CONTROLLERS

The ultimate goal of control systems engineering is to build real physical systems to perform some specified tasks. To accomplish this goal, design and physical implementation of a control strategy are required. The standard approach to design is this: Making necessary assumptions about various uncertain quantities on the dynamics of the system, a mathematical model is built. If the objective is well defined in precise mathematical terms, then control strategies can be derived mathematically (e.g., by optimizing some criterion of performance). This is the basis of all *model-based control* strategies. This approach is feasible when it is possible to specify the objective and the model mathematically. Many sophisticated methods based on model-based control approach will appear in later chapters of the book.

For motion control applications (position, and speed control systems), identification of mathematical models of systems close enough to reality is usually possible. However, for process control applications (pressure, flow, liquid-level, temperature, and composition control systems), identification of process dynamics precisely can be expensive if meaningful identification is possible at all. This is because industrial processes are relatively slow and complex. In process-control field, therefore, it is not uncommon to follow adhoc approach for controller development when high demands on control-system performance are not made. In the *adhoc approach*, we select a certain type of controller based on past experience with the process to be controlled, and then set controller parameters by experiment once the controller is installed. The 'experimental design' of controller settings has come to be known as *controller tuning*.

Many years of experience have shown that a PID controller is versatile enough to control a wide variety of industrial processes. The common practice is to interface a PID controller (with adjustment features) to the process and adjust the parameters of the controller online by trial-and-error to obtain acceptable performance. A number of *tuning methods* have been introduced to obtain fast convergence to control solution. These methods consist of the following two steps:

(i) experimental determination of the dynamic characteristic of the control loop; and
(ii) estimation of the controller tuning parameters that produce a desired response for the dynamic characteristics determined in first step.

It may be noted that for tuning purposes, simple experiments are performed to estimate important dynamic attributes of the process. The approximate models have proven to be quite useful for process control applications (For processes whose dynamics are precisely known, the use of trial-and-error tuning is not justified since many model-based methods to the design of PID controllers are available which predict the controller parameters fairly well at the design stage itself). The predicted parameter values based on approximate models simply provide initial trial values for the online trial-and-error approach. These trial values may turn out to be a poor guess. *Fine tuning* the controller parameters online is usually necessary to obtain acceptable control performance.

Some of the tuning methods which have been successfully used in process industry, will be described here. For more details, refer companion book [180].

Analog PID Controllers

Approximately 75% of feedback controllers in the process industry are PI controllers; most of the balance are PID controllers. Some applications require only P, or PD controllers, but these are few.

Proportional Controller

The equation that describes the proportional controller is:

$$u(t) = K_c e(t) \tag{3.44a}$$

or
$$U(s) = K_c E(s) \tag{3.44b}$$

where K_c is the *controller gain*, e is the error, and u is the perturbation in controller output signal from the bias or base value corresponding to the normal operating conditions; the base value on the controller is adjusted to produce zero error under the conditions of no disturbance and/or set-point change.

Some instrument manufacturers calibrate the controller gain as *proportional band* (*PB*). A 10% *PB* means that a 10% change in the controller input causes a full-scale (100%) change in controller output. The conversion relation is thus

$$K_c = \frac{100}{PB} \tag{3.45}$$

A proportional controller has only one adjustable or tuning parameter: K_c or *PB*.

A proportionally controlled process with no integration property will always exhibit error at steady-state in the presence of disturbances and changes in set-point. The error, of course, can be made negligibly small by increasing the gain of the proportional controller. However, as the gain is increased, the performance of the closed-loop system becomes more oscillatory and takes longer to settle down after being disturbed. Further, most process plants have a considerable amount of dead-time which severely restricts the value of the gain that can be used. In processes where the control within a band from the set-point is acceptable, proportional control is sufficient. However, in processes which require perfect control at the set-point, proportional controllers will not provide satisfactory performance.

Proportional-Integral Controller

To remove the steady-state offset in the controlled variable of a process, an extra amount of intelligence must be added to the proportional controller. This extra intelligence is the integral or reset action; consequently the controller becomes a PI controller. The equation describing a PI controller is as follows:

$$u(t) = K_c \left[e(t) + \frac{1}{T_I} \int_0^t e(t)\, dt \right] \tag{3.46a}$$

or
$$U(s) = K_c \left[1 + \frac{1}{T_I s} \right] E(s) \tag{3.46b}$$

where T_I is the *integral* or *reset time*.

A PI controller has thus two adjustable or tuning parameters: K_c (or *PB*) and T_I. The integral or reset action in this controller removes the steady-state offset in the controlled variable. However, the integral mode of control has a considerable destabilizing effect which, in most of the situations, can be compensated by adjusting the gain K_c.

Some instrument manufacturers calibrate the integral mode parameter as the *reset rate*, which is simply the reciprocal of the reset time.

Proportional-Integral-Derivative Controller

Sometimes a mode faster than the proportional mode is added to the PI controller. This new mode of control is the derivative action, also called the rate action, which responds to the rate of change of error with time. This speeds up the controller action. The equation describing the PID controller is as follows:

$$u(t) = K_c \left[e(t) + \frac{1}{T_I} \int_0^t e(t)\, dt + T_D \frac{de(t)}{dt} \right] \tag{3.47a}$$

or

$$U(s) = K_c \left[1 + \frac{1}{T_I s} + T_D s \right] E(s) \tag{3.47b}$$

where T_D is the *derivative* or *rate time*.

A PID controller has thus three adjustable or tuning parameters: K_c (or *PB*), T_I, and T_D. The derivative action anticipates the error, initiates an early corrective action, and tends to increase the stability of the system. It does not affect the steady-state error directly. A derivative control mode in isolation produces no corrective effort for any constant error, no matter how large, and would therefore allow uncontrolled steady-state errors. Thus, we cannot consider derivative modes in isolation; they will always be considered as augmenting some other mode.

The block diagram implementation of Eqn (3.47b) is sketched in Fig. 3.24a. The alternative form, Fig. 3.24b, is more commonly used because it avoids taking the rate of change of the set-point input to the controller, thus preventing the undesirable derivative 'kick' on set-point changes by the process operator.

Due to the noise-accentuating characteristics of derivative operations, the lowpass-filtered derivative $T_D s/(\alpha T_D s + 1)$ is actually preferred in practice (Fig. 3.24c). The value of the filter parameter α is not adjustable but is built into the design of the controller. It is usually of the order of 0.05 to 0.3.

The controller of Fig. 3.24 is considered to be *noninteracting* in that its derivative and integral modes operate independently of each other (although proportional gain affects all the three modes). Noninteraction is provided by the parallel functioning of integral and derivative modes. By contrast, many controllers have derivative and integral action applied serially to the controlled variable resulting in interaction between them. Many of the analog industrial controllers commercially available today realize the following *interacting* PID control action (it is claimed that this form of controller is easier to tune manually).

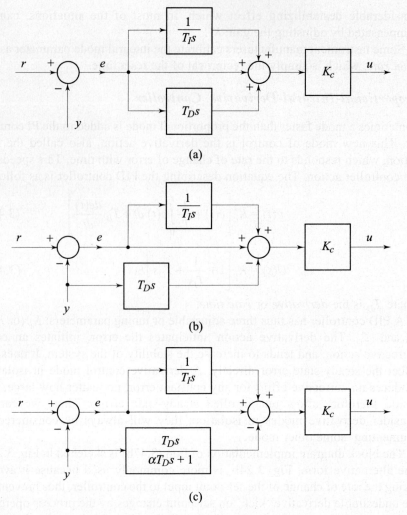

Fig. 3.24 Block diagram of PID controller

$$U(s) = K'_c \left[\frac{T'_D s + 1}{\alpha T'_D s + 1}\right]\left[1 + \frac{1}{T'_I s}\right] E(s) \qquad (3.48)$$

The first term in brackets is a derivative unit attached to the standard PI controller serially to create the PID controller (Fig. 3.25a). The derivative unit is installed on the controlled variable input to the controller in order to avoid the derivative kick (Fig. 3.25b).

From Figs 3.24a and 3.25a, the following equivalence can easily be established between the parameters of idealized versions of noninteracting PID controller and interacting PID controller ($\alpha = 0$):

$$F_{12} = 1 + (T'_D / T'_I) \qquad (3.49)$$

$$K_c = K'_c F_{12}; \quad T_I = T'_I F_{12}; \quad T_D = T'_D / F_{12}$$

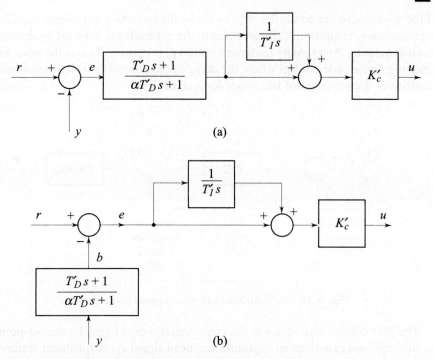

Fig. 3.25 Alternative realization schemes for PID control action

or
$$F_{21} = 0.5 + [0.25 - (T_D/T_I)]^{1/2} \tag{3.50}$$

$$K'_c = K_c F_{21}; \quad T'_I = T_I F_{21}; \quad T'_D = T_D/F_{21}$$

Adjustment Features in Industrial Controllers

A great number of manufacturers are now making available in the market process controllers (electronic, pneumatic, and computer-based) with features that permit adjusting the set-point, transferring between manual and automatic control modes, adjusting the output signal from the control-action unit (tuning the parameters K_c, T_I, and T_D), and displaying the controlled variable, set-point, and control signal. Figure 3.26 shows the basic structure of an industrial controller. The controller has been broken down into three main units:

1. The set-point control unit
2. The PID control unit
3. The manual/automatic control unit.

The set-point control unit receives the measurement y of controlled variable of the process together with the set-point r of the control. A switch gives an option of choosing between local and remote (external) set-point operation.

If the set-point to the controller is to be set by the operating personnel, then the local option r_L is chosen. If the set-point to the controller is to be set by another control module, then remote (external) option r_e is chosen. This is the case, for example, in cascade control where the drive of the controller in the major loop constitutes the set-point of the minor-loop controller.

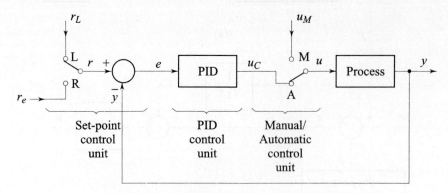

Fig. 3.26 Basic structure of an industrial controller

The PID control unit receives the error signal e developed by the set-point control unit and generates an appropriate control signal u_C. Adjustment features provided in the control unit for generating appropriate control signals include tuning of the three parameters K_c, T_I, and T_D.

The manual/automatic control unit has a switch which determines the mode of control action. When the switch is in the auto (A) position, the control signal u_C calculated by PID control unit is sent to the process (in such a case, the process is controlled in closed loop). When the switch is in the manual (M) position, the PID control unit 'freezes' its output. The control signal u_M can then be changed manually by the operating personnel (the process is then controlled in open loop).

The basic structure of a process controller shown in Fig. 3.26 is common for pneumatic, electronic, and computer-based controllers. These controllers are different in terms of realization of adjustment features.

Ziegler-Nichols Tuning Method Based on Ultimate Gain and Period

This pioneer method, also known as the closed-loop or on-line tuning method, was proposed by J.G. Ziegler and N.B. Nichols around 1940. In this method, the parameters by which the dynamic characteristics of the process are represented are the ultimate gain and period. These parameters are used in tuning the controller for a specified response: the quarter-decay ratio (QDR) response.

Determination of Ultimate Gain and Period

When the process is under closed-loop P control, the gain of the P controller at which the loop oscillates with constant amplitude has been defined as the *ultimate gain K_{cu}*. *Ultimate period T_u* is the period of these sustained oscillations. The ultimate gain is thus a measure of difficulty in controlling a process; the higher the ultimate gain, the easier it is to control the process loop. The ultimate period is, in turn, a measure of speed of response of the loop; the larger the period, the slower the loop.

By its definition, it can be deduced that the ultimate gain is the gain at which the loop is at the threshold of instability. At gains just below the ultimate, the loop signals will oscillate with decreasing amplitude, and at gains above the ultimate, the amplitude of the oscillations will grow with time.

For experimental determination of K_{cu} and T_u, the controller is set in 'auto' mode and the following procedure is followed (refer Fig. 3.26).

1. Remove the integral mode by setting the integral time to its highest value. Alternatively, if the PID controller allows for switching off the integral mode, switch it off.
2. Switch off the derivative mode or set the derivative time to its lowest value, usually zero.
3. Increase the proportional gain in steps. After each increase, disturb the loop by introducing a small step change in set-point and observe the response of the controlled variable, preferably on a trend recorder. The controlled variable should start oscillating as the gain is increased. When the amplitude of the oscillations remains approximately constant, the ultimate controller gain has been reached. Record it as K_{cu}.
4. Measure the period of the oscillations from the trend recording. This parameter is T_u.

The procedure just outlined is simple and requires a minimum upset to the process, just enough to be able to observe the oscillations. Nevertheless, the prospect of taking a process control loop to the verge of instability is not an attractive one from a process operation standpoint.

Tuning for Quarter-Decay Ratio Response

Ziegler and Nichols proposed that the parameters K_{cu} and T_u characterizing a process be used in tuning the controller for QDR response. The QDR response is illustrated in Fig. 3.27 for a step change in disturbance and for a step change in set-point. Its characteristic is that each oscillation has an amplitude that is one-fourth of the previous oscillation.

Empirical relations [7] for calculating the QDR tuning parameters of P, PI and PID controllers from the ultimate gain K_{cu} and period T_u are given in Table 3.2.

PI and PID tuning parameters that produce quarter-decay response are not unique. For each setting of the integral and derivative times, there will usually be a setting of the controller gain that produces quarter-decay response. The settings given in Table 3.2 are the figures based on experience; these settings have produced fast response for most industrial loops.

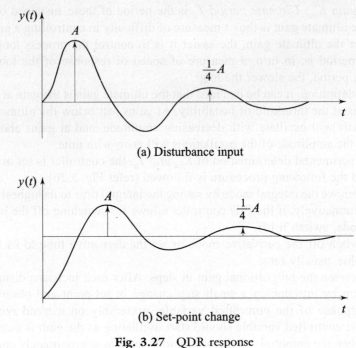

Fig. 3.27 QDR response

Table 3.2 QDR tuning formulas based on ultimate gain and period

Controller	Gain	Integral time	Derivative time
P	$K_c = 0.5\, K_{cu}$	—	—
PI	$K_c = 0.45\, K_{cu}$	$T_I = T_u/1.2$	—
PID (noninteracting)	$K_c = 0.75\, K_{cu}$	$T_I = T_u/1.6$	$T_D = T_u/10$
PID (interacting)	$K'_c = 0.6\, K_{cu}$	$T'_I = T_u/2$	$T'_D = T_u/8$

Ziegler-Nichols Tuning Method Based on Process Reaction Curve

Although the tuning method based on ultimate gain and period is simple and fast, other methods of characterizing the dynamic response of feedback control loops have been developed over the years. The need for these alternative methods is based on the fact that it is not always possible to determine the ultimate gain and period of a loop; some loops would not exhibit sustained oscillations with a proportional controller. Also, the ultimate gain and period do not give insight into which process or control system characteristics could be modified to improve the feedback controller performance. A more fundamental method of characterizing process dynamics is needed to guide such modifications. In the following, we present an open-loop method for characterizing the dynamic response of the process in the loop.

Process Reaction Curve

Process control is characterized by systems which are relatively slow and complex and which in many cases include an element of pure time delay (dead-time). Even where a dead-time element is not present, the complexity of the system which will typically contain several first-order sub-systems, will often result in a *process reaction curve* (dynamic response to a step change in input) which has the appearance of pure time delay.

Process reaction curve may be obtained by carrying out the following step-test procedure.

With the controller on 'manual', i.e., the loop opened (refer Fig. 3.26), a step change of magnitude Δu in the control signal $u(t)$ is applied to the process. The magnitude Δu should be large enough for the consequent change $\Delta y(t)$ in the process output variable to be measurable, but not so large that the response will be distorted by process nonlinearities. The process output is recorded for a period from the introduction of the step change in the input until the process reaches a new steady-state.

A typical process reaction curve is sketched in Fig. 3.28. The most common model used to characterize the process reaction curve is the following:

$$\frac{Y(s)}{U(s)} = G(s) = \frac{Ke^{-\tau_D s}}{\tau s + 1} \qquad (3.51)$$

where K = the process steady-state gain;
τ_D = the effective process dead-time; and
τ = the effective process time-constant.

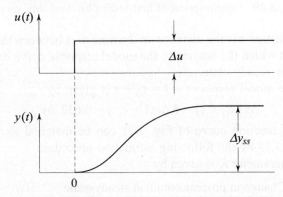

Fig. 3.28 A process reaction curve

This is a *first-order plus dead-time model*. The model response for a step change in the input signal of magnitude Δu, is given by

$$Y(s) = \frac{Ke^{-\tau_D s}}{\tau s + 1} \frac{\Delta u}{s} = K\Delta u \, e^{-\tau_D s} \left[\frac{1}{s} - \frac{\tau}{\tau s + 1} \right]$$

Inverting with the help of a Laplace transform table given in Appendix A, and applying the real translation theorem of Laplace transforms (Eqn. (A.30) in Appendix A), we get

$$\Delta y(t) = K\Delta u \left[1 - e^{-(t-\tau_D)/\tau} \right] \; ; \; t > \tau_D$$
$$= 0 \qquad\qquad\qquad\qquad ; \; t \leq \tau_D \tag{3.52}$$

The term Δy is the perturbation or change in the output from its initial value:

$$\Delta y(t) = y(t) - y(0)$$

Figure 3.29 shows the model response to a step change of magnitude Δu in the input signal. Δy_{ss} is the steady-state change in the process output (refer Eqn. (3.52)):

$$\Delta y_{ss} = \lim_{t \to \infty} \Delta y(t) = K\Delta u$$

At the point $t = \tau_D$ on the time axis, the process output variable leaves the initial steady-state with a maximum rate of change (refer Eqn. (3.52)):

$$\left. \frac{d}{dt} \Delta y(t) \right|_{t=\tau_D} = K\Delta u \left(\frac{1}{\tau} \right) = \frac{\Delta y_{ss}}{\tau}$$

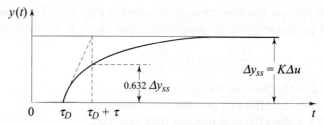

Fig. 3.29 Step-response of first-order plus dead-time model

The time-constant τ is the distance on the time axis between the point $t = \tau_D$, and the point at which the tangent to the model response curve drawn at $t = \tau_D$ crosses the new steady-state.

Note that the model response at $t = \tau_D + \tau$ is given by

$$\Delta y (\tau_D + \tau) = K\Delta u (1 - e^{-1}) = 0.632 \, \Delta y_{ss}$$

The process reaction curve of Fig. 3.28 can be matched to the model response of Fig. 3.29 by the following estimation procedure.

The model parameter K is given by

$$K = \frac{\text{Change in process output at steady-state}}{\text{Step change in process input}} = \frac{\Delta y_{ss}}{\Delta u} \tag{3.53a}$$

The estimation of the model parameters τ_D and τ can be done by at least three methods, each of which results in different values.

Tangent method

This method makes use of the line that is tangent to the process reaction curve at the point of maximum rate of change. The time-constant is then defined as the distance on the time axis between the point where the tangent crosses the initial steady-state of the output variable, and the point where it crosses the new steady-state value. The dead-time is the distance on the time axis between the occurrence of the input step change and the point where the tangent line crosses the initial steady-state. These estimates are indicated in Fig. 3.30a.

Tangent-and-point method

In this method, τ_D is determined in the same manner as in the earlier method, but the value of τ is the one that forces the model response to coincide with the actual response at $t = \tau_D + \tau$. Construction for this method is shown in Fig. 3.30b. The value of τ obtained by this method is usually less than that obtained by the earlier method, and the process reaction curve is usually closer to the response of the model obtained by this method compared to the one obtained by the earlier method.

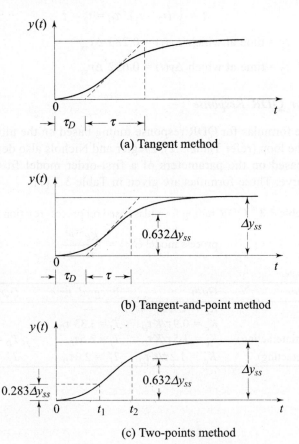

(a) Tangent method

(b) Tangent-and-point method

(c) Two-points method

Fig. 3.30 Estimation of model parameters

Two-points method

The least precise step in the determination of τ_D and τ by the previous two methods is the drawing of the line tangent to the process reaction curve at the point of maximum rate of change. To eliminate this dependence on the tangent line, it is proposed that the values of τ_D and τ be selected such that the model and the actual responses coincide at two points in the region of high rate of change. The two points recommended are $(\tau_D + \frac{1}{3}\tau)$ and $(\tau_D + \tau)$. To locate these points, we make use of Eqn. (3.52):

$$\Delta y(\tau_D + \tfrac{1}{3}\tau) = K\Delta u[1 - e^{-1/3}] = 0.283\,\Delta y_{ss}$$

$$\Delta y(\tau_D + \tau) = K\Delta u\,[1 - e^{-1}] = 0.632\,\Delta y_{ss}$$

These two points are labelled t_1 and t_2 in Fig. 3.30c. Knowing t_1 and t_2, we can obtain the values of τ_D and τ.

$$\tau_D + \tau = t_2; \quad \tau_D + \tfrac{1}{3}\tau = t_1$$

which reduces to

$$\tau = \tfrac{3}{2}(t_2 - t_1); \quad \tau_D = t_2 - \tau \qquad (3.53b)$$

where t_1 = time at which $\Delta y(t) = 0.283\,\Delta y_{ss}$

t_2 = time at which $\Delta y(t) = 0.632\,\Delta y_{ss}$

Tuning for QDR Response

Besides the formulas for QDR response tuning based on the ultimate gain and period of the loop (refer Table 3.2), Ziegler and Nichols also developed tuning formulas based on the parameters of a first-order model fit to the process reaction curve. These formulas are given in Table 3.3 [7].

Table 3.3 QDR tuning formulas based on process reaction curve:

$$\text{process model } G(s) = \frac{Ke^{-\tau_D s}}{\tau s + 1}$$

Controller	Gain	Integral time	Derivative time
P	$K_c = \tau/K\tau_D$	—	—
PI	$K_c = 0.9\tau/K\tau_D$	$T_I = 3.33\,\tau_D$	—
PID (noninteracting)	$K_c = 1.5\tau/K\tau_D$	$T_I = 2.5\tau_D$	$T_D = 0.4\tau_D$
PID (interacting)	$K'_c = 1.2\tau/K\tau_D$	$T'_I = 2.0\tau_D$	$T'_D = 0.5\tau_D$

Three major conclusions can be drawn from this table.
1. The controller gain is inversely proportional to the process gain K which represents the product of gain of all the elements in the loop other than the controller. It means that if the gain of any of the elements were to change because of recalibration, resizing, or nonlinearity, the response of the feedback loop will change unless the controller gain is readjusted.
2. The controller gain must be reduced when the ratio of the process dead-time to its time-constant increases. This means that the difficulty in controlling the loop increases when the ratio of the process dead-time to its time-constant increases. This ratio, which can be used as a measure of difficulty in controlling a process, will be called the *normalized dead-time* τ_{ND}.

$$\frac{\text{Apparent dead-time } \tau_D}{\text{Apparent time-constant } \tau} = \text{Normalized dead-time } \tau_{ND} \quad (3.54)$$

τ_{ND} can be estimated from the process reaction curve. Processes with small τ_{ND} are easy to control and processes with large τ_{ND} are difficult to control. The parameter τ_{ND} has been called the *controllability ratio* in the literature. To avoid confusion with the standard terminology of modern control theory (Chapter 5), the word normalized dead-time will be used here.

Notice that having a long dead-time parameter means that the loop is difficult to control only if the time-constant is short. In other words, a loop with a dead-time of several minutes would be just as difficult to control as one with a dead-time of a few seconds if the normalized dead-time for both the loops is the same.

3. The speed of response of the controller, which is determined by integral and derivative times, must match the speed of response of the process. The formulas in Table 3.3 match these response speeds by relating the integral and derivative times of the controller to the process dead-time.

In using the formulas of Table 3.3, we must keep in mind that they were developed empirically for the most common range of the normalized dead-time parameter, which is between 0.1 and 0.3, based on the fact that most processes do not exhibit significant transportation lag (rather, the dead-time is the result of several first-order lags in series).

As was pointed out in the earlier discussion on QDR tuning based on ultimate gain and period, the difficulty of the QDR performance specification for PI and PID controllers is that there is an infinite set of values of the controller parameters that can produce it; i.e., for each setting of the integral time on a PI controller and for each reset-derivative time combination on a PID controller, there is a setting of the gain that results in QDR response. The settings given in Table 3.3 are the figures based on experience; these settings have produced fast response for most industrial loops.

Digital PID Controllers

Most process industries today use computers to carry out the basic feedback control calculations. The formulas that are programmed to calculate the controller output are mostly the discrete versions of the analog controllers presented earlier in this section. This practice allows the use of established experience with analog controllers and, in principle, their well-known tuning rules could be applied.

As there is no extra cost in programming all the three modes of control, most computer-based algorithms contain all the three and then use flags and logic to allow the process engineer to specify any signal mode or a combination of two or three modes. Many computer controllers use the noninteracting version of PID control (refer Eqn. (3.47b)) and some computer control systems allow the option of either noninteracting or interacting (refer Eqn. (3.48)) version of PID control.

Noninteracting Position PID Algorithm

The equation describing the idealized noninteracting PID controller is as follows (refer Eqn. (3.47a)):

$$u(t) = K_c \left[e(t) + \frac{1}{T_I} \int_0^t e(t)dt + T_D \frac{de(t)}{dt} \right] \quad (3.55)$$

with parameters

K_c = controller gain;
T_I = integral time; and
T_D = derivative time.

For small sample times T, this equation can be turned into a difference equation by discretization. Various methods of discretization were presented in Section 2.12.

Approximating the derivative mode by the backward-difference approximation and the integral mode by backward integration rule, we obtain (refer Eqns (2.89))

$$u(k) = K_c \left[e(k) + \frac{1}{T_I} S(k) + \frac{T_D}{T}(e(k) - e(k-1)) \right] \quad (3.56)$$

$$S(k) = S(k-1) + Te(k)$$

where
$u(k)$ = the controller output at sample k;
$S(k)$ = the sum of the errors; and
T = the sampling interval.

This is a nonrecursive algorithm. For the formation of the sum, all past errors $e(\cdot)$ have to be stored.

Equation (3.56) is known as the 'absolute form' or 'position form' of the PID algorithm. It suffers from one particular disadvantage which is manifest when the process it is controlling is switched from manual to automatic control. The initial value of the control variable u will simply be

$$u(0) = K_c \left[1 + \frac{T}{T_I} + \frac{T_D}{T} \right] e(0)$$

Since the controller has no knowledge of the previous sample values, it is not likely that this output value will coincide with that previously available under manual control. As a result, the transfer of control will cause a 'bump' which may seriously disturb the plant operation. This can only be overcome by laboriously aligning the manual and computer outputs or by adding complexity to the controller so that it will automatically 'track' the manual controller.

Practical implementation of the PID algorithm includes the following additional features.
1. It is seldom desirable for the derivative mode of the controller to respond to set-point changes. This is because the set-point changes cause large changes in the error that last for only one sample; when the derivative mode acts on this error, undesirable pulses or 'derivative kicks' occur on the controller output right after the set-point is changed. These pulses, which last for one sampling interval, can be avoided by having the derivative mode act on the controlled variable rather than on the error.
2. A pure derivative term should not be implemented because it will give a very large amplification of the measurement noise. The gain of the derivative must thus be limited. This can be done by approximating the transfer function $T_D s$ as follows:

$$T_D s \cong \frac{T_D s}{\alpha T_D s + 1}$$

where α is the filter parameter whose value is not adjustable, but is built into the design of the controller. It is usually of the order of 0.05 to 0.3. The PID controller, therefore, takes the form (refer Fig. 3.24c):

$$U(s) = K_c \left[E(s) + \frac{1}{T_I s} E(s) - \frac{T_D s}{\alpha T_D s + 1} Y(s) \right] \qquad (3.57)$$

Discretization of this equation results in the following PID algorithm:

$$u(k) = K_c \left[e(k) + \frac{1}{T_I} S(k) + D(k) \right] \qquad (3.58)$$

$$S(k) = S(k-1) + Te(k)$$

$$D(k) = \frac{\alpha T_D}{\alpha T_D + T} D(k-1) - \frac{T_D}{\alpha T_D + T} [y(k) - y(k-1)]$$

Noninteracting Velocity PID Algorithm

This is a recursive algorithm characterized by the calculation of the current control variable $u(k)$ based on the previous control variable $u(k-1)$ and correction terms. To derive the recursive algorithm, we subtract from Eqn. (3.56)

$$u(k-1) = K_c\left[e(k-1) + \frac{1}{T_I}S(k-1) + \frac{T_D}{T}(e(k-1) - e(k-2))\right]$$

This gives

$$u(k) - u(k-1) = K_c\left[e(k) - e(k-1) + \frac{T}{T_I}e(k) + \frac{T_D}{T}[e(k) - 2e(k-1) + e(k-2)]\right] \qquad (3.59)$$

Now, only the current change in the control variable
$$\Delta u(k) = u(k) - u(k-1)$$
is calculated. This algorithm is known as the 'incremental form' or 'velocity form' of the PID algorithm. The distinction between the position and velocity algorithms is significant only for controllers with integral effect.

The velocity algorithm provides a simple solution to the requirement of bumpless transfer. The problem of bumps arises mainly from the need for an 'initial condition' on the integral, and the solution adopted is to externalize the integration as shown in Fig. 3.31. The external integration may take the form of an electronic integrator but frequently the type of actuating element is changed so that recursive algorithm is used with actuators which, by their very nature, contain integral action. Stepper motor (refer Section 3.9) is one such actuating element.

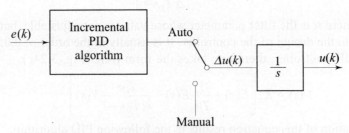

Fig. 3.31 Control scheme of bumpless transfer

Practical implementation of this algorithm includes the features of avoiding derivative kicks and filtering measurement noise. Using Eqn. (3.58) we obtain

$$\Delta u(k) = K_c\left[e(k) - e(k-1) + \frac{T}{T_I}e(k) - \frac{T}{\alpha T_D + T}D(k-1) \right.$$
$$\left. \frac{T_D}{\alpha T_D + T}(y(k) - y(k-1))\right] \qquad (3.60)$$

$$D(k) = \frac{\alpha T_D}{\alpha T_D + T} D(k-1) - \frac{T_D}{\alpha T_D + T}[y(k) - y(k-1)]$$

where $y(k)$ = controlled variable;

$\Delta u(k)$ = incremental control variable = $u(k) - u(k-1)$;

$e(k)$ = error variable; K_c = controller gain;

T_I = integral time; T_D = derivative time; and

T = sampling interval.

Interacting PID Algorithms

The interacting PID controller is described by the equation (refer Fig. 3.25b)

$$U(s) = K'_c \left[E(s) + \frac{1}{T'_I s} E(s) \right] \quad (3.61)$$

where $$E(s) = R(s) - B(s) = R(s) - \frac{T'_D s + 1}{\alpha T'_D s + 1} Y(s)$$

Discretization of this equation results in the following interacting PID algorithm in 'position form':

$$u(k) = K'_c \left[e(k) + \frac{1}{T'_I} S(k) \right] \quad (3.62)$$

where $S(k) = S(k-1) + Te(k)$; $e(k) = r(k) - b(k)$;

$$b(k) = \frac{\alpha T'_D}{\alpha T'_D + T} b(k-1) + \frac{T}{\alpha T'_D + T} y(k) + \frac{T'_D}{\alpha T'_D + T}[y(k) - y(k-1)]$$

The interacting PID algorithm in 'velocity form' is given by

$$\Delta u(k) = u(k) - u(k-1) = K'_c \left[e(k) - e(k-1) + \frac{T}{T'_I} z(k) \right] \quad (3.63)$$

where $e(k) = r(k) - b(k)$;

$$b(k) = \frac{\alpha T'_D}{\alpha T'_D + T} b(k-1) + \frac{T}{\alpha T'_D + T} y(k) + \frac{T'_D}{\alpha T'_D + T}[y(k) - y(k-1)]$$

Tuning Rules for Digital Controllers

Though tuning formulas that are specifically applicable to digital control algorithms have been developed [7, 89], the most popular and widely used tuning approach for digital PID controllers is to apply rules of Tables 3.2–3.3 with a simple correction to account for the effect of sampling. When a continuous-time signal is sampled at regular intervals of time and is then reconstructed by holding the sampled values constant for each sampling interval, the reconstructed signal is effectively delayed by approximately one half of the sampling interval, as shown in Fig. 3.32a (also refer Example 2.10). In the digital control configuration of Fig. 3.32b, the D/A converter holds the output of the digital controller constant between updates, thus adding one half the sampling time to the dead-time of the process components. The correction for sampling is then simply to add one half the sampling time to the dead-time obtained from the process reaction curve.

$$\tau_{CD} = \tau_D + \tfrac{1}{2} T \tag{3.64}$$

where τ_{CD} is the corrected dead-time, τ_D is the dead-time of the process, and T is the sampling interval.

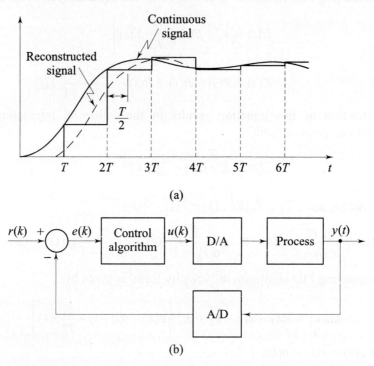

Fig. 3.32 Time delay introduced by sampling and reconstruction

The tuning formulas given in Table 3.3 can directly be used for digital PID controllers with τ_D replaced by τ_{CD}.

Notice that the online tuning method based on ultimate gain and period inherently incorporates the effect of sampling when the ultimate gain and period are determined with the digital controller included in the loop. Tuning rules of Table 3.2 can therefore be applied to digital control algorithms without any correction.

3.6 DIGITAL TEMPERATURE CONTROL SYSTEM

This section describes the hardware features of the design of a microprocessor-based controller for temperature control in an air-flow system.

Figure 3.33 shows the air-flow system, provided with temperature measurement and having a heater grid with controlled power input. Air drawn through a variable orifice by a centrifugal blower is driven past the heater grid and through a length of tubing to the atmosphere again. The temperature sensing element consists of a bead thermistor fitted to the end of a probe inserted into the air stream 30 cms from the heater. The task is to implement a controller, in the position shown by dotted box, to provide temperature control of the air stream. It is a practical process control problem in miniature, simulating the conditions found in furnaces, boilers, air-conditioning systems, etc.

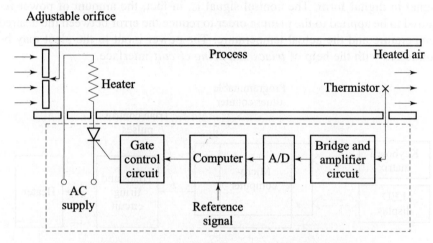

Fig. 3.33 Temperature control in an air-flow system

The functions within the control loop can be broken down as follows:
(a) sampling of the temperature measurement signal at an appropriate rate;
(b) transfer of the measurement signal into the computer;
(c) comparison of the measured temperature with a stored desired temperature to form an error signal;
(d) operation on the error signal by an appropriate algorithm to form an output signal; and
(e) transfer of the output signal through the interface to the power control unit.

Figure 3.34 gives hardware description of the temperature control system. Let us examine briefly the function of each block. The block labelled *keyboard matrix*, interfaced to the microcomputer through a programmable keyboard/display interface chip, enables the user to feed reference input to the temperature control system. The LED *display* unit provides display of the actual temperature of the heating chamber.

The temperature range for the system under consideration is 20 to 60°C. When a thermistor is used as temperature *transducer*, it is necessary to convert the change in its resistance to an equivalent analog voltage. This is accomplished with Wheatstone bridge; the thermistor exposed to the process air forms one arm of the bridge. The millivolt range of the bridge error voltage is *amplified* to the range required by A/D converter. The output of the A/D converter is the digital measurement of the actual temperature of the process air. This data is fed to the microcomputer through an input port. The microcomputer compares the actual temperature with the desired temperature at each sampling instant and generates an error signal. The error signal is then processed as per the control algorithm (to be given later), resulting in a control signal in digital form. The control signal is, in fact, the amount of power required to be applied to the plant in order to reduce the error between the desired temperature and the actual temperature. The power input to the plant may be controlled with the help of *triacs* and *firing circuit* interface.

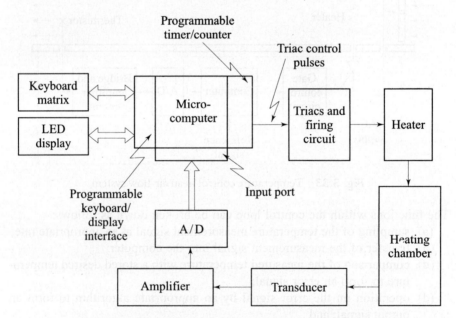

Fig. 3.34 Block diagram of the temperature control system

A basic circuit using a triac (bidirectional thyristor) which controls the flow of alternating current through the heater is shown in Fig. 3.35a. If the triac closes the circuit for t_p seconds out of T seconds, the average power applied to the plant over the sampling period T is

$$u = \frac{1}{T}\int_0^{t_p} \frac{V^2}{R} dt = \frac{V^2}{R}\frac{t_p}{T};$$

V = rms value of the voltage applied to the heater,
R = resistance of the heater.

This gives
$$t_p = \frac{u}{V^2/R} T \qquad (3.65)$$

Depending on the control signal u (power required to be applied to the plant), t_p is calculated in the microcomputer. A number is latched in a down counter (in the programmable timer/counter chip interfaced to the microcomputer) which is determined by the value of t_p and the counter's clock frequency. A pulse of required width t_p is thus available at each sampling instant from the programmable timer/counter chip. This, in fact, is a pulse width modulated (PWM) wave whose time period is constant and width is varied in accordance with the power required to be fed to the plant (Fig. 3.35b).

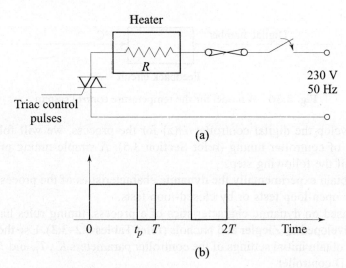

Fig. 3.35 Control scheme for the thermal process

The function of the triacs and firing circuit interface is thus to process the PWM output of the microcomputer such that the heater is ON when the PWM output is logic 1, and OFF when it is logic 0. Since the heater is operated off 230 V ac at 50 Hz, the firing circuit should also provide adequate isolation between the high voltage ac signals and the low voltage digital signals.

Control Algorithm

A model for the temperature control system under study is given by the block diagram of Fig. 3.36. A gain of unity in the feedback path corresponds to the design of feedback circuit (temperature transducer + amplifier + A/D converter) which enables us to interpret the magnitude of the digital output of A/D converter directly as temperature in °C. The temperature command is given

174 Digital Control and State Variable Methods

in terms of the digital number with magnitude equal to the desired temperature in °C. The error e (°C) is processed by the control algorithm with transfer function $D(z)$. The computer generates a PWM wave whose time period is equal to the sampling interval and width is varied in accordance with the control signal u(watts). The PWM wave controls the power input to the plant through the triacs and the firing circuit interface. Since the width of PWM remains constant over a sampling interval, we can use S/H to model the input-output relation of the triacs and the firing circuit interface.

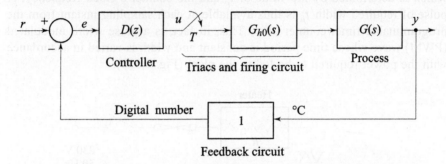

Fig. 3.36 A model for the temperature control system

To develop the digital controller $D(z)$ for the process, we will follow the approach of controller tuning (refer Section 3.5). A simple tuning procedure consists of the following steps:

(i) Obtain experimentally the dynamic characteristics of the process, either by open-loop tests or by closed-loop tests.

(ii) Based on dynamic characteristics of a process, tuning rules have been developed by Ziegler and Nichols (refer Tables 3.2–3.3). Use these rules to obtain initial settings of the controller parameters K_c, T_I, and T_D of the PID controller

$$D(s) = K_c \left[1 + \frac{1}{T_I s} + T_D s \right] \qquad (3.66)$$

(iii) Discretize the PID controller to obtain digital control algorithm for the temperature control process. Thumb rules given in Section 2.11 may be used for initial selection of sampling interval T.

In digital mode, the PID controller takes the form (refer Eqn. (2.107))

$$D(z) = K_c \left[1 + \frac{T}{2T_I}\left(\frac{z+1}{z-1}\right) + \frac{T_D}{T}\left(\frac{z-1}{z}\right)\right] = \frac{U(z)}{E(z)} \qquad (3.67)$$

(iv) Implement the digital PID controller. Figure 3.37 shows a realization scheme for the controller; the proportional, integral, and derivative terms are implemented separately and summed up at the output.

(v) Fine tune K_c, T_I, T_D and T to obtain acceptable performance.

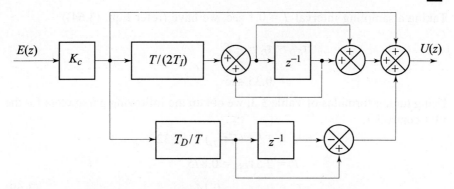

Fig. 3.37 A realization scheme for the PID controller

▲▲

An open-loop test was performed on the air-flow system (Fig. 3.33) to obtain its dynamic characteristics.

    Input   : heater power
    Output   : air temperature

The test was carried out with a dc input signal. A wattmeter on the input side measured the heater power and a voltmeter on the output side measured the output (in volts) of the bridge circuit, which is proportional to the air temperature in °C.

Figure 3.38 shows the response for a step input of 20 watts. This process reaction curve was obtained for a specific orifice setting.

Approximation of the process reaction curve by a first-order plus dead-time model is obtained as follows (refer Fig. 3.29):

The change in the process output at steady-state is found to be Δy_{ss} = 24.8 volts. Therefore the process gain

$$K = \frac{24.8}{20} = 1.24 \text{ volts/watt}$$

The line that is tangent to the process reaction curve at the point of maximum rate of change, gives τ_D = 0.3 sec. The time at which the response is $0.632\Delta y_{ss}$ is found to be 0.83 sec. Therefore $\tau + \tau_D$ = 0.83; which gives τ = 0.53 sec. (It may be noted that the response is oscillatory in nature; therefore a second-order model will give a better fit. However, for coarse tuning, we have approximated the response by a first-order plus dead-time model).

The process reaction curve of the air-flow system is thus represented by the model

$$G(s) = \frac{Ke^{-\tau_D s}}{\tau s + 1} = \frac{1.24 e^{-0.3s}}{0.53 s + 1} \tag{3.68}$$

Taking a sampling interval $T = 0.1$ sec, we have (refer Eqn. (3.64))

$$\tau_{CD} = \tau_D + \frac{1}{2}T$$

$$= 0.35 \text{ sec}$$

Using tuning formulas of Table 3.3, we obtain the following parameters for the PID controller.

$$K_c = 1.5\tau/(K\tau_{CD}) = 1.832$$
$$T_I = 2.5\tau_{CD} = 0.875$$
$$T_D = 0.4\tau_{CD} = 0.14 \tag{3.69}$$

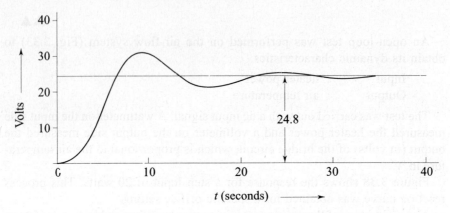

Fig. 3.38 Process reaction curve of the air-flow system

3.7 DIGITAL POSITION CONTROL SYSTEM

This section describes hardware features of the design of a microprocessor based controller for a position control system. The plant of our digital control system is an inertial load driven by an armature-controlled dc servomotor. The plant also includes a motor-drive circuit. The output of the drive circuit is fed to the armature of the motor which controls the position of the motor shaft. In addition, it also controls the direction of rotation of the motor shaft.

Figure 3.39 gives hardware description of the position control system. Let us examine briefly the function of each block.

The block labelled *digital signal generator*, interfaced to the microcomputer through an input port, enables the user to feed the desired position (set point) of the motor shaft. A keyboard matrix can be used for entering numerical commands into the digital system.

The microcomputer compares the actual position of the motor shaft with the desired position at each sampling instant and generates an error signal. The error signal is then processed as per the control algorithm (to be given later)

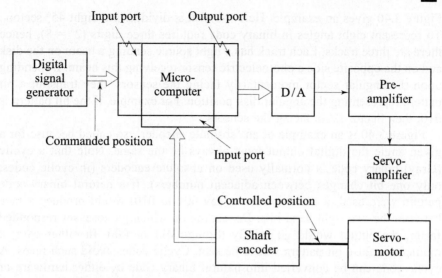

Fig. 3.39 Block diagram of the digital positioning system

resulting in a control signal in digital form. The digital control signal is converted to a bipolar (can be +ve or –ve) analog voltage in the D/A converter interfaced to the microcomputer. This bipolar signal is processed in preamplifier and servoamplifier (power amplifier), enabling the motor to be driven in one direction for a +ve voltage at preamplifier input and in opposite direction for a negative voltage.

With these units, the block diagram of Fig. 3.39 also shows a shaft encoder for digital measurement of shaft position/speed. We now examine in detail the principle of operation of this digital device.

Digital Measurement of Shaft Position/Speed

The digital measurement of shaft position requires conversion from the analog quantity 'shaft angle' to a binary number. One way of doing this would be to change shaft angle to a voltage using a potentiometer, and then to convert it to a binary number through an electronic A/D converter. This is perfectly feasible but is not sensible because:

(i) high quality potentiometers of good accuracy are expensive and subject to wear; and
(ii) the double conversion is certain to introduce more errors than a single conversion would do.

We can go straight from angle to number using an optical angular *absolute-position encoder*. It consists of a rotary disk made of a transparent material. The disk is divided into a number of equal angular sectors depending on the resolution required. Several tracks, which are transparent in certain sectors but opaque in others, are laid out. Each track represents one digit of a binary number. Detectors on these tracks sense whether the digit is a '1' or a '0'.

Figure 3.40 gives an example. Here the disk is divided into eight 45° sectors. To represent eight angles in binary code requires three digits ($2^3 = 8$), hence there are three tracks. Each track has a light source sending a beam on the disk and on the opposite side a photoelectric sensor receiving this beam. Depending upon the angular sector momentarily facing the sensors, they transmit a bit pattern representing the angular disk position. For example, if the bit pattern is 010, then Sector IV is facing the sensors.

Figure 3.40 is an example of an 'absolute encoder', so called because for a given angle the digital output must always be the same. Note that a cyclic (Gray) binary code is normally used on absolute encoders (in cyclic codes, only one bit changes between adjacent numbers). If a natural binary-code pattern were used, a transition from, say 001 to 010, would produce a race between the two right-hand bits. Depending on which photosensor responded faster, the output would go briefly through 011 or 000. In either case, a momentary false bit pattern would be sent. Cyclic codes avoid such races. A cyclic code can be converted into natural binary code by either hardware or computer software.

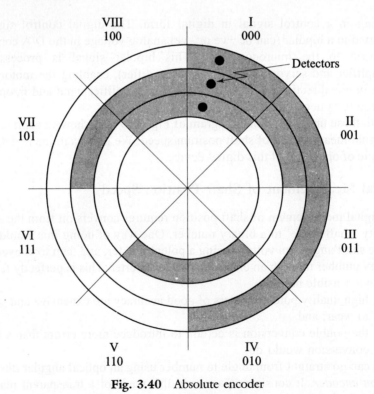

Fig. 3.40 Absolute encoder

Encoders similar to Fig. 3.40 have been widely used. But they have certain disadvantages.

(i) The resolution obtainable with these encoders is limited by the number of tracks on the encoder disk. The alignment of upto ten detectors and the laying out of ten tracks is still quite difficult and thus expensive.
(ii) The resulting digital measurement is in a cyclic code and must usually be converted to natural binary before use.
(iii) The large number of tracks and detectors inevitably increases the chance of mechanical and/or electrical failure.

For these reasons, another form of encoder is commonly used today and is known as the *incremental encoder*. The basis of an incremental encoder is a single track served by a single detector and laid out in equal segments of '0' and '1' as in Fig. 3.41. As the track moves relative to the detector, a pulse train is generated, and can be fed to a counter to record how much motion has occurred. With regard to this scheme of measurement, the following questions may be raised.
(i) How do we know which direction the motion was?
(ii) If we can record only the distance moved, how do we know where we were?

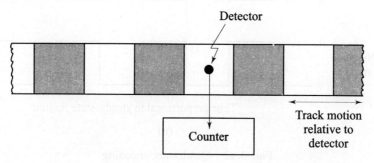

Fig. 3.41 Incremental encoder

The answer to the first question involves the addition of a second detector. Figure 3.42a shows two detectors, spaced one half of a segment apart. As the track moves relative to the detectors (we assume at a constant rate), the detector outputs vary with time as shown in the waveforms of Fig. 3.42b. We can see that the relative 'phasing' of the A and B signals depends upon the direction of motion and so gives us a means of detecting the direction.

For example, if signal B goes from '0' to '1' while signal A is at '1', the motion is positive. For the same direction, we see that B goes from '1' to '0' whilst A is at '0'. For negative motion, a similar but different pair of statements can be made. By application of some fairly simple logic, it is possible to control a reversible counter as is indicated in Fig. 3.43

This method of direction sensing is referred to as *quadrature encoding*. The detectors are one half of a segment apart, but reference to the waveforms of Fig. 3.42 shows that there are two segments to one cycle; so the detectors are one-quarter of a cycle apart, and hence the name.

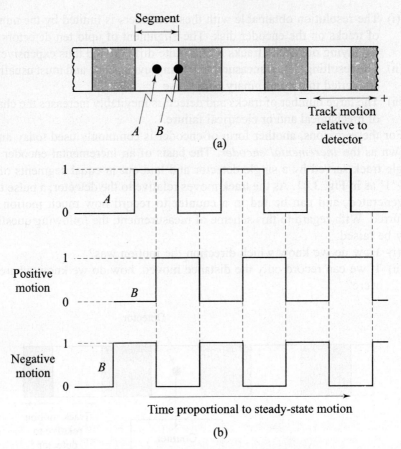

Fig. 3.42 Quadrature encoding

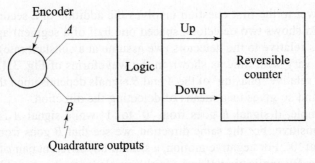

Fig. 3.43 Direction-detecting circuit for an incremental encoder

The solution to the second problem also requires an additional **detector** working on a datum track as shown in Fig. 3.44. The datum resets the **counter** every time it goes by.

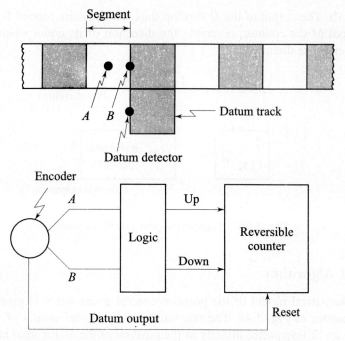

Fig. 3.44 Datum detector in an incremental encoder

We have thus three detectors in an incremental encoder. But this is still a lot less than on an absolute encoder.

In an analog system, speed is usually measured by a tachogenerator attached to the motor shaft. This is because the time differentiation of analog position signal presents practical problems.

In a digital system, however, it is relatively easy to carry out step-by-step calculation of the 'slope' of the position/time curve. We have the position data in digital form from the shaft encoder, so the rest is fairly straightforward.

Encoder Interface Implementation

In a position control system, the load is connected to the motor through a gear train. The encoder may be connected to the load shaft/motor shaft directly or through a pulley system.

To know the shaft position, the number of pulses obtained at the output of detector A or detector B have to be counted. To know the direction of rotation, the relative phasing of the outputs of detectors A and B has to be sensed. To implement the direction sensing, a negative edge-triggered D-flipflop may be used (Fig. 3.45). This flipflop has two inputs:

(i) clock input, derived from the output of detector A of the encoder; and
(ii) 'D' input, derived from the output of detector B of the encoder.

Every time the flipflop is triggered on the $1 \rightarrow 0$ transition of waveform A, the output of the D flipflop is either 1 or 0 depending on the direction of rotation of the shaft. The output of the D flipflop thus serves as the control for the up/down input of the counter, reversing the direction of its count whenever the shaft reverses its direction.

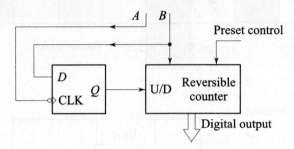

Fig. 3.45 A simple encoder interface circuit

Control Algorithm

The mathematical model of the position control under study is given by the block diagram of Fig. 3.46. The magnitude of the digital output of the shaft encoder can be interpreted directly as the position of the motor shaft in degrees by proper design of the encoder interface. Similarly, the magnitude of the digital reference input can be interpreted directly as reference input in degrees by proper design of the keyboard matrix interface. The error e (degrees) in position is processed by the control algorithm with transfer function $D(z)$. The control signal u (in volts) is applied to the preamplifier through the D/A converter. The plant (preamplifier + servoamplifier + dc motor + load) is described by the transfer function

$$\frac{\theta(s)}{V(s)} = G(s) = \frac{94}{s(0.3s+1)} \qquad (3.70)$$

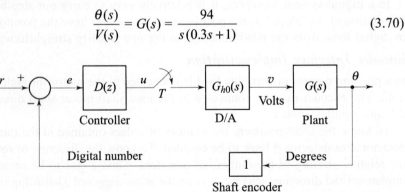

Fig. 3.46 Mathematical model for the position control system

To design the digital controller $D(z)$ for this plant, we will follow the approach of discretization of analog design (refer Section 2.12). The design requirements may be fixed as $\zeta = 0.7$ and $\omega_n \cong 10$. The first step is to find a proper analog controller $D(s)$ that meets the specifications. The transfer function

$$D(s) = K_c \frac{(s+3.33)}{s+\alpha}$$

cancels the plant pole at $s = -3.33$. The characteristic roots of

$$1 + D(s)G(s) = 0$$

give $\zeta = 0.7$ and $\omega_n = 10$ if we choose $K_c = 0.32$ and $\alpha = 14$.

The controller
$$D(s) = \frac{0.32(s+3.33)}{s+14} \tag{3.71}$$

gives the following steady-state behaviour:

$$K_v = \lim_{s \to 0} sG(s)D(s) = 7.15$$

This may be considered satisfactory.

The discretized version of the controller $D(s)$ is the proposed digital controller $D(z)$ for the control loop of Fig. 3.46. The $D(z)$ will perform as per the specifications if the lagging effect of zero-order hold is negligible. We take a small value for sampling interval T to satisfy this requirement. For a system with $\omega_n = 10$ rad/sec, a very 'safe' sample rate would be a factor of 20 faster than ω_n, yielding

$$\omega_s = 10 \times 20 = 200 \text{ rad/sec}$$

and
$$T = \frac{2\pi}{\omega_s} \cong 0.03 \text{ sec}$$

The dominant time constant of the plant is 0.3 sec. The sampling interval T is one-tenth of this value.

We use the bilinear transformation given by

$$s = \frac{2}{T}\left(\frac{z-1}{z+1}\right)$$

to digitize $D(s)$. This results in

$$D(z) = \frac{22.4z - 20.27}{80.67z - 52.67}$$

$$= \frac{0.278 - 0.25z^{-1}}{1 - 0.653z^{-1}} = \frac{U(z)}{E(z)} \tag{3.72a}$$

The control algorithm is, therefore, given by

$$u(k) = 0.653\ u(k-1) + 0.278\ e(k) - 0.25\ e(k-1) \tag{3.72b}$$

This completes the digital algorithm design.

184 Digital Control and State Variable Methods

3.8 STEPPING MOTORS AND THEIR CONTROL

The explosive growth of the computer industry in recent years has also meant an enormous growth for stepping motors because these motors provide the driving force in many computer peripheral devices. Stepping motors can be found, for example, driving the paper-feed mechanism in printers. These motors are also used exclusively in floppy disk drives, where they provide precise positioning of magnetic head on the disks. The X and Y coordinate pens in plotters are driven by stepping motors.

The stepping motor can be found performing countless tasks outside the computer industry as well. The most common application is probably in analog quartz watches where tiny stepping motors drive the hands. These motors are also popular in numerical-control applications (positioning of the workpiece and/or the tool in a machine according to previously specified numerical data).

A stepping motor is especially suited for applications mentioned above because essentially it is a device that serves to convert input information in *digital form* to an output that is mechanical. It thereby provides a natural interface with the digital computer. A stepping motor plus its associated drive electronics accepts a pulse-train input and produces an increment of rotary displacement for each pulse. We can control average speed by manipulating pulse rate, and motor position by controlling the total pulse count.

Two types of stepping motors are in common use—the permanent magnet (PM), and the variable reluctance (VR). Attention here is directed first to the PM motor.

Permanent Magnet Motor

Constructional features

A PM stepping motor in its simplest form is shown in Fig. 3.47. The motor has a permanent magnet rotor that, in this example, has two poles, though often many more poles are used. The stator is made of soft iron with a number of pole

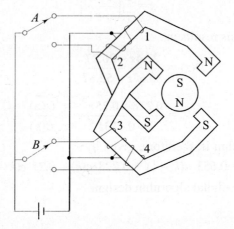

Fig. 3.47 A PM stepping motor (four phase)

pieces and associated windings. Only four windings (grouped into two sets of two windings each) are used in this example. These windings must be excited sequentially in a certain order. Although this is commonly done by solid-state switching circuits, mechanical switches are shown in the figure since their operation is easier to visualize.

Assume the switches to be in the positions shown. Windings 1 and 3 are energized and, as a result, the pole pieces have the polarities shown. The rotor is thus found in the position shown with its S pole centred between the two upper N pole pieces, and its N pole between the two lower S pole pieces. With the field maintained, if we try to twist the shaft away from its standstill (equilibrium) position, we feel a 'magnetic spring' restoring torque. However, a sufficiently large external torque can overcome the magnetic spring.

With the rotor energized and in equilibrium position, the torque required from an external source to break away the motor from this position is called the *holding torque*. The holding torque is a basic characteristic of the stepping motors and provides positional integrity under standstill conditions.

If we now imagine the position of switch A changed, then winding 2 is energized instead of winding 1. As a result the right upper pole piece becomes S instead of N, and the left lower one N, so that the rotor is forced to rotate 90° counterclockwise. Changing switch B produces the next 90° step, etc. The rotor is thus forced to realign itself continuously, according to the prevalent magnetic fields. If it is desired to reverse the direction of rotation, the order of changing the switch positions need only be reversed.

One characteristic feature of PM stepping motors is that they have a so-called *residual* or *detent torque* when power to the stator windings is cut off. It is the result of the permanent-magnet flux of the PM motor acting on residual flux on stator poles. The detent torque is naturally much lower than the holding torque produced when the stator if energized, but it does help in keeping the shaft from moving due to outside forces.

Many motors have more than four stator pole pieces—and possibly also more rotor poles—resulting in smaller step angles. Typical step angles for PM motors range from 90° to as low as 1.8°. Stator pole windings are connected in so-called *phases*, with all windings belonging to the same phase energized at the same time. Typically the phases can range from as low as two to as high as eight. The more phases the motor has, the smoother is its output torque.

Sequence logic and drive amplifier

Figure 3.48 shows a simple power drive scheme; each time the power transistors are switched as per the sequence given in the chart, the motor moves through a fixed angle referred to as the *step angle*. The chart is circular in the sense that the next entry after Step 4 is Step 1. To rotate the motor in a clockwise direction, the chart is traversed from top to bottom, and to rotate the motor in counterclockwise direction, the chart is traversed from bottom to top. Number of step movements/sec gives the *stepping rate*—a parameter that gives a measure of the speed of operation of the stepping motor. The stepping rate is controlled by changing the switching frequency of the transistors.

186 Digital Control and State Variable Methods

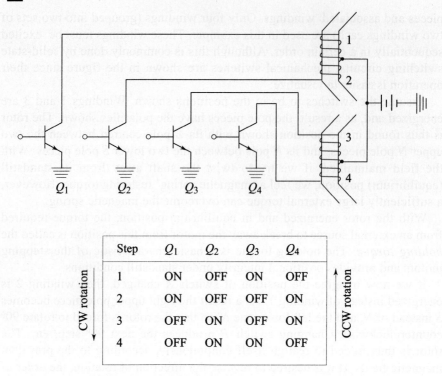

Fig. 3.48 A simple stepping motor drive scheme

From the foregoing description of the method of operation of a stepping motor, we observe that the stepping action of the motor is dependent on a specific switching sequence that serves to energize and de-energize the stator windings. In addition to the sequence requirement, the windings must be provided with sufficient current. These requirements are met by the stepping motor driver whose block diagram is shown in Fig. 3.49. The sequence-logic section of the motor driver accepts the pulse-train input, and also receives a binary direction signal indicating the direction in which the motor is to step. It then produces an appropriate switching sequence so that each phase of the motor is energized at the proper time. The drive-amplifier section consists of power transistors supplying sufficient current to drive the motor.

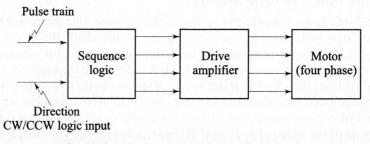

Fig. 3.49 Stepping motor driver (four phase)

Variable Reluctance Motor

Figure 3.50 illustrates a typical variable reluctance (VR) motor. The rotor is made of magnetic material, but it is not a permanent magnet, and it has a series of teeth (eight in this case) machined into it. As with the PM stepping motor, the stator consists of a number of pole pieces with windings connected in phases; all windings belonging to the same phase are energized at the same time. The stator in Fig. 3.50 is designed for 12 pole pieces with 12 associated windings arranged in three phases (labelled 1, 2, and 3 respectively). The figure shows a set of four windings for phase 1; the windings for the other two phases have been omitted for clarity.

The operating principle of the VR motor is straightforward. Let any phase of the windings be energized with a dc signal. The magnetomotive force set up will position the rotor such that the teeth of the rotor section in the neighbourhood of the excited phase of the stator are aligned opposite the pole pieces associated with the excited phase. This is the position of minimum reluctance and the motor is in a stable equilibrium. Figure 3.50 illustrates the rotor in the position it would assume when Phase 1 is energized. If we now de-energize Phase 1 and energize Phase 2, the rotor rotates counterclockwise so that the four rotor teeth nearest to the four pole pieces belonging to Phase 2 align them-

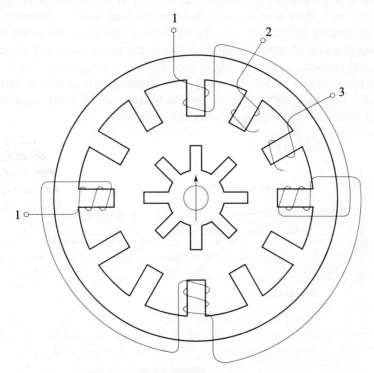

Fig. 3.50 A VR stepping motor (three phase)

selves with these. The step angle of the motor equals the difference in angular pitch between adjacent rotor teeth and adjacent pole pieces; in this cas 45 – 30 = 15°. Due to this difference relationship, VR motors can be designed to operate with considerably smaller step angles than PM motors. Other advantages of VR motors include faster dynamic response and the ability to accept higher pulse rates.

Among the drawbacks—their output torque is lower than that of a PM motor of similar size, and they do not provide any detent torque when not energized.

Torque-Speed Curves of a Stepping Motor

Torque *versus* speed curves of a stepping motor give the dynamic torque produced by the stepping motor at a given stepping rate on excitation under rated conditions. The dynamic torque of a motor is the most important data and it plays a major role in selection of a motor for a specified application. In a load-positioning application, for instance, the rotor would typically start from rest and accelerate the load to the desired position. To provide this type of motion, the motor must develop sufficient torque to overcome friction and to accelerate the total inertia. In accelerating the inertia, the motor may be required to develop a large amount of torque, particularly if the acceleration must be completed in a short time so as to position the load quickly. Inability of the motor to develop sufficient torque during motion may cause the motor to stall, resulting in a loss of synchronization between the motor steps and phase excitation, and consequently, resulting in incorrect positioning of the load.

A typical torque *versus* stepping rate characteristic graph is shown in Fig. 3.51, in which curve *a* gives pull-in torque *versus* rotor steps/sec and curve *b* gives pull-out torque *versus* rotor steps/sec.

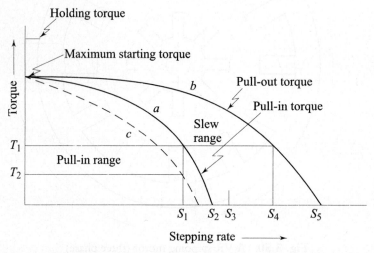

Fig. 3.51 Torque versus stepping rate characteristics

The *pull-in-range* (the area between axes and curve *a*) of the motor is the range of switching speeds at which the motor can start and stop without losing steps. For a frictional load requiring torque T_1 to overcome friction, the maximum *pull-in-rate* is S_1 steps per sec. S_2 is the *maximum pull-in-rate* at which the unloaded motor can start and stop without losing steps.

When the motor is running, the stepping rate can be increased above the maximum pull-in-rate and when this occurs the motor is operating in the *slew-range* region (the area between horizontal axis, and curves *a* and *b*). The slew-range gives the range of switching speeds within which the motor can run unidirectionally, but cannot be started or reversed (at shaft torque T_1, the motor cannot be started or reversed at step rate S_3). When the motor is running in the slew-range, it can follow changes in the stepping rate without losing steps but only with a certain acceleration limit.

For a frictional load requiring torque T_1 to overcome friction, the maximum *slewing rate* at which the motor can run is S_4. S_5 is the maximum slewing rate at which the unloaded motor can run without losing steps.

Curve *c* in Fig. 3.51 gives the pull-in torque with external inertia. It is obvious that if the external load results in a pull-in torque curve *c*, the torque developed by the motor at step rate S_1 is $T_2 < T_1$. The stepping motors are more sensitive to the inertia of the load than they are to its friction.

Interfacing of Stepping Motors to Microprocessors

In motion control technology, the rise of the stepping motors has in fact begun with the availability of easy-to-use integrated circuit chips to drive the stepping motors. These chips require as inputs a pulse train at the stepping frequency, a logic signal to specify CW and CCW rotation, and a logic signal for STOP/START operation. An adjustable frequency pulse train is readily obtained from another integrated circuit chip—a voltage-controlled oscillator.

The application of stepping motors has shot up with the availability of low-cost microprocessors. A simplified form of microprocessor-based stepping motor drive is shown in Fig. 3.52. The system requires an input port and an output port (this requirement is reduced to one port if a programmable I/O port is used). Output port handles the binary pattern applied to the stepping motor (which is assumed to be a four-phase motor). The excitation sequence is usually stored in a table of numbers. A pattern for four-phase motor is shown in the chart of Fig. 3.52. The chart is circular in the sense that the next entry after Step 4 is Step 1. To rotate the motor in a clockwise direction, the chart is traversed from top to bottom, and to rotate the motor in counter-clockwise direction, the chart is traversed from bottom to top. By controlling the number of bit-pattern changes and the speed at which they change, it is possible to control the angle through which the motor rotates and the speed of rotation. These controls can easily be realized through software.

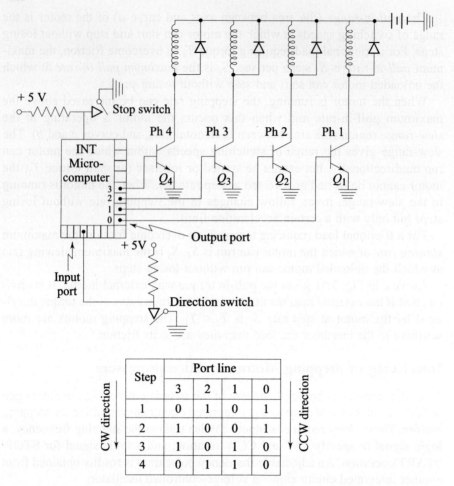

Fig. 3.52 Microprocessor-based stepping motor drive

The system operator has control over the direction of rotation of the motor by means of a DIRECTION switch, which is interfaced to the CPU through the input port. The operator is also provided with a STOP switch which is connected to an interrupt line of the CPU. The interrupt routine must stop the motor by sending out logic '0's on the data bus lines connected to the stepping motor windings through the output port.

Figure 3.52 also shows a simple drive circuit for the stepping motor. Power transistors Q_1-Q_4 act as switching elements.

When a power transistor is turned off, a high voltage builds up due to *di/dt*, which may damage the transistor. This surge in voltage can be suppressed by connecting a diode in parallel with each winding in the polarity shown in Fig. 3.52. Now there will be a flow of circulating current after the transistor is turned off and the current will decay with time.

Comments

Stepping motors present a number of pronounced advantages, as compared to conventional electric motors:

(i) Since the stepping-motor shaft angle bears an exact relation to the number of input pulses, the motor provides an accurate open-loop positioning system without the need for closing the loop with a position encoder, comparator, and servoamplifier, as is done in conventional closed-loop systems.

(ii) If the stepping motor receives a continuous train of pulses at constant frequency, it rotates at a constant speed provided neither the load torque nor the pulse frequency are excessive for the given motor. The stepping motor can thus take the place of a velocity servo, again, without the need for a closed-loop system. By changing pulse frequency, the motor speed can be controlled. Even low velocities can be maintained accurately, which is difficult to do with conventional dc motors.

(iii) By driving several motors from the same frequency source, synchronized motions at different points in a machine are easily obtained. Using standard frequency-divider chips, we can drive a motor at a precise fraction of another motor's speed, giving an electronic gear train.

(iv) If the motor stator is kept energized during standstill, the motor produces an appreciable holding torque. Thus, the load position can be locked without the need for clutch-brake arrangements. The motor can be stalled in this manner indefinitely without adverse effects.

There are, of course, also certain drawbacks.

(i) If the input pulse rate is too fast, or if the load is excessive, the motor will 'miss' steps, making the speed and position inaccurate.

(ii) If the motor is at rest, an external disturbing torque greater than the motor's holding torque can twist the motor shaft away from its commanded position by any number of steps.

(iii) With high load inertias, overshooting and oscillations can occur unless proper damping is applied, and under certain conditions, the stepping motor may become unstable.

(iv) The stepping motors are only available in low or medium hp ratings, up-to a couple of hp (in theory, larger stepping motors could be built, but the real problem lies with the controller—how to get large currents into and out of motor windings at a sufficiently high rate inspite of winding inductance).

(v) Stepping motors are inherently low-speed devices, more suited for low-speed applications because gearing is avoided. If high speeds are required, this of course becomes a drawback.

Since the cost and simplicity advantages of stepping-motor control systems erode when motion sensors and feedback loops are added, much effort has gone into improving the performance of open-loop systems:

(i) As explained earlier, in connection with Fig. 3.51, the permissible pulse rate for starting an inertia load (i.e., the pull-in rate) is much lower than the permissible pulse rate once the motor has reached maximum speed (pull-out rate). A good controller brings the motor upto its maximum speed gradually, a process called *ramping*[2], in such a manner that no pulses are lost. Similarly, a good controller controls deceleration when the motor is to be stopped.

(ii) Various schemes for improving damping to prevent overshooting and oscillations when the motor is to be stopped are available. Mechanical damping devices provide a simple solution, but these devices reduce the available motor torque and also mostly require a motor with a double-ended shaft. Therefore, electronic damping methods are usually preferred. A technique called *back-phase damping* consists of switching the motor into the reverse direction using the last few pulses of a move.

(iii) The more sophisticated controllers are able to provide so-called *microstepping*. This technique permits the motor shaft to be positioned at places other than the natural stable points of the motor. It is accomplished by proportioning the current in two adjacent motor windings. Instead of operating the winding in the on-off mode, the current in one winding is decreased slightly, but increased in the adjacent winding.

(iv) Complex drive circuits that offer good current build-up without loss at high stepping rates are used.

Although the advantages of stepping-motor drives in open-loop systems are most obvious, closed-loop applications also exist. A closed-loop stepping motor drive can be analysed using classical techniques employed for continuous-motion systems. For a detailed account of stepping motors, refer [51, 55–56].

3.9 PROGRAMMABLE LOGIC CONTROLLERS

A great deal of what has been said in this book so far about control systems seems exotic: algorithms for radar tracking, drives for rolling mills, filters for extracting information from noisy data, methods for numerical control of machine tools, fluid-temperature control in process plants etc. Underlying most of these are much more mudane tasks: turning equipment (pumps, conveyor belts, etc.) on and off; opening and closing of valves (pneumatic, hydraulic); checking sensors to be certain they are working; sending alarms when monitored signals go out of range; etc. Process control plants and manufacturing floors share this need for simple but important tasks.

---

2. Refer [88] for detailed description of hardware and software.

These so-called *logic control* functions can be implemented using one of the most ingenious devices ever devised to advance the field of *industrial automation*. So versatile are these devices that they are employed in the automation of almost every type of industry. The device, of course, is the *programmable controller*, and thousands of these devices go unrecognized in process plants and factory environments—quietly monitoring security, manipulating valves, and controlling machines and automatic production lines.

Industrial applications of logic control fall mainly into two types; those in which the control system is entirely based on logic principles and those that are mainly of a continuous feedback nature and use a 'relatively small' amount of logic in auxiliary functions such as start-up/shut-down, safety interlocks and overrides, and mode switching. Programmable controllers, originally intended for '100%' logic systems, have in recent years added the capability of conventional feedback control, making them very popular since one controller can now handle in an integrated way *all* aspects of operation of a practical system that includes both types of control problems. General purpose digital computers could also handle such situations but they are not as popular as the programmable controllers for the following reasons:

In theory, general-purpose computers can be programmed to perform most of the functions of programmable controllers. However, these machines are not built to operate reliably under industrial conditions, where they can be exposed to heat, humidity, corrosive atmosphere, mechanical shock and vibration, electromagnetic noise, unreliable ac power with dropping voltages, voltage spikes etc. A programmable controller is a special-purpose computer especially designed for industrial environments. A general-purpose computer is a complex machine, capable of executing several programs or tasks simultaneously, and in any order. By contrast, a programmable controller typically executes its tiny program continuously hundreds of millions of times before being interrupted to introduce a new program. General-purpose computers can be interfaced with external equipment with special circuit cards. In programmable controllers by comparison, the hardware interfaces for connecting the field devices are actually a part of the controller and are easily connected. The software of the controllers is designed for easy use by plant technicians. A programmable controller is thus a special-purpose device for industrial automation applications requiring logic control functions and simple PID control functions; it cannot compete with conventional computers when it comes to complex control algorithms and/or fast feedback loops requiring high program execution speeds.

Early devices were called 'programmable logic controllers (PLCs)', and were designed to accept on-off (binary logic) voltage inputs from sensors, switches, relay contacts, etc., and produce on-off voltage outputs to actuate motors, solenoids, control relays, lights, alarms, fans, heaters, and other electrical equipment. As many of today's 'programmable controllers', also accept analog data, perform simple arithmetic operations and even act as PID (proportional-integral-derivative) process controllers, the word 'logic' and the letter 'L' were dropped from the name long ago. This frequently causes confusion, since the letters 'PC' mean different things to different people; the most common usage of these letters being for 'Personal Computer'. To avoid this confusion there has been a tendency lately to restore the letter 'L' and revive the designation 'PLC'. We will follow this practice.

Before the era of PLCs, hardwired relay control panels were, in fact, the major type of systems and this historical development explains why the most modern, microprocessor-based PLCs still are usually programmed according to relay ladder diagrams. This feature has been responsible for much of the widespread and rapid acceptance of PLCs; the computer was forced to *learn* the already familiar human language rather than making the humans learn a new computer language. Originally cost-effective for only large-scale systems, small versions of PLCs are now available. However, electromechanical (relays having moving parts) type of design is still used for some applications. If non-PLC control without moving parts in desired, all the necessary elements are available as individual solid-state electronic devices.

A sequenced but brief presentation of building blocks of a PLC, ladder diagrams, and examples of industrial automation follows [25–28]. It is not appropriate here to teach the reader the internal details, performance specifications and programming details for any particular manufacturer's PLC. These aspects are described in manufacturers' literature.

Logic Controls for Industrial Automation

A definition of *logic controls* that adequately describes most applications is that they are controls that work with one-bit binary signals. That is, the system needs only to know that a signal is absent or present; its *exact* size is not important. This definition excludes the large and important field of digital computer control discussed so far in the book, since conventionally computer control also uses binary signals, but usually with many bits. The type of application and the analysis methods are quite different for logic controls and conventional computer controls, which is why we make the distinction.

Logic control systems can involve both *combinational* and *sequential* aspects. Combinational aspects are implemented by a proper interconnection of basic logic elements such as AND, OR, NOT, so as to provide a desired output or outputs when a certain combination of present inputs exists. Sequential effects use logic elements together with memory elements (counters, timers, etc.), to ensure that a chain of events occurs in some desired sequence. The present status of outputs depends both on the past and present status of inputs.

It is important to be able to distinguish between the nature of variables in a logic control system and those in a conventional feedback control system. To define the difference, we consider an example that employs both the control schemes.

Figure 3.53 shows a tank with a valve that controls flow of liquid into the tank, and another valve that controls flow out of the tank. A transducer is available to measure the level of the liquid in the tank. Also shown is the block diagram of a feedback control system whose *objective* is to maintain the level of the liquid in the tank at some preset or set-point value. We assume that the controller operates according to PID mode of control to regulate the level against variations induced from external influences. This is a *continuous variable* control system because both the level and the control valve setting can vary over a range to achieve the desired regulation.

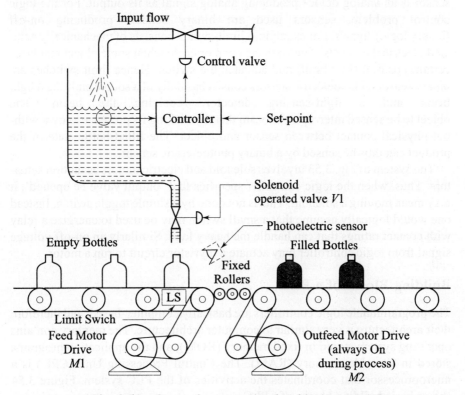

Fig. 3.53 Composite conventional and logic control

The liquid-level control system is a part of the continuous bottle filling process. Periodically, a bottle comes into position under the outlet valve. The level must be maintained at the set-point *while the outlet valve is opened and the bottle filled*. This requirement is necessary to assure a constant pressure head during bottle filling. Figure 3.53 shows a pictorial representation of process hardware for continuous bottle filling control. The objective is to fill bottles moving on a conveyor, from the constant-head tank. This is a typical logic control problem. We are to implement a control program that will detect the position of a bottle under the tank outlet, via a mechanically actuated *limit switch*, stop the feed motor $M1$ to stop the feed conveyor, open the solenoid-operated outlet valve

V1, and then fill the bottle until the photo sensor detects the filled position. After the bottle is filled, it will close the valve V1, and restart the conveyor to continue to the next bottle. The start and stop push buttons (PB) will be included for the outfeed motor and for the start of the bottle-filling process. Once the start PB is pushed, the outfeed motor M2 will be ON until the stop PB is pushed. The feed motor M1 is energized once the system starts (M2 ON) and is stopped when the limit switch detects the correct position of the bottle.

The sensors used for the logic control problem have characteristics different from those used for the regulator problem. For the regulator problem, the level sensor is an analog device producing analog signal as its output. For the logic control problem, sensors used are binary sensors producing on-off (binary logic) signals. For example, a limit-switch consists of mechanically actuated electrical contacts. The contacts open or close when some object reaches a certain position (i.e., limit) and actuates the switch. Hence limit switches are binary sensors. Photoelectric sensors consist basically of a source emitting a light beam and a light-sensing detector receiving the beam. The object to be sensed interrupts the beam thereby making its presence known without physical contact between sensor and object. The filled-bottle state of the product can thus be sensed by a binary photoelectric sensor.

The system of Fig. 3.53 involves solenoid and electric motors as motion actuators. Thus, when the logic controller specifies that 'output valve be opened', it may mean moving a solenoid. This is not done by a simple toggle switch. Instead one would logically assume that a small switch may be used to energize a relay with contact ratings that can handle the heavy load. Similarly an on-off voltage signal from logic controller may actuate a thyristor circuit to run a motor.

Building Blocks of a PLC

The programmable logic controllers are basically computer-based; and therefore their architecture is very similar to computer architecture. The memory contains operating system stored in fixed memory (ROM), and the application programs stored in alterable memory (RAM). The Central Processing Unit (CPU) is a microprocessor that coordinates the activities of the PLC system. Figure 3.54 shows basic building blocks of a PLC.

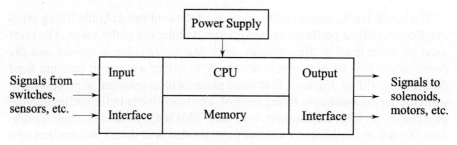

Fig. 3.54 Basic building blocks of a PLC

Input devices such as push-buttons, sensors, and limit switches are connected to the *input module*. This section gathers information from the outside environment and sends it on to the CPU. Output devices such as solenoids, motor controls, indicator lights and alarms are connected to the *output module*. This section is where the calculation results from the CPU are output to the outside environment. With the control application program (stored within the PLC memory) in execution, the PLC constantly monitors the state of the system through the field input devices; and based on the program logic, it then determines the course of action to be carried out at the field output devices. This process of sequentially reading the inputs, executing the program in memory, and updating the outputs is known as *scanning*.

Input Module

Intelligence of an automated system is greatly dependent on the ability of a PLC to read in the signals from various types of automatic sensing and manual input field devices. The most common class of input devices in an automated system is the binary type. These devices provide input signals that are ON/OFF, OPEN/CLOSED, or equivalent to a switch closure. To the input interface circuit, all binary input devices are essentially a switch that is either open or closed, signalling a 1(ON) or 0(OFF). Some of the binary input field devices alongwith their symbolic representation are listed in Fig. 3.55.

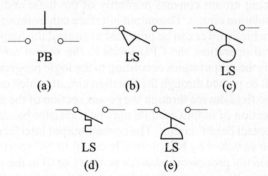

Fig. 3.55 (a) Push-button, (b) Position limit switch, (c) Level limit switch, (d) Temperature limit switch, (e) Pressure limit switch

As mentioned earlier, a switch is a symbolic representation of the field input device interfaced to the **input module** of the PLC. The device may be a manually operated push button, mechanically actuated limit switch (the contacts open/close when some object reaches a certain position and actuates the switch), proximity switch (device based on inductive/capacitive/magnetic effect which, with appropriate electronics, can sense the presence of an object without a physical contact with the object), photoelectric sensor, level sensor, temperature sensor, shaft encoder, etc. The main purpose of the input module is to condition the various signals received from the field devices to produce an output to be sensed by the CPU. The signal conditioning involves converting power-level signals from field devices to logic-level signals acceptable to the CPU, and providing *electrical* *isolation* so that there is no electrical connection between the field device (power) and the controller (logic). The coupling between the power and the logic sections in normally provided by an optical coupler.

During our discussion on PLC programming, it will be helpful if we keep in mind the relationship between the interface signals (ON/OFF) and their mapping and addressing used in the program. When in operation, if an input signal is energized (ON), the input interface circuit senses the field device's supplied voltage and converts it to a logic-level signal acceptable to the CPU to indicate the status of the device. The field status information provided to the standard input module is placed into the *input table* in memory through PLC instructions. The I/O address assignment document of the PLC manufacturer identifies each field device by an address. During scanning, the PLC reads the status of all field input devices and places this information at the corresponding address locations.

Output Module

An automation system is incomplete without means for interface to the field output devices. The *output module* provides connections between the CPU and output field devices. The output module receives from CPU logic-level signals (1 or 0).

The main purpose of the output interface circuit is to condition the signals received from CPU to produce outputs to actuate the output field devices. The signal conditioning circuit consists primarily of the logic and power sections, coupled by an isolation circuit. The output interface can be thought of as a simple switch through which power can be provided to control the output device.

During normal operation, the CPU sends to the *output table* at predefined address locations the output status according to the logic program. If the status is 1, ON signal will be passed through the isolation circuit, which in turn will switch the voltage to the field device through the power section of the module.

The power section of output module may be transistor based, triac based, or simply relay 'contact based' circuit. The contact output interface allows the output devices to be switched by NO (normally-open) or NC (normally-closed) relay contact. When the processor sends the status (1 or 0) to the module (through output table) during the output update, the state of the contact will change. If a 1 is sent to the module from the processor, a normally open contact will close, and a normally closed contact will change to an open position. If a 0 is sent, no change occurs to the normal state of the contacts. The contact output can be used to switch either ac or dc loads; switching small currents at low voltages. High power contact outputs are also available for switching of high currents.

Some of the output field devices alongwith their symbolic representation are given in Fig. 3.56.

I/O Rack Enclosures

Once we have the CPU programmed, we get information in and out of the PLC through the use of input and output modules. The input module terminals receive signals from wires connected to switches and other input information devices. The output module terminals provide output voltages to energize motors and valves, operate indicating devices, and so on.

Models of Digital Control Devices and Systems

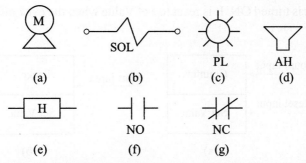

Fig. 3.56 (a) Motor, (b) Solenoid, (c) Pilot lamp, (d) Alarm horn, (e) Heater, (f) Relay contact (Normally-Open), (g) Relay contact (Normally-Closed)

For small PLC systems, the input and output terminals may be included on the same frame as the CPU. In large systems, the input and output modules are separate units; modules are placed in groups on racks, and the racks are connected to the CPU via appropriate connector cables.

Generally speaking, there are three categories of rack enclosures; the *master rack*, the *local rack*, and the *remote rack*. A master rack refers to the enclosure containing the CPU module. This rack may or may not have slots available for the insertion of I/O modules. The larger the PLC system, in terms of I/O, the less likely the master rack will have I/O housing capability or space. A local rack is an enclosure which is placed in the same location or area where the master rack is housed. If a master rack contains I/O, it can also be considered a local rack. In general, a local rack contains a local I/O processor which receives and sends data to and from the CPU.

As the name implies, remote racks are enclosures containing I/O modules located far away from the CPU. A remote rack contains an I/O processor which communicates I/O information just like the local rack.

Timers/Counters

Timers and counters play an important part in many industrial automation systems. The timers are used to initiate events at defined intervals. The counters, on the other hand, are used to count the occurrences of any defined event.

Basically the operation of both the timer and the counter is same as timer operates like a counter. The counter shown in Fig. 3.57a counts down from Set Value when its execution condition (count input) goes from OFF to ON. When the value reaches zero, the counter contact point is turned ON. It is reset with a reset input. The Set Value is decided by the programmer, and stored in internal register of the counter through control program instructions. The count input signal may refer to any event which may occur randomly.

When signal occurs at fixed frequency, i.e., after every fixed interval of time, the counter performs as timer. Now 10 pulses, i.e., counts, will mean an elapsed time of 5 seconds, if signal is occurring after a regular interval of 0.5 seconds. The timer shown Fig. 3.57b is activated when its execution condition goes ON and starts decreasing from the Set Value. When the value reaches zero, the timer contact point is turned ON. It is reset to Set Value when the execution condition goes OFF.

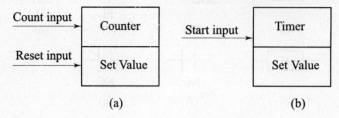

Fig. 3.57 (a) Counter, (b) Timer

Memory Map

It is unlikely that two different PLCs will have identical memory maps, but a generalization of memory organization is still valid in the light of the fact that all PLCs have similar storage requirements. In general, all PLCs must have memory allocated for the four items described below.

Executive: The executive software is a permanently stored collection of programs that are considered a part of the system itself. These programs direct system activities such as execution of the control program, communication with peripheral devices, and other housekeeping activities. The executive area of memory is not accessible to the user.

Scratch pad: It is a temporary storage used by the CPU to store a relatively small amount of data for interim calculations or control.

Data table: This area stores any data associated with the control program, such as timer/counter set values, and any other stored constants or variables that are used by the control program. This section also retains the status information of the system inputs once they have been read, and the system outputs once they have been set by the control program.

User program: This area provides storage for any programmed instructions entered by the user. The control program is stored in this area.

The Data Table and the User Program areas are accessible and are required by the user for control application. The Executive and Scratch Pad areas together are normally referred to as 'system memory', and Data Table and User Program areas together are labelled as 'application memory'.

Data Table Memory Area

The data table area of the PLC's application memory is composed of several sections described below.

Input table: The input table is an array of bits that stores the status of discrete inputs, which are connected to input interface circuits. The maximum number of bits in the input table is equal to the maximum number of field inputs that can be connected to the PLC. For instance, a controller with 128 inputs would require an input table of 128 bits. If the PLC system has 8 input modules, each with 16 terminal points, then the input table in PLC memory (assuming 16 bit word length) will look like that is Fig. 3.58.

Each terminal point on each of the input modules will have an *address* by which it is referenced. This address will be a pointer to a bit in the input table. Thus each connected input has a bit in the input table that corresponds exactly to the terminal to which the input is connected. The address of the input device can be interpreted as word location in the input table corresponding to the input module, and bit location in the word corresponding to the terminal of the input module to which the device is connected.

Several factors determine the address of the word location of each module. The type of module, input or output, determines the first number in the address from left to right (say, **0** for input, and **1** for output). The next two address numbers are determined by the rack number and the slot location where the module is placed. Figure 3.58 graphically illustrates a mapping of the input table and the modules placed in rack **0** (master rack). Note that the numbers associated with address assignment depend on the PLC model used. These addresses can be represented in octal, decimal, or hexadecimal. We have used decimal numbers.

The limit switch connected to the input interface (refer Fig. 3.58) has an address of **00012** for its corresponding bit in the input table. This address comes from the word location **000** and the bit number **12**, both of which are related to the rack position where the module is installed and the molule's terminal connected to the field device. If the limit switch is ON (closed), the corresponding bit **00012** will be 1; if the limit switch is OFF (open), its corresponding bit will be 0.

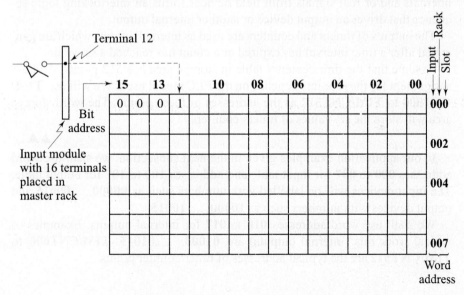

Fig. 3.58 An illustrative input table

During PLC operation, the processor will read the status of each input in the input modules, and then place this value (1 or 0) in the corresponding address in the input table. The input table is constantly changing to reflect the changes in the field devices connected to the input modules. These changes in the input table take place during the reading part of the processor scan.

Output table: The output table is an array of bits that controls the status of output devices, which are connected to the output interface circuits. The maximum number of bits available in the output table is equal to the maximum number of output field devices that can be interfaced to the PLC. For instance, a controller with a maximum number of 128 outputs would require an output table of 128 bits.

Each connected output has a bit in the output table that corresponds exactly to the terminal to which the output is connected. The bits in the output table are controlled (ON/OFF) by the processor as it interprets the control program logic. If a bit in the table is tuned On (logic 1), then the connected output is switched ON. If a bit is cleared or turned OFF (logic 0), the output is switched OFF. Remember that the turning ON or OFF the field devices occurs during the update of outputs after the end of the scan.

Storage area: This section of the data table may be subdivided in two parts consisting of a *work bit storage area* and a *word storage area*. The purpose of this data table section is to store data that can change whether it is a bit or a word (16 bits). Work bits are *internal outputs* which are normally used to provide interlocking logic. The internal outputs do not directly control output field devices. When the processor evaluates the control program, and any of these outputs is energized (logic 1), then this internal output in conjunction with other internals and/or real signals from field devices forms an interlocking logic sequence that drives an output device or another internal output.

The outputs of timers and counters are used as internal outputs which are generated after a time interval has expired or a count has reached a set value.

Assume that the timer/counter table in storage area has 512 points. Address assignment for these points depends on the PLC model used. We will use **TIM/CNT000** to **TIM/CNT512** as the addresses of these points. The word storage area will store the set values of timers/counters.

▲▲

In our application examples given in the next subsection, we shall use word addresses **000** to **007** for input table and addresses **100** to **107** for output table. The input devices will be labelled with numbers such as **00000**,..., **00015** and output devices with numbers such as **10000**, ..., **10015**.

We shall use word addresses **010** to **017** for internal outputs. Examples of typical work bits (internal outputs) are **01000**, ..., **01015**. **TIM/CNT000** to **TIM/CNT512** are the typical addresses of timer/counter points.

Ladder Diagrams

Although specialized functions are useful in certain situations, most logic control systems may be implemented with the three basic logic functions AND, OR, and NOT. These functions are used either singly or in combinations to form instructions that will determine if an output field device is to be switched ON or OFF. The most widely used language for implementing these instructions are *ladder diagrams*. Ladder diagrams are also called *contact symbology*, since its instructions, as we shall see, are relay-equivalent contact symbols shown in Figs 3.56f and 3.56g.

An AND device may have any number of inputs and one output. To turn the output ON, *all* the inputs must be ON. This function is most easily visualized in terms of switch arrangement of Fig. 3.59a, and timing chart of Fig. 3.59b. The corresponding ladder diagram is given in Fig. 3.59c. Figure 3.59d gives the *Boolean algebra* expression for the two-input AND, read as "A AND B equals C".

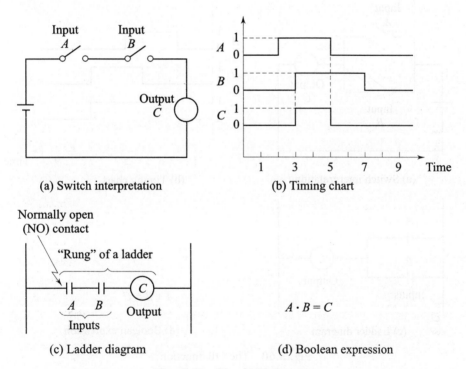

Fig. 3.59 The AND function

The timing chart in Fig. 3.59b is simply a series of graphs, each representing a logic variable, in which the horizontal axis is time and the vertical axis is logic state, that is, 0 or 1. The graphs are placed so that their time-axis are synchronized; in this way, a vertical line at any point on the graph describes a point in time, and all input and output variables can be evaluated at that point.

204 Digital Control and State Variable Methods

The graph of the output variable is determined by the structure of the logic system and of course the pattern of the input.

The input contacts in Fig. 3.59c are normally-open (NO) contacts (Do not confuse this symbol with the familiar electrical symbol for capacitors). If the status of the input A is '1', the contact A in ladder diagram will close and allow current to flow through the contact. If the status of the input A is '0', the contact will remain open and not allow current to flow through the contact.

The ladder diagram of Fig. 3.59c can be thought of as a circuit having many inputs. A circuit is known as the "rung" of the ladder. A complete PLC ladder diagram consists of several rungs; a rung either controls an output field device through output module or an internal output. The input to a rung can be logic commands from input modules or from output modules connected to field devices or from internal outputs.

Figure 3.60 gives similar details for logical OR operation and should be self-explanatory. The Boolean expression is read as "A OR B equals C".

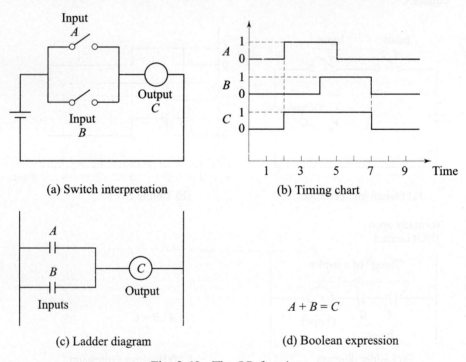

(a) Switch interpretation

(b) Timing chart

(c) Ladder diagram

(d) Boolean expression

Fig. 3.60 The OR function

The contact in Fig. 3.61 in normally closed (NC) contact. If the status of the input A is '0' the contact will remain closed, thus allowing current to flow through the contact. If the status of the input A is '1', the contact will open and *not* allow current to flow through the contact. This symbol permits the use of logic NOT operator. The Boolean expression is read as "NOT A equals B"; the overbar is used in general for applying NOT function.

Models of Digital Control Devices and Systems 205

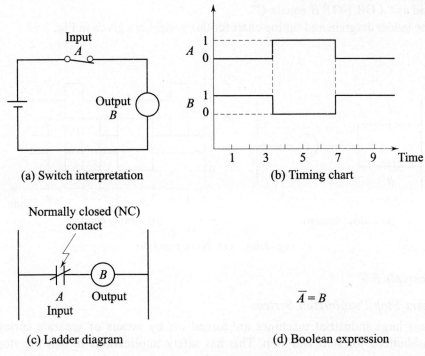

Fig. 3.61 The NOT function

Consider the logic system

$$A . \overline{B} = C$$

read as "A AND NOT B equals C".
The ladder diagram and timing chart for this system are given in Fig. 3.62.

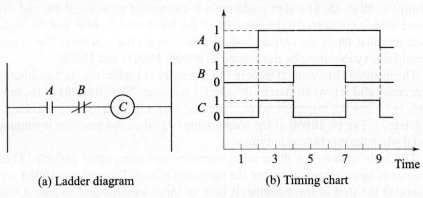

Fig. 3.62 AND NOT function

Consider now the logic system

$$A + \overline{B} = C$$

read as "*A* OR NOT *B* equals *C*".

The ladder diagram and timing chart for this system are given in Fig. 3.63.

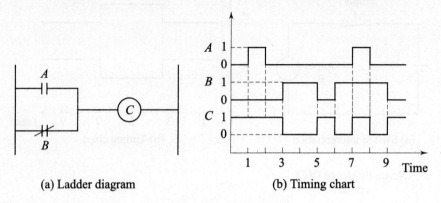

(a) Ladder diagram (b) Timing chart

Fig. 3.63 OR NOT function

Example 3.5

Start/Stop Pushbutton System

Most large industrial machines are turned off by means of separate spring pushbuttons for start and stop. This has safety implications in that the stop pushbutton can be given priority to shut down the machine in an emergency regardless of the status of the start pushbutton. The start/stop pushbutton system is a logic control system with three variables, each of which can take on two and only two values, or states. These variables and their states are defined as follows:

Assume that **000** is the word address of the input module and **100** is the word address of the output module of a PLC. Each module is assumed to have 16 terminals: **00** to **15**. The start pushbutton is connected to terminal **00**, and stop pushbutton is connected to terminal **01** of the input module **000**; and the signal from terminal **00** of the output module **100** controls the machine. The system variables may therefore be designated as **00000**, **00001**, and **10000**.

The bit **00000** of the input table in PLC memory is 1 when the start pushbutton is pressed, and is 0 when start pushbutton is released. The bit **00001** of the input table is 1 when the stop pushbutton is pressed, and is 0 when the stop pushbutton is released. The bit **10000** of the output table is 1 when the machine is running, and 0 when the machine is not running.

The logic system has three input variables and one output variable. There appears to be a contradiction, but the statement is true. The variable **10000**, representing the start of the machine, is both an input variable and an output variable. This makes sense because the current state of the machine may affect the future state.

Figure 3.64a illustrates a simple situation in which pushbutton **00000** turns ON machine **10000**. This of course would not be satisfactory pushbutton switch because as soon as pushbutton is released, the machine comes to OFF state.

Figure 3.64b adds an OR condition that keeps the machine ON if it is already ON. This is an improvement, but now there is a new problem; once turned ON, the output will never be turned OFF by the logic system. We add another input switch in Fig. 3.64c. Note that **00001** contact is normally closed. Input **00000** turns ON output **10000**; input **10000** keeps output **10000** ON until input **00001** turns it OFF.

The timing chart of the logic system is shown in Fig. 3.64d.

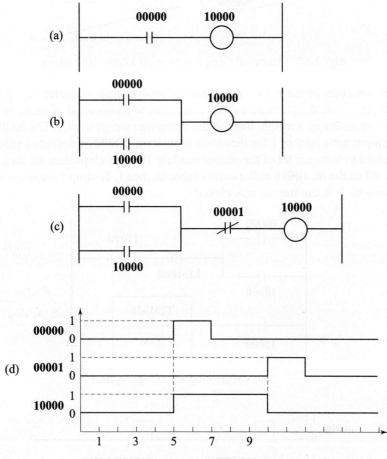

Fig. 3.64 Start/Stop pushbutton system

Example 3.6

Automatic Weigh Station

Consider the conveyer system of Fig. 3.65, in which an automatic weigh station activates a trap door or diverter in the event an overweight item passes over the weigh station. The trap door opens immediately and remains open for 4 seconds to allow sufficient time for the item to drop through to the overweight track (For the system to work properly, it is necessary for successive items on the conveyor to be separated by distances of at least 5 seconds or so).

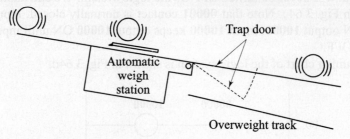

Fig. 3.65 Automatic weigh station on a conveyor system

The variables of the logic system are defined as follows (refer Fig. 3.66a). **00000** represents the pressure switch connected to terminal **00** of input module **000**. It senses the overweight item on the automatic weigh station. The bit **00000** in the input table latches 1 for the overweight item. **10000** represents a solenoid connected to terminal **00** of the output module **100**, which pushes the trap door open. When the bit **10000** in the output table latches 1, the trap door is open and when the bit is 0, the trap door is closed.

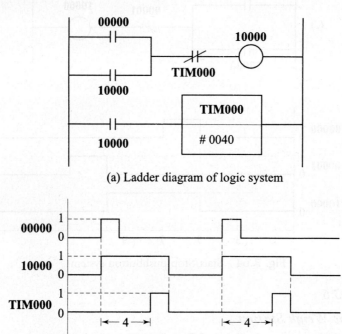

(a) Ladder diagram of logic system

(b) Timing chart

Fig. 3.66

For a 4 sec delay, the set value 0040 is stored in word storage area of memory. Count down of this number to 0000 will give an elapsed time of 4 seconds in our PLC system wherein we assume that the timer counts time-based intervals of 0.1 sec (40 counts will mean an elapsed time of 4 sec).

The timer **TIM000** is activated when its execution condition goes ON and starts decreasing from the set value. When the value reaches 0000, the timer contact point is tuned ON. It is reset to set value when the execution condition goes OFF. The timer contact point works as internal work bit (A work bit/internal output is for use of program only; it does not turn ON/OFF external field devices).

It is obvious from the ladder diagram of Fig. 3.66a that once an overweight item is detected, the trap door opens; it remains open for 4 sec, and thereafter it closes. Figure 3.66b shows the timing chart for the logic system.

Example 3.7

Packaging Line Control

Figure 3.67 shows a pictorial representation of process hardware of a conveyor system used for automatic packaging. The objective is to fill boxes moving on Box Conveyor from the Part Conveyor system.

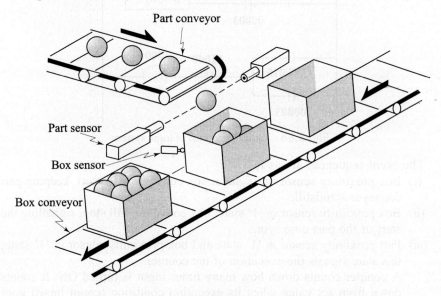

Fig. 3.67 Packaging line control

Input devices	: Box proximity sensor	00003
	Part proximity sensor	00002
	Stop pushbutton	00001
	Start pushbutton	00000
Output devices	: Part conveyor motor	10000
	Box conveyor motor	10001

Pushbutton **00000** starts the packaging line control which stops when pushbutton **00001** is pressed. Let us generate a work bit (internal output) **01000** which depends on the state of both the pushbuttons (refer Fig. 3.68). The work bit is 1 when packaging line control is ON, and it is 0 when packaging line control is OFF. The work bit is useful where the same combination of input signals appears repeatedly in the ladder diagram. We will shortly see that work bit **01000** is helpful in simplifying the ladder diagram.

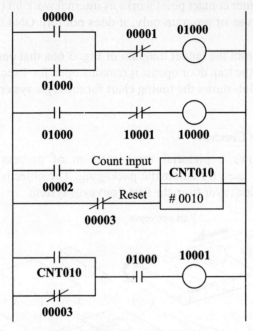

Fig. 3.68 Ladder diagram of logic system

The event sequences are as follows.
 (i) Box proximity sensor in '0' state; box conveyor will start, keeping part conveyor standstill.
 (ii) Box proximity sensor in '1' state; box conveyor will stop, signalling the start of the part conveyor.
 (iii) Part proximity sensor in '1' state and box proximity sensor in '1' state; this state signals the execution of the counter.
 A counter counts down how many times input is turned ON. It counts down from set value when its execution condition (count input) goes from OFF to ON. It decrements one count every time the input signal goes from OFF to ON. When the value reaches 0000, the counter contact point is turned ON. It is reset with a reset input. The counter contact point works as internal work bit.
 The execution signal for our counter is generated by part proximity sensor **00002** and reset signal is generated by box proximity sensor **00003**. For counting to occur, we need both the part proximity sensor and the box proximity sensor in '1' state.

(iv) Assume that set value 0010 of count has been loaded in word storage area of memory. When the value reaches 0000, the work bit **CNT010** takes '1' state which starts the box conveyor.

(v) Box proximity switch gets deactivated and count stops. The timing chart is shown in Fig. 3.69.

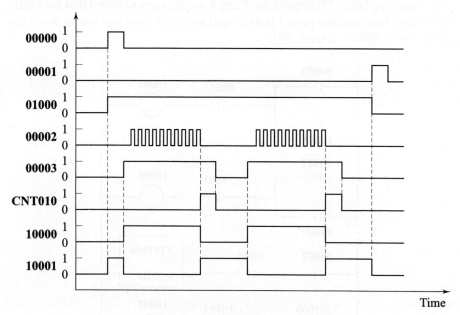

Fig. 3.69 Timing chart

Example 3.8: Automatic Bottle Filling Control

In this application (shown in Fig. 3.53), we are to implement a control program that will detect the position of a bottle via a limit switch, wait for 5 seconds, and then fill the bottle until the photo sensor detects the filled position. After the bottle is filled, it will wait for 7 seconds to continue to the next bottle. The start and stop circuits will also be included for the outfeed motor and for the start of the process. The I/O assignment follows:

Start-process Pushbutton	:	00000
Stop-process Pushbutton	:	00001
Limit switch (position detect)	:	00002
Photo sensor (level detect)	:	00003
Feed-motor drive	:	10000
Outfeed-motor drive	:	10001
Solenoid control	:	10002

The work bits may be assigned as follows:

Timer for 5 sec delay	:	**TIM000**
Timer for 7 sec delay	:	**TIM001**

212 Digital Control and State Variable Methods

Ladder diagram for automatic bottle filling controller is shown in Fig. 3.70. Rung 1 provides a start/stop latch for the system. The outfeed motor is always ON during process operation. Rung 2 drives the feed conveyor until a bottle is in position. Rung 3 introduces a time delay of 5 sec. The work bit TIM000 turns ON the valve solenoid (Rung 4). Rung 5 introduces a time delay of 7 sec after detecting bottle filled position. Rung 6 is necessary to detect that the bottle is full and 7 sec waiting period is over, and to restart the conveyor to move the bottle out (01000 is a work bit).

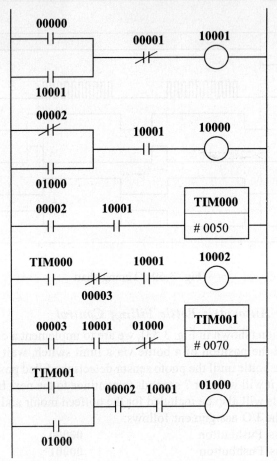

Fig. 3.70 Ladder diagram of logic system

PLC Programming

PLC programming methods vary from manufacturer to manufacturer, but the basic ladder diagram approach appears to be the standard throughout the industry. A CRT connected to the CPU of the PLC through a peripheral port, is perhaps the most common device used for programming the controller. A CRT is a self-contained video display unit with a keyboard and the necessary electronics to communicate with the CPU. The graphic display on the CRT screen appears as a ladder diagram. This ladder diagram takes form while the programmer builds it up using the keyboard. The keys themselves have symbols such as: —| |—; —|/|—, which are interpreted exactly as explained earlier in this section.

A limitation of CRT is that the device is not interchangeable from one manufacturer's PLC family to another. However, with the increasing number of products in the manufacturers' product lines and user standardization of products, these programming devices may be a good choice, especially if the user has standardized with one brand of PLCs.

At the other end of the spectrum of PLC programming devices is a Programming Console for programming small PLCs (upto 128 I/O). Physically, these devices resemble handheld calculators but have a larger display and somewhat different keyboard. The Programming Console uses keys with two- or three-letter abbreviations to write programs that bear some semblance to computer coding. The display at the top of the Console exhibits the PLC instruction located in the User Program memory area. As with CRTs, Programming Consoles are designed so that they are compatible with controllers of the product family.

Common usage of personal computer (PC) in our daily lives has led to a new breed of PLC programming devices. Due to the PC's general purpose architecture and de facto standard operating system, PLC manufacturers provide the necessary software to implement the ladder diagram entry, editing and real-time monitoring of the PLC's control program. PCs will soon be the programming device of choice not so much because of its PLC programming capabilities, but because these PCs may already be present at the location where the user may be performing the programming.

The programming device is connected to the CPU through a peripheral port. After the CPU has been programmed, the programming device is no longer required for CPU and process operation; it can be disconnected and removed. Therefore, we may need only one programming device for a number of operational PLCs. The programming device may be moved about in the plant as needed.

Programming details for any manufacturer's PLC are not included here. These aspects are described in manufacturers' literature.

Scope of Applications

Programmable logic controllers are available in many sizes, covering a wide spectrum of capability. On the low end are 'relay replacers' with minimum I/O and memory capability. At the high end are large supervisory controllers, which play an important role in distributed control systems by performing a variety of control and data acquisition functions. In between these two extremes are multifunctional controllers with communication capability which allows integration with various peripherals, and expansion capability which allows the product to grow as the application requirements change.

PLCs with analog input modules and analog output modules for driving analog valves and actuators using the PID control algorithms are being used in process industries.

Large PLCs are used for complicated control tasks that require analog control, data acquisition, data manipulation, numerical computations and reporting. The enhanced capabilities of these controllers allow them to be used effectively in applications where LAN (local area network) may be required.

Some PLCs offer the ability to program in other languages beside the conventional ladder language. An example is the BASIC programming language. Other manufacturers use what is called Boolean Mnemonics to program a controller. The Boolean language is really a method used to enter and explain the control logic which follows Boolean algebra.

3.10 REVIEW EXAMPLES

Review Example 3.1

We have so far used the z-transform technique to obtain system response at the sampling instants only. In ordinary cases, this information is adequate because if the sampling theorem is satisfied, then the output will not vary too much between any two consecutive sampling instants. In certain cases, however, we may need to find the response between consecutive sampling instants. This problem can be tackled by introducing a fictitious delay of ΔT seconds at the output of the system, where $0 \leq \Delta \leq 1$ and T is the sampling period. By varying Δ between 0 and 1, the output $y(t)$ at $t = kT - \Delta T$ (where $k = 1, 2, ...$) may be obtained.

In Example 3.1, unit-step response of the sampled-data system of Fig. 3.13a was obtained. The sampling period $T = 1$ sec, and the output at the sampling instants is given by

$$y(T) = 0.3679,\ y(2T) = 1,\ y(3T) = 1.3996,\ y(4T) = 1.3996,$$

$$y(5T) = 1.1469,\ y(6T) = 0.8944,\ y(7T) = 0.8015, ...$$

We now introduce at the output a fictitious delay of ΔT seconds with $\Delta = 0.5$ as shown in Fig. 3.71. The output $\hat{y}(kT)$ can be determined as follows:

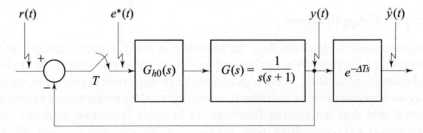

Fig. 3.71 Closed-loop discrete-time system for Review Example 3.1

$$\hat{y}(z) = \mathscr{Z}[G_{h0}(s)\, e^{-\Delta T s}]\, E(z)$$

$$E(z) = R(z) - Y(z)$$

$$Y(z) = \mathscr{Z}[G_{h0}(s)G(s)]\, E(z)$$

Therefore
$$E(z) = \frac{R(z)}{1 + \mathscr{Z}[G_{h0}(s)G(s)]}$$

$$\hat{Y}(z) = \frac{\mathscr{Z}[G_{h0}(s)G(s)e^{-\Delta T s}]}{1 + \mathscr{Z}[G_{h0}(s)G(s)]}\, R(z) \qquad (3.73)$$

Referring to Example 3.1, we have

$$\mathscr{Z}[G_{h0}(s)G(s)] = \frac{0.3679z + 0.2642}{z^2 - 1.3679z + 0.3679}$$

Referring to Table 3.1, we get

$$\mathscr{Z}[G_{h0}(s)G(s)e^{-\Delta T s}] = (1 - z^{-1})\left[\frac{1}{(z-1)^2} - \frac{0.5}{(z-1)} + \frac{0.6065}{(z - 0.3679)}\right]$$

$$= \frac{0.1065 z^2 + 0.4709 z + 0.0547}{z^3 - 1.3679 z^2 + 0.3679 z}$$

Referring to Eqn. (3.73) and noting that $R(z) = z/(z - 1)$, we have

$$\hat{Y}(z) = \frac{0.1065 z^{-1} + 0.4709 z^{-2} + 0.0547 z^{-3}}{1 - 2z^{-1} + 1.6321 z^{-2} - 0.6321 z^{-3}}$$

This equation can be expanded into an infinite series in z^{-1}:

$$\hat{Y}(z) = 0.1065\, z^{-1} + 0.6839\, z^{-2} + 1.2487\, z^{-3} + 1.4485\, z^{-4}$$
$$+ 1.2913\, z^{-5} + 1.0078\, z^{-6} + 0.8236\, z^{-7} + 0.8187\, z^{-8} + \cdots$$

Therefore

$$\hat{y}(T) = y(0.5T) = 0.1065$$
$$\hat{y}(2T) = y(1.5T) = 0.6839$$
$$\hat{y}(3T) = y(2.5T) = 1.2487$$
$$\hat{y}(4T) = y(3.5T) = 1.4485$$
$$\hat{y}(5T) = y(4.5T) = 1.2913$$
$$\hat{y}(6T) = y(5.5T) = 1.0078$$
$$\hat{y}(7T) = y(6.5T) = 0.8236$$
$$\hat{y}(8T) = y(7.5T) = 0.8187$$
$$\vdots$$

These values give the response at the midpoints between pairs of consecutive sampling points. Note that by varying the value of Δ between 0 and 1, it is possible to find the response at any point between two consecutive sampling points.

Review Example 3.2

Reconsider the sampled-data system of Example 3.2 (Fig. 3.15). The characteristic polynomial of the system is

$$\Delta(z) = 1 + G_{h0}G(z) = z^2 - az + b$$

where
$$a = 1 + e^{-2T} - 0.5K\,(T + 0.5e^{-2T} - 0.5)$$
$$b = e^{-2T} + 0.5K\,(0.5 - 0.5e^{-2T} - Te^{-2T})$$

The bilinear transformation

$$z = \frac{1+r}{1-r}$$

gives
$$\Delta(r) = \left(\frac{1+r}{1-r}\right)^2 - a\left(\frac{1+r}{1-r}\right) + b$$

On the r-plane, the characteristic equation becomes

$$r^2(1 + a + b) + 2r(1 - b) + 1 - a + b = 0$$

Applying the Routh criterion to this equation yields the array

$$\begin{array}{c|cc} r^2 & 1+a+b & 1-a+b \\ r^1 & 2(1-b) & \\ r^0 & 1-a+b & \end{array}$$

Thus for a stable system, $1 + a + b > 0$ \hfill (3.74a)

$$1 - b > 0 \hfill (3.74b)$$

$$1 - a + b > 0 \hfill (3.74c)$$

Substituting for a and b into (3.74a) and solving for K yields

$$K < \frac{4}{T - (1 - e^{-2T})/(1 + e^{-2T})}$$

From (3.74b), we get

$$K < \frac{2}{0.5 - (Te^{-2T})/(1 - e^{-2T})}$$

and from (3.74c),

$$e^{-2T} < 1$$

Table 3.4 indicates the values of K obtained for various values of T from (3.74a) and (3.74b). A sketch of T versus the boundary value of K for a stable system is shown in Fig. 3.72. Note that $\lim_{T \to 0} K = \infty$.

Table 3.4 Stability requirements for Review Example 3.2

T	Inequality (3.74a)	Inequality (3.74b)	Value of K for stability
0.01	$K < 3986844$	$K < 401.4$	$K < 401.4$
0.1	$K < 12042.7$	$K < 41.378$	$K < 41.378$
1.0	$K < 16.778$	$K < 5.8228$	$K < 5.8228$
1.0	$K < 4$	$K < 4$	$K < 4$

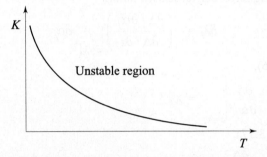

Fig. 3.72 T versus boundary value of K for stability

Review Example 3.3

Consider a digital control function

$$D(z) = \frac{N(z)}{\Delta(z)} = \frac{\beta_0 z^n + \beta_1 z^{n-1} + \cdots + \beta_n}{z^n + \alpha_1 z^{n-1} + \cdots + \alpha_n} \quad (3.75)$$

For direct realization of this control function, the computer must store the parameters $\alpha_1, \alpha_2, \ldots, \alpha_n, \beta_0, \beta_1, \ldots \beta_n$ (refer Fig. 3.18). If the machine uses fixed-point arithmetic, the parameter values will be rounded off to the accuracy of the machine. Thus a program designed to realize Eqn. (3.75), actually realizes[3]

$$\hat{D}(z) = \frac{(\beta_0 + \delta\beta_0)z^n + (\beta_1 + \delta\beta_1)z^{n-1} + \cdots + (\beta_n + \delta\beta_n)}{z^n + (\alpha_1 + \delta\alpha_1)z^{n-1} + \cdots + (\alpha_n + \delta\alpha_n)}$$

---

3. In addition to parameter quantization error, accuracy of a realization is affected by the error due to quantization of the input signal, and the error due to accumulation of round-off errors in arithmetic operations.

To study the effects of this realization on the dynamic response, we consider the characteristic equation and determine how a particular root changes when a particular parameter changes.

$$\Delta(z, \alpha_1, \alpha_2, ..., \alpha_n) = (z^n + \alpha_1 z^{n-1} + ... + \alpha_n) = 0 \quad (3.76)$$

is the characteristic equation with roots $\lambda_1, \lambda_2, ..., \lambda_n$:

$$\Delta(z, \alpha_1, \alpha_2, ..., \alpha_n) = (z - \lambda_1)(z - \lambda_2) ... (z - \lambda_n) \quad (3.77)$$

We shall consider the effect of parameter α_j on the root λ_k. By definition,

$$\Delta(\lambda_k, \alpha_j) = 0$$

If α_j is changed to $\alpha_j + \delta\alpha_j$, then λ_k also changes and the new polynomial is

$$\Delta(\lambda_k + \delta\lambda_k, \alpha_j + \delta\alpha_j) = \Delta(\lambda_k, \alpha_j) + \left.\frac{\partial \Delta}{\partial z}\right|_{z=\lambda_k} \delta\lambda_k + \left.\frac{\partial \Delta}{\partial \alpha_j}\right|_{z=\lambda_k} \delta\alpha_j + ... = 0$$

Neglecting the higher-order terms, we obtain

$$\delta\lambda_k = -\left(\frac{\partial \Delta / \partial \alpha_j}{\partial \Delta / \partial z}\right)\bigg|_{z=\lambda_k} \delta\alpha_j \quad (3.78)$$

From Eqn. (3.76),

$$\left.\frac{\partial \Delta}{\partial \alpha_j}\right|_{z=\lambda_k} = \lambda_k^{n-j}$$

and from Eqn. (3.77),

$$\left.\frac{\partial \Delta}{\partial z}\right|_{z=\lambda_k} = \prod_{i=k} (\lambda_k - \lambda_i)$$

Therefore Eqn. (3.78) gives

$$\delta\lambda_k = -\frac{\lambda_k^{n-j}}{\prod_{i \neq k}(\lambda_k - \lambda_i)} \delta\alpha_j$$

The sensitivity of the root λ_k to the parameter α_j is given by[4]

$$S_{\alpha_j}^{\lambda_k} = \frac{\partial \lambda_k}{\partial \alpha_j / \alpha_j}$$

$$= \frac{-\lambda_k^{n-j} \alpha_j}{\prod_{i \neq k}(\lambda_k - \lambda_i)} \quad (3.79)$$

Following observations are made from Eqn. (3.79).
(i) The numerator term in Eqn. (3.79) varies with j, the index number of the parameter whose variation is under consideration. For a stable system, $\lambda_k < 1$ and therefore the numerator term in Eqn. (3.79) is largest for $j = n$. Therefore, the most sensitive parameter in the characteristic Eqn. (3.76) is α_n.

---

4. Section 7.12 of reference [180].

(ii) The denominator in Eqn. (3.79) is the product of vectors from the characteristic roots to λ_k. Thus if all the roots are in a cluster, the sensitivity is high.

In the cascade and parallel realizations, the coefficients mechanized in the algorithm are poles themselves; these realizations are generally less sensitive than the direct realization.

PROBLEMS

3.1 Find $Y(z)/R(z)$ for the sampled-data closed-loop system of Fig. P3.1.

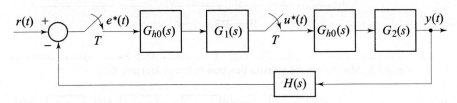

Fig. P3.1

3.2 For the sampled-data feedback system with digital network in the feedback path as shown in Fig. P3.2, find $Y(z)/R(z)$.

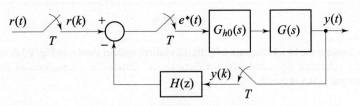

Fig. P3.2

3.3 Find $Y(z)$ for the sampled-data closed-loop system of Fig. P3.3.

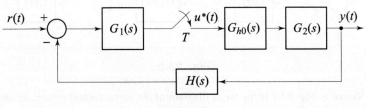

Fig. P3.3

3.4 Obtain the z-transform of the system output for the block diagram of Fig. P3.4.

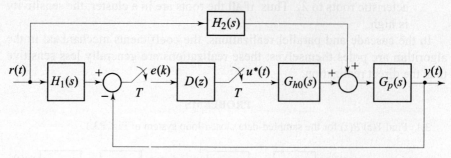

Fig. P3.4

3.5 Obtain the transfer function $Y(z)/R(z)$ of the closed-loop control system shown in Fig. P3.5. Also obtain the transfer function between $X(z)$ and $R(z)$.

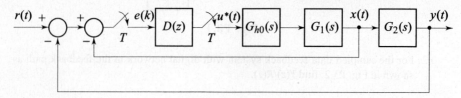

Fig. P3.5

3.6 Consider the block diagram of a digital control system shown in Fig. P3.6; $r(t)$ stands for reference input and $w(t)$ for disturbance. Obtain the z-transform of the system output when $r(t) = 0$.

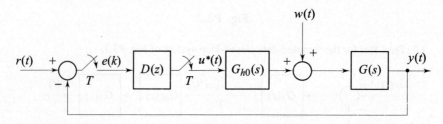

Fig. P3.6

3.7 Shown in Fig. P3.7 is the block diagram of the servo control system for one of the joints of a robot. With $D(z) = 1$, find the transfer function model of the closed-loop system. Sampling period $T = 0.25$ sec.

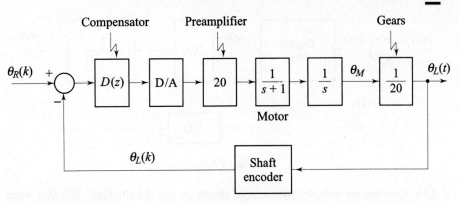

Fig. P3.7

3.8 The plant of the speed control system shown in Fig. P3.8 consists of load, armature-controlled dc motor and a power amplifier. Its transfer function is given by

$$\frac{\omega(s)}{V(s)} = \frac{185}{0.025s + 1}$$

Find the discrete-time transfer function $\omega(z)/\omega_r(z)$ for the closed-loop system. Sampling period $T = 0.05$ sec.

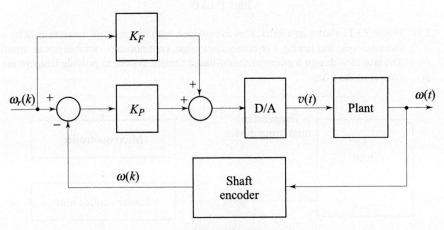

Fig. P3.8

3.9 For the system shown in Fig. P3.9, the computer solves the difference equation $u(k) = u(k-1) + 0.5\ e(k)$, where $e(k)$ is the filter input and $u(k)$ is the filter output. If the sampling rate $f_s = 5$ Hz, find $Y(z)/R(z)$.

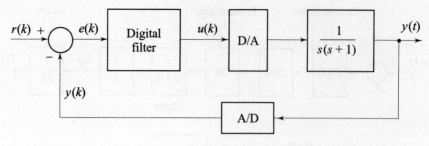

Fig. P3.9

3.10 Consider the sampled-data system shown in Fig. P3.10. Find $Y(z)/R(z)$ when (i) $\tau_D = 0.4$ sec, (ii) $\tau_D = 1.4$ sec.

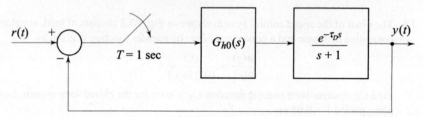

Fig. P3.10

3.11 Figure P3.11 shows an electrical oven provided with temperature measurement by a thermocouple and having a remotely controlled, continuously variable power input. The task is to design a microprocessor-based control system to provide temperature control of the oven.

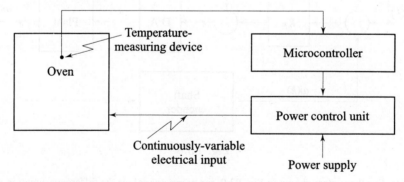

Fig. P3.11

The functions within the control loop can be broken down as follows:
   (i) sampling of the output of thermocouple;
   (ii) transfer of temperature signal into the computer;
   (iii) comparison of the measured temperature with the stored desired temperature to form an error signal;
   (iv) operation on the error signal by an appropriate algorithm to form a control signal; and
   (v) processing of the control signal and its transfer through the interface to the power control unit.

Suggest suitable hardware to implement these control-loop functions. Make a sketch of the system showing how the hardware is connected.

3.12 A mechanical manipulator arm has three degrees of freedom, each controlled by a permanent magnet stepping motor (Fig. P3.12). The base motor controls the rotation of the arm in the horizontal plane, the shoulder motor controls the raising and lowering of the arm in the vertical plane at the shoulder joint, and the elbow motor controls the bending of the arm in the vertical plane at the elbow joint. The stepping motors are coupled to the arm through gears and shafts. A solenoid is fitted at the end of the arm to lift magnetic materials. The task is to design a microprocessor-based control system to provide motion control of the manipulator arm.

The control functions can be broken down as follows:
 (i) suitable drive circuits for stepping motors and solenoid;
 (ii) interfacing drive circuits to the computer;
 (iii) interfacing a keyboard for applying motion commands:

Key 1: clockwise rotation of the base,
     2: counterclockwise rotation of the base,
     3: upward motion of the arm at the shoulder,
     4: downward motion of the arm at the shoulder,
     5: upward motion of the arm at the elbow,
     6: downward motion of the arm at the elbow,
     7: energize solenoid, and
     8: de-energize solenoid.

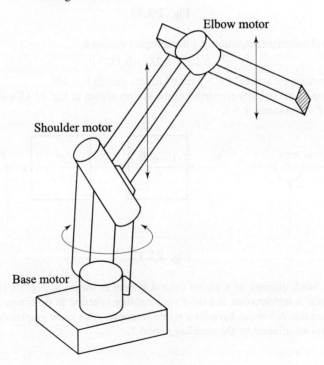

Fig. P3.12

(iv) operation on the motion commands by an appropriate sequence-generation program: and

(v) transfer of excitation sequences to the drive circuits through the interface.

Suggest suitable hardware to implement these control functions. Make a sketch of the system showing how the hardware is connected.

3.13 (a) A unity feedback system has the open-loop transfer function

$$G(s) = \frac{5}{s(s+1)(s+2)}$$

Using the Routh stability criterion, show that the closed-loop system is stable.

(b) A sampler and ZOH are now introduced in the forward path (Fig. P3.13). Show that the stable linear continuous-time system becomes unstable upon the introduction of sampler and ZOH.

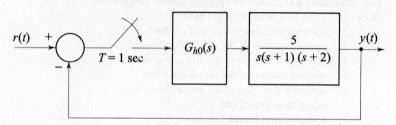

Fig. P3.13

3.14 The characteristic equation of a linear digital system is

$$z^3 - 0.1 z^2 + 0.2Kz - 0.1K = 0$$

Determine the values of $K > 0$ for which the system is stable.

3.15 Compare the stability properties of the system shown in Fig. P3.15 with (i) $T = 0.5$, (ii) $T = 1$. Assume $K > 0$.

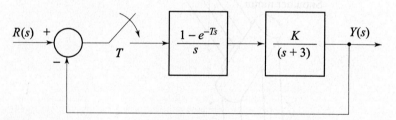

Fig. P3.15

3.16 The block diagram of a digital control system is shown in Fig. P3.16. Apply the bilinear transformation and the Routh stability criterion to determine the range of values that $K > 0$ can have for a stable response. Also show graphically how these values are affected by the sampling period T.

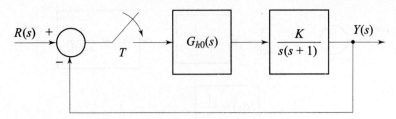

Fig P3.16

3.17 Consider the system shown in Fig. P3.17. Find the range of $K > 0$ for which the system is stable.

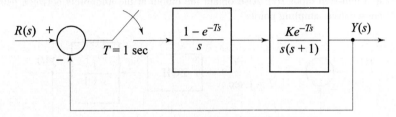

Fig. P3.17

3.18 (a) A unity feedback system has the open-loop transfer function
$$G(s) = \frac{Y(s)}{R(s)} = \frac{4500K}{s(s+361.2)}; K = 14.5$$
Find the response $y(t)$ of the system to a unit-step input.
(b) A sampler and ZOH are now introduced in the forward path (Fig. P3.18). For a unit-step input, determine the output $y(k)$ for first five sampling instants when (i) $T = 0.01$ sec, and (ii) $T = 0.001$ sec. Compare the result with that obtained earlier in part (a) above.

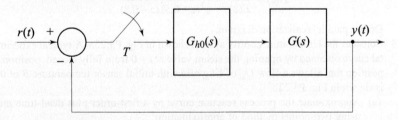

Fig. P3.18

3.19 For the sampled-data system shown in Fig. P3.19, find the output $y(k)$ for $r(t)$ = unit step.

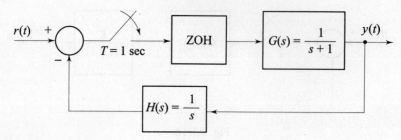

Fig. P3.19

3.20 For the sampled-data system of Fig. P3.20, find the response $y(kT)$; $k = 0, 1, 2, ...,$ to a unit-step input $r(t)$. Also obtain the output at the midpoints between pairs of consecutive sampling points.

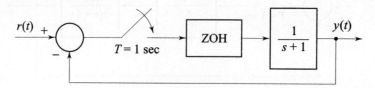

Fig. P3.20

3.21 Consider the digital controller defined by

$$D(z) = \frac{U(z)}{E(z)} = \frac{4(z-1)(z^2+1.2z+1)}{(z+0.1)(z^2-0.3z+0.8)}$$

Realize this digital controller in the cascade scheme and in parallel scheme. Use one first-order section and one second-order section.

3.22 Consider the digital controller defined by

$$D(z) = \frac{U(z)}{E(z)} = \frac{10(z^2+z+1)}{z^2(z-0.5)(z-0.8)}$$

Draw a parallel realization diagram.

3.23 Consider the temperature control system shown in Fig. P3.23a. A typical experimental curve obtained by opening the steam valve at $t = 0$ from fully closed position to a position that allows a flow Q_m of 1 kg/min with initial sensor temperature θ of 0 °C, is shown in Fig. P3.23b.
  (a) Approximate the process reaction curve by a first-order plus dead-time model using two-points method of approximation.
  (b) Calculate the QDR tuning parameters for a noninteracting PID controller. The PID control is to be carried out with a sampling period of 1 min on a computer control installation.

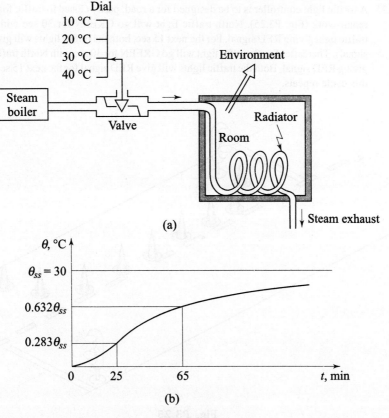

Fig. P3.23

3.24 Consider the liquid-level control system shown in Fig. 1.6. The digital computer was programmed to act as adjustable-gain proportional controller with a sampling period of $T = 10$ sec. The proportional gain was increased in steps. After each increase, the loop was disturbed by introducing a small change in set-point, and the response of the controlled variable (level in tank 2) was observed. The proportional gain of 4.75 resulted in oscillatory behaviour with amplitude of oscillations approximately constant. The period of oscillations measured from the response is 800 sec.

The PC implements the digital PI control algorithm. Determine tuning parameters for the controller:

$$\Delta u(k) = u(k) - u(k-1) = K_c \left[e(k) - e(k-1) + \frac{T}{T_I} e(k) \right]$$

where
  $u(k)$ = output of controller at kth sampling instant
  $\Delta u(k)$ = change in output of controller at kth sampling instant
  $e(k)$ = error at kth sampling instant
  T = sampling time
  T_I = integral time
  K_c = proportional gain

3.25 A traffic light controller is to be designed for a road, partly closed to traffic for urgent repair work (Fig. P3.25). North traffic light will go GREEN for 30 sec with South traffic light giving RED signal. For the next 15 sec, both the traffic lights will give RED signals. Thereafter South traffic light will go GREEN for 30 sec with North traffic light giving RED signal. Both the traffic lights will give RED signal for the next 15 sec. Then this cycle repeats.

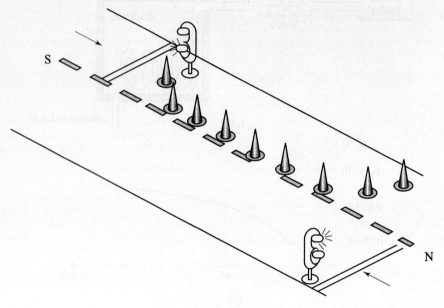

Fig. P3.25

Develop a PLC ladder diagram that accomplishes this objective.

3.26 Consider the tank system of Fig. P3.26. Valve V1 opens on pressing a pushbutton PB1 and liquid begins to fill the tank. At the same time, the stirring motor M starts operations. When the liquid level passes LL2 and reaches LL1, the valve V1 closes and the stirring motor stops. When PB1 is pressed again, the valve V2 opens and starts draining the liquid. When the liquid level drops below LL2, valve V2 closes. This cycle is repeated five times. A Buzzer will go high after 5 repetitions. The buzzer will be silenced by pressing push button PB2. The process will now be ready to take up another filling-stirring-draining operation under manual control. Develop a PLC ladder diagram that accomplishes this objective.

3.27 A control circuit is to be developed to detect and count the number of products being carried on an assembly line (Fig, 3.27). A sensor activates a counter as a product leaves the conveyor and enters the packaging section. When the counter counts five products, the circuit energizes a solenoid. The solenoid remains energized for a period of 2 seconds; the time being measured by a software timer. When the set-time has lapsed, the solenoid is deenergized, causing it to retract; and the control circuit is ready for the next cycle.

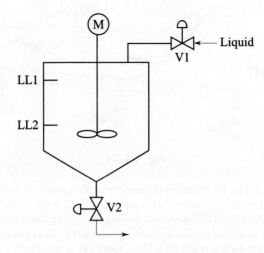

Fig. P3.26

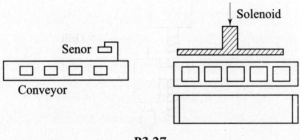

P3.27

Develop a suitable PLC ladder diagram.

3.28 In the system of Fig. P3.28, a PLC is used to start and stop the motors of a segmented conveyor belt. This allows only belt sections carrying an object to move. Motor M3 is kept ON during the operation. Position of a product is first detected by proximity switch S3 which switches on the motor M2. Sensor S2 switches on the motor M1 upon detection of the product. When the product moves beyond the range of sensor S2, a timer is activated and when the set-time of 20 sec has lapsed, motor M2 stops. Similarly, when the product moves beyond the range of sensor S1, another timer is activated and when the set-time of 20 sec (for unloading the product) has lapsed, motor M1 stops.

Develop a suitable ladder diagram for control.

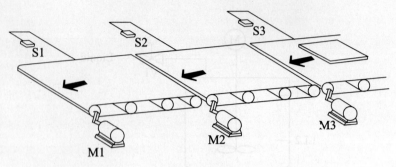

Fig. P3.28

3.29 The system of Fig. P3.29 has the objective of drilling a hole in workpiece moved on a carriage. When the start button PB1 is pushed and LS1 is ON (workpiece loaded), feed carriage motor runs in CW direction, moving the carriage from left to right. When the work comes exactly under the drill, which is sensed by limit switch LS2, the motor is cut-off and the work is ready for drilling operation. A timer with a set-time of 7 sec is activated. When timer set-value has lapsed, the motor reverses moving the carriage from right to left. When the work piece reaches LS1 position, the motor stops.

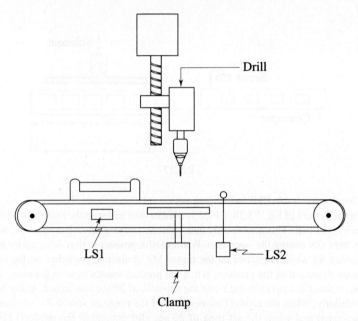

Fig. P3.29

The motor can be stopped by a stop pushbutton while in operation.
Develop a suitable ladder diagram for PLC control.

chapter 4
Design of Digital Control Algorithms

4.1 INTRODUCTION

During recent decades, the design procedures for analog control systems have been well formulated and a large body of knowledge accumulated. The analog-design methodology, based on conventional techniques of root locus and Bode plots or the tuning methods of Ziegler and Nichols, may be applied to designing digital control systems. The procedure would be to first design the analog form of the controller or compensator to meet a particular set of performance specifications. Having done this, the analog form can be transformed to a discrete-time formulation. This approach has already been introduced in Chapter 2 and is based on the fact that a digital system with a high sampling rate approximates to an analog system. The justification for using digital control under these circumstances must be that the practical limitations of the analog controller are overcome, the implementation cheaper, or that the supervisory control and communications more easily implemented.

However, the use of high sampling rates wastes computer power, and can lead to problems of arithmetic precision, etc. One is therefore driven to find methods of design which take account of the sampling process.

The alternative approach is to design controllers directly in the discrete-time domain based on the time-domain specifications of closed-loop system response. The controlled plant is represented by a discrete-time model which is a continuous-time system observed, analysed, and controlled at discrete intervals of time. Since the time response is the ultimate objective of the design, this approach, which we shall now consider, provides a direct path to the design of digital controllers. The features of direct digital design are that sample rates are generally lower than for discretized analog design, and the design is directly 'performance based'.

Figure 4.1 shows the basic structure of a digital control system. The design problem generally evolves around the choice of the control function $D(z)$ in order to impart a satisfactory form to the closed-loop transfer function. The choice is constrained by the form of the function $G_{h0}G(z)$ representing the fixed process elements.

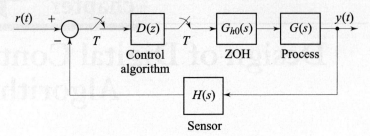

Fig. 4.1 Basic structure of a digital control system

A wide variety of digital-design procedures is available; these fall into two categories:
(i) direct synthesis procedures; and
(ii) iterative design procedures.

The direct synthesis procedures assume that the control function $D(z)$ is not restricted in any way by hardware or software limitations and can be allowed to take any form demanded by the nature of the fixed process elements and the specifications of the required system performance. This design approach has found wider applications in digital control systems than has the equivalent technique used with the analog systems. In a digital control system, realization of the required $D(z)$ may involve no more than programming a special purpose software procedure. With analog systems, the limitation was in terms of the complications involved in designing special purpose analog controllers.

The design obtained by a direct synthesis procedure will give perfect nominal performance. However, the performance may be inadequate in the field because of the sensitivity of the design to plant disturbances and modelling errors. It is important that a control system is robust in its behaviour with respect to the discrepancies between the model and the real process and uncertainties in disturbances acting on the process. Robustness property of some of the standard control structures, such as a three-term (PID) control algorithm, has been very well established. The design of such algorithms calls for an iterative design procedure where the choice of control function $D(z)$ is restricted to using a standard algorithm with variable parameters; the designer must then examine the effect of the choice of controller parameters on the system performance and make an appropriate final choice. The iterative design procedures for digital control systems are similar to the techniques evolved for analog system design using root locus and frequency response plots.

This chapter is devoted to the design of digital control algorithms, both by iterative design and direct synthesis procedures.

4.2 z-PLANE SPECIFICATIONS OF CONTROL SYSTEM DESIGN

Different properties that describe the performance of a feedback control system (Fig. 4.2) are listed below.
- Stability
- Steady-state accuracy
- Transient accuracy
- Disturbance rejection
- Insensitivity and robustness

We will discuss each of these in turn. Our attention will be focused on the unity feedback systems[1] of the form shown in Fig. 4.3.

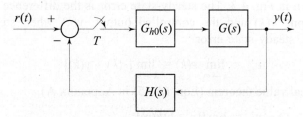

Fig. 4.2 A nonunity feedback discrete-time system

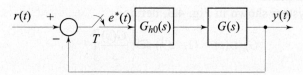

Fig. 4.3 A unity feedback discrete-time system

Stability

Stability is a very important property of closed-loop systems. Almost every working system is designed to be stable. We seek to improve the system performance within the constraints imposed by stability considerations.

Analytical study of stability of discrete-time systems can be carried out using the Jury stability criterion (Section 2.7) and/or the Routh stability criterion (Section 2.13).

---

1. It is assumed that the reader is familiar with the design of unity and nonunity feedback continuous-time systems. With this background, the results presented in this chapter for unity feedback discrete-time systems can easily be extended for the nonunity feedback case.

Steady-State Accuracy

Steady-state accuracy refers to the requirement that after all transients become negligible, the error between the reference input r and the controlled output y must be acceptably small. The specification on steady-state accuracy is often based on polynomial inputs of degree k: $r(t) = \dfrac{t^k}{k!}\mu(t)$. If $k = 0$, the input is a step of unit amplitude; if $k = 1$, the input is a ramp with unit slope; and if $k = 2$, the input is a parabola with unit second derivative. From the common problems of mechanical motion control, these inputs are called, respectively, *position*, *velocity*, and *acceleration* inputs.

For quantitative analysis, we consider the unity-feedback discrete-time system shown in Fig. 4.3. The steady-state error is the difference between the reference input $r(k)$ and the controlled output $y(k)$ when steady-state is reached, i.e., steady-state error

$$e_{ss}^* = \lim_{k \to \infty} e(k) = \lim_{k \to \infty} [r(k) - y(k)] \tag{4.1a}$$

Using the final value theorem (Eqn. (A.45) in Appendix A),

$$e_{ss}^* = \lim_{z \to 1} [(z - 1)E(z)] \tag{4.1b}$$

provided that $(z - 1) E(z)$ has no poles on the boundary and outside of the unit circle in the z-plane.

For the system shown in Fig. 4.3, define

$$G_{h0}G(z) = (1 - z^{-1}) \mathscr{Z}\left[\frac{G(s)}{s}\right]$$

Then we have
$$\frac{Y(z)}{R(z)} = \frac{G_{h0}G(z)}{1 + G_{h0}G(z)}$$

and
$$E(z) = R(z) - Y(z) = \frac{R(z)}{1 + G_{h0}G(z)} \tag{4.2}$$

By substituting Eqn. (4.2) into Eqn. (4.1b), we obtain

$$e_{ss}^* = \lim_{z \to 1} [(z - 1)E(z)] \tag{4.3a}$$

$$= \lim_{z \to 1} \left[(z - 1)\frac{R(z)}{1 + G_{h0}G(z)}\right] \tag{4.3b}$$

Thus, the steady-state error of a discrete-time system with unity feedback depends on the reference input signal $R(z)$, and the forward-path transfer function $G_{h0}G(z)$. By the nature of the limit in Eqns (4.3), we see that the result of the limit can be zero or can be a constant different from zero. Also the limit may not exist, in which case the final-value theorem does not apply. However, it is easy to see from basic definition (4.1a) that $e_{ss}^* = \infty$ in this case anyway because $E(z)$ will have a pole at $z = 1$ that is of order higher than one. Discrete-time systems having a finite nonzero steady-state error when the reference input is a zero-order polynomial input (a constant) are labelled 'Type-0'. Similarly, a system that has finite nonzero steady-state error to a first-order polynomial input (a ramp) is called a 'Type-1' system, and a system with finite nonzero steady-state error to a second-order polynomial input (a parabola) is called a 'Type-2' system.

Let the reference input to the system of Fig. 4.3 be a step function of magnitude unity. The z-transform of $r(t) = \mu(t)$ is

$$R(z) = \frac{z}{z-1} \tag{4.4a}$$

Substituting $R(z)$ into Eqn. (4.3b), we have

$$e_{ss}^* = \lim_{z \to 1} \frac{1}{1 + G_{h0}G(z)} = \frac{1}{1 + \lim_{z \to 1} G_{h0}G(z)}$$

In terms of the *position error constant* K_p, defined as

$$K_p = \lim_{z \to 1} G_{h0}G(z) \tag{4.4b}$$

the steady-state error to unit-step input becomes

$$e_{ss}^* = \frac{1}{1 + K_p} \tag{4.4c}$$

For a ramp input $r(t) = t\mu(t)$, the z-transform

$$R(z) = \frac{Tz}{(z-1)^2} \tag{4.5a}$$

Substituting into Eqn. (4.3b), we get

$$e_{ss}^* = \lim_{z \to 1} \frac{T}{(z-1)[1 + G_{h0}G(z)]} = \frac{1}{\lim_{z \to 1} \left[\frac{z-1}{T} G_{h0}G(z)\right]}$$

In terms of *velocity error constant* K_v, defined as

$$K_v = \frac{1}{T} \lim_{z \to 1} [(z-1)G_{h0}G(z)] \tag{4.5b}$$

the steady-state error to unit-ramp input becomes

$$e_{ss}^* = \frac{1}{K_v} \tag{4.5c}$$

For a parabolic input $r(t) = (t^2/2)\mu(t)$, the z-transform

$$R(z) = \frac{T^2 z(z+1)}{2(z-1)^3} \tag{4.6a}$$

Substituting into Eqn. (4.3b), we get

$$e_{ss}^* = \lim_{z \to 1} \frac{T^2}{(z-1)^2[1+G_{h0}G(z)]} = \frac{1}{\lim_{z \to 1}\left[\left(\frac{z-1}{T}\right)^2 G_{h0}G(z)\right]}$$

In terms of *acceleration error constant* K_a, defined as

$$K_a = \frac{1}{T^2}\lim_{z \to 1}[(z-1)^2 G_{h0}G(z)] \qquad (4.6b)$$

the steady-state error to unit-parabolic input becomes

$$e_{ss}^* = \frac{1}{K_a} \qquad (4.6c)$$

▲▲

As said earlier, discrete-time systems can be classified on the basis of their steady-state response to polynomial inputs. We can always express the forward-path transfer function $G_{h0}G(z)$ as

$$G_{h0}G(z) = \frac{K\prod_i(z-z_i)}{(z-1)^N \prod_j(z-p_j)}; p_j \neq 1, z_i \neq 1 \qquad (4.7)$$

$G_{h0}G(z)$ in Eqn. (4.7) involves the term $(z-1)^N$ in the denominator. As $z \to 1$, this term dominates in determining the steady-state error. Digital control systems are therefore classified in accordance with the number of poles at $z = 1$ in the forward-path transfer function as described below.

Type-0 system

If $N = 0$, the steady-state errors to various standard inputs obtained from Eqns (4.1)–(4.7) are

$$e_{ss}^* = \begin{cases} \dfrac{1}{1+K_p} & \text{in response to unit-step input; } K_p = \left.\dfrac{K\prod_i(z-z_i)}{\prod_j(z-p_j)}\right|_{z=1} \\ \infty & \text{in response to unit-ramp input} \\ \infty & \text{in response to unit-parabolic input} \end{cases} \qquad (4.8a)$$

Thus a system with $N = 0$, or no pole at $z = 1$ in $G_{h0}G(z)$, has a finite non-zero position error, infinite velocity and acceleration errors at steady-state.

Type-1 system

If $N = 1$, the steady-state errors to various standard inputs are

$$e_{ss}^* = \begin{cases} 0 & \text{in response to unit-step input} \\ \dfrac{1}{K_v} & \text{in response to unit-ramp input; } K_v = \left.\dfrac{\dfrac{K}{T}\Pi_i(z-z_i)}{\Pi_j(z-p_j)}\right|_{z=1} \\ \infty & \text{in response to unit-parabolic input} \end{cases} \quad (4.8b)$$

Thus a system with $N = 1$, or one pole at $z = 1$ in $G_{h0}G(z)$, has zero position error, a finite non-zero velocity error and infinite acceleration error at steady-state.

Type-2 system

If $N = 2$, the steady-state errors to various standard inputs are

$$e_{ss}^* = \begin{cases} 0 & \text{in response to unit-step input} \\ 0 & \text{in response to unit-ramp input} \\ \dfrac{1}{K_a} & \text{in response to unit-parabolic input; } K_a = \left.\dfrac{\dfrac{K}{T^2}\Pi_i(z-z_i)}{\Pi_j(z-p_j)}\right|_{z=1} \end{cases} \quad (4.8c)$$

Thus a system with $N = 2$, or two poles at $z = 1$ in $G_{h0}G(z)$, has zero position and velocity errors and a finite non-zero acceleration error at steady-state.

Steady-state errors for various inputs and systems are summarized in Table 4.1.

Table 4.1 Steady-state errors for various inputs and systems

Type of input	Steady-state error		
	Type-0 system	Type-1 system	Type-2 system
Unit-step	$\dfrac{1}{1+K_p}$	0	0
Unit-ramp	∞	$\dfrac{1}{K_v}$	0
Unit-parabolic	∞	∞	$\dfrac{1}{K_a}$
	$K_p = \lim\limits_{z\to 1} G_{h0}G(z);\ K_v = \dfrac{1}{T}\lim\limits_{z\to 1}[(z-1)G_{h0}G(z)];$		
	$K_a = \dfrac{1}{T^2}\lim\limits_{z\to 1}[(z-1)^2 G_{h0}G(z)]$		

238   Digital Control and State Variable Methods

The development above indicates that, in general, increased system gain K and/or addition of poles at $z = 1$ to the open-loop transfer function $G_{h0}G(z)$ tend to decrease steady-state errors. However, as will be seen later in this chapter, both large system gain and the poles at $z = 1$ in the loop transfer function have destabilizing effects on the system. Thus a control system design is usually a trade off between steady-state accuracy and acceptable relative stability.

Example 4.1

In the previous chapter we have shown that sampling usually has a deterimental effect on the transient response and the relative stability of a control system. It is natural to ask what the effect of sampling on the steady-state error of a closed-loop system will be? In other words, if we start out with a continuous-time system and then add S/H to form a digital control system, how would the steady-state errors of the two systems compare, when subject to the same type of input?

Let us first consider the system of Fig. 4.3 without S/H. Assume that the process $G(s)$ is represented by Type-1 transfer function:

$$G(s) = \frac{K(1+\tau_a s)(1+\tau_b s)\cdots(1+\tau_m s)}{s(1+\tau_1 s)(1+\tau_2 s)\cdots(1+\tau_n s)}$$

having more poles than zeros.

The velocity error constant

$$K_v = \lim_{s \to 0} sG(s) = K$$

The steady-state error of the system to unit-step input is zero, to unit-ramp input is $1/K$, and to unit-parabolic input is ∞.

We now consider the system of Fig. 4.3 with S/H:

$$G_{h0}G(z) = (1-z^{-1})\mathscr{Z}\left[\frac{K(1+\tau_a s)(1+\tau_b s)\cdots(1+\tau_m s)}{s^2(1+\tau_1 s)(1+\tau_2 s)\cdots(1+\tau_n s)}\right]$$

$$= (1-z^{-1})\mathscr{Z}\left[\frac{K}{s^2}+\frac{K_1}{s}+ \text{terms due to the non-zero poles}\right]$$

$$= (1-z^{-1})\left[\frac{KTz}{(z-1)^2}+\frac{K_1 z}{z-1}+ \text{terms due to the non-zero poles}\right]$$

It is important to note that the terms due to the non-zero poles do not contain the term $(z - 1)$ in the denominator. Thus, the velocity error constant is

$$K_v = \frac{1}{T}\lim_{z \to 1}[(z-1)G_{h0}G(z)] = K$$

The steady-state error of the discrete-time system to unit-step input is zero, to unit-ramp input is $1/K$, and to unit-parabolic input is ∞. Thus for a Type-1 system, the system with S/H has exactly the same steady-state error as the continuous-time system with the same process transfer function (this, in fact, is true for Type-0 and Type-2 systems also).

Equations (4.5b) and (4.6b) may purport to show that the velocity error constant and the acceleration error constant of a digital control system depend on the sampling period T. However, in the process of evaluation, T gets cancelled and the error depends only on the parameters of the process and the type of inputs.

Transient Accuracy

Figure 4.4 shows a typical unit-step response of a digital control system. Specifications of transient performance may be made in the time domain in terms of rise time, peak time, peak overshoot, settling time, etc. The use of root locus plots for the design of digital control systems necessitates the translation of time-domain performance specifications into desired locations of closed-loop poles in the z-plane. Whereas the use of frequency response plots necessitates the translation of time-domain specifications in terms of frequency response features such as bandwidth, phase margin, gain margin, resonance peak, resonance frequency, etc.

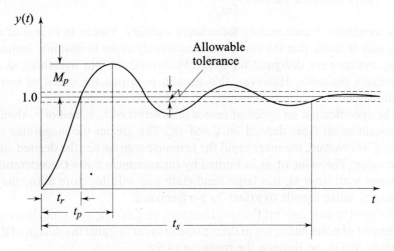

Fig. 4.4 Typical unit-step response of a digital control system

Specifications in terms of root locations in z-plane

Our approach is to first obtain the transient response specifications in terms of characteristic roots in the s-plane, and then use the relation

$$z = e^{sT} \tag{4.9}$$

to map the s-plane characteristic roots to the z-plane.

The transient response of Fig. 4.4 resembles the unit-step response of an underdamped second-order system

$$\frac{Y(s)}{R(s)} = \frac{\omega_n^2}{s^2 + 2\zeta\omega_n s + \omega_n^2} \tag{4.10}$$

where

 ζ = damping ratio, and
 ω_n = undamped natural frequency.

The transient response specifications in terms of rise time t_r, peak time t_p, peak overshoot M_p, and settling time t_s can be approximated to the parameters ζ and ω_n of the second-order system defined by Eqn. (4.10) using the following correlations[2]:

$$t_r(0\% \text{ to } 100\%) = \frac{\pi - \cos^{-1}\zeta}{\omega_n\sqrt{1-\zeta^2}} \qquad (4.11)$$

$$t_p = \frac{\pi}{\omega_n\sqrt{1-\zeta^2}} \qquad (4.12)$$

$$M_p = \exp(-\pi\zeta/\sqrt{1-\zeta^2}) \qquad (4.13)$$

$$t_s(2\% \text{ tolerance band}) = \frac{4}{\zeta\omega_n} \qquad (4.14)$$

Peak overshoot is used mainly for relative stability. Values in excess of about 40% may indicate that the system is dangerously close to absolute instability. Many systems are designed for 5 to 25% overshoot No overshoot at all is sometimes desirable. However, this usually penalizes the speed of response needlessly.

The specification on speed of response in terms of t_r, t_p and/or t_s should be consistent as all these depend on ζ and ω_n. The greater the magnitude of ω_n when ζ is constant, the more rapid the response approaches the desired steady-state value. The value of ω_n is limited by measurement noise considerations—a system with large ω_n has large bandwidth and will therefore allow the high frequency noise signals to affect its performance.

We need to now convert the specifications on ζ and ω_n into guidelines on the placement of poles and zeros in the z-plane in order to guide the design of digital controls. We do so through the mapping (4.9).

Figure 4.5 illustrates the translation of specifications on ζ and ω_n to the characteristic root locations in the z-plane (referring to Section 2.12 will be helpful). The s-plane poles

$$s_{1,2} = -\zeta\omega_n \pm j\omega_n\sqrt{1-\zeta^2} = -\zeta\omega_n \pm j\omega_d \qquad (4.15a)$$

for constant ζ, lie along a radial line in the s-plane (Fig. 4.5a). In the z-plane,

$$z_{1,2} = e^{-\zeta\omega_n T} e^{\pm j\omega_n T\sqrt{1-\zeta^2}} = r\, e^{\pm j\theta} \qquad (4.15b)$$

---

2. Chapter 6 of reference [180].

Design of Digital Control Algorithms   **241**

The magnitude of z (i.e., the distance to the origin) is $r = e^{-\zeta\omega_n T}$ and the angles with the positive real axis of the z-plane, measured positive in the counterclockwise direction, are $\theta = \omega_n T\sqrt{1-\zeta^2}$. It should be observed that the z-plane pole locations depend on the s-plane positions as well as the sampling interval T.

As ω_n increases for a constant-ζ, the magnitude of z decreases and the phase angle increases; constant-ζ locus is a logarithmic spiral in the z-plane (Fig. 4.5b). Increasing ω_n negatively, gives the mirror image.

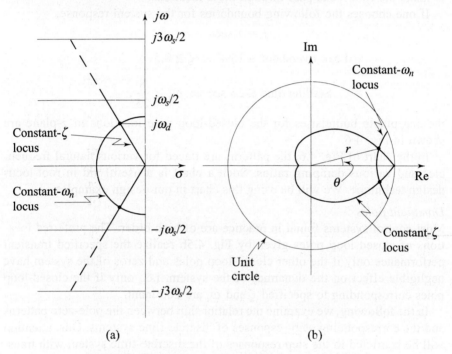

Fig. 4.5 Mapping of s-plane patterns on the z-plane

In Fig. 4.5a, the s-plane has been divided into strips of width ω_s, where $\omega_s = 2\pi/T$ is the sampling frequency. The primary strip extends from $\omega = -\omega_s/2$ to $+\omega_s/2$, and the complimentary strips extend from $-\omega_s/2$ to $-3\omega_s/2, \ldots$, for negative frequencies, and from $\omega_s/2$ to $3\omega_s/2,\ldots$, for positive frequencies. We will assume that the low-pass analog filtering characteristics of the continuous-time plant and the ZOH device attenuate the responses due to the poles in the complimentary strips; only the poles in the primary strip, generally, need be considered.

Figure 4.5 illustrates the mapping of constant-ζ locus in the primary strip of the s-plane to the z-plane. As the imaginary parts $\pm j\omega_d = \pm j\omega_n\sqrt{1-\zeta^2}$ of the s-plane poles move closer to the limit $\pm j\omega_s/2$ of the primary strip, the angles

$\theta = \pm \omega_d T = \pm \omega_n T\sqrt{1-\zeta^2}$ of the z-plane poles approach the direction of the negative real axis. The negative real axis in the z-plane thus corresponds to the boundaries of the primary strip in the s-plane. Figure 4.5 also shows the mapping of a constant-ω_n locus in the primary strip of the s-plane to the z-plane.

In the z-plane, the closed-loop poles must lie on the constant-ζ spiral to satisfy peak overshoot requirement, also the poles must lie on constant-ω_n curve to satisfy speed of response requirement. The intersection of the two curves (Fig. 4.5b) provides the preferred pole locations, and the design aim is to make the root locus pass through these locations.

If one chooses the following boundaries for the system response:

$$T = 1 \text{ sec}$$

$$\text{Peak overshoot} \leq 15\% \Rightarrow \zeta \geq 0.5$$

$$\text{Settling time} \leq 25 \text{ sec} \Rightarrow \omega_n \geq \frac{8}{25},$$

the acceptable boundaries for the closed-loop pole locations in z-plane are shown in Fig. 4.6.

In the chart of Fig. 4.6, the patterns are traced for various natural frequencies and various damping ratios. Such a chart is a useful aid in root-locus design technique. We will be using this chart in our design examples.

Dominant poles
Most control systems found in practice are of high order. The preferred locations of closed-loop poles given by Fig. 4.5b realize the specified transient performance only if the other closed-loop poles and zeros of the system have negligible effect on the dynamics of the system, i.e., only if the closed-loop poles corresponding to specified ζ and ω_n are dominant.

In the following, we examine the relationship between the pole-zero patterns and the corresponding step-responses of discrete-time systems. Our attention will be restricted to the step responses of the discrete-time system with transfer function

$$\frac{Y(z)}{R(z)} = \frac{K(z-z_1)(z-z_2)}{(z-p)(z-re^{j\theta})(z-re^{-j\theta})} = \frac{K(z-z_1)(z-z_2)}{(z-p)(z^2-2r\cos\theta z+r^2)} \quad (4.16)$$

for a selected set of values of the parameters K, z_1, z_2, p, r and θ.

We assume that the roots of the equation

$$z^2 - 2r\cos\theta z + r^2 = 0$$

are the preferred closed-loop poles corresponding to the specified values of ζ and ω_n. Complex-conjugate pole pairs corresponding to $\zeta = 0.5$ with $\theta = 18°$, 45° and 72° will be considered in our study. The pole pair with $\theta = 18°$ is shown in Fig. 4.7.

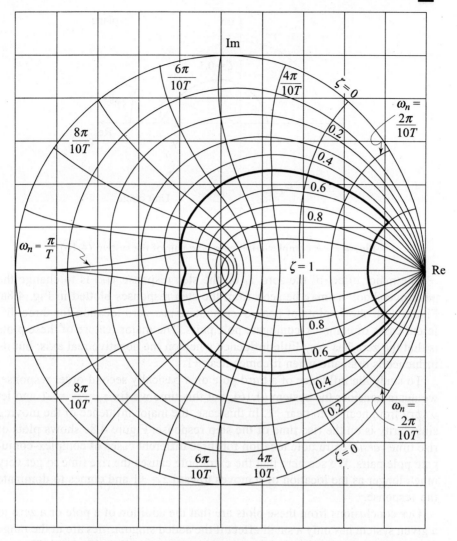

Fig. 4.6 Plot of an acceptable region for poles of a second-order system to satisfy dynamic response specifications

To study the effect of zero location, we let $z_2 = p$ and explore the effect of the (remaining) zero location z_1 on the transient performance. We take the gain K to be such that the steady-state output value equals the step size. For a unit-step input,

$$Y(z) = \left[\frac{K(z-z_1)}{z^2 - 2r\cos\theta\, z + r^2}\right]\left(\frac{z}{z-1}\right) \quad (4.17)$$

with

$$K = \frac{1 - 2r\cos\theta + r^2}{(1 - z_1)}$$

244 Digital Control and State Variable Methods

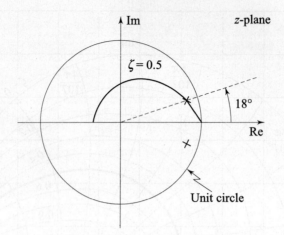

Fig. 4.7 A complex-conjugate pole pair of the system (4.16)

The major effect of the zero z_1 on the step response $y(k)$ is to change the peak overshoot, as may be seen from the step responses plotted in Fig. 4.8a. Figure 4.8b shows plots of peak overshoot *versus* zero location for three different cases of complex-conjugate pole pairs. The major feature of these plots is that the zero has very little influence when on the negative real axis, but its influence is dramatic when it comes near +1.

To study the influence of a third pole on a basically second-order response, we again consider the system (4.16), but this time we fix $z_1 = z_2 = -1$ and let p vary from near -1 to near $+1$. In this case, the major influence of the moving singularity is on the rise time of the step response. Figure 4.8c shows plots of rise time *versus* extra pole location for three different cases of complex-conjugate pole pairs. We see here that the extra pole causes the rise time to get very much longer as the location of p moves toward $z = +1$ and comes to dominate the response.

Our conclusions from these plots are that the addition of a pole or a zero to a given system has only a small effect if the added singularities are in the range 0 to -1. However, a zero moving toward $z = +1$ greatly increases the system overshoot. A pole placed toward $z = +1$ causes the response to slow down and thus primarily affects the rise time which is being progressively increased. The pole pair corresponding to specified ζ and ω_n is a dominant pole pair of the closed-loop system only if the influence of additional poles and zeros is negligibly small on the dynamic response of the system.

Specifications in terms of frequency response features

The translation of time-domain specifications into desired locations of pair of dominant closed-loop poles in z-plane is useful if the design is to be carried out by using root locus plots. The use of frequency response plots necessitates the translation of time-domain performance specifications in terms of frequency response features.

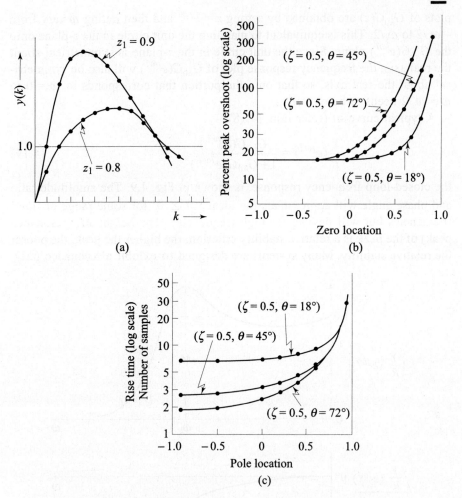

Fig. 4.8 (a) Effect of an extra zero on a discrete-time second-order system, $\zeta = 0.5$, $\theta = 18°$; (b) effect of an extra zero on a discrete-time second-order system; and (c) effect of an extra pole on a discrete-time second-order system

All the frequency-domain methods of continuous-time systems can be extended for the analysis and design of digital control systems. Consider the system shown in Fig. 4.3. The closed-loop transfer function of the sampled-data system is

$$\frac{Y(z)}{R(z)} = \frac{G_{h0}G(z)}{1 + G_{h0}G(z)} \qquad (4.18)$$

Just as in the case of continuous-time systems, the absolute and relative stability conditions of the closed-loop discrete-time system can be investigated by making the frequency response plots of $G_{h0}G(z)$. The frequency response plots of $G_{h0}G(z)$ are obtained by setting $z = e^{j\omega T}$ and then letting ω vary from $-\omega_s/2$ to $\omega_s/2$. This is equivalent to mapping the unit circle in the z-plane onto the $G_{h0}G(e^{j\omega T})$-plane. Since the unit circle in the z-plane is symmetrical about the real axis, the frequency response plot of $G_{h0}G(e^{j\omega T})$ will also be symmetrical about the real axis, so that only the portion that corresponds to $\omega = 0$ to $\omega = \omega_s/2$ needs to be plotted.

A typical curve of (refer Eqn. (4.18))

$$\frac{Y}{R}(e^{j\omega T}) = \frac{G_{h0}G(e^{j\omega T})}{1 + G_{h0}G(e^{j\omega T})}, \qquad (4.19)$$

the closed-loop frequency response, is shown in Fig. 4.9. The amplitude ratio and phase angle will approximate the ideal $1.0 \angle 0°$ for some range of 'low' frequencies but will deviate for high frequencies. The height M_r (resonance peak) of the peak is a relative stability criterion; the higher the peak, the poorer the relative stability. Many systems are designed to exhibit a resonance peak

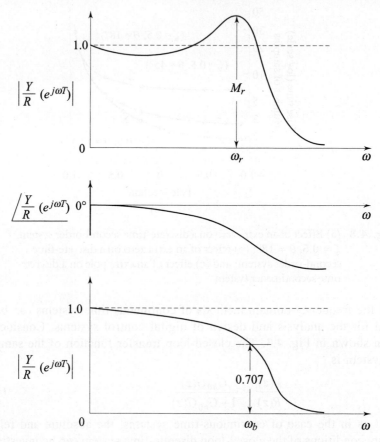

Fig. 4.9 Closed-loop frequency response criteria

in the range 1.2 to 1.4. The frequency ω_r (resonance frequency) at which this peak occurs is a speed of response criterion; the higher the ω_r, the faster the system. For systems that exhibit no peak (sometimes the case), the *bandwidth* ω_b is used for speed of response specifications. Bandwidth is the frequency at which amplitude ratio has dropped to $1/\sqrt{2}$ times its zero-frequency value. It can, of course, be specified even if there is a peak.

Two open-loop performance criteria are in common use to specify relative stability. These are gain margin *GM* and phase margin ΦM. A typical curve of $G_{h0}G(e^{j\omega T})$, the open-loop frequency response, is shown on the polar plane in Fig. 4.10. Gain margin is the multiplying factor by which the steady-state gain of $G_{h0}G(e^{j\omega T})$ could be increased so as to put the system on the edge of instability. Phase margin is the number of degrees of additional phase lag required to drive the system to the edge of instability.

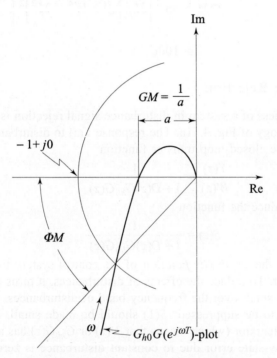

Fig. 4.10 Gain margin and phase margin

Both a good gain margin and a good phase margin are needed; neither is sufficient by itself. General numerical design goals for gain and phase margins cannot be given since systems that satisfy other specific performance criteria may exhibit a wide range of these margins. It is possible, however, to give useful lower bounds—the gain margin should usually exceed 2.5 and the phase margin should exceed 30°.

The translation of time-domain specifications in terms of frequency response features is carried out by using the explicit correlations for second-order system (4.10). The following correlations are valid approximations for higher-order systems dominated by a pair of complex conjugate poles[3].

$$M_r = \frac{1}{2\zeta\sqrt{1-\zeta^2}}; \quad \zeta \leq 0.707 \quad (4.20)$$

$$\omega_r = \omega_n\sqrt{1-2\zeta^2} \quad (4.21)$$

$$\omega_b = \omega_n\left[1 - 2\zeta^2 + \sqrt{(2-4\zeta^2+4\zeta^4)}\right]^{\frac{1}{2}} \quad (4.22)$$

$$\Phi M = \tan^{-1}\left\{2\zeta \Big/ \left[\sqrt{1+4\zeta^4} - 2\zeta^2\right]^{\frac{1}{2}}\right\}$$

$$\cong 100\zeta \quad (4.23)$$

Disturbance Rejection

The effectiveness of a system in disturbance signal rejection is readily studied with the topology of Fig. 4.11a. The response $Y(z)$ to disturbance $W(z)$ can be found from the closed-loop transfer function

$$\frac{Y(z)}{W(z)} = \frac{1}{1 + D(z)G_{h0}G(z)} \quad (4.24a)$$

We now introduce the function

$$S(z) = \frac{1}{1 + D(z)G_{h0}G(z)} \quad (4.24b)$$

which we call the *sensitivity function* of the control system for reasons to be explained later. To reduce the effects of disturbances, it turns out that $S(e^{j\omega T})$ must be made small over the frequency band of disturbances. If constant disturbances are to be suppressed, $S(1)$ should be made small. If $D(z)G_{h0}G(z)$ includes an integrator (which means that $D(z)$ or $G_{h0}G(z)$ has a pole at $z = 1$), then the steady-state error due to constant disturbance is zero. This may be seen as follows. Since for a constant disturbance of amplitude A, we have

$$W(z) = \frac{Az}{z-1},$$

the steady-state value of the output is given by

$$y_{ss} = \lim_{z \to 1}(z-1)Y(z) = \lim_{z \to 1}(z-1)S(z)W(z) = \lim_{z \to 1}AS(z)$$

which is equal to zero if $D(z)G_{h0}G(z)$ has a pole at $z = 1$.

---

3. Chapter 9 of reference [180].

Note that the point where the disturbance enters the system is very important in adjusting the gain of $D(z)G_{h0}G(z)$. For example, consider the system shown in Fig. 4.11b. The closed-loop transfer function for the disturbance is

$$\frac{Y(z)}{W(z)} = \frac{G_{h0}G(z)}{1+D(z)G_{h0}G(z)}$$

In this case, the steady-state error due to constant disturbance $W(z)$ is not equal to zero when $G_{h0}G(z)$ has a pole at $z = 1$. This may be seen as follows.

Let
$$G_{h0}G(z) = Q(z)/(z-1)$$

where $Q(z)$ is a rational polynomial of z such that $Q(1) \neq 0$ and $Q(1) \neq \infty$; and $D(z)$ is a controller which does not have pole at $z = 1$. Then

$$y_{ss} = \lim_{z \to 1} \frac{(z-1)\,G_{h0}G(z)}{1+D(z)\,G_{h0}G(z)}\,W(z) = \lim_{z \to 1} \frac{AzQ(z)}{z-1+D(z)Q(z)} = \frac{A}{D(1)}$$

Thus the steady-state error is non-zero; the magnitude of the error can be reduced by increasing the controller gain.

Figure 4.11c gives a block diagram of the situation where measurement noise $W_n(z)$ enters the system through feedback link. The closed-loop transfer function for this disturbance is

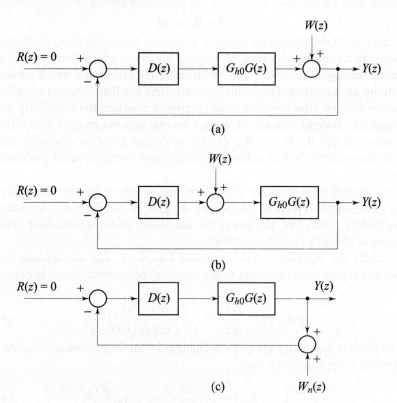

Fig. 4.11 Disturbance rejection

$$\frac{Y(z)}{W_n(z)} = \frac{D(z)G_{h0}G(z)}{1+D(z)G_{h0}G(z)} \qquad (4.25)$$

Thus the measurement noise is transferred to the output whenever $|D(z)G_{h0}G(z)| > 1$. Hence large gains of $D(z)G_{h0}G(z)$ will lead to large output errors due to measurement noise. This is in conflict with the disturbance-rejection property with respect to configurations of Figs 4.11a and 4.11b. To solve this problem, we can generally examine the measuring instrument and modify the filtering so that it satisfies the requirements of a particular control problem.

Insensitivity and Robustness

Finally, in our design we must take into account both the small and often the large differences between the derived process model and the real process behaviour. The differences may appear due to modelling approximations and the process behaviour changes with time during operation. If, for simplicity, it is assumed that the structure and order of the process model are chosen exactly and they do not change with time, then these differences are manifested as parameter errors.

Parameter changes with respect to nominal parameter vector $\boldsymbol{\theta}_n$ are assumed. The closed-loop behaviour for parameter vector

$$\boldsymbol{\theta} = \boldsymbol{\theta}_n + \Delta\boldsymbol{\theta}$$

is of interest. If the parameter changes are small, then sensitivity methods can be used. For controller design, both good control performance (steady-state accuracy, transient accuracy, and disturbance rejection) and small parameter sensitivity are required. The resulting controllers are then referred to as *insensitive controllers*. However, for large parameter changes, the sensitivity design is unsuitable. Instead, one has to assume several process models with different parameter vectors $\boldsymbol{\theta}_1, \boldsymbol{\theta}_2, ..., \boldsymbol{\theta}_M$, and try to design a *robust controller* which for all process models will maintain stability and certain control performance range.

As to design of insensitive controllers, the situation is very much like the disturbance-signal rejection. The larger the gain of the feedback loop around the offending parameter, the lower the sensitivity of the closed-loop transfer function to changes in that parameter.

Consider the digital control system of Fig. 4.11. The closed-loop input-output behaviour corresponding to the nominal parameter vector is described by

$$M(\boldsymbol{\theta}_n, z) = \frac{Y(z)}{R(z)} = \frac{D(z)G_{h0}G(\boldsymbol{\theta}_n, z)}{1+D(z)G_{h0}G(\boldsymbol{\theta}_n, z)} \qquad (4.26)$$

The process parameter vector now changes by an infinitesimal value $d\boldsymbol{\theta}$. For the control loop, it follows that

$$\left.\frac{\partial M(\boldsymbol{\theta}, z)}{\partial \boldsymbol{\theta}}\right|_{\boldsymbol{\theta}=\boldsymbol{\theta}_n} = \frac{D(z)}{[1+D(z)G_{h0}G(\boldsymbol{\theta}_n, z)]^2} \left.\frac{\partial G_{h0}G(\boldsymbol{\theta}, z)}{\partial \boldsymbol{\theta}}\right|_{\boldsymbol{\theta}=\boldsymbol{\theta}_n} \qquad (4.27)$$

From Eqns. (4.26)–(4.27), it follows that

$$\frac{dM(\theta_n,z)}{M(\theta_n,z)} = S(\theta_n, z)\frac{dG_{h0}G(\theta_n,z)}{G_{h0}G(\theta_n,z)} \quad (4.28a)$$

with the sensitivity function $S(\theta_n, z)$ of the feedback control given as

$$S^M_{G_{h0}G} = S(\theta_n, z) = \frac{1}{1+D(z)G_{h0}G(\theta_n,z)} \quad (4.28b)$$

This sensitivity function shows how relative changes of input/output behaviour of a closed loop depend on changes of the process transfer function. Small parameter-sensitivity of the closed-loop behaviour can be obtained by making $S(\theta_n, e^{j\omega T})$ small in the significant frequency range.

The Case for High-Gain Feedback

Control system design with high-gain feedback results in:
  (i) good steady-state tracking accuracy,
  (ii) good disturbance-signal rejection, and
  (iii) low sensitivity to process-parameter variations.

There are, however, factors limiting the gain:
  (i) High gain may result in instability problems.
  (ii) Input amplitudes limit the gain; excessively large magnitudes of control signals will drive the process to saturation region of its operation, and the control system design based on linear model of the plant will no longer give satisfactory performance.
  (iii) Measurement noise limits the gain; with high gain feedback, measurement noise appears unattenuated in the controlled output.

Therefore in design we are faced with trade-offs.

4.3 DIGITAL COMPENSATOR DESIGN USING FREQUENCY RESPONSE PLOTS

All the frequency response methods of continuous-time systems[4] are directly applicable for the analysis and design of digital control systems. For a system with closed-loop transfer function

$$\frac{Y(z)}{R(z)} = \frac{G_{h0}G(z)}{1+G_{h0}G(z)} \quad (4.29)$$

the absolute and relative stability conditions can be investigated by making the frequency response plots of $G_{h0}G(z)$. The frequency response plots of $G_{h0}G(z)$ can be obtained by setting

$$z = e^{j\omega T};\ T = \text{sampling interval} \quad (4.30)$$

and then letting the frequency ω vary from $-\omega_s/2$ to $\omega_s/2$; $\omega_s = 2\pi/T$. Computer assistance is normally required to make the frequency response plots.

---

4. Chapters 8–10 of reference [180].

Since the frequency appears in the form $z = e^{j\omega T}$, the discrete-time transfer functions are typically not rational functions and the simplicity of Bode's design technique is altogether lost in the z-plane. The simplicity can be regained by transforming the discrete-time transfer function in the z-plane to a different plane (called w) by the bilinear transformation (refer Eqn. (2.95))

$$z = \frac{1+wT/2}{1-wT/2} \qquad (4.31\text{a})$$

By solving Eqn. (4.31a) for w, we obtain the inverse relationship

$$w = \frac{2}{T}\frac{z-1}{z+1} \qquad (4.31\text{b})$$

Through the z-transformation and the w-transformation, the primary strip of the left half of the s-plane is first mapped into inside of the unit circle in the z-plane, and then mapped into the entire left half of the w-plane. The two mapping processes are depicted in Fig. 4.12. Notice that as s varies from 0 to $j\omega_s/2$ along the $j\omega$-axis in the s-plane, z varies from 1 to -1 along the unit circle in the z-plane, and w varies from 0 to ∞ along the imaginary axis in the w-plane. The bilinear transformation (4.31) does not have any physical significance in itself and therefore all w-plane quantities are fictitious quantities that correspond to the physical quantities of either the s-plane or the z-plane. The correspondence between the real frequency ω and the fictitious w-plane frequency, denoted as v, is obtained as follows:

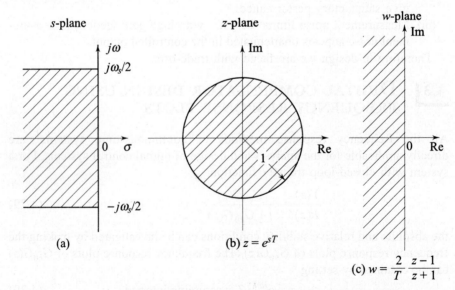

Fig. 4.12 Diagrams showing mappings from s-plane to z-plane and from z-plane to w-plane

From Eqn. (4.31b),

$$jv = \frac{2}{T}\frac{e^{j\omega T}-1}{e^{j\omega T}+1} = \frac{2}{T}\frac{e^{j\omega T/2}-e^{-j\omega T/2}}{e^{j\omega T/2}+e^{-j\omega T/2}} = \frac{2}{T}j\tan\frac{\omega T}{2}$$

or
$$v = \frac{2}{T}\tan\frac{\omega T}{2} \qquad (4.32)$$

Thus a nonlinear relationship or 'warping' exists between the two frequencies ω and v. As ω moves from 0 to $\omega_s/2$, v moves from 0 to ∞ (Fig. 4.13). Note that for relatively small $\frac{\omega T}{2}$ $\left(\frac{\omega T}{2} < 17°\text{ or about 0.3 rad}\right)$,

$$v \cong \frac{2}{T}\left(\frac{\omega T}{2}\right) \cong \omega$$

and the 'warping' effect on the frequency response is negligible.

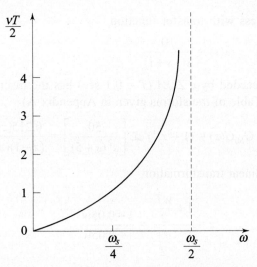

Fig. 4.13 Relationship between the fictitious frequency v and actual frequency ω

The distortion depicted in Fig. 4.13 may be taken into account in our design of digital compensation by frequency 'prewarping'. The idea of prewarping is simply to adjust the critical frequencies in our design; for example, if the closed-loop bandwidth is specified as ω_b, then the corresponding bandwidth on the w-plane is $v_b = \frac{2}{T}\tan\left(\frac{\omega_b T}{2}\right)$. Our design based on the frequency response plots of $G_{h0}G(jv)$ attempts to realize a closed-loop bandwidth equal to v_b.

The algebra in transforming $G_{h0}G(z)$ to r-plane by using the bilinear transformation (refer Eqn. (2.96))

$$z = \frac{1+r}{1-r} \quad (4.33a)$$

$$r = \frac{z-1}{z+1} \quad (4.33b)$$

is comparatively less tedious. The simplicity of Bode's design technique is regained by this transformation also. The transformed frequency ($\triangleq v'$) differs from the real frequency ω by $v' = \tan \omega T/2$.

Comparison of Eqns (4.31b) and (4.33b) reveals that the w-transformation and the r-transformation are identical except for the scale factor $2/T$. The presence of this scale factor in the transformation enables us to maintain the same error constants before and after the w-transformation (this means that the transfer function in the w-plane will approach that in the s-plane as T approaches zero, (see Example 4.2)). In our design examples, we will use the w-transformation given by Eqns (4.31).

Example 4.2

Consider a process with transfer function

$$G(s) = \frac{10}{s(\tfrac{1}{5}s + 1)} \quad (4.34)$$

which, when preceded by a ZOH ($T = 0.1$ sec) has the discrete-time transfer function (refer Table of transforms given in Appendix A)

$$G_{h0}G(z) = (1 - z^{-1}) \mathscr{Z}\left[\frac{50}{s^2(s+5)}\right] = \frac{0.215(z + 0.85)}{(z-1)(z-0.61)} \quad (4.35)$$

By use of the bilinear transformation

$$z = \frac{1 + \dfrac{wT}{2}}{1 - \dfrac{wT}{2}} = \frac{1 + 0.05w}{1 - 0.05w}$$

$G_{h0}G(z)$ can be transformed into $G_{h0}G(w)$ given below:

$$G_{h0}G(w) = \frac{10\left(1 - \dfrac{w}{20}\right)\left(1 + \dfrac{w}{246.67}\right)}{w\left(1 + \dfrac{w}{4.84}\right)} \quad (4.36)$$

Notice that the gain of $G_{h0}G(w)$ is precisely the same as that of $G(s)$—it is 10 in both the cases. This will always be true for a $G_{h0}G(w)$ computed using the bilinear transformation given by Eqns (4.31). The gain of 10 in Eqn. (4.36) is the K_v of the uncompensated system (4.35).

We also note that in Eqn. (4.36) the denominator looks very much similar to that of $G(s)$ and that the denominators will be the same as T approaches zero. This would also have been true for any zeros of $G_{h0}G(w)$ that corresponded to zeros of $G(s)$, but our example does not have any. Our example also shows the creation of a right-half plane zero of $G_{h0}G(w)$ at $2/T$ and the creation of a fast left-half plane zero when compared to the original $G(s)$. The transfer function $G_{h0}G(w)$ is thus a *nonminimum phase* function.

▲▲

To summarize, the w-transformation maps the inside of the unit circle in the z-plane into the left half of the w-plane. The magnitude and phase of $G_{h0}G(jv)$ are the magnitude and phase of $G_{h0}G(z)$ as z takes on values around the unit circle. Since $G_{h0}G(jv)$ is a rational function of v, we can apply all the standard straight line approximations to the log-magnitude and phase curves.

To obtain
$$G_{h0}G(w) = G_{h0}G(z)\bigg|_{z=\frac{1+wT/2}{1-wT/2}}$$

the following ready-to-use formula may be used.

$$G_{h0}G(z) = \frac{K \prod_{i=1}^{m}(z+a_i)}{(z-1)^l \prod_{j=1}^{n}(z+b_j)} \qquad (4.37a)$$

$$G_{h0}G(w) = \frac{K \prod_{i=1}^{m}(1+a_i)\left(1-\frac{w}{2/T}\right)^{l-m+n} \prod_{i=1}^{m}\left(1+\frac{w}{(2/T)[(1+a_i)/(1-a_i)]}\right)}{\prod_{j=1}^{n}(1+b_j)T^l w^l \prod_{j=1}^{n}\left(1+\frac{w}{(2/T)[(1+b_j)/(1-b_j)]}\right)}$$

$$(4.37b)$$

The design of analog control systems usually falls into one of the following categories: (1) lead compensation, (2) lag compensation, (3) lag-lead compensation. Other more complex schemes of course do exist, but knowing the effects of these three basic types of compensation gives a designer much insight into the design problem. With reference to the design of digital control systems by Bode plots, the basic forms of compensating network $D(w)$ have also been classified as lead, lag, and lag-lead. In the following paragraphs, we briefly review the fundamental frequency-domain features of these compensators.

Lead Compensation

A simple lead compensator model in the w-plane is described by the transfer function

$$D(w) = \frac{1+w\tau}{1+\alpha w\tau}; \ 0 < \alpha < 1, \ \tau > 0 \qquad (4.38)$$

The zero-frequency gain of the compensator is found by letting $w = 0$. Thus in Eqn. (4.38), we are assuming a unity zero-frequency gain for the compensator. Most of the designs require a compensator with a nonunity zero-frequency gain to improve steady-state response, disturbance rejection, etc. A nonunity zero-frequency gain is obtained by multiplying the right side of Eqn. (4.38) by a constant equal to the value of the desired zero-frequency gain. For the purpose of simplifying the design procedure, we normally add the required increase in gain to the plant transfer function, and design the unity zero-frequency gain compensator given by Eqn. (4.38) based on the new plant transfer function. Then the compensator is realized as the transfer function of (4.38) multiplied by the required gain factor.

The Bode plot of the unity zero-frequency gain lead compensator is shown in Fig. 4.14. The maximum phase lead ϕ_m of the compensator is given by the relation

$$\alpha = \frac{1 - \sin \phi_m}{1 + \sin \phi_m} \quad (4.39)$$

and it occurs at the frequency

$$v_m = \sqrt{\left(\frac{1}{\tau}\right)\left(\frac{1}{\alpha \tau}\right)} \quad (4.40)$$

The magnitude of $D(jv)$ at $v = v_m$ is $20 \log(1/\sqrt{\alpha})$.

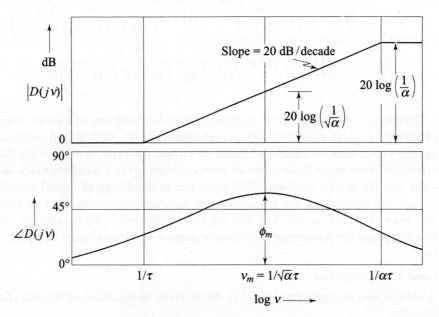

Fig. 4.14 Bode plot of lead compensator

The phase lead is introduced in the vicinity of the gain crossover frequency of the uncompensated system in order to increase the system's phase margin. Lead compensation increases the system gain at higher frequencies, thereby increasing the system bandwidth and hence the speed of response. However a system with large bandwidth may be subjected to high-frequency noise problems.

Lag Compensation

A simple lag compensator model in the w-plane is described by the transfer function

$$D(w) = \frac{1+w\tau}{1+\beta w\tau}; \quad \beta > 1, \tau > 0 \qquad (4.41)$$

The Bode plot of this unity zero-frequency gain compensator is shown in Fig. 4.15.

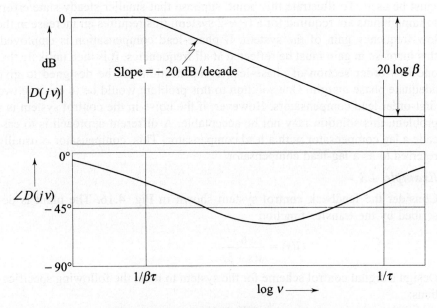

Fig. 4.15 Bode plot of lag compensator

Since the lag compensator reduces the system gain in the high frequency range without reducing the gain at low frequencies, the total system gain can be appreciably increased by a nonunity zero-frequency gain obtained by multiplying the right side of Eqn. (4.41) by a constant. This is equivalent to increasing the gain for the entire frequency range and then attenuating the magnitude curve in the high frequency region. This results in an appreciable increase in gain in the low frequency range of the lag-compensated system thereby improving steady-state accuracy.

In the design method using Bode plots, the attenuation property of lag compensator is utilized; the phase lag characteristic is of no consequence. The attenuation provided by the lag compensator in the high frequency range shifts the gain crossover frequency to a lower value and gives the system sufficient phase margin. So that a significant phase lag will not be contributed near the new gain crossover, the upper corner frequency $1/\tau$ of $D(w)$ is placed far below the new gain crossover.

With the reduction in system gain at high frequencies, the system bandwidth gets reduced and thus the system has a slower speed of response. This may be an advantage if high frequency noise is a problem.

Lag-lead Compensation

Equations (4.38) and (4.41) describe simple first-order compensators. In many system design problems, however, the system specifications cannot be satisfied by a first-order compensator. In these cases, higher-order compensators must be used. To illustrate this point, suppose that smaller steady-state errors to ramp inputs are required for a type-2 system; this requires an increase in the low-frequency gain of the system. If phase-lead compensation is employed, this increase in gain must be reflected at all frequencies. It is then unlikely that one first-order section of phase-lead compensation can be designed to give adequate phase margin. One solution to this problem would be to cascade two first-order lead compensators. However, if the noise in the control system is a problem, this solution may not be acceptable. A different approach is to cascade a lag compensator with a lead compensator. This compensator is usually referred to as a lag-lead compensator.

Example 4.3

Consider the feedback control system shown in Fig. 4.16. The plant is described by the transfer function

$$G(s) = \frac{K}{s(s+5)}$$

Design a digital control scheme for the system to meet the following specifications:

(i) the velocity error constant $K_v \geq 10$;
(ii) peak overshoot M_p to step input $\leq 25\%$, and
(iii) settling time t_s (2% tolerance band) ≤ 2.5 sec.

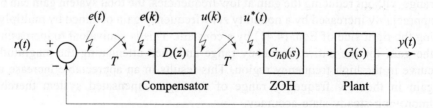

Fig. 4.16 A feedback control system with digital compensation

Solution
The design parameters are the sampling interval T, the system gain K, and the parameters of the unity zero-frequency gain compensator $D(z)$.

Let us translate the transient accuracy requirements to frequency response measures. $\zeta = 0.4$ corresponds to a peak overshoot of about 25% (Eqn. (4.13)), and a phase margin of about 40° (Eqn. (4.23)). The requirement of $t_s \cong 2.5$ sec corresponds to $\omega_n = 4$ rad/sec (Eqn. (4.14)) and closed-loop bandwidth $\omega_b \cong 5.5$ rad/sec (Eqn. (4.22)). Taking the sampling frequency about 10 times the bandwidth, we choose the sampling interval

$$T = \frac{2\pi}{10\omega_b} \cong 0.1 \text{ sec}$$

Our design approach is to first fix the system gain K to a value that results in the desired steady-state accuracy. A unity zero-frequency gain compensator is then introduced that satisfies the transient accuracy requirements without affecting the steady-state accuracy.

Since sampling does not affect the error constant of the system, we can relate K with K_v as follows for the system of Fig. 4.16 with $D(z) = 1$ (i.e., for uncompensated system):

$$K_v = \lim_{s \to 0} sG(s) = \frac{K}{5}$$

Thus $K = 50$ meets the requirements on steady-state accuracy.

For $T = 0.1$ and $K = 50$, we have

$$G_{h0}G(z) = \mathscr{Z}\left[\frac{1-e^{-Ts}}{s}\left(\frac{50}{s(s+5)}\right)\right] = \frac{0.215(z+0.85)}{(z-1)(z-0.61)} \tag{4.42}$$

$$G_{h0}G(w) = G_{h0}G(z)\bigg|_{z=\frac{1+\frac{wT}{2}}{1-\frac{wT}{2}}} = \frac{10\left(1-\frac{w}{20}\right)\left(1+\frac{w}{246.67}\right)}{w\left(1+\frac{w}{4.84}\right)} \tag{4.43}$$

$$G_{h0}G(jv) = G_{h0}G(w)\bigg|_{w=jv} = \frac{10\left(1-\frac{jv}{20}\right)\left(1+\frac{jv}{246.67}\right)}{jv\left(1+\frac{jv}{4.84}\right)} \tag{4.44}$$

The Bode plot of $G_{h0}G(jv)$ (i.e. the uncompensated system) is shown in Fig. 4.17. We find from this plot that the uncompensated system has gain crossover frequency $v_{c1} = 6.6$ rad/sec and phase margin $\Phi M_1 \cong 20°$. The magnitude versus phase angle curve of the uncompensated system is drawn in Fig. 4.18. The bandwidth[5] of the system is read as

---

5. The -3dB closed-loop gain contour of the Nichols chart has been used to determine bandwidth. The contour has been constructed using the following table obtained from the Nichols chart.

Degrees	−90	−100	−120	−140	−160	−180	−200	−220
dB	0	−1.5	−4.18	−6.13	−7.28	−7.66	−7.28	−6.13

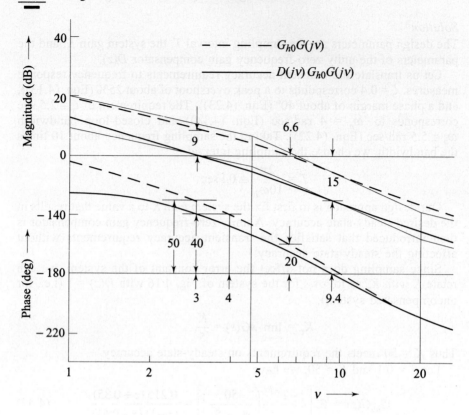

Fig. 4.17 Compensator design (Example 4.3)

$$v_{b1} = 11$$

In terms of the real frequency, the bandwidth (Eqn. (4.32))

$$\omega_{b1} = \frac{2}{T}\tan^{-1}\left(\frac{v_{b1}T}{2}\right) = 10 \text{ rad/sec}$$

It is desired to raise the phase margin to 40° without altering K_v. Also the bandwidth should not increase.

Obviously we should first try a lag compensator.

From the Bode plot of uncompensated system we observe that the phase margin of 40° is obtained if the gain crossover frequency is reduced to 4 rad/sec. The high frequency gain $-20 \log \beta$ of the lag compensator (Fig. 4.15) is utilized to reduce the gain crossover frequency. The upper corner frequency $1/\tau$ of the compensator is placed one octave to one decade below the new gain crossover, so that the phase lag contribution of the compensator in the vicinity of the new gain crossover is made sufficiently small. To nullify the small phase lag contribution which will still be present, the gain crossover frequency is reduced to a value v_{c2} where the phase angle of the uncompensated system is

$$\phi = -180° + \Phi M_s + \varepsilon;$$

ΦM_s is the specified phase margin and ε is allowed a value 5°–15°.

The uncompensated system (Fig. 4.17) has a phase angle

$$\phi = -180° + \Phi M_s + \varepsilon = -180° + 40° + 10° = -130°$$

at $v_{c2} = 3$ rad/sec. Placing the upper corner frequency of the compensator two octaves below v_{c2}, we have

$$\frac{1}{\tau} = \frac{v_{c2}}{(2)^2} = \frac{3}{4}$$

To bring the magnitude curve down to 0 dB at v_{c2}, the lag compensator must provide an attenuation of 9 dB (Fig. 4.17). Therefore

$$20 \log \beta = 9 \text{ or } \beta = 2.82$$

The lower corner frequency of the compensator is then fixed at

$$\frac{1}{\beta\tau} = 0.266$$

The transfer function of the lag compensator is then

$$D(w) = \frac{1+\tau w}{1+\beta\tau w} = \frac{1+1.33w}{1+3.76w}$$

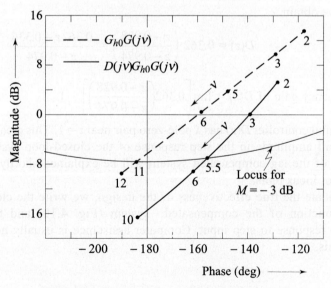

Fig. 4.18 Compensator design (Example 4.3)

Phase lag introduced by the compensator at $v_{c2} = \tan^{-1}(1.33\ v_{c2}) - \tan^{-1}(3.76\ v_{c2}) = 75.93° - 84.93° = -9°$. Therefore the safety margin of $\varepsilon = 10°$ is justified.

The open-loop transfer function of the compensated system becomes

$$D(w)G_{h0}G(w) = \frac{10\left(1 - \dfrac{w}{20}\right)\left(1 + \dfrac{w}{246.67}\right)\left(1 + \dfrac{w}{0.75}\right)}{w\left(1 + \dfrac{w}{4.84}\right)\left(1 + \dfrac{w}{0.266}\right)}$$

The Bode plot of $D(w)G_{h0}G(w)$ is shown in Fig. 4.17, from where the phase margin of the compensated system is found to be 40° and the gain margin is 15 dB. The magnitude versus phase angle curve of the compensated system is shown on Nichols chart in Fig. 4.18. The bandwidth of the compensated system is

$$v_{b2} = 5.5 \left(\omega_{b2} = \frac{2}{T} \tan^{-1}\left(\frac{v_{b2}T}{2}\right) = 5.36\ \text{rad/sec} \right)$$

Therefore the addition of the compensator has reduced the bandwidth from 10 rad/sec to 5.36 rad/sec. However, the reduced value lies in the acceptable range.

Substituting

$$w = \frac{2}{T}\frac{z-1}{z+1}$$

in $D(w)$, we obtain

$$D(z) = 0.362\left(\frac{z - 0.928}{z - 0.974}\right) = \frac{0.362z - 0.336}{z - 0.974}$$

Zero-frequency gain of $D(z) = \lim_{z \to 1}\left[0.362\left(\dfrac{z - 0.928}{z - 0.974}\right)\right] = 1$

The digital controller $D(z)$ has a pole-zero pair near $z = 1$. This creates a long tail of small amplitude in the step response of the closed-loop system. This behaviour of the lag-compensated system will be explained shortly with the help of root locus plots.

To evaluate the true effectiveness of the design, we write the closed-loop transfer function of the compensated system (Fig. 4.19) and therefrom obtain the response to step input. Computer assistance is usually needed for this analysis.

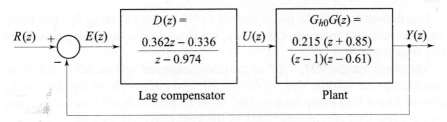

Fig. 4.19 Compensator design (Example 4.3)

Comment

We have obtained a digital control algorithm which meets the following objectives: $K_v \cong 10$, $M_p \cong 25\%$, $t_s \cong 2.5$ sec. We may attempt to improve upon this design to obtain $K_v > 10$, $M_p < 25\%$ and $t_s < 2.5$ sec. However, the scope of such an exercise is limited because the improvement in steady-state accuracy will be at the cost of stability margins and *vice versa*. Also the conflicting requirements of limiting the magnitudes of control signals to avoid saturation problems, limiting the bandwidth to avoid high-frequency noise problems, etc., have to be taken into consideration.

Example 4.4

Reconsider the feedback control system of Example 4.3 (Fig. 4.16). We now set the following goal for our design.

(i) $K_v \geq 10$
(ii) Phase margin $\cong 40°$
(iii) Bandwidth $\cong 12$ rad/sec.

Sampling interval $T = 0.1$ sec corresponds to a sampling frequency which is about five times the closed-loop bandwidth. A smaller value of T is more appropriate for the present design problem which requires higher speed of response; we will however take $T = 0.1$ sec to compare our results with those of Example 4.3.

Following the initial design steps of Example 4.3, we find that $K = 50$ meets the requirement on steady-state accuracy. For $K = 50$ and $T = 0.1$ sec, we have (refer Eqn. (4.44))

$$G_{h0}G(jv) = \frac{10\left(1 - \dfrac{jv}{20}\right)\left(1 + \dfrac{jv}{246.67}\right)}{jv\left(1 + \dfrac{jv}{4.84}\right)}$$

The uncompensated system has a gain crossover frequency $v_{c1} = 6.6$ rad/sec, phase margin $\Phi M_1 \cong 20°$ and bandwidth $v_{b1} = 11$ ($\omega_{b1} = 10$ rad/sec). This follows from Figs 4.20 and 4.21.

It is desired to raise the phase margin to 40° without altering K_v. The bandwidth should also increase. Obviously, we should try a lead compensator for this situation.

The phase margin $\Phi M_1 = 20°$ of the uncompensated system falls short of the specified phase margin $\Phi M_s = 40°$ by 20°. Additional phase margin can be provided by a lead compensator (Fig. 4.14) so placed that its corner frequencies $1/\tau$ and $1/\alpha\tau$ are on either side of the gain crossover frequency $v_{c1} = 6.6$ rad/sec. The compensator so placed will increase the system gain in the vicinity of v_{c1}; this will cause the gain crossover to shift to the right to some unknown value v_{c2}. The phase lead provided by the compensator at v_{c2} adds to the phase margin of the system.

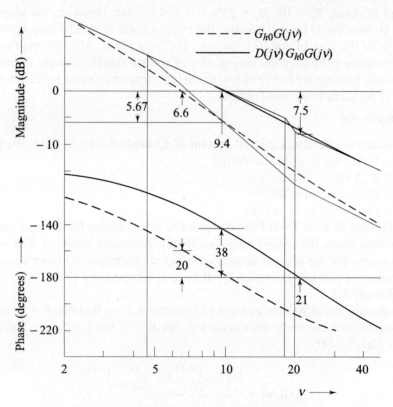

Fig. 4.20 Compensator design (Example 4.4)

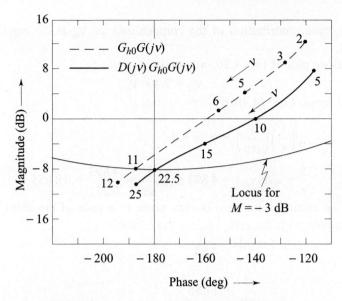

Fig. 4.21 Compensator design (Example 4.4)

Phase margin of the uncompensated system at v_{c1} is ΦM_1. At v_{c2}, which is expected to be close to v_{c1}, let us assume the phase margin of the uncompensated system to be $(\Phi M_1 - \varepsilon)$ where ε is allowed a value $5° - 15°$. The phase lead required at v_{c2} to bring the phase margin to the specified value ΦM_s is given by

$$\phi_l = \Phi M_s - (\Phi M_1 - \varepsilon) = \Phi M_s - \Phi M_1 + \varepsilon$$

In our design we will force the frequency v_m of the compensator to coincide with v_{c2} so that maximum phase lead ϕ_m of the compensator is added to the phase margin of the system. Thus we set

$$v_{c2} = v_m$$

Therefore $\phi_m = \phi_l$

The α-parameter of the compensator can then be computed from (refer Eqn. (4.39))

$$\alpha = \frac{1 - \sin\phi_m}{1 + \sin\phi_m}$$

Since at v_m, the compensator provides a dB-gain of $20\log(1/\sqrt{\alpha})$, the new crossover frequency $v_{c2} = v_m$ can be determined as that frequency at which the uncompensated system has a dB-gain of $-20\log(1/\sqrt{\alpha})$.

For the design problem under consideration,

$$\phi_l = 40° - 20° + 15° = 35°$$

Therefore $$\alpha = \frac{1 - \sin 35°}{1 + \sin 35°} = 0.271$$

The magnitude contribution of the compensator at v_m is $20\log(1/\sqrt{0.271})$ = 5.67dB.

From Bode plot of Fig. 4.20, we obtain
$$v_{c2} = 9.4 = v_m$$

Therefore (refer Eqn. (4.40))
$$\sqrt{\left(\frac{1}{\tau}\right)\left(\frac{1}{\alpha\tau}\right)} = v_m = 9.4$$

or $\quad \dfrac{1}{\tau} = \sqrt{\alpha}(v_m) = 4.893$ and $\dfrac{1}{\alpha\tau} = \dfrac{4.893}{0.271} = 18.055$

Since the compensator zero is very close to a pole of the plant, we may cancel the pole with the zero, i.e., we may choose
$$\frac{1}{\tau} = 4.84$$
$$\frac{1}{\alpha\tau} = 17.86$$

The transfer function of the lead compensator becomes
$$D(w) = \frac{1+\tau w}{1+\alpha\tau w} = \frac{1+0.21w}{1+0.056w}$$

Substituting
$$w = \frac{2}{T}\frac{z-1}{z+1}$$

in $D(w)$, we obtain
$$D(z) = \frac{2.45(z-0.616)}{z-0.057}$$

The open-loop transfer function of the compensated system is
$$D(w)G_{h0}G(w) = \frac{10\left(1-\dfrac{w}{20}\right)\left(1+\dfrac{w}{246.67}\right)}{w\left(1+\dfrac{w}{17.86}\right)}$$

The Bode plot of $D(w)G_{h0}G(w)$ is shown in Fig. 4.20, from where the phase margin of the compensated system is found to be 38°, and gain margin is 7.5 dB. The magnitude versus phase angle curve of the compensated system is shown in Fig. 4.21. The bandwidth of the compensated system is

$$v_{b2} = 22.5 \; \left(\omega_{b2} = \frac{2}{T}\tan^{-1}\left(\frac{v_{b2}T}{2}\right) = 16.9 \text{ rad/sec}\right)$$

Thus the addition of the lead compensator has increased the system bandwidth from 10 to 16.9 rad/sec. It may lead to noise problems if the control system is burdened with high frequency noise.

A solution to noise problems involves the use of a lag compensator cascaded with lead compensator. The lag compensation is employed to realize a part of the required phase margin, thus reducing the amount of lead compensation required.

4.4 DIGITAL COMPENSATOR DESIGN USING ROOT LOCUS PLOTS

Design of compensation networks using the root locus plots is a well established procedure in analog control systems. This is essentially a trial-and-error method whereby varying the controller parameters, the roots of the characteristic equation are relocated to favourable locations. In the present section, we shall consider the application of root locus method to the design of digital control systems.

The Root Locus on the z-plane

The characteristic equation of a discrete-time system can always be written in the form

$$1 + F(z) = 0 \qquad (4.45)$$

where $F(z)$ is a rational function of z.

From Eqn. (4.45), it is seen that the roots of the characteristic equation (i.e., the closed-loop poles of the discrete-time system) occur only for those values of z where

$$F(z) = -1 \qquad (4.46)$$

Since z is a complex variable, Eqn. (4.46) is converted into two conditions given below.

  (i) **Magnitude condition:** $|F(z)| = 1$ \qquad (4.47a)
  (ii) **Angle condition:** $\angle F(z) = \pm 180° (2q + 1)$; $q = 0, 1, 2, ...$ \qquad (4.47b)

In essence, the construction of the z-plane root loci is to find the points that satisfy these conditions. If we write $F(z)$ in the standard pole-zero form:

$$F(z) = \frac{K \prod_i (z - z_i)}{\prod_j (z - p_j)}; \quad K \geq 0 \qquad (4.48a)$$

then the two conditions given in Eqns (4.47) become

$$|F(z)| = \frac{K \prod_i |z - z_i|}{\prod_j |z - p_j|} = 1 \qquad (4.48b)$$

and

$$\angle F(z) = \sum_i \angle z - z_i - \sum_j \angle z - p_j$$

$$= \pm 180° (2q + 1); \quad q = 0, 1, 2, ... \qquad (4.48c)$$

Consequently, given the pole-zero configuration of $F(z)$, the construction of the root loci on the z-plane involves the following steps:

(i) A search for the points on the z-plane that satisfy the angle condition given by Eqn. (4.48c).

(ii) The value of K at a given point on a root locus is determined from the magnitude condition given by Eqn. (4.48b).

The root locus method developed for continuous-time systems can be extended to discrete-time systems without modifications, except that the stability boundary is changed from the $j\omega$ axis in the s-plane to the unit circle in the z-plane. The reason the root locus method can be extended to discrete-time systems is that the characteristic Eqn. (4.45) for the discrete-time system is of exactly the same form as the equation for root locus analysis in the s-plane. However, the pole locations for closed-loop systems in the z-plane must be interpreted differently from those in the s-plane.

We assume that the reader is already familiar with the s-plane root locus technique. We shall concentrate on the interpretation of the root loci in the z-plane with reference to the system performance, rather than the construction of root loci on the z-plane. Rules of construction of root loci are summarized, in Table 4.2 for ready reference.[6]

Table 4.2 Rules for construction of Root Locus Plot of $1 + F(z) = 0$;

$$F(z) = \frac{K \prod_{i=1}^{m}(z-z_i)}{\prod_{j=1}^{n}(z-p_j)}; K \geq 0, n \geq m$$

z_i: m open-loop zeros; p_j: n open-loop poles

Rules
(i) The root locus plot consists of n root loci as K varies from 0 to ∞. The loci are symmetric with respect to the real axis.
(ii) As K increases from zero to infinity, each root locus originates from an open-loop pole with $K = 0$ and terminates either on an open-loop zero or on infinity with $K = \infty$. The number of loci terminating on infinity equals the number of open-loop poles minus zeros.
(iii) The $(n - m)$ root loci which tend to infinity, do so along straight-line asymptotes radiating out from a single point $z = -\sigma_A$ on the real axis (called the centroid), where $$-\sigma_A = \frac{\Sigma \text{ (real parts of open-loop poles)} - \Sigma \text{ (real parts of open-loop zeros)}}{n-m}$$ These $(n - m)$ asymptotes have angles $$\phi_A = \frac{(2q+1)180°}{n-m}; q = 0, 1, 2, ..., (n-m-1)$$

contd.

---
6. Chapter 7 of reference [180].

(iv) A point on the real axis lies on the locus if the number of open-loop poles plus zeros on the real axis to the right of this point is odd. By use of this fact, the real axis can be divided into segments *on-locus* and *not-on-locus*; the dividing points being the real open-loop poles and zeros.

(v) The intersections (if any) of root loci with the imaginary axis can be determined by use of the Routh criterion.

(vi) The angle of departure, ϕ_p, of a root locus from a complex open-loop pole is given by
$$\phi_p = 180° + \phi$$
where ϕ is the net angle contribution at this pole of all other open-loop poles ans zeros.

(vii) The angle of arrival, ϕ_z, of a locus at a complex zero is given by
$$\phi_z = 180° - \phi$$
where ϕ is the net angle contribution at this zero of all other open-loop poles and zeros.

(viii) Points at which multiple roots of the characteristic equation occur (breakaway points of root loci) are the solutions of
$$\frac{dK}{dz} = 0$$
where
$$K = -\frac{\prod_{j=1}^{n}(z-p_j)}{\prod_{i=1}^{m}(z-z_i)}$$

(ix) The gain K at any point z_0 on a root locus is given by
$$K = \frac{\prod_{j=1}^{n}|z_0 - p_j|}{\prod_{i=1}^{m}|z_0 - z_i|}$$

$$= \frac{\text{[Product of phasor lengths (read to scale) from } z_0 \text{ to poles of } F(z)\text{]}}{\text{[Product of phasor lengths (read to scale) from } z_0 \text{ to zeros of } F(z)\text{]}}$$

Example 4.5

Consider a process with the transfer function

$$G(s) = \frac{K}{s(s+2)} \tag{4.49a}$$

which when preceded by a zero-order hold ($T = 0.2$ sec) has the discrete-time transfer function (refer Table of transforms given in Appendix A)

$$G_{h0}G(z) = (1 - z^{-1}) \mathscr{Z}\left[\frac{K}{s^2(s+2)}\right]$$

270 Digital Control and State Variable Methods

$$= \frac{K'(z-b)}{(z-a_1)(z-a_2)} \qquad (4.49b)$$

where $K' = 0.01758K$, $b = -0.876$, $a_1 = 0.67$, $a_2 = 1$.
The root locus plot of

$$1 + G_{h0}G(z) = 0 \qquad (4.50)$$

can be constructed using the rules given in Table 4.2. $G_{h0}G(z)$ has two poles at $z = a_1$ and $z = a_2$, and a zero at $z = b$. From rule (iv), the parts of the real axis between a_1 and a_2, and between $-\infty$ and b constitute sections of the loci. From rule (ii), the loci start from $z = a_1$ and $z = a_2$; one of the loci terminates at $z = b$, and the other locus terminates at $-\infty$. From rule (viii), the breakaway points (there are two) may be obtained by solving for the roots of

$$\frac{dK'}{dz} = 0$$

where
$$K' = -\frac{(z-a_1)(z-a_2)}{(z-b)}$$

However, we can show that for this simple two-pole and one zero configuration, the complex conjugate section of the root locus plot is a circle. The breakaway points are easily obtained from this result, which is proved as follows:

Let
$$z = x + jy$$

Equation (4.49b) becomes

$$G_{h0}G(z) = \frac{K'(x+jy-b)}{(x+jy-a_1)(x+jy-a_2)}$$

$$= \frac{K'(x-b+jy)}{(x-a_1)(x-a_2)-y^2 + jy(2x-a_1-a_2)}$$

On the root loci, z must satisfy Eqn. (4.50).
Therefore

$$\angle G_{h0}G(z) = \tan^{-1}\frac{y}{x-b} - \tan^{-1}\frac{y(2x-a_1-a_2)}{(x-a_1)(x-a_2)-y^2} = (2q+1)\,180°$$

Taking the tangent of both sides of this equation yields

$$\frac{\dfrac{y}{x-b} - \dfrac{y(2x-a_1-a_2)}{(x-a_1)(x-a_2)-y^2}}{1 + \dfrac{y}{x-b}\left[\dfrac{y(2x-a_1-a_2)}{(x-a_1)(x-a_2)-y^2}\right]} = 0$$

or
$$\frac{1}{x-b} - \frac{2x-a_1-a_2}{(x-a_1)(x-a_2)-y^2} = 0$$

Simplifying, we get
$$(x - b)^2 + y^2 = (b - a_1)(b - a_2) \qquad (4.51)$$
which is the equation of a circle with the centre at the open-loop zero $z = b$, and the radius equal to $[(b - a_1)(b - a_2)]^{1/2}$.

The root locus plot for the system given by Eqn. (4.49b) is constructed in Fig. 4.22. The limiting value of K for stability may be found by graphical construction, by the Jury stability test or the Routh stability criterion on the characteristic equation obtained by the bilinear transformation of the given z-domain characteristic equation. We illustrate the use of graphical construction.

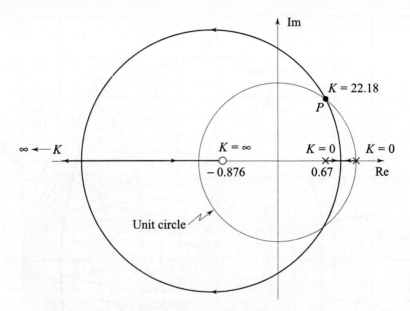

Fig. 4.22 Root locus plot for the system of Example 4.5

By rule (ix) of Table 4.2, the value of K' at point P where the root locus crosses the unit circle is given by

$$K' = \frac{\text{(Phasor length from } P \text{ to pole at } z = 1) \times}{\text{(Phasor length from } P \text{ to zero at } z = -0.876)}$$

$$= \frac{0.85 \times 0.78}{1.7} = 0.39 = 0.01758 K$$

Therefore $\qquad K = \dfrac{0.39}{0.01758} = 22.18$

The relative stability of the system can be investigated by superimposing the constant-ζ loci on the system root locus plot. This is shown in Fig. 4.23.

Inspection of this figure shows that the root locus intersects the $\zeta = 0.3$ locus at point Q. The value of K' at point Q is determined to be 0.1; the gain

$$K = K'/0.01758 = 5.7$$

The value of ω_n for $K' = 0.1$ may be obtained by superimposing constant-ω_n loci on the root locus plot and locating the constant-ω_n locus which passes through the point Q. From Fig. 4.23, we observe that none of the constant-ω_n loci on the standard chart passes through the point Q; we have to make a guess for the ω_n value. We can instead construct a constant-ω_d locus passing through the point Q and from there obtain ω_n more accurately.

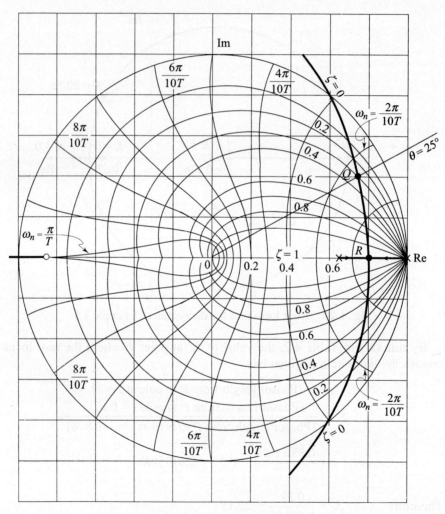

Fig 4.23 Relative stability analysis

are mapped to
$$s_{1,2} = -\zeta\omega_n \pm j\omega_n\sqrt{1-\zeta^2} = -\zeta\omega_n \pm j\omega_d$$
$$z_{1,2} = e^{-\zeta\omega_n T}e^{\pm j\omega_d T} = re^{\pm j\theta}$$
in the z-plane.

A constant-ω_d locus is thus a radial line passing through the origin at an angle $\theta = \omega_d T$ with the positive real axis of the z-plane measured positive in the counter clockwise direction.

The radial line passing through the point Q makes an angle $\theta = 25°$ with the real axis (Fig. 4.23). This is a constant-ω_d locus with ω_d given by

$$\omega_d T = \frac{25 \times \pi}{180} \text{ rad}$$

Therefore $\quad \omega_n T\sqrt{1-\zeta^2} = \dfrac{25\pi}{180}$

This gives $\omega_n = 2.29$ rad/sec.

The value of K' at the breakaway point R located at $z = 0.824$ is determined to be 0.01594. Therefore the gain $K = 0.01594/0.01758 = 0.9067$ results in critical damping ($\zeta = 1$) with the two closed-loop poles at $z = 0.824$.

A pole in the s-plane at $s = -a$ has a time constant of $\tau = 1/a$ and an equivalent z-plane location of $e^{-aT} = e^{-T/\tau}$. Thus for the critically damped case,

$$e^{-0.2/\tau} = 0.824$$

or $\quad \tau = 1.033 =$ time constant of the closed-loop poles.

In the frequency-response design procedure described in the previous section, we attempted to reshape the open-loop frequency response to achieve certain stability margins, steady-state response characteristics and so on. A different design technique is presented in this section—the root-locus procedure. In this procedure, we add poles and zeros through a digital controller so as to shift the roots of the characteristic equation to more appropriate locations in the z-plane. Therefore it is useful to investigate the effects of various pole-zero configurations of the digital controller on the root locus plots.

Lead Compensation

A simple lead compensator model in the w-plane is described by the transfer function (refer Eqn. (4.38))

$$D(w) = \frac{1+w\tau}{1+\alpha w\tau}; \; \alpha < 1, \tau > 0$$

The bilinear transformation

$$w = \frac{2}{T}\frac{z-1}{z+1}$$

transforms $D(w)$ into the following $D(z)$.

$$D(z) = \frac{1+2\tau/T}{1+2\alpha\tau/T}\left[\frac{z+(1-2\tau/T)/(1+2\tau/T)}{z+(1-2\alpha\tau/T)/(1+2\alpha\tau/T)}\right]$$

Since τ and α are both positive numbers and since $\alpha < 1$, the pole and zero of $D(z)$ always lie on the real axis inside the unit circle in the z-plane; the zero is always to the right of the pole. A typical pole-zero configuration of a lead compensator

$$D(z) = K_{c1}\frac{z-\alpha_1}{z-\alpha_2} \qquad (4.52)$$

is shown in Fig. 4.24a.

For the purpose of simplifying the design procedure, we normally associate the gain K_{c1} with the plant transfer function and design the lead compensator

$$D(z) = \frac{z-\alpha_1}{z-\alpha_2} \qquad (4.53a)$$

based on the new plant transfer function. It may be noted that $D(z)$ given by Eqn. (4.53a) is not a unity-gain model; the dc gain of $D(z)$ is given by

$$\lim_{z\to 1}\left(\frac{z-\alpha_1}{z-\alpha_2}\right) = \left(\frac{1-\alpha_1}{1-\alpha_2}\right) \qquad (4.53b)$$

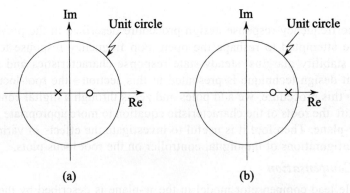

Fig. 4.24 Pole-zero configurations of compensators

To study the effect of a lead compensator on the root loci, we consider a unity feedback sampled-data system with open-loop transfer function

$$G_{h0}G(z) = \frac{K(z+0.368)}{(z-0.368)(z-0.135)}; \; T = 1 \text{ sec} \qquad (4.54)$$

The root locus plot of the uncompensated system is shown in Fig. 4.25a. The plot intersects the $\zeta = 0.5$ locus[7] at point P. The value of gain K at this point is determined to be 0.3823.

Constant-ω_d locus passing through point P is a radial line at an angle of 82° with the real axis (Fig. 4.25a). Therefore

$$\omega_d T = \omega_n T \sqrt{1-\zeta^2} = \frac{82\pi}{180}$$

This gives $\qquad \omega_n = 1.65$ rad/sec

Since $G_{h0}G(z)$ given by Eqn. (4.54) is a Type-0 system, we will consider position error constant K_p to study steady-state accuracy. For $K = 0.3823$,

$$K_p = \lim_{z \to 1} G_{h0}G(z) = \frac{0.3823(1+0.368)}{(1-0.368)(1-0.135)} = 0.957$$

We now cancel the pole of $G_{h0}G(z)$ at $z = 0.135$ by the zero of the lead compensator and add a pole at $z = -0.135$, i.e., we select

$$D(z) = \frac{z - 0.135}{z + 0.135}$$

---

7. For a given ζ, the constant-ζ curve may be constructed using Eqn. (4.15b). The following table gives the real and imaginary coordinates of points on some constant-ζ curves.

$\zeta = 0.3$	Re	0.932	0.735	0.360	0	−0.259	−0.380	−0.373
	Im	0.164	0.424	0.623	0.610	0.448	0.220	0
$\zeta = 0.4$	Re	0.913	0.689	0.317	0	−0.201	−0.276	−0.254
	Im	0.161	0.398	0.549	0.504	0.347	0.160	0
$\zeta = 0.5$	Re	0.891	0.640	0.273	0	−0.149	−0.191	−0.163
	Im	0.157	0.370	0.473	0.404	0.259	0.110	0
$\zeta = 0.6$	Re	0.864	0.585	0.228	0	−0.104	−0.122	−0.095
	Im	0.152	0.338	0.395	0.308	0.180	0.070	0
$\zeta = 0.7$	Re	0.830	0.519	0.179	0	−0.064	−0.067	−0.046
	Im	0.146	0.299	0.310	0.215	0.111	0.039	0
$\zeta = 0.8$	Re	0.780	0.431	0.124	0	−0.031	−0.026	−0.015
	Im	0.138	0.249	0.215	0.123	0.053	0.015	0

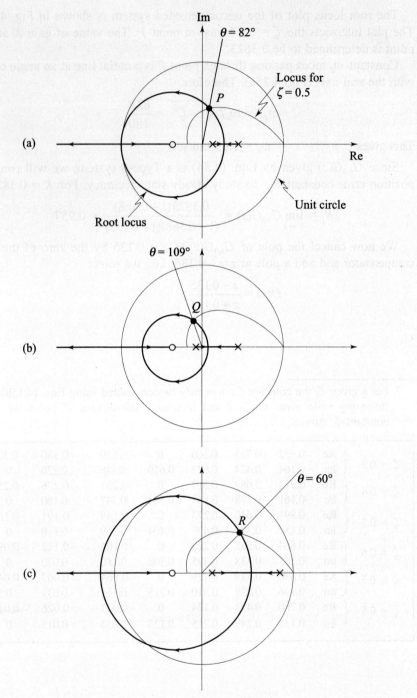

Fig. 4.25 Root locus plot for
(a) uncompensated;
(b) lead compensated; and
(c) lag compensated system

Figure 4.25b shows the root locus plot of lead compensated system. The modified locus has moved to the left, toward the more stable part of the plane. The intersection of the locus with the $\zeta = 0.5$ line is at point Q. The value of ω_n at this point is determined to be 2.2 rad/sec. The lead compensator has thus increased ω_n and hence the speed of response of the system. The gain K at point Q is determined to be 0.433. The position error constant of the lead compensated system is given by

$$K_p = \lim_{z \to 1} D(z) G_{h0} G(z) = \lim_{z \to 1} \frac{0.433(z + 0.368)}{(z - 0.368)(z + 0.135)} = 0.826$$

The lead compensator has thus given satisfactory dynamic response, but the position error constant is too low. We will shortly see how K_p can be increased by lag compensation.

The selection of the exact values of pole and zero of the lead compensator is done by experience and by trial-and-error. In general, the zero is placed in the neighbourhood of the desired dominant closed-loop poles, and the pole is located at a reasonable distance to the left of the zero location.

Lag Compensation

A simple lag compensator model in the w-plane is described by the transfer function (refer Eqn. (4.41))

$$D(w) = \frac{1 + w\tau}{1 + \beta w\tau}; \beta > 1, \tau > 0$$

The bilinear transformation

$$w = \frac{2}{T} \frac{z - 1}{z + 1}$$

transforms $D(w)$ into the following $D(z)$.

$$D(z) = \frac{1 + 2\tau/T}{1 + 2\beta\tau/T} \left[\frac{z + (1 - 2\tau/T)/(1 + 2\tau/T)}{z + (1 - 2\beta\tau/T)/(1 + 2\beta\tau/T)} \right]$$

Since τ and β are both positive numbers and since $\beta > 1$, the pole and zero of $D(z)$ always lie on the real axis inside the unit circle; the pole is always to the right of the zero. A typical pole-zero configuration of the lag compensator

$$D(z) = K_{c2} \frac{z - \beta_1}{z - \beta_2} \qquad (4.55)$$

is shown in Fig. 4.24b. Note that both the pole and the zero have been shown close to $z = 1$. This, as we shall see, gives better stability properties.

Again, we will associate the gain K_{c2} with the plant transfer function and design the lag compensator

$$D(z) = \frac{z - \beta_1}{z - \beta_2} \qquad (4.56)$$

based on the new plant transfer function. The dc gain of the lag compensator given by (4.56) is equal to

$$\lim_{z \to 1} \frac{z - \beta_1}{z - \beta_2} = \frac{1 - \beta_1}{1 - \beta_2} \qquad (4.57)$$

To study the effect of lag compensator on the root loci, we reconsider the system described by Eqn. (4.54):

$$G_{h0}G(z) = \frac{K(z + 0.368)}{(z - 0.368)(z - 0.135)}; \; T = 1 \text{ sec}$$

The root locus plot of the uncompensated system is shown in Fig. 4.25a. At point P, $\zeta = 0.5$, $\omega_n = 1.65$ and $K = 0.3823$ ($K_p = 0.957$).

We now cancel the pole of $G_{h0}G(z)$ at $z = 0.368$ by the zero of the lag compensator and add a pole at $z = 0.9$, i.e., we select

$$D(z) = \frac{z - 0.368}{z - 0.9}$$

Figure 4.25c shows the root locus plot of the lag compensated system. The intersection of the locus with $\zeta = 0.5$ line is at point R. The value of ω_n at this point is determined to be 1.2 rad/sec. The lag compensator has thus reduced ω_n and hence the speed of response. The value of the gain K at point R is determined to be 0.478. The position error constant of the lag compensated system is

$$K_p = \lim_{z \to 1} D(z) \, G_{h0}G(z) = \lim_{z \to 1} \frac{0.478(z + 0.368)}{(z - 0.135)(z - 0.9)} = 7.56$$

Thus we have been able to increase position error constant appreciably by lag compensation.

If both the pole and the zero of the lag compensator are moved close to $z = 1$, then the root locus plot of the lag compensated system moves back towards its uncompensated shape. Consider the root locus plot of the uncompensated system shown in Fig. 4.25a. The angle contributed at point P by additional pole-zero pair close to $z = 1$ (called a *dipole*) will be negligibly small; therefore the point P will continue to lie on the lag compensated root locus plot. However, the lag compensator

$$D(z) = \frac{z - \beta_1}{z - \beta_2}$$

will raise the system K_p (refer Eqn. (4.57)) by a factor of $(1 - \beta_1)/(1 - \beta_2)$.

▲▲

The following examples illustrate typical digital control system design problems carried out in the z-plane using the root locus technique. As we shall see, the design of digital compensation using root locus plots is essentially a trial-and-error method. The designer may rely on a digital computer to plot out a large number of root loci by scanning through a wide range of possible values of the compensator parameters, and select the best solution. However, one can still make proper and intelligent initial 'guesses' so that the amount of trial-and-error effort is kept to a minimum.

Example 4.6

Consider the feedback control system shown in Fig. 4.26. The plant is described by the transfer function

$$G(s) = \frac{K}{s(s+2)}$$

Design a digital control scheme for the system to meet the following specifications;
 (i) the velocity error constant $K_v = 6$;
 (ii) peak overshoot M_p to step input $\leq 15\%$, and
 (iii) settling time t_s (2% tolerance band) ≤ 5 sec.

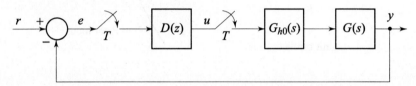

Fig. 4.26 A feedback system with digital compensation

Solution

The transient accuracy requirements correspond to $\zeta = 0.5$ and $\omega_n = 1.6$. We select $T = 0.2$ sec. Note that sampling frequency $\omega_s = 2\pi/T$ is about 20 times the natural frequency; therefore our choice of sampling period is satisfactory.

The transfer function $G_{h0}G(z)$ of the plant preceded by a ZOH can be obtained as follows:

$$G_{h0}G(z) = (1 - z^{-1}) \mathscr{Z}\left[\frac{K}{s^2(s+2)}\right]$$

$$= \frac{0.01758K(z+0.876)}{(z-1)(z-0.67)} = \frac{K'(z+0.876)}{(z-1)(z-0.67)} \quad (4.58)$$

The root locus plot of this system for $0 \leq K' < \infty$ was earlier constructed in Fig. 4.22. Complex conjugate sections of this plot are shown in Fig. 4.27. The plot intersects the $\zeta = 0.5$ locus at point P. At this point $\omega_n = 1.7$ rad/sec, $K' = 0.0546$.

Therefore the transient accuracy requirements ($\zeta = 0.5$, $\omega_n = 1.6$) are almost satisfied by gain adjustment only. Let us now examine the steady-state accuracy of the uncompensated system ($D(z) = 1$) with $K' = 0.0546$.

The velocity error constant K_v of the system is given by

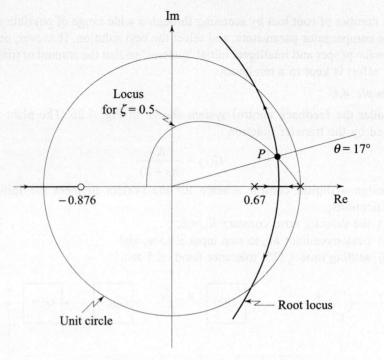

Fig. 4.27 Root locus plot for system (4.58)

$$K_v = \frac{1}{T} \lim_{z \to 1} (z-1)\, G_{h0}G(z) = \frac{5(0.0546)(1+0.876)}{(1-0.67)} = 1.55$$

The specified value of K_v is 6. Therefore an increase in K_v by a factor of 3.87 (= 6/1.55) is required. The objective before us now is to introduce a $D(z)$ that raises the system K_v by a factor of 3.87 without appreciably affecting the transient performance of the uncompensated system, i.e., without appreciably affecting the root locus plot in the vicinity of point P. This objective can be realized by a properly designed lag compensator, as is seen below.

We add the compensator pole and zero as shown in Fig. 4.28. Since both the pole and the zero are very close to $z = 1$, the scale in the vicinity of these points has been greatly expanded. The angle contributed by the compensator pole at point P is almost equal to the angle contributed by the compensator zero. Therefore, the addition of dipole near $z = 1$ does not appreciably disturb the root locus plot in the vicinity of point P. It only slightly reduces ω_n. The lag compensator

$$D(z) = \frac{z - 0.96}{z - 0.99}$$

raises the system K_v by a factor of $(1 - 0.96)/(1 - 0.99) = 4$.

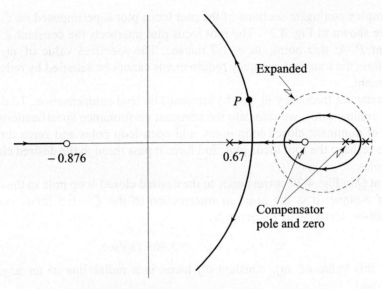

Fig. 4.28 Compensator design (Example 4.6)

Note that because of lag compensator, a third closed-loop pole has been added. This pole, as seen from Fig. 4.28, is a real pole lying close to $z = 1$. This pole, fortunately, does not disturb the dominance of the complex conjugate closed-loop poles. The reason is simple.

The closed-loop pole close to $z = 1$ has a long time constant. However, there is a zero close to this additional pole. The net effect is that the settling time will increase because of the third pole but the amplitude of the response term contributed by this pole will be very small. In system response, a long tail of small amplitude will appear which may not appreciably degrade the performance of the system.

Example 4.7

Reconsider the feedback control system of Example 4.6 (Fig. 4.26). We now set the following goal for our design.

(i) $K_v \geq 2.5$
(ii) $\zeta \cong 0.5$
(iii) t_s (2% tolerance band) ≤ 2 sec

The transient accuracy requirements correspond to $\zeta = 0.5$ and $\omega_n = 4$. For sampling interval $T = 0.2$ sec, the sampling frequency is about eight times the natural frequency. A smaller value of T is more appropriate for the present design problem which requires higher speed of response. We will however take $T = 0.2$ sec to compare our results with those of Example 4.6.

Following the initial design steps of Example 4.6, we find that

$$G_{h0}G(z) = \frac{0.01758K(z+0.876)}{(z-1)(z-0.67)} = \frac{K'(z+0.876)}{(z-1)(z-0.67)}$$

Complex conjugate sections of the root locus plot superimposed on $\zeta = 0.5$ line are shown in Fig. 4.27. The root locus plot intersects the constant-ζ locus at point P. At this point, $\omega_n = 1.7$ rad/sec. The specified value of ω_n is 4. Therefore, the transient accuracy requirements cannot be satisfied by only gain adjustment.

The natural frequency ω_n can be increased by lead compensation. To design a lead compensator, we translate the transient performance specifications into a pair of dominant closed-loop poles, add open-loop poles and zeros through $D(z)$ to reshape the root locus plot, and force it pass through the desired closed-locp poles.

Point Q in Fig. 4.29 corresponds to the desired closed-loop pole in the upper half of z-plane. It is the point of intersection of the $\zeta = 0.5$ locus and the constant-ω_d locus, with ω_d given by

$$\omega_d = \omega_n \sqrt{1 - \zeta^2} = 3.464 \text{ rad/sec}$$

For this value of ω_d, constant-ω_d locus is a radial line at an angle of $\omega_d T \left(\dfrac{180}{\pi} \right) = 39.7°$ with the real axis.

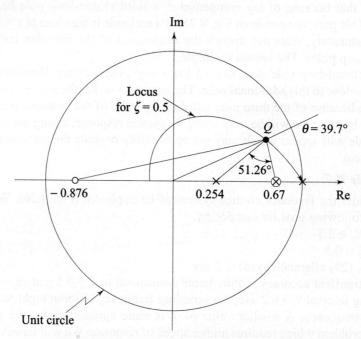

Fig. 4.29 Compensator design (Example 4.7)

If the point Q is to lie on the root locus plot of the compensated system, then the sum of the angles contributed by open-loop poles and zeros of the plant, and the pole and zero of the compensator at the point Q must be equal to $\pm(2q+1)180°$; $q = 0, 1, 2, \ldots$

The sum of the angle contributions due to open-loop poles and zero of the plant at plant Q is

$$17.10° - 138.52° - 109.84° = -231.26°$$

Hence the compensator $D(z)$ must provide $+51.26°$. The transfer function of the compensator may be assumed to be

$$D(z) = \frac{z - \alpha_1}{z - \alpha_2}$$

If we decide to cancel the pole at $z = 0.67$ by the zero of the compensator at $z = \alpha_1$, then the pole of the compensator can be determined (from the condition that the compensator must provide $+51.26°$) as a point at $z = 0.254$ ($\alpha_2 = 0.254$). Thus the transfer function of the compensator is obtained as

$$D(z) = \frac{z - 0.67}{z - 0.254}$$

The open-loop transfer function now becomes

$$D(z)G_{h0}G(z) = \frac{0.01758K(z+0.876)(z-0.67)}{(z-0.254)(z-1)(z-0.67)}$$

$$= \frac{0.01758K(z+0.876)}{(z-0.254)(z-1)} = \frac{K'(z+0.876)}{(z-0.254)(z-1)}$$

The value of K' at point Q, obtained form Fig. 4.29 by graphical construction, is 0.2227. Therefore $K = 12.67$.

The velocity error constant of the compensated system is given by

$$K_v = \frac{1}{T} \lim_{z \to 1} [(z-1)D(z)\, G_{h0}G(z)] = 2.8$$

It meets the specification on steady-state accuracy.

If it is required to have a large K_v, then we may include a lag compensator. The lag-lead compensator can satisfy the requirements of high steady-state accuracy and high speed of response.

Form the viewpoint of microprocessor implementation of the lag, lead, and lag-lead compensators, the lead compensators present the least coefficient quantization problems, because the locations of poles and zeros are widely separated and the numerical inaccuracies in realization of these compensators will result in only small deviations in expected system behaviour. However, in the case of lag compensators and lag-lead compensators, the lag section may result in considerable coefficient quantization problems because the locations of poles and zeros are usually close to each other (they are near the point $z = 1$). Numerical problems associated with realization of compensator coefficients may lead to significant deviations in expected system behaviour.

4.5 z-PLANE SYNTHESIS

Much of the style of the transform domain techniques we have been discussing in this chapter grew out of the limitations of technology which was available for realization of the compensators with pneumatic components or electric networks and amplifiers. In the digital computer, such limitations on realization are, of course, not relevant, and one can ignore these particular constraints. One design method which eliminates these constraints begins from the very direct point of view that we are given a process (plus hold) transfer function $G_{h0}G(z)$, that we want to construct a desired transfer function, $M(z)$, between input r and output y and that we have the computer transfer function, $D(z)$, to do the job as per the feedback control structure of Fig. 4.30.

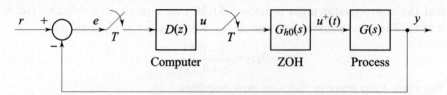

Fig. 4.30 A feedback system with digital compensation

The closed-loop transfer function is given by the formula

$$M(z) = \frac{D(z)G_{h0}G(z)}{1 + D(z)G_{h0}G(z)} \tag{4.59}$$

from which we get the design formula

$$D(z) = \frac{1}{G_{h0}G(z)} \left[\frac{M(z)}{1 - M(z)} \right] \tag{4.60}$$

As is seen from Eqn. (4.60), the controller transfer function consists of the inverse of the plant transfer function and the additional term which depends on the system closed-loop transfer function. Thus the design procedure outlined above looks for a $D(z)$ which will cancel the process effects and add whatever is necessary to give the desired performance.

For prescribing the required closed-loop transfer function $M(z)$, the following restrictions have to be noted.

Realizability of Digital Controller

Assume that a digital controller

$$D(z) = \frac{Q_v(z)}{P_\mu(z)} = \frac{q_0 z^v + q_1 z^{v-1} + \cdots + q_v}{z^\mu + p_1 z^{\mu-1} + \cdots + p_\mu} \tag{4.61}$$

is cascaded with the process

$$G_{h0}G(z) = \frac{B_m(z)}{A_n(z)} = \frac{b_0 z^m + b_1 z^{m-1} + \cdots + b_m}{z^n + a_1 z^{n-1} + \cdots + a_n}; \quad m \leq n \tag{4.62}$$

in the control loop given by Fig. 4.30.
For $D(z)$ to be physically realizable, $v \leq \mu$.
The closed-loop transfer function

$$M(z) = \frac{D(z)G_{h0}G(z)}{1+D(z)G_{h0}G(z)} = \frac{Q_v(z)B_m(z)}{P_\mu(z)A_n(z)+Q_v(z)B_m(z)} = \frac{N_{v+m}(z)}{D_{\mu+n}(z)}$$

The order of the numerator polynomial of $M(z)$ is $v+m$, and the order of the denominator polynomial of $M(z)$ is $\mu+n$. The pole excess[8] of $M(z)$ is therefore $\{(\mu-v)+(n-m)\}$.

This means that because of the condition of realizability of digital controller, the pole excess of the closed-loop transfer function $M(z)$ has to be greater than or equal to the pole excess of the process transfer function $G_{h0}G(z)$.

Cancellation of Poles and Zeros

If the digital controller $D(z)$ given by Eqn. (4.60) and the process $G_{h0}G(z)$ are in a closed loop, the poles and zeros of the process are cancelled by the zeros and poles of the controller. The cancellation is perfect if the process model $G_{h0}G(z)$ matches the process exactly. Since the process models used for design practically never describe the process behaviour exactly, the corresponding poles and zeros will not be cancelled exactly; the cancellation will be approximate. For poles and zeros of $G_{h0}G(z)$ which are sufficiently spread in the inner of the unit disc in the z-plane, the approximation in cancellation leads to only small deviations of the assumed behaviour $M(z)$ in general. However, one has to be careful if $G_{h0}G(z)$ has poles or zeros on or outside the unit circle. Imperfect cancellation may lead to weakly damped or unstable behaviour (refer Review Example 4.4). Therefore the design of digital controllers according to Eqn. (4.60) has to be restricted to cancellation of poles and zeros of $G_{h0}G(z)$ located inside the unit circle. This imposes certain restrictions on the desired transfer function $M(z)$ as is seen below.

Assume that $G_{h0}G(z)$ involves an unstable (or critically stable) pole at $z = \alpha$. Let us define

$$G_{h0}G(z) = \frac{G_1(z)}{z-\alpha}$$

where $G_1(z)$ does not include a term that cancels with $(z-\alpha)$. Then the closed-loop transfer function becomes

$$M(z) = \frac{D(z)\dfrac{G_1(z)}{z-\alpha}}{1+D(z)\dfrac{G_1(z)}{z-\alpha}} \qquad (4.63)$$

Since we require that no zero of $D(z)$ cancel the pole of $G_{h0}G(z)$ at $z = \alpha$, we must have

---

8. Pole excess of $M(z) = \{$number of finite poles of $M(z)$ – number of finite zeros of $M(z)\}$.

$$1 - M(z) = \frac{1}{1 + D(z)\dfrac{G_1(z)}{z-\alpha}} = \frac{z-\alpha}{z-\alpha + D(z)G_1(z)}$$

that is, $1 - M(z)$ must have $z = \alpha$ as a zero. This argument applies equally if $G_{h0}G(z)$ involves two or more unstable (or critically stable) poles.

Also note from Eqn. (4.63) that if poles of $D(z)$ do not cancel zeros of $G_{h0}G(z)$, then the zeros of $G_{h0}G(z)$ become zeros of $M(z)$.

Let us summarize what we have stated concerning cancellation of poles and zeros of $G_{h0}G(z)$.

  (i) Since the digital controller $D(z)$ should not cancel unstable (or critically stable) poles of $G_{h0}G(z)$, all such poles of $G_{h0}G(z)$ must be included in $1 - M(z)$ as zeros.
 (ii) Zeros of $G_{h0}G(z)$ that lie on or outside the unit circle should not be cancelled with poles of $D(z)$; all such zeros of $G_{h0}G(z)$ must be included in $M(z)$ as zeros.

The design procedure thus essentially involves the following three steps:

1. The closed-loop transfer function $M(z)$ of the final system is determined from the performance specifications, and the fixed parts of the system, i.e., $G_{h0}G(z)$.
2. The transfer function $D(z)$ of the digital controller is found using the design formula (4.60).
3. The digital controller $D(z)$ is synthesized.

Step 1 is certainly the most difficult one to satisfy. In order to pass step 1, a designer must fulfil the following requirements.

  (i) The digital controller $D(z)$ must be physically realizable.
 (ii) The poles and zeros of $G_{h0}G(z)$ on or outside the unit circle should not be cancelled by $D(z)$.
(iii) The system specifications on transient and steady-state accuracy be satisfied.

For common types of specifications (error constants, stability margins, speed of response, etc.), these requirements are usually easily satisfied. However, we can ask for more than what has been achieved earlier by compensation techniques based on root locus and frequency response plots. Design examples will best illustrate this point.

Example 4.8

The plant of sampled-data system of Fig. 4.30 is described by the transfer function

$$G(s) = \frac{1}{s(10s+1)} \qquad (4.64a)$$

The sampling period is 1 sec.

The problem is to design a digital controller $D(z)$ to realize the following specifications:

(i) $K_v \geq 1$
(ii) $\zeta = 0.5$, and
(iii) t_s (2% tolerance band) ≤ 8 sec.

The selection of a suitable $M(z)$ is described by the following steps.

(i) The z-transfer function of the plant is given by (refer Table of transforms given in Appendix A)

$$G_{h0}G(z) = (1 - z^{-1})\mathscr{Z}\left[\frac{1}{s^2(10s+1)}\right]$$

$$= 0.04837 \frac{(z+0.9672)}{(z-1)(z-0.9048)} \qquad (4.64b)$$

Since $G_{h0}G(z)$ has one more pole than zero, $M(z)$ must have a pole excess of at least one.

(ii) $G_{h0}G(z)$ has a pole at $z = 1$. This must be included in $1 - M(z)$ as zero, i.e.,

$$1 - M(z) = (z-1)F(z) \qquad (4.65)$$

where $F(z)$ is a ratio of polynomials of appropriate dimensions.

(iii) The transient accuracy requirements are specified as $\zeta = 0.5$, $\omega_n = 1$ ($t_s = 4/\zeta\omega_n$). With a sampling period $T = 1$ sec, this maps to a pair of dominant closed-loop poles in the z-plane with

$$|z_{1,2}| = e^{-\zeta\omega_n T} = 0.6065$$

$$\angle z_{1,2} = \pm \omega_n T\sqrt{1-\zeta^2} = \pm \frac{0.866 \times 180}{3.14} = \pm 49.64°$$

This corresponds to

$$z_{1,2} = 0.3928 \pm j\, 0.4621$$

The closed-loop transfer function $M(z)$ should have dominant poles at the roots of the equation

$$\Delta(z) = z^2 - 0.7856\, z + 0.3678 = 0 \qquad (4.66)$$

The steady-state accuracy requirements demand that steady-state error to unit-step input is zero, and steady-state error to unit-ramp input is less than $1/K_v$.

$$E(z) = R(z) - Y(z) = R(z)[1 - M(z)] = R(z)(z-1)F(z)$$

$$e_{ss}^*\Big|_{\text{unit step}} = \lim_{z \to 1} z(z-1)F(z) = 0$$

Thus, with the choice of $M(z)$ given by Eqn. (4.65), the steady-state error to unit step input is always zero.

$$e_{ss}^*\Big|_{\text{unit ramp}} = \lim_{z \to 1} (z-1)\frac{Tz}{(z-1)^2}(z-1)F(z) = T F(1) = 1/K_v$$

For $T = 1$ and $K_v = 1$,

$$F(1) = 1 \qquad (4.67)$$

From Eqns (4.65) and (4.66), we observe that

$$F(z) = \frac{z - \alpha}{z^2 - 0.7856z + 0.3678}$$

meets the requirements on realizability of $D(z)$, cancellation of poles and zeros of $G_{h0}G(z)$, and transient accuracy. The requirement on steady-state accuracy is also met if we choose α such that (refer Eqn. (4.67))

$$\frac{1-\alpha}{1-0.7856+0.3678} = 1$$

This gives $\quad\quad \alpha = 0.4178$

Therefore $\quad F(z) = \dfrac{z - 0.4178}{z^2 - 0.7856z + 0.3678}$

$$1 - M(z) = \frac{(z-1)(z-0.4178)}{z^2 - 0.7856z + 0.3678}$$

$$M(z) = \frac{0.6322z - 0.05}{z^2 - 0.7856z + 0.3678} \quad\quad (4.68)$$

Now turning to the basic design formula (4.60), we compute

$$D(z) = \frac{1}{G_{h0}G(z)}\left[\frac{M(z)}{1-M(z)}\right]$$

$$= \frac{(z-1)(z-0.9048)}{(0.04837)(z+0.9672)}\left[\frac{0.6322z - 0.05}{(z-1)(z-0.4178)}\right]$$

$$= 13.07 \frac{(z-0.9048)(z-0.079)}{(z+0.9672)(z-0.4178)} \quad\quad (4.69)$$

A plot of the step response of the resulting design is provided in Fig. 4.31, which also shows the control effort. We can see the oscillation of $u(k)$ which is associated with the pole of $D(z)$ at $z = -0.9672$, which is quite near the unit circle. Strong oscillations of $u(k)$ are often considered unsatisfactory even though the process is being controlled as was intended. In the literature, poles near $z = -1$ are often referred to as *ringing poles*.

To avoid the ringing effect, we could include the zero of $G_{h0}G(z)$ at $z = -0.9672$ in $M(z)$ as zero, so that this zero of $G_{h0}G(z)$ is not cancelled with pole of $D(z)$. $M(z)$ may have additional poles at $z = 0$, where the transient is as short as possible. The result will be a simpler $D(z)$ with a slightly more complicated $M(z)$.

The underdamped response shown in Fig. 4.31 settles within a two percent band of the desired value of unity in less than 8 sec. The steady-state error is zero. It may be noted that the response approaches the steady-state value of unity only as $k \to \infty$. It does not settle at the desired value of unity in finite time.

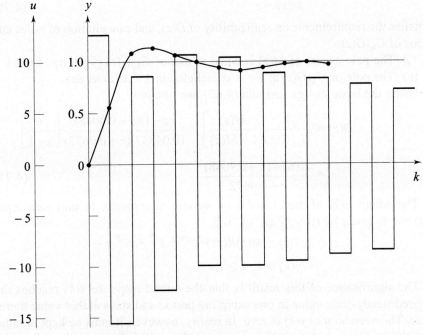

Fig. 4.31 Step response (Example 4.8)

The settling time of 8 sec corresponds to settling within the tolerance band and not settling at the desired steady-state value. Interpreting this behaviour in terms of impulse response, we can say that the impulse response of the system has an infinite number of non-zero samples, although their magnitudes become negligibly small as k increases. This may be seen by dividing the numerator polynomial of $M(z)$ in Eqn. (4.68) with its denominator polynomial:

$$M(z) = 0.6322\, z^{-1} + 0.4467\, z^{-2} + 0.1184\, z^{-3} - 0.0995\, z^{-5} - 0.0522\, z^{-6}$$
$$- 0.0044\, z^{-7} + 0.0157\, z^{-8} + \cdots$$

The closed-loop system is thus an IIR (Infinite Impulse Response) filter (refer Section 3.4).

As is obvious, the design principles used in the present example involve the extension of the design experience acquired in the design of analog control systems. However, since the digital controller has a great deal of flexibility in its configuration, we should be able to come up with independent methods not relying completely on the principles of design of analog control systems. In the next example we show that we can do better with independent methods of design of digital controls; the methods for which comparable analog methods do not exist.

Example 4.9

Reconsider the design problem of the previous example. We now select an FIR (Finite Impulse Response) filter as the desired closed-loop system. A simple model

290 Digital Control and State Variable Methods

$$M(z) = z^{-1} \tag{4.70}$$

satisfies the requirements on realizability of $D(z)$, and cancellation of poles and zeros of $G_{h0}G(z)$:

(i) The pole excess of $M(z)$ is equal to that of $G_{h0}G(z)$ given by (4.64b).
(ii) The pole of $G_{h0}G(z)$ at $z = 1$ is included in $1 - M(z)$ as zero.

Using the basic design formula (4.60), we obtain

$$D(z) = \frac{1}{G_{h0}G(z)} \left[\frac{M(z)}{1 - M(z)}\right] = \frac{(z-1)(z-0.9048)}{(0.04837)(z+0.9672)} \left[\frac{1}{z-1}\right]$$

$$= \frac{20.674\,(z-0.9048)}{z+0.9672} \tag{4.71}$$

The output $y(k)$ of the closed-loop system in response to unit step input $r(k) = 1$ is given by (refer Eqn. (4.70))

$$Y(z) = M(z)R(z) = z^{-1} + z^{-2} + z^{-3} + \cdots$$

$$y(k) = 1;\ k \geq 1$$

The significance of this result is that the output response $y(k)$ reaches the desired steady-state value in one sampling period and stays at that value thereafter. The overshoot of $y(k)$ is zero. In reality, however, it must be kept in mind that the true judgement on the system performance should be based on the behaviour of $y(t)$. In general, although $y(k)$ may exhibit little or no overshoot, the actual response $y(t)$ may have oscillations between the sampling instants.

Let us examine the intersample response of the present system. With reference to Fig. 4.30, we have

$$G_{h0}G(z)U(z) = Y(z) = M(z)R(z)$$

Therefore
$$U(z) = \frac{1}{G_{h0}G(z)}[M(z)R(z)] \tag{4.72}$$

$$= \frac{20.674\,(z-0.9048)(z-1)}{(z+0.9672)}\left[\frac{1}{z}\left(\frac{z}{z-1}\right)\right] = \frac{20.674 - 18.706\,z^{-1}}{1+0.9672\,z^{-1}}$$

$$= 20.674 - 38.702\,z^{-1} + 37.433\,z^{-2} - 36.205\,z^{-3} + \cdots$$

We can see the oscillation of $u(k)$ which is associated with the pole of $D(z)$ at $z = -0.9672$, which is quite near the unit circle.

Noting that $u^+(t)$, the output of the zero-order hold (refer Fig. 4.30), is a continuous function of time; a constant $y(t \geq T)$ requires that $u^+(t)$ also be constant for $t \geq T$. This obviously is not true in the present case and therefore the system will respond with intersample oscillations.

The intersample ripples may be eliminated by increasing the settling time from 1 second to 2 seconds. The $M(z)$ of Eqn. (4.70) is now modified to the following form:

$$M(z) = \alpha_1 z^{-1} + \alpha_2 z^{-2} \tag{4.73}$$

where α_1 and α_2 are constants with the following restrictions.

(i) The pole excess of $M(z)$ must be equal to or greater than the pole excess of $G_{h0}G(z)$. $M(z)$ given by Eqn. (4.73) satisfies this restriction.

(ii) The pole of $G_{h0}G(z)$ at $z = 1$ must be included in $1 - M(z)$ as zero, i.e.,

$$1 - M(z) = 1 - \alpha_1 z^{-1} - \alpha_2 z^{-2} = (1 - z^{-1})F(z)$$

where $F(z)$ is an FIR filter.

If we divide $(1 - \alpha_1 z^{-1} - \alpha_2 z^{-2})$ by $(1 - z^{-1})$, the quotient is $(1 + (1 - \alpha_1)z^{-1})$ and the remainder is $(1 - \alpha_1 - \alpha_2)z^{-2}$. Hence

$$F(z) = 1 + (1 - \alpha_1) z^{-1}$$

$$1 - \alpha_1 - \alpha_2 = 0 \tag{4.74}$$

(iii) The system must have zero steady-state error for specified reference input signal.

For a unit-step input, this requirement is met if

$$\lim_{z \to 1} (z - 1)Y(z) = 1$$

or

$$\lim_{z \to 1} (z - 1)M(z)R(z) = 1$$

Substituting for $M(z)$ and $R(z)$, we get

$$\alpha_1 + \alpha_2 = 1$$

This restriction is same as that given by Eqn. (4.74).

(iv) The response should not exhibit intersample ripples. Since the input is a unit-step function, we require that

$$y(t \geq 2T) = 1$$

This requirement is met if

$$u(t \geq 2T) = \text{constant}$$

Therefore, $U(z)$ must be of the following type of series in z^{-1}:

$$U(z) = u(0) + u(1)z^{-1} + u(2)[z^{-2} + z^{-3} + \cdots]$$

As the plant transfer function $G(s)$ given by Eqn. (4.64a) involves an integrator, $u(2)$ must be zero (otherwise the input cannot stay constant). Consequently, we have

$$U(z) = \beta_0 + \beta_1 z^{-1}$$

where β_0 and β_1 are the magnitudes of the control signal at $k = 0$ and $k = 1$ respectively.

Using the relation (4.72), we obtain

$$U(z) = \frac{20.674(z-1)(z-0.9048)}{z+0.9672} \left[\frac{\alpha_1 z + \alpha_2}{z^2} \left(\frac{z}{z-1} \right) \right]$$

$$= \frac{20.674(\alpha_1 z + \alpha_2)(z - 0.9048)}{z(z + 0.9672)}$$

$$= \frac{20.674(\alpha_1 + \alpha_2 z^{-1})(1 - 0.9048 z^{-1})}{1 + 0.9672 z^{-1}} \tag{4.75}$$

292 Digital Control and State Variable Methods

In order for $U(z)$ to be a series in z^{-1} with only two terms, $(\alpha_1 + \alpha_2 z^{-1})$ must be of the following form:
$$\alpha_1 + \alpha_2 z^{-1} = K_1(1 + 0.9672 z^{-1})$$
where K_1 is a constant.
This gives
$$\alpha_1 = K_1, \; \alpha_2 = 0.9672 K_1$$
Substituting in Eqn. (4.74), we obtain
$$1 - K_1 - 0.9672 K_1 = 0$$
or
$$K_1 = 1/1.9672$$
$$\alpha_1 = K_1 = 0.5083 \tag{4.76}$$
$$\alpha_2 = 0.9672 K_1 = 0.4917$$
Therefore from Eqns (4.73) and (4.76), we get
$$M(z) = \frac{0.5083 z + 0.4917}{z^2}$$
Using the basic design formula (4.60), $D(z)$ is determined as
$$D(z) = \frac{20.674 (z-1)(z-0.9048)}{z+0.9672} \left[\frac{0.5083z + 0.4917}{(z-1)(z+0.4917)} \right]$$
$$= \frac{10.5086 (z-0.9048)(z+0.9672)}{(z+0.9672)(z+0.4917)}$$
$$= \frac{10.5086 (z-0.9048)}{z+0.4917} \tag{4.77}$$

With the digital controller thus designed, the system output in response to a unit-step input is obtained as
$$Y(z) = M(z)R(z)$$
$$= (0.5083 z^{-1} + 0.4917 z^{-2}) \left(\frac{1}{1-z^{-1}} \right)$$
$$= 0.5083 z^{-1} + z^{-2} + z^{-3} + \cdots$$
Hence $\quad y(0) = 0, \; y(1) = 0.5083, \; y(i) = 1; \; i \geq 2$

From Eqns (4.75) and (4.76), we get
$$U(z) = 10.5094 - 9.5089 z^{-1}$$
Hence $\quad u(0) = 10.5094, \; u(1) = -9.5089, \; u(i) = 0; \; i \geq 2$

A plot of the step response is provided in Fig. 4.32, which also shows the control effort. The system reaches its steady-state value of unity in minimum time (two sampling periods) and there is no ripple in between the sampling instants. Systems with this type of response are generally referred to as *deadbeat control systems*.

It may be noted that the deadbeat-response design leads to a digital control system which gives an 'ideal' output response to an input it is designed for. It may give inferior or sometimes unacceptable performance to other inputs (disturbances). For example, our deadbeat response design gives the following value of K_v:

$$K_v = \frac{1}{T} \lim_{z \to 1} (z-1)D(z)G_{h0}G(z)$$

$$= \lim_{z \to 1} \frac{10.5086\,(z+0.9672)}{20.674\,(z+0.4917)} = 0.67$$

which may be unsatisfactory.

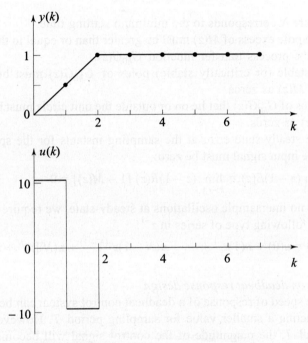

Fig. 4.32 Step response (Example 4.9)

Digital Controllers for Deadbeat Performance

Now let us look at the deadbeat response design as a general design tool. The deadbeat response is characterized by the following properties:

(i) The system must have zero steady-state error at the sampling instants for the specified reference input signal.

(ii) The settling time, defined as the number of samples N, required for the response to reach the steady-state and stay there for $k \geq N$, should be a minimum.

(iii) The system should not exhibit inter sampling ripples at steady-state.

The system configuration considered is shown in Fig. 4.30. The digital controller for deadbeat performance is obtained by using the formula

$$D(z) = \frac{1}{G_{h0}G(z)} \left[\frac{M(z)}{1 - M(z)}\right] \quad (4.78)$$

For prescribing the required closed-loop transfer function $M(z)$, the following restrictions have to be satisfied.

(i) The desired closed-loop system must be an FIR filter:
$$M(z) = \alpha_0 + \alpha_1 z^{-1} + \cdots + \alpha_N z^{-N} \quad (4.79a)$$
$$= \frac{\alpha_0 z^N + \alpha_1 z^{N-1} + \cdots + \alpha_N}{z^N} \quad (4.79b)$$
where N corresponds to the minimum settling time.

(ii) The pole excess of $M(z)$ must be greater than or equal to the pole excess of the process transfer function $G_{h0}G(z)$.

(iii) Unstable (or critically stable) poles of $G_{h0}G(z)$ must be included in $1 - M(z)$ as zeros.

(iv) Zeros of $G_{h0}G(z)$ that lie on or outside the unit circle must be included in $M(z)$ as zeros.

(v) The steady-state error at the sampling instants for the specified reference input signal must be zero:
$$\lim_{z \to 1} (z - 1)E(z) = \lim_{z \to 1} (z - 1)R(z)[1 - M(z)] = 0 \quad (4.80)$$

(vi) For no intersample oscillations at steady-state, we require $U(z)$ to be of the following type of series in z^{-1}:
$$U(z) = u(0) + u(1)z^{-1} + \cdots + u(N-1)z^{-(N-1)} + u(N)[z^{-N} + z^{-N-1} + \cdots] \quad (4.81)$$

Comments on deadbeat response design

(i) The speed of response of a deadbeat control system can be increased by selecting a smaller value for sampling period T. However, for a very small T, the magnitude of the control signal will become excessively large, with the result that saturation phenomena will take place in the system. Hence the sampling interval T should not be too small (recall that in the case where an analog controller is discretized, sampling interval T is required to be very small).

(ii) The deadbeat response design results in a multiple-order pole at $z = 0$ of the closed-loop transfer function $M(z)$. It was shown earlier in Review Example 3.3 that if the closed-loop poles are in a cluster, the sensitivity of the poles to changes in parameters of the system is high. Therefore, deadbeat response design is highly sensitive to system parameter variations.

(iii) Deadbeat response design can also be carried out by the method of pole assignment using state feedback. The state variable method will be developed in Chapter 7.

Example 4.10

In this example, a method for the design of deadbeat controllers for a class of digital control systems is described for which the resulting synthesis requires little calculations. The plant is assumed to be stable (all poles inside the unit circle except possibly a simple pole at $z = 1$). The design problem is to obtain a $D(z)$ that will give deadbeat response to a step input. The settling time is assumed to be equal to nT, where n is the order of the plant model (In Chapter 7, we will prove that nT is, in fact, the minimum settling time for deadbeat performance).

We reconsider the system of the previous example (refer Fig. 4.30). The plant model is given by

$$G(s) = \frac{1}{s(10s+1)}$$

The sampling period $T = 1$ sec.
Equation (4.64b) gives

$$\frac{Y(z)}{U(z)} = G_{h0}G(z) = \frac{0.04837(z+0.9672)}{(z-1)(z-0.9048)} \quad (4.82a)$$

$$= \frac{0.04837 + 0.04678\, z^{-1}}{1 - 1.9048\, z^{-1} + 0.9048\, z^{-2}} \quad (4.82b)$$

We note that $G_{h0}G(z)$ is a second-order model with a pole excess of one. A second-order FIR filter with a pole excess of one is of the form

$$M(z) = \alpha_1 z^{-1} + \alpha_2 z^{-2}$$

Assuming this filter to be the desired closed-loop transfer function, we have

$$\frac{Y(z)}{R(z)} = M(z) = \alpha_1 z^{-1} + \alpha_2 z^{-2} \quad (4.83)$$

$$E(z) = R(z) - Y(z) = R(z)[1 - M(z)]$$

$$\left. e_{ss}^* \right|_{\text{unit step}} = \lim_{z \to 1} (z-1)E(z)$$

The steady-state error to step change in input is zero if

$$\alpha_1 + \alpha_2 = 1 \quad (4.84)$$

Note that this equivalently means that $(1 - M(z))$ has a zero at $z = 1$. For no intersample oscillations at steady-state, we require that

$$U(z) = u(0) + u(1)\, z^{-1} + u(2)\, [z^{-2} + z^{-3} + \cdots]$$

$$= u(0) + u(1)\, z^{-1} + u(2)\, z^{-2} \left[\frac{1}{1-z^{-1}}\right]$$

Dividing $U(z)$ by $R(z) = 1/(1 - z^{-1})$, we obtain

$$\frac{U(z)}{R(z)} = u(0) + [u(1) - u(0)]\,z^{-1} + [u(2) - u(1)]\,z^{-2} \quad (4.85a)$$

$$= \beta_0 + \beta_1 z^{-1} + \beta_2 z^{-2} \quad (4.85b)$$

It follows from Eqns (4.83) and (4.85) that

$$G_{h0}G(z) = \frac{Y(z)}{U(z)} = \frac{Y(z)/R(z)}{U(z)/R(z)}$$

$$= \frac{\alpha_1 z^{-1} + \alpha_2 z^{-2}}{\beta_0 + \beta_1 z^{-1} + \beta_2 z^{-2}} = \frac{\dfrac{\alpha_1}{\beta_0} z^{-1} + \dfrac{\alpha_2}{\beta_0} z^{-2}}{1 + \dfrac{\beta_1}{\beta_0} z^{-1} + \dfrac{\beta_2}{\beta_0} z^{-2}} \quad (4.86)$$

Comparison of the coefficients in Eqns (4.82) and (4.86), and the steady-state error constraint given by Eqn. (4.84) yield the following values for the parameters of the closed-loop transfer function and the controller:

$$\alpha_1 = 0.5083,\ \alpha_2 = 0.4917$$

$$\beta_0 = 10.5097,\ \beta_1 = -20.0189,\ \beta_2 = 9.5092$$

Now (refer Eqn. (4.78))

$$D(z) = \frac{1}{G_{h0}G(z)}\left[\frac{M(z)}{1-M(z)}\right] = \frac{U(z)/R(z)}{1-Y(z)/R(z)} = \frac{\beta_0 + \beta_1 z^{-1} + \beta_2 z^{-2}}{1 - \alpha_1 z^{-1} - \alpha_2 z^{-2}}$$

$$= \frac{10.5097 - 20.0189\,z^{-1} + 9.5092\,z^{-2}}{1 - 0.5083\,z^{-1} - 0.4917\,z^{-2}} = \frac{10.5097 - 9.5092\,z^{-1}}{1 + 0.4917\,z^{-1}}$$

4.6 REVIEW EXAMPLES

Review Example 4.1

Consider the digital control system shown in Fig. 4.33. The transfer function of the plant is $G(s) = 1/[s(s + 1)]$. Design a lead compensator $D(z)$ in the w-plane such that the phase margin in 50°, the gain margin is at least 10 dB, and the velocity error constant K_v is 2. Assume that the sampling period is 0.2 sec.

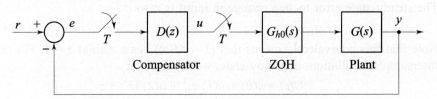

Fig. 4.33 A digital control configuration

Solution

The digital controller is assumed to be of the form

$$\frac{U(z)}{E(z)} = KD_1(z) = D(z)$$

To simplify the design procedure, we will associate the gain K of the controller with the plant model. The design problem is therefore to obtain compensator $D_1(z)$ for the plant

$$G(s) = \frac{K}{s(s+1)}$$

to meet the specifications on steady-state and transient performance.

We will fix the gain K to a value that realizes given K_v. A unity dc gain compensator $D_1(z)$ will then be introduced to meet the transient accuracy requirements without affecting the steady-state accuracy.

$$K_v = \lim_{s \to 0} sG(s) = K$$

Therefore $K = 2$ meets the requirement on steady-state accuracy.

For $T = 0.2$ and $K = 2$, we have (refer Table of transforms in Appendix A)

$$G_{h0}G(z) = (1 - z^{-1}) \mathscr{Z}\left[\frac{2}{s^2(s+1)}\right] = 0.03746\left[\frac{z + 0.9356}{(z-1)(z-0.8187)}\right]$$

By use of the bilinear transformation

$$z = \frac{1 + wT/2}{1 - wT/2} = \frac{1 + 0.1w}{1 - 0.1w}$$

$G_{h0}G(z)$ can be transformed to $G_{h0}G(w)$ given below (refer Eqn. (4.37)).

$$G_{h0}G(w) = \frac{2\left(1 - \dfrac{w}{10}\right)\left(1 + \dfrac{w}{300.6}\right)}{w\left(1 + \dfrac{w}{0.997}\right)}$$

The Bode plot of $G_{h0}G(jv)$ is shown in Fig. 4.34. The phase margin can be read from the Bode plot as 32° and the gain margin as 14.2 dB.

It is desired to raise the phase margin to 50° without altering K_v. Also the gain margin should at least be 10 dB. We now design a lead compensator

$$D_1(w) = \frac{1 + w\tau}{1 + \alpha w\tau}; \quad \alpha < 1, \tau > 0$$

to meet these objectives. We choose the zero of the compensator at 0.997 (This choice cancels a pole of $G_{h0}G(w)$). Addition of this zero shifts the gain crossover frequency of the uncompensated system to $v_c = 1.8$. The phase margin of the uncompensated system at v_c is $\Phi M_1 = 22°$. The phase lead required at v_c to bring the phase margin to the specified value $\Phi M_s = 50°$ is given by

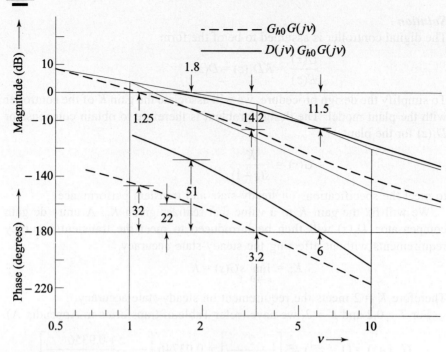

Fig. 4.34 Compensator design (Review Example 4.1)

$$\phi_l = \Phi M_s - \Phi M_1 + \varepsilon = 50° - 22° + 3° = 31°$$

By using Eqn. (4.39), we obtain

$$\alpha = \frac{1 - \sin 31°}{1 + \sin 31°} = 0.3$$

The phase lead of 31° is provided at the frequency (refer Eqn. (4.40))

$$v_m = \sqrt{\frac{1}{\tau}\left(\frac{1}{\alpha\tau}\right)} = \sqrt{0.997 \times 3.27} = 1.8$$

which is same as the gain crossover frequency.

Thus the compensator transfer function is

$$D_1(w) = \frac{1 + \dfrac{w}{0.997}}{1 + \dfrac{w}{3.27}} \qquad (4.87)$$

The magnitude and phase angle curves for the compensated open-loop transfer function are shown by solid curves in Fig. 4.34. From these curves we see that the phase margin is 51° and the gain margin is 11.5 dB.

The compensator transfer function given by Eqn. (4.87) will now be transformed back to the z-plane by the bilinear transformation

$$w = \frac{2}{T}\frac{z-1}{z+1} = 10\,\frac{z-1}{z+1}$$

This gives
$$D_1(z) = 2.718\left(\frac{z-0.8187}{z-0.5071}\right)$$

The system gain K was determined to be 2. Therefore, for the plant of the system of Fig. 4.33, the digital controller is given by

$$\frac{U(z)}{E(z)} = D(z) = 2.718\,K\left(\frac{z-0.8187}{z-0.5071}\right) = 5.436\left(\frac{z-0.8187}{z-0.5071}\right)$$

Review Example 4.2

Consider the digital control configuration shown in Fig. 4.33. The transfer function

$$G(s) = \frac{e^{-1.5s}}{s+1}$$

describes a process of temperature control via mixing (refer Example 3.3). In the following we design a digital compensator for the temperature control process; the sampling interval T is assumed to be 1 sec.

The transfer function $G_{h0}G(z)$ derived in Example 3.3 is repeated below.

$$G_{h0}G(z) = 0.3935\,\frac{z+0.6066}{z^2(z-0.3679)}$$

Since $G_{h0}G(z)$ is a type-0 transfer function, the system will have a steady-state error to a constant command or disturbance. If we assume that such a behaviour in steady-state is unacceptable, we can correct the problem by including integral control through the transfer function

$$D_1(z) = \frac{Kz}{z-1}$$

The effective plant transfer function is now

$$D_1(z)G_{h0}G(z) = 0.3935K\,\frac{(z+0.6066)}{z(z-1)(z-0.3679)} = \frac{K'(z+0.6066)}{z(z-1)(z-0.3679)}$$

The unity-feedback root locus plot for this transfer function is sketched in Fig. 4.35. The point P on the root locus corresponds to $\zeta = 0.5$,

$$\omega_n = 0.423\left(\theta = \omega_n T\sqrt{1-\zeta^2} = \frac{21\pi}{180}\right)$$

The natural frequency ω_n has to be raised to improve the speed of response. We employ a lead compensation which cancels the plant pole at $z = 0.3679$ and the plant zero at $z = -0.6066$. The open-loop transfer function of the lead-compensated system becomes

$$D_2(z)D_1(z)G_{h0}G(z) = \frac{z-0.3679}{z+0.6066}\left[\frac{K'(z+0.6066)}{z(z-1)(z-0.3679)}\right] = \frac{K'}{z(z-1)}$$

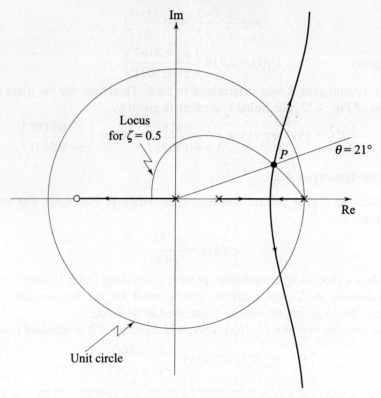

Fig. 4.35 Root locus plot of mixing flow plant with integral control

The root locus plot is sketched in Fig. 4.36. The point Q on the root locus corresponds to $\zeta = 0.5$, $\omega_n = 0.826$, $K_v = K' = 0.45$.

Suppose we wish to raise K_v to 1. A lag compensator

$$D_3(z) = \frac{z - 0.9}{z - 0.9545}$$

will raise K_v by a factor of $(1 - 0.9)/(1 - 0.9545)$. The lag pole-zero pair are very close to each other and do not change the root locus near the dominant roots significantly. However, the lag compensation does introduce a small but very slow transient, whose effect on dynamic response needs to be evaluated, especially in terms of the response to disturbances.

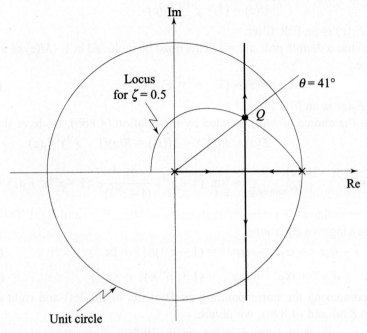

Fig. 4.36 Root locus plot of mixing flow plant with integral control and lead compensation

Review Example 4.3

Consider the digital control system of Fig. 4.33. The plant transfer function $G(s) = 1/s^2$. Design a digital controller $D(z)$ that meets the following control requirements:
  (i) Zero steady-state error to ramp input
  (ii) Finite settling time
  (iii) Sampling period $T = 0.1$ sec.

Solution

The z-transfer function of the plant is given by (refer Table of transforms given in Appendix A)

$$G_{h0}G(z) = (1 - z^{-1})\mathcal{Z}\left[\frac{1}{s^3}\right] = \frac{T^2}{2}\frac{z+1}{(z-1)^2} = 5 \times 10^{-3}\frac{z+1}{(z-1)^2}$$

The design requirement of finite settling time demands that the closed-loop transfer function $M(z)$ is an FIR filter:

$$M(z) = \alpha_0 + \alpha_1 z^{-1} + \alpha_2 z^{-2} + \cdots + \alpha_N z^{-N}$$

Since $G_{h0}G(z)$ has one more pole than zero, $M(z)$ must have pole excess of at least one. Therefore

$$\alpha_0 = 0$$

$G_{h0}G(z)$ has a zero at $z = -1$. This must be included in $M(z)$ as a zero, i.e.,

$$M(z) = (1 + z^{-1}) F_1(z) \tag{4.88a}$$

where $F_1(z)$ is an FIR filter.
$G_{h0}G(z)$ has a double pole at $z = 1$. This must be included in $1 - M(z)$ as a double zero, i.e.,

$$1 - M(z) = (1 - z^{-1})^2 F_2(z) \tag{4.88b}$$

where $F_2(z)$ is an FIR filter.

With the choice of $M(z)$ dictated by the relation (4.88b), we have the error
$$E(z) = R(z)(1 - M(z)) = R(z)(1 - z^{-1})^2 F_2(z)$$

Therefore $\left. e_{ss}^* \right|_{\text{unit ramp}} = \lim_{z \to 1} (1 - z^{-1}) \dfrac{Tz^{-1}}{(1-z^{-1})^2} (1 - z^{-1})^2 F_2(z) = 0$

We are looking for a minimal realization of $M(z)$, $F_1(z)$ and $F_2(z)$. This yields the following two equations:

$$1 - \alpha_1 z^{-1} - \alpha_2 z^{-2} - \alpha_3 z^{-3} = (1 - z^{-1})^2 (1 + \beta z^{-1}) \tag{4.89a}$$

$$\alpha_1 z^{-1} + \alpha_2 z^{-2} + \alpha_3 z^{-3} = (1 + z^{-1})(\gamma_1 + \gamma_2 z^{-1}) \tag{4.89b}$$

By comparing the corresponding coefficients of the left and right side of Eqns (4.89a) and (4.89b), we obtain

$$\alpha_1 = 1.25, \ \alpha_2 = 0.5, \ \alpha_3 = -0.75$$

$$\beta = 0.75, \ \gamma_1 = 1.25, \ \gamma_2 = -0.75$$

Thus $\quad M(z) = 1.25 z^{-1} + 0.5 z^{-2} - 0.75 z^{-3}$

The digital compensator

$$D(z) = \dfrac{1}{G_{h0}G(z)} \left[\dfrac{M(z)}{1 - M(z)} \right]$$

becomes $\quad D(z) = 2 \dfrac{125 - 75 z^{-1}}{1 + 0.75 z^{-1}}$

Review Example 4.4

A unity feedback sampled-data system with plant transfer function $G_{h0}G(z)$ and cascade controller $D(z)$, has the closed-loop transfer function

$$M(z) = \dfrac{D(z) G_{h0} G(z)}{1 + D(z) G_{h0} G(z)} \tag{4.90}$$

In the following, we study the effects of imperfect cancellation of poles and zeros of $G_{h0}G(z)$ with $D(z)$.

Consider a plant model of the form

$$G_{h0}G(z) = \left(\dfrac{z - z_1}{z - p_1} \right) F_1(z) \tag{4.91}$$

where z_1 is a zero on or outside the unit circle, and p_1 is a pole on or outside the unit circle; $F_1(z)$ represents a rational function with all poles and zeros inside the unit circle. Assuming that a digital controller is designed with poles and zeros that will cancel the zero z_1 and the pole p_1 of $G_{h0}G(z)$, and some poles and zeros of $F_1(z)$, $D(z)$ is given by

$$D(z) = \left(\frac{z - z_2}{z - p_2}\right) F_2(z) \tag{4.92}$$

where z_2 and p_2 are the zero and the pole which are to cancel p_1 and z_1 respectively. Substitution of Eqns (4.91) and (4.92) into the overall system function (4.90) yields

$$M(z) = \frac{\left(\dfrac{z - z_2}{z - p_2}\right)\left(\dfrac{z - z_1}{z - p_1}\right) F_1(z)F_2(z)}{1 + \left(\dfrac{z - z_2}{z - p_2}\right)\left(\dfrac{z - z_1}{z - p_1}\right) F_1(z)F_2(z)}$$

The characteristic equation of the system is

$$(z - p_2)(z - p_1) + (z - z_2)(z - z_1) F_1(z) F_2(z) = 0 \tag{4.93}$$

Let us assume that imperfect cancellation occurs between z_1 and p_2, so that

$$z_1 = p_2 - \Delta z_1 \tag{4.94}$$

but $p_1 = z_2$. Substituting Eqn. (4.94) into Eqn. (4.93) with $p_1 = z_2$, yields

$$1 + \frac{\Delta z_1 F_1(z) F_2(z)}{(z - p_2)[1 + F_1(z) F_2(z)]} = 0$$

The zeros of the last equation are the roots of the characteristic Eqn. (4.93). Using the root locus technique, we see that if Δz_1 is considered as a variable parameter, one of the root loci of the characteristic equation starts at $z = p_2$ which is located outside the unit circle. Therefore, for small values of Δz_1, the root will be very close to p_2, and the closed-loop system is unstable.

Similarly, if imperfect cancellation exists between p_1 and z_2, but $z_1 = p_2$, then it can be seen that one of the root loci of the characteristic equation starts at $z = z_2$ which is located outside the unit circle. Therefore, for small values of Δp_1, the closed-loop system is unstable.

Since Δp_1 and Δz_1 can have +ve or −ve values, imperfect cancellation of p_1 and z_1 located on the unit circle or just inside the unit circle may also lead to an unstable system.[9]

---

9. Study the root locus plot of $1 + KG_{h0}G(z) = 0$ for $-\infty < K < \infty$ to verify this statement [180].

PROBLEMS

4.1 For the system shown in Fig. P4.1, find
  (i) position error constant, K_p,
  (ii) velocity error constant, K_v, and
  (iii) acceleration error constant, K_a.
  Express the results in terms of K_1, K_2, J, and T.

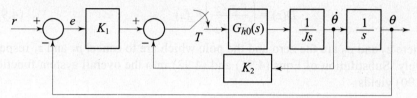

Fig. P4.1

4.2 Consider the analog control system shown in Fig. P4.2a. Show that the phase margin of the system is about 45°.

We wish to replace the analog controller by a digital controller as shown in Fig. P4.2b. First modify the analog controller to take into account the effect of the hold that must be included in the equivalent digital control system (the zero-order hold may be approximated by a pure time delay of one half the sampling period T (Fig. P4.2c) and then a lag compensator $D_1(s)$ may be designed to realize the phase margin of 45°). Then by using the bilinear transformation, determine the equivalent digital controller.

Compare the velocity error constants of the original analog system of Fig. P4.2a and the equivalent digital system of Fig. P4.2b.

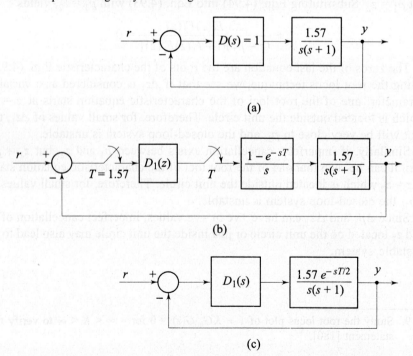

Fig. P4.2

4.3 A unity feedback system is characterized by the open-loop transfer function
$$G_{h0}G(z) = \frac{0.2385(z+0.8760)}{(z-1)(z-0.2644)}$$
The sampling period $T = 0.2$ sec.
Determine steady-state errors for unit-step, unit-ramp, and unit-acceleration inputs.

4.4 Predict the nature of the transient response of a discrete-time system whose characteristic equation is given by
$$z^2 - 1.9z + 0.9307 = 0$$
The sampling interval $T = 0.02$ sec.

4.5 The system of Fig. P4.5 contains a disturbance input $W(s)$, in addition to the reference input $R(s)$.
  (a) Express $Y(z)$ as a function of the two inputs.
  (b) Suppose that $D_2(z)$ and $D_3(z)$ are chosen such that $D_3(z) = D_2(z)G_{h0}G(z)$. Find $Y(z)$ as a function of the two inputs.
  (c) What is the advantage of the choice in part (b) if it is desired to minimize the response $Y(z)$ to the disturbance $W(s)$?

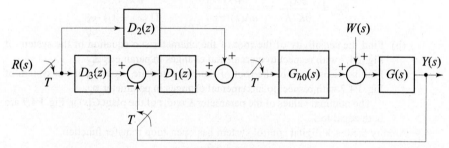

Fig. P4.5

4.6 Consider the system of Fig. P4.6. The design specifications for the system require that
  (i) the steady-state error to a unit-ramp reference input be less than 0.01,
  (ii) a constant disturbance w should not affect the steady-state value of the output.
  Show that these objectives can be met if $D(z)$ is a proportional-plus-integral compensator.

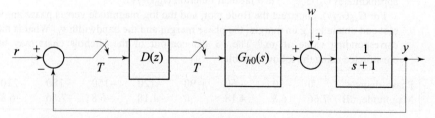

Fig. P4.6

4.7 Consider the feedback system shown in Fig. P4.7. The nominal values of the parameters K and τ of the plant $G(s)$ are both equal to 1. Find an expression for the sensitivity $S(z)$ of the closed-loop transfer function $M(z)$ with respect to incremental changes in open-loop transfer function $G_{h0}G(z)$. Plot $|S(e^{j\omega T})|$ for $0 \leq \omega \leq \omega_s/2$, where ω_s is the sampling frequency. Determine the bandwidth of the system if it is designed to have $|S(e^{j\omega T})| < 1$.

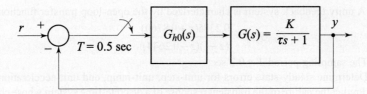

Fig. P4.7

4.8 (a) The characteristic polynomial of a digital system is of the form

$$\Delta(z) = A(z) + KB(z)$$
$$= (z - \lambda_1)(z - \lambda_2) \cdots (z - \lambda_n)$$

where K is a system parameter.

Prove that the sensitivity of the kth root of the characteristic equation with respect to incremental changes in K is given by

$$S_K^{\lambda_k} \triangleq \frac{\partial \lambda_k}{\partial K/K} = \left. \frac{A(z)}{d\Delta(z)/dz} \right|_{z=\lambda_k} = \frac{A(\lambda_k)}{\prod_{i \neq k}(\lambda_k - \lambda_i)}$$

(b) Find the sensitivity of the root of the characteristic equation of the system of Fig. P4.7 with respect to incremental changes in parameter K.

(c) Find the sensitivity of the root of the characteristic equation of the system of Fig. P4.7 with respect to incremental changes in parameter τ.

The nominal values of the parameter K and τ of the plant $G(s)$ in Fig. P4.7 are both equal to 1.

4.9 A unity feedback digital control system has open-loop transfer function

$$G_{h0}G(z) = \frac{0.368z + 0.264}{z^2 - 1.368z + 0.368}; \quad T = 1 \text{ sec}$$

The function $G_{h0}G(e^{j\omega T})$ may be used to obtain frequency response plots of the system. This function is, however, irrational. Prove that the relation

$$\omega = \frac{2}{T}\tan^{-1}\frac{vT}{2}$$

approximates $G_{h0}G(e^{j\omega T})$ to a rational function $G_{h0}G(jv)$.

For $G_{h0}G(jv)$, construct the Bode plot, and the log-magnitude versus phase angle plot and obtain the gain margin, the phase margin and the bandwidth v_b. What is the corresponding value of ω_b? The -3 dB contour of the Nichols chart may be constructed using the following table:

Phase, degrees	0	-30	-60	-90	-120	-150	-180	-210
Magnitude, dB	7.66	6.8	4.18	0	-4.18	-6.8	-7.66	-6.8

4.10 Consider the control system of Fig. P4.10 where the plant transfer function

$$G(s) = \frac{1}{s(s+2)}, \text{ and } T = 0.1 \text{ sec}$$

(a) Increase the plant gain to the value that results in $K_v = 5$. Then find the phase margin.

(b) Design a lead compensator that results in $55°$ phase margin with $K_v = 5$.

(c) Design a lag compensator that results in $55°$ phase margin with $K_v = 5$.

(d) Obtain the bandwidth realized by the three designs corresponding to parts (a), (b) and (c). Comment on the result.
(e) Is the selection of $T = 0.1$ sec justified from closed-loop bandwidth considerations?

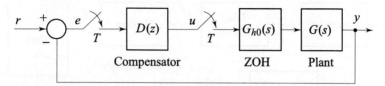

Fig. P4.10

4.11 Consider the control system of Fig. P4.10 where the plant transfer function is $G(s) = 1/s^2$, and $T = 0.1$ sec. Design a lead compensator such that the phase margin is 50° and the gain margin is at least 10 dB. Obtain the velocity error constant K_v of the compensated system.

Can the design be achieved using a lag compensator? Justify your answer.

4.12 Consider the control system of Fig. P4.10 where the plant transfer function is

$$G(s) = \frac{K}{s(s+5)}, \text{ and } T = 0.1 \text{ sec}$$

The performance specifications are given as
(i) velocity error constant $K_v \geq 10$,
(ii) phase margin $\Phi M \geq 60°$, and
(iii) bandwidth $\omega_b = 8$ rad/sec

(a) Find the value of K that gives $K_v = 10$. Determine the phase margin and the bandwidth of the closed-loop system.
(b) Show that if lead compensation is employed, the system bandwidth will increase beyond the specified value, and if lag compensation is attempted, the bandwidth will decrease sufficiently so as to fall short of the specified value.
(c) Design a lag section of a lag-lead compensator to provide partial compensation for the phase margin. Add a lead section to realize phase margin of 60°. Check the bandwidth of the lag-lead compensated system.
(d) Find the transfer function $D(z)$ of the lag-lead compensator and suggest a realization scheme.

4.13 Shown in Fig. P4.13a is a closed-loop temperature control system. Controlled electric heaters maintain the desired temperature of the liquid in the tank. The computer output controls electronic switches (triacs) to vary the effective voltage supplied to the heaters from 0 V to 230 V. The temperature is measured by a thermocouple whose output is amplified to give a voltage in the range required by A/D converter. A simplified block diagram of the system, showing perturbation dynamics, is given in Fig. P4.13b.

(a) Consider the analog control loop of Fig. P4.13c. Determine K that gives 2% steady-state error to a step input.
(b) Let $D(z) = K$ obtained in part (a). Is the sampled-data system of Fig. P4.13b stable for this value of $D(z)$?
(c) Design a lag compensator for the system of part (b), such that 2% steady-state error is realized, the phase margin is greater than 40° and the gain margin is greater than 6 dB. Give the total transfer function $D(z)$ of the compensator.

(d) Can the design of part (c) be achieved using a lead compensator? Justify your answer.

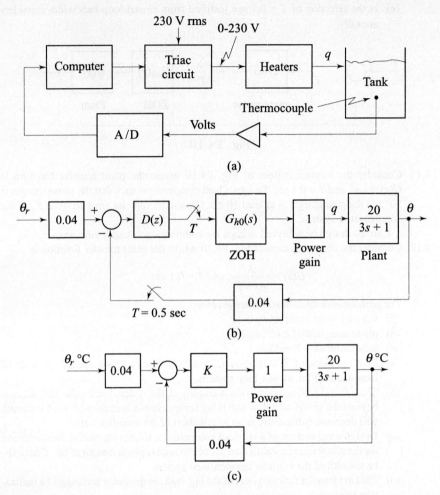

Fig. P4.13

4.14 (a) Consider a unity-feedback system with open-loop transfer function

$$G_{h0}G(z) = \frac{K(z-z_1)}{(z-p_1)(z-p_2)}; \ 0 \leq K < \infty$$

The poles and zero of this second-order transfer function lie on the real axis; the poles are adjacent or congruent with the zero to their left. Prove that the complex-conjugate section of the root locus plot is a circle with the centre at $z = z_1$ and the radius equal to $\sqrt{(z_1 - p_1)(z_1 - p_2)}$.

(b) Given $$G_{h0}G(z) = \frac{K(z - 0.9048)}{(z-1)^2}$$

Sketch the root locus plot for $0 \leq K < \infty$. Using the information in the root locus plot, determine the range of values of K for which the closed-loop system is stable. Also determine the value of K for which the system closed-loop poles are real and multiple.

4.15 A sampled-data feedback control system is shown in Fig. P4.15. The controlled process of the system is described by the transfer function

$$G(s) = \frac{K}{s(s+1)}; \quad 0 \le K < \infty$$

The sampling period $T = 1$ sec.
(a) Sketch the root locus plot for the system on the z-plane and from there obtain the value of K that results in marginal stability.
(b) Repeat part (a) for (i) $T = 2$ sec, (ii) $T = 4$ sec, and compare the stability properties of the system with different values of sampling interval.

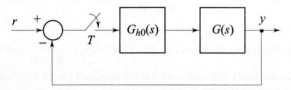

Fig. P4.15

4.16 The digital process of a unity feedback system is described by the transfer function

$$G_{h0}G(z) = \frac{K(z+0.717)}{(z-1)(z-0.368)}; \quad T = 1 \text{ sec}$$

Sketch the root locus plot for $0 \le K < \infty$ and from there obtain the following information:
(a) The value of K that results in marginal stability. Also find the frequency of oscillations.
(b) The value of K that results in $\zeta = 1$. What are the time constants of the closed-loop poles?
(c) The value of K that results in $\zeta = 0.5$. Also find the natural frequency ω_n for this value of K. You may use the following table to construct a constant-ζ locus on the z-plane corresponding to $\zeta = 0.5$.

Re	0.891	0.64	0.389	0.169	0	−0.113	−0.174	−0.188	−0.163
Im	0.157	0.37	0.463	0.464	0.404	0.310	0.207	0.068	0

4.17 The characteristic equation of a feedback control system is

$$z^2 + 0.2A\,z - 0.1\,A = 0$$

Sketch the root loci for $0 \le A < \infty$, and therefrom obtain the range of parameter A for which the system is stable.

4.18 The block diagram of a sampled-data system using a dc motor for speed control is shown in Fig. P4.18. The encoder senses the motor speed and the output of the encoder is compared with the speed command. Sketch the root locus plot for $0 \le K < \infty$.
(a) For $K = 1$, find the time constant of the closed-loop pole.
(b) Find the value of K which results in a closed-loop pole whose time constant is less than or equal to one fourth of the value found in part (a).

Use the parameter values:
$$K_m = 1, \tau_m = 1, T = 0.1 \text{ sec}, P = 60 \text{ pulses/revolution}.$$

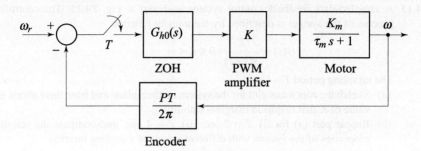

Fig. P4.18

4.19 Consider the system shown in Fig. P4.10 with $G(s) = \dfrac{1}{s(s+1)}$ and $T = 0.2$ sec.

(a) Design a lead compensator so that the dominant closed-loop poles of the system will have $\zeta = 0.5$ and $\omega_n = 4.5$.
(b) Obtain the velocity error constant K_v of the lead compensated system.
(c) Add a lag compensator in cascade so that K_v is increased by a factor of 3. What is the effect of the lag compensator on the transient response of the system?
(d) Obtain the transfer function $D(z)$ of the lag-lead compensator and suggest a realization scheme.

Use root locus method.

4.20 Consider the system shown in Fig. P4.10 with
$$G(s) = \dfrac{1}{(s+1)(s+2)}; T = 1 \text{ sec}$$

Design a compensator $D(z)$ that meets the following specifications on system performance:

(a) $\zeta = 0.5$
(b) $\omega_n = 1.5$, and
(c) $K_p \geq 7.5$

Use root locus method.

4.21 The block diagram of a digital control system is shown in Fig. P4.10. The controlled process is described by the transfer function
$$G(s) = \dfrac{K}{s^2}; T = 1 \text{ sec}$$

which may represent a pure inertial load.

(a) The dominant closed-loop poles of the system are required to have $\zeta = 0.7$, $\omega_n = 0.3$ rad/sec. Mark the desired dominant closed-loop pole locations in the z-plane. The root loci must pass through these points.
(b) Place the zero of the compensator $D(z)$ below the dominant poles and find the location of pole of $D(z)$ so that the angle criterion at the dominant poles is satisfied. Find the value of K so that the magnitude criterion at the dominant poles is satisfied.

(c) Find the acceleration error constant, K_a.
(d) Your design will result in specified values of ζ and ω_n for the closed-loop system response only if the dominance condition is satisfied. Find the third pole of the closed-loop system and comment on the effectiveness of your design.

4.22 The configuration of a commercial broadcast videotape positioning system is shown in Fig. P4.22. The relationship between the armature voltage (applied to drive motor) and tape speed at the recording and playback heads is approximated by the transfer function $G(s)$. The delay term involved accounts for the propagation of speed changes along the tape over the distance of physical separation of the tape drive mechanism and the recording and playback heads. The tape position is sensed by a recorded signal on the tape itself.

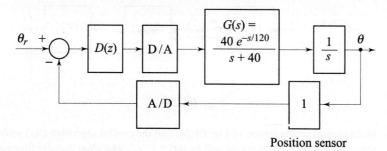

Fig. P4.22

Design the digital controller that should result in zero steady-state error to any step change in the desired tape position. The closed-loop poles of the system are required to lie within a circle of radius 0.56. Take the sampling interval $T = 1/120$ sec.

4.23 (a) Show that a necessary condition for a system to have a finite settling time, after the application of a step input, is that all the system poles should be located at the origin of the z-plane.
(b) Figure P4.23 shows the block diagram of a digital positioning system. A sampling period of 0.1 sec is to be used. Determine $D(z)$ such that the closed-loop step response settles in finite time with the output following one sampling interval behind the reference step input.

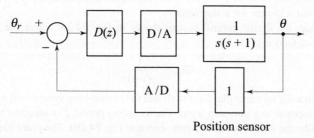

Fig. P4.23

4.24 Consider the sampled-data system shown in Fig. P4.24; the plant is known to have the transfer function

$$G(s) = \frac{1}{s(s+1)}$$

A sampling period of $T = 0.1$ sec is to be used.
(a) Design a digital controller to realize the following specifications:
  (i) $\zeta = 0.8$,
  (ii) $\omega_n = 2\pi/10T$, and
  (iii) $K_v \geq 5$.
(b) Design a digital controller so that the response to unit-step input is
$$y(k) = 0, 0.5, 1, 1, \ldots$$
Find the steady-state error to unit-ramp input.

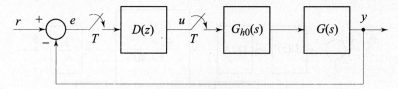

Fig. P4.24

4.25 In the control configuration of Fig. P4.24, find the control algorithm $D(z)$ so that the response to a unit-step function will be $y(t) = 1 - e^{-t}$. The plant transfer function is
$$G(s) = \frac{1}{10s+1}$$
Assume that the sampling interval $T = 2$ sec.

4.26 Consider the digital control system of Fig. P4.24. The plant transfer function $G(s) = 1/s^2$. Design a digital controller that meets the following control requirements:
  (i) zero steady-state error to ramp input,
  (ii) finite settling time, and
  (iii) sampling interval $T = 0.2$ sec.

4.27 The plant of the sampled-data system of Fig. P4.24 is given by
$$G(s) = \frac{2}{(s+1)(s+2)}$$
The sampling interval $T = 1$ sec. Find the digital compensator $D(z)$ which will give deadbeat response for a step input.

4.28 Consider the digital control system shown in Fig. P4.24, where the plant transfer function $G(s)$ is given by
$$G(s) = \frac{1}{s(s+1)}$$
Design a digital controller $D(z)$ such that the closed-loop system will exhibit a deadbeat response to a unit-step input. The sampling period T is assumed to be 1 sec.

4.29 Consider the digital control system shown in Fig. P4.29a. The plant transfer function involves a transportation lag of 5 sec. The desired output $y(t)$ in response to unit-step input is shown in Fig. P4.29b. Design a digital controller $D(z)$. Check the response of the system for intersampling ripples.

4.30 Consider the digital control system shown in Fig. P4.29a. It is desired that the system exhibit a deadbeat response to a unit-step input. Design a digital controller $D(z)$.

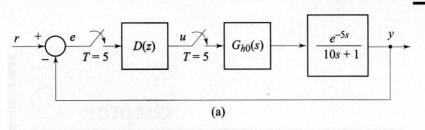

(a)

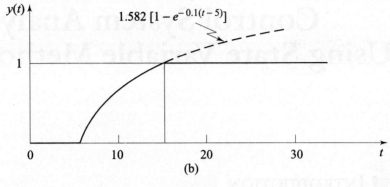

$1.582[1 - e^{-0.1(t-5)}]$

(b)

Fig. P4.29

chapter 5

Control System Analysis Using State Variable Methods

5.1 INTRODUCTION

In the preceding chapters, we have seen that the root-locus method and the frequency-response methods are quite powerful for the analysis and design of feedback control systems. The analysis and design are carried out using transfer functions, together with a variety of graphical tools such as root-locus plots, Nyquist plots, Bode plots, Nichols chart, etc. These techniques of the so called *classical control theory* have been greatly enhanced by the availability and low cost of digital computers for system analysis and simulation. The graphical tools can now be more easily used with computer graphics.

The classical design methods suffer from certain limitations due to the fact that the transfer function model is applicable only to linear time-invariant systems, and that there too it is generally restricted to single-input, single-output (SISO) systems. This is because the classical design approach becomes highly cumbersome for use in multi-input, multi-output (MIMO) systems. Another limitation of the transfer function technique is that it reveals only the system output for a given input and provides no information about the internal behaviour of the system. There may be situations where the output of a system is stable and yet some of the system elements may have a tendency to exceed their specified ratings. In addition to this, it may sometimes be necessary and advantageous to provide a feedback proportional to the internal variables of a system, rather than the output alone, for the purpose of stabilizing and improving the performance of a system.

The limitations of classical methods based on transfer function models have led to the development of state variable approach of analysis and design. It is a direct time-domain approach which provides a basis for *modern control theory*. It is a powerful technique for the analysis and design of linear and nonlinear, time-invariant or time-varying MIMO systems. The organization of the state variable approach is such that it is easily amenable to solution through digital computers.

It will be incorrect to conclude from the foregoing discussion that the state variable design methods can completely replace the classical design methods. In fact the classical control theory comprising a large body of use-tested knowledge is still going strong. State variable design methods prove their mettle in applications that are intractable by classical methods.

The state variable formulation contributes to the application areas of classical control theory in a different way. To compute the response of $G(s)$ to an input $R(s)$ requires the expansion of $\{G(s)R(s)\}$ into partial fractions; which, in turn, requires computation of all the poles of $\{G(s)R(s)\}$, or all the roots of a polynomial. The roots of a polynomial are very sensitive to their coefficients (refer Review Example 3.3). Furthermore, to develop a computer program to carry out partial fraction expansion is not simple. On the other hand, the response of state variable equations is easy to program. Its computation does not require the computation of roots or eigenvalues. Therefore, it is less sensitive to parameter variations. For these reasons, it is desirable to compute the response of $G(s)$ through state variable equations. State variable formulation is thus the most efficient form of system representation from the stand point of computer simulation. For this reason, many Computer-Aided-Design (CAD) packages, handling both the classical and the modern tools of control system design, use this notation. It is, therefore, helpful for the control engineer to be familiar with state variable methods of system representation and analysis.

Chapters 5–9 of this text present an introduction to a range of topics which fall within the domain of state variable analysis and design. Our approach is to build on and compliment the classical methods of analysis and design. State variable analysis and design methods use vector and matrix algebra and are, to some extent, different from those based on transfer functions. For this reason, we have not integrated the state variable approach with the frequency-domain approach based on transfer functions.

We have been mostly concerned with SISO systems in the text so far. In the remaining chapters also our emphasis will be on the control of SISO systems. However, many of the analysis and design methods based on state variable concepts are applicable to both SISO and MIMO systems with almost equal convenience; the only difference being the additional computational effort for MIMO systems, which is taken care of by CAD packages. A specific reference to such results will be made at appropriate places in these chapters.

5.2 VECTORS AND MATRICES

This section is intended to be a concise summary of facts about vectors and matrices. Having them all at hand will minimize the need to consult a book on matrix theory. It also serves to define the notation and terminology which are, regretably, not entirely standard.

No attempt has been made at proving every statement made in this section. The interested reader is urged to consult a suitable book (for example [32, 33]) for details of proofs.

Matrices[1]

Basic definitions and algebraic operations associated with matrices are given below.

Matrix

The matrix
$$\mathbf{A} = \begin{bmatrix} a_{11} & a_{12} & \cdots & a_{1m} \\ a_{21} & a_{22} & \cdots & a_{2m} \\ \vdots & \vdots & & \vdots \\ a_{n1} & a_{n2} & \cdots & a_{nm} \end{bmatrix} = [a_{ij}] \qquad (5.1)$$

is a rectangular array of nm elements. It has n rows and m columns. a_{ij} denotes (i, j)th element, i.e., the element located in ith row and jth column. $\mathbf{A}$ is said to be a *rectangular matrix* of order $n \times m$.

When $m = n$, i.e., the number of columns is equal to that of rows, the matrix is said to be a *square matrix* of order n.

An $n \times 1$ matrix, i.e., a matrix having only one column is called a *column matrix*. A $1 \times n$ matrix, i.e., a matrix having only one row is called a *row matrix*.

Diagonal matrix

A diagonal matrix is a square matrix whose elements off the *principal diagonal* are all zeros ($a_{ij} = 0$ for $i \neq j$). The following matrix is a diagonal matrix.

$$\Lambda = \begin{bmatrix} a_{11} & 0 & \cdots & 0 \\ 0 & a_{22} & \cdots & 0 \\ \vdots & \vdots & & \vdots \\ 0 & 0 & \cdots & a_{nn} \end{bmatrix} = \text{diag} \, [a_{11} \quad a_{12} \quad \cdots \quad a_{nm}] \qquad (5.2)$$

Unit (identity) matrix

A unit matrix $\mathbf{I}$ is a diagonal matrix whose diagonal elements are all equal to unity ($a_{ii} = 1$, $a_{ij} = 0$ for $i \neq j$).

$$\mathbf{I} = \begin{bmatrix} 1 & 0 & \cdots & 0 \\ 0 & 1 & \cdots & 0 \\ \vdots & \vdots & & \vdots \\ 0 & 0 & \cdots & 1 \end{bmatrix}$$

Whenever necessary, an $n \times n$ unit matrix will be denoted by $\mathbf{I}_n$.

---

1. We will use upper case bold letters to represent matrices and lower case bold letters to represent vectors.

Null (zero) matrix
A null matrix **0** is a matrix whose elements are all equal to zero.

$$\mathbf{0} = \begin{bmatrix} 0 & 0 & \cdots & 0 \\ 0 & 0 & \cdots & 0 \\ \vdots & \vdots & & \vdots \\ 0 & 0 & \cdots & 0 \end{bmatrix}$$

Whenever necessary, the dimensions of the null matrix will be indicated by two subscripts: $\mathbf{0}_{nm}$.

Lower-triangular matrix
A lower-triangular matrix **L** has all its elements *above* the principal diagonal equal to zero; $l_{ij} = 0$ if $i < j$ for $1 \leq i \leq n$ and $1 \leq j \leq m$.

$$\mathbf{L} = \begin{bmatrix} l_{11} & 0 & \cdots & 0 \\ l_{21} & l_{22} & \cdots & 0 \\ \vdots & \vdots & & \vdots \\ l_{n1} & l_{n2} & \cdots & l_{nm} \end{bmatrix}$$

Upper-triangular matrix
An upper-triangular matrix **U** has all its elements *below* the principal diagonal equal to zero; $u_{ij} = 0$ if $i > j$ for $1 \leq i \leq n$ and $1 \leq j \leq m$.

$$\mathbf{U} = \begin{bmatrix} u_{11} & u_{12} & \cdots & u_{1m} \\ 0 & u_{22} & \cdots & u_{2m} \\ \vdots & \vdots & & \vdots \\ 0 & 0 & \cdots & u_{nm} \end{bmatrix}$$

Matrix transpose
If the rows and columns of an $n \times m$ matrix **A** are interchanged, the resulting $m \times n$ matrix, denoted as $\mathbf{A}^T$, is called the *transpose* of the matrix **A**. Namely, if **A** is given by Eqn. (5.1), then

$$\mathbf{A}^T = \begin{bmatrix} a_{11} & a_{21} & \cdots & a_{n1} \\ a_{12} & a_{22} & \cdots & a_{n2} \\ \vdots & \vdots & & \vdots \\ a_{1m} & a_{2m} & \cdots & a_{nm} \end{bmatrix}$$

Some properties of the matrix transpose are
  (i) $(\mathbf{A}^T)^T = \mathbf{A}$
  (ii) $(k\mathbf{A})^T = k\mathbf{A}^T$, where k is a scalar
  (iii) $(\mathbf{A} + \mathbf{B})^T = \mathbf{A}^T + \mathbf{B}^T$
  (iv) $(\mathbf{AB})^T = \mathbf{B}^T \mathbf{A}^T$

Symmetric matrix
If a square matrix **A** is equal to its transpose;
$$\mathbf{A} = \mathbf{A}^T,$$

then the matrix **A** is called a symmetric matrix.

Skew-symmetric matrix
If a square matrix **A** is equal to the negative of its transpose;
$$\mathbf{A} = -\mathbf{A}^T,$$
then the matrix **A** is called a skew-symmetric matrix.

Conjugate matrix
If the complex elements of a matrix **A** are replaced by their respective conjugates, then the resulting matrix is called the conjugate of **A**.

Conjugate transpose
The conjugate transpose is the conjugate of the transpose of a matrix. Given a matrix **A**, the conjugate transpose is denoted by $\mathbf{A}^*$, and is equal to conjugate of $\mathbf{A}^T$.

Hermitian matrix
If a square matrix **A** is equal to its conjugate transpose;
$$\mathbf{A} = \mathbf{A}^*,$$
then the matrix **A** is called a Hermitian matrix. For matrices whose elements are all real (*real matrices*), symmetric and Hermitian mean the same thing.

Skew-Hermitian matrix
If a square matrix **A** is equal to the negative of its conjugate transpose;
$$\mathbf{A} = -\mathbf{A}^*,$$
then the matrix **A** is called a skew-Hermitian matrix. For real matrices, skew-symmetric and skew-Hermitian mean the same thing.

Determinant of a matrix
Determinants are defined for square matrices only. The determinant of the $n \times n$ matrix **A**, written as $|\mathbf{A}|$ or *det* **A**, is a scalar-valued function of **A**. It is found through the use of minors and cofactors.

The *minor* m_{ij} of the element a_{ij} is the determinant of a matrix of order $(n-1) \times (n-1)$ obtained from **A** by removing the row and the column containing a_{ij}.

The *cofactor* c_{ij} of the element a_{ij} is defined by the equation
$$c_{ij} = (-1)^{i+j} m_{ij}$$

Determinants can be evaluated by the method of *Laplace expansion*. If **A** is an $n \times n$ matrix, any arbitrary row k can be selected and $|\mathbf{A}|$ is then given by
$$|\mathbf{A}| = \sum_{j=1}^{n} a_{kj} c_{kj}$$

Similarly, Laplace expansion can be carried out with respect to any arbitrary column l, to obtain
$$|\mathbf{A}| = \sum_{i=1}^{n} a_{il} c_{il}$$

Laplace expansion reduces the evaluation of an $n \times n$ determinant down to the evaluation of a string of $(n-1) \times (n-1)$ determinants, namely the cofactors. Some properties of determinants are
  (i) $det\ \mathbf{AB} = (det\ \mathbf{A})(det\ \mathbf{B})$
  (ii) $det\ \mathbf{A}^T = det\ \mathbf{A}$
  (iii) $det\ k\mathbf{A} = k^n\ det\ \mathbf{A}$; $\mathbf{A}$ is $n \times n$ matrix and k is scalar
  (iv) The determinant of any diagonal or triangular matrix is the product of its diagonal elements.

Singular matrix
A square matrix is called singular if the associated determinant is zero.

Nonsingular matrix
A square matrix is called nonsingular if the associated determinant is non-zero.

Adjoint matrix
The adjoint matrix of a square matrix $\mathbf{A}$ is found by replacing each element a_{ij} of matrix $\mathbf{A}$ by its cofactor c_{ij} and then transposing.

$$adj\ \mathbf{A} = \mathbf{A}^+$$

$$= \begin{bmatrix} c_{11} & c_{21} & \cdots & c_{n1} \\ c_{12} & c_{22} & \cdots & c_{n2} \\ \vdots & \vdots & & \vdots \\ c_{1n} & c_{2n} & \cdots & c_{nn} \end{bmatrix} = [c_{ji}]$$

Note that

$$\mathbf{A}(adj\ \mathbf{A}) = (adj\ \mathbf{A})\mathbf{A} = |\mathbf{A}|\ \mathbf{I} \qquad (5.3)$$

Matrix inverse
The inverse of a square matrix is written as $\mathbf{A}^{-1}$, and is defined by the relation

$$\mathbf{A}^{-1}\mathbf{A} = \mathbf{A}\mathbf{A}^{-1} = \mathbf{I}$$

From Eqn. (5.3) and the definition of the inverse matrix, we have

$$\mathbf{A}^{-1} = \frac{adj\ \mathbf{A}}{|\mathbf{A}|} \qquad (5.4)$$

Some properties of matrix inverse are
  (i) $(\mathbf{A}^{-1})^{-1} = \mathbf{A}$
  (ii) $(\mathbf{A}^T)^{-1} = (\mathbf{A}^{-1})^T$
  (iii) $(\mathbf{AB})^{-1} = \mathbf{B}^{-1}\mathbf{A}^{-1}$
  (iv) $det\ \mathbf{A}^{-1} = \dfrac{1}{det\ \mathbf{A}}$
  (v) $det\ \mathbf{P}^{-1}\mathbf{AP} = det\ \mathbf{A}$
  (vi) Inverse of diagonal matrix given by Eqn. (5.2) is

$$\Lambda^{-1} = \begin{bmatrix} 1/a_{11} & 0 & \cdots & 0 \\ 0 & 1/a_{22} & \cdots & 0 \\ \vdots & \vdots & & \vdots \\ 0 & 0 & \cdots & 1/a_{nn} \end{bmatrix} = \text{diag} \begin{bmatrix} \dfrac{1}{a_{11}} & \dfrac{1}{a_{22}} & \cdots & \dfrac{1}{a_{nn}} \end{bmatrix}$$

Rank of a matrix

The rank $\rho(\mathbf{A})$ of a matrix $\mathbf{A}$ is the dimension of the largest array in $\mathbf{A}$ with a non-zero determinant. Some properties of rank are

(i) $\rho(\mathbf{A}^T) = \rho(\mathbf{A})$

(ii) The rank of a rectangular matrix cannot exceed the lesser of the number of rows or the number of columns. A matrix whose rank is equal to the lesser of the number of rows and number of columns is said to be of *full rank*.

$$\rho(\mathbf{A}) \leq \min(n, m); \, \mathbf{A} \text{ is } n \times m \text{ matrix}$$

(iii) The rank of a product of two matrices cannot exceed the rank of the either:

$$\rho(\mathbf{AB}) \leq \min[\rho(\mathbf{A}), \rho(\mathbf{B})]$$

Trace of a matrix

The trace of a square matrix $\mathbf{A}$ is the sum of the elements on the principal diagonal.

$$tr \, \mathbf{A} = \sum_i a_{ii} \tag{5.5}$$

Some properties of trace are

(i) $tr \, \mathbf{A}^T = tr \, \mathbf{A}$

(ii) $tr \, (\mathbf{A} + \mathbf{B}) = tr \, \mathbf{A} + tr \, \mathbf{B}$

(iii) $tr \, \mathbf{AB} = tr \, \mathbf{BA}; \, tr \, \mathbf{AB} \neq (tr \, \mathbf{A})(tr \, \mathbf{B})$

(iv) $tr \, \mathbf{P}^{-1} \mathbf{AP} = tr \, \mathbf{A}$

Partitioned matrix

A matrix can be partitioned into submatrices or vectors. Broken lines are used to show the partitioning when the elements of the submatrices are explicitly shown. For example

$$\mathbf{A} = \begin{bmatrix} a_{11} & a_{12} & \vdots & a_{13} \\ a_{21} & a_{22} & \vdots & a_{23} \\ \hdashline a_{31} & a_{32} & \vdots & a_{33} \end{bmatrix}$$

The broken lines indicating the partitioning are sometimes omitted when the context makes it clear that partitioned matrices are being considered. For example, the matrix $\mathbf{A}$ given above may be expressed as

$$\mathbf{A} = \begin{bmatrix} \mathbf{A}_{11} & \mathbf{A}_{12} \\ \mathbf{A}_{21} & \mathbf{A}_{22} \end{bmatrix}$$

We will be frequently using the following forms of partitioning.
(i) Matrix **A** partitioned into its columns:
$$\mathbf{A} = [\, \mathbf{a}_1 \quad \mathbf{a}_2 \quad \cdots \quad \mathbf{a}_m]$$
where
$$\mathbf{a}_i = \begin{bmatrix} a_{1i} \\ a_{2i} \\ \vdots \\ a_{ni} \end{bmatrix} = i\text{th column in } \mathbf{A}$$

(ii) Matrix **A** partitioned into its rows:
$$\mathbf{A} = \begin{bmatrix} \boldsymbol{\alpha}_1 \\ \boldsymbol{\alpha}_2 \\ \vdots \\ \boldsymbol{\alpha}_n \end{bmatrix}$$
where
$$\boldsymbol{\alpha}_i = [a_{i1} \quad a_{i2} \quad \cdots \quad a_{im}] = i\text{th row in } \mathbf{A}$$

(iii) A *block diagonal matrix* is a square matrix that can be partitioned so that the non-zero elements are contained only in square submatrices along the main diagonal,
$$\mathbf{A} = \begin{bmatrix} \mathbf{A}_1 & 0 & \cdots & 0 \\ 0 & \mathbf{A}_2 & \cdots & 0 \\ \vdots & \vdots & & \vdots \\ 0 & 0 & \cdots & \mathbf{A}_m \end{bmatrix}$$
$$= \text{diag}\,[\mathbf{A}_1 \quad \mathbf{A}_2 \quad \cdots \quad \mathbf{A}_m]$$

For this case
(i) $|\mathbf{A}| = |\mathbf{A}_1|\,|\mathbf{A}_2| \cdots |\mathbf{A}_m|$
(ii) $\mathbf{A}^{-1} = \text{diag}\,[\mathbf{A}_1^{-1}\,\mathbf{A}_2^{-1} \cdots \mathbf{A}_m^{-1}]$, provided that $\mathbf{A}^{-1}$ exists.

Vectors

We will be mostly concerned with vectors and matrices that have *real* elements. We therefore restrict our discussion to these cases only. An extension of the results to the situations where the vectors/matrices have complex elements is quite straightforward.

Scalar (inner) product of vectors

The scalar product of two $n \times 1$ constant vectors **x** and **y** is defined as
$$<\mathbf{x}, \mathbf{y}> = \mathbf{x}^T \mathbf{y}$$

$$= [x_1 \; x_2 \; \cdots \; x_n] \begin{bmatrix} y_1 \\ y_2 \\ \vdots \\ y_n \end{bmatrix}$$

$$= x_1 y_1 + x_2 y_2 + \cdots + x_n y_n$$
$$= \mathbf{y}^T \mathbf{x}$$

Vector norm

The concept of norm of a vector is a generalization of the idea of length. For the vector

$$\mathbf{x} = \begin{bmatrix} x_1 \\ x_2 \\ \vdots \\ x_n \end{bmatrix}$$

the *Euclidean vector norm* $\|\mathbf{x}\|$ is defined by

$$\|\mathbf{x}\| = (x_1^2 + x_2^2 + \cdots + x_n^2)^{1/2} = (\mathbf{x}^T \mathbf{x})^{1/2} \qquad (5.6a)$$

In two or three dimensions, it is easy to see that this definition for the length of $\mathbf{x}$ satisfies the conditions of Euclidean geometry. It is a generalization to n dimensions of the theorem of Pythagoras.

For any nonsingular matrix $\mathbf{P}$, the vector

$$\mathbf{y} = \mathbf{P}\mathbf{x}$$

has the Euclidean norm

$$\|\mathbf{y}\| = [(\mathbf{P}\mathbf{x})^T (\mathbf{P}\mathbf{x})]^{1/2}$$
$$= (\mathbf{x}^T \mathbf{P}^T \mathbf{P} \mathbf{x})^{1/2}$$

Letting $\mathbf{Q} = \mathbf{P}^T \mathbf{P}$, we write

$$\|\mathbf{y}\| = (\mathbf{x}^T \mathbf{Q} \mathbf{x})^{1/2}$$

or

$$\|\mathbf{x}\|_\mathbf{Q} = (\mathbf{x}^T \mathbf{Q} \mathbf{x})^{1/2} \qquad (5.6b)$$

We call $\|\mathbf{x}\|_\mathbf{Q}$ the norm of $\mathbf{x}$ with respect to $\mathbf{Q}$. It is, in fact, a generalization of the norm defined in (5.6a) in that it is a measure of the size of $\mathbf{x}$ 'weighted' by the matrix $\mathbf{Q}$.

Matrix norm

The norm of a matrix is a measure of the 'size' of the matrix (not its dimension). For the matrix

$$\mathbf{A} = \begin{bmatrix} a_{11} & a_{12} & \cdots & a_{1n} \\ a_{21} & a_{22} & \cdots & a_{2n} \\ \vdots & \vdots & & \vdots \\ a_{n1} & a_{n2} & \cdots & a_{nn} \end{bmatrix}$$

the *Euclidean matrix norm* $\|\mathbf{A}\|$ is defined by

$$\|\mathbf{A}\| = \left[\sum_{i,j=1}^{n} a_{ij}^2\right]^{1/2} = [tr\,(\mathbf{A}^T\mathbf{A})]^{1/2} \qquad (5.6c)$$

We can also describe the size of $\mathbf{A}$ by

$$\|\mathbf{A}\| = \max_{\mathbf{x}} \frac{\|\mathbf{A}\mathbf{x}\|}{\|\mathbf{x}\|}$$

i.e., the largest value of the ratio of the length $\|\mathbf{A}\mathbf{x}\|$ to the length $\|\mathbf{x}\|$. Using the Euclidean norm for vectors, we obtain

$$\|\mathbf{A}\| = \max_{\mathbf{x}} \frac{\left(\mathbf{x}^T\mathbf{A}^T\mathbf{A}\mathbf{x}\right)^{1/2}}{\left(\mathbf{x}^T\mathbf{x}\right)^{1/2}} = \max_{\mathbf{x}} \left(\frac{\mathbf{x}^T\mathbf{A}^T\mathbf{A}\mathbf{x}}{\mathbf{x}^T\mathbf{x}}\right)^{1/2}$$

The maximum value of the ratio in this expression can be determined in terms of the *eigenvalues*[2] of the matrix $\mathbf{A}^T\mathbf{A}$. The *real symmetric* matrix $\mathbf{A}^T\mathbf{A}$ has all real and non-negative eigenvalues and the maximum value of the ratio $(\mathbf{x}^T\mathbf{A}^T\mathbf{A}\mathbf{x})/(\mathbf{x}^T\mathbf{x})$ is equal to the maximum eigenvalue of $\mathbf{A}^T\mathbf{A}$ (for proof, refer [106]). Therefore,

$$\|\mathbf{A}\| = (\text{Maximum eigenvalue of } \mathbf{A}^T\mathbf{A})^{1/2} \qquad (5.6d)$$

This definition of the matrix norm is known as the *spectral norm*[3] of $\mathbf{A}$.

The square roots of the eigenvalues of $\mathbf{A}^T\mathbf{A}$ are called the *singular values* of $\mathbf{A}$. The spectral norm of $\mathbf{A}$ is equal to its largest singular value.

Singular values of a matrix are useful in numerical analysis. The ratio of the largest to the smallest singular values of $\mathbf{A}$, called the *condition number* of $\mathbf{A}$, is a measure of how close the matrix $\mathbf{A}$ comes to being singular. The matrix $\mathbf{A}$ is therefore ill-conditioned if its condition number is large.

Orthogonal vectors

Any two vectors which have a zero scalar product are said to be orthogonal vectors. Two $n \times 1$ vectors $\mathbf{x}$ and $\mathbf{y}$ are orthogonal if

$$\mathbf{x}^T\mathbf{y} = 0$$

A set of vectors is said to be orthogonal if, and only if, every two vectors from the set are orthogonal.

Unit vector

A unit vector $\hat{\mathbf{x}}$ is, by definition, a vector whose norm is unity; $\|\hat{\mathbf{x}}\| = 1$. Any non-zero vector $\mathbf{x}$ can be normalized to form unit vector.

$$\hat{\mathbf{x}} = \frac{\mathbf{x}}{\|\mathbf{x}\|}$$

---

2. The roots of the equation
$$|\lambda \mathbf{I} - \mathbf{A}| = 0$$
are called the eigenvalues of matrix $\mathbf{A}$. Detailed discussion will appear in Section 5.6.

3. Refer [102] for other valid vector and matrix norms.

Orthonormal vectors
A set of vectors is said to be orthonormal if, and only if, the set is orthogonal and each vector in this orthogonal set is a unit vector.

Orthogonal matrix
Suppose that $\{x_1, x_2, \ldots, x_n\}$ is an orthogonal set:

$$x_i^T x_i = 1, \quad \text{for all } i$$

and

$$x_i^T x_j = 0 \quad \text{for all } i \text{ and } j \text{ with } i \neq j.$$

If we form the $n \times n$ matrix

$$\mathbf{P} = [\, x_1 \; x_2 \; \cdots \; x_n \,]$$

it follows from partitioned multiplication that

$$\mathbf{P}^T \mathbf{P} = \mathbf{I}$$

That is,

$$\mathbf{P}^T = \mathbf{P}^{-1}$$

Such a matrix **P** is called an orthogonal matrix.

Linearly dependent vectors
Consider a set of m vectors $\{x_1, x_2, \ldots, x_m\}$, each of which has n components. If there exists a set of m scalars α_i, at least one of which is not zero, which satisfies

$$\alpha_1 x_1 + \alpha_2 x_2 + \cdots + \alpha_m x_m = 0,$$

then the set of vectors $\{x_i\}$ is said to be linearly dependent.

Linearly independent vectors
Any set of vectors $\{x_i\}$ which is not linearly dependent is said to be linearly independent. That is, if

$$\alpha_1 x_1 + \alpha_2 x_2 + \cdots + \alpha_m x_m = 0$$

implies that each $\alpha_i = 0$, then $\{x_i\}$ are linearly independent vectors.

Test for linear independence
Consider the set of m vectors $\{x_i\}$, each of which has n components, with $m \neq n$. Assume that this set is linearly dependent so that

$$\alpha_1 x_1 + \alpha_2 x_2 + \cdots + \alpha_m x_m = 0$$

with at least one non-zero α_i.

Premultiplying both sides of this equation by x_i^T gives a set of m simultaneous equations:

$$\alpha_1 x_i^T x_1 + \alpha_2 x_i^T x_2 + \cdots + \alpha_m x_i^T x_m = 0;$$
$$i = 1, 2, \ldots, m$$

These equations can be written in the matrix form as

$$\begin{bmatrix} \mathbf{x}_1^T\mathbf{x}_1 & \mathbf{x}_1^T\mathbf{x}_2 & \cdots & \mathbf{x}_1^T\mathbf{x}_m \\ \mathbf{x}_2^T\mathbf{x}_1 & \mathbf{x}_2^T\mathbf{x}_2 & \cdots & \mathbf{x}_2^T\mathbf{x}_m \\ \vdots & \vdots & & \vdots \\ \mathbf{x}_m^T\mathbf{x}_1 & \mathbf{x}_m^T\mathbf{x}_2 & \cdots & \mathbf{x}_m^T\mathbf{x}_m \end{bmatrix} \begin{bmatrix} \alpha_1 \\ \alpha_2 \\ \vdots \\ \alpha_m \end{bmatrix} = \begin{bmatrix} 0 \\ 0 \\ \vdots \\ 0 \end{bmatrix} \quad (5.7a)$$

or

$$\mathbf{G}\boldsymbol{\alpha} = \mathbf{0}$$

If the $m \times m$ matrix $\mathbf{G}$ has a non-zero determinant, then $\mathbf{G}^{-1}$ exists, and

$$\boldsymbol{\alpha} = \mathbf{G}^{-1}\mathbf{0} = \mathbf{0} \quad (5.7b)$$

This contradicts the assumption of at least one non-zero α_i. The matrix $\mathbf{G}$ is called the *Grammian matrix*.

A necessary and sufficient condition for the set $\{\mathbf{x}_i\}$ to be linearly dependent is that $|\mathbf{G}| = 0$.

Linear independence and rank

The column rank of a matrix $\mathbf{A}$ is equal to maximum number of linearly independent columns in $\mathbf{A}$.

The maximum number of linearly independent columns of a matrix is equal to the maximum number of linearly independent rows. Therefore, the column rank of $\mathbf{A}$ = the row rank of $\mathbf{A} = \rho(\mathbf{A})$, which is in turn equal to the order of the largest square array in $\mathbf{A}$ whose determinant does not vanish.

Vector Functions

In Section 5.8, we will require a test for the linear independence of the rows of a matrix whose elements are functions of time.
Consider a matrix

$$\mathbf{F}(t) = \begin{bmatrix} f_{11}(t) & f_{12}(t) & \cdots f_{1m}(t) \\ \vdots & \vdots & \vdots \\ f_{n1}(t) & f_{n2}(t) & \cdots f_{nm}(t) \end{bmatrix} = \begin{bmatrix} \mathbf{f}_1(t) \\ \vdots \\ \mathbf{f}_n(t) \end{bmatrix}$$

$\mathbf{f}_i(t)$; $i = 1, \ldots, n$ are the n row vectors of matrix $\mathbf{F}$; each vector has m components.

The scalar product of two $1 \times m$ vector functions $\mathbf{f}_i(t)$ and $\mathbf{f}_j(t)$ on $[t_0, t_1]$ is by definition,

$$< \mathbf{f}_i, \mathbf{f}_j > = \int_{t_0}^{t_1} \mathbf{f}_i(t)\mathbf{f}_j^T(t)\, dt$$

The set of n row-vector functions $\{\mathbf{f}_1(t), \ldots, \mathbf{f}_n(t)\}$ are linearly dependent if there exists a set of n scalars α_i, at least one of which is not zero, which satisfies

$$\alpha_1\mathbf{f}_1(t) + \alpha_2\mathbf{f}_2(t) + \cdots + \alpha_n\mathbf{f}_n(t) = \mathbf{0}_{1 \times m}$$

or
$$\alpha_1 \begin{bmatrix} f_{11}(t) \\ \vdots \\ f_{1m}(t) \end{bmatrix} + \alpha_2 \begin{bmatrix} f_{21}(t) \\ \vdots \\ f_{2m}(t) \end{bmatrix} + \cdots + \alpha_n \begin{bmatrix} f_{n1}(t) \\ \vdots \\ f_{nm}(t) \end{bmatrix} = \begin{bmatrix} 0 \\ \vdots \\ 0 \end{bmatrix}$$

Equivalently, the n rows $\mathbf{f}_i(t)$ are linearly dependent if

$$\boldsymbol{\alpha}^T \mathbf{F}(t) = \mathbf{0} \tag{5.8a}$$

for some

$$\boldsymbol{\alpha} = \begin{bmatrix} \alpha_1 \\ \vdots \\ \alpha_n \end{bmatrix} \neq \mathbf{0}$$

The Grammian matrix of functions $\mathbf{f}_i(t)$, $i = 1, \ldots, n$; where $\mathbf{f}_i(t)$ is the ith row of the matrix $\mathbf{F}(t)$, is given by (refer Eqns (5.7)),

$$\mathbf{W}(t_0, t_1) = \int_{t_0}^{t_1} \mathbf{F}(t) \mathbf{F}^T(t) \, dt \tag{5.8b}$$

The functions $\mathbf{f}_1(t), \ldots, \mathbf{f}_n(t)$, which are the rows of the matrix $\mathbf{F}(t)$, are linearly dependent on $[t_0, t_1]$ if and only if the $n \times n$ constant matrix $\mathbf{W}(t_0, t_1)$ is singular.

Quadratic Forms and Definiteness of Matrices

An expression such as

$$V(x_1, x_2, \ldots, x_n) = \sum_{i=1}^{n} \sum_{j=1}^{n} q_{ij} x_i x_j$$

involving terms of second degree in x_i and x_j is known as the *quadratic form* of n variables. Such scalar-valued functions are extensively used in stability analysis and modern control design.

In practice, one is usually concerned with quadratic forms $V(x_1, x_2, \ldots, x_n)$ that assume only *real* values. When x_i, x_j, and q_{ij} are all real, the value of V is real, and the quadratic form can be expressed in the vector-matrix notation as

$$V(\mathbf{x}) = \begin{bmatrix} x_1 & x_2 & \cdots & x_n \end{bmatrix} \begin{bmatrix} q_{11} & q_{12} & \cdots & q_{1n} \\ q_{21} & q_{22} & \cdots & q_{2n} \\ \vdots & \vdots & & \vdots \\ q_{n1} & q_{n2} & \cdots & q_{nn} \end{bmatrix} \begin{bmatrix} x_1 \\ x_2 \\ \vdots \\ x_n \end{bmatrix}$$

or
$$V(\mathbf{x}) = \mathbf{x}^T \mathbf{Q} \mathbf{x}$$

Any real square matrix $\mathbf{Q}$ may be written as the sum of a symmetric matrix $\mathbf{Q}_s$ and a skew-symmetric matrix $\mathbf{Q}_{sk}$, as shown below.

Let
$$Q = Q_s + Q_{sk}$$
Taking transpose of both sides,
$$Q^T = Q_s^T + Q_{sk}^T = Q_s - Q_{sk}$$
Solving for Q_s and Q_{sk}, we obtain
$$Q_s = \frac{Q + Q^T}{2}; \quad Q_{sk} = \frac{Q - Q^T}{2}$$
For a real matrix Q, the quadratic function $V(x)$ is therefore given by
$$V(x) = x^T Q x = x^T (Q_s + Q_{sk}) x$$
$$= x^T Q_s x + \tfrac{1}{2} x^T Q x - \tfrac{1}{2} x^T Q^T x$$
Since $x^T Q x = (x^T Q x)^T = x^T Q^T x$, we have
$$V(x) = x^T Q_s x$$

Thus, in quadratic function $V(x)$, only the symmetric portion of Q is of importance. We shall therefore tacitly assume that Q is symmetric.

It may be noted that real vector x and real matrix Q do not constitute necessary requirements for $V(x)$ to be real. $V(x)$ can be real when Q and x are possibly complex; it can easily be established that for a Hermitian matrix Q,
$$V(x) = x^* Q x$$
has real values.

Our discussion will be restricted to *real symmetric* matrices Q.

If for all $x \neq 0$,
  (i) $V(x) = x^T Q x \geq 0$,
    then $V(x)$ is called a positive semidefinite function and Q is called a *positive semidefinite matrix*;
  (ii) $V(x) = x^T Q x > 0$,
    then $V(x)$ is called a positive definite function and Q is called a *positive definite matrix*;
  (iii) $V(x) = x^T Q x \leq 0$,
    then $V(x)$ is called a negative semidefinite function and Q is called a *negative semidefinite matrix*;
  (iv) $V(x) = x^T Q x < 0$,
    then $V(x)$ is called a negative definite function and Q is called a *negative definite matrix*.

Tests for definiteness

1 Eigenvalues of Q and the nature of quadratic form

A real symmetric matrix Q has all real eigenvalues, and the signs of the eigenvalues of Q determine the nature of the quadratic form $x^T Q x$, as summarized in Table 5.1.

Table 5.1

Eigenvalues of $\mathbf{Q}$	Nature of quadratic form $\mathbf{x}^T\mathbf{Q}\mathbf{x}$
All $\lambda_i > 0$	Positive definite
All $\lambda_i \geq 0$	Positive semidefinite
All $\lambda_i < 0$	Negative definite
All $\lambda_i \leq 0$	Negative semidefinite
Some $\lambda_i \geq 0$, some $\lambda_j \leq 0$	Indefinite

II Sylvester's Criterion

The Sylvester's criterion states that the necessary and sufficient conditions for

$$V(\mathbf{x}) = \mathbf{x}^T\mathbf{Q}\mathbf{x} = [x_1 \; x_2 \; \cdots \; x_n] \begin{bmatrix} q_{11} & q_{12} & \cdots & q_{1n} \\ q_{21} & q_{22} & \cdots & q_{2n} \\ \vdots & \vdots & & \vdots \\ q_{n1} & q_{n2} & \cdots & q_{nn} \end{bmatrix} \begin{bmatrix} x_1 \\ x_2 \\ \vdots \\ x_n \end{bmatrix}$$

to be positive definite are that all the successive *principal minors* of $\mathbf{Q}$ be positive, i.e.,

$$q_{11} > 0; \; \begin{vmatrix} q_{11} & q_{12} \\ q_{21} & q_{22} \end{vmatrix} > 0; \; \begin{vmatrix} q_{11} & q_{12} & q_{13} \\ q_{21} & q_{22} & q_{23} \\ q_{31} & q_{32} & q_{33} \end{vmatrix} > 0; \; \cdots \; ; |\mathbf{Q}| > 0 \qquad (5.9)$$

The necessary and sufficient conditions for $V(\mathbf{x})$ to be positive semidefinite are that $\mathbf{Q}$ is singular and all the other principal minors of $\mathbf{Q}$ are non-negative.

$V(\mathbf{x})$ is negative definite if $[-V(\mathbf{x})]$ is positive definite. Similarly, $V(\mathbf{x})$ is negative semidefinite if $[-V(\mathbf{x})]$ is positive semidefinite.

5.3 STATE VARIABLE REPRESENTATION

We will be mostly concerned with SISO system configurations of the type shown in the block diagram of Fig. 5.1. The plant in the figure is a physical process characterized by the state variables $x_1, x_2, \ldots, x_n$, the output variable y and the input variable u.

State Variable Concepts

The modelling process of linear systems involves setting up a chain of cause–effect relationships, beginning from the input variable and ending at the output variable. This cause-effect chain includes a number of internal variables. These variables are eliminated both in the differential equation model and in the transfer function model, to obtain the final relationship between the input and the output. Analysis of systems with the input-output model will not give any

Control System Analysis Using State Variable Methods 329

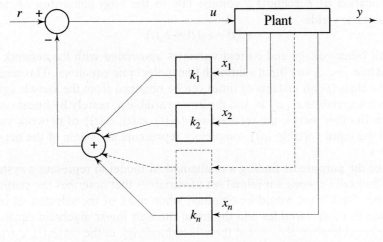

Fig. 5.1 A state-feedback control configuration

information about the behaviour of the internal variables for different operating conditions. For a better understanding of the system behaviour, its mathematical model should include the internal variables also. The state variable techniques of system representation and analysis make the internal variables an integral part of the system model, and thus provide more complete information about the system behaviour. In order to appreciate how the internal variables are included in the system representation, let us examine the modelling process by means of a simple example.

Consider the network shown in Fig. 5.2a. The set of voltages and currents associated with all the branches of the network at any time t represents the *status* of the network at that time. Application of Kirchhoff's current law at nodes 1 and 2 of the network gives the following equations:

$$\frac{de_1}{dt} + 2\frac{de_2}{dt} = \frac{u - e_1}{2}$$

$$2\frac{de_3}{dt} = 2\frac{de_2}{dt}$$

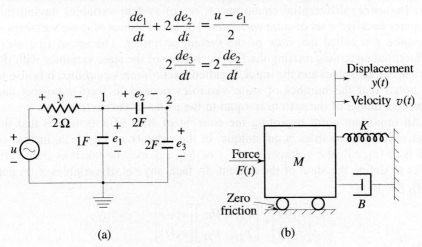

(a)　　　　　　　　　(b)

Fig. 5.2

330 Digital Control and State Variable Methods

Application of Kirchhoff's voltage law to the loop consisting of the three capacitors, yields

$$e_1(t) - e_2(t) = e_3(t)$$

All other voltage and current variables associated with the network are related to e_1, e_2, e_3 and input u through linear algebraic equations. This means that their values (at all instants of time) can be obtained from the knowledge of the network variables e_1, e_2, e_3 and the input variable u, merely by linear combinations. In other words, the reduced set $\{e_1(t), e_2(t), e_3(t)\}$ of network variables with the input variable $u(t)$ completely represents the status of the network at time t.

For the purpose of finding a mathematical model to represent a system, we will naturally choose a minimal set of variables that describes the status of the system. Such a set would be obtained when none of the selected variables is related to other variables and the input through linear algebraic equations. A little consideration shows that there is redundancy in the set $\{e_1(t), e_2(t), e_3(t)\}$ for the network of Fig. 5.2a; a set of two variables, say $\{e_1(t), e_2(t)\}$, with the input $u(t)$ represents the network completely at time t.

Manipulation of the network equations obtained earlier gives

$$\frac{de_1(t)}{dt} = -\tfrac{1}{4}e_1(t) + \tfrac{1}{4}u(t)$$

$$\frac{de_2(t)}{dt} = -\tfrac{1}{8}e_1(t) + \tfrac{1}{8}u(t)$$

This set of equations constitutes a mathematical model for the system. It is a set of two first-order differential equations. Its complete solution for any given $u(t)$ applied at $t = t_0$, will require a knowledge of the value of the selected variables $\{e_1, e_2\}$ at $t = t_0$. To put it differently, we can say that if the values of $\{e_1, e_2\}$ at $t = t_0$ are known, then the values of these variables at any time $t > t_0$, in response to a given input $u_{(t_0, t]}$, can be obtained by the solution of the two first-order differential equations. A set of system variables having this property is called a set of *state variables*. The set of values of these variables at any time t is called the *state* of the system at time t. The set of first-order differential equations relating the first derivative of the state variables with the variables themselves and the input, is called a set of *state equations*. It is also to be noted that the number of state variables needed to form a correct and complete model of the system is equal to the *order of the system*.

An important point regarding the concept of state of a system is that the choice of state variables is not unique. In the network of Fig. 5.2a, instead of voltages $\{e_1, e_2\}$, the voltages $\{e_2, e_3\}$ or $\{e_3, e_1\}$ may be taken as state variables to define the state of the system. In fact, any set of variables $x_1(t)$ and $x_2(t)$, given by

$$\begin{bmatrix} x_1(t) \\ x_2(t) \end{bmatrix} = \begin{bmatrix} p_{11} & p_{12} \\ p_{21} & p_{22} \end{bmatrix} \begin{bmatrix} e_1(t) \\ e_2(t) \end{bmatrix}$$

where p_{ij} are constants such that the matrix

$$\mathbf{P} = \begin{bmatrix} p_{11} & p_{12} \\ p_{21} & p_{22} \end{bmatrix}$$

is nonsingular, is qualified to describe the state of the system of Fig. 5.2a because we can express the capacitor voltages in terms of the selected variables $x_1(t)$ and $x_2(t)$. This brings out additional facts that there are *infinitely many choices of state variables for any given system, and that the selected state variables may not correspond to physically measurable quantities of the system*. Later in this chapter we will see that all the choices of state variables are not equally convenient. Usually, state variables are chosen so that they correspond to physically measurable quantities or lead to particularly simplified calculations.

For a particular goal of study of a given physical system, we may not be interested in the total information about the system at a particular time. We may be interested only in a part of the total information. This is called the *output of the system*, which can be obtained algebraically from the information of the system state and the input. The output is, by definition, a physical attribute of the system and is measurable. For example, in the electric network of Fig. 5.2a, the information of interest may be the voltage across the resistor. The output

$$y(t) = -e_1(t) + u(t)$$

As another example of the state of a system, consider the mechanical network shown in Fig. 5.2b. The force $F(t)$ is the input variable. Defining the displacement $y(t)$ of the mass as the output variable, we obtain the following input–output model for the system

$$M \frac{d^2 y(t)}{dt^2} + B \frac{dy(t)}{dt} + Ky(t) = F(t)$$

An alternative form of the input-output model is the transfer function model:

$$\frac{Y(s)}{F(s)} = \frac{1}{Ms^2 + Bs + K}$$

The set of forces, velocities, and displacements associated with all the elements of the mechanical network at any time t represents the status of the network at that time. A little consideration shows that values of all the system variables (at all instants of time) can be obtained from the knowledge of the system variables $y(t)$ and $v(t)$, and the input variable $F(t)$, merely by linear combinations. The dynamics of $y(t)$ and $v(t)$ are given by the following first-order differential equations:

$$\frac{dy(t)}{dt} = v(t)$$

$$\frac{dv(t)}{dt} = -\frac{K}{M} y(t) - \frac{B}{M} v(t) + \frac{1}{M} F(t)$$

The variables $\{y(t), v(t)\}$ are therefore the state variables of the system of Fig. 5.2b, and the two first-order differential equations given above are the state equations of the system. Using standard symbols for state variables and input variable, we can write the state equations as

$$\dot{x}_1 = x_2$$
$$\dot{x}_2 = -\frac{K}{M}x_1 - \frac{B}{M}x_2 + \frac{1}{M}u$$

$$x_1(t) \triangleq y(t)$$
$$x_2(t) \triangleq v(t)$$
$$u(t) \triangleq F(t)$$

Defining $y(t)$ as the output variable, the output equation becomes
$$y = x_1$$

We can now appreciate the following definitions.

State The *state* of a dynamic system is the smallest set of variables (called *state variables*) such that the knowledge of these variables at $t = t_0$, together with the knowledge of the input for $t \geq t_0$, completely determines the behaviour of the system for any time $t \geq t_0$.

State vector If n state variables $x_1, x_2, \ldots, x_n$, are needed to completely describe the behaviour of a given system, then these n state variables can be considered the n components of a vector **x**. Such a vector is called a *state vector*.

State space The n-dimensional space whose coordinate axes consist of the x_1-axis, x_2-axis, $\ldots$, x_n-axis, is called a *state space*.

State trajectory At any time t_0, the state vector (and hence the state of the system) defines a point in the state space. As time progresses and the system state changes, a set of points will be defined. This set of points, the locus of the tip of the state vector as time progresses, is called the *state trajectory* of the system.

State space and state trajectory in two-dimensional cases are referred to as the *phase plane* and *phase trajectory*, respectively.

State Variable Modelling

We know through our modelling experience that the application of physical laws to mechanical, electrical, thermal, liquid-level, and other physical processes results in a set of first-order and second-order differential equations.[4] Linear time-invariant differential equations can be rearranged in the following form:

---

4. Chapter 2 of reference [180].

Control System Analysis Using State Variable Methods

$$\dot{x}_1(t) = a_{11}x_1(t) + a_{12}x_2(t) + \cdots + a_{1n}x_n(t) + b_1 u(t)$$
$$\dot{x}_2(t) = a_{21}x_1(t) + a_{22}x_2(t) + \cdots + a_{2n}x_n(t) + b_2 u(t) \quad (5.10a)$$
$$\vdots$$
$$\dot{x}_n(t) = a_{n1}x_1(t) + a_{n2}x_2(t) + \cdots + a_{nn}x_n(t) + b_n u(t)$$

where the coefficients a_{ij} and b_i are constants. These n first-order differential equations are called *state equations* of the system.
Integration of Eqns (5.10a) gives

$$x_i(t) = x_i(t_0) + \int_{t_0}^{t} [a_{i1}x_1(t) + \cdots + a_{in}x_n(t) + b_i u(t)]dt;$$

$$i = 1, 2, \ldots, n$$

Thus, the n state variables and hence the state of the system can be determined uniquely at any $t > t_0$ if each state variable is known at $t = t_0$ and the control force $u(t)$ is known throughout the interval t_0 to t.

The output $y(t)$ at any time t will be a function of $x_1(t), x_2(t), \ldots, x_n(t)$. However, in some cases the output may also depend upon the instantaneous value of the input $u(t)$. For linear systems, the output is a linear combination of the state variables and the input:

$$y(t) = c_1 x_1(t) + c_2 x_2(t) + \cdots + c_n x_n(t) + d u(t) \quad (5.10b)$$

where c_i and d are constants. The algebraic Eqn. (5.10b) is called *output equation* of the system.

Since every real-world system has some nonlinearity, a mathematical model of the form (5.10) is an approximation to reality. Many real-world nonlinearities involve a 'smooth' curvelinear relation between independent and dependent variables. Nonlinear functions $f_i(.)$ and $g(.)$ of the form

$$\dot{x}_i(t) = f_i(x_1(t), x_2(t), \ldots, x_n(t), u(t)) \; ; \; x_i(t_0) \triangleq x_i^0 \quad (5.11a)$$
$$y(t) = g(x_1(t), x_2(t), \ldots, x_n(t), u(t)) \quad (5.11b)$$

may be linearized about a selected operating point using the multivariable form of the Taylor series:

$$f(x_1, x_2, x_3, \ldots) = f(x_{10}, x_{20}, \ldots) + \left[\frac{\partial f}{\partial x_1}\bigg|_{x_{10}, x_{20}, \ldots}\right](x_1 - x_{10})$$
$$+ \left[\frac{\partial f}{\partial x_2}\bigg|_{x_{10}, x_{20}, \ldots}\right](x_2 - x_{20}) + \cdots \quad (5.11c)$$

One of the advantages of state variable formulation is that an extremely compact vector-matrix notation can be used for the mathematical model. Using the laws of matrix algebra, it becomes much less cumbersome to manipulate the equations.

In the vector-matrix notation, we may write Eqns (5.10) as

$$\begin{bmatrix} \dot{x}_1(t) \\ \dot{x}_2(t) \\ \vdots \\ \dot{x}_n(t) \end{bmatrix} = \begin{bmatrix} a_{11} & a_{12} & \cdots & a_{1n} \\ a_{21} & a_{22} & \cdots & a_{2n} \\ \vdots & \vdots & & \vdots \\ a_{n1} & a_{n2} & \cdots & a_{nn} \end{bmatrix} \begin{bmatrix} x_1(t) \\ x_2(t) \\ \vdots \\ x_n(t) \end{bmatrix} + \begin{bmatrix} b_1 \\ b_2 \\ \vdots \\ b_n \end{bmatrix} u(t); \begin{bmatrix} x_1(t_0) \\ x_2(t_0) \\ \vdots \\ x_n(t_0) \end{bmatrix} \triangleq \begin{bmatrix} x_1^0 \\ x_2^0 \\ \vdots \\ x_n^0 \end{bmatrix}$$

(5.12a)

$$y(t) = \begin{bmatrix} c_1 & c_2 & \cdots & c_n \end{bmatrix} \begin{bmatrix} x_1(t) \\ x_2(t) \\ \vdots \\ x_n(t) \end{bmatrix} + d\, u(t) \qquad (5.12b)$$

In compact notation, Eqns (5.12) may be expressed as

$$\dot{\mathbf{x}}(t) = \mathbf{A}\mathbf{x}(t) + \mathbf{b}u(t); \mathbf{x}(t_0) \triangleq \mathbf{x}^0 : \textit{State equation} \qquad (5.13a)$$
$$y(t) = \mathbf{c}\mathbf{x}(t) + du(t) \qquad \qquad : \textit{output equation} \qquad (5.13b)$$

where

$\mathbf{x}(t) = n \times 1$ state vector of nth-order dynamic system
$u(t) =$ system input
$y(t) =$ defined output
$\mathbf{A} = n \times n$ matrix
$\mathbf{b} = n \times 1$ column matrix
$\mathbf{c} = 1 \times n$ row matrix
$d =$ scalar, representing direct coupling between input and output (direct coupling is rare in control systems, i.e., usually $d = 0$)

Example 5.1

Two very usual applications of dc motors are in speed and position control systems.

Figure 5.3 gives the basic block diagram of a speed control system. A separately excited dc motor drives the load. A dc tachogenerator is attached to the

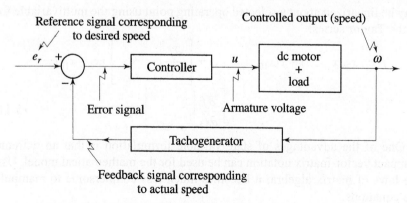

Fig. 5.3 Basic block diagram of a closed-loop speed control system

motor shaft; speed signal is fedback and the error signal is used to control the armature voltage of the motor.

In the following, we derive the plant model for the speed control system. A separately excited dc motor with armature voltage control is shown in Fig. 5.4.

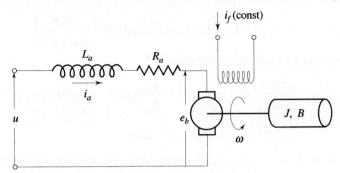

Fig. 5.4 Model of a separately excited dc motor

The voltage loop equation is

$$u(t) = L_a \frac{di_a(t)}{dt} + R_a i_a(t) + e_b(t) \tag{5.14a}$$

where

- L_a = inductance of armature winding (henrys)
- R_a = resistance of armature winding (ohms)
- i_a = armature current (amperes)
- e_b = back emf (volts)
- u = applied armature voltage (volts)

The torque balance equation is

$$T_M(t) = J \frac{d\omega(t)}{dt} + B\omega(t) \tag{5.14b}$$

where

- T_M = torque developed by the motor (newton-m)
- J = equivalent moment of inertia of motor and load referred to motor shaft (kg-m²)
- B = equivalent viscous friction coefficient of motor and load referred to motor shaft $\left(\dfrac{\text{newton-m}}{\text{rad/sec}}\right)$
- ω = angular velocity of motor shaft (rad/sec)

In servo applications, the dc motors are generally used in the linear range of the magnetization curve. Therefore, the air gap flux ϕ is proportional to the field current. For the armature controlled motor, the field current i_f is held constant. Therefore, the torque T_M developed by the motor, which is proportional to the product of the armature current and the air gap flux, can be expressed as

$$T_M(t) = K_T i_a(t) \tag{5.14c}$$

where

K_T = motor torque constant $\left(\dfrac{\text{newton-m}}{\text{amp}}\right)$

The counter electromotive force e_b, which is proportional to ϕ and ω, can be expressed as

$$e_b(t) = K_b\, \omega(t) \tag{5.14d}$$

where

$$K_b = \text{back emf constant}^5 \left(\frac{\text{volts}}{\text{rad/sec}}\right)$$

Equations (5.14) can be reorganized as

$$\frac{di_a(t)}{dt} = -\frac{R_a}{L_a} i_a(t) - \frac{K_b}{L_a}\omega(t) + \frac{1}{L_a} u(t)$$

$$\frac{d\omega(t)}{dt} = \frac{K_T}{J} i_a(t) - \frac{B}{J}\omega(t) \tag{5.15}$$

$x_1(t) = \omega(t)$, and $x_2(t) = i_a(t)$ is the obvious choice for state variables. The output variable is $y(t) = \omega(t)$.

The plant model of the speed control system organized into the vector-matrix notation is given below:

$$\begin{bmatrix} \dot{x}_1(t) \\ \dot{x}_2(t) \end{bmatrix} = \begin{bmatrix} -\dfrac{B}{J} & \dfrac{K_T}{J} \\ -\dfrac{K_b}{L_a} & -\dfrac{R_a}{L_a} \end{bmatrix} \begin{bmatrix} x_1(t) \\ x_2(t) \end{bmatrix} + \begin{bmatrix} 0 \\ \dfrac{1}{L_a} \end{bmatrix} u(t)$$

$$y(t) = x_1(t)$$

Let us assign numerical values to the system parameters.

For the parameters[6]

$$R_a = 1 \text{ ohm}, \; L_a = 0.1 \text{ H}, \; J = 0.1 \text{ kg-m}^2,$$

$$B = 0.1 \; \frac{\text{newton-m}}{\text{rad/sec}}, \; K_b = K_T = 0.1, \tag{5.16}$$

the plant model becomes

$$\dot{\mathbf{x}}(t) = \mathbf{A}\mathbf{x}(t) + \mathbf{b}u(t)$$
$$y(t) = \mathbf{c}\mathbf{x}(t) \tag{5.17}$$

where $\mathbf{A} = \begin{bmatrix} -1 & 1 \\ -1 & -10 \end{bmatrix}; \; \mathbf{b} = \begin{bmatrix} 0 \\ 10 \end{bmatrix}; \; \mathbf{c} = \begin{bmatrix} 1 & 0 \end{bmatrix}$

Example 5.2

Figure 5.5 gives the basic block diagram of a position control system. The controlled variable is now the angular position $\theta(t)$ of the motor shaft:

$$\frac{d\theta(t)}{dt} = \omega(t) \tag{5.18}$$

---

5. In MKS untis, $K_b = K_T$; Section 3.5 of reference [180].
6. These parameters have been chosen for computational convenience.

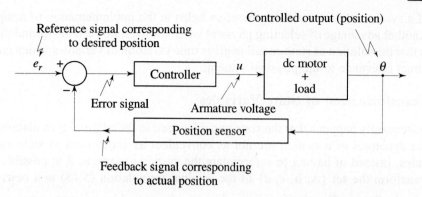

Fig. 5.5 Basic block diagram of a closed-loop position control system

We make the following choice for state and output variables.

$$x_1(t) = \theta(t), \quad x_2(t) = \omega(t), \quad x_3(t) = i_a(t), \quad y(t) = \theta(t)$$

For this choice, we obtain the following plant model from Eqns (5.15) and (5.18).

$$\begin{bmatrix} \dot{x}_1(t) \\ \dot{x}_2(t) \\ \dot{x}_3(t) \end{bmatrix} = \begin{bmatrix} 0 & 1 & 0 \\ 0 & -\dfrac{B}{J} & \dfrac{K_T}{J} \\ 0 & -\dfrac{K_b}{L_a} & -\dfrac{R_a}{L_a} \end{bmatrix} \begin{bmatrix} x_1(t) \\ x_2(t) \\ x_3(t) \end{bmatrix} + \begin{bmatrix} 0 \\ 0 \\ \dfrac{1}{L_a} \end{bmatrix} u(t)$$

$$y(t) = x_1(t)$$

For the system parameters given by (5.16), the plant model for position control system becomes

$$\dot{\mathbf{x}}(t) = \mathbf{A}\mathbf{x}(t) + \mathbf{b}u(t)$$
$$y(t) = \mathbf{c}\mathbf{x}(t) \tag{5.19}$$

where

$$\mathbf{A} = \begin{bmatrix} 0 & 1 & 0 \\ 0 & -1 & 1 \\ 0 & -1 & -10 \end{bmatrix}; \quad \mathbf{b} = \begin{bmatrix} 0 \\ 0 \\ 10 \end{bmatrix}; \quad \mathbf{c} = [1 \; 0 \; 0]$$

▲▲

In Examples 5.1 and 5.2 discussed above, the selected state variables are the physical quantities of the systems which can be measured.

We will see in Chapter 7 that in a physical system, in addition to output, other state variables could be utilized for the purpose of feedback. The implementation of design with state variable feedback becomes straightforward if the state variables are available for feedback. The choice of physical variables of a system as state variables therefore helps in the implementation of design. Another advantage of selecting physical variables for state variable formulation is that the solution of state equation gives time variation of variables which have direct relevance to the physical system.

Transformation of State Variables

It frequently happens that the state variables used in the original formulation of the dynamics of a system are not as convenient as another set of state variables. Instead of having to reformulate the system dynamics, it is possible to transform the set $\{\mathbf{A}, \mathbf{b}, \mathbf{c}, d\}$ of the original formulation (5.13) to a new set $\{\bar{\mathbf{A}}, \bar{\mathbf{b}}, \bar{\mathbf{c}}, \bar{d}\}$. The change of variables is represented by a linear transformation

$$\mathbf{x} = \mathbf{P}\bar{\mathbf{x}} \qquad (5.20a)$$

where $\bar{\mathbf{x}}$ is a state vector in the new formulation and $\mathbf{x}$ is the state vector in the original formulation. It is assumed that the transformation matrix $\mathbf{P}$ is a nonsingular $n \times n$ matrix, so that we can always write

$$\bar{\mathbf{x}} = \mathbf{P}^{-1}\mathbf{x} \qquad (5.20b)$$

We assume, moreover, that $\mathbf{P}$ is a constant matrix.

The original dynamics are expressed by

$$\dot{\mathbf{x}}(t) = \mathbf{A}\mathbf{x}(t) + \mathbf{b}u(t); \ \mathbf{x}(t_0) \triangleq \mathbf{x}^0 \qquad (5.21a)$$

and the output by

$$y(t) = \mathbf{c}\mathbf{x}(t) + du(t) \qquad (5.21b)$$

Substitution of $\mathbf{x}$ as given by Eqn. (5.20a) into these equations, gives

$$\mathbf{P}\dot{\bar{\mathbf{x}}}(t) = \mathbf{A}\mathbf{P}\bar{\mathbf{x}}(t) + \mathbf{b}u(t)$$
$$y(t) = \mathbf{c}\mathbf{P}\bar{\mathbf{x}}(t) + du(t)$$

or

$$\dot{\bar{\mathbf{x}}}(t) = \bar{\mathbf{A}}\bar{\mathbf{x}}(t) + \bar{\mathbf{b}}u(t); \ \bar{\mathbf{x}}(t_0) = \mathbf{P}^{-1}\mathbf{x}(t_0) \qquad (5.22a)$$
$$y(t) = \bar{\mathbf{c}}\bar{\mathbf{x}}(t) + \bar{d}u(t) \qquad (5.22b)$$

with

$$\bar{\mathbf{A}} = \mathbf{P}^{-1}\mathbf{A}\mathbf{P}, \ \bar{\mathbf{b}} = \mathbf{P}^{-1}\mathbf{b}, \ \bar{\mathbf{c}} = \mathbf{c}\mathbf{P}, \ \bar{d} = d$$

In the next section we will prove that both the linear systems (5.21) and (5.22) have identical output responses for the same input. The linear system (5.22) is said to be *equivalent* to the linear system (5.21), and $\mathbf{P}$ is called an *equivalence* or *similarity transformation*.

It is obvious that there exist an infinite number of equivalent systems since the transformation matrix $\mathbf{P}$ can be arbitrarily chosen. Some transformations have been extensively used for the purposes of analysis and design. Five of such special (*canonical*) transformations will be used in the present and the next two chapters.

Example 5.3
Example 5.1 revisited

For the system of Fig. 5.4, we have taken angular velocity $\omega(t)$ and armature current $i_a(t)$ as state variables:

$$\mathbf{x} = \begin{bmatrix} x_1 \\ x_2 \end{bmatrix} = \begin{bmatrix} \omega \\ i_a \end{bmatrix}$$

We now define new state variables as

$$\bar{x}_1 = \omega, \ \bar{x}_2 = -\omega + i_a$$

or

$$\bar{\mathbf{x}} = \begin{bmatrix} \bar{x}_1 \\ \bar{x}_2 \end{bmatrix} = \begin{bmatrix} x_1 \\ -x_1 + x_2 \end{bmatrix} = \begin{bmatrix} 1 & 0 \\ -1 & 1 \end{bmatrix} \begin{bmatrix} x_1 \\ x_2 \end{bmatrix}$$

We can express velocity $x_1(t)$ and armature current $x_2(t)$ in terms of the variables $\bar{x}_1(t)$ and $\bar{x}_2(t)$:

$$\mathbf{x} = \mathbf{P}\bar{\mathbf{x}} \tag{5.23}$$

with

$$\mathbf{P} = \begin{bmatrix} 1 & 0 \\ 1 & 1 \end{bmatrix}$$

Using Eqns (5.22) and (5.17), we obtain the following state variable model for the system of Fig. 5.4 in terms of the transformed state vector $\bar{\mathbf{x}}(t)$:

$$\begin{aligned} \dot{\bar{\mathbf{x}}}(t) &= \bar{\mathbf{A}}\bar{\mathbf{x}}(t) + \bar{\mathbf{b}}u(t) \\ y(t) &= \bar{\mathbf{c}}\,\bar{\mathbf{x}}(t) \end{aligned} \tag{5.24}$$

where

$$\bar{\mathbf{A}} = \mathbf{P}^{-1}\mathbf{A}\mathbf{P} = \begin{bmatrix} 1 & 0 \\ -1 & 1 \end{bmatrix}\begin{bmatrix} -1 & 1 \\ -1 & -10 \end{bmatrix}\begin{bmatrix} 1 & 0 \\ 1 & 1 \end{bmatrix} = \begin{bmatrix} 0 & 1 \\ -11 & -11 \end{bmatrix}$$

$$\bar{\mathbf{b}} = \mathbf{P}^{-1}\mathbf{b} = \begin{bmatrix} 1 & 0 \\ -1 & 1 \end{bmatrix}\begin{bmatrix} 0 \\ 10 \end{bmatrix} = \begin{bmatrix} 0 \\ 10 \end{bmatrix}$$

$$\bar{\mathbf{c}} = \mathbf{c}\mathbf{P} = \begin{bmatrix} 1 & 0 \end{bmatrix}\begin{bmatrix} 1 & 0 \\ 1 & 1 \end{bmatrix} = \begin{bmatrix} 1 & 0 \end{bmatrix}$$

$$\bar{x}_1(t_0) = x_1(t_0); \ \bar{x}_2(t_0) = -x_1(t_0) + x_2(t_0)$$

Equations (5.24) give an alternative state variable model of the system previously represented by Eqns (5.17). $\bar{\mathbf{x}}(t)$ and $\mathbf{x}(t)$ both qualify to be state vectors of the given system (the two vectors individually characterize the system completely at time t), and the output $y(t)$, as we shall see shortly, is uniquely determined from either of the models (5.17) and (5.24). State variable model (5.24) is thus *equivalent* to the model (5.17), and the matrix $\mathbf{P}$ given by Eqn. (5.23) is an *equivalence* or *similarity transformation*.

The state variable model given by Eqns (5.24) is in a *canonical* (special) form. In Chapter 7, we will use this form of models for pole-placement design by state feedback.

State Diagrams

An important advantage of state variable formulation is that it is a straightforward method to obtain a simulation diagram for the state equations. This is extremely useful if we wish to use computer simulation methods to study dynamic systems. In the following, we give an example of analog simulation diagram. Examples of digital simulation will appear in Chapter 6.

For brevity we consider a second-order system:

$$\begin{aligned} \dot{x}_1(t) &= a_{11} x_1(t) + a_{12} x_2(t) + b_1 u(t) \\ \dot{x}_2(t) &= a_{21} x_1(t) + a_{22} x_2(t) + b_2 u(t) \\ y(t) &= c_1 x_1(t) + c_2 x_2(t) \end{aligned} \qquad (5.25)$$

It is evident that if we knew $\dot{x}_1$ and $\dot{x}_2$, we could obtain x_1 and x_2 by simple integration. Hence $\dot{x}_1$ and $\dot{x}_2$ should be the inputs to two integrators. The corresponding integrator outputs are x_1 and x_2. This leaves only the problem of obtaining $\dot{x}_1$ and $\dot{x}_2$ for use as inputs to the integrators. In fact this is already specified by state equations. The completed state diagram is shown in Fig. 5.6. This diagram is essentially an analog-computer program for the given system.

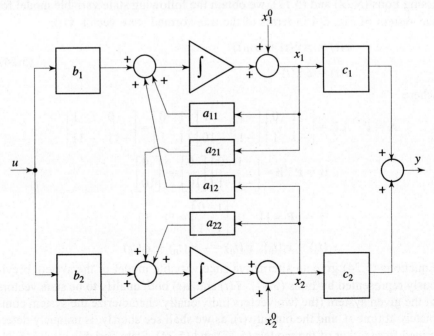

Fig. 5.6 State diagram for the system (5.25)

5.4 CONVERSION OF STATE VARIABLE MODELS TO TRANSFER FUNCTIONS

We shall derive the transfer function of a SISO system from the Laplace-transformed version of the state and output equations. Refer Section 5.2 for the vector and matrix operations used in the derivation.

Consider the state variable model (Eqns (5.13)):

$$\dot{\mathbf{x}}(t) = \mathbf{A}\mathbf{x}(t) + \mathbf{b}u(t); \quad \mathbf{x}(t_0) \triangleq \mathbf{x}^0 \quad (5.26)$$
$$y(t) = \mathbf{c}\mathbf{x}(t) + du(t)$$

Taking the Laplace transform of Eqns (5.26), we obtain

$$s\mathbf{X}(s) - \mathbf{x}^0 = \mathbf{A}\mathbf{X}(s) + \mathbf{b}U(s)$$
$$Y(s) = \mathbf{c}\mathbf{X}(s) + dU(s)$$

where

$$\mathbf{X}(s) \triangleq \mathscr{L}[\mathbf{x}(t)]; \quad U(s) \triangleq \mathscr{L}[u(t)]; \quad Y(s) \triangleq \mathscr{L}[y(t)]$$

Manipulation of these equations gives

$$(s\mathbf{I} - \mathbf{A})\mathbf{X}(s) = \mathbf{x}^0 + \mathbf{b}U(s); \quad \mathbf{I} \text{ is } n \times n \text{ identity matrix}$$

or
$$\mathbf{X}(s) = (s\mathbf{I} - \mathbf{A})^{-1}\mathbf{x}^0 + (s\mathbf{I} - \mathbf{A})^{-1}\mathbf{b}U(s) \quad (5.27a)$$
$$Y(s) = \mathbf{c}(s\mathbf{I} - \mathbf{A})^{-1}\mathbf{x}^0 + [\mathbf{c}(s\mathbf{I} - \mathbf{A})^{-1}\mathbf{b} + d]U(s) \quad (5.27b)$$

Equations (5.27) are algebraic equations. If $\mathbf{x}^0$ and $U(s)$ are knwon, $\mathbf{X}(s)$ and $Y(s)$ can be computed from these equations.

In the case of a zero initial state (i.e., $\mathbf{x}^0 = \mathbf{0}$), the input-output behaviour of the system (5.26) is determined entirely by the transfer function

$$\frac{Y(s)}{U(s)} = G(s) = \mathbf{c}(s\mathbf{I} - \mathbf{A})^{-1}\mathbf{b} + d \quad (5.28)$$

We can express the inverse of the matrix $(s\mathbf{I} - \mathbf{A})$ as

$$(s\mathbf{I} - \mathbf{A})^{-1} = \frac{(s\mathbf{I} - \mathbf{A})^+}{|s\mathbf{I} - \mathbf{A}|} \quad (5.29)$$

where

$|s\mathbf{I} - \mathbf{A}|$ = determinant of the matrix $(s\mathbf{I} - \mathbf{A})$
$(s\mathbf{I} - \mathbf{A})^+$ = adjoint of the matrix $(s\mathbf{I} - \mathbf{A})$

Using Eqn. (5.29), the transfer function $G(s)$ given by Eqn. (5.28) can be written as

$$G(s) = \frac{\mathbf{c}(s\mathbf{I} - \mathbf{A})^+\mathbf{b}}{|s\mathbf{I} - \mathbf{A}|} + d \quad (5.30)$$

For a general nth-order matrix

$$\mathbf{A} = \begin{bmatrix} a_{11} & a_{12} & \cdots & a_{1n} \\ a_{21} & a_{22} & \cdots & a_{2n} \\ \vdots & \vdots & & \vdots \\ a_{n1} & a_{n2} & \cdots & a_{nn} \end{bmatrix},$$

the matrix $(s\mathbf{I} - \mathbf{A})$ has the following appearance:

$$(s\mathbf{I} - \mathbf{A}) = \begin{bmatrix} s - a_{11} & -a_{12} & \cdots & -a_{1n} \\ -a_{21} & s - a_{22} & \cdots & -a_{2n} \\ \vdots & \vdots & & \vdots \\ -a_{n1} & -a_{n2} & \cdots & s - a_{nn} \end{bmatrix}$$

If we imagine calculating $det(s\mathbf{I} - \mathbf{A})$, we see that one of the terms will be product of diagonal elements of $(s\mathbf{I} - \mathbf{A})$:

$$(s - a_{11})(s - a_{22}) \cdots (s - a_{nn}) = s^n + \alpha_1' s^{n-1} + \cdots + \alpha_n',$$

a polynomial of degree n with the leading coefficient of unity. There will be other terms coming from the off-diagonal elements of $(s\mathbf{I} - \mathbf{A})$, but none will have a degree as high as n. Thus $|s\mathbf{I} - \mathbf{A}|$ will be of the following form:

$$|s\mathbf{I} - \mathbf{A}| = \Delta(s) = s^n + \alpha_1 s^{n-1} + \cdots + \alpha_n \tag{5.31}$$

where α_i are constant scalars.

This is known as the *characteristic polynomial* of the matrix $\mathbf{A}$. It plays a vital role in the dynamic behaviour of the system. The roots of this polynomial are called the *characteristic roots* or *eigenvalues* of matrix $\mathbf{A}$. These roots, as we shall see in Section 5.7, determine the essential features of the unforced dynamic behaviour of the system (5.26).

The adjoint of an $n \times n$ matrix is itself an $n \times n$ matrix whose elements are the cofactors of the original matrix. Each cofactor is obtained by computing the determinant of the matrix that remains when a row and a column of the original matrix are deleted. It thus follows that each element in $(s\mathbf{I} - \mathbf{A})^+$ is a polynomial in s of maximum degree $(n - 1)$. Adjoint of $(s\mathbf{I} - \mathbf{A})$ can therefore be expressed as

$$(s\mathbf{I} - \mathbf{A})^+ = \mathbf{Q}_1 s^{n-1} + \mathbf{Q}_2 s^{n-2} + \cdots + \mathbf{Q}_{n-1} s + \mathbf{Q}_n \tag{5.32}$$

where $\mathbf{Q}_i$ are constant $n \times n$ matrices.

We can express transfer function $G(s)$ given by Eqn. (5.30) in the following form:

$$G(s) = \frac{\mathbf{c}[\mathbf{Q}_1 s^{n-1} + \mathbf{Q}_2 s^{n-2} + \cdots + \mathbf{Q}_{n-1} s + \mathbf{Q}_n]\mathbf{b}}{s^n + \alpha_1 s^{n-1} + \cdots + \alpha_{n-1} s + \alpha_n} + d \tag{5.33}$$

$G(s)$ is thus a rational function of s. When $d = 0$, the degree of numerator polynomial of $G(s)$ is strictly less than the degree of the denominator polynomial and therefore the resulting transfer function is a *strictly proper transfer function*. When $d \neq 0$, the degree of numerator polynomial of $G(s)$ will be equal to the degree of the denominator polynomial, giving a *proper transfer function*. Further,

$$d = \lim_{s \to \infty} [G(s)] \tag{5.34}$$

From Eqns (5.31) and (5.33) we observe that the characteristic polynomial of matrix $\mathbf{A}$ of the system (5.26) is same as the denominator polynomial of the corresponding transfer function $G(s)$. If there are no cancellations between the numerator and denominator polynomials of $G(s)$ in Eqn. (5.33), the *eigenvalues* of matrix **A** are same as the *poles* of $G(s)$. We will take up in Section 5.9, this aspect of the correspondence between state variable models and transfer functions. It will be proved that for a completely controllable and completely observable state variable model, the eigenvalues of matrix **A** are same as the poles of the corresponding transfer function.

Invariance Property

It is recalled that the state variable model for a system is not unique, but depends on the choice of a set of state variables. A transformation

$$\mathbf{x}(t) = \mathbf{P}\bar{\mathbf{x}}(t); \ \mathbf{P} \text{ is a nonsingular matrix} \tag{5.35}$$

results in the following alternative state variable model (refer Eqns (5.22)) for the system (5.26):

$$\dot{\bar{\mathbf{x}}}(t) = \bar{\mathbf{A}}\bar{\mathbf{x}}(t) + \bar{\mathbf{b}}u(t); \ \bar{\mathbf{x}}(t_0) = \mathbf{P}^{-1}\mathbf{x}(t_0) \tag{5.36a}$$

$$y(t) = \bar{\mathbf{c}}\bar{\mathbf{x}}(t) + du(t) \tag{5.36b}$$

where $\bar{\mathbf{A}} = \mathbf{P}^{-1}\mathbf{A}\mathbf{P}, \ \bar{\mathbf{b}} = \mathbf{P}^{-1}\mathbf{b}, \ \bar{\mathbf{c}} = \mathbf{c}\mathbf{P}$

The definition of new set of internal state variables should evidently not affect the eigenvalues or input-output behaviour. This may be verified by evaluating the characteristic polynomial and the transfer function of the transformed system.

(i) $|s\mathbf{I} - \bar{\mathbf{A}}| = |s\mathbf{I} - \mathbf{P}^{-1}\mathbf{A}\mathbf{P}| = |s\mathbf{P}^{-1}\mathbf{P} - \mathbf{P}^{-1}\mathbf{A}\mathbf{P}|$
$$= |\mathbf{P}^{-1}(s\mathbf{I} - \mathbf{A})\mathbf{P}| = |\mathbf{P}^{-1}| \, |s\mathbf{I} - \mathbf{A}| \, |\mathbf{P}| = |s\mathbf{I} - \mathbf{A}| \tag{5.37}$$

(ii) System output in response to input $u(t)$ is given by the transfer function

$$\bar{G}(s) = \bar{\mathbf{c}}(s\mathbf{I} - \bar{\mathbf{A}})^{-1}\bar{\mathbf{b}} + d$$
$$= \mathbf{c}\mathbf{P}(s\mathbf{I} - \mathbf{P}^{-1}\mathbf{A}\mathbf{P})^{-1}\mathbf{P}^{-1}\mathbf{b} + d$$
$$= \mathbf{c}\mathbf{P}(s\mathbf{P}^{-1}\mathbf{P} - \mathbf{P}^{-1}\mathbf{A}\mathbf{P})^{-1}\mathbf{P}^{-1}\mathbf{b} + d$$
$$= \mathbf{c}\mathbf{P}[\mathbf{P}^{-1}(s\mathbf{I} - \mathbf{A})\mathbf{P}]^{-1}\mathbf{P}^{-1}\mathbf{b} + d$$
$$= \mathbf{c}\mathbf{P}\mathbf{P}^{-1}(s\mathbf{I} - \mathbf{A})^{-1}\mathbf{P}\mathbf{P}^{-1}\mathbf{b} + d$$
$$= \mathbf{c}(s\mathbf{I} - \mathbf{A})^{-1}\mathbf{b} + d = G(s) \tag{5.38}$$

(iii) System output in response to initial state $\bar{\mathbf{x}}(t_0)$ is given by (refer Eqn. (5.27b))

$$\bar{\mathbf{c}}(s\mathbf{I} - \bar{\mathbf{A}})^{-1}\bar{\mathbf{x}}(t_0) = \mathbf{c}\mathbf{P}(s\mathbf{I} - \mathbf{P}^{-1}\mathbf{A}\mathbf{P})^{-1}\mathbf{P}^{-1}\mathbf{x}(t_0)$$
$$= \mathbf{c}(s\mathbf{I} - \mathbf{A})^{-1}\mathbf{x}(t_0) \tag{5.39}$$

The input-output behaviour of the system (5.26) is thus *invariant* under the transformation (5.35).

Example 5.4

Consider the position control system of Example 5.2. The plant model of the system is reproduced below:

$$\dot{\mathbf{x}}(t) = \mathbf{A}\mathbf{x}(t) + \mathbf{b}u(t)$$
$$y(t) = \mathbf{c}\mathbf{x}(t) \tag{5.40}$$

with

$$\mathbf{A} = \begin{bmatrix} 0 & 1 & 0 \\ 0 & -1 & 1 \\ 0 & -1 & -10 \end{bmatrix}; \mathbf{b} = \begin{bmatrix} 0 \\ 0 \\ 10 \end{bmatrix}; \mathbf{c} = [1 \ 0 \ 0]$$

The characteristic polynomial of matrix **A** is

$$|s\mathbf{I} - \mathbf{A}| = \begin{vmatrix} s & -1 & 0 \\ 0 & s+1 & -1 \\ 0 & 1 & s+10 \end{vmatrix} = s(s^2 + 11s + 11)$$

The transfer function

$$G(s) = \frac{Y(s)}{U(s)} = \frac{\mathbf{c}(s\mathbf{I} - \mathbf{A})^+ \mathbf{b}}{|s\mathbf{I} - \mathbf{A}|}$$

$$= \frac{[1 \ 0 \ 0] \begin{bmatrix} s^2 + 11s + 11 & s+10 & 1 \\ 0 & s(s+10) & s \\ 0 & -s & s(s+1) \end{bmatrix} \begin{bmatrix} 0 \\ 0 \\ 10 \end{bmatrix}}{s(s^2 + 11s + 11)}$$

$$= \frac{10}{s(s^2 + 11s + 11)} \tag{5.41}$$

Alternatively, we can draw the state diagram of the plant model in signal flow graph form and from there obtain the transfer function using Mason's gain formula. For the plant model (5.40), the state diagram is shown in Fig. 5.7. Application of Mason's gain formula[7] yields

$$\frac{Y(s)}{U(s)} = G(s) = \frac{10s^{-3}}{1-(-10s^{-1}-s^{-1}-s^{-2})+10s^{-2}}$$

$$= \frac{10}{s^3 + 11s^2 + 11s} = \frac{10}{s(s^2 + 11s + 11)}$$

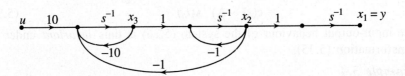

Fig. 5.7 State diagram for the system (5.40)

---

7. Section 3.4 of reference [180].

Resolvent Algorithm

The matrix

$$\Phi(s) = (s\mathbf{I} - \mathbf{A})^{-1} = \frac{(s\mathbf{I} - \mathbf{A})^+}{|s\mathbf{I} - \mathbf{A}|} \quad (5.42)$$

is known in mathematical literature as the *resolvent* of $\mathbf{A}$. Resolvent matrix $\Phi(s)$ can be expressed in the following form (refer Eqns (5.31) and (5.32)):

$$\Phi(s) = (s\mathbf{I} - \mathbf{A})^{-1} = \frac{\mathbf{Q}_1 s^{n-1} + \mathbf{Q}_2 s^{n-2} + \cdots + \mathbf{Q}_{n-1} s + \mathbf{Q}_n}{s^n + \alpha_1 s^{n-1} + \cdots + \alpha_{n-1} s + \alpha_n} \quad (5.43)$$

where $\mathbf{Q}_i$ are constant $(n \times n)$ matrices and α_j are constant scalars.

An interesting and useful relationship for the coefficient matrices $\mathbf{Q}_i$ of the adjoint matrix can be obtained by multiplying both sides of Eqn. (5.43) by $|s\mathbf{I} - \mathbf{A}|(s\mathbf{I} - \mathbf{A})$. The result is

$$|s\mathbf{I} - \mathbf{A}|\,\mathbf{I} = (s\mathbf{I} - \mathbf{A})(\mathbf{Q}_1 s^{n-1} + \mathbf{Q}_2 s^{n-2} + \cdots + \mathbf{Q}_{n-1} s + \mathbf{Q}_n)$$

or

$$s^n \mathbf{I} + \alpha_1 s^{n-1} \mathbf{I} + \cdots + \alpha_n \mathbf{I} = s^n \mathbf{Q}_1 + s^{n-1}(\mathbf{Q}_2 - \mathbf{A}\mathbf{Q}_1)$$
$$+ \cdots + s(\mathbf{Q}_n - \mathbf{A}\mathbf{Q}_{n-1}) - \mathbf{A}\mathbf{Q}_n$$

Equating the coefficients of s^i on both the sides, gives

$$\begin{aligned}
\mathbf{Q}_1 &= \mathbf{I} \\
\mathbf{Q}_2 &= \mathbf{A}\mathbf{Q}_1 + \alpha_1 \mathbf{I} \\
\mathbf{Q}_3 &= \mathbf{A}\mathbf{Q}_2 + \alpha_2 \mathbf{I} \\
&\vdots \\
\mathbf{Q}_n &= \mathbf{A}\mathbf{Q}_{n-1} + \alpha_{n-1} \mathbf{I} \\
0 &= \mathbf{A}\mathbf{Q}_n + \alpha_n \mathbf{I}
\end{aligned} \quad (5.44a)$$

We have thus determined that the leading coefficient of $(s\mathbf{I} - \mathbf{A})^+$ is the identity matrix, and that the subsequent coefficients can be obtained recursively. The last equation in (5.44a) is redundant, but can be used as a check when these recursion equations are used as the basis of a numerical algorithm.

An algorithm based on Eqns (5.44a) requires the coefficients α_i ($i = 1, 2, \ldots, n$) of the characteristic polynomial. Fortunately, the determination of these coefficients can be included in the algorithm, for it can be shown that[8]

$$\alpha_i = -\frac{1}{i} tr(\mathbf{A}\mathbf{Q}_i); \quad i = 1, 2, \ldots, n \quad (5.44b)$$

where $tr(\mathbf{M})$, the trace of $\mathbf{M}$, is the sum of all the diagonal elements of the matrix $\mathbf{M}$.

---

8. The proof of relation (5.44b) is quite involved and will not be presented here. Refer [115].

The algorithm given by Eqns (5.44), called the *resolvent algorithm*, is convenient for hand calculation and also easy to implement on a digital computer.

Example 5.5

Here we again compute $(s\mathbf{I} - \mathbf{A})^{-1}$ which appeared in Example 5.4, but this time using the resolvent algorithm (5.44).

$$\mathbf{Q}_1 = \mathbf{I},\ \alpha_1 = -tr(\mathbf{A}) = 11$$
$$\mathbf{Q}_2 = \mathbf{A} + \alpha_1 \mathbf{I}$$

$$= \begin{bmatrix} 11 & 1 & 0 \\ 0 & 10 & 1 \\ 0 & -1 & 1 \end{bmatrix};\ \alpha_2 = -\tfrac{1}{2} tr(\mathbf{AQ}_2) = 11$$

$$\mathbf{Q}_3 = \mathbf{AQ}_2 + \alpha_2 \mathbf{I}$$

$$= \begin{bmatrix} 11 & 10 & 1 \\ 0 & 0 & 0 \\ 0 & 0 & 0 \end{bmatrix};\ \alpha_3 = -\tfrac{1}{3} tr(\mathbf{AQ}_3) = 0$$

As a numerical check, we see that the relation

$$0 = \mathbf{AQ}_3 + \alpha_3 \mathbf{I}$$

is satisfied. Therefore,

$$(s\mathbf{I} - \mathbf{A})^{-1} = \mathbf{\Phi}(s) = \frac{\mathbf{Q}_1 s^2 + \mathbf{Q}_2 s + \mathbf{Q}_3}{s^3 + \alpha_1 s^2 + \alpha_2 s + \alpha_3}$$

$$= \frac{1}{s(s^2 + 11s + 11)} \begin{bmatrix} s^2 + 11s + 11 & s + 10 & 1 \\ 0 & s(s+10) & s \\ 0 & -s & s(s+1) \end{bmatrix}$$

Cayley-Hamilton Theorem

Using resolvent algorithm, we develop here a fundamental property of the characteristic equation. To this end, we write from Eqns (5.44a)

$$\mathbf{Q}_2 = \mathbf{A} + \alpha_1 \mathbf{I}$$
$$\mathbf{Q}_3 = \mathbf{AQ}_2 + \alpha_2 \mathbf{I} = \mathbf{A}^2 + \alpha_1 \mathbf{A} + \alpha_2 \mathbf{I}$$
$$\vdots$$
$$\mathbf{Q}_n = \mathbf{A}^{n-1} + \alpha_1 \mathbf{A}^{n-2} + \cdots + \alpha_{n-1} \mathbf{I}$$
$$\mathbf{AQ}_n = \mathbf{A}^n + \alpha_1 \mathbf{A}^{n-1} + \cdots + \alpha_{n-1} \mathbf{A} = -\alpha_n \mathbf{I}$$

Therefore,

$$\mathbf{A}^n + \alpha_1 \mathbf{A}^{n-1} + \cdots + \alpha_{n-1}\mathbf{A} + \alpha_n \mathbf{I} = 0 \qquad (5.45)$$

This is the well-known result known as the *Cayley-Hamilton theorem*. Note that this equation is same as the characteristic equation

$$s^n + \alpha_1 s^{n-1} + \cdots + \alpha_{n-1} s + \alpha_n = 0 \tag{5.46}$$

with the scalar s^i in the latter replaced by the matrix $\mathbf{A}^i$ ($i = 1, 2, \ldots, n$).

Thus another way of stating the Cayley-Hamilton theorem is: *Every matrix satisfies its own characteristic equation.*

Later we will use the resolvent algorithm and the Cayley-Hamilton theorem for evaluation of the state transition matrix required for the solution of the state equations.

5.5 CONVERSION OF TRANSFER FUNCTIONS TO CANONICAL STATE VARIABLE MODELS

In the last section, we studied the problem—finding the transfer function from the state variable model of a system. The converse problem—finding a state variable model from the transfer function of a system, is the subject of discussion in this section. This problem is quite important because of the following reasons:

(i) Quite often the system dynamics is determined experimentally using standard test signals like a step, impulse, or sinusoidal signal. A transfer function is conveniently fitted to the experimental data in some best possible manner.

There are, however, many design techniques developed exclusively for state variable models. In order to apply these techniques, experimentally obtained transfer function descriptions must be realized into state variable models.

(ii) Realization of transfer functions into state variable models is needed even if the control system design is based on frequency-domain design methods. In these cases the need arises for the purpose of transient response simulation. Many algorithms and numerical integration computer programs designed for solution of systems of first-order equations are available, but there is not much software for the numerical inversion of Laplace transforms. Thus, if a reliable method is needed for calculating the transient response of a system, one may be better off converting the transfer function of the system to state variable description and numerically integrating the resulting differential equations rather than attempting to compute the inverse Laplace transform by numerical methods.

We shall discuss here the problem of realization of transfer function into state variable models. Note the use of the term 'realization'. A state variable model that has a prescribed rational function $G(s)$ as its transfer function, is the *realization* of $G(s)$. The term 'realization' is justified by the fact that by using the state diagram corresponding to the state vari-

348 Digital Control and State Variable Methods

able model, the system with the transfer function $G(s)$ can be built in the real world by an op amp circuit.[9]

The following three problems are involved in the realization of a given transfer function into state variable models:

(i) Is it possible at all to obtain state variable description from the given transfer function?
(ii) If yes, is the state variable description unique for a given transfer function?
(iii) How do we obtain the state variable description from the given transfer function?

The answer to the first problem has been given in the last section. A rational function $G(s)$ is realizable by a finite dimensional linear time-invariant state model if and only if $G(s)$ is a proper rational function. A proper rational function will have state model of the form:

$$\dot{\mathbf{x}}(t) = \mathbf{A}\mathbf{x}(t) + \mathbf{b}u(t) \\ y(t) = \mathbf{c}\mathbf{x}(t) + du(t) \tag{5.47}$$

where $\mathbf{A}$, $\mathbf{b}$, $\mathbf{c}$ and d are constant matrices of appropriate dimensions. A strictly proper rational function will have state model of the form

$$\dot{\mathbf{x}}(t) = \mathbf{A}\mathbf{x}(t) + \mathbf{b}u(t) \\ y(t) = \mathbf{c}\mathbf{x}(t) \tag{5.48}$$

Let us now turn to the second problem. In the last section, we saw that there are innumerable systems that have the same transfer function. Hence, the representation of a transfer function in state variable form is obviously not unique. However, all these representations will be equivalent.

In the remaining part of this section, we deal with the third problem. We shall develop three standard, or 'canonical' representations of transfer functions.

A linear time-invariant SISO system is described by transfer function of the form

$$G(s) = \frac{\beta_0 s^m + \beta_1 s^{m-1} + \cdots + \beta_m}{s^n + \alpha_1 s^{n-1} + \cdots + \alpha_n} \quad ; m \leq n$$

where the coefficients α_i and β_i are real constant scalars. Note that there is no loss in generality to assume the coefficient of s^n to be unity.

In the following, we derive results for $m = n$; these results may be used for the case $m < n$ by setting appropriate β_i coefficients equal to zero. Therefore, our problem is to obtain a state variable model corresponding to the transfer function

$$G(s) = \frac{\beta_0 s^n + \beta_1 s^{n-1} + \cdots + \beta_n}{s^n + \alpha_1 s^{n-1} + \cdots + \alpha_n} \tag{5.49}$$

---

9. Section 2.9 of reference [180].

First Companion Form

Our development starts with a transfer function of the form

$$\frac{Z(s)}{U(s)} = \frac{1}{s^n + \alpha_1 s^{n-1} + \cdots + \alpha_n} \qquad (5.50)$$

which can be written as

$$(s^n + \alpha_1 s^{n-1} + \cdots + \alpha_n) Z(s) = U(s)$$

The corresponding differential equation is

$$p^n z(t) + \alpha_1 p^{n-1} z(t) + \cdots + \alpha_n z(t) = u(t)$$

where

$$p^k z(t) \triangleq \frac{d^k z(t)}{dt^k}$$

Solving for highest derivative of $z(t)$, we obtain

$$p^n z(t) = -\alpha_1 p^{n-1} z(t) - \alpha_2 p^{n-2} z(t) - \cdots - \alpha_n z(t) + u(t) \qquad (5.51)$$

Now consider a chain of n integrators as shown in Fig. 5.8. Suppose that the output of the last integrator is $z(t)$. Then the output of the just previous integrator is $pz = dz/dt$, and so forth. The output from the first integrator is $p^{n-1} z(t)$, and the input to this integrator is thus $p^n z(t)$. This leaves only the problem of obtaining $p^n z(t)$ for use as input to the first integrator. In fact, this is already specified by Eqn. (5.51). Realization of this equation is shown in Fig. 5.8.

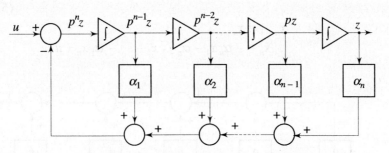

Fig. 5.8 Realization of the system (5.51)

Having developed a realization of the simple transfer function (5.50), we are now in a position to consider the more general transfer function (5.49). We decompose this transfer function into two parts, as shown in Fig. 5.9. The output $Y(s)$ can be written as

$$Y(s) = (\beta_0 s^n + \beta_1 s^{n-1} + \cdots + \beta_n) Z(s) \qquad (5.52a)$$

where $Z(s)$ is given by

$$\frac{Z(s)}{U(s)} = \frac{1}{s^n + \alpha_1 s^{n-1} + \cdots + \alpha_n} \qquad (5.52b)$$

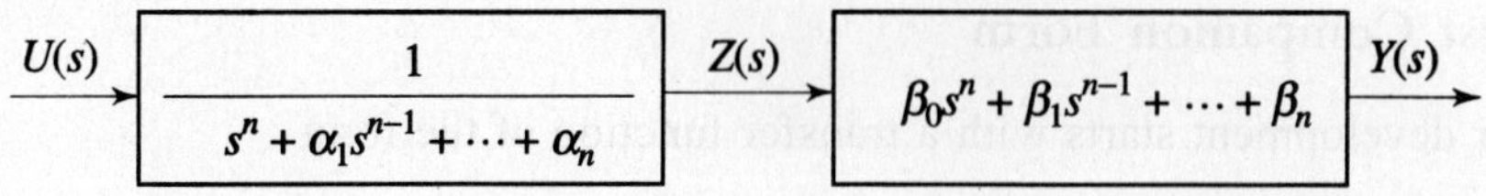

350 Digital Control and State Variable Methods

$$U(s) \to \boxed{\frac{1}{s^n + \alpha_1 s^{n-1} + \cdots + \alpha_n}} \xrightarrow{Z(s)} \boxed{\beta_0 s^n + \beta_1 s^{n-1} + \cdots + \beta_n} \to Y(s)$$

Fig. 5.9 Decomposition of the transfer function (5.49)

A realization of the transfer function (5.52b) has already been developed. Figure 5.8 shows this realization. The output of the last integrator is $z(t)$ and the inputs to the integrators in the chain from the last to the first are the n successive derivatives of $z(t)$.

Realization of the transfer function (5.52a) is now straightforward. The output

$$y(t) = \beta_0 p^n z(t) + \beta_1 p^{n-1} z(t) + \cdots + \beta_n z(t),$$

is nothing but the sum of the scaled versions of the inputs to the n integrators. Figure 5.10 shows complete realization of the transfer function (5.49). All that remains to be done is to write the corresponding differential equations.

To get one state variable model of the system, we identify the output of each integrator in Fig. 5.10 with a state variable, starting at the right and proceeding to the left. The corresponding differential equations using this identification of state variables are

$$\begin{aligned}
\dot{x}_1 &= x_2 \\
\dot{x}_2 &= x_3 \\
&\vdots \\
\dot{x}_{n-1} &= x_n \\
\dot{x}_n &= -\alpha_n x_1 - \alpha_{n-1} x_2 - \cdots - \alpha_1 x_n + u
\end{aligned} \qquad (5.53a)$$

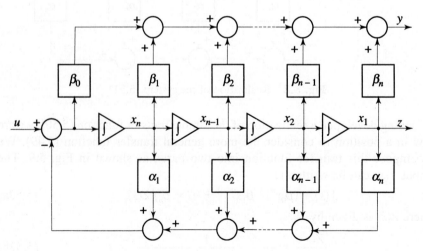

Fig. 5.10 Realization of the system (5.49)

The output equation is found by careful examination of the block diagram of Fig. 5.10. Note that there are two paths from the output of each integrator to the system output—one path upward through the box labelled β_i, and a second path down through the box labelled α_i and thence through the box labelled β_0. As a consequence,

$$y = (\beta_n - \alpha_n\beta_0)\, x_1 + (\beta_{n-1} - \alpha_{n-1}\beta_0)\, x_2 + \cdots$$
$$+ (\beta_1 - \alpha_1\beta_0)\, x_n + \beta_0 u \tag{5.53b}$$

The state and output Eqns (5.53), organized in vector-matrix form, are given below:

$$\dot{\mathbf{x}}(t) = \mathbf{A}\mathbf{x}(t) + \mathbf{b}u(t)$$
$$y(t) = \mathbf{c}\mathbf{x}(t) + du(t) \tag{5.54}$$

with

$$\mathbf{A} = \begin{bmatrix} 0 & 1 & 0 & \cdots & 0 \\ 0 & 0 & 1 & \cdots & 0 \\ \vdots & \vdots & \vdots & & \vdots \\ 0 & 0 & 0 & \cdots & 1 \\ -\alpha_n & -\alpha_{n-1} & -\alpha_{n-2} & \cdots & -\alpha_1 \end{bmatrix}; \mathbf{b} = \begin{bmatrix} 0 \\ 0 \\ \vdots \\ 0 \\ 1 \end{bmatrix}$$

$$\mathbf{c} = [\beta_n - \alpha_n\beta_0,\ \beta_{n-1} - \alpha_{n-1}\beta_0,\ \ldots,\ \beta_1 - \alpha_1\beta_0];\ d = \beta_0$$

If the direct path through β_0 is absent (refer Fig. 5.10), then the scalar d is zero and the row matrix $\mathbf{c}$ contains only the β_i coefficients.

The matrix $\mathbf{A}$ in Eqns (5.54) has a very special structure: the coefficients of the denominator of the transfer function preceded by minus signs form a string along the bottom row of the matrix. The rest of the matrix is zero except for the 'superdiagonal' terms which are all unity. In matrix theory, a matrix with this structure is said to be in *companion form*. For this reason, we identify the realization (5.54) as *companion-form realization* of the transfer function (5.49). We call this the *first companion form*; another companion form follows.

Second Companion Form

In the first companion form, the coefficients of the denominator of the transfer function appear in one of the rows of the $\mathbf{A}$ matrix. There is another companion form in which the coefficients appear in a column of the $\mathbf{A}$ matrix. This can be obtained by writing Eqn. (5.49) as

$$(s^n + \alpha_1 s^{n-1} + \cdots + \alpha_n)\, Y(s) = (\beta_0 s^n + \beta_1 s^{n-1} + \cdots + \beta_n)\, U(s)$$

or $\quad s^n\, [Y(s) - \beta_0 U(s)] + s^{n-1}\, [\alpha_1 Y(s) - \beta_1 U(s)] + \cdots$
$$+ [\alpha_n Y(s) - \beta_n U(s)] = 0$$

On dividing by s^n and solving for $Y(s)$, we obtain

$$Y(s) = \beta_0 U(s) + \frac{1}{s}\, [\beta_1 U(s) - \alpha_1 Y(s)] + \cdots + \frac{1}{s^n}\, [\beta_n U(s) - \alpha_n Y(s)] \tag{5.55}$$

Note that $1/s^n$ is the transfer function of a chain of n integrators. Realization of $\dfrac{1}{s^n}[\beta_n U(s) - \alpha_n Y(s)]$ requires a chain of n integrators with input $[\beta_n u - \alpha_n y]$ to the first integrator in the chain from left-to-right. Realization of $\dfrac{1}{s^{n-1}}[\beta_{n-1} U(s) - \alpha_{n-1} Y(s)]$ requires a chain of $(n-1)$ integrators with input $[\beta_{n-1} u - \alpha_{n-1} y]$ to the second integrator in the chain from left-to-right, and so forth. This immediately leads to the structure shown in Fig. 5.11. The signal y is fedback to each of the integrators in the chain and the signal u is fed forward. Thus the signal $[\beta_n u - \alpha_n y]$ passes through n integrators; the signal $[\beta_{n-1} u - \alpha_{n-1} y]$ passes through $(n-1)$ integrators and so forth to complete the realization of Eqn. (5.55). The structure retains the ladder-like shape of the first companion form, but the feedback paths are in different directions.

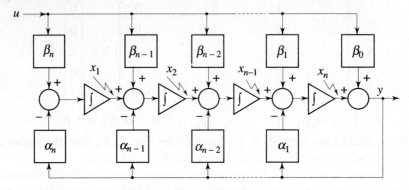

Fig. 5.11 Realization of Eqn. (5.55)

We can now write differential equations for the realization given by Fig. 5.11. To get one state variable model, we identify the output of each integrator in Fig. 5.11 with a state variable starting at the left and proceeding to the right. The corresponding differential equations are

$$\dot{x}_n = x_{n-1} - \alpha_1 (x_n + \beta_0 u) + \beta_1 u$$
$$\dot{x}_{n-1} = x_{n-2} - \alpha_2 (x_n + \beta_0 u) + \beta_2 u$$
$$\vdots$$
$$\dot{x}_2 = x_1 - \alpha_{n-1} (x_n + \beta_0 u) + \beta_{n-1} u$$
$$\dot{x}_1 = -\alpha_n (x_n + \beta_0 u) + \beta_n u$$

and the output equation is

$$y = x_n + \beta_0 u$$

The state and output equations organized in vector-matrix form are given below:

$$\dot{\mathbf{x}}(t) = \mathbf{A}\mathbf{x}^c(t) + \mathbf{b}u(t)$$
$$y(t) = \mathbf{c}\mathbf{x}(t) + du(t) \quad (5.56)$$

with

$$A = \begin{bmatrix} 0 & 0 & \cdots & 0 & -\alpha_n \\ 1 & 0 & \cdots & 0 & -\alpha_{n-1} \\ 0 & 1 & \cdots & 0 & -\alpha_{n-2} \\ \vdots & \vdots & & \vdots & \vdots \\ 0 & 0 & \cdots & 1 & -\alpha_1 \end{bmatrix}; \quad b = \begin{bmatrix} \beta_n - \alpha_n \beta_0 \\ \beta_{n-1} - \alpha_{n-1} \beta_0 \\ \vdots \\ \beta_1 - \alpha_1 \beta_0 \end{bmatrix}$$

$c = [0\ 0\ \cdots\ 0\ 1];\ d = \beta_0$

Compare A, b, and c matrices of the second companion form with that of the first. We observe that A, b, and c matrices of one companion form correspond to the transpose of the A, c, and b matrices, respectively, of the other. ▲▲

There are many benefits derived from the companion forms of state variable models. One obvious benefit is that both the companion forms lend themselves easily to simple analog computer models. Both the companion forms also play an important role in pole-placement design through state feedback. This will be discussed in Chapter 7.

Jordan Canonical Form

In the two canonical forms (5.54) and (5.56), the coefficients of the denominator of the transfer function appear in one of the rows or columns of matrix A. In another of the canonical forms, the poles of the transfer function form a string along the main diagonal of the matrix. This canonical form follows directly from the partial fraction expansion of the transfer function.

The general transfer function under consideration is (refer Eqn. (5.49))

$$G(s) = \frac{\beta_0 s^n + \beta_1 s^{n-1} + \cdots + \beta_n}{s^n + \alpha_1 s^{n-1} + \cdots + \alpha_n}$$

By long division, $G(s)$ can be written as

$$G(s) = \beta_0 + \frac{\beta'_1 s^{n-1} + \beta'_2 s^{n-2} + \cdots + \beta'_n}{s^n + \alpha_1 s^{n-1} + \cdots + \alpha_n} = \beta_0 + G'(s)$$

The results are simplest when the poles of the transfer function are all distinct. The partial fraction expansion of the transfer function then has the form (refer Eqns (A.14)-(A.16) in Appendix A):

$$G(s) = \frac{Y(s)}{U(s)} = \beta_0 + \frac{r_1}{s - \lambda_1} + \frac{r_2}{s - \lambda_2} + \cdots + \frac{r_n}{s - \lambda_n} \quad (5.57)$$

The coefficients $r_i (i = 1, 2, \ldots, n)$ are the residues of the transfer function $G'(s)$ at the corresponding poles at $s = \lambda_i$ $(i = 1, 2, \ldots, n)$. In the form of Eqn. (5.57), the transfer function consists of a direct path with gain β_0, and n first-order transfer functions in parallel. A block diagram representation of Eqn. (5.57) is shown in Fig. 5.12. The gains corresponding to the residues have been placed at the outputs of the integrators. This is quite arbitrary. They could have been located on the input side, or indeed split between the input and the output.

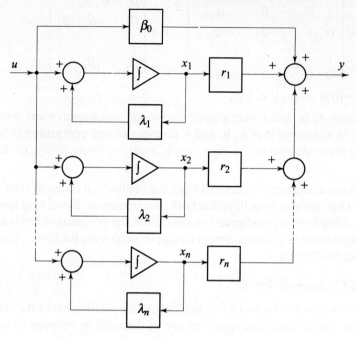

Fig. 5.12 Realization of $G(s)$ in Eqn. (5.57)

Identifying the outputs of the integrators with the state variables results in the following state and output equations:

$$\dot{\mathbf{x}}(t) = \mathbf{\Lambda}\mathbf{x}(t) + \mathbf{b}u(t)$$
$$y(t) = \mathbf{c}\mathbf{x}(t) + du(t) \qquad (5.58)$$

with

$$\mathbf{\Lambda} = \begin{bmatrix} \lambda_1 & 0 & \cdots & 0 \\ 0 & \lambda_2 & \cdots & 0 \\ \vdots & \vdots & & \vdots \\ 0 & 0 & \cdots & \lambda_n \end{bmatrix}; \mathbf{b} = \begin{bmatrix} 1 \\ 1 \\ \vdots \\ 1 \end{bmatrix}$$

$$\mathbf{c} = [r_1 \quad r_2 \quad \cdots \quad r_n]; d = \beta_0$$

It is observed that for this canonical state variable model, the matrix $\mathbf{\Lambda}$ is a diagonal matrix with the poles of $G(s)$ as its diagonal elements. The unique *decoupled* nature of the canonical model is obvious from Eqns (5.58); the n first-order differential equations are independent of each other:

$$\dot{x}_i(t) = \lambda_i x_i(t) + u(t); i = 1, 2, \ldots, n \qquad (5.59)$$

Control System Analysis Using State Variable Methods 355

This decoupling feature, as we shall see later in this chapter, greatly helps in system analysis.

The block diagram representation of Fig. 5.12 can be turned into hardware only if all the poles at $s = \lambda_1, \lambda_2, ..., \lambda_n$ are real. If they are complex, the feedback gains and the gains corresponding to the residues are complex. In this case, the representation must be considered as being purely conceptual; valid for theoretical studies, but not physically realizable. A realizable representation can be obtained by introducing an equivalence transformation.

Suppose that $s = \sigma + j\omega$, $s = \sigma - j\omega$ and $s = \lambda$ are the three poles of a transfer function. The residues at the pair of complex conjugate poles must be themselves complex conjugates. Partial fraction expansion of the transfer function with a pair of complex conjugate poles and a real pole has the form

$$G(s) = d + \frac{p+jq}{s-(\sigma+j\omega)} + \frac{p-jq}{s-(\sigma-j\omega)} + \frac{r}{s-\lambda}$$

A state variable model for this transfer function is given below (refer Eqns (5.58):

$$\dot{\mathbf{x}} = \mathbf{\Lambda}\mathbf{x} + \mathbf{b}u$$
$$y = \mathbf{c}\mathbf{x} + du$$
(5.60)

with

$$\mathbf{\Lambda} = \begin{bmatrix} \sigma+j\omega & 0 & 0 \\ 0 & \sigma-j\omega & 0 \\ 0 & 0 & \lambda \end{bmatrix}; \mathbf{b} = \begin{bmatrix} 1 \\ 1 \\ 1 \end{bmatrix}$$

$$\mathbf{c} = [p+jq \quad p-jq \quad r]$$

Introducing an equivalence transformation

$$\mathbf{x} = \mathbf{P}\bar{\mathbf{x}}$$

with

$$\mathbf{P} = \begin{bmatrix} 1/2 & -j1/2 & 0 \\ 1/2 & j1/2 & 0 \\ 0 & 0 & 1 \end{bmatrix}$$

we obtain (refer Eqns (5.22))

$$\dot{\bar{\mathbf{x}}}(t) = \bar{\mathbf{A}}\,\bar{\mathbf{x}}(t) + \bar{\mathbf{b}}u(t)$$
$$y(t) = \bar{\mathbf{c}}\,\bar{\mathbf{x}}(t) + du(t)$$
(5.61)

where

$$\bar{\mathbf{A}} = \mathbf{P}^{-1}\mathbf{\Lambda}\mathbf{P} = \begin{bmatrix} 1 & 1 & 0 \\ j & -j & 0 \\ 0 & 0 & 1 \end{bmatrix} \begin{bmatrix} \sigma+j\omega & 0 & 0 \\ 0 & \sigma-j\omega & 0 \\ 0 & 0 & \lambda \end{bmatrix} \begin{bmatrix} 1/2 & -j1/2 & 0 \\ 1/2 & j1/2 & 0 \\ 0 & 0 & 1 \end{bmatrix}$$

$$\overline{\mathbf{A}} = \mathbf{P}^{-1}\mathbf{AP} = \begin{bmatrix} \sigma & \omega & 0 \\ -\omega & \sigma & 0 \\ 0 & 0 & \lambda \end{bmatrix}$$

$$\overline{\mathbf{b}} = \mathbf{P}^{-1}\mathbf{b} = \begin{bmatrix} 2 \\ 0 \\ 1 \end{bmatrix}$$

$$\overline{\mathbf{c}} = \mathbf{cP} = [p \quad q \quad r]$$

▲▲

When the transfer function $G(s)$ has repeated poles, the partial fraction expansion will not be as simple as Eqn. (5.57). Assume that $G(s)$ has m distinct poles at $s = \lambda_1, \lambda_2, \ldots, \lambda_m$ of multiplicity $n_1, n_2, \ldots, n_m$ respectively; $n = n_1 + n_2 + \cdots + n_m$. That is, $G(s)$ is of the form

$$G(s) = \beta_0 + \frac{\beta'_1 s^{n-1} + \beta'_2 s^{n-2} + \cdots + \beta'_n}{(s-\lambda_1)^{n_1}(s-\lambda_2)^{n_2}\cdots(s-\lambda_m)^{n_m}} \tag{5.62}$$

The partial fraction expansion of $G(s)$ is of the form (refer Eqns (A.20)–(A.25) in Appendix A)

$$G(s) = \beta_0 + H_1(s) + \cdots + H_m(s) = \frac{Y(s)}{U(s)} \tag{5.63}$$

where

$$H_i(s) = \frac{r_{i1}}{(s-\lambda_i)^{n_i}} + \frac{r_{i2}}{(s-\lambda_i)^{n_i-1}} + \cdots + \frac{r_{in_i}}{(s-\lambda_i)} = \frac{Y_i(s)}{U(s)}$$

The first term in $H_i(s)$ can be synthesized as a chain of n_i identical, first-order systems, each having transfer function $1/(s - \lambda_i)$. The second term can be synthesized by a chain of $(n_i - 1)$ first-order systems, and so forth. The entire $H_i(s)$ can be synthesized by the system having the block diagram shown in Fig. 5.13.

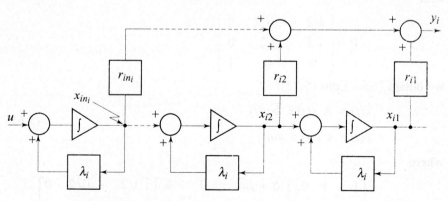

Fig. 5.13 Realization of $H_i(s)$ in Eqn. (5.63)

We can now write differential equations for the realization of $H_i(s)$ given by Fig. 5.13. To get one state variable formulation, we identify the output of each integrator with a state variable starting at the right and proceeding to the left. The corresponding differential equations are

$$\begin{aligned} \dot{x}_{i1} &= \lambda_i x_{i1} + x_{i2} \\ \dot{x}_{i2} &= \lambda_i x_{i2} + x_{i3} \\ &\vdots \\ \dot{x}_{in_i} &= \lambda_i x_{in_i} + u \end{aligned} \qquad (5.64a)$$

and the output is given by

$$y_i = r_{i1} x_{i1} + r_{i2} x_{i2} + \cdots + r_{in_i} x_{in_i} \qquad (5.64b)$$

If the state vector for the subsystem is defined by

$$\mathbf{x}_i = [x_{i1} \quad x_{i2} \quad \cdots \quad x_{in_i}]^T$$

then Eqns (5.64) can be written in the standard form

$$\begin{aligned} \dot{\mathbf{x}}_i &= \mathbf{\Lambda}_i \mathbf{x}_i + \mathbf{b}_i u \\ y_i &= \mathbf{c}_i \mathbf{x}_i \end{aligned} \qquad (5.65)$$

where

$$\mathbf{\Lambda}_i = \begin{bmatrix} \lambda_i & 1 & 0 & \cdots & 0 & 0 \\ 0 & \lambda_i & 1 & \cdots & 0 & 0 \\ \vdots & \vdots & \vdots & & \vdots & \vdots \\ 0 & 0 & 0 & \cdots & \lambda_i & 1 \\ 0 & 0 & 0 & \cdots & 0 & \lambda_i \end{bmatrix}; \quad \mathbf{b}_i = \begin{bmatrix} 0 \\ 0 \\ \vdots \\ 0 \\ 1 \end{bmatrix}$$

$$\mathbf{c}_i = [r_{i1} \quad r_{i2} \quad \cdots \quad r_{in_i}]$$

Note that matrix $\mathbf{\Lambda}_i$ has two diagonals—the principal diagonal has the corresponding characteristic root (pole) and the superdiagonal has all 1's. In matrix theory, a matrix having this structure is said to be in *Jordan form*. For this reason, we identify the realization (5.65) as *Jordan canonical form*.

According to Eqn. (5.63), the overall transfer function $G(s)$ consists of a direct path with gain β_0 and m subsystems, each of which is in the Jordan canonical form as shown in Fig. 5.14. The state vector of the overall system consists of the concatenation of the state vectors of each of the *Jordan blocks*:

$$\mathbf{x} = \begin{bmatrix} \mathbf{x}_1 \\ \mathbf{x}_2 \\ \vdots \\ \mathbf{x}_m \end{bmatrix} \qquad (5.66a)$$

Since there is no coupling between any of the subsystems, the $\mathbf{\Lambda}$ matrix of the overall system is 'block diagonal':

358 Digital Control and State Variable Methods

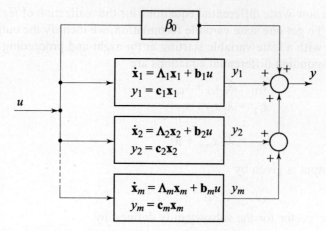

Fig. 5.14 Subsystems of Jordan canonical form combined into overall system

$$\Lambda = \begin{bmatrix} \Lambda_1 & 0 & \cdots & 0 \\ 0 & \Lambda_2 & \cdots & 0 \\ \vdots & \vdots & & \vdots \\ 0 & 0 & \cdots & \Lambda_m \end{bmatrix} \quad (5.66b)$$

where each of the submatrices Λ_i is in the Jordan canonical form (5.65). The **b** and **c** matrices of the o erall system are the concatenations of the $\mathbf{b}_i$ and $\mathbf{c}_i$ matrices respectively of each of the subsystems:

$$\mathbf{b} = \begin{bmatrix} \mathbf{b}_1 \\ \mathbf{b}_2 \\ \vdots \\ \mathbf{b}_m \end{bmatrix}; \mathbf{c} = [\mathbf{c}_1 \quad \mathbf{c}_2 \cdots \mathbf{c}_m]; d = \beta_0 \quad (5.66c)$$

The state variable model (5.58) derived for the case of distinct poles, is a special case of Jordan canonical form (5.66) where each Jordan block is of 1×1 dimension.

Example 5.6

In the following, we obtain three different realizations for the transfer function

$$G(s) = \frac{s+3}{s^3 + 9s^2 + 24s + 20} = \frac{Y(s)}{U(s)}$$

First companion form
Note that the given $G(s)$ is a strictly proper fraction; the realization will therefore be of the form (5.48), i.e., the parameter d in the realization $\{\mathbf{A}, \mathbf{b}, \mathbf{c}, d\}$ is zero.

The state variable formulation in the first companion form can be written just by inspection of the given transfer function. Referring to Eqns (5.54), we obtain

$$\begin{bmatrix} \dot{x}_1 \\ \dot{x}_2 \\ \dot{x}_3 \end{bmatrix} = \begin{bmatrix} 0 & 1 & 0 \\ 0 & 0 & 1 \\ -20 & -24 & -9 \end{bmatrix} \begin{bmatrix} x_1 \\ x_2 \\ x_3 \end{bmatrix} + \begin{bmatrix} 0 \\ 0 \\ 1 \end{bmatrix} u$$

$$y = \begin{bmatrix} 3 & 1 & 0 \end{bmatrix} \begin{bmatrix} x_1 \\ x_2 \\ x_3 \end{bmatrix}$$

Figure 5.15a shows the state diagram in signal flow graph form.

Second companion form

Referring to Eqns (5.56), we obtain

$$\begin{bmatrix} \dot{x}_1 \\ \dot{x}_2 \\ \dot{x}_3 \end{bmatrix} = \begin{bmatrix} 0 & 0 & -20 \\ 1 & 0 & -24 \\ 0 & 1 & -9 \end{bmatrix} \begin{bmatrix} x_1 \\ x_2 \\ x_3 \end{bmatrix} + \begin{bmatrix} 3 \\ 1 \\ 0 \end{bmatrix} u$$

$$y = x_3$$

Figure 5.15b shows the state diagram.

Jordan canonical form

The given transfer function $G(s)$ in the factored form:

$$G(s) = \frac{s+3}{(s+2)^2(s+5)}$$

Using partial fraction expansion, we obtain

$$G(s) = \frac{1/3}{(s+2)^2} + \frac{2/9}{s+2} + \frac{-2/9}{s+5}$$

A matrix of the state variable model in Jordan canonical form will be block-diagonal; consisting of two Jordan blocks (refer Eqns (5.65)):

$$\Lambda_1 = \begin{bmatrix} -2 & 1 \\ 0 & -2 \end{bmatrix},$$

$$\Lambda_2 = [-5]$$

The corresponding $\mathbf{b}_i$ and $\mathbf{c}_i$ vectors are (refer Eqns (5.65)):

$$\mathbf{b}_1 = \begin{bmatrix} 0 \\ 1 \end{bmatrix} ; \mathbf{c}_1 = [\tfrac{1}{3} \quad \tfrac{2}{9}]$$

$$\mathbf{b}_2 = [1] ; \mathbf{c}_2 = [-\tfrac{2}{9}]$$

The state variable model of the given $G(s)$ in Jordan canonical form is therefore given by (refer Eqns (5.66))

$$\begin{bmatrix} \dot{x}_1 \\ \dot{x}_2 \\ \dot{x}_3 \end{bmatrix} = \begin{bmatrix} -2 & 1 & 0 \\ 0 & -2 & 0 \\ 0 & 0 & -5 \end{bmatrix} \begin{bmatrix} x_1 \\ x_2 \\ x_3 \end{bmatrix} + \begin{bmatrix} 0 \\ 1 \\ 1 \end{bmatrix} u$$

$$y = \begin{bmatrix} \tfrac{1}{3} & \tfrac{2}{9} & \tfrac{-2}{9} \end{bmatrix} \begin{bmatrix} x_1 \\ x_2 \\ x_3 \end{bmatrix}$$

Figure 5.15c shows the state diagram. We note that Jordan canonical state variables are not completely decoupled. The decoupling is blockwise; state variables of one block are independent of state variables of all other blocks. However, the state variables of one block among themselves are coupled; the coupling is unique and simple.

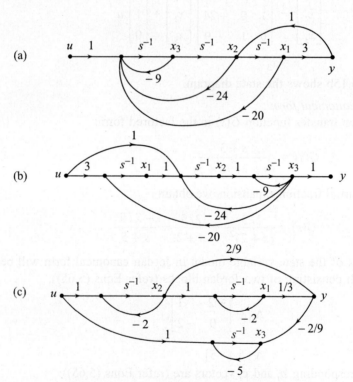

Fig. 5.15 Three realizations of given $G(s)$

5.6 EIGENVALUES AND EIGENVECTORS

The last section was concerned with the derivation of state variable models for a given transfer function. Out of infinitely many realizations possible for a given transfer function, we have derived the following three 'standard' or canonical forms:

  (i) First companion form

(ii) Second companion form
(iii) Jordan form

Consider now the situation where the system dynamics is already known in the form of a state variable model. For example, state equations representing the dynamics of a physical system may be obtained by the application of physical laws. However, state variables in such a formulation may not be as convenient as some other canonical state variables. Transformation of original state variable model to a canonical form may therefore be helpful in solving analysis and design problems.

In this section, we deal with the problem of transformation of a given state variable model to Jordan canonical form (transformation of a given model to other canonical forms will be taken up in Section 5.9, and to companion forms in Chapter 7).

Given state variable model:

$$\dot{\mathbf{x}}(t) = \mathbf{A}\mathbf{x}(t) + \mathbf{b}u(t) \qquad (5.67)$$
$$y(t) = \mathbf{c}\mathbf{x}(t) + du(t)$$

where $\mathbf{A}, \mathbf{b}, \mathbf{c}$ and d are constant matrices of dimensions $n \times n$, $n \times 1$, $1 \times n$ and 1×1 respectively.

The problem is to find an equivalence transformation

$$\mathbf{x} = \mathbf{P}\bar{\mathbf{x}} \qquad (5.68)$$

such that the equivalent model (refer Eqns (5.22))

$$\dot{\bar{\mathbf{x}}}(t) = \mathbf{P}^{-1}\mathbf{A}\mathbf{P}\bar{\mathbf{x}}(t) + \mathbf{P}^{-1}\mathbf{b}u(t) \qquad (5.69)$$
$$y(t) = \mathbf{c}\mathbf{P}\bar{\mathbf{x}}(t) + du(t)$$

is in Jordan canonical form.

Eigenvalues

For a general nth-order matrix

$$\mathbf{A} = \begin{bmatrix} a_{11} & a_{12} & \cdots & a_{1n} \\ a_{21} & a_{22} & \cdots & a_{2n} \\ \vdots & \vdots & & \vdots \\ a_{n1} & a_{n2} & \cdots & a_{nn} \end{bmatrix}$$

the determinant

$$|\lambda \mathbf{I} - \mathbf{A}| = \begin{vmatrix} \lambda - a_{11} & -a_{12} & \cdots & -a_{1n} \\ -a_{21} & \lambda - a_{22} & \cdots & -a_{2n} \\ \vdots & \vdots & & \vdots \\ -a_{n1} & -a_{n2} & \cdots & \lambda - a_{nn} \end{vmatrix}$$

On expanding the determinant we find that $|\lambda \mathbf{I} - \mathbf{A}|$, which is called the *characteristic polynomial* of the matrix $\mathbf{A}$, is a polynomial of degree n:

$$|\lambda \mathbf{I} - \mathbf{A}| = \Delta(\lambda) = \lambda^n + \alpha_1 \lambda^{n-1} + \cdots + \alpha_{n-1}\lambda + \alpha_n$$

where α_i are constant scalars.

The equation

$$\Delta(\lambda) = \lambda^n + \alpha_1 \lambda^{n-1} + \cdots + \alpha_{n-1}\lambda + \alpha_n = 0 \qquad (5.70)$$

is called the *characteristic equation* of the matrix $\mathbf{A}$, and its n roots are called *characteristic roots*, or *characteristic values*, or *eigenvalues* of the matrix $\mathbf{A}$. When $\mathbf{A}$ represents the dynamic matrix of a linear system, the eigenvalues determine the dynamic response of the system (the next section will establish this fact), and also turn out to be the poles of the corresponding transfer function (refer Eqn. (5.31)).

Eigenvalues of a matrix $\mathbf{A}$ are invariant under equivalence transformation (refer Eqn. (5.37)), i.e.,

$$|\lambda \mathbf{I} - \mathbf{A}| = |\lambda \mathbf{I} - \mathbf{P}^{-1}\mathbf{A}\mathbf{P}|$$

for any nonsingular matrix $\mathbf{P}$.

Eigenvectors

Consider an $n \times n$ matrix $\mathbf{A}$ with eigenvalues $\{\lambda_1, \lambda_2, ..., \lambda_n\}$. We start with the assumption of distinct eigenvalues; later we will relax this assumption.

Case I: *All eigenvalues are distinct*

State transformation to Jordan canonical form requires a transformation matrix $\mathbf{P}$ such that

$$\mathbf{P}^{-1}\mathbf{A}\mathbf{P} = \Lambda = \begin{bmatrix} \lambda_1 & 0 & \cdots & 0 \\ 0 & \lambda_2 & \cdots & 0 \\ \vdots & \vdots & & \vdots \\ 0 & 0 & \cdots & \lambda_n \end{bmatrix} \qquad (5.71)$$

Let the transformation matrix $\mathbf{P}$ required to transform $\mathbf{A}$ to Λ be of the form

$$\mathbf{P} = [\mathbf{v}_1 \quad \mathbf{v}_2 \quad \cdots \quad \mathbf{v}_n]; \qquad (5.72a)$$

$$\mathbf{v}_i = \begin{bmatrix} v_{1i} \\ v_{2i} \\ \vdots \\ v_{ni} \end{bmatrix} = i\text{th column of } \mathbf{P} \qquad (5.72b)$$

Equation (5.71) shows that

$$\mathbf{AP} = \mathbf{P}\Lambda$$

or

$$\mathbf{A}[\mathbf{v}_1 \quad \mathbf{v}_2 \quad \cdots \quad \mathbf{v}_n] = [\mathbf{v}_1 \quad \mathbf{v}_2 \quad \cdots \quad \mathbf{v}_n]\begin{bmatrix} \lambda_1 & 0 & \cdots & 0 \\ 0 & \lambda_2 & \cdots & 0 \\ \vdots & \vdots & & \vdots \\ 0 & 0 & \cdots & \lambda_n \end{bmatrix}$$

By equating the ith columns, we obtain

$$\mathbf{A}\mathbf{v}_i = \lambda_i \mathbf{v}_i$$

or
$$(\lambda_i \mathbf{I} - \mathbf{A})\mathbf{v}_i = \mathbf{0} \tag{5.73}$$

This is a set of n homogeneous equations in n unknowns $v_{1i}, v_{2i}, \ldots, v_{ni}$.
There are two questions of interest with regard to Eqn. (5.73):
 (i) whether a solution to Eqn. (5.73) exists; and
 (ii) if the answer to the first question is yes, how many linearly independent solutions occur?

We consider an example to answer these questions. Refer Section 5.2 for the basic definitions from linear algebra used in the sequel.

Example 5.7
The matrix

$$\mathbf{A} = \begin{bmatrix} -4 & 1 & 0 \\ 0 & -3 & 1 \\ 0 & 0 & -2 \end{bmatrix}$$

has the characteristic equation

$$|\lambda \mathbf{I} - \mathbf{A}| = \begin{vmatrix} \lambda+4 & -1 & 0 \\ 0 & \lambda+3 & -1 \\ 0 & 0 & \lambda+2 \end{vmatrix}$$

$$= (\lambda + 4)(\lambda + 3)(\lambda + 2) = 0$$

Therefore, the eigenvalues of $\mathbf{A}$ are $\lambda_1 = -2$, $\lambda_2 = -3$ and $\lambda_3 = -4$.
Consider a set of homogeneous equations

$$(\lambda_1 \mathbf{I} - \mathbf{A})\mathbf{v}_1 = \mathbf{0}$$

or
$$\begin{bmatrix} 2 & -1 & 0 \\ 0 & 1 & -1 \\ 0 & 0 & 0 \end{bmatrix} \begin{bmatrix} v_{11} \\ v_{21} \\ v_{31} \end{bmatrix} = \begin{bmatrix} 0 \\ 0 \\ 0 \end{bmatrix} \tag{5.74}$$

It is easy to check that rank of the matrix $(\lambda_1 \mathbf{I} - \mathbf{A})$ is two, i.e.,

$$\rho(\lambda_1 \mathbf{I} - \mathbf{A}) = 2$$

A highest-order array having nonvanishing determinant is

$$\begin{bmatrix} 2 & -1 \\ 0 & 1 \end{bmatrix},$$

which is obtained from $(\lambda_1 \mathbf{I} - \mathbf{A})$ by omitting the third row and the third column. Consequently, a set of linearly independent equations is

$$2v_{11} - v_{21} = 0$$
$$v_{21} = v_{31}$$

or
$$\begin{bmatrix} 2 & -1 \\ 0 & 1 \end{bmatrix} \begin{bmatrix} v_{11} \\ v_{21} \end{bmatrix} = \begin{bmatrix} 0 \\ v_{31} \end{bmatrix}$$

Therefore,
$$\begin{bmatrix} v_{11} \\ v_{21} \end{bmatrix} = \begin{bmatrix} 2 & -1 \\ 0 & 1 \end{bmatrix}^{-1} \begin{bmatrix} 0 \\ v_{31} \end{bmatrix} = \begin{bmatrix} v_{31}/2 \\ v_{31} \end{bmatrix}$$

There are three components in $\mathbf{v}_1$ and two equations governing them; therefore one of the three components can be arbitrarily chosen. For $v_{31} = 2$, a solution to Eqn. (5.74) is

$$\mathbf{v}_1 = \begin{bmatrix} 1 \\ 2 \\ 2 \end{bmatrix}$$

A different choice for v_{31} leads to a different solution to Eqn. (5.74). In fact, this set of equations has infinite solutions as demonstrated below:
For $v_{31} = 2\alpha$ (with α arbitrary), the solution

$$\mathbf{v}_1 = \alpha \begin{bmatrix} 1 \\ 2 \\ 2 \end{bmatrix}$$

Obviously this solution is non-unique. However, all non-trivial solutions have a unique direction and they differ only in terms of a scalar multiplier. There is, thus, only one independent solution.

Corresponding to the eigenvalue $\lambda_2 = -3$, a linearly independent solution to homogeneous equations

$$(\lambda_2 \mathbf{I} - \mathbf{A})\mathbf{v}_2 = \mathbf{0}$$

is given by

$$\mathbf{v}_2 = \begin{bmatrix} 1 \\ 1 \\ 0 \end{bmatrix}$$

And for $\lambda_3 = -4$, the equations

$$(\lambda_3 \mathbf{I} - \mathbf{A})\mathbf{v}_3 = \mathbf{0}$$

have a linearly independent solution

$$\mathbf{v}_3 = \begin{bmatrix} 2 \\ 0 \\ 0 \end{bmatrix}$$

In general, the number of equations that the vector $\mathbf{v}_i$ in (5.73) has to obey is equal to $\rho(\lambda_i \mathbf{I} - \mathbf{A})$ where $\rho(\mathbf{M})$ denotes the rank of matrix $\mathbf{M}$. There are n components in $\mathbf{v}_i$ (n = number of columns of $(\lambda_i \mathbf{I} - \mathbf{A})$); therefore $(n - \rho(\lambda_i \mathbf{I} - \mathbf{A}))$ components of $\mathbf{v}_i$ can be arbitrarily chosen. Thus the number of linearly independent solutions of the homogeneous Eqn. (5.73) $= [n - \rho(\lambda_i \mathbf{I} - \mathbf{A})] = \gamma(\lambda_i \mathbf{I} - \mathbf{A})$ where $\gamma(\mathbf{M})$ denotes the *nullity* of matrix $\mathbf{M}$.

We have the following answers to the two questions raised earlier with regard to Eqn. (5.73):

(i) For Eqn. (5.73) to have a non-trivial solution, rank of $(\lambda_i \mathbf{I} - \mathbf{A})$ must be less than n, or equivalently $det(\lambda_i \mathbf{I} - \mathbf{A}) = 0$. This condition is satisfied by virtue of the fact that λ_i is an eigenvalue.

(ii) The number of linearly independent solutions to Eqn. (5.73) is equal to nullity of $(\lambda_i \mathbf{I} - \mathbf{A})$.

The nullity of matrix $(\lambda_i \mathbf{I} - \mathbf{A})$ does not exceed the multiplicity of the eigenvalue λ_i (refer Lancaster and Tismenetsky [33] for proof of the result). Therefore, for distinct eigenvalue λ_i, there is one and only one linearly independent solution to Eqn. (5.73). This solution is called the *eigenvector* of $\mathbf{A}$ associated with the eigenvalue λ_i.

Theorem 5.1

Let $\mathbf{v}_1, \mathbf{v}_2, \ldots, \mathbf{v}_n$ be the eigenvectors associated with the distinct eigenvalues $\lambda_1, \lambda_2, \ldots, \lambda_n$, respectively, of matrix $\mathbf{A}$. The vectors $\mathbf{v}_1, \mathbf{v}_2, \ldots, \mathbf{v}_n$ are linearly independent and the nonsingular matrix

$$\mathbf{P} = [\mathbf{v}_1 \ \mathbf{v}_2 \ \cdots \ \mathbf{v}_n]$$

transforms matrix $\mathbf{A}$ into Jordan canonical form.

Proof
Let
$$\alpha_1 \mathbf{v}_1 + \alpha_2 \mathbf{v}_2 + \cdots + \alpha_n \mathbf{v}_n = \mathbf{0} \tag{5.75}$$

If it can be shown that this implies that $\alpha_1 = \alpha_2 = \cdots = \alpha_n = 0$, then the set $\{\mathbf{v}_i\}$ is linearly independent. Define

$$\mathbf{T}_i = \lambda_i \mathbf{I} - \mathbf{A}$$

Note that $\quad \mathbf{T}_i \mathbf{v}_i = \mathbf{0}$

and $\quad \mathbf{T}_i \mathbf{v}_j = (\lambda_i - \lambda_j) \mathbf{v}_j$ if $i \neq j$

Multiplying Eqn. (5.75) by $\mathbf{T}_1$ gives

$$\alpha_2 (\lambda_1 - \lambda_2) \mathbf{v}_2 + \alpha_3 (\lambda_1 - \lambda_3) \mathbf{v}_3 + \cdots + \alpha_n (\lambda_1 - \lambda_n) \mathbf{v}_n = \mathbf{0}$$

Multiplying this in turn by $\mathbf{T}_2, \mathbf{T}_3, \ldots, \mathbf{T}_{n-1}$, gives

$$\alpha_3 (\lambda_1 - \lambda_3)(\lambda_2 - \lambda_3) \mathbf{v}_3 + \cdots + \alpha_n (\lambda_1 - \lambda_n)(\lambda_2 - \lambda_n) \mathbf{v}_n = \mathbf{0}$$

$$\vdots$$

$$\alpha_{n-1} (\lambda_1 - \lambda_{n-1})(\lambda_2 - \lambda_{n-1}) \cdots (\lambda_{n-2} - \lambda_{n-1}) \mathbf{v}_{n-1}$$
$$+ \alpha_n (\lambda_1 - \lambda_n)(\lambda_2 - \lambda_n) \cdots (\lambda_{n-2} - \lambda_n) \mathbf{v}_n = \mathbf{0} \tag{5.76}$$

$$\alpha_n (\lambda_1 - \lambda_n)(\lambda_2 - \lambda_n) \cdots (\lambda_{n-2} - \lambda_n)(\lambda_{n-1} - \lambda_n) \mathbf{v}_n = \mathbf{0} \tag{5.77}$$

Since $\mathbf{v}_n \neq \mathbf{0}$ and $\lambda_n \neq \lambda_i$ for $i \neq n$, Eqn. (5.77) requires that $\alpha_n = 0$. This plus Eqn. (5.76) requires that $\alpha_{n-1} = 0$

Continuing this reasoning shows that Eqn. (5.75) requires $\alpha_i = 0$ for $i = 1, 2, \ldots, n$; so the eigenvectors $\mathbf{v}_i$ are linearly independent.

The matrix $\mathbf{P}$ constructed by placing the eigenvectors (columns) together is therefore a nonsingular matrix.

As per Eqns (5.71)–(5.73), $\mathbf{P}^{-1} \mathbf{A} \mathbf{P} = \mathbf{\Lambda}$.

Example 5.8
Consider the matrix

$$A = \begin{bmatrix} -4 & 1 & 0 \\ 0 & -3 & 1 \\ 0 & 0 & -2 \end{bmatrix}$$

for which we found in Example 5.7, the eigenvalues and eigenvectors to be

$$\lambda_1 = -2, \mathbf{v}_1 = \begin{bmatrix} 1 \\ 2 \\ 2 \end{bmatrix};$$

$$\lambda_2 = -3, \mathbf{v}_2 = \begin{bmatrix} 1 \\ 1 \\ 0 \end{bmatrix};$$

$$\lambda_3 = -4, \mathbf{v}_3 = \begin{bmatrix} 2 \\ 0 \\ 0 \end{bmatrix}$$

The transformation matrix

$$\mathbf{P} = \begin{bmatrix} 1 & 1 & 2 \\ 2 & 1 & 0 \\ 2 & 0 & 0 \end{bmatrix}$$

This gives

$$\mathbf{P}^{-1}\mathbf{AP} = \frac{1}{4}\begin{bmatrix} 0 & 0 & 2 \\ 0 & 4 & -4 \\ 2 & -2 & 1 \end{bmatrix}\begin{bmatrix} -4 & 1 & 0 \\ 0 & -3 & 1 \\ 0 & 0 & -2 \end{bmatrix}\begin{bmatrix} 1 & 1 & 2 \\ 2 & 1 & 0 \\ 2 & 0 & 0 \end{bmatrix}$$

$$= \begin{bmatrix} -2 & 0 & 0 \\ 0 & -3 & 0 \\ 0 & 0 & -4 \end{bmatrix} = \Lambda$$

which is the diagonal matrix (a special case of Jordan canonical form) with eigenvalues of $\mathbf{A}$ as its diagonal elements. In fact Λ could be written down directly without computing $\mathbf{P}^{-1}\mathbf{AP}$.

Computation of eigenvectors
The eigenvectors $\mathbf{v}_i$ which satisfy the equations

$$(\lambda_i \mathbf{I} - \mathbf{A})\mathbf{v}_i = \mathbf{0} \tag{5.78}$$

can be computed by solving the set of linear algebraic equations. The method of Gauss elimination is a straightforward and powerful procedure for reducing systems of linear equations to a simple *reduced form* easily solved by substitution (refer Noble and Daniel [32]). High quality software is available commercially; for example, the MATLAB system from the Math Works [177].

In the following, we give an analytical procedure of computing the eigenvectors. This procedure is quite useful for hand calculations.

Using the property (refer Eqn. (5.3))
$$M \, adj \, M = |M|I$$
and letting $M = (\lambda_i I - A)$ yields
$$(\lambda_i I - A) \, adj(\lambda_i I - A) = |\lambda_i I - A|I$$
Since $|\lambda_i I - A|$ is the characteristic polynomial and λ_i is an eigenvalue, this equation becomes
$$(\lambda_i I - A) \, adj(\lambda_i I - A) = \mathbf{0} \tag{5.79}$$
A comparison of Eqn. (5.78) with (5.79) shows that $\mathbf{v}_i$ is proportional to any non-zero column of $adj(\lambda_i I - A)$.

Example 5.9

Consider the state variable model

$$\dot{x} = Ax + bu$$
$$y = cx$$

with
$$A = \begin{bmatrix} -9 & 1 & 0 \\ -26 & 0 & 1 \\ -24 & 0 & 0 \end{bmatrix}; \, b = \begin{bmatrix} 2 \\ 5 \\ 0 \end{bmatrix}; \, c = \begin{bmatrix} 1 & 2 & -1 \end{bmatrix}$$

The characteristic equation
$$|\lambda I - A| = 0$$
yields the roots $\quad \lambda_1 = -2, \, \lambda_2 = -3, \, \text{and} \, \lambda_3 = -4.$

$$adj(\lambda I - A) = adj \begin{bmatrix} \lambda + 9 & -1 & 0 \\ 26 & \lambda & -1 \\ 24 & 0 & \lambda \end{bmatrix}$$

$$= \begin{bmatrix} \lambda^2 & \lambda & 1 \\ -26\lambda - 24 & \lambda^2 + 9\lambda & \lambda + 9 \\ -24\lambda & -24 & \lambda^2 + 9\lambda + 26 \end{bmatrix}$$

For $\quad \lambda_1 = -2,$
$$adj(\lambda_1 I - A) = \begin{bmatrix} 4 & -2 & 1 \\ 28 & -14 & 7 \\ 48 & -24 & 12 \end{bmatrix}; \, \mathbf{v}_1 = \begin{bmatrix} 1 \\ 7 \\ 12 \end{bmatrix}$$

For $\quad \lambda_2 = -3,$
$$adj(\lambda_2 I - A) = \begin{bmatrix} 9 & -3 & 1 \\ 54 & -18 & 6 \\ 72 & -24 & 8 \end{bmatrix}; \, \mathbf{v}_2 = \begin{bmatrix} 1 \\ 6 \\ 8 \end{bmatrix}$$

For $\quad \lambda_3 = -4,$
$$adj(\lambda_3 I - A) = \begin{bmatrix} 16 & -4 & 1 \\ 80 & -20 & 5 \\ 96 & -24 & 6 \end{bmatrix}; \, \mathbf{v}_3 = \begin{bmatrix} 1 \\ 5 \\ 6 \end{bmatrix}$$

In each case, the columns of $adj(\lambda_i \mathbf{I} - \mathbf{A})$ are linearly related. In practice, it is necessary to calculate only one (non-zero) column of the adjoint matrix.

The transformation matrix

$$\mathbf{P} = [\mathbf{v}_1 \; \mathbf{v}_2 \; \mathbf{v}_3] = \begin{bmatrix} 1 & 1 & 1 \\ 7 & 6 & 5 \\ 12 & 8 & 6 \end{bmatrix}$$

State transformation

$$\mathbf{x} = \mathbf{P}\bar{\mathbf{x}}$$

results in the following model (refer Eqns (5.22)):

$$\dot{\bar{\mathbf{x}}} = \mathbf{\Lambda}\bar{\mathbf{x}} + \bar{\mathbf{b}}u$$

$$y = \bar{\mathbf{c}}\bar{\mathbf{x}}$$

with

$$\mathbf{\Lambda} = \mathbf{P}^{-1}\mathbf{A}\mathbf{P} = -\frac{1}{2}\begin{bmatrix} -4 & 2 & -1 \\ 18 & -6 & 2 \\ -16 & 4 & -1 \end{bmatrix}\begin{bmatrix} -9 & 1 & 0 \\ -26 & 0 & 1 \\ -24 & 0 & 0 \end{bmatrix}\begin{bmatrix} 1 & 1 & 1 \\ 7 & 6 & 5 \\ 12 & 8 & 6 \end{bmatrix}$$

$$= \begin{bmatrix} -2 & 0 & 0 \\ 0 & -3 & 0 \\ 0 & 0 & -4 \end{bmatrix}$$

$$\bar{\mathbf{b}} = \mathbf{P}^{-1}\mathbf{b} = \begin{bmatrix} -1 \\ -3 \\ 6 \end{bmatrix}$$

$$\bar{\mathbf{c}} = \mathbf{c}\mathbf{P} = [3 \; 5 \; 5]$$

Case II: *Some eigenvalues are multiple roots of the characteristic equation*

For notational convenience, we assume that matrix $\mathbf{A}$ has an eigenvalue λ_1 of multiplicity n_1, and all other eigenvalues $\lambda_{n_1+1}, ..., \lambda_n$ are distinct, i.e.,

$$|\lambda \mathbf{I} - \mathbf{A}| = (\lambda - \lambda_1)^{n_1}(\lambda - \lambda_{n_1+1}) \cdots (\lambda - \lambda_n)$$

Recall the result stated earlier: the nullity γ of matrix $(\lambda_i \mathbf{I} - \mathbf{A})$ does not exceed the multiplicity of λ_i. Therefore,

$$1 \le \gamma(\lambda_1 \mathbf{I} - \mathbf{A}) \le n_1$$

$$\gamma(\lambda_{n_1+1} \mathbf{I} - \mathbf{A}) = 1$$

$$\vdots$$

$$\gamma(\lambda_n \mathbf{I} - \mathbf{A}) = 1$$

We know that the number of linearly independent eigenvectors associated with an eigenvalue λ_i is equal to the nullity γ of the matrix $(\lambda_i \mathbf{I} - \mathbf{A})$. Thus, when one or more eigenvalues is a repeated root of the characteristic equation, a full set of n linearly independent eigenvectors may or may not exist.

It is convenient to consider three sub-classifications for Case II.

Case II$_1$: *Nullity of $(\lambda_1 I - A) = n_1$*

In this case, the vector equation

$$(\lambda_1 I - A)v = 0$$

has n_1 linearly independent solutions, say, $v_1, v_2, \ldots, v_{n_1}$. We have thus a full set of n_1 eigenvectors associated with multiple eigenvalue λ_1.

The remaining $(n - n_1)$ eigenvectors are obtained from the vector equations

$$(\lambda_j I - A)v_j = 0, j = n_1 + 1, \ldots, n$$

Each of these vector equations has only one linearly independent solution. The matrix

$$P = [v_1 \quad v_2 \cdots v_{n_1} \quad v_{n_1+1} \cdots v_n]$$

gives

$$P^{-1}AP = \Lambda = \begin{bmatrix} \lambda_1 & 0 & \cdots & 0 & 0 & \cdots & 0 \\ 0 & \lambda_1 & \cdots & 0 & 0 & \cdots & 0 \\ \vdots & \vdots & & \vdots & \vdots & \cdots & \vdots \\ 0 & 0 & \cdots & \lambda_1 & 0 & \cdots & 0 \\ 0 & 0 & \cdots & 0 & \lambda_{n_1+1} & \cdots & 0 \\ \vdots & \vdots & & \vdots & \vdots & & \vdots \\ 0 & 0 & \cdots & 0 & 0 & \cdots & \lambda_n \end{bmatrix}$$

Case II$_2$: *Nullity of $(\lambda_1 I - A) = 1$*

For this case, there is only one eigenvector associated with λ_1 regardless of multiplicity n_1. This eigenvector is given by the linearly independent solution of the vector equation

$$(\lambda_1 I - A)v = 0$$

The solution to this equation may be found as in Case I.

We have seen in Cases I and II$_1$, that the transformation matrix P yields a diagonal matrix Λ if and only if P has a set of n linearly independent eigenvectors. When nullity of the matrix $(\lambda_1 I - A)$ is one, n linearly independent eigenvectors cannot be constructed and therefore the transformation to a diagonal matrix is not possible.

The simplest form to which matrix A having a multiple eigenvalue λ_1 of multiplicity n_1 with $\gamma(\lambda_1 I - A) = 1$ and all other distinct eigenvalues, can be reduced is the Jordan canonical form:

$$\Lambda = \begin{bmatrix} \Lambda_1 & 0 & \cdots & 0 \\ 0 & \Lambda_{n_1+1} & \cdots & 0 \\ \vdots & \vdots & & \vdots \\ 0 & 0 & \cdots & \Lambda_n \end{bmatrix}$$

where the Jordan blocks Λ_i are

$$\Lambda_1 = \begin{bmatrix} \lambda_1 & 1 & 0 & \cdots & 0 \\ 0 & \lambda_1 & 1 & \cdots & 0 \\ \vdots & \vdots & \vdots & & \vdots \\ 0 & 0 & 0 & \cdots & \lambda_1 \end{bmatrix}$$

$$\Lambda_{n_1+1} = [\lambda_{n_1+1}]$$
$$\vdots$$
$$\Lambda_n = [\lambda_n]$$

The transformation matrix $\mathbf{P}$ is given by

$$\mathbf{P} = [\mathbf{v}_1 \quad \mathbf{v}_2 \quad \cdots \quad \mathbf{v}_{n_1} \quad \mathbf{v}_{n_1+1} \quad \cdots \quad \mathbf{v}_n]$$

with $\mathbf{v}_1, \mathbf{v}_2, ..., \mathbf{v}_{n_1}$ determined as follows:

$$\mathbf{A}[\mathbf{v}_1 \quad \mathbf{v}_2 \quad \cdots \quad \mathbf{v}_{n_1}] = [\mathbf{v}_1 \quad \mathbf{v}_2 \quad \cdots \quad \mathbf{v}_{n_1}] \begin{bmatrix} \lambda_1 & 1 & 0 & \cdots & 0 \\ 0 & \lambda_1 & 1 & \cdots & 0 \\ \vdots & \vdots & \vdots & & \vdots \\ 0 & 0 & 0 & \cdots & \lambda_1 \end{bmatrix}$$

or
$$\mathbf{A}\mathbf{v}_1 = \lambda_1 \mathbf{v}_1$$
$$\mathbf{A}\mathbf{v}_2 = \mathbf{v}_1 + \lambda_1 \mathbf{v}_2$$
$$\vdots$$
$$\mathbf{A}\mathbf{v}_{n_1} = \mathbf{v}_{n_1-1} + \lambda_1 \mathbf{v}_{n_1}$$

Rearranging these equations, we obtain

$$(\lambda_1 \mathbf{I} - \mathbf{A})\mathbf{v}_1 = \mathbf{0}$$
$$(\lambda_1 \mathbf{I} - \mathbf{A})\mathbf{v}_2 = -\mathbf{v}_1$$
$$\vdots$$
$$(\lambda_1 \mathbf{I} - \mathbf{A})\mathbf{v}_{n_1} = -\mathbf{v}_{n_1-1}$$

It can easily be established that each of these vector equations gives one linearly independent solution and the solutions $\mathbf{v}_1, \mathbf{v}_2, ..., \mathbf{v}_{n_1}$ form a linearly independent set of vectors. We shall call the set of vectors $\{\mathbf{v}_1, ..., \mathbf{v}_{n_1}\}$ the chain of *generalized eigenvectors*. Note that the vector $\mathbf{v}_1$ in the chain is in fact the eigenvector associated with multiple eigenvalue λ_1.

Eigenvectors for the Jordan blocks $\Lambda_{n_1+1}, ..., \Lambda_n$ are given by the solution of the vector equations

$$(\lambda_j \mathbf{I} - \mathbf{A})\mathbf{v}_j = \mathbf{0}; j = n_1 + 1, ..., n$$

The eigenvectors corresponding to distinct eigenvalues and the chains of generalized eigenvectors corresponding to multiple eigenvalues form the transformation matrix $\mathbf{P}$.

Case II$_3$: $1 < \gamma(\lambda_1 \mathbf{I} - \mathbf{A}) < n_1$

For this case, there are γ eigenvectors associated with λ_1. There will be one Jordan block for each eigenvector; that is, λ_1 will have γ blocks associated with it. This case is just a combination of the Cases II$_1$ and II$_2$; there is only one ambiguity: the knowledge of n_1 and γ does not directly give the information about the dimension of each of the Jordan blocks associated with λ_1.

Assume that λ_1 is a fourth-order root of the characteristic equation and $\gamma(\lambda_1 \mathbf{I} - \mathbf{A}) = 2$. The two eigenvectors associated with λ_1 satisfy
$$(\lambda_1 \mathbf{I} - \mathbf{A})\mathbf{v}_a = \mathbf{0}, \ (\lambda_1 \mathbf{I} - \mathbf{A})\mathbf{v}_b = \mathbf{0}$$
To form the transformation matrix, we require two generalized eigenvectors but it is still uncertain whether the generalized eigenvectors are both associated with $\mathbf{v}_a$, or both with $\mathbf{v}_b$, or one with each. That is, the two Jordan blocks could take one of the following forms:

$$\Lambda_1 = \begin{bmatrix} \lambda_1 & 1 & 0 \\ 0 & \lambda_1 & 1 \\ 0 & 0 & \lambda_1 \end{bmatrix}, \ \Lambda_2 = [\lambda_1]$$

or
$$\Lambda_1 = \begin{bmatrix} \lambda_1 & 1 \\ 0 & \lambda_1 \end{bmatrix}, \ \Lambda_2 = \begin{bmatrix} \lambda_1 & 1 \\ 0 & \lambda_1 \end{bmatrix}$$

The first pair corresponds to the equations
$$(\lambda_1 \mathbf{I} - \mathbf{A})\mathbf{v}_1 = \mathbf{0}$$
$$(\lambda_1 \mathbf{I} - \mathbf{A})\mathbf{v}_2 = -\mathbf{v}_1$$
$$(\lambda_1 \mathbf{I} - \mathbf{A})\mathbf{v}_3 = -\mathbf{v}_2$$
$$(\lambda_1 \mathbf{I} - \mathbf{A})\mathbf{v}_4 = \mathbf{0}$$

The second pair corresponds to the equations
$$(\lambda_1 \mathbf{I} - \mathbf{A})\mathbf{v}_1 = \mathbf{0}$$
$$(\lambda_1 \mathbf{I} - \mathbf{A})\mathbf{v}_2 = -\mathbf{v}_1$$
$$(\lambda_1 \mathbf{I} - \mathbf{A})\mathbf{v}_3 = \mathbf{0}$$
$$(\lambda_1 \mathbf{I} - \mathbf{A})\mathbf{v}_4 = -\mathbf{v}_3$$

Ambiguities such as this can be resolved by the trial-and-error procedure.

▲▲

An n-dimensional SISO system with m distinct eigenvalues $\lambda_1, \lambda_2, ..., \lambda_m$, of multiplicity $n_1, n_2, ..., n_m$ respectively $\left(n = \sum_{i=1}^{m} n_i \right)$, has the following Jordan canonical representation.

$$\dot{\mathbf{x}} = \mathbf{\Lambda}\mathbf{x} + \mathbf{b}u$$
$$y = \mathbf{c}\mathbf{x} + du$$

where $\mathbf{\Lambda}$ is a block diagonal matrix with Jordan blocks $\Lambda_1, ..., \Lambda_m$ corresponding to the eigenvalues $\lambda_1, ..., \lambda_m$ respectively, on its principal diagonal; each Jordan block Λ_i corresponding to the eigenvalue λ_i is again block diagonal with $\gamma(i)$ sub-blocks on its principal diagonal; $\gamma(i)$ being the number of linearly independent eigenvectors associated with the eigenvalue λ_i:

$$\mathbf{\Lambda}_{(n \times n)} = \begin{bmatrix} \Lambda_1 & 0 & \cdots & 0 \\ 0 & \Lambda_2 & \cdots & 0 \\ \vdots & \vdots & & \vdots \\ 0 & 0 & \cdots & \Lambda_m \end{bmatrix}$$

$$\Lambda_i \atop (n_i \times n_i) = \begin{bmatrix} \Lambda_{1i} & 0 & \cdots & 0 \\ 0 & \Lambda_{2i} & \cdots & 0 \\ \vdots & \vdots & & \vdots \\ 0 & 0 & \cdots & \Lambda_{\gamma(i)i} \end{bmatrix} ; i = 1, 2, \ldots, m$$

$$\Lambda_{ki} = \begin{bmatrix} \lambda_i & 1 & 0 & \cdots & 0 \\ 0 & \lambda_i & 1 & \cdots & 0 \\ \vdots & \vdots & \vdots & & \vdots \\ 0 & 0 & 0 & \cdots & \lambda_i \end{bmatrix} ; k = 1, 2, \ldots, \gamma(i)$$

The topic of computation of eigenvectors and generalized eigenvectors for systems with multiple eigenvalues, is much too detailed and specialized for this book to treat (Refer Gopal [102] and Brogan [107]). Over the years, experts have developed excellent general-purpose computer programs for the efficient and accurate determination of eigenvectors and generalized eigenvectors. MATLAB [177] is one reference for commercially available software.

In this book, the usefulness of the transformation of state variable models to Jordan canonical form will be illustrated through system examples having distinct eigenvalues.

5.7 SOLUTION OF STATE EQUATIONS

In this section, we investigate the solution of the state equation

$$\dot{\mathbf{x}}(t) = \mathbf{A}\mathbf{x}(t) + \mathbf{b}u(t); \mathbf{x}(t_0) \triangleq \mathbf{x}^0 \qquad (5.80)$$

where $\mathbf{x}$ is $n \times 1$ state vector, u is a scalar input, $\mathbf{A}$ is $n \times n$ constant matrix, and $\mathbf{b}$ is $n \times 1$ constant vector.

Matrix Exponential

Functions of square matrices arise in connection with the solution of vector differential equations. Of immediate interest to us are matrix infinite series.

Consider the infinite series in a scalar variable x:

$$f(x) = \alpha_0 + \alpha_1 x + \alpha_2 x^2 + \cdots = \sum_{i=0}^{\infty} \alpha_i x^i \qquad (5.81a)$$

with the radius of convergence r.

We can define infinite series in a matrix variable $\mathbf{A}$ as

$$f(\mathbf{A}) = \alpha_0 \mathbf{I} + \alpha_1 \mathbf{A} + \alpha_2 \mathbf{A}^2 + \cdots = \sum_{i=0}^{\infty} \alpha_i \mathbf{A}^i \qquad (5.81b)$$

An important relation between the scalar power series (5.81a) and the matrix power series (5.81b) is that if the absolute values of eigenvalues of $\mathbf{A}$ are smaller than r, then the matrix power series (5.81b) converges (for proof, refer Lefschetz [36]).

Consider, in particular, the scalar power series

$$f(x) = 1 + x + \frac{1}{2!}x^2 + \cdots + \frac{1}{k!}x^k + \cdots = \sum_{i=0}^{\infty} \frac{1}{i!}x^i \qquad (5.82a)$$

It is well-known that this power series converges on to the exponential e^x for all finite x, so that

$$f(x) = e^x \qquad (5.82b)$$

It follows from this result that the matrix power series

$$f(\mathbf{A}) = \mathbf{I} + \mathbf{A} + \frac{1}{2!}\mathbf{A}^2 + \cdots + \frac{1}{k!}\mathbf{A}^k + \cdots = \sum_{i=0}^{\infty} \frac{1}{i!}\mathbf{A}^i$$

converges for all $\mathbf{A}$. By analogy with the power series in Eqns (5.82) for the ordinary exponential function, we adopt the following nomenclature:

If $\mathbf{A}$ is an $n \times n$ matrix, the *matrix exponential* of $\mathbf{A}$ is

$$e^{\mathbf{A}} \triangleq \mathbf{I} + \mathbf{A} + \frac{1}{2!}\mathbf{A}^2 + \cdots + \frac{1}{k!}\mathbf{A}^k + \cdots = \sum_{i=0}^{\infty} \frac{1}{i!}\mathbf{A}^i$$

The following matrix exponential will appear in the solution of state equations:

$$e^{\mathbf{A}t} = \mathbf{I} + \mathbf{A}t + \frac{1}{2!}\mathbf{A}^2 t^2 + \cdots + \frac{1}{k!}\mathbf{A}^k t^k + \cdots = \sum_{i=0}^{\infty} \frac{1}{i!}\mathbf{A}^i t^i \qquad (5.83)$$

It converges for all $\mathbf{A}$ and all finite t.

In the following we examine some of the properties of the matrix exponential.

1.
$$e^{\mathbf{A}0} = \mathbf{I} \qquad (5.84)$$

This is easily verified by setting $t = 0$ in Eqn. (5.83).

2.
$$e^{\mathbf{A}(t+\tau)} = e^{\mathbf{A}t}e^{\mathbf{A}\tau} = e^{\mathbf{A}\tau}e^{\mathbf{A}t} \qquad (5.85)$$

This is easily verified by multiplying out the first few terms for $e^{\mathbf{A}t}$ and $e^{\mathbf{A}\tau}$.

3.
$$(e^{\mathbf{A}t})^{-1} = e^{-\mathbf{A}t} \qquad (5.86)$$

Setting $\tau = -t$ in Eqn. (5.85), we obtain

$$e^{\mathbf{A}t}e^{-\mathbf{A}t} = e^{\mathbf{A}0} = \mathbf{I}$$

Thus the inverse of $e^{\mathbf{A}t}$ is $e^{-\mathbf{A}t}$.

Since the inverse of $e^{\mathbf{A}t}$ always exists, the matrix exponential is nonsingular for all finite values of t.

4.
$$\frac{d}{dt}e^{\mathbf{A}t} = \mathbf{A}e^{\mathbf{A}t} = e^{\mathbf{A}t}\mathbf{A} \qquad (5.87)$$

Term-by-term differentiation of Eqn. (5.83) gives

$$\frac{d}{dt}e^{\mathbf{A}t} = \mathbf{A} + \mathbf{A}^2 t + \frac{1}{2!}\mathbf{A}^3 t^2 + \cdots + \frac{1}{(k-1)!}\mathbf{A}^k t^{k-1} + \cdots$$

$$= \mathbf{A}[\mathbf{I} + \mathbf{A}t + \frac{1}{2!}\mathbf{A}^2 t^2 + \cdots + \frac{1}{(k-1)!}\mathbf{A}^{k-1} t^{k-1} + \cdots] = \mathbf{A} e^{\mathbf{A}t}$$

$$= [\mathbf{I} + \mathbf{A}t + \frac{1}{2!}\mathbf{A}^2 t^2 + \cdots + \frac{1}{(k-1)!}\mathbf{A}^{k-1} t^{k-1} + \cdots]\mathbf{A} = e^{\mathbf{A}t}\mathbf{A}$$

Solution of Homogeneous State Equation

The simplest form of the general differential Eqn. (5.80) is the homogeneous, i.e., unforced equation

$$\dot{\mathbf{x}}(t) = \mathbf{A}\mathbf{x}(t); \ \mathbf{x}(t_0) \triangleq \mathbf{x}^0 \tag{5.88}$$

We assume a solution $\mathbf{x}(t)$ of the form

$$\mathbf{x}(t) = e^{\mathbf{A}t}\mathbf{k} \tag{5.89}$$

where $e^{\mathbf{A}t}$ is the matrix exponential function defined in Eqn. (5.83) and $\mathbf{k}$ is a suitably chosen constant vector.

The assumed solution is, in fact, the true solution since it satisfies the differential Eqn. (5.88) as is seen below.

$$\dot{\mathbf{x}}(t) = \frac{d}{dt}[e^{\mathbf{A}t}\mathbf{k}] = \frac{d}{dt}[e^{\mathbf{A}t}]\mathbf{k}$$

Using property (5.87) of the matrix exponential, we obtain

$$\dot{\mathbf{x}}(t) = \mathbf{A}e^{\mathbf{A}t}\mathbf{k} = \mathbf{A}\mathbf{x}(t)$$

To evaluate the constant vector $\mathbf{k}$ in terms of the known initial state $\mathbf{x}(t_0)$, we substitute $t = t_0$ in Eqn. (5.89):

$$\mathbf{x}(t_0) = e^{\mathbf{A}t_0}\mathbf{k}$$

Using property (5.86) of the matrix exponential, we obtain

$$\mathbf{k} = (e^{\mathbf{A}t_0})^{-1}\mathbf{x}(t_0) = e^{-\mathbf{A}t_0}\mathbf{x}(t_0)$$

Thus the general solution to Eqn. (5.88) for the state $\mathbf{x}(t)$ at time t, given the state $\mathbf{x}(t_0)$ at time t_0, is

$$\mathbf{x}(t) = e^{\mathbf{A}t} e^{-\mathbf{A}t_0} \mathbf{x}(t_0) = e^{\mathbf{A}(t-t_0)}\mathbf{x}(t_0) \tag{5.90a}$$

We have used the property (5.85) of the matrix exponential to express the solution in this form.

If the initial time $t_0 = 0$, i.e., the initial state $\mathbf{x}^0$ is known at $t = 0$, we have from Eqn. (5.90a):

$$\mathbf{x}(t) = e^{\mathbf{A}t}\mathbf{x}(0) \tag{5.90b}$$

From Eqn. (5.90b) it is observed that the initial state $\mathbf{x}(0) \triangleq \mathbf{x}^0$ at $t = 0$ is driven to a state $\mathbf{x}(t)$ at time t. This transition in state is carried out by the matrix exponential $e^{\mathbf{A}t}$. Due to this property, $e^{\mathbf{A}t}$ is known as the *state transition matrix*, and is denoted by $\boldsymbol{\phi}(t)$.

Properties of state transition matrix

Properties of the matrix exponential, given earlier in Eqns (5.84)–(5.87), are restated below in terms of state transition matrix $\boldsymbol{\phi}(t)$.

1. $$\frac{d}{dt}\phi(t) = A\phi(t); \quad \phi(0) = I$$
2. $$\phi(t_2 - t_1)\phi(t_1 - t_0) = \phi(t_2 - t_0) \text{ for any } t_0, t_1, t_2$$

This property of the state transition matrix is important since it implies that a state transition process can be divided into a number of sequential transitions. The transition form t_0 to t_2:

$$x(t_2) = \phi(t_2 - t_0)x(t_0);$$

is equal to the transition from t_0 to t_1 and then from t_1 to t_2:

$$x(t_1) = \phi(t_1 - t_0)x(t_0)$$
$$x(t_2) = \phi(t_2 - t_1)x(t_1)$$

3. $$\phi^{-1}(t) = \phi(-t)$$
4. $\phi(t)$ is a nonsingular matrix for all finite t.

Evaluation of state transition matrix

The state transition matrix $\phi(t) = e^{At}$ of an $n \times n$ matrix A is given by the infinite series (5.83). The series converges for all A and all finite t. Hence, e^{At} can be evaluated within prescribed accuracy by truncating the series at, say, $i = N$. An algorithm for evaluation of matrix series is given in Section 6.3.

In the following, we discuss the commonly used methods for evaluating e^{At} in closed form.

Evaluation Using Inverse Laplace Transforms

Taking the Laplace transform on both sides of Eqn. (5.88) yields

$$sX(s) - x^0 = AX(s)$$

where

$$X(s) \triangleq \mathscr{L}[x(t)]$$
$$x^0 \triangleq x(0)$$

Solving for $X(s)$, we get

$$X(s) = (sI - A)^{-1}x^0$$

The state vector $x(t)$ can be obtained by inverse transforming $X(s)$:

$$x(t) = \mathscr{L}^{-1}[(sI - A)^{-1}]x^0$$

Comparing this equation with Eqn. (5.90b), we get

$$e^{At} = \phi(t) = \mathscr{L}^{-1}[(sI - A)^{-1}] \tag{5.91}$$

The matrix $(sI - A)^{-1} = \Phi(s)$ is known in mathematical literature as the resolvent of A. The entries of the resolvent matrix $\Phi(s)$ are rational functions of s. Resolvent matrix $\Phi(s)$ can be expressed in the following form (refer Eqn. (5.43)):

$$\Phi(s) = \frac{Q(s)}{\Delta(s)} = \frac{Q_1 s^{n-1} + Q_2 s^{n-2} + \cdots + Q_{n-1}s + Q_n}{s^n + \alpha_1 s^{n-1} + \cdots + \alpha_{n-1}s + \alpha_n} \tag{5.92a}$$

where $\mathbf{Q}_i$ are constant ($n \times n$) matrices and α_j are constant scalars. The coefficients of the scalar polynomial $\Delta(s)$ and the matrix polynomial $\mathbf{Q}(s)$ may be determined sequentially by *resolvent algorithm* (convenient for digital computer) given in Eqns (5.44).

The inverse transform
$$\mathscr{L}^{-1}[\mathbf{Q}(s)/\Delta(s)] = e^{\mathbf{A}t} \qquad (5.92b)$$
can be expressed as a power series in t.

Example 5.10
Consider the system
$$\dot{\mathbf{x}} = \begin{bmatrix} 0 & 0 & -2 \\ 0 & 1 & 0 \\ 1 & 0 & 3 \end{bmatrix} \mathbf{x}; \ \mathbf{x}(0) = \begin{bmatrix} 0 \\ 1 \\ 0 \end{bmatrix}$$

By direct computation, we have
$$(s\mathbf{I} - \mathbf{A})^{-1} = \begin{bmatrix} s & 0 & 2 \\ 0 & s-1 & 0 \\ -1 & 0 & s-3 \end{bmatrix}^{-1} = \frac{(s\mathbf{I} - \mathbf{A})^+}{|s\mathbf{I} - \mathbf{A}|}$$

$$|s\mathbf{I} - \mathbf{A}| = (s-1)^2 (s-2)$$

$$(s\mathbf{I} - \mathbf{A})^+ = \begin{bmatrix} (s-1)(s-3) & 0 & -2(s-1) \\ 0 & (s-1)(s-2) & 0 \\ (s-1) & 0 & s(s-1) \end{bmatrix}$$

$$e^{\mathbf{A}t} = \mathscr{L}^{-1}[(s\mathbf{I} - \mathbf{A})^{-1}]$$

$$= \mathscr{L}^{-1} \begin{bmatrix} \dfrac{(s-3)}{(s-1)(s-2)} & 0 & \dfrac{-2}{(s-1)(s-2)} \\ 0 & \dfrac{1}{(s-1)} & 0 \\ \dfrac{1}{(s-1)(s-2)} & 0 & \dfrac{s}{(s-1)(s-2)} \end{bmatrix}$$

$$= \begin{bmatrix} 2e^t - e^{2t} & 0 & 2e^t - 2e^{2t} \\ 0 & e^t & 0 \\ -e^t + e^{2t} & 0 & 2e^{2t} - e^t \end{bmatrix}$$

Consequently, the free response of the system is
$$\mathbf{x}(t) = e^{\mathbf{A}t}\mathbf{x}(0) = \begin{bmatrix} 0 \\ e^t \\ 0 \end{bmatrix}$$

Note that $\mathbf{x}(t)$ could be more easily computed by taking the inverse Laplace transform of
$$\mathbf{X}(s) = [(s\mathbf{I} - \mathbf{A})^{-1}\mathbf{x}(0)].$$

Evaluation Using Similarity Transformation

Suppose that $\mathbf{A}$ is an $n \times n$ nondiagonal matrix with distinct eigenvalues $\lambda_1, \lambda_2, ..., \lambda_n$. We define the diagonal matrix Λ as

$$\Lambda = \begin{bmatrix} \lambda_1 & 0 & 0 & \cdots & 0 \\ 0 & \lambda_2 & 0 & \cdots & 0 \\ \vdots & \vdots & \vdots & & \vdots \\ 0 & 0 & 0 & \cdots & \lambda_n \end{bmatrix}$$

$\mathbf{A}$ and Λ are similar matrices; there exists a nonsingular transformation matrix $\mathbf{P}$ such that (refer Eqns (5.22))
$$\Lambda = \mathbf{P}^{-1}\mathbf{A}\mathbf{P}$$

Now
$$\mathbf{P}^{-1}e^{\mathbf{A}t}\mathbf{P} = \mathbf{P}^{-1}[\mathbf{I} + \mathbf{A}t + \frac{1}{2!}\mathbf{A}^2 t^2 + \cdots]\mathbf{P}$$

$$= \mathbf{I} + \mathbf{P}^{-1}\mathbf{A}\mathbf{P}t + \frac{1}{2!}\mathbf{P}^{-1}\mathbf{A}^2\mathbf{P}t^2 + \cdots$$

$$= \mathbf{I} + \mathbf{P}^{-1}\mathbf{A}\mathbf{P}t + \frac{1}{2!}\mathbf{P}^{-1}\mathbf{A}\mathbf{P}\mathbf{P}^{-1}\mathbf{A}\mathbf{P}t^2 + \cdots$$

$$= \mathbf{I} + \Lambda t + \frac{1}{2!}\Lambda^2 t^2 + \cdots = e^{\Lambda t}$$

Thus the matrices $e^{\mathbf{A}t}$ and $e^{\Lambda t}$ are similar. Since Λ is diagonal, $e^{\Lambda t}$ is given by

$$e^{\Lambda t} = \begin{bmatrix} e^{\lambda_1 t} & 0 & 0 & \cdots & 0 \\ 0 & e^{\lambda_2 t} & 0 & \cdots & 0 \\ \vdots & \vdots & \vdots & & \vdots \\ 0 & 0 & 0 & \cdots & e^{\lambda_n t} \end{bmatrix}$$

The matrix exponential $e^{\mathbf{A}t}$ of matrix $\mathbf{A}$ with distinct eigenvalues $\lambda_1, \lambda_2, ..., \lambda_n$ may therefore be evaluated using the following relation:

$$e^{\mathbf{A}t} = \mathbf{P}e^{\Lambda t}\mathbf{P}^{-1}$$

$$= \mathbf{P}\begin{bmatrix} e^{\lambda_1 t} & 0 & \cdots & 0 \\ 0 & e^{\lambda_2 t} & \cdots & 0 \\ \vdots & \vdots & & \vdots \\ 0 & 0 & \cdots & e^{\lambda_n t} \end{bmatrix}\mathbf{P}^{-1} \quad (5.93)$$

where **P** is a transformation matrix that transforms **A** into the diagonal form. (For the general case wherein matrix **A** has multiple eigenvalues, refer [102]. Also refer Review Example 5.3 given at the end of this chapter).

Example 5.11
Consider the system

$$\dot{x} = \begin{bmatrix} 0 & 1 \\ -2 & -3 \end{bmatrix} x; \quad x(0) = \begin{bmatrix} 0 \\ 1 \end{bmatrix}$$

The characteristic equation for this system is

$$|\lambda I - A| = \begin{vmatrix} \lambda & -1 \\ 2 & (\lambda+3) \end{vmatrix} = 0$$

or
$$(\lambda + 1)(\lambda + 2) = 0$$

Therefore, the eigenvalues of system matrix **A** are

$$\lambda_1 = -1, \lambda_2 = -2$$

Eigenvectors v_1 and v_2 corresponding to the eigenvalues λ_1 and λ_2, respectively, can be determined from the adjoint matrix $(\lambda I - A)^+$ (refer Eqn. (5.79)).

$$(\lambda I - A)^+ = \begin{bmatrix} (\lambda+3) & 1 \\ -2 & \lambda \end{bmatrix}$$

For $\lambda = \lambda_1 = -1$,

$$(\lambda_1 I - A)^+ = \begin{bmatrix} 2 & 1 \\ -2 & -1 \end{bmatrix}; v_1 = \begin{bmatrix} 1 \\ -1 \end{bmatrix}$$

For $\lambda = \lambda_2 = -2$,

$$(\lambda_2 I - A)^+ = \begin{bmatrix} 1 & 1 \\ -2 & -2 \end{bmatrix}; v_2 = \begin{bmatrix} 1 \\ -2 \end{bmatrix}$$

The transformation matrix **P** that transforms **A** into diagonal form, is

$$P = \begin{bmatrix} 1 & 1 \\ -1 & -2 \end{bmatrix}$$

The matrix exponential

$$e^{At} = P \begin{bmatrix} e^{-t} & 0 \\ 0 & e^{-2t} \end{bmatrix} P^{-1}$$

$$= \begin{bmatrix} 1 & 1 \\ -1 & -2 \end{bmatrix} \begin{bmatrix} e^{-t} & 0 \\ 0 & e^{-2t} \end{bmatrix} \begin{bmatrix} 2 & 1 \\ -1 & -1 \end{bmatrix}$$

$$= \begin{bmatrix} 2e^{-t} - e^{-2t} & e^{-t} - e^{-2t} \\ -2e^{-t} + 2e^{-2t} & -e^{-t} + 2e^{-2t} \end{bmatrix}$$

Consequently, the free response of the system is

$$\mathbf{x}(t) = e^{\mathbf{A}t}\mathbf{x}(0) = \begin{bmatrix} e^{-t} - e^{-2t} \\ -e^{-t} + 2e^{-2t} \end{bmatrix}$$

Evaluation Using Cayley-Hamilton Technique

The state transition matrix may be evaluated using a technique based on the Cayley-Hamilton theorem. To begin with we restate the theorem proved earlier in Section 5.4 (refer Eqns (5.45)–(5.46)).

Every square matrix $\mathbf{A}$ satisfies its own characteristic equation.

Thus if we have, for an $n \times n$ matrix $\mathbf{A}$, the characteristic equation

$$\Delta(\lambda) = |\lambda \mathbf{I} - \mathbf{A}| = \lambda^n + \alpha_1 \lambda^{n-1} + \cdots + \alpha_{n-1} \lambda + \alpha_n = 0,$$

then according to this theorem

$$\Delta(\mathbf{A}) = \mathbf{A}^n + \alpha_1 \mathbf{A}^{n-1} + \cdots + \alpha_{n-1} \mathbf{A} + \alpha_n \mathbf{I} = \mathbf{0}$$

where $\mathbf{I}$ is an identity matrix and $\mathbf{0}$ is a null matrix.

This theorem provides a simple procedure for evaluating the function of a matrix. In the study of linear systems, we are mostly concerned with functions which can be represented as a series of the powers of a matrix. Consider the matrix polynomial

$$f(\mathbf{A}) = a_0 \mathbf{I} + a_1 \mathbf{A} + a_2 \mathbf{A}^2 + \cdots + a_n \mathbf{A}^n + a_{n+1} \mathbf{A}^{n+1} + \cdots \quad (5.94a)$$

This matrix polynomial, which is of degree higher than the order of $\mathbf{A}$, can be computed by consideration of the scalar polynomial

$$f(\lambda) = a_0 + a_1 \lambda + a_2 \lambda^2 + \cdots + a_n \lambda^n + a_{n+1} \lambda^{n+1} + \cdots \quad (5.94b)$$

Dividing $f(\lambda)$ by the characteristic polynomial $\Delta(\lambda)$, we get

$$\frac{f(\lambda)}{\Delta(\lambda)} = q(\lambda) + \frac{g(\lambda)}{\Delta(\lambda)} \quad (5.95a)$$

where $g(\lambda)$ is the remainder polynomial of the following form:

$$g(\lambda) = \beta_0 + \beta_1 \lambda + \cdots + \beta_{n-1} \lambda^{n-1} \quad (5.95b)$$

Equation (5.95a) may be written as

$$f(\lambda) = q(\lambda)\Delta(\lambda) + g(\lambda) \quad (5.96)$$

Assume that the $n \times n$ matrix $\mathbf{A}$ has n distinct eigenvalues $\lambda_1, \lambda_2, ..., \lambda_n$;

$$\Delta(\lambda_i) = 0; \quad i = 1, 2, ..., n$$

If we evaluate $f(\lambda)$ in Eqn. (5.96) at the eigenvalues $\lambda_1, \lambda_2, ..., \lambda_n$, we have

$$f(\lambda_i) = g(\lambda_i), \quad i = 1, 2, ..., n \quad (5.97)$$

The coefficients $\beta_0, \beta_1, ..., \beta_{n-1}$ in Eqn. (5.95b) can be computed by solving the set of n simultaneous equations obtained by successively substituting $\lambda_1, \lambda_2, ..., \lambda_n$ in Eqn. (5.97).

Substituting $\mathbf{A}$ for λ in Eqn. (5.96), we get

$$f(\mathbf{A}) = q(\mathbf{A})\Delta(\mathbf{A}) + g(\mathbf{A})$$

Since $\Delta(\mathbf{A})$ is identically zero, it follows that
$$f(\mathbf{A}) = g(\mathbf{A}) = \beta_0 \mathbf{I} + \beta_1 \mathbf{A} + \cdots + \beta_{n-1} \mathbf{A}^{n-1}$$

If $\mathbf{A}$ possesses an eigenvalue λ_k of multiplicity n_k, then only one independent equation can be obtained by substituting λ_k into Eqn. (5.97). The remaining $(n_k - 1)$ linear equations, which must be obtained in order to solve for β_i's, can be found by differentiating both sides of Eqn. (5.97). Since

$$\left[\frac{d^j}{d\lambda^j} \Delta(\lambda)\right]_{\lambda = \lambda_k} = 0 \, ; j = 0, 1, \ldots, (n_k - 1),$$

it follows that

$$\left[\frac{d^j}{d\lambda^j} f(\lambda)\right]_{\lambda = \lambda_k} = \left[\frac{d^j}{d\lambda^j} g(\lambda)\right]_{\lambda = \lambda_k} \, ; j = 0, 1, \ldots, (n_k - 1)$$

The formal procedure of evaluation of the matrix polynomial $f(\mathbf{A})$ is given below.

(i) Compute $\Delta(\lambda) \triangleq |\lambda \mathbf{I} - \mathbf{A}|$
(ii) Find the roots of $\Delta(\lambda) = 0$, say

$$\Delta(\lambda) = (\lambda - \lambda_1)^{n_1} (\lambda - \lambda_2)^{n_2} \cdots (\lambda - \lambda_m)^{n_m} \quad (5.98a)$$

where $n_1 + n_2 + \cdots + n_m = n$. In other words, $\Delta(\lambda)$ has root λ_i with multiplicity n_i. If λ_i is a complex number, then its complex conjugate is also a root of $\Delta(\lambda)$.

(iii) Form a polynomial $g(\lambda)$ of degree $(n-1)$; i.e.,

$$g(\lambda) = \beta_0 + \beta_1 \lambda + \cdots + \beta_{n-1} \lambda^{n-1} \quad (5.98b)$$

where the unknown parameters $\beta_0, \beta_1, \ldots, \beta_{n-1}$ are to be solved in Step (v).

(iv) Form the following n equations:

$$\left[\frac{d^j}{d\lambda^j} f(\lambda)\right]_{\lambda = \lambda_i} = \left[\frac{d^j}{d\lambda^j} g(\lambda)\right]_{\lambda = \lambda_i} \quad (5.98c)$$

$$j = 0, 1, \ldots, (n_i - 1)$$
$$i = 1, 2, \ldots, m$$

(v) Solve for the n unknown parameters $\beta_0, \beta_1, \ldots, \beta_{n-1}$ from the n equations in Step (iv).

Then

$$f(\mathbf{A}) = g(\mathbf{A}) = \beta_0 \mathbf{I} + \beta_1 \mathbf{A} + \cdots + \beta_{n-1} \mathbf{A}^{n-1} \quad (5.98d)$$

Example 5.12
Find $f(\mathbf{A}) = \mathbf{A}^{10}$ for

$$\mathbf{A} = \begin{bmatrix} 0 & 1 \\ -2 & -3 \end{bmatrix}$$

Solution

The characteristic polynomial is
$$\Delta(\lambda) = |\lambda \mathbf{I} - \mathbf{A}| = \begin{vmatrix} \lambda & -1 \\ 2 & \lambda + 3 \end{vmatrix} = (\lambda + 1)(\lambda + 2)$$

The roots of $\Delta(\lambda) = 0$ are $\lambda_1 = -1$, $\lambda_2 = -2$.

Since $\mathbf{A}$ is of second order, the polynomial $g(\lambda)$ will be of the following form:
$$g(\lambda) = \beta_0 + \beta_1 \lambda$$

The coefficients β_0 and β_1 are evaluated from equations
$$f(\lambda_1) = (\lambda_1)^{10} = g(\lambda_1) = \beta_0 + \beta_1 \lambda_1$$
$$f(\lambda_2) = (\lambda_2)^{10} = g(\lambda_2) = \beta_0 + \beta_1 \lambda_2$$

The result is
$$\beta_0 = -1022, \; \beta_1 = -1023$$

Therefore
$$f(\mathbf{A}) = \mathbf{A}^{10} = \beta_0 \mathbf{I} + \beta_1 \mathbf{A} = \begin{bmatrix} -1022 & -1023 \\ 2046 & 2047 \end{bmatrix}$$

▲▲

The Cayley-Hamilton technique allows us to solve the problem of evaluation of $e^{\mathbf{A}t}$, where $\mathbf{A}$ is a constant $n \times n$ matrix. Since the matrix power series
$$e^{\mathbf{A}t} = \mathbf{I} + \mathbf{A}t + \frac{\mathbf{A}^2 t^2}{2!} + \cdots + \frac{\mathbf{A}^n t^n}{n!} + \cdots$$
converges for all $\mathbf{A}$ and for all finite t, the matrix polynomial $f(\mathbf{A}) = e^{\mathbf{A}t}$ can be expressed as a polynomial $g(\mathbf{A})$ of degree $(n-1)$. This is illustrated below with the help of an example.

Example 5.13

Consider the system
$$\dot{\mathbf{x}} = \mathbf{A}\mathbf{x}$$
with
$$\mathbf{A} = \begin{bmatrix} 0 & 0 & -2 \\ 0 & 1 & 0 \\ 1 & 0 & 3 \end{bmatrix}; \; \mathbf{x}(0) = \begin{bmatrix} 0 \\ 1 \\ 0 \end{bmatrix}$$

In the following, we evaluate the function
$$f(\mathbf{A}) = e^{\mathbf{A}t}$$
using the Cayley-Hamilton technique.

The characteristic polynomial of matrix $\mathbf{A}$ is
$$\Delta(\lambda) = |\lambda \mathbf{I} - \mathbf{A}| = \begin{vmatrix} \lambda & 0 & 2 \\ 0 & (\lambda - 1) & 0 \\ -1 & 0 & (\lambda - 3) \end{vmatrix} = (\lambda - 1)^2 (\lambda - 2)$$

The characteristic equation $\Delta(\lambda) = 0$ has a second-order root at $\lambda_1 = 1$ and a simple root at $\lambda_2 = 2$.

Since $\mathbf{A}$ is of third order, the polynomial $g(\lambda)$ will be of the form
$$g(\lambda) = \beta_0 + \beta_1 \lambda + \beta_2 \lambda^2$$
The coefficients β_0, β_1, and β_2 are evaluated using the following relations:
$$f(\lambda_1) = g(\lambda_1)$$
$$\left.\frac{d}{d\lambda} f(\lambda)\right|_{\lambda = \lambda_1} = \left.\frac{d}{d\lambda} g(\lambda)\right|_{\lambda = \lambda_1}$$
$$f(\lambda_2) = g(\lambda_2)$$
These relations yield the following set of simultaneous equations.
$$e^t = \beta_0 + \beta_1 + \beta_2$$
$$te^t = \beta_1 + 2\beta_2$$
$$e^{2t} = \beta_0 + 2\beta_1 + 4\beta_2$$
Solving these equations, we obtain
$$\beta_0 = -2te^t + e^{2t}$$
$$\beta_1 = 3t\,e^t + 2e^t - 2e^{2t}, \text{ and}$$
$$\beta_2 = e^{2t} - e^t - te^t$$
Hence we have
$$e^{\mathbf{A}t} = g(\mathbf{A}) = \beta_0 \mathbf{I} + \beta_1 \mathbf{A} + \beta_2 \mathbf{A}^2$$
$$= (-2te^t + e^{2t})\mathbf{I} + (3te^t + 2e^t - 2e^{2t})\mathbf{A} + (e^{2t} - e^t - te^t)\mathbf{A}^2$$
$$= \begin{bmatrix} 2e^t - e^{2t} & 0 & 2e^t - 2e^{2t} \\ 0 & e^t & 0 \\ -e^t + e^{2t} & 0 & 2e^{2t} - e^t \end{bmatrix}$$
Consequently, the free response ($u(t) = 0$) of the system is
$$\mathbf{x}(t) = e^{\mathbf{A}t}\mathbf{x}(0) = \begin{bmatrix} 0 \\ e^t \\ 0 \end{bmatrix}$$
This result is identical to the one obtained earlier in Example 5.10.

Solution of Non-homogeneous State Equation

When an input $u(t)$ is present, the complete solution $\mathbf{x}(t)$ is obtained from the non-homogeneous Eqn. (5.80).

By writing Eqn. (5.80) as
$$\dot{\mathbf{x}}(t) - \mathbf{A}\mathbf{x}(t) = \mathbf{b}u(t)$$
and premultiplying both sides of this equation by $e^{-\mathbf{A}t}$, we obtain
$$e^{-\mathbf{A}t}[\dot{\mathbf{x}}(t) - \mathbf{A}\mathbf{x}(t)] = e^{-\mathbf{A}t}\mathbf{b}u(t) \qquad (5.99)$$

By applying the rule for the derivative of the product of two matrices, we can write (refer Eqn. (5.87))

$$\frac{d}{dt}[e^{-At}\mathbf{x}(t)] = e^{-At}\frac{d}{dt}(\mathbf{x}(t)) + \frac{d}{dt}(e^{-At})\mathbf{x}(t)$$

$$= e^{-At}\dot{\mathbf{x}}(t) - e^{-At}\mathbf{A}\mathbf{x}(t)$$

$$= e^{-At}[\dot{\mathbf{x}}(t) - \mathbf{A}\mathbf{x}(t)]$$

Use of this equality in Eqn. (5.99) gives

$$\frac{d}{dt}[e^{-At}\mathbf{x}(t)] = e^{-At}\mathbf{b}u(t)$$

Integrating both sides with respect to t between the limits 0 and t, we get

$$e^{-At}\mathbf{x}(t)\bigg|_0^t = \int_0^t e^{-A\tau}\mathbf{b}u(\tau)d\tau$$

or

$$e^{-At}\mathbf{x}(t) - \mathbf{x}(0) = \int_0^t e^{-A\tau}\mathbf{b}u(\tau)d\tau$$

Now premultiplying both sides by e^{At}, we have

$$\mathbf{x}(t) = e^{At}\mathbf{x}(0) + \int_0^t e^{A(t-\tau)}\mathbf{b}u(\tau)d\tau \qquad (5.100)$$

If the initial state is known at $t = t_0$, rather than $t = 0$, Eqn. (5.100) becomes

$$\mathbf{x}(t) = e^{A(t-t_0)}\mathbf{x}(t_0) + \int_{t_0}^t e^{A(t-\tau)}\mathbf{b}u(\tau)d\tau \qquad (5.101)$$

Equation (5.101) can also be written as

$$\mathbf{x}(t) = \phi(t-t_0)\mathbf{x}(t_0) + \int_{t_0}^t \phi(t-\tau)\mathbf{b}u(\tau)d\tau \qquad (5.102)$$

where

$$\phi(t) = e^{At}$$

Equation (5.102) is the solution of Eqn. (5.80). This equation is called the *state transition equation*. It describes the change of state relative to the initial conditions $\mathbf{x}(t_0)$ and the input $u(t)$.

Example 5.14

For the speed control system of Fig. 5.3, following plant model was derived in Example 5.1 (refer Eqns (5.17)):

$$\dot{\mathbf{x}} = \mathbf{A}\mathbf{x} + \mathbf{b}u$$

$$y = \mathbf{c}\mathbf{x}$$

with

$$\mathbf{A} = \begin{bmatrix} -1 & 1 \\ -1 & -10 \end{bmatrix}; \mathbf{b} = \begin{bmatrix} 0 \\ 10 \end{bmatrix}; \mathbf{c} = [1\ 0]$$

State variables x_1 and x_2 are the physical variables of the system:
$$x_1(t) = \omega(t), \text{ angular velocity of the motor shaft}$$
$$x_2(t) = i_a(t), \text{ armature current}$$

The output
$$y(t) = x_1(t) = \omega(t)$$

In the following, we evaluate the response of this system to a unit step input under zero initial conditions.

$$(s\mathbf{I} - \mathbf{A})^{-1} = \begin{bmatrix} s+1 & -1 \\ 1 & s+10 \end{bmatrix}^{-1}$$

$$= \frac{1}{s^2 + 11s + 11} \begin{bmatrix} s+10 & 1 \\ -1 & s+1 \end{bmatrix}$$

$$= \begin{bmatrix} \dfrac{s+10}{(s+a_1)(s+a_2)} & \dfrac{1}{(s+a_1)(s+a_2)} \\ \dfrac{-1}{(s+a_1)(s+a_2)} & \dfrac{s+1}{(s+a_1)(s+a_2)} \end{bmatrix}; a_1 = 1.1125, a_2 = 9.8875$$

$$e^{\mathbf{A}t} = \mathscr{L}^{-1}[(s\mathbf{I} - \mathbf{A})^{-1}]$$

$$= \begin{bmatrix} 1.0128e^{-a_1 t} - 0.0128e^{-a_2 t} & 0.114e^{-a_1 t} - 0.114e^{-a_2 t} \\ -0.114e^{-a_1 t} + 0.114e^{-a_2 t} & -0.0128e^{-a_1 t} + 1.0128e^{-a_2 t} \end{bmatrix}$$

$u(t) = 1; t \geq 0$

$\mathbf{x}(0) = \mathbf{0}$

Therefore,

$$\mathbf{x}(t) = \int_0^t e^{\mathbf{A}(t-\tau)} \mathbf{b} d\tau = \int_0^t \begin{bmatrix} 1.14\left(e^{-a_1(t-\tau)} - e^{-a_2(t-\tau)}\right) \\ 1.14\left(-0.1123e^{-a_1(t-\tau)} + 8.8842e^{-a_2(t-\tau)}\right) \end{bmatrix} d\tau$$

$$= \begin{bmatrix} 0.9094 - 1.0247e^{-a_1 t} + 0.1153e^{-a_2 t} \\ -0.0132 + 0.1151e^{-a_1 t} - 0.1019e^{-a_2 t} \end{bmatrix}$$

The output
$$y(t) = \omega(t) = 0.9094 - 1.0247\, e^{-1.1125t} + 0.1153 e^{-9.8875t};\ t \geq 0$$

5.8 CONCEPTS OF CONTROLLABILITY AND OBSERVABILITY

Controllability and observability are properties which describe structural features of a dynamic system. These properties play an important role in modern control system design theory; the conditions on controllability and observability often govern the control solution.

To illustrate the motivation of investigating controllability and observability properties, we consider the problem of the stabilization of an inverted pendulum on a motor-driven cart.

Example 5.15

Figure 5.16 shows an inverted pendulum with its pivot mounted on a cart. The cart is driven by an electric motor. The motor drives a pair of wheels of the cart; the whole cart and the pendulum become the 'load' on the motor. The motor at time t exerts a torque $T(t)$ on the wheels. The linear force applied to the cart is $u(t)$; $T(t) = Ru(t)$, where R is the radius of the wheels.

The pendulum is obviously unstable. It can, however, be kept upright by applying a proper control force $u(t)$. This somewhat artificial system example represents a dynamic model of a space booster on take off—the booster is balanced on top of the rocket engine thrust vector.

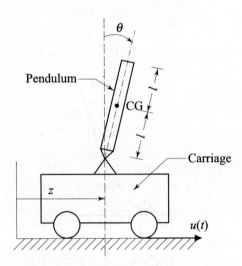

Fig. 5.16 Inverted pendulum system

From inspection of Fig. 5.16, we construct the differential equations describing the dynamics of the inverted pendulum and the cart. The horizontal displacement of the pivot on the cart is $z(t)$, while the rotational angle of the pendulum is $\theta(t)$. The parameters of the system are as follows.

$M =$ the mass of the cart
$L =$ the length of the pendulum $= 2l$
$m =$ the mass of the pendulum
$J =$ the moment of inertia of the pendulum with respect to centre of gravity (CG)

The horizontal and vertical positions of the CG of the pendulum are given by $(z + l\sin\theta)$ and $(l\cos\theta)$ respectively.

The forces exerted on the pendulum are the force mg on the centre of gravity, a horizontal reaction force H and a vertical reaction force V (Fig. 5.17a). Taking moments around CG of the pendulum, we get

$$J\frac{d^2\theta(t)}{dt^2} = V(t)\, l\, \sin\theta(t) - H(t)\, l\, \cos\theta(t) \qquad (5.103a)$$

Summing up all forces on the pendulum in vertical and horizontal directions, we obtain

$$m\frac{d^2}{dt^2}(l\cos\theta(t)) = V(t) - mg \qquad (5.103b)$$

$$m\frac{d^2}{dt^2}(z(t) + l\sin\theta(t)) = H(t) \qquad (5.103c)$$

Summing up all forces on the cart in the horizontal direction (Fig. 5.17b), we get

$$M\frac{d^2z(t)}{dt^2} = u(t) - H(t) \qquad (5.103d)$$

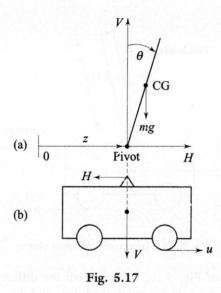

Fig. 5.17

In our problem, since the objective is to keep the pendulum upright, it seems reasonable to assume that $\dot\theta(t)$ and $\theta(t)$ will remain close to zero. In view of this, we can set with sufficient accuracy $\sin\theta \cong \theta$, $\cos\theta \cong 1$. With this approximation, we get from Eqns (5.103),

$$ml\,\ddot\theta(t) + (m + M)\ddot z(t) = u(t)$$
$$(J + ml^2)\ddot\theta(t) + ml\,\ddot z(t) - mgl\,\theta(t) = 0$$

These equations may be rearranged as

$$\ddot{\theta}(t) = \frac{ml(M+m)g}{\Delta}\theta(t) - \frac{ml}{\Delta}u(t) \qquad (5.104a)$$

$$\ddot{z}(t) = -\frac{m^2 l^2 g}{\Delta}\theta(t) + \frac{(J+ml^2)}{\Delta}u(t) \qquad (5.104b)$$

where

$$\Delta = (M+m)J + Mml^2$$

Suppose that the system parameters are $M = 1$ kg, $m = 0.15$ kg, $l = 1$ m. Recall that

$$g = 9.81 \text{ m/sec}^2$$
$$J = \tfrac{1}{3}mL^2 = \tfrac{4}{3}ml^2 = 0.2 \text{ kg-m}^2$$

For these parameters, we have from Eqns (5.104),

$$\ddot{\theta}(t) = 4.4537\,\theta(t) - 0.3947\,u(t) \qquad (5.105a)$$
$$\ddot{z}(t) = -0.5809\,\theta(t) + 0.9211\,u(t) \qquad (5.105b)$$

Choosing the states $x_1 = \theta$, $x_2 = \dot{\theta}$, $x_3 = z$, and $x_4 = \dot{z}$ we obtain the following state model for the inverted pendulum on moving cart.

$$\dot{\mathbf{x}} = \mathbf{A}\mathbf{x} + \mathbf{b}u \qquad (5.106)$$

with

$$\mathbf{A} = \begin{bmatrix} 0 & 1 & 0 & 0 \\ 4.4537 & 0 & 0 & 0 \\ 0 & 0 & 0 & 1 \\ -0.5809 & 0 & 0 & 0 \end{bmatrix}; \mathbf{b} = \begin{bmatrix} 0 \\ -0.3947 \\ 0 \\ 0.9211 \end{bmatrix}$$

The plant (5.106) is said to be *completely controllable* if every state $\mathbf{x}(t_0)$ can be affected or controlled to reach a desired state in finite time by some unconstrained control $u(t)$. Shortly we will see that the plant (5.106) satisfies this condition and therefore a solution exists to the following control problem:

Move the cart from one location to another without causing the pendulum to fall.

The solution to this control problem is not unique. We normally look for a feedback control scheme so that the destabilizing effects of disturbance forces (due to wind, for example) are filtered out. Figure 5.18a shows a state-feedback control scheme for stabilizing the inverted pendulum. The closed-loop system is formed by feeding back the state variables through a real constant matrix $\mathbf{k}$;

$$u(t) = -\mathbf{k}\mathbf{x}(t)$$

The closed-loop system is thus described by

$$\dot{\mathbf{x}}(t) = (\mathbf{A} - \mathbf{b}\mathbf{k})\mathbf{x}(t)$$

388 Digital Control and State Variable Methods

The design objective in this case is to find the feedback matrix **k** such that the closed-loop system is stable. The existence of solution to this design problem is directly based on the controllability property of the plant (5.106). This will be established in Chapter 7.

Implementation of the state-feedback control solution requires access to all the state variables of the plant model. In many control situations of interest, it is possible to install sensors to measure all the state variables. This may not be possible or practical in some cases. For example, if the plant model includes non-physical state variables, measurement of these variables using physical sensors is not possible. Accuracy requirements or cost considerations may prohibit the use of sensors for some physical variables also.

The input and the output of a system are always physical quantities, and are normally easily accessible to measurement. We therefore need a subsystem that performs the estimation of state variables based on the information received from the input $u(t)$ and the output $y(t)$. This subsystem is called an *observer* whose design is based on *observability property* of the controlled system.

The plant (5.106) is said to be *completely observable* if all the state variables in $\mathbf{x}(t)$ can be observed from the measurements of the output $y(t) = \theta(t)$ and the input $u(t)$. Shortly we will see that the plant (5.106) does not satisfy this condition, and therefore a solution to the observer-design problem does not exist when the inputs to the observer subsystem are $u(t)$ and $\theta(t)$.

Cart position $z(t)$ is easily accessible to measurement and, as we shall see, the observability condition is satisfied with this choice of input information to the observer subsystem. Figure 5.18b shows the block diagram of the closed-loop system with an observer that estimates the state vector from measurements of $u(t)$ and $z(t)$. The observed or estimated state vector, designated as $\hat{\mathbf{x}}$, is then used to generate the control u through the feedback matrix **k**.

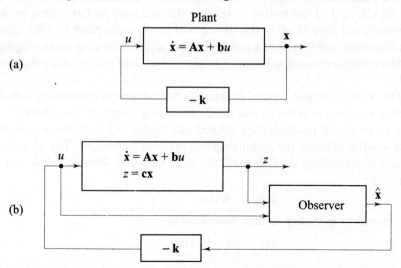

Fig. 5.18 Control system with state feedback

Control System Analysis Using State Variable Methods

A study of controllability and observability properties, presented in this section, provides a basis for the state-feedback design problems discussed in Chapter 7. Further, these properties establish the conditions for complete equivalence between the state variable and transfer function representations.

Definitions of Controllability and Observability

In this section, we study the controllability and observability of linear time-invariant systems described by state variable model of the following form:

$$\dot{\mathbf{x}}(t) = \mathbf{A}\mathbf{x}(t) + \mathbf{b}u(t) \tag{5.107a}$$
$$y(t) = \mathbf{c}\mathbf{x}(t) + du(t) \tag{5.107b}$$

where $\mathbf{A}, \mathbf{b}, \mathbf{c}$ and d are respectively $n \times n$, $n \times 1$, $1 \times n$ and 1×1 matrices, $\mathbf{x}(t)$ is $n \times 1$ state vector, $y(t)$ and $u(t)$ are respectively output and input variables.

Controllability

For the linear system given by Eqns (5.107), if there exists an input $u_{[0, t_1]}$ which transfers the initial state $\mathbf{x}(0) \triangleq \mathbf{x}^0$ to the state $\mathbf{x}^1$ in a finite time t_1, the state $\mathbf{x}^0$ is said to be controllable. If all initial states are controllable, the system is said to be *completely controllable*, or simply *controllable*. Otherwise, the system is said to be *uncontrollable*.

From Eqn. (5.100), the solution of Eqn. (5.107a) is

$$\mathbf{x}(t) = e^{\mathbf{A}t}\mathbf{x}^0 + \int_0^t e^{\mathbf{A}(t-\tau)} \mathbf{b}u(\tau)\, d\tau$$

To study the controllability property, we may assume without loss of generality that $\mathbf{x}^1 \equiv \mathbf{0}$. Therefore if the system (5.107) is controllable, there exists an input $u_{[0, t_1]}$ such that

$$-\mathbf{x}^0 = \int_0^{t_1} e^{-\mathbf{A}\tau} \mathbf{b}u(\tau)\, d\tau \tag{5.108}$$

From this equation, we observe that complete controllability of a system depends on $\mathbf{A}$ and $\mathbf{b}$, and is independent of output matrix $\mathbf{c}$. The controllability of the system (5.107) is frequently referred to as the controllability of the pair $\{\mathbf{A}, \mathbf{b}\}$.

It may be noted that according to the definition of controllability, there is no constraint imposed on the input or on the trajectory that the state should follow. Further, the system is said to be uncontrollable although it may be 'controllable in part'.

From the definition of controllability, we observe that by complete controllability of a plant, we mean that we can make the plant do whatever we please. Perhaps this definition is too restrictive in the sense that we are asking too much of the plant. But if we are able to show that system equations satisfy this definition, certainly there can be no intrinsic limitation on the design of the control system for the plant. However, if the system turns out to be uncontrollable, it does not necessarily mean that the plant can never be operated in a satisfactory manner. Provided that a control system will maintain the important variables in an acceptable region, the fact that the plant is not completely controllable is immaterial.

Another important point which the reader must bear in mind is that almost all physical systems are nonlinear in nature to a certain extent and a linear model is obtained after making certain approximations. Small perturbations of the elements of **A** and **b** may cause an uncontrollable system to become controllable. It may also be possible to increase the number of control variables and make the plant completely controllable (controllability of multi-input systems will be discussed in Section 5.10).

A common source of uncontrollable state variable models arises when redundant state variables are defined. No one would intentionally use more state variables than the minimum number needed to characterize the behaviour of the dynamic system. In a complex system with unfamiliar physics, one may be tempted to write down differential equations for everything in sight and in doing so, may write down more equations than are necessary. This will invariably result in an uncontrollable model for the system.

Observability

For the linear system given by Eqns (5.107), if the knowledge of the output y and the input u over a finite interval of time $[0, t_1]$ suffices to determine the state $x(0) \triangleq x^0$, the state x^0 is said to be observable. If all initial states are observable, the system is said to be *completely observable*, or simply *observable*. Otherwise, the system is said to be *unobservable*.

The output of the system (5.107) is given by

$$y(t) = c e^{At} x^0 + c \int_0^t e^{A(t-\tau)} b u(\tau) d\tau + du(t)$$

The output and the input can be measured and used, so that following signal $\eta(t)$ can be obtained from u and y.

$$\eta(t) \triangleq y(t) - c \int_0^t e^{A(t-\tau)} b u(\tau) d\tau - du(t) = c e^{At} x^0 \qquad (5.109)$$

Premultiplying by $e^{A^T t} c^T$ and integrating from 0 to t_1, gives

$$\left\{ \int_0^{t_1} e^{A^T t} c^T c e^{At} dt \right\} x^0 = \int_0^{t_1} e^{A^T t} c^T \eta(t) dt \qquad (5.110)$$

When the signal $\eta(t)$ is available over a time interval $[0, t_1]$, and the system (5.107) is observable, then the initial state x^0 can be uniquely determined from Eqn. (5.110).

From Eqn. (5.110) we see that complete observability of a system depends on **A** and **c**, and is independent of **b**. The observability of the system (5.107) is frequently referred to as the observability of the pair $\{\mathbf{A}, \mathbf{c}\}$.

Note that the system is said to be unobservable, although it may be 'observable in part'. Plants that are not completely observable can often be made observable by making more measurements (observability of multi-output systems will be discussed in Section 5.10). Alternately, one may examine feedback control schemes which do not require complete state feedback.

Controllability Tests

It is difficult to guess whether a system is controllable or not from the defining Eqn. (5.108). Some simple mathematical tests which answer the question of controllability have been developed. The following theorem gives two controllability tests.

Theorem 5.2

The necessary and sufficient condition for the system (5.107) to be completely controllable is given by any one of the following:

I.
$$\mathbf{W}(0, t_1) = \int_0^{t_1} e^{-\mathbf{A}t} \mathbf{b}\, \mathbf{b}^T e^{-\mathbf{A}^T t}\, dt \tag{5.111}$$

is nonsingular.

II. The $n \times n$ controllability matrix

$$\mathbf{U} \triangleq [\mathbf{b} \quad \mathbf{A}\mathbf{b} \quad \mathbf{A}^2\mathbf{b} \cdots \mathbf{A}^{n-1}\mathbf{b}] \tag{5.112}$$

has rank equal to n, i.e., $\rho(\mathbf{U}) = n$.

Since Test II can be computed without integration, it allows the controllability of a system to be easily checked.

Proof of Test I

Sufficiency: If $\mathbf{W}(0, t_1)$ given in Eqn. (5.111) is nonsingular, the input

$$u(t) = -\mathbf{b}^T e^{-\mathbf{A}^T t} \mathbf{W}^{-1}(0, t_1) \mathbf{x}^0 \tag{5.113}$$

can be applied to the system. This input satisfies the condition given in Eqn. (5.108):

$$\int_0^{t_1} e^{-\mathbf{A}t} \mathbf{b} u(t) dt = -\int_0^{t_1} e^{-\mathbf{A}t} \mathbf{b}\mathbf{b}^T e^{-\mathbf{A}^T t} \mathbf{W}^{-1}(0, t_1) \mathbf{x}^0 dt$$

$$= -\left\{\int_0^{t_1} e^{-\mathbf{A}t} \mathbf{b}\mathbf{b}^T e^{-\mathbf{A}^T t} dt\right\} \mathbf{W}^{-1}(0, t_1) \mathbf{x}^0 = -\mathbf{x}^0$$

Necessity: Assume that the system is controllable though $\mathbf{W}(0, t_1)$ is singular for any t_1. Then as per the results given in Eqns (5.8), the n rows of $e^{-\mathbf{A}t}\mathbf{b}$ are linearly dependent, i.e., there exists a nonzero $n \times 1$ vector $\boldsymbol{\alpha}$ such that

$$\boldsymbol{\alpha}^T e^{-\mathbf{A}t}\mathbf{b} = 0 \tag{5.114}$$

From the assumption of controllability, there exists an input u satisfying Eqn. (5.108); therefore from Eqns (5.108) and (5.114),

$$-\boldsymbol{\alpha}^T \mathbf{x}^0 = \int_0^{t_1} \boldsymbol{\alpha}^T e^{-\mathbf{A}t}\mathbf{b}\, u(t)\, dt = 0 \tag{5.115}$$

holds for any initial state $\mathbf{x}^0$. By choosing $\mathbf{x}^0 = \boldsymbol{\alpha}$, Eqn. (5.115) gives (refer Eqn. (5.6a)),

$$\boldsymbol{\alpha}^T \boldsymbol{\alpha} = [||\boldsymbol{\alpha}||]^2 = 0$$

This is true only for $\boldsymbol{\alpha} = \mathbf{0}$ which contradicts the nonzero property of $\boldsymbol{\alpha}$. Therefore the nonsingularity of $\mathbf{W}(0, t_1)$ is proved.

Proof of Test II

Sufficiency: It is first assumed that though $\rho(\mathbf{U}) = n$, the system is not controllable, and by showing that this is a contradiction, the controllability of the system is proved.

By the above assumption,

$$\rho(\mathbf{U}) = n \quad \text{and} \quad \mathbf{W}(0, t_1) \text{ is singular.}$$

Therefore Eqn. (5.114) holds, i.e.,

$$\boldsymbol{\alpha}^T e^{-\mathbf{A}t}\mathbf{b} = 0;\ t \geq 0,\ \boldsymbol{\alpha} \neq \mathbf{0}$$

Derivatives of the above equation at $t = 0$ yield (refer Eqn. (5.87)),

$$\boldsymbol{\alpha}^T \mathbf{A}^k \mathbf{b} = 0;\ k = 0, 1, \ldots, (n-1),$$

which is equivalent to

$$\boldsymbol{\alpha}^T [\mathbf{b}\ \ \mathbf{A}\mathbf{b}\ \cdots\ \mathbf{A}^{n-1}\mathbf{b}] = \boldsymbol{\alpha}^T \mathbf{U} = \mathbf{0}$$

Therefore, n rows of controllability matrix $\mathbf{U}$ are linearly dependent (refer Eqn. (5.8a)). This contradicts the assumption that $\rho(\mathbf{U}) = n$; hence the system is completely controllable.

Necessity: It is assumed that the system is completely controllable but $\rho(\mathbf{U}) < n$. From this assumption, there exists nonzero vector $\boldsymbol{\alpha}$ satisfying

$$\boldsymbol{\alpha}^T \mathbf{U} = \mathbf{0}$$

or

$$\boldsymbol{\alpha}^T \mathbf{A}^k \mathbf{b} = 0;\ k = 0, 1, \ldots, (n-1) \tag{5.116a}$$

Also from the Cayley-Hamilton theorem, $\mathbf{A}^{n+1}$ can be expressed as a linear combination of $\mathbf{I}, \mathbf{A}, \ldots, \mathbf{A}^{n-1}$ (refer Eqn. (5.98d)). Therefore $e^{-\mathbf{A}t}$ can be expressed as a linear combination of $\mathbf{I}, \mathbf{A}, \ldots, \mathbf{A}^{n-1}$:

$$e^{-\mathbf{A}t} = \beta_0 \mathbf{I} + \beta_1 \mathbf{A} + \cdots + \beta_{n-1} \mathbf{A}^{n-1} \tag{5.116b}$$

From Eqns (5.116a) and (5.116b), we obtain
$$\alpha^T e^{-At} \mathbf{b} = 0, \, t \geq 0, \, \boldsymbol{\alpha} \neq \mathbf{0}$$
and therefore (refer Eqns (5.8)),
$$\int_0^{t_1} \alpha^T e^{-At} \mathbf{b} \, \mathbf{b}^T e^{-A^T t} \alpha \, dt = \alpha^T \mathbf{W}(0, t_1) \alpha = 0$$

Since the system is completely controllable, $\mathbf{W}(0, t_1)$ should be nonsingular from Test I; this contradicts the assumption that α is nonzero. Therefore, $\rho(\mathbf{U}) = n$.

Example 5.16

Recall the inverted pendulum of Example 5.15, shown in Fig. 5.16, in which the object is to apply a force $u(t)$ so that the pendulum remains balanced in the vertical position. We found the linearized equations governing the system to be:
$$\dot{\mathbf{x}} = \mathbf{A}\mathbf{x} + \mathbf{b}u$$
where
$$\mathbf{x} = [\theta \; \dot{\theta} \; z \; \dot{z}]^T$$

$$\mathbf{A} = \begin{bmatrix} 0 & 1 & 0 & 0 \\ 4.4537 & 0 & 0 & 0 \\ 0 & 0 & 0 & 1 \\ -0.5809 & 0 & 0 & 0 \end{bmatrix}; \, \mathbf{b} = \begin{bmatrix} 0 \\ -0.3947 \\ 0 \\ 0.9211 \end{bmatrix}$$

$z(t)$ = horizontal displacement of the pivot on the cart
$\theta(t)$ = rotational angle of the pendulum.

To check the controllability of this system, we compute the controllability matrix $\mathbf{U}$:

$$\mathbf{U} = [\mathbf{b} \; \mathbf{Ab} \; \mathbf{A}^2\mathbf{b} \; \mathbf{A}^3\mathbf{b}] = \begin{bmatrix} 0 & -0.3947 & 0 & -1.7579 \\ -0.3947 & 0 & -1.7579 & 0 \\ 0 & 0.9211 & 0 & 0.2293 \\ 0.9211 & 0 & 0.2293 & 0 \end{bmatrix}$$

Since $|\mathbf{U}| = 2.3369$, $\mathbf{U}$ has full rank, and by Theorem 5.2, the system is completely controllable. Thus if the angle θ departs from equilibrium by a small amount, a control always exists which will drive it back to zero.[10] Moreover, a control also exists which will drive both θ and z, as well as their derivatives, to zero.

It may be noted that Eqn. (5.113) suggests a control law to prove the sufficiency of the controllability test. It does not necessarily give an acceptable solution to the control problem. The open-loop control given by Eqn. (5.113) is normally not acceptable. In Chapter 7, we will derive a state-feedback control law for the inverted pendulum. As we shall see, for such a control to exist, complete controllability of the plant is a *necessary requirement*.

---

10. This justifies the assumption that $\theta(t) \cong 0$, provided we choose an appropriate control strategy.

Example 5.17

Consider the electrical network shown in Fig. 5.19. Differential equations governing the dynamics of this network can be obtained by various standard methods. By use of nodal analysis, for example, we get

$$C_1 \frac{de_1}{dt} + \frac{e_1 - e_2}{R_3} + \frac{e_1 - e_0}{R_1} = 0$$

$$C_2 \frac{de_2}{dt} + \frac{e_2 - e_1}{R_3} + \frac{e_2 - e_0}{R_2} = 0$$

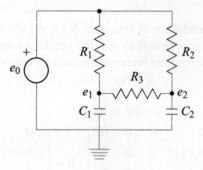

Fig. 5.19

The appropriate state variables for the network are the capacitor voltages e_1 and e_2. Thus, the state equations of the network are

$$\dot{\mathbf{x}} = \mathbf{A}\mathbf{x} + \mathbf{b}e_0$$

where
$$\mathbf{x} = [e_1 \ e_2]^T$$

$$\mathbf{A} = \begin{bmatrix} -\left(\dfrac{1}{R_1} + \dfrac{1}{R_3}\right)\dfrac{1}{C_1} & \dfrac{1}{R_3 C_1} \\ \dfrac{1}{R_3 C_2} & -\left(\dfrac{1}{R_2} + \dfrac{1}{R_3}\right)\dfrac{1}{C_2} \end{bmatrix}$$

$$\mathbf{b} = \begin{bmatrix} \dfrac{1}{R_1 C_1} \\ \dfrac{1}{R_2 C_2} \end{bmatrix}$$

The controllability matrix of the system is

$$\mathbf{U} = [\mathbf{b} \ \mathbf{A}\mathbf{b}] = \begin{bmatrix} \dfrac{1}{R_1 C_1} & -\dfrac{1}{(R_1 C_1)^2} + \dfrac{1}{R_3 C_1}\left(\dfrac{1}{R_2 C_2} - \dfrac{1}{R_1 C_1}\right) \\ \dfrac{1}{R_2 C_2} & -\dfrac{1}{(R_2 C_2)^2} + \dfrac{1}{R_3 C_2}\left(\dfrac{1}{R_1 C_1} - \dfrac{1}{R_2 C_2}\right) \end{bmatrix}$$

We see that under the condition
$$R_1 C_1 = R_2 C_2$$
$\rho(\mathbf{U}) = 1$ and the system becomes 'uncontrollable'. This condition is the one required to balance the bridge, and in this case, the voltage across the terminals of R_3 cannot be influenced by the input e_0.

Observability Tests

The following theorem gives two observability tests.

Theorem 5.3
The necessary and sufficient condition for the system (5.107) to be completely observable is given by any one of the following:

I. $\mathbf{M}(0, t_1) = \int_0^{t_1} e^{\mathbf{A}^T t} \mathbf{c}^T \mathbf{c} e^{\mathbf{A} t} dt$ (5.117)

is nonsingular.

II. The $n \times n$ observability matrix

$$\mathbf{V} \triangleq \begin{bmatrix} \mathbf{c} \\ \mathbf{cA} \\ \vdots \\ \mathbf{cA}^{n-1} \end{bmatrix} \quad (5.118)$$

has rank equal to n, i.e., $\rho(\mathbf{V}) = n$.

Proof
Using the observability Eqn. (5.110), this theorem can be proved in a manner similar to Theorem 5.2.

Example 5.18
We now return to the inverted pendulum of Example 5.16. Assuming that the only output variable for measurement is $\theta(t)$, the position of the pendulum, then the linearized equations governing the system are

$$\dot{\mathbf{x}} = \mathbf{A}\mathbf{x} + \mathbf{b}u$$
$$y = \mathbf{c}\mathbf{x}$$

where
$$\mathbf{A} = \begin{bmatrix} 0 & 1 & 0 & 0 \\ 4.4537 & 0 & 0 & 0 \\ 0 & 0 & 0 & 1 \\ -0.5809 & 0 & 0 & 0 \end{bmatrix}; \mathbf{b} = \begin{bmatrix} 0 \\ -0.3947 \\ 0 \\ 0.9211 \end{bmatrix}$$
$$\mathbf{c} = \begin{bmatrix} 1 & 0 & 0 & 0 \end{bmatrix}$$

The observability matrix

$$\mathbf{V} = \begin{bmatrix} \mathbf{c} \\ \mathbf{cA} \\ \mathbf{cA}^2 \\ \mathbf{cA}^3 \end{bmatrix} = \begin{bmatrix} 1 & 0 & 0 & 0 \\ 0 & 1 & 0 & 0 \\ 4.4537 & 0 & 0 & 0 \\ 0 & 4.4537 & 0 & 0 \end{bmatrix}$$

$|\mathbf{V}| = 0$, and therefore by Theorem 5.3, the system is not completely observable.

Consider now the displacement $z(t)$ of the cart as the output variable. Then
$$\mathbf{c} = [0 \ 0 \ 1 \ 0]$$
and the observability matrix

$$\mathbf{V} = \begin{bmatrix} 0 & 0 & 1 & 0 \\ 0 & 0 & 0 & 1 \\ -0.5809 & 0 & 0 & 0 \\ 0 & -0.5809 & 0 & 0 \end{bmatrix}$$

$|\mathbf{V}| = 0.3374 \neq 0$; the system is therefore completely observable. The values of $\dot{z}(t)$, $\theta(t)$ and $\dot{\theta}(t)$ can all be determined by observing $z(t)$ over an arbitrary time interval. Observer design for the inverted-pendulum system is given in Chapter 7.

Invariance Property

It is recalled that the state variable model for a system is not unique, but depends on the choice of a set of state variables. A transformation
    $\mathbf{x}(t) = \mathbf{P}\bar{\mathbf{x}}(t)$; $\mathbf{P}$ is a nonsingular constant matrix,
results in the following alternative state variable model (refer Eqns (5.22)) for the system (5.107):

$$\dot{\bar{\mathbf{x}}}(t) = \bar{\mathbf{A}}\bar{\mathbf{x}}(t) + \bar{\mathbf{b}}u(t); \ \bar{\mathbf{x}}(t_0) = \mathbf{P}^{-1}\mathbf{x}(t_0)$$
$$y(t) = \bar{\mathbf{c}}\bar{\mathbf{x}}(t) + du(t)$$

where
$$\bar{\mathbf{A}} = \mathbf{P}^{-1}\mathbf{A}\mathbf{P}, \ \bar{\mathbf{b}} = \mathbf{P}^{-1}\mathbf{b}, \ \bar{\mathbf{c}} = \mathbf{c}\mathbf{P}$$

The definition of new set of internal state variables should evidently not affect the controllability and observability properties. This may be verified by evaluating the controllability and observability matrices of the transformed system.

I.
$$\bar{\mathbf{U}} = [\bar{\mathbf{b}} \ \bar{\mathbf{A}}\bar{\mathbf{b}} \ \cdots \ (\bar{\mathbf{A}})^{n-1}\bar{\mathbf{b}}] \tag{5.119a}$$

$$\bar{\mathbf{b}} = \mathbf{P}^{-1}\mathbf{b}$$
$$\bar{\mathbf{A}}\bar{\mathbf{b}} = \mathbf{P}^{-1}\mathbf{A}\mathbf{P}\mathbf{P}^{-1}\mathbf{b} = \mathbf{P}^{-1}\mathbf{A}\mathbf{b}$$
$$(\bar{\mathbf{A}})^2\bar{\mathbf{b}} = \bar{\mathbf{A}}(\bar{\mathbf{A}}\bar{\mathbf{b}}) = \mathbf{P}^{-1}\mathbf{A}\mathbf{P}\mathbf{P}^{-1}\mathbf{A}\mathbf{b} = \mathbf{P}^{-1}\mathbf{A}^2\mathbf{b}$$
$$\vdots$$
$$(\bar{\mathbf{A}})^{n-1}\bar{\mathbf{b}} = \mathbf{P}^{-1}\mathbf{A}^{n-1}\mathbf{b}$$

Therefore,
$$\bar{\mathbf{U}} = [\mathbf{P}^{-1}\mathbf{b} \ \mathbf{P}^{-1}\mathbf{A}\mathbf{b} \ \cdots \ \mathbf{P}^{-1}\mathbf{A}^{n-1}\mathbf{b}] = \mathbf{P}^{-1}\mathbf{U}$$
where $\mathbf{U} = [\mathbf{b} \ \mathbf{A}\mathbf{b} \ \cdots \ \mathbf{A}^{n-1}\mathbf{b}]$ \tag{5.119b}

Since $\mathbf{P}^{-1}$ is nonsingular,
$$\rho(\bar{\mathbf{U}}) = \rho(\mathbf{U}) \tag{5.119c}$$

II. A similar relationship can be shown for the observability matrices.

Controllability and Observability of State Variable Model in Jordan Canonical Form.

If the system equations are known in Jordan canonical form, then one need not resort to controllability and observability tests given by Theorems 5.2 and 5.3. These properties can be determined almost by inspection of the system equations as will be shown below.

Consider a SISO system with distinct eigenvalues $\lambda_1, \lambda_2, ..., \lambda_n$. The Jordan canonical state model of this system is of the form

$$\dot{\mathbf{x}} = \mathbf{\Lambda}\mathbf{x} + \mathbf{b}u$$
$$y = \mathbf{c}\mathbf{x} + du \qquad (5.120)$$

with

$$\mathbf{\Lambda} = \begin{bmatrix} \lambda_1 & 0 & \cdots & 0 \\ 0 & \lambda_2 & \cdots & 0 \\ \vdots & \vdots & & \vdots \\ 0 & 0 & \cdots & \lambda_n \end{bmatrix}; \mathbf{b} = \begin{bmatrix} b_1 \\ b_2 \\ \vdots \\ b_n \end{bmatrix}; \mathbf{c} = [c_1 \ c_2 \ \cdots \ c_n]$$

Theorem 5.4

The system (5.120) is completely controllable if and only if none of the elements of the column matrix $\mathbf{b}$ is zero, and (5.120) is completely observable if and only if none of the elements of the row matrix $\mathbf{c}$ is zero.

Proof

The controllability matrix

$$\mathbf{U} = [\mathbf{b} \ \mathbf{\Lambda}\mathbf{b} \ \cdots \ \mathbf{\Lambda}^{n-1}\mathbf{b}]$$

$$= \begin{bmatrix} b_1 & b_1\lambda_1 & \cdots & b_1\lambda_1^{n-1} \\ b_2 & b_2\lambda_2 & \cdots & b_2\lambda_2^{n-1} \\ \vdots & \vdots & & \vdots \\ b_n & b_n\lambda_n & \cdots & b_n\lambda_n^{n-1} \end{bmatrix}; |\mathbf{U}| = b_1 \times b_2 \times \cdots \times b_n \begin{vmatrix} 1 & \lambda_1 & \cdots & \lambda_1^{n-1} \\ 1 & \lambda_2 & \cdots & \lambda_2^{n-1} \\ \vdots & \vdots & & \vdots \\ 1 & \lambda_n & \cdots & \lambda_n^{n-1} \end{vmatrix}$$

$$\neq 0 \text{ if } b_i \neq 0, \ i = 1, 2, ..., n.$$

This proves the first part of the theorem. The second part can be proved in a similar manner.[11]

---

11. Refer Gopal [102] for controllability and observability tests using Jordan canonical representation of systems with multiple eigenvalues.

5.9 EQUIVALENCE BETWEEN TRANSFER FUNCTION AND STATE VARIABLE REPRESENTATIONS

In frequency-domain analysis, it is tacitly assumed that the dynamic properties of a system are completely determined by the transfer function of the system. That this is not always the case is illustrated by the following examples.

Example 5.19
Consider the system

$$\dot{\mathbf{x}} = \mathbf{A}\mathbf{x} + \mathbf{b}u$$
$$y = \mathbf{c}\mathbf{x}$$
(5.121)

with $\mathbf{A} = \begin{bmatrix} -2 & 1 \\ 1 & -2 \end{bmatrix}; \mathbf{b} = \begin{bmatrix} 1 \\ 1 \end{bmatrix}; \mathbf{c} = [0 \ 1]$

The controllability matrix

$$\mathbf{U} = [\mathbf{b} \ \mathbf{A}\mathbf{b}] = \begin{bmatrix} 1 & -1 \\ 1 & -1 \end{bmatrix}$$

Since $\rho(\mathbf{U}) = 1$, the second-order system (5.121) is not completely controllable. The eigenvalues of matrix $\mathbf{A}$ are the roots of the characteristic equation

$$|s\mathbf{I} - \mathbf{A}| = \begin{vmatrix} s+2 & -1 \\ -1 & s+2 \end{vmatrix} = 0$$

The eigenvalues are obtained as $-1, -3$. The *modes* of the transient response are therefore e^{-t} and e^{-3t}.

The transfer function of the system (5.121) is calculated as

$$G(s) = \mathbf{c}(s\mathbf{I} - \mathbf{A})^{-1}\mathbf{b} = [0 \ 1] \begin{bmatrix} s+2 & -1 \\ -1 & s+2 \end{bmatrix}^{-1} \begin{bmatrix} 1 \\ 1 \end{bmatrix}$$

$$= [0 \ 1] \begin{bmatrix} \dfrac{s+2}{(s+1)(s+3)} & \dfrac{1}{(s+1)(s+3)} \\ \dfrac{1}{(s+1)(s+3)} & \dfrac{s+2}{(s+1)(s+3)} \end{bmatrix} \begin{bmatrix} 1 \\ 1 \end{bmatrix} = \dfrac{1}{s+1}$$

We find that because of pole-zero cancellation, both the eigenvalues of matrix $\mathbf{A}$ do not appear as poles in $G(s)$. The dynamic mode e^{-3t} of the system (5.121) does not show up in input-output characterization given by the transfer function $G(s)$. Note that the system under consideration is not a completely controllable system.

Example 5.20
Consider the system

$$\dot{\mathbf{x}} = \mathbf{A}\mathbf{x} + \mathbf{b}u$$
$$y = \mathbf{c}\mathbf{x}$$
(5.122)

with
$$A = \begin{bmatrix} -2 & 1 \\ 1 & -2 \end{bmatrix}; \quad b = \begin{bmatrix} 1 \\ 0 \end{bmatrix}; \quad c = \begin{bmatrix} 1 & -1 \end{bmatrix}$$

The observability matrix
$$V = \begin{bmatrix} c \\ cA \end{bmatrix} = \begin{bmatrix} 1 & -1 \\ -3 & 3 \end{bmatrix}$$

Since $\rho(V) = 1$, the second-order system (5.122) is not completely observable.

The eigenvalues of matrix A are $-1, -3$. The transfer function of the system (5.122) is calculated as

$$G(s) = c(sI - A)^{-1}b$$

$$= \begin{bmatrix} 1 & -1 \end{bmatrix} \begin{bmatrix} \dfrac{s+2}{(s+1)(s+3)} & \dfrac{1}{(s+1)(s+3)} \\ \dfrac{1}{(s+1)(s+3)} & \dfrac{s+2}{(s+1)(s+3)} \end{bmatrix} \begin{bmatrix} 1 \\ 0 \end{bmatrix} = \dfrac{1}{s+3}$$

The dynamic mode e^{-t} of the system (5.122) does not show up in the input-output characterization given by the transfer function $G(s)$. Note that the system under consideration is not a completely observable system.

In the following, we give two specific state transformations to reveal the underlying structure imposed upon a system by its controllability and observability properties (for proof, refer [102]). These results are then used to establish equivalence between transfer function and state variable representations.

Theorem 5.5

Consider an nth-order system
$$\dot{x} = Ax + bu$$
$$y = cx \tag{5.123a}$$

Assume that
$$\rho(U) = \rho[b \ Ab \ \cdots \ A^{n-1}b] = m < n$$

Consider the equivalence transformation
$$x = P\bar{x} = [P_1 \ P_2] \bar{x} \tag{5.123b}$$

where P_1 is composed of m linearly independent columns of U, and $(n - m)$ columns of P_2 are chosen arbitrarily so that matrix P is nonsingular.

The equivalence transformation (5.123b) transforms the system (5.123a) to the following form:

$$\begin{bmatrix} \dot{\bar{x}}_1 \\ \dot{\bar{x}}_2 \end{bmatrix} = \begin{bmatrix} \bar{A}_c & \bar{A}_{12} \\ 0 & \bar{A}_{22} \end{bmatrix} \begin{bmatrix} \bar{x}_1 \\ \bar{x}_2 \end{bmatrix} + \begin{bmatrix} \bar{b}_c \\ 0 \end{bmatrix} u$$

$$\dot{\bar{\mathbf{x}}} = \bar{\mathbf{A}}\bar{\mathbf{x}} + \bar{\mathbf{b}}u$$

$$y = [\bar{\mathbf{c}}_1 \ \bar{\mathbf{c}}_2] \begin{bmatrix} \bar{\mathbf{x}}_1 \\ \bar{\mathbf{x}}_2 \end{bmatrix} = \bar{\mathbf{c}}\bar{\mathbf{x}}$$

(5.123c)

where the m-dimensional subsystem

$$\dot{\bar{\mathbf{x}}}_1 = \bar{\mathbf{A}}_c \bar{\mathbf{x}}_1 + \bar{\mathbf{b}}_c u + \bar{\mathbf{A}}_{12} \bar{\mathbf{x}}_2$$

is controllable from u (the additional driving term $\bar{\mathbf{A}}_{12}\bar{\mathbf{x}}_2$ has no effect on controllability), and the $(n - m)$ dimensional subsystem

$$\dot{\bar{\mathbf{x}}}_2 = \bar{\mathbf{A}}_{22} \bar{\mathbf{x}}_2$$

is not affected by the input, and is therefore entirely uncontrollable. ▲▲

This theorem shows that any system which is not completely controllable can be decomposed into controllable and uncontrollable subsystems shown in Fig. 5.20. The state model (5.123c) is said to be in *controllability canonical form*.

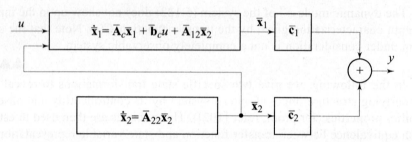

Fig. 5.20 The controllability canonical form of a state variable model

In Section 5.4, it was shown that the characteristic equations and transfer functions of equivalent systems are identical. Thus the set of eigenvalues of matrix $\mathbf{A}$ of system (5.123a) is same as the set of eigenvalues of matrix $\mathbf{A}$ of system (5.123c), which is a union of the subsets of eigenvalues of matrices $\bar{\mathbf{A}}_c$ and $\bar{\mathbf{A}}_{22}$. Also the transfer function of system (5.123a) must be the same as that of (5.123c). The transfer function of (5.123a) is calculated from Eqn. (5.123c) as[12]

---

12. $\begin{bmatrix} \mathbf{A}_1 & \mathbf{A}_2 \\ \mathbf{0} & \mathbf{A}_3 \end{bmatrix} \begin{bmatrix} \mathbf{B}_1 & \mathbf{B}_2 \\ \mathbf{B}_3 & \mathbf{B}_4 \end{bmatrix} = \begin{bmatrix} \mathbf{I} & \mathbf{0} \\ \mathbf{0} & \mathbf{I} \end{bmatrix}$

gives $\begin{bmatrix} \mathbf{B}_1 & \mathbf{B}_2 \\ \mathbf{B}_3 & \mathbf{B}_4 \end{bmatrix} = \begin{bmatrix} \mathbf{A}_1^{-1} & -\mathbf{A}_1^{-1}\mathbf{A}_2\mathbf{A}_3^{-1} \\ \mathbf{0} & \mathbf{A}_3^{-1} \end{bmatrix}$

$$G(s) = \begin{bmatrix} \bar{\mathbf{c}}_1 & \bar{\mathbf{c}}_2 \end{bmatrix} \begin{bmatrix} s\mathbf{I} - \bar{\mathbf{A}}_c & -\bar{\mathbf{A}}_{12} \\ 0 & s\mathbf{I} - \bar{\mathbf{A}}_{22} \end{bmatrix}^{-1} \begin{bmatrix} \bar{\mathbf{b}}_c \\ 0 \end{bmatrix}$$

$$= \begin{bmatrix} \bar{\mathbf{c}}_1 & \bar{\mathbf{c}}_2 \end{bmatrix} \begin{bmatrix} (s\mathbf{I} - \bar{\mathbf{A}}_c)^{-1} & (s\mathbf{I} - \bar{\mathbf{A}}_c)^{-1} \bar{\mathbf{A}}_{12} (s\mathbf{I} - \bar{\mathbf{A}}_{22})^{-1} \\ 0 & (s\mathbf{I} - \bar{\mathbf{A}}_{22})^{-1} \end{bmatrix} \begin{bmatrix} \bar{\mathbf{b}}_c \\ 0 \end{bmatrix}$$

$$= \bar{\mathbf{c}}_1 (s\mathbf{I} - \bar{\mathbf{A}}_c)^{-1} \bar{\mathbf{b}}_c$$

Therefore, the input-output relationship for the system is dependent only on the controllable part of the system. We will refer to the eigenvalues of $\bar{\mathbf{A}}_c$ as *controllable poles* and the eigenvalues of $\bar{\mathbf{A}}_{22}$ as *uncontrollable poles*.

Only the controllable poles appear in the transfer function model; the uncontrollable poles are cancelled by the zeros.

Theorem 5.6

Consider the nth order system

$$\dot{\mathbf{x}} = \mathbf{A}\mathbf{x} + \mathbf{b}u$$
$$y = \mathbf{c}\mathbf{x} \qquad (5.124a)$$

Assume that

$$\rho(\mathbf{V}) = \rho \begin{bmatrix} \mathbf{c} \\ \mathbf{c}\mathbf{A} \\ \vdots \\ \mathbf{c}\mathbf{A}^{n-1} \end{bmatrix} = l < n$$

Consider the equivalence transformation

$$\bar{\mathbf{x}} = \mathbf{Q}\mathbf{x} = \begin{bmatrix} \mathbf{Q}_1 \\ \mathbf{Q}_2 \end{bmatrix} \mathbf{x} \qquad (5.124b)$$

where $\mathbf{Q}_1$ is composed of l linearly independent rows of $\mathbf{V}$, $(n - l)$ rows of $\mathbf{Q}_2$ are chosen arbitrarily so that matrix $\mathbf{Q}$ is nonsingular.

The equivalence transformation (5.124b) transforms the system (5.124a) to the following form:

$$\begin{bmatrix} \dot{\bar{\mathbf{x}}}_1 \\ \dot{\bar{\mathbf{x}}}_2 \end{bmatrix} = \begin{bmatrix} \bar{\mathbf{A}}_0 & 0 \\ \bar{\mathbf{A}}_{21} & \bar{\mathbf{A}}_{22} \end{bmatrix} \begin{bmatrix} \bar{\mathbf{x}}_1 \\ \bar{\mathbf{x}}_2 \end{bmatrix} + \begin{bmatrix} \bar{\mathbf{b}}_1 \\ \bar{\mathbf{b}}_2 \end{bmatrix} u = \bar{\mathbf{A}}\bar{\mathbf{x}} + \bar{\mathbf{b}}u$$

$$y = \begin{bmatrix} \bar{\mathbf{c}}_0 & 0 \end{bmatrix} \begin{bmatrix} \bar{\mathbf{x}}_1 \\ \bar{\mathbf{x}}_2 \end{bmatrix} = \bar{\mathbf{c}}\bar{\mathbf{x}} \qquad (5.124c)$$

where the l-dimensional subsystem

$$\dot{\bar{\mathbf{x}}}_1 = \bar{\mathbf{A}}_0 \bar{\mathbf{x}}_1 + \bar{\mathbf{b}}_1 u$$
$$y = \bar{\mathbf{c}}_0 \bar{\mathbf{x}}_1$$

is observable from y, and the $(n - l)$-dimensional subsystem

$$\dot{\bar{x}}_2 = \bar{A}_{22} \bar{x}_2 + \bar{b}_2 u + \bar{A}_{21} \bar{x}_1$$

has no effect upon the output y, and is therefore entirely unobservable, i.e., nothing about $\bar{x}_2$ can be inferred from output measurement.

▲▲

This theorem shows that any system which is not completely observable can be decomposed into the observable and unobservable subsystems shown in Fig. 5.21. The state model (5.124c) is said to be in *observability canonical form*.

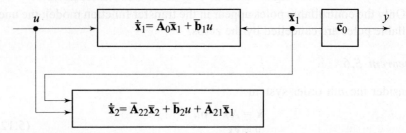

Fig. 5.21 The observability canonical form of a state variable model

Since systems (5.124a) and (5.124c) are equivalent, the set of eigenvalues of matrix A of system (5.124a) is same as the set of eigenvalues of matrix $\bar{A}$ of system (5.124c), which is a union of the subsets of eigenvalues of matrices $\bar{A}_0$ and $\bar{A}_{22}$. The transfer function of the system (5.124a) may be calculated from (5.124c) as follows:

$$G(s) = [\bar{c}_0 \quad 0] \begin{bmatrix} sI - \bar{A}_0 & 0 \\ -\bar{A}_{21} & sI - \bar{A}_{22} \end{bmatrix}^{-1} \begin{bmatrix} \bar{b}_1 \\ \bar{b}_2 \end{bmatrix}$$

$$= [\bar{c}_0 \quad 0] \begin{bmatrix} (sI - \bar{A}_0)^{-1} & 0 \\ (sI - \bar{A}_{22})^{-1} \bar{A}_{21} (sI - \bar{A}_0)^{-1} & (sI - \bar{A}_{22})^{-1} \end{bmatrix} \begin{bmatrix} \bar{b}_1 \\ \bar{b}_2 \end{bmatrix}$$

$$= \bar{c}_0 (sI - \bar{A}_0)^{-1} \bar{b}_1 \qquad (5.125)$$

which shows that the unobservable part of the system does not affect the input-output relationship. We will refer to the eigenvalues of $\bar{A}_0$ as *observable poles* and the eigenvalues of $\bar{A}_{22}$ as *unobservable poles*.

▲▲

We now examine the use of state variable and transfer function models of a system to study its dynamic properties.

We know that a system is asymptotically stable if all the eigenvalues of the characteristic matrix **A** of its state variable model are in the left-half of complex plane. Also we know that a system is (bounded-input bounded-output) BIBO stable if all the poles of its transfer function model are in the left-half of complex plane. Since, in general, the poles of the transfer function model of a system are a subset of the eigenvalues of the characteristic matrix **A** of the system, *asymptotic stability always implies BIBO stability*.

The reverse, however, may not always be true because the eigenvalues of the uncontrollable and/or unobservable part of the system are hidden from the BIBO stability analysis. These may lead to instability of a BIBO stable system. When a state variable model is both controllable and observable, all the eigenvalues of characteristic matrix **A** appear as poles in the corresponding transfer function. Therefore *BIBO stability implies asymptotic stability only for completely controllable and completely observable system*.

To conclude, we may say that the transfer function model of a system represents its complete dynamics only if the system is both controllable and observable.

5.10 MULTIVARIABLE SYSTEMS

Many of the analysis results developed in earlier sections of this chapter for SISO systems have obvious extensions for MIMO systems.

Consider a general MIMO system shown in the block diagram of Fig. 5.22. The input variables are represented by $u_1, u_2, ..., u_p$, and the output variables by $y_1, y_2, ..., y_q$. The state, as in the case of SISO systems, is represented by variables $x_1, x_2, ..., x_n$.

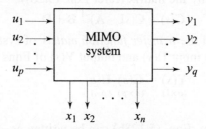

Fig. 5.22 A general MIMO system

The state variable model for a MIMO system takes the following form:

$$\dot{\mathbf{x}}(t) = \mathbf{A}\mathbf{x}(t) + \mathbf{B}\mathbf{u}(t); \quad \mathbf{x}(t_0) \triangleq \mathbf{x}^0 \quad (5.126a)$$

$$\mathbf{y}(t) = \mathbf{C}\mathbf{x}(t) + \mathbf{D}\mathbf{u}(t) \quad (5.126b)$$

where

$$\mathbf{A} = \begin{bmatrix} a_{11} & a_{12} & \cdots & a_{1n} \\ a_{21} & a_{22} & \cdots & a_{2n} \\ \vdots & \vdots & & \vdots \\ a_{n1} & a_{n2} & \cdots & a_{nn} \end{bmatrix}; \mathbf{B} = \begin{bmatrix} b_{11} & b_{12} & \cdots & b_{1p} \\ b_{21} & b_{22} & \cdots & b_{2p} \\ \vdots & \vdots & & \vdots \\ b_{n1} & b_{n2} & \cdots & b_{np} \end{bmatrix}$$

$$\mathbf{C} = \begin{bmatrix} c_{11} & c_{12} & \cdots & c_{1n} \\ c_{21} & c_{22} & \cdots & c_{2n} \\ \vdots & \vdots & & \vdots \\ c_{q1} & c_{q2} & \cdots & c_{qn} \end{bmatrix}, \mathbf{D} = \begin{bmatrix} d_{11} & d_{12} & \cdots & d_{1p} \\ d_{21} & d_{22} & \cdots & d_{2p} \\ \vdots & \vdots & & \vdots \\ d_{q1} & d_{q2} & \cdots & d_{qp} \end{bmatrix}$$

A, B, C and **D** are respectively $n \times n$, $n \times p$, $q \times n$, and $q \times p$ constant matrices; **x** is $n \times 1$ state vector, **u** is $p \times 1$ input vector, and **y** is $q \times 1$ output vector.
The solution of the state Eqn. (5.126a) is given by (refer Eqn. (5.100))

$$\mathbf{x}(t) = e^{\mathbf{A}t}\mathbf{x}(0) + \int_0^t e^{\mathbf{A}(t-\tau)} \mathbf{B}\mathbf{u}(\tau)\, d\tau \qquad (5.127a)$$

The output

$$\mathbf{y}(t) = \mathbf{C}\left[e^{\mathbf{A}t}\mathbf{x}(0) + \int_0^t e^{\mathbf{A}(t-\tau)}\mathbf{B}\mathbf{u}(\tau)d\tau \right] + \mathbf{D}\mathbf{u}(\tau) \qquad (5.127b)$$

In the transform domain, the input-output behaviour of the system (5.126) is determined entirely by the matrix (refer Eqn. (5.28))

$$\mathbf{G}(s) = \mathbf{C}(s\mathbf{I} - \mathbf{A})^{-1}\mathbf{B} + \mathbf{D} \qquad (5.128a)$$

This matrix is called the *transfer function matrix* of system (5.126) and it has the property that the input **U**(s) and output **Y**(s) of Eqns (5.126) are related by

$$\underset{(q \times 1)}{\mathbf{Y}(s)} = \underset{(q \times p)}{\mathbf{G}(s)}\; \underset{(p \times 1)}{\mathbf{U}(s)} \qquad (5.128b)$$

whenever $\mathbf{x}^0 = \mathbf{0}$.
In an expanded form, Eqn. (5.128b) can be written as

$$\begin{bmatrix} Y_1(s) \\ Y_2(s) \\ \vdots \\ Y_q(s) \end{bmatrix} = \begin{bmatrix} G_{11}(s) & G_{12}(s) & \cdots & G_{1p}(s) \\ G_{21}(s) & G_{22}(s) & \cdots & G_{2p}(s) \\ \vdots & \vdots & & \vdots \\ G_{q1}(s) & G_{q2}(s) & \cdots & G_{qp}(s) \end{bmatrix} \begin{bmatrix} U_1(s) \\ U_2(s) \\ \vdots \\ U_p(s) \end{bmatrix}$$

The (i, j)th element $G_{ij}(s)$ of **G**(s) is the transfer function relating the ith output to the jth input.

Example 5.21

The scheme of Fig. 5.23 describes a simple concentration control process. Two concentrated solutions of some chemical with constant concentrations C_1 and C_2 are fed with flow rates $Q_1(t) = \overline{Q}_1 + q_1(t)$, and $Q_2(t) = \overline{Q}_2 + q_2(t)$ respectively, and are continuously mixed in the tank. The outflow from the mixing tank is at a rate $Q(t) = \overline{Q} + q(t)$ with concentration $C(t) = \overline{C} + c(t)$. Let it be assumed that stirring causes perfect mixing so that the concentration of the solution in the tank is uniform throughout and equals that of the outflow. We shall also assume that the density remains constant.

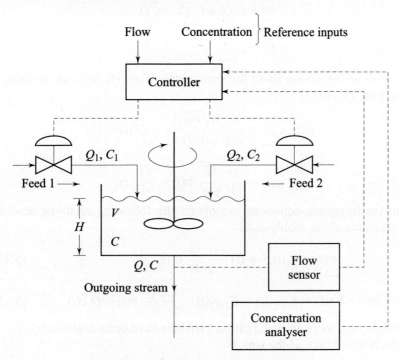

Fig. 5.23 A concentration control process

Let $V(t) = \overline{V} + v(t)$ be the volume of the fluid in the tank. The mass balance equations are

$$\frac{d}{dt}[\overline{V} + v(t)] = \overline{Q}_1 + q_1(t) + \overline{Q}_2 + q_2(t) - \overline{Q} - q(t) \tag{5.129a}$$

$$\frac{d}{dt}[\{\overline{C} + c(t)\}\{\overline{V} + v(t)\}]$$

$$= C_1[\overline{Q}_1 + q_1(t)] + C_2[\overline{Q}_2 + q_2(t)] - [\overline{C} + c(t)][\overline{Q} + q(t)] \tag{5.129b}$$

The flow $Q(t)$ is characterized by the turbulent flow relation

$$Q(t) = k\sqrt{H(t)} = k\sqrt{\frac{V(t)}{A}} \tag{5.130}$$

where $H(t) = \overline{H} + h(t)$ is the head of the liquid in the tank, A is the cross-sectional area of the tank and k is a constant.

The steady-state operation is described by the equations (obtained from Eqns (5.129) and (5.130))

$$0 = \overline{Q}_1 + \overline{Q}_2 - \overline{Q}$$

$$0 = C_1 \overline{Q}_1 + C_2 \overline{Q}_2 - \overline{C}\,\overline{Q}$$

$$\overline{Q} = k\sqrt{\frac{\overline{V}}{A}}$$

For small perturbations about the steady-state, Eqn. (5.130) can be linearized using Eqn. (5.11c):

$$Q(t) - \overline{Q} = \frac{k}{\sqrt{A}} \left.\frac{\partial \sqrt{V(t)}}{\partial V(t)}\right|_{V=\overline{V}} (V(t) - \overline{V})$$

or

$$q(t) = \frac{k}{2\overline{V}}\sqrt{\frac{\overline{V}}{A}}\, v(t) = \frac{\overline{Q}}{2\overline{V}}\, v(t)$$

From the foregoing equations, we obtain the following relations describing perturbations about steady-state:

$$\dot{v}(t) = q_1(t) + q_2(t) - \frac{1}{2}\frac{\overline{Q}}{\overline{V}}\, v(t) \tag{5.131a}$$

$$\overline{C}\dot{v}(t) + \overline{V}\dot{c}(t) = C_1 q_1(t) + C_2 q_2(t) - \frac{1}{2}\frac{\overline{C}\,\overline{Q}}{\overline{V}}\, v(t) - \overline{Q}\, c(t) \tag{5.131b}$$

(Second-order terms in perturbation variables have been neglected)

The hold-up time of the tank is

$$\tau = \frac{\overline{V}}{\overline{Q}}$$

Let us define

$$x_1(t) = v(t),\ x_2(t) = c(t),\ u_1(t) = q_1(t),\ u_2(t) = q_2(t),\ y_1(t) = q(t),$$

and $y_2(t) = c(t)$.

In terms of these variables, we get the following state model from Eqns (5.131):

$$\dot{\mathbf{x}}(t) = \begin{bmatrix} -\dfrac{1}{2\tau} & 0 \\ 0 & -\dfrac{1}{\tau} \end{bmatrix} \mathbf{x}(t) + \begin{bmatrix} \dfrac{1}{\overline{V}} & \dfrac{1}{\overline{V}} \\ \dfrac{C_1 - \overline{C}}{\overline{V}} & \dfrac{C_2 - \overline{C}}{\overline{V}} \end{bmatrix} \mathbf{u}(t) \tag{5.132a}$$

$$y(t) = \begin{bmatrix} \dfrac{1}{2\tau} & 0 \\ 0 & 1 \end{bmatrix} x(t) \tag{5.132b}$$

For the parameters

$\overline{Q}_1 = 10$ litres/sec

$\overline{Q}_2 = 20$ litres/sec

$C_1 = 9$ g-moles/litre

$C_2 = 18$ g-moles/litre

$\overline{V} = 1500$ litres

the state variable model becomes

$$\dot{x}(t) = Ax(t) + Bu(t) \tag{5.133a}$$
$$y(t) = Cx(t) \tag{5.133b}$$

with

$$A = \begin{bmatrix} -0.01 & 0 \\ 0 & -0.02 \end{bmatrix}; \quad B = \begin{bmatrix} 1 & 1 \\ -0.004 & 0.002 \end{bmatrix}$$

$$C = \begin{bmatrix} 0.01 & 0 \\ 0 & 1 \end{bmatrix}$$

In the transform domain, the input-output behaviour of the system is given by
$$Y(s) = G(s) U(s)$$
where
$$G(s) = C(sI - A)^{-1}B$$
For **A**, **B**, and **C** given by Eqns (5.133), we have

$$(sI - A) = \begin{bmatrix} s + 0.01 & 0 \\ 0 & s + 0.02 \end{bmatrix}$$

$$G(s) = C(sI - A)^{-1}B = \begin{bmatrix} 0.01 & 0 \\ 0 & 1 \end{bmatrix} \begin{bmatrix} \dfrac{1}{s+0.01} & 0 \\ 0 & \dfrac{1}{s+0.02} \end{bmatrix} \begin{bmatrix} 1 & 1 \\ -0.004 & 0.002 \end{bmatrix}$$

$$= \begin{bmatrix} \dfrac{0.01}{s+0.01} & \dfrac{0.01}{s+0.01} \\ \dfrac{-0.004}{s+0.02} & \dfrac{0.002}{s+0.02} \end{bmatrix} \tag{5.134}$$

Controllability Test

The necessary and sufficient condition for the system (5.126) to be completely controllable is that the $n \times np$ matrix

$$U \triangleq [B \; AB \; A^2B \; \cdots \; A^{n-1}B] \tag{5.135}$$

has rank equal to n, i.e., $\rho(U) = n$.

Observability Test

The necessary and sufficient condition for the system (5.126) to be completely observable is that the $nq \times n$ matrix

$$V \triangleq \begin{bmatrix} C \\ CA \\ \vdots \\ CA^{n-1} \end{bmatrix} \tag{5.136}$$

has rank equal to n, i.e., $\rho(V) = n$.

Controllability and Observability of State Variable Model in Jordan Canonical Form

The controllability and observability properties can be determined by the inspection of the system equations in Jordan canonical form. A MIMO system with distinct eigenvalues $\lambda_1, \lambda_2, ..., \lambda_n$ has the following Jordan canonical state model.

$$\dot{x} = \Lambda x + Bu \tag{5.137a}$$

$$y = Cx + Du \tag{5.137b}$$

with

$$\Lambda = \begin{bmatrix} \lambda_1 & 0 & \cdots & 0 \\ 0 & \lambda_2 & \cdots & 0 \\ \vdots & \vdots & & \vdots \\ 0 & 0 & \cdots & \lambda_n \end{bmatrix}; B = \begin{bmatrix} b_{11} & b_{12} & \cdots & b_{1p} \\ b_{21} & b_{22} & \cdots & b_{2p} \\ \vdots & \vdots & & \vdots \\ b_{n1} & b_{n2} & \cdots & b_{np} \end{bmatrix}$$

$$C = \begin{bmatrix} c_{11} & c_{12} & \cdots & c_{1n} \\ \vdots & \vdots & & \vdots \\ c_{q1} & c_{q2} & \cdots & c_{qn} \end{bmatrix}$$

The system (5.137) is completely controllable if and only if none of the rows of B matrix is a zero row, and (5.137) is completely observable if and only if none of the columns of C matrix is a zero column.

▲▲

We have been using Jordan canonical structure only for systems with distinct eigenvalues. Refer [102] for controllability and observability tests using Jordan canonical representation of systems with multiple eigenvalues.

Example 5.22

Consider the mixing-tank system discussed in Example 5.21. Suppose the feeds Q_1 and Q_2 have equal concentrations, i.e., $C_1 = C_2 = C_0$ (Fig. 5.23). Then the steady-state concentration $\bar{C}$ in the tank is also C_0 and from Eqn. (5.132a) we have

$$\dot{\mathbf{x}}(t) = \begin{bmatrix} -\dfrac{1}{2\tau} & 0 \\ 0 & -\dfrac{1}{\tau} \end{bmatrix} \mathbf{x}(t) + \begin{bmatrix} 1 & 1 \\ 0 & 0 \end{bmatrix} \mathbf{u}(t)$$

This state variable model is in Jordan canonical form. Since one row of the **B** matrix is a zero row, the system is not completely controllable. As is obvious from the Jordan canonical model, the input $\mathbf{u}(t)$ affects only the state variable $x_1(t)$, the incremental volume. The variable $x_2(t)$, the incremental concentration, has no connection with the input $\mathbf{u}(t)$.

If $C_1 \neq C_2$, the system is completely controllable.

5.11 REVIEW EXAMPLES

Review Example 5.1

A feedback system has a closed-loop transfer function

$$\frac{Y(s)}{R(s)} = \frac{10(s+4)}{s(s+1)(s+3)}$$

Construct three different state models for this system:
(a) One where the system matrix **A** is a diagonal matrix.
(b) One where **A** is in first companion form.
(c) One where **A** is in second companion form.

Solution

(a) The given transfer function can be expressed as

$$\frac{Y(s)}{R(s)} = \frac{10(s+4)}{s(s+1)(s+3)} = \frac{40/3}{s} + \frac{-15}{s+1} + \frac{5/3}{s+3}$$

Therefore,

$$Y(s) = \frac{40/3}{s} R(s) + \frac{-15}{s+1} R(s) + \frac{5/3}{s+3} R(s)$$

Let

$$X_1(s) = \frac{40/3}{s} R(s); \text{ this gives } \dot{x}_1 = \frac{40}{3} r$$

$$X_2(s) = \frac{-15}{s+1} R(s); \text{ this gives } \dot{x}_2 + x_2 = -15r$$

$$X_3(s) = \frac{5/3}{s+3} R(s); \text{ this gives } \dot{x}_3 + 3x_3 = \frac{5}{3} r$$

In terms of x_1, x_2 and x_3, the output $y(t)$ is given by
$$y(t) = x_1(t) + x_2(t) + x_3(t)$$
A state variable formulation for the given transfer function is defined by the following matrices:

$$\mathbf{A} = \begin{bmatrix} 0 & 0 & 0 \\ 0 & -1 & 0 \\ 0 & 0 & -3 \end{bmatrix}; \mathbf{b} = \begin{bmatrix} 40/3 \\ -15 \\ 5/3 \end{bmatrix}; \mathbf{c} = [1 \ \ 1 \ \ 1]; d = 0$$

Note that the coefficient matrix $\mathbf{A}$ is diagonal, and the state model is in Jordan canonical form.

We now construct two state models for the given transfer function in companion form. To do this, we express the transfer function as

$$\frac{Y(s)}{R(s)} = \frac{10(s+4)}{s(s+1)(s+3)} = \frac{10s+40}{s^3+4s^2+3s} = \frac{\beta_0 s^3 + \beta_1 s^2 + \beta_2 s + \beta_3}{s^3 + \alpha_1 s^2 + \alpha_2 s + \alpha_3};$$

$$\beta_0 = \beta_1 = 0, \ \beta_2 = 10, \ \beta_3 = 40, \ \alpha_1 = 4, \ \alpha_2 = 3, \ \alpha_3 = 0$$

(b) With reference to Eqns (5.54), we obtain the following state model in the first companion form:

$$\mathbf{A} = \begin{bmatrix} 0 & 1 & 0 \\ 0 & 0 & 1 \\ 0 & -3 & -4 \end{bmatrix}; \mathbf{b} = \begin{bmatrix} 0 \\ 0 \\ 1 \end{bmatrix}; \mathbf{c} = [40 \ \ 10 \ \ 0]; d = 0$$

(c) With reference to Eqns (5.56), the state model in second companion form becomes

$$\mathbf{A} = \begin{bmatrix} 0 & 0 & 0 \\ 1 & 0 & -3 \\ 0 & 1 & -4 \end{bmatrix}; \mathbf{b} = \begin{bmatrix} 40 \\ 10 \\ 0 \end{bmatrix}, \mathbf{c} = [0 \ \ 0 \ \ 1]; d = 0$$

Review Example 5.2

A linear time-invariant system is characterized by the homogeneous state equation

$$\begin{bmatrix} \dot{x}_1 \\ \dot{x}_2 \end{bmatrix} = \begin{bmatrix} 0 & 1 \\ 0 & -2 \end{bmatrix} \begin{bmatrix} x_1 \\ x_2 \end{bmatrix}$$

(a) Compute the solution of the homogeneous equation assuming the initial state vector

$$\mathbf{x}(0) = \begin{bmatrix} 1 \\ 0 \end{bmatrix}$$

Employ both the Laplace transform method and the canonical transformation method.

(b) Consider now that the system has a forcing function and is represented by the following non-homogeneous state equation:

$$\begin{bmatrix} \dot{x}_1 \\ \dot{x}_2 \end{bmatrix} = \begin{bmatrix} 0 & 1 \\ 0 & -2 \end{bmatrix} \begin{bmatrix} x_1 \\ x_2 \end{bmatrix} + \begin{bmatrix} 0 \\ 1 \end{bmatrix} u$$

where u is a unit-step input.

Compute the solution of this equation assuming initial conditions of part (a).

Solution

(a) Since

$$(s\mathbf{I} - \mathbf{A}) = \begin{bmatrix} s & -1 \\ 0 & s+2 \end{bmatrix}$$

we obtain

$$(s\mathbf{I} - \mathbf{A})^{-1} = \begin{bmatrix} \frac{1}{s} & \frac{1}{s(s+2)} \\ 0 & \frac{1}{s+2} \end{bmatrix}$$

Hence

$$e^{\mathbf{A}t} = \mathscr{L}^{-1}[(s\mathbf{I} - \mathbf{A})^{-1}] = \begin{bmatrix} 1 & \frac{1}{2}(1-e^{-2t}) \\ 0 & e^{-2t} \end{bmatrix}$$

To obtain the state transition matrix $e^{\mathbf{A}t}$ by the canonical transformation method, we compute the eigenvalues and eigenvectors of matrix $\mathbf{A}$. The roots of the characteristic equation

$$|\lambda \mathbf{I} - \mathbf{A}| = 0$$

are $\lambda_1 = 0$, and $\lambda_2 = -2$. These are the eigenvalues of matrix $\mathbf{A}$. Eigenvectors corresponding to the distinct eigenvalues λ_i may be obtained from the non-zero columns of $adj(\lambda_i \mathbf{I} - \mathbf{A})$.

For the given $\mathbf{A}$ matrix

$$adj(\lambda_i \mathbf{I} - \mathbf{A}) = \begin{bmatrix} \lambda_i + 2 & 1 \\ 0 & \lambda_i \end{bmatrix}$$

For $\lambda_1 = 0$,

$$adj(\lambda_1 \mathbf{I} - \mathbf{A}) = \begin{bmatrix} 2 & 1 \\ 0 & 0 \end{bmatrix}$$

The eigenvector $\mathbf{v}_1$ corresponding to the eigenvalue λ_1 is therefore given by

$$\mathbf{v}_1 = \begin{bmatrix} 1 \\ 0 \end{bmatrix}$$

For $\lambda_2 = -2$,

$$adj(\lambda_2 \mathbf{I} - \mathbf{A}) = \begin{bmatrix} 0 & 1 \\ 0 & -2 \end{bmatrix}$$

The eigenvector $\mathbf{v}_2$ corresponding to the eigenvalue λ_2 is given by

$$\mathbf{v}_2 = \begin{bmatrix} 1 \\ -2 \end{bmatrix}$$

The transformation matrix

$$\mathbf{P} = \begin{bmatrix} 1 & 1 \\ 0 & -2 \end{bmatrix}$$

gives

$$\mathbf{P}^{-1}\mathbf{A}\mathbf{P} = \mathbf{\Lambda} = \begin{bmatrix} 0 & 0 \\ 0 & -2 \end{bmatrix}$$

The state transition matrix (refer Eqn. (5.93))

$$e^{\mathbf{A}t} = \mathbf{P}e^{\mathbf{\Lambda}t}\mathbf{P}^{-1} = \begin{bmatrix} 1 & 1 \\ 0 & -2 \end{bmatrix} \begin{bmatrix} e^0 & 0 \\ 0 & e^{-2t} \end{bmatrix} \begin{bmatrix} 1 & \frac{1}{2} \\ 0 & -\frac{1}{2} \end{bmatrix} = \begin{bmatrix} 1 & \frac{1}{2}(1-e^{-2t}) \\ 0 & e^{-2t} \end{bmatrix}$$

$$\mathbf{x}(t) = e^{\mathbf{A}t}\mathbf{x}(0) = \begin{bmatrix} 1 \\ 0 \end{bmatrix}$$

(b) $$\mathbf{x}(t) = e^{\mathbf{A}t}\mathbf{x}(0) + \int_0^t e^{\mathbf{A}(t-\tau)}\mathbf{b}u(\tau)d\tau$$

Now

$$\int_0^t e^{\mathbf{A}(t-\tau)}\mathbf{b}u(\tau)d\tau = \begin{bmatrix} \frac{1}{2}\int_0^t [1-e^{-2(t-\tau)}]d\tau \\ \int_0^t e^{-2(t-\tau)}d\tau \end{bmatrix} = \begin{bmatrix} -\frac{1}{4} + \frac{1}{2}t + \frac{1}{4}e^{-2t} \\ \frac{1}{2}(1-e^{-2t}) \end{bmatrix}$$

Therefore,

$$x_1(t) = -\frac{1}{4} + \frac{1}{2}t + \frac{1}{4}e^{-2t}$$

$$x_2(t) = \frac{1}{2}(1-e^{-2t})$$

Review Example 5.3
Given

$$\underset{n \times n}{\Lambda} = \begin{bmatrix} \lambda_1 & 1 & 0 & \cdots & 0 \\ 0 & \lambda_1 & 1 & \cdots & 0 \\ \vdots & \vdots & \vdots & & \vdots \\ 0 & 0 & 0 & \cdots & 1 \\ 0 & 0 & 0 & \cdots & \lambda_1 \end{bmatrix}$$

Compute $e^{\Lambda t}$ using the Cayley-Hamilton technique.

Solution
Equations (5.98) outline the procedure of evaluation of matrix exponential using the Cayley-Hamilton technique.

The matrix Λ has n eigenvalues at $\lambda = \lambda_1$. To evaluate $f(\Lambda) = e^{\Lambda t}$, we define (refer Eqn. (5.98b)) the polynomial $g(\lambda)$ as

$$g(\lambda) = \beta_0 + \beta_1 \lambda + \cdots + \beta_{n-1} \lambda^{n-1}$$

This polynomial may be rearranged as

$$g(\lambda) = b_0 + b_1(\lambda - \lambda_1) + \cdots + b_{n-1}(\lambda - \lambda_1)^{n-1}$$

The coefficients $b_0, b_1, \ldots, b_{n-1}$ are given by the following equations (refer Eqns (5.98c)):

$$f(\lambda_1) = g(\lambda_1)$$

$$\left. \frac{d}{d\lambda} f(\lambda) \right|_{\lambda = \lambda_1} = \left. \frac{d}{d\lambda} g(\lambda) \right|_{\lambda = \lambda_1}$$

$$\vdots$$

$$\left. \frac{d^{n-1}}{d\lambda^{n-1}} f(\lambda) \right|_{\lambda = \lambda_1} = \left. \frac{d^{n-1}}{d\lambda^{n-1}} g(\lambda) \right|_{\lambda = \lambda_1}$$

Solving, we get

$$b_0 = e^{\lambda_1 t}$$

$$b_1 = \frac{t}{1!} e^{\lambda_1 t}$$

$$b_2 = \frac{t^2}{2!} e^{\lambda_1 t}$$

$$\vdots$$

$$b_{n-1} = \frac{t^{n-1}}{(n-1)!} e^{\lambda_1 t}$$

Therefore,
$$e^{\Lambda t} = b_0\mathbf{I} + b_1(\Lambda - \lambda_1\mathbf{I}) + \cdots + b_{n-1}(\Lambda - \lambda_1\mathbf{I})^{n-1}$$

$$(\Lambda - \lambda_1\mathbf{I}) = \begin{bmatrix} 0 & 1 & 0 & 0 & \cdots & 0 \\ 0 & 0 & 1 & 0 & \cdots & 0 \\ \vdots & \vdots & \vdots & \vdots & & \vdots \\ 0 & 0 & 0 & 0 & \cdots & 0 \end{bmatrix}$$

$$(\Lambda - \lambda_1\mathbf{I})(\Lambda - \lambda_1\mathbf{I}) = \begin{bmatrix} 0 & 0 & 1 & 0 & \cdots & 0 \\ 0 & 0 & 0 & 1 & \cdots & 0 \\ \vdots & \vdots & \vdots & \vdots & & \vdots \\ 0 & 0 & 0 & 0 & \cdots & 0 \end{bmatrix}$$

$$\vdots$$

$$e^{\Lambda t} = \begin{bmatrix} b_0 & b_1 & b_2 & \cdots & b_{n-1} \\ 0 & b_0 & b_1 & \cdots & b_{n-2} \\ \vdots & \vdots & \vdots & & \vdots \\ 0 & 0 & 0 & \cdots & b_0 \end{bmatrix}$$

$$= \begin{bmatrix} e^{\lambda_1 t} & te^{\lambda_1 t} & t^2 e^{\lambda_1 t}/2! & \cdots & t^{n-1}e^{\lambda_1 t}/(n-1)! \\ 0 & e^{\lambda_1 t} & te^{\lambda_1 t} & \cdots & t^{n-2}e^{\lambda_1 t}/(n-2)! \\ \vdots & \vdots & \vdots & & \vdots \\ 0 & 0 & 0 & \cdots & e^{\lambda_1 t} \end{bmatrix}$$

Review Example 5.4

The motion of a satellite in the equatorial (r, θ) plane is given by [109] the state equation

$$\begin{bmatrix} \dot{x}_1 \\ \dot{x}_2 \\ \dot{x}_3 \\ \dot{x}_4 \end{bmatrix} = \begin{bmatrix} 0 & 1 & 0 & 0 \\ 3\omega^2 & 0 & 0 & 2\omega \\ 0 & 0 & 0 & 1 \\ 0 & -2\omega & 0 & 0 \end{bmatrix} \begin{bmatrix} x_1 \\ x_2 \\ x_3 \\ x_4 \end{bmatrix} + \begin{bmatrix} 0 & 0 \\ 1 & 0 \\ 0 & 0 \\ 0 & 1 \end{bmatrix} \begin{bmatrix} u_1 \\ u_2 \end{bmatrix} = \mathbf{Ax + Bu}$$

where ω is the angular frequency of the satellite in circular, equatorial orbit, $x_1(t)$ and $x_3(t)$ are, respectively, the deviations in position variables $r(t)$ and $\theta(t)$ of the satellite, and $x_2(t)$ and $x_4(t)$ are, respectively, the deviations in velocity variables $\dot{r}(t)$ and $\dot{\theta}(t)$. The inputs $u_1(t)$ and $u_2(t)$ are the thrusts u_r and u_θ in the radial and tangential directions reapectively, applied by small rocket engines or gas jets ($\mathbf{u = 0}$ when $\mathbf{x = 0}$).

(a) Prove that the system is completely controllable.

(b) Suppose the tangential thruster becomes inoperable. Determine the controllability of the system with the radial thruster alone.
(c) Suppose that the radial thruster becomes inoperable. Determine the controllability of the system with the tangential thruster alone.
(d) Prove that the system is completely observable from radial ($x_1 = r$) and tangential ($x_3 = \theta$) position measurements.
(e) Suppose that the tangential measuring device becomes inoperable. Determine the observability of the system from radial position measurement alone.
(f) Suppose that the radial measurements are lost. Determine the observability of the system from tangential position measurement alone.

Solution

(a) The controllability matrix
$$\mathbf{U} = [\mathbf{B} \quad \mathbf{AB} \quad \mathbf{A}^2\mathbf{B} \quad \mathbf{A}^3\mathbf{B}]$$

Consider the matrix:
$$\mathbf{U}_1 = [\mathbf{B} \quad \mathbf{AB}] = \begin{bmatrix} 0 & 0 & 1 & 0 \\ 1 & 0 & 0 & 2\omega \\ 0 & 0 & 0 & 1 \\ 0 & 1 & -2\omega & 0 \end{bmatrix}$$

$$|\mathbf{U}_1| = -1$$

Therefore $\rho(\mathbf{U}_1) = \rho(\mathbf{U}) = 4$; the system is completely controllable.

(b) With $u_2 = 0$, the **B** matrix becomes
$$\mathbf{b} = \begin{bmatrix} 0 \\ 1 \\ 0 \\ 0 \end{bmatrix}$$

The controllability matrix
$$\mathbf{U} = [\mathbf{b} \quad \mathbf{Ab} \quad \mathbf{A}^2\mathbf{b} \quad \mathbf{A}^3\mathbf{b}] = \begin{bmatrix} 0 & 1 & 0 & -\omega^2 \\ 1 & 0 & -\omega^2 & 0 \\ 0 & 0 & -2\omega & 0 \\ 0 & -2\omega & 0 & 2\omega^3 \end{bmatrix}$$

$$|\mathbf{U}| = - \begin{vmatrix} 1 & 0 & -\omega^2 \\ 0 & -2\omega & 0 \\ -2\omega & 0 & 2\omega^3 \end{vmatrix} = -[-2\omega(2\omega^3 - 2\omega^3)] = 0$$

Therefore, $\rho(\mathbf{U}) < 4$, and the system is not completely controllable with u_1 alone.

(c) With $u_1 = 0$, the **B** matrix becomes

$$\mathbf{b} = \begin{bmatrix} 0 \\ 0 \\ 0 \\ 1 \end{bmatrix}$$

The controllability matrix

$$\mathbf{U} = \begin{bmatrix} 0 & 0 & 2\omega & 0 \\ 0 & 2\omega & 0 & -2\omega^3 \\ 0 & 1 & 0 & -4\omega^2 \\ 1 & 0 & -4\omega^2 & 0 \end{bmatrix}$$

$$|\mathbf{U}| = 2\omega \begin{vmatrix} 0 & 2\omega & -2\omega^3 \\ 0 & 1 & -4\omega^2 \\ 1 & 0 & 0 \end{vmatrix} = -12\omega^4 \neq 0$$

Therefore, $\rho(\mathbf{U}) = 4$, and the system is completely controllable with u_2 alone.

(d) The observability matrix

$$\mathbf{V} = \begin{bmatrix} \mathbf{C} \\ \mathbf{CA} \\ \mathbf{CA}^2 \\ \mathbf{CA}^3 \end{bmatrix}$$

Taking radial and tangential position measurements as the outputs, we have

$$y_1 = x_1$$
$$y_2 = x_3$$

or
$$\mathbf{y} = \mathbf{Cx} = \begin{bmatrix} 1 & 0 & 0 & 0 \\ 0 & 0 & 1 & 0 \end{bmatrix} \mathbf{x}$$

Consider the matrix

$$\mathbf{V}_1 = \begin{bmatrix} \mathbf{C} \\ \mathbf{CA} \end{bmatrix} = \begin{bmatrix} 1 & 0 & 0 & 0 \\ 0 & 0 & 1 & 0 \\ 0 & 1 & 0 & 0 \\ 0 & 0 & 0 & 1 \end{bmatrix}$$

$$|\mathbf{V}_1| \neq 0$$

Therefore, $\rho(\mathbf{V}_1) = \rho(\mathbf{V}) = 4$, and the system is completely observable.

(e) With $x_3 = 0$, the **C** matrix becomes
$$\mathbf{c} = [1\ 0\ 0\ 0]$$
The observability matrix
$$\mathbf{V} = \begin{bmatrix} \mathbf{c} \\ \mathbf{cA} \\ \mathbf{cA}^2 \\ \mathbf{cA}^3 \end{bmatrix} = \begin{bmatrix} 1 & 0 & 0 & 0 \\ 0 & 1 & 0 & 0 \\ 3\omega^2 & 0 & 0 & 2\omega \\ 0 & -\omega^2 & 0 & 0 \end{bmatrix}$$
$$|\mathbf{V}| = 0$$
Therefore, $\rho(\mathbf{V}) < 4$, and the system is not completely observable from $y_1 = x_1$ alone.

(f) With $x_1 = 0$, the **C** matrix becomes
$$\mathbf{c} = [0\ 0\ 1\ 0]$$
The observability matrix
$$\mathbf{V} = \begin{bmatrix} 0 & 0 & 1 & 0 \\ 0 & 0 & 0 & 1 \\ 0 & -2\omega & 0 & 0 \\ -6\omega^3 & 0 & 0 & -4\omega^2 \end{bmatrix}$$
$$|\mathbf{V}| = -12\omega^4 \neq 0$$
Therefore, $\rho(\mathbf{V}) = 4$, and the system is completely observable from $y_2 = x_3$ alone.

PROBLEMS

5.1 Figure P5.1 shows a control scheme for controlling the azimuth angle of a rotating antenna. The plant consists of an armature-controlled dc motor with dc generator used as an amplifier. The parameters of the plant are given below:

Motor torque constant, $K_T = 1.2$ N-m/amp
Motor back emf constant, $K_b = 1.2$ V/(rad/sec)
Generator gain constant, $K_g = 100$ V/amp
Motor to load gear ratio, $n = (\dot{\theta}_L/\dot{\theta}_M) = 1/2$

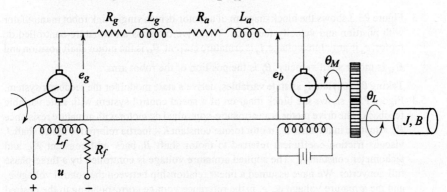

Fig. P5.1

$R_f = 21\ \Omega$, $L_f = 5$H, $R_g = 9\ \Omega$, $L_g = 0.06$ H, $R_a = 10\ \Omega$, $L_a = 0.04$ H, $J = 1.6$ N-m/(rad/sec^2), $B = 0.04$ N-m/(rad/sec), motor inertia and friction are negligible.

Taking physically meaningful and measurable variables as state variables, derive a state model for the system.

5.2 Figure P5.2 shows a position control system with state variable feedback. The plant consists of a field-controlled dc motor with a dc amplifier. The parameters of the plant are given below:

Amplifier gain, $K_A = 50$ volt/volt
Motor field resistance, $R_f = 99\ \Omega$
Motor field inductance, $L_f = 20$ H
Motor torque constant, $K_T = 10$ N-m/amp
Moment of inertia of load, $J = 0.5$ N-m/(rad/sec^2)
Coefficient of viscous friction of load, $B = 0.5$ N-m/(rad/sec)
Motor inertia and friction are negligible.

Taking $x_1 = \theta$, $x_2 = \dot{\theta}$, and $x_3 = i_f$ as the state variables, $u = e_f$ as the input, and $y = \theta$ as the output, derive a state variable model for the plant.

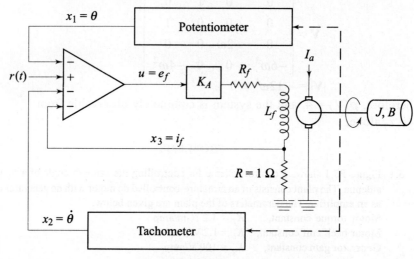

Fig. P5.2

5.3 Figure P5.3 shows the block diagram of a motor-driven single-link robot manipulator with position and velocity feedback. The drive motor is an armature-controlled dc motor; e_a is armature voltage, i_a is armature current, θ_M is the motor shaft position and $\dot{\theta}_M$ is motor shaft velocity. θ_L is the position of the robot arm.

Taking θ_M, $\dot{\theta}_M$ and i_a as state variables, derive a state model for the feedback system.

5.4 Figure P5.4 shows the block diagram of a speed control system with state variable feedback. The drive motor is an armature-controlled dc motor with armature resistance R_a, armature inductance L_a, motor torque constant K_T, inertia referred to motor shaft J, viscous friction coefficient referred to motor shaft B, back emf constant K_b, and tachometer constant K_t. The applied armature voltage is controlled by a three-phase full-converter. We have assumed a linear relationship between the control voltage e_c and the armature voltage e_a, e_r is the reference voltage corresponding to the desired speed.

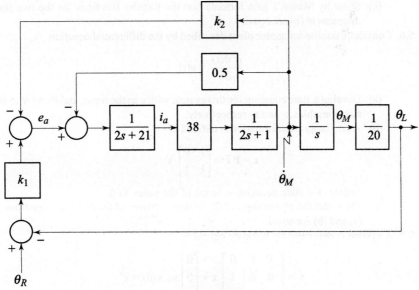

Fig. P5.3

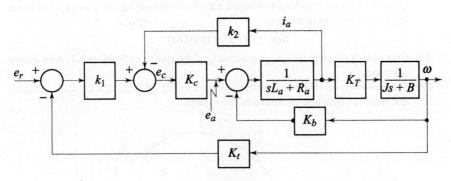

Fig. P5.4

Taking $x_1 = \omega$ (speed) and $x_2 = i_a$ (armature current) as the state variables, $u = e_r$ as the input, and $y = \omega$ as the output, derive a state variable model for the feedback system.

5.5 Consider the system

$$\dot{\mathbf{x}} = \begin{bmatrix} -3 & 1 \\ -2 & 0 \end{bmatrix} \mathbf{x} + \begin{bmatrix} 0 \\ 1 \end{bmatrix} u$$

$$y = \begin{bmatrix} 1 & 0 \end{bmatrix} \mathbf{x}$$

A similarity transformation is defined by

$$\mathbf{x} = \mathbf{P}\bar{\mathbf{x}} = \begin{bmatrix} 2 & -1 \\ -1 & 1 \end{bmatrix} \bar{\mathbf{x}}$$

(a) Express the state model in terms of the states $\bar{\mathbf{x}}(t)$.
(b) Draw state diagrams in signal flow graph form for the state models in $\mathbf{x}(t)$ and $\bar{\mathbf{x}}(t)$.

(c) Show by Mason's gain formula that the transfer functions for the two state diagrams in (b) are equal.

5.6 Consider a double-integrator plant described by the differential equation

$$\frac{d^2\theta(t)}{dt^2} = u(t)$$

(a) Develop a state equation for this system with u as the input, and θ and $\dot{\theta}$ as the state variables x_1 and x_2 respectively.
(b) A similarity transformation is defined as

$$\mathbf{x} = \mathbf{P}\bar{\mathbf{x}} = \begin{bmatrix} 1 & 0 \\ 1 & 1 \end{bmatrix} \bar{\mathbf{x}}$$

Express the state equation in terms of the states $\bar{\mathbf{x}}(t)$.
(c) Show that the eigenvalues of the system matrices of the two state equations in (a) and (b) are equal.

5.7 A system is described by the state equation

$$\dot{\mathbf{x}} = \begin{bmatrix} 0 & 1 & 0 \\ 0 & 0 & 1 \\ -1 & 0 & -3 \end{bmatrix} \mathbf{x} + \begin{bmatrix} 0 \\ 0 \\ 1 \end{bmatrix} u; \quad \mathbf{x}(0) = \mathbf{x}^0$$

Using the Laplace transform technique, transform the state equation into a set of linear algebraic equations in the form

$$\mathbf{X}(s) = \mathbf{G}(s)\mathbf{x}^0 + \mathbf{H}(s)U(s)$$

5.8 Give a block diagram for the programming of the system of Problem 5.7 on an analog computer.

5.9 The state diagram of a linear system is shown in Fig. P5.9. Assign the state variables and write the dynamic equations of the system.

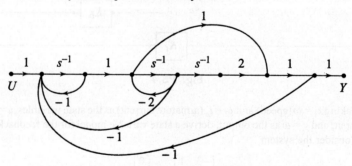

Fig. P5.9

5.10 Construct a state model for the system of Fig. P5.10.
5.11 Derive transfer functions corresponding to the following state models:

(a) $\dot{\mathbf{x}} = \begin{bmatrix} 0 & 1 \\ -2 & -3 \end{bmatrix} \mathbf{x} + \begin{bmatrix} 1 \\ 0 \end{bmatrix} u; \quad y = \begin{bmatrix} 1 & 0 \end{bmatrix} \mathbf{x}$

(b) $\dot{\mathbf{x}} = \begin{bmatrix} -3 & 1 \\ -2 & 0 \end{bmatrix} \mathbf{x} + \begin{bmatrix} 0 \\ 1 \end{bmatrix} u; \quad y = \begin{bmatrix} 1 & 0 \end{bmatrix} \mathbf{x}$

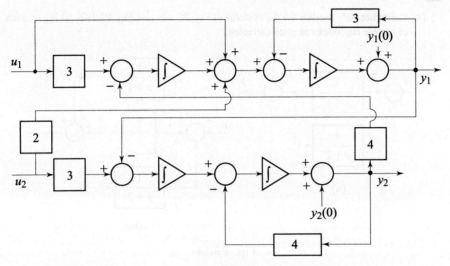

Fig. P5.10

5.12 Derive the transfer function matrix corresponding to the following state model, using resolvent algorithm.

$$\dot{\mathbf{x}} = \begin{bmatrix} 2 & -1 & 0 \\ 1 & 1 & 2 \\ -1 & 0 & 1 \end{bmatrix} \mathbf{x} + \begin{bmatrix} -1 & 0 \\ 1 & 0 \\ 0 & 2 \end{bmatrix} \mathbf{u}$$

$$\mathbf{y} = \begin{bmatrix} 1 & 1 & 0 \\ 1 & 0 & 1 \end{bmatrix} \mathbf{x}$$

5.13 Figure P5.13 shows the block diagram of a control system with state variable feedback and integral control. The plant model is

$$\begin{bmatrix} \dot{x}_1 \\ \dot{x}_2 \end{bmatrix} = \begin{bmatrix} -3 & 2 \\ 4 & -5 \end{bmatrix} \begin{bmatrix} x_1 \\ x_2 \end{bmatrix} + \begin{bmatrix} 1 \\ 0 \end{bmatrix} u$$

$$y = \begin{bmatrix} 0 & 1 \end{bmatrix} \mathbf{x}$$

(a) Derive a state model of the feedback system.
(b) Derive the transfer function $Y(s)/R(s)$.

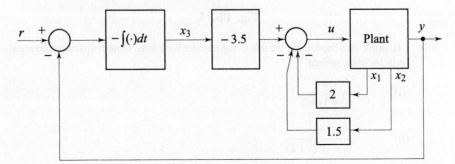

Fig. P5.13

5.14 Construct state models for the systems of Fig. P5.14a and Fig. P5.14b, taking outputs of simple lag blocks as state variables.

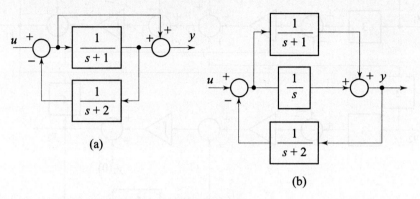

Fig. P5.14

5.15 Derive a state model for the two-input, two-output feedback control system shown in Fig. P5.15. Take outputs of simple lags as state variables.

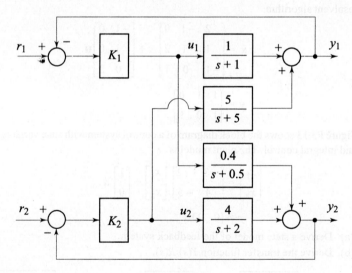

Fig. P5.15

5.16 Construct state models for the following transfer functions. Obtain different canonical form for each system.

(i) $\dfrac{s+3}{s^2+3s+2}$

(ii) $\dfrac{5}{(s+1)^2(s+2)}$

(iii) $\dfrac{s^3 + 8s^2 + 17s + 8}{(s+1)(s+2)(s+3)}$

Give block diagrams for the analog computer simulation of these transfer functions.

5.17 Construct state models for the following differential equations. Obtain a different canonical form for each system.
   (i) $\dddot{y} + 3\ddot{y} + 2\dot{y} = \dot{u} + u$
   (ii) $\dddot{y} + 6\ddot{y} + 11\dot{y} + 6y = u$
   (iii) $\dddot{y} + 6\ddot{y} + 11\dot{y} + 6y = \dddot{u} + 8\ddot{u} + 17\dot{u} + 8u$

5.18 Derive two state models for the system with transfer function
$$\frac{Y(s)}{U(s)} = \frac{50(1 + s/5)}{s(1 + s/2)(1 + s/50)}$$
   (a) One for which the system matrix is a companion matrix.
   (b) One for which the system matrix is diagonal.

5.19 (a) Obtain state variable model in Jordan canonical form for the system with transfer function
$$\frac{Y(s)}{U(s)} = \frac{2s^2 + 6s + 5}{(s+1)^2 (s+2)}$$
   (b) Find the response $y(t)$ to a unit-step input using the state variable model in (a).
   (c) Give a block diagram for analog computer simulation of the transfer function.

5.20 Find the eigenvalues and eigenvectors for the following matrices.

(i) $\begin{bmatrix} 1 & 1 \\ 0 & 2 \end{bmatrix}$ (ii) $\begin{bmatrix} -3 & 2 \\ -1 & 0 \end{bmatrix}$ (iii) $\begin{bmatrix} 0 & 1 & 0 \\ 3 & 0 & 2 \\ -12 & -7 & -6 \end{bmatrix}$

5.21 (a) If $\lambda_1, \lambda_2, \cdots, \lambda_n$ are distinct eigenvalues of
$$\mathbf{A} = \begin{bmatrix} 0 & 1 & 0 & \cdots & 0 \\ 0 & 0 & 1 & \cdots & 0 \\ \vdots & \vdots & \vdots & & \vdots \\ 0 & 0 & 0 & \cdots & 1 \\ -\alpha_n & -\alpha_{n-1} & -\alpha_{n-2} & \cdots & -\alpha_1 \end{bmatrix}$$
prove that the matrix
$$\mathbf{P} = \begin{bmatrix} 1 & 1 & \cdots & 1 \\ \lambda_1 & \lambda_2 & \cdots & \lambda_n \\ \lambda_1^2 & \lambda_2^2 & \cdots & \lambda_n^2 \\ \vdots & \vdots & & \vdots \\ \lambda_1^{n-1} & \lambda_2^{n-1} & \cdots & \lambda_n^{n-1} \end{bmatrix}$$
transforms $\mathbf{A}$ into Jordan canonical form.

(b) Using the result in (a), find the eigenvalues and eigenvectors of the following matrix:
$$\mathbf{A} = \begin{bmatrix} 0 & 1 & 0 \\ 0 & 0 & 1 \\ -24 & -26 & -9 \end{bmatrix}$$

5.22 Consider the matrix

$$A = \begin{bmatrix} 0 & 1 & 0 \\ 0 & 0 & 1 \\ -2 & -4 & -3 \end{bmatrix}$$

(a) Suggest a transformation matrix P such that $\Lambda = P^{-1}AP$ is in Jordan canonical form.
(b) Matrix Λ in (a) has complex elements. Real arithmetic is often preferable, and can be achieved by further transformation. Suggest a transformation matrix Q such that $Q^{-1}AQ$ has all real elements.

5.23 Given the system

$$\dot{x} = \begin{bmatrix} -4 & 3 \\ -6 & 5 \end{bmatrix} x = Ax$$

Determine eigenvalues and eigenvectors of matrix A, and use these results to find the state transition matrix.

5..24 Using Laplace transform method, find the matrix exponential e^{At} for

(a) $A = \begin{bmatrix} 0 & -3 \\ 1 & -4 \end{bmatrix}$ (b) $A = \begin{bmatrix} 0 & 1 \\ -3 & -4 \end{bmatrix}$

5.25 Using the Cayley-Hamilton technique, find e^{At} for

(a) $A = \begin{bmatrix} 0 & 1 \\ -6 & -5 \end{bmatrix}$ (b) $A = \begin{bmatrix} 0 & 2 \\ -2 & -4 \end{bmatrix}$

5.26 Given the system

$$\dot{x} = \begin{bmatrix} -2 & 1 \\ 1 & -2 \end{bmatrix} x + \begin{bmatrix} 1 \\ 1 \end{bmatrix} u$$

(a) Obtain a state diagram in signal flow graph form.
(b) From the signal flow graph, determine the state equation in the form
$$X(s) = G(s)x(0) + H(s)U(s)$$
(c) Using inverse Laplace transformation, obtain the
 (i) zero-input response to initial condition
$$x(0) = [x_1^0 \ x_2^0]^T;$$
 (ii) zero-state response to unit-step input.

5.27 A linear time-invariant system is described by the following state model:

$$\dot{x} = \begin{bmatrix} 0 & 1 & 0 \\ 0 & 0 & 1 \\ -6 & -11 & -6 \end{bmatrix} x + \begin{bmatrix} 0 \\ 0 \\ 2 \end{bmatrix} u$$

$$y = [1 \ 0 \ 0] x$$

Diagonalize the coefficient matrix of the state model using a similarity transformation, and from there obtain the explicit solutions for the state vector and output when the control force u is a unit-step function and the initial state vector is
$$x(0) = [0 \ 0 \ 2]^T$$

5.28 Consider the system

$$\dot{x} = \begin{bmatrix} 0 & 1 \\ -2 & -3 \end{bmatrix} x + \begin{bmatrix} 0 \\ 1 \end{bmatrix} u; \; x(0) = \begin{bmatrix} 1 \\ 1 \end{bmatrix}$$

$$y = \begin{bmatrix} 1 & 0 \end{bmatrix} x$$

(a) Determine the stability of the system.
(b) Find the output response of the system to unit-step input.

5.29 Find the response of the system

$$\dot{x} = \begin{bmatrix} 0 & 1 \\ -2 & -3 \end{bmatrix} x + \begin{bmatrix} 2 & 1 \\ 0 & 1 \end{bmatrix} u; \; x(0) = \begin{bmatrix} 0 \\ 0 \end{bmatrix}$$

$$y = \begin{bmatrix} 1 & 0 \\ 1 & 1 \end{bmatrix} x$$

to the following input:

$$\mathbf{u}(t) = \begin{bmatrix} u_1(t) \\ u_2(t) \end{bmatrix} = \begin{bmatrix} \mu(t) \\ e^{-3t} \mu(t) \end{bmatrix} ; \; \mu(t) \text{ is unit-step function.}$$

5.30 Figure P5.30 shows the block diagram of a control system with state variable feedback and feedforward control. The plant model is

$$\dot{x} = \begin{bmatrix} -3 & 2 \\ 4 & -5 \end{bmatrix} x + \begin{bmatrix} 1 \\ 0 \end{bmatrix} u$$

$$y = \begin{bmatrix} 0 & 1 \end{bmatrix} x$$

(a) Derive a state model for the feedback system.
(b) Find the output $y(t)$ of the feedback system to a unit-step input $r(t)$; the initial state is assumed to be zero.

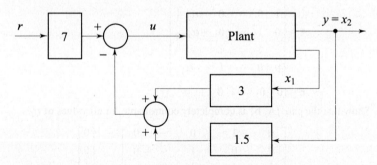

Fig. P5.30

5.31 Consider the state equation

$$\dot{x} = \begin{bmatrix} 0 & 1 \\ -1 & -2 \end{bmatrix} x$$

Find a set of states $x_1(1)$ and $x_2(1)$ such that $x_1(2) = 2$.

5.32 Consider the system
$$\dot{\mathbf{x}} = \begin{bmatrix} 0 & 1 & 0 \\ 3 & 0 & 2 \\ -12 & -7 & -6 \end{bmatrix} \mathbf{x}$$

(a) Find the modes of the system.
(b) Find the initial condition vector $\mathbf{x}(0)$ which will only excite the mode corresponding to the eigenvalue with the most negative real part.

5.33 Consider the system
$$\dot{\mathbf{x}}(t) = \begin{bmatrix} 0 & 1 \\ 2 & 1 \end{bmatrix} \mathbf{x}(t)$$

$$y(t) = \begin{bmatrix} 1 & 2 \end{bmatrix} \mathbf{x}(t)$$

(a) Show that the system modes are e^{-t} and e^{2t}.
(b) Find a set of initial conditions such that the mode e^{2t} is suppressed in $y(t)$.

5.34 The following facts are known about the linear system
$$\dot{\mathbf{x}}(t) = \mathbf{A}\mathbf{x}(t).$$

If $\mathbf{x}(0) = \begin{bmatrix} 1 \\ -2 \end{bmatrix}$, then $\mathbf{x}(t) = \begin{bmatrix} e^{-2t} \\ -2e^{-2t} \end{bmatrix}$

If $\mathbf{x}(0) = \begin{bmatrix} 1 \\ -1 \end{bmatrix}$, then $\mathbf{x}(t) = \begin{bmatrix} e^{-t} \\ -e^{-t} \end{bmatrix}$

Find $e^{\mathbf{A}t}$ and hence $\mathbf{A}$.

5.35 Show that the pair $\{\mathbf{A}, \mathbf{c}\}$ is completely observable for all values of α_i's.
$$\mathbf{A} = \begin{bmatrix} 0 & 0 & \cdots & 0 & -\alpha_n \\ 1 & 0 & \cdots & 0 & -\alpha_{n-1} \\ 0 & 1 & \cdots & 0 & -\alpha_{n-2} \\ \vdots & \vdots & & \vdots & \vdots \\ 0 & 0 & \cdots & 1 & -\alpha_1 \end{bmatrix}$$

$$\mathbf{c} = \begin{bmatrix} 0 & 0 & \cdots & 0 & 1 \end{bmatrix}$$

5.36 Show that the pair $\{\mathbf{A}, \mathbf{b}\}$ is completely controllable for all values of α_i's.
$$\mathbf{A} = \begin{bmatrix} 0 & 1 & 0 & \cdots & 0 \\ 0 & 0 & 1 & \cdots & 0 \\ \vdots & \vdots & \vdots & & \vdots \\ 0 & 0 & 0 & \cdots & 1 \\ -\alpha_n & -\alpha_{n-1} & -\alpha_{n-2} & \cdots & -\alpha_1 \end{bmatrix}; \mathbf{b} = \begin{bmatrix} 0 \\ 0 \\ \vdots \\ 0 \\ 1 \end{bmatrix}$$

5.37 Given the system
$$\dot{\mathbf{x}} = \begin{bmatrix} -1 & 0 & 0 \\ 0 & -2 & 0 \\ 0 & 0 & -3 \end{bmatrix} \mathbf{x} + \begin{bmatrix} 0 & 1 \\ 2 & 0 \\ 0 & 1 \end{bmatrix} \mathbf{u}$$

$$y = \begin{bmatrix} 0 & 1 & 2 \\ 0 & 1 & 0 \end{bmatrix} \mathbf{x}$$

What can we say about controllability and observability without making any further calculations?

5.38 Determine the controllability and observability properties of the following systems:

(i) $\mathbf{A} = \begin{bmatrix} -2 & 1 \\ 1 & -2 \end{bmatrix}; \mathbf{b} = \begin{bmatrix} 1 \\ 0 \end{bmatrix}; \mathbf{c} = [1 \ -1]$

(ii) $\mathbf{A} = \begin{bmatrix} -1 & 0 \\ 0 & -2 \end{bmatrix}; \mathbf{b} = \begin{bmatrix} 2 \\ 5 \end{bmatrix}; \mathbf{c} = [0 \ 1]$

(iii) $\mathbf{A} = \begin{bmatrix} -1 & 0 & 0 \\ 0 & -2 & 0 \\ 0 & 0 & -3 \end{bmatrix}; \mathbf{B} = \begin{bmatrix} 1 & 0 \\ 1 & 2 \\ 2 & 1 \end{bmatrix}; \mathbf{C} = \begin{bmatrix} 1 & 1 & 2 \\ 3 & 1 & 5 \end{bmatrix}$

(iv) $\mathbf{A} = \begin{bmatrix} 0 & 1 & 0 \\ 0 & 0 & 1 \\ 0 & -2 & -3 \end{bmatrix}; \mathbf{b} = \begin{bmatrix} 0 \\ 0 \\ 1 \end{bmatrix}; \mathbf{c} = [10 \ 0 \ 0]$

(v) $\mathbf{A} = \begin{bmatrix} 0 & 0 & 0 \\ 1 & 0 & -3 \\ 0 & 1 & -4 \end{bmatrix}; \mathbf{b} = \begin{bmatrix} 40 \\ 10 \\ 0 \end{bmatrix}; \mathbf{c} = [0 \ 0 \ 1]$

5.39 The following models realize the transfer function $G(s) = \dfrac{1}{s+1}$.

(i) $\mathbf{A} = \begin{bmatrix} -2 & 1 \\ 1 & -2 \end{bmatrix}; \mathbf{b} = \begin{bmatrix} 1 \\ 1 \end{bmatrix}; \mathbf{c} = [0 \ 1]$

(ii) $\mathbf{A} = \begin{bmatrix} -1 & 0 \\ 0 & -3 \end{bmatrix}; \mathbf{b} = \begin{bmatrix} 1 \\ 1 \end{bmatrix}; \mathbf{c} = [1 \ 0]$

(iii) $\mathbf{A} = \begin{bmatrix} -2 & 0 \\ 0 & -1 \end{bmatrix}; \mathbf{b} = \begin{bmatrix} 0 \\ 1 \end{bmatrix}; \mathbf{c} = [0 \ 1]$

Investigate the controllability and observability properties of these models. Find a state variable model for the given transfer function which is both controllable and observable.

5.40 Consider the systems

(i) $\mathbf{A} = \begin{bmatrix} 0 & -2 \\ 1 & -3 \end{bmatrix}; \mathbf{b} = \begin{bmatrix} 1 \\ 1 \end{bmatrix}; \mathbf{c} = [0 \ 1]$

(ii) $\mathbf{A} = \begin{bmatrix} 0 & 1 & 0 \\ 0 & 0 & 1 \\ -6 & -11 & -6 \end{bmatrix}; \mathbf{b} = \begin{bmatrix} 0 \\ 0 \\ 1 \end{bmatrix}; \mathbf{c} = [4 \ 5 \ 1]$

Determine the transfer function in each case. What can we say about controllability and observability properties without making any further calculations?

5.41 Consider the system

$$\dot{x} = \begin{bmatrix} 1 & 1 & 0 \\ 0 & -2 & 1 \\ 0 & 0 & -1 \end{bmatrix} x + \begin{bmatrix} 0 \\ 1 \\ -2 \end{bmatrix} u; \ y = \begin{bmatrix} 1 & 0 & 0 \end{bmatrix} x$$

(a) Find the eigenvalues of **A** and from there determine the stability of the system.
(b) Find the transfer function model and from there determine the stability of the system.
(c) Are the two results same? If not, why?

5.42 Given a transfer function

$$G(s) = \frac{10}{s(s+1)} = \frac{Y(s)}{U(s)}$$

Construct three different state models for this system:
(a) One which is both controllable and observable.
(b) One which is controllable but not observable.
(c) One which is observable but not controllable.

5.43 Prove that the transfer function

$$G(s) = Y(s)/U(s)$$

of the system

$$\dot{x}(t) = Ax(t) + bu(t)$$
$$y(t) = cx(t) + du(t)$$

is invariant under state transformation $x(t) = P\bar{x}(t)$; **P** is a constant nonsingular matrix.

chapter 6

State Variable Analysis of Digital Control Systems

6.1 INTRODUCTION

In the previous chapter of this book, we treated in considerable detail the analysis of linear continuous-time systems using state variable methods. In this chapter we give a condensed review of the same methods for linear discrete-time systems. Since the theory of linear discrete-time systems very closely parallels the theory of linear continuous-time systems, many of the results are similar. For this reason, the comments in this chapter are brief, except in those cases where the results for discrete-time systems deviate markedly from the continuous-time situation. For the same reason, many proofs are omitted.

We will be mostly concerned with single-input, single-output (SISO) system configurations of the type shown in the block diagram of Fig. 6.1. The plant in the figure is a physical process characterized by continuous-time input and output variables. A digital computer is used to control the continuous-time plant. The interface system that takes care of the communication between the digital computer and the continuous-time plant consists of analog-to-digital (A/D) converter and digital-to-analog (D/A) converter. In order to analyse such a system, it is often convenient to represent the continuous-time plant together with the D/A converter and the A/D converter by an equivalent discrete-time system.

The discrete-time systems we will come across can therefore be classified into two types.
1. Inherently discrete-time systems (digital processors), where it makes sense to consider the system at discrete instants of time only, and what happens in between is irrelevant.
2. Discrete-time systems that result from considering continuous-time systems at discrete instants of time only.

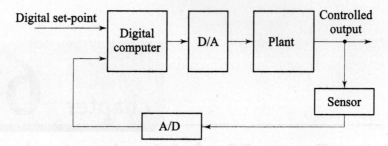

Fig. 6.1 Basic structure of digital control systems

6.2 STATE DESCRIPTIONS OF DIGITAL PROCESSORS

A discrete-time system is a transformation or operator that maps a given input sequence $u(k)$ into an output sequence $y(k)$. Classes of discrete-time systems are defined by placing constraints on the transformation. As they are relatively easy to characterize mathematically, and as they can be designed to perform useful signal processing functions, the class of linear time-invariant systems will be studied here.

In the control structure of Fig. 6.1, the digital computer transforms an input sequence into a form which is in some sense more desirable. Therefore, the discrete-time systems we consider here are in fact *computer programs*. Needless to say, digital computers can do many things other than control dynamic systems; it is our purpose to examine their characteristics when doing this elementary control task.

State variable model of a SISO discrete-time system consists of a set of first-order difference equations relating state variables $x_1(k), x_2(k), ..., x_n(k)$ of the discrete-time system to the input $u(k)$; the output $y(k)$ is algebraically related to the state variables and the input. Assuming that the input is switched on to the system at $k = 0$ ($u(k) = 0$ for $k < 0$), then the initial state is given by

$$\mathbf{x}(0) \triangleq \mathbf{x}^0; \text{ a specified } n \times 1 \text{ vector}$$

The dynamics of a linear time-invariant system is described by equations of the form:

$$\mathbf{x}(k+1) = \mathbf{F}\mathbf{x}(k) + \mathbf{g}u(k); \; \mathbf{x}(0) \triangleq \mathbf{x}^0 \tag{6.1a}$$

$$y(k) = \mathbf{c}\mathbf{x}(k) + du(k) \tag{6.1b}$$

where $\mathbf{x}(k) = \begin{bmatrix} x_1(k) \\ x_2(k) \\ \vdots \\ x_n(k) \end{bmatrix} = n \times 1$ state vector of nth-order system

$u(k)$ = system input

$y(k)$ = defined output

$$\mathbf{F} = \begin{bmatrix} f_{11} & f_{12} & \cdots & f_{1n} \\ f_{21} & f_{22} & \cdots & f_{2n} \\ \vdots & \vdots & & \vdots \\ f_{n1} & f_{n2} & \cdots & f_{nn} \end{bmatrix} = n \times n \text{ constant matrix}$$

$$\mathbf{g} = \begin{bmatrix} g_1 \\ g_2 \\ \vdots \\ g_n \end{bmatrix} = n \times 1 \text{ constant column matrix}$$

$\mathbf{c} = [c_1 \quad c_2 \quad \cdots \quad c_n] = 1 \times n$ constant row matrix

d = scalar, representing direct coupling between input and output

Equation (6.1a) is called the *state equation* of the system, Eqn. (6.1b) is called the *output equation*; the two equations together give the *state variable model* of the system.

Conversion of State Variable Models to Transfer Functions

In the study of linear time-invariant discrete-time equations, we may also apply the z-transform techniques. Taking the z-transform of Eqns (6.1), we obtain (refer Appendix A):

$$z\mathbf{X}(z) - z\mathbf{x}^0 = \mathbf{F}\mathbf{X}(z) + \mathbf{g}U(z)$$

$$Y(z) = \mathbf{c}\mathbf{X}(z) + dU(z)$$

where
$$\mathbf{X}(z) \triangleq \mathscr{Z}[\mathbf{x}(k)]$$
$$U(z) \triangleq \mathscr{Z}[u(k)]$$
$$Y(z) \triangleq \mathscr{Z}[y(k)]$$

Manipulation of these equations gives

$$(z\mathbf{I} - \mathbf{F})\mathbf{X}(z) = z\mathbf{x}^0 + \mathbf{g}U(z); \mathbf{I} \text{ is } n \times n \text{ identity matrix}$$

or
$$\mathbf{X}(z) = (z\mathbf{I} - \mathbf{F})^{-1} z\mathbf{x}^0 + (z\mathbf{I} - \mathbf{F})^{-1} \mathbf{g}U(z) \quad (6.2a)$$

$$Y(z) = \mathbf{c}(z\mathbf{I} - \mathbf{F})^{-1} z\mathbf{x}^0 + [\mathbf{c}(z\mathbf{I} - \mathbf{F})^{-1} \mathbf{g} + d] U(z) \quad (6.2b)$$

Equations (6.2) are algebraic equations. If $\mathbf{x}^0$ and $U(z)$ are known, $\mathbf{X}(z)$ can be computed from these equations.

In the case of zero initial state (i.e., $\mathbf{x}^0 = \mathbf{0}$), the input-output behaviour of the system (6.1) is determined entirely by the transfer function

$$\frac{Y(z)}{U(z)} = G(z) = \mathbf{c}(z\mathbf{I} - \mathbf{F})^{-1}\mathbf{g} + d \qquad (6.3a)$$

$$= \mathbf{c}\frac{(z\mathbf{I} - \mathbf{F})^{+}\mathbf{g}}{|z\mathbf{I} - \mathbf{F}|} + d \qquad (6.3b)$$

where $(z\mathbf{I} - \mathbf{F})^{+}$ = adjoint of the matrix $(z\mathbf{I} - \mathbf{F})$

$|z\mathbf{I} - \mathbf{F}|$ = determinant of the matrix $(z\mathbf{I} - \mathbf{F})$

$|\lambda\mathbf{I} - \mathbf{F}|$ is the *characteristic polynomial* of matrix $\mathbf{F}$. The roots of this polynomial are the *characteristic roots* or *eigenvalues* of matrix $\mathbf{F}$.

From Eqn. (6.3b) we observe that the characteristic polynomial of matrix $\mathbf{F}$ of the system (6.1) is same as the denominator polynomial of the corresponding transfer function $G(z)$. If there are no cancellations between the numerator and denominator polynomials of $G(z)$ in Eqn. (6.3b), the eigenvalues of matrix $\mathbf{F}$ are same as the poles of $G(z)$.

In a later section, we shall see that for a completely controllable and observable state variable model, the eigenvalues of matrix $\mathbf{F}$ are same as the poles of the corresponding transfer function.

Conversion of Transfer Functions to Canonical State Variable Models

In Chapters 2–4, we have seen that transform-domain design techniques yield digital control algorithms in the form of transfer functions of the form

$$D(z) = \frac{\beta_0 z^n + \beta_1 z^{n-1} + \cdots + \beta_{n-1} z + \beta_n}{z^n + \alpha_1 z^{n-1} + \cdots + \alpha_{n-1} z + \alpha_n} \qquad (6.4)$$

where the coefficients α_i and β_i are real constant scalars. Equation (6.4) represents an nth-order digital controller. Several different structures for realization of this controller using delay elements, adders, and multipliers were presented in Section 3.4. Each of these realizations is a dynamic system with n first-order dynamic elements—the unit delayers. We know that output of a first-order dynamic element represents the state of that element. Therefore each realization of Eqn. (6.4) is, in fact, a *state diagram*; by labelling the unit-delayer outputs as state variables, we can obtain the state variable model.

In what follows, we shall use two of the structures presented in Section 3.4 for obtaining canonical state variable models corresponding to the general transfer function

$$G(z) = \frac{Y(z)}{U(z)} = \frac{\beta_0 z^n + \beta_1 z^{n-1} + \cdots + \beta_{n-1} z + \beta_n}{z^n + \alpha_1 z^{n-1} + \cdots + \alpha_{n-1} z + \alpha_n} \qquad (6.5)$$

Revisiting Section 3.4 at this stage will be helpful in our discussion.

First Companion Form

A direct realization structure for the system described by Eqn. (6.5) is shown in Fig. 6.2. Notice that n delay elements have been used in this realization. The coefficients $\alpha_1, \alpha_2, \ldots, \alpha_n$ appear as feedback elements, and the coefficients $\beta_0, \beta_1, \ldots, \beta_n$ appear as feedforward elements. To get one state variable model, we identify the output of each unit delayer with a state variable, starting at the right and proceeding to the left. The corresponding difference equations are

$$x_1(k+1) = x_2(k)$$
$$x_2(k+1) = x_3(k) \tag{6.6a}$$
$$\vdots$$
$$x_{n-1}(k+1) = x_n(k)$$
$$x_n(k+1) = -\alpha_n x_1(k) - \alpha_{n-1} x_2(k) - \cdots - \alpha_1 x_n(k) + u(k)$$

Careful examination of Fig. 6.2 reveals that there are two paths from the output of each unit delayer to the system output: one path upward through the box labelled β_i and a second path down through the box labelled α_i and thence through the box labelled β_0. As a consequence

$$y(k) = (\beta_n - \alpha_n \beta_0) x_1(k) + (\beta_{n-1} - \alpha_{n-1} \beta_0) x_2(k)$$
$$+ \cdots + (\beta_1 - \alpha_1 \beta_0) x_n(k) + \beta_0 u(k) \tag{6.6b}$$

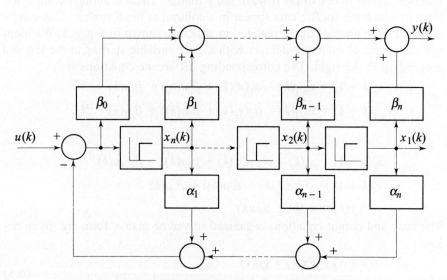

Fig. 6.2 A direct realization structure for system given by Eqn. (6.5)

The state and output Eqns (6.6), organized in vector-matrix form, are given below.

$$\mathbf{x}(k+1) = \mathbf{F}\mathbf{x}(k) + \mathbf{g}u(k)$$
$$y(k) = \mathbf{c}\mathbf{x}(k) + du(k) \tag{6.7}$$

with
$$\mathbf{F} = \begin{bmatrix} 0 & 1 & 0 & \cdots & 0 \\ 0 & 0 & 1 & \cdots & 0 \\ \vdots & \vdots & \vdots & & \vdots \\ 0 & 0 & 0 & \cdots & 1 \\ -\alpha_n & -\alpha_{n-1} & -\alpha_{n-2} & \cdots & -\alpha_1 \end{bmatrix}; \mathbf{g} = \begin{bmatrix} 0 \\ 0 \\ \vdots \\ 0 \\ 1 \end{bmatrix}$$

$$\mathbf{c} = [\beta_n - \alpha_n\beta_0,\ \beta_{n-1} - \alpha_{n-1}\beta_0,\ \ldots,\ \beta_1 - \alpha_1\beta_0];\ d = \beta_0$$

The matrix $\mathbf{F}$ in Eqns (6.7) has a very special structure—the coefficients of the denominator of the transfer function preceded by minus signs form a string along the bottom row of the matrix. The rest of the matrix is zero except for the 'superdiagonal' terms which are all unity. A matrix with this structure is said to be in *companion form*. We call the state variable model (6.7) the *first companion form*[1] state model for the transfer function (6.5); another companion form follows.

Second Companion Form

In the first companion form, the coefficients of the denominator of the transfer function appear in one of the rows of the $\mathbf{F}$ matrix. There is another companion form in which the coefficients appear in a column of the $\mathbf{F}$ matrix. This can be obtained from another direct realization structure shown in Fig. 6.3. We identify the output of each unit delayer with a state variable starting at the left and proceeding to the right. The corresponding difference equations are

$$x_n(k+1) = x_{n-1}(k) - \alpha_1(x_n(k) + \beta_0 u(k)) + \beta_1 u(k)$$
$$x_{n-1}(k+1) = x_{n-2}(k) - \alpha_2(x_n(k) + \beta_0 u(k)) + \beta_2 u(k)$$
$$\vdots$$
$$x_2(k+1) = x_1(k) - \alpha_{n-1}(x_n(k) + \beta_0 u(k)) + \beta_{n-1} u(k)$$
$$x_1(k+1) = -\alpha_n(x_n(k) + \beta_0 u(k)) + \beta_n u(k)$$
$$y(k) = x_n(k) + \beta_0 u(k)$$

The state and output equations organized in vector-matrix form are given below.

$$\begin{aligned} \mathbf{x}(k+1) &= \mathbf{F}\mathbf{x}(k) + \mathbf{g}u(k) \\ y(k) &= \mathbf{c}\mathbf{x}(k) + du(k) \end{aligned} \quad (6.8)$$

with
$$\mathbf{F} = \begin{bmatrix} 0 & 0 & \cdots & 0 & -\alpha_n \\ 1 & 0 & \cdots & 0 & -\alpha_{n-1} \\ 0 & 1 & \cdots & 0 & -\alpha_{n-2} \\ \vdots & \vdots & & \vdots & \vdots \\ 0 & 0 & \cdots & 1 & -\alpha_1 \end{bmatrix};\ \mathbf{g} = \begin{bmatrix} \beta_n - \alpha_n\beta_0 \\ \beta_{n-1} - \alpha_{n-1}\beta_0 \\ \vdots \\ \beta_1 - \alpha_1\beta_0 \end{bmatrix}$$

---
1. The pair $(\mathbf{F}, \mathbf{g})$ of Eqns (6.7) is completely controllable for all values of α_i's (Refer Problem 5.36).

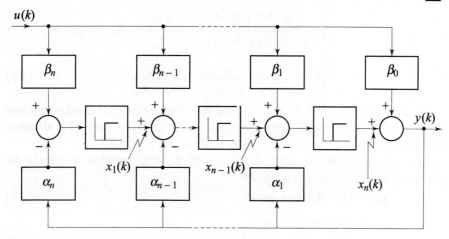

Fig. 6.3 An alternative direct realization structure for the system given by Eqn. (6.5)

$$\mathbf{c} = [0 \quad 0 \quad \cdots \quad 0 \quad 1]; \; d = \beta_0$$

Comparing the **F**, **g** and **c** matrices of the *second companion form*[2] with that of the first, we observe that **F**, **g**, and **c** matrices of one companion form correspond to the transpose of **F**, **c**, and **g** matrices respectively of the other.

Both the companion forms of state variable models play an important role in pole-placement design through state feedback. This will be discussed in Chapter 7.

Jordan Canonical Form

In the two canonical forms (6.7) and (6.8), the coefficients of the denominator of the transfer function appear in one of the rows or columns of matrix **F**. In another of the canonical forms, the poles of the transfer function form a string along the main diagonal of the matrix. The canonical form follows directly from the parallel realization structure of transfer function. We shall first discuss the case where all poles are distinct. Then we shall consider the case where multiple poles are involved.

Case I: *The transfer function involves distinct poles only*

Assume that $z = \lambda_i$ ($i = 1, 2, ..., n$) are the distinct poles of the given transfer function (6.5). Partial-fraction expansion of the transfer function gives

$$\frac{Y(z)}{U(z)} = G(z) = \frac{\beta_0 z^n + \beta_1 z^{n-1} + \cdots + \beta_{n-1} z + \beta_n}{z^n + \alpha_1 z^{n-1} + \cdots + \alpha_{n-1} z + \alpha_n}$$

$$= \beta_0 + \frac{\beta'_1 z^{n-1} + \beta'_2 z^{n-2} + \cdots + \beta'_n}{z^n + \alpha_1 z^{n-1} + \cdots + \alpha_n}$$

---

2. The pair {**F**, **c**} of Eqns (6.8) is completely observable for all values of α_i's (Refer Problem 5.35).

$$= \beta_0 + \frac{\beta_1' z^{n-1} + \beta_2' z^{n-2} + \cdots + \beta_n'}{(z-\lambda_1)(z-\lambda_2)\cdots(z-\lambda_n)} = \beta_0 + G'(z)$$

$$= \beta_0 + \frac{r_1}{z-\lambda_1} + \frac{r_2}{z-\lambda_2} + \cdots + \frac{r_n}{z-\lambda_n} \tag{6.9}$$

The coefficients $r_i(i = 1, 2, \ldots, n)$ are the residues of the transfer function $G'(z)$ at the corresponding poles at $z = \lambda_i(i = 1, 2, \ldots, n)$. A parallel realization structure of the transfer function (6.9) is shown in Fig. 6.4.

Identifying the outputs of the delayers with the state variables results in the following state and output equations:

$$\mathbf{x}(k+1) = \mathbf{\Lambda}\mathbf{x}(k) + \mathbf{g}u(k) \tag{6.10}$$

$$y(k) = \mathbf{c}\mathbf{x}(k) + du(k)$$

with
$$\mathbf{\Lambda} = \begin{bmatrix} \lambda_1 & 0 & \cdots & 0 \\ 0 & \lambda_2 & \cdots & 0 \\ \vdots & \vdots & & \vdots \\ 0 & 0 & \cdots & \lambda_n \end{bmatrix}; \mathbf{g} = \begin{bmatrix} 1 \\ 1 \\ \vdots \\ 1 \end{bmatrix}$$

$$\mathbf{c} = [r_1 \; r_2 \; \cdots \; r_n]; \; d = \beta_0$$

It is observed that for this canonical state variable model, the matrix $\mathbf{\Lambda}$ is a diagonal matrix with the poles of $G(z)$ as its diagonal elements.

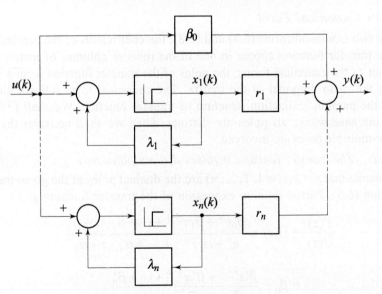

Fig. 6.4 A parallel structure realization for the system given by Eqn. (6.9)

Case II: The transfer function involves multiple poles

When the transfer function $G(z)$ involves multiple poles, the partial fraction expansion will not be as simple as (6.9). In the discussion that follows, we assume that $G(z)$ involves a multiple pole of order m at $z = \lambda_1$, and that all other poles are distinct. Performing the partial-fraction expansion for this case, we get

$$\frac{Y(z)}{U(z)} = G(z) = \frac{\beta_0 z^n + \beta_1 z^{n-1} + \cdots + \beta_{n-1} z + \beta_n}{z^n + \alpha_1 z^{n-1} + \cdots + \alpha_{n-1} z + \alpha_n}$$

$$= \beta_0 + \frac{\beta'_1 z^{n-1} + \beta'_2 z^{n-2} + \cdots + \beta'_n}{z^n + \alpha_1 z^{n-1} + \cdots + \alpha_n}$$

$$= \beta_0 + \frac{\beta'_1 z^{n-1} + \beta'_2 z^{n-2} + \cdots + \beta'_n}{(z - \lambda_1)^m (z - \lambda_{m+1}) \cdots (z - \lambda_n)}$$

$$= \beta_0 + H_1(z) + H_{m+1}(z) + \cdots + H_n(z) \quad (6.11a)$$

where
$$H_{m+1}(z) = \frac{r_{m+1}}{z - \lambda_{m+1}}, \ldots, H_n(z) = \frac{r_n}{z - \lambda_n}, \quad (6.11b)$$

and
$$H_1(z) = \frac{r_{11}}{(z - \lambda_1)^m} + \frac{r_{12}}{(z - \lambda_1)^{m-1}} + \cdots + \frac{r_{1m}}{(z - \lambda_1)} \quad (6.11c)$$

A realization of $H_1(z)$ is shown in Fig. 6.5. Other terms of Eqn. (6.11a) may be realized as per Fig. 6.4.

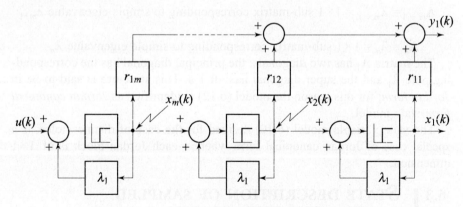

Fig. 6.5 A realization of $H_1(z)$ given by Eqn. (6.11c)

Identifying the outputs of the delayers with the state variables results in the following state and output equations:

$$\begin{aligned} \mathbf{x}(k+1) &= \mathbf{\Lambda}\mathbf{x}(k) + \mathbf{g}u(k) \\ y(k) &= \mathbf{c}\mathbf{x}(k) + du(k) \end{aligned} \quad (6.12)$$

with $m \times m$ Jordan block

$$\Lambda = \begin{bmatrix} \lambda_1 & 1 & 0 & \cdots & 0 & 0 & \cdots & 0 \\ 0 & \lambda_1 & 1 & \cdots & 0 & 0 & \cdots & 0 \\ \vdots & \vdots & \vdots & & \vdots & \vdots & & \vdots \\ 0 & 0 & 0 & \cdots & \lambda_1 & 0 & \cdots & 0 \\ \hline 0 & 0 & 0 & \cdots & 0 & \lambda_{m+1} & \cdots & 0 \\ \vdots & \vdots & \vdots & & \vdots & \vdots & & \vdots \\ 0 & 0 & 0 & \cdots & 0 & 0 & \cdots & \lambda_n \end{bmatrix}; \mathbf{g} = \begin{bmatrix} 0 \\ 0 \\ \vdots \\ 1 \\ \hline 1 \\ \vdots \\ 1 \end{bmatrix}$$

$$\mathbf{c} = [r_{11} \quad r_{12} \quad \cdots \quad r_{1m} \mid r_{m+1} \quad \cdots \quad r_n]; d = \beta_0$$

Note that the Λ matrix in Eqns (6.12) is block diagonal:

$$\Lambda = \begin{bmatrix} \Lambda_1 & 0 & \cdots & 0 \\ 0 & \Lambda_{m+1} & \cdots & 0 \\ \vdots & \vdots & & \vdots \\ 0 & 0 & \cdots & \Lambda_n \end{bmatrix}$$

with $\quad \Lambda_1 = \begin{bmatrix} \lambda_1 & 1 & 0 & \cdots & 0 \\ 0 & \lambda_1 & 1 & \cdots & 0 \\ \vdots & \vdots & \vdots & & \vdots \\ 0 & 0 & 0 & \cdots & \lambda_1 \end{bmatrix} = m \times m$ sub-matrix corresponding to eigenvalue λ_1 of multiplicity m

$\Lambda_{m+1} = \lambda_{m+1} = 1 \times 1$ sub-matrix corresponding to simple eigenvalue λ_{m+1}

$\vdots$

$\Lambda_n = \lambda_n = 1 \times 1$ sub-matrix corresponding to simple eigenvalue λ_n

The matrix Λ_1 has two diagonals: the principal diagonal has the corresponding pole λ_1 and the super diagonal has all 1's. This structure is said to be in *Jordan form*; for this reason the model (6.12) is identified as *Jordan canonical form* state model.

The state variable model (6.10) derived for the case of distinct poles, is a special case of Jordan canonical form wherein each Jordan block is of 1×1 dimension.

6.3 STATE DESCRIPTION OF SAMPLED CONTINUOUS-TIME PLANTS

Systems that consist of an interconnection of a discrete-time system and a continuous-time system are frequently encountered. An example of particular interest occurs when a digital computer is used to control a continuous-time plant. Whenever such interconnections exist, there must be some type of *interface system* that takes care of the communication between the discrete-time and continuous-time systems. In the system of Fig. 6.1, the interface function is performed by D/A and A/D converters.

Simple models of the interface actions of D/A and A/D converters have been developed in Chapter 2. A brief review is in order here.

A simple model of A/D converter is shown in Fig. 6.6. A continuous-time function $f(t)$, $t \geq 0$, is the input and the sequence of real numbers $f(k)$, $k = 0, 1, 2, \ldots$, is the output; the following relation holds between input and output:

$$f(k) = f(t = kT); \; T \text{ is the time interval between samples} \qquad (6.13a)$$

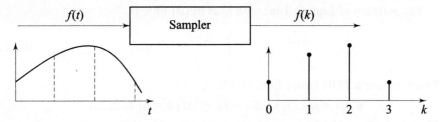

Fig. 6.6 A model of an A/D converter

A simple model of D/A converter is shown in Fig. 6.7. A sequence of numbers $f(k)$, $k = 0, 1, 2, \ldots$, is the input and the continuous-time function $f^+(t)$, $t \geq 0$, is the output; the following relation holds between input and output:

$$f^+(t) = f(k); \; kT \leq t < (k + 1)T \qquad (6.13b)$$

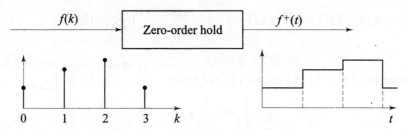

Fig. 6.7 A model of a D/A converter

Figure 6.8 illustrates a typical example of an interconnection of discrete-time and continuous-time systems. In order to analyse such a system, it is often convenient to represent the continuous-time system together with the zero-order hold (ZOH) and the sampler by an *equivalent discrete-time system*.

We assume that the continuous-time system of Fig. 6.8 is a linear system with state variable model

$$\dot{\mathbf{x}}(t) = \mathbf{A}\mathbf{x}(t) + \mathbf{b}u^+(t) \qquad (6.14a)$$

$$y(t) = \mathbf{c}\mathbf{x}(t) + du^+(t) \qquad (6.14b)$$

where $\mathbf{x}(t)$ is $n \times 1$ state vector, $u^+(t)$ is scalar input, $y(t)$ is scalar output; $\mathbf{A}$, $\mathbf{b}$, $\mathbf{c}$, and d are, respectively, $n \times n$, $n \times 1$, $1 \times n$, and 1×1 real constant matrices.

440 Digital Control and State Variable Methods

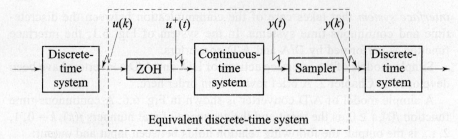

Fig. 6.8 Interconnection of discrete-time and continuous-time systems

The solution of Eqn. (6.14a) with t_0 as initial time is

$$\mathbf{x}(t) = e^{\mathbf{A}(t-t_0)} \mathbf{x}(t_0) + \int_{t_0}^{t} e^{\mathbf{A}(t-\tau)} \mathbf{b} u^+(\tau) \, d\tau \qquad (6.15)$$

Since we use a ZOH (refer Eqn. (6.13b)),

$$u^+(t) = u(kT); \, kT \leq t < (k+1)T; \, k = 0, 1, 2, \ldots$$

Then from Eqn. (6.15) we can write,

$$\mathbf{x}(t) = e^{\mathbf{A}(t-kT)} \mathbf{x}(kT) + \left[\int_{kT}^{t} e^{\mathbf{A}(t-\tau)} \mathbf{b} \, d\tau \right] u(kT); \, kT \leq t < (k+1)T \qquad (6.16)$$

In response to the input $u(kT)$, the state settles to the value $\mathbf{x}((k+1)T)$ prior to the application of the input $u((k+1)T)$, where

$$\mathbf{x}((k+1)T) = e^{\mathbf{A}T} \mathbf{x}(kT) + \left[\int_{kT}^{(k+1)T} e^{\mathbf{A}[(k+1)T-\tau]} \mathbf{b} \, d\tau \right] u(kT)$$

$$= \mathbf{F}\mathbf{x}(kT) + \mathbf{g}u(kT) \qquad (6.17)$$

Letting $\sigma = (\tau - kT)$ in Eqn. (6.17), we have

$$\mathbf{g} = \int_{0}^{T} e^{\mathbf{A}(T-\sigma)} \mathbf{b} \, d\sigma$$

With $\theta = T - \sigma$, we get

$$\mathbf{g} = \int_{0}^{T} e^{\mathbf{A}\theta} \mathbf{b} \, d\theta$$

If we are interested in the value of $\mathbf{x}(t)$ (or $y(t)$) between sampling instants, we first solve for $\mathbf{x}(kT)$ for any k using Eqn. (6.17) and then use Eqn. (6.16) to determine $\mathbf{x}(t)$ for $kT \leq t < (k+1)T$.

Since we have a sampler in configuration of Fig. 6.8 (refer Eqn. (6.13a)), we have from Eqn. (6.14b),

$$y(kT) = \mathbf{c}\mathbf{x}(kT) + du(kT)$$

State description of the equivalent discrete-time system of Fig. 6.8 is, therefore, of the form:

$$\mathbf{x}(k+1) = \mathbf{F}\mathbf{x}(k) + \mathbf{g}u(k) \tag{6.18a}$$

$$y(k) = \mathbf{c}\mathbf{x}(k) + du(k) \tag{6.18b}$$

where
$$\mathbf{F} = e^{\mathbf{A}T} \tag{6.18c}$$

$$\mathbf{g} = \int_0^T e^{\mathbf{A}\theta}\mathbf{b}\,d\theta \tag{6.18d}$$

Machine Computation of $e^{\mathbf{A}T}$

There are several methods available for computing $e^{\mathbf{A}T}$. Some of these methods have been discussed in the earlier chapter. Standard computer programs based on these methods are available.

In the following, we present an alternative technique of computing $e^{\mathbf{A}T}$. The virtues of this technique are its simplicity and the ease of programming.

The infinite series expansion for $\mathbf{F} = e^{\mathbf{A}T}$ is

$$\mathbf{F} = e^{\mathbf{A}T} = \mathbf{I} + \mathbf{A}T + \frac{1}{2!}\mathbf{A}^2 T^2 + \frac{1}{3!}\mathbf{A}^3 T^3 + \cdots$$

$$= \sum_{i=0}^{\infty} \frac{\mathbf{A}^i T^i}{i!}; \quad \mathbf{A}^0 = \mathbf{I} \tag{6.19}$$

For a finite T, this series is uniformly convergent (Section 5.7). It is therefore possible to evaluate $\mathbf{F}$ within prescribed accuracy. If the series is truncated at $i = N$, then we may write the finite series sum as

$$\overline{\mathbf{F}} = \sum_{i=0}^{N} \frac{\mathbf{A}^i T^i}{i!} \tag{6.20}$$

which represents the infinite series approximation. The larger the N, the better is the approximation. We evaluate $\overline{\mathbf{F}}$ by a series in the form

$$\overline{\mathbf{F}} = \mathbf{I} + \mathbf{A}T\left(\mathbf{I} + \frac{\mathbf{A}T}{2}\left\{\mathbf{I} + \frac{\mathbf{A}T}{3}\left[\mathbf{I} + \cdots + \frac{\mathbf{A}T}{N-1}\left(\mathbf{I} + \frac{\mathbf{A}T}{N}\right)\cdots\right]\right\}\right) \tag{6.21}$$

which has better numerical properties than the direct series of powers. Starting with the innermost factor, this nested product expansion lends itself easily to digital programming. The empirical relation giving the number of terms, N, is

$$N = \min\{3\|\mathbf{A}T\| + 6, 100\} \tag{6.22}$$

where $\|\mathbf{A}T\|$ is a norm of the matrix $\mathbf{A}T$. There are several different forms of matrix norms commonly used. Any one of them may be used in Eqn. (6.22). Two forms of matrix norms are defined in Section 5.2.

The relation (6.22) assumes that no more than 100 terms are included. The series e^{AT} will be accurate to at least six significant figures.

The integral in Eqn. (6.18d) can be evaluated term by term to give

$$\mathbf{g} = \left[\int_0^T \left(\mathbf{I} + \mathbf{A}\theta + \frac{1}{2!}\mathbf{A}^2\theta^2 + \cdots\right)d\theta\right]\mathbf{b} = \sum_{i=0}^{\infty} \frac{\mathbf{A}^i T^{i+1}}{(i+1)!}\mathbf{b} \quad (6.23)$$

$$= \sum_{i=0}^{\infty} \frac{\mathbf{A}^i T^i}{(i+1)!} T\mathbf{b}$$

$$= \left(\mathbf{I} + \frac{\mathbf{A}T}{2!} + \frac{\mathbf{A}^2 T^2}{3!} + \cdots\right)T\mathbf{b} = (e^{\mathbf{A}T} - \mathbf{I})\mathbf{A}^{-1}\mathbf{b} \quad (6.24)$$

The transition from Eqn. (6.23) to (6.24) is possible only for a nonsingular matrix $\mathbf{A}$. For a singular $\mathbf{A}$, we may evaluate $\mathbf{g}$ from Eqn. (6.23) by the approximation technique described above. Since the series expansion for $\mathbf{g}$ converges faster than that for $\mathbf{F}$, it suffices to determine N for $\mathbf{F}$ from Eqn. (6.22) and apply the same value for $\mathbf{g}$.

Example 6.1

Figure 6.9 shows the block diagram of a digital positioning system. Defining the state variables as

$$x_1(t) = \theta(t), \; x_2(t) = \dot{\theta}(t),$$

the state variable model of the plant becomes

$$\dot{\mathbf{x}}(t) = \mathbf{A}\mathbf{x}(t) + \mathbf{b}u^+(t)$$
$$y(t) = \mathbf{c}\mathbf{x}(t) \quad (6.25)$$

with

$$\mathbf{A} = \begin{bmatrix} 0 & 1 \\ 0 & -5 \end{bmatrix}; \; \mathbf{b} = \begin{bmatrix} 0 \\ 1 \end{bmatrix}; \; \mathbf{c} = \begin{bmatrix} 1 & 0 \end{bmatrix}$$

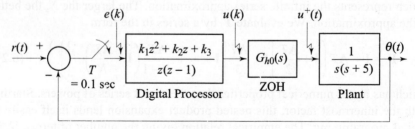

Fig. 6.9 A digital positioning system

We apply here the Cayley-Hamilton technique to evaluate state transition matrix $e^{\mathbf{A}t}$.

Eigenvalues of matrix $\mathbf{A}$ are given by

$$|\lambda \mathbf{I} - \mathbf{A}| = \begin{bmatrix} \lambda & -1 \\ 0 & \lambda + 5 \end{bmatrix} = 0$$

Therefore $\lambda_1 = 0, \lambda_2 = -5$

Since $\mathbf{A}$ is of second order, the polynomial $g(\lambda)$ will be of the form (refer Eqns (5.98)),

$$g(\lambda) = \beta_0 + \beta_1 \lambda$$

The coefficients β_0 and β_1 are evaluated from the following equations:

$$1 = \beta_0$$
$$e^{-5t} = \beta_0 - 5\beta_1$$

The result is
$$\beta_0 = 1$$
$$\beta_1 = \tfrac{1}{5}(1 - e^{-5t})$$

Hence
$$e^{\mathbf{A}t} = \beta_0 \mathbf{I} + \beta_1 \mathbf{A} = \begin{bmatrix} 1 & \tfrac{1}{5}(1-e^{-5t}) \\ 0 & e^{-5t} \end{bmatrix}$$

The equivalent discrete-time plant with input $u(k)$ and output $\theta(k)$ (refer Fig. 6.9) is described by the equations

$$\mathbf{x}(k+1) = \mathbf{F}\mathbf{x}(k) + \mathbf{g}u(k)$$
$$y(k) = \mathbf{c}\mathbf{x}(k)$$
(6.26)

where
$$\mathbf{F} = e^{\mathbf{A}T} = \begin{bmatrix} 1 & \tfrac{1}{5}(1-e^{-5T}) \\ 0 & e^{-5T} \end{bmatrix}$$

$$\mathbf{g} = \int_0^T e^{\mathbf{A}\theta} \mathbf{b}\, d\theta$$

$$= \begin{bmatrix} \int_0^T \tfrac{1}{5}(1-e^{-5\theta})d\theta \\ \int_0^T e^{-5\theta} d\theta \end{bmatrix} = \begin{bmatrix} \tfrac{1}{5}(T - \tfrac{1}{5} + \tfrac{1}{5}e^{-5T}) \\ \tfrac{1}{5}(1 - e^{-5T}) \end{bmatrix}$$

For $T = 0.1$ sec,

$$\mathbf{F} = \begin{bmatrix} 1 & 0.0787 \\ 0 & 0.6065 \end{bmatrix};\ \mathbf{g} = \begin{bmatrix} 0.0043 \\ 0.0787 \end{bmatrix}$$

Consider now the digital processor. The input-output model of the processor is

$$\frac{U(z)}{E(z)} = \frac{k_1 z^2 + k_2 z + k_3}{z(z-1)}$$

444 Digital Control and State Variable Methods

Direct digital realization of the processor is shown in Fig. 6.10. Taking outputs of unit delayers as state variables, we get the following state description for the processor dynamics (refer Eqns (6.8)):

$$\begin{bmatrix} x_3(k+1) \\ x_4(k+1) \end{bmatrix} = \begin{bmatrix} 0 & 0 \\ 1 & 1 \end{bmatrix} \begin{bmatrix} x_3(k) \\ x_4(k) \end{bmatrix} + \begin{bmatrix} k_3 \\ k_2 + k_1 \end{bmatrix} e(k) \qquad (6.27)$$

$$u(k) = x_4(k) + k_1 e(k)$$

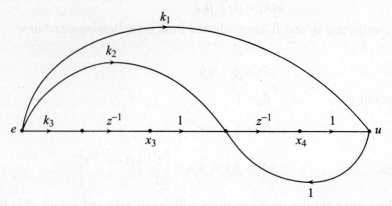

Fig. 6.10 Realization of digital processor of positioning system shown in Fig. 6.9

The processor input is derived from the reference input and the position feedback (Fig. 6.9):

$$e(k) = r(k) - x_1(k) \qquad (6.28)$$

From Eqns (6.26)–(6.28), we get the following state variable model for the feedback system of Fig. 6.9.

$$\begin{bmatrix} x_1(k+1) \\ x_2(k+1) \\ x_3(k+1) \\ x_4(k+1) \end{bmatrix} = \begin{bmatrix} 1-0.0043k_1 & 0.0787 & 0 & 0.0043 \\ -0.0787k_1 & 0.6065 & 0 & 0.0787 \\ -k_3 & 0 & 0 & 0 \\ -(k_2+k_1) & 0 & 1 & 1 \end{bmatrix} \begin{bmatrix} x_1(k) \\ x_2(k) \\ x_3(k) \\ x_4(k) \end{bmatrix}$$

$$+ \begin{bmatrix} 0.0043\,k_1 \\ 0.0787k_1 \\ k_3 \\ k_2+k_1 \end{bmatrix} r(k) \qquad (6.29)$$

$$y(k) = \begin{bmatrix} 1 & 0 & 0 & 0 \end{bmatrix} \begin{bmatrix} x_1(k) \\ x_2(k) \\ x_3(k) \\ x_4(k) \end{bmatrix}$$

6.4 STATE DESCRIPTION OF SYSTEMS WITH DEAD-TIME

Consider a state equation of a single-input system which includes delay in control action:

$$\dot{\mathbf{x}}(t) = \mathbf{A}\mathbf{x}(t) + \mathbf{b}u^+(t - \tau_D) \tag{6.30}$$

where $\mathbf{x}$ is $n \times 1$ state vector, u^+ is scalar input, τ_D is the dead-time, and $\mathbf{A}$ and $\mathbf{b}$ are respectively $n \times n$ and $n \times 1$ real constant matrices.
The solution of Eqn. (6.30) with t_0 as initial time is

$$\mathbf{x}(t) = e^{\mathbf{A}(t-t_0)} \mathbf{x}(t_0) + \int_{t_0}^{t} e^{\mathbf{A}(t-\tau)} \mathbf{b}u^+(\tau - \tau_D) d\tau$$

If we let $t_0 = kT$ and $t = kT + T$, we obtain

$$\mathbf{x}(kT + T) = e^{\mathbf{A}T} \mathbf{x}(kT) + \int_{kT}^{kT+T} e^{\mathbf{A}(kT + T - \tau)} \mathbf{b}u^+(\tau - \tau_D) d\tau$$

With $\sigma = kT + T - \tau$, we get

$$\mathbf{x}(kT + T) = e^{\mathbf{A}T} \mathbf{x}(kT) + \int_{0}^{T} e^{\mathbf{A}\sigma} \mathbf{b}u^+(kT + T - \tau_D - \sigma) d\sigma \tag{6.31}$$

If N is the largest integer number of sampling periods in τ_D, we can write

$$\tau_D = NT + \Delta T; \; 0 \leq \Delta < 1 \tag{6.32a}$$

Substituting in Eqn. (6.31), we get

$$\mathbf{x}(kT + T) = e^{\mathbf{A}T} \mathbf{x}(kT) + \int_{0}^{T} e^{\mathbf{A}\sigma} \mathbf{b}u^+(kT + T - NT - \Delta T - \sigma) d\sigma$$

We introduce a parameter m such that

$$m = 1 - \Delta \tag{6.32b}$$

Then

$$\mathbf{x}(kT + T) = e^{\mathbf{A}T} \mathbf{x}(kT) + \int_{0}^{T} e^{\mathbf{A}\sigma} \mathbf{b}u^+(kT - NT + mT - \sigma) d\sigma \tag{6.33}$$

Since we use a ZOH, u^+ is piecewise constant. The nature of the integral in Eqn. (6.33) with respect to variable σ becomes clear from the sketch of the piecewise constant input u^+ over a segment of time axis near $t = kT - NT$ (Fig. 6.11). The integral runs for σ from 0 to T which corresponds to t from $kT - NT + mT$ backward to $kT - NT - T + mT$. Over this period, the control first takes on the value $u(kT - NT)$ and then the value $u(kT - NT - T)$. Therefore we can break the integral in Eqn. (6.33) into two parts as follows:

446 Digital Control and State Variable Methods

$$\mathbf{x}(kT + T) = e^{\mathbf{A}T}\mathbf{x}(kT) + \left[\int_0^{mT} e^{\mathbf{A}\sigma}\mathbf{b}d\sigma\right]u(kT - NT)$$

$$+ \left[\int_{mT}^{T} e^{\mathbf{A}\sigma}\mathbf{b}d\sigma\right]u(kT - NT - T)$$

$$= \mathbf{F}\mathbf{x}(kT) + \mathbf{g}_1 u(kT - NT - T) + \mathbf{g}_2 u(kT - NT) \tag{6.34a}$$

where
$$\mathbf{F} = e^{\mathbf{A}T} \tag{6.34b}$$

$$\mathbf{g}_1 = \int_{mT}^{T} e^{\mathbf{A}\sigma}\mathbf{b}d\sigma \tag{6.34c}$$

$$\mathbf{g}_2 = \int_0^{mT} e^{\mathbf{A}\sigma}\mathbf{b}d\sigma \tag{6.34d}$$

Setting $\theta = \sigma - mT$ in Eqn. (6.34c), we get

$$\mathbf{g}_1 = \int_0^{\Delta T} e^{\mathbf{A}(mT + \theta)}\mathbf{b}d\theta = e^{\mathbf{A}mT}\int_0^{\Delta T} e^{\mathbf{A}\theta}\mathbf{b}d\theta \tag{6.34e}$$

The matrices/vectors $\mathbf{F}$, $\mathbf{g}_1$ and $\mathbf{g}_2$ can be evaluated by series truncation method discussed in the earlier section.

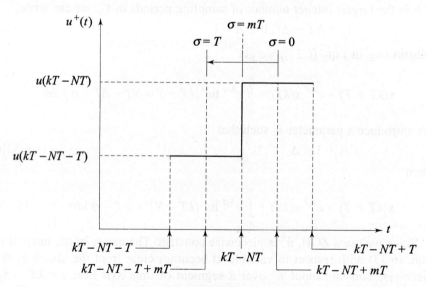

Fig. 6.11

Equation (6.34a) can be expressed in the standard state variable format. To do this, we consider first the case of $N = 0$. For this case, Eqn. (6.34a) becomes

$$\mathbf{x}(k+1) = \mathbf{F}\mathbf{x}(k) + \mathbf{g}_1 u(k-1) + \mathbf{g}_2 u(k)$$

We must eliminate $u(k-1)$ from the right-hand side, which we do by defining a new state

$$x_{n+1}(k) = u(k-1)$$

The augmented state equation is given by

$$\begin{bmatrix} \mathbf{x}(k+1) \\ x_{n+1}(k+1) \end{bmatrix} = \begin{bmatrix} \mathbf{F} & \mathbf{g}_1 \\ \mathbf{0} & 0 \end{bmatrix} \begin{bmatrix} \mathbf{x}(k) \\ x_{n+1}(k) \end{bmatrix} + \begin{bmatrix} \mathbf{g}_2 \\ 1 \end{bmatrix} u(k) \qquad (6.35)$$

For $N > 0$, Eqn. (6.34a) can be expressed as

$$\mathbf{x}(k+1) = \mathbf{F}\mathbf{x}(k) + \mathbf{g}_1 u(k-N-1) + \mathbf{g}_2 u(k-N)$$

Let us introduce $(N+1)$ new states defined below:

$$x_{n+1}(k) = u(k-N-1)$$

$$x_{n+2}(k) = u(k-N)$$

$$\vdots$$

$$x_{n+N+1}(k) = u(k-1)$$

The augmented state equation now becomes

$$\begin{bmatrix} \mathbf{x}(k+1) \\ x_{n+1}(k+1) \\ x_{n+2}(k+1) \\ \vdots \\ x_{n+N}(k+1) \\ x_{n+N+1}(k+1) \end{bmatrix} = \begin{bmatrix} \mathbf{F} & \mathbf{g}_1 & \mathbf{g}_2 & 0 & \cdots & 0 \\ \mathbf{0} & 0 & 1 & 0 & \cdots & 0 \\ \mathbf{0} & 0 & 0 & 1 & \cdots & 0 \\ \vdots & \vdots & \vdots & \vdots & & \vdots \\ \mathbf{0} & 0 & 0 & 0 & \cdots & 1 \\ \mathbf{0} & 0 & 0 & 0 & \cdots & 0 \end{bmatrix} \begin{bmatrix} \mathbf{x}(k) \\ x_{n+1}(k) \\ x_{n+2}(k) \\ \vdots \\ x_{n+N}(k) \\ x_{n+N+1}(k) \end{bmatrix} + \begin{bmatrix} \mathbf{0} \\ 0 \\ 0 \\ \vdots \\ 0 \\ 1 \end{bmatrix} u(k) \qquad (6.36)$$

Example 6.2

In the following, we reconsider the tank fluid temperature control system discussed in Example 3.3 (see Fig. 3.17). The differential equation governing the tank fluid temperature was found to be

$$\dot{x}_1(t) = -x_1(t) + u(t-1.5) \qquad (6.37)$$

where

$x_1(t) = \theta(t) =$ tank fluid temperature

$u(t) = \theta_i(t) =$ temperature of the incoming fluid (control temperature)

$\tau_D = 1.5$ sec

Assume that the system is sampled with period $T = 1$ sec. From Eqn. (6.32), we have

$$N = 1, \Delta = 0.5, m = 0.5$$

Equations (6.34b), (6.34d), and (6.34e) give

$$\mathbf{F} = e^{-1} = 0.3679$$

$$g_2 = \int_0^{0.5} e^{-\sigma} d\sigma = 1 - e^{-0.5} = 0.3935$$

$$g_1 = e^{-0.5} \int_0^{0.5} e^{-\theta} d\theta = e^{-0.5} - e^{-1} = 0.2387$$

The discrete-time model of the tank fluid temperature control system becomes (refer Eqn. (6.34a))

$$x_1(k+1) = 0.3679\, x_1(k) + 0.2387\, u(k-2) + 0.3935\, u(k-1) \quad (6.38)$$

Let us introduce two new states defined below.

$$x_2(k) = u(k-2)$$
$$x_3(k) = u(k-1)$$

The augmented state equation becomes

$$\mathbf{x}(k+1) = \mathbf{F}\mathbf{x}(k) + \mathbf{g}u(k)$$
$$y(k) = \mathbf{c}\mathbf{x}(k) \quad (6.39)$$

with

$$\mathbf{F} = \begin{bmatrix} 0.3679 & 0.2387 & 0.3935 \\ 0 & 0 & 1 \\ 0 & 0 & 0 \end{bmatrix}; \mathbf{g} = \begin{bmatrix} 0 \\ 0 \\ 1 \end{bmatrix}$$

$$\mathbf{c} = \begin{bmatrix} 1 & 0 & 0 \end{bmatrix}$$

From Eqns (6.39), the transfer function model is given as follows:

$$G(z) = \frac{Y(z)}{U(z)} = \mathbf{c}(z\mathbf{I} - \mathbf{F})^{-1} \mathbf{g}$$

$$= \begin{bmatrix} 1 & 0 & 0 \end{bmatrix} \begin{bmatrix} z - 0.3679 & -0.2387 & -0.3935 \\ 0 & z & -1 \\ 0 & 0 & z \end{bmatrix}^{-1} \begin{bmatrix} 0 \\ 0 \\ 1 \end{bmatrix}$$

$$= \frac{1}{z^2(z-0.3679)} \begin{bmatrix} 1 & 0 & 0 \end{bmatrix} \begin{bmatrix} z^2 & 0.2387z & 0.2387 + 0.3935z \\ 0 & z(z-0.3679) & z - 0.3679 \\ 0 & 0 & z(z-0.3679) \end{bmatrix} \begin{bmatrix} 0 \\ 0 \\ 1 \end{bmatrix}$$

$$= \frac{0.2387 + 0.3935z}{z^2(z-0.3679)} = \frac{0.3935(z+0.6066)}{z^2(z-0.3679)} \quad (6.40)$$

Note that the same result was obtained earlier in Example 3.3.

6.5 SOLUTION OF STATE DIFFERENCE EQUATIONS

In this section, we investigate the solution of the state equation

$$\mathbf{x}(k+1) = \mathbf{F}\mathbf{x}(k) + \mathbf{g}u(k); \quad \mathbf{x}(0) \triangleq \mathbf{x}^0 \qquad (6.41)$$

where $\mathbf{x}$ is $n \times 1$ state vector, u is a scalar input, $\mathbf{F}$ is $n \times n$ real constant matrix, and $\mathbf{g}$ is $n \times 1$ real constant vector.

In general, discrete-time equations are easier to solve than differential equations because the former can be solved easily by means of a recursion procedure. The recursion procedure is quite simple and convenient for digital computations.

The solution of Eqn. (6.41) for any positive integer k may be obtained directly by recursion as follows.

From $\mathbf{x}(0)$ and $u(0)$, $\mathbf{x}(1)$ can be calculated:

$$\mathbf{x}(1) = \mathbf{F}\mathbf{x}(0) + \mathbf{g}u(0) \qquad (6.42a)$$

Then using $\mathbf{x}(1)$ and $u(1)$:

$$\mathbf{x}(2) = \mathbf{F}\mathbf{x}(1) + \mathbf{g}u(1) \qquad (6.42b)$$

From $\mathbf{x}(2)$ and $u(2)$:

$$\mathbf{x}(3) = \mathbf{F}\mathbf{x}(2) + \mathbf{g}u(2) \qquad (6.42c)$$

$$\vdots$$

From $\mathbf{x}(k-1)$ and $u(k-1)$:

$$\mathbf{x}(k) = \mathbf{F}\mathbf{x}(k-1) + \mathbf{g}u(k-1)$$

Closed-Form Solution

In the following, we obtain the closed-form solution of state equation (6.41).
From Eqns (6.42a)–(6.42b), we obtain

$$\mathbf{x}(2) = \mathbf{F}[\mathbf{F}\mathbf{x}(0) + \mathbf{g}u(0)] + \mathbf{g}u(1)$$
$$= \mathbf{F}^2\mathbf{x}(0) + \mathbf{F}\mathbf{g}u(0) + \mathbf{g}u(1) \qquad (6.43)$$

From Eqns (6.43) and (6.42c), we get

$$\mathbf{x}(3) = \mathbf{F}[\mathbf{F}^2\mathbf{x}(0) + \mathbf{F}\mathbf{g}u(0) + \mathbf{g}u(1)] + \mathbf{g}u(2)$$
$$= \mathbf{F}^3\mathbf{x}(0) + \mathbf{F}^2\mathbf{g}u(0) + \mathbf{F}\mathbf{g}u(1) + \mathbf{g}u(2)$$

By repeating this procedure, we obtain

$$\mathbf{x}(k) = \mathbf{F}^k\mathbf{x}(0) + \mathbf{F}^{k-1}\mathbf{g}u(0) + \mathbf{F}^{k-2}\mathbf{g}u(1) + \cdots + \mathbf{F}^0\mathbf{g}u(k-1); \quad \mathbf{F}^0 = \mathbf{I}$$

$$= \mathbf{F}^k\mathbf{x}(0) + \sum_{i=0}^{k-1} \mathbf{F}^{k-1-i}\mathbf{g}u(i) \qquad (6.44)$$

Clearly $\mathbf{x}(k)$ consists of two parts, one representing the contribution of the initial state $\mathbf{x}(0)$, and the other the contribution of the input $u(i)$; $i = 0, 1, 2, ..., (k-1)$.

State Transition Matrix

Notice that it is possible to write the solution of the homogeneous state equations

$$\mathbf{x}(k+1) = \mathbf{F}\mathbf{x}(k); \quad \mathbf{x}(0) \triangleq \mathbf{x}^0 \tag{6.45a}$$

as
$$\mathbf{x}(k) = \mathbf{F}^k\mathbf{x}(0) \tag{6.45b}$$

From Eqn. (6.45b) it is observed that the initial state $\mathbf{x}(0)$ at $k = 0$ is driven to the state $\mathbf{x}(k)$ at the sampling instant k. This transition in state is carried out by the matrix $\mathbf{F}^k$. Due to this property, $\mathbf{F}^k$ is known as the *state transition matrix*, and is denoted by $\boldsymbol{\phi}(k)$:

$$\boldsymbol{\phi}(k) = \mathbf{F}^k; \quad \boldsymbol{\phi}(0) = \mathbf{I} \text{ (Identity matrix)} \tag{6.46}$$

In the following, we discuss commonly used methods for evaluating state transition matrix in closed form.

Evaluation Using Inverse z-Transforms

Taking the z-transform on both sides of Eqn. (6.45a) yields

$$z\mathbf{X}(z) - z\mathbf{x}(0) = \mathbf{F}\mathbf{X}(z)$$

where $\mathbf{X}(z) \triangleq \mathscr{Z}[\mathbf{x}(k)]$

Solving for $\mathbf{X}(z)$, we get

$$\mathbf{X}(z) = (z\mathbf{I} - \mathbf{F})^{-1}z\mathbf{x}(0)$$

The state vector $\mathbf{x}(k)$ can be obtained by inverse transforming $\mathbf{X}(z)$:

$$\mathbf{x}(k) = \mathscr{Z}^{-1}[(z\mathbf{I} - \mathbf{F})^{-1}z]\,\mathbf{x}(0)$$

Comparing this equation with Eqn. (6.45b), we get

$$\mathbf{F}^k = \boldsymbol{\phi}(k) = \mathscr{Z}^{-1}[(z\mathbf{I} - \mathbf{F})^{-1}z] \tag{6.47}$$

Example 6.3

Consider the matrix $\mathbf{F} = \begin{bmatrix} 0 & 1 \\ -0.16 & -1 \end{bmatrix}$

For this $\mathbf{F}$, $(z\mathbf{I} - \mathbf{F})^{-1} = \begin{bmatrix} z & -1 \\ 0.16 & z+1 \end{bmatrix}^{-1}$

$$= \begin{bmatrix} \dfrac{z+1}{(z+0.2)(z+0.8)} & \dfrac{1}{(z+0.2)(z+0.8)} \\ \dfrac{-0.16}{(z+0.2)(z+0.8)} & \dfrac{z}{(z+0.2)(z+0.8)} \end{bmatrix}$$

$$= \begin{bmatrix} \dfrac{4/3}{z+0.2} + \dfrac{-1/3}{z+0.8} & \dfrac{5/3}{z+0.2} + \dfrac{-5/3}{z+0.8} \\ \dfrac{-0.8/3}{z+0.2} + \dfrac{0.8/3}{z+0.8} & \dfrac{-1/3}{z+0.2} + \dfrac{4/3}{z+0.8} \end{bmatrix}$$

Therefore $\phi(k) = \mathbf{F}^k = \mathscr{Z}^{-1}[(z\mathbf{I} - \mathbf{F})^{-1}z]$

$$= \mathscr{Z}^{-1}\begin{bmatrix} \frac{4}{3}\frac{z}{z+0.2} - \frac{1}{3}\frac{z}{z+0.8} & \frac{5}{3}\frac{z}{z+0.2} - \frac{5}{3}\frac{z}{z+0.8} \\ \frac{-0.8}{3}\frac{z}{z+0.2} + \frac{0.8}{3}\frac{z}{z+0.8} & \frac{-1}{3}\frac{z}{z+0.2} + \frac{4}{3}\frac{z}{z+0.8} \end{bmatrix}$$

$$= \begin{bmatrix} \frac{4}{3}(-0.2)^k - \frac{1}{3}(-0.8)^k & \frac{5}{3}(-0.2)^k - \frac{5}{3}(-0.8)^k \\ \frac{-0.8}{3}(-0.2)^k + \frac{0.8}{3}(-0.8)^k & \frac{-1}{3}(-0.2)^k + \frac{4}{3}(-0.8)^k \end{bmatrix}$$

Evaluation Using Similarity Transformation

Suppose that $\mathbf{F}$ is an $n \times n$ non-diagonal matrix with distinct eigenvalues $\lambda_1, \lambda_2, ..., \lambda_n$. We define the diagonal matrix

$$\mathbf{\Lambda} = \begin{bmatrix} \lambda_1 & 0 & \cdots & 0 \\ 0 & \lambda_2 & \cdots & 0 \\ \vdots & \vdots & & \vdots \\ 0 & 0 & \cdots & \lambda_n \end{bmatrix}$$

$\mathbf{F}$ and $\mathbf{\Lambda}$ are similar matrices; there exists a nonsingular matrix $\mathbf{P}$ such that (refer Eqns (5.22))

$$\mathbf{\Lambda} = \mathbf{P}^{-1}\mathbf{F}\mathbf{P}$$

Now $\quad \mathbf{P}^{-1}\mathbf{F}^k\mathbf{P} = \mathbf{P}^{-1}[\mathbf{F}\mathbf{F} \cdots \mathbf{F}]\mathbf{P}$

$$= \mathbf{P}^{-1}[(\mathbf{P}\mathbf{\Lambda}\mathbf{P}^{-1})(\mathbf{P}\mathbf{\Lambda}\mathbf{P}^{-1}) \cdots (\mathbf{P}\mathbf{\Lambda}\mathbf{P}^{-1})]\mathbf{P} = \mathbf{\Lambda}^k$$

Thus the matrices $\mathbf{F}^k$ and $\mathbf{\Lambda}^k$ are similar. Since $\mathbf{\Lambda}$ is diagonal, $\mathbf{\Lambda}^k$ is given by

$$\mathbf{\Lambda}^k = \begin{bmatrix} \lambda_1^k & 0 & \cdots & 0 \\ 0 & \lambda_2^k & \cdots & 0 \\ \vdots & \vdots & & \vdots \\ 0 & 0 & \cdots & \lambda_n^k \end{bmatrix}$$

The state transition matrix $\mathbf{F}^k$ of matrix $\mathbf{F}$ with distinct eigenvalues $\lambda_1, \lambda_2, ..., \lambda_n$ may therefore be evaluated using the following relation:

$$\mathbf{F}^k = \mathbf{P}\mathbf{\Lambda}^k\mathbf{P}^{-1} = \mathbf{P}\begin{bmatrix} \lambda_1^k & 0 & \cdots & 0 \\ 0 & \lambda_2^k & \cdots & 0 \\ \vdots & \vdots & & \vdots \\ 0 & 0 & \cdots & \lambda_n^k \end{bmatrix}\mathbf{P}^{-1} \qquad (6.48)$$

where $\mathbf{P}$ is a transformation matrix that transforms $\mathbf{F}$ into the diagonal form (For the general case where matrix $\mathbf{F}$ has multiple eigenvalues, refer [102]; also refer Review Example 6.5 given at the end of this chapter).

Example 6.4
Consider the matrix
$$\mathbf{F} = \begin{bmatrix} 0 & 1 \\ -2 & -3 \end{bmatrix}$$

The characteristic equation of the system is
$$|\lambda \mathbf{I} - \mathbf{F}| = \lambda^2 + 3\lambda + 2 = 0$$
which yields $\lambda_1 = -1$ and $\lambda_2 = -2$ as the eigenvalues of $\mathbf{F}$.

Since the matrix $\mathbf{F}$ is in companion form, the eigenvectors[3] and hence the transformation matrix can easily be obtained (refer Problem 5.21).
$$\mathbf{P} = \begin{bmatrix} 1 & 1 \\ \lambda_1 & \lambda_2 \end{bmatrix} = \begin{bmatrix} 1 & 1 \\ -1 & -2 \end{bmatrix}$$

gives the diagonalized matrix
$$\mathbf{\Lambda} = \mathbf{P}^{-1} \mathbf{F} \mathbf{P} = \begin{bmatrix} -1 & 0 \\ 0 & -2 \end{bmatrix}$$

From Eqn. (6.48), we may write
$$\mathbf{F}^k = \mathbf{P} \begin{bmatrix} (-1)^k & 0 \\ 0 & (-2)^k \end{bmatrix} \mathbf{P}^{-1}$$

$$= \begin{bmatrix} 1 & 1 \\ -1 & -2 \end{bmatrix} \begin{bmatrix} (-1)^k & 0 \\ 0 & (-2)^k \end{bmatrix} \begin{bmatrix} 2 & 1 \\ -1 & -1 \end{bmatrix}$$

$$= \begin{bmatrix} 2(-1)^k - (-2)^k & (-1)^k - (-2)^k \\ -2(-1)^k + 2(-2)^k & -(-1)^k + 2(-2)^k \end{bmatrix}$$

Evaluation Using Cayley-Hamilton Technique

The Cayley-Hamilton technique has already been explained in the earlier chapter. We illustrate the use of this technique for evaluation of $\mathbf{F}^k$ by an example.

Example 6.5
Consider the matrix
$$\mathbf{F} = \begin{bmatrix} 0 & 1 \\ -1 & -2 \end{bmatrix}$$

Let us evaluate $f(\mathbf{F}) = \mathbf{F}^k$.
Matrix $\mathbf{F}$ has two eigenvalues at $\lambda_1 = \lambda_2 = -1$.

---

3 Refer Section 5.6 for methods of determination of eigenvectors for a given general matrix $\mathbf{F}$.

Since **F** is of second order, the polynomial $g(\lambda)$ will be of the form (refer Eqns (5.98)):
$$g(\lambda) = \beta_0 + \beta_1 \lambda$$
The coefficients β_0 and β_1 are evaluated from the following equations:
$$f(-1) = (-1)^k = \beta_0 - \beta_1$$
$$\left.\frac{d}{d\lambda} f(\lambda)\right|_{\lambda=-1} = k(-1)^{k-1} = \left.\frac{d}{d\lambda} g(\lambda)\right|_{\lambda=-1} = \beta_1$$

The result is
$$\beta_0 = (1-k)(-1)^k$$
$$\beta_1 = -k(-1)^k$$

Hence
$$f(\mathbf{F}) = \mathbf{F}^k = \beta_0 \mathbf{I} + \beta_1 \mathbf{F}$$
$$= (1-k)(-1)^k \begin{bmatrix} 1 & 0 \\ 0 & 1 \end{bmatrix} - k(-1)^k \begin{bmatrix} 0 & 1 \\ -1 & -2 \end{bmatrix}$$
$$= (-1)^k \begin{bmatrix} (1-k) & -k \\ k & (1+k) \end{bmatrix}$$

State Transition Equation

The solution of the non-homogeneous state difference Eqn. (6.41) is given by Eqn. (6.44). In terms of the state transition matrix $\phi(k)$, Eqn. (6.44) can be written in the form

$$\mathbf{x}(k) = \phi(k) \mathbf{x}(0) + \sum_{i=0}^{k-1} \phi(k-1-i) g u(i) \qquad (6.49)$$

This equation is called the *state transition equation*; it describes the change of state relative to the initial conditions $\mathbf{x}(0)$ and the input $u(k)$.

Example 6.6

Consider the system
$$\begin{bmatrix} x_1(k+1) \\ x_2(k+1) \end{bmatrix} = \begin{bmatrix} 0 & 1 \\ -2 & -3 \end{bmatrix} \begin{bmatrix} x_1(k) \\ x_2(k) \end{bmatrix} + \begin{bmatrix} 0 \\ 1 \end{bmatrix} (-1)^k$$
$$x_1(0) = 1 = x_2(0)$$
$$y(k) = x_1(k)$$

Find $y(k)$ for $k \geq 1$.

Solution

For the given state equation, we have
$$\mathbf{F} = \begin{bmatrix} 0 & 1 \\ -2 & -3 \end{bmatrix}$$

For this **F**, $\phi(k) = \mathbf{F}^k$ was evaluated in Example 6.4:

$$\phi(k) = \mathbf{F}^k = \begin{bmatrix} 2(-1)^k - (-2)^k & (-1)^k - (-2)^k \\ -2(-1)^k + 2(-2)^k & -(-1)^k + 2(-2)^k \end{bmatrix}$$

The state $\quad \mathbf{x}(k) = \mathbf{F}^k \mathbf{x}(0) + \sum_{i=0}^{k-1} \mathbf{F}^{k-1-i} \mathbf{g} u(i)$

With $\quad \mathbf{g} = \begin{bmatrix} 0 \\ 1 \end{bmatrix}, \mathbf{x}(0) = \begin{bmatrix} 1 \\ 1 \end{bmatrix},$

and $\quad u(k) = (-1)^k,$

we get

$$y(k) = x_1(k) = 3(-1)^k - 2(-2)^k + \sum_{i=0}^{k-1} [(-1)^{k-1-i} - (-2)^{k-1-i}](-1)^i$$

$$= 3(-1)^k - 2(-2)^k + k(-1)^{k-1} - (-2)^{k-1} \sum_{i=0}^{k-1} \left(\tfrac{1}{2}\right)^i$$

Since[4]

$$\sum_{i=0}^{k-1} \left(\tfrac{1}{2}\right)^i = \frac{1 - \left(\tfrac{1}{2}\right)^k}{1 - \tfrac{1}{2}} = -2\left[\left(\tfrac{1}{2}\right)^k - 1\right], \text{ we have}$$

$$y(k) = 3(-1)^k - 2(-2)^k - k(-1)^k + (-2)^k \left[1 - \left(\tfrac{1}{2}\right)^k\right]$$

$$= 3(-1)^k - 2(-2)^k - k(-1)^k + (-2)^k - (-1)^k = (2-k)(-1)^k - (-2)^k$$

6.6 CONTROLLABILITY AND OBSERVABILITY

In this section, we study the controllability and observability properties of linear time-invariant systems described by state variable model of the following form:

$$\mathbf{x}(k+1) = \mathbf{F}\mathbf{x}(k) + \mathbf{g}u(k) \quad (6.50a)$$

$$y(k) = \mathbf{c}\mathbf{x}(k) + du(k) \quad (6.50b)$$

where **F**, **g**, **c** and d are respectively $n \times n$, $n \times 1$, $1 \times n$, and 1×1 matrices. **x** is $n \times 1$ state vector, and y and u are respectively output and input variables.

---

4. $\sum_{j=0}^{k} a^j = \dfrac{1 - a^{k+1}}{1 - a}; a \neq 1.$

Controllability

For the linear system given by Eqns (6.50), if there exists an input $u(k)$; $k \in [0, N-1]$ with N a finite positive integer, which transfers the initial state $x(0) \triangleq x^0$ to the state x^1 at $k = N$, the state x^0 is said to be controllable. If all initial states are controllable, the system is said to be *completely controllable* or simplify *controllable*. Otherwise, the system is said to be *uncontrollable*.

▲▲

The following theorem gives a simple controllability test.

Theorem 6.1

The necessary and sufficient condition for the system (6.50) to be completely controllable is that the $n \times n$ controllability matrix,

$$\mathbf{U} \triangleq [\mathbf{g} \quad \mathbf{Fg} \quad \mathbf{F}^2\mathbf{g} \cdots \mathbf{F}^{n-1}\mathbf{g}] \tag{6.51}$$

has rank equal to n, i.e., $\rho(\mathbf{U}) = n$.

Proof

Solution of Eqn. (6.50a) is

$$\mathbf{x}(k) = \mathbf{F}^k \mathbf{x}(0) + \sum_{i=0}^{k-1} \mathbf{F}^{k-1-i} \mathbf{g}\, u(i)$$

Letting $\mathbf{x}(0) \triangleq \mathbf{x}^0$ and $\mathbf{x}(n) \triangleq \mathbf{x}^1$, we obtain

$$\mathbf{x}^1 - \mathbf{F}^n \mathbf{x}^0 = \mathbf{F}^{n-1}\mathbf{g}u(0) + \mathbf{F}^{n-2}\mathbf{g}u(1) + \cdots + \mathbf{g}u(n-1)$$

or

$$\mathbf{x}^1 - \mathbf{F}^n \mathbf{x}^0 = [\mathbf{g} \quad \mathbf{Fg} \cdots \mathbf{F}^{n-1}\mathbf{g}] \begin{bmatrix} u(n-1) \\ u(n-2) \\ \vdots \\ u(0) \end{bmatrix} \tag{6.52}$$

Since $\mathbf{g}$ is an $n \times 1$ matrix, we find that each of the matrices $\mathbf{g}, \mathbf{Fg}, \ldots, \mathbf{F}^{n-1}\mathbf{g}$ is an $n \times 1$ matrix. Therefore

$$\mathbf{U} = [\mathbf{g} \quad \mathbf{Fg} \cdots \mathbf{F}^{n-1}\mathbf{g}]$$

is an $n \times n$ matrix. If the rank of $\mathbf{U}$ is n, then for arbitrary states $\mathbf{x}^0$ and $\mathbf{x}^1$, there exists a sequence of unconstrained control signals $u(0), u(1), \ldots, u(n-1)$ that satisfies Eqn. (6.52). Hence the condition that the rank of the controllability matrix is n gives a sufficient condition for complete controllability.

To prove that the condition $\rho(\mathbf{U}) = n$ is also a necessary condition for complete controllability, we assume that

$$\rho[\mathbf{g} \quad \mathbf{Fg} \cdots \mathbf{F}^{n-1}\mathbf{g}] < n$$

The matrix $\mathbf{U}$ is therefore singular and for arbitrary $\mathbf{x}^0$ and $\mathbf{x}^1$, a solution $\{u(0), u(1), \ldots, u(n-1)\}$ satisfying Eqn. (6.52) does not exist.

456 Digital Control and State Variable Methods

Let us attempt a solution of the form $\{u(0), u(1), ..., u(N-1)\}$; $N > n$. This will amount to adding columns $\mathbf{F}^n\mathbf{g}, \mathbf{F}^{n+1}\mathbf{g}, ..., \mathbf{F}^{N-1}\mathbf{g}$ in the $\mathbf{U}$ matrix. But by Cayley-Hamilton theorem, $f(\mathbf{F}) = \mathbf{F}^j$; $j \geq n$ is a linear combination of $\mathbf{F}^{n-1}, ..., \mathbf{F}^1, \mathbf{F}^0$ (refer Eqn. (5.98d)) and therefore columns $\mathbf{F}^n\mathbf{g}, \mathbf{F}^{n+1}\mathbf{g}, ..., \mathbf{F}^{N-1}\mathbf{g}$ add no new rank. Thus if a state cannot be transferred to some other state in n sampling intervals, no matter how long the input sequence $\{u(0), u(1), ..., u(N-1)\}$; $N > n$ is, it still cannot be achieved. Consequently we find that the rank condition given by Eqn. (6.51) is necessary and sufficient condition for complete controllability.

Observability

For the linear system given by Eqns (6.50), if the knowledge of the input $u(k)$; $k \in [0, N-1]$ and the output $y(k)$; $k \in [0, N-1]$ with N a finite positive integer, suffices to determine the state $\mathbf{x}(0) \triangleq \mathbf{x}^0$, the state $\mathbf{x}^0$ is said to be observable. If all initial conditions are observable, the system is said to be *completely observable*, or simply *observable*. Otherwise, the system is said to be *unobservable*.

▲▲

The following theorem gives a simple *observability test*.

Theorem 6.2

The necessary and sufficient condition for the system (6.50) to be completely observable is that the $n \times n$ *observability matrix*

$$\mathbf{V} = \begin{bmatrix} \mathbf{c} \\ \mathbf{cF} \\ \mathbf{cF}^2 \\ \vdots \\ \mathbf{cF}^{n-1} \end{bmatrix} \quad (6.53)$$

has rank equal to n, i.e., $\rho(\mathbf{V}) = n$.

Proof

The solution of Eqns (6.50) is

$$y(k) = \mathbf{cF}^k\mathbf{x}(0) + \left[\sum_{i=0}^{k-1} \mathbf{cF}^{k-1-i}\mathbf{g}u(i)\right] + du(k)$$

This gives
$$y(0) = \mathbf{cx}(0) + du(0)$$
$$y(1) = \mathbf{cFx}(0) + \mathbf{cg}u(0) + du(1)$$
$$\vdots$$
$$y(n-1) = \mathbf{cF}^{n-1}\mathbf{x}(0) + \mathbf{cF}^{n-2}\mathbf{g}u(0) + \mathbf{cF}^{n-3}\mathbf{g}u(1)$$
$$+ \cdots + \mathbf{cg}u(n-2) + du(n-1)$$

From these equations, we may write

$$\begin{bmatrix} y(0) - du(0) \\ y(1) - \mathbf{cg}u(0) - du(1) \\ \vdots \\ y(n-1) - \mathbf{cF}^{n-2}\mathbf{g}u(0) - \mathbf{cF}^{n-3}\mathbf{g}u(1) - \cdots - \mathbf{cg}u(n-2) - du(n-1) \end{bmatrix}$$

$$= \begin{bmatrix} \mathbf{c} \\ \mathbf{cF} \\ \mathbf{cF}^2 \\ \vdots \\ \mathbf{cF}^{n-1} \end{bmatrix} \mathbf{x}^0 = \underset{n \times n}{\mathbf{V}} \underset{n \times 1}{\mathbf{x}^0} \qquad (6.54)$$

If the rank of $\mathbf{V}$ is n, then there exists a unique solution $\mathbf{x}^0$ of Eqn. (6.54). Hence the condition that the rank of the observability matrix is n gives a sufficient condition for complete observability.

It can easily be proved (refer proof of Theorem 6.1) that the condition $\rho(\mathbf{V}) = n$ is also a necessary condition for complete observability.

Controllability and Observability of State Variable Model in Jordan Canonical Form

The following result for discrete-time systems easily follows from the corresponding result for continuous-time systems given in the earlier chapter.

Consider a SISO system with distinct eigenvalues[5] $\lambda_1, \lambda_2, ..., \lambda_n$.
The Jordan canonical state model of the system is of the form

$$\mathbf{x}(k+1) = \mathbf{\Lambda}\mathbf{x}(k) + \mathbf{g}u(k) \qquad (6.55)$$
$$y(k) = \mathbf{c}\mathbf{x}(k) + du(k)$$

with $\mathbf{\Lambda} = \begin{bmatrix} \lambda_1 & 0 & \cdots & 0 \\ 0 & \lambda_2 & \cdots & 0 \\ \vdots & \vdots & & \vdots \\ 0 & 0 & \cdots & \lambda_n \end{bmatrix}; \mathbf{g} = \begin{bmatrix} g_1 \\ g_2 \\ \vdots \\ g_n \end{bmatrix}; \mathbf{c} = [c_1 \; c_2 \; \cdots \; c_n]$

The system (6.55) is completely controllable if and only if none of the elements of the column matrix $\mathbf{g}$ is zero. The system (6.55) is completely observable if and only if none of the elements of the row matrix $\mathbf{c}$ is zero.

---

5. Refer Gopal [102] for the case of multiple eigenvalues.

Equivalence Between Transfer Function and State Variable Representations

The following result for discrete-time systems easily follows from the corresponding result for continuous-time systems given in the earlier chapter.

The general state variable model of nth-order linear time-invariant discrete-time system is given by Eqns (6.50):

$$\mathbf{x}(k+1) = \mathbf{F}\mathbf{x}(k) + \mathbf{g}u(k); \ \mathbf{x}(0) \triangleq \mathbf{x}^0$$
$$y(k) = \mathbf{c}\mathbf{x}(k) + du(k) \tag{6.56}$$

The corresponding transfer function model is

$$\frac{Y(z)}{U(z)} = \frac{\mathbf{c}(z\mathbf{I} - \mathbf{F})^+ \mathbf{g} + d\,|z\mathbf{I} - \mathbf{F}|}{|z\mathbf{I} - \mathbf{F}|} \tag{6.57}$$

The uncontrollable and unobservable modes of the state variable model (6.56) do not show up in the corresponding transfer function representation (6.57); the poles of the transfer function are therefore a subset of the eigenvalues of matrix $\mathbf{F}$, and the asymptotic stability of the system always implies bounded-input, bounded-output (BIBO) stability. The reverse however may not be true because the eigenvalues of uncontrollable and/or unobservable parts of the system are hidden from the BIBO stability analysis. When the state variable model (6.56) is both controllable and observable, all the eigenvalues of $\mathbf{F}$ appear as poles in the transfer function (6.57), and therefore BIBO stability implies asymptotic stability only for controllable and observable systems.

Conclusion

The transfer function model of a system represents its complete dynamics only if the system is both controllable and observable.

Loss of Controllability and Observability due to Sampling

Sampling of a continuous-time system gives a discrete-time system with system matrices that depend on the sampling period. How will that influence the controllability and observability of the sampled system? To get a controllable sampled system, it is necessary that the continuous-time system also be controllable, because the allowable control signals for the sampled system—piecewise constant signals—are a subset of allowable control signals for the continuous-time system. However, it may happen that the controllability is lost for some sampling periods.

The conditions for unobservability are more restricted in the continuous-time case because the output has to be zero over a time interval, while the sampled system output has to be zero only at the sampling instants. This means that the continuous output may oscillate between the sampling times and be zero at the sampling instants. This condition is sometimes called *hidden oscillations*. The sampled system can thus be unobservable even if the corresponding continuous-time system is observable.

The harmonic oscillator can be used to illustrate the preceding discussion. The transfer function model of the oscillator system is

$$\frac{Y(s)}{U(s)} = \frac{\omega^2}{s^2 + \omega^2} \tag{6.58}$$

From this model, we have

$$\ddot{y} + \omega^2 y = \omega^2 u$$

Define

$$x_1 = y$$

$$x_2 = \frac{1}{\omega} \dot{y}$$

This gives the following state variable representation of the oscillator system:

$$\begin{bmatrix} \dot{x}_1 \\ \dot{x}_2 \end{bmatrix} = \begin{bmatrix} 0 & \omega \\ -\omega & 0 \end{bmatrix} \begin{bmatrix} x_1 \\ x_2 \end{bmatrix} + \begin{bmatrix} 0 \\ \omega \end{bmatrix} u$$

$$y = \begin{bmatrix} 1 & 0 \end{bmatrix} \begin{bmatrix} x_1 \\ x_2 \end{bmatrix} \tag{6.59}$$

The discrete-time state variable representation of the system is obtained as follows. Noting that

$$\mathbf{A} = \begin{bmatrix} 0 & \omega \\ -\omega & 0 \end{bmatrix}, \mathbf{b} = \begin{bmatrix} 0 \\ \omega \end{bmatrix}$$

we have

$$\mathbf{F} = e^{\mathbf{A}T} = \mathcal{L}^{-1}\left[(s\mathbf{I} - \mathbf{A})^{-1}\right]\big|_{t=T} = \mathcal{L}^{-1}\left(\begin{bmatrix} s & -\omega \\ \omega & s \end{bmatrix}^{-1}\right)\bigg|_{t=T}$$

$$= \mathcal{L}^{-1}\begin{bmatrix} \frac{s}{s^2+\omega^2} & \frac{\omega}{s^2+\omega^2} \\ \frac{-\omega}{s^2+\omega^2} & \frac{s}{s^2+\omega^2} \end{bmatrix}\bigg|_{t=T} = \begin{bmatrix} \cos\omega T & \sin\omega T \\ -\sin\omega T & \cos\omega T \end{bmatrix}$$

and

$$\mathbf{g} = \left[\int_0^T e^{\mathbf{A}\theta}\,d\theta\right]\mathbf{b} = \left[\int_0^T \begin{pmatrix} \cos\omega\theta & \sin\omega\theta \\ -\sin\omega\theta & \cos\omega\theta \end{pmatrix} d\theta\right]\begin{bmatrix} 0 \\ \omega \end{bmatrix} = \begin{bmatrix} 1-\cos\omega T \\ \sin\omega T \end{bmatrix}$$

Hence, the discrete-time state variable representation of the oscillator system becomes

$$\begin{bmatrix} x_1(k+1) \\ x_2(k+1) \end{bmatrix} = \begin{bmatrix} \cos\omega T & \sin\omega T \\ -\sin\omega T & \cos\omega T \end{bmatrix} \begin{bmatrix} x_1(k) \\ x_2(k) \end{bmatrix} + \begin{bmatrix} 1-\cos\omega T \\ \sin\omega T \end{bmatrix} u(k)$$

(6.60)

$$y(k) = \begin{bmatrix} 1 & 0 \end{bmatrix} \begin{bmatrix} x_1(k) \\ x_2(k) \end{bmatrix}$$

The determinants of the controllability and observability matrices are

$$|\mathbf{U}| = |[\mathbf{g} \quad \mathbf{Fg}]| = -2\sin\omega T(1-\cos\omega T)$$

$$|\mathbf{V}| = \left| \begin{bmatrix} \mathbf{c} \\ \mathbf{cF} \end{bmatrix} \right| = \sin\omega T$$

Both controllability and observability are lost for $\omega T = n\pi$, $n = 1, 2, ...$ (i.e., when the sampling interval is half the period of oscillation of the harmonic oscillator or an integer multiple of that period), although the corresponding continuous-time system given by Eqns (6.59) is both controllable and observable.

Loss of controllability and/or observability due to sampling occurs only when the continuous-time system has oscillatory modes and the sampling interval is half the period of oscillation of an oscillatory mode or an integer multiple of that period. This implies that controllability and observability properties of a continuous-time system are preserved after introduction of sampling if, and only if, for every eigenvalue of the characteristic equation, the relation

$$\text{Re } \lambda_i = \text{Re } \lambda_j$$

(6.61)

implies $\quad\text{Im } (\lambda_i - \lambda_j) \neq \dfrac{2n\pi}{T}$

where T is the sampling period and $n = \pm 1, \pm 2, ...$.

We know that controllability and/or observability is lost when the transfer function corresponding to a state model has common poles and zeros. The poles and zeros are functions of sampling interval. This implies that if the choice of sampling interval does not satisfy the condition given by (6.61), pole-zero cancellation will occur in passing from the continuous-time to the discrete-time case; the pole-zero cancellation will not take place if the continuous-time system does not contain complex poles.

It is very unlikely that the sampling interval chosen for a plant control system would be precisely one resulting in loss of controllability and/or observability. In fact the rules of thumb for the choice of sampling interval given in Section 2.11 imply a sampling interval of about one-tenth of the period of oscillation of an oscillatory mode, and not just half.

6.7 MULTIVARIABLE SYSTEMS

The state variable model of the multi-input, multi-output (MIMO) system takes the following form (refer Eqns (2.14)–(2.15)):

$$\mathbf{x}(k+1) = \mathbf{F}\mathbf{x}(k) + \mathbf{G}\mathbf{u}(k); \; \mathbf{x}(0) \triangleq \mathbf{x}^0 \quad (6.62a)$$

$$\mathbf{y}(k) = \mathbf{C}\mathbf{x}(k) + \mathbf{D}\mathbf{u}(k) \quad (6.62b)$$

F, **G**, **C**, and **D** are respectively $n \times n$, $n \times p$, $q \times n$ and $q \times p$ constant matrices, **x** is $n \times 1$ state vector, **u** is $p \times 1$ input vector, and **y** is $q \times 1$ output vector.

Many of the analysis results developed in earlier sections of this chapter for SISO systems have obvious extensions for the system (6.62).

The solution of the state Eqn. (6.62a) is given by (refer Eqn. (6.44))

$$\mathbf{x}(k) = \mathbf{F}^k \mathbf{x}(0) + \sum_{i=0}^{k-1} \mathbf{F}^{k-1-i} \mathbf{G}\mathbf{u}(i) \quad (6.63a)$$

The output

$$\mathbf{y}(k) = \mathbf{C}\left[\mathbf{F}^k \mathbf{x}(0) + \sum_{i=0}^{k-1} \mathbf{F}^{k-1-i} \mathbf{G}\mathbf{u}(i)\right] + \mathbf{D}\mathbf{u}(k) \quad (6.63b)$$

▲▲

In the transform domain, the input-output behaviour of the system (6.62) is determined entirely by the *transfer function matrix* (refer Eqns (6.3))

$$\mathbf{G}(z) = \mathbf{C}(z\mathbf{I} - \mathbf{F})^{-1}\mathbf{G} + \mathbf{D} \quad (6.64a)$$

The output $\quad \underset{q \times 1}{\mathbf{Y}(z)} = \underset{q \times p}{\mathbf{G}(z)} \; \underset{p \times 1}{\mathbf{U}(z)} \quad (6.64b)$

▲▲

The necessary and sufficient condition for the system (6.62) to be completely controllable is that the $n \times np$ matrix

$$\mathbf{U} \triangleq [\mathbf{G} \quad \mathbf{FG} \quad \mathbf{F}^2\mathbf{G} \cdots \mathbf{F}^{n-1}\mathbf{G}] \quad (6.65)$$

has rank equal to n.

The necessary and sufficient condition for the system (6.62) to be completely observable is that the $nq \times n$ matrix

$$\mathbf{V} \triangleq \begin{bmatrix} \mathbf{C} \\ \mathbf{CF} \\ \vdots \\ \mathbf{CF}^{n-1} \end{bmatrix} \quad (6.66)$$

has rank equal to n.

A MIMO system with distinct eigenvalues[6] $\lambda_1, \lambda_2, ..., \lambda_n$ has the following Jordan canonical state model.

---

6. Refer Gopal [102] for the case of multiple eigenvalues.

$$\mathbf{x}(k+1) = \mathbf{\Lambda x}(k) + \mathbf{Gu}(k)$$
$$\mathbf{y}(k) = \mathbf{Cx}(k) + \mathbf{Du}(k) \tag{6.67}$$

with

$$\mathbf{\Lambda} = \begin{bmatrix} \lambda_1 & 0 & \cdots & 0 \\ 0 & \lambda_2 & \cdots & 0 \\ \vdots & \vdots & & \vdots \\ 0 & 0 & \cdots & \lambda_n \end{bmatrix}; \mathbf{G} = \begin{bmatrix} g_{11} & g_{12} & \cdots & g_{1p} \\ g_{21} & g_{22} & \cdots & g_{2p} \\ \vdots & \vdots & & \vdots \\ g_{n1} & g_{n2} & \cdots & g_{np} \end{bmatrix}$$

$$\mathbf{C} = \begin{bmatrix} c_{11} & c_{12} & \cdots & c_{1n} \\ \vdots & \vdots & & \vdots \\ c_{q1} & c_{q2} & \cdots & c_{qn} \end{bmatrix}$$

The system (6.67) is completely controllable if and only if none of the rows of $\mathbf{G}$ matrix is a zero row, and (6.67) is completely observable if and only if none of the columns of $\mathbf{C}$ matrix is a zero column.

Example 6.7

The scheme of Fig. 5.23 (refer Example 5.21) describes a simple concentration control process. Mathematical model of the plant, given by Eqns (5.133), is reproduced below:

$$\dot{\mathbf{x}} = \mathbf{Ax} + \mathbf{Bu}$$
$$\mathbf{y} = \mathbf{Cx} \tag{6.68}$$

with $\mathbf{A} = \begin{bmatrix} -0.01 & 0 \\ 0 & -0.02 \end{bmatrix}; \mathbf{B} = \begin{bmatrix} 1 & 1 \\ -0.004 & -0.002 \end{bmatrix}; \mathbf{C} = \begin{bmatrix} 0.01 & 0 \\ 0 & 1 \end{bmatrix}$

The state, input, and output variables are deviations from steady-state values:

x_1 = incremental volume of fluid in the tank (litres)
x_2 = incremental outgoing concentration (g-moles/litre)
u_1 = incremental feed 1 (litres/sec)
u_2 = incremental feed 2 (litres/sec)
y_1 = incremental outflow (litres/sec)
y_2 = incremental outgoing concentration (g-moles/litre)

Matrix $\mathbf{A}$ in Eqns (6.68) is in diagonal form; none of the rows of $\mathbf{B}$ matrix is a zero row, and none of the columns of $\mathbf{C}$ matrix is a zero column. The state model (6.68) is therefore completely controllable and observable.

With initial values of x_1 and x_2 equal to zero at $t = 0$, a step of 2 litres/sec in feed 1 results in

$$\mathbf{y}(t) = \mathbf{C}\left[\int_0^t e^{\mathbf{A}(t-\tau)}\mathbf{Bu}(\tau)d\tau\right]$$

with
$$e^{At} = \begin{bmatrix} e^{-0.01t} & 0 \\ 0 & e^{-0.02t} \end{bmatrix}$$

and
$$\mathbf{u}(\tau) = \begin{bmatrix} 2 \\ 0 \end{bmatrix}$$

Solving for $\mathbf{y}(t)$, we get

$$\mathbf{y}(t) = \mathbf{C} \begin{bmatrix} \int_0^t 2e^{-0.01(t-\tau)} d\tau \\ -\int_0^t 0.008 e^{-0.02(t-\tau)} d\tau \end{bmatrix} = \mathbf{C} \begin{bmatrix} \frac{2}{0.01}(1-e^{-0.01t}) \\ -0.4(1-e^{-0.02t}) \end{bmatrix}$$

Therefore
$$y_1(t) = 2(1 - e^{-0.01t}) \tag{6.69a}$$
$$y_2(t) = -0.4(1 - e^{-0.02t}) \tag{6.69b}$$

Suppose that the plant (6.68) forms part of a process commanded by a process control computer. As a result, the valve settings change at discrete instants only and remain constant in between. Assuming that these instants are separated by time period $T = 5$ sec, we derive the discrete-time description of the plant.

$$\mathbf{x}(k+1) = \mathbf{F}\mathbf{x}(k) + \mathbf{G}u(k) \tag{6.70a}$$
$$\mathbf{y}(k) = \mathbf{C}\mathbf{x}(k) \tag{6.70b}$$

$$\mathbf{F} = e^{AT} = \begin{bmatrix} e^{-0.01T} & 0 \\ 0 & e^{-0.02T} \end{bmatrix} = \begin{bmatrix} 0.9512 & 0 \\ 0 & 0.9048 \end{bmatrix} \tag{6.70c}$$

$$\mathbf{G} = \int_0^T e^{A\theta} \mathbf{B} d\theta = \begin{bmatrix} \int_0^T e^{-0.01\theta} d\theta & \int_0^T e^{-0.01\theta} d\theta \\ -0.004 \int_0^T e^{-0.02\theta} d\theta & 0.002 \int_0^T e^{-0.02\theta} d\theta \end{bmatrix}$$

$$= \begin{bmatrix} 4.88 & 4.88 \\ -0.019 & 0.0095 \end{bmatrix} \tag{6.70d}$$

Matrix $\mathbf{F}$ in Eqns (6.70) is in diagonal form; none of the rows of $\mathbf{G}$ matrix is a zero row, and none of the columns of $\mathbf{C}$ matrix is a zero column. The state model (6.70) is therefore completely controllable and observable.

With initial values of x_1 and x_2 equal to zero at $k = 0$, a step of 2 litres/sec in feed 1 results in

$$\mathbf{y}(k) = \mathbf{C} \left[\sum_{i=0}^{k-1} \mathbf{F}^{k-1-i} \mathbf{G}u(i) \right]$$

with
$$\mathbf{F}^k = \begin{bmatrix} (0.9512)^k & 0 \\ 0 & (0.9048)^k \end{bmatrix}$$

and
$$\mathbf{u}(i) = \begin{bmatrix} 2 \\ 0 \end{bmatrix}$$

Solving for y(k), we get
$$y_1(k) = 0.01 x_1(k)$$
$$= 0.01 \sum_{i=0}^{k-1} 9.76(0.9512)^{k-1-i}$$

Since (refer footnote 4)

$$\sum_{i=0}^{k-1} \left(\frac{1}{0.9512}\right)^i = \frac{1-\left(\frac{1}{0.9512}\right)^k}{1-\frac{1}{0.9512}} = \frac{0.9512}{-0.0488}[1-(0.9512)^{-k}]$$

we have
$$y_1(k) = 2[1-(0.9512)^k] \qquad (6.71a)$$

$$y_2(k) = x_2(k) = -0.038 \sum_{i=0}^{k-1} (0.9048)^{k-1-i}$$
$$= -0.4[1-(0.9048)^k] \qquad (6.71b)$$

Comparison of $y_1(k)$ and $y_2(k)$, with $y_1(t)$ and $y_2(t)$ shows that the two sets of responses match exactly at the sampling instants.

6.8 REVIEW EXAMPLES

Review Example 6.1

Give three different canonical state variable models corresponding to the transfer function

$$G(z) = \frac{4z^3 - 12z^2 + 13z - 7}{(z-1)^2(z-2)}$$

Solution

The given transfer function is

$$G(z) = \frac{Y(z)}{U(z)} = \frac{4z^3 - 12z^2 + 13z - 7}{z^3 - 4z^2 + 5z - 2} = \frac{\beta_0 z^3 + \beta_1 z^2 + \beta_2 z + \beta_3}{z^3 + \alpha_1 z^2 + \alpha_2 z + \alpha_3}$$

The controllable canonical state model (first companion form) follows directly from Eqns (6.5) and (6.7):

$$\begin{bmatrix} x_1(k+1) \\ x_2(k+1) \\ x_3(k+1) \end{bmatrix} = \begin{bmatrix} 0 & 1 & 0 \\ 0 & 0 & 1 \\ 2 & -5 & 4 \end{bmatrix} \begin{bmatrix} x_1(k) \\ x_2(k) \\ x_3(k) \end{bmatrix} + \begin{bmatrix} 0 \\ 0 \\ 1 \end{bmatrix} u(k)$$

$$y(k) = \begin{bmatrix} 1 & -7 & 4 \end{bmatrix} \begin{bmatrix} x_1(k) \\ x_2(k) \\ x_3(k) \end{bmatrix} + 4u(k)$$

The observable canonical state model (second companion form) follows directly from Eqns (6.5) and (6.8):

$$\begin{bmatrix} \bar{x}_1(k+1) \\ \bar{x}_2(k+1) \\ \bar{x}_3(k+1) \end{bmatrix} = \begin{bmatrix} 0 & 0 & 2 \\ 1 & 0 & -5 \\ 0 & 1 & 4 \end{bmatrix} \begin{bmatrix} \bar{x}_1(k) \\ \bar{x}_2(k) \\ \bar{x}_3(k) \end{bmatrix} + \begin{bmatrix} 1 \\ -7 \\ 4 \end{bmatrix} u(k)$$

$$y(k) = \begin{bmatrix} 0 & 0 & 1 \end{bmatrix} \begin{bmatrix} \bar{x}_1(k) \\ \bar{x}_2(k) \\ \bar{x}_3(k) \end{bmatrix} + 4u(k)$$

The state variable model in Jordan canonical form follows from Eqns (6.11) and (6.12):

$$\frac{Y(z)}{U(z)} = G(z) = \frac{4z^3 - 12z^2 + 13z - 7}{z^3 - 4z^2 + 5z - 2} = 4 + \frac{4z^2 - 7z + 1}{(z-1)^2(z-2)}$$

$$= 4 + \frac{2}{(z-1)^2} + \frac{1}{(z-1)} + \frac{3}{(z-2)}$$

$$\begin{bmatrix} \hat{x}_1(k+1) \\ \hat{x}_2(k+1) \\ \hat{x}_3(k+1) \end{bmatrix} = \begin{bmatrix} 1 & 1 & 0 \\ 0 & 1 & 0 \\ 0 & 0 & 2 \end{bmatrix} \begin{bmatrix} \hat{x}_1(k) \\ \hat{x}_2(k) \\ \hat{x}_3(k) \end{bmatrix} + \begin{bmatrix} 0 \\ 1 \\ 1 \end{bmatrix} u(k)$$

$$y(k) = \begin{bmatrix} 2 & 1 & 3 \end{bmatrix} \begin{bmatrix} \hat{x}_1(k) \\ \hat{x}_2(k) \\ \hat{x}_3(k) \end{bmatrix} + 4u(k)$$

Review Example 6.2

Prove that a discrete-time system obtained by zero-order-hold sampling of an asymptotically stable continuous-time system is also asymptotically stable.

Solution

Consider an asymptotically stable continuous-time system

$$\dot{\mathbf{x}}(t) = \mathbf{A}\mathbf{x}(t) \tag{6.72}$$

We assume for simplicity that the eigenvalues $\lambda_1, \lambda_2, ..., \lambda_n$ of matrix $\mathbf{A}$ are all distinct. Let $\mathbf{P}$ be a transformation matrix such that

$$\mathbf{P}^{-1}\mathbf{A}\mathbf{P} = \mathbf{\Lambda} = \begin{bmatrix} \lambda_1 & 0 & \cdots & 0 \\ 0 & \lambda_2 & \cdots & 0 \\ \vdots & \vdots & & \vdots \\ 0 & 0 & \cdots & \lambda_n \end{bmatrix}$$

This gives (refer Eqn. (5.93))

$$\mathbf{P}^{-1}e^{\mathbf{A}t}\mathbf{P} = e^{\mathbf{\Lambda}t}$$

$$= \begin{bmatrix} e^{\lambda_1 t} & 0 & \cdots & 0 \\ 0 & e^{\lambda_2 t} & \cdots & 0 \\ \vdots & \vdots & & \vdots \\ 0 & 0 & \cdots & e^{\lambda_n t} \end{bmatrix}$$

The zero-order-hold sampling of the continuous-time system (6.72) results in a discrete-time system

$$\mathbf{x}(k+1) = \mathbf{F}\mathbf{x}(k)$$

where
$$\mathbf{F} = e^{\mathbf{A}T}; \; T = \text{sampling interval}$$

The characteristic polynomial of the system is

$$|z\mathbf{I} - \mathbf{F}| = |z\mathbf{I} - e^{\mathbf{A}T}| = |\mathbf{P}^{-1}| \, |z\mathbf{I} - e^{\mathbf{A}T}| \, |\mathbf{P}|$$
$$= |z\mathbf{P}^{-1}\mathbf{P} - \mathbf{P}^{-1}e^{\mathbf{A}T}\mathbf{P}| = |z\mathbf{I} - e^{\mathbf{\Lambda}T}|$$
$$= (z - e^{\lambda_1 T})(z - e^{\lambda_2 T}) \cdots (z - e^{\lambda_n T})$$

Notice that the eigenvalues of $\mathbf{F}$ are given by $z_i = e^{\lambda_i T}$; $i = 1, 2, ..., n$. We see the equivalence of $\text{Re}\,\lambda_i < 0$ and $|z_i| < 1$. Thus, the discrete-time system obtained by zero-order-hold sampling of an asymptotically stable continuous-time system is also asymptotically stable.

The proof for the case where matrix $\mathbf{A}$ has multiple eigenvalues follows on identical lines.

Review Example 6.3

Consider a unity feedback system with the plant

$$\dot{\mathbf{x}} = \mathbf{A}\mathbf{x} + \mathbf{b}u$$
$$y = \mathbf{c}\mathbf{x}$$

where
$$\mathbf{A} = \begin{bmatrix} 0 & 1 \\ 0 & -2 \end{bmatrix}; \; \mathbf{b} = \begin{bmatrix} 0 \\ K \end{bmatrix}; \; \mathbf{c} = \begin{bmatrix} 1 & 0 \end{bmatrix}$$

(a) Find the range of values of K for which the closed-loop system is stable.
(b) Introduce now a sampler and zero-order hold in the forward path of the closed-loop system. Show that sampling has a destabilizing effect on the stability of the closed-loop system. To establish this result, you may find the range of values of K for which the closed-loop digital system is stable when (i) $T = 0.4$ sec, and (ii) $T = 3$ sec; and then compare with that obtained in (a) above.

Solution

Consider the feedback system of Fig. 6.12a. Substituting
$$u = r - y = r - x_1$$
in the plant model, we get the following state variable description of the closed-loop system:
$$\begin{bmatrix} \dot{x}_1 \\ \dot{x}_2 \end{bmatrix} = \begin{bmatrix} 0 & 1 \\ -K & -2 \end{bmatrix} \begin{bmatrix} x_1 \\ x_2 \end{bmatrix} + \begin{bmatrix} 0 \\ K \end{bmatrix} r$$

The characteristic equation of the closed-loop system is
$$\lambda^2 + 2\lambda + K = 0$$
The closed-loop system is stable for all values of $K > 0$.

Figure 6.12b shows a block diagram of the closed-loop digital system. The discrete-time description of the plant is obtained as follows:
$$\mathbf{x}(k+1) = \mathbf{F}\mathbf{x}(k) + \mathbf{g}u(k)$$
$$y(k) = \mathbf{c}\mathbf{x}(k)$$

where
$$\mathbf{F} = e^{\mathbf{A}T}; \text{ and } \mathbf{g} = \int_0^T e^{\mathbf{A}\theta}\mathbf{b}\,d\theta$$

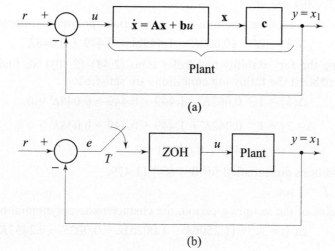

Fig. 6.12

Now
$$e^{At} = \mathscr{L}^{-1}[(s\mathbf{I}-\mathbf{A})^{-1}] = \mathscr{L}^{-1}\left(\begin{bmatrix} s & -1 \\ 0 & s+2 \end{bmatrix}^{-1}\right)$$

$$= \mathscr{L}^{-1}\begin{bmatrix} \dfrac{1}{s} & \dfrac{1}{s(s+2)} \\ 0 & \dfrac{1}{s+2} \end{bmatrix} = \begin{bmatrix} 1 & \tfrac{1}{2}(1-e^{-2t}) \\ 0 & e^{-2t} \end{bmatrix}$$

Therefore
$$\mathbf{F} = e^{AT} = \begin{bmatrix} 1 & \tfrac{1}{2}(1-e^{-2T}) \\ 0 & e^{-2T} \end{bmatrix}$$

$$\mathbf{g} = \int_0^T \begin{bmatrix} \dfrac{K}{2}(1-e^{-2\theta}) \\ Ke^{-2\theta} \end{bmatrix} d\theta = \tfrac{1}{2}K \begin{bmatrix} T-\tfrac{1}{2}+\tfrac{1}{2}e^{-2T} \\ 1-e^{-2T} \end{bmatrix}$$

The state variable description of the closed-loop digital system becomes

$$\mathbf{x}(k+1) = \begin{bmatrix} 1-\tfrac{1}{2}K(T-\tfrac{1}{2}+\tfrac{1}{2}e^{-2T}) & \tfrac{1}{2}(1-e^{-2T}) \\ -\tfrac{1}{2}K(1-e^{-2T}) & e^{-2T} \end{bmatrix} \mathbf{x}(k)$$

$$+ \tfrac{1}{2}K \begin{bmatrix} T-\tfrac{1}{2}+\tfrac{1}{2}e^{-2T} \\ 1-e^{-2T} \end{bmatrix} r(k)$$

The characteristic equation is given by

$$\lambda^2 + [-(1+e^{-2T}) + \tfrac{1}{2}K(T-\tfrac{1}{2}+\tfrac{1}{2}e^{-2T})]\lambda$$
$$+ e^{-2T} + \tfrac{1}{2}K(\tfrac{1}{2} - \tfrac{1}{2}e^{-2T} - Te^{-2T}) = 0$$

Case I: $T = 0.4$ sec

For this value of sampling period, the characteristic polynomial becomes

$$\Delta(\lambda) = \lambda^2 + (0.062K - 1.449)\lambda + 0.449 + 0.048K$$

Applying the Jury stability test (refer Eqns (2.48)–(2.50)) we find that the system is stable if the following conditions are satisfied:

$$\Delta(1) = 1 + 0.062K - 1.449 + 0.449 + 0.048K > 0$$

$$\Delta(-1) = 1 - 0.062K + 1.449 + 0.449 + 0.048K > 0$$

$$|0.449 + 0.048K| < 1$$

These conditions are satisfied for $0 < K < 11.479$.

Case II: $T = 3$ sec

For this value of the sampling period, the characteristic polynomial becomes

$$\Delta(\lambda) = \lambda^2 + (1.2506K - 1.0025)\lambda + 0.0025 + 0.2457K$$

The system is found to be stable for $0 < K < 1.995$.

Thus the system which is stable for all $K > 0$ when $T = 0$ (continuous-time system) becomes unstable for $K > 11.479$, when $T = 0.4$ sec. When T is increased to 3 sec, it becomes unstable for $K > 1.995$. It means that increasing the sampling period (or decreasing the sampling rate) reduces the margin of stability.

Review Example 6.4

A closed-loop computer control system is shown in Fig. 6.13. The digital controller is described by the difference equation

$$e_2(k+1) + ae_2(k) = be_1(k)$$

The state variable model of the plant is given below:

$$\dot{\mathbf{x}} = \mathbf{A}\mathbf{x} + \mathbf{b}u$$
$$y = \mathbf{c}\mathbf{x}$$

with
$$\mathbf{A} = \begin{bmatrix} 0 & 1 \\ 0 & -1 \end{bmatrix}; \mathbf{b} = \begin{bmatrix} 0 \\ 1 \end{bmatrix}; \mathbf{c} = \begin{bmatrix} 1 & 0 \end{bmatrix}$$

Obtain discrete-time state description for the closed-loop system.

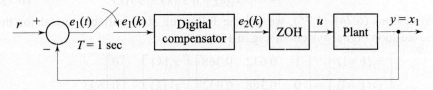

Fig. 6.13

Solution
Given
$$\mathbf{A} = \begin{bmatrix} 0 & 1 \\ 0 & -1 \end{bmatrix}$$

$$e^{\mathbf{A}t} = \mathscr{L}^{-1}[(s\mathbf{I} - \mathbf{A})^{-1}] = \mathscr{L}^{-1} \begin{bmatrix} \dfrac{1}{s} & \dfrac{1}{s(s+1)} \\ 0 & \dfrac{1}{s+1} \end{bmatrix} = \begin{bmatrix} 1 & 1-e^{-t} \\ 0 & e^{-t} \end{bmatrix}$$

The discretized state equation of the plant is
$$\mathbf{x}(k+1) = \mathbf{F}\mathbf{x}(k) + \mathbf{g}u(k) \qquad (6.73a)$$

where
$$\mathbf{F} = e^{\mathbf{A}T} = \begin{bmatrix} 1 & 1-e^{-T} \\ 0 & e^{-T} \end{bmatrix}$$

$$\mathbf{g} = \int_0^T e^{\mathbf{A}\theta} \mathbf{b}\, d\theta = \begin{bmatrix} \int_0^T (1-e^{-\theta})\, d\theta \\ \int_0^T e^{-\theta}\, d\theta \end{bmatrix} = \begin{bmatrix} T-1+e^{-T} \\ 1-e^{-T} \end{bmatrix}$$

For $T = 1$ sec, we have

$$\mathbf{F} = \begin{bmatrix} 1 & 0.632 \\ 0 & 0.368 \end{bmatrix};\ \mathbf{g} = \begin{bmatrix} 0.368 \\ 0.632 \end{bmatrix} \tag{6.73b}$$

Consider now the feedback system of Fig. 6.13 with the plant described by the equation (refer Eqns 6.73)):

$$\begin{bmatrix} x_1(k+1) \\ x_2(k+1) \end{bmatrix} = \begin{bmatrix} 1 & 0.632 \\ 0 & 0.368 \end{bmatrix} \begin{bmatrix} x_1(k) \\ x_2(k) \end{bmatrix} + \begin{bmatrix} 0.368 \\ 0.632 \end{bmatrix} e_2(k) \tag{6.74}$$

$e_2(k)$ may be taken as the third state variable $x_3(k)$ whose dynamics are given by

$$\begin{aligned} x_3(k+1) &= -a\, x_3(k) + b\, e_1(k) \\ &= -a\, x_3(k) + b\, (r(k) - x_1(k)) \\ &= -b\, x_1(k) - a\, x_3(k) + b\, r(k) \end{aligned} \tag{6.75}$$

From Eqns (6.74)–(6.75), we get the following state variable model for the closed-loop digital system of Fig. 6.13:

$$\begin{bmatrix} x_1(k+1) \\ x_2(k+1) \\ x_3(k+1) \end{bmatrix} = \begin{bmatrix} 1 & 0.632 & 0.368 \\ 0 & 0.368 & 0.632 \\ -b & 0 & -a \end{bmatrix} \begin{bmatrix} x_1(k) \\ x_2(k) \\ x_3(k) \end{bmatrix} + \begin{bmatrix} 0 \\ 0 \\ b \end{bmatrix} r(k)$$

$$y(k) = \begin{bmatrix} 1 & 0 & 0 \end{bmatrix} \begin{bmatrix} x_1(k) \\ x_2(k) \\ x_3(k) \end{bmatrix}$$

Review Example 6.5

Given

$$\underset{n \times n}{\Lambda} = \begin{bmatrix} \lambda_1 & 1 & 0 & \cdots & 0 \\ 0 & \lambda_1 & 1 & \cdots & 0 \\ \vdots & \vdots & \vdots & & \vdots \\ 0 & 0 & 0 & \cdots & 1 \\ 0 & 0 & 0 & \cdots & \lambda_1 \end{bmatrix}$$

Compute $\mathbf{\Lambda}^k$ using the Cayley-Hamilton technique.

Solution

Equations (5.98) outline the procedure of evaluation of functions of a matrix using the Cayley-Hamilton technique.

The matrix $\mathbf{\Lambda}$ has n eigenvalues at $\lambda = \lambda_1$. To evaluate $f(\mathbf{\Lambda}) = \mathbf{\Lambda}^k$, we define (refer Eqn. (5.98b)) the polynomial $g(\lambda)$ as

$$g(\lambda) = \beta_0 + \beta_1 \lambda + \cdots + \beta_{n-1} \lambda^{n-1}$$

This polynomial may be rearranged as

$$g(\lambda) = b_0 + b_1(\lambda - \lambda_1) + \cdots + b_{n-1}(\lambda - \lambda_1)^{n-1}$$

The coefficients $b_0, b_1, \ldots, b_{n-1}$ are given by the following equations (refer Eqns (5.98c)):

$$f(\lambda_1) = g(\lambda_1)$$

$$\left.\frac{d}{d\lambda} f(\lambda)\right|_{\lambda=\lambda_1} = \left.\frac{d}{d\lambda} g(\lambda)\right|_{\lambda=\lambda_1}$$

$$\vdots$$

$$\left.\frac{d^{n-1}}{d\lambda^{n-1}} f(\lambda)\right|_{\lambda=\lambda_1} = \left.\frac{d^{n-1}}{d\lambda^{n-1}} g(\lambda)\right|_{\lambda=\lambda_1}$$

Solving, we get

$$b_0 = \lambda_1^k$$

$$b_1 = \frac{k}{1!} \lambda_1^{k-1}$$

$$b_2 = \frac{k(k-1)}{2!} \lambda_1^{k-2}$$

$$\vdots$$

$$b_{n-1} = \frac{k(k-1)(k-2)\cdots(k-n+2)}{(n-1)!} \lambda_1^{k-n+1}$$

$$= \frac{k(k-1)(k-2)\cdots(k-n+2)(k-n+1)(k-n)\cdots 1}{(k-n+1)(k-n)\cdots 1} \left[\frac{1}{(n-1)!}\right] \lambda_1^{k-n+1}$$

$$= \frac{k!}{(k-n+1)!(n-1)!} \lambda_1^{k-n+1}$$

Therefore (refer Review Example 5.3)

$$\mathbf{\Lambda}^k = b_0 \mathbf{I} + b_1(\mathbf{\Lambda} - \lambda_1 \mathbf{I}) + \cdots + b_{n-1}(\mathbf{\Lambda} - \lambda_1 \mathbf{I})^{n-1}$$

$$= \begin{bmatrix} \lambda_1^k & \dfrac{k}{1!}\lambda_1^{k-1} & \dfrac{k(k-1)}{2!}\lambda_1^{k-2} & \cdots & \dfrac{k!\,\lambda_1^{k-n+1}}{(k-n+1)!(n-1)!} \\ 0 & \lambda_1^k & \dfrac{k}{1!}\lambda_1^{k-1} & \cdots & \bullet \\ 0 & 0 & \lambda_1^k & \cdots & \bullet \\ \vdots & \vdots & \vdots & & \vdots \\ 0 & 0 & 0 & \cdots & \lambda_1^k \end{bmatrix}$$

PROBLEMS

6.1 A system is described by the state equation

$$\mathbf{x}(k+1) = \begin{bmatrix} -3 & 1 & 0 \\ -4 & 0 & 1 \\ -1 & 0 & 0 \end{bmatrix} \mathbf{x}(k) + \begin{bmatrix} -3 \\ -7 \\ 0 \end{bmatrix} u(k); \; \mathbf{x}(0) = \mathbf{x}^0$$

Using the z-transform technique, transform the state equation into a set of linear algebraic equations in the form

$$\mathbf{X}(z) = \mathbf{G}(z)\mathbf{x}^0 + \mathbf{H}(z)U(z)$$

6.2 Give a block diagram for digital realization of the state equation of Problem 6.1.

6.3 Obtain the transfer function description for the following system:

$$\begin{bmatrix} x_1(k+1) \\ x_2(k+1) \end{bmatrix} = \begin{bmatrix} 2 & -5 \\ \tfrac{1}{2} & -1 \end{bmatrix} \begin{bmatrix} x_1(k) \\ x_2(k) \end{bmatrix} + \begin{bmatrix} 1 \\ 0 \end{bmatrix} u(k)$$

$$y(k) = 2x_1(k)$$

6.4 A second-order multivariable system is described by the following equations:

$$\begin{bmatrix} x_1(k+1) \\ x_2(k+1) \end{bmatrix} = \begin{bmatrix} 2 & -5 \\ \tfrac{1}{2} & -1 \end{bmatrix} \begin{bmatrix} x_1(k) \\ x_2(k) \end{bmatrix} + \begin{bmatrix} 1 & -2 & 0 \\ 0 & 1 & 3 \end{bmatrix} \begin{bmatrix} u_1(k) \\ u_2(k) \\ u_3(k) \end{bmatrix}$$

$$\begin{bmatrix} y_1(k) \\ y_2(k) \end{bmatrix} = \begin{bmatrix} 2 & 0 \\ 1 & -1 \end{bmatrix} \begin{bmatrix} x_1(k) \\ x_2(k) \end{bmatrix} + \begin{bmatrix} 0 & 4 & 0 \\ 0 & 0 & -2 \end{bmatrix} \begin{bmatrix} u_1(k) \\ u_2(k) \\ u_3(k) \end{bmatrix}$$

Convert the state variable model into a transfer function matrix.

6.5 The state diagram of a linear system is shown in Fig. P6.5. Assign the state variables and write the dynamic equations of the system.

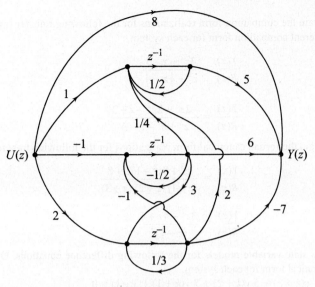

Fig. P6.5

6.6 Set up a state variable model for the system of Fig. P6.6.

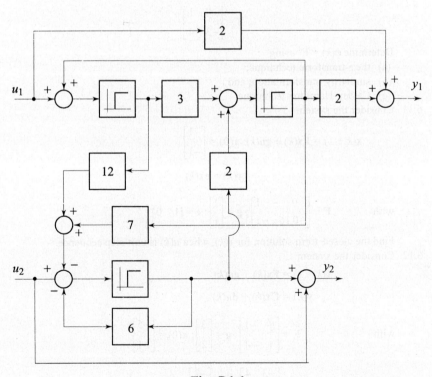

Fig. P6.6

6.7 Obtain the companion form realizations for the following transfer functions. Obtain different companion form for each system.

(i) $$\frac{Y(z)}{R(z)} = \frac{3z^2 - z - 3}{z^2 + \frac{1}{3}z - \frac{2}{3}}$$

(ii) $$\frac{Y(z)}{R(z)} = \frac{-2z^3 + 2z^2 - z + 2}{z^3 + z^2 - z - \frac{3}{4}}$$

6.8 Obtain the Jordan canonical form realizations for the following transfer functions.

(i) $$\frac{Y(z)}{R(z)} = \frac{z^3 + 8z^2 + 17z + 8}{(z+1)(z+2)(z+3)}$$

(ii) $$\frac{Y(z)}{R(z)} = \frac{3z^3 - 4z + 6}{(z - \frac{1}{3})^3}$$

6.9 Find state variable models for the following difference equations. Obtain different canonical form for each system.
   (i) $y(k+3) + 5 y(k+2) + 7 y(k+1) + 3 y(k) = 0$
   (ii) $y(k+2) + 3 y(k+1) + 2 y(k) = 5 r(k+1) + 3 r(k)$
   (iii) $y(k+3) + 5 y(k+2) + 7 y(k+1) + 3 y(k) = r(k+1) + 2 r(k)$

6.10 Given
$$\mathbf{F} = \begin{bmatrix} 0 & 1 \\ -3 & 4 \end{bmatrix}$$

Determine $\phi(k) = \mathbf{F}^k$ using
(a) the z-transform technique;
(b) similarity transformation; and
(c) Cayley-Hamilton technique.

6.11 Consider the system
$$\mathbf{x}(k+1) = \mathbf{F}\mathbf{x}(k) + \mathbf{g}u(k); \; \mathbf{x}(0) = \begin{bmatrix} 1 \\ -1 \end{bmatrix}$$

$$y(k) = \mathbf{c}\mathbf{x}(k)$$

with $\mathbf{F} = \begin{bmatrix} 0 & 1 \\ -0.16 & -1 \end{bmatrix}; \; \mathbf{g} = \begin{bmatrix} 1 \\ 1 \end{bmatrix}; \; \mathbf{c} = \begin{bmatrix} 1 & 0 \end{bmatrix}$

Find the closed-form solution for $y(k)$, when $u(k)$ is unit-step sequence.

6.12 Consider the system
$$\mathbf{x}(k+1) = \mathbf{F}\mathbf{x}(k) + \mathbf{g}u(k)$$
$$y(k) = \mathbf{C}\mathbf{x}(k) + \mathbf{d}u(k)$$

with $\mathbf{F} = \begin{bmatrix} \frac{3}{2} & -1 \\ 1 & -1 \end{bmatrix}; \; \mathbf{g} = \begin{bmatrix} 3 \\ 2 \end{bmatrix}; \; \mathbf{x}(0) = \begin{bmatrix} -5 \\ 1 \end{bmatrix}$

$\mathbf{C} = \begin{bmatrix} -3 & 4 \\ -1 & 1 \end{bmatrix}; \; \mathbf{d} = \begin{bmatrix} -2 \\ 0 \end{bmatrix}; \; u(k) = \left(\frac{1}{2}\right)^k, \; k \geq 0$

Find the response $y(k)$, $k \geq 0$.

6.13 Consider the system
$$\mathbf{x}(k+1) = \mathbf{F}\mathbf{x}(k)$$
with
$$\mathbf{F} = \begin{bmatrix} -1 & 1 & 0 \\ 0 & -1 & 0 \\ 0 & 0 & -2 \end{bmatrix}$$

(a) Find the modes of the free response.
(b) Find $\mathbf{x}(k)$ for
$$\mathbf{x}(0) = [0 \quad 1 \quad 1]^T$$

6.14 Consider the continuous-time system
$$G_a(s) = \frac{Y(s)}{R(s)} = \frac{1}{s(s+2)}$$

Insert sample-and-hold devices and determine the vector difference state model for digital simulation of the continuous-time system when the computation interval is $T = 1$ sec. Use the following methods to obtain the simulation model.
(a) Obtain $G(z)$ by taking the z transform of $G_a(s)$ when it is preceded by a sampler-and-hold; convert $G(z)$ into a vector difference state model.
(b) Obtain a continuous-time state model for the given $G_a(s)$; insert sample-and-hold and discretize the model.

6.15 Consider a continuous-time system
$$\dot{\mathbf{x}}(t) = \begin{bmatrix} -2 & 2 \\ 1 & -3 \end{bmatrix} \mathbf{x}(t) + \begin{bmatrix} -1 \\ 5 \end{bmatrix} u(t)$$
$$y(t) = [2 \quad -4] \, \mathbf{x}(t) + 6 \, u(t)$$

Insert sample-and-hold devices and determine the vector difference state model for digital simulation of the continuous-time system when the computation interval $T = 0.2$.

6.16 The plant of a single-input, single-output digital control system is shown in the block diagram of Fig. P6.16, where $u(t)$ is the control input and $w(t)$ is a unit-step load disturbance. Obtain the state difference equations of the plant. Sampling period $T = 0.1$ second.

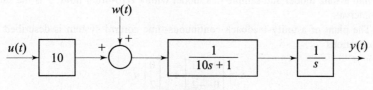

Fig. P6.16

6.17 The mathematical model of the plant of a two-input, two-output temperature control system is given below:
$$\dot{\mathbf{x}} = \mathbf{A}\mathbf{x} + \mathbf{B}\mathbf{u}$$
$$\mathbf{y} = \mathbf{C}\mathbf{x}$$

$$A = \begin{bmatrix} -0.1 & 0 \\ 0.1 & -0.1 \end{bmatrix}; B = \begin{bmatrix} 100 & 0 \\ 0 & 100 \end{bmatrix}; C = \begin{bmatrix} 1 & 0 \\ 0 & 1 \end{bmatrix}$$

For the computer control of this system, obtain the discrete-time model of the plant. Sampling period $T = 3$ seconds.

6.18 Consider the closed-loop control system shown in Fig. P6.18.
 (a) Obtain the z-transform of the feedforward transfer function.
 (b) Obtain the closed-loop transfer function, and convert it into a state variable model for digital simulation.

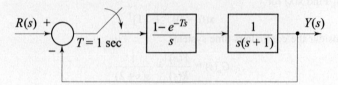

Fig. P6.18

6.19 The mathematical model of the plant of a control system is given below:

$$\frac{Y(s)}{U(s)} = G_a(s) = \frac{e^{-0.4s}}{s+1}$$

For digital simulation of the plant, obtain a vector difference state model with $T = 1$ sec as the sampling period. Use the following methods to obtain the plant model:
 (a) Sample $G_a(s)$ with a zero-order hold and convert the resulting discrete-time transfer function into a state model.
 (b) Convert the given $G_a(s)$ into a state model and sample this model with a zero-order hold.

6.20 Determine zero-order hold sampling of the process

$$\dot{x}(t) = -x(t) + u(t - 2.5)$$

with sampling interval $T = 1$.

6.21 Convert the transfer function

$$\frac{Y(s)}{U(s)} = G_a(s) = \frac{e^{-s\tau_D}}{s^2}; \; 0 \le \tau_D < T$$

into a state model and sample this model with a zero-order hold; T is the sampling interval.

6.22 The plant of a unity-feedback continuous-time control system is described by the equations

$$\dot{\mathbf{x}} = \begin{bmatrix} 0 & 1 \\ 0 & -2 \end{bmatrix} \mathbf{x} + \begin{bmatrix} 0 \\ 2 \end{bmatrix} u$$

$$y = x_1$$

 (a) Show that the continuous-time closed-loop system is stable.
 (b) A sampler and zero-order hold are now introduced in the forward loop. Show that the stable linear continuous-time system becomes unstable upon the introduction of a sampler and a zero-order hold with sampling period $T = 3$ sec.

6.23 The block diagram of a sampled-data system is shown in Fig. P6.23.

(a) Obtain a discrete-time state model for the system.
(b) Obtain the equation for intersample response of the system.

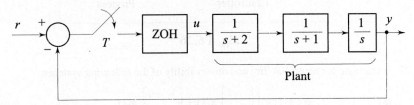

Fig. P6.23

6.24 The block diagram of a sampled-data system is shown in Fig. P6.24. Obtain the discrete-time state model of the system.

Given
$$\mathbf{A} = \begin{bmatrix} 0 & 1 \\ -2 & -3 \end{bmatrix}; \mathbf{b} = \begin{bmatrix} 0 \\ 1 \end{bmatrix}; \mathbf{c} = \begin{bmatrix} 1 & 0 \end{bmatrix}$$

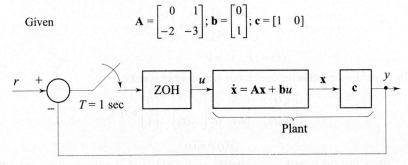

Fig. P6.24

6.25 A closed-loop computer control system is shown in Fig. P6.25. The digital compensator is described by the difference equation

$$e_2(k+1) + 2e_2(k) = e_1(k)$$

The state model of the plant is as given in Problem 6.24. Obtain the discrete-time state model for the system.

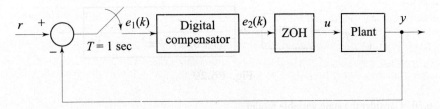

Fig. P6.25

6.26 Consider the closed-loop analog control system shown in Fig. P6.26. For computer control of the process, transform the controller transfer function into a difference equation using backward-difference approximation of the derivative.
Sample the process model with a zero-order hold and obtain the state variable model of the closed-loop computer-controlled system. Take $T = 0.1$ sec as sampling interval.

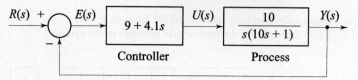

Fig. P6.26

6.27 Investigate the controllability and observability of the following systems:

(a) $$\mathbf{x}(k+1) = \begin{bmatrix} 1 & -2 \\ 1 & -1 \end{bmatrix} \mathbf{x}(k) + \begin{bmatrix} 1 & -1 \\ 0 & 0 \end{bmatrix} \mathbf{u}(k)$$

$$\mathbf{y}(k) = \begin{bmatrix} 1 & 0 \\ 0 & 1 \end{bmatrix} \mathbf{x}(k)$$

(b) $$\mathbf{x}(k+1) = \begin{bmatrix} -1 & 1 \\ 0 & -1 \end{bmatrix} \mathbf{x}(k) + \begin{bmatrix} 0 \\ 1 \end{bmatrix} u(k)$$

$$y(k) = \begin{bmatrix} 1 & 1 \end{bmatrix} \mathbf{x}(k)$$

6.28 Consider the following continuous-time control system:

$$\begin{bmatrix} \dot{x}_1(t) \\ \dot{x}_2(t) \end{bmatrix} = \begin{bmatrix} 0 & 1 \\ -1 & 0 \end{bmatrix} \begin{bmatrix} x_1(t) \\ x_2(t) \end{bmatrix} + \begin{bmatrix} 0 \\ 1 \end{bmatrix} u(t)$$

$$y(t) = x_1(t)$$

Show that the system is completely controllable and completely observable. The control signal $u^+(t)$ is now generated by processing the signal $u(t)$ through a sampler and a zero-order hold. Study the controllability and observability properties of the system under this condition. Determine the values of the sampling period for which the discretized system may exhibit hidden oscillations.

6.29 For the digital system shown in Fig. P6.29, determine what values of T must be avoided so that the system will be assured of complete controllability and observability.

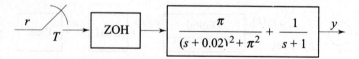

Fig. P6.29

6.30 Consider the state variable model

$$\mathbf{x}(k+1) = \mathbf{F}\mathbf{x}(k) + \mathbf{g}r(k)$$

$$y(k) = \mathbf{c}\mathbf{x}(k)$$

with $$\mathbf{F} = \begin{bmatrix} 0 & 1 \\ -\frac{1}{8} & \frac{3}{4} \end{bmatrix}; \mathbf{g} = \begin{bmatrix} 0 \\ 1 \end{bmatrix}; \mathbf{c} = \begin{bmatrix} -\frac{1}{2} & 1 \end{bmatrix}$$

(a) Find the eigenvalues of matrix **F**.
(b) Find the transfer function $G(z) = Y(z)/R(z)$ and determine the poles of the transfer function.
(c) Comment upon the controllability and observability properties of the given system without making any further calculations.

chapter 7

Pole-Placement Design and State Observers

7.1 INTRODUCTION

The design techniques presented in the preceding chapters are based on either frequency response or the root locus. These transfer function-based methods have been referred to as *classical control design*. The goal of this chapter is to solve the identical problem using different techniques which are based on state variable formulation. The use of the state-space approach has often been referred to as *modern control design*. However, since the state-space method of description for differential equations is over 100 years old and was introduced in the control design in the late 1950s, it seems somewhat misleading to refer to it as 'modern'. We prefer to refer to the two approaches to design as state variable methods and transform methods.

The transform methods of design are powerful methods of practical design. Most control systems are designed using variations of these methods. An important property of these methods is robustness. The resultant closed-loop system characteristics tend to be insensitive to small inaccuracies in the system model. This property is very important because of the difficulty in finding an accurate linear model of a physical system and also because many systems have significant nonlinear operations.

The state variable methods appear to be much more dependent on having an accurate system model for the design process. An advantage of these methods is that the system representation provides a complete (internal) description of the system, including possible internal oscillations or instabilities that might be hidden by inappropriate cancellations in the transfer function (input/output) description. The power of state variable techniques is especially apparent when we design controllers for systems with more than one control input or sensed output. However, in this chapter, we will illustrate the state variable design methods using single-input, single-output (SISO) systems. Methods for multi-input, multi-output (MIMO) design are discussed in Chapter 9.

Pole-Placement Design and State Observers

In this chapter, we present a design method known as *pole placement* or *pole assignment*. This method is similar to the root-locus design in that the closed-loop poles may be placed in desired locations. However, pole-placement design allows all closed-loop poles to be placed in desirable locations, whereas the root-locus design procedure allows only the two dominant poles to be placed. There is a cost associated with placing all closed-loop poles, however, because placing all closed-loop poles requires measurement and feedback of all the state variables of the system.

It many applications, all the state variables cannot be measured because of cost considerations or because of the lack of suitable transducers. In these cases, those state variables that cannot be measured must be estimated from the ones that are measured. Fortunately, we can separate the design into two phases. During the first phase, we design the system as though all states of the system will be measured. The second phase is concerned with the design of the state estimator. In this chapter, we consider both phases of the design process, and the effects that the state estimator has on closed-loop system operation.

Figure 7.1 shows how the state-feedback control law and the state estimator fit together and how the combination takes the place of what we have been previously referring to as dynamic *compensation*. We will see in this chapter that the estimator-based dynamic compensators are very similar to the classical compensators of Chapter 4, inspite of the fact that they are arrived at by entirely different means.

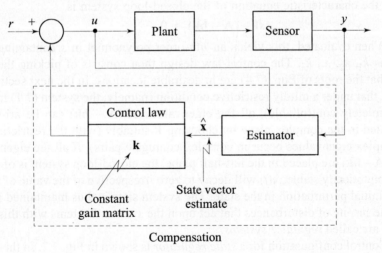

Fig. 7.1 Schematic diagram of a state-feedback control system

7.2 STABILITY IMPROVEMENT BY STATE FEEDBACK

An important aspect of the feedback system design is the stability of the control system. Whatever we want to achieve with the control system, its stability must be assured. Sometimes the main goal of a feedback design is actually to stabilize a system if it is initially unstable, or to improve its stability if transient phenomena do not die out sufficiently fast.

The purpose of this section is to investigate how the stability properties of linear systems can be improved by state feedback.

Consider the single-input linear time-invariant system with nth-order state differential equation

$$\dot{\mathbf{x}}(t) = \mathbf{A}\mathbf{x}(t) + \mathbf{b}u(t) \tag{7.1}$$

If we suppose that all the n state variables $x_1, x_2, ..., x_n$ can be accurately measured at all times, it is possible to implement a linear control law of the form

$$u(t) = -k_1 x_1(t) - k_2 x_2(t) - \cdots - k_n x_n(t) = -\mathbf{k}\mathbf{x}(t) \tag{7.2}$$

where $\quad \mathbf{k} = [k_1 \; k_2 \; \cdots \; k_n]$

is a constant state-feedback gain matrix. If this state-feedback control law is connected to the system (7.1), the closed-loop system is described by the state differential equation

$$\dot{\mathbf{x}}(t) = (\mathbf{A} - \mathbf{b}\mathbf{k})\,\mathbf{x}(t) \tag{7.3}$$

and the characteristic equation of the closed-loop system is

$$|s\mathbf{I} - (\mathbf{A} - \mathbf{b}\mathbf{k})| = 0 \tag{7.4}$$

When evaluated, this yields an nth-order polynomial in s containing the n gains $k_1, k_2, ..., k_n$. The control-law design then consists of picking the gains so that the roots of Eqn. (7.4) are in desirable locations. In the next section, we find that under a mildly restrictive condition (namely, the system (7.1) must be completely controllable), all the eigenvalues of $(\mathbf{A} - \mathbf{b}\mathbf{k})$ can be arbitrarily located in the complex plane by choosing $\mathbf{k}$ suitably (with the restriction that complex eigenvalues occur in complex-conjugate pairs). If all the eigenvalues of $(\mathbf{A} - \mathbf{b}\mathbf{k})$ are placed in the left-half plane, the closed-loop system is of course asymptotically stable; $\mathbf{x}(t)$ will decay to zero irrespective of the value of $\mathbf{x}(0)$—the initial perturbation in the state. The system state is thus maintained at zero value inspite of disturbances that act upon the system. Systems with this property are called *regulator systems*.

Control configuration for a *state regulator* is shown in Fig. 7.2. In this structure, there is no command input ($r = 0$). Control systems in which the output must follow the command signals (called *servo systems*) will be considered later.

Selection of desirable locations for the closed-loop poles requires some iteration by the designer. Some of the issues in their selection will be discussed later in this chapter. For now, we will assume that the desired locations are known, say,

$$s = \lambda_1, \lambda_2, \ldots, \lambda_n$$

Then the desired characteristic equation is

$$(s - \lambda_1)(s - \lambda_2) \cdots (s - \lambda_n) = 0 \qquad (7.5)$$

The required elements of **k** are obtained by matching coefficients in Eqns (7.4) and (7.5), thus forcing the system characteristic equation to be identical with the desired equation. An example should clarify this pole placement idea.

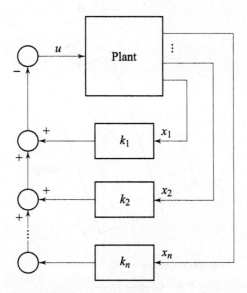

Fig. 7.2 Control configuration for a state regulator

Example 7.1

Consider the problem of designing an attitude control system for a rigid satellite. Satellites usually require attitude control so that antennas, sensors, and solar panels are properly oriented. For example, antennas are usually pointed towards a particular location on the earth while solar panels need to be oriented towards the sun for maximum power generation. To gain an insight into the full three-axis attitude-control system, we often consider one axis at a time. Figure 7.3 depicts this case. The angle θ that describes the satellite orientation must be measured with respect to an 'inertial' reference, that is, a reference that has no angular acceleration. The control signal comes from the reaction jets that produce a torque $T(t)$ $(= Fd)$ about the mass centre.

The satellite is assumed to be in frictionless environment. If $T(t)$ is the system input and $\theta(t)$ is the system output, we have

$$T(t) = J \frac{d^2 \theta(t)}{dt^2}$$

where J is the moment of inertia of the satellite. Normalizing, we define

$$u = T(t)/J$$

and obtain

$$\ddot{\theta} = u \text{ or } \frac{\theta(s)}{U(s)} = \frac{1}{s^2}$$

This is a reasonably accurate model of a rigid satellite in a frictionless environment, and is useful in examples because of its simplicity.

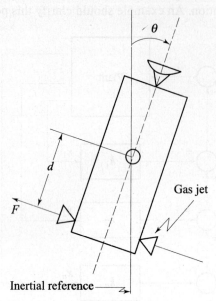

Fig. 7.3 Satellite control schematic

Choosing $x_1 = \theta$ and $x_2 = \dot{\theta}$ as state variables, we obtain the following state equation for the system.

$$\dot{\mathbf{x}} = \mathbf{A}\mathbf{x} + \mathbf{b}u = \begin{bmatrix} 0 & 1 \\ 0 & 0 \end{bmatrix} \mathbf{x} + \begin{bmatrix} 0 \\ 1 \end{bmatrix} u \quad (7.6)$$

To stabilize the system, the input signal is chosen to be of the form

$$u(t) = -k_1 x_1(t) - k_2 x_2(t) = -\mathbf{k}\mathbf{x}(t)$$

The state equation for the closed-loop system (Fig. 7.4), then becomes

$$\dot{\mathbf{x}} = (\mathbf{A} - \mathbf{b}\mathbf{k})\mathbf{x} = \left(\begin{bmatrix} 0 & 1 \\ 0 & 0 \end{bmatrix} - \begin{bmatrix} 0 \\ 1 \end{bmatrix} [k_1 \ k_2] \right) \mathbf{x} = \begin{bmatrix} 0 & 1 \\ -k_1 & -k_2 \end{bmatrix} \mathbf{x}$$

The characteristic equation of the closed-loop system is

$$|s\mathbf{I} - (\mathbf{A} - \mathbf{b}\mathbf{k})| = s^2 + k_2 s + k_1 = 0 \quad (7.7)$$

Suppose that the design specifications for this system require $\zeta = 0.707$ with a settling time of 1 sec $\left(\frac{4}{\zeta \omega_n} = 1 \text{ or } \zeta \omega_n = 4 \right)$. The closed-loop pole locations needed are at $s = -4 \pm j4$. The desired characteristic equation is

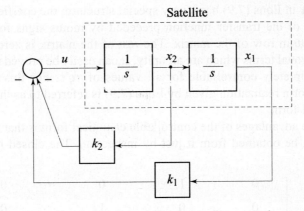

Fig. 7.4 Simulation diagram for the satellite control system

$$(s + 4 + j4)(s + 4 - j4) = s^2 + 8s + 32 \qquad (7.8)$$

Equating the coefficients with like powers of s in Eqns (7.7) and (7.8) yields

$$k_1 = 32, \ k_2 = 8$$

▲▲

The calculation of the gains using the technique illustrated in this example becomes rather tedious when the order of the system is larger than three. There are however 'canonical' forms of the state variable equations where the algebra for finding the gains is especially simple. On such canonical form useful in control-law design is the *controllable canonical form*. Consider a system represented by the transfer function

$$\frac{Y(s)}{U(s)} = \frac{\beta_1 s^{n-1} + \beta_2 s^{n-2} + \cdots + \beta_n}{s^n + \alpha_1 s^{n-1} + \cdots + \alpha_n}$$

A companion-form realization of this transfer function is given below (refer Eqns (5.54)):

$$\dot{\mathbf{x}} = \mathbf{A}\mathbf{x} + \mathbf{b}u$$
$$y = \mathbf{c}\mathbf{x} \qquad (7.9)$$

where

$$\mathbf{A} = \begin{bmatrix} 0 & 1 & 0 & \cdots & 0 \\ 0 & 0 & 1 & \cdots & 0 \\ \vdots & \vdots & \vdots & & \vdots \\ 0 & 0 & 0 & \cdots & 1 \\ -\alpha_n & -\alpha_{n-1} & -\alpha_{n-2} & \cdots & -\alpha_1 \end{bmatrix}; \ \mathbf{b} = \begin{bmatrix} 0 \\ 0 \\ \vdots \\ 0 \\ 1 \end{bmatrix}$$

$$\mathbf{c} = [\beta_n \ \beta_{n-1} \ \cdots \ \beta_2 \ \beta_1]$$

The matrix **A** in Eqns (7.9) has a very special structure: the coefficients of the denominator of the transfer function preceded by minus signs form a string along the bottom row of the matrix. The rest of the matrix is zero except for the superdiagonal terms which are all unity. It can easily be proved that the pair (**A**,**b**) is completely controllable for all values of α_i's. For this reason, the companion-form realization given by Eqns (7.9) is referred to as the controllable canonical form.

One of the advantages of the controllable canonical form is that the controller gains can be obtained from it just by inspection. The closed-loop system matrix

$$\mathbf{A} - \mathbf{bk} = \begin{bmatrix} 0 & 1 & 0 & \cdots & 0 \\ 0 & 0 & 1 & \cdots & 0 \\ \vdots & \vdots & \vdots & & \vdots \\ 0 & 0 & 0 & \cdots & 1 \\ -\alpha_n - k_1 & -\alpha_{n-1} - k_2 & -\alpha_{n-2} - k_3 & \cdots & -\alpha_1 - k_n \end{bmatrix}$$

has the characteristic equation

$$s^n + (\alpha_1 + k_n)s^{n-1} + \cdots + (\alpha_{n-2} + k_3)s^2 + (\alpha_{n-1} + k_2)s + \alpha_n + k_1 = 0$$

and the controller gains can be found by comparing the coefficients of this characteristic equation with Eqn. (7.5).

We now have the basis for a design procedure. Given an arbitrary state variable model and a desired characteristic polynomial, we transform the model to controllable canonical form and solve for the controller gains by inspection. Since these gains are for the state in the controllable canonical form, we must transform the gains back to the original state. We will develop this pole-placement design procedure in the subsequent sections.

7.3 NECESSARY AND SUFFICIENT CONDITIONS FOR ARBITRARY POLE-PLACEMENT

Consider the linear time-invariant system (7.1) with state feedback control law (7.2); the resulting closed-loop system is given by Eqn. (7.3). In the following, we shall prove that a necessary and sufficient condition for arbitrary placement of closed-loop eigenvalues in the complex plane (with the restriction that complex eigenvalues occur in complex-conjugate pairs) is that the system (7.1) is completely controllable. We shall first prove the sufficient condition, i.e., if the system (7.1) is completely controllable, all the eigenvalues of (**A** − **bk**) in Eqn. (7.3) can be arbitrarily placed.

In proving the sufficient condition on arbitrary pole-placement, it is convenient to transform the state Eqn. (7.1) into the controllable canonical form (7.9). Let us assume that such a transformation exists and is given by

$$\bar{\mathbf{x}} = \mathbf{Px} \qquad (7.10)$$

$$= \begin{bmatrix} p_{11} & p_{12} & \cdots & p_{1n} \\ p_{21} & p_{22} & \cdots & p_{2n} \\ \vdots & \vdots & & \vdots \\ p_{n1} & p_{n2} & \cdots & p_{nn} \end{bmatrix} \mathbf{x} = \begin{bmatrix} \mathbf{p}_1 \\ \mathbf{p}_2 \\ \vdots \\ \mathbf{p}_n \end{bmatrix} \mathbf{x}$$

$$\mathbf{p}_i = [p_{i1}\ p_{i2}\ \cdots\ p_{in}];\ i = 1, 2, \ldots, n$$

Under the transformation (7.10), system (7.1) is transformed to the following controllable canonical model:

$$\dot{\bar{\mathbf{x}}} = \overline{\mathbf{A}}\bar{\mathbf{x}} + \overline{\mathbf{b}}u \tag{7.11}$$

where

$$\overline{\mathbf{A}} = \mathbf{P}\mathbf{A}\mathbf{P}^{-1} = \begin{bmatrix} 0 & 1 & 0 & \cdots & 0 \\ 0 & 0 & 1 & \cdots & 0 \\ \vdots & \vdots & \vdots & & \vdots \\ 0 & 0 & 0 & & 1 \\ -\alpha_n & -\alpha_{n-1} & -\alpha_{n-2} & \cdots & -\alpha_1 \end{bmatrix}$$

$$\overline{\mathbf{b}} = \mathbf{P}\mathbf{b} = \begin{bmatrix} 0 \\ 0 \\ \vdots \\ 0 \\ 1 \end{bmatrix}$$

$$|s\mathbf{I} - \overline{\mathbf{A}}| = s^n + \alpha_1 s^{n-1} + \cdots + \alpha_{n-1} s + \alpha_n = |s\mathbf{I} - \mathbf{A}|$$

(Characteristic polynomial is invariant under equivalence transformation)

The first equation in the set (7.10) is given by

$$\bar{x}_1 = p_{11} x_1 + p_{12} x_2 + \cdots + p_{1n} x_n = \mathbf{p}_1 \mathbf{x}$$

Taking the derivative on both sides of this equation, we get

$$\dot{\bar{x}}_1 = \mathbf{p}_1 \dot{\mathbf{x}} = \mathbf{p}_1 \mathbf{A}\mathbf{x} + \mathbf{p}_1 \mathbf{b} u$$

But $\dot{\bar{x}}_1 (= \bar{x}_2)$ is a function of $\mathbf{x}$ only as per the canonical model (7.11). Therefore

$$\mathbf{p}_1 \mathbf{b} = 0 \text{ and } \bar{x}_2 = \mathbf{p}_1 \mathbf{A}\mathbf{x}$$

Taking derivative on both sides once again, we get

$$\mathbf{p}_1 \mathbf{A}\mathbf{b} = 0 \text{ and } \bar{x}_3 = \mathbf{p}_1 \mathbf{A}^2 \mathbf{x}$$

Continuing the process, we obtain

$$\mathbf{p}_1 \mathbf{A}^{n-2} \mathbf{b} = 0 \text{ and } \bar{x}_n = \mathbf{p}_1 \mathbf{A}^{n-1} \mathbf{x}$$

488 Digital Control and State Variable Methods

Taking derivative once again, we obtain
$$\mathbf{p}_1 \mathbf{A}^{n-1}\mathbf{b} = 1$$
Thus
$$\bar{\mathbf{x}} = \mathbf{P}\mathbf{x} = \begin{bmatrix} \mathbf{p}_1 \\ \mathbf{p}_1\mathbf{A} \\ \vdots \\ \mathbf{p}_1\mathbf{A}^{n-1} \end{bmatrix} \mathbf{x}$$

where $\mathbf{p}_1$ must satisfy the conditions
$$\mathbf{p}_1\mathbf{b} = \mathbf{p}_1\mathbf{A}\mathbf{b} = \cdots = \mathbf{p}_1\mathbf{A}^{n-2}\mathbf{b} = 0, \mathbf{p}_1\mathbf{A}^{n-1}\mathbf{b} = 1$$

From Eqn. (7.11), we have
$$\mathbf{Pb} = \begin{bmatrix} 0 \\ 0 \\ \vdots \\ 0 \\ 1 \end{bmatrix} = \begin{bmatrix} \mathbf{p}_1\mathbf{b} \\ \mathbf{p}_1\mathbf{A}\mathbf{b} \\ \vdots \\ \mathbf{p}_1\mathbf{A}^{n-2}\mathbf{b} \\ \mathbf{p}_1\mathbf{A}^{n-1}\mathbf{b} \end{bmatrix}$$

or
$$\mathbf{p}_1[\mathbf{b} \quad \mathbf{A}\mathbf{b} \quad \cdots \quad \mathbf{A}^{n-2}\mathbf{b} \quad \mathbf{A}^{n-1}\mathbf{b}] = [0 \quad 0 \quad \cdots \quad 0 \quad 1]$$

This gives
$$\mathbf{p}_1 = [0 \quad 0 \quad \cdots \quad 0 \quad 1]\mathbf{U}^{-1}$$
where
$$\mathbf{U} = [\mathbf{b} \quad \mathbf{A}\mathbf{b} \quad \cdots \quad \mathbf{A}^{n-1}\mathbf{b}]$$
is the controllability matrix, which is nonsingular because of the assumption of controllability of the system (7.1).

Therefore, the controllable state model (7.1) can be transformed to the canonical form (7.11) by the transformation
$$\bar{\mathbf{x}} = \mathbf{P}\mathbf{x} \qquad (7.12)$$

where
$$\mathbf{P} = \begin{bmatrix} \mathbf{p}_1 \\ \mathbf{p}_1\mathbf{A} \\ \vdots \\ \mathbf{p}_1\mathbf{A}^{n-1} \end{bmatrix}; \mathbf{p}_1 = [0 \quad 0 \quad \cdots \quad 0 \quad 1]\mathbf{U}^{-1}$$

Under the equivalence transformation (7.12), the state-feedback control law (7.2) becomes
$$u = -\mathbf{k}\mathbf{x} = -\bar{\mathbf{k}}\bar{\mathbf{x}} \qquad (7.13)$$

where
$$\bar{\mathbf{k}} = \mathbf{k}\mathbf{P}^{-1} = [\bar{k}_1 \ \bar{k}_2 \ \cdots \ \bar{k}_n]$$

With this control law, system (7.11) becomes

$$\dot{\bar{\mathbf{x}}} = (\bar{\mathbf{A}} - \bar{\mathbf{b}}\bar{\mathbf{k}})\bar{\mathbf{x}}$$

$$= \begin{bmatrix} 0 & 1 & 0 & \cdots & 0 & 0 \\ 0 & 0 & 1 & \cdots & 0 & 0 \\ \vdots & \vdots & \vdots & \cdots & \vdots & \vdots \\ 0 & 0 & 0 & \cdots & 0 & 1 \\ \vdots & \vdots & \vdots & & \vdots & \vdots \\ -\alpha_n - \bar{k}_1 & -\alpha_{n-1} - \bar{k}_2 & -\alpha_{n-2} - \bar{k}_3 & \cdots & -\alpha_2 - \bar{k}_{n-1} & -\alpha_1 - \bar{k}_n \end{bmatrix} \bar{\mathbf{x}}$$

(7.14)

$$|s\mathbf{I} - (\bar{\mathbf{A}} - \bar{\mathbf{b}}\bar{\mathbf{k}})| = s^n + (\alpha_1 + \bar{k}_n)s^{n-1} + (\alpha_2 + \bar{k}_{n-1})s^{n-2} +$$
$$\cdots + (\alpha_{n-1} + \bar{k}_2)s + (\alpha_n + \bar{k}_1)$$

(7.15)

Since the coefficients $\bar{k}_i$ are arbitrarily chosen real numbers, the coefficients of the characteristic polynomial of $(\mathbf{A} - \mathbf{bk})$ can be given any desired values. Hence, the closed-loop poles can be placed at any desired locations in the complex plane (subject to conjugate pairing: coefficients of a characteristic polynomial will be real only if the complex poles are present in conjugate pairs).

Assume that the desired characteristic polynomial of $(\mathbf{A} - \mathbf{bk})$, and hence $(\bar{\mathbf{A}} - \bar{\mathbf{b}}\bar{\mathbf{k}})$, is

$$s^n + a_1 s^{n-1} + \cdots + a_n$$

From Eqn. (7.15), it is obvious that this requirement is met if $\bar{\mathbf{k}}$ is chosen as

$$\bar{\mathbf{k}} = [a_n - \alpha_n \quad a_{n-1} - \alpha_{n-1} \quad \cdots \quad a_1 - \alpha_1]$$

Transforming the feedback controller (7.13) to the original coordinates, we obtain

$$\mathbf{k} = \bar{\mathbf{k}}\mathbf{P} = [a_n - \alpha_n \quad a_{n-1} - \alpha_{n-1} \quad \cdots \quad a_1 - \alpha_1]\mathbf{P} \quad (7.16)$$

This proves that if (7.1) is controllable, the closed-loop poles can be arbitrarily assigned (sufficient condition).

We now derive the necessary condition by proving that if the system (7.1) is not completely controllable, then there are eigenvalues of $(\mathbf{A} - \mathbf{bk})$ that cannot be controlled by state-feedback.

It was shown in Section 5.9 that an uncontrollable system can be transformed into controllability canonical from (Eqn. (5.123c))

$$\begin{bmatrix} \dot{\bar{\mathbf{x}}}_1 \\ \dot{\bar{\mathbf{x}}}_2 \end{bmatrix} = \begin{bmatrix} \bar{\mathbf{A}}_c & \bar{\mathbf{A}}_{12} \\ 0 & \bar{\mathbf{A}}_{22} \end{bmatrix} \begin{bmatrix} \bar{\mathbf{x}}_1 \\ \bar{\mathbf{x}}_2 \end{bmatrix} + \begin{bmatrix} \bar{\mathbf{b}}_c \\ 0 \end{bmatrix} u = \bar{\mathbf{A}}\bar{\mathbf{x}} + \bar{\mathbf{b}}u$$

where the pair $(\overline{\mathbf{A}}_c, \overline{\mathbf{b}}_c)$ is completely controllable.

The set of eigenvalues of $\overline{\mathbf{A}}$ is the union of the sets of eigenvalues of $\overline{\mathbf{A}}_c$ and $\overline{\mathbf{A}}_{22}$. In view of the form of $\overline{\mathbf{b}}$, it is obvious that the matrix $\overline{\mathbf{A}}_{22}$ is not affected by the introduction of any state-feedback of the form $u = -\overline{\mathbf{k}}\,\overline{\mathbf{x}}$. Therefore the eigenvalues of $\overline{\mathbf{A}}_{22}$ cannot be controlled. This proves the necessary condition.

7.4 STATE REGULATOR DESIGN

Consider the nth-order, single-input linear time-invariant system
$$\dot{\mathbf{x}}(t) = \mathbf{A}\mathbf{x}(t) + \mathbf{b}u(t) \qquad (7.17)$$
with state-feedback control law
$$u(t) = -\mathbf{k}\mathbf{x}(t) \qquad (7.18)$$
The resulting closed-loop system is
$$\dot{\mathbf{x}}(t) = (\mathbf{A} - \mathbf{b}\mathbf{k})\,\mathbf{x}(t) \qquad (7.19)$$
The eigenvalues of $(\mathbf{A} - \mathbf{b}\mathbf{k})$ can be arbitrarily placed in the complex plane (with the restriction that complex eigenvalues occur in complex-conjugate pairs) by choosing $\mathbf{k}$ suitably if and only if the system (7.17) is completely controllable.

This important result on pole placement was proved in the previous section. Following design steps for pole placement emerge from the proof.

Step 1: From the characteristic polynomial of matrix $\mathbf{A}$:
$$|s\mathbf{I} - \mathbf{A}| = s^n + \alpha_1 s^{n-1} + \cdots + \alpha_{n-1} s + \alpha_n \qquad (7.20)$$
determine the values of $\alpha_1, \alpha_2, \ldots, \alpha_{n-1}, \alpha_n$.

Step 2: Determine the transformation matrix $\mathbf{P}$ that transforms the system (7.17) into controllable canonical form:
$$\mathbf{P} = \begin{bmatrix} \mathbf{p}_1 \\ \mathbf{p}_1 \mathbf{A} \\ \vdots \\ \mathbf{p}_1 \mathbf{A}^{n-1} \end{bmatrix}; \quad \begin{array}{l} \mathbf{p}_1 = [0\ 0 \cdots 0\ 1]\mathbf{U}^{-1} \\ \mathbf{U} = [\mathbf{b}\ \ \mathbf{A}\mathbf{b} \cdots \mathbf{A}^{n-1}\mathbf{b}] \end{array} \qquad (7.21)$$

Step 3: Using the desired eigenvalues (desired closed-loop poles) $\lambda_1, \lambda_2, \ldots, \lambda_n$, write the desired characteristic polynomial:
$$(s - \lambda_1)(s - \lambda_2) \cdots (s - \lambda_n) = s^n + a_1 s^{n-1} + \cdots + a_{n-1} s + a_n \qquad (7.22)$$
and determine the values of $a_1, a_2, \ldots, a_{n-1}, a_n$.

Step 4: The required state-feedback gain matrix is determined from the following equation:
$$\mathbf{k} = [a_n - \alpha_n \quad a_{n-1} - \alpha_{n-1} \quad \cdots \quad a_1 - \alpha_1]\,\mathbf{P} \qquad (7.23)$$

Pole-Placement Design and State Observers

There are other approaches also for the determination of the state-feedback gain matrix **k**. In what follows, we shall present a well-known formula, known as the *Ackermann's formula*, which is convenient for computer solution.

From Eqns (7.23) and (7.21), we get

$$\mathbf{k} = [a_n - \alpha_n \quad a_{n-1} - \alpha_{n-1} \quad \cdots \quad a_1 - \alpha_1] \begin{bmatrix} (0 & 0 & \cdots & 0 & 1)\mathbf{U}^{-1} \\ (0 & 0 & \cdots & 0 & 1)\mathbf{U}^{-1}\mathbf{A} \\ & & \vdots & & \\ (0 & 0 & \cdots & 0 & 1)\mathbf{U}^{-1}\mathbf{A}^{n-1} \end{bmatrix}$$

$$= [0 \quad 0 \cdots 0 \quad 1]\mathbf{U}^{-1}[(a_1 - \alpha_1)\mathbf{A}^{n-1} +$$
$$(a_2 - \alpha_2)\mathbf{A}^{n-2} + \cdots + (a_n - \alpha_n)\mathbf{I}] \quad (7.24)$$

The characteristic polynomial of matrix **A** is (Eqn. (7.20))

$$|s\mathbf{I} - \mathbf{A}| = s^n + \alpha_1 s^{n-1} + \alpha_2 s^{n-2} + \cdots + \alpha_{n-1} s + \alpha_n$$

Since the Cayley-Hamilton theorem states that a matrix satisfies its own characteristic equation, we have

$$\mathbf{A}^n + \alpha_1 \mathbf{A}^{n-1} + \alpha_2 \mathbf{A}^{n-2} + \cdots + \alpha_{n-1}\mathbf{A} + \alpha_n \mathbf{I} = \mathbf{0}$$

Therefore $\quad \mathbf{A}^n = -\alpha_1 \mathbf{A}^{n-1} - \alpha_2 \mathbf{A}^{n-2} - \cdots - \alpha_{n-1}\mathbf{A} - \alpha_n \mathbf{I}$ (7.25)

From Eqns (7.24) and (7.25), we get

$$\mathbf{k} = [0 \quad 0 \cdots 0 \quad 1]\mathbf{U}^{-1}\phi(\mathbf{A}) \quad (7.26a)$$

where

$$\phi(\mathbf{A}) = \mathbf{A}^n + a_1 \mathbf{A}^{n-1} + a_2 \mathbf{A}^{n-2} + \cdots + a_{n-1}\mathbf{A} + a_n \mathbf{I} \quad (7.26b)$$

$$\mathbf{U} = [\mathbf{b} \quad \mathbf{Ab} \quad \cdots \quad \mathbf{A}^{n-1}\mathbf{b}] \quad (7.26c)$$

Equations (7.26) describe the Ackermann's formula for the determination of the state-feedback gain matrix **k**.

Example 7.2

Recall the inverted pendulum of Example 5.15, shown in Fig. 5.16, in which the object is to apply a force $u(t)$ so that the pendulum remains balanced in the vertical position. We found the linearized equations governing the system to be:

$$\dot{\mathbf{x}} = \mathbf{A}\mathbf{x} + \mathbf{b}u$$

where $\quad \mathbf{x} = [\theta \quad \dot{\theta} \quad z \quad \dot{z}]^T$

$$\mathbf{A} = \begin{bmatrix} 0 & 1 & 0 & 0 \\ 4.4537 & 0 & 0 & 0 \\ 0 & 0 & 0 & 1 \\ -0.5809 & 0 & 0 & 0 \end{bmatrix}; \quad \mathbf{b} = \begin{bmatrix} 0 \\ -0.3947 \\ 0 \\ 0.9211 \end{bmatrix}$$

$z(t)$ = horizontal displacement of the pivot on the cart
$\theta(t)$ = rotational angle of the pendulum

It is easy to verify that the characteristic polynomial of matrix **A** is
$$|s\mathbf{I} - \mathbf{A}| = s^4 - 4.4537s^2$$
Since there are poles at 0, 0, 2.11, and -2.11, the system is quite unstable, as one would expect from physical reasoning.

Suppose we require a feedback control of the form
$$u(t) = -\mathbf{kx} = -k_1 x_1 - k_2 x_2 - k_3 x_3 - k_4 x_4,$$
such that the closed-loop system has the stable pole configuration given by multiple poles at -1. We verified in Example 5.16 that the system under consideration is a controllable system; therefore such a feedback gain matrix **k** does exist. We will determine the required **k** by using the design Eqns (7.17)–(7.23).

The controllability matrix
$$\mathbf{U} = [\mathbf{b} \quad \mathbf{Ab} \quad \mathbf{A}^2\mathbf{b} \quad \mathbf{A}^3\mathbf{b}] = \begin{bmatrix} 0 & -0.3947 & 0 & -1.7579 \\ -0.3947 & 0 & -1.7579 & 0 \\ 0 & 0.9211 & 0 & 0.2293 \\ 0.9211 & 0 & 0.2293 & 0 \end{bmatrix}$$

$$\mathbf{U}^{-1} = \begin{bmatrix} 0 & 0.15 & 0 & 1.1499 \\ 0.15 & 0 & 1.1499 & 0 \\ 0 & -0.6025 & 0 & -0.2582 \\ -0.6025 & 0 & -0.2582 & 0 \end{bmatrix}$$

Therefore
$$\mathbf{P}_1 = [-0.6025 \quad 0 \quad -0.2582 \quad 0]$$

$$\mathbf{P} = \begin{bmatrix} \mathbf{p}_1 \\ \mathbf{p}_1\mathbf{A} \\ \mathbf{p}_1\mathbf{A}^2 \\ \mathbf{p}_1\mathbf{A}^3 \end{bmatrix} = \begin{bmatrix} -0.6025 & 0 & -0.2582 & 0 \\ 0 & -0.6025 & 0 & -0.2582 \\ -2.5334 & 0 & 0 & 0 \\ 0 & -2.5334 & 0 & 0 \end{bmatrix}$$

$$|s\mathbf{I} - \mathbf{A}| = s^4 + \alpha_1 s^3 + \alpha_2 s^2 + \alpha_3 s + \alpha_4$$
$$= s^4 + 0s^3 - 4.4537s^2 + 0s + 0$$
$$|s\mathbf{I} - (\mathbf{A} - \mathbf{bk})| = s^4 + a_1 s^3 + a_2 s^2 + a_3 s + a_4 = (s+1)^4$$
$$= s^4 + 4s^3 + 6s^2 + 4s + 1$$

$$\mathbf{k} = [a_4 - \alpha_4 \quad a_3 - \alpha_3 \quad a_2 - \alpha_2 \quad a_1 - \alpha_1]\mathbf{P}$$
$$= [1 \quad 4 \quad 10.4537 \quad 4]\mathbf{P}$$
$$= [-27 \quad -12.544 \quad -0.258 \quad -1.033]$$

This feedback control law yields a stable closed-loop system so that the entire state vector, when disturbed from the zero state, returns asymptotically to this state. This means that not only is the pendulum balanced ($\theta \to 0$), but that the cart returns to its origin as well ($z \to 0$).

Example 7.3

Let us use Ackermann's formula to the state-regulator design problem of Example 7.1. The plant model is given by (Eqn. (7.6))

$$\dot{\mathbf{x}} = \mathbf{A}\mathbf{x} + \mathbf{b}u = \begin{bmatrix} 0 & 1 \\ 0 & 0 \end{bmatrix} \mathbf{x} + \begin{bmatrix} 0 \\ 1 \end{bmatrix} u$$

The desired characteristic polynomial is (Eqn. (7.8))

$$s^2 + a_1 s + a_2 = s^2 + 8s + 32$$

To use Ackermann's formula (7.26) to calculate the gain matrix $\mathbf{k}$, we first evaluate $\mathbf{U}^{-1}$ and $\phi(\mathbf{A})$:

$$\mathbf{U} = [\mathbf{b} \quad \mathbf{Ab}] = \begin{bmatrix} 0 & 1 \\ 1 & 0 \end{bmatrix}$$

$$\mathbf{U}^{-1} = \begin{bmatrix} 0 & 1 \\ 1 & 0 \end{bmatrix}$$

$$\phi(\mathbf{A}) = \mathbf{A}^2 + a_1 \mathbf{A} + a_2 \mathbf{I}$$

$$= \begin{bmatrix} 0 & 1 \\ 0 & 0 \end{bmatrix} \begin{bmatrix} 0 & 1 \\ 0 & 0 \end{bmatrix} + 8 \begin{bmatrix} 0 & 1 \\ 0 & 0 \end{bmatrix} + 32 \begin{bmatrix} 1 & 0 \\ 0 & 1 \end{bmatrix} = \begin{bmatrix} 32 & 8 \\ 0 & 32 \end{bmatrix}$$

Now using Eqn. (7.26a), we obtain

$$\mathbf{k} = [0 \quad 1] \mathbf{U}^{-1} \phi(\mathbf{A}) = [0 \quad 1] \begin{bmatrix} 0 & 1 \\ 1 & 0 \end{bmatrix} \begin{bmatrix} 32 & 8 \\ 0 & 32 \end{bmatrix} = [32 \quad 8]$$

The solution is seen to be the same as that obtained in Example 7.1.

▲▲

Comments

1. Through the pole-placement design procedure described in the present section, it is always possible to stabilize a completely controllable system by state feedback, or to improve its stability by assigning the closed-loop poles to locations in the left-half complex plane. The design procedure, however, gives no guidance as to where in the left-half plane the closed-loop poles should be located.

It appears that we can choose the magnitude of the real part of the closed-loop poles arbitrarily large, making the system response arbitrarily fast. However, to increase the rate at which the plant responds, the input signal to the plant must become larger, requiring large values of gains. As the magnitudes of the signals in a system increase, the likelihood of the system entering nonlinear regions of operation increases. For very large signals, this nonlinear operation will occur for almost every physical system. Hence, the linear model that is used in design no longer accurately models the physical system.

Thus, the selection of desired closed-loop poles requires a proper balance of bandwidth, overshoot, sensitivity, control effort, and other design requirements. If the system is of second-order, then the system dynamics (response characteristics) can be precisely correlated to the locations of the desired closed-loop poles. For higher-order systems, the location of the closed-loop poles and the response characteristics are not easily correlated. Hence, in determining the state-feedback gain matrix **k** for a given system, it is desirable to examine, by computer simulations, the response characteristics for several different matrices **k** (based on several different characteristic equations), and choose the one that gives the best overall performance.

2. For the case of single-input systems, the gain matrix **k** which places the closed-loop poles at the desired locations is unique.

If the dynamic system under consideration

$$\dot{x} = Ax + Bu$$

has more than one input, that is, **B** has more than one column, then the gain matrix **K** in the control law

$$u = -Kx$$

has more than one row. Since each row of **K** furnishes n gains (n is the order of the system) that can be adjusted, it is clear that in a controllable system there will be more gains available than are needed to place all of the closed-loop poles. This is a benefit: the designer has more flexibility in the design; it is possible to specify all the closed-loop poles and still be able to satisfy other requirements. How should these other requirements be specified? The answer to this question may well depend on the circumstances of the particular application. A number of results using the design freedom in multi-input systems to improve robustness of the control system have appeared in the literature. We will not be able to accommodate these results in this book.

The non-uniqueness in the design of state-feedback control law for multi-input systems is removed by optimal control theory which is discussed in Chapter 9.

7.5 DESIGN OF STATE OBSERVERS

The pole-placement design procedure introduced in the preceding sections results in control law of the form

$$u(t) = -kx(t) \qquad (7.27)$$

which requires the ability to directly measure the entire state vector $x(t)$. Full state-feedback control for many second-order systems requires feedback of position and rate variables which can easily be measured. However, for most of the higher-order systems, full state measurements are not practical. Thus, either a new approach that directly accounts for the non-availability of the entire state vector (Chapter 9) is to be devised, or a suitable approximation of the state vector must be determined. The latter approach is much simpler in many situations.

The purpose of this section is to demonstrate the estimation of all the state variables of a system from the measurements that can be made on the system. If the estimate of the state vector is denoted by $\hat{\mathbf{x}}$, it would be nice if the true state in the control law given by Eqn. (7.27) could be replaced by its estimate:

$$u(t) = - \mathbf{k}\hat{\mathbf{x}}(t) \tag{7.28}$$

This indeed is possible, as we shall see in the next section.

A device (or a computer program) that estimates the state variables is called a *state observer*, or simply an *observer*. If the state observer estimates all the state variables of the system, regardless of whether some state variables are available for direct measurement, it is called a *full-order state observer*. However, if accurate measurements of certain states are possible, we may estimate only the remaining states, and the accurately measured signals are then used directly for feedback. The resulting observer is called a *reduced-order state observer*.

Full-Order State Observer

Consider a process described by the state equation

$$\dot{\mathbf{x}}(t) = \mathbf{A}\mathbf{x}(t) + \mathbf{b}u(t) \tag{7.29a}$$

where $\mathbf{A}$ and $\mathbf{b}$ are, respectively, $n \times n$ and $n \times 1$ real constant matrices. The measurement $y(t)$ is related to the state by the equation

$$y(t) = \mathbf{c}\mathbf{x}(t) \tag{7.29b}$$

where $\mathbf{c}$ is $1 \times n$ real constant matrix. Without loss of generality, the direct transmission part has been assumed to be zero.

One method of estimating all the state variables that we may consider is to construct a model of the plant dynamics:

$$\dot{\hat{\mathbf{x}}}(t) = \mathbf{A}\hat{\mathbf{x}}(t) + \mathbf{b}u(t) \tag{7.30}$$

where $\hat{\mathbf{x}}$ is the estimate of the actual state $\mathbf{x}$. We know $\mathbf{A}$, $\mathbf{b}$ and $u(t)$, and hence this estimator is satisfactory if we can obtain the correct initial condition $\mathbf{x}(0)$ and set $\hat{\mathbf{x}}(0)$ equal to it. Figure 7.5 depicts this 'open-loop' estimator. However, it is precisely the lack of information on $\mathbf{x}(0)$ that requires the construction of an estimator. If $\hat{\mathbf{x}}(0) \neq \mathbf{x}(0)$, the estimated state $\hat{\mathbf{x}}(t)$ obtained from the open-loop scheme of Fig. 7.5 would have a continually growing error or an error that goes to zero too slowly to be of any use. Furthermore, small errors in our knowledge of the system $(\mathbf{A}, \mathbf{b})$, and the disturbances that enter the system but not the model would also cause the estimate to slowly diverge from the true state.

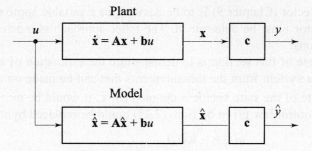

Fig. 7.5 Open-loop estimator

In order to speed up the estimation process and provide a useful state estimate, we feedback the difference between the measured and the estimated outputs and correct the model continuously with this error signal. This scheme, commonly known as 'Luenberger state observer', is shown in Fig. 7.6, and the equation for it is

$$\dot{\hat{\mathbf{x}}}(t) = \mathbf{A}\hat{\mathbf{x}}(t) + \mathbf{b}u(t) + \mathbf{m}(y(t) - \hat{y}(t)) \qquad (7.31)$$

where $\mathbf{m}$ is an $n \times 1$ real constant gain matrix.

The state error vector

$$\tilde{\mathbf{x}}(t) = \mathbf{x}(t) - \hat{\mathbf{x}}(t) \qquad (7.32)$$

Differentiating both sides, we get

$$\dot{\tilde{\mathbf{x}}}(t) = \dot{\mathbf{x}}(t) - \dot{\hat{\mathbf{x}}}(t)$$

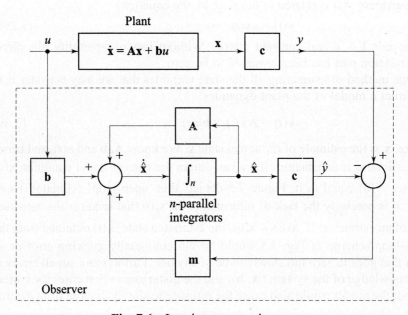

Fig. 7.6 Luenberger state observer

Substituting for $\dot{\mathbf{x}}(t)$ and $\dot{\hat{\mathbf{x}}}(t)$ from Eqns (7.29) and (7.31) respectively, we get

$$\dot{\tilde{\mathbf{x}}}(t) = \mathbf{A}\mathbf{x}(t) + \mathbf{b}u(t) - \mathbf{A}\hat{\mathbf{x}}(t) - \mathbf{b}u(t) - \mathbf{mc}(\mathbf{x}(t) - \hat{\mathbf{x}}(t))$$

$$= (\mathbf{A} - \mathbf{mc})\tilde{\mathbf{x}}(t) \tag{7.33}$$

The characteristic equation of the error is given by

$$|s\mathbf{I} - (\mathbf{A} - \mathbf{mc})| = 0 \tag{7.34a}$$

If $\mathbf{m}$ can (we hope) be chosen so that $(\mathbf{A} - \mathbf{mc})$ has stable and reasonably fast roots, $\tilde{\mathbf{x}}(t)$ will decay to zero irrespective of $\tilde{\mathbf{x}}(0)$. This means that $\hat{\mathbf{x}}(t)$ will converge to $\mathbf{x}(t)$ regardless of the value of $\hat{\mathbf{x}}(0)$, and furthermore, the dynamics of the error can be chosen to be faster than the open-loop dynamics. Note that Eqn. (7.33) is independent of applied control. This is a consequence of assuming $\mathbf{A}$, $\mathbf{b}$ and $\mathbf{c}$ to be identical in the plant and the observer. Therefore, the estimation error $\tilde{\mathbf{x}}$ converges to zero and remains there, independent of any known forcing function $u(t)$ on the plant and its effect on the state $\mathbf{x}(t)$. If we do not have a very accurate model of the plant $(\mathbf{A}, \mathbf{b}, \mathbf{c})$, the dynamics of the error are no longer governed by Eqn. (7.33). However, $\mathbf{m}$ can typically be chosen so that the error system is stable and the error is acceptably small, even with small modelling errors and disturbance inputs.

The selection of $\mathbf{m}$ can be approached in exactly the same fashion as the selection of $\mathbf{k}$ is in the control law design. If we specify the desired location of the observer error roots as

$$s = \lambda_1, \lambda_2, \ldots, \lambda_n,$$

the desired observer characteristic equation is

$$(s - \lambda_1)(s - \lambda_2) \cdots (s - \lambda_n) = 0 \tag{7.34b}$$

and one can solve for $\mathbf{m}$ by comparing coefficients in Eqns (7.34a) and (7.34b). However, as we shall see shortly, this can be done only if the system (7.29) is completely observable.

The calculation of the gains using this simple technique becomes rather tedious when the order of the system is larger than three. As in the controller design, there is an *observable canonical form* for which the observer design equations are particularly simple. Consider a system represented by the transfer function

$$\frac{Y(s)}{U(s)} = \frac{\beta_1 s^{n-1} + \beta_2 s^{n-2} + \cdots + \beta_n}{s^n + \alpha_1 s^{n-1} + \cdots + \alpha_n}$$

A companion-form realization of this transfer function is given below (refer Eqns (5.56)):

$$\begin{aligned}\dot{\mathbf{x}} &= \mathbf{A}\mathbf{x} + \mathbf{b}u \\ y &= \mathbf{c}\mathbf{x}\end{aligned} \tag{7.35}$$

where

$$A = \begin{bmatrix} 0 & 0 & \cdots & 0 & -\alpha_n \\ 1 & 0 & \cdots & 0 & -\alpha_{n-1} \\ 0 & 1 & \cdots & 0 & -\alpha_{n-2} \\ \vdots & \vdots & & \vdots & \vdots \\ 0 & 0 & \cdots & 1 & -\alpha_1 \end{bmatrix}; \quad b = \begin{bmatrix} \beta_n \\ \beta_{n-1} \\ \beta_{n-2} \\ \vdots \\ \beta_1 \end{bmatrix}$$

$$c = [0 \ 0 \ \cdots \ 0 \ 1]$$

It can easily be proved that the pair (A,c) is completely observable for all values of α_i's. For this reason, the companion form realization given by Eqn. (7.35) is referred to as observable canonical form.

One of the advantages of the observable canonical form is that the observer gains m can be obtained from it just by inspection. The observer-error matrix is

$$(A - mc) = \begin{bmatrix} 0 & 0 & \cdots & 0 & -\alpha_n - m_1 \\ 1 & 0 & \cdots & 0 & -\alpha_{n-1} - m_2 \\ 0 & 1 & \cdots & 0 & -\alpha_{n-2} - m_3 \\ \vdots & \vdots & & \vdots & \vdots \\ 0 & 0 & \cdots & 1 & -\alpha_1 - m_n \end{bmatrix}$$

which has the characteristic equation

$$s^n + (\alpha_1 + m_n) s^{n-1} + \cdots + (\alpha_{n-2} + m_3)s^2 + (\alpha_{n-1} + m_2) s + \alpha_n + m_1 = 0$$

and the observer gains can be found by comparing the coefficients of this equation with Eqn. (7.34b).

A procedure for observer design, therefore, consists of transforming the given state variable model to observable canonical form, solving for the observer gains, and transforming the gains back to the original state.

We can, however, directly use the equations of the control-law design for computing the observer gain matrix m if we examine the resemblance between the estimation and control problems. In fact, the two problems are mathematically equivalent. This property is called *duality*. The design of a full-order observer requires the determination of the gain matrix m such that $(A - mc)$ has desired eigenvalues λ_i; $i = 1, 2, ..., n$. This is mathematically equivalent to designing a full state-feedback controller for the 'transposed auxiliary system',

$$\dot{\zeta}(t) = A^T \zeta(t) + c^T \eta(t) \qquad (7.36a)$$

with feedback

$$\eta(t) = - m^T \zeta(t) \qquad (7.36b)$$

so that the closed-loop auxiliary system

$$\dot{\zeta}(t) = (\mathbf{A}^T - \mathbf{c}^T\mathbf{m}^T)\zeta(t) \qquad (7.37)$$

has eigenvalues λ_i; $i = 1, 2, ..., n$.
Since

$$det\ \mathbf{W} = det\ \mathbf{W}^T,$$

one obtains

$$det\ [s\mathbf{I} - (\mathbf{A}^T - \mathbf{c}^T\mathbf{m}^T)] = det\ [s\mathbf{I} - (\mathbf{A} - \mathbf{mc})]$$

i.e., the eigenvalues of $(\mathbf{A}^T - \mathbf{c}^T\mathbf{m}^T)$ are same as the eigenvalues of $(\mathbf{A} - \mathbf{mc})$.

By comparing the characteristic equation of the closed-loop system (7.19) and that of the auxiliary system (7.37), we obtain the duality relations given in Table 7.1 between the control and estimation problems. The Ackermann's control-design formula given by Eqns (7.26) becomes the observer-design formula if the substitutions of Table 7.1 are made.

Table 7.1 Duality between control and estimation

Control	Estimation
A	$\mathbf{A}^T$
b	$\mathbf{c}^T$
k	$\mathbf{m}^T$

A necessary and sufficient condition for determination of the observer gain matrix $\mathbf{m}$ for the desired eigenvalues of $(\mathbf{A} - \mathbf{mc})$ is that the auxiliary system (7.36) be completely controllable. The controllability condition for this system is that the rank of

$$[\mathbf{c}^T\ \mathbf{A}^T\mathbf{c}^T\ \cdots\ (\mathbf{A}^T)^{n-1}\ \mathbf{c}^T]$$

is n. This is the condition for complete observability of the original system defined by Eqns (7.29). This means that a necessary and sufficient condition for estimation of the state of the system defined by Eqns (7.29) is that the system be completely observable.

Again by duality, we can say that for the case of single-output systems, the gain matrix $\mathbf{m}$ which places the observer poles at desired locations is unique. In the multi-output case, the same pole configuration can be achieved by various feedback gain matrices. This non-uniqueness is removed by optimal control theory which is discussed in Chapter 9.

Example 7.4

We will consider the satellite control system of Example 7.3. The state equation of the plant is

$$\dot{\mathbf{x}} = \mathbf{A}\mathbf{x} + \mathbf{b}u$$

with

$$\mathbf{A} = \begin{bmatrix} 0 & 1 \\ 0 & 0 \end{bmatrix};\ \mathbf{b} = \begin{bmatrix} 0 \\ 1 \end{bmatrix}$$

$x_1 = \theta$, the orientation of the satellite

$x_2 = \dot{\theta}$

We assume that the orientation θ can be accurately measured from the antenna signal; therefore

$$y = \mathbf{cx}(t)$$

with
$$\mathbf{c} = [1 \quad 0]$$

Let us design a state observer for the system. We choose the observer to be critically damped with a settling time of 0.4 sec $\left(\dfrac{4}{\zeta\omega_n} = 0.4; \zeta\omega_n = 10\right)$. To satisfy these specifications, the observer poles will be placed at $s = -10, -10$.

The transposed auxiliary system is given by

$$\dot{\boldsymbol{\zeta}} = \mathbf{A}^T\boldsymbol{\zeta} + \mathbf{c}^T\eta$$

$$\eta = -\mathbf{m}^T\boldsymbol{\zeta}$$

The desired characteristic equation of the closed-loop auxiliary system is

$$s^2 + a_1 s + a_2 = (s + 10)(s + 10) = s^2 + 20s + 100$$

To apply Ackermann's formula given by Eqns (7.26), we compute

$$\mathbf{U}^{-1} = [\mathbf{c}^T \quad \mathbf{A}^T\mathbf{c}^T]^{-1} = \begin{bmatrix} 1 & 0 \\ 0 & 1 \end{bmatrix}$$

$$\phi(\mathbf{A}^T) = (\mathbf{A}^T)^2 + a_1 \mathbf{A}^T + a_2 \mathbf{I}$$

$$= \begin{bmatrix} 0 & 0 \\ 1 & 0 \end{bmatrix}\begin{bmatrix} 0 & 0 \\ 1 & 0 \end{bmatrix} + 20\begin{bmatrix} 0 & 0 \\ 1 & 0 \end{bmatrix} + 100\begin{bmatrix} 1 & 0 \\ 0 & 1 \end{bmatrix} = \begin{bmatrix} 100 & 0 \\ 20 & 100 \end{bmatrix}$$

The observer gain matrix is given by the equation

$$\mathbf{m}^T = [0 \quad 1]\begin{bmatrix} 100 & 0 \\ 20 & 100 \end{bmatrix} = [20 \quad 100]$$

Therefore $\mathbf{m} = \begin{bmatrix} 20 \\ 100 \end{bmatrix}$

Example 7.5

Consider once again the inverted-pendulum system of Example 7.2. Suppose that the only output available for measurement is $z(t)$, the position of the cart. The linearized equations governing this system are

$$\dot{\mathbf{x}} = \mathbf{Ax} + \mathbf{b}u$$

$$y = \mathbf{cx}$$

where
$$\mathbf{A} = \begin{bmatrix} 0 & 1 & 0 & 0 \\ 4.4537 & 0 & 0 & 0 \\ 0 & 0 & 0 & 1 \\ -0.5809 & 0 & 0 & 0 \end{bmatrix}; \mathbf{b} = \begin{bmatrix} 0 \\ -0.3947 \\ 0 \\ 0.9211 \end{bmatrix}$$

$$\mathbf{c} = [0 \ 0 \ 1 \ 0]$$

We verified in Example 5.18 that this system is completely observable. In the following, we design a full-order observer for this system. We choose the observer pole locations as $-2, -2 \pm j1, -3$. The corresponding characteristic equation is

$$s^4 + 9s^3 + 31s^2 + 49s + 30 = 0$$

The transposed auxiliary system is given by

$$\dot{\boldsymbol{\zeta}}(t) = \mathbf{A}^T \boldsymbol{\zeta}(t) + \mathbf{c}^T \eta(t)$$

$$\eta(t) = -\mathbf{m}^T \boldsymbol{\zeta}(t)$$

We will determine the gain matrix $\mathbf{m}$ using the design Eqns (7.17)–(7.23).

The controllability matrix

$$\mathbf{U} = [\mathbf{c}^T \ \mathbf{A}^T\mathbf{c}^T \ (\mathbf{A}^T)^2\mathbf{c}^T \ (\mathbf{A}^T)^3\mathbf{c}^T]$$

$$= \begin{bmatrix} 0 & 0 & -0.5809 & 0 \\ 0 & 0 & 0 & -0.5809 \\ 1 & 0 & 0 & 0 \\ 0 & 1 & 0 & 0 \end{bmatrix}$$

$$\mathbf{U}^{-1} = \begin{bmatrix} 0 & 0 & 1 & 0 \\ 0 & 0 & 0 & 1 \\ -1.7215 & 0 & 0 & 0 \\ 0 & -1.7215 & 0 & 0 \end{bmatrix}$$

Therefore

$$\mathbf{p}_1 = [0 \ -1.7215 \ 0 \ 0]$$

$$\mathbf{P} = \begin{bmatrix} \mathbf{p}_1 \\ \mathbf{p}_1(\mathbf{A}^T) \\ \mathbf{p}_1(\mathbf{A}^T)^2 \\ \mathbf{p}_1(\mathbf{A}^T)^3 \end{bmatrix} = \begin{bmatrix} 0 & -1.7215 & 0 & 0 \\ -1.7215 & 0 & 0 & 0 \\ 0 & 0 & -7.667 & 0 & 1 \\ -7.667 & 0 & 1 & 0 \end{bmatrix}$$

$$|s\mathbf{I} - \mathbf{A}^T| = s^4 + \alpha_1 s^3 + \alpha_2 s^2 + \alpha_3 s + \alpha_4$$

$$= s^4 + 0s^3 - 4.4537s^2 + 0s + 0$$

$$|s\mathbf{I} - (\mathbf{A}^T - \mathbf{c}^T\mathbf{m}^T)| = s^4 + a_1 s^3 + a_2 s^2 + a_3 s + a_4$$
$$= s^4 + 9s^3 + 31s^2 + 49s + 30$$
$$\mathbf{m}^T = [a_4 - \alpha_4 \quad a_3 - \alpha_3 \quad a_2 - \alpha_2 \quad a_1 - \alpha_1]\,\mathbf{P}$$
$$= [30 \quad 49 \quad 35.4537 \quad 9]\,\mathbf{P}$$
$$= [-153.3569 \quad -323.4701 \quad 9 \quad 35.4544]$$

Therefore
$$\mathbf{m} = \begin{bmatrix} -153.3569 \\ -323.4701 \\ 9 \\ 35.4544 \end{bmatrix}$$

With this $\mathbf{m}$, the observer

$$\dot{\hat{\mathbf{x}}} = (\mathbf{A} - \mathbf{mc})\hat{\mathbf{x}} + \mathbf{b}u + \mathbf{m}y$$

will process the cart position $y(t) = z(t)$, and input $u(t)$ to continuously provide an estimate $\hat{\mathbf{x}}(t)$ of the state vector $\mathbf{x}(t)$, and any errors in the estimate will decay at least as fast as e^{-2t}.

Reduced-Order State Observer

The observer developed in the previous sub-section reconstructs the entire state vector. However, usually the measurements available are some of the states of the plant. For example, for the satellite considered in the previous sub-section, the measurement is orientation of the satellite, which is $x_1(t)$. The measurement of a state in general will be more accurate than any estimate of the state based on the measurement. Hence it is not logical in most cases to estimate states that we are measuring. One possible exception is the case in which a measurement is very noisy. The state observer for this case may furnish some beneficial noise filtering.

Since we will not usually want to estimate any state that we are measuring, we prefer to design an observer that estimates only those states that are not measured. This type of observer is called a *reduced-order state observer*. We develop design equations for such an observer in this sub-section. We consider only the case of one measurement. It is assumed that the state variables are always chosen such that the state measured is $x_1(t)$; we can do this without loss of generality. The output equation then is given by

$$y(t) = x_1(t) = \mathbf{c}\mathbf{x}(t)$$

where $\mathbf{c} = [1 \quad 0 \quad 0 \cdots 0]$

To derive the reduced-order observer, we partition the state vector into two parts: one part is x_1 which is directly measured and the other part is $\mathbf{x}_e$, representing the state variables that need to be estimated:

$$\mathbf{x}(t) = \begin{bmatrix} x_1(t) \\ \mathbf{x}_e(t) \end{bmatrix}$$

If we partition the system matrices accordingly, the complete description of the system is given by

$$\begin{bmatrix} \dot{x}_1 \\ \dot{\mathbf{x}}_e \end{bmatrix} = \begin{bmatrix} a_{11} & \mathbf{a}_{1e} \\ \mathbf{a}_{e1} & \mathbf{A}_{ee} \end{bmatrix} \begin{bmatrix} x_1 \\ \mathbf{x}_e \end{bmatrix} + \begin{bmatrix} b_1 \\ \mathbf{b}_e \end{bmatrix} u \quad (7.38a)$$

$$y = \begin{bmatrix} 1 & \mathbf{0} \end{bmatrix} \begin{bmatrix} x_1 \\ \mathbf{x}_e \end{bmatrix} \quad (7.38b)$$

The dynamics of the unmeasured state variables are given by

$$\dot{\mathbf{x}}_e = \mathbf{A}_{ee} \mathbf{x}_e + \underbrace{\mathbf{a}_{e1} x_1 + \mathbf{b}_e u}_{\text{known input}} \quad (7.39)$$

where the rightmost two terms are known and can be considered as an input into the $\mathbf{x}_e$ dynamics. Since $x_1 = y$, measured dynamics are given by the scalar equation

$$\dot{x}_1 = \dot{y} = a_{11} y + \mathbf{a}_{1e} \mathbf{x}_e + b_1 u \quad (7.40)$$

If we collect the known terms of Eqn. (7.40) on one side, we get

$$\underbrace{\dot{y} - a_{11} y - b_1 u}_{\text{known measurement}} = \mathbf{a}_{1e} \mathbf{x}_e \quad (7.41)$$

Note that Eqns (7.39) and (7.41) have the same relationship to the state $\mathbf{x}_e$ that the original Eqns (7.38) had to the entire state $\mathbf{x}$. Following this line of reasoning, we can establish the following substitutions in the original observer-design equations to obtain an (reduced-order) observer of $\mathbf{x}_e$:

$$\begin{aligned}
\mathbf{x} &\leftarrow \mathbf{x}_e \\
\mathbf{A} &\leftarrow \mathbf{A}_{ee} \\
\mathbf{b}u &\leftarrow \mathbf{a}_{e1} y + \mathbf{b}_e u \\
y &\leftarrow \dot{y} - a_{11} y - b_1 u \\
\mathbf{c} &\leftarrow \mathbf{a}_{1e}
\end{aligned} \quad (7.42)$$

Making these substitutions into the equations for full-order observer (Eqn. (7.31)), we obtain the equations of the reduced-order observer:

$$\dot{\hat{\mathbf{x}}}_e = \mathbf{A}_{ee} \hat{\mathbf{x}}_e + \underbrace{\mathbf{a}_{e1} y + \mathbf{b}_e u}_{\text{input}}$$

$$+ \mathbf{m} \, (\underbrace{\dot{y} - a_{11} y - b_1 u}_{\text{measurement}} - \mathbf{a}_{1e} \hat{\mathbf{x}}_e) \quad (7.43)$$

If we define the estimation error as
$$\tilde{\mathbf{x}}_e = \mathbf{x}_e - \hat{\mathbf{x}}_e \tag{7.44}$$
the dynamics of error are given by subtracting Eqn. (7.43) from Eqn. (7.39):
$$\dot{\tilde{\mathbf{x}}}_e = (\mathbf{A}_{ee} - \mathbf{ma}_{1e})\tilde{\mathbf{x}}_e \tag{7.45}$$
Its characteristic equation is given by
$$|s\mathbf{I} - (\mathbf{A}_{ee} - \mathbf{ma}_{1e})| = 0 \tag{7.46}$$

We design the dynamics of this observer by selecting $\mathbf{m}$ so that Eqn. (7.46) matches a desired reduced-order characteristic equation. To carry out the design using state regulator results, we form a 'transposed auxiliary system'
$$\dot{\boldsymbol{\zeta}}(t) = \mathbf{A}_{ee}^T \boldsymbol{\zeta}(t) + \mathbf{a}_{1e}^T \boldsymbol{\eta}(t)$$
$$\boldsymbol{\eta}(t) = - \mathbf{m}^T \boldsymbol{\zeta}(t) \tag{7.47}$$

Use of Ackermann's formula given by Eqns (7.26) for this auxiliary system gives the gains $\mathbf{m}$ of the reduced-order observer. We should point out that the conditions for the existence of the reduced-order observer are the same as for the full-order observer—namely observability of the pair $(\mathbf{A}, \mathbf{c})$.

Let us now look at the implementational aspects of the reduced-order observer given by Eqn. (7.43). This equation can be rewritten as
$$\dot{\hat{\mathbf{x}}}_e = (\mathbf{A}_{ee} - \mathbf{ma}_{1e})\hat{\mathbf{x}}_e + (\mathbf{a}_{e1} - \mathbf{m}a_{11})y + (\mathbf{b}_e - \mathbf{m}b_1)u + \mathbf{m}\dot{y} \tag{7.48}$$

The fact that the reduced-order observer requires the derivative of $y(t)$ as an input appears to present a practical difficulty. It is known that differentiation amplifies noise, so if y is noisy, the use of $\dot{y}$ is unacceptable. To get around this difficulty, we define the new state as
$$\mathbf{x}'_e \triangleq \hat{\mathbf{x}}_e - \mathbf{m}y \tag{7.49a}$$
Then, in terms of this new state, the implementation of the reduced-order observer is given by
$$\dot{\mathbf{x}}'_e = (\mathbf{A}_{ee} - \mathbf{ma}_{1e})\hat{\mathbf{x}}_e + (\mathbf{a}_{e1} - \mathbf{m}a_{11})y + (\mathbf{b}_e - \mathbf{m}b_1)u \tag{7.49b}$$
and $\dot{y}$ no longer appears directly. A block-diagram representation of the reduced-order observer is shown in Fig. 7.7.

Example 7.6

In Example 7.4, a second-order observer for the satellite control system was designed with the observer poles at $s = -10, -10$. We now design a reduced-order (first-order) observer for the system with observer pole at $s = -10$.
The plant equations are
$$\begin{bmatrix} \dot{x}_1 \\ \dot{x}_2 \end{bmatrix} = \begin{bmatrix} 0 & 1 \\ 0 & 0 \end{bmatrix} \begin{bmatrix} x_1 \\ x_2 \end{bmatrix} + \begin{bmatrix} 0 \\ 1 \end{bmatrix} u$$
$$y = \begin{bmatrix} 1 & 0 \end{bmatrix} \begin{bmatrix} x_1 \\ x_2 \end{bmatrix}$$

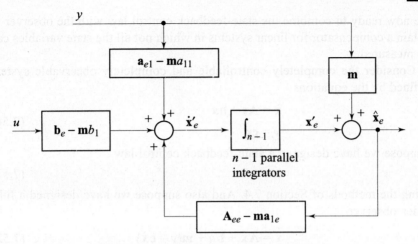

Fig. 7.7 Reduced-order observer structure

The partitioned matrices are

$$\left[\begin{array}{c|c} a_{11} & a_{1e} \\ \hline a_{e1} & A_{ee} \end{array}\right] = \left[\begin{array}{c|c} 0 & 1 \\ \hline 0 & 0 \end{array}\right] = \left[\begin{array}{c} b_1 \\ b_e \end{array}\right] = \left[\begin{array}{c} 0 \\ 1 \end{array}\right]$$

From Eqn. (7.46), we find the characteristic equation in terms of m:

$$s - (0 - m) = 0$$

We compare it with the desired equation

$$s + 10 = 0$$

which yields,

$$m = 10$$

The observer equations are (refer Eqns (7.49))

$$\dot{x}'_2 = -10\hat{x}_2 + u$$

$$\hat{x}_2 = x'_2 + 10y$$

This completes the design of the reduced-order observer which estimates the angular velocity of the satellite from the measurement of the angular position.

7.6 COMPENSATOR DESIGN BY THE SEPARATION PRINCIPLE

In Sections 7.2–7.4 we studied the design of control laws for systems in which the state variables are all accessible for measurement. We promised to overcome the difficulty of not being able to measure all the state variables by the use of an observer to estimate those state variables that cannot be measured. Then in Section 7.5, we studied the design of observers for systems with known inputs, but not when the state estimate is used for the purpose of control. We are now ready to combine the state-feedback control law with the observer to obtain a compensator for linear systems in which not all the state variables can be measured.

Consider the completely controllable and completely observable system defined by the equations

$$\dot{\mathbf{x}} = \mathbf{A}\mathbf{x} + \mathbf{b}u$$
$$y = \mathbf{c}\mathbf{x}$$
(7.50)

Suppose we have designed a state-feedback control law

$$u = -\mathbf{k}\mathbf{x}$$
(7.51)

using the methods of Section 7.4. And also suppose we have designed a full-order observer

$$\dot{\hat{\mathbf{x}}} = \mathbf{A}\hat{\mathbf{x}} + \mathbf{b}u + \mathbf{m}(y - \mathbf{c}\hat{\mathbf{x}})$$
(7.52)

using the methods of Section 7.5.

For the state-feedback control based on the observed state $\hat{\mathbf{x}}$,

$$u = -\mathbf{k}\hat{\mathbf{x}}$$
(7.53)

The control system based on combining the state-feedback control law and state observer, has the configuration shown in Fig. 7.8. Note that the number of state variables in the compensator is equal to the order of the embedded observer and hence is equal to the order of the plant. Thus, the order of the overall closed-loop system, when a full-order observer is used in the compensator, is $2n$ for a plant of order n. We are interested in the dynamic behaviour of the $2n$th-order system comprising the plant and the compensator. With the control law (7.53) used, the plant dynamics become

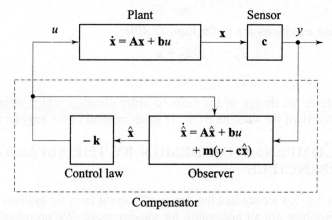

Fig. 7.8 Combined state-feedback control and state estimation

$$\dot{\mathbf{x}} = \mathbf{A}\mathbf{x} - \mathbf{b}\mathbf{k}\hat{\mathbf{x}} = (\mathbf{A} - \mathbf{b}\mathbf{k})\mathbf{x} + \mathbf{b}\mathbf{k}(\mathbf{x} - \hat{\mathbf{x}}) \tag{7.54}$$

The difference between the actual state $\mathbf{x}$ and observed state $\hat{\mathbf{x}}$ has been defined as the error $\tilde{\mathbf{x}}$:

$$\tilde{\mathbf{x}} = \mathbf{x} - \hat{\mathbf{x}}$$

Substitution of the error vector into Eqn. (7.54) gives

$$\dot{\mathbf{x}} = (\mathbf{A} - \mathbf{b}\mathbf{k})\mathbf{x} + \mathbf{b}\mathbf{k}\tilde{\mathbf{x}} \tag{7.55}$$

Note that the observer error was given by Eqn. (7.33), repeated here:

$$\dot{\tilde{\mathbf{x}}} = (\mathbf{A} - \mathbf{m}\mathbf{c})\tilde{\mathbf{x}} \tag{7.56}$$

Combining Eqns (7.55) and (7.56) we obtain

$$\begin{bmatrix} \dot{\mathbf{x}} \\ \dot{\tilde{\mathbf{x}}} \end{bmatrix} = \begin{bmatrix} \mathbf{A} - \mathbf{b}\mathbf{k} & \mathbf{b}\mathbf{k} \\ 0 & \mathbf{A} - \mathbf{m}\mathbf{c} \end{bmatrix} \begin{bmatrix} \mathbf{x} \\ \tilde{\mathbf{x}} \end{bmatrix} \tag{7.57}$$

Equation (7.57) describes the dynamics of the $2n$-dimensional system of Fig. 7.8. The characteristic equation for the system is

$$|s\mathbf{I} - (\mathbf{A} - \mathbf{b}\mathbf{k})| \, |s\mathbf{I} - (\mathbf{A} - \mathbf{m}\mathbf{c})| = 0$$

In other words, the poles of the combined system consist of the union of control and observer roots. This means that the design of the control law and the observer can be carried out independently. Yet, when they are used together the roots are unchanged. This is a special case of the *separation principle* which holds in much more general contexts and allows for the separate design of control law and estimator in certain stochastic cases.

To compare the state-variable method of design with the transform methods discussed in earlier chapters, we obtain the transfer function model of the compensator used in the control system of Fig. 7.8. The state variable model for this compensator is obtained by including the feedback law $u = -\mathbf{k}\hat{\mathbf{x}}$ (since it is part of the compensator) in the observer Eqn. (7.52).

$$\dot{\hat{\mathbf{x}}} = (\mathbf{A} - \mathbf{b}\mathbf{k} - \mathbf{m}\mathbf{c})\hat{\mathbf{x}} + \mathbf{m}y \tag{7.58}$$

$$u = -\mathbf{k}\hat{\mathbf{x}}$$

The formula for conversion of state variable model to the transfer function model is given by Eqn. (5.28). Applying this result to the model given by Eqn. (7.58), we obtain

$$\frac{U(s)}{-Y(s)} = D(s) = \mathbf{k}(s\mathbf{I} - \mathbf{A} + \mathbf{b}\mathbf{k} + \mathbf{m}\mathbf{c})^{-1}\mathbf{m} \tag{7.59}$$

Figure 7.9 shows the block diagram representation of the system with observer-based controller.

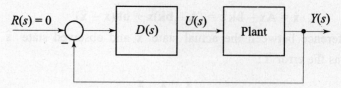

Fig. 7.9 Block diagram representation of a system with observer-based controller

Note that the poles of $D(s)$ in Eqn. (7.59) were never specified nor used during the state-variable design process. It may even happen that $D(s)$ has one or more poles in the right-half plane; the compensator, in other words, could turn out to be unstable. But the closed-loop system, if so designed, would be stable. There is however one problem if the compensator is unstable. The open-loop poles of the system are the poles of the plant and the poles of the compensator. If the latter are in the right-half plane, then the closed-loop poles may be in the right-half plane when the loop gain becomes too small. Robustness considerations put certain restrictions on the use of unstable compensators to stabilize a system.

Example 7.7

In this example, we study the closed-loop system obtained by implementing the state-feedback control law of Example 7.3 and state-observer design of Examples 7.4 and 7.6 for the attitude control of a satellite. The plant model is given by

$$\dot{x} = Ax + bu$$
$$y = cx$$

with

$$A = \begin{bmatrix} 0 & 1 \\ 0 & 0 \end{bmatrix}; \; b = \begin{bmatrix} 0 \\ 1 \end{bmatrix}; \; c = \begin{bmatrix} 1 & 0 \end{bmatrix}$$

In Example 7.3, the gain matrix required to place the closed-loop poles at $s = -4 \pm j4$ was calculated to be

$$k = [32 \quad 8]$$

If both the state variables are available for feedback, the control law becomes

$$u = -kx = -[32 \quad 8]x$$

resulting in the closed-loop system

$$\dot{x} = (A - bk)x = \begin{bmatrix} 0 & 1 \\ -32 & -8 \end{bmatrix} x$$

Figure 7.10a shows the response of the system to an initial condition $x(0) = [1 \quad 0]^T$. Assume now that the state-feedback control law is implemented

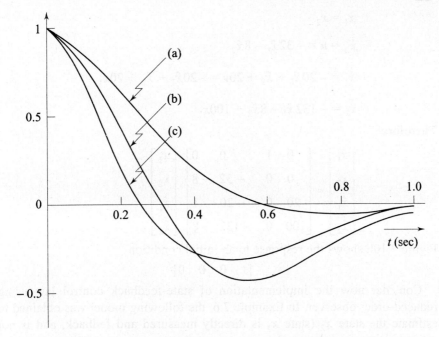

Fig. 7.10 Initial condition response; (a) without observer (b) with full-order observer, and (c) with reduced-order observer

using a full-order observer. In Example 7.4, the observer gain matrix was calculated to be

$$\mathbf{m} = \begin{bmatrix} 20 \\ 100 \end{bmatrix}$$

The state variable model of the compensator obtained by cascading the state-feedback control law and the state observer is obtained as (refer Eqns (7.58))

$$\dot{\hat{\mathbf{x}}} = (\mathbf{A} - \mathbf{bk} - \mathbf{mc})\hat{\mathbf{x}} + \mathbf{m}y = \begin{bmatrix} -20 & 1 \\ -132 & -8 \end{bmatrix} \hat{\mathbf{x}} + \begin{bmatrix} 20 \\ 100 \end{bmatrix} y$$

$$u = -\mathbf{k}\hat{\mathbf{x}} = -[32 \quad 8]\hat{\mathbf{x}}$$

The compensator transfer function is (refer Eqn. (7.59))

$$D(s) = \frac{U(s)}{-Y(s)} = \mathbf{k}(s\mathbf{I} - \mathbf{A} + \mathbf{bk} + \mathbf{mc})^{-1}\mathbf{m} = \frac{1440s + 3200}{s^2 + 28s + 292}$$

The state variable model of the closed-loop system can be constructed as follows:

$$\dot{x}_1 = x_2$$

$$\dot{x}_2 = u = -32\hat{x}_1 - 8\hat{x}_2$$

$$\dot{\hat{x}}_1 = -20\hat{x}_1 + \hat{x}_2 + 20y = -20\hat{x}_1 + \hat{x}_2 + 20x_1$$

$$\dot{\hat{x}}_2 = -132\hat{x}_1 - 8\hat{x}_2 + 100x_1$$

Therefore

$$\begin{bmatrix} \dot{x}_1 \\ \dot{x}_2 \\ \dot{\hat{x}}_1 \\ \dot{\hat{x}}_2 \end{bmatrix} = \begin{bmatrix} 0 & 1 & 0 & 0 \\ 0 & 0 & -32 & -8 \\ 20 & 0 & -20 & 1 \\ 100 & 0 & -132 & -8 \end{bmatrix} \begin{bmatrix} x_1 \\ x_2 \\ \hat{x}_1 \\ \hat{x}_2 \end{bmatrix}$$

Figure 7.10b shows the response to an initial condition

$$[1 \ 0 \ 0 \ 0]^T$$

Consider now the implementation of state-feedback control law using reduced-order observer. In Example 7.6, the following model was obtained to estimate the state x_2 (state x_1 is directly measured and fedback, and is not estimated using an observer):

$$\hat{x}_2 = x'_2 + 10y$$

$$\dot{x}'_2 = -10\hat{x}_2 + u$$

The control law is given by

$$u = -32x_1 - 8\hat{x}_2$$

From these equations, the following transfer function model of the compensator is obtained:

$$\frac{U(s)}{-Y(s)} = \frac{112(s+2.86)}{s+18}$$

The reduced-order compensator is precisely the lead network; this is a pleasant discovery, as it shows that the classical and state variable methods can result in exactly the same type of compensation.

The state variable model of the closed-loop system with the reduced-order compensator is derived below.

$$\dot{x}_1 = x_2$$

$$\dot{x}_2 = u = -32x_1 - 8\hat{x}_2 = -32x_1 - 8(x'_2 + 10x_1)$$

$$= -112x_1 - 8x'_2$$

$$\dot{x}'_2 = -10\hat{x}_2 + u = -10\hat{x}_2 - 32x_1 - 8\hat{x}_2$$

$$= -18(x'_2 + 10x_1) - 32x_1 = -18x'_2 - 212x_1$$

Therefore

$$\begin{bmatrix} \dot{x}_1 \\ \dot{x}_2 \\ \dot{x}'_2 \end{bmatrix} = \begin{bmatrix} 0 & 1 & 0 \\ -112 & 0 & -8 \\ -212 & 0 & -18 \end{bmatrix} \begin{bmatrix} x_1 \\ x_2 \\ x'_2 \end{bmatrix}$$

Figure 7.10c shows the response to an initial condition

$$[1 \quad 0 \quad 0]^T$$

Comments

1. Underlying the separation principle is a critical assumption, namely that the observer includes an exact dynamic model of the plant—the process under control. This assumption is almost never valid in reality. In practical systems, the precise dynamic model is rarely known. Even that which is known about the real process dynamics is often too complicated to include in the observer. Thus, the observer must in practice be configured to use only an approximate model of the plant. This encounter with the real world does not vitiate the separation principle, but it means that the effect of an inaccurate plant model must be considered. If the design achieved through use of the separation principle is *robust*, it will be able to tolerate uncertainty of the plant dynamics. Doyle and Stein [142] have proposed a 'design adjustment procedure' to improve robustness with observers.

2. One of the considerations in the design of a gain matrix **k** in the state-feedback control law is that the resulting control signal u not be too large: the use of large control effort increases the likelihood of the system entering nonlinear regions of operation. Since the function of the observer is only to process data, there is no limitation on the size of the gain matrix **m** for its realization. (Nowadays, it is all but certain that the entire compensator would be realized by a digital computer. With floating-point numerics, a digital computer would be capable of handling variables of any reasonable dynamic range). Though the realization of observer may impose no limitation on the observer dynamics, it may nevertheless be desirable to limit the observer speed of response (bandwidth). Remember that real sensors are noisy, and much of the noise occurs at relatively high frequencies. By limiting the bandwidth of the observer, we can attenuate and smooth the noise contribution to the compensator output—which is the control signal.

3. The desired closed-loop poles to be generated by state-feedback are chosen to satisfy the performance requirements. The poles of the observer are usually chosen so that the observer response is much faster than the system response. A rule of thumb is to choose an observer response at least two to five times faster than the system response. This is to ensure a faster decay of estimation errors compared with the desired dynamics, thus causing the closed-loop poles generated by state-feedback to dominate the total response. If the sensor noise is large enough to be a major concern, one may decide to choose the observer poles to be slower than two times the system poles, which would

yield a system with lower bandwidth and more noise-smoothing. However the total system response in this case will be strongly influenced by the observer poles. Doyle and Stein [142] have shown that the commonly suggested approach of 'speeding-up' observer dynamics will not work in all cases. They have suggested that procedures which drive some observer poles toward stable plant zeros and the rest toward infinity achieve the desired objective.

4. A final comment concerns the reduced-order observer. Due to the presence of a direct transmission term (see Fig. 7.7), the reduced-order observer has much higher bandwidth from sensor to control compared with the full-order observer. Therefore, if sensor noise is a significant factor, the reduced-order observer is less attractive, since the potential savings in complexity is more than offset by the increased sensitivity to noise.

7.7 SERVO DESIGN: INTRODUCTION OF THE REFERENCE INPUT BY FEEDFORWARD CONTROL

In the state regulator design studied in Section 7.4, the characteristic equation of the closed-loop system is chosen so as to give satisfactory transients to disturbances. However, no mention is made of a reference input or of design considerations to yield good transient response with respect to command changes. In general, these considerations should be taken into account in the design of a control system. This can be done by proper introduction of the reference input into the system equations.

Consider the completely controllable SISO linear time-invariant system with nth-order state variable model

$$\dot{\mathbf{x}}(t) = \mathbf{A}\mathbf{x}(t) + \mathbf{b}u(t)$$
$$y(t) = \mathbf{c}\mathbf{x}(t) \quad (7.60)$$

We assume that all the n state variables can be accurately measured at all times. Implementation of appropriately designed control law of the form

$$u(t) = -\mathbf{k}\mathbf{x}(t)$$

results in a state regulator system: any perturbation in the system state will asymptotically decay to the equilibrium state $\mathbf{x} = \mathbf{0}$.

Let us now assume that for the system given by Eqns (7.60), the desired steady-state value of the controlled variable $y(t)$ is a constant reference input r. For this servo system, the desired equilibrium state $\mathbf{x}_s$ is a constant point in state space and is governed by the equations

$$\mathbf{c}\mathbf{x}_s = r \quad (7.61)$$

We can formulate this command-following problem as a 'shifted regulator problem', by shifting the origin of the state space to the equilibrium point $\mathbf{x}_s$. Formulation of the shifted regulator problem is as follows.

Let u_s be the needed input to maintain $\mathbf{x}(t)$ at the equilibrium point $\mathbf{x}_s$, i.e. (refer Eqn. (7.60)),

$$\mathbf{0} = \mathbf{A}\mathbf{x}_s + \mathbf{b}u_s \quad (7.62)$$

Assuming for the present that a u_s exists that satisfies Eqns (7.61)–(7.62), we define shifted input, shifted state, and shifted controlled variable as

$$\tilde{u}(t) = u(t) - u_s$$
$$\tilde{\mathbf{x}}(t) = \mathbf{x}(t) - \mathbf{x}_s \qquad (7.63)$$
$$\tilde{y}(t) = y(t) - r$$

The shifted variables satisfy the equations

$$\dot{\tilde{\mathbf{x}}} = \mathbf{A}\tilde{\mathbf{x}} + \mathbf{b}\tilde{u}$$
$$\tilde{y} = \mathbf{c}\tilde{\mathbf{x}} \qquad (7.64)$$

This system possesses a time-invariant asymptotically stable control law

$$\tilde{u} = -\mathbf{k}\tilde{\mathbf{x}} \qquad (7.65)$$

The application of this control law ensures that

$$\tilde{\mathbf{x}} \to \mathbf{0} \quad (\mathbf{x}(t) \to \mathbf{x}_s, \ y(t) \to r)$$

In terms of the original state variables, total control effort

$$u(t) = -\mathbf{k}\mathbf{x}(t) + u_s + \mathbf{k}\mathbf{x}_s \qquad (7.66)$$

Manipulation of Eqn. (7.62) gives

$$(\mathbf{A} - \mathbf{b}\mathbf{k})\mathbf{x}_s + \mathbf{b}(u_s + \mathbf{k}\mathbf{x}_s) = 0 \quad \text{or} \quad \mathbf{x}_s = -(\mathbf{A} - \mathbf{b}\mathbf{k})^{-1}\mathbf{b}(u_s + \mathbf{k}\mathbf{x}_s)$$

or

$$\mathbf{c}\mathbf{x}_s = r = -\mathbf{c}(\mathbf{A} - \mathbf{b}\mathbf{k})^{-1}\mathbf{b}(u_s + \mathbf{k}\mathbf{x}_s)$$

This equation has a unique solution for $(u_s + \mathbf{k}\mathbf{x}_s)$:

$$(u_s + \mathbf{k}\mathbf{x}_s) = Nr$$

where N is a scalar feedforward gain, given by

$$(N)^{-1} = -\mathbf{c}(\mathbf{A} - \mathbf{b}\mathbf{k})^{-1}\mathbf{b} \qquad (7.67)$$

The control law (7.66), therefore takes the form

$$u(t) = -\mathbf{k}\mathbf{x}(t) + Nr \qquad (7.68)$$

The block diagram of Fig. 7.11 shows the configuration of feedback control system with feedforward compensation for non-zero equilibrium state.

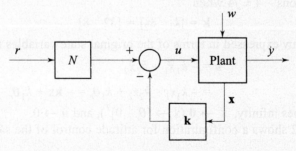

Fig. 7.11 Control configuration of a servo system

Example 7.8

The system considered in this example is the attitude control system for a rigid satellite. The plant equations are (refer Example 7.1)

$$\dot{\mathbf{x}} = \mathbf{A}\mathbf{x} + \mathbf{b}u$$

$$y = \mathbf{c}\mathbf{x}$$

where

$$\mathbf{A} = \begin{bmatrix} 0 & 1 \\ 0 & 0 \end{bmatrix}; \mathbf{b} = \begin{bmatrix} 0 \\ 1 \end{bmatrix}; \mathbf{c} = \begin{bmatrix} 1 & 0 \end{bmatrix}$$

$$x_1(t) = \text{position } \theta(t)$$

$$x_2(t) = \text{velocity } \dot{\theta}(t)$$

The reference input $r = \theta_r$, is a step function. The desired steady-state is

$$\mathbf{x}_s = [\theta_r \quad 0]^T,$$

which is a non-null state.

As the plant has integrating property, the steady-state value u_s of the input must be zero (otherwise the output cannot stay constant). For this case, the shifted regulator problem may be formulated as follows:

$$\tilde{x}_1 = x_1 - \theta_r$$

$$\tilde{x}_2 = x_2$$

Shifted state variables satisfy the equations

$$\dot{\tilde{\mathbf{x}}} = \mathbf{A}\tilde{\mathbf{x}} + \mathbf{b}u$$

The state-feedback control

$$u = -\mathbf{k}\tilde{\mathbf{x}}$$

results in dynamics of $\tilde{\mathbf{x}}$ given by

$$\dot{\tilde{\mathbf{x}}} = (\mathbf{A} - \mathbf{b}\mathbf{k})\tilde{\mathbf{x}}$$

In Example 7.1, we found that the eigenvalues of $(\mathbf{A} - \mathbf{b}\mathbf{k})$ are placed at the desired locations $-4 \pm j4$ when

$$\mathbf{k} = [k_1 \quad k_2] = [32 \quad 8]$$

The control law expressed in terms of the original state variables is given as

$$u = -k_1 \tilde{x}_1 - k_2 \tilde{x}_2$$
$$= -k_1 x_1 - k_2 x_2 + k_1 \theta_r = -\mathbf{k}\mathbf{x} + k_1 \theta_r$$

As t approaches infinity, $\tilde{\mathbf{x}} \to \mathbf{0}$ ($\mathbf{x} \to [\theta_r \quad 0]^T$), and $u \to 0$.

Figure 7.12 shows a configuration for attitude control of the satellite.

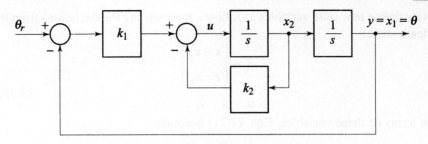

Fig. 7.12 Attitude control of a satellite

In fact, control configuration of the form shown in Fig. 7.12 may be used for any SISO plant with integrating property.

7.8 STATE FEEDBACK WITH INTEGRAL CONTROL

Control configuration of Fig. 7.12 produces a generalization of proportional and derivative feedback but it does not include integral control unless special steps are taken in the design process. One way to introduce integral control is to augment the state vector with the desired integral. More specifically, for the system (7.60), we can feedback the state **x** as well as the integral of the error in output by augmenting the plant state **x** with the extra 'integral state' z, defined by the equation

$$z(t) = \int_0^t (y(t) - r) dt \tag{7.69a}$$

where r is the constant reference input of the system. Since $z(t)$ satisfies the differential equation

$$\dot{z}(t) = y(t) - r = \mathbf{c}\mathbf{x}(t) - r \tag{7.69b}$$

it is easily included by augmenting the original system (7.60) as follows:

$$\begin{bmatrix} \dot{\mathbf{x}} \\ \dot{z} \end{bmatrix} = \begin{bmatrix} \mathbf{A} & \mathbf{0} \\ \mathbf{c} & 0 \end{bmatrix} \begin{bmatrix} \mathbf{x} \\ z \end{bmatrix} + \begin{bmatrix} \mathbf{b} \\ 0 \end{bmatrix} u + \begin{bmatrix} \mathbf{0} \\ -1 \end{bmatrix} r \tag{7.70}$$

Since r is constant, in the steady-state $\dot{\mathbf{x}} = \mathbf{0}$, $\dot{z} = 0$, provided that the system is stable. This means that the steady-state solutions $\mathbf{x}_s$, z_s and u_s must satisfy the equation

$$\begin{bmatrix} \mathbf{0} \\ -1 \end{bmatrix} r = -\begin{bmatrix} \mathbf{A} & \mathbf{0} \\ \mathbf{c} & 0 \end{bmatrix} \begin{bmatrix} \mathbf{x}_s \\ z_s \end{bmatrix} - \begin{bmatrix} \mathbf{b} \\ 0 \end{bmatrix} u_s$$

Substituting this for the last term in Eqn. (7.70) gives

$$\begin{bmatrix} \dot{\mathbf{x}} \\ \dot{z} \end{bmatrix} = \begin{bmatrix} \mathbf{A} & \mathbf{0} \\ \mathbf{c} & 0 \end{bmatrix} \begin{bmatrix} \mathbf{x} - \mathbf{x}_s \\ z - z_s \end{bmatrix} + \begin{bmatrix} \mathbf{b} \\ 0 \end{bmatrix} (u - u_s) \tag{7.71}$$

Now define new state variables as follows, representing the deviations from the steady-state:

$$\tilde{x} = \begin{bmatrix} x - x_s \\ z - z_s \end{bmatrix}$$

$$\tilde{u} = u - u_s \qquad (7.72a)$$

In terms of these variables, Eqn. (7.71) becomes

$$\dot{\tilde{x}} = \overline{A}\tilde{x} + \overline{b}\tilde{u} \qquad (7.72b)$$

$$\overline{A} = \begin{bmatrix} A & 0 \\ c & 0 \end{bmatrix}, \quad \overline{b} = \begin{bmatrix} b \\ 0 \end{bmatrix}$$

The significance of this result is that by defining the deviations from steady-state as state and control variables, the design problem has been reformulated to be the standard regulator problem with $\tilde{x} = 0$ as the desired state. We assume that an asymptotically stable solution to this problem exists and is given by

$$\tilde{u} = -k\tilde{x}$$

Partitioning k appropriately and using Eqns (7.72a) yields

$$k = [k_p \quad k_i]$$

$$u - u_s = -[k_p \quad k_i]\begin{bmatrix} x - x_s \\ z - z_s \end{bmatrix} = -k_p(x - x_s) - k_i(z - z_s)$$

The steady-state terms must balance, therefore

$$u = -k_p x - k_i z = -k_p x - k_i \int_0^t (y(t) - r)dt \qquad (7.73)$$

The control, thus, consists of proportional state feedback and integral control of output error. At steady-state, $\dot{\tilde{x}} = 0$; therefore

$$\lim_{t \to \infty} \dot{z}(t) \to 0 \quad \text{or} \quad \lim_{t \to \infty} y(t) \to r$$

Thus, by integrating action, the output y is driven to the no-offset condition. This will be true even in the presence of constant disturbances acting on the plant. Block diagram of Fig. 7.13 shows the configuration of feedback control system with proportional state feedback and integral control of output error.

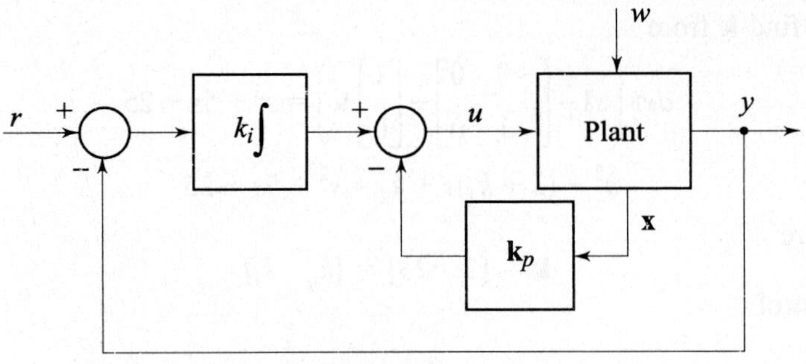

Fig. 7.13 State feedback with integral control

Example 7.9

Suppose the system is given by

$$\frac{Y(s)}{U(s)} = \frac{1}{s+3}$$

with a constant reference command signal. We wish to have integral control with closed-loop poles at $\omega_n = 5$ and $\zeta = 0.5$, which is equivalent to asking for a desired characteristic equation

$$s^2 + 5s + 25 = 0$$

The plant model is

$$\dot{x} = -3x + u$$
$$y = x$$

Augmenting the plant state x with the integral state z defined by the equation

$$z(t) = \int_0^t (y(t) - r)dt$$

we obtain

$$\begin{bmatrix} \dot{x} \\ \dot{z} \end{bmatrix} = \begin{bmatrix} -3 & 0 \\ 1 & 0 \end{bmatrix} \begin{bmatrix} x \\ z \end{bmatrix} + \begin{bmatrix} 1 \\ 0 \end{bmatrix} u + \begin{bmatrix} 0 \\ -1 \end{bmatrix} r$$

In terms of state and control variables representing deviations from the steady-state:

$$\tilde{\mathbf{x}} = \begin{bmatrix} x - x_s \\ z - z_s \end{bmatrix},$$

$$\tilde{u} = u - u_s$$

the state equation becomes

$$\dot{\tilde{\mathbf{x}}} = \begin{bmatrix} -3 & 0 \\ 1 & 0 \end{bmatrix} \tilde{\mathbf{x}} + \begin{bmatrix} 1 \\ 0 \end{bmatrix} \tilde{u}$$

We can find **k** from

$$\det\left(s\mathbf{I} - \begin{bmatrix} -3 & 0 \\ 1 & 0 \end{bmatrix} + \begin{bmatrix} 1 \\ 0 \end{bmatrix}\mathbf{k}\right) = s^2 + 5s + 25$$

or
$$s^2 + (3 + k_1)s + k_2 = s^2 + 5s + 25$$

Therefore
$$\mathbf{k} = [2 \quad 25] = [k_p \quad k_i]$$

The control

$$u = -k_p x - k_i z = -2x - 25\int_0^t (y(t) - r)dt$$

The control configuration is shown in Fig. 7.14, along with a disturbance input w. This system will behave according to the desired closed-loop roots ($\omega_n = 5$, $\zeta = 0.5$) and will exhibit the characteristics of integral control: zero steady-state error to a step r and zero steady-state error to a constant disturbance w.

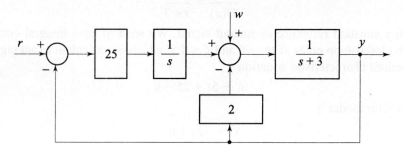

Fig. 7.14 Integral control example

7.9 DIGITAL CONTROL SYSTEMS WITH STATE FEEDBACK

This section covers the key results on the pole-placement design and state observers for discrete-time systems. Our discussion will be brief because of the strong analogy between the discrete-time and continuous-time cases. Consider the discretized model of the given plant:

$$\mathbf{x}(k+1) = \mathbf{F}\mathbf{x}(k) + \mathbf{g}u(k)$$
$$y(k) = \mathbf{c}\mathbf{x}(k) \tag{7.74}$$

where **x** is the $n \times 1$ state vector, u is the scalar input, y is the scalar output; **F**, **g**, and **c** are, respectively, $n \times n$, $n \times 1$ and $1 \times n$ real constant matrices; and $k = 0, 1, 2, ...$

We will carry out the design of digital control system for the plant (7.74) in two steps. One step assumes that we have all the elements of the state vector at our disposal for feedback purposes. The next step is to design a state observer which estimates the entire state vector when provided with the measurements of the system indicated by the output equation in (7.74).

Pole-Placement Design and State Observers

The final step will consist of combining the control law and the observer, where the control law calculations are based on the estimated state variables rather than the actual state.

State Regulator Design

Consider the nth-order, single-input, linear time-invariant system (7.74) with state-feedback control law

$$u(k) = -\mathbf{k}\mathbf{x}(k) \tag{7.75}$$

where
$$\mathbf{k} = [k_1 \quad k_2 \quad \cdots \quad k_n]$$

The resulting closed-loop system is

$$\mathbf{x}(k+1) = (\mathbf{F} - \mathbf{g}\mathbf{k})\mathbf{x}(k) \tag{7.76}$$

If all the eigenvalues of $(\mathbf{F} - \mathbf{g}\mathbf{k})$ are placed inside the unit circle in the complex plane, the state $\mathbf{x}(k)$ will decay to the equilibrium state $\mathbf{x} = \mathbf{0}$ irrespective of the value of $\mathbf{x}(0)$—the initial perturbation in the state.

A necessary and sufficient condition for arbitrary placement of closed-loop eigenvalues (with the restriction that complex eigenvalues occur in conjugate pairs) is that the system (7.74) is completely controllable.

The characteristic equation of the closed-loop system is

$$|z\mathbf{I} - (\mathbf{F} - \mathbf{g}\mathbf{k})| = 0 \tag{7.77a}$$

Assuming that the desired characteristic equation is

$$(z - \lambda_1)(z - \lambda_2) \cdots (z - \lambda_n) = z^n + a_1 z^{n-1} + \cdots + a_{n-1} z + a_n = 0 \tag{7.77b}$$

the required elements of $\mathbf{k}$ are obtained by matching coefficients in Eqns (7.77a) and (7.77b).

The calculation of the gains using this method becomes rather tedious when the order of the system is greater than three. The algebra for finding the gains becomes especially simple when the state variable equations are in controllable canonical form. A design procedure based on the use of controllable canonical state variable model, is given below (refer Eqns (7.20)–(7.23)):

Step 1: From the characteristic polynomial of matrix $\mathbf{F}$:

$$|z\mathbf{I} - \mathbf{F}| = z^n + \alpha_1 z^{n-1} + \cdots + \alpha_{n-1} z + \alpha_n \tag{7.78}$$

determine the values of $\alpha_1, \alpha_2, \ldots, \alpha_n$.

Step 2: Determine the transformation matrix $\mathbf{P}$ that transforms the system (7.74) into controllable canonical form:

$$\mathbf{P} = \begin{bmatrix} \mathbf{p}_1 \\ \mathbf{p}_1 \mathbf{F} \\ \vdots \\ \mathbf{p}_1 \mathbf{F}^{n-1} \end{bmatrix} ; \quad \begin{aligned} \mathbf{p}_1 &= [0 \quad 0 \cdots 0 \quad 1]\mathbf{U}^{-1} \\ \mathbf{U} &= [\mathbf{g} \quad \mathbf{F}\mathbf{g} \cdots \mathbf{F}^{n-1}\mathbf{g}] \end{aligned} \tag{7.79}$$

Step 3: Using the desired eigenvalues (desired closed-loop poles) $\lambda_1, \lambda_2, \ldots, \lambda_n$, write the desired characteristic polynomial:

$$(z - \lambda_1)(z - \lambda_2) \cdots (z - \lambda_n) = z^n + a_1 z^{n-1} + \cdots + a_{n-1} z + a_n, \tag{7.80}$$

and determine the values of $a_1, a_2, \ldots, a_{n-1}, a_n$.

Step 4: The required state-feedback gain matrix is determined from the following equation:

$$\mathbf{k} = [a_n - \alpha_n \quad a_{n-1} - \alpha_{n-1} \quad \cdots \quad a_1 - \alpha_1]\mathbf{P} \qquad (7.81)$$

▲▲

The Ackermann's formula given below is more convenient for computer solution (refer Eqns (7.26)).

$$\mathbf{k} = [0 \quad 0 \quad \cdots \quad 0 \quad 1\,]\mathbf{U}^{-1}\phi(\mathbf{F}) \qquad (7.82)$$

where

$$\phi(\mathbf{F}) = \mathbf{F}^n + a_1\mathbf{F}^{n-1} + \cdots + a_{n-1}\mathbf{F} + a_n\mathbf{I}$$

$$\mathbf{U} = [\mathbf{g} \quad \mathbf{Fg} \quad \cdots \quad \mathbf{F}^{n-1}\mathbf{g}]$$

Example 7.10

Consider the problem of attitude control of a rigid satellite. A state variable model of the plant is (refer Eqn. (7.6))

$$\dot{\mathbf{x}} = \mathbf{A}\mathbf{x} + \mathbf{b}u = \begin{bmatrix} 0 & 1 \\ 0 & 0 \end{bmatrix}\mathbf{x} + \begin{bmatrix} 0 \\ 1 \end{bmatrix}u$$

where $x_1 = \theta$ is the attitude angle and u is the system input.

The discrete-time description of the plant (assuming that the input u is applied through a zero-order hold (ZOH)) is given below (refer Section 6.3):

$$\mathbf{x}(k+1) = \mathbf{F}\mathbf{x}(k) + \mathbf{g}u(k) \qquad (7.83)$$

where

$$\mathbf{F} = e^{\mathbf{A}T} = \begin{bmatrix} 1 & T \\ 0 & 1 \end{bmatrix}$$

$$\mathbf{g} = \int_0^T e^{\mathbf{A}\tau}\mathbf{b}d\tau = \begin{bmatrix} T^2/2 \\ T \end{bmatrix}$$

The characteristic equation of the open-loop system is

$$|z\mathbf{I} - \mathbf{F}| = \begin{vmatrix} z-1 & -T \\ 0 & z-1 \end{vmatrix} = (z-1)^2 = 0$$

With the control law

$$u(k) = -\mathbf{k}\mathbf{x}(k) = -[k_1 \quad k_2]\mathbf{x}(k)$$

the closed-loop system becomes

$$\mathbf{x}(k+1) = (\mathbf{F} - \mathbf{g}\mathbf{k})\mathbf{x}(k)$$

The characteristic equation of the closed-loop system is

$$|z\mathbf{I} - (\mathbf{F} - \mathbf{g}\mathbf{k})|$$
$$= z^2 + (Tk_2 + (T^2/2)k_1 - 2)z + (T^2/2)k_1 - Tk_2 + 1 = 0 \qquad (7.84a)$$

We assume that $T = 0.1$ sec, and the desired characteristic roots of the closed-loop system are $z_{1,2} = 0.875 \angle \pm 17.9°$.

Note that these roots correspond to $\zeta = 0.5$, and $\omega_n = 3.6$ (refer Eqns (4.15)):

$$z_{1,2} = e^{-\zeta \omega_n T} e^{\pm j\omega_n T \sqrt{1-\zeta^2}}$$

The desired characteristic equation is then (approximately)

$$z^2 - 1.6z + 0.70 = 0 \qquad (7.84b)$$

Matching coefficients in Eqns (7.84a) and (7.84b), we obtain

$$k_1 = 10, \ k_2 = 3.5$$

Design of State Observers

The control law designed in the last sub-section assumed that all states were available for feedback. Since, typically, not all states are measured, the purpose of this sub-section is to show how to determine algorithms which will reconstruct all the states, given measurements of a portion of them. If the state is $\mathbf{x}$, then the estimate is $\hat{\mathbf{x}}$ and the idea is to let $u = -\mathbf{k}\hat{\mathbf{x}}$; replacing the true states by their estimates in the control law.

Prediction Observer

An estimation scheme employing a full-order observer is shown in Fig. 7.15, and the equation for it is

$$\hat{\mathbf{x}}(k+1) = \mathbf{F}\hat{\mathbf{x}}(k) + \mathbf{g}u(k) + \mathbf{m}(y(k) - \mathbf{c}\hat{\mathbf{x}}(k)) \qquad (7.85)$$

where $\mathbf{m}$ is an $n \times 1$ real constant gain matrix. We will call this a *prediction observer* because the estimate $\hat{\mathbf{x}}(k+1)$ is one sampling period ahead of the measurement $y(k)$.

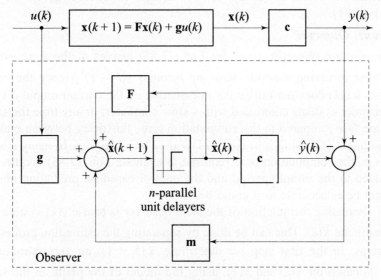

Fig. 7.15 Prediction observer

A difference equation describing the behaviour of the error is obtained by subtracting Eqn. (7.85) from Eqn. (7.74):

$$\tilde{\mathbf{x}}(k+1) = (\mathbf{F} - \mathbf{mc})\,\tilde{\mathbf{x}}(k) \tag{7.86}$$

where

$$\tilde{\mathbf{x}} = \mathbf{x} - \hat{\mathbf{x}}$$

The characteristic equation of the error is given by

$$|z\mathbf{I} - (\mathbf{F} - \mathbf{mc})| = 0 \tag{7.87a}$$

Assuming that the desired characteristic equation is

$$(z - \lambda_1)(z - \lambda_2)\cdots(z - \lambda_n) = 0, \tag{7.87b}$$

the required elements of $\mathbf{m}$ are obtained by matching coefficients in Eqns (7.87a) and (7.87b). A necessary and sufficient condition for the arbitrary assignment of eigenvalues of $(\mathbf{F} - \mathbf{mc})$ is that the system (7.74) is completely observable.

The problem of designing a full-order observer is mathematically equivalent to designing a full state-feedback controller for the 'transposed auxiliary system'

$$\boldsymbol{\zeta}(k+1) = \mathbf{F}^T \boldsymbol{\zeta}(k) + \mathbf{c}^T \boldsymbol{\eta}(k) \tag{7.88a}$$

with feedback

$$\boldsymbol{\eta}(k) = -\mathbf{m}^T \boldsymbol{\zeta}(k) \tag{7.88b}$$

so that the closed-loop auxiliary system

$$\boldsymbol{\zeta}(k+1) = (\mathbf{F}^T - \mathbf{c}^T \mathbf{m}^T)\boldsymbol{\zeta}(k) \tag{7.88c}$$

has eigenvalues $\lambda_i;\ i = 1, 2, \ldots, n$.

This duality principle may be used to design full-order state observers by Ackermann's formula (7.82) or by design procedure given in Eqns (7.78)–(7.81).

Current Observer

The prediction observer given by Eqn. (7.85) arrives at the state estimate $\hat{\mathbf{x}}(k)$ after receiving measurements up through $y(k-1)$. Hence the control $u(k) = -\mathbf{k}\hat{\mathbf{x}}(k)$ does not utilize the information on the current output $y(k)$. For higher-order systems controlled with a slow computer, or any time the sample rates are fast compared to the computation time, this delay between making a measurement and using it in control law may be a blessing. In many systems, however, the computation time required to evaluate Eqn. (7.85) is quite short compared to the sample period and the control based on prediction observer may not be as accurate as it could be.

An alternative formulation of the state observer is to use $y(k)$ to obtain the state estimate $\hat{\mathbf{x}}(k)$. This can be done by separating the estimation process into two steps. In the first step we determine $\bar{\mathbf{x}}(k+1)$, an approximation of $\mathbf{x}(k+1)$ based on $\hat{\mathbf{x}}(k)$ and $u(k)$, using the model of the plant. In the second step we use $y(k+1)$ to improve $\bar{\mathbf{x}}(k+1)$. The improved $\bar{\mathbf{x}}(k+1)$ is $\hat{\mathbf{x}}(k+1)$.

The state observer based on this formulation is called the *current observer*. The current observer equations are given by

$$\bar{\mathbf{x}}(k+1) = \mathbf{F}\hat{\mathbf{x}}(k) + \mathbf{g}u(k) \tag{7.89a}$$

$$\hat{\mathbf{x}}(k+1) = \bar{\mathbf{x}}(k+1) + \mathbf{m}[y(k+1) - \mathbf{c}\bar{\mathbf{x}}(k+1)] \tag{7.89b}$$

In practice, the current observer cannot be implemented exactly because it is impossible to sample, perform calculations, and output with absolutely no time elapsing. However, the errors introduced due to computational delays will be negligible if the computation time is quite short compared to the sample period.

The error equation for the current observer is similar to the error equation for the prediction observer that was given in (7.86). The current-estimate error equation is obtained by subtracting Eqns (7.89) from (7.74).

$$\begin{aligned}\tilde{\mathbf{x}}(k+1) &= \mathbf{x}(k+1) - \hat{\mathbf{x}}(k+1) \\ &= \mathbf{F}\mathbf{x}(k) + \mathbf{g}u(k) - \mathbf{F}\hat{\mathbf{x}}(k) - \mathbf{g}u(k) - \mathbf{mc}[\mathbf{x}(k+1) - \bar{\mathbf{x}}(k+1)] \\ &= \mathbf{F}\tilde{\mathbf{x}}(k) - \mathbf{mcF}\tilde{\mathbf{x}}(k) = (\mathbf{F} - \mathbf{mcF})\tilde{\mathbf{x}}(k) \end{aligned} \tag{7.90}$$

Therefore, the gain matrix $\mathbf{m}$ is obtained exactly as before, except that $\mathbf{c}$ is replaced by $\mathbf{cF}$.

Reduced-Order Observer

The observers discussed so far are designed to reconstruct the entire state vector, given measurements of some of the states. To pursue an observer for only the unmeasured states, we partition the state vector into two parts: one part is x_1 which is directly measured, and the other part is $\mathbf{x}_e$, representing the state variables that need to be estimated. If we partition the system matrices accordingly, the complete description of the system (7.74) is given by

$$\begin{bmatrix} x_1(k+1) \\ \mathbf{x}_e(k+1) \end{bmatrix} = \begin{bmatrix} f_{11} & \mathbf{f}_{1e} \\ \mathbf{f}_{e1} & \mathbf{F}_{ee} \end{bmatrix} \begin{bmatrix} x_1(k) \\ \mathbf{x}_e(k) \end{bmatrix} + \begin{bmatrix} g_1 \\ \mathbf{g}_e \end{bmatrix} u(k) \tag{7.91a}$$

$$y(k) = \begin{bmatrix} 1 & \mathbf{0} \end{bmatrix} \begin{bmatrix} x_1(k) \\ \mathbf{x}_e(k) \end{bmatrix} \tag{7.91b}$$

The portion describing the dynamics of unmeasured states is

$$\mathbf{x}_e(k+1) = \mathbf{F}_{ee}\mathbf{x}_e(k) + \underbrace{\mathbf{f}_{e1}x_1(k) + \mathbf{g}_e u(k)}_{\text{known input}} \tag{7.92}$$

The measured dynamics are given by the scalar equation

$$\underbrace{y(k+1) - f_{11}y(k) - g_1 u(k)}_{\text{known measurement}} = \mathbf{f}_{1e}\mathbf{x}_e(k) \tag{7.93}$$

Equations (7.92) and (7.93) have the same relationship to the state $\mathbf{x}_e$ that the original Eqns (7.74) had to the entire state $\mathbf{x}$. Following this reasoning, we arrive at the desired observer by making the following substitutions into the observer equations:

$$\mathbf{x} \leftarrow \mathbf{x}_e$$

$$\mathbf{F} \leftarrow \mathbf{F}_{ee}$$

$$\mathbf{g}u(k) \leftarrow \mathbf{f}_{e1} y(k) + \mathbf{g}_e u(k) \tag{7.94}$$

$$y(k) \leftarrow y(k+1) - f_{11}y(k) - g_1 u(k)$$

$$\mathbf{c} \leftarrow \mathbf{f}_{1e}$$

Thus, the reduced-order observer equations are

$$\hat{\mathbf{x}}_e(k+1) = \mathbf{F}_{ee}\hat{\mathbf{x}}_e(k) + \underbrace{\mathbf{f}_{e1}y(k) + \mathbf{g}_e u(k)}_{\text{input}}$$

$$+ \mathbf{m}\underbrace{(y(k+1) - f_{11}y(k) - g_1 u(k) - \mathbf{f}_{1e}\hat{\mathbf{x}}_e(k))}_{\text{measurement}} \tag{7.95}$$

Subtracting Eqn. (7.95) from (7.92) yields the error equation

$$\tilde{\mathbf{x}}_e(k+1) = (\mathbf{F}_{ee} - \mathbf{m}\mathbf{f}_{1e})\tilde{\mathbf{x}}_e(k) \tag{7.96}$$

where

$$\tilde{\mathbf{x}}_e = \mathbf{x}_e - \hat{\mathbf{x}}_e$$

The characteristic equation is given by

$$|z\mathbf{I} - (\mathbf{F}_{ee} - \mathbf{m}\mathbf{f}_{1e})| = 0 \tag{7.97}$$

We design the dynamics of this observer by selecting $\mathbf{m}$ so that Eqn. (7.97) matches a desired reduced-order characteristic equation. The design may be carried out directly or by using duality principle.

Compensator Design by the Separation Principle

If we take the state-feedback control law and implement it using an estimated state vector, the control system can be completed. A schematic of such a scheme using a prediction observer[1] is shown in Fig. 7.16. Note that by the separation principle, the control law and the state observer can be designed separately, and yet used together.

The portion within the dotted line in Fig. 7.16 corresponds to dynamic compensation. The state variable model of the compensator is obtained by including the state-feedback control (since it is a part of the compensator) in the observer equations, yielding

$$\hat{\mathbf{x}}(k+1) = (\mathbf{F} - \mathbf{g}\mathbf{k} - \mathbf{m}\mathbf{c})\hat{\mathbf{x}}(k) + \mathbf{m}y(k) \tag{7.98}$$

$$u(k) = -\mathbf{k}\hat{\mathbf{x}}(k)$$

---

1. We will design the compensator only for the prediction observer case. The other observers give very similar results.

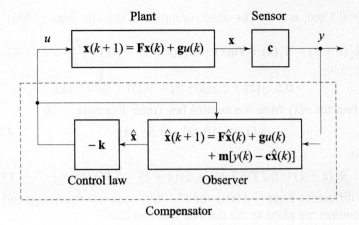

Fig. 7.16 Combined state-feedback control and state estimation

The formula for conversion of a discrete-time state variable model to the transfer function model is given by Eqn. (6.3). Applying this result to the model (7.98), we obtain

$$\frac{U(z)}{-Y(z)} = D(z) = \mathbf{k}(z\mathbf{I} - \mathbf{F} + \mathbf{g}\mathbf{k} + \mathbf{m}\mathbf{c})^{-1}\mathbf{m} \qquad (7.99)$$

Example 7.11

As an example of complete design, we will add a state observer to the satellite attitude control considered in Example 7.10. The system equations of motion are (refer Eqn. (7.83))

$$\mathbf{x}(k+1) = \mathbf{F}\mathbf{x}(k) + \mathbf{g}u(k) = \begin{bmatrix} 1 & T \\ 0 & 1 \end{bmatrix} \mathbf{x}(k) + \begin{bmatrix} T^2/2 \\ T \end{bmatrix} u(k)$$

We assume that the position state x_1 is measured and the velocity state x_2 is to be estimated; the measurement equation is therefore

$$y(k) = \mathbf{c}\mathbf{x}(k) = \begin{bmatrix} 1 & 0 \end{bmatrix} \mathbf{x}(k)$$

We will design a first-order observer for the state $x_2(k)$.

The partitioned matrices are

$$\begin{bmatrix} f_{11} & f_{1e} \\ f_{e1} & F_{ee} \end{bmatrix} = \begin{bmatrix} 1 & T \\ 0 & 1 \end{bmatrix}; \quad \begin{bmatrix} g_1 \\ g_e \end{bmatrix} = \begin{bmatrix} T^2/2 \\ T \end{bmatrix}$$

From Eqn. (7.97), we find the characteristic equation in terms of m:

$$z - (1 - mT) = 0$$

For the observer to be about four times faster than the control, we place the observer pole at

$$z = 0.5 \; (\cong (0.835)^4); \text{ therefore}$$

$$1 - mT = 0.5$$

For: $T = 0.1$ sec, $m = 5$. The observer equation is (refer Eqn. (7.95))

$$\hat{x}_2(k+1) = \hat{x}_2(k) + Tu(k) + m(y(k+1) - y(k) - \frac{T^2}{2} u(k) - T\hat{x}_2(k))$$

$$= 0.5\hat{x}_2(k) + 5(y(k+1) - y(k)) + 0.075u(k)$$

Substituting for $u(k)$ from the control law (refer Example 7.10)

$$u(k) = -10y(k) - 3.5\hat{x}_2(k), \quad (7.100a)$$

we obtain

$$\hat{x}_2(k+1) = 0.2375\hat{x}_2(k) + 5y(k+1) - 5.75y(k) \quad (7.100b)$$

The two difference Eqns (7.100a) and (7.100b) complete the design and can be used to control the plant to the desired specifications.

To relate the observer-based state-feedback design to a classical design, one needs to compute the z-transform of Eqns (7.100a) and (7.100b) obtaining

$$\frac{U(z)}{-Y(z)} = \frac{27.5(z - 0.818)}{z - 0.2375}$$

The compensation looks very much like the classical lead compensation that would be used for $1/s^2$ plant.

Servo Design

Let us assume that for the system given by Eqns (7.74), the desired steady-state value for the controlled variable $y(k)$ is a constant reference input r. For this servo system, the desired equilibrium state $\mathbf{x}_s$ is a constant point in state space, and is governed by the equations

$$\mathbf{c}\mathbf{x}_s = r \quad (7.101a)$$

We formulate this command following problem as a 'shifted regulator problem' by shifting the origin of the state space to the equilibrium point $\mathbf{x}_s$. Let u_s be the needed input to maintain $\mathbf{x}(k)$ at the equilibrium point $\mathbf{x}_s$, i.e. (refer Eqns (7.74)),

$$\mathbf{x}_s = \mathbf{F}\mathbf{x}_s + \mathbf{g}u_s \quad (7.101b)$$

Assuming for the present that a u_s exists that satisfies Eqns (7.101a)–(7.101b), we define shifted input, shifted state, and shifted controlled variable as

$$\tilde{u}(k) = u(k) - u_s$$
$$\tilde{\mathbf{x}}(k) = \mathbf{x}(k) - \mathbf{x}_s \quad (7.102)$$
$$\tilde{y}(k) = y(k) - r$$

The shifted variables satisfy the equations

$$\tilde{\mathbf{x}}(k+1) = \mathbf{F}\tilde{\mathbf{x}}(k) + \mathbf{g}\tilde{u}(k)$$
$$\tilde{y}(k) = \mathbf{c}\tilde{\mathbf{x}}(k) \quad (7.103)$$

This system possesses a time-invariant asymptotically stable control law (assuming {**F**, **g**} is controllable)

$$\tilde{u} = -\mathbf{k}\tilde{\mathbf{x}}$$

The application of this control law ensures that

$$\tilde{\mathbf{x}}(k) \to \mathbf{0} \quad (\mathbf{x}(k) \to \mathbf{x}_s; y(k) \to r)$$

In terms of the original state variables, total control effort

$$u(k) = -\mathbf{k}\mathbf{x}(k) + u_s + \mathbf{k}\mathbf{x}_s \tag{7.104}$$

Manipulation of Eqn. (7.101b) gives

$$(\mathbf{F} - \mathbf{g}\mathbf{k} - \mathbf{I})\mathbf{x}_s + \mathbf{g}(u_s + \mathbf{k}\mathbf{x}_s) = \mathbf{0}$$

or

$$\mathbf{x}_s = -(\mathbf{F} - \mathbf{g}\mathbf{k} - \mathbf{I})^{-1}\mathbf{g}(u_s + \mathbf{k}\mathbf{x}_s)$$

or

$$\mathbf{c}\mathbf{x}_s = r = -\mathbf{c}(\mathbf{F} - \mathbf{g}\mathbf{k} - \mathbf{I})^{-1}\mathbf{g}(u_s + \mathbf{k}\mathbf{x}_s)$$

This equation has a unique solution for $(u_s + \mathbf{k}\mathbf{x}_s)$:

$$(u_s + \mathbf{k}\mathbf{x}_s) = Nr$$

where N is a scalar feedforward gain, given by

$$(N)^{-1} = -\mathbf{c}(\mathbf{F} - \mathbf{g}\mathbf{k} - \mathbf{I})^{-1}\mathbf{g} \tag{7.105}$$

The control law (7.104), therefore, takes the form

$$u(k) = -\mathbf{k}\mathbf{x}(k) + Nr \tag{7.106}$$

State Feedback with Integral Control

In the following we study a control scheme for the system (7.74), where we feedback the state **x** as well as the integral of the error in the output.

One way to introduce an integrator is to augment the plant state vector **x** with the 'integral state' v that integrates the difference between the output $y(k)$ and the constant reference input r. The 'integral state' v is defined by

$$v(k) = v(k-1) + y(k) - r \tag{7.107a}$$

This equation can be rewritten as follows:

$$v(k+1) = v(k) + y(k+1) - r = v(k) + \mathbf{c}[\mathbf{F}\mathbf{x}(k) + \mathbf{g}u(k)] - r$$

$$= \mathbf{c}\mathbf{F}\mathbf{x}(k) + v(k) + \mathbf{c}\mathbf{g}u(k) - r \tag{7.107b}$$

From Eqns (7.74) and (7.107b), we obtain

$$\begin{bmatrix} \mathbf{x}(k+1) \\ v(k+1) \end{bmatrix} = \begin{bmatrix} \mathbf{F} & \mathbf{0} \\ \mathbf{c}\mathbf{F} & 1 \end{bmatrix} \begin{bmatrix} \mathbf{x}(k) \\ v(k) \end{bmatrix} + \begin{bmatrix} \mathbf{g} \\ \mathbf{c}\mathbf{g} \end{bmatrix} u(k) + \begin{bmatrix} \mathbf{0} \\ -1 \end{bmatrix} r \tag{7.108}$$

Since r is constant, in the steady-state $\mathbf{x}(k+1) = \mathbf{x}(k)$ and $v(k+1) = v(k)$, provided that the system is stable. This means that the steady-state solutions $\mathbf{x}_s$, v_s and u_s must satisfy the equation

$$\begin{bmatrix} \mathbf{0} \\ -1 \end{bmatrix} r = \begin{bmatrix} \mathbf{x}_s \\ v_s \end{bmatrix} - \begin{bmatrix} \mathbf{F} & \mathbf{0} \\ \mathbf{c}\mathbf{F} & 1 \end{bmatrix} \begin{bmatrix} \mathbf{x}_s \\ v_s \end{bmatrix} - \begin{bmatrix} \mathbf{g} \\ \mathbf{c}\mathbf{g} \end{bmatrix} u_s$$

Substituting this for the last term in Eqn. (7.108) gives

$$\tilde{\mathbf{x}}(k+1) = \overline{\mathbf{F}}\tilde{\mathbf{x}}(k) + \overline{\mathbf{g}}\tilde{u}(k) \qquad (7.109)$$

where

$$\tilde{\mathbf{x}} = \begin{bmatrix} \mathbf{x} - \mathbf{x}_s \\ v - v_s \end{bmatrix}$$

$$\tilde{u} = u - u_s$$

$$\overline{\mathbf{F}} = \begin{bmatrix} \mathbf{F} & 0 \\ \mathbf{cF} & 1 \end{bmatrix}, \overline{\mathbf{g}} = \begin{bmatrix} \mathbf{g} \\ \mathbf{cg} \end{bmatrix}$$

The significance of this result is that by defining the deviations from steady-state as state and control variables, the design problem has been reformulated to be the standard regulator problem with $\tilde{\mathbf{x}} = 0$ as the desired state. We assume that an asymptotically stable solution to this problem exists and is given by

$$\tilde{u}(k) = -\mathbf{k}\tilde{\mathbf{x}}(k)$$

Partitioning $\mathbf{k}$ appropriately and using Eqn. (7.109) yields

$$\mathbf{k} = [\mathbf{k}_p \quad k_i]$$

$$u - u_s = -[\mathbf{k}_p \quad k_i] \begin{bmatrix} \mathbf{x} - \mathbf{x}_s \\ v - v_s \end{bmatrix}$$

$$= -\mathbf{k}_p(\mathbf{x} - \mathbf{x}_s) - k_i(v - v_s)$$

The steady-state terms must balance, therefore

$$u(k) = -\mathbf{k}_p \mathbf{x}(k) - k_i v(k) \qquad (7.110)$$

At steady-state, $\tilde{\mathbf{x}}(k+1) - \tilde{\mathbf{x}}(k) = 0$; therefore

$$v(k+1) - v(k) = 0 = y(k) - r, \text{ i.e., } y(k) \to r$$

The block diagram of Fig. 7.17 shows the control configuration.

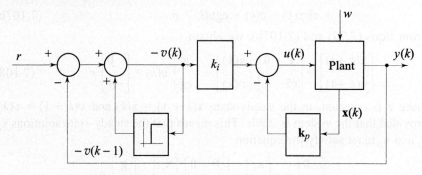

Fig. 7.17 State feedback with integral control

Example 7.12

Consider the problem of digital control of a plant described by the transfer function

$$G(s) = \frac{1}{s+3}$$

Discretization of the plant model gives

$$G_{h0}G(z) = \frac{Y(z)}{U(z)} = \mathscr{Z}\left[\left(\frac{1-e^{-sT}}{s}\right)\left(\frac{1}{s+3}\right)\right]$$

$$= (1 - z^{-1})\,\mathscr{Z}\left[\frac{1}{s(s+3)}\right] = \frac{1}{3}\left(\frac{1-e^{-3T}}{z-e^{-3T}}\right)$$

For a sampling interval $T = 0.1$ sec,

$$G_{h0}G(z) = \frac{0.0864}{z - 0.741}$$

The difference equation model of the plant is

$$y(k + 1) = 0.741y(k) + 0.0864u(k)$$

The plant has a constant reference command signal. We wish to design a PI control algorithm that results in system response characteristics: $\zeta = 0.5$, $\omega_n = 5$. This is equivalent to asking for the closed-loop poles at

$$z_{1,2} = e^{-\zeta\omega_n T}e^{\pm j\omega_n T\sqrt{1-\zeta^2}} = 0.7788 \angle \pm 24.82° = 0.7068 \pm j0.3269$$

The desired characteristic equation is therefore

$$(z - 0.7068 - j0.3269)(z - 0.7068 + j0.3269)$$
$$= z^2 - 1.4136z + 0.6065 = z^2 + a_1 z + a_2 = 0$$

Augmenting the plant state $y(k)$ with the 'integral state' $v(k)$ defined by

$$v(k) = v(k - 1) + y(k) - r,$$

we obtain

$$\begin{bmatrix} y(k+1) \\ v(k+1) \end{bmatrix} = \begin{bmatrix} 0.741 & 0 \\ 0.741 & 1 \end{bmatrix}\begin{bmatrix} y(k) \\ v(k) \end{bmatrix} + \begin{bmatrix} 0.0864 \\ 0.0864 \end{bmatrix} u(k) + \begin{bmatrix} 0 \\ -1 \end{bmatrix} r$$

In terms of state variables representing deviations from the steady-state:

$$\tilde{\mathbf{x}} = \begin{bmatrix} y - y_s \\ v - v_s \end{bmatrix},$$

the state equation becomes

$$\tilde{\mathbf{x}}(k + 1) = \overline{\mathbf{F}}\,\tilde{\mathbf{x}}(k) + \overline{\mathbf{g}}\,\tilde{u}(k)$$

where
$$\overline{\mathbf{F}} = \begin{bmatrix} 0.741 & 0 \\ 0.741 & 1 \end{bmatrix}; \overline{\mathbf{g}} = \begin{bmatrix} 0.0864 \\ 0.0864 \end{bmatrix}$$

By Ackermann's formula (7.82),
$$\mathbf{k} = \begin{bmatrix} 0 & 1 \end{bmatrix} \mathbf{U}^{-1} \phi(\overline{\mathbf{F}})$$
where
$$\phi(\overline{\mathbf{F}}) = \overline{\mathbf{F}}^2 + a_1 \overline{\mathbf{F}} + a_2 \mathbf{I} = \begin{bmatrix} 0.108 & 0 \\ 0.2425 & 0.1929 \end{bmatrix}$$

$$\mathbf{U}^{-1} = [\overline{\mathbf{g}} \quad \overline{\mathbf{F}}\overline{\mathbf{g}}]^{-1} = \begin{bmatrix} 0.0864 & 0.064 \\ 0.0864 & 0.15 \end{bmatrix}^{-1}$$

$$= \frac{1}{7.43 \times 10^{-3}} \begin{bmatrix} 0.15 & -0.064 \\ -0.0864 & 0.0864 \end{bmatrix}$$

This gives
$$\mathbf{k} = [1.564 \quad 2.243]$$

The control algorithm is given by
$$u(k) = -1.564 y(k) - 2.243 v(k)$$

7.10 DEADBEAT CONTROL BY STATE FEEDBACK AND DEADBEAT OBSERVERS

A completely controllable and observable SISO system of order n is considered.
$$\mathbf{x}(k + 1) = \mathbf{F}\mathbf{x}(k) + \mathbf{g}u(k)$$
$$y(k) = \mathbf{c}\mathbf{x}(k) \qquad (7.111)$$

With the state-feedback control law
$$u(k) = -\mathbf{k}\mathbf{x}(k) \qquad (7.112a)$$

the closed-loop system becomes
$$\mathbf{x}(k + 1) = (\mathbf{F} - \mathbf{g}\mathbf{k})\mathbf{x}(k) \qquad (7.112b)$$

with the characteristic equation
$$|z\mathbf{I} - (\mathbf{F} - \mathbf{g}\mathbf{k})| = 0 \qquad (7.112c)$$

The control-law design consists of picking the gains $\mathbf{k}$ so that Eqn. (7.112c) matches the desired characteristic equation
$$z^n + a_1 z^{n-1} + \cdots + a_{n-1} z + a_n = 0$$

A case of special interest occurs when $a_1 = a_2 = \cdots = a_{n-1} = a_n = 0$, that is, desired characteristic equation is
$$z^n = 0 \qquad (7.113)$$

By the Cayley-Hamilton theorem (a matrix satisfies its own characteristic equation),

$$(\mathbf{F} - \mathbf{gk})^n = \mathbf{0}$$

This result implies that the force-free response of closed-loop system (7.112b),

$$\mathbf{x}(k) = (\mathbf{F} - \mathbf{gk})^k \mathbf{x}(0) = \mathbf{0} \text{ for } k \geq n$$

In other words, any initial state $\mathbf{x}(0)$ is driven to the equilibrium state $\mathbf{x} = \mathbf{0}$ in (at most) n steps. The feedback control law that assigns all the closed-loop poles to origin is, therefore, a *deadbeat control law* (refer Section 4.5). A state observer defined by the equation

$$\hat{\mathbf{x}}(k+1) = \mathbf{F}\hat{\mathbf{x}}(k) + \mathbf{g}u(k) + \mathbf{m}[y(k) - \mathbf{c}\hat{\mathbf{x}}(k)] \quad (7.114)$$

gives an estimate $\hat{\mathbf{x}}(k)$ of the state $\mathbf{x}(k)$; the observer design procedure consists of picking the gains $\mathbf{m}$ so that the error system

$$\widetilde{\mathbf{x}}(k+1) = (\mathbf{F} - \mathbf{mc})\widetilde{\mathbf{x}}(k) \quad (7.115)$$

has the desired characteristic equation. A case of special interest occurs when all the observer poles (i.e., eigenvalues of $(\mathbf{F} - \mathbf{mc})$) are zero. In analogy with the deadbeat control law, we refer to observers with this property as *deadbeat observers*.

Comments

The concept of deadbeat performance was introduced earlier in Section 4.5. This concept is unique to discrete-time systems. By deadbeat control, any nonzero error vector will be driven to zero in atmost n sampling periods if the magnitude of the scalar control $u(k)$ is unbounded. The settling time depends on the sampling period T. If T is chosen very small, the settling time will also be very small, which implies that the control signal must have an extremely large magnitude. The designer must choose the sampling period so that an extremely large control magnitude is not required in normal operation of the system. Thus, in deadbeat control, the sampling period is the only design parameter.

Example 7.13

The system considered in this example is the attitude control system for a rigid satellite. The plant equations are (refer Example 7.10)

$$\mathbf{x}(k+1) = \mathbf{F}\mathbf{x}(k) + \mathbf{g}u(k)$$

where

$$\mathbf{F} = \begin{bmatrix} 1 & T \\ 0 & 1 \end{bmatrix}; \; \mathbf{g} = \begin{bmatrix} T^2/2 \\ T \end{bmatrix}$$

$x_1(k)$ = position state θ

$x_2(k)$ = velocity state ω

The reference input $r = \theta_r$, a step function. The desired steady-state

$$\mathbf{x}_s = [\theta_r \quad 0]^T$$

which is a non-null state.

As the plant has integrating property, the steady-state value u_s of the input must be zero (otherwise the output cannot stay constant). For this case, the shifted regulator problem may be formulated as follows:

$$\tilde{x}_1 = x_1 - \theta_r$$
$$\tilde{x}_2 = x_2$$

Shifted state variables satisfy the equations

$$\tilde{\mathbf{x}}(k+1) = \mathbf{F}\tilde{\mathbf{x}}(k) + \mathbf{g}u(k)$$

The state-feedback control

$$u(k) = -\mathbf{k}\tilde{\mathbf{x}}(k)$$

results in the dynamics of $\tilde{\mathbf{x}}$ given by

$$\tilde{\mathbf{x}}(k+1) = (\mathbf{F} - \mathbf{g}\mathbf{k})\tilde{\mathbf{x}}(k)$$

We now determine the gain matrix $\mathbf{k}$ such that the response to an arbitrary initial condition is deadbeat. The desired characteristic equation is

$$z^2 = 0$$

Using Ackermann's formula (7.82), we obtain

$$\mathbf{k} = \begin{bmatrix} 0 & 1 \end{bmatrix} \mathbf{U}^{-1} \phi(\mathbf{F})$$

where

$$\phi(\mathbf{F}) = \mathbf{F}^2 = \begin{bmatrix} 1 & 2T \\ 0 & 1 \end{bmatrix}$$

$$\mathbf{U}^{-1} = [\mathbf{g} \quad \mathbf{F}\mathbf{g}]^{-1} = \begin{bmatrix} -\dfrac{1}{T^2} & \dfrac{3}{2T} \\ \dfrac{1}{T^2} & -\dfrac{1}{2T} \end{bmatrix}$$

This gives

$$\mathbf{k} = \begin{bmatrix} \dfrac{1}{T^2} & \dfrac{3}{2T} \end{bmatrix}$$

For $T = 0.1$ sec,
$$\mathbf{k} = \begin{bmatrix} 100 & 15 \end{bmatrix}$$

The control law expressed in terms of original state variables is given as

$$u(k) = -k_1 \tilde{x}_1(k) - k_2 \tilde{x}_2(k) = -100(x_1(k) - \theta_r) - 15 x_2(k)$$

Example 7.14

Reconsider the problem of attitude control of a satellite. For implementation of the design of the previous example, we require the states $x_1(k)$ and $x_2(k)$ to be measurable. Assuming that the output $y(k) = x_1(k)$ is the only state variable that can be measured, we design a state observer for the system. It is desired that the error vector exhibit deadbeat response. The measurement equation is

$$y(k) = \mathbf{c}\mathbf{x}(k) = \begin{bmatrix} 1 & 0 \end{bmatrix} \mathbf{x}(k)$$

The prediction observer for the system is given as
$$\hat{\mathbf{x}}(k+1) = \mathbf{F}\hat{\mathbf{x}}(k) + \mathbf{g}u(k) + \mathbf{m}(y(k) - \mathbf{c}\hat{\mathbf{x}}(k))$$
The gains $\mathbf{m}$ may be calculated by solving the state regulator design problem for the 'transposed auxiliary system'
$$\boldsymbol{\zeta}(k+1) = \mathbf{F}^T\boldsymbol{\zeta}(k) + \mathbf{c}^T\boldsymbol{\eta}(k)$$
$$\boldsymbol{\eta}(k) = -\mathbf{m}^T\boldsymbol{\zeta}(k)$$
The desired characteristic equation is
$$z^2 = 0$$
Using Ackermann's formula, we obtain
$$\mathbf{m}^T = [0 \quad 1]\mathbf{U}^{-1}\boldsymbol{\phi}(\mathbf{F}^T)$$
where
$$\boldsymbol{\phi}(\mathbf{F}^T) = (\mathbf{F}^T)^2 = \begin{bmatrix} 1 & 0 \\ 2T & 1 \end{bmatrix}$$
$$\mathbf{U}^{-1} = [\mathbf{c}^T \quad \mathbf{F}^T\mathbf{c}^T]^{-1} = \begin{bmatrix} 1 & -1/T \\ 0 & 1/T \end{bmatrix}$$
This gives
$$\mathbf{m}^T = [2 \quad 1/T]$$
For $T = 0.1$ sec,
$$\mathbf{m} = \begin{bmatrix} 2 \\ 10 \end{bmatrix}$$

7.11 INTRODUCTION TO SYSTEM IDENTIFICATION AND ADAPTIVE CONTROL

System Identification

The types of models that are needed for the design methods presented in this and earlier chapters can be grouped into two categories: transfer function model and state variable model. If we have a transfer function description, we can obtain an equivalent state variable description and vice versa. These equivalent models are described by certain *parameters*—the elements of $\mathbf{F}, \mathbf{g}, \mathbf{c}$ matrices of the state model
$$\mathbf{x}(k+1) = \mathbf{F}\mathbf{x}(k) + \mathbf{g}u(k)$$
$$y(k) = \mathbf{c}\mathbf{x}(k)$$
$\mathbf{x}(k)$: $n \times n$ state vector
$u(k)$: scalar input
$y(k)$: scalar output

or the parameters α_i and β_j of the transfer function

$$G(z) = \frac{Y(z)}{U(z)} = \frac{\beta_1 z^m + \beta_2 z^{m-1} + \cdots + \beta_{m+1}}{z^n + \alpha_1 z^{n-1} + \cdots + \alpha_n}$$

The category of such models gives us the *parametric description* of the plant. The other category of models such as frequency-response curves (Bode plots, polar plots, etc.), time-response curves, etc., gives *non-parametric description* of the plant.

Plant models can be obtained from the first principles of physics. In many cases, however, it is not possible to make a complete model only from physical knowledge. In these circumstances, the designer may turn to the other source of information about plant dynamics, which is the data taken from experiments directly conducted to excite the plant and to measure the response. The process of constructing models from experimental data is called *system identification*. In this section, we restrict our attention to identification of discrete parameteric models, which includes the following steps:
1. Experimental planning
2. Selection of model structure
3. Parameter estimation

Experimental Planning: The choice of experimental conditions for parameter estimation is of considerable importance. It is clear that the best experimental conditions are those that account for the final application of the model. This may occur naturally in some cases; e.g., in adaptive control (discussed later in this section), the model is adjusted under normal operating conditions. Many 'classic' methods depend strongly on having specific input, e.g., sinusoid or impulse. There could be advantages in contriving such an artificial experiment if it subjects the system to a rich and informative set of conditions in the shortest possible time. A requirement on the input signal is that it should excite all the modes of the process sufficiently.

One broad distinction in identification methods is between *on-line* and *off-line* experimentation. The on-line methods give estimates recursively as the measurements are obtained and are the only alternative if the identification is going to be used in an adaptive controller.

Selection of Model Structure: The model structures are derived from prior knowledge of the plant. In some cases the only *a priori* knowledge is that the plant can be described as a linear system in a particular range. It is then natural to use general representations of linear systems.

Consider a SISO dynamic system with input $\{u(t)\}$ and output $\{y(t)\}$. The sampled values of these signals can be related through the linear difference equation

$$y(k+n) + \alpha_1 y(k+n-1) + \cdots + \alpha_n y(k) = \beta_1 u(k+m) + \beta_2 u(k+m-1)$$
$$+ \cdots + \beta_{m+1} u(k); \quad n \geq m \tag{7.116}$$

α_i and β_j are constant (unknown) parameters.

The number of parameters to be identified depends on the order of the selected model, i.e., n in Eqn. (7.116). The calculations can be arranged so that it is possible to make a recursion in the number of parameters in the model. The methods of model-order selection are usually developed for the off-line solution.

The unknown process is not completely a black box. Some information about its dynamic behaviour is known from basic principles and/or plant experience. Therefore, some estimate of the model's order and some initial values for the unknown parameters will be available. The more we know about the process, the more effective the postulated model will be. Consequently, we should use all available information for its development. The order of the postulated model is a very important factor. Remember that complex models of high order will not necessarily produce better controller designs and may burden the computational effort without tangible results.

Equation (7.116) may be expressed as

$$y(k) + \alpha_1 y(k-1) + \cdots + \alpha_n y(k-n)$$
$$= \beta_1 u(k+m-n) + \beta_2 u(k+m-1-n) + \cdots + \beta_{m+1} u(k-n)$$

or

$$y(k) + \alpha_1 y(k-1) + \cdots + \alpha_n y(k-n)$$
$$= \beta_1 u(k-d) + \beta_2 u(k-1-d) + \cdots + \beta_{n-d+1} u(k-n) \qquad (7.117)$$

$d = n - m \geq 0$ is the relative degree or control delay.

We shall use operator notation for conveniently writing linear difference equations. Let z^{-1} be the backward shift (or delay) operator:

$$z^{-1} y(k) = y(k-1) \qquad (7.118)$$

Then Eqn. (7.117) can be written as

$$A(z^{-1}) y(k) = B(z^{-1}) u(k)$$

where $A(z^{-1})$ and $B(z^{-1})$ are polynomials in the delay operator:

$$A(z^{-1}) = 1 + \alpha_1 z^{-1} + \cdots + \alpha_n z^{-n}$$
$$B(z^{-1}) = z^{-d} (\beta_1 + \beta_2 z^{-1} + \cdots + \beta_{n-d+1} z^{-(n-d)})$$

We shall present the parameter-estimation algorithms for the case of $d = 1$ without any loss of generality; the results for any value of d easily follow.

For $d = 1$, we get the input-output model structure

$$A(z^{-1}) y(k) = B(z^{-1}) u(k) \qquad (7.119)$$

where

$$A(z^{-1}) = 1 + \alpha_1 z^{-1} + \cdots + \alpha_n z^{-n}$$
$$B(z^{-1}) = \beta_1 z^{-1} + \beta_2 z^{-2} + \cdots + \beta_n z^{-n}$$

In the presence of disturbance, model (7.119) takes the form

$$A(z^{-1}) y(k) = B(z^{-1}) u(k) + \varepsilon(k) \qquad (7.120)$$

where $\varepsilon(k)$ is some disturbance of unspecified character.

The model (7.120) describes the dynamic relationship between the input and output signals, expressed in terms of the parameter vector

$$\theta = [\alpha_1 \cdots \alpha_n \ \beta_1 \cdots \beta_n]^T \quad (7.121)$$

Introducing the vector of lagged input-output data,

$$\phi(k) = [-y(k-1) - \cdots - y(k-n) \ u(k-1) \cdots u(k-n)]; \quad (7.122)$$

Eqn. (7.120) can be rewritten as

$$y(k) = \phi(k)\theta + \varepsilon(k) \quad (7.123)$$

A model structure should be selected (i) that has a minimal set of parameters and is yet equivalent to the assumed plant description; (ii) whose parameters are uniquely determined by the observed data; and (iii) which will make subsequent control design simple.

Parameter Estimation: The dynamic relationship between the input and output of a scalar system is given by the model (7.123). Ignoring random effects $\varepsilon(k)$ on data collection, we have

$$y(k) = \phi(k)\theta \quad (7.124)$$

where $\phi(k)$ is given by Eqn. (7.122) and θ is given by Eqn. (7.121).
Using the observations

$$\{u(0), u(1), ..., u(N), y(0), y(1), ..., y(N)\}$$

we wish to compute the values of α_i and β_j in parameter vector θ, which will fit the observed data.

Thus, solving the parameter-estimation problem requires techniques for selecting a parameter estimate which best represents the given data. For this we require some idea of the goodness of fit of a proposed value of θ to the true θ°. Because, by the very nature of the problem, θ° is unknown, it is unrealistic to define a direct parameter error between θ and θ°. We must define the error in a way that can be computed from $\{u(k)\}$ and $\{y(k)\}$.

Let $e(k,\theta)$ be the *equation error* comprising the extent to which the equations of motion (7.124) fail to be true for a specific value of θ when used with the specific actual data:

$$e(k,\theta) = y(k) - \phi(k)\theta \quad (7.125)$$

A simple criterion representing the goodness of fit of a proposed value of θ is given by

$$J(\theta) = \sum_{k=1}^{N} e^2(k,\theta) \quad (7.126)$$

The method called the *Least Squares Method*, based on minimizing the sum of the squares of the error, is a very simple and effective method of parameter estimation.

Since $y(k)$ depends on past data upto n periods earlier, the first error we can form is $e(n,\theta)$; the subsequent errors being $e(n + 1,\theta)$, ..., $e(N,\theta)$:

$$e(n,\theta) = y(n) - \phi(n)\theta$$
$$e(n + 1,\theta) = y(n + 1) - \phi(n + 1)\theta$$
$$\vdots$$
$$e(N,\theta) = y(N) - \phi(N)\theta$$

In vector-matrix notation,

$$\mathbf{e}(N,\theta) = \mathbf{y}(N) - \Phi(N)\theta \qquad (7.127)$$

where

$$\mathbf{e}(N,\theta) = [e(n,\theta)\ e(n + 1,\theta)\ \cdots\ e(N,\theta)]^T$$
$$\Phi(N) = [\phi^T(n)\ \phi^T(n + 1)\ \cdots\ \phi^T(N)]^T$$
$$\mathbf{y}(N) = [y(n)\ y(n + 1)\ \cdots\ y(N)]^T$$

Note that $\mathbf{e}$ is $(N - n + 1) \times 1$ vector, $\mathbf{y}$ is $(N - n + 1) \times 1$ vector, Φ is $(N - n + 1) \times 2n$ matrix and θ is $2n \times 1$ vector.

The principle of least squares says that the parameters should be selected in such a way that the performance measure

$$J(\theta) = \sum_{k=n}^{N} e^2(k,\theta) = \mathbf{e}^T(N,\theta)\mathbf{e}(N,\theta) \qquad (7.128)$$

is minimized.

The performance measure $J(\theta)$ can be written as

$$J(\theta) = [\mathbf{y}(N) - \Phi(N)\theta]^T [\mathbf{y}(N) - \Phi(N)\theta]$$
$$= \mathbf{y}^T(N)\mathbf{y}(N) - \theta^T \Phi^T(N)\mathbf{y}(N) - \mathbf{y}^T(N) \Phi(N)\theta + \theta^T \Phi^T(N) \Phi(N)\theta$$
$$= \mathbf{y}^T(N)\ \mathbf{y}(N) - \theta^T \Phi^T(N)\mathbf{y}(N) - \mathbf{y}^T(N) \Phi(N)\theta + \theta^T \Phi^T(N) \Phi(N)\theta$$
$$+ \mathbf{y}^T(N)\ \Phi(N)(\Phi^T(N)\Phi(N))^{-1}\ \Phi^T(N)\mathbf{y}(N)$$
$$- \mathbf{y}^T(N)\Phi(N)(\Phi^T(N)\Phi(N))^{-1}\Phi^T(N)\mathbf{y}(N)$$

(Note that we have simply added and subtracted the same terms under the assumption that $[\Phi^T(N)\Phi(N)]$ is invertible).
Hence

$$J(\theta) = \mathbf{y}^T(N)[I - (\Phi(N)(\Phi^T(N)\Phi(N))^{-1}\Phi^T(N)]\mathbf{y}(N)$$
$$+ (\theta - (\Phi^T(N)\Phi(N))^{-1}\Phi^T(N)\mathbf{y}(N))^T\Phi^T(N)\Phi(N) \times$$
$$(\theta - (\Phi^T(N)\Phi(N))^{-1}\Phi^T(N)\mathbf{y}(N))$$

The first term in this equation is independent of θ, so we cannot reduce J via this term. Hence, to get the smallest value of J, we choose θ so that the second term is zero. Denoting the value of θ that achieves the minimization of J by $\hat{\theta}$, we notice that

$$\hat{\boldsymbol{\theta}} = [\boldsymbol{\Phi}^T(N)\boldsymbol{\Phi}(N)]^{-1}\boldsymbol{\Phi}^T(N)\mathbf{y}(N) \qquad (7.129a)$$
$$= \mathbf{P}(N)\boldsymbol{\Phi}^T(N)\mathbf{y}(N) \qquad (7.129b)$$

where
$$\mathbf{P}(N) = [\boldsymbol{\Phi}^T(N)\boldsymbol{\Phi}(N)]^{-1}$$

The least squares calculation for $\hat{\boldsymbol{\theta}}$ given by (7.129) is a 'batch' calculation since one has a batch of data from which the matrix $\boldsymbol{\Phi}$ and vector $\mathbf{y}$ are composed according to (7.127). In many cases, the observations are obtained sequentially. If the least squares problem has been solved for N observations, it seems to be a waste of computational resources to start from scratch when a new observation is obtained. Hence, it is desirable to arrange the computations in such a way that the results obtained for N observations can be used in order to get the estimates for $(N + 1)$ observations. The algorithm for calculating the least-squares estimate recursively is discussed below.

Let $\hat{\boldsymbol{\theta}}(N)$ denote the least-squares estimate based on N measurements. Then from (7.129)
$$\hat{\boldsymbol{\theta}}(N) = [\boldsymbol{\Phi}^T(N)\boldsymbol{\Phi}(N)]^{-1}\boldsymbol{\Phi}^T(N)\mathbf{y}(N)$$

It is assumed that the matrix $[\boldsymbol{\Phi}^T(N)\boldsymbol{\Phi}(N)]$ is nonsingular for all N. When an additional measurement is obtained, a row is added to the matrix $\boldsymbol{\Phi}$ and an element is added to the vector $\mathbf{y}$. Hence

$$\boldsymbol{\Phi}(N+1) = \begin{bmatrix} \boldsymbol{\Phi}(N) \\ \boldsymbol{\phi}(N+1) \end{bmatrix}; \; \mathbf{y}(N+1) = \begin{bmatrix} \mathbf{y}(N) \\ y(N+1) \end{bmatrix}$$

The estimate $\hat{\boldsymbol{\theta}}(N+1)$ based on $N+1$ measurements can then be written as

$$\hat{\boldsymbol{\theta}}(N+1) = [\boldsymbol{\Phi}^T(N+1)\boldsymbol{\Phi}(N+1)]^{-1}\boldsymbol{\Phi}^T(N+1)\mathbf{y}(N+1)$$
$$= [\boldsymbol{\Phi}^T(N)\boldsymbol{\Phi}(N) + \boldsymbol{\phi}^T(N+1)\boldsymbol{\phi}(N+1)]^{-1}[\boldsymbol{\Phi}^T(N)\mathbf{y}(N)$$
$$+ \boldsymbol{\phi}^T(N+1)y(N+1)]$$

$$\mathbf{P}(N+1) = [\boldsymbol{\Phi}^T(N+1)\boldsymbol{\Phi}(N+1)]^{-1} \qquad (7.130)$$

Then from (7.130), we obtain
$$\mathbf{P}(N+1) = [\mathbf{P}^{-1}(N) + \boldsymbol{\phi}^T(N+1)\boldsymbol{\phi}(N+1)]^{-1} \qquad (7.131)$$

We now need the inverse of a sum of two matrices. We will use the well-known *matrix inversion lemma*[2] for this purpose.

---

2   Matrix inversion lemma is proved below.
$$[\mathbf{A} + \mathbf{BCD}]\{\mathbf{A}^{-1} - \mathbf{A}^{-1}\mathbf{B}\,[\mathbf{C}^{-1} + \mathbf{DA}^{-1}\mathbf{B}]^{-1}\mathbf{DA}^{-1}\}$$
$$= \mathbf{I} + \mathbf{BCDA}^{-1} - \mathbf{B}[\mathbf{C}^{-1} + \mathbf{DA}^{-1}\mathbf{B}]^{-1}\mathbf{DA}^{-1} - \mathbf{BCDA}^{-1}\mathbf{B}[\mathbf{C}^{-1} + \mathbf{DA}^{-1}\mathbf{B}]^{-1}\mathbf{DA}^{-1}$$
$$= \mathbf{I} + \mathbf{BCDA}^{-1} - \mathbf{BC}[\mathbf{C}^{-1} - \mathbf{DA}^{-1}\mathbf{B}]\,[\mathbf{C}^{-1} - \mathbf{DA}^{-1}\mathbf{B}]^{-1}\mathbf{DA}^{-1}$$
$$= \mathbf{I} + \mathbf{BCDA}^{-1} - \mathbf{BCDA}^{-1}$$
$$= \mathbf{I}$$

Let $\mathbf{A}$, $\mathbf{C}$ and $\mathbf{C}^{-1} + \mathbf{D}\mathbf{A}^{-1}\mathbf{B}$ be nonsingular square matrices; then
$$[\mathbf{A} + \mathbf{B}\mathbf{C}\mathbf{D}]^{-1} = \mathbf{A}^{-1} - \mathbf{A}^{-1}\mathbf{B}[\mathbf{C}^{-1} + \mathbf{D}\mathbf{A}^{-1}\mathbf{B}]^{-1}\mathbf{D}\mathbf{A}^{-1} \quad (7.132)$$

To apply (7.132) to (7.131), we make the associations
$$\mathbf{A} = \mathbf{P}^{-1}(N)$$
$$\mathbf{B} = \boldsymbol{\phi}^T(N+1)$$
$$\mathbf{C} = 1$$
$$\mathbf{D} = \boldsymbol{\phi}(N+1)$$

Now, the following result can easily be established.
$$\mathbf{P}(N+1) = \mathbf{P}(N) - \mathbf{P}(N)\boldsymbol{\phi}^T(N+1)[1 + \boldsymbol{\phi}(N+1)\mathbf{P}(N) \times$$
$$\boldsymbol{\phi}^T(N+1)]^{-1}\boldsymbol{\phi}(N+1)\mathbf{P}(N)$$

Substituting the expression for $\mathbf{P}(N+1)$ into (7.130), we obtain
$$\hat{\boldsymbol{\theta}}(N+1) = \{\mathbf{P}(N) - \mathbf{P}(N)\boldsymbol{\phi}^T(N+1)[1 + \boldsymbol{\phi}(N+1)\mathbf{P}(N)\boldsymbol{\phi}^T(N+1)]^{-1} \times$$
$$\boldsymbol{\phi}(N+1)\,\mathbf{P}(N)\}\,[\boldsymbol{\Phi}^T(N)\mathbf{y}(N) + \boldsymbol{\phi}^T(N+1)y(N+1)]$$
$$= \mathbf{P}(N)\boldsymbol{\Phi}^T(N)\mathbf{y}(N) + \mathbf{P}(N)\boldsymbol{\phi}^T(N+1)\,y(N+1) - \mathbf{P}(N)\boldsymbol{\phi}^T(N+1) \times$$
$$[1 + \boldsymbol{\phi}(N+1)\mathbf{P}(N)\boldsymbol{\phi}^T(N+1)]^{-1}\boldsymbol{\phi}(N+1)\mathbf{P}(N)\boldsymbol{\Phi}^T(N)\,\mathbf{y}(N) - \mathbf{P}(N) \times$$
$$\boldsymbol{\phi}^T(N+1)[1 + \boldsymbol{\phi}(N+1)\mathbf{P}(N)\boldsymbol{\phi}^T(N+1)]^{-1}\boldsymbol{\phi}(N+1)\mathbf{P}(N) \times$$
$$\boldsymbol{\phi}^T(N+1)y(N+1)$$
$$= \hat{\boldsymbol{\theta}}(N) + \mathbf{P}(N)\boldsymbol{\phi}^T(N+1)[1 + \boldsymbol{\phi}(N+1)\mathbf{P}(N)\boldsymbol{\phi}^T(N+1)]^{-1} \times$$
$$[1 + \boldsymbol{\phi}(N+1)\mathbf{P}(N)\boldsymbol{\phi}^T(N+1)]y(N+1) - \mathbf{P}(N)\boldsymbol{\phi}^T(N+1) \times$$
$$[1 + \boldsymbol{\phi}(N+1)\mathbf{P}(N)\boldsymbol{\phi}^T(N+1)]^{-1}\boldsymbol{\phi}(N+1)\hat{\boldsymbol{\theta}}(N) - \mathbf{P}(N) \times$$
$$\boldsymbol{\phi}^T(N+1)[1 + \boldsymbol{\phi}(N+1)\mathbf{P}(N)\boldsymbol{\phi}^T(N+1)]^{-1}\boldsymbol{\phi}(N+1)\mathbf{P}(N) \times$$
$$\boldsymbol{\phi}^T(N+1)y(N+1)$$

This gives
$$\hat{\boldsymbol{\theta}}(N+1) = \hat{\boldsymbol{\theta}}(N) + \mathbf{K}(N)[y(N+1) - \boldsymbol{\phi}(N+1)\hat{\boldsymbol{\theta}}(N)] \quad (7.133a)$$
$$\mathbf{K}(N) = \mathbf{P}(N)\boldsymbol{\phi}^T(N+1)[1 + \boldsymbol{\phi}(N+1)\mathbf{P}(N)\boldsymbol{\phi}^T(N+1)]^{-1} \quad (7.133b)$$
$$\mathbf{P}(N+1) = [1 - \mathbf{K}(N)\boldsymbol{\phi}(N+1)]\mathbf{P}(N) \quad (7.133c)$$

Equations (7.133) give the recursive least squares algorithm. Notice that for a single-output system, no matrix inversion is required.

The equations (7.133) have a strong intuitive appeal. The estimate $\hat{\boldsymbol{\theta}}(N+1)$ is obtained by adding a correction to the previous estimate $\hat{\boldsymbol{\theta}}(N)$. The correction is proportional to $y(N+1) - \boldsymbol{\phi}(N+1)\hat{\boldsymbol{\theta}}(N)$ where the term $\boldsymbol{\phi}(N+1)\hat{\boldsymbol{\theta}}(N)$ is the expected output at the time $N+1$, based on the previous data $\boldsymbol{\phi}(N+1)$ and the previous estimate $\hat{\boldsymbol{\theta}}(N)$. Thus, the next estimate of $\boldsymbol{\theta}$ is given by the old estimate corrected by a term linear in error between the observed output $y(N+1)$ and the predicted output $\phi(N+1)\hat{\theta}(N)$. The components of the vector $\mathbf{K}(N)$ are weighting factors that show how the correction and the previous estimate should be combined.

Replacing N by recursive parameter k in Eqns (7.133), we rewrite the recursive least squares (RLS) algorithm as

$$\hat{\theta}(k+1) = \hat{\theta}(k) + \mathbf{K}(k)[y(k+1) - \phi(k+1)\hat{\theta}(k)] \qquad (7.134a)$$

$$\mathbf{K}(k) = \mathbf{P}(k)\phi^T(k+1)[1 + \phi(k+1)\mathbf{P}(k)\phi^T(k+1)]^{-1} \qquad (7.134b)$$

$$\mathbf{P}(k+1) = [1 - \mathbf{K}(k)\phi(k+1)]\mathbf{P}(k) \qquad (7.134c)$$

Any recursive algorithm requires some initial value to be started up. In (7.133), we require $\hat{\theta}(N)$ and $\mathbf{P}(N)$ (equivalently, in (7.134) we require $\hat{\theta}(k)$ and $\mathbf{P}(k)$). We may collect a batch of $N > 2n$ data values and solve the batch formula once for $\mathbf{P}(N)$ and $\hat{\theta}(N)$.

However, it is more common to start the recursion at $k = 0$ with $\mathbf{P}(0) = \alpha \mathbf{I}$ and $\hat{\theta}(0) = \mathbf{0}$, where α is some large constant. You may pick $\mathbf{P}(0) = \alpha \mathbf{I}$ but choose $\hat{\theta}(0)$ to be the best guess that you have at what the parameter values are.

We have presented the least squares method, ignoring random effects on data collection, i.e., $\varepsilon(k)$ in Eqn. (7.123) has been neglected. If $\varepsilon(k)$ is white noise, the least squares estimate given by (7.134) converges to the desired value. However, if $\varepsilon(k)$ is coloured noise, the least squares estimation usually gives a biased (wrong mean value) estimate. This can be overcome by using various extensions of the least squares estimation.

Adaptive Control

When we use a model of the plant as the basis of a control system design, we are tacitly assuming that this model is a reasonable representation of the plant. Although the design model almost always differs from the true plant in some details, we are confident that these details are not important enough to invalidate the design.

There are many applications, however, for which a design model cannot be developed with a reasonable degree of confidence. Moreover, most dynamic processes change with time. Parameters may vary because of normal wear, aging, breakdown, and changes in the environment in which the process operates. The feedback mechanism provides some degree of immunity to discrepancies between the physical plant and the model that is used for the design of the control system. But sometimes this is not enough. A control system designed on the basis of a nominal design model may not behave as well as expected, because the design model does not adequately represent the process in its operating environment.

How can we deal with processes that are prone to large changes, or for which adequate design models are not available? One approach is brute force, i.e., high loop-gain: as the loop-gain becomes infinite, the output of the process tracks the input with vanishing error. Brute force rarely works, however, for well-known reasons: such as dynamic instability, control saturation, and susceptibility to noise and other extraneous inputs.

In robust control-design methods, model uncertainties are captured in a family of perturbed plant models, where each member of the family may represent the nominal plant but it remains unknown which member does. A robust controller satisfies the design-requirements in connection with all the members of the family. Robust control design techniques are sophisticated and make it possible for the control system design to tolerate substantial variation in the model. But the price of achieving immunity to model uncertainties may be a sacrifice in performance. Moreover robust control design techniques are not applicable to processes for which no (uncertainty) design model is available.

The adaptive control theory provides another approach to the design of uncertain systems. Unlike the fixed-parameter controller, adaptive controllers adjust their behaviour on-line, in real-time, to the changing properties of the controlled processes.

The concept of controlling a process that is not well understood, or one in which the parameters are subject to wide variations, has a history that predates the beginning of modern control theory. The early theory was empirical and was developed before digital computer techniques could be used for extensive performance simulations. Prototype testing was one of the few techniques available for testing adaptive control. At least one early experiment had disastrous consequences. As the more mathematically rigorous areas of control theory were developed starting in the 1960s, interest in adaptive control faded for a time, only to be reawakened in the late 1970s with the discovery of mathematically rigorous proofs of the convergence of some popular adaptive-control algorithms. This interest continues unabated [145–157].

Many apparently different approaches to adaptive control have been proposed in the literature. Two schemes in particular have attracted much interest: Self-Tunning Regulator (STR), and Model Reference Adaptive Control (MRAC). These two approaches actually turn out to be special cases of a more general design philosophy.

Self-Tuning Regulator: If the plant is imperfectly known, perhaps because of random time-varying parameters or because of the effects of environmental changes on the plant's dynamic characteristics, then the initial plant model and the resulting control design will not be sufficient to obtain an acceptable performance for all time. It then becomes necessary to carry out plant-identification and control-design procedures continuously or at intervals of time, depending on how fast the plant parameters change. This "self-design" property of the system to compensate for unpredictable changes in the plant is the aspect of performance that is usually considered in defining an adaptive-control system.

The identification of the dynamic characteristics of the plant should be accomplished without affecting the normal operation of the system. To identify the plant model, we must impose a control signal on the plant and analyze the system response. Identification may be made from normal operating data of the plant or by use of test signals, such as sinusoidal ones of small amplitude. The plant should be in normal operation during the test; the test signals imposed should not unduly disturb normal outputs. Furthermore, inputs and system noise should not confuse the test. Normal inputs are ideal as test signals since no difficulties with undesired outputs or confusing inputs will arise. However, identification with normal inputs is only possible when they have adequate signal characteristics (bandwidth, amplitude, and so on) for proper identification.

Once the plant has been identified, a decision must be made as to how the adjustable parameters (controller characteristics) should be varied to maintain acceptable performance. The control signals are then modified according to the plant identification and control decision. The three functions:
  (i) plant identification,
  (ii) control design based on the identification results, and
  (iii) actuation based on the control design
can easily be implemented using a digital computer. Figure 7.18 shows a block diagram representation of the adaptive control scheme. The system obtained is called a *self-tuning regulator* (STR) because it has facilities for tuning its own parameters. The regulator can be thought of as being composed of two loops:
1. The inner loop is the conventional feedback control loop consisting of the plant and the regulator.
2. The parameters of the regulator are adjusted on-line by the outer loop, which is composed of the recursive-parameter estimator and design calculations.

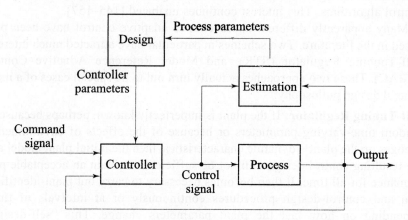

Fig. 7.18 Block diagram of a self-tuning regulator

A self-tuning regulator, therefore, consists of a recursive parameter estimator (plant identifier) coupled with a control-design procedure, such that the currently estimated parameter values are used to provide feedback-controller coefficients. At each sampling, an updated parameter estimate is generated and a controller is designed assuming that the current parameter estimate is actually the true value. The approach of using the estimates as if they were the true parameters for the purpose of design is called *certainty equivalence adaptive control*.

From the block diagram of Fig. 7.18, one may jump to the false conclusion that such regulators can be switched on and used blindly without any *a priori* considerations; the only requirement being a recursive parameter estimation scheme and a design procedure. We have no doubt an array of parameter-estimation schemes and an array of controller-design methods for plants with known parameters. However, all the possible combinations may not have a self-tuning property, which requires that the performance of the regulator coincides with the performance that would be obtained if the system parameters were known exactly. Before using a combination, important theoretical problems such as stability and convergence have to be tackled. There are cases wherein self-tuning regulators have been used profitably though some of their properties have not yet been fully understood theoretically; on the other hand, bad choices have been disastrous in some other cases.

So far only a small number of available combinations have been explored from the stability, convergence, and performance points of view. It is a striking fact, uncovered by Astrom and Wittenmark [150], that in some circumstances a combination of simple least-squares estimation and minimum-variance control has a self-tuning property. The same is true for some classes of pole-shifting regulators. Computer-based controllers incorporating these concepts are now commercially available.

Consider a first-order system with a plant model of the form

$$y(k + 1) = fy(k) + gu(k); \quad y(0) \triangleq y^0 \qquad (7.135)$$

where u is the control variable and y is the measured state (output); f and g are unknown coefficients.

An especially simple adaptive controller results by combining the least squares method of parameter estimation with the pole-placement controller. The least squares parameter-estimation algorithm requires relatively small computational effort and has a reliable convergence, but is applicable only for small noise-signal ratios. Several applications have shown that the combination of least squares parameter estimation with pole-placement control gives good results.

Let us assume that for the system given by Eqn. (7.135), the desired steady-state value for the controlled variable $y(k)$ is a constant reference input r. With the control algorithm (refer Eqns (7.60)–(7.68)),

$$u(k) = - Ky(k) + Nr, \qquad (7.136)$$

the feedback system becomes

$$y(k+1) = (f - gK)\, y(k) + gNr \quad (7.137)$$

The characteristic equation

$$z - f + gK = 0$$

when equated to the desired characteristic equation

$$z - \lambda = 0;\ \lambda = 0.2$$

gives

$$K = \frac{f - \lambda}{g} \quad (7.138a)$$

With this value of K, and

$$N = \frac{1 - \lambda}{g} \quad (7.138b)$$

we get the desired steady-state value of output y, as seen from Eqn. (7.137).

Therefore, the pole-placement control law for the tracking problem is

$$u(k) = -\frac{f - \lambda}{g}\, y(k) + \frac{1 - \lambda}{g} r;\ \lambda = 0.2 \quad (7.139)$$

If the system parameters were known, the feedback controller should take the form (7.139). Since the parameters are assumed to be unknown, the least squares error estimates will be used in place of the true values f^o and g^o of the parameters f and g. The parameter estimates $\hat{f}$ and $\hat{g}$ are derived from the input-output measurements. For example, if data is available at times $k = 0, 1, \ldots, N$, then writing out the system equations (7.135) and combining them into a matrix form yields

$$\begin{bmatrix} y(1) \\ y(2) \\ \vdots \\ y(N) \end{bmatrix} = \begin{bmatrix} y(0) & u(0) \\ y(1) & u(1) \\ \vdots & \vdots \\ y(N-1) & u(N-1) \end{bmatrix} \begin{bmatrix} f \\ g \end{bmatrix} = \begin{bmatrix} \phi(1) \\ \phi(2) \\ \vdots \\ \phi(N) \end{bmatrix} \begin{bmatrix} f \\ g \end{bmatrix}$$

or more compactly

$$\mathbf{y} = \mathbf{\Phi \theta}$$

The parameter estimate (refer Eqns (7.129))

$$\hat{\mathbf{\theta}} = \begin{bmatrix} \hat{f} \\ \hat{g} \end{bmatrix} = (\mathbf{\Phi}^T \mathbf{\Phi})^{-1} \mathbf{\Phi}^T \mathbf{y} \quad (7.140)$$

To simulate the system, we first collected a batch of 5 sets of data values to solve the formula (7.140) once for $\hat{\mathbf{\theta}}$. The data values were obtained from Eqn. (7.135) assuming the true parameters

$f^o = 1.1052;\ g^o = 0.0526$

and sampling interval $T = 0.1$ sec.

With the initial estimate $\hat{\theta}(0) = \begin{bmatrix} \hat{f}(0) \\ \hat{g}(0) \end{bmatrix}$, we implement the pole-placement control law. This gives us the first set of data values for the recursive least squares estimation:

$$u(0) = -\frac{\hat{f}(0) - 0.2}{\hat{g}(0)} y(0) + \frac{1 - 0.2}{\hat{g}(0)} r$$

$$y(1) = 1.1052\, y(0) + 0.0536\, u(0)$$

Discard the first set of data values from the batch of 5 sets of data and include the new set generated on-line to calculate the new parameter estimate $\hat{\theta}(1)$. Based on the new estimate,

$$u(1) = -\frac{\hat{f}(1) - 0.2}{\hat{g}(1)} y(1) + \frac{1 - 0.2}{\hat{g}(1)} r$$

$$y(2) = 1.1052\, y(1) + 0.0526\, u(1)$$

(Alternatively, we could use Eqns (7.134) to generate the new parameter estimate)

The plot of Fig. 7.19 was generated using this procedure. The input signal is a square wave with amplitude 50. The closed-loop system is close to the desired behaviour after a few transients.

Model Reference Adaptive Control: Another popular adaptive control strategy is MRAC. The basic principle is illustrated in Fig. 7.20. The desired performance is given in terms of a reference model (a computer program), which in turn gives the desired response to the command signal. The inner control loop is a conventional feedback loop composed of the plant and the controller; the parameters of the controller are adjusted by the outer loop in such a way that the error between the plant and model outputs becomes small. The key problem is to determine the adjustment mechanism so that a stable system which brings the error to zero is obtained.

From the block diagram of Fig. 7.20, one may jump to the false conclusion that MRAC has an answer to all control problems with uncertain plants. Before using such a scheme, important theoretical problems such as stability and convergence have to be considered. Since adaptive control schemes are both time-varying and nonlinear, stability and performance analysis becomes very difficult. Many advances have been made in proving stability under certain (sometimes limited) conditions. However, not much ground has been gained in proving performance bounds.

An example of Model Reference Adaptive System will appear in Section 8.7.

546 Digital Control and State Variable Methods

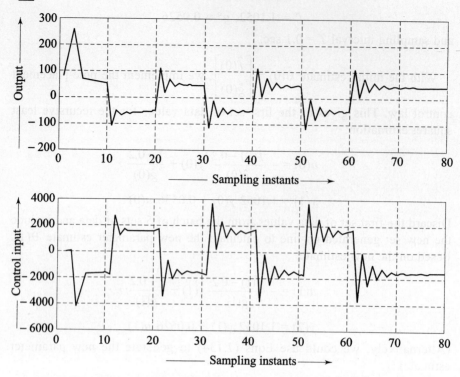

Fig. 7.19 Simulation results

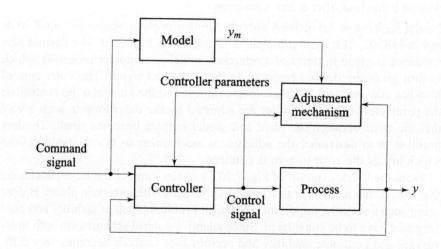

Fig. 7.20 Block diagram of model-reference adaptive system

7.12 REVIEW EXAMPLES

Review Example 7.1

DC motors are widely used in speed-control drives. In most applications, the armature voltage of the motor is controlled in a closed-loop feedback system. Figure 7.21a shows a plant model of a speed control system.

The state variables of the plant can be chosen as the motor shaft velocity $\omega(t)$, and the armature current $i_a(t)$. If both the state variables are used in feedback, then two voltages proportional, respectively, to these two state variables must be generated. The generation of the voltage proportional to $\omega(t)$ can be achieved by use of a tachometer. A voltage proportional to $i_a(t)$ can be generated by inserting a sampling resistor R_s in the armature circuit, as shown in Fig. 7.21a. It may, however, be noted that if R_s is very small, the voltage across R_s may consist largely of noise; and if R_s is large, the voltage is more accurate, but a considerable amount of power is wasted in R_s and the efficiency of the system is reduced.

In modern speed-control drives, thyristor rectifier is used as a power amplifier. The thyristor rectifier is supplied by an external single-phase or three-phase ac power, and it amplifies its input voltage u to produce an output voltage e_a, which is supplied to the armature of the dc motor. The state-feedback control, requiring the feedback of both the motor-shaft velocity and the armature current can, in fact, be effectively used to provide current-limiting protective feature to prevent damage to the thyristors.

The voltage u is fed to the driver of the thyristor rectifier. The driver produces time-gate pulses that control the conduction of the thyristors in the rectifier module. The rectified output voltage e_a depends on the firing angle of the pluses relative to the ac supply waveform. A linear relationship between the input voltage u and the output voltage e_a can be obtained when a proper firing control scheme is used. The time constants associated with the rectifier are negligibly small. Neglecting the dynamics of the rectifier, we get

$$e_a(t) = K_r\, u(t)$$

where K_r is the gain of the rectifier.

Figure 7.21b shows the functional block diagram of the plant with

B = viscous-friction coefficient of motor and load,
J = moment of inertia of motor and load,
K_T = motor torque constant,
K_b = motor back-emf constant,
T_L = constant load torque,
L_a = armature inductance, and
$R_a + R_s$ = armature resistance.

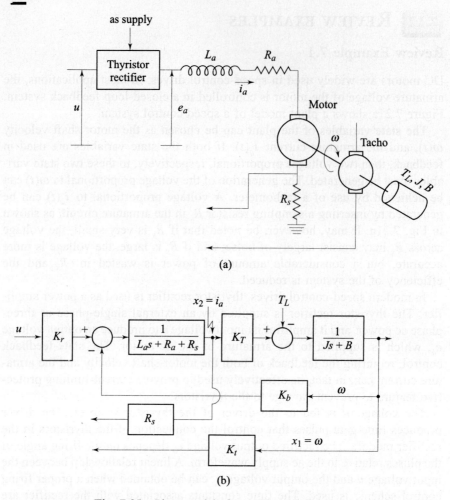

Fig. 7.21 Plant model of a speed-control system

As seen from Fig. 7.21b, the plant is a type-0 system. A control law of the form

$$u(t) = -\mathbf{k}\mathbf{x} + Nr$$

can shape the dynamics of the state variables $x_1(t) = \omega(t)$ and $x_2(t) = i_a(t)$ with zero steady-state error in $\omega(t)$ to constant reference input r. The closed-loop system will, however, be a type-0 system resulting in steady-state errors to constant disturbances. We assume that steady-state performance specifications require a type-1 system. Hence we employ state feedback with integral control. A block diagram of the control configuration is shown in Fig. 7.22.

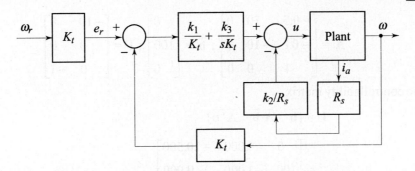

Fig. 7.22 Control configuration for a speed-control system

The state equations of the plant are

$$J\dot{x}_1 + Bx_1 = K_T x_2 - T_L$$

$$L_a \dot{x}_2 + (R_a + R_s)x_2 = K_r u - K_b x_1$$

or
$$\dot{x} = Ax + bu + \gamma T_L$$

where

$$A = \begin{bmatrix} -\dfrac{B}{J} & \dfrac{K_T}{J} \\ -\dfrac{K_b}{L_a} & -\dfrac{(R_a + R_s)}{L_a} \end{bmatrix} ; \; b = \begin{bmatrix} 0 \\ \dfrac{K_r}{L_a} \end{bmatrix} ; \; \gamma = \begin{bmatrix} -\dfrac{1}{J} \\ 0 \end{bmatrix}$$

Let the parameter values be such that these matrices become

$$A = \begin{bmatrix} -0.5 & 10 \\ -0.1 & -10 \end{bmatrix} ; \; b = \begin{bmatrix} 0 \\ 100 \end{bmatrix} ; \; \gamma = \begin{bmatrix} -10 \\ 0 \end{bmatrix}$$

We define an additional state variable x_3 as

$$x_3 = \int_0^t (\omega - r)\, dt,$$

i.e.,
$$\dot{x}_3 = \omega - r = x_1 - r$$

Augmenting this state variable with the plant equations, we obtain

$$\dot{\bar{x}} = \bar{A}\bar{x} + \bar{b}u + \Gamma w$$

where
$$\bar{x} = [x_1 \; x_2 \; x_3]^T; \; w = [T_L \; r]^T$$

$$\overline{A} = \begin{bmatrix} -0.5 & 10 & 0 \\ -0.1 & -10 & 0 \\ 1 & 0 & 0 \end{bmatrix}; \overline{b} = \begin{bmatrix} 0 \\ 100 \\ 0 \end{bmatrix}; \Gamma = \begin{bmatrix} -10 & 0 \\ 0 & 0 \\ 0 & -1 \end{bmatrix}$$

The controllability matrix

$$U = [\overline{b} \quad \overline{A}\overline{b} \quad \overline{A}^2\overline{b}]$$

$$= \begin{bmatrix} 0 & 1{,}000 & -10{,}500 \\ 100 & -1{,}000 & 9{,}900 \\ 0 & 0 & 1{,}000 \end{bmatrix}$$

The determinant of U is nonzero. The pair $(\overline{A}, \overline{b})$ is, therefore, completely controllable and the conditions for pole placement by state feedback and integral control are satisfied.

The characteristic polynomial of the closed-loop system is given by

$$|sI - (\overline{A} - \overline{b}\,\overline{k})| = \begin{vmatrix} s+0.5 & -10 & 0 \\ 0.1+100k_1 & s+10+100k_2 & 100k_3 \\ -1 & 0 & s \end{vmatrix}$$

$$= s^3 + (10.5 + 100k_2)s^2 + (6 + 50k_2 + 1{,}000k_1)s + 1{,}000k_3 \quad (7.141a)$$

Let the desired characteristic polynomial be

$$s^3 + 87.5\, s^2 + 5{,}374.5\, s + 124{,}969$$

$$= (s + 35.4)(s + 26.05 + j53.4)(s + 26.05 - j53.4) \quad (7.141b)$$

The quadratic term has a natural frequency $\omega_n = 59.39$ rad/sec, and a damping ratio $\zeta = 0.44$.

Matching the corresponding coefficients of Eqns (7.141a) and (7.141b), we obtain

$$k_1 = 5.33,\ k_2 = 0.77,\ k_3 = 124.97$$

With these values of the feedback gains, the state variable model of the closed-loop system becomes

$$\begin{bmatrix} \dot{x}_1 \\ \dot{x}_2 \\ \dot{x}_3 \end{bmatrix} = \begin{bmatrix} -0.5 & 10 & 0 \\ -533.1 & -87 & -12497 \\ 1 & 0 & 0 \end{bmatrix} \begin{bmatrix} x_1 \\ x_2 \\ x_3 \end{bmatrix} + \begin{bmatrix} -10 & 0 \\ 0 & 0 \\ 0 & -1 \end{bmatrix} \begin{bmatrix} T_L \\ r \end{bmatrix}$$

At steady-state $\dot{x} = 0$, and therefore the motor velocity $x_1 = \omega(t)$ will approach the constant reference set point r as t approaches infinity, independent of the disturbance torque T_L.

Review Example 7.2

One of the most common uses of feedback control is to position an inertia load using an electric motor. The inertia load may consist of a very large, massive object such as a radar antenna or a small object such as a precision instrument. Armature-controlled dc motors are used in many applications for positioning the load.

We consider here a motor-driven inertia system described by the following equations (refer Eqns (5.14)).

$$u(t) = R_a\, i_a(t) + K_b\, \omega(t) = R_a\, i_a(t) + K_b\, \frac{d\theta(t)}{dt}$$

$$K_T\, i_a(t) = J\, \frac{d\omega(t)}{dt} = J\, \frac{d^2\theta(t)}{dt^2}$$

where

u = applied armature voltage,
R_a = armature resistance,
i_a = armature current,
θ = angular position of the motor shaft,
ω = angular velocity of the motor shaft,
K_b = back emf constant,
K_T = motor torque constant, and
J = moment of inertia referred to the motor shaft.

Taking $x_1 = \theta$, and $x_2 = \dot{\theta} = \omega$ as the state variables, we obtain the following state variable equations for the system.

$$\dot{x}_1 = x_2$$

$$\dot{x}_2 = -\frac{K_T K_b}{JR_a} x_2 + \frac{K_T}{JR_a} u = -\alpha x_2 + \beta u$$

Assume that the physical parameters of the motor and the load yield $\alpha = 1$, $\beta = 1$. Then

$$\dot{\mathbf{x}} = \mathbf{Ax} + \mathbf{b}u$$

where

$$\mathbf{A} = \begin{bmatrix} 0 & 1 \\ 0 & -1 \end{bmatrix};\ \mathbf{b} = \begin{bmatrix} 0 \\ 1 \end{bmatrix}$$

The discrete-time description of this system, with sampling period $T = 0.1$ sec, is given by the following equations (refer Section 6.3).

$$\mathbf{x}(k+1) = \mathbf{F}\mathbf{x}(k) + \mathbf{g}u(k) \qquad (7.142)$$

where

$$\mathbf{F} = e^{\mathbf{A}T} = \begin{bmatrix} 1 & 0.0952 \\ 0 & 0.905 \end{bmatrix}$$

$$\mathbf{g} = \int_0^T e^{\mathbf{A}\tau} \mathbf{b} d\tau = \begin{bmatrix} 0.00484 \\ 0.0952 \end{bmatrix}$$

In this model, $x_1(k)$ is the shaft position and $x_2(k)$ is the shaft velocity. We assume that $x_1(k)$ and $x_2(k)$ can easily be measured using shaft encoders.

We choose the control configuration of Fig. 7.23 for digital positioning of the load; θ_r is a constant reference command. In terms of the error variables

$$\tilde{x}_1(k) = x_1(k) - \theta_r$$
$$\tilde{x}_2(k) = x_2(k) \qquad (7.143)$$

the control signal

$$u(k) = -k_1 \tilde{x}_1(k) - k_2 \tilde{x}_2(k) = -\mathbf{k}\tilde{\mathbf{x}}(k) \qquad (7.144)$$

where the gain matrix

$$\mathbf{k} = [k_1 \quad k_2]$$

The dynamics of the error-vector $\tilde{\mathbf{x}}(k)$ are given by the equations

$$\tilde{x}_1(k+1) = x_1(k+1) - \theta_r$$
$$= x_1(k) + 0.0952\, x_2(k) + 0.00484\, u(k) - \theta_r$$
$$= \tilde{x}_1(k) + 0.0952\, \tilde{x}_2(k) + 0.00484\, u(k)$$

$$\tilde{x}_2(k+1) = 0.905\, \tilde{x}_2(k) + 0.0952\, u(k)$$

or $\qquad \tilde{\mathbf{x}}(k+1) = \mathbf{F}\tilde{\mathbf{x}}(k) + \mathbf{g}u(k)$

where $\mathbf{F}$ and $\mathbf{g}$ are given by Eqn. (7.142).

Substituting for $u(k)$ from Eqn. (7.144), we obtain the following closed-loop model of the error dynamics:

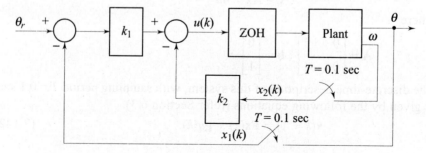

Fig. 7.23 Control configuration for a digital positioning system

$$\tilde{\mathbf{x}}(k+1) = (\mathbf{F} - \mathbf{gk})\tilde{\mathbf{x}}(k)$$

$$= \begin{bmatrix} 1 - 0.00484k_1 & 0.0952 - 0.00484k_2 \\ -0.0952k_1 & 0.905 - 0.0952k_2 \end{bmatrix} \tilde{\mathbf{x}}(k) \quad (7.145)$$

The characteristic equation is

$$|z\mathbf{I} - (\mathbf{F} - \mathbf{gk})| = z^2 + (0.00484k_1 + 0.0952k_2 - 1.905)z$$
$$+ 0.00468k_1 - 0.0952k_2 + 0.905 = 0 \quad (7.146a)$$

We choose the desired characteristic-equation zero locations to be

$$z_{1,2} = 0.888 \pm j0.173 = 0.905 \angle \pm 11.04°$$

Note that this corresponds to $\zeta = 0.46$ and $\omega_n = 2.17$ (refer Eqns (4.15)):

$$z_{1,2} = e^{-\zeta\omega_n T} e^{\pm j\omega_n T\sqrt{1-\zeta^2}}$$

The desired characteristic equation is given by

$$(z - 0.888 - j0.173)(z - 0.888 + j0.173) = z^2 - 1.776z + 0.819 = 0 \quad (7.146b)$$

Equating coefficients in Eqns (7.146a) and (7.146b) yields the equations

$$0.00484k_1 + 0.0952k_2 = -1.776 + 1.905$$

$$0.00468k_1 - 0.0952k_2 = 0.819 - 0.905$$

These equations are linear in k_1 and k_2 and upon solving yield

$$k_1 = 4.52, \; k_2 = 1.12$$

The control law is therefore given by

$$u(k) = -k_1\tilde{x}_1(k) - k_2\tilde{x}_2(k) = -4.52(x_1(k) - \theta_r) - 1.12x_2(k)$$

The implementation of this control law requires the feedback of the states $x_1(k)$ and $x_2(k)$. If we measure $x_1(k)$ using a shaft encoder and estimate $x_2(k)$ using a state observer, the control configuration will take the form shown in Fig. 7.24.

The state-feedback control has been designed for $\zeta = 0.46$, $\omega_n = 2.17$; $\zeta\omega_n \cong 1$ sec. The reduced-order observer for estimating velocity $x_2(k)$ from measurements of position $x_1(k)$ is a first-order system; we choose the time constant of this system to be 0.5 sec. Hence the desired pole location[3] in the observer design problem is

$$z = e^{-T/\tau} = e^{-0.1/0.5} = 0.819$$

The observer characteristic equation is then

$$z - 0.819 = 0 \quad (7.147)$$

From the plant state Eqn. (7.142), and Eqns (7.91), the partitioned matrices are seen to be

$$f_{11} = 1, \; f_{1e} = 0.0952, \; f_{e1} = 0, \; F_{ee} = 0.905,$$
$$g_1 = 0.00484, \; g_e = 0.0952$$

---

[3] The pole at $s = -1/\tau$ is mapped to $z = e^{-T/\tau}$; T = sampling interval.

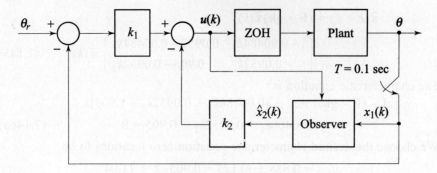

Fig. 7.24 Observer-based digital positioning system

The observer equation is (refer Eqn. (7.95))

$$\hat{x}_2(k+1) = 0.905\hat{x}_2(k) + 0.0952u(k)$$
$$+ m(\theta(k+1) - \theta(k) - 0.00484u(k) - 0.0952\hat{x}_2(k))$$
$$= (0.905 - m(0.0952))\hat{x}_2(k) + m\theta(k+1) - m\theta(k)$$
$$+ (0.0952 - m(0.00484))u(k) \qquad (7.148)$$

The characteristic equation is given by

$$z - (0.905 - 0.0952m) = 0$$

Comparing the coefficients with those of Eqn. (7.147), we obtain

$$m = 0.903$$

Substituting in Eqn. (7.148), we get

$$\hat{x}_2(k+1) = 0.819\hat{x}_2(k) + 0.903\theta(k+1) - 0.903\theta(k) + 0.0908u(k)$$

The control system is implemented as follows. A measurement $\theta(k)$ is made at $t = kT$. The observer state is calculated from

$$\hat{x}_2(k) = 0.819\hat{x}_2(k-1) + 0.903\theta(k) - 0.903\theta(k-1) + 0.0908u(k-1)$$

Then the control input is calculated, using

$$u(k) = -4.52(\theta(k) - \theta_r) - 1.12\hat{x}_2(k)$$

PROBLEMS

7.1 Consider an nth-order, single-input system

$$\dot{\mathbf{x}} = \mathbf{A}\mathbf{x} + \mathbf{b}u$$

and assume that we are using feedback of the form

$$u = -\mathbf{k}\mathbf{x} + r$$

where r is the reference input signal.
Show that the zeros of the system are invariant under state-feedback.

7.2 A regulator system has the plant

$$\dot{x} = \begin{bmatrix} 0 & 1 & 0 \\ 0 & 0 & 1 \\ -6 & -11 & -6 \end{bmatrix} x + \begin{bmatrix} 0 \\ 0 \\ 1 \end{bmatrix} u$$

$$y = [1 \quad 0 \quad 0] x$$

(a) Design a state-feedback controller which will place the closed-loop poles at $-2 \pm j3.464, -5$. Give a block diagram of the control configuration.

(b) Design a full-order state observer; the observer-error poles are required to be located at $-2 \pm j3.464, -5$. Give all the relevant observer equations and a block diagram description of the observer structure.

(c) The state variable x_1 (which is equal to y) is directly measurable and need not be observed. Design a reduced-order state observer for the plant; the observer-error poles are required to be located at $-2 \pm j3.464$. Give all the relevant observer equations.

7.3 A regulator system has the plant

$$\dot{x} = Ax + bu$$

$$y = cx$$

with

$$A = \begin{bmatrix} 0 & 0 & -6 \\ 1 & 0 & -11 \\ 0 & 1 & -6 \end{bmatrix}; b = \begin{bmatrix} 1 \\ 0 \\ 0 \end{bmatrix}; c = [0 \quad 0 \quad 1]$$

(a) Compute k so that the control law $u = -kx$ places the closed-loop poles at $-2 \pm j3.464, -5$. Give the state variable model of the closed-loop system.

(b) For the estimation of the state vector x, we use an observer defined by

$$\dot{\hat{x}} = (A - mc)\hat{x} + bu + my$$

Compute m so that the eigenvalues of $(A - mc)$ are located at $-2 \pm j3.464, -5$.

(c) The state variable x_3 (which is equal to y) is directly measurable and need not be observed. Design a reduced-order observer for the plant; the observer-error poles are required to be located at $-2 \pm j3.464$. Give all the relevant observer equations.

7.4 Consider the system

$$\dot{x} = Ax + Bu$$

$$y = cx + du$$

where

$$A = \begin{bmatrix} -2 & -1 \\ 1 & 0 \end{bmatrix}; B = \begin{bmatrix} 1 & 0 \\ 1 & 1 \end{bmatrix}; c = [0 \quad 1]; d = [2 \quad 0]$$

Design a full-order state observer so that the estimation error will decay in less than 4 seconds.

7.5 Consider the system

$$\dot{x} = \begin{bmatrix} 1 & 0 \\ 0 & 0 \end{bmatrix} x + \begin{bmatrix} 1 \\ 1 \end{bmatrix} u$$

$$y = [2 \quad -1] x$$

Design a reduced-order state observer that makes the estimation error to decay at least as fast as e^{-10t}.

7.6 Consider the system with the transfer function

$$\frac{Y(s)}{U(s)} = \frac{9}{s^2 - 9}$$

(a) Find $(\mathbf{A}, \mathbf{b}, \mathbf{c})$ for this system in observable canonical form.
(b) Compute $\mathbf{k}$ so that the control law $u = -\mathbf{k}\mathbf{x}$ places the closed-loop poles at $-3 \pm j3$.
(c) Design a full-order observer such that the observer-error poles are located at $-6 \pm j6$. Give all the relevant observer equations.
(d) Suppose the system has a zero such that

$$\frac{Y(s)}{U(s)} = \frac{9(s+1)}{s^2 - 9}$$

Prove that if $u = -\mathbf{k}\mathbf{x} + r$, there is a feedback matrix $\mathbf{k}$ such that the system is unobservable.

7.7 The equation of motion of an undamped oscillator with frequency ω_0 is

$$\ddot{y} + \omega_0^2 y = u$$

(a) Write the equations of motion in the state variable form with $x_1 = y$ and $x_2 = \dot{y}$ as the state variables.
(b) Find k_1 and k_2 such that $u = -k_1 x_1 - k_2 x_2$ gives closed-loop characteristic roots with $\omega_n = 2\omega_0$ and $\zeta = 1$.
(c) Design a second-order observer that estimates x_1 and x_2, given measurements of x_1. Pick the characteristic roots of the state-error equation with $\omega_n = 10\omega_0$ and $\zeta = 1$. Give a block diagram of the observer-based state-feedback control system.
(d) Design a first-order observer that estimates x_2, given measurements of x_1. The characteristic root of the state-error equation is required to be located at $-10\omega_0$. Give a block diagram of the observer-based state-feedback control system.

7.8 A regulator system has the plant

$$\dot{\mathbf{x}} = \begin{bmatrix} 0 & 1 \\ 20.6 & 0 \end{bmatrix} \mathbf{x} + \begin{bmatrix} 0 \\ 1 \end{bmatrix} u$$

$$y = \begin{bmatrix} 1 & 0 \end{bmatrix} \mathbf{x}$$

(a) Design a control law $u = -\mathbf{k}\mathbf{x}$ so that the closed-loop system has eigenvalues at $-1.8 \pm j2.4$.
(b) Design a full-order state observer to estimate the state vector. The observer matrix is required to have eigenvalues at $-8, -8$.
(c) Find the transfer function of the compensator obtained by combining (a) and (b).
(d) Find the state variable model of the complete observer-based state-feedback control system.

7.9 A regulator system has the double integrator plant

$$\frac{Y(s)}{U(s)} = \frac{1}{s^2}$$

(a) Taking $x_1 = y$ and $x_2 = \dot{y}$ as state variables, obtain the state variable model of the plant.

(b) Compute **k** such that $u = -\mathbf{kx}$ gives closed-loop characteristic roots with $\omega_n = 1$, $\zeta = \sqrt{2}/2$.

(c) Design a full-order observer that estimates x_1 and x_2, given measurements of x_1. Pick the characteristic roots of the state-error equation with $\omega_n = 5$, $\zeta = 0.5$.

(d) Find the transfer function of the compensator obtained by combining (b) and (c).

(e) Design a reduced-order observer that estimates x_2 given measurements of x_1; place the single observer pole at $s = -5$.

(f) Find the transfer function of the compensator obtained by combining (b) and (e).

7.10 A servo system has the type-1 plant described by the equation

$$\dot{\mathbf{x}} = \mathbf{A}\mathbf{x} + \mathbf{b}u$$

$$y = \mathbf{cx}$$

where

$$\mathbf{A} = \begin{bmatrix} 0 & 1 & 0 \\ 0 & -1 & 1 \\ 0 & 0 & -2 \end{bmatrix}; \mathbf{b} = \begin{bmatrix} 0 \\ 0 \\ 1 \end{bmatrix}; \mathbf{c} = \begin{bmatrix} 1 & 0 & 0 \end{bmatrix}$$

(a) If $u = -\mathbf{kx} + Nr$, compute **k** and N so that the closed-loop poles are located at $-1 \pm j1, -2$; and $y(\infty) = r$, a constant reference input.

(b) For the estimation of the state vector **x**, we use a full-order observer

$$\dot{\hat{\mathbf{x}}} = (\mathbf{A} - \mathbf{mc})\hat{\mathbf{x}} + \mathbf{b}u + \mathbf{m}y$$

Compute **m** so that observer-error poles are located at $-2 \pm j2, -4$.

(c) Replace the control law in (a) by $u = -\mathbf{k}\hat{\mathbf{x}} + Nr$, and give a block diagram of the observer-based servo system.

7.11 A plant is described by the equation

$$\dot{\mathbf{x}} = \begin{bmatrix} -1 & 0 \\ 0 & -2 \end{bmatrix} \mathbf{x} + \begin{bmatrix} 1 \\ 1 \end{bmatrix} u$$

$$y = \begin{bmatrix} 1 & 3 \end{bmatrix} \mathbf{x}$$

Add to the plant equations an integrator $\dot{z} = y - r$ (r is a constant reference input) and select gains **k**, k_i so that if $u = -\mathbf{kx} - k_i z$, the closed-loop poles are at $-2, -1 \pm j\sqrt{3}$. Give a block diagram of the control configuration.

7.12 Figure P7.12 shows the block diagram of a position control system employing a dc motor in armature control mode; θ (rad) is the motor shaft position, $\dot{\theta}$ (rad/sec) is the motor shaft velocity, i_a (amps) is the armature current, and K_P (volts/rad) is the sensitivity of the potentiometer. Find k_1, k_2 and k_3 so that the dominant poles of the closed-loop system are characterized by $\zeta = 0.5$, $\omega_n = 2$; non-dominant pole is at $s = -10$, and the steady-state error to constant reference input is zero.

7.13 A dc motor in armature control mode has been used in speed control system of Fig. P7.13 employing state-feedback with integral control; ω (rad/sec) is the motor shaft velocity, i_a (amps) is the armature current and K_t (volts/(rad/sec)) is the tachometer constant. Find k_1, k_2 and k_3 so that the closed-loop poles of the system are placed at $-1 \pm j\sqrt{3} - 10$; and the steady-state error to constant reference input is zero.

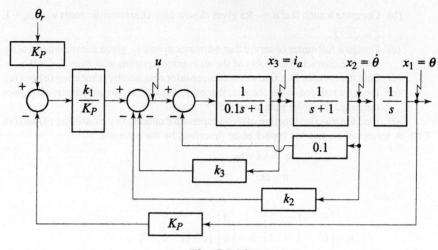

Fig. P7.12

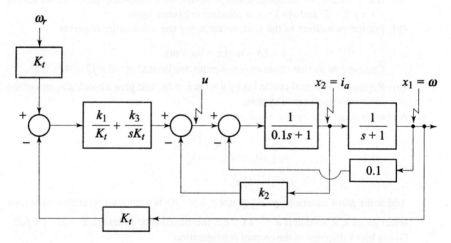

Fig. P7.13

7.14 The control law $u = -\mathbf{kx} - k_1\theta_r$ for position control system of Fig. P7.12 is to be replaced by $u = -\mathbf{k\hat{x}} + k_1\theta_r$, where $\hat{\mathbf{x}}$ is the estimate of the state vector $\mathbf{x}$ given by the observer system

$$\dot{\hat{\mathbf{x}}} = (\mathbf{A} - \mathbf{mc})\hat{\mathbf{x}} + \mathbf{b}u + \mathbf{m}\theta$$

Find the gain matrix $\mathbf{m}$ which places the eigenvalues of $(\mathbf{A} - \mathbf{mc})$ at $-3 \pm j\sqrt{3}, -10$. Give a block diagram of the observer-based position control system.

7.15 Consider the position control system of Fig. P7.15 employing a dc motor in armature control mode with state variables defined on the diagram. Full state-feedback is employed with position feedback being obtained from a potentiometer, rate feedback from a tachometer and current feedback from a voltage sample across a resistance in the armature circuit. K_A is the amplifier gain. Find the adjustable parameters $K_A, k_2,$ and k_3 so that the closed-loop poles of the system are placed at $-3 \pm j3, -20$.

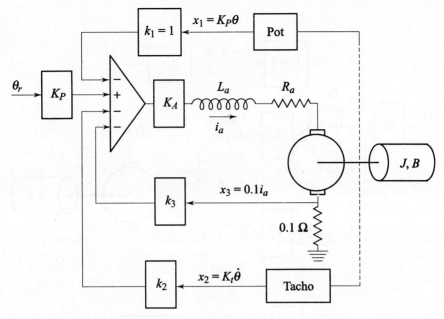

Fig. P7.15

Given:
Potentiometer sensitivity, $\quad K_P = 1$ volt/rad
Tachometer constant, $\quad K_t = 1$ volt/(rad/sec)
Armature inductance, $\quad L_a = 0.005$ H
Armature resistance, $\quad R_a = 0.9\ \Omega$
Moment of inertia of motor and load, $J = 0.02$ N-m/(rad/sec^2)
Viscous-friction coefficient of motor and load, $B = 0$
Back emf constant, $\quad K_b = 1$ volt/(rad/sec)
Motor torque constant, $\quad K_T = 1$ N-m/amp

7.16 Consider the position control system of Fig. P7.16 employing a dc motor in the field control mode, with state variables defined on the diagram. Full state-feedback is employed with position feedback being obtained from a potentiometer, rate feedback from a tachometer and current feedback from a voltage sample across a resistor connected in the field circuit. K_A is the amplifier gain.

Find the adjustable parameters K_A, k_2, and k_3 so that the closed-loop system has dominant poles characterized by $\zeta = 0.5$, $\omega_n = 2$, and the third pole at $s = -10$.

Given:
Potentiometer sensitivity, $\quad K_P = 1$ volt/rad
Tachometer constant, $\quad K_t = 1$ volt/(rad/sec)
Field inductance, $\quad L_f = 20$ H
Field resistance, $\quad R_f = 99\ \Omega$
Moment of inertia of motor and load, $J = 0.5$ N-m/(rad/sec^2)
Viscous-friction coefficient of motor and load, $B = 0.5$ N-m/(rad/sec)
Motor torque constant, $K_T = 10$ N-m/amp

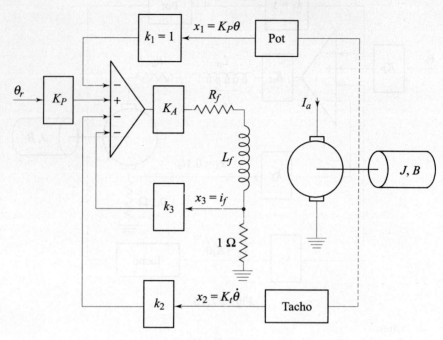

Fig. P7.16

7.17 Figure P7.17 shows control configuration of a type-1 servo system. Both the state variables x_1 and x_2 are assumed to be measurable. It is desired to regulate the output y to a constant value $r = 5$. Find the values of k_1, k_2 and N so that
   (i) $y(\infty) = r = 5$
   (ii) The closed-loop characteristic equation is
$$s^2 + a_1 s + a_2 = 0$$

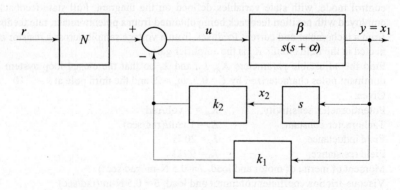

Fig. P7.17

7.18 A speed control system, employing a dc motor in the armature control mode, is described by the following state equations:

$$\frac{d\omega(t)}{dt} = -\frac{B}{j}\omega(t) + \frac{K_T}{J}i_a(t) - \frac{1}{J}T_L$$

$$\frac{di_a(t)}{dt} = -\frac{K_b}{L}\omega(t) - \frac{R}{L}i_a(t) + \frac{1}{L}u(t)$$

where

$i_a(t)$ = armature current, amps
$u(t)$ = armature applied voltage, volts
$\omega(t)$ = motor velocity, rad/sec
B = viscous-friction coefficient of motor and load = 0
J = moment of inertia of motor and load = 0.02 N-m/(rad/sec^2)
K_T = motor torque constant = 1 N-m/amp
K_b = motor back emf constant = 1 volt/(rad/sec)
T_L = constant disturbance torque (magnitude not known)
L = armature inductance = 0.005 H
R = armature resistance = 1 Ω

The design problem is to find the control $u(t)$ such that

(i) $\lim\limits_{t \to \infty} \dfrac{di_a(t)}{dt} = 0$ and $\lim\limits_{t \to \infty} \dfrac{d\omega(t)}{dt} = 0$, and

(ii) $\lim\limits_{t \to \infty} \omega(t)$ = constant set-point r.

Show that the control law of the form

$$u(t) = -k_1\omega(t) - k_2 i_a(t) - k_3 \int_0^t (\omega(t) - r)dt$$

can meet these objectives. Find k_1, k_2, and k_3 so that the closed-loop poles are placed at $-10 \pm j10, -300$. Suggest a suitable scheme for implementation of the control law.

7.19 Figure P7.19 shows a process consisting of two interconnected tanks. h_1 and h_2 represent deviations in tank levels from their steady-state values $\overline{H}_1$ and $\overline{H}_2$ respectively; q represents deviation in the flow rate from its steady-state value $\overline{Q}$. The flow rate q is controlled by signal u via valve and actuator. A disturbance flow rate w enters the first tank *via* a returns line from elsewhere in the process. The differential equations for levels in the tanks are given by

$$\dot{h}_1 = -3h_1 + 2h_2 + u + w$$
$$\dot{h}_2 = 4h_1 - 5h_2$$

(a) Compute the gains k_1 and k_2 so that the control law $u = -k_1 h_1(t) - k_2 h_2(t)$ places the closed-loop poles at $-4, -7$.
(b) Show that the steady-state error in the output $y(t) = h_2(t)$ in response to constant disturbance input w is non-zero.
(c) Add to the plant equations an integrator $\dot{z}(t) = y(t)$ and select gains k_1, k_2 and k_3 so that the control law $u = -k_1 h_1(t) - k_2 h_2(t) - k_3 z(t)$ places the closed-loop poles at $-1, -2, -7$. Find the steady-state value of the output in response to constant disturbance w. Give a block diagram depicting the control configuration.

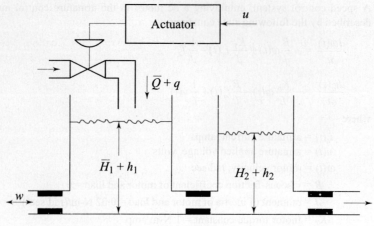

Fig. P7.19

7.20 The plant of a servo system is described by the equations

$$\dot{\mathbf{x}} = \mathbf{A}\mathbf{x} + \mathbf{b}u + \mathbf{b}w$$
$$y = \mathbf{c}\mathbf{x}$$

where

$$\mathbf{A} = \begin{bmatrix} -3 & 2 \\ 4 & -5 \end{bmatrix}; \mathbf{b} = \begin{bmatrix} 1 \\ 0 \end{bmatrix}; \mathbf{c} = \begin{bmatrix} 0 & 1 \end{bmatrix}$$

w is a disturbance input to the system.
A control law of the form $u = -\mathbf{k}\mathbf{x} + Nr$ is proposed; r is a constant reference input.
(a) Compute $\mathbf{k}$ so that the eigenvalues of $(\mathbf{A} - \mathbf{b}\mathbf{k})$ are $-4, -7$.
(b) Choose N so that the system has zero steady-state error to reference input, i.e., $y(\infty) = r$
(c) Show that the steady-state error to a constant disturbance input w is non-zero for the above choice of N.
(d) Add to the plant equation, an integrator equation ($z(t)$ being the state of the integrator):

$$\dot{z}(t) = y(t) - r$$

and select gains k_1, k_2 and k_3 so that the control law $u = -k_1 x_1(t) - k_2 x_2(t) - k_3 z(t)$ places the eigenvalues of closed-loop system matrix at $-1, -2, -7$.
(e) Draw a block diagram of the control scheme employing integral control and show that the steady-state error to constant disturbance input is zero.

7.21 Consider a plant consisting of a dc motor, the shaft of which has the angular velocity $\omega(t)$ and which is driven by an input voltage $u(t)$. The describing equation is

$$\dot{\omega}(t) = -0.5\,\omega(t) + 100\,u(t) = A\omega(t) + bu(t)$$

It is desired to regulate the angular velocity at the desired value $\omega^0 = r$.
(a) Use control law of the form $u = -K\omega(t) + Nr$. Choose K that results in closed-loop pole with time constant 0.1 sec. Choose N that guarantees zero steady-state error, i.e., $\omega(\infty) = r$.
(b) Show that if A changes to $A + \delta A$ subject to $(A + \delta A - bK)$ being stable, then the above choice of N will no longer make $\omega(\infty) = r$. Therefore, the system is not robust under changes in system parameters.

(c) The system can be made robust by augmenting it with an integrator

$$\dot{z} = \omega - r$$

where z is the state of the integrator. To see this, first use a feedback of the form $u = -K_1\omega(t) - K_2 z(t)$ and select K_1 and K_2 so that the characteristic polynomial of the closed-loop system becomes $\Delta(s) = s^2 + 11s + 50$. Show that the resulting system will have $\omega(\infty) = r$ no matter how the matrix A changes so long as the closed-loop system remains asymptotically stable.

7.22 A discrete-time regulator system has the plant

$$\mathbf{x}(k+1) = \begin{bmatrix} 0 & 1 & 0 \\ 0 & 0 & 1 \\ -4 & -2 & -1 \end{bmatrix} \mathbf{x}(k) + \begin{bmatrix} 0 \\ 0 \\ 1 \end{bmatrix} u(k)$$

Design a state-feedback controller which will place the closed-loop poles at $-\frac{1}{2} \pm j\frac{1}{2}, 0$. Give a block diagram of the control configuration.

7.23 Consider a plant defined by the following state variable model:

$$\mathbf{x}(k+1) = \mathbf{F}\mathbf{x}(k) + \mathbf{G}\mathbf{u}(k)$$

$$y(k) = \mathbf{c}\mathbf{x}(k) + \mathbf{d}\mathbf{u}(k)$$

where

$$\mathbf{F} = \begin{bmatrix} \frac{1}{2} & 1 & 0 \\ -1 & 0 & 1 \\ 0 & 0 & 0 \end{bmatrix}; \mathbf{G} = \begin{bmatrix} 1 & 4 \\ 0 & 0 \\ -3 & 2 \end{bmatrix}$$

$$\mathbf{c} = \begin{bmatrix} 1 & 0 & 0 \end{bmatrix}; \mathbf{d} = \begin{bmatrix} 0 & 4 \end{bmatrix}$$

Design a prediction observer for the estimation of the state vector $\mathbf{x}$; the observer-error poles are required to lie at $-\frac{1}{2} \pm j\frac{1}{4}, 0$. Give all the relevant observer equations and a block diagram description of the observer structure.

7.24 Consider the system defined by

$$\mathbf{x}(k+1) = \begin{bmatrix} 0 & 1 & 0 \\ 0 & 0 & 1 \\ -0.5 & -0.2 & 1.1 \end{bmatrix} \mathbf{x}(k) + \begin{bmatrix} 0 \\ 0 \\ 1 \end{bmatrix} u(k)$$

Determine the state-feedback gain matrix $\mathbf{k}$ such that when the control signal is given by $u(k) = -\mathbf{k}\mathbf{x}(k)$, the closed-loop system will exhibit the deadbeat response to any initial state $\mathbf{x}(0)$. Give the state variable model of the closed-loop system.

7.25 Consider the system

$$\mathbf{x}(k+1) = \begin{bmatrix} 0 & 1 \\ -0.16 & -1 \end{bmatrix} \mathbf{x}(k) + \begin{bmatrix} 0 \\ 1 \end{bmatrix} u(k)$$

$$y(k) = \begin{bmatrix} 1 & 1 \end{bmatrix} \mathbf{x}(k)$$

Design a current observer for the system; the response to the initial observer error is required to be deadbeat. Give all the relevant observer equations.

7.26 Consider the plant defined in Problem 7.24. Assuming that only $y(k) = x_2(k)$ is measurable, design a reduced-order observer such that the response to the observer error is deadbeat. Give all the relevant observer equations.

7.27 A discrete-time regulator system has the plant

$$\mathbf{x}(k+1) = \begin{bmatrix} 2 & -1 \\ -1 & 1 \end{bmatrix} \mathbf{x}(k) + \begin{bmatrix} 4 \\ 3 \end{bmatrix} u(k)$$

$$y(k) = [1 \quad 1] \mathbf{x}(k) + 7u(k)$$

(a) Design a state-feedback control algorithm $u(k) = -\mathbf{kx}(k)$ which places the closed-loop characteristic roots at $\pm j\frac{1}{2}$.

(b) Design a prediction observer for deadbeat response. Give the relevant observer equations.

(c) Combining (a) and (b), give a block diagram of the control configuration. Also obtain state variable model of the observer-based state-feedback control system.

7.28 A regulator system has the plant with transfer function

$$\frac{Y(z)}{U(z)} = \frac{z^{-2}}{(1+0.8z^{-1})(1+0.2z^{-1})}$$

(a) Find $(\mathbf{F}, \mathbf{g}, \mathbf{c})$ for the plant in controllable canonical form.

(b) Find k_1 and k_2 such that $u(k) = -k_1 x_1(k) - k_2 x_2(k)$ gives closed-loop characteristic roots at $0.6 \pm j0.4$.

(c) Design a first-order observer that estimates x_2, given measurements of x_1; the response to initial observer error is required to be deadbeat.

(d) Give a z-domain block diagram of the closed-loop system.

7.29 Consider the system

$$\mathbf{x}(k+1) = \mathbf{Fx}(k) + \mathbf{g}u(k)$$

$$y(k) = \mathbf{cx}(k)$$

where

$$\mathbf{F} = \begin{bmatrix} 0.16 & 2.16 \\ -0.16 & -1.16 \end{bmatrix}; \mathbf{g} = \begin{bmatrix} -1 \\ 1 \end{bmatrix}; \mathbf{c} = [1 \quad 1]$$

(a) Design a state-feedback control algorithm which gives closed-loop characteristic roots at $0.6 \pm j\, 0.4$.

(b) Design a reduced-order observer for deadbeat response.

(c) Find the transfer function of the compensator obtained by combining (a) and (b). Give a block diagram of the closed-loop system showing the compensator in the control loop.

7.30 A double integrator plant is to be controlled by a digital computer employing state-feedback. Figure P7.30 shows a model of the control scheme. Both the state variables x_1 and x_2 are assumed to be measurable.

(a) Obtain the discrete-time state variable model of the plant.

(b) Compute k_1 and k_2 so that the response $y(t)$ of the closed-loop system has the parameters: $\zeta = 0.5$, $\omega_n = 4$.

(c) Assume now that only x_1 is measurable. Design a prediction observer to estimate the state vector $\mathbf{x}$; the estimation error is required to decay in a deadbeat manner.

(d) Find the transfer function of the compensator obtained by combining (b) and (c).

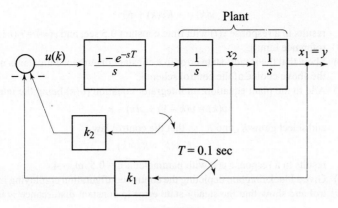

Fig. P.730

7.31 Figure P7.31 shows the block diagram of a digital positioning system. The plant is a dc motor driving inertia load. Both the position θ and velocity $\dot{\theta}$ are measurable.
(a) Obtain matrices (**F, g, c**) of the discrete-time state variable model of the plant.
(b) Compute k_1 and k_2 so that the closed-loop system positions the load in a deadbeat manner in response to any change in step command θ_r.
(c) Assume now that the position θ is measured by a shaft encoder and a second-order state observer is used to estimate the state vector **x** from plant input u and measurements of θ. Design a deadbeat observer. Give a block diagram of the observer-based digital positioning system.
(d) Design a first-order deadbeat observer to estimate velocity ω from measurements of position θ.

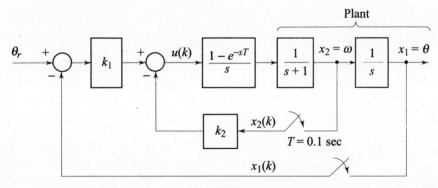

Fig. P7.31

7.32 A continuous-time plant described by the equation
$$\dot{y} = -y + u + w$$
is to be controlled by a digital computer; y is the output, u is the input, and w is the disturbance signal. Sampling interval $T = 1$ sec.
(a) Obtain a discrete-time state variable model of the plant.
(b) Compute K and N so that the control law

$$u(k) = -Ky(k) + Nr$$

results in a response $y(t)$ with time constant 0.5 sec, and $y(\infty) = r$ (r is a constant reference input).

(c) Show that the steady-state error to a constant disturbance input w is non-zero for the above choice of the control scheme.

(d) Add to the plant equation, an integrator equation ($v(k)$ being the integral state)

$$v(k) = v(k-1) + y(k) - r$$

and select gains K_1 and K_2 so that the control law

$$u = -K_1 y(k) - K_2 v(k)$$

results in a response $y(t)$ with parameters: $\zeta = 0.5$, $\omega_n = 4$.

(e) Give a block diagram depicting the control configuration employing integral control and show that the steady-state error to constant disturbance w is zero.

chapter 8

Lyapunov Stability Analysis

8.1 INTRODUCTION

For a given control system, stability is usually the most important attribute to be determined. If the system is linear and time-invariant, many stability criteria are applicable. Among them are the Routh stability criterion[1], and the Nyquist stability criterion[2]. However, if the system is nonlinear or linear but time-varying, then such stability criteria do not apply.

The tools available for analysis of nonlinear systems are, unfortunately, quite limited in nature, in number, in applicability, and in effectiveness. The classical tools[3] are

(i) the phase plane method, and
(ii) the describing function method.

The phase plane analysis applies primarily to second-order systems only. The describing function method is applicable to a special group of nonlinear systems; also the stability analysis using this approach is only approximate. Much of the work must be done by computer simulation. The describing-function analysis merely provides the background for the intelligent planning of the simulations.

In this chapter, the intention is to present a new concept of viewing system stability—the concept introduced by Russian mathematician A.M. Lyapunov. The Lyapunov's method of stability analysis is, in principle, the most general method for determination of stability of nonlinear and/or time-varying systems. The major drawback which seriously limits its use in practice is the difficulty often associated with the construction of the *Lyapunov function* required by the method.

---

1. Reference [180], Chapter 5.
2. *Ibid*, Chapter 8.
3. Discussed later in Chapter 10.

No new results are obtained by the use of the Lyapunov's method for the stability analysis of linear time-invariant systems. Simple and powerful methods such as the Nyquist stability criterion and the Routh stability criterion are adequate for such systems. However, as we shall see in Chapter 9, Lyapunov functions supply certain performance indices and synthesis data of linear time-invariant systems. The use of Lyapunov functions in adaptive control of linear time-invariant systems will be demonstrated in the present chapter.

This chapter aims to introduce the concept of Lyapunov stability and the role it plays in control system design. For detailed study, specialized books on nonlinear control systems [116–122] should be consulted. A moderate treatment of the subject is also available in reference [102].

8.2 BASIC CONCEPTS

The general state equation for a nonlinear system can be expressed as

$$\dot{\mathbf{x}} = \mathbf{f}(\mathbf{x}(t), \mathbf{u}(t), t); \; \mathbf{x}(t_0) \triangleq \mathbf{x}^0 \tag{8.1}$$

where $\mathbf{x}$ is the $n \times 1$ state vector, $\mathbf{u}$ is the $p \times 1$ input vector, and

$$\mathbf{f}(.) = \begin{bmatrix} f_1(.) \\ f_2(.) \\ \vdots \\ f_n(.) \end{bmatrix}$$

is the $n \times 1$ function vector.

Assuming that a suitable model of the form (8.1) is available, essentially any nonlinear system can be studied using computer simulation. It should, however, be clearly understood that simulation studies are simply controlled experiments performed on a computer model. Therefore adequate analysis results only if the experiments are carefully planned. All simulation studies are trial-and-error procedures.

Simulation of Eqn. (8.1) generates a set of plots of n state variables as a function of time for various initial states and the forcing functions. From the data so generated we can obtain state-space trajectories which provide a convenient visualization of the behaviour of nonlinear dynamical systems. At any time t, the state vector $\mathbf{x}(t)$ defines a point in the n-dimensional state space with components of the state vector as its coordinate axes. As time progresses and the system state changes, a set of points will be defined. This set of points, the locus of the tip of the state vector as time progresses, is called the *state trajectory* of the system. Time t is an implicit function along the trajectory. The family of all possible trajectories gives a *state-portrait* which defines the dynamical behaviour of the system.

Suppose that all the states of the system (8.1) settle to constant values (not necessarily zero values) for a constant input vector $\mathbf{u}(t) = \mathbf{u}^c$. The system is then said to be in an *equilibrium state* corresponding to the input $\mathbf{u}^c$. The state

Lyapunov Stability Analysis

trajectories converge to a point in state space called the *equilibrium point*. At this point, no states vary with time. Thus, we have the following definition of equilibrium point (equilibrium state).

If for any constant input vector $\mathbf{u}(t) = \mathbf{u}^c$, there exists a point $\mathbf{x}(t) = \mathbf{x}^e =$ constant, in state space such that at this point $\dot{\mathbf{x}}(t) = \mathbf{0}$ for all t, then this point is called an *equilibrium point* of the system corresponding to the input $\mathbf{u}^c$. Applying this definition to the system (8.1), any equilibrium point must satisfy
$$\mathbf{f}(\mathbf{x}^e, \mathbf{u}^c, t) = \mathbf{0} \quad \text{for all } t \tag{8.2}$$

The number of solutions depends entirely upon the nature of $\mathbf{f}(.)$ and no general statement is possible.

Example 8.1
Consider the nonlinear system described by the state equations:
$$\dot{x}_1 = x_2$$
$$\dot{x}_2 = -x_1 - x_1^2 - x_2$$

The equilibrium states of this system are given by the solutions of the following set of equations (refer Eqn. (8.2)):
$$\dot{x}_1^e = x_2^e = 0$$
$$\dot{x}_2^e = -x_1^e - (x_1^e)^2 - x_2^e = 0$$

From the first equation, x_2^e is equal to zero. From the second equation,
$$(x_1^e)^2 + x_1^e = x_1^e(x_1^e + 1) = 0$$

which has the solutions $x_1^e = 0$ and $x_1^e = -1$. Thus, there are two equilibrium states given by
$$\mathbf{x}^{e1} = \begin{bmatrix} 0 \\ 0 \end{bmatrix}, \quad \mathbf{x}^{e2} = \begin{bmatrix} -1 \\ 0 \end{bmatrix}$$

In the stability analysis of a system, we are usually concerned with the following two notions of stability:

(i) when a relaxed system ($\mathbf{x}(t_0) = \mathbf{0}$) is excited by a bounded input, the output must be bounded; and

(ii) in an unforced system ($\mathbf{u} = \mathbf{0}$) with arbitrary initial conditions, the system state must tend towards the equilibrium point in state space.

We have seen earlier in Chapter 5 that the two notions of stability defined above are essentially equivalent for linear time-invariant systems.

Unfortunately in nonlinear systems, there is no definite correspondence between the two notions. For a nonlinear system which is stable as per the second notion above, there is no guarantee that the output will be bounded whenever the system is excited by a bounded input. Also, if the output is bounded for a particular input, it may not be bounded for other inputs. Most of the important results obtained thus far concern the stability of nonlinear *autonomous*[4] systems:

$$\dot{\mathbf{x}}(t) = \mathbf{f}(\mathbf{x}(t)); \quad \mathbf{x}(t_0) \triangleq \mathbf{x}^0 \tag{8.3}$$

in the sense of second notion above.

It may be noted that even for this class of systems, the concept of stability is not clear cut. The linear autonomous systems have only one equilibrium state (the origin of the state space) and their behaviour about the equilibrium state completely determines the qualitative behaviour in the entire state space. In nonlinear systems, on the other hand, system behaviour for small deviations about the equilibrium point may be different from that for large deviations. Therefore, *local stability* does not imply stability in the overall state space and the two concepts should be considered separately.

Secondly, the set of nonlinear equations (refer Eqns (8.2)–(8.3)),

$$\mathbf{f}(\mathbf{x}^e) = \mathbf{0} \tag{8.4}$$

may result in a number of solutions (equilibrium points). Due to the possible existence of multiple equilibrium states, the system trajectories may move away from one equilibrium state to the other as time progresses. Thus, it appears that in the case of nonlinear systems, it is simpler to speak of system stability relative to the equilibrium state rather than using the general term 'stability of a system'.

Another important point to be kept in mind is that in linear autonomous systems, when oscillations occur, the amplitude of the oscillations is not fixed. It changes with the size of the initial conditions. Slight changes in system parameters (shifting the eigenvalues from the imaginary axis of the complex plane) will destroy the oscillations. In nonlinear systems, on the other hand, there can be oscillations that are independent of the size of initial conditions and these oscillations (*limit cycles*) are usually much less sensitive to parameter variations. Limit cycles of fixed amplitude and period can be sustained over a finite range of system parameters.

We shall confine our attention to nonlinear autonomous systems described by state equation of the form

$$\dot{\mathbf{x}}(t) = \mathbf{f}(\mathbf{x}(t)); \quad \mathbf{f}(\mathbf{0}) = \mathbf{0}; \quad \mathbf{x}(0) \triangleq \mathbf{x}^0 \tag{8.5}$$

Note that the origin of the state space has been taken as the equilibrium state of the system. There is no loss in generality in this assumption since any nonzero equilibrium state can be shifted to the origin by appropriate transformation. Further, we have taken $t_0 = 0$ in Eqn. (8.5), which is a convenient choice for time-invariant systems.

For nonlinear autonomous systems, local stability may be investigated through linearization in the neighbourhood of the equilibrium point. The validity of determining the stability of the unperturbed solution near the equilibrium

---

4. An unforced (i.e., $\mathbf{u} = \mathbf{0}$) and time-invariant system is called an autonomous system.

points from the linearized equations was developed independently by Poincaré and Lyapunov in 1892. Lyapunov designated this as the *first method*. This stability determination is applicable only in a small region near the equilibrium point and results in *stability-in-the-small*.

The region of validity of local stability is generally not known. In some cases, the region may be too small to be of any use practically; while in others the region may be much larger than the one assumed by the designer giving rise to systems that are too conservatively designed. We therefore need information about the domain of stability. We present, in this chapter, the 'second method of Lyapunov' (also called the 'direct method of Lyapunov') which is used to determine *stability in-the-large*.

8.3 STABILITY DEFINITIONS

The concept of stability formulated by Russian mathematician A.M. Lyapunov is concerned with the following question:

If a system with zero input is perturbed from the equilibrium point $\mathbf{x}^e$ at $t = 0$, will the state $\mathbf{x}(t)$ return to $\mathbf{x}^e$, remain 'close' to $\mathbf{x}^e$, or diverge from $\mathbf{x}^e$?

Lyapunov stability analysis is thus concerned with the boundedness of the free (unforced) response of a system. The free response of a system is said to be *stable in the sense of Lyapunov* at the equilibrium point $\mathbf{x}^e$ if, for every initial state $\mathbf{x}(t_0)$ which is sufficiently close to $\mathbf{x}^e$, $\mathbf{x}(t)$ remains near $\mathbf{x}^e$ for all t. It is *asymptotically stable* at $\mathbf{x}^e$ if $\mathbf{x}(t)$, in fact, approaches $\mathbf{x}^e$ as $t \to \infty$.

In the following, we give mathematically precise definitions of different types of stability with respect to the system described by Eqn. (8.5).

Stability in the Sense of Lyapunov

The system described by Eqn. (8.5) is *stable in the sense of Lyapunov* at the origin if, for every real number $\varepsilon > 0$, there exists a real number $\delta(\varepsilon) > 0$ such that $\|\mathbf{x}(0)\| < \delta$ results in $\|\mathbf{x}(t)\| < \varepsilon$ for all $t \geq 0$.

This definition uses the concept of the vector norm. The Euclidean norm for a vector with n components $x_1, x_2, \ldots, x_n$ is (refer Eqn. (5.6a))

$$\|\mathbf{x}\| = (x_1^2 + x_2^2 + \cdots + x_n^2)^{1/2}$$

$\|\mathbf{x}\| \leq R$ defines a hyper-spherical region $S(R)$ of radius R surrounding the equilibrium point $\mathbf{x}^e = \mathbf{0}$. In terms of the Euclidean norm, the above definition of stability implies that for any $S(\varepsilon)$ that we may designate, the designer must produce $S(\delta)$ so that the system state initially in $S(\delta)$ will never leave $S(\varepsilon)$. This is illustrated in Fig. 8.1a.

Note that this definition of stability permits the existence of continuous oscillation about the equilibrium point. The state-space trajectory for such an oscillation is a closed path called a *limit cycle*. The performance specifications for a control system must be used to determine whether or not a limit cycle can be permitted. The amplitude and frequency of the oscillation may influence whether it represents acceptable performance.

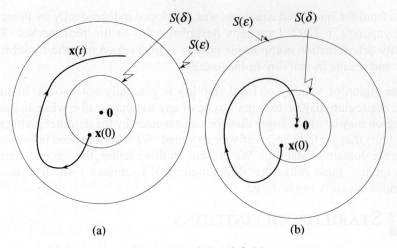

Fig. 8.1 Stability definitions

Example 8.2
Consider a linear oscillator described by the differential equation

$$\ddot{y}(t) + \omega^2 y(t) = 0$$

where ω is the frequency of oscillations.
Define the state variables as

$$x_1(t) = y(t), \; x_2(t) = \dot{y}(t)$$

This gives the state equations

$$\dot{x}_1(t) = x_2(t)$$
$$\dot{x}_2(t) = -\omega^2 x_1(t)$$

From these equations, we obtain the following equation for state trajectory:

$$\frac{dx_2}{dx_1} = -\omega^2 \frac{x_1}{x_2} \quad \text{or} \quad x_2^2 + \omega^2 x_1^2 = c^2; \quad c = \text{constant}$$

Several state trajectories for various values of c, corresponding to various initial conditions of x_1 and x_2, are shown in Fig. 8.2. For a specified value of ε, we can find a closed state trajectory whose maximum distance from the origin is ε. We then select a value of δ which is less than the minimum distance from that curve to the origin. The $\delta(\varepsilon)$ so chosen will satisfy the conditions that guarantee stability in the sense of Lyapunov.

Asymptotic Stability
The system (8.5) is asymptotically stable at the origin if
   (a) it is stable in the sense of Lyapunov, i.e., for each $S(\varepsilon)$ there is a region $S(\delta)$ such that trajectories starting within $S(\delta)$ do not leave $S(\varepsilon)$ as $t \to \infty$, and

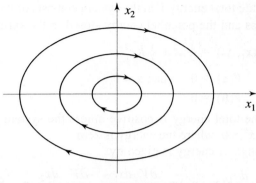

Fig. 8.2

(b) each trajectory starting within $S(\delta)$ converges to the origin as $t \to \infty$ (Fig. 8.1b).

Local and Global Stability

The definitions of asymptotic stability and stability in the sense of Lyapunov apply in a *local* sense (*stability in-the-small*) if the region $S(\delta)$ is small. When the region $S(\delta)$ includes the entire state space, the definitions of asymptotic stability and stability in the sense of Lyapunov are said to apply in a *global* sense (*stability in-the-large*).

8.4 STABILITY THEOREMS

The Lyapunov stability analysis is based upon the concept of energy and the relation of stored energy with system stability. We first give an example to motivate the discussion.

Consider the spring-mass-damper system of Fig. 8.3. The governing equation of the system is

$$\ddot{x}_1 + B\dot{x}_1 + Kx_1 = 0$$

A corresponding state variable model is

$$\begin{aligned} \dot{x}_1 &= x_2 \\ \dot{x}_2 &= -Kx_1 - Bx_2 \end{aligned} \tag{8.6}$$

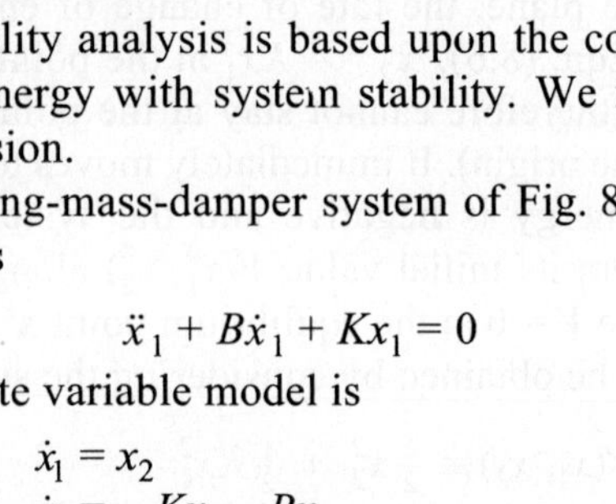

Fig. 8.3 A spring-mass-damper system

At any instant, the total energy V in the system consists of the kinetic energy of the moving mass and the potential energy stored in the spring.

$$V(x_1, x_2) = \tfrac{1}{2} x_2^2 + \tfrac{1}{2} K x_1^2 \tag{8.7}$$

Thus $\quad V(\mathbf{x}) > 0 \quad$ when $\mathbf{x} \neq 0$
$\qquad V(\mathbf{0}) = 0$

This means that the total energy is positive unless the system is at rest at the equilibrium point $\mathbf{x}^e = \mathbf{0}$, where the energy is zero.

The rate of change of energy is given by

$$\dot{V}(x_1, x_2) = \frac{d}{dt} V(x_1, x_2) = \frac{\partial V}{\partial x_1} \frac{dx_1}{dt} + \frac{\partial V}{\partial x_2} \frac{dx_2}{dt} = -Bx_2^2 \tag{8.8}$$

Case I: Positive damping ($B > 0$)

Let (x_1^0, x_2^0) be an arbitrary initial state of the system of Fig. 8.3. The solution of the differential Eqns (8.6) corresponding to this initial state gives the state trajectory $\mathbf{x}(t)$ for $t > 0$. Since the linear system (8.6) is stable under the condition of positive damping, $\mathbf{x}(t) \to \mathbf{0}$ as $t \to \infty$.

Let us study the relation of stored energy with system stability. The initial energy in the system is

$$V(x_1^0, x_2^0) = \tfrac{1}{2}(x_2^0)^2 + \tfrac{1}{2} K(x_1^0)^2$$

As per Eqn. (8.8), the rate of change of energy is negative and therefore system energy $V(x_1, x_2)$ continually decreases along the trajectory $\mathbf{x}(t)$, $t > 0$. There is only one exception; when the representative point $\mathbf{x}(t)$ of the trajectory reaches $x_2 = 0$ points in the state plane, the rate of change of energy becomes zero. However, as seen from Eqn. (8.6), $\dot{x}_2 = -Kx_1$ at the points where $x_2 = 0$. The representative point $\mathbf{x}(t)$ therefore cannot stay at the points in the state plane where $x_2 = 0$ (except at the origin). It immediately moves to the points at which the rate of change of energy is negative and the system energy therefore continually decreases from its initial value $V(x_1^0, x_2^0)$ along the trajectory $\mathbf{x}(t)$, $t > 0$ till it reaches a value $V = 0$ at the equlibrium point $\mathbf{x}^e = \mathbf{0}$.

A visual analogy may be obtained by considering the surface

$$V(x_1, x_2) = \tfrac{1}{2} x_2^2 + \tfrac{1}{2} K x_1^2 \tag{8.9}$$

This is paraboloid surface as shown in Fig. 8.4. The value $V(x_1, x_2) = k_i$ (a constant) is represented by the intersection of the surface $V(x_1, x_2)$ and the plane $z = k_i$. The projection of this intersection on the (x_1, x_2)-plane is a closed curve, an oval, around the origin. There is a family of such closed curves in the (x_1, x_2)-plane for different values of k_i. The closed curve corresponding to $V(x_1, x_2) = k_1$ lies entirely inside the closed curve corresponding to $V(x_1, x_2) = k_2$ if $k_1 < k_2$. The value $V(x_1, x_2) = 0$ is the point at the origin. It is the innermost curve of the family of closed curves representing different levels on the paraboloid for $V(x_1, x_2) = k_i$.

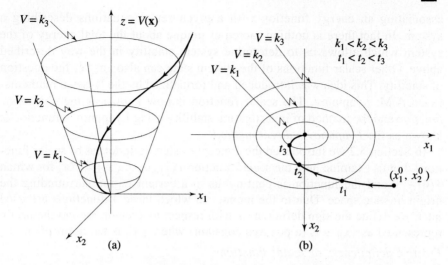

Fig. 8.4 Constant-V curves

If one plots a state-plane trajectory starting from the point (x_1^0, x_2^0), the representative point $\mathbf{x}(t)$ crosses the ovals for successively smaller values of $V(x_1, x_2)$ and moves towards the point corresponding to $V(x_1, x_2) = 0$, which is the equilibrium point. Figure 8.4 shows a typical trajectory.

Note also that $V(\mathbf{x})$ given by Eqn. (8.9) is *radially unbounded*[5], i.e., $V(\mathbf{x}) \to \infty$ as $\|\mathbf{x}\| \to \infty$. The ovals on the (x_1, x_2)-plane extend over the entire state plane and therefore for any initial state $\mathbf{x}^0$ in the entire state plane, the system energy continually decreases from the value $V(\mathbf{x}^0)$ to zero.

Case II: Zero damping $(B = 0)$

Under the condition of zero damping, Eqns (8.6) become
$$\dot{\mathbf{x}} = \mathbf{A}\mathbf{x}$$
with
$$\mathbf{A} = \begin{bmatrix} 0 & 1 \\ -K & 0 \end{bmatrix}$$

The eigenvalues of $\mathbf{A}$ lie on the imaginary axis in the complex plane; the system response is therefore oscillatory in nature.

From Eqn. (8.8) we observe that when $B = 0$, the rate of change of energy $\dot{V}(x_1, x_2)$ vanishes identically along any trajectory; the system energy $V(x_1, x_2) = V(x_1^0, x_2^0)$ for all $t \geq 0$. The representative point $\mathbf{x}(t)$ cannot cross the V-contours in Fig. 8.4; it simply moves along one of these contours and the system remains in a limit cycle.

▲▲

In the example given above, it was easy to associate the energy function V with the given system. However, in general, there is no obvious way of

---

5. Use of the norm definition given by Eqn. (5.6b) immediately proves this result.

associating an energy function with a given set of equations describing a system. In fact there is nothing sacred or unique about the total energy of the system which allows us to determine system stability in the way described above. Other scalar functions of the system state can also answer the question of stability. This idea was introduced and formalized by the Russian mathematician A.M. Lyapunov. The scalar function is now known as the *Lyapunov function* and the method of investigating stability using Lyapunov's function is known as the *Lyapunov's direct method*.

In Section 5.2 we introduced the concept of sign definiteness of scalar functions. Let us examine here the scalar function $V(x_1, x_2, \ldots, x_n) \triangleq V(\mathbf{x})$ for which $V(\mathbf{0}) = 0$ and the function is continuous in a certain region surrounding the origin in state space. Due to the manner in which these V-functions are used later, we define the sign definiteness with respect to a region around the origin represented as $\|\mathbf{x}\| \leq K$ (a positive constant) where $\|\mathbf{x}\|$ is the norm of $\mathbf{x}$.

Positive definiteness of scalar functions

A scalar function $V(\mathbf{x})$ is said to be *positive definite* in the region $\|\mathbf{x}\| \leq K$ (which includes the origin of the state space) if $V(\mathbf{x}) > 0$ at all points of the region except at the origin, where it is zero.

Negative definiteness of scalar functions

A scalar function $V(\mathbf{x})$ is said to be *negative definite* if $[-V(\mathbf{x})]$ is positive definite.

Positive semidefiniteness of scalar functions

A scalar function $V(\mathbf{x})$ is said to be *positive semidefinite* in the region $\|\mathbf{x}\| < K$, if its value is positive at all points of the region except at finite number of points, including origin, where it is zero.

Negative semidefiniteness of scalar functions

A scalar function $V(\mathbf{x})$ is said to be *negative semidefinite* if $[-V(\mathbf{x})]$ is positive semidefinite.

Indefiniteness of scalar functions

A scalar function $V(\mathbf{x})$ is said to be indefinite in the region $\|\mathbf{x}\| < K$ if it assumes both positive and negative values, within this region.

Examples

For all $\mathbf{x}$ in the state plane,
(i) $V(\mathbf{x}) = x_1^2 + x_2^2$ is positive definite,
(ii) $V(\mathbf{x}) = (x_1 + x_2)^2$ is positive semidefinite,
(iii) $V(\mathbf{x}) = -x_1^2 - (x_1 + x_2)^2$ is negative definite,
(iv) $V(\mathbf{x}) = x_1 x_2 + x_2^2$ is indefinite.

An important class of scalar functions is a quadratic form:
$$V(\mathbf{x}) = \mathbf{x}^T \mathbf{P} \mathbf{x}$$
where $\mathbf{P}$ is a real, symmetric constant matrix. In this form the definiteness of V is usually attributed to $\mathbf{P}$. We speak of the positive (negative) definite and the

positive (negative) semidefinite **P** depending upon the definiteness of $V(\mathbf{x}) = \mathbf{x}^T\mathbf{Px}$.

Tests for checking definiteness of a matrix were described in Section 5.2. We consider an example here:

$$\mathbf{P} = \begin{bmatrix} 10 & 1 & -2 \\ 1 & 4 & -1 \\ -2 & -1 & 1 \end{bmatrix}$$

As per Sylvester's test, the necessary and sufficient condition for **P** to be positive definite is that all the successive principal minors of **P** be positive.

Applying Sylvester's test to the given **P**, we obtain

$$10 > 0$$

$$\begin{vmatrix} 10 & 1 \\ 1 & 4 \end{vmatrix} > 0$$

$$\begin{vmatrix} 10 & 1 & -2 \\ 1 & 4 & -1 \\ -2 & -1 & 1 \end{vmatrix} > 0$$

Since all the successive principal minors of the matrix **P** are positive, $V(\mathbf{x}) = \mathbf{x}^T\mathbf{Px}$ is a positive definite function.

▲▲

In the following, we state, without proof, the basic Lyapunov stability results. For proof, refer [102].

Theorem 8.1

For the autonomous system (8.5), sufficient conditions of stability are as follows:

Suppose that there exists a scalar function $V(\mathbf{x})$ which, for some real number $\varepsilon > 0$, satisfies the following properties for all **x** in the region $\|\mathbf{x}\| \leq \varepsilon$:

(1) $V(\mathbf{x}) > 0; \mathbf{x} \neq \mathbf{0}$
(2) $V(\mathbf{0}) = 0$ (i.e., $V(\mathbf{x})$ is positive definite function)
(3) $V(\mathbf{x})$ has continuous partial derivatives with respect to all components of **x**.

Then the equilibrium state $\mathbf{x}^e = \mathbf{0}$ of the system (8.5) is

(4a) *asymptotically stable* if $\dot{V}(\mathbf{x}) < 0, \mathbf{x} \neq \mathbf{0}$, i.e., $\dot{V}(\mathbf{x})$ is a negative definite function;

(4b) *asymptotically stable in-the-large* if $\dot{V}(\mathbf{x}) < 0, \mathbf{x} \neq \mathbf{0}$, and in addition $V(\mathbf{x}) \to \infty$ as $\|\mathbf{x}\| \to \infty$.

Example 8.3

Consider a nonlinear system described by the equations

$$\begin{aligned} \dot{x}_1 &= x_2 - x_1(x_1^2 + x_2^2) \\ \dot{x}_2 &= -x_1 - x_2(x_1^2 + x_2^2) \end{aligned} \qquad (8.10)$$

Clearly, the origin is the only equilibrium state.

Let us choose the following positive definite scalar function as a possible Lyapunov function:

$$V(\mathbf{x}) = x_1^2 + x_2^2 \qquad (8.11)$$

Time derivative of $V(\mathbf{x})$ along any trajectory is given by

$$\dot{V}(\mathbf{x}) = \frac{dV(x_1, x_2)}{dt} = \frac{\partial V}{\partial x_1}\frac{dx_1}{dt} + \frac{\partial V}{\partial x_2}\frac{dx_2}{dt}$$

$$= 2x_1\dot{x}_1 + 2x_2\dot{x}_2 = -2(x_1^2 + x_2^2)^2 \qquad (8.12)$$

which is negative definite. This shows that $V(\mathbf{x})$ is continually decreasing along any trajectory; hence $V(\mathbf{x})$ is a Lyapunov function. By Theorem 8.1, the equilibrium state (at the origin) of the system (8.10) is asymptotically stable.

Further, $V(\mathbf{x}) \to \infty$ as $\|\mathbf{x}\| \to \infty$, i.e., $V(\mathbf{x})$ becomes infinite with infinite deviation from the equilibrium state. Therefore as per condition (4b) of Theorem 8.1, the equilibrium state of the system (8.10) is asymptotically stable in-the-large.

▲▲

Although Theorem 8.1 is a basic theorem of Lyapunov stability analysis, it is somewhat restrictive because $\dot{V}(\mathbf{x})$ must be negative definite. This requirement can be relaxed to $\dot{V}(\mathbf{x}) \leq 0$ (a negative semidefinite $\dot{V}(\mathbf{x})$) under proper conditions. This relaxed requirement is sufficient if it can be shown that no trajectory can stay forever at the points or on the line, other than the origin, at which $\dot{V}(\mathbf{x}) = 0$. This is the case for the system of Fig. 8.3 as described at the beginning of this section.

If, however, there exists a positive definite function $V(\mathbf{x})$ such that $\dot{V}(\mathbf{x})$ is identically zero along a trajectory, the system will remain in a limit cycle. The equilibrium state at the origin in this case is said to be stable in the sense of Lyapunov.

Theorem 8.2

For the autonomous system (8.5), sufficient conditions of stability are as follows.

Suppose that there exists a scalar function $V(\mathbf{x})$ which, for some real number $\varepsilon > 0$, satisfies the following properties for all $\mathbf{x}$ in the region $\|\mathbf{x}\| \leq \varepsilon$:

(1) $V(\mathbf{x}) > 0;\ \mathbf{x} \neq \mathbf{0}$
(2) $V(\mathbf{0}) = 0$ } (i.e, $V(\mathbf{x})$ is positive definite function)

(3) $V(\mathbf{x})$ has continuous partial derivatives with respect to all components of $\mathbf{x}$.

Then the equilibrium state $\mathbf{x}^e = \mathbf{0}$ of the system (8.5) is

(4a) *asymptotically stable* if $\dot{V}(\mathbf{x}) < 0$, $\mathbf{x} \neq \mathbf{0}$, i.e., $\dot{V}(\mathbf{x})$ is a negative definite function; or if $\dot{V}(\mathbf{x}) \leq 0$ (i.e., $\dot{V}(\mathbf{x})$ is negative semidefinite) and no trajectory can stay forever at the points or on the line, other than the origin, at which $\dot{V}(\mathbf{x}) = 0$;

(4b) *asymptotically stable in-the-large* if conditions (4a) are satisfied, and in addition $V(\mathbf{x}) \to \infty$ as $\|\mathbf{x}\| \to \infty$;

(4c) *stable in the sense of Lyapunov* if $\dot{V}(\mathbf{x})$ is identically zero along a trajectory.

Example 8.4

Consider the linear feedback system shown in Fig. 8.5 with $r(t) = 0$. We know that the closed-loop system will exhibit sustained oscillations.

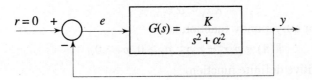

Fig. 8.5 Linear feedback System

The differential equation for the error signal is

$$\ddot{e} + \alpha^2 e = Ky = -Ke$$

Taking e and $\dot{e}$ as state variables x_1 and x_2 respectively, we obtain the following state equations:

$$\begin{aligned}\dot{x}_1 &= x_2 \\ \dot{x}_2 &= -(K + \alpha^2)x_1\end{aligned} \qquad (8.13)$$

Let us choose the following scalar positive definite function as a possible Lyapunov function:

$$V(\mathbf{x}) = x_1^2 + x_2^2 \qquad (8.14)$$

Then $\dot{V}(\mathbf{x})$ becomes

$$\dot{V}(\mathbf{x}) = 2x_1\dot{x}_1 + 2x_2\dot{x}_2 = 2[1 - (K + \alpha^2)]x_1 x_2$$

$\dot{V}(\mathbf{x})$ is indefinite. This implies that $V(\mathbf{x})$ given by Eqn. (8.14) is not a Lyapunov function and stability cannot be determined by its use (the system is known to be stable in the sense of Lyapunov as per the stability definition given in Section 8.3).

We now test

$$V(\mathbf{x}) = p_1 x_1^2 + p_2 x_2^2; \; p_1 > 0, \, p_2 > 0$$

for Lyapunov properties. Conditions (1)–(3) of Theorem 8.2 are obviously satisfied.

$$\dot{V}(\mathbf{x}) = 2p_1 x_1 x_2 - 2p_2(K + \alpha^2)x_1 x_2$$

If we set $p_1 = p_2(K + \alpha^2)$, $\dot{V}(\mathbf{x}) = 0$ and as per Theorem 8.2, the equilibrium state of the system (8.13) is stable in the sense of Lyapunov.

Example 8.5

Reconsider the system of Fig. 8.5 with

$$G(s) = \frac{K}{s(s + \alpha)}$$

If the reference variable $r(t) = 0$, then the differential equation for the actuating error will be

$$\ddot{e} + \alpha\dot{e} + Ke = 0$$

Taking e and $\dot{e}$ as state variables x_1 and x_2 respectively, we obtain the following state equations:

$$\dot{x}_1 = x_2$$
$$\dot{x}_2 = -Kx_1 - \alpha x_2 \quad (8.15)$$

A candidate for a Lyapunov function is

$$V(\mathbf{x}) = p_1 x_1^2 + p_2 x_2^2;\ p_1 > 0,\ p_2 > 0,$$

which is a positive definite function.

Its derivative is

$$\dot{V}(\mathbf{x}) = 2(p_1 x_1 \dot{x}_1 + p_2 x_2 \dot{x}_2) = 2(p_1 - p_2 K)x_1 x_2 - 2p_2 \alpha x_2^2$$

If we take $p_1 = Kp_2$ with $K > 0$, $\alpha > 0$, we obtain

$$\dot{V}(\mathbf{x}) = -2\alpha p_2 x_2^2$$

which is negative semidefinite.

The condition $\dot{V}(\mathbf{x}) = 0$ exists along the x_1-axis where $x_2 = 0$. A way of showing that $\dot{V}(\mathbf{x})$ being negative semidefinite is sufficient for asymptotic stability is to show that x_1-axis is not a trajectory of the system differential Eqn. (8.15). The first equation yields $\dot{x}_1 = 0$ or $x_1 = c$. The x_1-axis can be a trajectory only if $x_2 = 0$ and $\dot{x}_2 = 0$. Since on x_1-axis, $\dot{x}_2 = -Kc \neq 0$, x_1-axis is not a trajectory, and the equilibrium state at the origin of the system (8.15) is asymptotically stable.

Further, since $V(\mathbf{x}) \to \infty$ as $\|\mathbf{x}\| \to \infty$, the equilibrium state is asymptotically stable in-the-large.

This result, obtained by Lyapunov's direct method, is readily recognized as being correct either from the Routh stability criterion or from the root locus.

Example 8.6

Consider the system described by the state equations

$$\dot{x}_1 = x_2$$
$$\dot{x}_2 = -x_1 - x_2$$

Let us choose,

$$V(\mathbf{x}) = x_1^2 + x_2^2$$

which is a positive definite function; $V(\mathbf{x}) \to \infty$ as $\|\mathbf{x}\| \to \infty$.
This gives

$$\dot{V}(\mathbf{x}) = 2x_1\dot{x}_1 + 2x_2\dot{x}_2 = -2x_2^2$$

which is negative semidefinite. As per the procedure described in the earlier example, it can be established that $\dot{V}(\mathbf{x})$ vanishes identically only at the origin. Hence, by Theorem 8.2, the equilibrium state at the origin is asymptotically stable in-the-large.

To show that a different choice of a Lyapunov function yields the same stability information, let us choose the following positive definite function as another possible Lyapunov function:

$$V(\mathbf{x}) = \tfrac{1}{2}[(x_1 + x_2)^2 + 2x_1^2 + x_2^2]$$

Then $\dot{V}(\mathbf{x})$ becomes

$$\dot{V}(\mathbf{x}) = (x_1 + x_2)(\dot{x}_1 + \dot{x}_2) + 2x_1\dot{x}_1 + x_2\dot{x}_2$$
$$= (x_1 + x_2)(x_2 - x_1 - x_2) + 2x_1 x_2 + x_2(-x_1 - x_2) = -(x_1^2 + x_2^2)$$

which is negative definite. Since $V(\mathbf{x}) \to \infty$ as $\|\mathbf{x}\| \to \infty$, by Theorem 8.2, the equilibrium state at the origin is asymptotically stable in-the-large.

Instability

It may be noted that instability in a nonlinear system can be established by direct recourse to the instability theorem of the direct method. The basic instability theorem is presented below.

Theorem 8.3

For the autonomous system (8.5), sufficient conditions for instability are as follows:

Suppose that there exists a scalar function $W(\mathbf{x})$ which, for some real number $\varepsilon > 0$, satisfies the following properties for all $\mathbf{x}$ in the region $\|\mathbf{x}\| \leq \varepsilon$:

(1) $W(\mathbf{x}) > 0; \mathbf{x} \neq \mathbf{0}$
(2) $W(\mathbf{0}) = 0$
(3) $W(\mathbf{x})$ has continuous partial derivatives with respect to all components of $\mathbf{x}$.

Then the equilibrium state $\mathbf{x}^e = \mathbf{0}$ of the system (8.5) is *unstable* if $\dot{W}(\mathbf{x}) > 0$, $\mathbf{x} \neq \mathbf{0}$, i.e., $\dot{W}(\mathbf{x})$ is a positive definite function. ▲▲

Note that it requires as much ingenuity to devise a suitable W function as to devise a Lyapunov function V. In the stability analysis of nonlinear systems it is valuable to establish conditions for which the system is unstable. Then the regions of asymptotic stability need not be sought for such conditions and the analyst is saved of this fruitless effort.

8.5 LYAPUNOV FUNCTIONS FOR NONLINEAR SYSTEMS

The determination of stability through Lyapunov's direct method centres around the choice of a positive definite function $V(\mathbf{x})$ called the *Lyapunov function*. Unfortunately, there is no universal method for selecting the Lyapunov function which is unique for a given nonlinear system. Some Lyapunov functions may provide better answers than others. Several techniques have been devised for the systematic construction of Lyapunov functions; each is applicable to a particular class of systems.

In addition, if a Lyapunov function cannot be found, it in no way implies that the system is unstable (stability theorems presented in the earlier section merely provide *sufficient conditions* for stability). It only means that our attempt in trying to establish the stability of an equilibrium state of the system has failed. Also, if certain Lyapunov function provides stability for a specified parameter region, this does not necessarily mean that leaving that region will result in system instability. Another choice of Lyapunov function may lead to a larger stability region.

Further, for a given V-function, there is no general method which will allow us to ascertain whether it is positive definite. However, if $V(\mathbf{x})$ is in quadratic form in x_i's, we can use simple tests given in Section 5.2 to ascertain definiteness of the function.

Inspite of all these limitations, Lyapunov's direct method is the most powerful technique available today for stability analysis of nonlinear systems. A host of methods for constructing Lyapunov functions are available in the literature, and many refinements have been suggested to enlarge the region in which the equilibrium state is found to be stable. Since this treatise is meant as a first exposure of the student to Lyapunov's direct method, we restrict the choice of $V(\mathbf{x})$ to a quadratic form (note that the simplest positive definite function is quadratic form) in our examples on stability analysis.

Example 8.7

Consider a nonlinear system governed by the equations

$$\dot{x}_1 = -x_1 + 2x_1^2 x_2$$
$$\dot{x}_2 = -x_2$$

Note that $\mathbf{x} = \mathbf{0}$ is the equilibrium point.

A candidate for a Lyapunov function is

$$V = p_{11} x_1^2 + p_{22} x_2^2 \,;\, p_{11} > 0,\, p_{22} > 0$$

which is a positive definite function.
Then

$$\frac{dV}{dt} = 2p_{11} x_1 \dot{x}_1 + 2p_{22} x_2 \dot{x}_2$$
$$= 2p_{11} x_1 (-x_1 + 2x_1^2 x_2) + 2p_{22} x_2 (-x_2)$$
$$= -2p_{11} x_1^2 (1 - 2x_1 x_2) - 2p_{22} x_2^2$$

dV/dt is negative definite if

$$1 - 2x_1 x_2 > 0 \tag{8.16}$$

Therefore, for asymptotic stability we require that the condition (8.16) is satisfied. The region of state space where this condition is not satisfied is *possibly* the region of instability. Let us concentrate on the region of state space where this condition is satisfied. The limiting condition for such a region is

$$1 - 2x_1 x_2 = 0$$

The dividing lines lie in the first and the third quadrants and are rectangular hyperbolas as shown in Fig. 8.6. In the second and the fourth quadrants, the inequality is satisfied for all values of x_1 and x_2. Figure 8.6 shows the regions of stability and possible instability. Since the choice of the Lyapunov function is not unique, it may be possible to choose another Lyapunov function for the system under consideration which yields a larger region of stability.

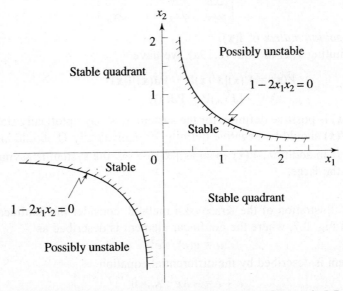

Fig. 8.6 Stability regions for the nonlinear system of Example 8.7

Example 8.8

In this example, we describe the *Krasovskii Method* of constructing Lyapunov functions for nonlinear systems [102].

Consider the nonlinear autonomous system

$$\dot{\mathbf{x}} = \mathbf{f}(\mathbf{x}); \quad \mathbf{f}(0) = 0 \qquad (8.17)$$
$$\mathbf{f} = [f_1 \ f_2 \ \cdots \ f_n]^T; \quad \mathbf{x} = [x_1 \ x_2 \ \cdots \ x_n]^T$$

We assume that $\mathbf{f}(\mathbf{x})$ has continuous first partial derivatives.

We define a Lyapunov function as

$$V(\mathbf{x}) = \mathbf{f}^T(\mathbf{x})\mathbf{P}\,\mathbf{f}(\mathbf{x}) \qquad (8.18)$$

where $\mathbf{P}$ = a symmetric positive definite matrix.

Now,

$$\dot{V}(\mathbf{x}) = \dot{\mathbf{f}}^T(\mathbf{x})\mathbf{P}\,\mathbf{f}(\mathbf{x}) + \mathbf{f}^T(\mathbf{x})\mathbf{P}\,\dot{\mathbf{f}}(\mathbf{x}) \qquad (8.19a)$$

where

$$\dot{\mathbf{f}}(\mathbf{x}) = \frac{\partial \mathbf{f}(\mathbf{x})}{\partial \mathbf{x}} \frac{d\mathbf{x}}{dt} = \mathbf{J}(\mathbf{x})\mathbf{f}(\mathbf{x});$$

$$\mathbf{J(x)} = \begin{bmatrix} \dfrac{\partial f_1}{\partial x_1} & \dfrac{\partial f_1}{\partial x_2} & \cdots & \dfrac{\partial f_1}{\partial x_n} \\ \dfrac{\partial f_2}{\partial x_1} & \dfrac{\partial f_2}{\partial x_2} & \cdots & \dfrac{\partial f_2}{\partial x_n} \\ \vdots & \vdots & & \vdots \\ \dfrac{\partial f_n}{\partial x_1} & \dfrac{\partial f_n}{\partial x_2} & \cdots & \dfrac{\partial f_n}{\partial x_n} \end{bmatrix} \quad (8.19b)$$

is the *Jacobian matrix* of $\mathbf{f(x)}$.

Substituting $\dot{\mathbf{f}}(\mathbf{x})$ in Eqn. (8.19a), we have

$$\dot{V}(\mathbf{x}) = \mathbf{f}^T(\mathbf{x})[\mathbf{J}^T(\mathbf{x})\mathbf{P} + \mathbf{PJ(x)}]\mathbf{f(x)} \quad (8.20a)$$

Let
$$\mathbf{Q} = -[\mathbf{J}^T(\mathbf{x})\mathbf{P} + \mathbf{PJ(x)}] \quad (8.20b)$$

Since $V(\mathbf{x})$ is positive definite, for the system to be asymptotically stable at the origin, $\dot{V}(\mathbf{x})$ should be negative definite, or equivalently $\mathbf{Q}$ should be positive definite. If in addition, $V(\mathbf{x}) \to \infty$ as $||\mathbf{x}|| \to \infty$, the system is asymptotically stable in-the-large.

As an illustration of the Krasovskii method, consider the nonlinear system shown in Fig. 8.7, where the nonlinear element is described as

$$u = g(e) = e^3$$

The system is described by the differential equation

$$\ddot{e} + \dot{e} = -Ke^3; \; K > 0$$

Defining $x_1 = e$ and $x_2 = \dot{e}$, we get the following state equations:

$$\begin{aligned} \dot{x}_1 &= f_1(\mathbf{x}) = x_2 \\ \dot{x}_2 &= f_2(\mathbf{x}) = -x_2 - Kx_1^3 \end{aligned} \quad (8.21)$$

The equilibrium point lies at the origin.

Now

$$\mathbf{J(x)} = \begin{bmatrix} \dfrac{\partial f_1}{\partial x_1} & \dfrac{\partial f_1}{\partial x_2} \\ \dfrac{\partial f_2}{\partial x_1} & \dfrac{\partial f_2}{\partial x_2} \end{bmatrix} = \begin{bmatrix} 0 & 1 \\ -3Kx_1^2 & -1 \end{bmatrix}$$

Let
$$\mathbf{P} = \begin{bmatrix} p_{11} & p_{12} \\ p_{12} & p_{22} \end{bmatrix}$$

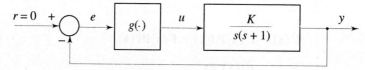

Fig. 8.7 A nonlinear system

For **P** to be positive definite,
$$p_{11} > 0 \tag{8.22a}$$
$$p_{11} p_{22} - p_{12}^2 > 0 \tag{8.22b}$$

The matrix
$$\mathbf{Q} = -[\mathbf{J}^T(\mathbf{x})\mathbf{P} + \mathbf{P}\mathbf{J}(\mathbf{x})]$$

$$= -\left\{ \begin{bmatrix} 0 & -3Kx_1^2 \\ 1 & -1 \end{bmatrix} \begin{bmatrix} p_{11} & p_{12} \\ p_{12} & p_{22} \end{bmatrix} + \begin{bmatrix} p_{11} & p_{12} \\ p_{12} & p_{22} \end{bmatrix} \begin{bmatrix} 0 & 1 \\ -3Kx_1^2 & -1 \end{bmatrix} \right\}$$

$$= -\begin{bmatrix} -6p_{12}Kx_1^2 & p_{11} - p_{12} - 3p_{22}Kx_1^2 \\ p_{11} - p_{12} - 3p_{22}Kx_1^2 & 2(p_{12} - p_{22}) \end{bmatrix}$$

For the system (8.21) to be asymptotically stable at the origin, **Q** should be positive definite, i.e.,
$$6p_{12} Kx_1^2 > 0 \tag{8.22c}$$
$$-12p_{12} Kx_1^2(p_{12} - p_{22}) - (p_{11} - p_{12} - 3p_{22} Kx_1^2)^2 > 0$$
or $\quad 12p_{12} Kx_1^2(p_{22} - p_{12}) > (p_{11} - p_{12} - 3p_{22}Kx_1^2)^2 \tag{8.22d}$

Choose $p_{12} > 0$. Inequality (8.22c) then yields the condition
$$x_1^2 > 0, \text{ which is always met.}$$

Choose, $p_{11} = p_{12}$, and $p_{22} = \beta p_{12}$ with $\beta > 1$. Inequalities (8.22a) and (8.22b) are satisfied, and inequality (8.22d) gives the condition

$$12(\beta - 1) > 9\beta^2 Kx_1^2 \quad \text{or} \quad x_1^2 < \frac{4}{3K}\left(\frac{1}{\beta} - \frac{1}{\beta^2}\right)$$

It can easily be shown that the largest value of x_1 occurs when $\beta = 2$. Therefore,

$$x_1^2 < \frac{1}{3K} \quad \text{or} \quad -\frac{1}{\sqrt{3K}} < x_1 < \frac{1}{\sqrt{3K}}$$

This region of asymptotic stability is illustrated in Fig. 8.8.

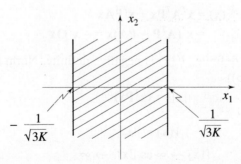

Fig. 8.8 Stability region of the nonlinear system of Fig. 8.7

8.6 LYAPUNOV FUNCTIONS FOR LINEAR SYSTEMS

As has been said earlier, the Lyapunov theorems give only sufficient conditions on the stability of the equilibrium state of a nonlinear system and, furthermore, there is no unique way of constructing a Lyapunov function. In case of linear systems, the direct method of Lyapunov provides a simple approach to stability analysis. For a linear system, a Lyapunov function can always be constructed and both the necessary and sufficient conditions on stability established.

Consider a linear autonomous system described by the state equation

$$\dot{\mathbf{x}} = \mathbf{A}\mathbf{x} \qquad (8.23)$$

where $\mathbf{A}$ is $n \times n$ real constant matrix.

Theorem 8.4

The linear system (8.23) is globally asymptotically stable at the origin if, and only if, for any given symmetric positive definite matrix $\mathbf{Q}$, there exists a symmetric positive definite matrix $\mathbf{P}$ that satisfies the matrix equation

$$\mathbf{A}^T\mathbf{P} + \mathbf{P}\mathbf{A} = -\mathbf{Q} \qquad (8.24)$$

Proof

Let us first prove the sufficiency of the result. Assume that a symmetric positive definite matrix $\mathbf{P}$ exists which is the unique solution of Eqn. (8.24). Consider the scalar function

$$V(\mathbf{x}) = \mathbf{x}^T \mathbf{P} \mathbf{x}$$

Note that

$$V(\mathbf{x}) > 0 \text{ for } \mathbf{x} \neq \mathbf{0} \quad \text{and} \quad V(\mathbf{0}) = 0$$

The time derivative of $V(\mathbf{x})$ is

$$\dot{V}(\mathbf{x}) = \dot{\mathbf{x}}^T \mathbf{P} \mathbf{x} + \mathbf{x}^T \mathbf{P} \dot{\mathbf{x}}$$

Using Eqns (8.23) and (8.24), we get

$$\dot{V}(\mathbf{x}) = \mathbf{x}^T \mathbf{A}^T \mathbf{P} \mathbf{x} + \mathbf{x}^T \mathbf{P} \mathbf{A} \mathbf{x}$$
$$= \mathbf{x}^T (\mathbf{A}^T \mathbf{P} + \mathbf{P} \mathbf{A}) \mathbf{x} = -\mathbf{x}^T \mathbf{Q} \mathbf{x}$$

Since $\mathbf{Q}$ is positive definite, $\dot{V}(\mathbf{x})$ is negative definite. Norm of $\mathbf{x}$ may be defined as (Eqn. (5.6b))

$$||\mathbf{x}|| = (\mathbf{x}^T \mathbf{P} \mathbf{x})^{1/2}$$

Then

$$V(\mathbf{x}) = ||\mathbf{x}||^2$$

$$V(\mathbf{x}) \to \infty \text{ as } ||\mathbf{x}|| \to \infty$$

Therefore, as per Theorem 8.1, the system is globally asymptotically stable at the origin.

To prove the necessity of the result, the reader is advised to refer [102] where the proof has been developed in two parts:

(i) If (8.23) is asymptotically stable, then for any $\mathbf{Q}$ there exists a matrix $\mathbf{P}$ satisfying (8.24).
(ii) If $\mathbf{Q}$ is positive definite, then $\mathbf{P}$ is also positive definite.

▲▲

Comments

(i) The implication of Theorem 8.4 is that if $\mathbf{A}$ is asymptotically stable and $\mathbf{Q}$ is positive definite, then the solution $\mathbf{P}$ of Eqn. (8.24) must be positive definite. Note that it does not say that if $\mathbf{A}$ is asymptotically stable and $\mathbf{P}$ is positive definite, then $\mathbf{Q}$ computed from Eqn. (8.24) is positive definite. For an arbitrary $\mathbf{P}$, $\mathbf{Q}$ may be positive definite (semi-definite) or negative definite (semi-definite).
(ii) Since matrix $\mathbf{P}$ is known to be symmetric, there are only $n(n + 1)/2$ independent equations in (8.24) rather than n^2.
(iii) In very simple cases, Eqn. (8.24), called the *Lyapunov equation*, can be solved analytically, but usually numerical solution is required. A number of computer programs for this purpose are available [176–179].
(iv) Since Theorem 8.4 holds for any positive definite symmetric matrix $\mathbf{Q}$, the matrix $\mathbf{Q}$ in Eqn. (8.24) is often chosen to be a unit matrix.
(v) If $\dot{V}(\mathbf{x}) = -\mathbf{x}^T\mathbf{Q}\mathbf{x}$ does not vanish identically along any trajectory, then $\mathbf{Q}$ may be chosen to be positive semi-definite. This follows from Theorem 8.2.

A necessary and sufficient condition that $\dot{V}(\mathbf{x})$ does not vanish identically along any trajectory (meaning that $\dot{V}(\mathbf{x}) = 0$ only at $\mathbf{x} = \mathbf{0}$) is that

$$\rho \begin{bmatrix} \mathbf{H} \\ \mathbf{H}\mathbf{A} \\ \vdots \\ \mathbf{H}\mathbf{A}^{n-1} \end{bmatrix} = n; \quad \mathbf{Q} = \mathbf{H}^T\mathbf{H} \tag{8.25}$$

where $\rho(.)$ stands for rank of a matrix.

This can be proved as follows. Since $\dot{V}(\mathbf{x})$ can be written as

$$\dot{V}(\mathbf{x}) = -\mathbf{x}^T\mathbf{Q}\mathbf{x} = -\mathbf{x}^T\mathbf{H}^T\mathbf{H}\mathbf{x},$$

$\dot{V}(\mathbf{x}) = 0$ means that

$$\mathbf{H}\mathbf{x} = \mathbf{0}$$

Differentiating with respect to t, gives

$$\mathbf{H}\dot{\mathbf{x}} = \mathbf{H}\mathbf{A}\mathbf{x} = \mathbf{0}$$

Differentiating once again, we get

$$\mathbf{H}\mathbf{A}\dot{\mathbf{x}} = \mathbf{H}\mathbf{A}^2\mathbf{x} = \mathbf{0}$$

Repeating the differentiation process and combining the equations, we obtain

$$\begin{bmatrix} \mathbf{H} \\ \mathbf{H}\mathbf{A} \\ \vdots \\ \mathbf{H}\mathbf{A}^{n-1} \end{bmatrix} \mathbf{x} = \mathbf{0}$$

A necessary and sufficient condition for $\mathbf{x} = \mathbf{0}$ to be the only solution of this equation is given by (8.25).

Example 8.9

Let us determine the stability of the system described by the following equation:

$$\dot{\mathbf{x}} = \mathbf{A}\mathbf{x}$$

with

$$\mathbf{A} = \begin{bmatrix} -1 & -2 \\ 1 & -4 \end{bmatrix}$$

We will first solve Eqn. (8.24) for $\mathbf{P}$ for an arbitrary choice of real symmetric positive definite matrix $\mathbf{Q}$. We may choose $\mathbf{Q} = \mathbf{I}$, the identity matrix. Equation (8.24) then becomes

$$\mathbf{A}^T \mathbf{P} + \mathbf{P} \mathbf{A} = -\mathbf{I}$$

or
$$\begin{bmatrix} -1 & 1 \\ -2 & -4 \end{bmatrix} \begin{bmatrix} p_{11} & p_{12} \\ p_{12} & p_{22} \end{bmatrix} + \begin{bmatrix} p_{11} & p_{12} \\ p_{12} & p_{22} \end{bmatrix} \begin{bmatrix} -1 & -2 \\ 1 & -4 \end{bmatrix} = \begin{bmatrix} -1 & 0 \\ 0 & -1 \end{bmatrix} \quad (8.26)$$

Note that we have taken $p_{12} = p_{21}$. This is because the solution matrix $\mathbf{P}$ is known to be a positive definite real symmetric matrix for a stable system. From Eqn. (8.26), we get

$$-2p_{11} + 2p_{12} = -1$$
$$-2p_{11} - 5p_{12} + p_{22} = 0$$
$$-4p_{12} - 8p_{22} = -1$$

Solving for p_{ij}'s, we obtain

$$\mathbf{P} = \begin{bmatrix} p_{11} & p_{12} \\ p_{12} & p_{22} \end{bmatrix} = \begin{bmatrix} \frac{23}{60} & -\frac{7}{60} \\ -\frac{7}{60} & \frac{11}{60} \end{bmatrix}$$

Using Sylvester's test (Section 5.2), we find that $\mathbf{P}$ is positive definite. Therefore, the system under consideration is asymptotically stable in-the-large.

In order to illustrate the arbitrariness in the choice of $\mathbf{Q}$, consider

$$\mathbf{Q} = \begin{bmatrix} 0 & 0 \\ 0 & 1 \end{bmatrix} \quad (8.27)$$

This is a positive semidefinite matrix. This choice of $\mathbf{Q}$ is permissible since it satisfies the condition (8.25) as is seen below.

$$\mathbf{Q} = \begin{bmatrix} 0 & 0 \\ 0 & 1 \end{bmatrix} = \begin{bmatrix} 0 \\ 1 \end{bmatrix} \begin{bmatrix} 0 & 1 \end{bmatrix} = \mathbf{H}^T \mathbf{H}$$

$$\rho \begin{bmatrix} \mathbf{H} \\ \mathbf{H}\mathbf{A} \end{bmatrix} = \rho \begin{bmatrix} 0 & 1 \\ 1 & -4 \end{bmatrix} = 2$$

It can easily be verified that with the choice of **Q** given by Eqn. (8.27), we derive the same conclusion about the stability of the system as obtained earlier with **Q = I**.

8.7 A MODEL REFERENCE ADAPTIVE SYSTEM

This section is concerned with the use of Lyapunov functions in designing the parameter adjustment mechanism of a simple *Model Reference Adaptive System* (MRAS). Detailed treatment of the subject is beyond the scope of the book (refer [150, 155–157]).

A process may vary with time, and therefore the real process model may differ from the model used for designing the controller. The variation of the process may be caused by a change in the operating conditions or the environment. For such circumstances, an adaptive control strategy can be considered. An adaptive control system is one in which the controller parameters are adjusted automatically to compensate for the changing process conditions or environment (refer Section 7.11).

A natural way to compensate for the changing process parameters is to estimate the system parameters on-line and to change the controller parameters accordingly. Conceptually, the design of an adaptive control system is a technique for combining a parameter estimation method with a control law. This approach is often referred to as *self-tuning control* (refer Fig. 8.9).

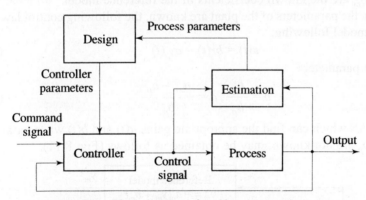

Fig. 8.9 Block diagram of a self-tuning regulator

Another popular adaptive control strategy is MRAS. The basic principle is illustrated in Fig. 8.10. The desired performance is given in terms of a reference model, which in turn gives the desired response to the command signal. The inner control loop is an ordinary feedback loop composed of the plant and the controller; the parameters of the controller are adjusted by the outer loop in such a way that the error between the plant and model outputs becomes small. The key problem is to determine the adjustment mechanism so that a stable system which brings the error to zero is obtained.

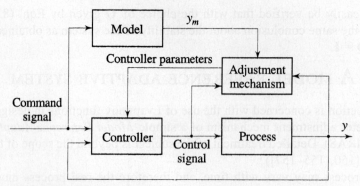

Fig. 8.10 Block diagram of a model reference adaptive system

This is called the *model-following* problem.

Consider a first-order system with the plant model of the form

$$\dot{y}_p = -a_p y_p + b_p u; \; y_p(0) \triangleq y_p^0 \tag{8.28}$$

where u is the control variable and y_p is the measured state (output); a_p and b_p are unknown coefficients.

Assume that it is desired to obtain a closed-loop system described by

$$\dot{y}_m = -a_m y_m + b_m r, \; y_m(0) \triangleq y_m^0 \tag{8.29}$$

a_m and b_m are the known coefficients of the reference model.

When the parameters of the plant are known, the following control law gives perfect model following:

$$u(t) = br(t) - ay_p(t) \tag{8.30}$$

with the parameters

$$b = \frac{b_m}{b_p}, \; a = \frac{a_m - a_p}{b_p} \tag{8.31}$$

An MRAS which can find the appropriate gains $a(t)$ and $b(t)$ when parameters a_p and b_p are not known, may be obtained as follows (Fig. 8.11).

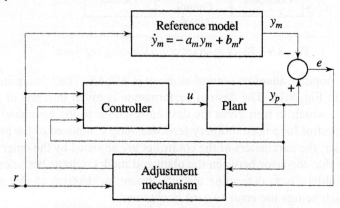

Fig. 8.11 A model reference adaptive system

Introduce the error variable
$$e(t) = y_p(t) - y_m(t) \tag{8.32a}$$
The rate of change of the error is given by
$$\frac{de(t)}{dt} = [-a_p y_p(t) + b_p u(t)] - [-a_m y_m(t) + b_m r(t)]$$
$$= -a_m e(t) + [a_m - a_p - b_p a(t)] y_p(t) + [b_p b(t) - b_m] r(t) \tag{8.32b}$$

Notice that the error goes to zero if the parameters $a(t)$ and $b(t)$ are equal to the ones given by (8.31). We will now attempt to construct a parameter adjustment mechanism that will drive the parameters $a(t)$ and $b(t)$ to appropriate values such that the resulting control law (8.30) forces the plant output $y_p(t)$ to follow the model output $y_m(t)$. For this purpose, we introduce the Lyapunov function,

$$V(e, a, b)$$

$$= \frac{1}{2}\left[e^2(t) + \frac{1}{b_p \gamma}(b_p a(t) + a_p - a_m)^2 + \frac{1}{b_p \gamma}(b_p b(t) - b_m)^2 \right]$$

where $\gamma > 0$.

This function is zero when $e(t)$ is zero and the controller parameters $a(t)$ and $b(t)$ are equal to the optimal values given by (8.31). The derivative of V is

$$\frac{dV}{dt} = e(t)\frac{de(t)}{dt} + \frac{1}{\gamma}[b_p a(t) + a_p - a_m]\frac{da(t)}{dt} + \frac{1}{\gamma}[b_p b(t) - b_m]\frac{db(t)}{dt}$$

$$= -a_m e^2(t) + \frac{1}{\gamma}[b_p a(t) + a_p - a_m]\left[\frac{da(t)}{dt} - \gamma y_p(t) e(t)\right]$$

$$+ \frac{1}{\gamma}[b_p b(t) - b_m]\left[\frac{db(t)}{dt} + \gamma r(t) e(t)\right]$$

If the parameters are updated as

$$\frac{db(t)}{dt} = -\gamma r(t)\, e(t)$$
$$\frac{da(t)}{dt} = \gamma y_p(t)\, e(t), \tag{8.33}$$

we get

$$\frac{dV}{dt} = -a_m e^2(t)$$

The function V will thus decrease as long as the error e is different from zero. It can thus be concluded that the error will go to zero.

Figure 8.12 shows the simulation of MRAS with $a_p = 1$, $b_p = 0.5$, $a_m = 2$, and $b_m = 2$. The input signal is a square wave with amplitude 1. The adaptation gain $\gamma = 2$. The closed-loop system is close to the desired behaviour after only

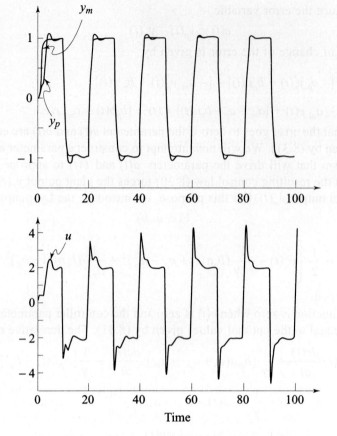

Fig. 8.12 Simulation of the MRAS of Fig. 8.11

a few transients. The convergence rate depends critically on the choice of the parameter γ.

Plots in Fig. 8.12 were generated by simulating the following sets of equations.

(i) $$\dot{y}_m(t) = -2y_m(t) + 2r(t);\ y_m(0) = 0$$

This gives $y_m(t)$.

(ii) $$\dot{y}_p(t) = -y_p(t) + 0.5u(t);\ y_p(0) = 0$$

$$u(t) = b(t)r(t) - a(t)y_p(t)$$

$$\frac{db(t)}{dt} = -2r(t)e(t);\ b(0) = 0.5$$

$$\frac{da(t)}{dt} = 2y_p(t)e(t);\ a(0) = 1$$

$$e(t) = y_p(t) - y_m(t)$$

From this set of equations, we obtain $u(t)$ and $y_p(t)$.

8.8 DISCRETE-TIME SYSTEMS

In this section, we extend the Lyapunov stability analysis presented thus far to autonomous discrete-time systems:

$$x(k + 1) = f(x(k)); \quad f(0) = 0 \tag{8.34}$$

Our discussion will be brief because of the strong analogy between the discrete-time and continuous-time cases.

Lyapunov's Stability Theorem

Theorem 8.5

For the autonomous system (8.34), sufficient conditions of stability are as follows.

Suppose that there exists a scalar function $V(x(k))$ which, for some real number $\varepsilon > 0$, satisfies the following properties for all x in the region $||x|| \leq \varepsilon$:

(1) $V(x) > 0; \, x \neq 0$
(2) $V(0) = 0$
$\left.\right\}$ (i.e., $V(x)$ is positive definite function)

(3) $V(x)$ is continuous for all x.

Then the equilibrium state $x^e = 0$ of the system (8.34) is

(4a) *asymptotically stable*, if the difference $\Delta V(x(k)) = [V(x(k + 1)) - V(x(k))] < 0, \, x \neq 0$, i.e., $\Delta V(x(k))$ is a negative definite function; or if $\Delta V(x(k)) \leq 0$ (i.e., $\Delta V(x(k))$ is negative semidefinite) and no trajectory can stay forever at the points or on the line, other than the origin, at which $\Delta V(x(k)) = 0$;

(4b) *asymptotically stable in-the-large*, if conditions (4a) are satisfied, and in addition $V(x) \to \infty$ as $||x|| \to \infty$;

(4c) *stable in the sense of Lyapunov*, if $\Delta V(x(k))$ is identically zero along a trajectory.

Stability of Linear Systems

Consider a linear autonomous system described by the state equation

$$x(k + 1) = Fx(k) \tag{8.35}$$

where F is $n \times n$ real constant matrix.

Theorem 8.6

The linear system (8.35) is globally asymptotically stable at the origin if and only if, for any given symmetric positive definite matrix Q, there exists a symmetric positive definite matrix P that satisfies the matrix equation

$$F^T PF - P = -Q \tag{8.36}$$

Proof

Let us first prove the sufficiency of the result. Assume that a symmetric positive definite matrix P exists which is the unique solution of Eqn. (8.36). Consider the scalar function

$$V(\mathbf{x}) = \mathbf{x}^T \mathbf{P} \mathbf{x}$$

Note that

$$V(\mathbf{x}) > 0 \quad \text{for } \mathbf{x} \neq \mathbf{0} \quad \text{and } V(\mathbf{0}) = 0$$

The difference

$$\Delta V(\mathbf{x}) = V(\mathbf{x}(k+1)) - V(\mathbf{x}(k))$$
$$= \mathbf{x}^T(k+1)\mathbf{P}\mathbf{x}(k+1) - \mathbf{x}^T(k)\mathbf{P}\mathbf{x}(k)$$

Using Eqns (8.35)–(8.36), we get

$$\Delta V(\mathbf{x}) = \mathbf{x}^T(k)\mathbf{F}^T\mathbf{P}\mathbf{F}\mathbf{x}(k) - \mathbf{x}^T(k)\mathbf{P}\mathbf{x}(k)$$
$$= \mathbf{x}^T(k)[\mathbf{F}^T\mathbf{P}\mathbf{F} - \mathbf{P}]\mathbf{x}(k) = -\mathbf{x}^T(k)\mathbf{Q}\mathbf{x}(k)$$

Since $\mathbf{Q}$ is positive definite, $\Delta V(\mathbf{x})$ is negative definite. Further $V(\mathbf{x}) \to \infty$ as $\|\mathbf{x}\| \to \infty$. Therefore, as per Theorem 8.5, the system is globally asymptotically stable at the origin.

The proof of necessity is analogous to that of continuous-time case (refer [102]).

Comments

(i) In very simple cases, Eqn. (8.36), called the *discrete Lyapunov equation*, can be solved analytically, but usually a numerical solution is required. A number of computer programs for this purpose are available [176–179].

(ii) If $\Delta V(\mathbf{x}(k)) = -\mathbf{x}^T(k)\mathbf{Q}\mathbf{x}(k)$ does not vanish identically along any trajectory, then $\mathbf{Q}$ may be chosen to be positive semidefinite. This follows from Theorem 8.5.

A necessary and sufficient condition that $\Delta V(\mathbf{x}(k))$ does not vanish identically along any trajectory (meaning that $\Delta V(\mathbf{x}(k)) = 0$ only at $\mathbf{x} = \mathbf{0}$) is that

$$\rho \begin{bmatrix} \mathbf{H} \\ \mathbf{H}\mathbf{F} \\ \vdots \\ \mathbf{H}\mathbf{F}^{n-1} \end{bmatrix} = n; \quad \mathbf{Q} = \mathbf{H}^T\mathbf{H} \tag{8.37}$$

where $\rho(.)$ stands for rank of a matrix.

Example 8.10

Let us determine the stability of the system described by the following equation:

$$\mathbf{x}(k+1) = \mathbf{F}\mathbf{x}(k)$$

with

$$\mathbf{F} = \begin{bmatrix} -1 & -2 \\ 1 & -4 \end{bmatrix}$$

We will first solve Eqn. (8.36) for **P** for an arbitrary choice of real symmetric positive definite matrix **Q**. We may choose **Q** = **I**, the identity matrix. Equation (8.36) then becomes

$$\mathbf{F}^T\mathbf{P}\mathbf{F} - \mathbf{P} = -\mathbf{I}$$

or

$$\begin{bmatrix} -1 & 1 \\ -2 & -4 \end{bmatrix} \begin{bmatrix} p_{11} & p_{12} \\ p_{12} & p_{22} \end{bmatrix} \begin{bmatrix} -1 & -2 \\ 1 & -4 \end{bmatrix} - \begin{bmatrix} p_{11} & p_{12} \\ p_{12} & p_{22} \end{bmatrix} = \begin{bmatrix} -1 & 0 \\ 0 & -1 \end{bmatrix}$$

or

$$-2p_{12} + p_{22} = -1$$
$$2p_{11} + p_{12} - 4p_{22} = 0$$
$$4p_{11} + 16p_{12} + 15p_{22} = -1$$

Solving for p_{ij}'s, we obtain

$$\mathbf{P} = \begin{bmatrix} p_{11} & p_{12} \\ p_{12} & p_{22} \end{bmatrix} = \begin{bmatrix} -\frac{43}{60} & \frac{11}{30} \\ \frac{11}{30} & -\frac{4}{15} \end{bmatrix}$$

Using Sylvester's test (Section 5.2) we find that **P** is negative definite. Therefore, the system under consideration is unstable.

8.9 REVIEW EXAMPLES

Review Example 8.1

Consider the nonlinear system

$$\dot{x}_1 = -x_1 - x_2^2$$
$$\dot{x}_2 = -x_2$$

Investigate the stability of the equilibrium points.

Solution

The given system is

$$\dot{\mathbf{x}} = \mathbf{f}(\mathbf{x})$$

or

$$\dot{x}_1 = f_1(\mathbf{x}) = -x_1 - x_2^2$$
$$\dot{x}_2 = f_2(\mathbf{x}) = -x_2$$

Clearly, **x** = **0** is the only equilibrium point.

In the following, we apply Krasovskii method to determine sufficient conditions for asymptotic stability in the vicinity of the equilibrium point.

A candidate for a Lyapunov function is

$$V(\mathbf{x}) = \mathbf{f}^T(\mathbf{x})\mathbf{P}\mathbf{f}(\mathbf{x})$$

Selecting **P** = **I** may lead to a successful determination of the conditions for asymptotic stability in the vicinity of the equilibrium point.

With this choice of Lyapunov function, we have (refer Eqns (8.20))

$$\dot{V}(\mathbf{x}) = \mathbf{f}^T(\mathbf{x})[\mathbf{J}^T(\mathbf{x}) + \mathbf{J}(\mathbf{x})]\mathbf{f}(\mathbf{x})$$

where

$$J(x) = \begin{bmatrix} \dfrac{\partial f_1}{\partial x_1} & \dfrac{\partial f_1}{\partial x_2} \\ \dfrac{\partial f_2}{\partial x_1} & \dfrac{\partial f_2}{\partial x_2} \end{bmatrix} = \begin{bmatrix} -1 & -2x_2 \\ 0 & -1 \end{bmatrix}$$

The matrix

$$Q = -[J^T(x) + J(x)] = \begin{bmatrix} 2 & 2x_2 \\ 2x_2 & 2 \end{bmatrix}$$

Using Sylvester's criterion (Section 5.2) we find that the matrix Q is positive definite if

$$4 - 4x_2^2 > 0 \quad \text{or} \quad |x_2| < 1$$

The shaded region in Fig. 8.13 is the region of asymptotic stability. It is, however, not the largest region. Another choice of Lyapunov function for the system under consideration may lead to a larger region of asymptotic stability in the vicinity of the equilibrium point.

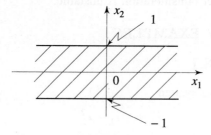

Fig. 8.13

Review Example 8.2

Using the Lyapunov equation, determine the stability range for the gain K of the system shown in Fig. 8.14.

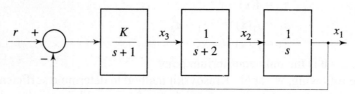

Fig. 8.14

Solution

The state equation of the system is

$$\begin{bmatrix} \dot x_1 \\ \dot x_2 \\ \dot x_3 \end{bmatrix} = \begin{bmatrix} 0 & 1 & 0 \\ 0 & -2 & 1 \\ -K & 0 & -1 \end{bmatrix} \begin{bmatrix} x_1 \\ x_2 \\ x_3 \end{bmatrix} + \begin{bmatrix} 0 \\ 0 \\ K \end{bmatrix} r$$

For the investigation of asymptotic stability, we consider the autonomous system

$$\dot{\mathbf{x}} = \mathbf{A}\mathbf{x}$$

with

$$\mathbf{A} = \begin{bmatrix} 0 & 1 & 0 \\ 0 & -2 & 1 \\ -K & 0 & -1 \end{bmatrix}$$

Clearly the equilibrium state is the origin.

Let us choose a Lyapunov function

$$V(\mathbf{x}) = \mathbf{x}^T \mathbf{P} \mathbf{x}$$

where $\mathbf{P}$ is to be determined from the Lyapunov equation

$$\mathbf{A}^T \mathbf{P} + \mathbf{P}\mathbf{A} = -\mathbf{Q}$$

The matrix $\mathbf{Q}$ could be chosen as identity matrix. However, we make the following choice for $\mathbf{Q}$.

$$\mathbf{Q} = \begin{bmatrix} 0 & 0 & 0 \\ 0 & 0 & 0 \\ 0 & 0 & 1 \end{bmatrix}$$

This is a positive semidefinite matrix which satisfies the condition (8.25) as is seen below.

$$\mathbf{Q} = \begin{bmatrix} 0 & 0 & 0 \\ 0 & 0 & 0 \\ 0 & 0 & 1 \end{bmatrix} = \begin{bmatrix} 0 \\ 0 \\ 1 \end{bmatrix} \begin{bmatrix} 0 & 0 & 1 \end{bmatrix} = \mathbf{H}^T \mathbf{H}$$

$$\rho \begin{bmatrix} \mathbf{H} \\ \mathbf{H}\mathbf{A} \\ \mathbf{H}\mathbf{A}^2 \end{bmatrix} = \rho \begin{bmatrix} 0 & 0 & 1 \\ -K & 0 & -1 \\ K & -K & 1 \end{bmatrix} = 3$$

With this choice of $\mathbf{Q}$, as we shall see, manipulation of the Lyapunov equation for its analytical solution becomes easier.

Now let us solve the Lyapunov equation

$$\mathbf{A}^T \mathbf{P} + \mathbf{P}\mathbf{A} = -\mathbf{Q}$$

or

$$\begin{bmatrix} 0 & 0 & -K \\ 1 & -2 & 0 \\ 0 & 1 & -1 \end{bmatrix} \begin{bmatrix} p_{11} & p_{12} & p_{13} \\ p_{12} & p_{22} & p_{23} \\ p_{13} & p_{23} & p_{33} \end{bmatrix} + \begin{bmatrix} p_{11} & p_{12} & p_{13} \\ p_{12} & p_{22} & p_{23} \\ p_{13} & p_{23} & p_{33} \end{bmatrix} \begin{bmatrix} 0 & 1 & 0 \\ 0 & -2 & 1 \\ -K & 0 & -1 \end{bmatrix}$$

$$= \begin{bmatrix} 0 & 0 & 0 \\ 0 & 0 & 0 \\ 0 & 0 & -1 \end{bmatrix}$$

Solving this equation for p_{ij}'s, we obtain

$$\mathbf{P} = \begin{bmatrix} \dfrac{K^2 + 12K}{12 - 2K} & \dfrac{6K}{12 - 2K} & 0 \\ \dfrac{6K}{12 - 2K} & \dfrac{3K}{12 - 2K} & \dfrac{K}{12 - 2K} \\ 0 & \dfrac{K}{12 - 2K} & \dfrac{6}{12 - 2K} \end{bmatrix}$$

For **P** to be positive definite, it is necessary and sufficient that
$$(12 - 2K) > 0 \text{ and } K > 0 \quad \text{or} \quad 0 < K < 6$$
Thus for $0 < K < 6$, the system is asymtotically stable.

Review Example 8.3

Consider the system described by the equations
$$x_1(k + 1) = 2x_1(k) + 0.5x_2(k) - 5$$
$$x_2(k + 1) = 0.8x_2(k) + 2$$

Investigate the stability of the equilibrium state. Use the direct method of Lyapunov.

Solution

The equilibrium state $\mathbf{x}^e = \begin{bmatrix} x_1^e \\ x_2^e \end{bmatrix}$ can be determined from the equations

$$x_1^e = 2x_1^e + 0.5x_2^e - 5$$
$$x_2^e = 0.8x_2^e + 2$$

Solving, we get

$$\begin{bmatrix} x_1^e \\ x_2^e \end{bmatrix} = \begin{bmatrix} 0 \\ 10 \end{bmatrix}$$

Define
$$\tilde{x}_1(k) = x_1(k) - x_1^e$$
$$\tilde{x}_2(k) = x_2(k) - x_2^e$$

In terms of the shifted variables, the system equations become

$$\begin{bmatrix} \tilde{x}_1(k+1) \\ \tilde{x}_2(k+1) \end{bmatrix} = \begin{bmatrix} 2 & 0.5 \\ 0 & 0.8 \end{bmatrix} \begin{bmatrix} \tilde{x}_1(k) \\ \tilde{x}_2(k) \end{bmatrix}$$

or
$$\tilde{\mathbf{x}}(k+1) = \mathbf{F}\tilde{\mathbf{x}}(k) \tag{8.38}$$

Clearly $\tilde{\mathbf{x}} = \mathbf{0}$ is the equilibrium state of this autonomous system.
Let us choose a Lyapunov function
$$V(\tilde{\mathbf{x}}) = \tilde{\mathbf{x}}^T \mathbf{P} \tilde{\mathbf{x}}$$

where **P** is to be determined from the Lyapunov equation
$$\mathbf{F}^T \mathbf{P} \mathbf{F} - \mathbf{P} = -\mathbf{I}$$

or
$$\begin{bmatrix} 2 & 0 \\ 0.5 & 0.8 \end{bmatrix} \begin{bmatrix} p_{11} & p_{12} \\ p_{12} & p_{22} \end{bmatrix} \begin{bmatrix} 2 & 0.5 \\ 0 & 0.8 \end{bmatrix} - \begin{bmatrix} p_{11} & p_{12} \\ p_{12} & p_{22} \end{bmatrix} = \begin{bmatrix} -1 & 0 \\ 0 & -1 \end{bmatrix}$$

Solving for p_{ij}'s, we get

$$\mathbf{P} = \begin{bmatrix} -\frac{1}{3} & \frac{5}{9} \\ \frac{5}{9} & \frac{1225}{324} \end{bmatrix}$$

By applying the Sylvester's test for positive definiteness, we find that the matrix **P** is not positive definite. Therefore the origin of the system (8.38) is not asymptotically stable.

In terms of the original state variables, we can say that the equilibrium state $\mathbf{x}^e = [0 \quad 10]^T$ of the given system is not asymptotically stable.

PROBLEMS

8.1 Consider the linear autonomous system

$$\dot{\mathbf{x}} = \begin{bmatrix} 0 & 1 \\ -1 & -2 \end{bmatrix} \mathbf{x}$$

Using direct method of Lyapunov, determine the stability of the equilibrium state.

8.2 Using direct method of Lyapunov, determine the stability of the equilibrium state of the system

$$\dot{\mathbf{x}} = \mathbf{A}\mathbf{x}$$

with

$$\mathbf{A} = \begin{bmatrix} 0 & 1 \\ -1 & 1 \end{bmatrix}$$

8.3 Consider the system described by the equations

$$\dot{x}_1 = x_2$$
$$\dot{x}_2 = -x_1 - x_2 + 2$$

Investigate the stability of the equilibrium state. Use the direct method of Lyapunov.

8.4 A linear autonomous system is described by the state equation

$$\dot{\mathbf{x}} = \mathbf{A}\mathbf{x}$$

where

$$\mathbf{A} = \begin{bmatrix} -4K & 4K \\ 2K & -6K \end{bmatrix}$$

Using the direct method of Lyapunov, find restrictions on the parameter K to guarantee the stability of the system.

8.5 Consider the system of Fig. P8.5. Find the restrictions on the parameter K to guarantee system stability. Use the Lyapunov's direct method and the Routh criterion. Compare the results.

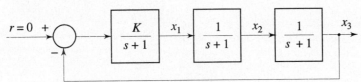

Fig. P8.5

8.6. Consider the linear autonomous system

$$\mathbf{x}(k+1) = \begin{bmatrix} 0.5 & 1 \\ -1 & -1 \end{bmatrix} \mathbf{x}(k)$$

Using the direct method of Lyapunov, determine the stability of the equilibrium state.

8.7. Using direct method of Lyapunov, determine the stability of the equilibrium state of the system

$$\mathbf{x}(k+1) = \mathbf{F}\mathbf{x}(k)$$

with

$$\mathbf{F} = \begin{bmatrix} 0 & 0.5 \\ -0.5 & -1 \end{bmatrix}$$

8.8. Consider the nonlinear system described by the equations

$$\dot{x}_1 = x_2$$
$$\dot{x}_2 = -(1 - |x_1|)x_2 - x_1$$

Find the region in the state plane for which the equilibrium state of the system is asymptotically stable.

8.9. Check the stability of the equilibrium state of the system described by

$$\dot{x}_1 = x_2$$
$$\dot{x}_2 = -x_1 - x_1^2 x_2$$

8.10. Consider a nonlinear system described by the equations

$$\dot{x}_1 = -3x_1 + x_2$$
$$\dot{x}_2 = x_1 - x_2 - x_2^3$$

Using the Krasovskii method for constructing the Lyapunov function with **P** as identity matrix, investigate the stability of the equilibrium state.

chapter 9
Linear Quadratic Optimal Control

9.1 PARAMETER OPTIMIZATION AND OPTIMAL CONTROL PROBLEMS

In previous chapters, we encountered various methods for designing feedback control laws, ranging from root-locus and Bode-plot techniques to pole-placement by state feedback and state estimation. In each case, the designer was left with decisions regarding the locations of closed-loop poles. We have given a fairly complete treatment of these design techniques for linear time-invariant single-variable systems.

Here in this chapter, a somewhat different approach to design is taken. The performance of the system is measured with a single scalar quantity—the performance index. A configuration of the controller is selected and free parameters of the controller that optimize (minimize or maximize as the case may be) the performance index are determined. In most industrial control problems, the nature of the performance index is such that the design process requires its minimization.

The design approach based on parameter optimization consists of the following steps:

(i) Compute the performance index J as a function of the free parameters $k_1, k_2, ..., k_n$, of the system with fixed configuration:

$$J = J(k_1, k_2, ..., k_n) \tag{9.1}$$

(ii) Determine the solution set k_i of the equations

$$\frac{\partial J}{\partial k_i} = 0; \ i = 1, 2, ..., n \tag{9.2}$$

Equations (9.2) give the necessary conditions for J to be minimum. From the solution set of these equations, find the subset that satisfies the sufficient conditions which require that the *Hessian matrix* given below is positive definite.

$$\mathbf{H} = \begin{bmatrix} \dfrac{\partial^2 J}{\partial k_1^2} & \dfrac{\partial^2 J}{\partial k_1 \partial k_2} & \cdots & \dfrac{\partial^2 J}{\partial k_1 \partial k_n} \\ \dfrac{\partial^2 J}{\partial k_2 \partial k_1} & \dfrac{\partial^2 J}{\partial k_2^2} & \cdots & \dfrac{\partial^2 J}{\partial k_2 \partial k_n} \\ \vdots & \vdots & & \vdots \\ \dfrac{\partial^2 J}{\partial k_n \partial k_1} & \dfrac{\partial^2 J}{\partial k_n \partial k_2} & \cdots & \dfrac{\partial^2 J}{\partial k_n^2} \end{bmatrix} \qquad (9.3)$$

Since $\dfrac{\partial^2 J}{\partial k_i \partial k_j} = \dfrac{\partial^2 J}{\partial k_j \partial k_i}$,

the matrix **H** is always symmetric.

(iii) If there are two or more sets of k_i satisfying the necessary as well as sufficient conditions of minimization of J, then compute the corresponding J for each set. The set that gives the smallest J is the optimum set.

Selection of an appropriate performance index is as much a part of the design process as the minimization of the index. We know that the performance of a control system can be adequately specified in terms of settling time, peak overshoot, and steady-state error. The performance index could then be chosen as

$$J \triangleq K_1(\text{settling time}) + K_2(\text{peak overshoot}) + K_3(\text{steady-state error})$$

where the K_i are weighing factors.

Although the criterion seems reasonable, it is not trackable analytically. A compromise must be made between specifying a performance index which includes all the desired system characteristics, and a performance index which can be minimized with a reasonable amount of computation.

In the following, we present several performance indices which include the desired system characteristics and in addition have good mathematical trackability. These indices often involve integrating some function of system error over some time interval when the system is subjected to a standard command or disturbance such as step. A common example is the integral of absolute error (IAE) defined by

$$\text{IAE} \triangleq \int_0^\infty |e(t)|\, dt$$

If the index is to be computed numerically, the infinite upper limit can be replaced by the limit t_f, where t_f is large enough so that $e(t)$ is negligible for $t > t_f$. This index is not unreasonable since both the fast but highly oscillatory systems and the sluggish systems will give large IAE value (see Fig. 9.1). Minimization of IAE by adjusting system parameters will provide acceptable relative stability and speed of response. Also a finite value of IAE implies that the steady-state error is zero.

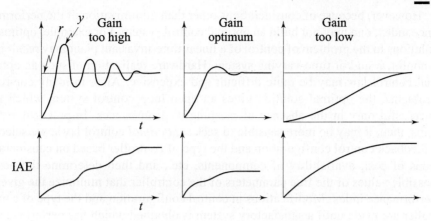

Fig. 9.1 The IAE optimal response criterion

Another similar index is the integral of time multiplied by absolute error (ITAE) which exhibits the additional useful features that the initial large error (unavoidable for a step input) is not heavily weighted, whereas errors that persist are more heavily weighted.

$$\text{ITAE} \triangleq \int_0^\infty t|e(t)|\,dt$$

The integral of square error (ISE) and integral of time multiplied by square error (ITSE) indices are analogous to IAE and ITAE criteria except that the square of the error is employed for three reasons: (i) in some applications, the squared error represents the system's power consumption, (ii) squaring the error weighs large errors more heavily than small errors, and (iii) the squared error is much easier to handle analytically.

$$\text{ISE} \triangleq \int_0^\infty e^2(t)\,dt$$

$$\text{ITSE} \triangleq \int_0^\infty te^2(t)\,dt$$

The system whose design minimizes (or maximizes) the selected performance index with no constraints on controller configuration is, by definition, *optimal*.

The difference between parameter optimization and optimal control problems is that no constraint on controllers is imposed on the later. In optimal design, the designer is permitted to use controllers of any degree and any configuration, whereas in parameter optimization the configuration and the type of controllers are predetermined. Since there is no constraint imposed on controllers, optimal design results in a better system, i.e., lower value of the performance index.

However, because of considerations other than minimization of the performance index, one may not build an optimal control system. For example, optimal solutions to the problem of control of a linear time-invariant plant may result in a nonlinear and/or time-varying system. Hardware realization of such an optimal control law may be quite difficult and expensive. Also, in many control problems, the optimal solution gives an open-loop control system which is successful only in the absence of meaningful disturbances. In practical systems, then, it may be more sensible to seek *suboptimal* control laws: we select a feedback control configuration and the type of controller based on considerations of cost, availability of components, etc., and then determine the best possible values of the free parameters of the controller that minimize the given performance index. Modifications in control configuration and the type of controller are made until a satisfactory system is obtained which has performance characteristics close to the optimal control system we have worked out in theory.

There exists an important class of optimal control problems for which quite general results have been obtained. It involves control of linear systems with the objective of minimizing the integral of a quadratic performance index. An important feature of this class of problems is that optimal control is possible by feedback controllers. For linear time-invariant plants, the optimal control results in a linear time-invariant closed-loop system. The implementation of optimal control is therefore simple and less expensive. Many problems of industrial control belong to this class of problems—*linear quadratic optimal control problems*.

As we shall see in this chapter, the linear quadratic optimal control laws have some computational advantage, and a number of useful properties. The task of the designer shifts to one of specifying various parameters in the performance index.

In the previous chapters, we have been mostly concerned with the design of single-variable systems. Extensions of the root-locus method and the Bode/Nyquist-plot design to multivariable cases have been reported in the literature. However, the design of multivariable systems using these techniques is much more complicated than the single-variable cases. Design of multivariable systems through pole-placement can also be carried out; the computations required are however highly complicated.

The optimal control theory provides a simple and powerful tool for designing multivariable systems. Indeed the equations and computations required in the design of optimal single-variable systems and those in the design of optimal multivariable systems are almost identical. We will therefore use in this chapter, the multi-input, multi-output (MIMO) state variable model in the formulation of optimal control problem.

The objective set for this chapter is the presentation of simple and analytically solvable optimal control and parameter optimization problems. This will provide insight into optimal and suboptimal structures and algorithms that may be applied in practical cases. For detailed study, specialized books on optimal control [123–127] should be consulted. A moderate treatment of the subject is also available in reference [102].

Commercially available software [176–179] may be used for solving complex optimal/suboptimal control problems.

9.2 QUADRATIC PERFORMANCE INDEX

A commonly used performance criterion is the integral square error (ISE):

$$J = \int_0^\infty [y(t) - y_r]^2 \, dt \tag{9.4a}$$

$$= \int_0^\infty e^2(t) \, dt \tag{9.4b}$$

where y_r is the command or set-point value of the output, $y(t)$ is the actual output, $e(t) = y(t) - y_r$ is the error of the system.

This criterion, which has good mathematical trackability properties, is acceptable in practice as a measure of system performance. The criterion penalizes positive and negative errors equally. It penalizes heavily on large errors; hence a small J usually results in a system with small overshoot. Since the integration is carried out over $[0, \infty)$, a small J limits the effect of small error lasting for long time and thus results in a small settling time. Also, a finite J implies that the steady-state error is zero.

The optimal design obtained by minimizing the performance index given by Eqns (9.4) may be unsatisfactory because it may lead to excessively large magnitudes of control signals. A more realistic solution to the problem is reached if the performance index is modified to account for physical constraints like saturation in physical devices. Therefore, a more realistic performance index is of the form:

$$J = \int_0^\infty e^2(t) \, dt \tag{9.5a}$$

subject to the following constraint on control signal $u(t)$.

$$\max |u(t)| \leq M \tag{9.5b}$$

for some constant M. The constant M is determined by the linear range of the plant.

Although the criterion expressed in (9.5) can be used in the design, it is not convenient to work with. In a number of problems, $u^2(t)$ is a measure of the instantaneous rate of expenditure of energy. To minimize energy expenditure, we minimize

$$\int_0^\infty u^2(t) \, dt \tag{9.6}$$

We would very much like to replace the performance criterion given by (9.5) by the following *quadratic performance index*:

$$J = \int_0^\infty [e^2(t) + u^2(t)]\, dt$$

To allow greater generality, we can insert a real positive constant λ to obtain

$$J = \int_0^\infty [e^2(t) + \lambda u^2(t)]\, dt \tag{9.7}$$

By adjusting the *weighting factor* λ, we can weigh the relative importance of the system error and the expenditure of energy. By increasing λ, i.e., by giving sufficient weight to control effort, the amplitude of the control signal which minimizes the overall performance index may be kept within practical bounds, although at the expense of the increased system error. Note that as $\lambda \to 0$, the performance index reduces to the integral square error criterion. In this case, the magnitude of $u(t)$ will be very large and the constraint given by (9.5b) may be violated. If $\lambda \to \infty$, the performance index reduces to the one given by Eqn. (9.6), and the optimal system that minimizes this J is one with $u = 0$. From these two extreme cases, we conclude that if λ is properly chosen, then the constraint of Eqn. (9.5b) will be satisfied.

Example 9.1

For the system of Fig. 9.2a, let us compute the value of K that minimizes ISE for the unit-step input.

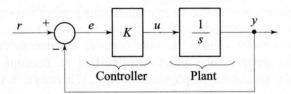

Fig. 9.2a Control system for parameter optimization

For the system under consideration,

$$\frac{E(s)}{R(s)} = \frac{s}{s+K}$$

For unit-step input,

$$E(s) = \frac{1}{s+K}$$

Therefore,
$$e(t) = e^{-Kt}$$

$$\text{ISE} = \int_0^\infty e^2(t)\, dt = \frac{1}{2K}$$

Obviously, the minimum value of ISE is obtained as $K \to \infty$. This is an impractical solution resulting in excessive strain on the physical components of the system.

Sound engineering judgement tells us that we must include in our performance index, the 'cost' of the control effort. The quadratic performance index

$$J = \int_0^\infty [e^2(t) + u^2(t)] \, dt$$

may serve the objective.

From Fig. 9.2a,

$$u(t) = K\, e(t) = K\, e^{-Kt}$$

Therefore,

$$J = \frac{1}{2K} + \frac{K}{2}$$

The minimum value of J is obtained when

$$\frac{\partial J}{\partial K} = -\frac{1}{2K^2} + \frac{1}{2} = 0 \text{ or } K = 1$$

Note that

$$\frac{\partial^2 J}{\partial K^2} = \frac{1}{K^3} > 0$$

The minimum value of J is 1.

This solution, which weighs error and control effort equally, seems to be acceptable.

The following performance index assigns larger weight to error minimization:

$$J = \int_0^\infty [e^2(t) + \lambda u^2(t)] \, dt; \; \lambda = 0.5$$

For the system under consideration,

$$J = \frac{1}{2K} + \lambda \frac{K}{2}$$

$$\frac{\partial J}{\partial K} = 0 \text{ gives}$$

$$K = \sqrt{2}, \; J_{\min} = 0.707$$

When λ is greater than unity, it means that more importance is given to the constraint on amplitude of $u(t)$ compared to the performance of the system. A suitable value of λ is chosen so that relative importance of the system performance is contrasted with the importance of the limit on control effort. Figure 9.2b gives a plot of the performance index *versus* K for various values of λ.

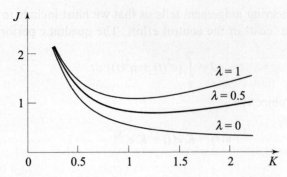

Fig. 9.2b Performance index versus gain for the system shown in Fig. 9.2a

Example 9.2

Consider the liquid-level system shown in Fig. 9.3. h represents the deviation of liquid head from the steady-state value $\bar{H}$.

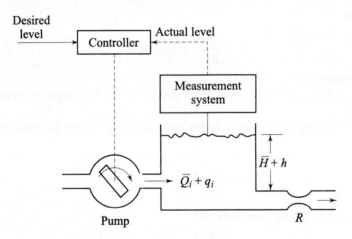

Fig. 9.3 A liquid-level system

The pump controls the liquid head h by supplying liquid at a rate $(\bar{Q}_i + q_i)$ m³/sec to the tank. We shall assume that the flow rate q_i is proportional to the error in liquid level (desired level − actual level). Under these assumptions, the system equations are [180]:

(i) $$A\frac{dh}{dt} = q_i - \frac{\rho g h}{R}$$

where A = area of cross-section of the tank;

R = total resistance offered by the tank outlet and pipe ($R \triangleq$ incremental change in pressure across the restriction/incremental change in flow through the restriction);

ρ = density of the liquid, and

g = acceleration due to gravity.

(ii)
$$q_i = Ke$$

where e = error in liquid level
K = gain constant

Let $A = 1$, and $\dfrac{R}{\rho g} = 1$. Then

$$\frac{H(s)}{Q_i(s)} = \frac{1}{s+1}$$

The block diagram representation is given in Fig. 9.4. The output $y(t) = h(t)$ is the deviation in liquid head from steady-state value. Therefore, the output $y(t)$ is itself the error which is to be minimized. Let us pose the problem of computing the value of K that minimizes the ISE for the initial condition $y(0) = 1$.

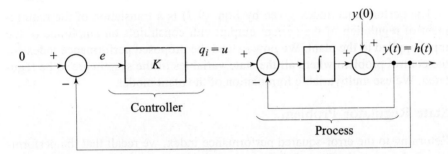

Fig. 9.4 Block diagram representation of the system shown in Fig. 9.3

From Fig. 9.4, we get

$$Y(s) = \frac{s}{s+1+K}\left(\frac{y(0)}{s}\right) = \frac{1}{s+1+K}$$

Therefore,

$$y(t) = e^{-(1+K)t}$$

$$\text{ISE} = \int_0^\infty y^2(t)\, dt = \frac{1}{2(1+K)}$$

Obviously, the minimum value of ISE is obtained as $K \to \infty$.

This is an impractical solution resulting in excessive strain on the physical components of the system. Increasing the gain means, in effect, increasing the pump size.

Consider now the problem of minimization of

$$J = \int_0^\infty [y^2(t) + u^2(t)]\, dt$$

From Fig. 9.4, we have

$$u(t) = -Ky(t)$$

Therefore,
$$u(t) = -Ke^{-(1+K)t}$$
$$J = \frac{1}{2(1+K)} + \frac{K^2}{2(1+K)}$$
$$\frac{\partial J}{\partial K} = 0 \text{ gives } K = \left(\sqrt{2}-1\right)$$

Note that
$$\frac{\partial^2 J}{\partial K^2} = \frac{2}{(1+K)^3} > 0$$

The minimum value of J is $\left(\sqrt{2}-1\right)$.

▲▲

The performance index given by Eqn. (9.7) is a translation of the requirement of regulation of the system output with constraints on amplitude of the input applied to the plant. We now extend the proposed performance index for the control problem where all the state variables of the system are to be regulated. We use multivariable formulation of the plant model.

State Regulator Problem

Returning to the error-squared performance index, we recall that the performance index given by Eqn. (9.4) for a single variable $e(t) = y(t) - y_r$, ensures that the design based on its minimization maintains system output $y(t)$ near the desired value y_r for all time.

Consider now the control problem where the objective is to maintain the system state given by the $n \times 1$ state vector $\mathbf{x}(t)$, near the desired state $\mathbf{x}_d$ (which, in many cases, is the equilibrium point of the system) for all time.

Relative to the desired state $\mathbf{x}_d$, $(\mathbf{x}(t) - \mathbf{x}_d)$ can be viewed as the instantaneous system error. If we transform the system coordinates such that the desired state becomes the origin of the state space, then the new state $\mathbf{x}(t)$ is itself the error.

One measure of the magnitude of the state vector $\mathbf{x}(t)$ (or of its distance from the origin) is the norm $\|\mathbf{x}(t)\|$ defined by
$$\|\mathbf{x}(t)\|^2 = \mathbf{x}^T(t)\mathbf{x}(t)$$
Therefore,
$$J = \int_0^\infty [\mathbf{x}^T(t)\mathbf{x}(t)]\, dt = \int_0^\infty [x_1^2(t) + x_2^2(t) + \cdots + x_n^2(t)]\, dt$$
is a reasonable measure of the system transient response.

In practical systems, the control of all the states of the system is not equally important. To be more general,
$$J = \int_0^\infty [\mathbf{x}^T(t)\mathbf{Q}\mathbf{x}(t)]dt \qquad (9.8)$$

with $\mathbf{Q}$ as $n \times n$ real, symmetric, positive definite (or positive semidefinite) constant matrix, can be used as a performance measure. The simplest form of $\mathbf{Q}$ one can use is the diagonal matrix:

$$\mathbf{Q} = \begin{bmatrix} q_1 & 0 & \cdots & 0 \\ 0 & q_2 & \cdots & 0 \\ \vdots & \vdots & & \vdots \\ 0 & 0 & \cdots & q_n \end{bmatrix}$$

The ith entry of $\mathbf{Q}$ represents the weight the designer places on the constraint on the state variable $x_i(t)$. The larger the value of q_i relative to the other values of q, the more control effort is spent to regular $x_i(t)$.

The design obtained by minimizing the performance index of the form (9.8) may be unsatisfactory in practice. A more realistic solution is obtained if the performance index is modified by adding a penalty term for physical constraints on the $p \times 1$ control vector $\mathbf{u}(t)$. One of the ways of accomplishing this is to introduce the following quadratic control term in the performance index:

$$J = \int_0^\infty [\mathbf{u}^T(t)\mathbf{R}\mathbf{u}(t)] \, dt \qquad (9.9)$$

where $\mathbf{R}$ is $p \times p$ real, symmetric, positive definite,[1] constant matrix.

By giving sufficient weight to control terms, the amplitudes of control signals which minimize overall performance index may be kept within practical bounds, although at the expense of increased error in $\mathbf{x}(t)$.

For the state regulator problem, a useful performance measure is therefore[2]

$$J = \tfrac{1}{2} \int_0^\infty [\mathbf{x}^T(t)\mathbf{Q}\mathbf{x}(t) + \mathbf{u}^T(t)\mathbf{R}\mathbf{u}(t)] \, dt \qquad (9.10)$$

Output Regulator Problem

In the state regulator problem, we are concerned with maintaining the $n \times 1$ state vector $\mathbf{x}(t)$ near the origin of the state space for all time. In the output regulator problem, on the other hand, we are concerned with maintaining the $q \times 1$ output vector $\mathbf{y}(t)$ near origin for all time. A useful performance measure for the output regulator problem is

$$J = \tfrac{1}{2} \int_0^\infty [\mathbf{y}^T(t)\mathbf{Q}\mathbf{y}(t) + \mathbf{u}^T(t)\mathbf{R}\mathbf{u}(t)] \, dt \qquad (9.11a)$$

---

1. As we shall see in Section 9.5, positive definiteness of $\mathbf{R}$ is a necessary condition for the existence of the optimal solution to the control problem.
2. Note that multiplication by 1/2 does not affect the minimization problem. The constant helps us in mathematical manipulations as we shall see later in Section 9.5.

where **Q** is a $q \times q$ positive definite (or positive semidefinite) real, symmetric constant matrix and **R** is a $p \times p$ positive definite, real, symmetric, constant matrix.

Substituting $\mathbf{y} = \mathbf{Cx}$ in Eqn. (9.11a), we get

$$J = \tfrac{1}{2} \int_0^\infty (\mathbf{x}^T \mathbf{C}^T \mathbf{Q}\mathbf{C}\mathbf{x} + \mathbf{u}^T \mathbf{R}\mathbf{u}) \, dt \qquad (9.11b)$$

Comparing Eqn. (9.11b) with Eqn. (9.10) we observe that the two indices are identical in form; **Q** in Eqn. (9.10) is replaced by $\mathbf{C}^T\mathbf{Q}\mathbf{C}$ in Eqn. (9.11b). If we assume that the plant is completely observable, then **C** cannot be zero; $\mathbf{C}^T\mathbf{Q}\mathbf{C}$ will be positive definite (or positive semidefinite) whenever **Q** is positive definite (or positive semidefinite).

Thus, the solution to the output regulator problem directly follows from that of the state regulator problem.

9.3 CONTROL CONFIGURATIONS

State Regulator

Consider the plant represented by linear state equations of the form

$$\dot{\mathbf{x}}(t) = \mathbf{A}\mathbf{x}(t) + \mathbf{B}\mathbf{u}(t); \; \mathbf{x}(0) \triangleq \mathbf{x}^0 \qquad (9.12a)$$

$$\mathbf{y}(t) = \mathbf{C}\mathbf{x}(t) \qquad (9.12b)$$

where $\mathbf{x}(t)$ is the $n \times 1$ state vector, $\mathbf{u}(t)$ is the $p \times 1$ input vector, $\mathbf{y}(t)$ is the $q \times 1$ output vector; **A**, **B** and **C** are, respectively, $n \times n$, $n \times p$ and $q \times n$ real constant matrices. We will assume that the null state $\mathbf{x} = \mathbf{0}$ is the desired state; $\mathbf{x}(t)$ is thus system-error vector at time t.

We shall be interested in selecting the controls $\mathbf{u}(t)$ which quickly move the system state $\mathbf{x}(t)$ to the null state $\mathbf{x} = \mathbf{0}$ for any initial perturbation $\mathbf{x}^0$. The control problem is, thus, to determine $\mathbf{u}(t)$ which minimizes performance index of the form

$$J = \tfrac{1}{2} \int_0^\infty [\mathbf{x}^T(t)\mathbf{Q}\mathbf{x}(t) + \mathbf{u}^T(t)\mathbf{R}\mathbf{u}(t)] \, dt \qquad (9.13)$$

where **Q** is a positive definite (or positive semidefinite), real, symmetric constant matrix, and **R** is a positive definite, real, symmetric constant matrix.

An important feature of this class of problems is that optimal control is possible by feedback control law of the form

$$\mathbf{u}(t) = -\mathbf{K}\mathbf{x}(t) \qquad (9.14a)$$

where **K** is a $p \times n$ constant matrix, or

$$\begin{bmatrix} u_1 \\ u_2 \\ \vdots \\ u_p \end{bmatrix} = - \begin{bmatrix} k_{11} & k_{12} & \cdots & k_{1n} \\ k_{21} & k_{22} & \cdots & k_{2n} \\ \vdots & \vdots & & \vdots \\ k_{p1} & k_{p2} & \cdots & k_{pn} \end{bmatrix} \begin{bmatrix} x_1 \\ x_2 \\ \vdots \\ x_n \end{bmatrix} \qquad (9.14b)$$

If the unknown elements of the matrix **K** are determined so as to minimize the performance index given by Eqn. (9.13), then the control law given by Eqns (9.14) is optimal. The configuration of the optimal closed-loop control system is represented by the block diagram of Fig. 9.5. As we shall see later in this chapter, controllability of the plant (9.12) and positive definiteness of matrix **Q** in the performance index (9.13) are sufficient conditions for the existence of asymptotically stable ($\mathbf{x}(t) \to 0$ as $t \to \infty$) optimal solution to the control problem.

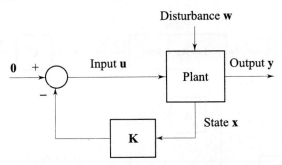

Fig. 9.5 Control configuration of a state regulator

It may be noted that the optimal solution obtained by minimizing the performance index (9.13) may not be the best solution in all circumstances. For example, all the elements of the matrix **K** may not be free; some gains are fixed by the physical constraints of the system and are therefore relatively inflexible. Similarly, if all the states $\mathbf{x}(t)$ are not accessible for feedback, one has to go for a state observer whose complexity is comparable to that of the system itself. It is natural to seek a procedure that relies on the use of feedback from the accessible state variables only, constraining the gain elements of matrix **K** corresponding to the inaccessible state variables to have zero value. Thus, whether one chooses an *optimal* or *suboptimal* solution depends on many factors in addition to the performance required out of the system.

State Observer

Implementation of the optimal control law given by Eqns (9.14) requires the ability to directly measure the entire state vector $\mathbf{x}(t)$. For many systems, full state measurements are not practical. In Section 7.5, we found that the state vector of an observable linear system can be estimated using a state observer which operates on input and output measurements. We assumed that all inputs can be specified exactly and all outputs can be measured with unlimited precision. The dynamic behaviour of the observer was assumed to be specified in terms of its characteristic equation.

Here we are concerned with the optimal design of the state observer for the multivariable system given by Eqns (9.12).

We postulate the existence of an observer of the form

$$\dot{\hat{\mathbf{x}}}(t) = \mathbf{A}\hat{\mathbf{x}}(t) + \mathbf{B}\mathbf{u}(t) + \mathbf{M}[\mathbf{y}(t) - \mathbf{C}\hat{\mathbf{x}}(t)] \qquad (9.15)$$

where $\hat{\mathbf{x}}$ is the estimate of state $\mathbf{x}$ and $\mathbf{M}$ is an $n \times q$ real constant gain matrix. The observer structure is shown in Fig. 9.6, which is of the same form as that considered in Section 7.5. The estimation error is given by

$$\tilde{\mathbf{x}}(t) = \mathbf{x}(t) - \hat{\mathbf{x}}(t) \tag{9.16a}$$

From Eqns (9.12) and (9.15), we have

$$\dot{\tilde{\mathbf{x}}}(t) = (\mathbf{A} - \mathbf{MC})\tilde{\mathbf{x}}(t) \tag{9.16b}$$

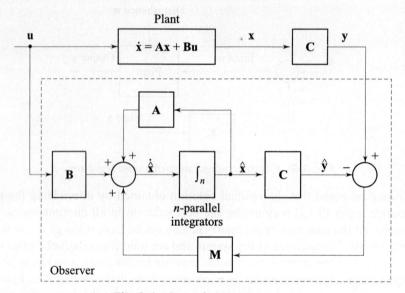

Fig. 9.6 State-observer structure

To design gain matrix $\mathbf{M}$, we may use the duality between control and estimation problems, developed in Section 7.5. (refer Table 7.1). As per the duality principle, the problem of determination of gain matrix $\mathbf{M}$ for the optimal state observer is mathematically equivalent to designing optimal state regulator for the 'transposed auxiliary system' (refer Eqns (7.36))

$$\dot{\boldsymbol{\zeta}}(t) = \mathbf{A}^T \boldsymbol{\zeta}(t) + \mathbf{C}^T \boldsymbol{\eta}(t) \tag{9.17a}$$

The design problem is to determine $n \times q$ gain matrix $\mathbf{M}$ such that

$$\boldsymbol{\eta}(t) = -\mathbf{M}^T \boldsymbol{\zeta}(t) \tag{9.17b}$$

minimizes a quadratic performance index of the form

$$J = \tfrac{1}{2} \int_0^\infty (\boldsymbol{\zeta}^T \mathbf{Q}_0 \boldsymbol{\zeta} + \boldsymbol{\eta}^T \mathbf{R}_0 \boldsymbol{\eta}) \, dt \tag{9.18}$$

where $\mathbf{Q}_0$ is a positive definite (or positive semidefinite), real, symmetric constant matrix, and $\mathbf{R}_0$ is a positive definite, real, symmetric constant matrix.

The solution to this problem exists if the auxiliary system (9.17) is completely controllable. This condition is met, if the original system (9.12) is completely observable.

The *separation principle* (refer Section 7.6) allows for the separate designs of state-feedback control law and state observer; the control law and the observer are then combined as per the configuration of Fig. 9.7. The weighting matrices $\mathbf{Q}_0$ and $\mathbf{R}_0$ for the observer design can be assumed to be equal to the weighting matrices $\mathbf{Q}$ and $\mathbf{R}$, respectively, of the control-law design. Generally, however, one would design a faster observer in comparison with the regulator, i.e., for $\mathbf{Q}_0 = \mathbf{Q}$, the elements of $\mathbf{R}_0$ are chosen smaller than those of $\mathbf{R}$.

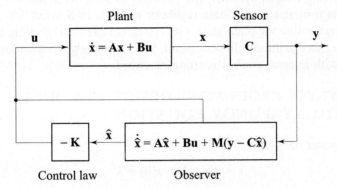

Fig. 9.7 Combined state-feedback control and state estimation

A state variable model of a dynamic system which is more accurate than (9.12) is

$$\dot{\mathbf{x}}(t) = \mathbf{A}\mathbf{x}(t) + \mathbf{B}\mathbf{u}(t) + \mathbf{w}(t)$$
$$\mathbf{y}(t) = \mathbf{C}\mathbf{x}(t) + \mathbf{v}(t)$$
(9.19)

where $\mathbf{w}(t)$ is the $n \times 1$ state excitation noise vector and $\mathbf{v}(t)$ is $q \times 1$ observation noise vector. The state estimator which is optimum with respect to the state excitation noise and observation noise is commonly known as *Kalman filter* [102].

Servo Systems

The control configuration of Fig. 9.5 implicitly assumes that the null state $\mathbf{x} = \mathbf{0}$ is the desired equilibrium state of the system. It is a state regulator with zero command input.

In servo systems where the output $\mathbf{y}(t)$ is required to track a constant command input, the equilibrium state is a constant point (other than the origin) in state space. This servo problem can be formulated as a 'shifted regulator problem' by shifting the origin of the state space to the equilibrium point. Formulation of the shifted regulator problem for single-input systems was given in Section 7.7. Extension of the formulation to the multi-input case is straightforward.

State-Feedback with Integral Control

In a state-feedback control system (which is a generalization of proportional plus derivative feedback), it is usually required that the system have one or more integrators within the closed loop. This will lead to zero steady-state error when the command input and disturbance have constant steady-state values. Unless the plant to be controlled has integrating property, it is generally necessary to add one or more integrators within the loop.

For the system (9.12), we can feedback the state **x** as well as the integral of the error in output by augmenting the plant state **x** with the extra 'integral state'. For single-input systems, the problem of state-feedback with integral control was formulated as a state regulator problem in Section 7.8. This was done by augmenting the plant state with 'integral state', and shifting the origin of the state space to the equilibrium point. Multivariable generalization of state-feedback with integral control is straightforward.

9.4 STATE REGULATOR DESIGN THROUGH THE LYAPUNOV EQUATION

Let us consider the system:

$$\dot{\mathbf{x}} = \mathbf{A}\mathbf{x} + \mathbf{B}\mathbf{u}; \; \mathbf{x}(0) \triangleq \mathbf{x}^0 \qquad (9.20)$$

where **x** is the $n \times 1$ state vector, and **u** is the $p \times 1$ input vector. **A** and **B** are, respectively, $n \times n$ and $n \times p$ real constant matrices.

It is desired to minimize the following performance index:

$$J = \tfrac{1}{2} \int_0^\infty (\mathbf{x}^T \mathbf{Q}\mathbf{x} + \mathbf{u}^T \mathbf{R}\mathbf{u}) \, dt \qquad (9.21)$$

where **Q** is $n \times n$ positive definite (or positive semidefinite), real, symmetric, constant matrix, and **R** is $p \times p$ positive definite, real, symmetric, constant matrix.

The present section is devoted to constrained optimization leading to suboptimal solution to the control problem. Unconstrained optimization, leading to optimal solution to the control problem, will be discussed in the next section.

We shall obtain a direct relationship between Lyapunov functions and quadratic performance measures, and solve the parameter-optimization problem using this relationship. We select the feedback control configuration of Fig. 9.5 described by the control law

$$\mathbf{u} = -\mathbf{K}\mathbf{x} \qquad (9.22)$$

where **K** is $p \times n$ matrix which involves adjustable parameters. With this control law, the closed-loop system becomes

$$\dot{\mathbf{x}} = \mathbf{A}\mathbf{x} - \mathbf{B}\mathbf{K}\mathbf{x} = (\mathbf{A} - \mathbf{B}\mathbf{K})\mathbf{x} \qquad (9.23)$$

All the parameters of the matrix **K** may not be available for adjustments. Some of them may have fixed values (including zero values). We will assume that a matrix **K** satisfying the imposed constraints on its parameters exists such that $(\mathbf{A} - \mathbf{B}\mathbf{K})$ is a stable matrix.

The optimization problem is to determine the values of free parameters of the matrix **K** so as to minimize the performance index given by Eqn. (9.21).

Substituting the control vector **u** from Eqn. (9.22) in the performance index J of Eqn. (9.21), we have

$$J = \tfrac{1}{2} \int_0^\infty (\mathbf{x}^T \mathbf{Q} \mathbf{x} + \mathbf{x}^T \mathbf{K}^T \mathbf{R} \mathbf{K} \mathbf{x}) dt$$

$$= \tfrac{1}{2} \int_0^\infty \mathbf{x}^T (\mathbf{Q} + \mathbf{K}^T \mathbf{R} \mathbf{K}) \mathbf{x}\, dt \tag{9.24}$$

Let us assume a Lyapunov function

$$V(\mathbf{x}(t)) = \tfrac{1}{2} \int_t^\infty \mathbf{x}^T (\mathbf{Q} + \mathbf{K}^T \mathbf{R} \mathbf{K}) \mathbf{x}\, dt$$

Note that the value of the performance index for system trajectory starting at $\mathbf{x}(0)$ is $V(\mathbf{x}(0))$.

The time derivative of the Lyapunov function is

$$\dot{V}(\mathbf{x}) = \tfrac{1}{2} \mathbf{x}^T (\mathbf{Q} + \mathbf{K}^T \mathbf{R} \mathbf{K}) \mathbf{x} \Big|_t^\infty$$

$$= \tfrac{1}{2} \mathbf{x}^T(\infty) [\mathbf{Q} + \mathbf{K}^T \mathbf{R} \mathbf{K}] \mathbf{x}(\infty) - \tfrac{1}{2} \mathbf{x}^T(t) [\mathbf{Q} + \mathbf{K}^T \mathbf{R} \mathbf{K}] \mathbf{x}(t)$$

Assuming that the matrix $(\mathbf{A} - \mathbf{B}\mathbf{K})$ is stable, we have from Eqn. (9.23),

$$\mathbf{x}(\infty) \to \mathbf{0}$$

Therefore,

$$\dot{V}(\mathbf{x}) = -\tfrac{1}{2} \mathbf{x}^T (\mathbf{Q} + \mathbf{K}^T \mathbf{R} \mathbf{K}) \mathbf{x} \tag{9.25}$$

Since $\dot{V}(\mathbf{x})$ is quadratic in **x** and the plant equation is linear, let us assume that $V(\mathbf{x})$ is also given by the quadratic form:

$$V(\mathbf{x}) = \tfrac{1}{2} \mathbf{x}^T \mathbf{P} \mathbf{x} \tag{9.26}$$

where **P** is a positive definite real, symmetric, constant matrix. Therefore,

$$\dot{V}(\mathbf{x}) = \tfrac{1}{2} (\dot{\mathbf{x}}^T \mathbf{P} \mathbf{x} + \mathbf{x}^T \mathbf{P} \dot{\mathbf{x}})$$

Substituting for $\dot{\mathbf{x}}$ from Eqn. (9.23), we get

$$\dot{V}(\mathbf{x}) = \tfrac{1}{2} \mathbf{x}^T [(\mathbf{A} - \mathbf{B}\mathbf{K})^T \mathbf{P} + \mathbf{P}(\mathbf{A} - \mathbf{B}\mathbf{K})] \mathbf{x}$$

Comparison of this result with Eqn. (9.25) gives

$$\tfrac{1}{2} \mathbf{x}^T [(\mathbf{A} - \mathbf{B}\mathbf{K})^T \mathbf{P} + \mathbf{P}(\mathbf{A} - \mathbf{B}\mathbf{K})] \mathbf{x} = -\tfrac{1}{2} \mathbf{x}^T (\mathbf{Q} + \mathbf{K}^T \mathbf{R} \mathbf{K}) \mathbf{x}$$

Since the above equality holds for arbitrary $\mathbf{x}(t)$, we have

$$(\mathbf{A} - \mathbf{B}\mathbf{K})^T \mathbf{P} + \mathbf{P}(\mathbf{A} - \mathbf{B}\mathbf{K}) + \mathbf{K}^T \mathbf{R} \mathbf{K} + \mathbf{Q} = 0 \tag{9.27}$$

This equation is of the form of Lyapunov equation defined in Section 8.6. In Eqn. (9.27) we have n^2 nonliner algebraic equations. However, since $n \times n$

matrix **P** is symmetric, we need to solve only $\frac{n(n+1)}{2}$ equations for the elements p_{ij} of the matrix **P**. The solution will give p_{ij} as functions of the feedback matrix **K**.

As pointed out earlier, $V(\mathbf{x}(0))$ is the value of the performance index for the system trajectory starting at $\mathbf{x}(0)$. From Eqn. (9.26), we get,

$$J = \tfrac{1}{2}\mathbf{x}^T(0)\mathbf{P}\mathbf{x}(0) \qquad (9.28)$$

A suboptimal control law may be obtained by minimizing J with respect to the available elements k_{ij} of **K**, i.e., by setting

$$\frac{\partial [\mathbf{x}^T(0)\mathbf{P}\mathbf{x}(0)]}{\partial k_{ij}} = 0 \qquad (9.29)$$

If for the suboptimal solution thus obtained, the matrix $(\mathbf{A} - \mathbf{B}\mathbf{K})$ is stable, then the minimization of J as per the procedure described above gives the correct result. From Eqn. (9.25) we observe that for a positive definite **Q**, $\dot{V}(\mathbf{x}) < 0$ for all $\mathbf{x} \neq \mathbf{0}$ (note that $\mathbf{K}^T\mathbf{R}\mathbf{K}$ is non-negative definite). Also Eqn. (9.26) shows that $V(\mathbf{x}) > 0$ for all $\mathbf{x} \neq \mathbf{0}$ if **P** is positive definite. Therefore minimization of J with respect to k_{ij} (Eqn. (9.29)) will lead to a stable closed-loop system if the optimal k_{ij} result in a positive definite matrix[3] **P**.

We now study the effect of choosing a positive semidefinite **Q** in the performance index J. If **Q** is positive semidefinite and in addition the following rank condition is satisfied:

$$\rho \begin{bmatrix} \mathbf{H} \\ \mathbf{H}\mathbf{A} \\ \vdots \\ \mathbf{H}\mathbf{A}^{n-1} \end{bmatrix} = n; \; \mathbf{Q} = \mathbf{H}^T\mathbf{H}, \qquad (9.30)$$

then $\dot{V}(\mathbf{x}) < 0$ for all $\mathbf{x} \neq \mathbf{0}$.

We prove this result by contradiction: the rank condition is satisfied but $\dot{V}(\mathbf{x}) = 0$ for some $\mathbf{x} \neq \mathbf{0}$. Substituting $\mathbf{Q} = \mathbf{H}^T\mathbf{H}$ in Eqn. (9.25), we obtain (refer Eqns (5.6))

$$\dot{V}(\mathbf{x}) = -\tfrac{1}{2}(\mathbf{x}^T\mathbf{H}^T\mathbf{H}\mathbf{x} + \mathbf{x}^T\mathbf{K}^T\mathbf{R}\mathbf{K}\mathbf{x}) = -\tfrac{1}{2}[\|\mathbf{H}\mathbf{x}\|^2 + \|\mathbf{K}\mathbf{x}\|_\mathbf{R}^2]$$

Therefore $\mathbf{H}\mathbf{x} = \mathbf{0}$ and $\mathbf{K}\mathbf{x} = \mathbf{0}$ should be simultaneously satisfied. $\mathbf{K}\mathbf{x} = \mathbf{0}$ reduces the closed-loop system (9.23) to the open-loop system.

---

3. One would like to examine the existence of a solution to the optimization problem before actually starting the optimization procedure. For the problem under consideration, existence of a **K** that minimizes J is ensured if, and only if, there exists a $\mathbf{K}_0$ satisfying the imposed constraints on its parameters, such that $(\mathbf{A} - \mathbf{B}\mathbf{K}_0)$ is asymptotically stable. The question of existence of a $\mathbf{K}_0$ has, as yet, no straightforward answer.

From the condition
$$\dot{\mathbf{x}} = \mathbf{A}\mathbf{x}$$
$$\mathbf{H}\mathbf{x} = \mathbf{0}$$
we obtain
$$\mathbf{H}\dot{\mathbf{x}} = \mathbf{H}\mathbf{A}\mathbf{x} = \mathbf{0}$$
Continuing the process of taking derivative, we get
$$\mathbf{H}\mathbf{A}^2\mathbf{x} = \mathbf{0}$$
$$\vdots$$
$$\mathbf{H}\mathbf{A}^{n-1}\mathbf{x} = \mathbf{0}$$

or
$$\begin{bmatrix} \mathbf{H} \\ \mathbf{H}\mathbf{A} \\ \vdots \\ \mathbf{H}\mathbf{A}^{n-1} \end{bmatrix} \mathbf{x} = \mathbf{0} = \mathbf{V}\mathbf{x}$$

Since $\rho[\mathbf{V}] = n$, $\mathbf{V}\mathbf{x} = \mathbf{0}$ only when $\mathbf{x} = \mathbf{0}$. This proves the result.

Let us see an alternative interpretation of the rank condition (9.30). The rank condition implies that the system
$$\dot{\mathbf{x}} = \mathbf{A}\mathbf{x} + \mathbf{B}\mathbf{u}$$
with the 'auxiliary output'
$$\zeta(t) = \mathbf{H}\mathbf{x}(t)$$
is completely observable. Since the performance index
$$J = \tfrac{1}{2}\int_0^\infty (\mathbf{x}^T\mathbf{Q}\mathbf{x} + \mathbf{u}^T\mathbf{R}\mathbf{u})\,dt = \tfrac{1}{2}\int_0^\infty (\mathbf{x}^T\mathbf{H}^T\mathbf{H}\mathbf{x} + \mathbf{u}^T\mathbf{R}\mathbf{u})\,dt$$
$$= \tfrac{1}{2}\int_0^\infty (\zeta^T\zeta + \mathbf{u}^T\mathbf{R}\mathbf{u})\,dt,$$
the observability of the pair $(\mathbf{A}, \mathbf{H})$ implies that all the modes of the state trajectories are reflected in the performance index. A finite value of J, therefore, ensures that unstable modes (if any) have been stabilized[4] by the control $\mathbf{u} = -\mathbf{K}\mathbf{x}$.

The observability condition is always satisfied when the matrix $\mathbf{Q}$ is positive definite.

▲▲

---

4. Observability canonical form for a state model which is not completely observable is given in Eqn. (5.124c). It decomposes the model into two parts: the observable part and the unobservable part. If the unobservable part is stable, then the model is said to be *detectable*.

In the optimization problem under consideration, the observability of the pair $(\mathbf{A}, \mathbf{H})$ is not a necessary condition for the existence of stable solution. If the pair $(\mathbf{A}, \mathbf{H})$ is detectable, then the modes of state trajectories not reflected in J are stable and a finite value of J will ensure asymptotic stability of the closed-loop system.

In the special case where the performance index is independent of control $\mathbf{u}$, we have

$$J = \tfrac{1}{2}\int_0^\infty \mathbf{x}^T\mathbf{Q}\mathbf{x}\, dt \qquad (9.31a)$$

In this case, the matrix $\mathbf{P}$ is obtained from Eqn. (9.27) by putting $\mathbf{R} = \mathbf{0}$, resulting in the modified Lyapunov equation

$$(\mathbf{A} - \mathbf{BK})^T\mathbf{P} + \mathbf{P}(\mathbf{A} - \mathbf{BK}) + \mathbf{Q} = \mathbf{0} \qquad (9.31b)$$

Even though $\mathbf{R}$ is originally assumed to be positive definite, substituting $\mathbf{R} = \mathbf{0}$ is a valid operation here as the positive definiteness of $\mathbf{R}$ has not been used in the derivation so far.

The reader is reminded here that the matrix $\mathbf{K}$ corresponding to the sub-optimal solution has to satisfy the further constraint that the closed-loop system be asymptotically stable. The question of existence of $\mathbf{K}$ satisfying this constraint has, as yet, no straightforward answer. We now turn to the problem of finding such a $\mathbf{K}$, assuming that one exists.

The design procedure is summarized below:
- *Plant model*

$$\dot{\mathbf{x}} = \mathbf{A}\mathbf{x} + \mathbf{B}\mathbf{u};\ \mathbf{x}(0) \triangleq \mathbf{x}^0 \qquad (9.32a)$$

- *Performance index*

$$J = \tfrac{1}{2}\int_0^\infty (\mathbf{x}^T\mathbf{Q}\mathbf{x} + \mathbf{u}^T\mathbf{R}\mathbf{u})\, dt \qquad (9.32b)$$

- *Feedback control law*

$$\mathbf{u} = -\mathbf{K}\mathbf{x} \qquad (9.32c)$$

Step 1: Identify $\mathbf{A}$, $\mathbf{B}$, $\mathbf{Q}$ and $\mathbf{R}$ matrices and solve the Lyapunov equation

$$(\mathbf{A} - \mathbf{BK})^T\mathbf{P} + \mathbf{P}(\mathbf{A} - \mathbf{BK}) + \mathbf{K}^T\mathbf{R}\mathbf{K} + \mathbf{Q} = \mathbf{0} \qquad (9.32d)$$

for $\mathbf{P}$ as a function of the free parameters k_{ij} of the feedback matrix $\mathbf{K}$.

Step 2: Compute the performance index J as a function of the free parameters k_{ij} of the matrix $\mathbf{K}$:

$$J = \tfrac{1}{2}\mathbf{x}^T(0)\mathbf{P}\mathbf{x}(0) \qquad (9.32e)$$

Step 3: Determine the solution of the equations

$$\frac{\partial J}{\partial k_{ij}} = 0 \qquad (9.32f)$$

From the solution set of these equations, find the subset that satisfies the sufficient conditions for minimization. This subset is the optimum parameter set.

Step 4: Examine $(\mathbf{A} - \mathbf{BK})$ for stability.

Example 9.3

Reconsider the liquid-level system of Fig. 9.3. In Example 9.2, we designed a parameter-optimized controller for this process by direct minimization of the performance index. In the following, we use Lyapunov equation for designing a regulator for the process.

The state equation of the process is

$$\frac{dy}{dt} = -y + u; \quad y(0) = 1$$

where

$y = h$ = deviation of the liquid head from the steady-state; $u = q_i$ = rate of liquid inflow, and the performance index

$$J = \int_0^\infty (y^2 + u^2)\, dt$$

Comparison of the state equation and the performance index with those given in Eqns (9.32a) and (9.32b), gives

$$A = -1,\ B = 1,\ Q = 2,\ R = 2$$

The control law:

$$u = -Ky$$

The Lyapunov Eqn. (9.32d) reduces to the following:

$$(-1 - K)P + P(-1 - K) + 2K^2 + 2 = 0$$

which gives

$$P = \frac{K^2 + 1}{K + 1}$$

For $K > 0$, P is positive definite.

The performance index

$$J = \tfrac{1}{2} y^T(0) P y(0) = \frac{1 + K^2}{2(K+1)} [y(0)]^2$$

Minimization of J gives

$$K = \left(\sqrt{2} - 1\right);\ J_{min} = \left(\sqrt{2} - 1\right) [y(0)]^2$$

Note that optimum value of K is independent of initial conditions.

Example 9.4

Reconsider the system of Fig. 9.2a. In Example 9.1, we designed a parameter-optimized controller for this system by direct minimization of the performance index. In the following, we use Lyapunov equation for designing a regulator for the system.

The state equation for the plant is

$$\dot{y} = u;\ y(0) = 0$$

The performance index

$$J = \int_0^\infty (e^2 + 0.5u^2)\, dt$$

$$e = r - y$$

In terms of the equivalent state variable

$$\tilde{x} = y - r$$

the state equation becomes

$$\dot{\tilde{x}} = u; \ \tilde{x}(0) = -1$$

and the performance index

$$J = \int_0^\infty (\tilde{x}^2 + 0.5u^2)\, dt$$

Comparison of the transformed state equation and performance index with those given in Eqns (9.32a) and (9.32b), gives

$$A = 0,\ B = 1,\ Q = 2,\ R = 1$$

The control law:

$$u = -K\tilde{x}$$

The Lyapunov Eqn. (9.32d) reduces to the following:

$$-KP - PK + K^2 + 2 = 0$$

which gives

$$P = \frac{1 + 0.5K^2}{K}$$

For $K > 0$, P is positive definite.

The performance index

$$J = \tfrac{1}{2}\tilde{x}^T(0)P\tilde{x}(0) = \frac{1 + 0.5K^2}{2K}[\tilde{x}(0)]^2$$

Minimization of J gives

$$K = \sqrt{2},\ J_{\min} = 0.707\,[\tilde{x}(0)]^2$$

Note that optimum value of K is independent of $\tilde{x}(0)$.

Example 9.5

Consider the problem of attitude control of a rigid satellite which was discussed in Example 7.1. An attitude control system for the satellite that utilizes rate feedback is shown in Fig. 9.8; $\theta(t)$ is the actual attitude, $\theta_r(t)$ is the reference attitude which is a step function, and $u(t)$ is the torque developed by the thrusters.

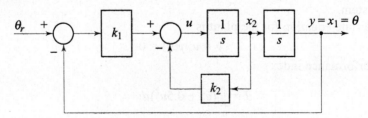

Fig. 9.8 Attitude control of a satellite

Linear Quadratic Optimal Control

The state variable model of the system is

$$\dot{\mathbf{x}} = \mathbf{A}\mathbf{x} + \mathbf{B}u; \mathbf{x}(0) = 0$$
$$y = \mathbf{C}\mathbf{x} \qquad (9.33)$$

with
$$\mathbf{A} = \begin{bmatrix} 0 & 1 \\ 0 & 0 \end{bmatrix}; \mathbf{B} = \begin{bmatrix} 0 \\ 1 \end{bmatrix}; \mathbf{C} = \begin{bmatrix} 1 & 0 \end{bmatrix}$$

The feedback control law is
$$u = -k_1(x_1 - \theta_r) - k_2 x_2; x_1 = \theta, x_2 = \dot{\theta} \qquad (9.34)$$

with the constraint
$$k_1 = 1$$

The problem is to obtain the optimum value of the parameter k_2 that minimizes the performance index

$$J = \int_0^\infty (\theta_r - \theta)^2 \, dt \qquad (9.35)$$

for a unit-step input θ_r.

The following state transformation converts the above servo problem to a standard regulator problem where null state is the desired state.

$$\tilde{x}_1 = x_1 - \theta_r$$
$$\tilde{x}_2 = x_2$$

The transformed state equation is of the form:

$$\dot{\tilde{\mathbf{x}}} = \mathbf{A}\tilde{\mathbf{x}} + \mathbf{B}u; \tilde{\mathbf{x}}(0) = [-1 \quad 0]^T$$

where $\mathbf{A}$ and $\mathbf{B}$ are given by Eqns (9.33)
The problem is to find k_2 such that

$$u = -[1 \quad k_2]\tilde{\mathbf{x}}$$

minimizes the performance index

$$J = \int_0^\infty (\tilde{x}_1)^2 \, dt$$

Comparison of the transformed state equation and performance index with those given in Eqns (9.32a) and (9.32b), gives

$$\mathbf{A} = \begin{bmatrix} 0 & 1 \\ 0 & 0 \end{bmatrix}; \mathbf{B} = \begin{bmatrix} 0 \\ 1 \end{bmatrix}; \mathbf{Q} = \begin{bmatrix} 2 & 0 \\ 0 & 0 \end{bmatrix}; R = 0$$

The Lyapunov Eqn. (9.32d) reduces to the following:

$$\begin{bmatrix} 0 & -1 \\ 1 & -k_2 \end{bmatrix} \begin{bmatrix} p_{11} & p_{12} \\ p_{12} & p_{22} \end{bmatrix} + \begin{bmatrix} p_{11} & p_{12} \\ p_{12} & p_{22} \end{bmatrix} \begin{bmatrix} 0 & 1 \\ -1 & -k_2 \end{bmatrix} + \begin{bmatrix} 2 & 0 \\ 0 & 0 \end{bmatrix} = \begin{bmatrix} 0 & 0 \\ 0 & 0 \end{bmatrix}$$

Solving we get

$$\mathbf{P} = \begin{bmatrix} \dfrac{1+k_2^2}{k_2} & 1 \\ 1 & \dfrac{1}{k_2} \end{bmatrix}$$

For $k_2 > 0$, $\mathbf{P}$ is positive definite.

$$J = \tfrac{1}{2}\tilde{\mathbf{x}}^T(0)\mathbf{P}\tilde{\mathbf{x}}(0) = \frac{1+k_2^2}{2k_2}$$

The optimum value of k_2 is obtained by setting

$$\frac{\partial J}{\partial k_2} = \frac{k_2^2 - 1}{2k_2^2} = 0$$

This gives

$$k_2 = 1$$

Note that

$$\frac{\partial^2 J}{\partial k_2^2} = \frac{1}{k_2^3} > 0$$

The suboptimal control law is therefore,

$$u = -\tilde{x}_1 - \tilde{x}_2 = -x_1 - x_2 + \theta_r$$

The state equation for the closed-loop system becomes

$$\dot{\mathbf{x}} = (\mathbf{A} - \mathbf{BK})\mathbf{x} + \mathbf{B}\theta_r$$

where

$$(\mathbf{A} - \mathbf{BK}) = \begin{bmatrix} 0 & 1 \\ -1 & -1 \end{bmatrix}; \mathbf{B} = \begin{bmatrix} 0 \\ 1 \end{bmatrix}$$

The characteristic equation of the closed-loop system is:

$$det[\lambda \mathbf{I} - (\mathbf{A} - \mathbf{BK})] = \begin{bmatrix} \lambda & -1 \\ 1 & \lambda+1 \end{bmatrix} = \lambda^2 + \lambda + 1$$

The closed-loop system is therefore asymptotically stable.

A curve of the performance measure as a function of k_2 is shown in Fig. 9.9. It is clear that the system is not very sensitive to changes in k_2 and will maintain a near minimum performance index if the k_2 parameter is altered some percentage.

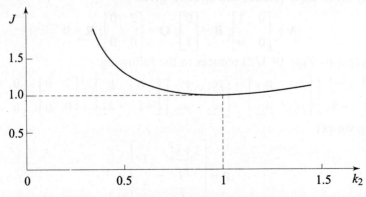

Fig. 9.9 Performance measure versus gain for the system shown in Fig. 9.8

We define the sensitivity of an optimum system as

$$S_k^{opt} = \frac{\Delta J/J}{\Delta k/k} \qquad (9.36)$$

where k is the design parameter.

For this example, we have $k = k_2$ and therefore

$$S_{k_2}^{opt} \cong \frac{0.00556/1}{0.1/1} = 0.0556$$

More on Parameter Optimization

(i) If the feedback matrix **K** is unconstrained, i.e., all the parameters k_{ij} are free, then the control law is optimal and will be independent of initial conditions. The optimal k_{ij}'s can be obtained from the equations

$$\frac{\partial P}{\partial k_{ij}} = 0 \quad \text{for all } i, j \qquad (9.37)$$

However, this procedure for obtaining the optimal unconstrained feedback matrix **K** gets very complicated for higher-order systems. An alternative route to the solution is through the matrix Riccati equation discussed in the next section.

(ii) When the configuration of the controller is constrained, i.e., some k_{ij}'s are not available for adjustment, a solution which is independent of initial conditions can no longer be found. If a system is to operate satisfactorily for a range of initial disturbances, it may not be clear which is the most suitable for optimization.

The dependence on initial conditions can be avoided by averaging the performance obtained for a linearly independent set of initial conditions. This is equivalent to assuming the initial state $\mathbf{x}(0)$ to be a random variable uniformly distributed on the surface of the n-dimensional unit sphere (refer [102] and [129] for details).

(iii) We have carefully selected simple optimization problems as our design examples; we could solve these problems analytically. The need for mathematical programming techniques is obvious. Reference [102] describes a numerical algorithm based on Fletcher-Powell method for the minimization of functions.

9.5 OPTIMAL STATE REGULATOR THROUGH THE MATRIX RICCATI EQUATION

The optimal control problem for a linear multivariable system with the quadratic criterion function is one of the most common problems in linear system theory. It is defined below:

Given the completely controllable plant

$$\dot{\mathbf{x}} = \mathbf{A}\mathbf{x} + \mathbf{B}\mathbf{u} \qquad (9.38)$$

where **x** is the $n \times 1$ state vector, **u** is the $p \times 1$ input vector; **A** and **B** are, respectively, $n \times n$ and $n \times p$ real constant matrices, and the null state $\mathbf{x} = \mathbf{0}$ is the desired steady-state.

Find the control law

$$\mathbf{u} = -\mathbf{K}\mathbf{x}(t) \tag{9.39}$$

where **K** is $p \times n$ real constant unconstrained gain matrix, that minimizes the following performance index subject to the initial conditions $\mathbf{x}(0) \triangleq \mathbf{x}^0$:

$$J = \tfrac{1}{2}\int_0^\infty (\mathbf{x}^T\mathbf{Q}\mathbf{x} + \mathbf{u}^T\mathbf{R}\mathbf{u})\, dt \tag{9.40}$$

where **Q** is $n \times n$ positive definite, real, symmetric, constant matrix, and **R** is $p \times p$ positive definite, real, symmetric, constant matrix.

There are several ways to solve this optimal control problem. We will use the Laypunov function approach developed in the previous section. Substituting Eqn. (9.39) into Eqn. (9.38), we obtain

$$\dot{\mathbf{x}} = \mathbf{A}\mathbf{x} - \mathbf{B}\mathbf{K}\mathbf{x} = (\mathbf{A} - \mathbf{B}\mathbf{K})\mathbf{x} \tag{9.41}$$

Since the (**A**,**B**) pair is completely controllable, there exists a feedback matrix **K** such that $(\mathbf{A} - \mathbf{B}\mathbf{K})$ is a stable matrix (refer Section 7.3). The controllability[5] of the given plant, thus, ensures the existence of a **K** that minimizes J.

Following the discussion given in the previous section, we obtain the following design equations for the optimal control problem.

The time derivative of the Lyapunov function is

$$\dot{V}(\mathbf{x}) = -\tfrac{1}{2}\mathbf{x}^T(\mathbf{Q} + \mathbf{K}^T\mathbf{R}\mathbf{K})\mathbf{x} \tag{9.42}$$

The Lyapunov function

$$V(\mathbf{x}) = \tfrac{1}{2}\mathbf{x}^T\mathbf{P}\mathbf{x} \tag{9.43}$$

where **P** is a positive definite, real, symmetric, constant matrix, and

$$(\mathbf{A} - \mathbf{B}\mathbf{K})^T\mathbf{P} + \mathbf{P}(\mathbf{A} - \mathbf{B}\mathbf{K}) + \mathbf{K}^T\mathbf{R}\mathbf{K} + \mathbf{Q} = 0 \tag{9.44}$$

The performance index

$$J = \tfrac{1}{2}\mathbf{x}^T(0)\mathbf{P}\mathbf{x}(0) \tag{9.45}$$

---

5. The controllability of the (**A**,**B**) pair is not a necessary condition for the existence of the optimal solution. If the (**A**,**B**) pair is not completely controllable, we can transform the plant model to controllability canonical form given in Eqn. (5.123c). It decomposes the model into two parts: The controllable part and the uncontrollable part. If the uncontrollable part is stable, then the model is said to be *stabilizable*. Stabilizability of the (**A**,**B**) pair is a necessary condition for the existence of optimal solution.

Since feedback matrix **K** is unconstrained, the optimum value of J is independent of initial conditions. The optimal k_{ij}'s are obtained from the equations

$$\frac{\partial \mathbf{P}}{\partial k_{ij}} = \mathbf{0} \tag{9.46}$$

for all i, j.

Since **R** has been assumed to be a positive definite matrix, we can write

$$\mathbf{R} = \mathbf{\Gamma}^T \mathbf{\Gamma}$$

where $\mathbf{\Gamma}$ is a nonsingular matrix. Then Eqn. (9.44) can be written as

$$(\mathbf{A}^T - \mathbf{K}^T\mathbf{B}^T)\mathbf{P} + \mathbf{P}(\mathbf{A} - \mathbf{BK}) + \mathbf{Q} + \mathbf{K}^T\mathbf{\Gamma}^T\mathbf{\Gamma}\mathbf{K} = \mathbf{0}$$

which can be rewritten as

$$\mathbf{A}^T\mathbf{P} + \mathbf{P}\mathbf{A} + [\mathbf{\Gamma}\mathbf{K} - (\mathbf{\Gamma}^T)^{-1}\mathbf{B}^T\mathbf{P}]^T[\mathbf{\Gamma}\mathbf{K} - (\mathbf{\Gamma}^T)^{-1}\mathbf{B}^T\mathbf{P}]$$
$$- \mathbf{PBR}^{-1}\mathbf{B}^T\mathbf{P} + \mathbf{Q} = \mathbf{0} \tag{9.47}$$

The condition (9.46) for unconstrained minimization of J leads to the following equations:

$$\frac{\partial}{\partial k_{ij}} [(\mathbf{\Gamma}\mathbf{K} - (\mathbf{\Gamma}^T)^{-1}\mathbf{B}^T\mathbf{P})^T(\mathbf{\Gamma}\mathbf{K} - (\mathbf{\Gamma}^T)^{-1}\mathbf{B}^T\mathbf{P})] = 0$$

Since the matrix within the brackets is non-negative definite, the minimum occurs when it is zero, or when

$$\mathbf{\Gamma}\mathbf{K} = (\mathbf{\Gamma}^T)^{-1}\mathbf{B}^T\mathbf{P}$$

Hence

$$\mathbf{K} = \mathbf{\Gamma}^{-1}(\mathbf{\Gamma}^T)^{-1}\mathbf{B}^T\mathbf{P} = \mathbf{R}^{-1}\mathbf{B}^T\mathbf{P} \tag{9.48}$$

Equation (9.48) gives the optimal gain matrix **K**. Thus the optimal control law is given by

$$\mathbf{u}(t) = -\mathbf{K}\mathbf{x}(t) = -\mathbf{R}^{-1}\mathbf{B}^T\mathbf{P}\mathbf{x}(t) \tag{9.49}$$

The matrix **P** in Eqn. (9.49) must satisfy Eqn. (9.47), or the following reduced equation:

$$\mathbf{A}^T\mathbf{P} + \mathbf{P}\mathbf{A} - \mathbf{PB}\,\mathbf{R}^{-1}\mathbf{B}^T\mathbf{P} + \mathbf{Q} = \mathbf{0} \tag{9.50}$$

Equation (9.50) is called the *matrix Riccati equation*.

We have assumed **Q** to be a positive definite matrix. For this choice of **Q**, $\dot{V}(\mathbf{x})$ in Eqn. (9.42) is always negative definite; therefore the optimal feedback system is asymptotically stable. Controllability of the (**A**,**B**) pair and positive definiteness of **Q** are, thus, sufficient conditions for the existence of asymptotically stable optimal solution to the control problem.

The design steps may be stated as follows:
1. Solve the matrix Riccati Eqn. (9.50) for the positive definite matrix **P**.
2. Substitute this matrix **P** into Eqn. (9.49); the resulting equation gives optimal control law.

▲▲

This is a basic and well-known result in the theory of optimal control. Once the designer has specified **Q** and **R**, representing his/her assessment of the relative importance of various terms in the performance index, the solution of Eqn. (9.50) specifies the optimal control law (9.49). This yields the optimal closed-loop system. If the resulting transient response is unsatisfactory, the designer may alter the weighting matrices **Q** and **R** and try again.

Comments

(i) The matrix **R** has been assumed to be positive definite. This is a necessary condition for the existence of the optimal solution to the control problem as seen from Eqn. (9.49).

(ii) We have assumed that the plant (9.38) is completely controllable, and the matrix **Q** in performance index J given by Eqn. (9.40) is positive definite. These are sufficient conditions for the existence of asymptotically stable optimal solution to the control problem. The requirement on matrix **Q** may be relaxed to a positive semidefinite matrix with the pair (**A**,**H**) completely observable, where $\mathbf{Q} = \mathbf{H}^T\mathbf{H}$.

(iii) It is important to be able to find out whether the sought-after solution exists or not before we start working out the solution. This is possible only if necessary conditions for the existence of asymptotically stable optimal solution are established. A discussion on this subject entails not only controllability and observability but also the concepts of *stabilizability* and *detectability*. Basic ideas about these concepts have been given in footnotes of this chapter; a detailed discussion is beyond the scope of this book.

(iv) Equation (9.50) is a set of n^2 nonlinear algebraic equations. Since **P** is a symmetric matrix, we need to solve only $\dfrac{n(n+1)}{2}$ equations.

(v) The solution of Eqn. (9.50) is not unique. Of the several possible solutions, the desired answer is obtained by enforcing the requirement that **P** be positive definite. The positive definite solution of Eqn. (9.50) is unique.

(vi) In very simple cases, the Riccati equation can be solved analytically, but usually a numerical solution is required. A number of computer programs for the purpose are available [176–179]. Some methods for the numerical solution of the Riccati equation are given in reference [102].

Example 9.6

Consider the problem of attitude control of a rigid satellite, which was discussed in Example 9.5. An attitude control system for the satellite is shown in Fig. 9.8; $\theta(t)$ is the actual attitude, $\theta_r(t)$ is the reference attitude which is a step function and $u(t)$ is the torque developed by the thrusters.

The state variable model of the system is
$$\dot{x} = Ax + Bu$$
$$y = Cx \qquad (9.51)$$
with
$$A = \begin{bmatrix} 0 & 1 \\ 0 & 0 \end{bmatrix}; \quad B = \begin{bmatrix} 0 \\ 1 \end{bmatrix}; \quad C = \begin{bmatrix} 1 & 0 \end{bmatrix}$$

The problem is to obtain the optimal control law
$$u = -k_1(x_1 - \theta_r) - k_2 x_2; \quad x_1 = \theta, \; x_2 = \dot{\theta}$$
that minimizes the performance index
$$J = \int_0^\infty [(\theta_r - \theta)^2 + u^2]\, dt \qquad (9.52)$$

In terms of the shifted state variables
$$\tilde{x}_1 = x_1 - \theta_r$$
$$\tilde{x}_2 = x_2$$
the state equation becomes
$$\dot{\tilde{x}} = A\tilde{x} + Bu \qquad (9.53)$$
where A and B are given by Eqns (9.51).

Now the problem is to find optimal values of the parameters k_1 and k_2 such that the control law
$$u = -k_1 \tilde{x}_1 - k_2 \tilde{x}_2$$
minimizes the performance index
$$J = \int_0^\infty (\tilde{x}_1^2 + u^2)\, dt \qquad (9.54)$$

The Q and R matrices are
$$Q = \begin{bmatrix} 2 & 0 \\ 0 & 0 \end{bmatrix}; \quad R = 2$$

Note that Q is positive semidefinite matrix.
$$Q = H^T H = \begin{bmatrix} \sqrt{2} \\ 0 \end{bmatrix} \begin{bmatrix} \sqrt{2} & 0 \end{bmatrix}$$

The pair (A, H) is completely observable. Also the pair (A, B) is completely controllable. Therefore, sufficient conditions for the existence of asymptotically stable optimal solution are satisfied.

The matrix Riccati equation is
$$A^T P + PA - PBR^{-1}B^T P + Q = 0$$

or
$$\begin{bmatrix} 0 & 0 \\ 1 & 0 \end{bmatrix} \begin{bmatrix} p_{11} & p_{12} \\ p_{12} & p_{22} \end{bmatrix} + \begin{bmatrix} p_{11} & p_{12} \\ p_{12} & p_{22} \end{bmatrix} \begin{bmatrix} 0 & 1 \\ 0 & 0 \end{bmatrix}$$
$$- \begin{bmatrix} p_{11} & p_{12} \\ p_{12} & p_{22} \end{bmatrix} \begin{bmatrix} 0 \\ 1 \end{bmatrix} [\tfrac{1}{2}] \begin{bmatrix} 0 & 1 \end{bmatrix} \begin{bmatrix} p_{11} & p_{12} \\ p_{12} & p_{22} \end{bmatrix} + \begin{bmatrix} 2 & 0 \\ 0 & 0 \end{bmatrix} = \begin{bmatrix} 0 & 0 \\ 0 & 0 \end{bmatrix}$$

Upon simplification, we get

$$\frac{-p_{12}^2}{2} + 2 = 0$$

$$p_{11} - \frac{p_{12}p_{22}}{2} = 0$$

$$\frac{-p_{22}^2}{2} + 2p_{12} = 0$$

Solving these three simultaneous equations for $p_{11}, p_{12},$ and p_{22}, requiring **P** to be positive definite, we obtain

$$\mathbf{P} = \begin{bmatrix} 2\sqrt{2} & 2 \\ 2 & 2\sqrt{2} \end{bmatrix}$$

The optimal control law is given by

$$u = -\mathbf{R}^{-1}\mathbf{B}^T\mathbf{P}\tilde{\mathbf{x}}(t) = -[\tfrac{1}{2}] \begin{bmatrix} 0 & 1 \end{bmatrix} \begin{bmatrix} 2\sqrt{2} & 2 \\ 2 & 2\sqrt{2} \end{bmatrix} \begin{bmatrix} \tilde{x}_1 \\ \tilde{x}_2 \end{bmatrix}$$

$$= -\tilde{x}_1(t) - \sqrt{2}\,\tilde{x}_2(t) = -(x_1 - \theta_r) - \sqrt{2}\,x_2$$

It can easily be verified that the closed-loop system is asymptotically stable.

Example 9.7

Consider the liquid-level system of Fig. 9.3. In Example 9.3, we designed an optimal regulator for this process using the Lyapunov equation. In the following, we use the Riccati equation for designing the optimal regulator. The state equation of the process is

$$\frac{dy}{dt} = -y + u \tag{9.55}$$

where

y = deviation of the liquid head from the steady-state
u = rate of liquid inflow

The performance index

$$J = \int_0^\infty (y^2 + u^2)\, dt$$

For this design problem

$$A = -1,\ B = 1,\ Q = 2,\ R = 2$$

The matrix Riccati equation is

$$A^T P + PA - PBR^{-1}B^T P + Q = 0$$

or
$$-P - P - \frac{P^2}{2} + 2 = 0$$

Solving for P, requiring it to be positive definite, we get
$$P = 2(\sqrt{2} - 1)$$

The optimal control law is
$$u = -R^{-1}B^T P y(t) = -(\sqrt{2} - 1)\, y(t)$$

Substituting in Eqn. (9.55), we get the following equation for the closed-loop system:
$$\frac{dy}{dt} = -\sqrt{2}\, y$$

Obviously $y(t) \to 0$ for any initial displacement $y(0)$.

Assume now that a constant disturbance due to the pump enters the system as shown in Fig. 9.10. This type-0 regulator system cannot reject the disturbance; there will be a steady-state offset in the liquid head y. Let us introduce integral control to eliminate this offset.

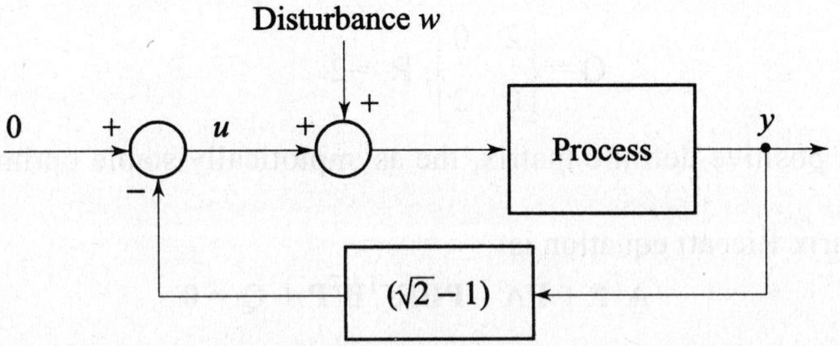

Fig. 9.10 Optimal control of the liquid-level process shown in Fig. 9.3

Defining the integral state z by
$$\dot{z}(t) = y(t),$$
we get the following augmented system:
$$\begin{bmatrix} \dot{y} \\ \dot{z} \end{bmatrix} = \begin{bmatrix} -1 & 0 \\ 1 & 0 \end{bmatrix} \begin{bmatrix} y \\ z \end{bmatrix} + \begin{bmatrix} 1 \\ 0 \end{bmatrix} u \qquad (9.56)$$

Now the design problem is to obtain the control law
$$u = -k_1 y(t) - k_2 z(t)$$
that minimizes
$$J = \int_0^\infty (y^2 + u^2)\, dt$$

The **Q** and **R** matrices are
$$\mathbf{Q} = \begin{bmatrix} 2 & 0 \\ 0 & 0 \end{bmatrix},\ \mathbf{R} = 2$$

The state Eqn. (9.56) is completely controllable, satisfying one of the sufficient conditions for the existence of the optimal solution.

The matrix $\mathbf{Q}$ is positive semidefinite;

$$\mathbf{Q} = \mathbf{H}^T\mathbf{H} = \begin{bmatrix} \sqrt{2} \\ 0 \end{bmatrix}[\sqrt{2} \quad 0]$$

The pair $\left(\begin{bmatrix} -1 & 0 \\ 1 & 0 \end{bmatrix}, [\sqrt{2} \quad 0]\right)$ is not completely observable. Therefore, the other sufficient condition for the existence of the asymptotically stable optimal solution is not satisfied. It can easily be verified that a positive definite solution to the matrix Riccati equation does not exist in this case; the chosen matrix $\mathbf{Q}$ cannot give a closed-loop stable optimal system.

We not modify the performance index to the following:

$$J = \int_0^\infty (y^2 + z^2 + u^2)\,dt$$

The $\mathbf{Q}$ and $\mathbf{R}$ matrices are

$$\mathbf{Q} = \begin{bmatrix} 2 & 0 \\ 0 & 2 \end{bmatrix}, \mathbf{R} = 2$$

Since $\mathbf{Q}$ is positive definite matrix, the asymptotically stable optimal solution exists.

The matrix Riccati equation is

$$\mathbf{A}^T\mathbf{P} + \mathbf{PA} - \mathbf{PBR}^{-1}\mathbf{B}^T\mathbf{P} + \mathbf{Q} = 0$$

or

$$\begin{bmatrix} -1 & 1 \\ 0 & 0 \end{bmatrix}\begin{bmatrix} p_{11} & p_{12} \\ p_{12} & p_{22} \end{bmatrix} + \begin{bmatrix} p_{11} & p_{12} \\ p_{12} & p_{22} \end{bmatrix}\begin{bmatrix} -1 & 0 \\ 1 & 0 \end{bmatrix}$$

$$-\begin{bmatrix} p_{11} & p_{12} \\ p_{12} & p_{22} \end{bmatrix}\begin{bmatrix} 1 \\ 0 \end{bmatrix}[\tfrac{1}{2}][1 \quad 0]\begin{bmatrix} p_{11} & p_{12} \\ p_{12} & p_{22} \end{bmatrix} + \begin{bmatrix} 2 & 0 \\ 0 & 2 \end{bmatrix} = \begin{bmatrix} 0 & 0 \\ 0 & 0 \end{bmatrix}$$

From this equation, we obtain the following three simultaneous equations:

$$-2p_{11} + 2p_{12} - \tfrac{1}{2}p_{11}^2 + 2 = 0$$

$$-p_{12} + p_{22} - \frac{p_{11}p_{12}}{2} = 0$$

$$-\frac{p_{12}^2}{2} + 2 = 0$$

Solving for p_{11}, p_{12} and p_{22}, requiring $\mathbf{P}$ to be positive definite, we obtain

$$\mathbf{P} = \begin{bmatrix} 2 & 2 \\ 2 & 4 \end{bmatrix}$$

The gain matrix

$$\mathbf{K} = [k_1 \quad k_2] = \mathbf{R}^{-1}\mathbf{B}^T\mathbf{P} = [\tfrac{1}{2}] \, [1 \quad 0] \begin{bmatrix} 2 & 2 \\ 2 & 4 \end{bmatrix} = [1 \quad 1]$$

Therefore,

$$u = -y(t) - z(t) = -y(t) - \int_0^\infty y(t)\,dt$$

The block diagram of Fig. 9.11 shows the configuration of the optimal control system employing state-feedback and integral control. It is a Type-1 regulator system.

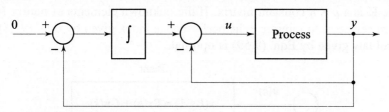

Fig. 9.11 Optimal control system employing state-feedback and integral control

Since at steady-state

$$\dot{z}(t) \to 0,$$

therefore,

$$y(t) \to 0,$$

and there will be no steady-state offset in the liquid head y even in the presence of constant disturbances acting on the plant.

9.6 OPTIMAL DIGITAL CONTROL SYSTEMS

This section covers the key results on the design of optimal controllers for discrete-time systems. Our discussion will be brief because of the strong analogy between the discrete-time and continuous-time cases.

Consider the discretized model of the given plant:

$$\mathbf{x}(k+1) = \mathbf{F}\mathbf{x}(k) + \mathbf{G}\mathbf{u}(k); \quad \mathbf{x}(0) \triangleq \mathbf{x}^0$$
$$\mathbf{y}(k) = \mathbf{C}\mathbf{x}(k) \tag{9.57}$$

where $\mathbf{x}$ is the $n \times 1$ state vector, $\mathbf{u}$ is the $p \times 1$ input vector, $\mathbf{y}$ is the $q \times 1$ output vector; $\mathbf{F}$, $\mathbf{G}$, and $\mathbf{C}$ are, respectively, $n \times n$, $n \times p$, and $q \times n$ real constant matrices; and $k = 0, 1, 2, \ldots$ We will assume that the null state $\mathbf{x} = \mathbf{0}$ is the desired steady-state; $\mathbf{x}(k)$ is thus the system-error vector at $t = kT$ where T is the sampling interval. The state variable model (9.57) is assumed to be completely controllable and observable.

We shall be interested in selecting the controls $\mathbf{u}(k)$; $k = 0, 1, \ldots,$ which minimize a performance index of the form

$$J = \tfrac{1}{2} \sum_{k=0}^{\infty} [\mathbf{x}^T(k)\mathbf{Q}\mathbf{x}(k) + \mathbf{u}^T(k)\mathbf{R}\mathbf{u}(k)] \qquad (9.58)$$

where $\mathbf{Q}$ is an $n \times n$ positive definite, real, symmetric, constant matrix (or a positive semidefinite, real, symmetric, constant matrix with the restriction that the pair $(\mathbf{F},\mathbf{H})$ is observable, where $\mathbf{H}^T\mathbf{H} = \mathbf{Q}$), and $\mathbf{R}$ is a $p \times p$ positive definite, real, symmetric, constant matrix. This criterion is the discrete analog of that given by Eqn. (9.10); a summation replaces integration.

We assume a control configuration of the form shown in Fig. 9.12, described by the state-feedback control law

$$\mathbf{u}(k) = -\mathbf{K}\mathbf{x}(k) \qquad (9.59)$$

where $\mathbf{K}$ is a $p \times n$ constant matrix. If the unknown elements of matrix $\mathbf{K}$ are determined so as to minimize the performance index given by Eqn. (9.58), the control law given by Eqn. (9.59) is optimal.

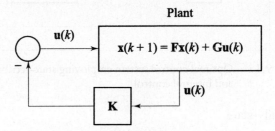

Fig. 9.12 Control configuration of the state regulator for discrete-time systems

The control problem stated above, as we know, is a state regulator design problem. The equations developed below for the state regulator design can also be used for servo design by an appropriate transformation of the state variables (refer Section 7.9 for details). State regulator design equations can also be used for state-feedback control schemes with integral control. This is done by the augmentation of the plant state with integral state and appropriate transformation of the state variables (refer Section 7.9 for details).

In the following, we develop state regulator design equations using the Lyapunov function approach.

With the linear feedback control law (9.59), the closed-loop system is described by

$$\mathbf{x}(k+1) = (\mathbf{F} - \mathbf{GK})\mathbf{x}(k) \qquad (9.60)$$

We will assume that a matrix $\mathbf{K}$ exists such that $(\mathbf{F} - \mathbf{GK})$ is a stable matrix. The controllability of the model (9.57) is sufficient to ensure this if all the elements k_{ij} of matrix $\mathbf{K}$ are available for adjustments.

Substituting for the control vector $\mathbf{u}(k)$ from Eqn. (9.59) in the performance index J given by Eqn. (9.58), we get

$$J = \tfrac{1}{2} \sum_{k=0}^{\infty} \mathbf{x}^T(k)[\mathbf{Q} + \mathbf{K}^T\mathbf{R}\mathbf{K}]\mathbf{x}(k) \qquad (9.61)$$

Let us assume a Lyapunov function

$$V(\mathbf{x}(k)) = \tfrac{1}{2} \sum_{i=k}^{\infty} \mathbf{x}^T(i)[\mathbf{Q} + \mathbf{K}^T\mathbf{R}\mathbf{K}]\mathbf{x}(i) \qquad (9.62)$$

Note that the value of the performance index for system trajectory starting at $\mathbf{x}(0)$ is $V(\mathbf{x}(0))$. The difference

$$V(\mathbf{x}(k+1)) - V(\mathbf{x}(k)) = \Delta V(\mathbf{x}(k)) = -\tfrac{1}{2}\mathbf{x}^T(k)[\mathbf{Q} + \mathbf{K}^T\mathbf{R}\mathbf{K}]\mathbf{x}(k) \qquad (9.63)$$

(Note that $\mathbf{x}(\infty)$ has been taken as zero under the assumption of asymptotic stability of the closed-loop system).

Since $\Delta V(\mathbf{x}(k))$ is quardratic in $\mathbf{x}(k)$ and the plant equation is linear, let us assume that $V(\mathbf{x}(k))$ is also given by the quadratic form:

$$V(\mathbf{x}(k)) = \tfrac{1}{2}\mathbf{x}^T(k)\,\mathbf{P}\mathbf{x}(k) \qquad (9.64)$$

where $\mathbf{P}$ is a positive definite, real, symmetric, constant matrix.
Therefore

$$\Delta V(\mathbf{x}(k)) = \tfrac{1}{2}\mathbf{x}^T(k+1)\mathbf{P}\mathbf{x}(k+1) - \tfrac{1}{2}\mathbf{x}^T(k)\mathbf{P}\mathbf{x}(k)$$

Substituting for $\mathbf{x}(k+1)$ from Eqn. (9.60), we get

$$\Delta V(\mathbf{x}(k)) = \tfrac{1}{2}\mathbf{x}^T(k)\,[(\mathbf{F}-\mathbf{G}\mathbf{K})^T\mathbf{P}(\mathbf{F}-\mathbf{G}\mathbf{K}) - \mathbf{P}]\mathbf{x}(k)$$

Comparing this result with Eqn. (9.63), we obtain

$$(\mathbf{F}-\mathbf{G}\mathbf{K})^T\mathbf{P}(\mathbf{F}-\mathbf{G}\mathbf{K}) - \mathbf{P} + \mathbf{K}^T\mathbf{R}\mathbf{K} + \mathbf{Q} = 0 \qquad (9.65)$$

This equation is of the form of Lyapunov equation defined in Section 8.8.

Since $V(\mathbf{x}(0))$ is the value of the performance index, we have

$$J = \tfrac{1}{2}\mathbf{x}^T(0)\mathbf{P}\mathbf{x}(0) \qquad (9.66)$$

Optimum values of the parameters k_{ij} may be obtained from the equations

$$\frac{\partial[\mathbf{x}^T(0)\mathbf{P}\mathbf{x}(0)]}{\partial k_{ij}} = 0 \qquad (9.67)$$

If the matrix $\mathbf{K}$ thus obtained results in a stable $(\mathbf{F}-\mathbf{G}\mathbf{K})$ matrix, then the minimization of J as per the procedure described above gives the correct result. If $\mathbf{K}$ is unconstrained (all k_{ij} elements are adjustable), then controllability of $(\mathbf{F}, \mathbf{G})$ pair and the positive definiteness of matrix $\mathbf{Q}$ (or the positive semidefiniteness of $\mathbf{Q}$ under the restriction that the pair $(\mathbf{F}, \mathbf{H})$ is observable; $\mathbf{H}^T\mathbf{H} = \mathbf{Q}$) are sufficient conditions for the existence of the asymptotically stable optimal solution. However, if $\mathbf{K}$ is constrained (some of the k_{ij} elements have fixed values, including zero values), then nothing in general can be said about the existence of $\mathbf{K}$ that minimizes J.

We will select simple design examples and solve the parameter optimization problems analytically. Reference [102] describes a numerical algorithm for parameter optimization.

When the feedback matrix **K** is unconstrained, the optimum value of J is independent of initial conditions, and the optimal k_{ij}'s may be obtained from the equations

$$\frac{\partial \mathbf{P}}{\partial k_{ij}} = \mathbf{0} \quad \text{for all } i, j \qquad (9.68)$$

However, this procedure for obtaining the optimal unconstrained feedback matrix **K** gets very much involved for higher-order systems. An alternative route to the solution is through the matrix Riccati equation derived below.

The Lyapunov Eqn. (9.65) can be written as

$$\mathbf{Q} + \mathbf{F}^T\mathbf{PF} - \mathbf{P} + \mathbf{K}^T(\mathbf{R} + \mathbf{G}^T\mathbf{PG})\mathbf{K} - (\mathbf{K}^T\mathbf{G}^T\mathbf{PF} + \mathbf{F}^T\mathbf{PGK}) = \mathbf{0} \qquad (9.69)$$

Since $(\mathbf{R} + \mathbf{G}^T\mathbf{PG})$ is a positive definite matrix, we can write

$$\mathbf{R} + \mathbf{G}^T\mathbf{PG} = \mathbf{\Gamma}^T\mathbf{\Gamma}$$

where $\mathbf{\Gamma}$ is a nonsingular matrix. Then Eqn. (9.69) can be rewritten as

$$\mathbf{Q} + \mathbf{F}^T\mathbf{PF} - \mathbf{P} + [\mathbf{\Gamma K} - (\mathbf{\Gamma}^T)^{-1}\mathbf{G}^T\mathbf{PF}]^T [\mathbf{\Gamma K} - (\mathbf{\Gamma}^T)^{-1}\mathbf{G}^T\mathbf{PF}]$$
$$- \mathbf{F}^T\mathbf{PG}(\mathbf{R} + \mathbf{G}^T\mathbf{PG})^{-1}\mathbf{G}^T\mathbf{PF} = \mathbf{0} \qquad (9.70)$$

The condition (9.68) for unconstrained minimization of J leads to the following equations:

$$\frac{\partial}{\partial k_{ij}}[(\mathbf{\Gamma K} - (\mathbf{\Gamma}^T)^{-1}\mathbf{G}^T\mathbf{PF})^T(\mathbf{\Gamma K} - (\mathbf{\Gamma}^T)^{-1}\mathbf{G}^T\mathbf{PF})] = \mathbf{0}$$

Since the matrix within the brackets is non-negative definite, the minimum occurs when it is zero, or when

$$\mathbf{\Gamma K} = (\mathbf{\Gamma}^T)^{-1}\mathbf{G}^T\mathbf{PF}$$

Hence

$$\mathbf{K} = \mathbf{\Gamma}^{-1}(\mathbf{\Gamma}^T)^{-1}\mathbf{G}^T\mathbf{PF} = (\mathbf{R} + \mathbf{G}^T\mathbf{PG})^{-1}\mathbf{G}^T\mathbf{PF} \qquad (9.71)$$

Equation (9.71) gives the optimal gain matrix **K**. Thus the optimal control law is given by

$$\mathbf{u}(k) = -\mathbf{Kx}(k) = -(\mathbf{R} + \mathbf{G}^T\mathbf{PG})^{-1}\mathbf{G}^T\mathbf{PFx}(k) \qquad (9.72)$$

The matrix **P** in Eqn. (9.72) must satisfy Eqn. (9.70), or the following reduced equation:

$$\mathbf{P} = \mathbf{Q} + \mathbf{F}^T\mathbf{PF} - \mathbf{F}^T\mathbf{PG}(\mathbf{R} + \mathbf{G}^T\mathbf{PG})^{-1}\mathbf{G}^T\mathbf{PF} \qquad (9.73)$$

Equation (9.73) is called the *discrete matrix Riccati equation*.

The discrete matrix Riccati equation given in (9.73) is one of the many equivalent forms which satisfy the optimal regulator design. The analytical solution of the discrete Riccati equation is possible only for very simple cases. Some methods for the numerical solution of the equation are given in reference [102]. A number of computer programs for the solution of the discrete Riccati equation are available [176–179].

Example 9.8

Figure 9.13a illustrates a typical sampled-data system. The transfer functions $G(s)$ and $G_{h0}(s)$ of the controlled plant and the hold circuit, respectively, are known. The data-processing unit $D(z)$ which operates on the sampled error signal $e(k)$ is to be designed.

Assuming the processing unit $D(z)$ to be simply an amplifier of gain K, let us find K so that the sum square error

$$J = \sum_{k=0}^{\infty} e^2(k)$$

is minimized.

From Fig. 9.13a, we have

$$G_{h0}(s)G(s) = \frac{1-e^{-sT}}{s^2}$$

Therefore

$$G_{h0}G(z) = (1-z^{-1})\mathscr{Z}\left[\frac{1}{s^2}\right] = \frac{1}{z-1}$$

Figure 9.13b shows an equivalent block diagram of the sampled-data system. From this figure, we obtain the following state variable model:

$$\begin{aligned} y(k+1) &= y(k) + u(k) \\ u(k) &= -K[y(k) - r] \end{aligned} \qquad (9.74)$$

In terms of the shifted state variable,

$$\tilde{x}(k) = y(k) - r,$$

the state equation becomes

$$\tilde{x}(k+1) = \tilde{x}(k) + u(k) \qquad (9.75)$$

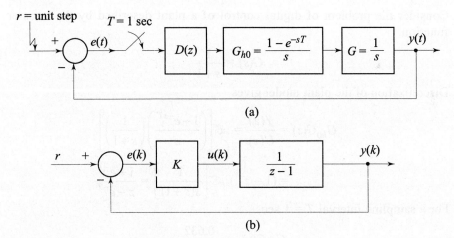

Fig. 9.13 A sampled-data system

The problem is to obtain the optimal control sequence

$$u(k) = -K\tilde{x}(k)$$

that minimizes the performance index

$$J = \sum_{k=0}^{\infty} \tilde{x}^2(k)$$

For this problem,

$$F = 1,\ G = 1,\ Q = 2,\ R = 0$$

The Lyapunov equation is (refer Eqn. (9.65))

$$(F - GK)^T P(F - GK) - P + Q = 0$$

or

$$(1 - K)^2 P - P + 2 = 0$$

This gives

$$P = \frac{2}{2K - K^2}$$

P is positive definite for $0 < K < 2$.

$$J = \tfrac{1}{2}\tilde{x}^T(0) P \tilde{x}(0) = \frac{1}{2K - K^2}[\tilde{x}(0)]^2$$

The necessary condition for J to be minimum is

$$\frac{\partial J}{\partial K} = \frac{-(2 - 2K)}{(2K - K^2)^2}[\tilde{x}(0)]^2 = 0 \quad \text{or} \quad K = 1$$

The sufficient condition $\dfrac{\partial^2 J}{\partial K^2} > 0$ is satisfied for $K = 1$. Therefore $K = 1$ is the optimum value of the amplifier gain; it is independent of $\tilde{x}(0)$.

Example 9.9

Consider the problem of digital control of a plant described by the transfer function

$$G(s) = \frac{1}{s+1}$$

Discretization of the plant model gives

$$G_{h0}G(z) = \frac{Y(z)}{U(z)} = \mathscr{Z}\left[\left(\frac{1 - e^{-sT}}{s}\right)\left(\frac{1}{s+1}\right)\right]$$

$$= (1 - z^{-1})\mathscr{Z}\left[\frac{1}{s(s+1)}\right] = \frac{1 - e^{-T}}{z - e^{-T}}$$

For a sampling interval $T = 1$ sec,

$$G_{h0}G(z) = \frac{0.632}{z - 0.368}$$

Linear Quadratic Optimal Control

The difference equation model of the plant is

$$y(k+1) = 0.368\, y(k) + 0.632\, u(k) \qquad (9.76)$$

The design specifications are given below:

(i) Minimize

$$J = \tfrac{1}{2} \sum_{k=0}^{\infty} [\tilde{y}^2(k) + \tilde{u}^2(k)]$$

where $\tilde{y}$ and $\tilde{u}$ are, respectively, the deviations of the output and the control signal from their steady-state values.

(ii) For a constant y_r, $y(\infty) = y_r$, i.e., there is zero steady-state error.

For this design problem, we select the feedback plus feedforward control scheme shown in Fig. 9.14. The feedback gain K is obtained from the solution of the shifted regulator problem as is seen below.

Let y_s and u_s be the steady-state values of the output and the control signal, respectively. Equation (9.76) at steady-state becomes

$$y_s = 0.368\, y_s + 0.632\, u_s$$

The state Eqn. (9.76) may equivalently be expressed as

$$\tilde{y}(k+1) = 0.368\, \tilde{y}(k) + 0.632\, \tilde{u}(k)$$

where

$$\tilde{y} = y - y_s$$
$$\tilde{u} = u - u_s$$

In terms of this equivalent formulation, the optimal control problem is to obtain

$$\tilde{u}(k) = -K\tilde{y}(k)$$

so that

$$J = \tfrac{1}{2} \sum_{k=0}^{\infty} [\tilde{y}^2(k) + \tilde{u}^2(k)]$$

is minimized.

For this shifted regulator problem,

$$F = 0.368,\ G = 0.632,\ Q = 1,\ R = 1$$

The Riccati Eqn. (9.73) gives

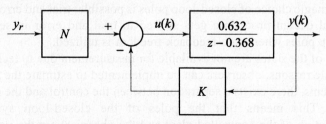

Fig. 9.14 Feedback plus feedforward control scheme

$$P = Q + F^TPF - F^TPG(R + G^TPG)^{-1}G^TPF = 1 + (0.368)^2 P$$
$$- (0.368)P(0.632)[1 + (0.632)^2 P]^{-1}(0.632)P(0.368)$$
$$= 1 + 0.135P - \frac{0.054 P^2}{1 + 0.4 P}$$

Solving for P, requiring it to be positive definite, we obtain,
$$P = 1.11$$

Feedback gain (refer Eqn. (9.71))
$$K = (R + G^TPG)^{-1}G^TPF = 0.18$$

The feedforward gain (refer Eqn. (7.105)) N is given by
$$\frac{1}{N} = -C(F - GK - I)^{-1}G = -[0.368 - 0.632(0.18) - 1]^{-1}(0.632)$$

or
$$N = \frac{0.746}{0.632} = 1.18$$

The optimal control sequence (refer Eqn. (7.106))
$$u(k) = -Ky(k) + Ny_r = -0.18\, y(k) + 1.18\, y_r$$

Substituting in Eqn. (9.76), we obtain.
$$y(k+1) = 0.254\, y(k) + 0.746\, y_r$$

At steady-state,
$$y(\infty) = y_s = \frac{0.746}{0.746} y_r = y_r$$

9.7 CRITIQUE OF LINEAR QUADRATIC CONTROL

In Chapter 7, we observed that the state feedback gives the designer the option of relocating all closed-loop poles. This is in contrast with classical design, whereby the designer can only hope to achieve a pair of complex conjugate poles that are dominant. This is because all other poles and zeros may fall anywhere; meeting the design specifications becomes a matter of trial and error. With the freedom of choice rendered by state feedback comes the responsibility of selecting closed-loop poles judiciously. An obvious temptation is to push the poles too far into the left half plane to get a faster system. This, however, is costlier because a faster system may require a more accurate sensor and a larger or stronger actuator (such as a motor) to perform the task. Also the system bandwidth increases, and the system becomes sensitive to noise. No magic choice of closed-loop poles is possible; trial and error involved in classical design procedure gets shifted to trial and error in selection of closed-loop poles when state-feedback freedom is utilized.

If some of the states are not available for measurement due to technological or economic reasons, observers can be implemented to estimate the states. For linear systems, there exists a separation between the control and the estimation problems. This means that the poles of the closed-loop system (the interconnection of the controlled plant and the observer) are the union of the poles of the controlled plant and the observer. A major limitation of observer-based controllers is the lack of guaranteed stability margins. The methods rely heavily on the plant models. As we rarely have accurate models of our plants, adequate stability margins are required to protect against these model uncertainties. In some situations we may design a controller that works perfectly under computer simulations but turns out to be unstable in practice. Therefore, any design must be tested thoroughly to prevent disastrous results.

For multi-input, multi-output (MIMO) regulators and tracking systems, the assignment of the poles (the eigenvalue spectrum) alone does not lead to a unique feedback matrix. The assignment of the entire eigenstructure, consisting of both the eigenvalues and the eigenvectors, leads to a unique feedback matrix. The method of the entire eigenstructure assignment identifies the subspaces in which the eigenvectors associated with each assigned eigenvalue must be located. The ability to select from many possible eigenvectors for a given eigenvalue provides the means for adjusting the magnitude of each mode which appears in the output. This method of shaping the output provides the potential for achieving the 'best possible' response characteristics. Here again, there exists a gap between theory and practice. Additional techniques need to be developed on the proper guidelines for selection of eigenvectors to satisfy specifications on output response.

In linear quadratic optimal control, the task of the designer shifts from selection of eigenvalues and eigenvectors to one of specifying various parameters in the performance index. A linear quadratic control design behaves quite well from a classical control point of view. It always results in an asymptotically stable system. It can easily be established that the linear quadratic control solution, in the single-input, single-output (SISO) case, has at least a 60 degree phase margin and an infinite gain margin. This is to be compared with the state-feedfack pole-placement technique where stability margins are not known or guaranteed ahead of time. Care must be exercised in using linear quadratic control solution because excessive stability margins may affect the noise suppression properties of the optimal system.

The linear quadratic control requires that all states be available for measurement. If all the states are not available for feedback, an optimal observer (or Kalman filter) can be designed. The combination of the observer and the linear quardratic controller results in an optimal closed-loop system. The design consists of the selection of the state and control weights **Q** and **R**, selection of the corresponding observer-design weights $\mathbf{Q}_0$ and $\mathbf{R}_0$, and solution of the two Reccati equations. This design paradigm, however, fails to meet the main objectives of control-system designers. The major problem with the observer-based optimal design is the lack of robustness. In a series of papers, researchers showed that these designs can become unstable in practice as more realism is added to the plant model. The same kinds of failures were also observed in industrial experiments. It became apparent that too much emphasis on optimality, and not enough attention to the model uncertainty issue, was the main culprit. During the 1980s, much of the attention was shifted back to feedback properties and frequency domain techniques (which are the main features of classical control), and their generalization to multivariable systems.

Loosp transfer recovery (LTR) is a modification of the observer-based linear quadratic design; it allows recovery of the linear quadratic control stability margins. One begins with the selection of the **Q** and **R** parameters of the linear quadratic design until the transfer function of the feedback system with full state feedback (also called the *target feedback loop*) has desirable time and/or frequency domain properties. The recovery step involves iterating on the observer design parameters $\mathbf{Q}_0$ and $\mathbf{R}_0$ so that the resulting loop transfer function is as close as possible to the target feedback loop. The LTR methodology converts the linear quadratic design technique from a rigid time-domain method to a flexible frequency-domain design technique. The computations are still based on state-space techniques [137].

H_∞ *control* is the latest tool for control-system design. It is a computer-aided frequency domain method for design of multivariable systems. The exogenous inputs (disturbances, command inputs, sensor noise) are collected into one vector; the regulated outputs (control signals, errors) are collected into another vector. The objective is to maintain the peak in the closed-loop frequency response of the system below a specified value γ. The optimal solution can be obtained by iterating on γ. The solution involves selecting weights (possibly frequency-dependent weights) and solving two Riccati equations [141]. The name 'H_∞ control' is somewhat unfortunate. H_∞ is one member of the family of spaces introduced by the mathematician Hardy. It is the space of functions on the complex plane that are analytic and bounded in the right-half plane. The space plays an important role in the deeper mathematics needed to solve the control problems.

A common feature of the latest control techniques for control system design is that they are frequency-domain techniques that use state-space machinery for computation.

9.8 REVIEW EXAMPLES

Review Example 9.1

Consider the second-order system of Fig. 9.15. Determine the value of the damping ratio $\zeta > 0$ so that when the system is subjected to a unit-step input r, the following performance index is minimized.

$$J = \int_0^\infty (e^2 + \dot{e}^2) dt$$

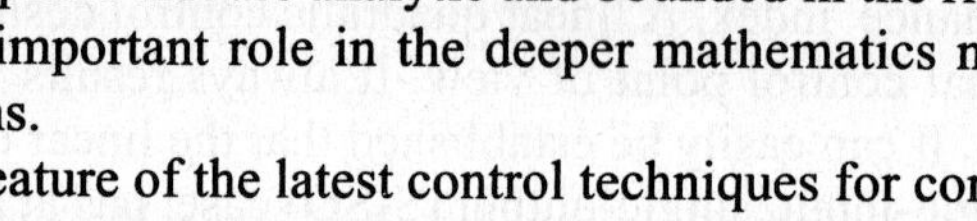

Fig. 9.15

The system is assumed to be at rest initially.

Solution

From Fig. 9.15, we have
$$\frac{Y(s)}{R(s)} = \frac{1}{s^2 + 2\zeta s + 1}$$

or $\quad \ddot{y} + 2\zeta \dot{y} + y = r$

In terms of the error signal e, we obtain
$$\ddot{e} + 2\zeta \dot{e} + e = 0$$
$$e = r - y;\ e(0) = 1,\ \dot{e}(0) = 0$$

Now define the state variables as
$$x_1 = e,\ x_2 = \dot{e}$$

The state equation becomes
$$\dot{\mathbf{x}} = \mathbf{A}\mathbf{x}$$

where $\quad \mathbf{A} = \begin{bmatrix} 0 & 1 \\ -1 & -2\zeta \end{bmatrix}$

The performance index J can be written as
$$J = \int_0^\infty (x_1^2 + x_2^2)\,dt = \tfrac{1}{2}\int_0^\infty \mathbf{x}^T \mathbf{Q}\mathbf{x}\,dt;\ \mathbf{Q} = \begin{bmatrix} 2 & 0 \\ 0 & 2 \end{bmatrix}$$

Since $\mathbf{A}$ is a stable matrix, referring to Eqns (9.31), the value of J can be given by
$$J = \tfrac{1}{2}\mathbf{x}^T(0)\mathbf{P}\mathbf{x}(0)$$

where $\mathbf{P}$ is determined from
$$\mathbf{A}^T \mathbf{P} + \mathbf{P}\mathbf{A} + \mathbf{Q} = \mathbf{0}$$

or $\begin{bmatrix} 0 & -1 \\ 1 & -2\zeta \end{bmatrix}\begin{bmatrix} p_{11} & p_{12} \\ p_{12} & p_{22} \end{bmatrix} + \begin{bmatrix} p_{11} & p_{12} \\ p_{12} & p_{22} \end{bmatrix}\begin{bmatrix} 0 & 1 \\ -1 & -2\zeta \end{bmatrix} + \begin{bmatrix} 2 & 0 \\ 0 & 2 \end{bmatrix} = \begin{bmatrix} 0 & 0 \\ 0 & 0 \end{bmatrix}$

This equation results in the following three equations:
$$-2p_{12} + 2 = 0$$
$$p_{11} - 2\zeta p_{12} - p_{22} = 0$$
$$2p_{12} - 4\zeta p_{22} + 2 = 0$$

Solving for p_{ij}, we obtain
$$\mathbf{P} = \begin{bmatrix} \dfrac{1}{\zeta} + 2\zeta & 1 \\ 1 & \dfrac{1}{\zeta} \end{bmatrix}$$

$\mathbf{P}$ is positive definite for $\zeta > 0$.

$$J = \tfrac{1}{2}\mathbf{x}^T(0)\mathbf{P}\mathbf{x}(0) = \frac{P_{11}}{2} = \zeta + \frac{1}{2\zeta}$$

To minimize J with respect to ζ, we set

$$\frac{\partial J}{\partial \zeta} = 1 - \frac{1}{2\zeta^2} = 0$$

This yields $\zeta = \dfrac{1}{\sqrt{2}}$

Note that $\partial^2 J/\partial \zeta^2 > 0$ for this value of ζ. Thus the optimal value of $\zeta = \dfrac{1}{\sqrt{2}} = 0.707$.

Review Example 9.2

Referring to the block diagram of Fig. 9.16, consider that $G(s) = 100/s^2$ and $R(s) = 1/s$. Determine the optimal values of parameters k_1 and k_2 such that

$$J = \int_0^\infty [e^2(t) + 0.25u^2(t)]dt$$

is minimized.

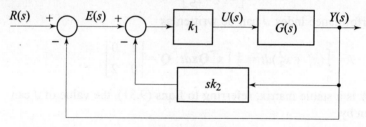

Fig. 9.16

Solution

From Fig. 9.16, we obtain

$$\ddot{y}(t) = 100u(t); \; y(0) = \dot{y}(0) = 0$$
$$u(t) = k_1[r - y(t) - k_2\dot{y}(t)]$$

In terms of the state variables

$$\tilde{x}_1(t) = y(t) - r$$
$$\tilde{x}_2(t) = \dot{y}(t),$$

the state variable model of the system becomes

$$\dot{\tilde{\mathbf{x}}} = \begin{bmatrix} 0 & 1 \\ 0 & 0 \end{bmatrix} \tilde{\mathbf{x}} + \begin{bmatrix} 0 \\ 100 \end{bmatrix} u; \; \tilde{\mathbf{x}}(0) = \begin{bmatrix} -1 \\ 0 \end{bmatrix}$$

$$u = -k_1\tilde{x}_1 - k_1k_2\tilde{x}_2 = -\mathbf{K}\tilde{\mathbf{x}} \tag{9.77}$$

where

$$\mathbf{K} = [k_1 \quad k_1k_2]$$

The optimization problem is to find **K** such that

$$J = \int_0^\infty (\tilde{x}_1^2 + 0.25 u^2) dt$$

is minimized.
Note that

$$\mathbf{A} = \begin{bmatrix} 0 & 1 \\ 0 & 0 \end{bmatrix}; \quad \mathbf{B} = \begin{bmatrix} 0 \\ 100 \end{bmatrix}; \quad \mathbf{Q} = \begin{bmatrix} 2 & 0 \\ 0 & 0 \end{bmatrix}; \quad R = 0.5$$

The Lyapunov equation is

$$(\mathbf{A} - \mathbf{BK})^T \mathbf{P} + \mathbf{P}(\mathbf{A} - \mathbf{BK}) + \mathbf{K}^T R \mathbf{K} + \mathbf{Q} = 0$$

or

$$\begin{bmatrix} 0 & -100k_1 \\ 1 & -100k_1k_2 \end{bmatrix} \begin{bmatrix} p_{11} & p_{12} \\ p_{12} & p_{22} \end{bmatrix} + \begin{bmatrix} p_{11} & p_{12} \\ p_{12} & p_{22} \end{bmatrix} \begin{bmatrix} 0 & 1 \\ -100k_1 & -100k_1k_2 \end{bmatrix}$$

$$+ 0.5 \begin{bmatrix} k_1^2 & k_1^2 k_2 \\ k_1^2 k_2 & k_1^2 k_2^2 \end{bmatrix} + \begin{bmatrix} 2 & 0 \\ 0 & 0 \end{bmatrix} = \begin{bmatrix} 0 & 0 \\ 0 & 0 \end{bmatrix}$$

Solving, we get

$$-200 k_1 p_{12} + 2 + 0.5 k_1^2 = 0$$

$$-100 k_1 p_{22} + p_{11} - 100 k_1 k_2 p_{12} + 0.5 k_1^2 k_2 = 0$$

$$2 p_{12} - 200 k_1 k_2 p_{22} + 0.5 k_1^2 k_2^2 = 0$$

$$p_{12} = \frac{2 + 0.5 k_1^2}{200 k_1}$$

$$p_{22} = \frac{1}{100 k_1 k_2} \left[\frac{2 + 0.5 k_1^2}{200 k_1} + 0.25 k_1^2 k_2^2 \right]$$

$$p_{11} = \frac{2 + 0.5 k_1^2}{200 k_1 k_2} + k_2$$

$$J = \tfrac{1}{2} \tilde{\mathbf{x}}^T(0) \mathbf{P} \tilde{\mathbf{x}}(0)$$

$$= \tfrac{1}{2} [-1 \quad 0] \begin{bmatrix} p_{11} & p_{12} \\ p_{12} & p_{22} \end{bmatrix} \begin{bmatrix} -1 \\ 0 \end{bmatrix}$$

$$= \tfrac{1}{2} p_{11}$$

$$= \frac{1 + 100 k_1 k_2^2 + 0.25 k_1^2}{200 k_1 k_2}$$

$$\frac{\partial J}{\partial k_i} = 0 \text{ for } i = 1, 2, \text{ gives}$$

$$0.25 k_1^2 = 1$$

$$100 k_1 k_2^2 - 1 - 0.25 k_1^2 = 0$$

Let us choose
$$k_1 = 2, \; k_2 = 0.1$$
as the solution. This gives positive definite **P**.
The Hessian matrix

$$\mathbf{H} = \begin{bmatrix} \dfrac{\partial^2 J}{\partial k_1^2} & \dfrac{\partial^2 J}{\partial k_1 \partial k_2} \\ \dfrac{\partial^2 J}{\partial k_1 \partial k_2} & \dfrac{\partial^2 J}{\partial k_2^2} \end{bmatrix} = \begin{bmatrix} \dfrac{1}{100 k_1^3 k_2} & \dfrac{1 - 0.25 k_1^2}{200 k_1^2 k_2^2} \\ \dfrac{1 - 0.25 k_1^2}{200 k_1^2 k_2^2} & \dfrac{1 + 0.25 k_1^2}{100 k_1 k_2^3} \end{bmatrix}$$

For $k_1 = 2, \; k_2 = 0.1$,

$$\mathbf{H} = \begin{bmatrix} \tfrac{1}{80} & 0 \\ 0 & 10 \end{bmatrix} \text{ is positive definite}$$

Therefore $k_1 = 2, \; k_2 = 0.1$ satisfy the necessary and sufficient conditions for J to be minimum. For these values of gain, the matrix

$$(\mathbf{A} - \mathbf{BK}) = \begin{bmatrix} 0 & 1 \\ -200 & -20 \end{bmatrix}$$

The eigenvalues of this matrix are in the left half of complex plane and therefore the closed-loop system is asymptotically stable.

We have determined k_1 and k_2 for a specific set of initial conditions. Since no constraints have been imposed on the elements of the gain matrix, we expect that the optimum values of k_1 and k_2 are independent of initial conditions. This, in fact, is true as will be seen in the next example.

Review Example 9.3

Reconsider the system of Fig. 9.16. Determine the optimal values of the parameters k_1 and k_2 through the Riccati equation.

Solution

The matrix Riccati equation is
$$\mathbf{A}^T \mathbf{P} + \mathbf{PA} - \mathbf{PBR}^{-1} \mathbf{B}^T \mathbf{P} + \mathbf{Q} = \mathbf{0}$$
or

$$\begin{bmatrix} 0 & 0 \\ 1 & 0 \end{bmatrix} \begin{bmatrix} p_{11} & p_{12} \\ p_{12} & p_{22} \end{bmatrix} + \begin{bmatrix} p_{11} & p_{12} \\ p_{12} & p_{22} \end{bmatrix} \begin{bmatrix} 0 & 1 \\ 0 & 0 \end{bmatrix}$$

$$- \begin{bmatrix} p_{11} & p_{12} \\ p_{12} & p_{22} \end{bmatrix} \begin{bmatrix} 0 \\ 100 \end{bmatrix} [\tfrac{1}{0.5}] \, [0 \; \; 100] \begin{bmatrix} p_{11} & p_{12} \\ p_{12} & p_{22} \end{bmatrix} + \begin{bmatrix} 2 & 0 \\ 0 & 0 \end{bmatrix} = \begin{bmatrix} 0 & 0 \\ 0 & 0 \end{bmatrix}$$

Solving for p_{11}, p_{12}, and p_{22}, requiring **P** to be positive definite, we obtain

$$\mathbf{P} = \begin{bmatrix} 2 \times 10^{-1} & 10^{-2} \\ 10^{-2} & 10^{-3} \end{bmatrix}$$

The feedback gain matrix
$$K = R^{-1}B^T P$$
$$= \tfrac{1}{0.5}[0 \quad 100]\begin{bmatrix} 2\times 10^{-1} & 10^{-2} \\ 10^{-2} & 10^{-3} \end{bmatrix} = [2 \quad 0.2]$$

From Eqn. (9.77), we obtain
$$[k_1 \quad k_1 k_2] = [2 \quad 0.2] \quad \text{or} \quad k_1 = 2,\ k_2 = 0.1$$

As expected, this result is the same as obtained by the Lyapunov equation approach in the previous example.

Review Example 9.4

Figure 9.17 shows the optimal control configuration of a position servo system. Both the state variables—angular position θ and angular velocity $\dot{\theta}$, are assumed to be measurable.

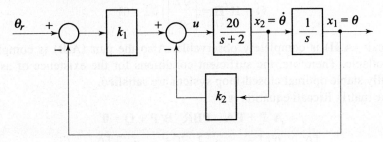

Fig. 9.17 A position servo system

It is desired to regulate the angular position to a unit-step function θ_r. Find the optimum values of the gains k_1 and k_2 that minimize
$$J = \int_0^\infty [(x_1 - \theta_r)^2 + u^2]dt$$

Solution

The state variable description of the system, obtained from Fig. 9.17, is given by
$$\dot{x} = \begin{bmatrix} 0 & 1 \\ 0 & -2 \end{bmatrix} x + \begin{bmatrix} 0 \\ 20 \end{bmatrix} u$$
$$y = x_1$$

In terms of the shifted state variables
$$\tilde{x}_1 = x_1 - \theta_r$$
$$\tilde{x}_2 = x_2,$$

the state variable model becomes
$$\dot{\tilde{x}} = A\tilde{x} + Bu \tag{9.78}$$

with
$$\mathbf{A} = \begin{bmatrix} 0 & 1 \\ 0 & -2 \end{bmatrix},\ \mathbf{B} = \begin{bmatrix} 0 \\ 20 \end{bmatrix}$$

The design problem is to determine optimal control
$$u = -k_1 \tilde{x}_1 - k_2 \tilde{x}_2 \qquad (9.79)$$
that minimizes
$$J = \int_0^\infty (\tilde{x}_1^2 + u^2)\, dt$$

For this J,
$$\mathbf{Q} = \begin{bmatrix} 2 & 0 \\ 0 & 0 \end{bmatrix};\ R = 2$$

The matrix $\mathbf{Q}$ is positive semidefinite.
$$\mathbf{Q} = \mathbf{H}^T \mathbf{H} = \begin{bmatrix} \sqrt{2} \\ 0 \end{bmatrix} [\sqrt{2}\ \ 0]$$

The pair $(\mathbf{A},\mathbf{H})$ is completely observable. Also the pair $(\mathbf{A},\mathbf{B})$ is completely controllable. Therefore, the sufficient conditions for the existence of asymptotically stable optimal closed-loop system are satisfied.

The matrix Riccati equation is
$$\mathbf{A}^T \mathbf{P} + \mathbf{P}\mathbf{A} - \mathbf{P}\mathbf{B}\mathbf{R}^{-1}\mathbf{B}^T\mathbf{P} + \mathbf{Q} = 0$$

or
$$\begin{bmatrix} 0 & 0 \\ 1 & -2 \end{bmatrix} \begin{bmatrix} p_{11} & p_{12} \\ p_{12} & p_{22} \end{bmatrix} + \begin{bmatrix} p_{11} & p_{12} \\ p_{12} & p_{22} \end{bmatrix} \begin{bmatrix} 0 & 1 \\ 0 & -2 \end{bmatrix}$$
$$- \begin{bmatrix} p_{11} & p_{12} \\ p_{12} & p_{22} \end{bmatrix} \begin{bmatrix} 0 \\ 20 \end{bmatrix} [\tfrac{1}{2}] [0\ \ 20] \begin{bmatrix} p_{11} & p_{12} \\ p_{12} & p_{22} \end{bmatrix} + \begin{bmatrix} 2 & 0 \\ 0 & 0 \end{bmatrix} = \begin{bmatrix} 0 & 0 \\ 0 & 0 \end{bmatrix}$$

or
$$\begin{bmatrix} 0 & p_{11} - 2p_{12} \\ p_{11} - 2p_{12} & 2p_{12} - 4p_{22} \end{bmatrix} - \begin{bmatrix} 200 p_{12}^2 & 200 p_{12} p_{22} \\ 200 p_{12} p_{22} & 200 p_{22}^2 \end{bmatrix}$$
$$+ \begin{bmatrix} 2 & 0 \\ 0 & 0 \end{bmatrix} = \begin{bmatrix} 0 & 0 \\ 0 & 0 \end{bmatrix}$$

or
$$2 - 200\, p_{12}^2 = 0$$
$$2p_{12} - 4p_{22} - 200\, p_{22}^2 = 0$$
$$p_{11} - 2p_{12} - 200 p_{12} p_{22} = 0$$

The positive definite solution of the Riccati equation is
$$\mathbf{P} = \begin{bmatrix} 0.664 & 0.1 \\ 0.1 & 0.0232 \end{bmatrix}$$

Linear Quadratic Optimal Control

The optimal gain matrix
$$\mathbf{K} = \mathbf{R}^{-1}\mathbf{B}^T\mathbf{P} = [1 \quad 0.232]$$
The minimum value of J for an initial condition $\tilde{\mathbf{x}}(0) = [-1 \quad 0]^T$, is
$$J = \tfrac{1}{2}\tilde{\mathbf{x}}^T(0)\mathbf{P}\tilde{\mathbf{x}}(0) = \frac{p_{11}}{2} = 0.332$$
It can easily be verified that the optimal closed-loop system is stable.

Review Example 9.5

Reconsider the position servo system of Fig. 9.17. Assume that only the angular position θ is available for feedback. It is desired to regulate the angular position to a unit-step function θ_r. Find the optimum values of the free parameters so that

$$J = \int_0^\infty [(x_1 - \theta_r)^2 + u^2]\, dt$$

is minimized.

Solution

The constraint of partial state feedback can be met by setting
$$k_2 = 0$$
in the optimization problem.

For the problem formulation given by Eqns (9.78)–(9.79), the Lyapunov equation is

$$(\mathbf{A} - \mathbf{B}\mathbf{K})^T\mathbf{P} + \mathbf{P}(\mathbf{A} - \mathbf{B}\mathbf{K}) + \mathbf{K}^T\mathbf{R}\mathbf{K} + \mathbf{Q} = \mathbf{0}$$

or

$$\begin{bmatrix} 0 & -20k_1 \\ 1 & -2 \end{bmatrix}\begin{bmatrix} p_{11} & p_{12} \\ p_{12} & p_{22} \end{bmatrix} + \begin{bmatrix} p_{11} & p_{12} \\ p_{12} & p_{22} \end{bmatrix}\begin{bmatrix} 0 & 1 \\ -20k_1 & -2 \end{bmatrix}$$
$$+ \begin{bmatrix} 2k_1^2 & 0 \\ 0 & 0 \end{bmatrix} + \begin{bmatrix} 2 & 0 \\ 0 & 0 \end{bmatrix} = \begin{bmatrix} 0 & 0 \\ 0 & 0 \end{bmatrix}$$

or

$$\begin{bmatrix} -40k_1p_{12} & p_{11} - 2p_{12} - 20k_1p_{22} \\ p_{11} - 2p_{12} - 20k_1p_{22} & 2p_{12} - 4p_{22} \end{bmatrix} + \begin{bmatrix} 2k_1^2 & 0 \\ 0 & 0 \end{bmatrix}$$
$$+ \begin{bmatrix} 2 & 0 \\ 0 & 0 \end{bmatrix} = \begin{bmatrix} 0 & 0 \\ 0 & 0 \end{bmatrix}$$

Solving for $\mathbf{P}$, we get

$$\mathbf{P} = \begin{bmatrix} \dfrac{(1+5k_1)(1+k_1^2)}{10k_1} & \dfrac{1+k_1^2}{20k_1} \\ \dfrac{1+k_1^2}{20k_1} & \dfrac{1+k_1^2}{40k_1} \end{bmatrix}$$

$\mathbf{P}$ is positive definite for $k_1 > 0$.

For the initial condition $\tilde{\mathbf{x}}(0) = [-1 \ \ 0]^T$, the performance index J has the value

$$J = \tfrac{1}{2}\tilde{\mathbf{x}}^T(0)\mathbf{P}\tilde{\mathbf{x}}(0) = \frac{p_{11}}{2} = \frac{5k_1^3 + k_1^2 + 5k_1 + 1}{20k_1}$$

$$\frac{\partial J}{\partial k_1} = 0 \text{ gives } k_1 = 0.43$$

For this value of k_1, $\dfrac{\partial^2 J}{\partial k_1^2} > 0$; and the matrix $(\mathbf{A} - \mathbf{BK})$ is a stable matrix.

$$J_{\min} = 0.434$$

It is observed that optimal cost $J_{\min}$ for constrained optimization (suboptimal control with partial state feedback) is only slightly higher than the optimal cost J_0 for unconstrained optimization (optimal control with full state feedback) determined in the previous example.

$J_{\min}$ corresponding to suboptimal control = 0.434
J_0 corresponding to optimal control = 0.332

Review Example 9.6

Consider the sampled-data system shown in Fig. 9.13. The processing unit $D(z)$ is simply a gain K.

Find the optimal value of K so that

$$J = \sum_{k=0}^{\infty} [e^2(k) + 0.75u^2(k)]$$

is minimized.

Solution

From Fig. 9.13, we obtain the following state variable model:

$$y(k+1) = y(k) + u(k)$$
$$u(k) = -K[y(k) - r]$$

In terms of the shifted state variable

$$\tilde{x}(k) = y(k) - r$$

the state equation becomes

$$\tilde{x}(k+1) = \tilde{x}(k) + u(k)$$

The problem is to obtain optimal control sequence

$$u(k) = -K\tilde{x}(k)$$

that minimizes the performance index

$$J = \sum_{k=0}^{\infty} [\tilde{x}^2(k) + 0.75u^2(k)]$$

For this problem,

$$F = 1, \ G = 1, \ Q = 2, \ R = 1.5$$

The Riccati equation is (refer Eqn. (9.73))

$$P = Q + F^T PF - F^T PG(R + G^T PG)^{-1} G^T PF$$

$$= 2 + P - \frac{P^2}{1.5 + P}$$

Solving for P, requiring it to be positive definite, we get

$$P = 3$$

The optimal control (refer Eqn. (9.72))

$$u(k) = -K\tilde{x}(k)$$

where

$$K = (R + G^T PG)^{-1} G^T PF = \frac{P}{1.5 + P} = \frac{2}{3}.$$

PROBLEMS

9.1 Consider the second-order system of Fig. P9.1. Determine the value of the damping ratio $\zeta > 0$ which minimizes the integral square error

$$J = \int_0^\infty e^2(t) \, dt$$

for the initial conditions $y(0) = 1$, $\dot{y}(0) = 0$. What is the minimum value of the performance index?

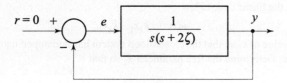

Fig. P9.1

9.2 A unity feedback system has the forward path transfer function

$$G(s) = \frac{K}{s(s + \alpha)}$$

The reference input is a unit-step function, and the system is at rest initially.
  (i) Taking K as constant, determine the value of α which minimizes the integral square error.
  (ii) Taking α as constant, determine the value of K which minimizes the integral square error.

9.3 Consider a system described by the vector differential equation

$$\dot{\mathbf{x}} = \begin{bmatrix} 0 & 1 \\ 0 & 0 \end{bmatrix} \mathbf{x} + \begin{bmatrix} 0 \\ 1 \end{bmatrix} u$$

The control signal is set as $u(t) = -x_1 - kx_2$. Design gain k so that the performance index

$$J = \tfrac{1}{2}\int_0^\infty \mathbf{x}^T\mathbf{x}\, dt$$

is minimized for $\mathbf{x}(0) = [1 \quad 1]^T$. Evaluate the minimum value of the performance index. Determine the sensitivity of the performance to a change in k.

9.4 Referring to the block diagram of Fig. P9.4, consider that $G(s) = 100/s^2$; the reference input is a unit-step function and the system is at rest initially. Determine the optimum value of the parameter K such that

$$J = \int_0^\infty e^2(t)\, dt$$

is minimized. Determine the sensitivity of J to a change in K.

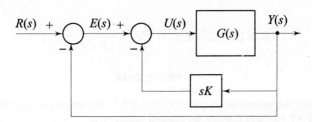

Fig. P9.4

9.5 Consider the system

$$\dot{\mathbf{x}} = \begin{bmatrix} 0 & 1 \\ 0 & 0 \end{bmatrix}\mathbf{x} + \begin{bmatrix} 0 \\ 1 \end{bmatrix}u$$

Assume the linear control law

$$u(t) = -k_1 x_1 - k_1 x_2$$

Fix the value of k_1 so that the closed-loop system has undamped natural frequency of 2 rad/sec. Determine the free parameter k_2 so that

$$J = \tfrac{1}{2}\int_0^\infty (x_1^2 + x_2^2)\, dt$$

is minimized for $\mathbf{x}(0) = [1 \quad 0]^T$. Find the minimum value of the performance index.

9.6 Consider the system

$$\dot{\mathbf{x}} = \begin{bmatrix} 0 & 1 \\ 0 & 0 \end{bmatrix}\mathbf{x} + \begin{bmatrix} 0 \\ 1 \end{bmatrix}u$$

Find the value of k such that

$$u(t) = -k(x_1 + x_2)$$

minimizes

$$J = \int_0^\infty \mathbf{x}^T\mathbf{x}\, dt$$

for $\mathbf{x}(0) = [1 \quad 0]^T$

9.7 Reconsider the system of Problem 9.6. The performance index is now set as

$$J = \int_0^\infty (x_1^2 + x_2^2 + u^2)\,dt$$

Design the control $u(t) = -k(x_1 + x_2)$ so that J is minimized for $\mathbf{x}(0) = [1 \quad 0]^T$. Determine the sensitivity of the performance to a change in k.

9.8 Consider the system shown in Fig. P9.8. Determine the optimal feedback gain matrix $\mathbf{K}$ such that the following performance index is minimized:

$$J = \frac{1}{2}\int_0^\infty (\mathbf{x}^T\mathbf{Q}\mathbf{x} + 2u^2)\,dt;\ \mathbf{Q} = \begin{bmatrix} 2 & 0 \\ 0 & 2 \end{bmatrix}$$

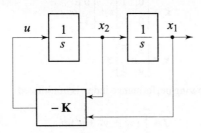

Fig. P9.8

9.9 The matrix $\mathbf{Q}$ in Problem 9.8 is replaced by the following positive semi-definite matrix:

$$\mathbf{Q} = \begin{bmatrix} 2 & 0 \\ 0 & 0 \end{bmatrix}$$

Show that sufficient conditions for the existence of the asymptotically stable optimal control solution are satisfied. Find the optimal feedback matrix $\mathbf{K}$.

9.10 Test whether sufficient conditions for the existence of the asymptotically stable optimal control solution for the plant

$$\dot{\mathbf{x}} = \begin{bmatrix} 0 & 0 \\ 0 & 1 \end{bmatrix}\mathbf{x} + \begin{bmatrix} 1 \\ 1 \end{bmatrix}u$$

with the performance index

$$J = \int_0^\infty (x_1^2 + u^2)\,dt$$

are satisfied. Find the optimal closed-loop system and determine its stability.

9.11 Consider the plant

$$\dot{\mathbf{x}} = \begin{bmatrix} -1 & 0 \\ 1 & 0 \end{bmatrix}\mathbf{x} + \begin{bmatrix} 1 \\ 0 \end{bmatrix}u$$

with the performance index

$$J = \int_0^\infty (x_1^2 + u^2)\,dt$$

Test whether an asymptotically stable optimal solution exists for this control problem.

9.12 Consider the system described by the state model:

$$\dot{\mathbf{x}} = \begin{bmatrix} 0 & 1 \\ 0 & -2 \end{bmatrix} \mathbf{x} + \begin{bmatrix} 0 \\ 20 \end{bmatrix} u$$

$$y = \begin{bmatrix} 1 & 0 \end{bmatrix} \mathbf{x}$$

Find the optimal control law that minimizes

$$J = \tfrac{1}{2} \int_0^\infty [(y(t) - 1)^2 + u^2] \, dt$$

9.13 Determine the optimal control law for the system

$$\dot{\mathbf{x}} = \begin{bmatrix} 0 & 1 \\ 0 & 0 \end{bmatrix} \mathbf{x} + \begin{bmatrix} 0 \\ 1 \end{bmatrix} u$$

$$\mathbf{y} = \begin{bmatrix} 1 & 0 \\ 0 & 2 \end{bmatrix} \mathbf{x}$$

such that the following performance index is minimized.

$$J = \int_0^\infty (y_1^2 + y_2^2 + u^2) \, dt$$

9.14 Consider the plant

$$\dot{\mathbf{x}} = \mathbf{A}\mathbf{x} + \mathbf{B}u$$
$$y = \mathbf{C}\mathbf{x}$$

$$\mathbf{A} = \begin{bmatrix} 0 & 1 \\ 0 & -1 \end{bmatrix}, \mathbf{B} = \begin{bmatrix} 0 \\ 1 \end{bmatrix}, \mathbf{C} = \begin{bmatrix} 1 & 0 \end{bmatrix}$$

with the performance index

$$J = \int_0^\infty (x_1^2 + x_2^2 + u^2) \, dt$$

Choose a control law that minimizes J. Design a state observer for implementation of the control law; both the poles of the state observer are required to lie at $s = -3$.

9.15 Figure P9.15 shows the optimal control configuration of a position servo system. Both the state variables—angular position θ and angular velocity $\dot{\theta}$, are assumed to be measurable.
It is desired to regulate the angular position to a constant value $\theta_r = 5$. Find optimum values of the gains k_1 and k_2 that minimize

$$J = \int_0^\infty [(x_1 - \theta_r)^2 + \tfrac{1}{2} u^2] \, dt$$

What is the minimum value of J?

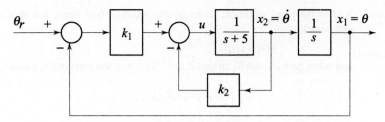

Fig. P9.15

9.16 Consider now that for the position servo system of Problem 9.15, the performance index is

$$J = \int_0^\infty [(x_1 - \theta_r)^2 + \rho u^2] dt$$

For $\rho = \frac{1}{10}, \frac{1}{100}$, and $\frac{1}{1000}$, find the optimal control law that minimizes the given J. Determine closed-loop poles for various values of ρ and comment on your result.

9.17 Reconsider the position servo system of Fig. P9.15. Assume that only the angular position θ is available for feedback.

It is desired to regulate the angular position to a constant value $\theta_r = 5$. Find optimum values of the free parameters so that

$$J = \int_0^\infty [(x_1 - \theta_r)^2 + \tfrac{1}{2} u^2] dt$$

is minimized. What is the minimum value of J?

9.18 In the control scheme of Fig. P9.18, the control law of the form $u = -Ky + Ny_r$ has been used; y_r is the constant command input.

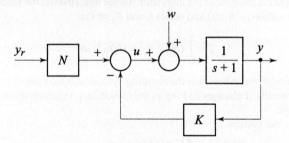

Fig. P9.18

(a) Find K such that

$$J = \int_0^\infty (\tilde{y}^2 + \tilde{u}^2) dt$$

is minimized; $\tilde{y}$ and $\tilde{u}$ are, respectively, the deviations of the output and the control signal from their steady-state values.

(b) Choose N so that the system has zero steady-state error, i.e., $y(\infty) = y_r$.

(c) Show that the steady-state error to a constant disturbance input w is non-zero for the above choice of N.

(d) Add to the plant equation, an integrator equation ($z(t)$ being the state of the integrator);
$$\dot{z}(t) = y(t) - y_r$$
and select gains K and K_1 so that if $u = -Ky - K_1 z$, the performance index
$$J = \int_0^\infty (\tilde{y}^2 + \tilde{z}^2 + \tilde{u}^2) dt$$
is minimized.

(e) Draw a block diagram of the control scheme employing integral control and show that the steady-state error to constant disturbance w is zero.

9.19 Consider a plant consisting of a dc motor, the shaft of which has the angular velocity $\omega(t)$ and which is driven by the input voltage $u(t)$. The describing equation is
$$\dot{\omega}(t) = -0.5\omega(t) + 100u(t) = A\omega(t) + Bu(t)$$
It is desired to regulate the angular velocity at the desired value $\omega^0 = r$.

(a) Use the control law of the form $u(t) = -K\omega(t) + Nr$.

Choose K that minimizes $J = \int_0^\infty (\tilde{\omega}^2 + 100\tilde{u}^2) dt$; $\tilde{\omega}$ and $\tilde{u}$ are, respectively, the deviations of the output and the control signal from their steady-state values. Choose N that guarantees zero steady-state error, i.e.,
$$\omega(\infty) = \omega^0 = r.$$

(b) Show that if A changes to $A + \delta A$, subject to $(A + \delta A - BK)$ being stable, then the above choice of N will no longer make $\omega(\infty) = r$. Therefore the system is not robust under changes in system parameters.

(c) The system can be made robust by augmenting it with an integrator:
$$\dot{z} = \omega - r$$
where z is the state of the integrator. To see this, first use the feedback of the form $u = -K\omega(t) - K_1 z(t)$ and select K and K_1 so that
$$J = \int_0^\infty (\tilde{\omega}^2 + \tilde{z}^2 + 100\tilde{u}^2) dt$$
is minimized. Show that the resulting system will have $\omega(\infty) = r$ no matter how the matrix A changes so long as the closed-loop system remains asymptotically stable.

9.20 Consider the system
$$x(k+1) = x(k) + u(k)$$
Using the Lyapunov equation, find the control sequence
$$u(k) = -Kx(k)$$
that minimizes the performance index
$$J = \sum_{k=0}^{\infty} [x^2(k) + 0.75u^2(k)]$$
Also determine the sensitivity of J to a change in K for $x(0) = 1$.

9.21 Consider the system
$$x(k+1) = 0.368\, x(k) + 0.632\, u(k)$$

Using the discrete matrix Riccati equation, find the control sequence
$$u(k) = -Kx(k)$$
that minimizes the performance index
$$J = \sum_{k=0}^{\infty} [x^2(k) + u^2(k)]$$

9.22 Consider the sampled-date system shown in Fig. P9.22.
  (a) Find K so that
$$J = \tfrac{1}{2} \sum_{k=0}^{\infty} [\tilde{y}^2(k) + \tilde{u}^2(k)]$$
   is minimized; $\tilde{y}$ and $\tilde{u}$ are, respectively, the deviations of the output and the control signal from their steady-state values.
  (b) Find the steady-state value of the output.
  (c) To eliminate steady-state error, introduce a feedforward controller. The control scheme now becomes $u(k) = -Ky(k) + Nr$. Find the value of N so that $y(\infty) = r$.

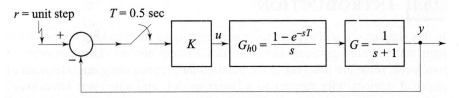

Fig. P9.22

9.23 A plant is described by the state equation
$$x(k+1) = 0.5x(k) + 2u(k) = Fx(k) + Gu(k)$$
  (a) Find K such that if $u(k) = -Kx(k) + Nr$, the performance index
$$J = \tfrac{1}{2} \sum_{k=0}^{\infty} [\tilde{x}^2(k) + \tilde{u}^2(k)]$$
   is minimized; r is a constant reference input, and $\tilde{x}$ and $\tilde{u}$ are, respectively, the deviations in state and control signal from their steady-state values.
  (b) Find N so that $x(\infty) = r$, i.e., there is no steady-state error.
  (c) Show that the property of zero steady-state error is not robust with respect to changes in F.
  (d) In order to obtain robust steady-state accuracy with respect to changes in F, we may use integral control in addition to state feedback. Describe through block diagram, the structure of such a control scheme.

chapter 10
Nonlinear Control Systems

10.1 INTRODUCTION

In this chapter, we will explore tools and techniques for attacking control problems that contain significant nonlinearities. System nonlinearities can occur in two ways: some are *inherent* in the plant model, representing imperfections of physical devices with respect to a linear model; and some are *intentional*, which are deliberately inserted into a loop to improve performance. We will examine control design in the presence of inherent nonlinearities, as well as introduce nonlinearities into control designs to improve performance. Unlike linear systems, no general techniques are available that apply for all nonlinear systems; each nonlinear system must generally be considered as a separate design problem. A multitude of techniques have been developed to facilitate the solution to a particular problem. An extensive literature exists on the subject [116–112]. This chapter is intended to introduce some of these techniques.

Analysis Techniques

System design contains two phases. The *design* phase encompasses finding the best structure and making trade offs between the goals and objectives for the complete system. The *analysis* phase includes examination of a fixed structure to determine such properties as signal sizes, stability and dynamic response.

Simulation

Perhaps the single most valuable asset to the field of engineering is the simulation tool—constructing a model of the proposed or actual system and using a numerical solution of the model to reveal the behaviour of the system. Simulation is the only general method of analysis applicable to finding solutions of linear and nonlinear differential and difference equations. Of course, simulation finds *specific* solutions; that is, solutions to the equations with specific inputs, initial conditions, and parametric conditions. It is for this reason that simulation is not a substitute for other forms of analysis. Important properties such as stability and conditional stability are not *proven* with simulations. When the complexity of a system precludes the use of any analytical approach to establish proof of stability, simulations will be the only way to obtain necessary information for design purposes.

A partial list of the simulation programs available today is contained in references [174–179].

Lyapunov stability analysis

The Lyapunov's method of stability analysis (refer Chapter 8) is, in principle, the most general method of determination of stability of nonlinear systems. The intuitive idea behind Lyapunov's method is that a physical system can store only finite energy, and thus if we can show that energy is *always* being dissipated except at the equilibrium point, then the system must finally reach equilibrium when the energy is gone. The mathematical representation of system 'energy' is in the Lyapunov function. The major drawback which seriously limits the use of Lyapunov's method of stability analysis is the difficulty often associated with the construction of the *Lyapunov function* required by the method (refer Section 8.5).

Describing function analysis

For the so-called separable systems, which comprise a linear part defined by its transfer function, and a nonlinear part defined by a time-independent relationship between its input and output variables, the describing function method is most practically useful. The method usually gives sufficiently accurate information about stability and limit cycles (A periodic oscillation in a linear time-invariant system is sinusoidal with the amplitude of oscillation a function of both the system excitation and the initial conditions. Slight changes in system parameters will shift the system poles from imaginary axis of the complex plane and destroy the oscillations. Nonlinear systems may exhibit periodic oscillations that are independent of the size of applied excitation and are usually much less sensitive to parameter variations. A periodic oscillation in a nonlinear system is called a *limit cycle*. In general, limit cycles are nonsinusoidal. Limit cycles of fixed amplitude and period can be sustained over a finite range of system parameters).

Design Techniques

If the system is only mildly nonlinear, the simplest approach might be to ignore the nonlinearity in designing the controller (i.e., to omit the nonlinearity in the design model), but to include its effect in evaluating the system performance. The inherent robustness of the control law designed for the approximating nonlinear system is relied upon to carry it over to the nonlinear system.

Linearization based on first-order approximation

When a system is significantly nonlinear, it is traditionally dealt with by linearization (refer Eqns (5.11)) about a selected operating point using Taylor series. We design a linear controller based on first-order approximation. If the controller works effectively, the perturbations in actual state about the equilibrium state will be small; if the perturbations are small, the neglected higher-order terms will be small and can be regarded as a disturbance. Since the controller is designed to counteract the effects of disturbances, the presence of higher-order terms should cause no problem. This reasoning cannot be justified rigorously. But nevertheless it usually works. Needless to say, it may not always work; so it is necessary to test the design that emerges for stability and performance—analytically, by Lyapunov's stability analysis for example, and/or by simulation.

Feedback linearization

In many systems, the nonlinearity inherent in the plant is so dominant that the linearization approach described above can hardly meet the stringent requirements on systems performance. This reality inevitably promotes the endeavour to develop control approaches that will more or less incorporate the nonlinear dynamics into the design process. One such approach is *feedback linearization*. Unlike the first-order approximation approach, wherein the higher-order terms of the plant are ignored, this approach utilizes the feedback to render the given system a linear input-output dynamics. On the basis of the linear system thus obtained, linear control techniques can be applied to address design issues. Figure 10.1 gives block diagram description of such a scheme. The results for such problems are couched in the terminology of *differential geometry* [116].

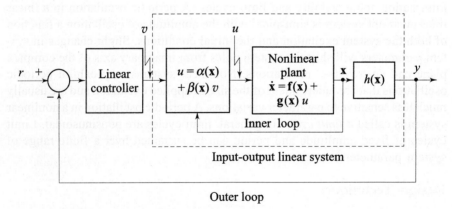

Fig. 10.1 Feedback linearization

Extended linearization

The roughness of the linearization approach based on first-order approximation can be viewed from two perspectives. First, it neglects all the higher-order terms. Second, the linear terms depend on the equilibrium point. These two uncertainties may explain why this linearization approach is incapable of dealing with the situation where the system operates over wide dynamic regions. Although the feedback linearization may overcome the first drawback, its applicability is limited mainly because it rarely leads to a design that guarantees the system performance over the whole dynamic regime. This is because feedback linearization is often performed locally around a specific equilibrium point. The resulting controller is of local nature. Another remedy to linearization based on first-order approximation is to design several control laws corresponding to several operating points that cover the whole dynamics of the system. Then these linear controllers are pieced together to obtain a nonlinear control law. This approach is often called *gain scheduling*. Though this approach does not account for the higher-order terms, it does accommodate the variations of the first-order terms with respect to the equilibrium points.

The extended linearization approach can be viewed as an analytic formulation of gain scheduling [116].

Adaptive control

Adaptive control theory provides an effective tool for the design of uncertain systems. Unlike *fixed-parameter controllers* (e.g. H_∞ theory based robust controller), adaptive controllers adapt (adjust) their behaviour on-line to the changing properties of the controlled processes.

If a fixed-parameter automatic control system is used, the plant–parameter variations directly affect the capability of the design to meet the performance specifications under all operating conditions. If an adaptive controller is used, the plant–parameter variations are accounted for at the price of increased complexity of the controller. Adaptive control is certainly more complex than fixed-parameter control and carries with it more complex failure mechanisms. In addition, adaptive control is both time-varying and nonlinear, increasing the difficulty of stability and performance analysis. It is this tradeoff of complexity versus performance that must be examined carefully in choosing the control structure.

Revisit Sections 7.11 and 8.7 for an introduction to adaptive control.

Variable structure control

The main distinctive feature of variable structure systems (VSS), setting them apart as an independent class of control systems, is that changes can occur in the structure of the system during the transient process. The structure of a VSS is changed intentionally in accordance with some law of structure change; the times at which these changes occur (and the type of structure formed) are determined not by a fixed program but in accordance with the current value of the states of the system.

There are two basis steps in the design of variable structure control [120]:
1. The design of a switching surface in system's state space so that the behaviour of the system has certain prescribed properties on the surface. For example, the switching surface will be designed so that the system is asymptotically stable on the surface.
2. The design of the control strategy to steer the system to the switching surface and to maintain it there.

Our discussion in this chapter will be limited to second-order systems on the phase plane. Emphasis will be on bang-bang control strategy to steer the system to the desired switching curve on the phase plane.

10.2 A CLASS OF NONLINEAR SYSTEMS: SEPARABLE NONLINEARITIES

In this chapter, we shall be concerned primarily with the so-called *separable* systems, which comprise a linear part defined by its transfer function, and a nonlinear part defined by a time-independent relationship between its input and output variables. Emphasis will be on the *piecewise linear systems* characterized by the existence of a number of regions of linear operation. Such systems often constitute a valid approximation which simplifies the investigation of nonlinear phenomena.

In the following paragraphs we shall discuss, in brief, the basic features of commonly encountered nonlinearities.

Saturation

This is perhaps the most common of all nonlinearities. All devices when driven by sufficiently large signals, exhibit the phenomenon of *saturation* due to limitations of their physical capabilities. Saturation in the output of electronic, rotating, and flow (hydraulic and pneumatic) amplifiers, speed and torque saturation in electric and hydraulic motors, saturation in the output of sensors for measuring position, velocity, temperature, etc., are the well-known examples. Figure 10.2 shows a linear-segmented approximation of saturation nonlinearity.

Deadzone

Many physical devices do not respond to small signals, i.e., if the input amplitude is less than some small value, there will be no output. This is true of many sensors and amplifiers. These devices are said to have a *deadzone*. Figure 10.3 shows linear-segmented approximation of deadzone nonlinearity.

Nonlinear Friction

In any system where there is relative motion between contacting surfaces, there are several types of friction; all of them nonlinear except the viscous components. The nonlinear effect is variously called *dry friction, nonlinear friction*, and *Coulomb friction*. It is, in essence, a drag (reaction) force which opposes motion, but is essentially constant in magnitude regardless of velocity (Fig. 10.4). The common example is an electric motor in which we find Coulomb friction drag due to the rubbing contact between the brushes and the commutator.

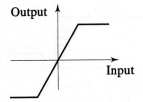 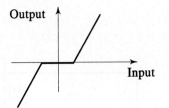

Fig. 10.2 Saturation nonlinearity   **Fig. 10.3** Deadzone nonlinearity

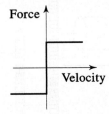

Fig. 10.4 Nonlinear furiction

On-off Controllers

Here we consider the class of controllers having a limited number of states rather than a continuous output. The most common is the two-state controller, usually called an on–off controller.

Electromechanical relays are frequently used in control systems where the control strategy requires control signals with only two or three states. Figure 10.5 shows a simple servo system using a bidirectional (polarized) relay. The error signal $e(t)$ controls the current through the relay field winding which moves the armature so as to make the contact in one direction or the other.

In linear servo systems, amount of power applied by the servomotor to the output member is normally proportional to the instantaneous error $e(t)$ of the system. For linear systems with cascaded compensation elements, the power applied to the output member may depend on the error e, its derivative, and its integral, in some linear manner. In the system shown in Fig. 10.5, the motor always applies full power to the output member, i.e., maximum available power is utilized at all times. This class of systems is often referred to as *bang-bang control systems*. As we shall see later in this chapter, by properly designing the switching of the relay, a bang-bang type of system is capable of driving the output to the desired value in minimum possible time.

Figure 10.6a shows the ideal characteristic of a bidirectional relay. In practice, the relay will not respond instantaneously as the error changes sign. Figure 10.6b shows the characteristic of a relay with switching time lag. For input currents between the two switching instants, the relay may be in one position

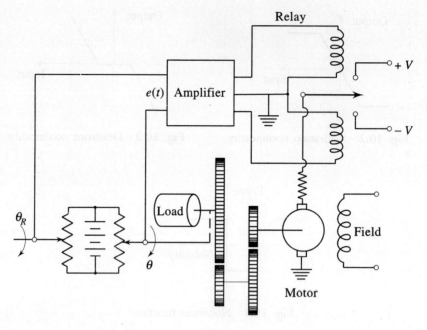

Fig. 10.5 A bang-bang control system

or the other depending upon the previous history of the input. This characteristic, called *on–off with hysteresis*, has thus inherent memory and is referred to as *memory-type nonlinearity*. We may also classify the two characteristics of Figs 10.6a and 10.6b as, respectively, a *single-valued nonlinearity* and a *double-valued nonlinearity*.

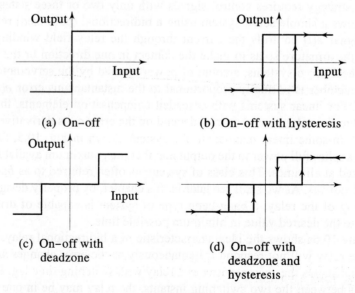

Fig. 10.6

A relay also has a definite amount of deadzone in practice. The deadzone is caused by the fact that the relay field winding requires a finite amount of current to move the armature. Figure 10.6c shows the characteristic of relay with deadzone, while Fig. 10.6d combines both the deadzone and the hysteresis characteristics.

Friction-controlled Backlash

An extremely complex form of nonlinear behaviour to model is the backlash, commonly found in mechanical linkages where coupling is not perfect. An example is the free play between teeth of the drive gear and those of the driven gear in a gear train. In situations frequently met in practice, the driven gear is subject to dry friction force sufficiently large that it *instantly* stops whenever the drive gear breaks contact, and it follows the input member *perfectly* whenever the contact is made. No elasticity is present and there is no 'bouncing' at the instants when input and output members make and break contact. Another example of backlash is the process control valve hysteresis.

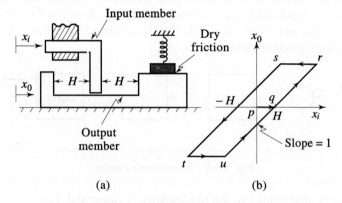

Fig. 10.7 Friction-controlled backlash

A physical model of backlash is shown in Fig. 10.7a. The transfer characteristic between input and output displacements is shown in Fig. 10.7b. As the input member is driven forward from the position shown in Fig. 10.7a, no output motion takes place until the input member travels a distance H. This situation corresponds to the segment pq of Fig. 10.7b. After the contact is made, the output member perfectly follows the input member; this is illustrated by the line segment qr of Fig. 10.7b. As the input motion is reversed, the contact between the input and output members is lost. The output motion therefore ceases till the input member has travelled a distance $2H$ in the reverse direction as illustrated by the segment rs. After the input member establishes contact with the output member, the output member moves in the reverse direction (segment *st*). As the input motion is reversed, the output member is again at standstill for the segment *tu* and then follows the input member along *uq*.

As is obvious, backlash is a double-valued (memory-type) nonlinearity.

10.3 FILTERED NONLINEAR SYSTEM: THE DESCRIBING FUNCTION ANALYSIS

Of all the analytical methods developed over the years for nonlinear systems, the describing function method is generally agreed upon as being the most practically useful. It is an approximate method but experience with real systems and computer simulation results shows adequate accuracy in many cases. The method predicts whether limit cycle oscillations will exist or not and gives numerical estimates of oscillation frequency and amplitude when limit cycles are predicted. Basically the method is an approximate extension of frequency-response methods (including Nyquist stability criterion) to nonlinear systems.

To discuss the basic concept underlying the describing function analysis, let us consider the block diagram of a nonlinear system shown in Fig. 10.8, where the blocks $G_1(s)$ and $G_2(s)$ represent the linear elements, while the block N represents the nonlinear element.

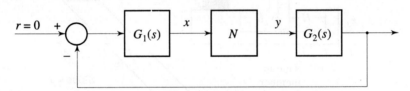

Fig. 10.8 A nonlinear system

Let us assume that input x to the nonlinearity is sinusoidal, i.e.,

$$x = X \sin \omega t \tag{10.1}$$

With such an input, the output y of the nonlinear element will in general be a nonsinusoidal periodic function which may be expressed in terms of Fourier series as follows (refer Eqns (A.6)–(A.7) in Appendix A):

$$y = Y_0 + A_1 \cos \omega t + B_1 \sin \omega t + A_2 \cos 2\omega t + B_2 \sin 2\omega t + \cdots \tag{10.2}$$

The nonlinear characteristics listed in the previous section are all odd-symmetrical/odd half-wave symmetrical; the mean value Y_0 for all such cases is zero and therefore the output

$$y = A_1 \cos \omega t + B_1 \sin \omega t + A_2 \cos 2\omega t + B_2 \sin 2\omega t + \cdots \tag{10.3}$$

In the absence of an external input (i.e., $r = 0$ in Fig. 10.8), the output y of the nonlinear element N is fedback to its input through the linear elements $G_2(s)$ and $G_1(s)$ in tandem. If $G_2(s)G_1(s)$ has lowpass characteristics (this is usually the case in control systems), it can be assumed to a good degree of approximation that all the higher harmonics of y are filtered out in the process, and the input x to the nonlinear element N is mainly contributed by the fundamental component (first harmonic) of y, i.e., x remains sinusoidal. Under such conditions, the second and higher harmonics of y can be thrown away for the purpose of analysis and the fundamental component of y, i.e.,

$$y_1 = A_1 \cos \omega t + B_1 \sin \omega t \qquad (10.4)$$

need only be considered.

The above procedure heuristically linearizes the nonlinearity since for a sinusoidal input, only a sinusoidal output of the same frequency is now assumed to be produced. This type of linearization, called the *first harmonic approximation*, is valid for large signals as well, so long as the filtering condition is satisfied.

We can write $y_1(t)$ in the form

$$y_1(t) = A_1 \sin(\omega t + 90°) + B_1 \sin \omega t = Y_1 \sin(\omega t + \phi_1) \qquad (10.5a)$$

where, by using phasors,

$$Y_1 \angle \phi_1 = B_1 + jA_1 = \sqrt{B_1^2 + A_1^2} \angle \tan^{-1}(A_1/B_1) \qquad (10.5b)$$

The coefficients A_1 and B_1 of the Fourier series are given by (refer Eqns (A.6) in Appendix A)

$$A_1 = \frac{1}{\pi} \int_0^{2\pi} y \cos \omega t \, d(\omega t) \qquad (10.5c)$$

$$B_1 = \frac{1}{\pi} \int_0^{2\pi} y \sin \omega t \, d(\omega t) \qquad (10.5d)$$

As we shall see shortly, the amplitude Y_1 and the phase shift ϕ_1 are both functions of X, but independent of ω. We may combine the amplitude ratio and the phase shift in a *complex equivalent gain* $N(X)$ such that

$$N(X) = \frac{Y_1(X)}{X} \angle \phi_1(X) = \frac{B_1 + jA_1}{X} \qquad (10.6)$$

Under first-harmonic approximation, the nonlinear element is completely characterized by the function $N(X)$; this function is usually referred to as the *describing function* of the nonlinearity.

The describing function differs from a linear system transfer function in that its numerical value will vary with input amplitude X. Also, it does not depend on frequency ω (there are, however, a few situations in which the describing function for the nonlinearity is a function of both the input amplitude X and the frequency ω (refer [121–122]). When embedded in an otherwise linear system (Fig. 10.9), the describing function can be combined with the 'ordinary' sinusoidal transfer function of the rest of the system to obtain the complete open-loop function; however, we will get a different open-loop function for every different amplitude X. We can check all of these open-loop functions for closed-loop stability using Nyquist stability criterion.

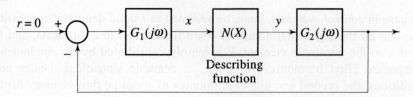

Fig. 10.9 Nonlinear system with nonlinearity replaced by describing function

It is important to remind ourselves here that the simplicity in analysis of nonlinear systems using describing functions has been achieved at the cost of certain limitations, the foremost being the assumption that in traversing the path through the linear parts of the system from nonlinearity output back to nonlinearity input, the higher harmonics will have been effectively lowpass filtered, relative to the first harmonic. When the linear part of the system does indeed provide a sufficiently strong filtering effect, then the predictions of describing function analysis usually are a good approximation to actual behaviour. Filtering characteristics of the linear part of the system improve as the order of the system goes up.

The 'lowpass filtering' requirement is never completely satisfied; for this reason the describing function method is mainly used for stability analysis and is not directly applied to the optimization of system design. Usually the describing function analysis will correctly predict the existence and characteristics of limit cycles. However, false indications cannot be ruled out; therefore the results must be verified by simulation. Simulation, in fact, is an almost indispensable tool for analysis and design of nonlinear systems; describing function and other analytical methods provide the background for intelligent planning of the simulations.

We will limit our discussion to separable nonlinear systems with reference input $r = 0$, and with symmetrical nonlinearities (listed in Section 10.2) in the loop. Refer [121–122] for situations wherein dissymmetrical nonlinearities are present, and/or the reference input is non-zero.

10.4 DESCRIBING FUNCTIONS OF COMMON NONLINEARITIES

Before coming to the stability study by the describing function method, it is worthwhile to derive the describing functions of some common nonlinearities. Our first example is an on–off controller with a deadzone as in Fig. 10.10. If X is less than deadzone Δ, then the controller produces no output; the first harmonic component of the Fourier series is of course zero, and the describing function is also zero. If $X > \Delta$, the controller produces the 'square wave' output y. One cycle of this periodic function of period 2π is described as follows:

Nonlinear Control Systems

$$y = \begin{cases} 0 & ; 0 \le \omega t < \alpha \\ M & ; \alpha \le \omega t < (\pi - \alpha) \\ 0 & ; (\pi - \alpha) \le \omega t < (\pi + \alpha) \\ -M & ; (\pi + \alpha) \le \omega t < (2\pi - \alpha) \\ 0 & ; (2\pi - \alpha) \le \omega t \le 2\pi \end{cases} \quad (10.7)$$

where $X \sin\alpha = \Delta$; or $\alpha = \sin^{-1}(\Delta/X)$

This periodic function has odd symmetry:

$$y(\omega t) = -y(-\omega t)$$

Therefore, the fundamental component of y is given by (refer Eqn. (A.7d) in Appendix A)

$$y_1 = B_1 \sin\omega t$$

where
$$B_1 = \frac{1}{\pi} \int_0^{2\pi} y \sin\omega t \, d(\omega t)$$

Due to the symmetry of y (refer Fig. 10.10), the coefficient B_1 can be calculated as follows:

$$B_1 = \frac{4}{\pi} \int_0^{\pi/2} y \sin\omega t \, d(\omega t) = \frac{4M}{\pi} \int_\alpha^{\pi/2} \sin\omega t \, d(\omega t) = \frac{4M}{\pi} \cos\alpha \quad (10.8)$$

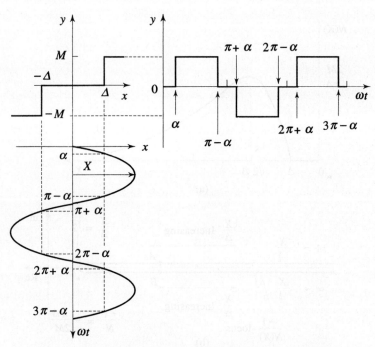

Fig. 10.10 Fourier-series analysis of an on–off controller with deadzone

Since A_1 (the Fourier series cosine coefficient) is zero, the first harmonic component of y is exactly in phase with $X\sin\omega t$, and the describing function $N(X)$ is given by (refer Eqns (10.6)–(10.8))

$$N(X) = \begin{cases} 0 & ; X < \Delta \\ \dfrac{4M}{\pi X}\sqrt{1-\left(\dfrac{\Delta}{X}\right)^2} & ; X \geq \Delta \end{cases} \qquad (10.9)$$

For a given controller, M and Δ are fixed and the describing function is a function of input amplitude X, which is graphed in Fig. 10.11a together with peak location and value found by standard calculus maximization procedure. Note that for a given X, $N(X)$ is just a pure real positive number, and thus plays the role of a steady-state gain in a block diagram of the form shown in Fig. 10.9. However, this gain term is unusual in that it changes when X changes.

A describing function $N(X)$ may be equivalently represented by a plot of

$$-\frac{1}{N(X)} = \left|-\frac{1}{N(X)}\right| \angle(-1/N(X)) \qquad (10.10)$$

as a function of X on the polar plane. We will use this form of representation in the next section for stability analysis.

Rearrangement of Eqn. (10.9) gives

$$-\frac{1}{N(X)} = -\frac{\pi\Delta}{4M}\frac{(X/\Delta)^2}{\sqrt{(X/\Delta)^2-1}} \qquad (10.11)$$

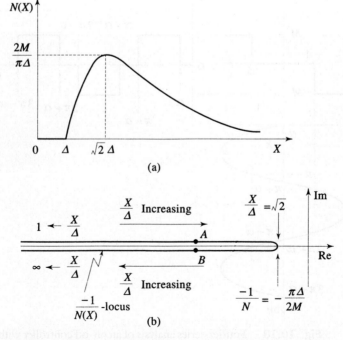

Fig. 10.11 Describing function of an on–off controller with deadzone

Figure 10.11b gives the representation on the polar plane of the describing function for an on–off controller with deadzone. It may be noted that though the points A and B lie at the same place on the negative real axis, they belong to different values of X/Δ.

We choose as another example the friction-controlled backlash since its behaviour brings out certain features not encountered in our earlier example. The characteristics of backlash nonlinearity and its response to sinusoidal input are shown in Fig. 10.12. The output y is again a periodic function of period 2π; one cycle of this function is described as follows:

$$y = \begin{cases} x - H\,; & 0 \leq \omega t < \pi/2 \\ X - H\,; & \pi/2 \leq \omega t < (\pi - \beta) \\ x + H\,; & (\pi - \beta) \leq \omega t < 3\pi/2 \\ -X + H\,; & 3\pi/2 \leq \omega t < (2\pi - \beta) \\ x - H\,; & (2\pi - \beta) \leq \omega t \leq 2\pi \end{cases} \quad (10.12)$$

where $X \sin\beta = X - 2H$; or $\beta = \sin^{-1}\left(1 - \dfrac{2H}{X}\right)$.

The periodic function does not possess odd symmetry:

$$y(\omega t) \neq -y(-\omega t),$$

but possesses odd half-wave symmetry:

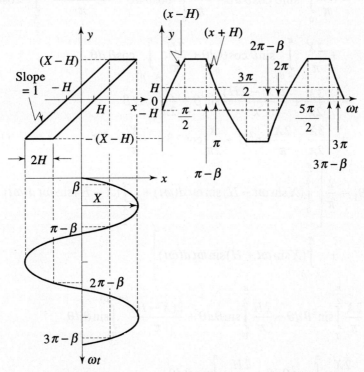

Fig. 10.12 Fourier-series analysis of friction-controlled backlash

$$y(\omega t \pm \pi) = -y(\omega t)$$

Therefore, the fundamental component of y is given by (refer Eqn. (A.7e) in Appendix A)

$$y_1 = A_1 \cos\omega t + B_1 \sin\omega t$$

where
$$A_1 = \frac{1}{\pi} \int_0^{2\pi} y \cos\omega t \, d(\omega t)$$

$$B_1 = \frac{1}{\pi} \int_0^{2\pi} y \sin\omega t \, d(\omega t)$$

Due to the symmetry of y, only the positive half wave need be considered (Fig. 10.12):

$$A_1 = \frac{2}{\pi}\left[\int_0^{\pi/2}(X\sin\omega t - H)\cos\omega t \, d(\omega t) + \int_{\pi/2}^{(\pi-\beta)}(X - H)\cos\omega t \, d(\omega t)\right.$$

$$\left. + \int_{(\pi-\beta)}^{\pi}(X\sin\omega t + H)\cos\omega t \, d(\omega t)\right]$$

$$= \frac{2X}{\pi}\int_0^{\pi/2}\sin\theta\cos\theta \, d\theta - \frac{2H}{\pi}\int_0^{\pi/2}\cos\theta \, d\theta + \frac{2(X-H)}{\pi}\int_{\pi/2}^{(\pi-\beta)}\cos\theta \, d\theta$$

$$+ \frac{2X}{\pi}\int_{(\pi-\beta)}^{\pi}\sin\theta\cos\theta \, d\theta + \frac{2H}{\pi}\int_{(\pi-\beta)}^{\pi}\cos\theta \, d\theta$$

$$= -\frac{3X}{2\pi} + \frac{2(X-2H)}{\pi}\sin\beta + \frac{X}{2\pi}\cos 2\beta$$

$$= -\frac{3X}{2\pi} + \frac{2X}{\pi}\sin^2\beta + \frac{X}{2\pi}\cos 2\beta = -\frac{X}{\pi}\cos^2\beta \qquad (10.13a)$$

$$B_1 = \frac{2}{\pi}\left[\int_0^{\pi/2}(X\sin\omega t - H)\sin\omega t \, d(\omega t) + \int_{\pi/2}^{(\pi-\beta)}(X - H)\sin\omega t \, d(\omega t)\right.$$

$$\left. + \int_{(\pi-\beta)}^{\pi}(X\sin\omega t + H)\sin\omega t \, d(\omega t)\right]$$

$$= \frac{2X}{\pi}\int_0^{\pi/2}\sin^2\theta \, d\theta - \frac{2H}{\pi}\int_0^{\pi/2}\sin\theta \, d\theta + \frac{2(X-H)}{\pi}\int_{\pi/2}^{(\pi-\beta)}\sin\theta \, d\theta$$

$$+ \frac{2X}{\pi}\int_{(\pi-\beta)}^{\pi}\sin^2\theta \, d\theta + \frac{2H}{\pi}\int_{(\pi-\beta)}^{\pi}\sin\theta \, d\theta$$

$$= \frac{X}{\pi}\left[\frac{\pi}{2} + \beta\right] + \frac{2(X-2H)}{\pi}\cos\beta - \frac{X}{2\pi}\sin 2\beta$$

$$= \frac{X}{\pi}\left[\frac{\pi}{2} + \beta\right] + \frac{2X}{\pi}\sin\beta\cos\beta - \frac{X}{2\pi}\sin 2\beta = \frac{X}{\pi}\left[\frac{\pi}{2} + \beta + \frac{1}{2}\sin 2\beta\right]$$
(10.13b)

It is clear that the fundamental component of y will have a phase shift with respect to $X\sin\omega t$ (a feature not present in our earlier example). The describing function $N(X)$ is given by (refer Eqns (10.6), (10.12), (10.13))

$$N(X) = \frac{1}{X}(B_1 + jA_1) = \frac{1}{\pi}\left[\frac{\pi}{2} + \beta + \frac{1}{2}\sin 2\beta - j\cos^2\beta\right]$$
(10.14)

$$\beta = \sin^{-1}\left(1 - \frac{2H}{X}\right)$$

Note that $N(X)$ is a function of the non-dimensional ratio H/X; we can thus tabulate or plot a single table or graph of $N(X)$ that will be usable for any numerical value of H (see Table 10.1, and Fig. 10.13).

Table 10.1 Describing function for friction-controlled backlash

H/X	$|-1/N(X)|$	$\angle(-1/N(X))$
0.000	1.000	180.0
0.050	1.017	183.5
0.125	1.066	188.5
0.200	1.134	193.4
0.300	1.259	199.7
0.400	1.435	206.0
0.500	1.687	212.5
0.600	2.072	219.3
0.700	2.720	226.7
0.800	4.024	235.1
0.850	5.330	239.9
0.900	7.946	245.6
0.925	10.560	248.9
0.950	15.800	252.8
0.975	31.500	257.9

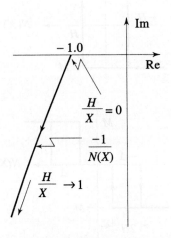

Fig. 10.13 Describing function for friction-controlled backlash

We have so far given illustrative derivations of describing functions for on–off controller with deadzone, and friction-controlled backlash. By similar procedures, the describing functions of other common nonlinearities can be derived; some of these are tabulated in Table 10.2.

Table 10.2 Describing functions of common nonlinearities

Nonlinearity	Describing function (input = $X\sin\omega t$)
1. (ideal relay)	$N(X) = \dfrac{4M}{\pi X}$
2. (relay with dead zone)	$N(X) = \begin{cases} 0 & ; X < \Delta \\ \dfrac{4M}{\pi X}\sqrt{1-\left(\dfrac{\Delta}{X}\right)^2} & ; X \geq \Delta \end{cases}$
3. (relay with hysteresis)	$N(X) = \begin{cases} 0 & ; X < H \\ \dfrac{4M}{\pi X}\left[\sqrt{1-\left(\dfrac{H}{X}\right)^2} - j\dfrac{H}{X}\right] & ; X \geq H \end{cases}$
4. (relay with dead zone and hysteresis)	$N(X) = \begin{cases} 0 & ; X < \Delta + H \\ \dfrac{2M}{\pi X}\left[\sqrt{1-\left(\dfrac{\Delta - H}{X}\right)^2} \\ \quad + \sqrt{1-\left(\dfrac{\Delta + H}{X}\right)^2} - j\dfrac{2H}{X}\right] & ; X \geq \Delta + H \end{cases}$
5. (saturation, Slope = K)	$N(X) = \begin{cases} K & ; X < S \\ \dfrac{2K}{\pi}\left[\sin^{-1}\dfrac{S}{X} + \dfrac{S}{X}\sqrt{1-\left(\dfrac{S}{X}\right)^2}\right] & ; X \geq S \end{cases}$

(Contd.)

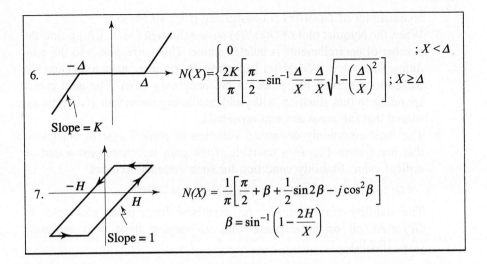

10.5 STABILITY ANALYSIS BY THE DESCRIBING FUNCTION METHOD

Consider the linear system of Fig. 10.14a. Application of Nyquist stability criterion[1] to this system involves the following steps.

1. Define the Nyquist contour in the s-plane that encloses the entire right-hand side (unstable region) of the s-plane (Fig. 10.14b).
2. Sketch the Nyquist plot, which is the locus of $KG(s)H(s)$ when s takes on values along the Nyquist contour (Fig. 10.14c).
3. The characteristic equation of the system is
$$1 + KG(s)H(s) = 0$$
or
$$KG(s)H(s) = -1 \qquad (10.15)$$
The stability of the closed-loop system is determined by investigating the behaviour of the Nyquist plot of $KG(s)H(s)$ with respect to the *critical point* $(-1 + j0)$ in the $KG(s)H(s)$-plane.

For the predominant case of systems wherein open-loop transfer function $KG(s)H(s)$ has no poles in the right-half of s-plane, the Nyquist stability criterion it stated below:

If the Nyquist plot of the open-loop transfer function $KG(s)H(s)$ corresponding to the Nyquist contour in the s-plane does not encircle the critical point $(-1 + j0)$, the closed-loop system is stable.

4. The characteristic Eqn. (10.15) may be rearranged as follows:
$$G(s)H(s) = -1/K \qquad (10.16)$$
For the linear system with open-loop transfer function $KG(s)H(s)$, we can count the number of encirclements of $(-1/K + j0)$ point if the

---

1. Chapter 8 of reference [180].

Nyquist plot of $G(s)H(s)$ is constructed (Fig. 10.14d).

5. When the Nyquist plot of $G(s)H(s)$ passes through $(-1/K + j0)$ point, the number of encirclements is indeterminate. This corresponds to the condition where $1 + KG(s)H(s)$ has zeros on the imaginary axis (i.e., the closed-loop system has poles on the imaginary axis). The gain corresponding to this situation will yield oscillatory behaviour (we have assumed that the zeros are non-repeated).

6. The most commonly occurring situation in control system designs is that the system becomes unstable if the gain increases past a certain critical value. Stability condition for such systems becomes

$$|G(j\omega)H(j\omega)| < 1/K \text{ at } \angle G(j\omega)H(j\omega) = -180°$$

The stability may therefore be examined from polar plot (plot of $G(j\omega)H(j\omega)$ on polar plane with ω varying from 0 to ∞) only (Fig. 10.14e).

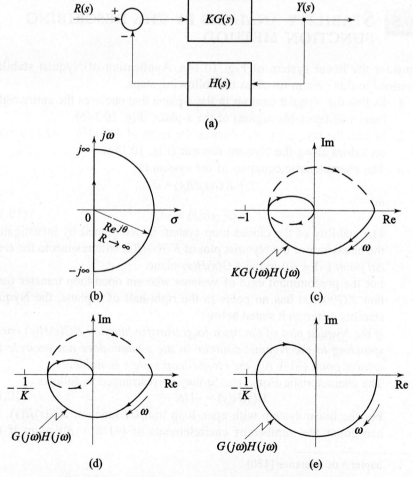

Fig. 10.14 Application of the Nyquist stability criterion

Consider now a nonlinear system of Fig. 10.15; $N(X)$ is the describing function of the nonlinear element and $G(s)$ is the transfer function of the linear part of the system. $G(s)$ is assumed to have no poles in the right half s-plane.

The validity of the block diagram shown in Fig. 10.15 is based on the assumption that the input to the nonlinearity is a pure sinusoid $x = X\sin\omega t$. This necessarily requires that r is zero, since non-zero values of the input usually result in the nonlinearity input signal containing components in addition to the assumed sine wave. So the describing function approach is applicable when the input r is zero and the system is excited by some initial conditions. For different values of the initial conditions, a signal of the form $x = X\sin\omega t$ will be generated at the input of an odd-symmetrical/odd half-wave symmetrical nonlinearity, with X varying from 0 to ∞. This is true only if the linear part of the system possesses the required lowpass characteristics (for situations where the nonlinearity input signal contains components in addition to $X\sin\omega t$ (such as r not being zero), the method of dual-input describing functions may be useful (refer [121–122]).

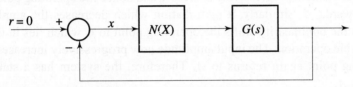

Fig. 10.15 A nonlinear system

For a given X, $N(X)$ in Fig. 10.15 is just a real/complex number; the condition (10.16) therefore becomes

$$G(s) = -1/N(X) \tag{10.17}$$

This modified condition differs from the condition (10.16) in the fact that the critical point $(-1/K + j0)$ now becomes the critical locus $-1/N(X)$ as a function of X. The stability analysis can be carried out by examining the relative position of the following plots on polar plane.

1. Plot of $G(j\omega)$ with ω varying from 0 to ∞, called the polar plot of $G(j\omega)$ (note that the Nyquist plot is the plot of $G(j\omega)$ with ω varying from $-\infty$ to $+\infty$).
2. Plot of $-1/N(X)$ with X varying from 0 to ∞.

When the critical points of $-1/N(X)$ lie to the left of the polar plot of $G(j\omega)$ (or are not encircled by the Nyquist plot of $G(j\omega)$), the closed-loop system is stable; any disturbances which appear in the system will tend to die out. Conversely, if any part of the $-1/N(X)$ locus lies to the right of the polar plot of $G(j\omega)$ (or is enclosed by the Nyquist plot of $G(j\omega)$), it implies that any disturbances which are characterized by the values of X corresponding to the enclosed critical points will provide unstable operations. The intersection of $G(j\omega)$ and $-1/N(X)$ loci corresponds to the possibility of a periodic oscillation (limit cycle) characterized by the value of X on the $-1/N(X)$ locus and the value of ω on the $G(j\omega)$ locus.

Figure 10.16a shows a $G(j\omega)$ plot superimposed on a $-1/N(X)$ locus. The values of X for which the $-1/N(X)$ locus lies in the region to the right of an observer traversing the polar plot of $G(j\omega)$ in the direction of increasing ω, correspond to unstable conditions. Similarly, the values of X for which the $-1/N(X)$ locus lies in the region to the left of an observer traversing the polar plot of $G(j\omega)$ in the direction of increasing ω, correspond to the stable conditions. The locus of $-1/N(X)$ and the polar plot of $G(j\omega)$ intersect at the point $A(\omega = \omega_2, X = X_2)$ which corresponds to the condition of limit cycle. The system is unstable for $X < X_2$ and is stable for $X > X_2$. The stability of the limit cycle can be judged by the perturbation technique described below.

Suppose that the system is originally operating at A under the state of a limit cycle. Assume that a slight perturbation is given to the system so that the input to the nonlinear element increases to X_3, i.e., the operating point is shifted to B. Since B is in the range of stable operation, the amplitude of the input to the nonlinear element progressively decreases and hence the operating point moves back towards A. Similarly, a perturbation which decreases the amplitude of input to the nonlinearity shifts the operating point to C which lies in the range of unstable operation. The input amplitude now progressively increases and the operating point again returns to A. Therefore, the system has a stable limit cycle at A.

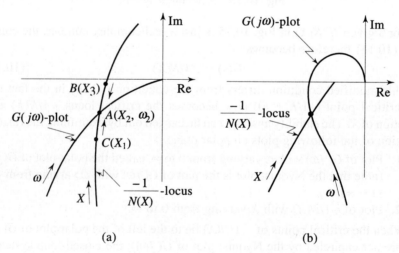

Fig. 10.16 Prediction and stability of limit cycles

Figure 10.16b shows the case of an unstable limit cycle. For systems having $G(j\omega)$ plots and $-1/N(X)$ loci as shown in Figs 10.17a and 10.17b, there are two limit cycles; one stable and the other unstable.

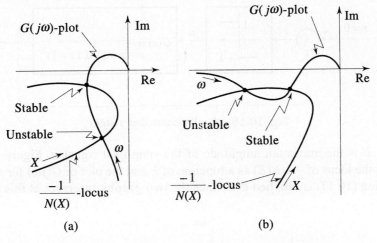

Fig. 10.17 Prediction and stability of limit cycles

Describing function method usually gives sufficiently accurate information about stability and limit cycles. This analysis is invariably followed by a simulation study.

The transfer function $G(s)$ in Fig. 10.15, when converted to state variable formulation, takes the form

$$\dot{\mathbf{x}}(t) = \mathbf{A}\mathbf{x}(t) + \mathbf{b}u(t); \quad \mathbf{x}(0) \triangleq \mathbf{x}^0$$
$$y(t) = \mathbf{c}\mathbf{x}(t)$$

where

$\mathbf{x}(t) = n \times 1$ state vector for nth order $G(s)$

$u(t)$ = input to $G(s)$

$y(t)$ = output of $G(s)$

$\mathbf{A} = n \times n$ matrix

$\mathbf{b} = n \times 1$ column matrix

$\mathbf{c} = 1 \times n$ row matrix

In Appendix B we use MATLAB Software SIMULINK to obtain response of nonlinear systems of the form given in Fig. 10.15 with zero reference input and initial state $\mathbf{x}^0$.

Example 10.1

Let us investigate the stability of a relay-controlled system shown in Fig. 10.18. Using the describing function of an ideal relay given in Table 10.2, we have

$$-\frac{1}{N(E)} = -\frac{\pi E}{4} \qquad (10.18)$$

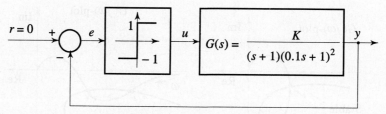

Fig. 10.18 A relay-controlled system

where E is the maximum amplitude of the sinusoidal signal e. Figure 10.19 shows the locus of $-1/N(E)$ as a function of E and the plot of $G(j\omega)$ for $K = 5$. Equation (10.17) is satisfied at A since the two graphs intersect at this point.

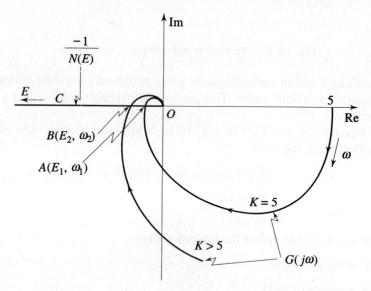

Fig. 10.19 Stability analysis of the system in Fig. 10.18

The point of intersection on the $G(j\omega)$ plot gives a numerical value ω_1 for the frequency of the limit cycle; whereas the same point on the $-1/N(E)$ locus gives us the predicted amplitude E_1 of the oscillation. As an observer traverses the $G(j\omega)$ plot in the direction of increasing ω, the portion O-A of the $-1/N(E)$ locus lies to its right and the portion A-C lies to its left. Using the arguments presented previously, we can conclude that the limit cycle is a stable one.

Since $-1/N(E)$ is a negative real number, it is clear that intersection occurs at $-180°$ phase angle. The frequency ω_1 that gives $\angle G(j\omega_1) = -180°$ is 10.95 rad/sec. Furthermore, at point A

$$|G(j\omega_1)| = \left| -\frac{1}{N(E_1)} \right|$$

At $\omega_1 = 10.95$, $|G(j\omega_1)| = 0.206$ and therefore (refer Eqn. (10.18)).

$$\left| -\frac{1}{N(E_1)} \right| = \frac{\pi E_1}{4} = 0.206$$

This gives $E_1 = 0.262$.

The describing function analysis thus predicts a limit cycle (sustained oscillation)

$$y(t) = -e(t) = -0.262 \sin 10.95 t$$

For $K > 5$, the intersection point shifts to B (Fig. 10.19) resulting in a limit cycle of amplitude $E_2 > E_1$ and frequency $\omega_2 = \omega_1$. It should be observed that the system has a limit cycle for all positive values of gain K.

To gain some further insight into on–off control behaviour and describing function analysis, let us modify the system of Fig. 10.18 by letting the linear portion be of second-order with

$$G(s) = \frac{5}{(s+1)(0.1s+1)}$$

Figure 10.20 shows the plot of $G(j\omega)$ superimposed on the locus of $-1/N(E)$. The intersection of the two graphs is now impossible since the phase angle of neither $G(j\omega)$ nor $-1/N(E)$ can be more lagging than $-180°$. Describing function analysis thus seems to predict no limit cycling, whereas the fact that the control signal u must be either $+1.0$ or -1.0 dictates that the system oscillate. One possible interpretation to this analysis would be that the second-order linear system provides less of the lowpass filtering assumed in the describing function method than did the third-order system and thus the approximation has become inaccurate to the point of predicting no limit cycle when actually one occurs. Another interpretation would be that the curves actually do 'intersect' at the origin, predicting a limit cycle of infinite frequency and infinitesimal amplitude. This latter interpretation, even though it predicts a physically impossible result, agrees with the rigorous mathematical solution of the differential equations: for some non-zero initial value of y, we find that $y(t)$ oscillates about zero with ever-decreasing amplitude and ever-increasing frequency. We will examine this solution on the phase plane in a later section.

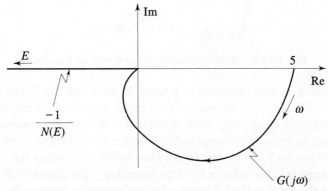

Fig. 10.20 Describing function analysis of a relay-controlled system with second-order plant

Let us now modify the system of Fig. 10.18 by giving the controller a deadzone Δ, as shown in Fig. 10.21. The $-1/N(E)$ locus for this type of relay controller is given by Fig. 10.11b. Plots of $G(j\omega)$ for different values of K, superimposed on $-1/N(E)$ locus, are shown in Fig. 10.22. From this figure, we observe that for $K = K_1$, the $G(j\omega)$ plot crosses the negative real axis at a point to the right of $-\pi\Delta/2$ such that no intersection takes place between the graphs of $G(j\omega)$ and $-1/N(E)$, and therefore no limit cycle results. With such a gain, the $-1/N(E)$ locus lies entirely to the left of the $G(j\omega)$ plot, the system is therefore stable, i.e., it has effectively positive damping.

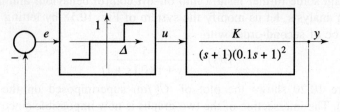

Fig. 10.21 A system controlled by relay with deadzone

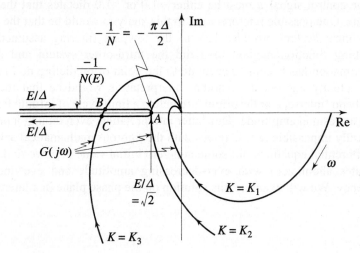

Fig. 10.22 Stability analysis of the system of Fig. 10.21

If the gain K is now increased to a value K_2 such that the $G(j\omega)$ plot intersects the $-1/N(E)$ locus at the point A (i.e., on the negative real axis at $-\pi\Delta/2$), then there exists a limit cycle. Now suppose that the system is operating at the point A. Any increase in the amplitude of E takes the operating point to the left so that it is not enclosed by the $G(j\omega)$ plot, which means that the system has positive damping. This reduces E till the operating point comes back to A. Any decrease in the amplitude of E again takes the operating point to the left of the $G(j\omega)$ plot, i.e., the system has positive damping which further reduces E, finally driving the system to rest. Since random disturbances are always present in any system, the system under discussion cannot remain at A. Therefore, the limit cycle represented by A is unstable.

When the gain K is further increased to K_3, the graphs of $G(j\omega)$ and $-1/N(E)$ intersect at two points B and C. By arguments similar to those advanced earlier, it can be shown that the point B represents an unstable limit cycle and C represents a stable limit cycle. It may be noted that though the points B and C lie at the same place on the negative real axis, they belong to different values of E/Δ.

It is also clear that limit cycling is predicted only for deadzone Δ smaller than the value given by

$$\frac{\pi \Delta}{2} = |G(j\omega_1)|$$

where ω_1 is the frequency at which the plot $G(j\omega)$ intersects the negative real axis. A deadzone in relay controllers appears to be a desirable feature to avoid limit cycling. However, as we shall see later in this chapter, a large value of Δ would cause the steady-state performance of the system to deteriorate.

Example 10.2

Figure 10.23a shows a block diagram for a servo system consisting of an amplifier, a motor, a gear train, and a load (gear 2 shown in the diagram includes the load element). It is assumed that the inertia of the gears and load element is negligible compared with that of the motor, and backlash exists between gear 1 and gear 2. The gear ratio between gear 1 and gear 2 is unity.

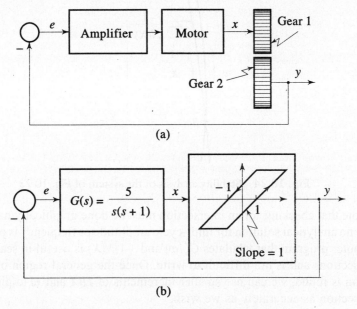

Fig. 10.23 A servo system with backlash in gears

The transfer function of the amplifier-motor combination is given by $5/s(s + 1)$ and the backlash amplitude is given as unity ($H = 1$).

From the problem statement, the block diagram of the system may be redrawn as shown in Fig. 10.23b. Let us investigate the stability of this system. The $-1/N(X)$ locus for the backlash nonlinearity is given by Fig. 10.13 (Table 10.1). Plot of $G(j\omega)$ superimposed on $-1/N(X)$ locus is shown in Fig. 10.24. As seen from this figure, there are two intersections of the two loci. Applying the stability test for the limit cycle reveals that point A corresponds to a stable limit cycle and point B corresponds to an unstable limit cycle. The stable limit cycle has a frequency of 1.6 rad/sec and an amplitude of 2 (the unstable limit cycle cannot physically occur). To avoid limit-cycle behaviour, the gain of the amplifier must be decreased sufficiently so that the entire $G(j\omega)$ plot lies to the left of $-1/N(X)$ locus.

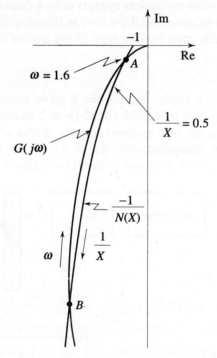

Fig. 10.24 Stability analysis of the system of Fig. 10.23

Note that checking for an intersection must be done graphically/numerically since no analytical solution for limit-cycle amplitude or frequency is possible. A computer program that tabulates $G(j\omega)$ and $-1/N(X)$ is useful in searching for intersections and is not difficult to write. Once the general region of an intersection is found, we can use smaller increments of H/X and ω to pinpoint the intersection as accurately as we wish.

10.6 NONLINEAR SAMPLED-DATA SYSTEMS

Figure 10.25 shows the block diagram of a simple nonlinear sampled-data system. The describing function method can be used to investigate the possibility of limit cycles in an analogous manner to the continuous-time case. The problem, however, is different in two ways; first, limit cycles normally occur at subharmonics of the sampling frequency, and secondly, the describing function is a function of the sampling phase, that is, the phase angle at which the sampler in Fig. 10.25 samples the assumed sinusoidal input, as well as the amplitude of the sinusoid.

Two describing function approaches are possible, one which uses a continuous describing function model for sampler, hold and nonlinearity combination, and the other which uses a discrete describing function model for the nonlinearity alone. The reader is advised to refer [84, 170] for application examples and MATLAB-based software.

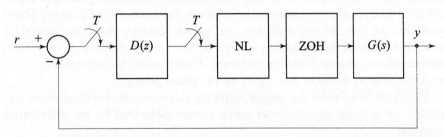

Fig. 10.25 Block diagram of a nonlinear sampled-data system

10.7 SECOND-ORDER NONLINEAR SYSTEM ON THE PHASE PLANE

The free motion of any second-order nonlinear system can always be described by an equation of the form

$$\ddot{y} + g(y,\dot{y})\dot{y} + h(y,\dot{y})y = 0 \tag{10.19}$$

The state of the system at any moment can be represented by a point of coordinates $(y,\dot{y})$ in a system of rectangular coordinates. Such a coordinate plane is called a 'phase plane'.

In terms of the state variables

$$x_1 = y, \ x_2 = \dot{y}, \tag{10.20a}$$

our second-order system is equivalent to the following canonical set of state equations:

$$\dot{x}_1 = x_2 \tag{10.20b}$$
$$\dot{x}_2 = -g(x_1,x_2)x_2 - h(x_1,x_2)x_1$$

If we eliminate time t by division, we obtain a first-order differential equation relating the variables x_2 and x_1:

$$\frac{dx_2}{dx_1} = -\frac{g(x_1, x_2)x_2 + h(x_1, x_2)x_1}{x_2} \qquad (10.21)$$

For a given set of initial conditions $\{x_1(0), x_2(0)\}$, the solution to Eqn. (10.21) may be represented by a single curve in the phase plane for which the coordinates are x_1 and x_2. The curve traced out by the state point $\{x_1(t), x_2(t)\}$ as time t is varied from 0 to ∞ is called the *phase trajectory*, and the family of all possible curves for different initial conditions is called the *phase portrait*. Normally, a finite number of trajectories, defined in a finite region, is considered a portrait.

For systems described by simple piecewise linear differential equations, closed-form solutions for phase trajectories may be obtained. However, the cases where closed-form solution is not possible are frequently met in practice, and we rely on numerical solutions to obtain the phase trajectories as needed.

One may obviously raise the question that when time solutions $x_1(t)$ and $x_2(t)$ as time t is varied from 0 to ∞, may be obtained by direct integration of Eqns. (10.20b) analytically or numerically, where is the necessity of drawing phase portraits? In fact, as we shall see, the phase portraits provide a powerful qualitative aid for investigating system behaviour and the design of system parameters to achieve a desired response. Furthermore, the existence of limit cycles is sharply brought into focus by the phase portrait.

Figure 10.26a shows the output response and the corresponding phase trajectory for a linear second-order servo system described by the differential equation

$$\ddot{y} + 2\zeta\dot{y} + y = 0;\ y(0) = y^0,\ \dot{y}(0) = 0,\ 0 < \zeta < 1$$

In terms of the state variables $x_1 = y$ and $x_2 = \dot{y}$, the system model is given by the equations

$$\dot{x}_1 = x_2;\ \dot{x}_2 = -2\zeta x_2 - x_1;\ x_1(0) = y^0,\ x_2(0) = 0$$

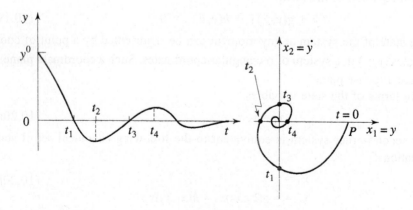

Fig. 10.26a A second-order linear system on the phase plane

The origin of the phase plane ($x_1 = 0$, $x_2 = 0$) is the *equilibrium point* of the system since at this point the derivatives $\dot{x}_1$ and $\dot{x}_2$ are zero (the system continues to lie at the equilibrium point unless otherwise disturbed). The nature of the transient can be readily inferred from the phase trajectory of Fig. 10.26a; starting from the point P, i.e., with initial deviation but no initial velocity, the system returns to rest, i.e., to the origin, with damped oscillatory behaviour. Consider now the well-known *Van der Pol's differential equation*

$$\ddot{y} - \mu(1 - y^2)\dot{y} + y = 0$$

which describes physical situations in many nonlinear systems. It terms of the state variables $x_1 = y$ and $x_2 = \dot{y}$, we obtain

$$\dot{x}_1 = x_2;\ \dot{x}_2 = \mu(1 - x_1^2)x_2 - x_1$$

Origin of the phase plane is the equilibrium point of the system. Figure 10.26b shows phase portraits for (i) $\mu > 0$, (ii) $\mu < 0$. In the case of $\mu > 0$, we observe that for large values of $x_1(0)$, the system response is damped and the amplitude of $x_1(t) = y(t)$ decreases till the system state enters the limit cycle as shown by the outer trajectory. On the other hand, if initially $x_1(0)$ is small, the damping is negative, hence the amplitude of $x_1(t) = y(t)$ increases till the system state enters the limit cycle as shown by the inner trajectory. The limit cycle is a stable one, since the paths in its neighbourhood converge toward the limit cycle. Figure 10.26b shows an unstable limit cycle for $\mu < 0$.

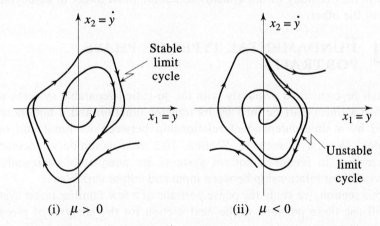

Fig. 10.26b A second-order nonlinear system on the phase plane

The *phase plane* for second-order systems is indeed a special case of *phase space* or *state space* defined for nth-order systems. Much work has been done to extend this approach of analysis to third-order systems. Though a phase trajectory for a third-order system can be graphically visualized through its projections on two planes, say (x_1, x_2) and (x_2, x_3) planes, this complicacy causes the technique to lose its major power of quick graphical visualization of the total system response. The phase trajectories are therefore generally restricted to second-order systems only.

For time-invariant systems, the entire phase plane is covered with trajectories with one and only one curve passing through each point of the plane except for certain critical points through which either infinite number or none of the trajectories pass. Such points (called *singular points*) are discussed in the next section.

If the parameters of a system vary with time, or if a time-varying driving function is imposed, two or more trajectories may pass through a single point in a phase plane. In such cases, the phase portrait becomes complex and more difficult to work with and interpret. Therefore, the use of phase-plane analysis is restricted to second-order systems with constant parameters and constant or zero input. However, it may be mentioned that investigators have made fruitful use of the phase-plane method in investigating second-order time-invariant systems under simple time-varying inputs such as ramp. Some simple time-varying systems have also been analysed by this method. Our discussion will be limited to second-order time-invariant systems with constant or zero input.

From the above discussion, we observe that the phase-plane analysis applies primarily to systems described by second-order differential equations. In the case of feedback control systems, systems of order higher than the second are likely to be well filtered and tractable by the describing-function method discussed earlier in this chapter. The two methods of the phase plane and of the describing function are therefore complementary to a large extent, each being available for the study of the systems which are most likely to be beyond the scope of the other.

10.8 FUNDAMENTAL TYPES OF PHASE PORTRAITS

We shall be concerned primarily with the so-called separable systems which comprise a linear part defined by its transfer function and a nonlinear part defined by a time-independent relationship between its input and output variables. We have seen in Section 10.2 that commonly encountered nonlinearities in feedback control systems are adequately represented by piecewise linear relationship between input and output variables.

In this section, we study the phase portraits of a few familiar linear systems. We will use these portraits in the next section for the analysis of piecewise linear systems.

Type-2 System

Consider the linear system with the transfer function

$$\frac{Y(s)}{R(s)} = \frac{1}{Js^2}; \qquad (10.22a)$$

$y(t)$ represents the motion of a purely inertial load subjected to a torque proportional to $r(t)$. For a constant input $r(t) = A$, we obtain the equation

$$J\ddot{y} = A \tag{10.22b}$$

In terms of the state variables $x_1 = y$ and $x_2 = \dot{y}$, the state equations become

$$\dot{x}_1 = x_2; \quad \dot{x}_2 = \frac{A}{J} \tag{10.22c}$$

Elimination of t by division yields the equation of the trajectories:

$$\frac{dx_2}{dx_1} = \frac{A}{Jx_2} \tag{10.23}$$

or
$$Jx_2 \, dx_2 = A \, dx_1$$

This equation is easily integrated; the general solution is

$$x_1(t) = \frac{Jx_2^2(t)}{2A} + C \tag{10.24a}$$

where C is a constant of integration and is determined by initial conditions, i.e.,

$$C = x_1(0) - \frac{Jx_2^2(0)}{2A} \tag{10.24b}$$

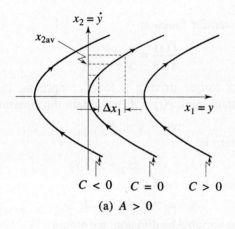

(a) $A > 0$

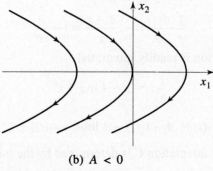

(b) $A < 0$

Fig. 10.27 Phase potraits for a type-2 system

For an initial state point $(x_1(0), x_2(0))$, the trajectory is a parabola passing through the point $x_1 = C$ on the x_1-axis where C is defined by Eqn. (10.24b).

A family of parabolas in the (x_1, x_2) plane is shown in Fig. 10.27a for $A > 0$. As time t increases, each trajectory is described in the clockwise direction as indicated by the arrows. The direction of the phase trajectories is dictated by the relationship $\dot{x}_1 = x_2$: x_1 increases with time in the upper half of the phase plane and the state point therefore moves from left to right ($\rightarrow$); in the lower half of the phase plane, x_1 decreases with time and the state point must therefore move from right to left ($\leftarrow$).

The time interval between two points of a trajectory is given by $\Delta t = \Delta x_1 / x_{2av}$. The trajectories may be provided with a time scale by means of this equation. This operation is, however, often unnecessary since the phase portrait is mainly used to display the general features of the system transients.

The phase portrait for $A < 0$ is shown in Fig. 10.27b. In the special case of $A = 0$ (no driving torque), the integration of the trajectory equation (10.23) gives $x_2(t) = x_2(0)$. The trajectories are therefore straight lines parallel to x_1-axis.

Type-1 System

Consider now the transfer function

$$\frac{Y(s)}{R(s)} = \frac{1}{s(\tau s + 1)} \qquad (10.25a)$$

corresponding to a torque driving a load comprising inertia and viscous friction. For a constant input $r(t) = A$, we obtain the equation

$$\tau \ddot{y} + \dot{y} = A \qquad (10.25b)$$

The equivalent system is

$$\dot{x}_1 = x_2; \quad \tau \dot{x}_2 = A - x_2 \qquad (10.25c)$$

Let us take a new variable z such that

$$A - x_2 = z; \quad dx_2 = -dz$$

Eliminating the time variable by division, we obtain

$$\frac{1}{\tau} \frac{dx_1}{dz} = -\frac{A-z}{z} = 1 - \frac{A}{z}$$

This first-order equation is readily integrated.

$$\frac{1}{\tau} x_1 = z - A \ln z + C$$

or

$$\frac{1}{\tau} x_1(t) = A - x_2(t) - A \ln(A - x_2(t)) + C \qquad (10.26a)$$

where the constant of integration C is determined by the initial conditions, i.e.,

$$C = \frac{1}{\tau} x_1(0) - A + x_2(0) + A \ln(A - x_2(0)) \qquad (10.26b)$$

Therefore, the trajectory equation becomes

$$\frac{1}{\tau}(x_1 - x_1(0)) = -(x_2 - x_2(0)) - A \ln\left(\frac{A - x_2}{A - x_2(0)}\right) \quad (10.26c)$$

The phase portrait is shown in Fig. 10.28a for $A > 0$.

For the case of initial state point at the origin ($x_1(0) = x_2(0) = 0$), Eqn. (10.26c) reads

$$\frac{1}{\tau}x_1 = -x_2 - A \ln\left(\frac{A - x_2}{A}\right) \quad (10.26d)$$

The phase trajectory described by this equation is shown in Fig. 10.28a as the curve Γ_0. It is seen that the trajectory is asymptotic to the line $x_2 = A$, which is the final velocity.

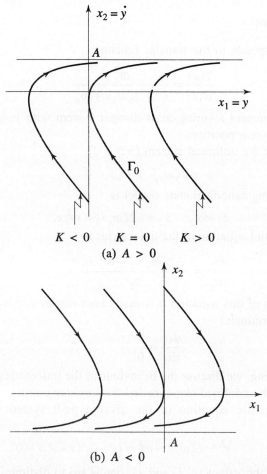

Fig. 10.28 Phase portraits for a type-1 system

For an initial state point $(x_1(0), x_2(0))$, the trajectory will have the same shape as the curve Γ_0, except that it is shifted horizontally so that it passes through the point $(x_1(0), x_2(0))$. This is obvious from Eqn. (10.26c) which can be written as

$$\frac{1}{\tau}(x_1 - K) = -x_2 - A \ln\left(\frac{A - x_2}{A}\right)$$

where $$K = x_1(0) + \tau x_2(0) + \tau A \ln\left(\frac{A - x_2(0)}{A}\right)$$

For an initial state point $(x_1(0), x_2(0))$, the trajectory is Γ_0 shifted horizontally by K units. The phase portrait for $A < 0$ is shown in Fig. 10.28b. In the special case of $A = 0$, the phase portrait consists of a family of straight lines of slope $-1/\tau$.

Type-0 System

This case corresponds to the transfer function

$$\frac{Y(s)}{R(s)} = \frac{\omega_n^2}{s^2 + 2\zeta\omega_n s + \omega_n^2} \tag{10.27}$$

which might represent a spring-mass-damper system with $r(t)$ as the driving force and y the linear position.

Consider first the unforced system ($r(t) = 0$):

$$\ddot{y} + 2\zeta\omega_n \dot{y} + \omega_n^2 y = 0 \tag{10.28a}$$

The corresponding canonical state model is

$$\dot{x}_1 = x_2; \quad \dot{x}_2 = -2\zeta\omega_n x_2 - \omega_n^2 x_1 \tag{10.28b}$$

and the differential equation of the trajectories is

$$\frac{dx_2}{dx_1} = \frac{-2\zeta\omega_n x_2 - \omega_n^2 x_1}{x_2} \tag{10.29}$$

By inspection of this equation it is easily seen that at $x_1 = x_2 = 0$, the slope dx_2/dx_1 is indeterminate:

$$\frac{dx_2}{dx_1} = \frac{0}{0} \tag{10.30}$$

In the following, we discuss the behaviour of the trajectories in the vicinity of this point with undefined slope, called the *singular point*.

The characteristic equation of the given type-0 system is (refer Eqn. (10.27))

$$s^2 + 2\zeta\omega_n s + \omega_n^2 = (s - \lambda_1)(s - \lambda_2) = 0 \tag{10.31}$$

According to the values of λ_1 and λ_2, one is led to distinguish between the six types of singular points shown in Fig. 10.29. Let us examine each of these cases in detail.

1. **Stable system with complex roots**

$$\lambda_1 = -\alpha + j\beta, \quad \lambda_2 = -\alpha - j\beta; \quad \alpha > 0, \beta > 0$$

The response $\quad y(t) = C_1 e^{-\alpha t} \sin(\beta t + C_2)$ (10.32)

where the constants C_1 and C_2 are determined by the initial conditions.

Using Eqn. (10.32), we can construct a phase portrait on the (x_1, x_2)-plane with $x_1 = y$ and $x_2 = \dot{y}$. A typical phase trajectory is shown in Fig. 10.29a which is a logarithmic spiral into the singular point. This type of singular point is called a *stable focus*.

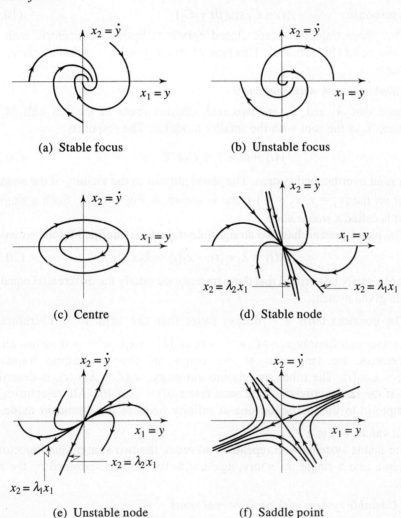

(a) Stable focus (b) Unstable focus

(c) Centre (d) Stable node

(e) Unstable node (f) Saddle point

Fig. 10.29 Phase portraits for type-0 systems

2. Unstable system with complex roots

$$\lambda_1 = \alpha + j\beta, \quad \lambda_2 = \alpha - j\beta; \quad \alpha > 0, \beta > 0$$

The response
$$y(t) = C_1 e^{\alpha t} \sin(\beta t + C_2) \tag{10.33}$$

The transient is an exponentially increasing sinusoid; the phase trajectory on the $(x_1 = y, x_2 = \dot{y})$-plane is a logarithmic spiral expanding out of the singular point (Fig. 10.29b). This type of singular point is called an *unstable focus*.

3. Marginally stable system with complex roots

$$\lambda_1 = j\beta, \quad \lambda_2 = -j\beta; \quad \beta > 0$$

The response
$$y(t) = C_1 \sin(\beta t + C_2) \tag{10.34}$$

The phase trajectories are closed curves (elliptical) concentric with the singular point (Fig. 10.29c). This type of singular point is called a *centre*, or a *vortex*.

4. Stable system with real roots

Assume that λ_1 and λ_2 are two real, distinct roots in the left half of the s-plane; λ_1 is the root with the smaller modulus. The response

$$y(t) = C_1 e^{\lambda_1 t} + C_2 e^{\lambda_2 t} \tag{10.35}$$

It is an overdamped system. The phase portrait in the vicinity of the singular point on the $(x_1 = y, x_2 = \dot{y})$-plane is shown in Fig. 10.29d. Such a singular point is called a *stable node*.

The phase portrait has two straightline trajectories defined by the equations

$$x_2(t) = \lambda_1 x_1(t); \quad x_2(t) = \lambda_2 x_1(t) \tag{10.36}$$

It can easily be verified that these trajectories satisfy the differential equation of the given system.

The transient term $e^{\lambda_2 t}$ decays faster than the term $e^{\lambda_1 t}$. Therefore as t increases indefinitely, $x_1 \to C_1 e^{\lambda_1 t} \to 0$, and $x_2 \to \lambda_1 C_1 e^{\lambda_1 t} \to 0$, so that all the trajectories are tangential at the origin to the straightline trajectory $x_2(t) = \lambda_1 x_1(t)$. The other straightline trajectory, $x_2(t) = \lambda_2 x_1(t)$, is described only if the initial conditions are such that $x_2(0) = \lambda_2 x_1(0)$. All trajectories are asymptotic to this particular one at infinity because the dominant mode for small values of t is $e^{\lambda_2 t}$.

For stable systems with repeated real roots, the two straightline trajectories coalesce into a single trajectory, again with the slope determined by the root value.

5. Unstable system with positive real roots

Assume that λ_1 and λ_2 are two real distinct roots in the right half of tne s-plane; λ_1 is the smaller root. The phase portrait in the vicinity of the singular point on the $(x_1 = y, x_2 = \dot{y})$-plane is shown in Fig. 10.29e. Such a singular point is called an *unstable node*.

All trajectories emerge from the singular point and go to infinity. The trajectories are tangential at the origin to the straightline trajectory, $x_2(t) = \lambda_1 x_1(t)$, and asymptotic at infinity to the other straightline trajectory.

If the roots are repeated, the two trajectories coalesce into a single trajectory.

6. **Unstable system with one negative real root and one positive real root**

The phase portrait in the vicinity of the singular point on the $(x_1 = y, x_2 = \dot{y})$-plane is shown in Fig. 10.29f. Such a singular point is called a *saddle*.

There are two straightline trajectories with slopes defined by the root values. The straightline due to the negative root provides a trajectory that *enters* the singular point, while the straightline trajectory due to the positive root *leaves* the singular point. All other trajectories approach the singular point adjacent to the incoming straight line, then curve away and leave the vicinity of the singular point, eventually approaching the second straight line asymptotically.

So far we have considered the case of zero input. Let us now return to the general case and make r constant. With $r = A$, we have from Eqn. (10.27):

$$\ddot{y} + 2\zeta\omega_n \dot{y} + \omega_n^2 y = A\omega_n^2 \qquad (10.37a)$$

If we write this equation in the form

$$\frac{d^2}{dt^2}(y - A) + 2\zeta\omega_n \frac{d}{dt}(y - A) + \omega_n^2(y - A) = 0 \qquad (10.37b)$$

and compare it with Eqn. (10.28a), we see that the trajectories are derived from those of Fig. 10.29 by shifting the singular point from the origin to the point $y = A$ on the y-axis.

10.9 SYSTEM ANALYSIS ON THE PHASE PLANE

Consider a time-invariant second-order system with constant or zero input, described by equations of the form

$$\dot{x}_1 = x_2; \quad \dot{x}_2 = f(x_1, x_2) \qquad (10.38)$$

Elimination of t by division yields the equation of the trajectories:

$$\frac{dx_2}{dx_1} = \frac{f(x_1, x_2)}{x_2} \qquad (10.39)$$

Integration of this equation, analytically or numerically (refer Appendix B), for various initial conditions, yields a family of phase trajectories which displays the general features of the system transients.

Example 10.3

Consider the nonlinear system shown in Fig. 10.30. The nonlinear element is a relay with deadzone whose characteristics are shown in the figure.

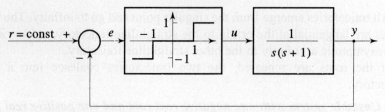

Fig. 10.30 A system controlled by relay with deadzone

The differential equation governing the dynamics of the system is given by

$$\ddot{y} + \dot{y} = u; \quad \text{or} \quad \ddot{e} + \dot{e} = -\phi(e) \quad (10.40)$$

where $e = r - y$; r is constant, and

$$\phi(e) = \begin{cases} +1; & e > 1 \\ 0; & -1 < e < 1 \\ -1; & e < -1 \end{cases}$$

Choosing the state variables $x_1 = e$ and $x_2 = \dot{e}$, we obtain the following first-order equations:

$$\dot{x}_1 = x_2; \quad \dot{x}_2 = -x_2 - \phi(x_1)$$

The phase plane may be divided into three regions:

1. *Region I* (defined by $x_1 > 1$): The trajectories in this region are given by the equation (refer Eqn. (10.26c): $\tau = 1$, $A = -1$)

$$x_1 - x_1(0) = -(x_2 - x_2(0)) + \ln\left(\frac{1 + x_2}{1 + x_2(0)}\right) \quad (10.41a)$$

The trajectories are asymptotic to the ordinate -1.

2. *Region II* (defined by $-1 < x_1 < 1$): The trajectories in this region are given by the equation (refer Eqn. (10.26c): $\tau = 1$, $A = 0$)

$$x_1 - x_1(0) = -(x_2 - x_2(0)) \quad (10.41b)$$

3. *Region III* (defined by $x_1 < -1$): The trajectories in this region are given by the equation (refer Eqn. (10.26c): $\tau = 1$, $A = 1$)

$$x_1 - x_1(0) = -(x_2 - x_2(0)) - \ln\left(\frac{1 - x_2}{1 - x_2(0)}\right) \quad (10.41c)$$

The trajectories are asymptotic to the ordinate $+1$.

For a step input $r = 3$ and zero initial conditions, the initial point of the phase trajectory is located at P in Fig. 10.31. The figure also shows a phase trajectory, constructed using Eqns (10.41).

It is important to note that a small deadzone region is not always undesirable in relay controllers. Let us investigate the behaviour of the system of Fig. 10.30 using ideal relay (no deadzone) as a controller. For such a controller, the width of region II (corresponding to deadzone) in the phase plane reduces to zero. The phase trajectory of such a system with $r = 3$ is shown in Fig. 10.32 : $e(t)$ oscillates about the origin with ever-decreasing amplitude and ever-increasing frequency.

Comparison of Figs 10.31 and 10.32 reveals that deadzone in relay characteristic helps to reduce system oscillations, thereby reducing settling time. However the relay controller with deadzone drives the system to a point within the deadzone width. A large deadzone would of course cause the steady-state performance of the system to deteriorate.

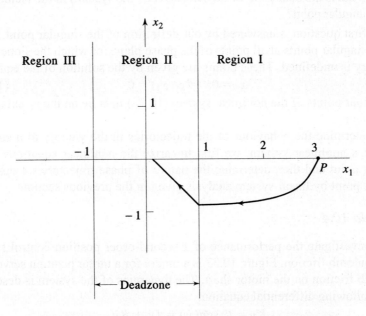

Fig. 10.31 A typical trajectory for the system in Fig. 10.30

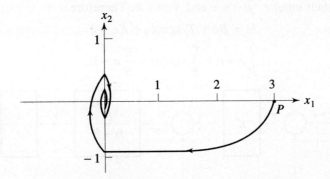

Fig. 10.32 Phase trajectory for the system in Fig. 10.30 when the deadzone is absent

Singular Points of a Nonlinear System

Every point (x_1, x_2) of the phase plane has associated with it the slope of the trajectory which passes through that point. The only exception are the *singular points* at which the trajectory slope is indeterminate (refer Eqn. (10.39)):

$$\frac{dx_2}{dx_1} = \frac{0}{0} = \frac{f(x_1, x_2)}{x_2}$$

We need to know the following:
1. Where will the singular points be and how many will there be?
2. What is the behaviour of trajectories (i.e., the system) in the vicinity of a singular point?

The first question is answered by our definition of the singular point. There will be singular points at all points of the phase plane for which the slope of the trajectory is undefined. These points are given by the solution of the equations

$$x_2 = 0; \quad f(x_1, x_2) = 0 \tag{10.42}$$

Singular points of the nonlinear system (10.38) thus lie on the x_1-axis of the phase plane.

To determine the behaviour of the trajectories in the vicinity of a singular point of a nonlinear system, we first linearize the nonlinear equations at the singular point and then determine the nature of phase trajectories around the singular point by linear system analysis given in the previous section.

Example 10.4

Let us investigate the performance of a second-order position control system with Coulomb friction. Figure 10.33 is a model for a motor position servo with Coulomb friction on the motor shaft. The dynamics of the system is described by the following differential equation:

$$Ke - T_c \operatorname{sgn}(\dot{y}) = J\ddot{y} + B\dot{y}$$

where T_c is the Coulomb frictional torque.

For constant input r, $\dot{y} = -\dot{e}$ and $\ddot{y} = -\ddot{e}$. Therefore

$$J\ddot{e} + B\dot{e} + T_c \operatorname{sgn}(\dot{e}) + Ke = 0$$

or

$$\frac{J}{B}\ddot{e} + \dot{e} + \frac{T_c}{B} \operatorname{sgn}(\dot{e}) + \frac{K}{B}e = 0$$

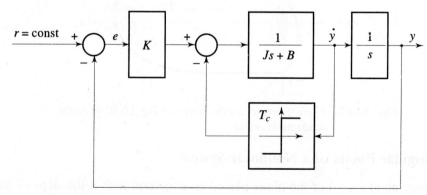

Fig. 10.33 A motor position servo with Coulomb friction on the motor shaft

Letting $J/B = \tau$, we get

$$\tau\ddot{e} + \dot{e} + \frac{T_c}{B}\text{sgn}(\dot{e}) + \frac{K}{B}e = 0 \quad (10.43)$$

In terms of state variables

$$x_1 = e;\ x_2 = \dot{e}$$

we get the following description of the system:

$$\dot{x}_1 = x_2;\quad \tau\dot{x}_2 = -\frac{K}{B}x_1 - x_2 - \frac{T_c}{B}\text{sgn}(x_2) \quad (10.44)$$

The singular points are given by the solution of the equations (refer Eqns (10.42))

$$0 = x_2;\quad 0 = -\frac{K}{B}x_1 - x_2 - \frac{T_c}{B}\text{sgn}(x_2)$$

The solution gives

$$x_1 = -\frac{T_c}{K}\text{sgn}(x_2)$$

Thus, there are two singular points. Their location can be interpreted physically—they are at a value of $e = x_1$, such that $|Ke| = |T_c|$, i.e., the drive torque is exactly equal to the Coulomb-friction torque. We note that both the singular points are on the x_1-axis ($x_2 \equiv 0$) and that the singular point given by $x_1 = T_c/K$ is related to the lower-half phase plane (x_2 negative) and the singular point given by $x_1 = -T_c/K$ is related to the upper-half phase plane (x_2 positive).

Let us now investigate the stability of the singular points. For $\dot{e} > 0$, Eqn. (10.43) may be expressed as

$$\tau\frac{d^2}{dt^2}\left(e + \frac{T_c}{K}\right) + \frac{d}{dt}\left(e + \frac{T_c}{K}\right) + \frac{K}{B}\left(e + \frac{T_c}{K}\right) = 0 \quad (10.45)$$

This is a linear second-order system with the singular point at $(-T_c/K, 0)$ on the $(e, \dot{e})$-plane. The characteristic equation of this system is given by

$$\lambda^2 + \frac{1}{\tau}\lambda + \frac{K}{\tau B} = 0$$

Let us assume the following parameter values for the system under consideration:

$$(K/B) = 5,\ \tau = 4 \quad (10.46)$$

With these parameters, the roots of the characteristic equation are complex conjugate with negative real parts; the singular point is therefore a stable focus (refer Fig. 10.29).

Let us now investigate the system behaviour when large inputs are applied. Phase trajectories may be obtained by solving the following second-order differential equations for given initial state points (refer Eqns (10.31)–(10.32)).

Region I (defined by $x_2 > 0$):

$$4\ddot{z} + \dot{z} + 5z = 0;\ z = x_1 + \frac{T_c}{K};\ x_2 = \dot{x}_1$$

Region II (defined by $x_2 < 0$):

$$4\ddot{z} + \dot{z} + 5z = 0;\ z = x_1 - \frac{T_c}{K};\ x_2 = \dot{x}_1$$

Figure 10.34 shows a few phase trajectories. It is observed that for small as well as large inputs, the resulting trajectories terminate on a line along the x_1-axis from $-T_c/K$ to $+T_c/K$, i.e., the line joining the singular points. Therefore, the system with Coulomb friction is stable; however, there is a possibility of large steady-state error.

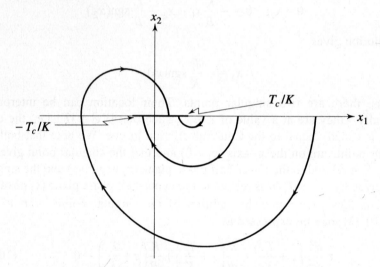

Fig. 10.34 Phase portrait for the system in Fig. 10.33

10.10 OPTIMAL SWITCHING IN BANG-BANG CONTROL SYSTEMS

Consider the system of Fig. 10.35, in which an ideal relay is actuated by a signal which is a function $\phi(e, \dot{e})$ of the error and its first derivative. The differential equation describing the dynamics of the system is given by

$$\ddot{y} = u \quad \text{or} \quad \ddot{e} = -u$$

where $e = r - y$; r is constant, and

$$u = \begin{cases} +1;\ \phi(e, \dot{e}) > 0 \\ -1;\ \phi(e, \dot{e}) < 0 \end{cases}$$

Choosing $x_1 = e$ and $x_2 = \dot{e}$ as state variables, we have

$$\dot{x}_1 = x_2;\ \dot{x}_2 = -u$$

The trajectories corresponding to $u = -1$ are given by (refer Eqns (10.24))

$$x_1(t) = \tfrac{1}{2}x_2^2(t) + x_1(0) - \tfrac{1}{2}x_2^2(0) \qquad (10.47a)$$

and the trajectories corresponding to $u = +1$ are given by

$$x_1(t) = -\tfrac{1}{2}x_2^2(t) + x_1(0) + \tfrac{1}{2}x_2^2(0) \qquad (10.47b)$$

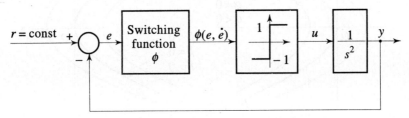

Fig. 10.35 An on–off control system

Equations (10.47a) and (10.47b) each define a family of parabolas shown in Fig. 10.36. One trajectory of each family goes through the origin. Segments A–O and B–O of these two trajectories terminating at the origin form the curve shown by the thick line in Fig. 10.37.

The following strategy seems to give an optimal control performance.
1. If the initial state point lies at P_1 on the segment A–O (Fig. 10.37), the state point $(x_1(t), x_2(t))$ is driven to the origin along the trajectory P_1–O, a segment of a parabola corresponding to $u = +1$.
2. If the initial point lies at P_2 on the segment B–O (Fig. 10.37), the state point $(x_1(t), x_2(t))$ is driven to the origin along the trajectory P_2–O, a segment of a parabola corresponding to $u = -1$.
3. If the initial state point lies above or below the curve A–O–B, then the relay has to switch only once to drive the state point to the origin. Consider the initial state point at P_3 which is above the curve A–O–B (Fig. 10.37). The state point $(x_1(t), x_2(t))$ follows a parabola corresponding to $u = +1$ till it reaches the segment B–O. This is followed by switching of the control to -1 and driving of the state point to the origin along B–O with $u = -1$.
4. Consider the initial point at P_4 which is below the curve A–O–B (Fig. 10.37). The state point $(x_1(t), x_2(t))$ follows a parabola corresponding to $u = -1$ till it reaches the segment A–O. This is followed by switching of the control to $+1$ and driving of the state point to the origin along A–O with $u = +1$.

It is also clear from Fig. 10.37 that for all initial conditions, the state point is driven to the origin along the shortest-time path with no oscillations (the output reaches the final value in minimum time with no ripples, and stays there; this type of response is commonly called a *deadbeat response*). Such bang-bang control systems provide optimal control (*minimum-time control*) [102].

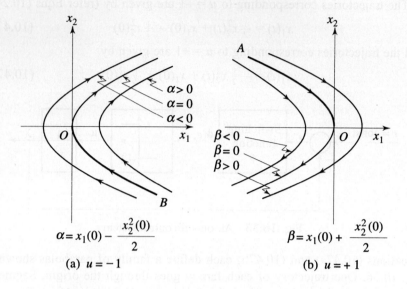

Fig. 10.36 Phase trjectories for the system of Fig. 10.35

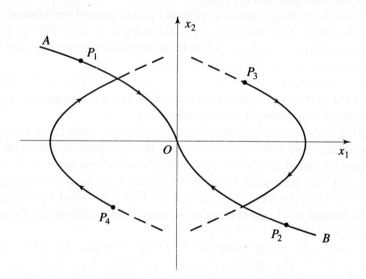

Fig. 10.37 Optimal switching curve

The equation of *optimum switching curve A–O–B* can be obtained from Eqns (10.47) by setting (refer Fig. 10.36)

$$x_1(0) - \frac{x_2^2(0)}{2} = x_1(0) + \frac{x_2^2(0)}{2} = 0$$

This gives

$$x_1(t) = -\tfrac{1}{2} x_2(t) |x_2(t)| \tag{10.48a}$$

Let us define the switching function $\phi(e, \dot{e}) = \phi(x_1, x_2)$ as

$$\phi(x_1, x_2) = x_1(t) + \tfrac{1}{2} x_2(t) |x_2(t)| \quad (10.48b)$$

1. $\phi(x_1, x_2) > 0$ implies that the state point (x_1, x_2) lies above the curve A–O–B.
2. $\phi(x_1, x_2) = 0$ and $x_2 > 0$ implies that the state point (x_1, x_2) lies on the segment A–O.
3. $\phi(x_1, x_2) = 0$ and $x_2 < 0$ implies that the state point (x_1, x_2) lies on the segment B–O.
4. $\phi(x_1, x_2) < 0$ implies that the state point (x_1, x_2) lies below the segment A–O–B.

In terms of the *optimal switching function* $\phi(x_1, x_2)$, the control law becomes

$$u(t) = \begin{cases} +1 & \text{when } \phi(x_1, x_2) > 0 \\ -1 & \text{when } \phi(x_1, x_2) = 0 \text{ and } x_2(t) < 0 \\ -1 & \text{when } \phi(x_1, x_2) < 0 \\ +1 & \text{when } \phi(x_1, x_2) = 0 \text{ and } x_2(t) > 0 \end{cases} \quad (10.49)$$

The optimal switching may be realized by a computer. It accepts the state point (x_1, x_2) and computes the switching function given by Eqn. (10.48b). It then controls the ideal relay to produce the optimal control components according to Eqn. (10.49).

10.11 REVIEW EXAMPLES

Review Example 10.1

Figure 10.38 shows the input-output waveforms of an on–off controller with deadzone and hysteresis. If X is less than $\Delta + H$, then the controller produces no output; the first harmonic component of the Fourier series is of course zero, and the describing function is also zero. If $X > \Delta + H$, the controller output y is equal to M, and it remains at this level until X becomes less than $\Delta - H$. One cycle of the output y, which is a periodic function of period 2π, is described as follows:

$$y = \begin{cases} 0; & 0 \leq \omega t < \alpha \\ M; & \alpha \leq \omega t < (\pi - \beta) \\ 0; & (\pi - \beta) \leq \omega t < (\pi + \alpha) \\ -M; & (\pi + \alpha) \leq \omega t < (2\pi - \beta) \\ 0; & (2\pi - \beta) \leq \omega t \leq 2\pi \end{cases}$$

where $X \sin\alpha = \Delta + H$, $X \sin\beta = \Delta - H$

This periodic function does not possess odd symmetry:
$$y(\omega t) \neq -y(-\omega t),$$
but possesses odd half-wave symmetry:
$$y(\omega t \pm \pi) = -y(\omega t)$$

Therefore, the fundamental component of y is given by (refer Eqn. (A.7e) in Appendix A)

$$y_1 = A_1\cos\omega t + B_1\sin\omega t$$

where $\quad A_1 = \dfrac{1}{\pi}\displaystyle\int_0^{2\pi} y\cos\omega t\, d(\omega t); \quad B_1 = \dfrac{1}{\pi}\displaystyle\int_0^{2\pi} y\sin\omega t\, d(\omega t)$

Due to the symmetry of y, only the positive half-wave need be considered (Fig. 10.38):

$$A_1 = \dfrac{2}{\pi}\left[\int_\alpha^{\pi-\beta} M\cos\omega t\, d(\omega t)\right] = \dfrac{2M}{\pi}(\sin\beta - \sin\alpha) = -\dfrac{4HM}{\pi X}$$

$$B_1 = \dfrac{2}{\pi}\left[\int_\alpha^{\pi-\beta} M\sin\omega t\, d(\omega t)\right] = \dfrac{2M}{\pi}(\cos\beta + \cos\alpha)$$

$$= \dfrac{2M}{\pi}\left[\sqrt{1-\left(\dfrac{\Delta-H}{X}\right)^2} + \sqrt{1-\left(\dfrac{\Delta+H}{X}\right)^2}\right]$$

$$N(X) = \dfrac{1}{X}(B_1 + jA_1)$$

$$= \dfrac{2M}{\pi X}\left[\sqrt{1-\left(\dfrac{\Delta-H}{X}\right)^2} + \sqrt{1-\left(\dfrac{\Delta+H}{X}\right)^2} - j\dfrac{2H}{X}\right]; \quad X \geq \Delta+H$$

$$= 0 \qquad\qquad\qquad\qquad\qquad\qquad\qquad ; \quad X < \Delta+H \qquad (10.50)$$

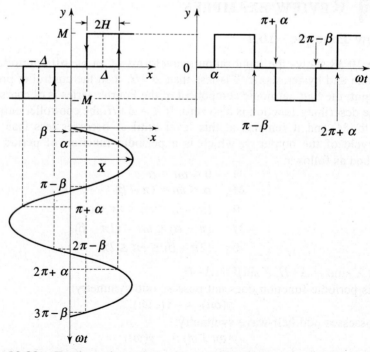

Fig. 10.38 Fourier-series analysis of an on–off controller with deadzone and hysteresis

Review Example 10.2

Figure 10.39 shows the input-output waveforms of a saturating element or a limiter. For small input signals ($X < S$), the output is proportional to the input. However, if the input amplitude is sufficiently large to cause saturation ($X > S$), the output is a clipped sine wave. One cycle of the output, which is a periodic function of period 2π, is described as follows:

$$y = \begin{cases} Kx; & 0 \leq \omega t < \alpha \\ KS; & \alpha \leq \omega t < (\pi - \alpha) \\ Kx; & (\pi - \alpha) \leq \omega t < (\pi + \alpha) \\ -KS; & (\pi + \alpha) \leq \omega t < (2\pi - \alpha) \\ Kx; & (2\pi - \alpha) \leq \omega t \leq 2\pi \end{cases}$$

where $\alpha = \sin^{-1}(S/X)$

This periodic function has odd symmetry:

$$y(\omega t) = -y(-\omega t)$$

Therefore, the fundamental component of y is given by (refer Eqn. (A.7d) in Appendix A)

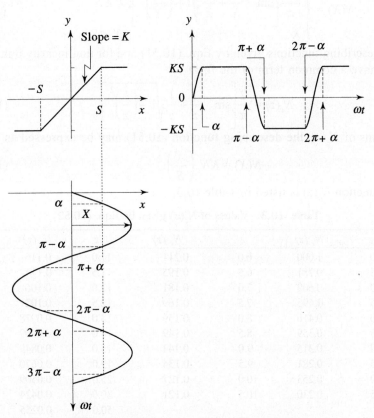

Fig. 10.39 Fourier-series analysis of a saturating element

$$y_1 = B_1 \sin \omega t$$

where $B_1 = \dfrac{1}{\pi} \displaystyle\int_0^{2\pi} y \sin \omega t \, d(\omega t)$

Due to symmetry of y (refer Fig. 10.39), the coefficient B_1 can be calculated as follows:

$$B_1 = \frac{4}{\pi} \int_0^{\pi/2} y \sin\theta \, d\theta = \frac{4K}{\pi} \left[\int_0^\alpha X\sin^2\theta \, d\theta + \int_\alpha^{\pi/2} S\sin\theta \, d\theta \right]$$

$$= \frac{4K}{\pi} \left[\frac{X}{2}(\alpha - \sin\alpha\cos\alpha) + S\cos\alpha \right]$$

$$\frac{B_1}{X} = \frac{2K}{\pi} \left[\sin^{-1}\frac{S}{X} - \frac{S}{X}\sqrt{1-\left(\frac{S}{X}\right)^2} + \frac{2S}{X}\sqrt{1-\left(\frac{S}{X}\right)^2} \right]$$

Therefore

$$N(X) = \begin{cases} \dfrac{2K}{\pi}\left[\sin^{-1}\dfrac{S}{X} + \dfrac{S}{X}\sqrt{1-\left(\dfrac{S}{X}\right)^2}\right]; & X \geq S \\ K; & X < S \end{cases} \qquad (10.51)$$

The describing functions given by Eqn. (10.51) and for nonlinearity 6 in Table 10.2 have a common term of the form

$$N_c(z) = \frac{2}{\pi}\left[\sin^{-1}\frac{1}{z} + \frac{1}{z}\sqrt{1-\left(\frac{1}{z}\right)^2}\right] \qquad (10.52)$$

In terms of $N_c(z)$, the describing function (10.51) may be expressed as

$$N(X) = KN_c\left(\frac{X}{S}\right) \qquad (10.53)$$

The function $N_c(z)$ is listed in Table 10.3.

Table 10.3 Values of $N_c(z)$ given by Eqn. (10.52)

z	$N_c(z)$	z	$N_c(z)$	z	$N_c(z)$
1.0	1.000	6.0	0.211	11.0	0.116
1.5	0.781	6.5	0.195	11.5	0.111
2.0	0.609	7.0	0.181	12.0	0.106
2.5	0.495	7.5	0.169	12.5	0.102
3.0	0.416	8.0	0.159	13.0	0.0978
3.5	0.359	8.5	0.149	14.0	0.0909
4.0	0.315	9.0	0.141	15.0	0.0848
4.5	0.281	9.5	0.134	19.0	0.0670
5.0	0.253	10.0	0.127	25.0	0.0509
5.5	0.230	10.5	0.121	30.0	0.0424
				50.0	0.0255
				100.0	0.0127

Review Example 10.3

Consider the nonlinear system of Fig. 10.40a, with a saturating amplifier having gain K in its linear region. Determine the largest value of gain K for the system to stay stable. What would be the frequency, amplitude and nature of the limit cycle for a gain $K = 3$?

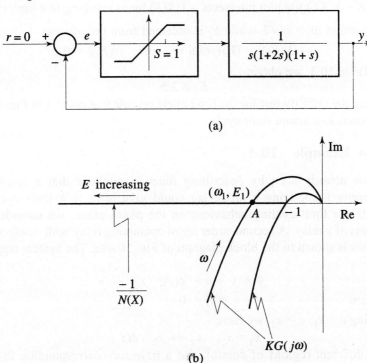

Fig. 10.40 Nonlinear system with saturating amplifier

Solution
It is convenient to regard the amplifier to have unit gain and the gain K to be attached to the linear part. From Eqn. (10.53), we obtain for $S = 1$ and $K = 1$,

$$N(E) = N_c(E);$$

the function $N_c(E)$ is listed in Table 10.3.

The locus of $-1/N(E)$ thus starts from $(-1 + j0)$ and travels along the negative real axis for increasing E, as shown in Fig. 10.40b. Now, for the equation

$$KG(j\omega) = -1/N(X)$$

to be satisfied, $G(j\omega)$ must have an angle of $-180°$:

$$\angle G(j\omega) = -90° - \tan^{-1} 2\omega - \tan^{-1} \omega = -180°$$

This gives

$$\frac{2\omega + \omega}{1 - 2\omega^2} = \tan 90° = \infty \quad \text{or} \quad \omega = 1/\sqrt{2} \text{ rad/sec.}$$

The largest value of K for stability is obtained when $KG(j\omega)$ passes through $(-1 + j0)$, i.e.,

$$|KG(j\omega)|_{\omega = 1/\sqrt{2}} = 1 \quad \text{or} \quad \frac{K}{\left(\dfrac{1}{\sqrt{2}}\right)(\sqrt{3})\left(\dfrac{\sqrt{3}}{\sqrt{2}}\right)} = 1 \text{ or } K = 3/2$$

For $K = 3$, $KG(j\omega)$ plot intersects $-1/N(X)$ locus resulting in a limit cycle at (ω_1, E_1) where $\omega_1 = 1/\sqrt{2}$ while E_1 is obtained from the relation

$$|-1/N(E_1)| = |3G(j\omega_1)| = 2 \quad \text{or} \quad |N(E_1)| = 0.5$$

From Table 10.3, we obtain

$$E_1 \cong 2.5$$

Applying the stability test for the limit cycle reveals that point A in Fig. 10.40b corresponds to a stable limit cycle.

Review Example 10.4

We have already seen by describing function analysis that a two-valued nonlinearity introducing a phase lag could give birth to a limit cycle. To illustrate the limit-cycling behaviour on the phase plane, we consider here hysteresis of a relay. A second-order servo containing relay with deadzone and hysteresis is shown in the block diagram of Fig. 10.41a. The system equations are

$$\ddot{y} + \dot{y} = u = \phi(e); \quad e = r - y$$

Therefore $\quad \ddot{e} + \dot{e} + \phi(e) = 0$

Assuming $e = x_1$, $\dot{e} = x_2$, we have

$$\dot{x}_1 = x_2; \quad \dot{x}_2 = -x_2 - \phi(x_1)$$

The different regions of control, and a trajectory corresponding to initial conditions $e(0) = 0.65$, $\dot{e}(0) = 0$, are shown in Fig. 10.41b (revisiting Example 10.3 will be helpful). It is found that there exists a limit cycle whose amplitude is 0.2. Time period of the limit-cycle may be obtained by determining graphically the time along the limit-cycle trajectory. We can, however, obtain the parameters of the limit-cycle more accurately from the computer solution of the system equations.

Review Example 10.5

Consider the servomechanism of Fig. 10.42a. The velocity of the output member is limited to the value 0.4. The region of linear proportional operation is thus limited in the phase plane by the two straight lines of ordinates 0.4 and -0.4.

The system equations in the linear range are

$$\ddot{y} + \dot{y} = e$$

or $\quad \ddot{e} + \dot{e} + e = 0 \hfill (10.54)$

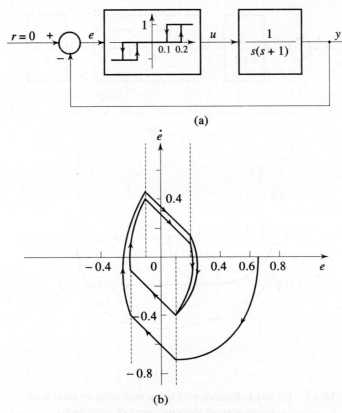

Fig. 10.41 (a) Block diagram of a servo containing relay controller with deadzone and hysteresis (b) phase portrait showing limit cycle

On the $(e, \dot{e})$-plane, the origin is a stable focus (revisiting Example 10.4 will be helpful). For an initial error amplitude of 0.74, the trajectory is tangent to the ordinate -0.4 (Fig. 10.42b). For a larger initial amplitude, e.g., 1.6, the representative point first follows a portion of the trajectory described by Eqn. (10.54) and then moves along the straight line of ordinate -0.4 until it meets the trajectory tangent to the ordinate (Fig. 10.42b).

It is obvious that the transient is slowed down as a result of saturation, since the velocity along the continuous trajectory of Fig. 10.42b is always smaller than that along the dotted trajectory.

Review Example 10.6

Let us reconsider the system of Fig. 10.35. For this system, optimal switching function $\phi(e, \dot{e})$ was described in Section 10.10.

In this example, we discuss a suboptimal method of switching the relay in Fig. 10.35. The advantage of the method described below is that implementation of the switching function is simple. The cost paid is in terms of increase in settling time compared to the optimal solution.

710 Digital Control and State Variable Methods

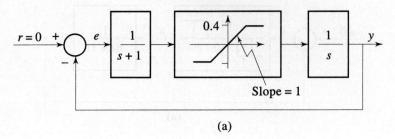

(a)

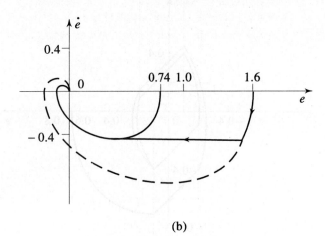

(b)

Fig. 10.42 (a) Block diagram for a servo with velocity saturation
(b) Phase portrait showing effect of saturation

Consider the following suboptimal switching function:
$$\phi(e, \dot{e}) = e + K_D \dot{e}$$
The system equations now become
$$\ddot{e} = -u; \quad u = \text{sgn}(e + K_D \dot{e})$$
In the state variable form ($x_1 = e$, $x_2 = \dot{e}$),
$$\dot{x}_1 = x_2; \quad \dot{x}_2 = -\text{sgn}(x_1 + K_D x_2)$$
The phase plane is divided into two regions by the switching line
$$x_1 + K_D x_2 = 0 \tag{10.55}$$
The trajectory equation for the region defined by $x_1 + K_D x_2 < 0$, is (refer Eqn. (10.47a))
$$x_1(t) = \tfrac{1}{2} x_2^2(t) + x_1(0) - \tfrac{1}{2} x_2^2(0)$$
and the trajectory equation for the region defined by $x_1 + K_D x_2 > 0$, is (refer Eqn. (10.47b))
$$x_1(t) = -\tfrac{1}{2} x_2^2(t) + x_1(0) + \tfrac{1}{2} x_2^2(0)$$

In each half of the phase plane separated by the switching line (10.55), the system trajectories would be parabolas. Assume that the system under consideration starts with initial conditions corresponding to point A in Fig. 10.43. The relay switches when the representative point reaches B. By geometry of the situation, we see that the trajectory resulting from the reversal of the drive at point B will bring the representative point on a parabola passing much closer to the origin. This will continue until the trajectory intersects the switching line at a point closer to the origin than the points A_1 and A_2 which are points of intersection of the switching line with parabolas passing through the origin. In Fig. 10.43, point C corresponds to this situation. Here, an instant after the relay is switched, the system trajectory will recross the switching line and the relay must switch back. The relay will thus chatter while the system stays on the switching line. In a second-order system, the chattering frequency will be infinite and amplitude will be zero; the representative point thus slides along the switching line. It can easily be verified that with $K_D = 0$ (relay switching on the vertical axis), the system always enters into a limit cycle. The switching process given by the switching line (10.55) has thus converted the oscillating system into an asymptotically stable one; though the goal has been achieved in a suboptimal way.

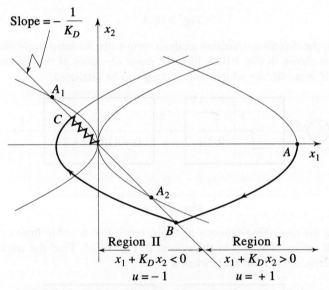

Fig. 10.43 Phase trajectories for a bang-bang control system

PROBLEMS

10.1 For a sinusoidal input $x = X \sin \omega t$, find the output waveforms for each of the nonlinearities listed in Table 10.2. By Fourier-series analysis of the output waveforms, derive the describing function for each entry of the table.

10.2 Consider the system shown in Fig. P10.2. Using the describing-function analysis, show that a stable limit cycle exists for all values of $K > 0$. Find the amplitude and frequency of the limit cycle when $K = 4$, and plot $y(t)$ versus t.

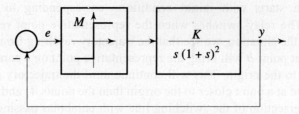

Fig. P10.2

10.3 Consider the system shown in Fig. P10.3. Use the describing function technique to investigate the possibility of limit cycles in this system. If a stable limit cycle is predicted, determine its amplitude and frequency.

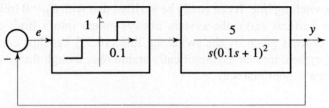

Fig. P10.3

10.4 Using the describing-function analysis, prove that no limit cycle exists in the system shown in Fig. P10.4. Find the range of values of the deadzone of the on–off controller for which limit cycling will be predicted.

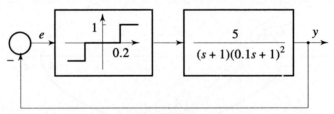

Fig. P10.4

10.5 Using the describing-function analysis, prove that a stable limit cycle exists for all values of $K > 0$ in the system of Fig. P10.5. Find the amplitude and frequency of the limit cycle when $K = 10$.

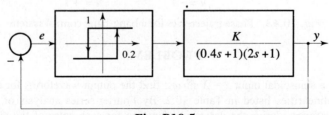

Fig. P10.5

10.6 Consider the system shown in Fig. P10.6. Using the describing-function technique, show that a stable limit cycle cannot exist in this system for any $K > 0$.

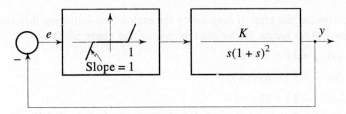

Fig. P10.6

10.7 Consider the system shown in Fig. P10.7. Using the describing-function analysis, investigate the possibility of a limit cycle in the system. If a limit cycle is predicted, determine its amplitude and frequency, and investigate its stability.

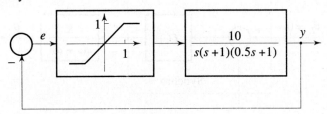

Fig. P10.7

10.8 An instrument servo system used for positioning a load may be adequately represented by the block diagram in Fig. P10.8a. The backlash characteristic is shown in Fig. P10.8b. Show that the system is stable for $K = 1$. If the value of K is now raised to 2, show that limit cycles exist. Investigate the stability of these limit cycles. Determine the amplitude and frequency of the stable limit cycle.
Given:

H/X	0	0.1	0.2	0.5	0.7	0.8	0.9	0.95	1.0
$\|N(X)\|$	1	0.954	0.882	0.593	0.367	0.249	0.126	0.063	0
$\angle N(X)$	0	$-6.7°$	$-13.4°$	$-32.5°$	$-46.6°$	$-55.1°$	$-65.6°$	$-72.8°$	$-90°$

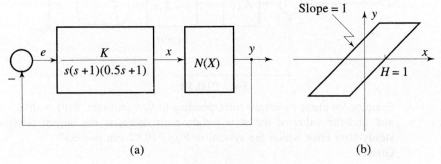

(a)   (b)

Fig. P10.8

10.9 Determine the kind of singularity for each of the following differential equations. Also locate the singular points on the phase plane.

(a) $\ddot{y} + 3\dot{y} + 2y = 0$

(b) $\ddot{y} + 5\dot{y} + 6y = 6$

(c) $\ddot{y} - 8\dot{y} + 17y = 34$

10.10 Consider the position control system shown in Fig. P10.10. The feedback action is provided by synchros which generate the actuating signal $e = \sin(\theta_R - \theta)$. Prove that this nonlinear system has multiple singular points. Linearize the system around the singular points (0,0) and $(\pi,0)$, and identify the kind of singularity at each point.

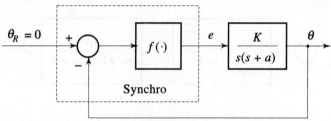

Fig. P10.10

10.11 A linear second-order servo is described by the equation
$$\ddot{y} + 2\zeta\omega_n \dot{y} + \omega_n^2 y = \omega_n^2$$
where $\omega_n = 1, y(0) = 2.0, \dot{y}(0) = 0$

Determine the singular points when (i) $\zeta = 0$, (ii) $\zeta = 0.15$. Construct the phase trajectory in each case.

10.12 The position control system shown in Fig. P10.12 has Coulomb friction $T_c \, \text{sgn}(\dot{\theta})$ at the output shaft. Prove that the phase trajectories on $(e, \dot{e}/\omega_n)$-plane are semicircles with the centre on horizontal axis at $+T_c/K$ for $\dot{e} < 0$ and $-T_c/K$ for $\dot{e} > 0$.

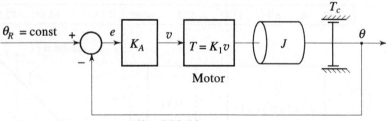

Fig. P10.12

Examine the phase trajectory corresponding to θ_R = unit step, $\dot{\theta}(0) = \theta(0) = 0$; and find the value of the steady-state error. What is the largest possible steady-state error which the system in Fig. P10.12 can possess?

Given:
$$\omega_n = \sqrt{K/J} = 1.2 \text{ rad/sec}; \quad T_c/K = 0.3 \text{ rad}$$
where $K = K_A K_1$.

10.13 Consider the nonlinear system with deadzone shown in Fig. P10.13. Construct phase trajectories and show that the closed-loop system is stable with equilibrium zone defined by

$$-1 \leq x_1 \leq 1; \quad \dot{x}_1 = 0$$

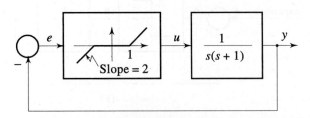

Fig. P10.13

10.14 Consider the system shown in Fig. P10.14 in which the nonlinear element is a power amplifier with gain equal to 1.0, which saturates for error magnitudes greater than 1.0. Given the initial condition: $e(0) = 2$, $\dot{e}(0) = 0$, plot phase trajectories with and without saturation, and comment upon the effect of saturation on the transient behaviour of the system.

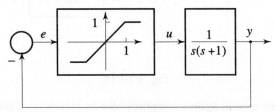

Fig. P10.14

10.15 A second-order servo is shown in the block diagram of Fig. P10.15. The relay characteristic possesses hysteresis with an overlap of 0.4. Given the initial condition: $e(0) = 1.4$, $\dot{e}(0) = 0$, obtain the phase trajectories with and without hysteresis, and comment upon the effect of hysteresis on the transient behaviour of the system.

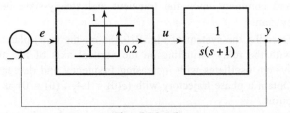

Fig. P10.15

10.16 (a) A plant with model $G(s) = 1/s^2$ is placed in a feedback configuration as in Fig. P10.16a. Construct a trajectory on the $(e, \dot{e})$ plane with $r = 2$ and $y(0) = \dot{y}(0) = 0$. Show that the system response is a limit cycle.

(b) To the feedback control system of Fig. P10.16a, we add a derivative feedback with gain K_D as in Fig. P10.16b. Show that the limit cycle gets eliminated by the introduction of derivative-control term.

(c) Show that if K_D is large, the trajectory may slide along the switching line towards the origin.

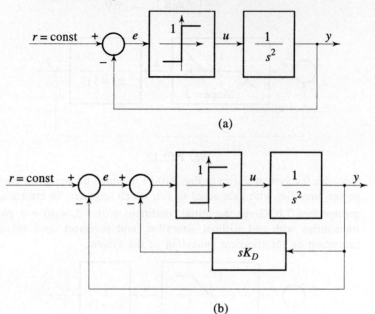

Fig. P10.16

10.17 A position control system comprises of a dc servomotor, potentiometer error detector, a relay amplifier, and a tachogenerator coupled to the motor shaft. The following equations describe the system:

Reaction torque $= \ddot{\theta} + 0.5\dot{\theta}$

Drive torque $= 2\,\text{sgn}(e + 0.5\,\dot{e})$; $e = \theta_R - \theta$

 (a) Make a sketch of the system showing how the hardware is connected.
 (b) Construct a phase trajectory on $(e, \dot{e})$-plane with $e(0) = 2$ and $\dot{e}(0) = 0$, and comment upon the transient and steady-state behaviour of the system.

10.18 (a) Describe the system of Fig. P10.18a on the $(e, \dot{e})$-plane, and show that with the relay switching on the vertical axis of the phase plane, the system oscillates with increasing frequency and decreasing amplitude. Obtain a phase trajectory with $(e(0) = 1.4,\ \dot{e}(0) = 0)$ as the initial state point.
 (b) Introduce now a deadzone of ± 0.2 in the relay characteristic. Obtain a phase trajectory for the modified system with $(e(0) = 1.4,\ \dot{e}(0) = 0)$ as the initial state point and comment upon the effect of deadzone.
 (c) The relay with deadzone is now controlled by the signal $\left(e + \tfrac{1}{3}\dot{e}\right)$, combining proportional and derivative control (Fig. P10.18b). Draw the switching line on the phase plane and construct a phase trajectory with $(e(0) = 1.4,\ \dot{e}(0) = 0)$ as the initial state point. What is the effect of the derivative-control action?

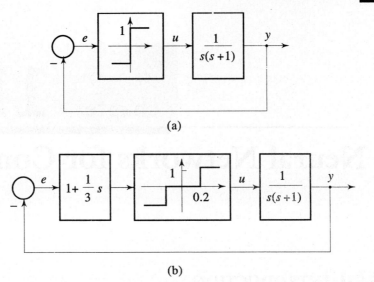

(a)

(b)

Fig. P10.18

10.19 Consider the second-order system

$$\dot{x}_1 = x_2; \quad \dot{x}_2 = -u$$

It is desired to transfer the system to the origin in minimum time from an arbitrary itinial state. Use the bang-bang control strategy with $|u| = 1$. Derive an expression for the optimum switching curve. Construct a phase portrait showing a few typical minimum-time trajectories.

10.20 A plant with model $G(s) = \dfrac{1}{s(s+1)}$ is placed in a feedback configuration as shown in Fig. P10.20. It is desired to transfer the system from any initial state to the origin in minimum time. Derive an expression for optimum switching curve and construct a phase portrait on the $(e, \dot{e})$-plane showing a few typical minimum-time trajectories.

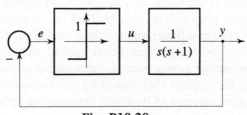

Fig. P10.20

chapter 11

Neural Networks for Control

11.1 INTRODUCTION

In engineering practice, nonlinear systems are omnipresent. However, because of the difficulty in handling nonlinearity, a nonlinear control system is traditionally dealt with by the Jacobian linearization approach. That is, designing a linear controller based on the linearized plant model around a certain equilibrium. When the requirements on the system performance are mild and the system operates in a small dynamic regime, such a controller is often acceptable. However, in many modern systems such as flight control, weapon control, robotics, and process control, the nonlinearity inherent in the plant is so dominant that the approach can hardly meet the stringent requirements of system performance. This reality inevitably promotes the endeavour to develop control approaches that will more or less incorporate nonlinear dynamics into the design process.

One such approach is (input–output) feedback linearization. Unlike the Jacobian linearization approach, which ignores all higher-order terms of the plant, this approach only utilizes the feedback and transforms coordinates to render the given system a linear input–output dynamics. Then on the basis of the linear system thus transformed, linear control techniques can be further applied to address design issues. Finally, the resulting control law can be transformed back into the original coordinates for implementation [116].

To give the basic concepts of feedback linearization technique, we consider here a single-input, single-output system

$$\begin{bmatrix} \dot{x}_1(t) \\ \vdots \\ \dot{x}_n(t) \end{bmatrix} = \begin{bmatrix} f_1(x_1(t),\ldots,x_n(t)) \\ \vdots \\ f_n(x_1(t),\ldots,x_n(t)) \end{bmatrix} + \begin{bmatrix} g_1(x_1(t),\ldots,x_n(t)) \\ \vdots \\ g_n(x_1(t),\ldots,x_n(t)) \end{bmatrix} u(t) \quad (11.1)$$

$$y(t) = h(x_1(t), \ldots, x_n(t))$$

In vector notation, we may write

$$\dot{\mathbf{x}} = \mathbf{f}(\mathbf{x}) + \mathbf{g}(\mathbf{x})u$$
$$y = h(\mathbf{x}) \qquad (11.2)$$

where $\mathbf{f}(.)$ and $\mathbf{g}(.)$ are smooth vector functions, and $h(.)$ is a smooth scalar function. Many nonlinear systems occur naturally in this form. Moreover, it is often possible to transform general nonlinear systems to this form [159].

Differentiating y with respect to time, we obtain

$$\dot{y} = L_f h(\mathbf{x}) + L_g h(\mathbf{x})u \qquad (11.3a)$$

Here $L_f h$ and $L_g h$ stand for the *Lie derivatives* of h with respect to $\mathbf{f}$ and $\mathbf{g}$ respectively:

$$L_f h(\mathbf{x}) \triangleq \sum_{i=1}^{n} \frac{\partial h(\mathbf{x})}{\partial x_i} f_i(\mathbf{x}) = \frac{\partial h(\mathbf{x})}{\partial \mathbf{x}} \mathbf{f}(\mathbf{x}) \qquad (11.3b)$$

$$L_g h(\mathbf{x}) \triangleq \sum_{i=1}^{n} \frac{\partial h(\mathbf{x})}{\partial x_i} g_i(\mathbf{x}) = \frac{\partial h(\mathbf{x})}{\partial \mathbf{x}} \mathbf{g}(\mathbf{x}) \qquad (11.3c)$$

If $L_g h(\mathbf{x}) \neq 0$ for all $\mathbf{x}$, then the control law of the form $\alpha(\mathbf{x}) + \beta(\mathbf{x})v$, namely

$$u = \frac{1}{L_g h(\mathbf{x})} [-L_f h(\mathbf{x}) + v] \qquad (11.4a)$$

yields the linear system

$$\dot{y} = v \qquad (11.4b)$$

In the instance that $L_g h(\mathbf{x}) \equiv 0$, we differentiate y further to obtain

$$\ddot{y} = L_f^2 h(\mathbf{x}) + L_g L_f h(\mathbf{x}) u \qquad (11.5a)$$

Here $L_f^2 h$ stands for $L_f(L_f h)$ and $L_g L_f h$ stands for $L_g(L_f h)$. As berore, if $L_g L_f h(\mathbf{x}) \neq 0$ for all $\mathbf{x}$, the law

$$u = \frac{1}{L_g L_f h(\mathbf{x})} [-L_f^2 h(\mathbf{x}) + v] \qquad (11.5b)$$

linearizes the system to yield

$$\ddot{y} = v \qquad (11.5c)$$

More generally, if γ is the smallest integer such that $L_g L_f^i h(\mathbf{x}) \equiv 0$ for $i = 0, \ldots, \gamma - 2$ ($L_f^0 h \triangleq h$), and $L_g L_f^{\gamma-1} h(\mathbf{x}) \neq 0$ for all $\mathbf{x}$, then the control law

$$u = \frac{1}{L_g L_f^{\gamma-1} h(\mathbf{x})} [-L_f^\gamma h(\mathbf{x}) + v] \qquad (11.6)$$

yields

$$y^{(\gamma)} = v \qquad (11.7)$$

Once linearization has been achieved, any further control objective, such as pole-placement, may be easily met. The theory is considerably more complicated if $L_g L_f^{\gamma-1} h = 0$ for some values of $\mathbf{x}$.

There are number of applications of this technique. The chief drawback of the technique, however, is that it relies on exact cancellation of nonlinear terms in order to get linear input–output behaviour (This is equivalent to cancellation of the nonlinearity with its inverse). Consequently, if there are errors or uncertainty in the model of the nonlinear terms, the cancellation is no longer exact. Therefore the applicability of such model-based approaches to feedback control of actual systems is quite limited because they rely on the exact mathematical models of system nonlinearities.

Many of the abilities one possesses as a human have been learned from examples. Thus it is only natural to try to carry this "didactic principle" over to a computer program to make it learn how to output the desired answer for a given input. In a sense, the artificial neural network is one such computer program; it is a mathematical formula with several adjustable parameters (*weights*) that are tuned from a set of examples. These examples represent what the network should output when it is shown a particular input.

Of fundamental importance in neural network (NN) closed-loop control applications is the *function approximation* property of NN. Consider a continuous, multivariable nonlinear function $\mathbf{f}(\mathbf{x})$ where $\mathbf{x} = [x_1, x_2, ..., x_n]^T$ is the input vector. It is desired to approximate function $\mathbf{f}(\mathbf{x})$ by a neural network (computer program) with adjustable parameter vector $\mathbf{w} = [w_1, w_2, ..., w_m]^T$. The learning task is to find $\mathbf{w}$ that produces the best possible approximation of $\mathbf{f}(\mathbf{x})$ based on the set of *exemplar* input/output pairs $\{\mathbf{x}, \mathbf{f}(\mathbf{x})\}$.

The basic approximation result says that any smooth function $\mathbf{f}(\mathbf{x})$ can be arbitrarily and closely approximated using a neural network. The result however does not indicate how to determine the required parameters. The issue of finding the parameters such that a neural network does indeed approximate a given function $\mathbf{f}(\mathbf{x})$ closely enough is not an easy one. In the subsequent sections we shall show how to accomplish this.

If the function approximation is to be carried out in the context of a dynamic closed-loop control scheme, the issue is thornier. As a simple example, consider i/o feedback linearization as a controller design technique for the system (11.2) with

$$\mathbf{f}(\mathbf{x}) = \begin{bmatrix} x_2 \\ f(x_1, x_2) \end{bmatrix}; \; \mathbf{g}(\mathbf{x}) = \begin{bmatrix} 0 \\ g(x_1, x_2) \end{bmatrix}; \; h(\mathbf{x}) = x_1 \qquad (11.8)$$

The control law

$$u = \frac{1}{g(\mathbf{x})}[-f(\mathbf{x}) + v] \qquad (11.9)$$

Linearizes the system to yield

$$\ddot{y} = v \tag{11.10}$$

Standard linear-system techniques can now be used to design a tracking controller for the feedback linearized system that causes $y(t) = x_1(t)$ to follow a desired trajectory $y_r(t)$ which is prescribed by the user. For instance, one possibility is the proportional-plus-derivative (PD) tracking control

$$v \equiv \ddot{y}_r + K_D \dot{e} + K_P e \tag{11.11}$$

where the tracking error is defined as

$$e(t) \equiv y_r(t) - y(t) \tag{11.12}$$

Substituting this control $v(t)$ into (11.10) yields the closed-loop system

$$\ddot{e} + K_D \dot{e} + K_P e = 0, \tag{11.13}$$

or equivalently, in state–space form

$$\frac{d}{dt}\begin{bmatrix} e \\ \dot{e} \end{bmatrix} = \begin{bmatrix} 0 & 1 \\ -K_P & -K_D \end{bmatrix} \begin{bmatrix} e \\ \dot{e} \end{bmatrix} \tag{11.14}$$

As long as the PD gains are positive, the tracking error converges to zero. The PD gains should allow for a suitable percentage overshoot and rise time.

The complete control scheme is shown in Fig. 11.1.

$$\mathbf{y}_r = \begin{bmatrix} y_r \\ \dot{y}_r \end{bmatrix}; \; \mathbf{e} = \begin{bmatrix} e \\ \dot{e} \end{bmatrix}; \; \mathbf{y} = \begin{bmatrix} y \\ \dot{y} \end{bmatrix} \tag{11.15}$$

Since the functions $f(\mathbf{x})$ and $1/g(\mathbf{x})$ are not exactly known, we construct their estimates $\hat{f}(\mathbf{x})$ and $1/\hat{g}(\mathbf{x})$ by two neural networks, NN1 and NN2 respectively.

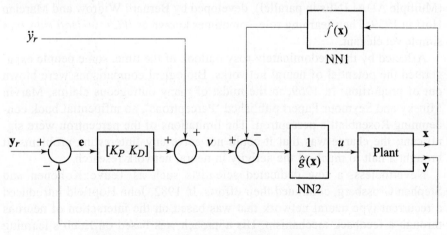

Fig. 11.1 A neural-network based control scheme

Despite the fact that "learning from examples" sounds easy, the issue is not as simple as it looks; learning the unknown nonlinear functions in the context of dynamic closed-loop control scheme of Fig. 11.1 is a complex task. The tuning of the adjustable parameters (weights) of NN1 and NN2 would have been relatively simple if the exemplar input/output pairs $\{\mathbf{x}, f(\mathbf{x})\}$ and $\{\mathbf{x}, 1/g(\mathbf{x})\}$ were known. Since a dynamic model of the plant is unknown, such exemplar pairs cannot be generated. The objective here is to create computer programs NN1 and NN2 that are able to learn from online experience interpreted as knowledge about how certain inputs affect the closed-loop system, and the experimental measurements of error and other variables. Such thornier issues will also be discussed in the subsequent sections.

Historically, research in artificial neural networks was inspired by the desire to produce artificial systems capable of sophisticated 'intelligent' processing similar to the human brain. The science of artificial neural networks made its first significant appearance in 1943 when Warren McCulloch and Walter Pitts published their study in this field. They suggested a simple neuron model (known today as *MP artificial neural model*) and implemented it as an electrical circuit. In 1949, Donald Hebb highlighted the connection between psychology and physiology, pointing out that a neural pathway is reinforced each time it is used. *Hebb's learning rule*, as it is sometimes known, is still used and quoted today. Improvements in hardware and software in the 1950s ushered in the age of computer simulation. It became possible to test theories about nervous system functions. Research expanded; neural network terminology came into its own.

The *perceptron* is the earliest of the neural network paradigms. Frank Rosenblatt built this learning machine device in hardware in 1958 and caused quite a stir.

The perceptron has been a fundamental building block for more powerful models, such as the ADALINE (ADAptive LINear Elements) and MEDALINE (Multiple ADALINEs in parallel), developed by Bernard Widrow and Marcian Hoff in 1959. Their learning rule, sometimes known as *Widrow–Hoff rule*, was simple yet elegant.

Affected by the predominately rosy outlook of the time, some people exaggerated the potential of neural networks. Biological comparisons were blown out of proportion. In 1969, in the midst of many outrageous claims, Marvin Minsky and Seymour Papert published 'Perceptrons', an influential book condemning Rosenblatt's perceptron. The limitations of the perceptron were significant; the charge was that it could not solve any 'interesting' problems. It brought a halt to much of the activity in neural network research.

Nevertheless, a few dedicated scientists such as Teuvo Kohonen and Stephen Grossberg, continued their efforts. In 1982, John Hopfield introduced a recurrent-type neural network that was based on the interaction of neurons through a feedback mechanism. His approach was based on Hebb's learning rule. The backpropagation learning rule arrived on the neural-network scene at approximately the same time from several independent sources (Werbos; Parker; Rumelhart, Hinton and Williams). Essentially a refinement of the

Widrow–Hoff learning rule, the backpropagation learning rule provided a systematic means for training multilayer networks, thereby overcoming the limitations presented by Minsky. Minsky's appraisal has proven excessively pessimistic; networks now routinely solve many of the problems that he posed in his book.

Research in the 1980s triggered the present boom in the scientific community. New and better models are being proposed and the limitations of some of the 'old' models are being chipped away. A number of today's technological problems are in areas where neural-network technology has demonstrated potential: speech processing, image processing and pattern recognition, time-series predictions, real-time control and others.

As the research on neural networks evolved, more and more types of networks are being introduced while still less emphasis is placed on the connection to the biological neural network. In fact, the neural networks that are most popular today have very little resemblance to the brain and one might argue that it would be more fair to regard them simply as a discipline under statistics.

The application of artificial neural networks in closed-loop control has recently been rigorously studied. One property of these networks, central to most control applications, is that of function approximation. Such networks can generate input/output maps which can approximate any continuous function with the required degree of accuracy. This emerging technology will give us control design techniques that do not depend on parametrized mathematical models. Neural networks will be used to estimate the unknown nonlinear functions; the controller formulation will use these estimated results.

When neural networks are used for control of systems, it is important that results and claims are based on firm analytical foundations. This is especially important when these control systems are to be used in areas where the cost of failure is very high, for example when human life is threatened, as in aircrafts, nuclear plants, etc. It is also true that without a good theoretical framework, it is unlikely that the research in the discipline will progress very far, as intuitive invention and tricks cannot be counted on to provide good solutions to controlling complex systems under a high degree of uncertainty. Strong theoretical results guaranteeing control system properties such as stability are still to come, although promising results have been reported recently of progress in special cases. The potential of neural networks in control systems clearly needs to be further explored and both theory and applications need to be further developed.

The rest of the chapter gives a gentle introduction to the application of neural networks in control systems. A single chapter can in no way do justice to the multitude of interesting neural network results that have appeared in the literature. Not only would space be required, but in the time required to detail current results, new results would certainly arise. Instead of trying to cover a large spectrum of such a vast field, we will focus on what is generally regarded as the core of the subject. This chapter is meant to be a stepping-stone that could lead interested readers on to other books for additional information on the current status and future trends of the subject.

11.2 NEURON MODELS

A discussion of anthropomorphism to introduce neural network technology may be worthwhile as it helps explain the terminology of neural networks. However, anthropomorphism can lead to misunderstanding when the metaphor is carried too far. We give here a brief description of how the brain works; a lot of details of the complex electrical and chemical processes that go on in the brain have been ignored. A pragmatic justification for such a simplification is that by starting with a simple model of the brain, scientists have been able to achieve very useful results.

Biological Neuron

To the extent a human brain is understood today, it seems to operate as follows:

Bundles of neurons, or nerve fibres, form nerve structures. There are many different types of neurons in the nerve structure, each having a particular shape, size and length depending upon its function and utility in the nervous system. While each type of neuron has its own unique features needed for specific purposes, all neurons have two important structural components in common. These may be seen in the typical biological neuron shown in Fig. 11.2. At one end of the neuron are a multitude of tiny, filament-like appendages called *dendrites*, which come together to form larger branches and trunks where they attach to *soma*, the body of the nerve cell. At the other end of the neuron is a single filament leading out of the soma, called an *axon*, which has extensive branching on its far end. These two structures have special electrophysiological properties which are basic to the function of neurons as *information processors*, as we shall see next.

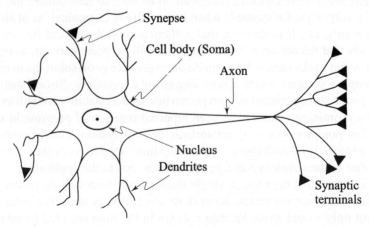

Fig. 11.2 A typical biological neuron

Neurons are connected to each other via their axons and dendrites. Signals are sent through the axon of one neuron to the dendrites of other neurons. Hence dendrites may be represented as the inputs to the neuron, and the axon as its output. Note that each neuron has many inputs through its multiple dendrites, whereas it has only one output through its single axon. The axon of each neuron forms connections with the dendrites of many other neurons, with each branch of the axon meeting exactly one dendrite of another cell at what is called a *synapse*. Actually, the axon terminals do not quite touch the dendrites of the other neurons, but are separated by a very small distance of between 50 and 200 angstroms. This separation is called the *synaptic gap*.

A conventional computer is typically a single processor acting on explicitly programmed instructions. Programmers break tasks into tiny components, to be performed in sequence rapidly. On the other hand, the brain is composed of ten billion or so neurons. Each nerve cell can interact directly with up to 200,000 other neurons (though 1000 to 10,000 is typical). In place of explicit rules that are used by a conventional computer, in the human brain it is the pattern of connections between the neurons that seems to embody the 'knowledge' required for carrying out various information-processing tasks. In human brain, there is no equivalent of a CPU that is in overall control of the actions of all the neurons.

The brain is organized into different regions, each responsible for different functions. The largest parts of the brain are the cerebral hemispheres, which occupy most of the interior of the skull. They are layered structures, the most complex being the outer layer, known as the *cerebral cortex*, where the nerve cells are extremely densely packed to allow greater interconnectivity. Interaction with the environment is through the visual, auditory and motion control (muscles and glands) parts of the cortex.

In essence, neurons are tiny electrophysiological information processing units which communicate with each other through electrical signals. The synaptic activity produces a voltage pulse on the dendrite which is then conducted into the soma. Each dendrite may have many synapses acting on it, allowing massive interconnectivity to be achieved. In the soma, the dendrite potentials are added. Note that neurons are able to perform more complex functions than simple addition on the inputs they receive, but considering a simple summation is a reasonable approximation.

When the soma potential rises above a critical threshold, the axon will fire an electrical signal. This sudden burst of electrical energy along the axon is called axon potential and has the form of an electrical impulse or spike that lasts about 1 msec. The magnitude of the axon potential is constant and is not related to the electrical stimulus (soma potential). However, neurons typically respond to a stimulus by firing not just one but a barrage of successive axon potentials. What varies is the frequency of axonal activity. Neurons can fire between 0 to 1500 times per second. Thus, information is encoded in the nerve signals as the instantaneous frequency of axon potentials and the mean frequency of the signal.

A synapse couples the axon with the dendrite of another cell. The synapse releases chemicals called *neurotransmitters* when its potential is raised sufficiently by the axon potential. It may take the arrival of more than one spike before the synapse is triggered. The neurotransmitters that are released by the synapse diffuse across the gap and chemically activate gates on the dendrites, which when open, allow charged ions to flow. It is this flow of ions that alters the dendritic potential and provides voltage pulse on the dendrite, which is then conducted into the neighbouring neuron body. At the synaptic junction, the number of gates open on the dendrite depends upon the number of neurotransmitters released. It also appears that some synapses excite the dendrites they affect, whilst others serve to inhibit it. This corresponds to altering the local potential of the dendrite in a positive or negative direction.

Synaptic junctions alter the effectiveness with which the signal is transmitted; some synapses are good junctions and pass a large signal across, whilst others are very poor, and allow very little through.

Essentially each neuron receives signals from a large number of other neurons. These are the inputs to the neuron which are 'weighted'. That is, some signals are stronger than others. Some signals excite (are positive), and others inhibit (are negative). The effects of all weighted inputs are summed. If the sum is equal to or greater than the *threshold* for the neuron, the neuron *fires* (gives output). This is an "all-or-nothing" situation. Because the neuron either fires or doesn't fire, the rate of firing, not the amplitude, conveys the magnitude of information.

The ease of transmission of signals is altered by activity in the nervous system. The neural pathway between two neurons is susceptible to fatigue, oxygen deficiency, and agents like anaesthetics. These events create a resistance to the passage of impulses. Other events may increase the rate of firing. This ability to adjust signals is a mechanism for *learning*.

After carrying a pulse, an axon fibre is in a state of complete non-excitability for a certain time called the *refractory period*. For this time interval, the nerve does not conduct any signals, regardless of the intensity of excitation. Thus, we may divide the time scale into consecutive intervals, each equal to the length of the refractory period. This will enable a discrete-time description of the neurons' performance in terms of their states at discrete-time instances.

Artificial Neuron

Artificial neurons bear only a modest resemblance to real things. They model approximately three of the processes that biological neurons perform (there are at least 150 processes performed by neurons in the human brain).

An artificial neuron
  (i) evaluates the input signals, determining the strength of each one;
  (ii) calculates a total for the combined input signals and compares that total to some threshold level; and
  (iii) determines what the output should be.

Inputs and outputs: Just as there are many inputs (stimulation levels) to a biological neuron, there should be many input signals to our artificial neuron (AN). All of them should come to our AN simultaneously. In response, a biological neuron either 'fires' or 'doesn't fire' depending upon some *threshold* level. Our AN will be allowed a single output signal, just as is present in a biological neuron: many inputs, one output (Fig. 11.3).

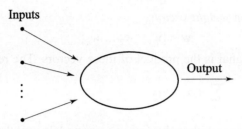

Fig. 11.3 Many inputs, one output model of a neuron

Weighting factors: Each input will be given a relative weighting, which will affect the impact of that input (Fig. 11.4). This is something like varying synaptic strengths of the biological neurons—some inputs are more important than others in the way they combine to produce an impulse. Weights are adaptive coefficients within the network that determine the intensity of the input signal. In fact, this adaptability of connection strength is precisely what provides neural networks with their ability to learn and store information, and, consequently, is an essential element of all neuron models.

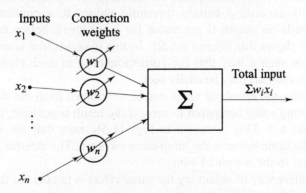

Fig. 11.4 A neuron with weighted inputs

Excitatory and inhibitory inputs are represented simply by positive or negative connection weights respectively. Positive inputs promote the firing of the neuron, while negative inputs tend to keep the neuron from firing.

Mathematically, we could look at the inputs and the weights on the inputs as vectors.

The *input vector*

$$\mathbf{x} = \begin{bmatrix} x_1 \\ x_2 \\ \vdots \\ x_n \end{bmatrix} \qquad (11.16a)$$

and the connection *weight vector*

$$\mathbf{w} = [w_1 \quad w_2 \ \ldots \ w_n] \qquad (11.16b)$$

The total input signal is the product of these vectors. The result is a scalar

$$\sum_{i=1}^{n} w_i x_i = \mathbf{w}\mathbf{x} \qquad (11.16c)$$

Activation functions: Although most neuron models sum their input signals in basically the same manner, as described above, they are not all identical in terms of how they produce an output response from this input. Artificial neurons use an *activation function*, often called a *transfer function*, to compute their activation as a function of total input stimulus. Several different functions may be used as activation functions, and in fact the most distinguishing feature between existing neuron models is precisely which transfer function they employ.

We will shortly take a closer look at the activation functions. We first build a neuron model, assuming that the transfer function has a threshold behaviour, which is in fact the type of response exhibited by biological neurons: when the total stimulus exceeds a certain threshold value θ, a constant output is produced, while no output is generated for input levels below the threshold. Figure 11.5a shows this neuron model. In this diagram, the neuron has been represented in such a way that the correspondence of each element with its biological counterpart may be easily seen.

Equivalently, the threshold value can be subtracted from the weighted sum and the resulting value compared to zero; if the result is positive, then output a 1, else output a 0. This is shown in Fig. 11.5b; note that the shape of the function is the same but now the jump occurs at zero. The threshold effectively adds an offset to the weighted sum.

An alternative way of achieving the same effect is to take the threshold out of the body of the model neuron and connect it to an extra input value that is fixed to be 'on' all the time. In this case, rather than subtracting the threshold value from the weighted sum, the extra input of +1 is multiplied by a weight and added in a manner similar to other inputs—this is known as *biasing* the neuron. Figure 11.5c shows a neuron model with a bias term. Note that we have taken constant input '1' with an adaptive weight '*b*' in our model.

The first formal definition of a synthetic neuron model based on the highly simplified considerations of the biological neuron was formulated by

Neural Networks for Control

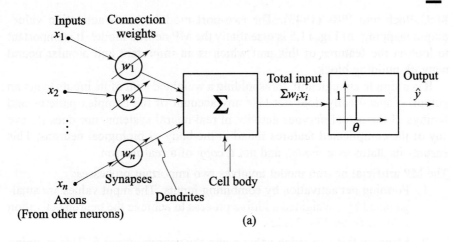

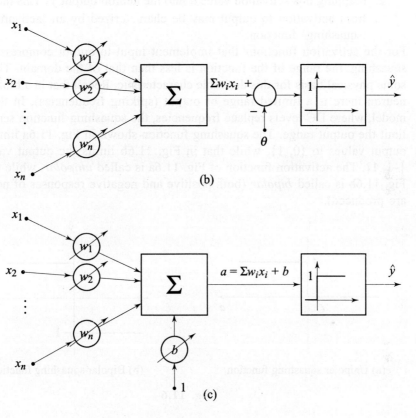

Fig. 11.5 A neuron model

McCulloch and Pitts (1943). The two-port model (inputs-activation value-output mapping) of Fig. 11.5 is essentially the MP neuron model. It is important to look at the features of this unit which is an important and popular neural network building block.

It is a simple enough unit, thresholding a weighted sum of its inputs to get an output. It specifically does not take any account of the complex patterns and timings of the actual nervous activity in real neural systems, nor does it have any of the complicated features found in the body of biological neurons. This ensures its status as a *model*, and not a *copy* of a real neuron.

The MP artificial neuron model involves two important processes:
1. Forming net activation by combining inputs. The input values are amalgamated by a weighted additive process to achieve the neuron activation value a (refer Fig. 11.5c).
2. Mapping this activation value a into the neuron output $\hat{y}$. This mapping from activation to output may be characterized by an 'activation' or 'squashing' function.

For the activation functions that implement input-to-output compression or squashing, the range of the function is less than that of the domain. There is some physical basis for this desirable characteristic. Recall that in a biological neuron there is a limited range of output (spiking frequencies). In the MP model, where DC levels replace frequencies, the squashing function serves to limit the output range. The squashing function shown in Fig. 11.6a limits the output values to $\{0, 1\}$, while that in Fig. 11.6b limits the output value to $\{-1, 1\}$. The activation function of Fig. 11.6a is called *unipolar*, while that in Fig. 11.6b is called *bipolar* (both positive and negative responses of neurons are produced).

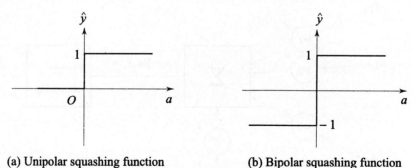

(a) Unipolar squashing function   (b) Bipolar squashing function

Fig. 11.6

Mathematical Model

From the above discussion, it is evident that the artificial neuron is really nothing more than a simple mathematical equation for calculating an output value from a set of input values. From now onwards, we will be more on mathematical footing; the reference to biological similarities will be reduced. Therefore

names like a *processing element*, a *unit*, a *node*, a *cell!*, etc., may be used for the neuron. A neuron model (a processing element/a unit/a node/a cell of our neural network) will be represented as follows:
The input vector

$$\mathbf{x} = [x_1 \ x_2 \ ... \ x_n]^T;$$

the connection weight vector

$$\mathbf{w} = [w_1 \ w_2 \ ... \ w_n];$$

the unity-input weight b (bias term), and the output $\hat{y}$ of the neuron are related by the following equation:

$$\hat{y} = \sigma(\mathbf{wx} + b) = \sigma\left(\sum_{i=1}^{n} w_i x_i + b\right) \quad (11.17)$$

where $\sigma(.)$ is the activation function (transfer function) of the neuron.

The weights are always adaptive. We can simplify our diagram as in Fig. 11.7a; adaptation need not be specifically shown in the diagram.

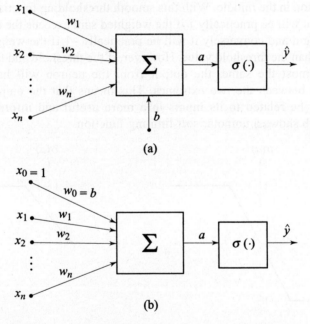

Fig. 11.7 Mathematical model of a neuron (perceptron)

The bias term may be absorbed in the input vector itself as shown in Fig. 11.7b.

$$\hat{y} = \sigma(a)$$
$$= \sigma\left(\sum_{i=0}^{n} w_i x_i\right); \ w_0 = b, \ x_0 = 1 \quad (11.18a)$$

$$= \sigma\left(\sum_{i=1}^{n} w_i x_i + w_0\right) = \sigma(\mathbf{w}\mathbf{x} + w_0) \qquad (11.18b)$$

In the literature, this model of an artificial neuron is also referred to as a *perceptron* (the name was given by Rosenblatt in 1958).

The expressions for the neuron output $\hat{y}$ are referred to as the *cell recall mechanism*. They describe how the output is reconstructed from the input signals and the values of the cell parameters.

The artificial neural systems under investigation and experimentation today employ a variety of activation functions that have more diversified features than the one presented in Fig. 11.6. Below, we introduce the main activation functions that will be used later in this chapter.

The MP neuron model shown in Fig. 11.5 used the *hard-limiting activation function*. When artificial neurons are cascaded together in layers (discussed in the next section), it is more common to use a *soft-limiting activation function*. Figure 11.8a shows a possible bipolar soft-limiting semilinear activation function. This function is more or less the ON-OFF type, as before, but has a sloping region in the middle. With this smooth thresholding function, the value of the output will be practically 1 if the weighted sum exceeds the threshold by a huge margin and conversely it will be practically −1 if the weighted sum is much less than the threshold value. However, if the threshold and the weighted sum are almost the same, the output from the neuron will have a value somewhere between the two extremes. This means that the output from the neuron can be related to its inputs in a more useful and informative way. Figure 11.8b shows a unipolar soft-limiting function.

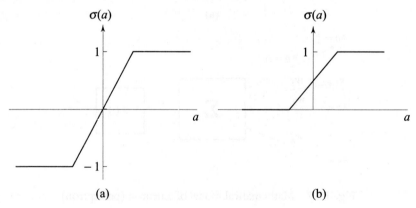

Fig. 11.8 Self-limiting activation functions

For many training algorithms (discussed in later sections), the derivative of the activation function is needed; therefore the activation function selected must be differentiable. The *logistic* or *sigmoid* function, which satisfies this requirement, is the most commonly used soft-limiting activation function. The sigmoid function (Fig. 11.9a):

$$\sigma(a) = \frac{1}{1+e^{-\lambda a}} \qquad (11.19)$$

is continuous and varies monotonically from 0 to 1 as a varies from $-\infty$ to ∞. The gain of the sigmoid, λ, determines the steepness of the transition region. Note that as the gain approaches infinity, the sigmoid approaches a hard-limiting nonlinearity. One of the advantages of the sigmoid is that it is *differentiable*. This property had a significant impact historically because it made it possible to derive a gradient search learning algorithm for networks with multiple layers (discussed in later sections).

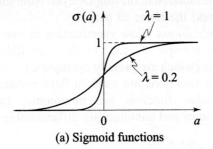

(a) Sigmoid functions

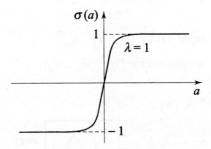

(b) Hyperbolic tangent function

Fig. 11.9

The sigmoid function is unipolar. A bipolar function with similar characteristics is a hyperbolic tangent (Fig. 11.9b):

$$\sigma(a) = \tanh(a) = \frac{1-e^{-\lambda a}}{1+e^{-\lambda a}} \qquad (11.20)$$

The biological basis of these activation functions can easily be established. It is known that neurons located in different parts of the nervous system have different characteristics. The neurons of the ocular motor system have a sigmoid characteristic while those located in the visual area have a Gaussian characteristic. As we said earlier, anthropomorphism can lead to misunderstanding when the metaphor is carried too far. It is now a well-known result in neural network theory that a 2-layer neural network is capable of solving any classification problem. It has also been shown that a 2-layer network is capable of solving any nonlinear function approximation problem [160, 161]. This result does not require the use of sigmoid nonlinearity. The proof assumes only that nonlinearity is a continuous, smooth, monotonically increasing function that is bounded above and below. Thus, numerous alternatives to sigmoid could be used, without a biological justification. In addition, the above result does not require that the nonlinearity be present in the second (output) layer. It is quite common to use linear output nodes since this tends to make learning easier. In other words

$$\sigma(a) = \lambda a;\ \lambda > 0 \qquad (11.21)$$

is used as an activation function in the output layer. Note that this function does not "squash" (compress) the range of output.

Many neuron models do not allow visualization in two ports (input-activation value-output). An example of such a model is the RBF model, which uses Radial Basis Functions (which are radially symmetric).

Typically Gaussian functions are used in RBF networks. Although both halves of the Gaussian function are monotonic, the curve itself is nonmonotonic, continuous and continuously differentiable (Fig. 11.10):

$$\sigma(\mathbf{x}) = e^{-\left(\|\mathbf{x}-\mathbf{w}^T\|b\right)^2} \qquad (11.22)$$

where

$$\mathbf{x} = \text{input vector } [x_1\ x_2\ ...x_n]^T$$

$$\mathbf{w} = \text{weight vector } [w_1\ w_2\ ...w_n]$$

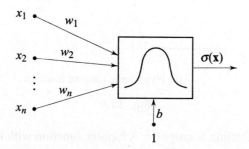

Fig. 11.10 Gaussian function in RBF neuron model

Our focus in this chapter will be on two-layer perceptron networks with the first (hidden) layer having *log-sigmoid*

$$\sigma(a) = \frac{1}{1+e^{-a}} \qquad (11.23a)$$

or *tan-sigmoid*

$$\sigma(a) = \frac{1-e^{-a}}{1+e^{-a}} \qquad (11.23b)$$

activation function, and the second (output) layer having *linear* activation function

$$\sigma(a) = a \tag{11.24}$$

The log-sigmoid function has historically been a very popular choice, but since it is related to the tan-sigmoid by the simple transformation

$$\sigma_{\text{log-sigmoid}} = (\sigma_{\text{tan-sigmoid}} + 1)/2 \tag{11.25}$$

it makes no difference which of these is used.

11.3 NETWORK ARCHITECTURES

In the biological brain, a huge number of neurons are interconnected to form the network and perform advanced intelligent activities. The artificial neural network is built by neuron models. Many different types of artificial neural networks have been proposed, just as there are many theories on how biological neural processing works. We may classify the organization of the neural networks largely into two types: a feedforward net and a recurrent net. The feedforward net has a hierarchical structure that consists of several layers without interconnection between neurons in each layer, and signals flow from input to output layer in one direction. In the recurrent net, multiple neurons in a layer are interconnected to organize the network. In the following, we give typical characteristics of the feedforward net and the recurrent net, respectively.

Feedforward Networks

A feedforward network consists of a set of *input terminals* which feed the input patterns to a layer or subgroup of neurons. The layer of neurons makes independent computations on data that it receives and passes the results to another layer. The next layer may in turn make its independent computations and pass on the results to yet another layer. Finally, a subgroup of one or more neurons determines the output from the network. This last layer of the network is the *output layer*. The layers that are placed between the input terminals and the output layer are called *hidden layers*.

Some authors refer to the input terminals as the input layer of the network. We do not use that convention since we wish to avoid ambiguity. Note that each neuron in a network makes its computation based on the weighted sum of its inputs. There is one exception to this rule: the role of the 'input layer' is somewhat different in that units in this layer are used only to hold input data and to distribute the data to units in the next layer. Thus the 'input layer' units perform no function other than serving as a buffer, fanning out the inputs to the next layer. These units do not perform any computation on the input data, and their weights, strictly speaking, do not exist.

The network outputs are generated from the output layer units. The output layer makes the network information available to the outside world. The hidden layers are internal to the network and have no direct contact with the external environment. There may be from zero to several hidden layers. The network is said to be *fully connected* if every output from a single node is channeled to every node in the next layer.

The number of input and output nodes needed for a network will depend on the nature of the data presented to the network and the type of the output desired from it, respectively. The number of neurons to use in a hidden layer, and the number of hidden layers required for a processing task is less obvious. Further comments on this question will appear in a later section.

A layer of neurons: A one-layer network with n inputs and q neurons is shown in Fig. 11.11. In the network, each input x_i; $i = 1, 2, ..., n$ is connected to the jth neuron input through the weight w_{ji}; $j = 1, 2, ..., q$. The jth neuron has a summer that gathers its weighted inputs to form its own scalar output

$$\sum_{i=1}^{n} w_{ji} x_i + w_{j0}; j = 1, 2, ..., q.$$

Finally, the jth neuron outputs $\hat{y}_j$ through its activation function $\sigma(.)$:

$$\hat{y}_j = \sigma\left(\sum_{i=1}^{n} w_{ji} x_i + w_{j0}\right); j = 1, 2, ..., q \qquad (11.26a)$$

$$= \sigma(\mathbf{w}_j \mathbf{x} + w_{j0}); j = 1, 2, ..., q \qquad (11.26b)$$

where weight vector $\mathbf{w}_j$ is defined as

$$\mathbf{w}_j = [w_{j1} \quad w_{j2} ... w_{jn}] \qquad (11.26c)$$

Note that it is common for the number of inputs to be different from the number of neurons (i.e., $n \neq q$). A layer is not constrained to have the number of its inputs equal to the number of its neurons.

In vector-matrix notation, the layer shown in Fig. 11.11 has $q \times 1$ output vector

$$\hat{\mathbf{y}} = \begin{bmatrix} \hat{y}_1 \\ \hat{y}_2 \\ \vdots \\ \hat{y}_q \end{bmatrix}, \qquad (11.27a)$$

$q \times n$ weight matrix

$$\mathbf{W} = \begin{bmatrix} w_{11} & w_{12} & \cdots & w_{1n} \\ w_{21} & w_{22} & \cdots & w_{2n} \\ \vdots & \vdots & & \vdots \\ w_{q1} & w_{q2} & \cdots & w_{qn} \end{bmatrix} = \begin{bmatrix} \mathbf{w}_1 \\ \mathbf{w}_2 \\ \mathbf{w}_3 \\ \mathbf{w}_q \end{bmatrix} \qquad (11.27b)$$

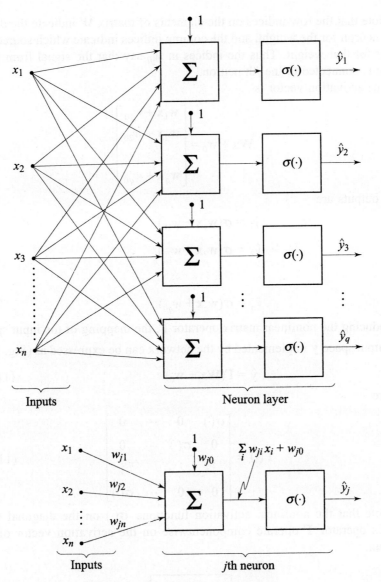

Fig. 11.11 A one-layer network

and $q \times 1$ bias vector

$$\mathbf{w}_0 = \begin{bmatrix} w_{10} \\ w_{20} \\ \vdots \\ w_{q0} \end{bmatrix} \qquad (11.27c)$$

Note that the row indices on the elements of matrix **W** indicate the destination neuron for the weight, and the column indices indicate which source is the input for that weight. Thus the indices in w_{ji} say that the signal from the ith input is connected to the jth neuron.

The activation vector is

$$\mathbf{Wx} + \mathbf{w}_0 = \begin{bmatrix} \mathbf{w}_1 \mathbf{x} + w_{10} \\ \mathbf{w}_2 \mathbf{x} + w_{20} \\ \vdots \\ \mathbf{w}_q \mathbf{x} + w_{q0} \end{bmatrix}$$

The outputs are

$$\hat{y}_1 = \sigma(\mathbf{w}_1 \mathbf{x} + w_{10})$$
$$\hat{y}_2 = \sigma(\mathbf{w}_2 \mathbf{x} + w_{20})$$
$$\vdots$$
$$\hat{y}_q = \sigma(\mathbf{w}_q \mathbf{x} + w_{q0})$$

Introducing the nonlinear matrix operator Γ, the mapping of the input space $\mathbf{x}$ to output space $\hat{\mathbf{y}}$ implemented by the network can be expressed as (Fig. 11.12)

$$\hat{\mathbf{y}} = \Gamma(\mathbf{Wx} + \mathbf{w}_0) \qquad (11.28a)$$

where

$$\Gamma(\cdot) \triangleq \begin{bmatrix} \sigma(\cdot) & 0 & \cdots & 0 \\ 0 & \sigma(\cdot) & \cdots & 0 \\ \vdots & \vdots & \cdots & \vdots \\ 0 & 0 & \cdots & \sigma(\cdot) \end{bmatrix} \qquad (11.28b)$$

Note that the nonlinear activation functions $\sigma(\cdot)$ on the diagonal of the matrix operator Γ operate componentwise on the activation vector of each neuron.

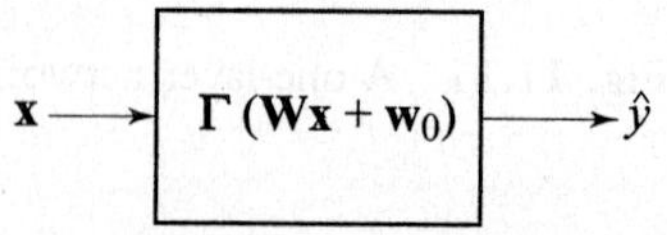

$$\mathbf{x} \longrightarrow \boxed{\Gamma(\mathbf{Wx} + \mathbf{w}_0)} \longrightarrow \hat{\mathbf{y}}$$

Fig. 11.12 Input–Output map of a one-layer network

The input and output vectors $\mathbf{x}$ and $\hat{\mathbf{y}}$ are often called *input* and *output patterns*, respectively. The mapping of the input pattern to an output pattern as given by (11.28) is of the feedforward and instantaneous type since it involves no time delay between input $\mathbf{x}$ and the output $\hat{\mathbf{y}}$.

Multiple layers of neurons: A two-layer NN, depicted in Fig. 11.13, has n inputs and two layers of neurons, with the first layer having m neurons that feed into the second layer having q neurons. The first layer is known as the *hidden layer*, with m the number of *hidden-layer neurons*; the second layer is known as the output layer, with q the number of *output-layer neurons*. It is common for different layers to have different numbers of neurons. Note that the outputs of the hidden layer are inputs to the following layer (output layer); and the network is fully connected. Neural networks with multiple layers are called *multilayer perceptrons*; their computing power is significantly enhanced over the one-layer NN.

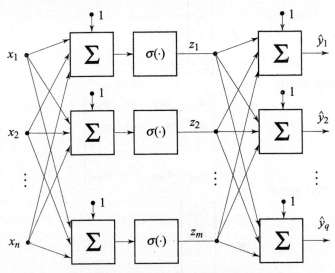

Fig. 11.13 A two-layer network

All continuous functions (exhibiting certain smoothness) can be approximated to any desired accuracy with a network of one hidden layer of sigmoidal hidden units and a layer of linear output units [159]. Does it mean that there is no need to use more than one hidden layer and/or mix different types of activation functions? This is not quite true. It may be that the accuracy can be improved using a more sophisticated network architecture. In particular when the complexity of the mapping to be learned is high, it is likely that the performance can be improved. However, since implementation and training of the network become more complicated, it is customary to apply only a single hidden layer of similar activation functions and an output layer of linear units. Our focus is on two-layer feedforward neural networks with hyperbolic tangent hidden units and linear output units. This is probably the most commonly used network architecture as it works quite well in many practical applications.

Defining the hidden-layer outputs z_l allows one to write

$$z_l = \sigma\left(\sum_{i=1}^{n} w_{li} x_i + w_{l0}\right); \quad l = 1, 2, \ldots, m \quad (11.29)$$

$$= \sigma(\mathbf{w}_l \mathbf{x} + w_{l0})$$

where

$$\mathbf{w}_l \triangleq [w_{l1} \quad w_{l2} \quad \ldots \quad w_{ln}]$$

In vector-matrix notation, the hidden layer in Fig. 11.13 has $m \times 1$ output vector

$$\mathbf{z} = \begin{bmatrix} z_1 \\ z_2 \\ \vdots \\ z_m \end{bmatrix}, \quad (11.30a)$$

$m \times n$ weight matrix

$$\mathbf{W} = \begin{bmatrix} w_{11} & w_{12} & \cdots & w_{1n} \\ w_{21} & w_{22} & \cdots & w_{2n} \\ \vdots & \vdots & & \vdots \\ w_{m1} & w_{m2} & \cdots & w_{mn} \end{bmatrix} \quad (11.30b)$$

and $m \times 1$ bias vector

$$\mathbf{w}_0 = \begin{bmatrix} w_{10} \\ w_{20} \\ \vdots \\ w_{m0} \end{bmatrix} \quad (11.30c)$$

The output

$$\mathbf{z} = \Gamma(\mathbf{W}\mathbf{x} + \mathbf{w}_0) \quad (11.31a)$$

where

$$\Gamma(\cdot) \triangleq \begin{bmatrix} \sigma(\cdot) & 0 & \cdots & 0 \\ 0 & \sigma(\cdot) & \cdots & 0 \\ \vdots & \vdots & & \vdots \\ 0 & 0 & \cdots & \sigma(\cdot) \end{bmatrix} \quad (11.31b)$$

Defining the second-layer weight matrix as

$$\mathbf{V} = \begin{bmatrix} v_{11} & v_{12} & \cdots & v_{1m} \\ v_{21} & v_{22} & \cdots & v_{2m} \\ \vdots & \vdots & & \vdots \\ v_{q1} & v_{q2} & \cdots & v_{qm} \end{bmatrix} \quad (11.32a)$$

and bias vector as

$$\mathbf{v}_0 = \begin{bmatrix} v_{10} \\ v_{20} \\ \vdots \\ v_{q0} \end{bmatrix}, \tag{11.32b}$$

one may write the NN output as

$$\hat{y}_j = \Gamma\left(\sum_{l=1}^{m} v_{jl} z_l + v_{j0}\right); j = 1, 2, \ldots, q \tag{11.33}$$

$$= \mathbf{v}_j \mathbf{z} + v_{j0}$$

where

$$\mathbf{v}_j \triangleq [v_{j1} \ v_{j2} \ \ldots \ v_{jm}]$$

The output vector

$$\hat{\mathbf{y}} = \begin{bmatrix} \hat{y}_1 \\ \hat{y}_2 \\ \vdots \\ \hat{y}_q \end{bmatrix} \tag{11.34a}$$

is given by the expression

$$\hat{\mathbf{y}} = \mathbf{V}\mathbf{z} + \mathbf{v}_0$$

$$= \mathbf{V}(\mathbf{\Gamma}(\mathbf{W}\mathbf{x} + \mathbf{w}_0) + \mathbf{v}_0) \tag{11.34b}$$

Figure 11.14 shows the input–output map.

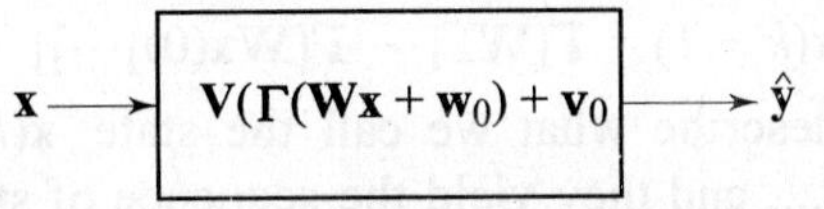

$$\mathbf{x} \longrightarrow \boxed{\mathbf{V}(\mathbf{\Gamma}(\mathbf{W}\mathbf{x} + \mathbf{w}_0) + \mathbf{v}_0)} \longrightarrow \hat{\mathbf{y}}$$

Fig. 11.14 Input–output map of a two-layer network

Recurrent Networks

Look at the directionality in the connections of multilayer feedforward network of Fig. 11.13; the output of a neuron scaled by the value of the connection weight is fed forward to provide a portion of the activation for the neurons in the next higher layer. No neuron output can be an input for a neuron in the same layer or preceding layer. A *feedback network* would allow outputs to be the input to the preceding layers, and feedlateral connections would send some outputs to other nodes in the same layer. This class of networks, wherein outputs are directed back as inputs to neurons in the previous or the same layer are referred to as *recurrent networks* because they incorporate feedback and thus are inherently recursive [160, 162].

Figure 11.15a shows a recurrent network architecture. The feedback loops in the network enable control of the output x_i through outputs x_j for $i, j = 1, 2, ..., n$. Such a control is especially meaningful if the present output, say $\mathbf{x}(t)$, controls the output at the following instant, $\mathbf{x}(t + \Delta)$. The time Δ elapsed between t and $t + \Delta$ is introduced by the delay elements in the feedback loops as shown in Fig. 11.15a. Here the time delay Δ has a symbolic meaning: it is an analogy to the refractory period of an elementary biological neuron model. Using the notation introduced for feedforward networks, the mapping of $\mathbf{x}(t)$ into $\mathbf{x}(t + \Delta)$ can be written as

$$\mathbf{x}(t + \Delta) = \Gamma[\mathbf{W}\mathbf{x}(t)] \qquad (11.35a)$$

This is represented by the block diagram shown in Fig. 11.15b. Note that the input $\mathbf{x}(0)$ is needed to initialize the network.

Assume that we consider time as a discrete variable and decide to observe the network performance at discrete time instants $\Delta, 2\Delta, 3\Delta, ...$ For notational convenience, the time step may be equated to unity, and the instants indexed by positive integers. Symbol Δ thus has the meaning of unit delay. Equation (11.35a) may therefore be expressed as (Fig. 11.15c)

$$\mathbf{x}(k + 1) = \Gamma[\mathbf{W}\mathbf{x}(k)]; \ k = 0, 1, 2,... \qquad (11.35b)$$

where k is the instant number. The response of the network at the $(k + 1)$th instant depends on the entire history of the network starting at $k = 0$. Indeed we have from Eqn. (11.35b), a series of nested solutions as follows:

$$\mathbf{x}(1) = \Gamma[\mathbf{W}\mathbf{x}(0)]$$

$$\mathbf{x}(2) = \Gamma[\mathbf{W}\Gamma[\mathbf{W}\mathbf{x}(0)]] \qquad (11.36)$$

$$\vdots$$

$$\mathbf{x}(k + 1) = \Gamma[\mathbf{W}\Gamma[\cdots \Gamma[\mathbf{W}\mathbf{x}(0)]\cdots]]$$

Equations (11.36) describe what we call the state $\mathbf{x}(k)$ of the network at instants $k = 0, 1, 2, ...$, and they yield the sequence of state transitions.

Figure 11.16 shows the architecture of a *generalized feedback network*. Examples of the $H(z)$ functions are:

1. $H(z) = z^{-1}$; the simplest situation where the linear dynamic transfer function $H(z)$ is a pure unit delay.
2. $H(z) = z^{-d}$; pure delay of d units.
3. $H(z) = \sum_{i=1}^{d} \alpha_i z^{-i}$.
4. $H(z)$ is a stable rational transfer function, e.g.,

$$H(z) = (z + 0.8)/(z^2 - 0.5z + 0.5)$$

Neural Networks for Control 743

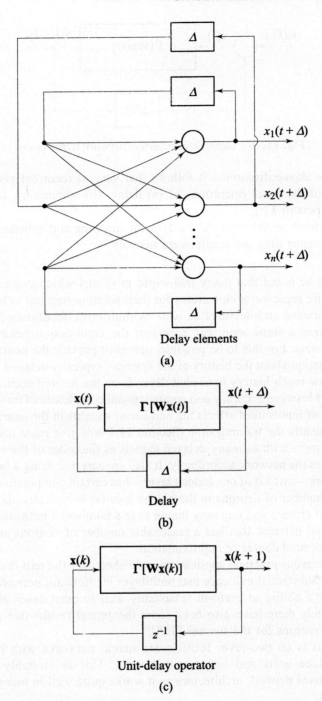

Fig. 11.15 A recurrent network architecture

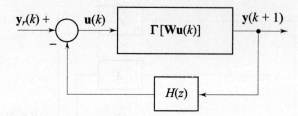

Fig. 11.16 Generalized recurrent network architecture

From the above discussion it follows that general recurrent networks are composed of the basic operations of (a) delay, (b) summation, and (c) the nonlinear operator $\Gamma[\cdot]$.

The recurrent neural network is a dynamic mapping and is better suited for dynamic systems than the feedforward network.

▲▲

It should be noted that many real-world problems which one might think would require recurrent architectures for their solution turn out to be solvable with feedforward architectures as well. A multilayer feedforward network, which realizes a static map, can represent the input/output behaviour of a dynamic system. For this to be possible, one must provide the neural network with information about the history of the system—typically delayed inputs and outputs. How much history is needed depends on the desired accuracy. There is a trade off between accuracy and computational complexity of training, since the number of inputs used affects the number of weights in the neural network and subsequently the training time (Section 11.7 will give more details). One sometimes starts with as many delayed signals as the order of the system and then modifies the network accordingly. It also appears that using a two hidden-layer network—instead of one hidden layer—has certain computational advantages. The number of neurons in the hidden layer(s) is typically chosen based on empirical criteria and one may iterate over a number of networks to determine a neural network that has a reasonable number of neurons and accomplishes the desired degree of approximation.

From numerous practical applications published over the past decade, there seems to be substantial evidence that multilayer feedforward networks possess an impressive ability to perform reasonably well in most cases of practical interest. Lately there have also been some theoretical results that attempt to explain the reasons for the success [158].

Our focus is on two-layer feedforward neural networks with hyperbolic tangent hidden units and linear output units. This is probably the most commonly used network architecture as it works quite well in many practical applications.

11.4 LEARNING IN NEURAL NETWORKS

We are accustomed to solving computational problems by breaking them down into steps and solving each small step. Neural networks don't work by developing an algorithm for each individual processing element. The knowledge is in the state of the whole network, not any one piece of it. Consequently, we have to think in overall terms of inputs, transformations, and outputs of the network.

Neural networks deal only with numeric input data. Therefore, we must convert or encode information from the external environment to numeric data form. Additionally it is often necessary to scale data. Inhibitory inputs are just as important as excitatory inputs. The input scheme should adequately allow for both the types. A provision is also usually made for constant-source input to serve as an offset or bias term for the transfer or activation function.

The numeric output data of a neural network will likewise require decoding and scaling to make it compatible with the external environment.

Important characteristics of the network depend on:

(i) the transfer or activation functions of the processing elements;
(ii) the structure of the network (number of neurons, layers and interconnections); and
(iii) the learning mechanism of the network.

We have seen in Section 11.2 that several different functions may be used as activation functions, and in fact, one of the distinguishing features between existing neural network models is precisely which transfer function they employ. Other important distinguishing feature is the network structure. Two major classes of networks, feedforward and recurrent networks, were introduced in the last section. In this section, we present basic learning features of neural networks.

The Basic Learning Mechanism

Each processing element (neuron) in a neural network has a number of inputs (x_i), each of which must store a connection weight (w_{ji}). The element sums up the weighted input ($w_{ji}x_i$) and computes one and only one activation signal (a_j). This signal is a function ($\sigma(\cdot)$) of the weighted sum. Figure 11.17 summarizes how a processing element works.

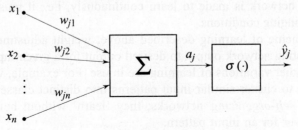

Fig. 11.17 A processing element

The function $\sigma(\cdot)$ remains fixed for the life of the processing element. It is generally decided upon as part of the design, and it cannot be changed dynamically. In other words, the transfer function currently cannot be adjusted or modified during the operation or running of the network.

However the weights (w_{ji}) are variables. They can be dynamically adjusted to produce a given output (y_j). This dynamic modification of the variable weights is the very essence of *learning*. At the level of a single processing element, this self-adjustment is a very simple thing. When many processing elements do it collectively, we say it resembles 'intelligence'. The meaningful information is in the modified weights. The ability of an entire neural network to adapt itself (change the w_{ji} values) to achieve a given output (y_j) is its uniqueness (of course, uniqueness for a nonliving entity).

In traditional programming, the programmer will be given the inputs, some type of processing requirements (what to do with the inputs), and the desired output. The programmer's job is to apply the necessary step-by-step instructions to develop the required relationship between the input and the output.

In contrast, neural networks do not require any instructions, rules or processing requirements about how to process the input data. In fact, neural networks determine the relationship between input and output by looking at examples of many input-output pairs.

Pairs of inputs and outputs are applied to the neural network. These pairs of data are used to *teach* or *train* the network, and as such are referred to as the *training set*. Knowing what output is expected from each input, the network automatically adjusts or adapts the strengths of the connections between processing elements. The method used for the adjusting process is called the *learning rule*.

The training set needs to be fairly large; it must also contain all the necessary information if the network is to 'learn' the features and relationships that are important. Also, note that the approach of teaching the network one thing at a time does not work. All the weights set so meticulously for one fact could be drastically altered in learning the next fact. The network has to learn everything together, finding the best weight settings for the total set of facts.

After training, the network is ready for use. Depending on the task to be done, the network may have its currently learned weights 'frozen' thus disabling its learning laws. This network will give reasonable outputs for the data it has not seen before. For some applications such as real-time process control, the network is made to learn continuously, i.e., it must continue to adjust to changing conditions.

In the scheme of learning described above, weight adjustment is correctional: adjusting network output to desired output for a given input pattern. A number of other variations of learning are in use. For example, weights could be modified to cluster similar input patterns into distinct classes. This is the property of *self-organizing* networks; they 'learn' without being given the correct answer for an input pattern.

Let us look at some of the characteristics of neural-network technology which set it apart from conventional computing.

Inherent parallelism: Processing sequence in neural networks is parallel and simultaneous. The processing elements in one layer all operate in concert. Computation is distributed over more than one processing element and is done simultaneously.

Although digital computers have to simulate this parallelism, true neural network hardware will really perform the operations in parallel. Very fast decisions made in real time will be possible.

Storing knowledge: Knowledge within a neural network is not stored in specific memory locations, as it is in conventional computing. Knowledge is distributed throughout the system; it is the dynamic response to the inputs and the network architecture. Because knowledge is distributed, the system uses many connections to retrieve solutions to particular problems.

Not only is the memory in a neural network distributed, it may also be *associative*, like in biological neural systems. For example, if we hum a few bars of a tune and it is one we have heard before, we may be able to 'name that tune'. This capability of looking at partial patterns and guessing the whole is an important characteristic of the neural networks because most of the data in this world is inexact.

If the partial pattern leads to the complete version, the term *auto-associative* is used to describe the network. On the other hand, if the identification is based on distorted pattern, the term *hetro-associative* applies. In a conventional digital computer, each fact is stored in a unique location. A fact is retrieved by providing the address, even though there is no particular relationship between the fact and the address. Neural networks don't work this way at all. The associative memories of neural networks are *content-addressable*; the address may be a subpart of the memory content.

Fault tolerance: Neural networks are extremely fault tolerant and degrade gracefully. They can learn from and make decisions based on incomplete data. Because the knowledge is distributed throughout the system rather than residing in a single memory location, a percentage of the nodes can be inoperative without significantly changing the overall system behaviour. Resistance to hardware failure is therefore much greater in a neural network than in conventional computers, where hardware failure is usually catastrophic.

Generalization: Generalization is the ability of the network to respond to input that is has not seen before. The input may be partial or incomplete. Generalization takes the ability of the neural network to learn and self-adjust a step further. The system can 'hypothesize' a response.

Learning Rules

As we have seen in the preceding paragraphs, learning in neural networks refers to the process of acquiring a desired behaviour by changing the connection weights. The main rules to produce these weight changes may be classified into three important categories, each of which borrow concepts from behavioural theories of learning in biological systems.

Supervised learning: These rules compute the necessary change in the connection weights by presenting the network given input pattern, comparing the obtained response with a desired response known *a priori* and then changing the weights in the direction of decreasing error. More clearly, in the supervised learning mode, a neural network is supplied with a sequence of examples $(\mathbf{x}^{(1)}, \mathbf{y}^{(1)})$, $(\mathbf{x}^{(2)}, \mathbf{y}^{(2)})$, ..., $(\mathbf{x}^{(p)}, \mathbf{y}^{(p)})$, ..., of desired input-output pairs. When each input $\mathbf{x}^{(p)}$ is fed into the neural network, the corresponding desired output $\mathbf{y}^{(p)}$ is also supplied to the neural network. As shown in Fig. 11.18a, the difference between the actual NN output $\hat{\mathbf{y}}^{(p)}$ and the desired output $\mathbf{y}^{(p)}$ is measured in the error-signal generator which then produces error signals for the NN to correct its weights in such a way that the actual output will move closer to the desired output.

With supervised learning, it is necessary to 'train' the neural network before it becomes operational. Training consists of, as we have discussed earlier, presenting to the network sets of input patterns and desired responses. During the training phase, connection weights are adapted to learn the desired input-output behaviour. During the operation phase, the network works with fixed weights to produce outputs in response to new patterns.

Reinforcement learning: In supervised learning it is assumed that the correct 'target' output values are known for each input pattern. But in some situations only less detailed information is available. For example, the NN may only be told that its current actual output is "too high", or "50% correct". In the extreme case, there is only a single bit of feedback information indicating whether the output is *right* or *wrong*. Learning based on this kind of *critic* information is called *reinforcement learning*, and the feedback information is called *reinforcement signal*. As shown in Fig. 11.18b, reinforcement learning is a form of supervised learning because the network still receives some feedback from its environment. But the feedback (i.e., the reinforcement signal) is only *evaluative* (*critic*) rather than instructive. That is, it just says how good or how bad a particular output is and provides no hint as to what the right answer should be. The external reinforcement signal is usually processed by the critic-signal generator (Fig. 11.18b) to produce a more informative critic signal for the NN to adjust its weights properly with the hope of getting better critic feedback in the future. Reinforcement learning is also called *learning with a critic* as opposed to *learning with a teacher*, which describes supervised learning.

Unsupervised learning: In unsupervised learning, there is no teacher to provide any feedback information (see Fig. 11.18c). There is no feedback from the environment to say what the outputs should be or whether they are correct. The network must discover for itself patterns, features, regularities, correlations, or categories in the input data and code for them in the output. While dis-

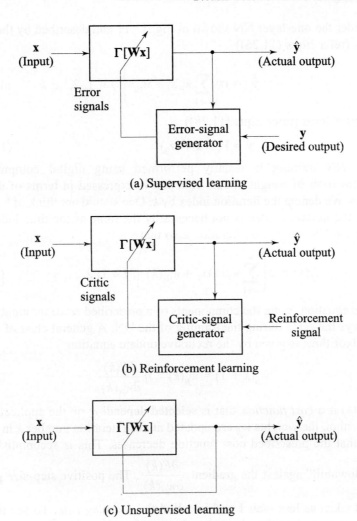

Fig. 11.18

covering these features, the network undergoes changes in its parameters; this process is called *self-organizing*. A typical example is making an unsupervised classification of objects without providing information about the actual classes. Proper clusters are formed by discovering the similarities and dissimilarities among the objects.

11.5 TRAINING THE MULTILAYER NEURAL NETWORK–BACKPROPAGATION TUNING

Gradient descent is a basic technique that plays an important role in learning algorithm for neural networks. Application of this technique to a single-layer network will be discussed as a prelude to backpropagation learning algorithm for multilayer feedforward neural networks.

Consider the one-layer NN shown in Fig. 11.11 and described by the recall equation (refer Eqns (11.26))

$$\hat{y}_j = \sigma\left(\sum_{i=1}^{n} w_{ji} x_i + w_{j0}\right); j = 1, 2, \ldots, q \qquad (11.37a)$$

or in matrix form (refer Eqns (11.28))

$$\hat{\mathbf{y}} = \Gamma(\mathbf{W}\mathbf{x} + \mathbf{w}_0) \qquad (11.37b)$$

Since NN training is usually performed using digital computers, a convenient form of weight update equation is expressed in terms of discrete iterations. We denote the iteration index by k. One should not think of k as time index as the iteration index is not necessarily the same as the time index. Let $w_{ji}(k)$ be the NN weights on iteration k so that

$$\hat{y}_j(k) = \sigma\left(\sum_{i=1}^{n} w_{ji}(k) x_i + w_{j0}(k)\right); j = 1, 2, \ldots, q \qquad (11.38)$$

In this equation, x_i are the components of a prescribed *constant* input vector $\mathbf{x}$ that stays the same during the training of the NN. A general class of weight update algorithms is given by the recursive update equation

$$w_{ji}(k+1) = w_{ji}(k) - \eta \frac{\partial E(k)}{\partial w_{ji}(k)} \qquad (11.39)$$

where $E(k)$ is a *cost function* that is selected depending on the application. In this algorithm, the weights w_{ji} are updated at each iteration number k in such a manner that the prescribed cost function decreases. This is accomplished by going "downhill" against the gradient $\dfrac{\partial E(k)}{\partial w_{ji}(k)}$. The positive step-size parameter η is taken as less than 1 and is called the *learning rate*. To see that the gradient descent algorithm decreases the cost function, note that

$$\Delta w_{ji}(k) \equiv w_{ji}(k+1) - w_{ji}(k) \qquad (11.40a)$$

and, to first order

$$\Delta E(k) \equiv E(k+1) - E(k)$$

$$\approx \sum_{j,i} \frac{\partial E(k)}{\partial w_{ji}(k)} \Delta w_{ji}(k)$$

$$\approx -\eta \sum_{j,i} \left(\frac{\partial E(k)}{\partial w_{ji}(k)}\right)^2 \qquad (11.40b)$$

Let a prescribed vector $\mathbf{x}$ be input to the NN and desired target output associated with $\mathbf{x}$ be $\mathbf{y}$. Then at iteration index k, jth component of the output error is

$$e_j(k) = y_j - \hat{y}_j(k) \tag{11.41}$$

where y_j is the desired output and $\hat{y}_j(k)$ is the actual output with input **x**. We define the cost function as least-squares NN output error:

$$E(k) = \frac{1}{2}\sum_{j=1}^{q} e_j^2(k) = \frac{1}{2}\sum_{j=1}^{q}\left(y_j - \hat{y}_j(k)\right)^2 \tag{11.42}$$

Note that the components x_i of the input **x** and the desired NN output components y_j are not functions of the iteration index k.

To derive the gradient descent algorithm with the least-squares output-error cost, the gradients with respect to the weights and biases are computed using the product rule and the chain rule as

$$\frac{\partial E(k)}{\partial w_{ji}(k)} = -e_j(k)\frac{\partial \hat{y}_j(k)}{\partial w_{ji}(k)} = -e_j(k)\sigma'\left(\sum_{i=1}^{n} w_{ji}(k)x_i + w_{j0}(k)\right)x_i \tag{11.43a}$$

$$\frac{\partial E(k)}{\partial w_{j0}(k)} = -e_j(k)\sigma'\left(\sum_{i=1}^{n} w_{ji}(k)x_i + w_{j0}(k)\right) \tag{11.43b}$$

where Eqns (11.42) and (11.38) were used. The notation $\sigma'(\cdot)$ denotes the derivative of the activation function evaluated at the argument. The argument is the activation value a_j:

$$a_j = \sum_{i=1}^{n} w_{ji}(k)x_i + w_{j0}(k)$$

$$\sigma'(a_j) = \frac{d\sigma}{da_j}$$

For a linear activation function, $\sigma(a_j) = a_j$ and $\sigma'(a_j) = 1$. The gradient descent algorithm becomes

$$w_{ji}(k+1) = w_{ji}(k) + \eta e_j(k) x_i \tag{11.44a}$$

$$w_{j0}(k+1) = w_{j0}(k) + \eta e_j(k) \tag{11.44b}$$

For a sigmoidal activation function

$$\sigma(a_j) = \frac{1}{1+e^{-a_j}}$$

$$\sigma'(a_j) = -\frac{1}{\left(1+e^{-a_j}\right)^2}\left(-e^{-a_j}\right)$$

$$= \frac{1}{1+e^{-a_j}}\left(\frac{e^{-a_j}}{1+e^{-a_j}}\right) = \sigma(a_j)[1-\sigma(a_j)] \tag{11.45}$$

The gradient descent algorithm becomes

$$w_{ji}(k+1) = w_{ji}(k) + \eta e_j(k)\hat{y}_j(1-\hat{y}_j)x_i \qquad (11.46a)$$

$$w_{j0}(k+1) = w_{j0}(k) + \eta e_j(k)\hat{y}_j(1-\hat{y}_j) \qquad (11.46b)$$

In terms of matrices, algorithm (11.44) may be expressed as

$$\mathbf{W}(k+1) = \mathbf{W}(k) + \eta \mathbf{e}(k)\mathbf{x}^T \qquad (11.47a)$$

where

$$\mathbf{e}(k) = \mathbf{y} - \hat{\mathbf{y}}(k)$$

The bias vector $\mathbf{w}_0$ is updated according to

$$\mathbf{w}_0(k+1) = \mathbf{w}_0(k) + \eta \mathbf{e}(k) \qquad (11.47b)$$

Similar expressions may be written down for the algorithm (11.46).

We have just discussed NN weight training procedure when one input-vector/desired output-vector pair $(\mathbf{x}, \mathbf{y})$ is given. In practical situations, there will be multiple input vectors prescribed by the user, each with an associated desired output vector. Suppose there are presecibed P desired input/output pairs $\left(\mathbf{x}^{(1)},\mathbf{y}^{(1)}\right)$, $\left(\mathbf{x}^{(2)},\mathbf{y}^{(2)}\right)$, ..., $\left(\mathbf{x}^{(P)},\mathbf{y}^{(P)}\right)$ for the NN. In such situations, the NN must be trained to associate *each* input vector with its prescribed output vector. There are many strategies for training the net in this scenario; at the two extremes are *incremental updating* and *batch updating*. For this discussion we shall use matrix updates. Define for $p = 1, 2, ..., P$ the quantities

$$\hat{\mathbf{y}}^{(p)}(k) = \sigma(\mathbf{W}(k)\mathbf{x}^{(p)} + \mathbf{w}_0)$$

$$\mathbf{e}^{(p)}(k) = \mathbf{y}^{(p)} - \hat{\mathbf{y}}^{(p)}(k) \qquad (11.48)$$

$$E^{(p)}(k) = \frac{1}{2}\left(\mathbf{e}^{(p)}(k)\right)^T \mathbf{e}^{(p)}(k)$$

In incremental updating, the vectors $(\mathbf{x}^{(p)}, \mathbf{y}^{(p)})$ are sequentially presented to the NN. At each presentation, one step of the training algorithm is performed, so that (refer (11.47))

$$\mathbf{W}(k+1) = \mathbf{W}(k) + \eta \mathbf{e}^{(p)}(k)(\mathbf{x}^{(p)})^T; \, p = 1, 2, ..., P \qquad (11.49a)$$

$$\mathbf{w}_0(k+1) = \mathbf{w}_0(k) + \eta \mathbf{e}^{(p)}(k); \, p = 1, 2, ..., P \qquad (11.49b)$$

which updates both the weights and biases. An *epoch* is defined as one complete run through all the P associated pairs. When one epoch has been completed, the pair $\left(\mathbf{x}^{(1)},\mathbf{y}^{(1)}\right)$ is presented again and another run through all the P pairs is performed. It is hoped that after many epochs, the output error will be small enough.

In batch updating, all P pairs are presented to the NN (one at a time) and a cumulative error is computed after all have been presented. At the end of this procedure, the NN weights and biases are updated once. The result is

$$\mathbf{W}(k+1) = \mathbf{W}(k) + \eta \sum_{p=1}^{P} \mathbf{e}^{(p)}(k)(\mathbf{x}^{(p)})^T \qquad (11.50a)$$

$$\mathbf{w}_0(k+1) = \mathbf{w}_0(k) + \eta \sum_{p=1}^{P} \mathbf{e}^{(p)}(k) \qquad (11.50b)$$

In batch updating, the iteration index corresponds to the number of times the set of P patterns in presented and the cumulative error computed. That is, k corresponds to the epoch number.

Backpropagation Algorithm

We shall drive the backpropagation algorithm for the two-layer NN shown in Fig. 11.13. We shall conserve simplicity of notation by dispensing with the iteration index k. Let there be a prescribed input vector $\mathbf{x}$ and an associated desired output vector $\mathbf{y}$ for the network. Define the least-squares NN output error as

$$E = \frac{1}{2} \sum_{j=1}^{q} e_j^2$$

$$e_j = y_j - \hat{y}_j$$

where $\hat{y}_j$ is evaluated using the equations

$$\hat{y}_j = \sum_{\ell=0}^{m} v_{j\ell} z_\ell; \quad z_0 \equiv 1$$

$$z_\ell = \sigma\left(\sum_{i=0}^{n} w_{\ell i} x_i\right); \quad x_0 \equiv 1$$

with the components x_i of the input vector.

We first consider the output layer of the network. Weights and biases in the layer are updated according to

$$v_{j\ell} = v_{j\ell} - \eta \frac{\partial E}{\partial v_{j\ell}}$$

With linear activation functions in the output layer, the update rule becomes (refer Eqns 11.44))

$$v_{j\ell} = v_{j\ell} + \eta e_j z_\ell$$

We now consider the hidden layer of the network. Unlike the output nodes, the desired output of the hidden nodes is unknown. For a given input/output pair, $(\mathbf{x}, \mathbf{y})$, the backpropagation algorithm performs two phases of data flow. First, the input pattern $\mathbf{x}$ is propagated from the input terminals to the output layer and, as a result of the forward flow of data, it produces an actual output

$\hat{\mathbf{y}}$. Then the error signals resulting from the difference between $\hat{\mathbf{y}}$ and $\mathbf{y}$ are *backpropagated* from the output layer to the previous layers, to update their weights. Error backpropagation may be computed by expanding the error derivative using the chain rule as follows:

$$z_\ell = \sigma(a_\ell)$$

$$a_\ell = \sum_{i=0}^{n} w_{\ell i} x_i; \; x_0 \equiv 1$$

$$w_{\ell i} = w_{\ell i} - \eta \frac{\partial E}{\partial w_{\ell i}}$$

$$\frac{\partial E}{\partial w_{\ell i}} = \frac{\partial E}{\partial a_\ell} \frac{\partial a_\ell}{\partial w_{\ell i}} = \left[\sum_{j=1}^{q} \frac{\partial E}{\partial \hat{y}_j} \frac{\partial \hat{y}_j}{\partial z_\ell} \frac{\partial z_\ell}{\partial a_\ell} \right] \frac{\partial a_\ell}{\partial w_{\ell i}}$$

$$\frac{\partial E}{\partial \hat{y}_j} = -e_j$$

$$\frac{\partial \hat{y}_j}{\partial z_\ell} = v_{j\ell}$$

$$\frac{\partial z_\ell}{\partial a_\ell} = \sigma'(a_\ell) = z_\ell(1 - z_\ell)$$

$$\frac{\partial a_\ell}{\partial w_{\ell i}} = x_i$$

Therefore

$$\frac{\partial E}{\partial w_{\ell i}} = -x_i \left[z_\ell(1 - z_\ell) \sum_{j=1}^{q} v_{j\ell} e_j \right]$$

and the update rule becomes

$$w_{\ell i} = w_{\ell i} + \eta \, x_i z_\ell(1 - z_\ell) \sum_{j=1}^{q} v_{j\ell} e_j$$

The backpropagation algorithm consists of repeating the following iterative procedure until the NN output error has become sufficiently small. Incremental or batch processing may be used.

Forward Recursion to Compute NN Output:

Present input vector $\mathbf{x}$ to the NN and compute the NN output using

$$z_\ell = \sigma\left(\sum_{i=0}^{n} w_{\ell i} x_i \right); \; \ell = 1, 2, \ldots, m \qquad (11.51\text{a})$$

$$\hat{y}_j = \left(\sum_{\ell=0}^{m} w_{\ell i} x_i\right) v_{j\ell} z_\ell; j = 1, 2, \ldots, q \qquad (11.51b)$$

with $x_0 = 1$ and $z_0 = 1$, where **y** is the desired output vector.

Backward Recursion for Backpropagated Errors:

$$e_j = y_j - \hat{y}_j; j = 1, 2, \ldots, q \qquad (11.51c)$$

$$\delta_\ell = z_\ell(1 - z_\ell) \sum_{j=1}^{q} v_{j\ell} e_j; \ell = 1, 2, \ldots, m \qquad (11.51d)$$

Computation of the NN Weights and Bias Updates:

$$v_{j\ell} = v_{j\ell} + \eta z_\ell e_j; j = 1, 2, \ldots, q; \ell = 0, 1, \ldots, m \qquad (11.51e)$$

$$w_{\ell i} = w_{\ell i} + \eta x_i \delta_\ell; \ell = 1, 2, \ldots, m; i = 0, 1, \ldots, n \qquad (11.51f)$$

Improvements on Gradient Descent

There are many sorts of training algorithms for NN; the basic type we have discussed in the previous subsection is the Backpropagation Training Algorithm. Though the backpropagation algorithm enjoys great success, one must remember that it is a gradient-based technique, so that the usual coveats associated with step sizes, local minima and so on must be kept in mind while using it.

The NN weights and biases are typically initialized to small random (positive and negative) values. A typical error surface graph in 1-D is shown in Fig. 11.19, which shows a local minimum and a global minimum. If the weight is initialized as shown in Case 1, there is a possibility that the gradient descent might find the local minimum. Several authors have determined better techniques to initialize the weights than the random selection, particularly for the multilayer NN. Among these are Nguyan and Widrow, whose techniques are used, for instance, in MATLAB. Such improved initialization techniques can also significantly speed up convergence of the weights to their final values.

An improved version of gradient decent is given by *Momentum Gradient Algorithm*. Momentum allows a network to respond not only to the local gradient but also to recent trends in error surface. The learning rule with the inclusion of a momentum term can be written as (refer Eqn. (11.39))

$$\Delta w_{ji}(k) = -\eta \frac{\partial E(k)}{\partial w_{ji}(k)} + \alpha \Delta w_{ji}(k-1); 0 \leq |\alpha| \leq 1 \qquad (11.52)$$

Without momentum, a network may get stuck in a shallow local minimum; adding mementum can help the NN "ride through" local minima. (Case 1 in Fig. 11.19 may not get stuck in local minimum while learning with momentum). In the MATLAB Neural Network Toolbox are some examples showing

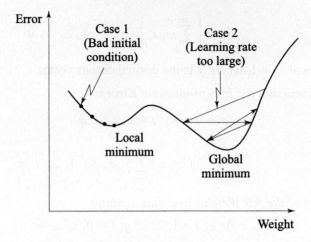

Fig. 11.19 Typical 1-D NN error surface

that learning with momentum can significantly speed up and improve the performance of backpropagation.

Only small learning constants η guarantee a true gradient descent. The price of this guarantee is an increased total number of learning steps that need to be made to reach a satisfactory solution. It is desirable to monitor the progress of learning so that η can be increased at appropriate stages of training to speed up the minimum seeking.

When broad minima yield small gradient values, then a larger value of η will result in a more rapid convergence. However, for problems with steep and narrow minima, if the learning rate η is too large, then the NN can overshoot the minimum cost value, jumping back and forth over the minimum and failing to converge, as shown in Fig. 11.19, Case 2. Adapting the learning rates can significantly speed up the convergence of the weights.

All the refinements: selecting better initial conditions, using learning with momentum, and using an adaptive learning rate are available in the MATLAB NN Toolbox.

In practice the gradient method is quite slow. Other methods are available which converge much faster. In most applications, it is therefore difficult to justify using the gradient method. Nevertheless, the method has gained a remarkable popularity in the neural network community. The primary properties in favour of the method are the simplicity at which it is implemented, and the modest requirement to data storage. In most situations, the drawback associated with slow convergence motivates the use of more sophisticated methods.

The category of fast algorithms uses standard numerical optimization techniques. Three types of numerical optimization techniques for neural network training have been incorporated in MATLAB: Conjugate gradient; quasi-Newton; and Levenberg-Marquardt.

The reader is advised to refer the literature [158–161] for details on improvements suggested above.

11.6 FUNCTION APPROXIMATION WITH NEURAL NETWORKS

Of fundamental importance in NN closed-loop control applications is the *universal function approximation* property of NNs having at least two layers (one-layer NNs do not generally have a universal approximation capability).

The basic universal approximation result says [159] that any smooth function $\mathbf{f(x)}$ can be approximated arbitrarily closely on a compact set using a two-layer NN with appropriate weights. This result has been shown using sigmoid activations, RBF activations, and others. Specifically, let $\mathbf{f(x)}$ be a smooth function; $\mathbf{x} = [x_1\ x_2\ ...\ x_n]^T$, $\mathbf{f}(\cdot) = [f_1(\cdot)\ f_2(\cdot)\ ...\ f_q(\cdot)]^T$, $\mathbf{S}$ be a compact set in n-dimensional state space and ε_N be a positive number. There exists a two-layer NN (Eqn. (11.34b)) such that

$$\mathbf{f(x)} = \mathbf{V}(\Gamma(\mathbf{Wx} + \mathbf{w}_0) + \mathbf{v}_0) + \boldsymbol{\varepsilon} \quad (11.53)$$

with $\|\boldsymbol{\varepsilon}\| < \varepsilon_N$ for all $\mathbf{x} \in \mathbf{S}$, for some (sufficiently large) number m of hidden-layer neurons. The value $\boldsymbol{\varepsilon}$ (generally a function of $\mathbf{x}$) is called the *NN function approximation error*, and it decreases as the hidden-layer size m increases. We say that on the compact set $\mathbf{S}$, $\mathbf{f(x)}$ is 'within ε_N of the NN functional range'. Approximation results have also been shown for smooth functions with a finite number of discontinuities.

Note that in this result the activation functions are not needed on the NN output layer (i.e., the output layer activation functions are linear). It also happens that the bias terms v_{j0} on the output layers are not needed, though the hidden layer bias terms $w_{\ell 0}$ are required.

Note further that though the result says 'there exists an NN that approximates $\mathbf{f(x)}$', it does not show how to determine the required number of units in the hidden layer. The issue of finding the required number of units in the hidden layer such that an NN does indeed approximate a given function $\mathbf{f(x)}$ closely enough is not an easy one (If the function approximation is to be carried out in the context of a dynamic closed-loop feedback control scheme, the issue is thornier and is discussed in subsequent sections). This issue has been addressed in the literature [158], and a significant result has been derived about the approximation capabilities of two-layer networks when the function to be approximated exhibits a certain smoothness. Unfortunately, the result is difficult to apply for selecting the number of hidden units. The guidelines to select the appropriate number of hidden neurons are rather empirical at the moment. To avoid large number of neurons and the corresponding inhibitively large training times, the smaller number of hidden layer neurons are often used in the first trial. One increases accuracy by adding more hidden neurons. Note that the number of inputs and outputs in the neural network are determined, respectively, by the number of the data presented to it and the type of the output desired from it.

Because of the above-mentioned results, one might think that there is no need for using more than one hidden layer and/or mixing different types of activation functions. This is not quite true: it may be that accuracy can be improved using a more sophisticated network architecture. In particular, when the complexity of the mapping to be learned is high (e.g., functions with discontinuities), it is likely that the performance can be improved. Experimental evidence tends to show that using a two hidden-layer network for continuous functions has sometimes advantages over a one-layer network as the former requires shorter training times.

An illustration of the NN function approximation in the context of closed-loop control was given in Section 11.1. Refer to [159] for a detailed account of controller design for nonlinear dynamic systems; the neural control having the structure shown in Fig. 11.1.

11.7 SYSTEM IDENTIFICATION WITH NEURAL NETWORKS

The main goal of the present chapter is to describe approaches to neural-network-based control that are found to be practically applicable to a reasonably wide class of unknown nonlinear systems. Systems identification is an integral part of such a control system design and consequently it calls for considerable attention as well. The system identification is necessary to establish a model based on which the controller can be designed, and it is useful for tuning and simulation before applying the controller to the real system. In this section, attention is drawn to identification of neural network models for nonlinear dynamic systems from a series of measurements on the systems.

We give here a generic working procedure for system identification with neural networks. One should be aware, of course, that such a procedure must always be squeezed a little here and there to conform to the application under consideration.

The multilayer feedforward network is straightforward to employ for the discrete-time modeling of dynamic systems for which there is a nonlinear relationship between the system's input and output. Let k count the multiple sampling periods so that $y(k)$ specifies the present output while $y(k-1)$ signifies the output observed at the previous sampling instant, etc. It is assumed that the output of the dynamic system at discrete-time instances can be described as a function of number of past inputs and outputs:

$$y(k) = f(y(k-1), ..., y(k-n), u(k-1), ..., u(k-m)) \quad (11.54)$$

A multilayer network can be used for approximating $f(\cdot)$ if the inputs to the network are chosen as the n past outputs and m past inputs of the dynamic system.

When attempting to identify a model of a dynamic system, it is a common practice to follow the procedure depicted in Fig. 11.20.

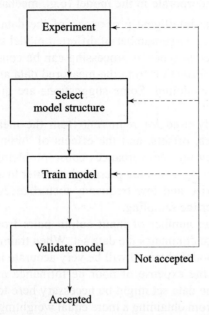

Fig. 11.20 System identification procedure

Experiment

The primary purpose of an experiment is to produce a set of examples of how the dynamic system to be identified responds to various control inputs. (These examples can later be used to train neural network to model the system). The experiment is particularly important in relation to nonlinear modeling; one must be extremely careful to collect a set of data that describes how the system behaves over its entire range of operaion. The following issues must be considered in relation to acquisition of data (For detailed information, see [151]).

Sampling frequency: The sampling frequency should be chosen in accordance with the desired dynamics of the closed-loop system consisting of controller and the system. A high sampling frequency permits a rapid reference tracking and a smoother control signal, but the problems with numerical ill-conditioning will become more pronounced. Consequently, the sampling frequency should be selected as a sensible compromise.

Input signals: While for identification of linear systems, it is sufficient to apply a signal containing a finite number of frequencies, a nonlinear system demands, roughly speaking, that all combinations of frequencies and amplitudes in the system's operating range are represented in the signal. As a consequence, the necessary size of the data set increases dramatically with the number of inputs and outputs. Unfortunately, there is no obvious remedy to this curse of dimensionality.

Before an input signal is selected, it is important to identify the operating range of the system. Special care must be taken not to excite dynamics that one does not intend to incorporate in the model (e.g., mechanical resonances).

Processing the data: Intelligent processing of the data is often much more important than trying a large number of different model structures and training schemes. Many different types of processing can be considered for extracting the most valuable information from the measured data and to make it suitable for neural-network modeling. Some suggestions are given in the following paragraphs.

Filtering is widely used for removing from the measured signals noise, periodic disturbances, offsets, and the effects of "uninteresting" dynamics. When high-frequency noise/disturbances cause problems, it is recommended to remove them by using an analog presampling filter to avoid an aliasing phenomenon. Offset, drift, and low-frequency disturbances can be removed by filtering the data after the sampling.

Sometimes a large number of input-output pairs from a small regime of entire operating range dominates the data set. When training on such a data set, it is likely that the model obtained will be very accurate in the regime that was over-represented at the expense of poor performance outside the regime. A little "surgery" on the data set might be necessary here to eliminate redundant information. Apart from obtaining a more equal weighting of the information, a reduction of the data set size also has the benefit that training times will be reduced.

It is also recommended to remove outliers from the data set (or alternatively insert interpolated values of the output signal). Outliers will often have a fatal impact on the training model.

Before training, it is often useful to scale all the signals so that they always fall within a specified range, say $[-1, 1]$. Another approach for scaling is to normalize the mean and standard deviation of the training set, e.g., to zero mean and unity standard deviation. The signals are likely to be measured in different physical units and without scaling there is a tendency that the signal of largest magnitude will be too dominating. Moreover, scaling makes the training algorithm numerically robust and leads to faster convergence.

Model Structure Selection

The model structure selection is basically concerned with the following two issues:
- Selecting an internal network architecture
- Selecting the inputs to the network

An often-used approach is to let the internal architecture be feedforward multilayer network. Probably the most commonly used network architecture is a two-layer feedforward network with hyperbolic tangent hidden units and linear output units. This architecture works quite well in many practical applications. In our presentation, we use this architecture. However, the reader is referred to more fundamental textbooks/research papers for a treatment of other types of neural networks in the control loop.

The input structure we use here consists of a number of past inputs and outputs (Refer Fig. 11.21):

$$\hat{y}(k|\theta) = \sum_{\ell=1}^{M} v_\ell \sigma\left(\sum_{i=1}^{N} w_{\ell i}\phi_i(k) + w_{\ell 0}\right) \quad (11.55a)$$

where $\hat{y}$ is the predicted value of the output y at sampling instant $t = kT$ (T = sampling interval), $\theta = \{v_\ell \; w_{\ell i}\}$ is the vector containing the adjustable parameters in the neural network (*weights*), ϕ is the *regression vector* which contains past outputs and past inputs (regressors's dependency on the weights is ignored):

$$\phi(k) = [y(k-1) \cdots y(k-n) \; u(k-1) \cdots u(k-m)]^T \quad (11.55b)$$
$$= [\phi_1(k) \; \phi_2(k) \cdots \phi_N(k)]^T$$

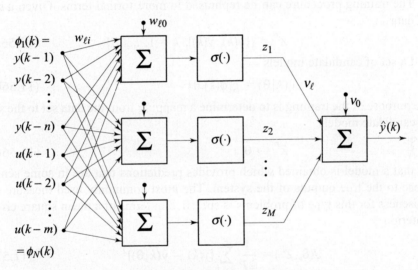

Fig. 11.21 Input structure

Often it is of little importance that the network architecture has selected vector θ too small or too large. However, a wrong choice of *lag space*, i.e., the number of delayed signals used as regressors, may have a disastrous impact on some control applications. Too small obviously implies that essential dynamics will not be modeled but too large can also be a problem. From the theory of linear systems it is known that too large a lag space may manifest itself as common factors in the identified transfer function. An equivalent behaviour must be expected in the nonlinear case. Although it is not always a problem, common factors (corresponding to hidden modes) may lead to difficulties in some of the controller designs.

It is necessary to determine both a sufficiently large lag space and an adequate number of hidden units. While it is difficult to apply physical insight towards the determination of number of hidden units, it can often guide the proper lag space. If the lag space is properly determined, the model structure selection problem is substantially reduced. If one has no idea regarding the lag space, it is sometimes possible to determine it empirically.

Training

Assume now that a data set has been acquired and that some model structure has been selected. According to the identification procedure in Fig. 11.20, the next step is to apply the data set to pick "the best" model among the candidates contained in the model structure. This is the *training stage*. The training can be computationally intensive, but it is generally one of the easiest stages in the identification. It is not very difficult to implement a training algorithm in a computer, but one might as well resort to one of the many available software packages, e.g., MATLAB.

The training procedure can be rephrased in more formal terms. Given a set of data

$$Z^P = \{[u(k), y(k)],\ k = 1, ..., P\} \quad (11.56a)$$

and a set of candidate models

$$\hat{y}(k|\boldsymbol{\theta}) = g[\boldsymbol{\phi}(k), \boldsymbol{\theta}] \quad (11.56b)$$

the purpose of the training is to determine a mapping from the data set to the set of candidate models

$$Z^P \rightarrow \hat{\boldsymbol{\theta}} \quad (11.56c)$$

so that a model is obtained which provides predictions that are in some sense close to the true outputs of the system. The most commonly used measure of closeness for this type of problems is specified in terms of a mean square error criterion

$$J(\boldsymbol{\theta}, Z^P) = \frac{1}{2P} \sum_{k=1}^{P} [y(k) - \hat{y}(k|\boldsymbol{\theta})]^2 \quad (11.57)$$

The most appealing feature of mean square error criterion is the simplicity with which a weight update rule can be derived. The principle of the gradient (descent) iterative search method is that at each iteration, the weights are modified along the opposite direction of the gradient. That is the search direction is selected as $-\dfrac{\partial J}{\partial \boldsymbol{\theta}}$.

$$\boldsymbol{\theta}^{(i+1)} = \boldsymbol{\theta}^{(i)} - \eta^{(i)} \frac{\partial J}{\partial \boldsymbol{\theta}^{(i)}} \quad (11.58)$$

When applying the gradient method to the training of multilayer feedforward networks, it is useful to order the computations in a fashion that utilizes the particular structure of the network. The method, called the *backpropagation algorithm*, was discussed in Section 11.5. Batch method of the backpropagation algorithm refers to the fact that each iteration on the parameter vector requires an evaluation of the entire data set, Z^P.

It is sometimes useful to identify a system online, simultaneously with the acquirement of measurements. Adaptive control is an example of such an application. In this case, a model must be identified and a control system designed online because the dynamics of the system to be controlled vary with time. Obviously *batch methods* are unsuitable in such applications as the amount of computation required in each iteration might exceed the time available within one sampling interval. Moreover, old data will be obsolete when the system to be identified is time dependent.

In a *recursive algorithm*, one input-output pair from the training set, $[\phi(k), y(k)]$, is evaluated at a time and used for updating the weights. In the neural network community, this is frequently referred to as *incremental* or *online* backpropagation (refer to Section 11.5).

Improving generalization: The NN modeling error may be separated into two contributions:
1. *The bias error:* The portions of the error that are due to insufficient model structure (number of hidden units, lag space), and an insufficient sample time.
2. *The variance error:* The portion of the error that is due to the fact that the function implemented by the network on a specific data set deviates from the average function.

The bias error is in practice inevitable; there will always be a certain error even with large data sets. However, typically one will find that the bias error decreases as more weights are added since the network will be able to describe the system more accurately. The reason for not just selecting a gigantic network architecture is that one must expect the variance error to work in the opposite direction: as more weights are estimated on the same data set, the variance on the estimated weights will increase. This quandary is often referred to as the *bias/variance dilemma* [158]. If a network has more hidden units, the variance error dominates. To describe this situation, it is common to use the expression *overfitting*. Overfitting means that the network not only models the features of the system, but to an undesired extent also the noise in the training set. The bias error on the training set is driven to a very small value with large number of hidden units, but when new data a presented to the network, the error is large. The network has memorized the training examples, but it has not learnt to *generalize* to new situations. Likewise, the expression *underfitting* is used when the bias error dominates.

One method for improving network generalization is to use a network which is just large enough to provide an adequate fit. However, it is difficult to know beforehand how large a network should be selected for a specific application. There are two other methods for improving generalization: regularization and early stopping.

Regularization: This means to augment the mean square error criterion with a regularization (or complexity) term. The most commonly used augmentation is the so-called *simple weight decay* term:

$$J = \frac{1}{2P} \sum_{k=1}^{P} [y(k) - \hat{y}(k|\theta)]^2 + \frac{1}{2P} \alpha \theta^T \theta \qquad (11.59)$$

where α denotes the *weight decay*.

Early stopping: An effect similar to regularization can be accomplished by stopping the training session before the minimum of the criterion has been reached. Error reduces in the beginning; after some iterations it reaches a minimum and then starts growing. Loosely speaking, the network initially captures the essential features of the system but after a while it adapts mostly to the noise.

Validation

In the validation stage, the trained model is evaluated to clarify if it represents the underlying system adequately. Ideally, the validation should be performed in accordance with the intended use of the model. As it turns out, this is often rather difficult. For instance, if the intention is to use the model for designing a control system, the validation ought to imply that a controller was designed and its performance tested in practice. For most applications, this level of ambition is somewhat high, and it is common to apply a series of simple "standard" tests instead that concentrate on investigating particular properties of the model. Although this is less than ideal, it is good as a preliminary validation to quickly exclude really poor models.

Most of the tests require a set of data that was not used during training. Such a data set is commonly known as *test* or *validation* set. It is desirable that the test set satisfies the same demands as the training set regarding representation of the entire operating range.

A very important part of the validation is to simply inspect the plot comparing observed outputs to predictions. Unless the signal-to-noise ratio is very poor, it can show the extent of overfitting as well as possible systematic errors.

If the sampling frequency is high compared to the dynamics of the system, a visual inspection of the predictions will not reveal possible problems. Some scalar quantities (correlation functions) to measure the accuracy of the predictions have been suggested. Reliable estimates of the average generalization error are also useful for validation purpose, but their primary application is for model structure selection. The estimates are good for rapidly comparing different model structures to decide which one is likely to be the best.

11.8 CONTROL WITH NEURAL NETWORKS

Neural-network-based control constitutes a very large research field, and it is difficult to give a clear overview of the entire field. Here in this section, an attempt has been made to outline a feasible path through the "jungle" of neural network solutions. A completely automatic procedure for control system design is not realistic; the emphasis is on the guidelines for working solutions.

It is believed that one of the most important lessons to be learnt from the numerous automatic control applications developed over the past half century is that *simple solutions* actually solve most problems quite well. Regardless of the fact that all systems to some extent exhibit a nonlinear behaviour, it turns out that they can often be controlled satisfactorily with simple linear controllers. When neural networks are introduced as a tool for improving the performance of control systems for a general class of unknown nonlinear systems, it should be done in the same spirit. A consequence of this philosophy is that our focus is on simple control structures that yield good performance in practice.

Inverse Model of the System used as Controller

When neural networks originally were proposed for controlling unknown nonlinear systems, one of the first methods being reported was on training a network to act as the inverse of the system and use this as a controller. Explained in brief, the basis principle is as follows:

Assume that the system to be controlled can be described by

$$y(k) = f_1[y(k-1), ..., y(k-n), u(k-1), ..., u(k-m)] \quad (11.60)$$

Perform an experiment on the system to collect a set of data that describes how the system behaves over its entire range of operation:

$$\{[u(k), y(k)], k = 1, ..., P\} \quad (11.61)$$

Using identification procedures described in the earlier section, we can infer a neural network model of the system using this data set.

An inverse model of the system can be inferred from the data set

$$\{[y(k), u(k)], k = 1, ..., P\} \quad (11.62)$$

The output of the inverse model is $u(k)$:

$$u(k) = f_2(y(k+1), y(k), ..., y(k-n+1), u(k-1), ..., u(k-m+1)) \quad (11.63)$$

The inverse model can be used as controller for the system. Let the "desired" closed-loop system behave as

$$\frac{Y(z)}{R(z)} = M(z) = z^{-1} \quad (11.64)$$

Substitute the output $y(k + 1)$ by the desired output: the reference, $r(k)$. If the network represents the exact inverse, the control input produced by it will drive the system output at time $k + 1$ to $r(k)$. The principle is illustrated in Fig. 11.22.

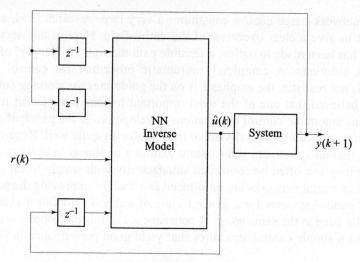

Fig. 11.22 Inverse model of the system as a controller

The most straightforward way of training a network as the inverse of a system is to approach the problem as a system-identification problem analogous to the one considered in the previous section: an experiment is performed, a network architecture is selected, and the network is trained off-line. The difference from system identification lies in the choice of regressors and network output. They are now selected as shown in a functional relation (11.63). The network is then trained to minimize the criterion

$$J = \frac{1}{2P} \sum_{k=1}^{P} [u(k) - \hat{u}(k|\boldsymbol{\theta})]^2 \qquad (11.65)$$

We will call this procedure, the *general training* procedure for an inverse model.

The practical relevance of using an inverse model of the system as a controller is limited due to a number of serious inconveniences. The control scheme will typically result in a poor robustness with a high sensitivity to noise and high-frequency disturbances (corresponding to unity forward-path transfer function in the linear case). In addition, one will often encounter a very active control signal, which may adversely affect the system/actuators. If the system is linear, this occurs when its zeros are situated close to the unit circle. In the nonlinear case, there is no unique set of zeros, but of course a similar phenomenon exists.

If the inverse model is unstable (corresponding to zeros of the system outside the unit circle in the linear case), one must anticipate that the closed-loop system becomes unstable. Unfortunately, this situation occurs quite frequently in practice. Discretization of linear continuous-time models under quite common circumstances can result in zeros outside the unit circle regardless that the continuous-time model has no zeros, or all zeros are in the left half of the plane. In fact, for a model with a pole excess of at least two, one or more zeros in the discretized model will converge to the unit circle or even outside as the sampling frequency is increased. It must be expected that a similar behaviour also can be found in discrete models of nonlinear systems.

Another problem with the design arises when the system to be controlled is not one-to-one, since then a unique inverse model does not exist. If this non-uniqueness is not reflected in the training set, one can, in principle, yield a particular inverse which might be adequate for controlling the system. Most often, however, one will end up with a useless, incorrect inverse model.

Feedforward-Feedback Control

Many of the problems mentioned in the previous subsection can be taken care of by employing a control structure of the form shown in Fig. 11.23. The feedforward control is used for improving the reference tracking while feedback is used for stabilizing the system and for suppressing disturbances.

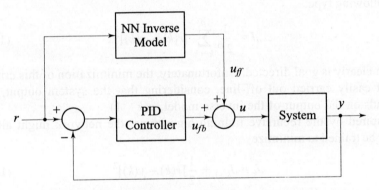

Fig. 11.23 Feedforward-feedback control structure

An inverse model is trained as discussed earlier (refer Eqn. (11.63)):

$$u(k) = f[y(k + 1), y(k), ..., y(k - n + 1), u(k - 1), ..., u(k - m + 1)] \quad (11.66)$$

The feedforward component of the control input is then composed by substituting all system outputs for corresponding reference values:

$$u_{ff}(k) = f[r(k + 1), ..., r(k - n + 1), u_{ff}(k - 1), ..., u_{ff}(k - m + 1)] \quad (11.67)$$

If the complete reference trajectory is known in advance, implementation of the scheme is particularly easy. It is then possible to compute the contribution from the feedforward controller beforehand and store the entire sequence of control inputs $\{u_{ff}\}$ for use in the computer program implementing the control system.

Although a neural network feedforward can be useful for optimizing many control systems, one must be careful not to use it uncritically. An inaccurate feedforward control may actually harm rather than enhance performance.

Model Reference Adaptive System

In the context of training inverse models, which are to be used as controllers, the trained inverse model somehow ought to be validated in terms of performance of the final closed-loop system. This points out a serious weakness associated with the general training procedure for an inverse model: the criterion (11.65) expresses the objective to minimize the discrepancy between the network output and a sequence of "true" control inputs. This is not really a relevant objective. In practice, it is not possible to achieve zero generalization error and consequently the trained network will have certain inaccuracies. Although these are reasonably small in terms of the network output being close to the ideal control signal, there may be large deviations between the reference and the output of the system when the network is applied as controller for the system. The weakness lies in the fact that the training procedure is not *goal directed*. The goal is that in some sense the system output should follow the reference signal closely. It would be more desirable to minimize a criterion of the following type:

$$J = \frac{1}{2P} \sum_{k=1}^{P} [r(k) - y(k)]^2 \qquad (11.68)$$

which clearly is goal directed. Unfortunately, the minimization of this criterion is not easily carried out off-line, considering that the system output, $y(k)$, depends on the output of the inverse model, $u(k-1)$.

Inspired by the recursive training algorithms, the network might alternatively be trained to minimize

$$J_k = J_{k-1} + \frac{1}{2}[r(k) - y(k)]^2 \qquad (11.69)$$

This is an on-line approach and, therefore, the scheme constitutes an *adaptive controller*.

Assuming that J_{k-1} has already been minimized, the weights at time k are adjusted according to

$$\hat{\theta}(k) = \hat{\theta}(k-1) - \eta \frac{de^2(k)}{d\theta} \qquad (11.70a)$$

where

$$e(k) = r(k) - y(k) \qquad (11.70b)$$

and

$$\frac{de^2(k)}{d\theta} = -\frac{dy(k)}{d\theta} e(k) \qquad (11.70c)$$

By application of the chain rule, the gradient $\dfrac{dy(k)}{d\theta}$ can be calculated:

$$\frac{dy(k)}{d\theta} = \frac{\partial y(k)}{\partial u(k-1)} \frac{du(k-1)}{d\theta} \qquad (11.71a)$$

Jacobians of the system, $\dfrac{\partial y(k)}{\partial u(k-1)}$, are required. These are generally unknown since the system is unknown. To overcome this problem, a *forward model* of the system is identified to provide estimates of the Jacobians:

$$\frac{\partial y(k)}{\partial u(k-1)} \approx \frac{\partial \hat{y}(k)}{\partial u(k-1)} \qquad (11.71b)$$

The forward model is obtained by the system identification procedure described in the earlier section.

Fortunately, inaccuracies in the forward model need not have a harmful impact on the training. The Jacobian is a scalar factor and in the simplified algorithm will only change the step-size of the algorithm. Thus, as long as the Jacobians have the correct sign, the algorithm will converge if the step-size parameter is sufficiently small.

The deadbeat character appearing when inverse models are used directly as controllers will often result in an unnecessarily fast response to reference changes. An active control signal may even harm the system or the actuators. Consequently, it might be desirable to train the network to achieve some prescribed low-pass behaviour of the closed-loop system. Say, have the closed-loop system following the model:

$$y_m(k) = \frac{B_m(z^{-1})}{A_m(z^{-1})} r(k) \qquad (11.72)$$

The polynomials A_m and B_m are selected arbitrarily by the designer.

The control design is, in this case, related to Model Reference Adaptive System (MRAS); a popular type of adaptive controller.

Since this *specialized training* is an on-line approach, the combination of having many weights to adjust and having only the slow convergence of a gradient method will often be disastrous. Before the weights are properly adjusted, the system may have been driven outside the operating range with possibly serious consequences. Often *general training* can be used to provide a decent initialization of the network so that the specialized training is only used for "fine tuning" of the controller.

The simplified specialized training is quite easily implemented with the backpropagation algorithm. This algorithm is applied on the inverse model:

$$u(k-1) = f(y(k+1), y(k), ..., y(k-n+1), u(k-2), ..., u(k-m))$$

by assuming the following "virtual" error on the output of the controller:

$$e_u(k) = \frac{\partial \hat{y}(k)}{\partial u(k-1)} e(k) \qquad (11.73)$$

The regressor's dependency on the weights in the network is ignored during training:

$$\frac{du(k-1)}{d\theta} = \frac{\partial u(k-1)}{\partial \theta}$$

For a multilayer feedforward network with one hidden layer of sigmoid units and a linear output (Fig. 11.24),

$$e(k) = [y_m(k) - y(k)]; \quad E(k) = \frac{1}{2}e^2(k)$$

$$e_u(k) = \frac{\partial \hat{y}(k)}{\partial u(k-1)} e(k)$$

Refer Eqns (11.70–11.73) and Fig. 11.21:

$$\hat{\theta}(k) = \hat{\theta}(k-1) - \eta \frac{\partial E}{\partial \theta} = \hat{\theta}(k-1) + \eta e_u(k) \frac{\partial u(k-1)}{\partial \theta}$$

$$\hat{y}(k) = \sum_{\ell=1}^{M} v_\ell \sigma \left(\sum_{i=1}^{N} w_{\ell i} \phi_i(k) + w_{\ell 0} \right) + v_0 \qquad (11.74c)$$

$$\phi(k) = [y(k-1), ..., y(k-n), u(k-1), ..., u(k-m)] \qquad (11.74b)$$

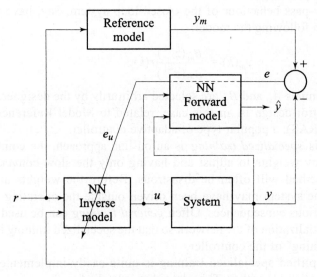

Fig. 11.24 Specialized training

The derivative of the output with respect to the regressor $\phi_i(k)$ is given by

$$\frac{\partial \hat{y}(k)}{\partial \phi_i(k)} = \sum_{\ell=1}^{M} v_\ell w_{\ell i} \sigma'(a) \quad (11.75a)$$

$$= \sum_{\ell=1}^{M} v_\ell w_{\ell i} \sigma(a)[1 - \sigma(a)]$$

$$a = \sum_{i=1}^{N} w_{\ell i} \phi_i(k) + w_{\ell 0} \quad (11.75b)$$

Feedback Linearization

In the area of nonlinear control, feedback linearization is a principle which has drawn much attention. The application is restricted to certain classes of systems, but these are actually not uncommon in practice. The advantage of feedback linearization is that the design can be used generically, in the sense that the same principle can be used on all systems of the right type.

A large number of publications exists on how to use neural networks for feedback linearization. The treatment given in the following has the objective of only demonstrating the principle (refer Fig. 11.1).

It is assumed that the system can be modeled as

$$\mathbf{x}(k+1) = \begin{bmatrix} x_1(k+1) \\ x_2(k+1) \\ \vdots \\ x_n(k+1) \end{bmatrix} = \begin{bmatrix} x_2(k) \\ x_3(k) \\ \vdots \\ f(\cdot) + g(\cdot)u(k) \end{bmatrix} \quad (11.76a)$$

$$y(k) = x_n(k) \quad (11.76b)$$

$$y(k+1) = f(\cdot) + g(\cdot)u(k)$$

$$= f[y(k), \ldots, y(k-n+1), u(k-1), \ldots, u(k-m+1)] \\ + g[y(k), \ldots, y(k-n+1), u(k-1), \ldots, u(k-m+1)]u(k)$$

$$(11.76c)$$

$$u(k) = \frac{1}{g(\cdot)}[-f(\cdot) + v(k)] \quad (11.77a)$$

will give the linear system

$$y(k+1) = v(k) \quad (11.77b)$$

In case the system is unknown, a model can be induced from data by letting two separate neural networks approximate the functions f and g.

$$\hat{y}(k) = \hat{f}[y(k-1), ..., y(k-n), u(k-2), ..., u(k-m), \boldsymbol{\theta}^f]$$
$$+ \hat{g}[y(k-1), ..., y(k-n), u(k-2), ..., u(k-m), \boldsymbol{\theta}^g]u(k-1)$$

(11.78)

where $\boldsymbol{\theta}^f$ and $\boldsymbol{\theta}^g$ are parameters (i.e., weights and biases) of the neural networks that approximate $f(\cdot)$ and $g(\cdot)$ respectively.

Both the networks $\hat{f}$ and $\hat{g}$ can be trained individually. The weight updation can be obtained as follows:

$$e(k) = y(k) - \hat{y}(k)$$
$$= y(k) - [\hat{f}(\cdot) + \hat{g}(\cdot)u(k-1)]$$
$$E(k) = \frac{1}{2}e^2(k)$$
$$\boldsymbol{\theta}^f(k+1) = \boldsymbol{\theta}^f(k) - \eta e(k)\frac{de(k)}{d\boldsymbol{\theta}^f}$$
$$= \boldsymbol{\theta}^f(k) - \eta e(k)\frac{d}{d\boldsymbol{\theta}^f}[y(k) - (\hat{f}(\cdot) + \hat{g}(\cdot)u(k-1))]$$
$$= \boldsymbol{\theta}^f(k) + \eta e(k)\frac{d\hat{f}(\cdot)}{d\boldsymbol{\theta}^f} \quad (11.79)$$

Similarly

$$\boldsymbol{\theta}^g(k+1) = \boldsymbol{\theta}^g(k) + \eta e(k)\frac{d\hat{g}(\cdot)}{d\boldsymbol{\theta}^g}u(k-1) \quad (11.80)$$

The backpropagation algorithm (Section 11.5) can be used for training the nets.

11.9 REVIEW EXAMPLES

Review Example 11.1

A high performance drive system consists of a motor and a controller integrated to perform a precise mechanical maneuver. This requires the shaft speed and/or position of the motor to clearly follow a specified trajectory regardless of unknown load variations and other parameter uncertainties.

A backpropagation neural network can be trained to emulate the unknown nonlinear plant dynamics by presenting a suitable set of input/output patterns generated by the plant. Once system dynamics have been identified using a neural network, many conventional control techniques can be applied to achieve the desired objective of trajectory tracking.

In this example, we study a neural-network-based identification and control strategy for trajectory control of a dc motor.

DC Motor Model

Although it is not mandatory to obtain a motor model if a neural network (NN) is used in the motor-control system, it may be worth doing so from the analytical perspective, in order to establish the foundation of the NN structure. We will use input/output patterns generated by simulation of this model for training of NN (In a real life situation, experimentally generated input/output patterns will be used for training).

The dc motor dynamics are given by the following equations (refer Fig. 11.25):

$$v_a(t) = R_a i_a(t) + L_a \frac{di_a}{dt} + e_b(t) \tag{11.81}$$

$$e_b(t) = K_b \omega(t) \tag{11.82}$$

$$T_M(t) = K_T i_a(t) \tag{11.83}$$

$$= J \frac{d\omega(t)}{dt} + B\omega(t) + T_L(t) + T_F \tag{11.84}$$

where

$v_a(t)$ = applied armature voltage (volts);
$e_b(t)$ = back emf (volts);
$i_a(t)$ = armature current (amps);
R_a = armature winding resistance (ohms);
L_a = armature winding inductance (henrys);
$\omega(t)$ = angular velocity of the motor rotor (rad/sec);

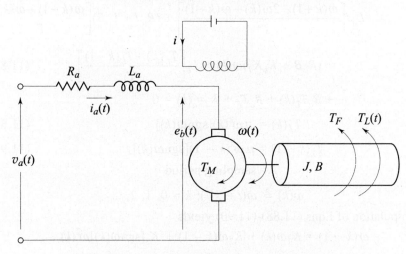

Fig. 11.25 Armature-controlled dc motor with load

$T_M(t)$ = torque developed by the motor (Newton-m);
K_T = torque constant (Newton-m/amp);
K_b = back emf constant (volts/(rad/sec));
J = moment of inertia of the motor rotor with attached mechanical load (kg-m^2);
B = viscous-friction coefficient of the motor rotor with attached mechanical load ((Newton-m)/(rad/sec));
$T_L(t)$ = disturbance load torque (Newton-m); and
T_F = frictional torque (Newton-m).

The load torque $T_L(t)$ can be expressed as

$$T_L(t) = \psi(\omega) \tag{11.85}$$

where the function $\psi(\cdot)$ depends on the nature of the load.

For most propeller driven or fan type loads, the function $\psi(\cdot)$ takes the following form:

$$T_L(t) = \mu \omega^2(t)[\text{sgn}\,\omega(t)] \tag{11.86}$$

where μ is a constant.

DC motor drive system can be expressed as single-input, single-output system by combining Eqns (11.81)-(11.84):

$$L_a J \frac{d^2\omega(t)}{dt^2} + (R_a J + L_a B)\frac{d\omega(t)}{dt} + (R_a B + K_b K_T)\omega(t)$$

$$+ L_a \frac{dT_L(t)}{dt} + R_a[T_L(t) + T_F] + K_T v_a(t) = 0 \tag{11.87}$$

The discrete-time model is derived by replacing all continuous differentials with finite differences.

$$L_a J \left[\frac{\omega(k+1) - 2\omega(k) + \omega(k-1)}{T^2}\right] + (R_a J + L_a B)\left[\frac{\omega(k+1) - \omega(k)}{T}\right]$$

$$+ (R_a B + K_b K_T)\omega(k) + L_a\left[\frac{T_L(k) - T_L(k-1)}{T}\right] \tag{11.88}$$

$$+ R_a T_L(k) + R_a T_F + K_T v_a(k) = 0$$

$$T_L(k) = \mu \omega^2(k)[\text{sgn}\,\omega(k)] \tag{11.89}$$

$$T_L(k-1) = \mu \omega^2(k-1)[\text{sgn}\,\omega(k)] \tag{11.90}$$

T = sampling period

$\omega(k) \triangleq \omega(t = kT);\ k = 0, 1, 2, \ldots$

Manipulation of Eqns (11.88)-(11.90) yields

$$\omega(k+1) = K_1 \omega(k) + K_2 \omega(k-1) + K_3[\text{sgn}\,\omega(k)]\omega^2(k)$$
$$+ K_4[\text{sgn}\,\omega(k)]\omega^2(k-1) + K_5 v_a(k) + K_6 \tag{11.91}$$

where

$$K_1 = \frac{2L_a J + T(R_a J + L_a B) - T^2(R_a B + K_b K_T)}{L_a J + T(R_a J + L_a B)}$$

$$K_2 = -\frac{L_a J}{L_a J + T(R_a J + L_a B)} \tag{11.92}$$

$$K_3 = -\frac{T(\mu L_a + \mu R_a T)}{L_a J + T(R_a J + L_a B)}$$

$$K_4 = \frac{T\mu L_a}{L_a J + T(R_a J + L_a B)}$$

$$K_5 = \frac{K_T T^2}{L_a J + T(R_a J + L_a B)}$$

$$K_6 = -\frac{T_F R_a T^2}{L_a J + T(R_a J + L_a B)}$$

The following parameter values are associated with the dc motor.

$J = 0.068$ kg-m^2
$B = 0.03475$ N-m/(rad/sec)
$R_a = 7.56$ Ω
$L_a = 0.055$ H
$K_T = 3.475$ N-m/amp (11.93)
$K_b = 3.475$ volts/(rad/sec)
$\mu = 0.0039$ N-m/(rad/sec)^2
$T_F = 0.212$ N-m
$T = 40$ msec $= 0.04$ sec

With these motor parameters, the constants K_1, K_2, K_3, K_4, K_5 and K_6 become

$K_1 = 0.34366$
$K_2 = -0.1534069$
$K_3 = -2.286928 \times 10^{-3}$ (11.94)
$K_4 = 3.5193358 \times 10^{-4}$
$K_5 = 0.2280595$
$K_6 = -0.105184$

Identification of Inverse Dynamics

Equation (11.91) can be manipulated to obtain the inverse dynamic model of the drive system as

$$v_a(k) = f[\omega(k+1), \omega(k), \omega(k-1)] \tag{11.95}$$

The right hand side of Eqn. (11.95) is a nonlinear function of the speed ω and is given by

$$f(\omega(k+1), \omega(k), \omega(k-1)) = \frac{1}{K_5}[\omega(k+1) - K_1\omega(k) - K_2\omega(k-1)$$
$$- K_3\{\mathrm{sgn}\,\omega(k)\}\omega^2(k) - K_4\{\mathrm{sgn}\,\omega(k)\}\omega^2(k-1) - K_6] \quad (11.96)$$

which is assumed to be unknown (It is assumed that the only available qualitative *a priori* knowledge about the plant is a rough estimate of the order of the plant). A neural network is trained to emulate the unknown function $f(\cdot)$. The values $\omega(k+1)$, $\omega(k)$ and $\omega(k-1)$, which are the independent variables of $f(\cdot)$, are selected as the inputs to the NN. The corresponding target $f(\omega(k+1), \omega(k), \omega(k-1))$ is given by Eqn. (11.96). This quantity is also equal to the armature voltage $v_a(k)$ as seen from Eqn. (11.95). Randomly generated input patterns of $[\omega(k+1), \omega(k), \omega(k-1)]$ and the corresponding target $v_a(k)$ are used for off-line training. The training data is generated within the constrained operating space. In conforming with the mechanical and electrical hardware limitations of the motor, and with a hypothetical operating scenario in mind, the following constrained operating space is defined:

$$-30 < \omega(k) < 30 \text{ rad/sec}$$
$$|\omega(k-1) - \omega(k)| < 1.0 \text{ rad/sec} \quad (11.97)$$
$$|v_a(k)| < 100 \text{ volts}$$

The estimated motor armature voltage given by the NN identifier is

$$\hat{v}_a(k-1) = N(\omega(k), \omega(k-1), \omega(k-2)) \quad (11.98)$$

Trajectory Control of DC Motor Using Trained NN

The objective of the control system is to drive the motor so that its speed $\omega(k)$ follows a reference (prespecified) trajectory $\omega_r(k)$. A controller topology is presented in Fig. 11.26. The NN trained to emulate inverse dynamics of the dc motor, is used to estimate the motor armature voltage $\hat{v}_a(k)$, which enables accurate trajectory control of the shaft speed $\omega(k)$. Refer Appendix C for realization of the controller.

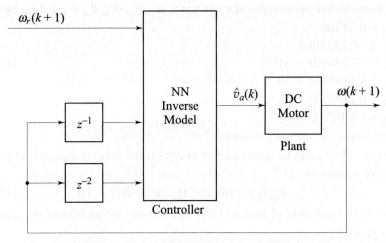

Fig. 11.26 A structure for NN-based speed control

Review Example 11.2

In this example we study a neural-network–based identification and control strategy for temperature control of a water bath.

The plant consists of a laboratory 7-litre water bath as depicted in Fig. 11.27. A personal computer reads the temperature of the water bath through a link consisting of a diode-based temperature sensor module (SM) and an 8-bit A/D converter. The plant input produced by the computer is limited between 0 and 5 volts, and controls the duty cycle for a 1.3 kW heater via a pulse-width-modulation (PWM) scheme.

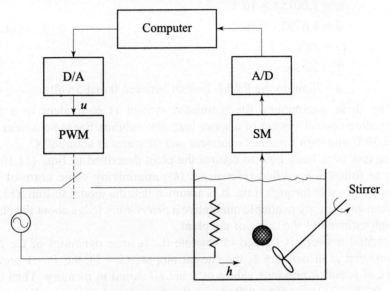

Fig. 11.27 Water bath control system

The temperature of water in a stirred tank is described by the equation

$$C\frac{dy(t)}{dt} = \frac{Y_0 - y(t)}{R} + h(t) \tag{11.99}$$

where $y(t)$ is the temperature of water in the tank (°C), $h(t)$ is the heat flowing into the tank through the base heater (watts), Y_0 is the temperature of the surroundings (assumed constant, for simplicity), C denotes the tank thermal capacity (Joules/°C), and R is the thermal resistance between tank borders and surroundings. Assuming R and C as essentially constant, we can obtain discrete-time description of the thermal system as follows:

$$x(t) \triangleq y(t) - Y_0 \tag{11.100}$$

$$\dot{x}(t) = -\frac{1}{RC}x(t) + \frac{1}{C}h(t) = -\alpha\, x(t) + \beta\, h(t)$$

The discrete-time state equation (sampling period = T):

$$x(k+1) = F\, x(k) + g h(k) \tag{11.101}$$

where

$$F = e^{-\alpha T}; \quad g = \beta \int_0^T e^{-\alpha \tau} d\tau = \frac{\beta}{\alpha}[1 - e^{-\alpha T}]$$

We modify this model to include a saturating nonlinearity so that the water temperature cannot exceed some limitation. The nonlinear plant model then becomes (obtained from real plant by experimentation)

$$y(k+1) = F\, y(k) + \frac{g}{1 + \exp[0.5 y(k) - 40]} u(k) + (1 - F) Y_0 \quad (11.102)$$

$\alpha = 1.00151 \times 10^{-4}$
$\beta = 8.67973 \times 10^{-3}$ \hfill (11.103)
$Y_0 = 25°C$
$T = 30$ sec
$u = $ input to the PWM, limited between 0 and 5 volts.

With these parameters, the simulated system is equivalent to a SISO temperature control system of a water bath that exhibits linear behaviour upto about 70°C and then becomes nonlinear and saturates at about 80°C.

The task is to learn how to control the plant described in Eqn. (11.102) in order to follow a specified reference $y_r(k)$, minimizing some norm of error $e(k) = y_r(k) - y(k)$ through time. It is assumed that the model in Eqn. (11.102) is unknown; the only available qualitative *a priori* knowledge about the plant is a rough estimate of the order of the plant.

A neural network is trained to emulate the inverse dynamics of the plant. Assume that at instant $k + 1$, the current output $y(k + 1)$, the $P - 1$ previous values of y, and P previous values of u are all stored in memory. Then the P pairs $(\mathbf{x}^T(k - i), u(k - i)); i = 0, 1, ..., P - 1, \mathbf{x}^T(k) = [y(k + 1), y(k)]$, can be used as patterns for training the NN at time $k + 1$. A train of pulses is applied to the plant and the corresponding input/output pairs are recorded. The NN is then trained with reasonably large sets of data chosen from the experimentally obtained data bank in order to span a considerable region of the control space (We will use input/output patterns generated by simulation of the plant model for training the NN).

A controller topology is presented in Fig. 11.28. It is assumed that the complete reference trajectory $y_r(k)$ is known in advance. The feedforward component of the control input is then composed by substituting all system outputs for corresponding reference values. Refer Appendix C for realization of the controller.

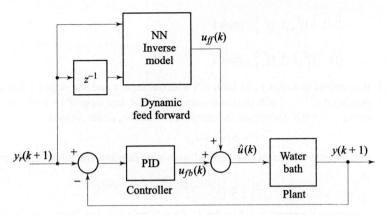

Fig. 11.28 A structure for NN-based temperature control

PROBLEMS

11.1 Four patterns in a two-dimensional pattern space need to be classified according to their membership in sets as follows:

$$\{[0,0]^T, [0,1]^T, [1,0]^T\} : \text{class 1}$$

$$\{[1,1]^T\} : \text{class 2}$$

Design a neuron with linear activation function:

$$\hat{y} = w_1 x_1 + w_2 x_2 + w_0$$

to solve this classification problem.
(*Hint*: $w_1 x_1 + w_2 x_2 + w_0 = 0$ specifies the decision boundary)

11.2 Six patterns in a two-dimensional pattern space need to be classified according to the membership in sets, as follows:

$$\{[0,0]^T, [-0.5,-1]^T, [-1,-2]^T\} : \text{class 1}$$

$$\{[2,0]^T, [1.5,-1]^T, [1,-2]^T\} : \text{class 2}$$

Design a neuron with linear activation function:

$$\hat{y} = w_1 x_1 + w_2 x_2 + w_0$$

to solve this classification problem.

11.3 A two-layer network with unipolar sigmoid (log-sigmoid) activation function in a hidden layer has the recall equation

$$\hat{y} = \mathbf{V}\sigma(\mathbf{W}\mathbf{x} + \mathbf{w}_0) + v_0$$

$$\mathbf{W} = \begin{bmatrix} -2.69 & -2.80 \\ -3.39 & -4.56 \end{bmatrix}; \mathbf{w}_0 = \begin{bmatrix} -2.21 \\ 4.76 \end{bmatrix}$$

$$\mathbf{V} = [-4.91 \quad 4.95]. \; v_0 = -2.28$$

Show that this network classifies four patterns in a two-dimensional pattern space according to their membership in sets as follows:

$$\left\{[-1,-1]^T, [1,1]^T\right\} : \text{class 1}$$

$$\left\{[1,-1]^T, [-1,1]^T\right\} : \text{class 2}$$

11.4 It is desired to design a one-layer NN with one input x and one output $\hat{y}$ that associates input $x^{(1)} = -3$ with the target output $y^{(1)} = 0.4$, and input $x^{(2)} = 2$ with the target output $y^{(2)} = 0.8$. Determine the parameters w and w_0 of the network

$$\hat{y} = \sigma(wx + w_0)$$

with unipolar sigmoidal (log-sigmoid) activation function, that minimize the error

$$E = \frac{1}{2}\left[\left(y^{(1)} - \hat{y}^{(1)}\right)^2 + \left(y^{(2)} - \hat{y}^{(2)}\right)^2\right]$$

11.5 Streamline the notation in Chapter 11 for a 3-layer NN. For instance define $\mathbf{W}^{h1}$ as weights of hidden layer 1 with m nodes; $\mathbf{W}^{h2}$ as weights of hidden layer 2 with p nodes; and $\mathbf{V}$ as weights of output layer with q nodes.

    Input variables   : $x_i; i = 1, ..., n$

    Outputs of hidden layer 1 : $z_\ell; \ell = 1, ..., m$

    Outputs of hidden layer 2 : $t_r; r = 1, ..., p$

    Outputs of output layer   : $\hat{y}_j; j = 1, ..., q$

    Desired outputs   : y_j

    Learning constant   : η

Derive the backpropagation algorithm for the 3-layer network assuming the output layer has linear activation and the two hidden layers have unipolar sigmoidal activations.

11.6 Consider a four-input single-node perceptron with a bipolar sigmoidal function (tan-sigmoid)

$$\sigma(a) = \frac{2}{1 + e^{-a}} - 1$$

where 'a' is the activation value for the node.
(a) Derive the weight update rule for $\{w_i\}$ for all i. The learning rate $\eta = 0.1$. Input variables: $x_i; i = 1, 2, 3, 4$. Desired output is y.
(b) Use the rule in part (a) to update the perceptron weights incrementally for one epoch. The set of input and desired output patterns is as follows:

$$\mathbf{x}^{(1)} = [1 \quad -2 \quad 0 \quad -1]^T, y^{(1)} = -1$$

$$\mathbf{x}^{(2)} = [0 \quad 1.5 \quad -0.5 \quad -1]^T, y^{(2)} = -1$$

$$\mathbf{x}^{(3)} = [-1 \quad 1 \quad 0.5 \quad -1]^T, y^{(3)} = 1$$

The initial weight vector is chosen as

$$\mathbf{w}_0 = [1 \quad -1 \quad 0 \quad 0.5]$$

The perceptron does not possess bias term.
(c) Use the training data and initial weights given in part (b) and update the perceptron weights for one epoch in batch mode.

11.7 We are given the two-layer backpropagation network shown in Fig. P11.7.
(a) Derive the weight update rules for $\{v_\ell\}$ and $\{w_{\ell i}\}$ for all i and ℓ. Assume that activation function for all the nodes is a unipolar sigmoid function

$$\sigma(a) = \frac{1}{1+e^{-a}}$$

where 'a' represents the activation value for the node. The learning constant $\eta = 0.1$. The desired output is y.

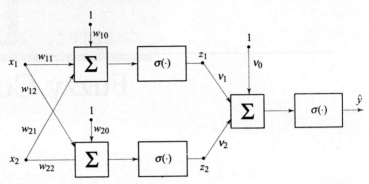

Fig. P11.7

(b) Use the equations derived in part (a) to update the weights in the network for one step with input vector $\mathbf{x} = [1 \ \ 0]^T$, desired output $y = 1$, and the initial weights:

$w_{10} = 1, w_{11} = 3, w_{12} = 4, w_{20} = -6, w_{21} = 6, w_{22} = 5$

$v_0 = -3.92, v_1 = 2$, and $v_2 = 4$

(c) As a check, compute the error with the same input for initial weights and updated weights and verify that the error has decreased.

11.8 We are given the two-layer backpropagation network in Fig. P11.8.
Derive the weight update rules for $\{v_\ell\}$ and $\{w_\ell\}$ for all ℓ. Assume that activation function for all the nodes is a bipolar sigmoid function

$$\sigma(a) = \frac{2}{1+e^{-a}} - 1$$

where 'a' is the activation value for the node. The learning constant is $\eta = 0.4$. The desired output is y.

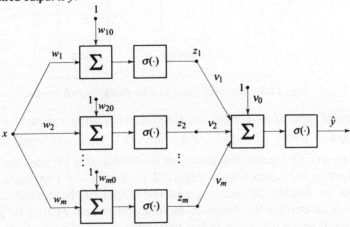

Fig. P11.8

chapter 12

Fuzzy Control

12.1 INTRODUCTION

In a man–machine system, there arises the problem of processing information with the "vagueness" that is characteristic of man. We consider here a real-life situation in process control.

The basic structure of a feedback control system is shown in Fig. 12.1. G represents the system to be controlled (*plant* or *process*). The purpose of the *controller D* is to guarantee a desired response of the *output y*. The process of keeping the output y close to the *set-point (reference input)* y_r, despite the presence of disturbances, fluctuations of the system parameters, and noisy measurements, is called *regulation*. The law governing corrective action of the controller is called the *control algorithm*. The output of the controller is the *control action u*.

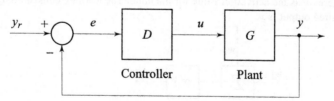

Fig. 12.1 Basic structure of a feedback control system

The general form of the control law (implemented using a digital computer) is

$$u(k) = f(e(k), e(k-1), ..., e(k-m), u(k-1), ..., u(k-m)) \qquad (12.1)$$

providing a control action that describes the relationship between the input and the output of the controller. In Eqn. (12.1), $e = y_r - y$ represents the error between the desired set-point y_r and the output of the controlled system; parameter m defines the order of the controller; and $f(\cdot)$ is, in general, a nonlinear function. k is an index representing sampling instant; T is the sampling interval used for digital implementation (Fig. 12.2). To distinguish control law (12.1) from the control schemes based on fuzzy logic/neural networks, we shall call this *conventional control law*.

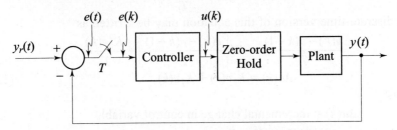

Fig. 12.2 Basic structure of a digital control system

A common feature of conventional control is that the control algorithm is analytically described by equations—algebraic, difference, differential, and so on. In general, the synthesis of such control algorithms requires a formalized analytical description of the controlled system by a mathematical model. The concept of analyticity is one of the main paradigms of conventional control theory. We will also refer to conventional control as *model-based* control.

When the underlying assumptions are satisfied, many of the model-based control techniques provide good stability, robustness to model uncertainties and disturbances, and speed of response. However, there are many practical deficiencies of these control algorithms. It is generally difficult to accurately represent a complex process by a mathematical model. If the process model has parameters whose values are partially known, ambiguous or vague, the control algorithms that are based on such *incomplete information* will not usually give satisfactory results. The environment with which the process interacts may not be completely predictable, and it is normally not possible for a model-based algorithm to accurately respond to a condition that it did not anticipate. Skilled human operators are, however, controlling complex plants quite successfully on the basis of their experience, without having quantitative models.

Regulatory control objectives, typical of many industrial applications, are
- to remove any significant errors in process output $y(t)$ by appropriate adjustment of the controller output $u(k)$,
- to prevent process output from exceeding some user-specified constraint y_c, i.e., for all t, $y(t)$ should be less than or equal to y_c, and
- to produce smooth control action near the set-point, i.e., minor fluctuations in the process output are not passed further to the controller.

A conventional PI controller uses an analytical expression of the following form to compute the control action:

$$u(t) = K'_c \left[e(t) + \frac{1}{T_I} \int e(\tau) d\tau \right] \tag{12.2}$$

where K'_c is the controller gain, and T_I is integral or reset time.

When this expression is differentiated, we obtain

$$\dot{u}(t) = K'_c \dot{e}(t) + \frac{K'_c}{T_I} e(t)$$

The discrete-time version of this equation may be written as

$$\frac{u(k) - u(k-1)}{T} = K'_c \left[\frac{e(k) - e(k-1)}{T} \right] + \frac{K'_c e(k)}{T_I}$$

or
$$\Delta u(k) = K_c v(k) + K_I e(k) \qquad (12.3)$$
where

$\Delta u(k)$ = incremental change in control variable
= $u(k) - u(k-1)$;

$e(k)$ = error variable
= $y_r - y(k)$;

$v(k)$ = time rate of change of error[1]
= $\dfrac{e(k) - e(k-1)}{T}$

The control objectives listed earlier would require variable gains when the process output is in different regions around the set-point. Figure 12.3 illustrates the type of control action desired; Δu should be "near zero" in the set-point region, very large in the constraint region, and normal in between.

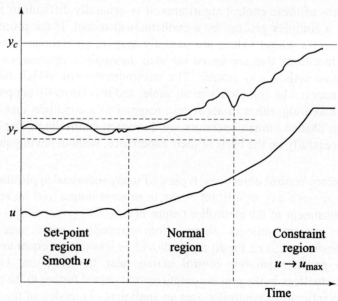

Fig. 12.3 Type of control action desired in different regions around the set-point

---

1. A PD controller in position form is
$u(k) = K_c e(k) + K_D v(k)$
We see that PD controller in position form is structurally related to PI controller in incremental form.

A simple PI controller is inherently incapable of achieving all of the above control objectives and has to be implemented with additional (nonlinear) control laws for set-point and constraint regions, making the control scheme a complex adaptive control scheme which would allow the desired gain modification when required.

On the other hand, an experienced process operator can easily meet all the three control objectives. An expert operator employs, consciously or subconsciously, a set of IF-THEN rules to control a process (Fig. 12.4). He estimates the *error e(k)* and *time rate of change of error v(k)* at a specific time instant, and based on this information he changes the control by $\Delta u(k)$.

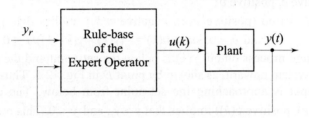

Fig. 12.4 A man–machine control system

A typical *production rule* of the rule-base in Fig. 12.4 is of the form:

$$\text{IF (process state) THEN (control action)} \qquad (12.4)$$

instead of an analytical expression defining the control variable as a function of process state. The "process state" part of the rule is called the rule *premise* (or *antecedent*) and contains a description of the process state at the kth sampling instant. This description is done in terms of particular values of error $e(k)$, velocity (time rate of change of error) $v(k)$, and the constraint. The "control action", part of the rule is called the *conclusion* (or *consequent*) and contains a description of the control variable which should be produced given the particular process state in the rule antecedent. This description is in terms of the value of the change-in-control, $\Delta u(k)$.

Negative values of $e(k)$ mean that the current process output $y(k)$ has a value above the set-point y_r, since $e(k) = y_r - y(k) < 0$. The magnitude of a negative value describes the magnitude of the difference $y_r - y$. On the other hand, positive values of $e(k)$ express the knowledge that the current value of the process output $y(k)$ is below the set-point y_r. The magnitude of such a positive value is the magnitude of the difference $y_r - y$.

Negative values of $v(k)$ mean that the current process output $y(k)$ has increased compared with its previous value $y(k-1)$, since $v(k) = -(y(k) - y(k-1))/T < 0$. The magnitude of such a negative value describes the magnitude of this increase. Positive values of $v(k)$ express the knowledge that $y(k)$ has decreased its value when compared to $y(k-1)$. The magnitude of such a value is the magnitude of the decrease.

Positive values of $\Delta u(k)$ mean that the value of the control $u(k-1)$ has to be increased to obtain the value of the control for the current sampling time k. A value with a negative sign means a decrease in the value of $u(k-1)$. The magnitude of such a value is the magnitude of increase/decrease in the value $u(k-1)$.

The possible combinations of positive/negative values of $e(k)$ and $v(k)$ are as follows:
- positive e, negative v
- negative e, positive v
- negative e, negative v
- positive e, positive v

The combination (positive $e(k)$, negative $v(k)$) implies that $y < y_r$, since $e(k) = y_r - y(k) > 0$; and $\dot{y} > 0$, since $v(k) = -(y(k) - y(k-1))/T < 0$. This means that the current process output $y(k)$ is below the set-point and the controller is driving the system upward, as shown by point D in Fig. 12.5. Thus, the current process output is approaching the set-point from below. The combination (negative $e(k)$, positive $v(k)$) implies that $y > y_r$, and $\dot{y} < 0$. This means that the current process output is above the set-point and the controller is driving the system downward, as shown by point B in Fig. 12.5. Thus the current process output is approaching the set-point from above. The combination (negative $e(k)$, negative $v(k)$) implies that $y > y_r$ and $\dot{y} > 0$. This means that the current process output $y(k)$ is above the set-point and the controller is driving the system upward, as shown by point C in Fig. 12.5. Thus the process output is moving further away from the set-point and approaching overshoot. The combination

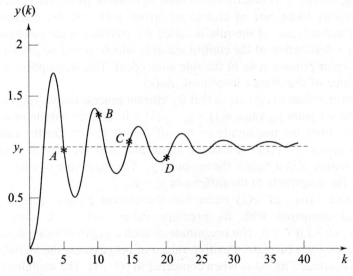

Fig. 12.5 Process output deviations from the set-point

(positive $e(k)$, positive $v(k)$) implies that $y < y_r$ and $\dot{y} < 0$. This means that the current process output is below the set-point and the controller is driving the system downward, as shown by point A in Fig. 12.5. Thus the process output is moving further away from the set-point and approaching undershoot.

In a man-machine control system of the type shown in Fig. 12.4, experience-based knowledge of the process operator and/or control engineer is instrumental in changing the control by $\Delta u(k)$, for a given estimate of error $e(k)$ and time rate of change of error $v(k)$.

(i) If both $e(k)$ and $v(k)$ (positive or negative) are small (or zero), it means that the current value of the process output variable $y(k)$ has deviated from the set-point but is still close to it. The amount of change $\Delta u(k)$ in the previous control $u(k-1)$ should also be small (or zero) in magnitude, which is intended to correct small deviations from the set-point.

(ii) Consider the situation when $e(k)$ has large negative value, which implies that $y(k)$ is significantly above the set-point. If $v(k)$ is positive at the same time, this means that y is moving towards the set-point. The amount of change Δu to be introduced is intended to either speed up or slow down the approach to the set-point. For example, if $y(k)$ is much above the set-point ($e(k)$ has a large negative value) and it is moving towards the set-point with a small step ($v(k)$ has small positive value), then the magnitude of this step has to be significantly increased ($\Delta u(k) \rightarrow$ large negative value).

(iii) $e(k)$ has either a small value (positive, negative, zero) or a large positive value, which implies that $y(k)$ is either close to the set-point or significantly below it. If $v(k)$ is positive at the same time, this means that y is moving away from the set-point. Then a positive change $\Delta u(k)$ in the previous control $u(k-1)$ is required to reverse this trend and make y to start moving towards it, instead of moving away from the set-point.

(iv) Consider a situation when $e(k)$ has large positive value (which implies that $y(k)$ is significantly below the set-point) and $v(k)$ is negative (which implies that y is moving towards the set-point). The amount of change Δu to be introduced is intended to either speed up or slow down the approach to the set-point. For example, if $y(k)$ is much below the set-point ($e(k)$ has large positive value), and it is moving towards the set-point with somewhat large step ($v(k)$ has large negative value), then the magnitude of this step need not be changed ($\Delta u(k) \rightarrow 0$) or only slightly enlarged ($\Delta u(k) \rightarrow$ small positive value).

(v) $e(k)$ has either a small value (positive, negative, zero) or a large negative value, and this implies that $y(k)$ is either close to the set-point or significantly above it. If $v(k)$ is negative at the same time, y is moving away from the set-point. Thus a negative change $\Delta u(k)$ in the previous control $u(k-1)$ is required to reverse this trend and make y to start moving towards it instead of moving away from the set-point.

The variables e, v and Δu are described as consisting of a finite number of verbally expressed values which these variables can take. Values are expressed as tuples of the form {value sign, value magnitude}. The 'value sign' component of such a tuple takes on either one of the two values: positive or negative. The 'value magnitude' component can take on any number of magnitudes, e.g., {zero, small, medium, big}, or {zero, small, big}, or {zero, very small, small, medium, big, very big}, etc.

The tuples of values may therefore look like: Negative Big (NB), Negative Medium (NM), Negative Small (NS), Zero (ZO), Positive Small (PS), Positive Medium (PM), Positive Big (PB) or an enhanced set/subset of these values.

We consider here a simple rule-based controller which employs only three values of the variables e, v, and Δu: Negative (N), Near Zero (NZ), Positive (P), for e and v; and Negative (N), Zero (Z), Positive (P) for Δu. A typical production rule of the rule-base in Fig. 12.4 is

$$\text{IF } e(k) \text{ is Positive \textbf{and} } v(k) \text{ is Positive THEN } \Delta u(k) \text{ is Positive} \qquad (12.5)$$

Let us see now what such a rule actually means. A positive $e(k)$ implies that $y(k)$ is below the set-point. If $v(k)$ is positive at the same time, it means that $y(k)$ is moving away from the set-point. Thus a positive change $\Delta u(k)$ in the previous control $u(k-1)$ is required to reverse this trend.

Consider another rule:

$$\text{IF } e(k) \text{ is Positive \textbf{and} } v(k) \text{ is Negative THEN } \Delta u(k) \text{ is Zero} \qquad (12.6)$$

This rule says that if $y(k)$ is below the set-point, and is moving towards the set-point, then no change in control is required.

We will present the rule-base in table format, shown in Fig. 12.6. The cell defined by the intersection of the third row and third column represents the rule given in (12.5), and the cell defined by the intersection of the third row and first column represents the rule given in (12.6).

e \ v	N	NZ	P
N	N	N	Z
NZ	N	Z	P
P	Z	P	P

Fig. 12.6 The rule-base for system of Fig. 12.4

The rule-base shown in Fig. 12.6 is designed to remove any significant errors in process output by appropriate adjustment of the controller output. Note that the rule

$$\text{IF } e(k) \text{ is Near Zero \textbf{and} } v(k) \text{ is Near Zero THEN } \Delta u(k) \text{ is Zero} \qquad (12.7)$$

ensures smooth action near the set-point, i.e., minor fluctuations in the process output are not passed further to the controller.

The rule-base of Fig. 12.6 is thus effective for control action in the set-point region and the normal region in Fig. 12.3. However, we require additional rules for the constraint region. The following three rules prescribe a control action when the error is in the constraint region, approaching it or leaving it.
1. IF $e(k)$ is in constraint region THEN value of $\Delta u(k)$ is drastic change. This rule specifies the magnitude of additional $\Delta U(k)$ to be added to the one already determined by the rules of Fig. 12.6 when $e(k)$ is in the constraint region.
2. IF $e(k)$ enters constraint region THEN start summing up the values of $\Delta u(k)$ determined by constraint rule 1.
3. IF $e(k)$ leaves constraint region THEN subtract the total value of $\Delta u(k)$ determined by constraint rule 2.

While the value of $e(k)$ is in the constraint region, the latter rule is necessary to prevent the drastic change in the value of the control variable from dominating the value of the control variable computed for the region outside of the constraint. This is similar to reset wind-up protection.

The man-machine control system of Fig. 12.4 has the capability of representing and manipulating data that is not precise, but rather fuzzy. The error variable is 'near zero', change in control is 'drastic', etc., are the type of linguistic information which the expert controller is required to handle. But what is a 'drastic change' in control? The property 'drastic' is inherently *vague*, meaning that the set of signals it is applied to has no sharp boundaries between 'drastic' and 'not drastic'. The fuzziness of a property lies in the lack of well-defined boundaries of the set of objects to which the property applies.

Problems featuring *uncertainty* and *ambiguity* have been successfully addressed subconsciously by humans. Humans can adapt to unfamiliar situations and they are able to gather information in an efficient manner and discard irrelevant details. The information gathered need not be complete and precise and could be *general, qualitative* and *vague* because humans can *reason, infer* and *deduce* new information and knowledge. They can *learn, perceive* and improve their skills through experience.

How can humans reason about complex systems, when the complete description of such a system often requires more detailed data than a human could ever hope to recognize simultaneously and assimilate with understanding? The answer is that humans have the capacity to reason approximately. In reasoning about a complex system, humans reason approximately about its behaviour, thereby maintaining only a generic understanding about the problem. Fortunately, this generality and ambiguity are sufficient for human comprehension of complex systems.

The seminal work by Dr. Lotfi Zadeh (1965) on system analysis based on the theory of fuzzy sets, has provided a mathematical strength to capture the uncertainties associated with human cognitive processes, such as thinking and reasoning. The conventional approaches to knowledge representation lack the means for representing the meaning of fuzzy concepts. As a consequence, the approaches based on classical logic and probability theory do not provide an appropriate conceptual framework for dealing with the representation of commonsense knowledge, since such knowledge is by its nature both lexically imprecise and non-categorical. The development of fuzzy logic was motivated in large measure by the need for a conceptual framework which can address the issue of uncertainty and lexical imprecision. Fuzzy logic provides an inference morphology that enables approximate human reasoning capabilities to be applied to knowledge-based systems.

Since the publication of Zadeh's seminal work "Fuzzy Sets" in 1965, the subject has been the focus of many independent research investigations by mathematicians, scientists and engineers from around the world. Fuzzy logic has rapidly become one of the most successful of technologies today for developing sophisticated control systems. With its aid, complex requirements may be implemented in amazingly simple, easily maintained and inexpensive controllers. Of course, fuzzy logic is not the best approach for every control problem. As designers look at its power and expressiveness, they must decide where to apply it.

Our focus in this chapter is on the essential ideas and tools necessary for the construction of the fuzzy knowledge-based models that have been successful in the development of intelligent controllers. Fuzzy control and modeling use only a small portion of the fuzzy mathematics that is available; this portion is also mathematically quite simple and conceptually easy to understand. This chapter begins with an introduction to some essential concepts, terminology, notations and arithmetic of fuzzy sets and fuzzy logic. We include only a minimum though adequate amount of fuzzy mathematics necessary for understanding fuzzy control and modeling. To facilitate easy reading, this background material is presented in a rather informal manner with simple and clear notation as well as explanation. Whenever possible, excessively rigorous mathematics is avoided. This material is intended to serve as an introductory foundation for the reader to understand not only the fuzzy controllers presented later in this chapter but also others in the literature. We point to references [161, 163–167] as further sources on fuzzy set theory.

12.2 FUZZY QUANTIFICATION OF KNOWLEDGE

Up to this point we have only quantified, in an abstract way, the knowledge that the human expert has about how to control the plant. Next we will show how to use fuzzy logic to fully quantify the meaning of linguistic descriptions so that we may automate in the fuzzy controller, the control rules specified by the expert.

What is Fuzzy Logic?

Knowledge is structured information and knowledge acquisition is done through learning and experience, which are forms of high-level processing of information. Knowledge representation and processing are the keys to any intelligent system. In *logic*, knowledge is represented by propositions and is processed through reasoning by the application of various laws of logic, including an appropriate *rule of inference*.

Fuzzy logic focusses on linguistic variables in natural language and aims to provide foundations for *approximate reasoning* with imprecise propositions.

In *classical logic*, a proposition is either TRUE, denoted by 1, or FALSE, denoted by 0. Consider the following proposition p:

"Team member is female"

Let X be a collection of 10 people: $x_1, x_2, ..., x_{10}$, who form a project team. The entire object of discussion is

$$X = \{x_1, x_2, ..., x_{10}\}$$

In general, the entire object of discussion is called a "universe of discourse", and each constituent member x is called an "element" of the universe (the fact that x is an element of X is written as $x \in X$).

If x_1, x_2, x_3 and x_4 are female members in the project team, then the proposition p on the universe of discourse X is equally well represented by the *crisp* (non-fuzzy) set A defined below:

$$A = \{x_1, x_2, x_3, x_4\}$$

The fact that A is a subset of X is denoted as $A \subset X$.

The proposition can also be expressed by a mapping μ_A from X into the binary space $\{0, 1\}$.

$$\mu_A : X \to \{0, 1\}$$

such that

$$\mu_A = \begin{cases} 0 ; x = x_5, x_6, x_7, x_8, x_9, x_{10} \\ 1 ; x = x_1, x_2, x_3, x_4 \end{cases}$$

That is to say, the value $\mu_A(x) = 1$ when the element x satisfies the attributes of set A; 0 when it does not. μ_A is called the *characteristic function* of A.

Next, supposing that within X only x_1 and x_2 are below age 20; we may call them "minors". Then

$$B = \{x_1, x_2\}$$

consists of minor team members. In this case

$$\mu_B(x) = \begin{cases} 1 ; x = x_1, x_2 \\ 0 ; \text{otherwise} \end{cases}$$

B is obviously a subset of A; we write $B \subset A$.

We have considered the "set of females A", and the "set of minors B" in X. Is it also possible to consider a "set of young females C"? If for convenience we consider the attribute "young" to be same as "minor", then $C = B$; but in this case we have created a sharp boundary, under which x_2 who is 19 is still young ($\mu_C(x_2) = 1$), but x_3 who just turned 20 today is no longer young ($\mu_C(x_3) = 0$).

In just one day, the value changed from yes (1) to no (0), and x_3 is now an old maid.

However, is it not possible that a young woman becomes an old maid over a period of 10 to 15 years, so that we ought to be patient with her? Prof. Zadeh admitted values such as 0.8 and 0.9 that are intermediate between 0 and 1, thus creating the concept of a "fuzzy set". Whereas a crisp set is defined by the characteristic function that can assume only the two values {0, 1}, a fuzzy set is defined by a "membership function" that can assume an infinite number of values; any real number in the closed interval [0, 1].

With this definition, the concept of "young women" in X can be expressed flexibly in terms of membership function (Fuzzy sets are denoted in this book by a set symbol with a tilde understrike).

$$\mu_{\underset{\sim}{C}} : X \to [0, 1]$$

such that

$$\mu_{\underset{\sim}{C}} = \begin{cases} 1; & x = x_1, x_2 \\ 0.9; & x = x_3 \\ 0.2; & x = x_4 \\ 0; & \text{otherwise} \end{cases}$$

The significance of such terms as "patient" and "flexibly" in the above description may be explained as follows. For example, we have taken $\mu_{\underset{\sim}{C}}(x_3) = 0.9$, but suppose that x_3 objects that "you are being unfair; I really ought to be a 1 but if you insist we can compromise on 0.95". There is a good amount of subjectivity in the choice of membership values. A great deal of research is being done on the question of assignment of membership values. However, even with this restriction, it has become possible to deal with many problems that could not be handled with only crisp sets.

Since [0, 1] incorporates {0, 1}, the concept of fuzzy set can be considered as an extended concept, which incorporates the concept of crisp set. For example, the crisp set B of 'minors' can be regarded as a fuzzy set $\underset{\sim}{B}$ with the membership function:

$$\mu_{\underset{\sim}{B}}(x) = \begin{cases} 1; & x = x_1, x_2 \\ 0; & \text{otherwise} \end{cases}$$

Example 12.1

One of the most commonly used examples of a fuzzy set is the set of tall people. In this case the universe of discourse is potential heights (the real line), say from 3 feet to 9 feet. If the set of tall people is given the well-defined boundary of a crisp set, we might say all people taller than 6 feet are officially considered tall. The characteristic function of the set $A = \{\text{tall men}\}$ then is

$$\mu_A(x) = \begin{cases} 1 \text{ for } 6 \leq x \\ 0 \text{ for } 3 \leq x < 6 \end{cases}$$

Such a condition is expressed by a Venn diagram shown in Fig. 12.7a and a characteristic function shown in Fig. 12.8a.

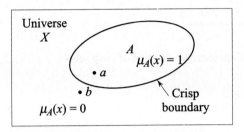

(a) The crisp set A and the universe of discourse

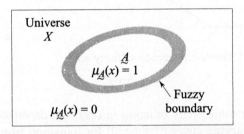

(b) The fuzzy set $\underline{A}$ and the universe of discourse

Fig. 12.7

For our example of universe X of heights of people, the crisp set A of all people with $x \geq 6$ has a sharp boundary: individual, 'a' corresponding to $x = 6$ is a member of the crisp set A, and individual 'b' corresponding to $x = 5.9$ is unambiguously *not* a member of set A. Is it not an absurd statement for the situation under consideration? A 0.1" reduction in the height of a person has changed μ_A from 1 to 0, and the person is no more tall.

It may make sense to consider the crisp set of all real numbers greater than 6 because the numbers belong to an abstract plane, but when we want to talk about real people, it is unreasonable to call one person short and another one tall when they differ in height by the width of a hair. But if this kind of distinction is unworkable, than what is the right way to define the set of all people? Much as with our example of "set of young females", the word 'tall' would correspond to a curve that defines the degree to which any person is tall. Figure 12.8b shows a possible membership function of this fuzzy set $\underline{A}$; the curve defines the transition from not tall to tall. Two people with membership values 0.9 and 0.3 are tall to some degree, but one significantly less than the other.

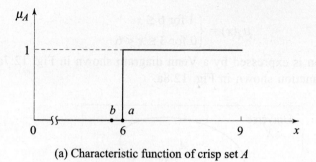

(a) Characteristic function of crisp set A

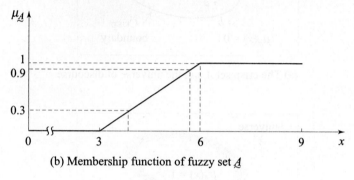

(b) Membership function of fuzzy set $\underset{\sim}{A}$

Fig. 12.8

Note that there is subjectivity inherent in fuzzy set description. Figure 12.9 shows a smoothly varying curve (S-shaped) for transition from not tall to tall. Compared to Fig. 12.8b, the membership values are lower for heights close to 3' and are higher for heights close to 6'. This looks more reasonable; however the price paid is in terms of a more complex function, which is more difficult to handle.

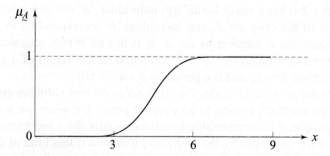

Fig. 12.9 A smoothly varying membership function for fuzzy set $\underset{\sim}{A}$

Figure 12.7b shows the representation of a fuzzy set by a Venn diagram. In the central (unshaded) region of the fuzzy set, $\mu_{\underset{\sim}{A}}(x) = 1$. Outside the boundary region of fuzzy set, $\mu_{\underset{\sim}{A}}(x) = 0$. On the boundary region, $\mu_{\underset{\sim}{A}}(x)$ assumes an intermediate value in the interval (0, 1). Presumably, the membership value of an x in fuzzy set $\underset{\sim}{A}$ approaches a value of 1 as it moves closer to the central (unshaded) region; it approaches a value of 0 as it moves closer to leaving the boundary region of $\underset{\sim}{A}$.

▲▲

Thus far we have discussed the *representation* of knowledge in logic. We have seen that the concept of fuzzy sets makes it possible to describe vague information (knowledge). But description alone will not lead to the development of any useful products. Indeed, a good deal of time passed after fuzzy sets were first proposed until they were applied at the industrial level. However, eventually it became possible to apply them in the form of "fuzzy inference", and fuzzy logic theory has now become legitimized as one component of applied high technology.

In fuzzy logic theory, nothing is done at random or haphazardly. Information containing a certain amount of vagueness is expressed as faithfully as possible, without the distortion produced by forcing it into a "crisp" mold, and it is then processed by applying an appropriate rule of inference.

'Approximate reasoning' is the best known from the fuzzy logic processing and covers a variety of inference rules.

Fuzziness is often confused with probability. The fundamental difference between them is that fuzziness deals with deterministic plausibility, while probability concerns the likelihood of non-deterministic (stochastic) events. Fuzziness is one aspect of uncertainty. It is the ambiguity (vagueness) found in the definition of a concept or the meaning of a term. However, the uncertainty of probability generally relates to the occurrence of phenomena, not the vagueness of phenomena. For example "There is a 50-50 chance that he will be there" has the uncertainty of randomness. "He is a young man", has the uncertainty in definition of "young man". Thus fuzziness describes the ambiguity of an event, whereas randomness describes the uncertainty in the occurrence of an event.

We can now give a formal definition to fuzzy sets.

Fuzzy Sets

A *universe of discourse*, X, is a collection of objects all having the same characteristics. The individual elements in the universe X will be denoted as x.

A universe of discourse and a membership function that spans the universe completely define a *fuzzy set*. Consider a universe of discourse X with x representing its generic element. A fuzzy set $\underset{\sim}{A}$ in X has the membership function $\mu_{\underset{\sim}{A}}(x)$ which maps the elements of the universe onto numerical values in the interval [0, 1]:

$$\mu_{\underset{\sim}{A}}(x) : X \to [0, 1] \tag{12.8a}$$

Every element x in X has a membership function $\mu_{\underset{\sim}{A}}(x) \in [0, 1]$. $\underset{\sim}{A}$ is then defined by the set of ordered pairs:

$$\underset{\sim}{A} = \left\{ \left(x, \mu_{\underset{\sim}{A}}(x) \right) \middle| x \in X, \mu_{\underset{\sim}{A}}(x) \in [0, 1] \right\} \tag{12.8b}$$

A membership value of zero implies that the corresponding element is definitely *not* an element of the fuzzy set $\underset{\sim}{A}$. A membership function of unity means that the corresponding element is definitely an element of fuzzy set $\underset{\sim}{A}$. A grade of membership greater than zero and less than unity corresponds to a noncrisp (or fuzzy) membership of the fuzzy set $\underset{\sim}{A}$. Classical sets can be considered as special case of fuzzy sets with all membership grades equal to unity.

A fuzzy set $\underset{\sim}{A}$ is formally given by its membership function $\mu_{\underset{\sim}{A}}(x)$. We will identify any fuzzy set with its membership function and use these two terms interchangeably.

Membership functions characterize the fuzziness in a fuzzy set. However, the shape of the membership functions used to describe the fuzziness has very few restrictions indeed. It might be claimed that the rules used to describe fuzziness are also fuzzy. Just as there are an infinite number of ways to characterize fuzziness, there are an infinite number of ways to graphically depict the membership functions that describe fuzziness. Although the selection of membership functions is *subjective*, *it cannot be arbitrary*; it should be *plausible*.

To avoid unjustified complications, $\mu_A(x)$ is usually constructed without a high degree of precision. It is advantageous to deal with membership functions involving a small number of parameters. Indeed, one of the key issues in the theory and practice of fuzzy sets is how to define the proper membership functions. Primary approaches include (1) asking the control expert to define them; (2) using data from the system to be controlled to generate them; and (3) making them in a trial-and-error manner. In more than 25 years of practice, it has been found that the third approach, though *adhoc*, works effectively and efficiently in many real-world applications.

Numerous applications in control have shown that only four types of membership functions are needed in most circumstances: trapezoidal, triangular (a special case of trapezoidal), Gaussian, and bell-shaped. Figure 12.10 shows an example of each type. Among the four, the first two are more widely used. All these fuzzy sets are continuous, normal and convex.

A fuzzy set is said to be *continuous* if its membership function is continuous.

A fuzzy set is said to be *normal* if its *height* is one (The largest membership value of a fuzzy set is called the height of the fuzzy set).

The *convexity* property of fuzzy sets is viewed as a generalization of the classical concept of crisp sets. Consider the universe X to be a set of real num-

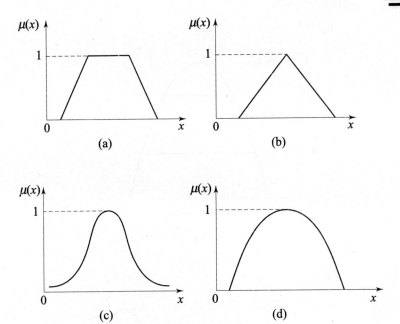

Fig. 12.10 Examples of fuzzy sets;
(a) trapezoidal, (b) triangular
(c) Gaussian, and (d) bell-shaped

bers $\Re$. A subset A of $\Re$ is said to be *convex* if and only if, for all $x_1, x_2 \in A$ and for every real number λ satisfying $0 \leq \lambda \leq 1$, we have

$$\lambda x_1 + (1 - \lambda)x_2 \in A \tag{12.9}$$

It can easily be established that any set defined by a single interval of real numbers is convex; any set defined by more than one interval that does not contain some points between the intervals is not convex.

An *alpha-cut* of a fuzzy set $\underset{\sim}{A}$ is a crisp set A_α that contains all the elements of the universal set X that have a membership grade in $\underset{\sim}{A}$ greater than or equal to α (refer Fig. 12.11). The convexity property of fuzzy sets is viewed as a generalization of the classical concept of convexity of crisp sets. In order to make the generalized convexity consistent with the classical definition of convexity, it is required that α-cuts of a *convex fuzzy set* be convex for all $\alpha \in (0, 1]$ in the classical sense (0-cut is excluded here since it is always equal to $\Re$ in this sense and thus includes $-\infty$ to $+\infty$). Figure 12.11a shows a fuzzy set that is convex. Two of the α-cuts shown in this figure are clearly convex in the classical sense, and it is easy to see that any other α-cuts for $\alpha > 0$ are convex as well. Figure 12.11b illustrates a fuzzy set that is not convex. The lack of convexity of this fuzzy set can be demonstrated by identifying some of its α-cuts ($\alpha > 0$) that are not convex.

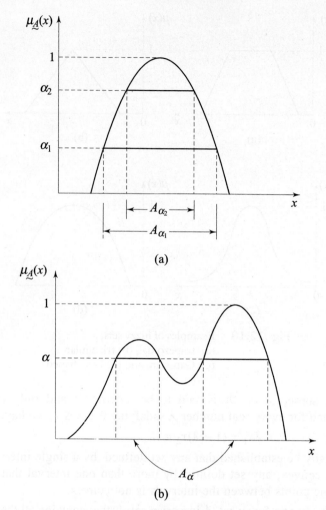

Fig. 12.11 (a) Convex fuzzy set
(b) Nonconvex fuzzy set

The support of a fuzzy set $\underset{\sim}{A}$ is the crisp set of all $x \in X$ such that $\mu_{\underset{\sim}{A}}(x) > 0$. That is

$$\mathrm{supp}(\underset{\sim}{A}) = \left\{ x \in X \mid \mu_{\underset{\sim}{A}}(x) > 0 \right\} \tag{12.10}$$

The element $x \in X$ at which $\mu_{\underset{\sim}{A}}(x) = 0.5$ is called the *crosspoint*.

A fuzzy set $\underset{\sim}{A}$ whose support is a single point in X with $\mu_{\underset{\sim}{A}}(x) = 1$ is referred to as a *fuzzy singleton*.

Example 12.2

Consider the fuzzy set described by membership function depicted in Fig. 12.12, where the universe of discourse is

$$X = [32°F, 104°F]$$

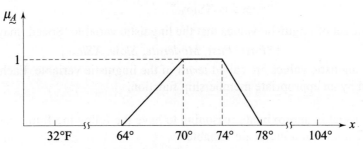

Fig. 12.12 A fuzzy set: linguistic 'warm'

This fuzzy set $\underset{\sim}{A}$ is linguistic 'warm' with membership function

$$\mu_{\underset{\sim}{A}}(x) = \begin{cases} 0 & ; x < 64° \\ (x - 64°)/6 & ; 64° \leq x < 70° \\ 1 & ; 70° < x \leq 74° \\ (78° - x)/4 & ; 74 < x \leq 78° \\ 0 & ; x > 78° \end{cases}$$

The support of $\underset{\sim}{A}$ is the crisp set

$$\{x \mid 64° < x < 78°\}$$

Example 12.3

Consider an *atomic primitive*, expressed in a natural language form:

"Speed sensor output is very large"

The formal, symbolic translation of this natural language expression in terms of linguistic variables proceeds as follows:

1. An abbreviation 'Speed' may be chosen to denote the physical variable "Speed sensor output".
2. An abbreviation 'XFast' (i.e., extra fast) may be chosen to denote the particular value "very large" of speed.
3. The above natural language expression is rewritten as "Speed is XFast".

Such an expression is an *atomic fuzzy proposition*. The 'meaning' of the atomic proposition is then defined by a fuzzy set $\underset{\sim}{XFast}$ or a membership function $\mu_{\underset{\sim}{XFast}}(x)$, defined on the physical domain $X = [0 \text{ mph}, 100 \text{ mph}]$ of the physical variable 'Speed'.

Many atomic propositions may be associated with a linguistic variable, e.g.,

"Speed is Fast"

"Speed is Moderate"

"Speed is Slow"

"Speed is XSlow"

Thus the set of linguistic values that the linguistic variable 'Speed' may take is
$$\{XFast, Fast, Moderate, Slow, XSlow\}$$
These linguistic values are called *terms* of the linguistic variable. Each term is defined by an appropriate membership function.

▲▲

It is usual in approximate reasoning to have the following frame associated with the notion of a linguistic variable:

$$\left\langle \underset{\sim}{A}, L\underset{\sim}{A}, X, \mu_{L\underset{\sim}{A}} \right\rangle \tag{12.11}$$

where $\underset{\sim}{A}$ denotes the symbolic name of a linguistic variable, e.g., speed, temperature, level, error, change-of-error, etc. $L\underset{\sim}{A}$ is the set of linguistic values that $\underset{\sim}{A}$ can take, i.e. $L\underset{\sim}{A}$ is the term set of $\underset{\sim}{A}$. X is the actual physical domain over which linguistic variable $\underset{\sim}{A}$ takes its quantitative (crisp) values, and $\mu_{L\underset{\sim}{A}}$ is a membership function which gives a meaning to the linguistic value in terms of the quantitative elements of X.

Example 12.4

Consider *speed*, interpreted as a linguistic variable with $X = [0\text{mph}, 100\text{mph}]$; i.e., $x = $ 'speed'. Its term set could be

$$\{Slow, Moderate, Fast\}$$

Slow = the fuzzy set for "a speed below about 40 miles per hour (mph)", with membership function μ_{Slow}.

Moderate = the fuzzy set for "a speed close to 55 mph" with membership function $\mu_{Moderate}$.

Fast = the fuzzy set for "a speed above about 70 mph" with membership function μ_{Fast}.

The frame of speed is

$$\left\langle \underset{\sim}{Speed}, L\underset{\sim}{Speed}, X, \mu_{L\underset{\sim}{Speed}} \right\rangle$$

where

$$L\text{ Speed} = \{Slow, Moderate, Fast\}$$

$$X = [0, 100] \text{ mph}$$

$\mu_{Slow}, \mu_{Moderate}, \mu_{Fast}$ are given in Fig. 12.13.

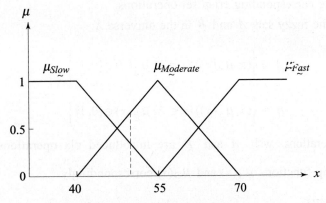

Fig. 12.13 Terms of linguistic variable 'speed'

The frame of speed helps us to decide the degree to which an atomic proposition associated with 'speed' is satisfied given a specific physical value of speed. For example, for crisp input

$$Speed = 50 \text{ mph},$$

$$\mu_{Slow}(50) = 1/3$$

$$\mu_{Moderate}(50) = 2/3$$

$$\mu_{Fast}(50) = 0$$

Therefore, the proposition "Speed is Slow" is satisfied to a degree of 1/3, the proposition "Speed is Moderate" is satisfied to a degree of 2/3, and the proposition "Speed is Fast" is not satisfied.

Fuzzy Operations

There are a variety of fuzzy set theories which differ from one another by the set operations (complement, intersection, union) they employ. The fuzzy complement, intersection and union are not unique operations, contrary to their crisp counterparts; different functions may be appropriate to represent these operations in different contexts. That is, not only membership functions of fuzzy sets but also operations on fuzzy sets are context-dependent. The capability to determine appropriate membership functions and meaningful fuzzy operations in the context of each particular application is crucial for making fuzzy set theory practically useful.

The intersection and union operations on fuzzy sets are often referred to as *triangular norms* (*t*-norms), and *triangular conorms* (*t*-conorms; also called *s*-norms), respectively. The reader is advised to refer [161] for the axioms which *t*-norms, *t*-conorms, and the complements of fuzzy sets are required to satisfy.

In the following, we define *standard fuzzy operations*, which are generalizations of the corresponding crisp set operations.

Consider the fuzzy sets $\underset{\sim}{A}$ and $\underset{\sim}{B}$ in the universe X.

$$\underset{\sim}{A} = \left\{(x, \mu_{\underset{\sim}{A}}(x)) \mid x \in X; \mu_{\underset{\sim}{A}}(x) \in [0,1]\right\} \quad (12.12)$$

$$\underset{\sim}{B} = \left\{(x, \mu_{\underset{\sim}{B}}(x)) \mid x \in X; \mu_{\underset{\sim}{B}}(x) \in [0,1]\right\} \quad (12.13)$$

The operations with $\underset{\sim}{A}$ and $\underset{\sim}{B}$ are introduced via operations on their membership functions $\mu_{\underset{\sim}{A}}(x)$ and $\mu_{\underset{\sim}{B}}(x)$ correspondingly.

Complement

The standard complement, $\overline{\underset{\sim}{A}}$, of fuzzy set $\underset{\sim}{A}$ with respect to the universal set X is defined for all $x \in X$ by the equation

$$\mu_{\overline{\underset{\sim}{A}}}(x) \triangleq 1 - \mu_{\underset{\sim}{A}}(x) \; \forall x \in X \quad (12.14)$$

Intersection

The standard operation, $\underset{\sim}{A} \cap \underset{\sim}{B}$ is defined for all $x \in X$ by the equation

$$\mu_{\underset{\sim}{A} \cap \underset{\sim}{B}}(x) \triangleq \min[\mu_{\underset{\sim}{A}}(x), \mu_{\underset{\sim}{B}}(x)] \equiv \mu_{\underset{\sim}{A}}(x) \wedge \mu_{\underset{\sim}{B}}(x) \; \forall x \in X \quad (12.15)$$

where $\wedge$ indicates the **min** operation.

Union

The standard union, $\underset{\sim}{A} \cup \underset{\sim}{B}$, is defined for all $x \in X$ by the equation

$$\mu_{\underset{\sim}{A} \cup \underset{\sim}{B}}(x) \triangleq \max[\mu_{\underset{\sim}{A}}(x), \mu_{\underset{\sim}{B}}(x)] \equiv \mu_{\underset{\sim}{A}}(x) \vee \mu_{\underset{\sim}{B}}(x) \; \forall x \in X \quad (12.16)$$

where $\vee$ indicates the **max** operation.

Fuzzy Relations

Consider two universes (crisp sets) X and Y. The *Cartesian product* (or *cross product*) of two sets X and Y (in this order) is the set of all ordered pairs such that the first element in each pair is a member of X, and the second element is a member of Y. Formally,

$$X \times Y = \{(x, y); x \in X, y \in Y\} \qquad (12.17)$$

where $X \times Y$ denotes the Cartesian product.

A *fuzzy relation* on $X \times Y$, denoted by $\underset{\sim}{R}$, or $\underset{\sim}{R}(X, Y)$ is defined as the set

$$\underset{\sim}{R} = \left\{ \left((x, y), \mu_{\underset{\sim}{R}}(x, y)\right) \middle| (x, y) \in X \times Y, \mu_{\underset{\sim}{R}}(x, y) \in [0, 1] \right\} \qquad (12.18)$$

where $\mu_{\underset{\sim}{R}}(x, y)$ is a function in two variables, called membership function of

the fuzzy relation. It gives the degree of membership of the ordered pair (x, y) in $\underset{\sim}{R}$ associating with each pair (x, y) in $X \times Y$ a real number in the interval $[0, 1]$. The degree of membership indicates the degree to which x is in relation with y. It is clear that a fuzzy relation is basically a fuzzy set.

Example 12.5

Consider an example of fuzzy sets: the set of people with *normal* weight. In this case the universe of discourse appears to be all potential weights (the real line). However, the reader might have realized by now that knowledge representation in terms of this universe is not useful. The normal weight of a person is a function of his/her height.

$$\text{Body Mass Index (BMI)} = \frac{\text{Weight. kg}}{(\text{Height, m})^2}$$

Normal BMI for males is 20–25, and for females is 19–24. Values between 25 to 27 in men and 24–27 in women indicate overweight; and those over 27 indicate obesity. Of course, values below 20 for man and below 19 for women indicate underweight.

The universe of discourse for this fuzzy set is more appropriately the Cartesian product of two universal sets: X, the set of all potential heights, and Y, the set of all potential weights. The Cartesian product space $X \times Y$ is a universal set which is a set of ordered pairs (x, y), for each $x \in X$ and each $y \in Y$.

A subset of the Cartesian product $X \times Y$, satisfying the knowledge attribute "normal weight" is set a of (height, weight) pairs. This is called a relation $\underset{\sim}{R}$.

The membership value for each element of $\underset{\sim}{R}$ depends on BMI. For men, a BMI of 27 and more could be given a membership value of 0, and a BMI of less than 18 could also be given a membership value of 0; and membership value between 0 and 1 for BMI between 18 and 27.

Example 12.6

Because fuzzy relations, in general, are fuzzy sets, we can define the Cartesian product to be a relation between two or more fuzzy sets. Let $\underset{\sim}{A}$ be a fuzzy set on universe X and $\underset{\sim}{B}$ be a fuzzy set on universe Y; then the Cartesian product between fuzzy sets $\underset{\sim}{A}$ and $\underset{\sim}{B}$ will result in a fuzzy relation $\underset{\sim}{R}$, which is contained within the full Cartesian product space, or

$$\underset{\sim}{A} \times \underset{\sim}{B} = \underset{\sim}{R} \subset X \times Y \qquad (12.19a)$$

where the fuzzy relation $\underset{\sim}{R}$ has membership function

$$\mu_{\underset{\sim}{R}}(x, y) = \mu_{\underset{\sim}{A} \times \underset{\sim}{B}}(x, y) = \min[\mu_{\underset{\sim}{A}}(x), \mu_{\underset{\sim}{B}}(y)] \quad \forall x \in X, \forall y \in Y \qquad (12.19b)$$

Note that the **min** combination applies here because each element (x, y) in the Cartesian product is formed by taking both elements x, y together, not just the one or the other.

As an example of the Cartesian product of fuzzy sets, we consider premise quantification. Atomic fuzzy propositions do not usually make a knowledge base in real-life situations. Many propositions connected by logical connectives may be needed. A set of such compound propositions connected by IF-THEN rules makes a knowledge base.

Consider two propositions defined by

$$p \triangleq x \text{ is } \underset{\sim}{A}$$

$$q \triangleq y \text{ is } \underset{\sim}{B}$$

where $\underset{\sim}{A}$ and $\underset{\sim}{B}$ are the fuzzy sets

$$\underset{\sim}{A} = \left\{ \left(x, \mu_{\underset{\sim}{A}}(x)\right) \middle| x \in X \right\}$$

$$\underset{\sim}{B} = \left\{ \left(y, \mu_{\underset{\sim}{B}}(y)\right) \middle| y \in Y \right\}$$

The meaning of the linguistic terms "x is $\underset{\sim}{A}$", and "y is $\underset{\sim}{B}$" is quantified via the membership functions $\mu_{\underset{\sim}{A}}(x)$ and $\mu_{\underset{\sim}{B}}(y)$, respectively. Now we seek to quantify the linguistic premise "x is $\underset{\sim}{A}$ and y is $\underset{\sim}{B}$" of the rule:

$$\text{IF } x \text{ is } \underset{\sim}{A} \text{ and } y \text{ is } \underset{\sim}{B} \text{ THEN } z \text{ is } \underset{\sim}{C} \qquad (12.20a)$$

The main item to focus on is how to quantify the logical "and" operation that combines the meaning of two linguistic terms. As said earlier, there are actually several ways (t-norms) to define this quantification. In the following, we use **min** operation:

$$\mu_{premise}(x,y) = \min[\mu_{\underset{\sim}{A}}(x), \mu_{\underset{\sim}{B}}(y)] \quad \forall x \in X, \forall y \in Y \quad (12.20b)$$

Does this quantification make sense? Notice that this way of quantifying the "and" operation in the premise indicates that you can be no more certain about the conjunction of two statements than you are about the individual terms that make them up.

The *conjunction operator* (logical connective **and**), implemented as Cartesian product, is described in Fig. 12.14.

$$\mu_{premise}(x,y) = \mu_{\underset{\sim}{A} \times \underset{\sim}{B}}(x,y) = \min[\mu_{\underset{\sim}{A}}(x), \mu_{\underset{\sim}{B}}(y)] \quad \forall x \in X, \forall y \in Y \quad (12.20c)$$

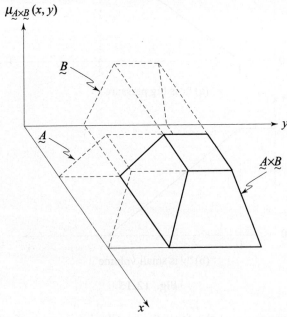

Fig. 12.14 The conjunction operator implemented as Cartesian product

Example 12.7

We consider here quantification of "implication" operator via fuzzy logic. Consider the implication statement

IF pressure is high THEN volume is small.

The membership function of the fuzzy set $\underset{\sim}{A}$ = "big pressure",

$$\mu_{\underset{\sim}{A}}(x) = \begin{cases} 1 & ; x \geq 5 \\ 1-(5-x)/4 & ; 1 \leq x \leq 5 \\ 0 & ; \text{otherwise} \end{cases}$$

is shown in Fig. 12.15a. The membership function of the fuzzy set $\underset{\sim}{B}$ = "small volume",

$$\mu_{\underset{\sim}{B}}(y) = \begin{cases} 1 & ; y \le 1 \\ 1-(y-1)/4 & ; 1 \le y \le 5 \\ 0 & ; \text{otherwise} \end{cases}$$

is shown in Fig. 12.15b.

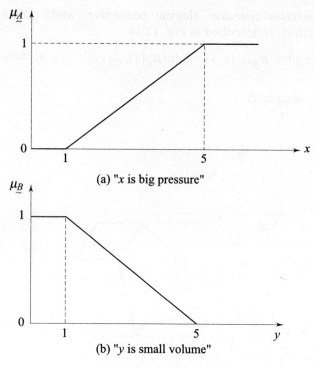

Fig. 12.15

If p is a proposition of the form "x is $\underset{\sim}{A}$" where $\underset{\sim}{A}$ is a fuzzy set on the universe X, e.g. "big pressure", and q is a proposition of the form "y is $\underset{\sim}{B}$" where $\underset{\sim}{B}$ is a fuzzy set on the universe Y, e.g., "small volume", then one encounters the following problem:

How does one define the membership function of the fuzzy implication $\underset{\sim}{A} \to \underset{\sim}{B}$?

There are different important classes of fuzzy implication operators based on t-norm and t-conorm. In many practical applications, one uses *Mamdani's implication operator* to model causal relationship between fuzzy variables:

$$\mu_{\underset{\sim}{A} \to \underset{\sim}{B}}(x, y) = \min[\mu_{\underset{\sim}{A}}(x), \mu_{\underset{\sim}{B}}(y)] \qquad \forall x \in X, \forall y \in Y \qquad (12.21)$$

The fuzzy implication $\underset{\sim}{A} \rightarrow \underset{\sim}{B}$ is a fuzzy relation in the Cartesian product space $X \times Y$.

Note that Mamdani's implication operator gives a relation which is symmetric with respect to $\underset{\sim}{A}$ and $\underset{\sim}{B}$. This is not intuitively satisfying because "implication" is not a commutative operation. In practice, however, the method provides good, robust results. The justification for the use of the **min** operator to represent the implication is that *we can be no more certain about our consequent than our premise.*

12.3 FUZZY INFERENCE

Problems featuring uncertainty and ambiguity have been successfully addressed subconsciously by humans. Humans can adapt to unfamiliar situations and they are able to gather information in an efficient manner and discard irrelevant details. The information gathered need not be complete and precise and could be general, qualitative and vague, because humans can reason, infer and deduce new information and knowledge. They can learn, perceive and improve their skills through experience.

How can humans reason about complex systems when the complete description of such a system often requires more detailed data than a human could ever hope to recognize simultaneously and assimilate with understanding? The answer is that humans have the capacity to reason approximately. In reasoning about a complex system, humans reason approximately about its behaviour, thereby maintaining only a generic understanding about the problem.

The fuzzy set theory has provided a mathematical strength to capture the uncertainties associated with human congnitive processes, such as thinking and reasoning. Fuzzy logic provides an *inference* morphology that enables approximate human reasoning capabilities to be applied to knowledge-based systems.

Fuzzy conditional or fuzzy IF-THEN production rules are symbolically expressed as

$$\{\text{IF (premise } i\text{) THEN (consequent } i\text{)}\}_{i=1}^{N}$$

Here N is the number of rules.

Two major types of fuzzy rules exist: Mamdani fuzzy rules, and Takagi-Sugeno fuzzy rules.

Mamdani Fuzzy Rules

In Mamdani fuzzy rules, both the premises and the consequents are fuzzy propositions (atomic/compound). Consider first the case of a rule with atomic propositions. For example:

$$\text{"IF } x \text{ is } \underset{\sim}{A} \text{ THEN } y \text{ is } \underset{\sim}{B}\text{"} \qquad (12.22a)$$

If we let X be the premise universe of discourse, and Y the consequent universe of discourse, then the relation between the premise $\underset{\sim}{A}$ and consequent $\underset{\sim}{B}$ can be described using fuzzy sets on the Cartesian product space $X \times Y$. Using Mamdani's implication rule,

$$\underset{\sim}{R} = \underset{\sim}{A} \to \underset{\sim}{B}$$

$$\mu_{\underset{\sim}{R}}(x, y) = \mu_{\underset{\sim}{A} \to \underset{\sim}{B}}(x, y)$$

$$= \min[\mu_{\underset{\sim}{A}}(x), \mu_{\underset{\sim}{B}}(y)] \quad \forall x \in X, \forall y \in Y \quad (12.22b)$$

When the rule premise or rule consequent are compound fuzzy propositions, then first the membership function corresponding to each such compound proposition is determined. Finally the above operation is applied to represent IF-THEN relation. Quite often in control applications, we come across logical connective **and** (conjunction operation on atomic propositions), which, as we have seen in Example 12.6, may be implemented by Cartesian product.

The rules of inference in fuzzy logic govern the deduction of final conclusion from IF-THEN rules for known inputs (Fig. 12.16). Consider the statements:

$$\begin{array}{lll} \text{rule} & : & \text{IF } x \text{ is } \underset{\sim}{A} \text{ THEN } y \text{ is } \underset{\sim}{B} \\ \text{input} & : & x \text{ is } \underset{\sim}{A}' \\ \hline \text{inference} & : & y \text{ is } \underset{\sim}{B}' \end{array} \quad (12.23)$$

Here the propositions "x is $\underset{\sim}{A}$", "x is $\underset{\sim}{A}'$", "y is $\underset{\sim}{B}$" and "y is $\underset{\sim}{B}'$" are characterized by fuzzy sets $\underset{\sim}{A}, \underset{\sim}{A}', \underset{\sim}{B},$ and $\underset{\sim}{B}'$ respectively.

$$\underset{\sim}{A} = \left\{ (x, \mu_{\underset{\sim}{A}}(x)) \mid x \in X; \mu_{\underset{\sim}{A}} \in [0,1] \right\}$$

$$\underset{\sim}{A}' = \left\{ (x, \mu_{\underset{\sim}{A}'}(x)) \mid x \in X; \mu_{\underset{\sim}{A}'} \in [0,1] \right\} \quad (12.24)$$

$$\underset{\sim}{B} = \left\{ (y, \mu_{\underset{\sim}{B}}(y)) \mid y \in Y; \mu_{\underset{\sim}{B}} \in [0,1] \right\}$$

$$\underset{\sim}{B}' = \left\{ (y, \mu_{\underset{\sim}{B}'}(y)) \mid y \in Y; \mu_{\underset{\sim}{B}'} \in [0,1] \right\}$$

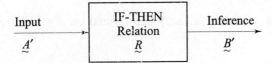

Fig. 12.16 Inference mechanism

Fuzzy sets $\underset{\sim}{A}$ and $\underset{\sim}{A}'$ are close but not equal, and same is valid for the sets $\underset{\sim}{B}$ and $\underset{\sim}{B}'$.

Inference mechanism is based on matching of two fuzzy sets $\underset{\sim}{A}'$ and $\underset{\sim}{R}$, and determining membership function of $\underset{\sim}{B}'$ according to the result. Note that X denotes the space in which the input $\underset{\sim}{A}'$ is defined, and it is subspace of the space $X \times Y$ in which the rule-base relation $\underset{\sim}{R}$ is defined. It is therefore not possible to take the intersection of $\underset{\sim}{A}'$ and $\underset{\sim}{R}$; an operation required for matching the two sets to incorporate the knowledge of the membership functions of both the input and the rule base. But when $\underset{\sim}{A}'$ is extended to $X \times Y$, this is possible.

Cylindrical extension of $\underset{\sim}{A}'$ (a fuzzy set defined on X) on $X \times Y$ is the set of all tuples $(x, y) \in X \times Y$ with membership degree equal to $\mu_{\underset{\sim}{A}'}(x)$, i.e.,

$$\mu_{ce(\underset{\sim}{A}')}(x, y) = \mu_{\underset{\sim}{A}'}(x) \text{ for every } y \in Y \qquad (12.25)$$

Now the intersection operation to incorporate the knowledge of membership functions of input and rule base is possible. It is given by

$$ce\left(\underset{\sim}{A}'\right) \cap \underset{\sim}{R}$$

In terms of membership functions, this operation may be expressed as

$$\mu_{ce(\underset{\sim}{A}')}(x, y) \wedge \mu_{\underset{\sim}{R}}(x, y) = \min[\mu_{ce(\underset{\sim}{A}')}(x, y), \mu_{\underset{\sim}{R}}(x, y)] \quad \forall x \in X, \forall y \in Y$$

$$\mu_{\underset{\sim}{R}}(x, y) = \mu_{\underset{\sim}{A} \to \underset{\sim}{B}}(x, y) = \min[\mu_{\underset{\sim}{A}}(x), \mu_{\underset{\sim}{B}}(y)]$$

$$\mu_{ce(\underset{\sim}{A}')}(x, y) = \mu_{\underset{\sim}{A}'}(x)$$

Therefore

$$\mu_{\underset{\sim}{S}}(x, y) = \mu_{ce(\underset{\sim}{A}')}(x, y) \wedge \mu_{\underset{\sim}{R}}(x, y) = \min(\mu_{\underset{\sim}{A}'}(x, y), \min(\mu_{\underset{\sim}{A}}(x), \mu_{\underset{\sim}{B}}(y))) \qquad (12.26)$$

By projecting this matched fuzzy set (defined on $X \times Y$) over to the inference subspace Y, we can determine the membership function $\mu_{\underset{\sim}{B}'}(y)$ of the fuzzy set $\underset{\sim}{B}'$ (defined on Y).

Projection of $\mu_{\underset{\sim}{S}}(x, y)$ (a fuzzy set defined on $X \times Y$) on Y is a set of all $y \in Y$ with membership grades equal to $\max_x \{\mu_{\underset{\sim}{S}}(x, y)\}$; max means maximum with respect to x while y is considered fixed, i.e.,

$$\mu_{proj(\underset{\sim}{S})}(y) = \max_x \{\mu_{\underset{\sim}{S}}(x, y)\} \qquad (12.27)$$

Projection on Y means that y_i is assigned the highest membership degree from the tuples $(x_1, y_i), (x_2, y_i), (x_3, y_i), \ldots$, where $x_1, x_2, x_2, \ldots \in X$ and $y_i \in Y$. The rationale for using the **max** operation on the membership functions of $\underset{\sim}{S}$ should be clear in view of the fact that we have a many-to-one mapping.

The combination of fuzzy sets with the aid of cylindrical extension and projection is called *composition*. It is denoted by ∘.

If $\underset{\sim}{A}'$ is a fuzzy set defined on X and $\underset{\sim}{R}$ is a fuzzy relation defined on $X \times Y$, then the composition of $\underset{\sim}{A}'$ and $\underset{\sim}{R}$ resulting in a fuzzy set $\underset{\sim}{B}'$ defined on Y is given by

$$\underset{\sim}{B}' = \underset{\sim}{A}' \circ \underset{\sim}{R} = proj\left(ce(\underset{\sim}{A}') \cap \underset{\sim}{R}\right) \text{ on } Y \qquad (12.28)$$

Note that, in general, intersection is given by a *t*-norm, and projection by a *t*-conorm, resulting in many definitions of composition operator. In our applications, we will restrict our use to **min** operator for *t*-norm and **max** operator for *t*-conorm. Therefore, we have the following *compositional rule of inference*:

$$\mu_{\underset{\sim}{B}'}(y) = \max_x \left\{ \min\left(\mu_{\underset{\sim}{A}'}(x), \min\left(\mu_{\underset{\sim}{A}}(x), \mu_{\underset{\sim}{B}}(y) \right) \right) \right\} \qquad (12.29)$$

This inference rule based on *max-min composition*, uses Mamdani's rule for implication operator.

In control applications, as we shall see later, the fuzzy set $\underset{\sim}{A}'$ is fuzzy singleton, i.e.,

$$\mu_{\underset{\sim}{A}'}(x) = \begin{cases} 1 & \text{for } x = x_0 \in X \\ 0 & \text{for all other } x \in X \end{cases} \qquad (12.30)$$

This results in a simple inference procedure, as is seen below.

$$\mu_{\underset{\sim}{B}'}(y) = \begin{cases} \min\left(\mu_{\underset{\sim}{A}}(x), \mu_{\underset{\sim}{B}}(y) \right) & \text{for } x = x_0, \forall y \in Y \\ 0 & \text{for all other } x, \forall y \in Y \end{cases} \qquad (12.31)$$

Graphical representation of the procedure is shown in Fig. 12.17.

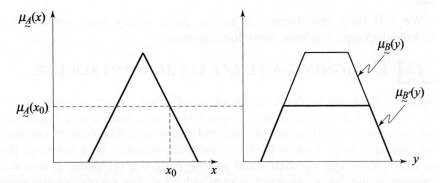

Fig. 12.17 Inference procedure for singleton fuzzy system

Singleton fuzzy system is most widely used because of its simplicity and lower computational requirements. However, this kind of fuzzy system may not always be adequate, especially in cases where noise is present in the data. *Nonsingleton fuzzy system* becomes necessary to account for uncertainty in the data.

Takagi-Sugeno Fuzzy Rules

Unlike Mamdani fuzzy rules, Takagi-Sugeno rules are functions of input variables on the rule consequent:

$$\text{IF } x_1 \text{ is } \underset{\sim}{A} \text{ and } x_2 \text{ is } \underset{\sim}{B} \text{ THEN } u = f(x_1, x_2) \quad (12.32a)$$

where $f(\cdot)$ is a real function.

In theory, $f(\cdot)$ can be any real function, linear or nonlinear. It seems to be appealing to use nonlinear functions; rules are more general and can potentially be more powerful. Unfortunately, the idea is impractical, for properly choosing or determining the mathematical formalism of nonlinear functions for every fuzzy rule is extremely difficult, if not impossible. For this reason, linear functions have been employed exclusively in theoretical research and practical development of Takagi-Sugeno fuzzy controllers and models:

$$\text{IF } x_1 \text{ is } \underset{\sim}{A} \text{ and } x_2 \text{ is } \underset{\sim}{B} \text{ THEN } u = \alpha x_1 + \beta x_2 \quad (12.32b)$$

A Mamdani fuzzy controller uses fuzzy sets as rule consequent. Hence fuzzy rules are more intuitive and can more easily be extracted from expert knowledge and experience. The Takagi-Sugeno rule scheme, on the other hand, possesses little linguistic meaning and hence is not intuitive. Also Takagi-Sugeno controllers have more design parameters and the number grows exponentially with the increase of the number of input variables. In theory, these parameters provide a means for designing and achieving desired local as well as global control action, possibly resulting in a superior control performance. To a large extent, the power of Takagi-Sugeno rule scheme lies in these parameters. In practice, however, tuning of these parameters is often very difficult [163].

We will limit our discussion to the more widely used controllers—the Mamdani type singleton fuzzy logic systems.

12.4 DESIGNING A FUZZY LOGIC CONTROLLER

Figure 12.18 shows the basic configuration of a fuzzy logic controller (FLC), which comprises four principal components: a rule base, a fuzzy inference system, an input fuzzification interface and an output defuzzification interface. The rule base holds a set of IF-THEN rules that quantify the knowledge that human experts have amassed about solving particular problems. It acts as a resource to the fuzzy inference system, which makes successive decisions about which rules are most relevant to the current situation and applies the actions indicated by these rules. The input fuzzifier takes the crisp numeric inputs and, as its name implies, converts them into the fuzzy form needed by the fuzzy inference system. At the output, the defuzzification interface combines the conclusions reached by the fuzzy inference system and converts them into crisp numeric values as control actions.

We will illustrate the FLC methodology, step by step, on a water-heating system.

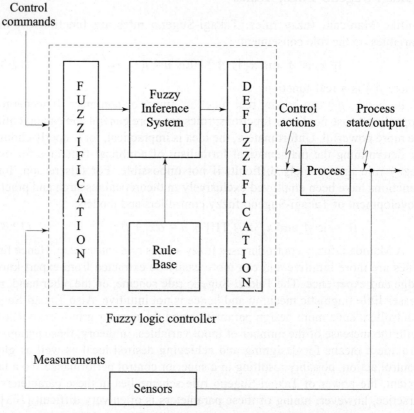

Fig. 12.18 A simple fuzzy logic control system block diagram

Consider a simple water-heating system shown in Fig. 12.19. The water heater has a knob (HeatKnob) to control the steam for circulation through the radiator. The higher the setting of the *HeatKnob*, the hotter the water gets with the value of '0' indicating the steam supply is turned off, and the value of '10' indicating the maximum possible steam supply. There is a sensor (*TempSense*) in the outlet pipe to tell us the temperature of the outflowing water, which varies from 0°C to 125°C. Another sensor (*LevelSense*) tells us the level of the water in the tank, which varies from 0 (= empty) to 10 (= full). We assume that there is an automatic flow control that determines how much cold water flows into the tank from the main water supply; whenever the level of the tank gets below 4, the flow control turns ON, and turns OFF when the level of the water gets above 9.5.

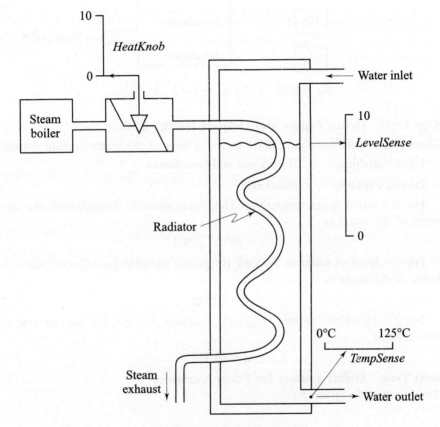

Fig. 12.19 Water heating system

Figure 12.20 shows a FLC diagram for the water-heating system.
The design objective can be stated as:
Keep the water temperature as close to 80°C as possible, inspite of changes in the hot water flowing out of the tank, and the cold water flowing into the tank.

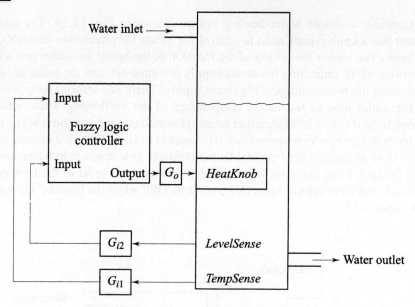

Fig. 12.20 Fuzzy control of a water heater

Step One: Define Inputs and Outputs for the FLC
Three fuzzy variables characterize the behaviour of the water-heating system.

    Input Variables: *TempSense* and *LevelSense*

    Output Variable: *HeatKnob*

For x = outlet water temperature (linguistic variable *TempSense*), the universe of discourse is

$$X = [0°C, 125°C]$$

For y = level of water in the tank (linguistic variable *LevelSense*), the universe of discourse is

$$Y = [0, 10]$$

For z = HeatKnob setting (linguistic variable *HeatKnob*), the universe of discourse is

$$Z = [0, 10]$$

Step Two: Define Frames for Fuzzy Variables
The frame of *TempSense* is

$$\left\langle \underset{\sim}{TempSense}, \mathcal{L}\underset{\sim}{TempSense}\ X\ ,\ \mu_{\mathcal{L}\underset{\sim}{TempSense}} \right\rangle$$

where $\mathcal{L}\underset{\sim}{TempSense}$ is the set of linguistic values that *TempSense* can take. We may use the following fuzzy subsets to describe the linguistic values:

XSmall (XS)
Small (S)
Medium (M)
Large (L)
XLarge (XL)

i.e.,

$$\mathcal{L}Temp\underset{\sim}{S}ense = \{XSmall,\ Small,\ Medium,\ Large,\ XLarge\}$$

The frame of *LevelSense* is

$$\left\langle Level\underset{\sim}{S}ense, \mathcal{L}Level\underset{\sim}{S}ense, Y, \mu_{\mathcal{L}Level\underset{\sim}{S}ense} \right\rangle$$

$$\mathcal{L}Level\underset{\sim}{S}ense = \{XSmall,\ Small,\ Medium,\ Large,\ XLarge\}$$

In our system, we have just one output which is the *HeatKnob*. We take the following frame for this linguistic variable.

$$\left\langle Heat\underset{\sim}{K}nob, \mathcal{L}Heat\underset{\sim}{K}nob, Z, \mu_{\mathcal{L}Heat\underset{\sim}{K}nob} \right\rangle$$

$$\mathcal{L}Heat\underset{\sim}{K}nob = \{VeryLittle,\ ALittle,\ AGoodAmount,\ ALot,\ AWholeLot\}$$

Step Three: Assign Membership Values to Fuzzy Variables

Since the membership function essentially embodies all fuzziness for a particular fuzzy set, its description is the essence of a fuzzy property or operation. Because of the importance of the "shape" of the membership function, a great deal of attention has been focussed on development of these functions. Many ways to develop membership functions, i.e., to assign membership values to fuzzy variables, have been reported in the literature—methods based on Inference, Neural Networks, Genetic Algorithms, Inductive Reasoning, etc. The assignment process can be intuitive, or it can be based on some algorithmic or logical operations. We shall rely on intuition in our application examples.

The input variables *TempSense* and *LevelSense*, as well as the output variable *HeatKnob*, are restricted to positive values. In Table 12.1 and Fig. 12.21, we show a possible assignment for ranges and triangular membership functions for *TempSense*. Similarly, we assign ranges and fuzzy membership functions for *LevelSense* in Table 12.2 and Fig. 12.22; and *HeatKnob* in Table 12.3 and Fig. 12.23. The optimization of these assignments is often done through trial and error for achieving optimum performance of FLC.

Table 12.1 Fuzzy variable ranges for *TempSense*

Crisp Input Range	Fuzzy Variable
0–20	XSmall
10–35	Small
30–75	Medium
60–95	Large
85–125	XLarge

Table 12.2 Fuzzy variable ranges for *LevelSense*

Crisp Input Range	Fuzzy Variable
0–2	XSmall
1.5–4	Small
3–7	Medium
6–8.5	Large
7.5–10	XLarge

Table 12.3 Fuzzy variable ranges for *HeatKnob*

Crisp Input Range	Fuzzy Variable
0–2	VeryLittle
1.5–4	ALittle
3–7	AGoodAmount
6–8.5	ALot
7.5–10	AWholeLot

The following guidelines were kept in mind while determining range of fuzzy variables as related to the crisp inputs.

1. Symmetrically distribute the fuzzified values across the universe of discourse.
2. Use an odd number of fuzzy sets for each variable so that some set is assured to be in the middle. The use of 5 to 7 sets is fairly typical.
3. Overlap adjacent sets (by 15% to 25% typically).

Step Four: Create a Rule Base

Now that we have the inputs and the outputs in terms of fuzzy variables, we need only specify what actions to take under what conditions; i.e., we need to construct a set of rules that describe the operation of the FLC. These rules usually take the form of IF-THEN rules and can be obtained from a human expert (heuristics).

The rule-base matrix for our example is given in Table 12.4. Our heuristic guidelines in determining this matrix are the following:

1. When the temperature is low, the *HeatKnob* should be set higher than when the temperature is high.
2. When the volume of water is low, the *HeatKnob* does not need to be as high as when the volume of water is high.

Fuzzy Control   817

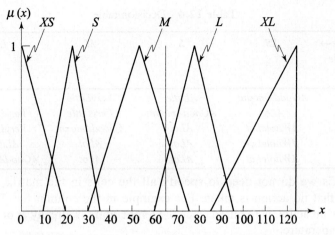

Fig. 12.21 Fuzzy membership functions for *TempSense*

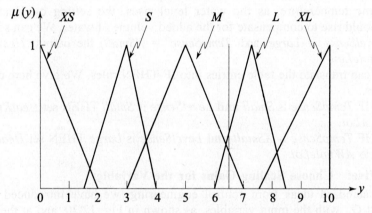

Fig. 12.22 Fuzzy membership functions for *LevelSense*

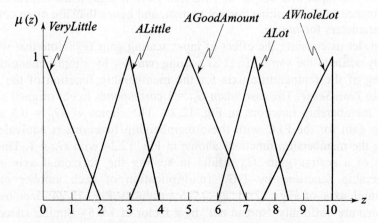

Fig. 12.23 Fuzzy membership functions for the output *HeatKnob*

Table 12.4 Decision table

TempSense → LevelSense ↓	XS	S	M	L	XL
XS	AGoodAmount	ALittle	VeryLittle		
S	ALot	AGoodAmount	VeryLittle	VeryLittle	
M	AWholeLot	ALot	AGoodAmount	VeryLittle	
L	AWholeLot	ALot	ALot	ALittle	
XL	AWholeLot	ALot	ALot	AGoodAmount	

In FLCs we do not need to specify all the cells in the matrix. No entry signifies that no action is taken. For example in the column for *TempSense* = *XLarge*, no action is required since the temperature is already at or above the target temperature.

Let us examine a couple of typical entries in the table: For *LevelSense* = *Small*, and *TempSense* = *XSmall*, the output is *HeatKnob* = *ALot*. Now for the same temperature, as the water level rises, the setting on HeatKnob also should rise to compensate for the added volume of water. We can see that for *LevelSense* = *Large* and *TempSense* = *XSmall*, the output *HeatKnob* = *AWholeLot*.

We can translate the table entries into IF-THEN rules. We give here couple of rules.

- IF *TempSense* is *Small* **and** *LevelSense* is *Small* THEN set *HeatKnob* to *ALot*.
- IF *TempSense* is *XSmall* **and** *LevelSense* is *Large* THEN set *HeatKnob* to *AWholeLot*.

Step Five: Choose Scaling Gains for the Variables

Using standard ideas from control engineering, we have introduced gains G_{i1} and G_{i2} with the input variables, as shown in Fig. 12.20, and at the same time we also put a gain G_o between FLC and the plant. Change in the *scaling gains* at the input and output of FLC can have a significant impact on the performance of the resulting control system, and hence they are often convenient parameters for *tuning*.

First, let us consider the effect of input scaling gain G_{i1}. Note that we can actually achieve the same effect as scaling *via* G_{i1} by simply changing the labeling of the temperature axis for the membership function of the input variable *TempSense*. The case when $G_{i1} = 1$ corresponds to our original choice of the membership functions in Fig. 12.21. The choice of $G_{i1} = 0.5$ as the scaling gain for the FLC with these membership functions is equivalent to having the membership functions shown in Fig. 12.24 with $G_{i1} = 1$. Thus the choice of a scaling gain G_{i1} results in scaling the horizontal axis of the membership functions by $1/G_{i1}$ (multiplication of each number on the horizontal axis of Fig. 12.21 by 1/0.5 produces Fig. 12.24; membership functions are uniformly "spread out" by a factor of 1/0.5). Similar statements can be made about G_{i2} (Fig. 12.25).

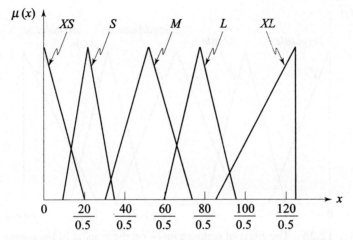

Fig. 12.24 Scaled membership functions for *TempSense*

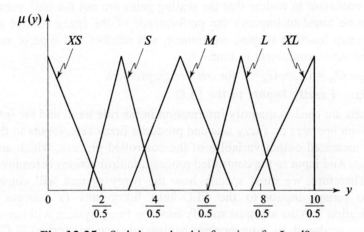

Fig. 12.25 Scaled membership functions for *LevelSense*

Figure 12.23 shows our choice of output membership functions with $G_o = 1$. There is a proportional effect between the scaling gain G_o and the output membership functions as shown in Fig. 12.26 for $G_o = 2$.

If for the process under consideration, the effective universes of discourse for all inputs and output are common, say, [0, 1], then we may say that the FLC is *normalized*. Clearly, scaling gains can be used to normalize the given FLC. Denormalization of the output of such a FLC will yield the required control action.

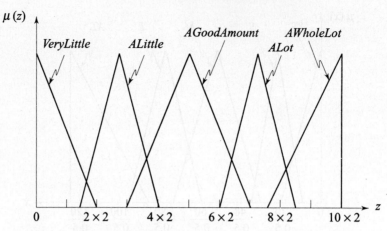

Fig. 12.26 The effect of scaling gain G_o on the spacing of the output membership functions

It is important to realize that the scaling gains are not the only parameters that can be tuned to improve the performance of the fuzzy control system. Membership function shapes, positioning, and number and type of rules are often the other parameters to tune.

We set $G_{i1} = G_{i2} = G_o = 1$ for our design problem.

Step Six: Fuzzify Inputs to the FLC

Fuzzy sets are used to quantify information in the rule base, and the inference mechanism operates on fuzzy sets and produces fuzzy sets. Inputs to the FLC are the measured output variables of the controlled process, which are crisp variables. And input to the controlled process (control action) is required to be crisp. Therefore, we must specify how the fuzzy system will convert the numeric (crisp) inputs to the FLC into fuzzy sets (a process called "fuzzification"). Also we must specify how the fuzzy system will convert the fuzzy sets produced by inference mechanism into numeric (crisp) FLC output (a process called "defuzzification"), which is the input to the controlled process.

Fuzzification can be defined as a mapping from an observed input space to fuzzy sets in a specified universe of discourse. A natural and simple fuzzification approach is to convert a crisp measurement into a fuzzy singleton within the specified universe of discourse. This approach is based on the assumption that the observed data is crisp, and not corrupted by random noise.

To understand fuzzification, we consider an example. Assume that at a particular point in time, *LevelSense* = 6.5 and *TempSense* = 65°C. These are the crisp inputs directly from the sensors. Figures 12.21 and 12.22 show the membership functions for the input variables and indicate with vertical lines the measured values of *LevelSense* and *TempSense*. These vertical lines are, in fact, graphical representation of the two fuzzy singletons obtained by the fuzzification process.

Step Seven: Determine which Rules Fire

We see that with singleton fuzzification, combining the fuzzy sets that were created by the fuzzification process to represent the inputs, with the premise membership functions for the rules, is particularly simple. It simply reduces to computing the membership values of the input fuzzy sets for the given inputs.

From Fig. 12.21 we find that for input *TempSense* = 65°C, $\mu_M(65) = 0.45$, $\mu_L(65) = 0.28$ and all other membership functions are off (i.e., their values are zero). Therefore the proposition "*TempSense* is *Medium*" is satisfied to a degree of 0.45, the proposition "*TempSense* is *Large*" is satisfied to a degree of 0.28; all other atomic propositions associated with *TempSense* are not satisfied.

From Fig. 12.22 we find that for input *LevelSense* = 6.5, $\mu_M(6.5) = 0.25$, $\mu_L(6.5) = 0.38$; all other membership functions are off.

We next form membership values of premises of all the rules. From the induced decision table (Table 12.5) we observe that the rules that have the premise terms:

1. *TempSense* is *Medium* **and** *LevelSense* is *Medium*
2. *TempSense* is *Large* **and** *LevelSense* is *Medium*
3. *TempSense* is *Medium* **and** *LevelSense* is *Large*
4. *TempSense* is *Large* **and** *LevelSense* is *Large*

have $\mu_{\text{premise}} > 0$. For all other rules, $\mu_{\text{premise}} = 0$.

Table 12.5 Induced Decision table

TempSense→	$\mu_{XS}=0$	$\mu_S=0$	$\mu_M=0.45$	$\mu_L=0.28$	$\mu_{XL}=0$
LevelSense ↓					
$\mu_{XS}=0$	0	0	0	0	0
$\mu_S=0$	0	0	0	0	0
$\mu_M=0.25$	0	0	AGoodAmount	VeryLittle	0
$\mu_L=0.38$	0	0	ALot	ALittle	0
$\mu_{XL}=0$	0	0	0	0	0

Determining applicability of each rule is called "firing". We say that, a rule fires at time t, if its premise membership value at time t is greater than zero. The inference mechanism seeks to determine which rules fire to find out which rules are relevant to the current situation. The inference mechanism combines the recommendations of all the rules to come up with a single conclusion.

For crisp input *TempSense* = 65°C, and *LevelSense* = 6.5, four rules fire. μ_{premise} for the four rules (refer Table 12.5), which amounts to firing strength in each case, can be calculated as follows.

1. $\mu_{Temp\tilde{S}ense \times Level\tilde{S}ense} = \min(0.45, 0.25) = 0.25$

2. $\mu_{Temp\tilde{S}ense \times Level\tilde{S}ense} = \min(0.28, 0.25) = 0.25$

3. $\mu_{Temp\tilde{S}ense \times Level\tilde{S}ense} = \min(0.45, 0.38) = 0.38$

4. $\mu_{Temp\tilde{S}ense \times Level\tilde{S}ense} = \min(0.28, 0.38) = 0.28$

Step Eight: Infer the Output Recommended by Each Rule

From the induced decision table (Table 12.5), we observe that only four cells contain nonzero terms. Let us call these cells *active*. The active cells correspond to the following rules.

1. *TempSense* is *Medium* **and** *LevelSense* is *Medium* : p_1
   Set *HeatKnob* to *AGoodAmount* : q_1
   IF p_1 THEN q_1

   $\mu_{premise(1)} = 0.25$

   $\mu_{inference(1)}$ is obtained by "chopping off" the top of $\mu_{A\tilde{G}oodAmount}$ function of the output variable *HeatKnob*, as shown in Fig. 12.27a.

2. *TempSense* is *Large* **and** *LevelSense* is *Medium* : p_2
   Set *HeatKnob* to *VeryLittle* : q_2
   IF p_2 THEN q_2

   $\mu_{premise(2)} = 0.25$

   $\mu_{inference(2)}$ is shown in Fig. 12.27b.

3. *TempSense* is *Medium* **and** *LevelSense* is *Large* : p_3
   Set *HeatKnob* to *ALot* : q_3
   IF p_3 THEN q_3

   $\mu_{premise(3)} = 0.38$

   $\mu_{inference(3)}$ is shown in Fig. 12.27c.

4. *TempSense* is *Large* **and** *LevelSense* is *Large* : p_4
   Set *HeatKnob* to *ALittle* : q_4
   IF p_4 THEN q_4

$\mu_{premise(4)} = 0.28$

$\mu_{inference(4)}$ is shown in Fig. 12.27d.

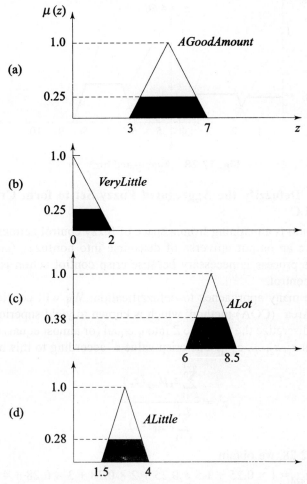

Fig. 12.27 Inference for each rule

The reader should note that for different crisp measurements *TempSense* and *LevelSense*, there will be different values of $\mu_{premise}$ and, hence, different $\mu_{inference}$ functions will be obtained.

Step Nine: Aggregate the Fuzzy Outputs Recommended by Each Rule
In the previous step, we noticed that the input to the inference process is the set of rules that fire; its output is the set of fuzzy sets that represent the inference reached by all the rules that fire. We now combine all the recommendations of all the rules to determine the control action. This is done by aggregating

the inferred fuzzy sets. Aggregated fuzzy set, obtained by drawing all the inferred fuzzy sets on one axis, is shown in Fig. 12.28. This fuzzy set represents the desired control action.

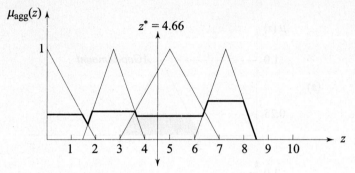

Fig. 12.28 Aggregated fuzzy set

Step Ten: Defuzzify the Aggregated Fuzzy Set to form Crisp Output from the FLC

Defuzzification is a mapping from a space of fuzzy control actions defined by fuzzy sets on an output universe of discourse into nonfuzzy (crisp) control actions. This process is necessary because crisp control action is required to actuate the control.

There are many approaches to defuzzification. We will consider here the "Centre of Area" (COA) method, which is known to yield superior results.

We may discretize the universe Z into q equal (or almost equal) subintervals by the points $z_1, z_2, ..., z_{q-1}$. The crisp value z^* according to this method is

$$z^* = \frac{\sum_{k=1}^{q-1} z_k \mu_{agg}(z_k)}{\sum_{k=1}^{q-1} \mu_{agg}(z_k)} \qquad (12.33)$$

From Fig. 12.28, we obtain

$\Sigma z_k \mu_{agg} = 1 \times 0.25 + 1.5 \times 0.25 + 2 \times 0.28 + 3 \times 0.28 + 4 \times 0.25 +$
$\qquad 5 \times 0.25 + 6 \times 0.25 + 7 \times 0.38 + 8 \times 0.38 = 11.475$

$\Sigma \mu_{agg} = 0.25 + 0.25 + 0.28 + 0.28 + 0.25 + 0.5 + 0.25 +$
$\qquad 0.38 + 0.38 = 2.57$

Therefore

$$z^* = \frac{11.475}{2.57} = 4.46$$

The physical interpretation of z^* is that if the area is cut of a thin piece of metal or wood, the centre of the area will be the centre of gravity.

In fact, there is hardly any need of discretization of the universe for situations like the one shown in Fig. 12.28; we can split up geometry into pieces and place a straight edge (centroid) through the figure to have it perfectly balanced with equal area of the figure on its either side. Analytical expression for z^* is

$$z^* = \frac{\int_z \mu_{agg}(z)z\,dz}{\int_z \mu_{agg}(z)\,dz} \qquad (12.34)$$

This completes the design for the simple example we chose.

12.5 GENETIC ALGORITHMS

A genetic algorithm (GA) uses the principles of evolution, natural selection and genetics from natural biological systems in a computer algorithm to simulate evolution. Essentially, the genetic algorithm is an optimization technique that performs a parallel, stochastic, but directed search to evolve the fittest population. In this section, we will introduce the genetic algorithm and explain how it can be used for design and tuning of fuzzy systems.

Introduction

In the 1950s and 1960s, several computer scientists independently studied evolutionary systems with the idea that evolution could be used as an optimization tool for engineering problems. The idea in all these systems was to evolve a population of candidate solutions to a given problem, using operators inspired by natural genetic variation and natural selection.

Why use evolution as an inspiration for solving computational problems? To evolutionary computation researchers, the mechanisms of evolution are well suited for some of the most pressing computational problems in many fields. Many computational problems require searching through a huge number of possibilities for solutions. An example is the problem of searching for proteins, in which an algorithm is sought that will search among the vast number of possible amino acid sequences for a protein with specified properties. Such search problems can often benefit from an effective use of parallelism, in which many different possibilities are explored simultaneously in an efficient way. For example, in searching for proteins with specified properties, rather than evaluate one amino acid sequence at a time, it would be much faster to evaluate many simultaneously. What is needed is both computational parallelism and an intelligent strategy for choosing the next set of sequences to evaluate.

Biological evolution is an appealing source of inspiration for addressing such problems. Evolution is, in effect, a method of searching among an enormous number of possibilities for "solutions". In biology, the enormous set of possibilities is the set of possible genetic sequences, and the desired "solutions" are highly fit organisms—organisms well able to survive and reproduce in their environments. Of course the fitness of a biological organism depends on many factors—for example, how well it can weather the physical characteristics of the environment and how well it can compete with or cooperate with the other organisms around it. The fitness criteria continually changes as creatures evolve; so evolution is searching a constantly changing set of possibilities. Searching for solutions in the face of changing conditions is precisely what is required for adaptive computer programs. Furthermore, evolution is a massively parallel search method: rather than work on one species at a time, evolution tests and changes millions of species in parallel. Finally, the "rules" of evolution are remarkably simple: species evolve by means of random variation (via mutation, recombination and other parameters), followed by natural selection in which the fittest tend to survive and reproduce, thus propagating their genetic material to future generations. Yet these simple rules are thought to be responsible, in large part, for the extraordinary variety and complexity we seen in the biosphere.

Knowledge of biological terminology, though not necessary, but may help better appreciation of genetic algorithms. All living organisms consist of cells, and each cell contains the same set of one or more *chromosomes*—strings of DNA (deoxyribonucleic acid). A chromosome can be conceptually divided into *genes*—functional blocks of DNA, each of which encodes a particular protein. Very roughly, one can think of a gene as encoding a *trait*, such as eye colour. The different possible "settings" for a trait (e.g., blue, brown, hazel) are called *alleles*. Each gene is located at a particular *locus* (position) on the chromosome.

Many organisms have multiple chromosomes in each cell. The complete collection of genetic material (all chromosomes taken together) is called the organism's *genome*. The term *genotype* refers to the particular set of genes contained in a genome. The genotype gives rise, under foetus and later development, to the organism's *phenotype*—its physical and mental characteristics, such as eye colour, height, brain size and intelligence.

Organisms whose chromosomes are arrayed in pairs are called *diploid*; organisms whose chromosomes are unpaired are called *haploid*. In nature, most sexually reproducing species are diploid, including human beings. In diploid sexual reproduction, *recombination* (or *crossover*) occurs: in each parent, genes are exchanged between each pair of chromosomes to form a *gemete* (a single chromosome), and then gemetes from the two parents pair up to create a full set of diploid chromosomes. In haploid sexual reproduction, genes are exchanged between the two parents' single-strand chromosomes. Offspring are subject to *mutation*, in which single nucleotides (elementary bits of DNA) are changed from parent to offspring; mutation may cause the chromosomes of children to be different from those of their biological parents. The *fitness* of an organism is typically defined as the probability that the organism will live to reproduce (*viability*) or as a function of the number of offspring the organism has (*fertility*).

The basic idea of a genetic algorithm is very simple. The term *chromosome* typically refers to a candidate solution to a problem, typically stored as strings of binary digits (1s and 0s) in the computer's memory. The 'genes' are short blocks of adjacent bits that encode a particular element of the candidate solution (e.g., in the context of multi-parameter function optimization, the bits encoding a particular parameter might be considered to be a gene). An 'allele' in a bit string is either 0 or 1. Crossover typically consists of exchanging genetic material between two single-chromosome haploid parents. Mutation consists of flipping the bit at a randomly chosen locus.

Most applications of genetic algorithms employ haploid individuals, particularly, single-chromosome individuals. The genotype of an individual in a genetic algorithm using bit strings is simply the configuration of bits in that individual's chromosome. Genetic algorithms with more complex representation, including diploid and multiple chromosomes, and higher cardinality alphabets, have also been developed. However, the binary case is both the simplest and the most general.

How are Genetic Algorithms Different from Traditional Methods?

The current literature identifies three main types of search methods: calculus-based, enumerative and random. Calculus-based methods have been studied extensively. These subdivide into two main classes: indirect and direct. Indirect methods seek local extrema by solving the usually nonlinear set of equations resulting from setting the gradient of the objective function equal to zero. Given a smooth, unconstrained function, finding a possible peak starts by restricting search to those points with slopes of zero in all directions. On the other hand, direct (search) methods seek local optima by hopping on the function and moving in a direction related to the local gradient. This is simply the notion of *hill climbing*: to find the local best, climb the function in the steepest permissible direction.

Both the calculus-based methods are local in scope: the optima they seek are the best in a neighbourhood of the current point. Clearly, starting the search procedures in the neighbourhood of the lower peak will cause us to miss the main event (the higher peak). Furthermore, once the lower peak is reached, further improvement must be sought through random restart or other trickery. Another problem with calculus-based methods is that they depend upon the existence of derivatives (well-defined slope values). Even if we allow numerical approximation of derivatives, this is a severe shortcoming. The real world of search is fraught with discontinuities and vast multi-modal (i.e., consisting of many 'hills') noisy search spaces; methods depending upon restrictive requirements of continuity and derivative existence are unsuitable for all but a very limited problem domain.

Enumerative schemes have been considered in many shapes and sizes. The idea is fairly straightforward: within a finite search space, the search algorithm starts looking at objective function values at every point in the space, one at a time. Although the simplicity of the type of algorithm is attractive, and enumeration is a very human kind of search, such schemes have applications wherein the number of possibilities is small. Even the highly touted enumerative scheme *dynamic programming* breaks down on problems of moderate size and complexity.

Random walks and random schemes that search and save the best, in the long run, can be expected to do not better than enumerative schemes. We must be careful to separate the strictly random search methods from randomized techniques. The genetic algorithm is an example of a search procedure that uses random choice as a tool to guide a highly exploitative search through a coding of parameter space. Using random choice as a tool in a directed search process seems strange at first, but nature contains many examples.

The traditional schemes have been used successfully in many applications; however as more complex problems are attacked, other methods will be necessary. We shall soon see how genetic algorithms help attack complex problems [168].

The GA literature describes a large number of successful applications, but there are also many cases in which GAs perform poorly. Given a potential application, how do we know if a GA is good method to use? There is no rigorous answer, though many researchers share the intuitions that if the space to be searched is large, is known not to be perfectly smooth and unimodal, or is not well understood; or if the fitness function is noisy; and if the task does not require a global optimum to be found—i.e., if quickly finding a sufficiently good solution is enough—a GA will have a good chance of being competitive or surpassing other methods. If the space is not large, it can be searched exhaustively by enumerative search methods, and one can be sure that the best possible solution has been found, whereas a GA might give only a 'good' solution. If the space is smooth and unimodal, a gradient ascent algorithm will be much more efficient than a GA. If the space is well understood, search methods using domain-specific heuristics can often be designed to outperform any general-purpose method such as a GA. If the fitness function is noisy, a one-candidate-solution-at-a-time search method such as simple hill climbing might be irrecoverably led astray by the noise; but GAs, since they work by accumulating fitness statistics over many generations, are thought to outperform robustly in the presence of small amounts of noise.

These intuitions, of course, do not rigorously predict when a GA will be an effective search procedure competitive with other procedures. It would be useful to have a mathematical characterization of how the genetic algorithm works, that is predictive. Research on this aspect of genetic algorithms has not yet produced definite answers.

Basics of Genetic Algorithms

Encoding: Simple genetic algorithms require the natural parameter set of the problem to be coded as a finite-length string of binary bits 0 and 1. For example, given a set of two-dimensional data ((x, y) data points), we want to fit a linear curve (straight line) through the data. To get a linear fit, we encode the parameter set for a line $y = \theta_1 x + \theta_2$ by creating independent bit strings for the two unknown constants θ_1 and θ_2 (parameter set describing the line) and then joining them (concatenating the strings). A bit string is a combination of 0s and 1s, which represents the value of a number in binary form. An n-bit string can accommodate all integers upto the value $2^n - 1$.

For problems that are solved by the genetic algorithm, it is usually known that the parameters that are manipulated by the algorithm will lie in a certain fixed range, say $\{\theta_{min}, \theta_{max}\}$. A bit string may then be mapped to the value of a parameter, say θ_i, by the mapping

$$\theta_i = \theta_{min\,i} + \frac{b}{2^L - 1} (\theta_{max\,i} - \theta_{min\,i}) \quad (12.35)$$

where "b" is the number in decimal form that is being represented in binary form (e.g., 152 may be represented in binary form as 10011000), L is the length of the bit string (i.e., the number of bits in each string), and θ_{max} and θ_{min} are user-specified constants, which depend on the problem in hand.

The length of the bit strings is based on the handling capacity of the computer being used, i.e., how long a string (strings of each parameter are concatenated to make one long string representing the whole parameter set) the computer can manipulate at an optimum speed.

Let us consider the data set in Table 12.6. For performing a line ($y = \theta_1 x + \theta_2$) fit, as mentioned earlier, we encode the parameter set (θ_1, θ_2) in the form of binary strings. We take the string length to be 12 bits. The first six bits encode the parameter θ_1, and the next six bits encode the parameter θ_2.

Table 12.6 Data set through which a line fit is required

Data number	x	y
1	1.0	1.0
2	2.0	2.0
3	4.0	4.0
4	6.0	6.0

The strings (000000, 000000) and (111111, 111111) represent the points (θ_{min1}, θ_{min2}) and (θ_{max1}, θ_{max2}), respectively, in the parameter space for the parameter set (θ_1, θ_2). Decoding of (000000) and (111111) to decimal form gives 0 and 63 respectively. However, problem specification may impose different values of minimum and maximum for θ_i. We assume that the minimum value to which we would expect θ_1 or θ_2 to go would be -2, and the maximum would be 5. Therefore

$$\theta_{min\,i} = -2 \text{ and } \theta_{max\,i} = 5$$

Consider a string (a concatenation of two substrings)

$$000111 \; 010100 \quad (12.36)$$

representing a point in the parameter space for the set (θ_1, θ_2). The decimal value of the substring (000111) is 7 and that of (010100) is 20. This, however, does not give the value of the parameter set (θ_1, θ_2) corresponding to the string in (12.36). The mapping (12.35) gives the value:

$$\theta_1 = \theta_{min1} + \frac{b}{2^L - 1}(\theta_{max1} - \theta_{min1})$$

$$= -2 + \frac{7}{2^6 - 1}(5 - (-2)) = -1.22$$

$$\theta_2 = \theta_{min2} + \frac{b}{2^L - 1}(\theta_{max2} - \theta_{min2})$$

$$= -2 + \frac{20}{2^6 - 1}(5 - (-2)) = 0.22$$

Fitness function: A fitness function takes a chromosome (binary string) as an input and returns a number that is a measure of the chromosome's performance on the problem to be solved. Fitness function plays the same role in GAs as the environment plays in natural evolution. The interaction of an individual with its environment provides a measure of fitness to reproduce. Similarly, the interaction of a chromosome with a fitness function provides a measure of fitness that the GA uses when carrying out reproduction. Genetic algorithm is a maximization routine; the fitness function must be a non-negative figure of merit.

It is often necessary to map the underlying natural objective function to a fitness function form through one or more mappings. If the optimization problem is to minimize cost function $\bar{J}(\boldsymbol{\theta})$, where $\boldsymbol{\theta}$ denotes the parameter set, then the following cost-to-fitness transformation may be used:

$$J(\boldsymbol{\theta}) = \frac{1}{\bar{J}(\boldsymbol{\theta}) + \varepsilon} \qquad (12.37)$$

where ε is a small positive number. Maximization of J can be achieved by minimization of $\bar{J}$; so the desired effect is achieved.

Another way to define the fitness function is to let

$$J(\boldsymbol{\theta}(k)) = -\bar{J}(\boldsymbol{\theta}(k)) + \max_{\boldsymbol{\theta}(k)}\{\bar{J}(\boldsymbol{\theta}(k))\} \qquad (12.38)$$

The minus sign in front of the $\bar{J}(\boldsymbol{\theta}(k))$ term turns the minimization problem into a maximization problem and $\max_{\boldsymbol{\theta}(k)}\{\bar{J}(\boldsymbol{\theta}(k))\}$ term is needed to shift the function up so that $\bar{J}(\boldsymbol{\theta}(k))$ is always positive; k is the iteration index.

A fitness function can be any nonlinear, non-differentiable, discontinuous, positive function because the algorithm only needs a fitness value assigned to each string.

For the problem in hand (fit a line through a given data set), let us choose a fitness function. Using decoded values of θ_1 and θ_2 of a chromosome, and the four data values of x given in the data table, calculate

$$\hat{y}_i = \theta_1 x + \theta_2; \; i = 1, 2, 3, 4$$

These computed values of $\hat{y}_i$ are compared with the correct values y_i given in the data table, and square of errors in estimating the y's is calculated for each string. The summation of the square of errors is subtracted from a large number (400 in this problem) to convert the problem into a maximization problem:

$$J(\boldsymbol{\theta}) = 400 - \sum_i (\hat{y}_i - y_i)^2 ; \; \boldsymbol{\theta} = [\theta_1 \; \theta_2] \qquad (12.39)$$

The fitness value of the string (12.36) is calculated as follows:

$$\theta_1 = -1.22, \; \theta_2 = 0.22$$

For $\quad x = 1.0, \; \hat{y}_1 = \theta_1 x + \theta_2 = -1.00$

For $\quad x = 2.0, \; \hat{y}_2 = -2.22$

For $\quad x = 4.0, \; \hat{y}_3 = -4.66$

For $\quad x = 6.0, \; \hat{y}_4 = -7.10$

$$J(\boldsymbol{\theta}) = 400 - \sum_{i=1}^{4} (\hat{y}_i - y_i)^2 = 131.586$$

Initialization of population: The basic element processed by a GA is the string formed by concatenating substrings each of which is a binary coding of a parameter of the search space. If there are N decision variables in an optimization problem, and each decision variable is encoded as an n-digit binary number, then a chromosome is a string of $n \times N$ binary digits. We start with a randomly selected initial population of such chromosomes; each chromosome in the population represents a point in the search space, and hence a possible solution to the problem. Each string is then decoded to obtain its fitness value which determines the probability of the chromosome being acted on genetic operators. The population then evolves and a new generation is created through the application of genetic operators (The total number of strings included in a population is kept unchanged throughout generations for computational economy and efficiency). The new generation is expected to perform better than the previous generation (better fitness values). The new set of strings is again decoded and evaluated, and another generation is created using the basic genetic operators. This process is continued until convergence is achieved within a population.

Let $\theta^j(k)$ be a single parameter in chromosome j of generation k. Chromosome j is composed of N of these parameters:

$$\boldsymbol{\theta}^j(k) = [\theta^j_1(k), \; \theta^j_2(k), \; ..., \; \theta^j_N(k)] \qquad (12.40)$$

The population of chromosome in generation k:

$$P(k) = \{\boldsymbol{\theta}^j(k) | \; j = 1, 2, ..., S\} \qquad (12.41)$$

where S represents the number of chromosomes in the population. We want to pick S to be big enough so that the population elements can cover the search space. However, we do not want S to be too big since this increases the number of computations we have to perform.

For the problem in hand, Table 12.7 gives an initial population of 4 strings, the corresponding decoded values of θ_1 and θ_2, and the fitness value for each string.

Table 12.7 Initial population

String no.	String	θ_1	θ_2	J
1.	000111 010100	−1.22	0.22	131.586
2.	010010 001100	0.00	−0.67	323.784
3.	010101 101010	0.33	2.67	392.41
4.	100100 001001	2.00	−1.00	365.00
			ΣJ	1212.8
			Av.J	303.2
			Max.J	392.41

Evolution occurs as we go from generation k to the next generation $k + 1$. Genetic operations of *selection*, *crossover* and *mutation* are used to produce a new generation.

Selection: Basically, according to Darwin the most qualified (fittest) creatures survive to mate. Fitness is determined by a creature's ability to survive predators, pestilence, and other obstacles to adulthood and subsequent reproduction. In our unabashedly artificial setting, we quantify "most qualified" via a chromosome's fitness $J(\theta^j(k))$. The fitness function is the final arbiter of the string-creature's life or death. Selecting strings according to their fitness values means that the strings with a higher value have a higher probability of contributing one or more offspring in the next generation.

Selection is a process in which good-fit strings in the population are selected to form a mating pool, which we denote by

$$M(k) = \{\mathbf{m}^j(k) | j = 1, 2, ..., S\} \tag{12.42}$$

The mating pool is the set of chromosomes that are selected for mating. A chromosome is selected for mating pool according to the probability proportional to its fitness value. The probability for selecting the ith string is

$$p_i = \frac{J(\theta^i(k))}{\sum_{j=1}^{S} J(\theta^j(k))} \tag{12.43}$$

For the initial population of four strings in Table 12.7, the probability for selecting each string is calculated as follows:

$$p_1 = \frac{131.586}{131.586 + 323.784 + 392.41 + 365} = \frac{131.586}{1212.8} = 0.108$$

$$p_2 = \frac{323.784}{1212.8} = 0.267$$

$$p_3 = \frac{392.41}{1212.8} = 0.324$$

$$p_4 = \frac{365.00}{1212.8} = 0.301$$

To clarify the meaning of the formula and hence the selection strategy, Goldberg [168] uses the analogy of spinning a unit-circumference roulette wheel; the wheel is cut like a pie into S regions where the ith region is associated with the ith element of $P(k)$. Each pie-shaped region has a portion of the circumference that is given by p_i in Eqn. (12.43).

The roulette wheel for the problem in hand is shown in Fig. 12.29. String 1 has solution probability of 0.108. As a result, string 1 is given 10.8% slice of the roulette wheel. Similarly, string 2 is given 26.7% slice, string 3 is given 32.4% slice and string 4 is given 30.1% of the roulette wheel.

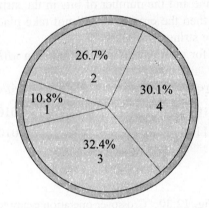

Fig. 12.29 Roulette Wheel

You spin the wheel, and if the pointer points at region i when the wheel stops, then you place θ^i into the mating pool $M(k)$. You spin the wheel S times so that S strings end up in the mating pool. Clearly, the strings which are more fit will end up with more copies in the mating pool; hence, chromosomes with larger-than-average fitness will embody a greater portion of the next generation. At the same time, due to the probabilistic nature of the selection process, it is possible that some relatively unfit strings may end up in the mating pool.

For the problem in hand, the four spins might choose strings 3, 3, 4 and 2 as parents (String 1 also may be selected in the process of roulette wheel spin; it is just the luck of the draw. If the roulette wheel were spun many times, the average results would be closer to the expected values).

Reproduction phase; crossover: We think of crossover as mating in biological terms, which at fundamental biological level involves the process of combining chromosomes. The crossover operation operates on the mating pool $M(k)$. First, specify the "crossover probability" p_c (usually chosen to be near one since when mating occurs in biological systems, genetic material is swapped between the parents).

The procedure for crossover consists of the following steps:
1. Randomly pair off the strings in the mating pool $M(k)$ (i.e., form pairs to mate by the flip of a coin). If there are an odd number of strings in $M(k)$, then, for instance, simply take the last string and pair it off with another string which has already been paired off.
2. Consider chromosome pair ($\boldsymbol{\theta}^j$, $\boldsymbol{\theta}^i$) that was formed in step 1. Generate a random number $r \in [0, 1]$.
   (a) If $r < p_c$, than crossover $\boldsymbol{\theta}^j$ and $\boldsymbol{\theta}^i$. To crossover these chromosomes, select at random a "cross site" and exchange all bits to the right of the cross site of one string with those of the other. This process is pictured in Fig. 12.30. In this example, the cross site is position four on the string, and hence we swap the last eight bits between the two strings. Clearly, the cross site is a random number between one and the number of bits in the string minus one.
   (b) If $r > p_c$, then the crossover will not take place; hence, we do not modify the strings.
3. Repeat step 2 for each pair of strings that is in $M(k)$.

Fig. 12.30 Crossover operation example

For the problem in hand, Table 12.8 shows the power of crossover. The first column shows the four strings selected for mating pool. We randomly pair off the strings. Suppose that random choice of mates has selected the first string in the mating pool to be mated with the fourth. With a cross site 4, the two strings cross and yield two new strings as shown in Table 12.8. The remaining two strings in the mating pool are crossed at site 9; the resulting strings are given in the table.

Table 12.8 Selection and crossover processes

String number	Mating pool	Couples	After crossover	θ_1	θ_2	J
3	010101 101010	0101 | 01 101010	010110 001100	0.44	−0.67	370.574
3	010101 101010	0100 | 10 001100	010001 101010	−0.11	2.67	378.311
4	100100 001001	010101 101 | 010	010101 101001	0.33	2.56	392.794
2	010010 001100	100100 001 | 001	100100 001010	2.00	−0.89	362.972
					ΣJ	1504.7
					Av.J	376.163
					Max.J	392.794

In nature, an offspring inherits genes from both the parents. The crossover process creates children strings from the parent strings. The children strings thus produced may or may not have combination of good substrings from parents strings, but we don't worry about this too much because if good strings are not created by crossover, they will not survive too long because of the selection operator. If good strings are created by crossover, there will be more copies of it in the next mating pool generated by the selection operator.

Besides the fact that crossover helps to model the mating part of the evolution process, why should the genetic algorithm perform crossover? Basically, the crossover operation perturbs the parameters near good positions to try to find better solutions to the optimization problem. It tends to help perform a localized search around the more fit strings (since on average the strings in the generation k mating pool are more fit than the ones in the generation k population).

Reproduction phase; mutation: Selection according to fitness combined with crossover gives genetic algorithms the bulk of their processing power. Mutation plays a secondary role in the operation of GAs. Mutation is needed because occasionally chromosomes may lose some potentially useful genetic material. In artificial genetic systems, mutation is realized by inverting a randomly chosen bit in a string. This is illustrated in Fig. 12.31.

Fig. 12.31 Mutation operation example

Besides the fact that this helps to model mutation in a biological system, why should the genetic algorithm perform mutation? Basically, it provides random excursions into new parts of the search space. It is possible that we will get lucky and mutate to a good solution. It is the mechanism that tries to make sure that we do not get stuck at a local maxima and that we seek to explore other areas of the search space to help find a global maximum for $J(\boldsymbol{\theta})$. Usually, the mutation probability p_m is chosen to be quite small (e.g., less than 0.01) since this will help guarantee that all the strings in the mating pool are not mutated so that any search progress that was made is lost (i.e., we keep it relatively low to avoid degradation to exhaustive search via a random walk in the search space).

After mutation we get a modified mating pool $M(k)$. To form the next generation for the population, we let

$$P(k+1) = M(k) \qquad (12.44)$$

where this $M(k)$ is the one that was formed by selection and modified by crossover and mutation. Then the above steps repeat, successive generations are produced, and we thereby model evolution (of course, it is a very crude model).

Terminal conditions: While the biological evolutionary process continues, perhaps indefinitely, we would like to terminate our artificial one and find the following:

- The population string—say, $\boldsymbol{\theta}^*(k)$—that best maximizes the fitness function. Notice that to determine this we also need to know the generation number k where the most fit string existed (it is not necessarily in the last generation). A computer code implementing the genetic algorithm keeps track of the highest J value, and the generation number and string that achieved this value of J.
- The value of the fitness function $J(\boldsymbol{\theta}^*(k))$.

There is then the question of how to terminate the genetic algorithm. There are many ways to terminate a genetic algorithm, many of them similar to termination conditions used for conventional optimization algorithms. To introduce a few of these, let $\varepsilon > 0$ be a small number and $n_1 > 0$ and $n_2 > 0$ be integers. Consider the following options for terminating the GA:

- Stop the algorithm after generating generation $P(n_1)$—that is after n_1 generations.
- Stop the algorithm after at least n_2 generations have occurred and at least n_1 steps have occurred when the maximum (or average) value of J for all population members has increased by no more than ε.
- Stop the algorithm once J takes on a value above some fixed value.

The above possibilities are easy to implement on a computer but, sometimes, you may want to watch the parameters evolve and decide yourself when to stop the algorithm.

Working parameters: A set of parameters is predefined to guide the genetic algorithm, such as:

- the length of each decision variable encoded as a binary string,
- the number of chromosomes to be generated and operated in each generation, i.e., population size,
- the crossover probability p_c,
- the mutation probability p_m,
- and the stopping criterion.

GA for Fuzzy System Design and Tuning

The genetic algorithm can be used in the computer-aided design of control systems since it can artificially evolve an appropriate controller that meets the performance specifications to the greatest extent possible. To do this, the genetic algorithm maintains a population of strings each representing a different controller (bits on the strings characterize parameters of the controller), and it uses a fitness function that characterizes the closed-loop specifications. Suppose, for instance, that the closed-loop specifications indicate that we want, for a step input, a (stable) response with a settling time of t_s^* and a percent overshoot of M_p^*. We need to define the fitness function so that it measures how close each string in the population at time k (i.e., each controller candidate) is meeting these specifications. Suppose that we let t_s and M_p denote the settling time and overshoot, respectively, for a given string (we compute these for a string in the population by performing a simulation of the closed-loop system with the candidate controller and a model of the plant). Given these values, we let (for each string and every time step k)

$$\bar{J} = w_1(t_s - t_s^*)^2 + w_2(M_p - M_p^*)^2$$

where $w_i > 0$; $i = 1, 2$ are weighting factors. The function $\bar{J}$ characterizes how well the candidate controller meets the closed-loop specifications where if $\bar{J} = 0$ it meets the specifications perfectly. The weighting factors can be used to prioritize the importance of meeting the various specifications (e.g., a high value of w_2 relative to w_1 indicates that the percent overshoot specification is more important than the settling time specification).

To minimize $\bar{J}$ with the genetic algorithm, we can choose the fitness function

$$J = \frac{1}{\bar{J} + \varepsilon}$$

where $\varepsilon > 0$ is a small positive number.

This completes the description of how to use a genetic algorithm for computer-aided control system design. Note that the above approach depends in no way on whether the controller that is evolved is a conventional controller (e.g., a PID controller) or a fuzzy system or a neural network. For instance you could use a fuzzy system for the controller and let the genetic algorithm tune the appropriate parameters (e.g., scaling factors).

Example 12.8

Consider the problem of maximizing the function

$$J(\theta) = \theta^2 \qquad (12.45)$$

where θ is permitted to vary between 0 and 31.

To use a GA, we must first code the decision variables of our problem as some finite length string. For this problem, we will code the variable θ simply as a binary unsigned integer of length 5. With a 5-bit unsigned integer, we can obtain numbers between 0 (00000) and 31 (11111). The fitness function is simply defined as the function $J(\theta)$.

To start off, we select an initial population at random. We select a population of size 4. Table 12.9 gives the selected initial population, decoded θ values, and the fitness function values $J(\theta)$. As an illustration of the calculations done, let's take a look at the third string of the initial population, string 01000. Decoding this string gives $\theta = 8$, and the fitness $J(\theta) = 64$. Other θ and $J(\theta)$ values are obtained similarly.

The mating pool of the next generation may be selected by spinning a roulette wheel. Alternatively, the roulette-wheel technique may be implemented using a computer algorithm:

1. Sum the fitness of all the population members, and call this result the total fitness ΣJ.
2. Generate r, a random number between 0 and total fitness.
3. Return the first population member whose fitness, added to the fitness of the preceding population members (running total) is greater than or equal to r.

We generate numbers randomly from the interval [0, 1170] (refer Table 12.9). For each number, we choose first chromosome for which the running total of fitness is greater than or equal to the random number. Four randomly generated numbers are 233, 9, 508, 967; string 1 and string 4 give one copy to the mating pool, string 2 gives two copies, and string 3 gives no copies.

With the above active pool of strings looking for mates, simple crossover proceeds in two steps: (1) strings are mated randomly, (2) mated-strings couples crossover. We take the crossover probability $p_c = 1$. Looking at Table 12.10, we find that random choice of mates has selected the second string in the mating pool to be mated with the first. With a crossing site of 4, the two strings 01101 and 11000 cross and yield two new strings 01100 and 11001. The remaining two strings in the mating pool are crossed at site 2; the resulting strings are given in Table 12.10.

Table 12.9 Selection process

String No.	Initial population	θ	$J(\theta)$	Running Total
1	01101	13	169	169
2	11000	24	576	745
3	01000	8	64	809
4	10011	19	361	1170
		ΣJ	1170	
		Av.J	293	
		Max.J	576	

Table 12.10 Crossover process

Mating pool	New Population	θ	$J(\theta)$
0110 | 1	01100	12	144
1100 | 0	11001	25	625
11 | 000	11011	27	729
10 | 011	10000	16	256
		ΣJ	1754
		Av.J	439
		Max.J	729

The last operator, mutation, is performed on a bit-by-bit basis. We assume that the probability of mutation in this test is 0.001. With 20 transferred bit positions, we should expect 20 × 0.001 = 0.02 bits to undergo mutation during a given generation. Simulation of this process indicates that no bits undergo mutation for this probability value. As a result, no bit positions are changed from 0 to 1 or vice versa during this generation.

Following selection, crossover and mutation, the new population is ready to be tested. To do this we simply decode the new strings created by the simple GA and calculate the fitness function values from the θ values thus decoded. The results are shown in Table 12.10. While drawing concrete conclusions from a single trial of a stochastic process is, at best, a risky business, we start to see how GAs combine high-performance notions to achieve better performance. Both the maximal and average performance have improved in the new population. The population average fitness has improved from 293 to 439 in one generation. The maximum fitness has increased from 576 to 729 during that same period.

12.6 REVIEW EXAMPLES

Review Example 12.1

We consider here the simplest fuzzy PI control scheme for a servo motor with the control model (refer Fig. 12.32a)

$$\frac{Y(s)}{U(s)} = G(s) = \frac{1}{s(s+3.6)} \qquad (12.46)$$

The objective of the fuzzy controller is to control angular position $y(t)$ of the servo motor to achieve a given set-point y_r within desired accuracy.

The discretized model for the plant (refer Chapter 3) is

$$G_{h0}G(z) = \frac{Y(z)}{U(z)} = \frac{0.0237z^{-1} + 0.0175z^{-2}}{1 - 1.407z^{-1} + 0.407z^{-2}}$$

$$y(k) = 1.407y(k-1) - 0.407y(k-2) + 0.0237u(k-1) + 0.0175u(k-2)$$

(12.47)

The proposed fuzzy controller (refer Fig. 12.32b) has two input variables:

$e(k)$ = error between the set-point and actual position of the shaft,

$v(k)$ = rate of change of error;

and one output variable:

$\Delta u(k)$ = incremental voltage signal to the driver circuit of the motor.

Universe of discourse for $e(k)$ = $\{-L_e, L_e\}$

Universe of discourse for $v(k)$ = $\{-L_v, L_v\}$

Universe of discourse for $\Delta u(k)$ = $\{-H_{\Delta u}, H_{\Delta u}\}$

Clockwise and counterclockwise rotations are defined as positive and negative, respectively.

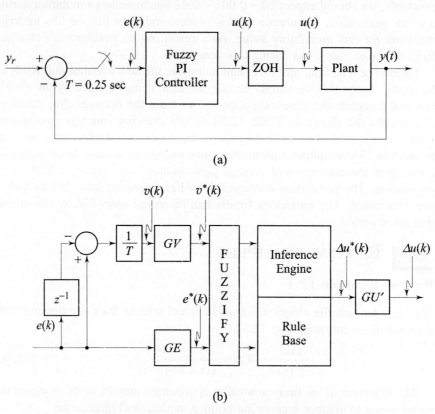

Fig. 12.32 A structure for fuzzy controller

The two input variables are quantized to two fuzzy subsets: Positive ($\underset{\sim}{P}$), Negative ($\underset{\sim}{N}$); and one output variable is quantized to three fuzzy subsets: Positive ($\underset{\sim}{P}$), Zero ($\underset{\sim}{Z}$), Negative ($\underset{\sim}{N}$). Triangular membership functions are used.

The scaling factors GE (gain for error variable), and GV (gain for velocity variable) describe input normalization:

$$e^*(k) = GE \times e(k) \; ; \; GE = L/L_e$$
$$v^*(k) = GV \times v(k) \; ; \; GV = L/L_v$$

where $\qquad e^*, v^* \in \{-L, L\}$

The output Δu^* of the fuzzy controller is denormalized to Δu by the relation

$$\Delta u(k) = GU' \times \Delta u^*(k) \; ; \; GU' = H_{\Delta u}/H$$

where $\qquad \Delta u^* \in \{-H, H\}$

Without loss of generality, we take $L = H = 1$ (refer Fig. 12.33).

The fuzzy PI controller uses the following four fuzzy rules:

IF $e^*(k)$ is $\underset{\sim}{P}$ **and** $v^*(k)$ is $\underset{\sim}{P}$ THEN $\Delta u^*(k)$ is $\underset{\sim}{P}$

IF $e^*(k)$ is $\underset{\sim}{P}$ **and** $v^*(k)$ is $\underset{\sim}{N}$ THEN $\Delta u^*(k)$ is $\underset{\sim}{Z}$

IF $e^*(k)$ is $\underset{\sim}{N}$ **and** $v^*(k)$ is $\underset{\sim}{P}$ THEN $\Delta u^*(k)$ is $\underset{\sim}{Z}$

IF $e^*(k)$ is $\underset{\sim}{N}$ **and** $v^*(k)$ is $\underset{\sim}{N}$ THEN $\Delta u^*(k)$ is $\underset{\sim}{N}$

The initial value of the system output and the initial velocity are set to zero, as is the initial output of the fuzzy PI controller.

The scaling factors GE, GV and GU' of the fuzzy controller may be tuned by trial and error. Refer Appendix C for realization of the controller.

Review Example 12.2

Although applied in many complex industrial processes, fuzzy logic–based expert systems experience a deficiency in knowledge acquisition, and rely to a great extent on empirical and heuristic knowledge which, in many cases, cannot be elicited objectively. Fuzziness describes event ambiguity. It measures the degree to which an event occurs, not whether it occurs. Fuzzy controller design involves the determination of the linguistic state space, definition of membership grades of each linguistic term, and the derivation of the control rules. The information on the above aspects can be gathered by interviewing process operators, process knowledge experts, and other sources of domain knowledge and theory.

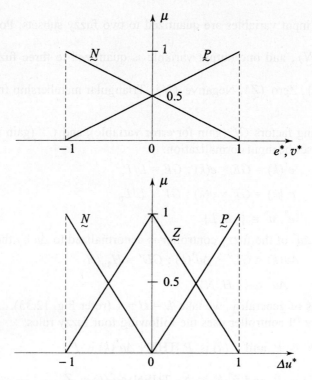

Fig. 12.33 Membership functions for input and output variables of the fuzzy controller

The choice of a FLC depends more on the intuition of the designer, and its effectiveness depends on the following parameters:
1. Selection of rule set.
2. Number, shape and size of the membership functions of the input and output variables.
3. Value of the normalizing factors to the FLC for the input variables.
4. Value of the denormalizing factors to the FLC for the output variables.

Genetic Algorithm (GA) has a capability to guide in poorly understood, irregular spaces. In the following, we illustrate the use of GA in designing a FLC for the thermal system described in Review Example 11.2. We design FLC by tuning only the normalizing and denormalizing factors.

The proposed fuzzy controller has two input variables (Refer Figs 12.32):

$e(k)$ – error between set-point and actual temperature of the tank,

$v(k)$ – rate of change of error,

and one output variable:

$\Delta u(k)$ – incremental heat input to the tank.

The universe of discourse for all the three variables may be taken as $[-1, 1]$. Proposed membership functions are shown in Fig. 12.34.

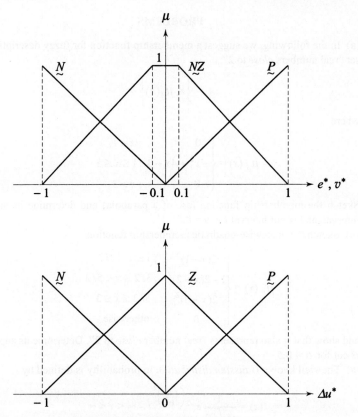

Fig. 12.34 Membership functions for the input and output variables of the fuzzy controller

The selected rules are

Rate of change of error→ Error ↓	N	NZ	P
N	N	N	Z
NZ	N	Z	P
P	Z	P	P

The initial value of the system output is Y_0. The initial velocity, and the initial output of the Fuzzy PI controller are set to zero.

The scaling factors GE, GV and GU' of the fuzzy controller may be tuned using genetic algorithm. Refer Appendix C for realization of the controller.

PROBLEMS

12.1 (a) In the following, we suggest a membership function for fuzzy description of the set "real numbers *close* to 2":

$$\underset{\sim}{A} = \left\{ x, \mu_{\underset{\sim}{A}}(x) \right\}$$

where

$$\mu_{\underset{\sim}{A}}(x) = \begin{cases} 0 & ; x < 1 \\ -x^2 + 4x - 3 & ; 1 \leq x \leq 3 \\ 0 & ; x > 3 \end{cases}$$

Sketch the membership function (arc of a parabola) and determine its supporting interval, and α-cut interval for $\alpha = 0.5$.

(b) Sketch the piecewise-quadratic membership function

$$\mu_{\underset{\sim}{B}}(x) = \begin{cases} 2(x-1)^2 & ; 1 \leq x < 3/2 \\ 1 - 2(x-2)^2 & ; 3/2 \leq x < 5/2 \\ 2(x-3)^2 & ; 5/2 \leq x \leq 3 \\ 0 & ; \text{otherwise} \end{cases}$$

and show that it also represents "real number *close* to 2". Determine its support, and α-cut for $\alpha = 0.5$.

12.2 (a) The well known *Gaussian distribution* in probability is defined by

$$f(x) = \frac{1}{\sigma\sqrt{2\pi}} e^{-\frac{1}{2}\left(\frac{x-\mu}{\sigma}\right)^2} ; -\infty < x < \infty$$

where μ is the mean and σ is the standard deviation of the distribution. Construct a normal, convex membership function from this distribution (select parameters μ and σ) that represents "real numbers *close* to 2". Find its support, and α-cut for $\alpha = 0.5$. Show that the membership function

$$\mu_{\underset{\sim}{A}}(x) = \frac{1}{1 + (x-2)^2}$$

also represents "real numbers *close* to 2". Find its support, and α-cut for $\alpha = 0.5$.

12.3 Consider the piecewise-quadratic function

$$f(x) = \begin{cases} 0 & ; x < a \\ 2\left(\dfrac{x-a}{b-a}\right)^2 & ; a \leq x < \dfrac{a+b}{2} \\ 1 - 2\left(\dfrac{x-b}{b-a}\right)^2 & ; \dfrac{a+b}{2} \leq x < b \\ 1 & ; b \leq x < c \end{cases}$$

Construct a normal, convex membership function from $f(x)$ (select parameters a, b and c) that represents the set "tall men" on the universe $\{3, 9\}$. Determine the crosspoints and support of the membership function.

12.4 (a) Write an analytical expression for the membership function $\mu_{\underset{\sim}{A}}(x)$ with supporting interval $[-1, 9]$ and α-cut interval for $\alpha = 1$ given as $[4, 5]$.

(b) Define what we mean by a normal membership function and a convex membership function. Is the function described in (a) above (i) normal, (ii) convex?

12.5 (a) Let the fuzzy set A be the linguitic 'warm' with membership function

$$\mu_A(x) = \begin{cases} 0 & ; x < a_1 \\ \dfrac{x - a_1}{b_1 - a_1} & ; a_1 \le x \le b_1 \\ 1 & ; b_1 \le x \le b_2 \\ \dfrac{x - a_2}{b_2 - a_2} & ; b_2 \le x \le a_2 \\ 0 & ; x \ge a_2 \end{cases}$$

$a_1 = 64°F$, $b_1 = 70°F$, $b_2 = 74°F$, $a_2 = 78°F$

(i) Is A a normal fuzzy set?

(ii) Is A a convex fuzzy set?

(iii) Is A a singleton fuzzy set?

If answer to one or more to these is 'no', then give an example of such a set.

(b) For fuzzy set A described in part (a), assume that $b_1 = b_2 = 72°F$.

Sketch the resulting membership function and determine its support, crosspoints and α-cuts for $\alpha = 0.2$ and 0.4.

12.6 Consider two fuzzy sets A and B; membership functions $\mu_A(x)$ and $\mu_B(x)$ are shown in Fig. P12.6. The fuzzy variable x is temperature.

Sketch the graph of $\mu_{\bar{A}}(x)$, $\mu_{A \cap B}(x)$ and $\mu_{A \cup B}(x)$.

Which t-norm and t-conorm have you used?

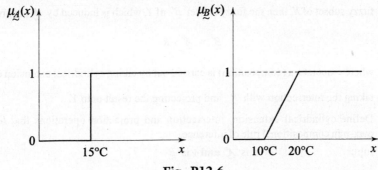

Fig. P12.6

12.7 Consider the fuzzy relation R on the universe $X \times Y$, given by the membership function

$$\mu_R(x, y) = \dfrac{1}{[1 + 100(x - 3y)^4]},$$

vaguely representing the crisp relation $x = 3y$. All elements satisfying $x = 3y$ have unity grade of membership; elements satisfying, for example, $x = 3.1y$ have membership grades less than 1. The farther away the elements are from the straight line, the lower are the membership grades.

Give a graphical representation of the fuzzy relation $\underset{\sim}{R}$.

12.8 Assume the membership function of the fuzzy set $\underset{\sim}{A}$, *big pressure*, is

$$\mu_{\underset{\sim}{A}}(x) = \begin{cases} 1 & ; x \geq 5 \\ 1 - \dfrac{5-x}{4} & ; 1 \leq x \leq 5 \\ 0 & ; \text{otherwise} \end{cases}$$

Assume the membership function of the fuzzy set $\underset{\sim}{B}$, *small volume*, is

$$\mu_{\underset{\sim}{B}}(y) = \begin{cases} 1 & ; y \leq 1 \\ 1 - \dfrac{y-1}{4} & ; 1 \leq y \leq 5 \\ 0 & ; \text{otherwise} \end{cases}$$

Find the truth values of the following propositions:
(i) 4 is big pressure,
(ii) 3 is small volume,
(iii) 4 is big pressure **and** 3 is small volume,
(iv) 4 is big pressure → 3 is small volume.

Explain the conjunction and implication operations you have used for this purpose.

12.9 Consider the following statements:

Input   : $\underset{\sim}{A}'$ is very small

Rule   :   IF $\underset{\sim}{A}'$ is small THEN $\underset{\sim}{B}'$ is large

Inference   : $\underset{\sim}{B}'$ is very large

If $\underset{\sim}{R}$ is a fuzzy relation from X to Y representing the implication rule, and $\underset{\sim}{A}'$ is a fuzzy subset of X, then the fuzzy subset $\underset{\sim}{B}'$ of Y, which is induced by $\underset{\sim}{A}'$, is given by

$$\underset{\sim}{B}' = \underset{\sim}{A}' \circ \underset{\sim}{R}$$

where ∘ operation (composition) is carried out by taking cylindrical extension of $\underset{\sim}{A}'$, taking the intersection with $\underset{\sim}{R}$, and projecting the result onto Y.

Define cylindrical extension, intersection and projection operations that lead to max-min compositional rule of inference.

12.10 Input   : x is $\underset{\sim}{A}'$ **and** y is $\underset{\sim}{B}'$

Rule 1   :   IF x is $\underset{\sim}{A}_1$ **and** y is $\underset{\sim}{B}_1$ THEN z is $\underset{\sim}{C}_1$

Rule 2   :   IF x is $\underset{\sim}{A}_2$ **and** y is $\underset{\sim}{B}_2$ THEN z is $\underset{\sim}{C}_2$

Inference   : z is $\underset{\sim}{C}'$

Taking arbitrary membership functions for $\underset{\sim}{A}_1, \underset{\sim}{B}_1, \underset{\sim}{C}_1, \underset{\sim}{A}_2, \underset{\sim}{B}_2$ and $\underset{\sim}{C}_2$, outline the procedure of determining $\underset{\sim}{C}'$ corresponding to the crisp inputs $x = x_0$ and $y = y_0$. Use

t-norm 'min' for conjunction operation, Mamdani's implication operation and max-min compositional rule of inference.

12.11 Fig. P12.11 shows the fuzzy output of a certain control problem. Defuzzify by using the centre of area method to obtain the value of crisp control action.

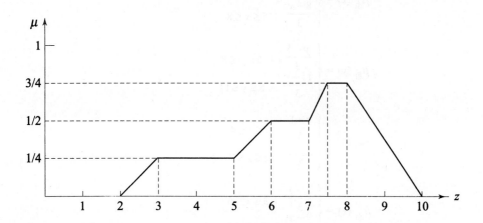

Fig. P12.11

12.12 Consider the fuzzy system concerning the terminal voltage and speed of an electric motor, described by the membership functions

x	100	150	200	250	300
$\mu_A(x)$	1	0.8	0.5	0.2	0.1
y	1600	1800	2000	2200	2400
$\mu_B(y)$	1	0.9	0.7	0.3	0

Input   : Voltage is rather small (x is A')

Rule   : IF voltage is small (x is A) THEN speed is small (y is B)

Inference   : Speed is rather small (y is B')

Assume that the input fuzzy set A' is a singleton at $x_0 = 125$. Determine the inference fuzzy set B' of the fuzzy system. Defuzzify this set to obtain crisp value for speed. Use piecewise continuous approximations of graphs of $\mu_A(x)$ and $\mu_B(y)$ to describe your solution.

12.13 Consider the two-input, one output fuzzy system:

Input   : x is A' and y is B'

Rule 1   : IF x is A_1 and y is B_1 THEN z is C_1

Rule 2   : IF x is A_2 and y is B_2 THEN z is C_2

Inference   : z is C'

The fuzzy sets A_i, B_i and C_i; $i = 1, 2$, have the membership functions

$$\mu_{A_1}(x) = \begin{cases} \dfrac{x-2}{3} & ; 2 \leq x \leq 5 \\ \dfrac{8-x}{3} & ; 5 \leq x \leq 8 \end{cases}$$

$$\mu_{B_1}(y) = \begin{cases} \dfrac{y-5}{3} & ; 5 \leq y \leq 8 \\ \dfrac{11-y}{3} & ; 8 \leq y \leq 11 \end{cases}$$

$$\mu_{C_1}(z) = \begin{cases} \dfrac{z-1}{3} & ; 1 \leq z \leq 4 \\ \dfrac{7-z}{3} & ; 4 \leq z \leq 7 \end{cases}$$

$$\mu_{A_2}(x) = \begin{cases} \dfrac{x-3}{3} & ; 3 \leq x \leq 6 \\ \dfrac{9-x}{3} & ; 6 \leq x \leq 9 \end{cases}$$

$$\mu_{B_2}(y) = \begin{cases} \dfrac{y-4}{3} & ; 4 \leq y \leq 7 \\ \dfrac{10-y}{3} & ; 7 \leq y \leq 10 \end{cases}$$

$$\mu_{C_2}(z) = \begin{cases} \dfrac{z-3}{3} & ; 3 \leq z \leq 6 \\ \dfrac{9-z}{3} & ; 6 \leq z \leq 9 \end{cases}$$

Assume fuzzy sets A' and B' are singletons at $x_0 = 4$ and $y_0 = 8$. Determine the inference fuzzy set C' of the fuzzy system. Defuzzify C'.

12.14 The control objective is to design an automatic braking system for motor cars. We need two analog signals: vehicle speed (V), and a measure of distance (D) from the vehicle in the front. A fuzzy logic control system will process these, giving a single output, braking force (B), which controls the brakes.
Term set for each of the variables (V, D, and B) is of the form:
{PS (positive small), PM (positive medium), PL (positive large)}
Membership functions for each term-set are given in Fig. P12.14.
Suppose that for the control problem, two rules have to be fired:
Rule 1:   IF $D = PS$ **and** $V = PM$ THEN $B = PL$
Rule 2:   IF $D = PM$ **and** $V = PL$ THEN $B = PM$

For the sensor readings of $V = 55$ km/hr, and $D = 27$ m from the car in front, find graphically
 (i) the firing strengths of the two rules,
 (ii) the aggregated output, and
 (iii) defuzzified control action.

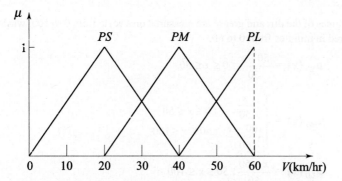

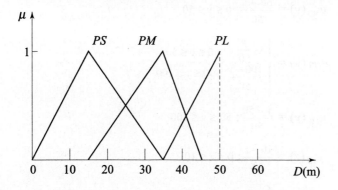

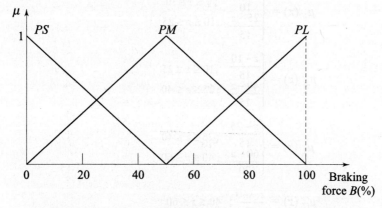

Fig. P12.14

12.15 The control objective is to automate the *wash time* when using a washing machine. Experts select for inputs *dirt* and *grease* of the clothes to be washed, and for output parameter the *wash time*, as follows:

$$Dirt \triangleq \{ SD \text{ (small dirt)}, MD \text{ (medium dirt)}, LD \text{ (large dirt)} \}$$

$$Grease \triangleq \{ NG \text{ (no grease)}, MG \text{ (medium grease)}, LG \text{ (large grease)} \}$$

$$Washtime \triangleq \{ VS \text{ (very short)}, S \text{ (short)}, M \text{ (medium)}, L \text{ (long)}, VL \text{ (very long)} \}$$

The degrees of the dirt and grease are measured on a scale from 0 to 100; washtime is measured in minutes from 0 to 60.

$$\mu_{\underset{\sim}{SD}}(x) = \frac{50-x}{50}; \; 0 \leq x \leq 50$$

$$\mu_{\underset{\sim}{MD}}(x) = \begin{cases} \dfrac{x}{50} & ; 0 \leq x \leq 50 \\ \dfrac{100-x}{50} & ; 50 \leq x \leq 100 \end{cases}$$

$$\mu_{\underset{\sim}{LD}}(x) = \frac{x-50}{50}; \; 50 \leq x \leq 100$$

$$\mu_{\underset{\sim}{NG}}(y) = \frac{50-y}{50}; \; 0 \leq y \leq 50$$

$$\mu_{\underset{\sim}{MG}}(y) = \begin{cases} \dfrac{y}{50} & ; 0 \leq y \leq 50 \\ \dfrac{100-y}{50} & ; 50 \leq y \leq 100 \end{cases}$$

$$\mu_{\underset{\sim}{LG}}(y) = \frac{y-50}{50}; \; 50 \leq y \leq 100$$

$$\mu_{\underset{\sim}{VS}}(z) = \frac{10-z}{10}; \; 0 \leq z \leq 10$$

$$\mu_{\underset{\sim}{S}}(z) = \begin{cases} \dfrac{z}{10} & ; 0 \leq z \leq 10 \\ \dfrac{25-z}{15} & ; 10 \leq z \leq 25 \end{cases}$$

$$\mu_{\underset{\sim}{M}}(z) = \begin{cases} \dfrac{z-10}{15} & ; 10 \leq z \leq 25 \\ \dfrac{40-z}{15} & ; 25 \leq z \leq 40 \end{cases}$$

$$\mu_{\underset{\sim}{L}}(z) = \begin{cases} \dfrac{z-25}{15} & ; 25 \leq z \leq 40 \\ \dfrac{60-z}{20} & ; 40 \leq z \leq 60 \end{cases}$$

$$\mu_{\underset{\sim}{VL}}(z) = \frac{z-40}{20}; \; 40 \leq z \leq 60$$

The selected rules are:

Grease→ Dirt ↓	NG	MG	LG
SD	VS	M	L
MD	S	M	L
LD	M	L	VL

Find a crisp control output for the following sensor readings:
$$Dirt = 60; Grease = 70$$

12.16 The objective is to minimize the function:

$$f(x_1, x_2) = (x_1^2 + x_2 - 11)^2 + (x_2^2 + x_1 - 7)^2$$

in the interval $0 \leq x_1, x_2 \leq 6$. The true solution to the problem is $[3, 2]^T$ having a function value equal to zero.

Take up this problem to explain the steps involved in GA: maximizing the function

$$F(x_1, x_2) = \frac{1.0}{1.0 + f(x_1, x_2)}; 0 \leq x_1, x_2 \leq 6.$$

Step 1: Take 10 bits to code each variable. With 10 bits, what is the solution accuracy in the interval (0, 6)?

Step 2: Take population size equal to total string length, i.e., 20. Create a random population of strings.

Step 3: Consider the first string of the initial random population. Decode the two substrings and determine the corresponding parameter values. What is the fitness function value corresponding to this string? Similarly for other strings, calculate the fitness values.

Step 4: Select good strings in the population to form the mating pool.

Step 5: Perform crossover on random pairs of strings (the crossover probability is 0.8).

Step 6: Perform bitwise mutation with probability 0.05 for every bit.

The resulting population is the new population. This completes one iteration of GA and the generation count is incremented by 1.

Appendix A

Mathematical Background

A.1 INTRODUCTION

In an attempt to make the book reasonably self-contained, we present in this appendix the background material that is needed for the analysis and design of control systems. Our treatment will be fairly rapid, as we shall assume that the reader has been exposed to some of the mathematical notions we need. Also, our treatment will be incomplete in the sense that mathematical rigour will be missing. The appendix should serve as a working reference.

A.2 FOURIER SERIES AND FOURIER TRANSFORMS

Consider a sinusoidal signal

$$y(t) = A\cos(\omega_0 t + \phi)$$
$$= \frac{A}{2}e^{j\phi}e^{j\omega_0 t} + \frac{A}{2}e^{-j\phi}e^{-j\omega_0 t} \quad (A.1)$$

with Fig. A.1a being its phasor diagram. In the phasor description, this signal consists of two well defined frequency components: one at $+\omega_0$ and the other at $-\omega_0$, corresponding to two directions of rotation. Each frequency component is completely specified by two parameters, the amplitude and the phase. Since these parameters are independent of time, they can be represented in the frequency domain as shown in Fig. A.1b, where at the frequency $\omega = +\omega_0$ we associate an amplitude $A/2$ and a phase $+\phi$, and at $\omega = -\omega_0$, we associate an amplitude $A/2$ and a phase angle $-\phi$. Figure A.1b, called a *line spectrum*, completely represents the time function of Eqn. (A.1) since forming and summing the indicated phasors yields $y(t)$. This frequency domain representation has *negative frequencies* because there are two rotational directions and it takes a pair of complex-conjugate phasors to give a real sum.

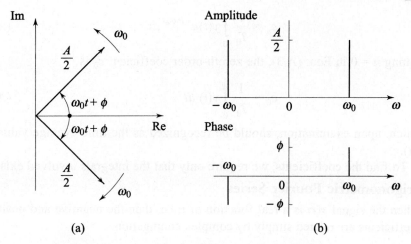

Fig. A.1 (a) Phasor diagram (b) Line spectrum

Exponential Fourier Series

Let $y(t)$ be a non-sinusoidal periodic signal with fundamental period T_0 (Fig. A.2). It can be expanded as a linear combination of phasors through the exponential Fourier series:

$$y(t) = \sum_{n=-\infty}^{\infty} c_n e^{jn\omega_0 t} \tag{A.2a}$$

where ω_0 is the fundamental angular frequency:

$$\omega_0 = \frac{2\pi}{T_0} \tag{A.2b}$$

and the coefficients c_n are given by

$$c_n = \frac{1}{T_0} \int_{T_0} y(t) e^{-jn\omega_0 t} \, dt \tag{A.2c}$$

The symbol $\int_{T_0}$ stands for integration over any period $t_1 \leq t \leq t_1 + T_0$ with t_1 being arbitrary. Letting $t_1 = 0$, we obtain

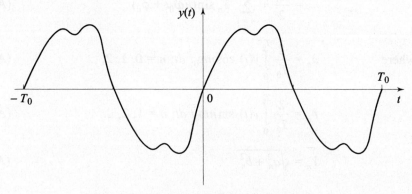

Fig. A.2 A non-sinusoidal periodic signal

$$c_n = \frac{1}{T_0} \int_0^{T_0} y(t) e^{-jn\omega_0 t} \, dt \tag{A.3}$$

Setting $n = 0$ in Eqn. (A.3), the zeroth-order coefficient c_0 is

$$c_0 = \frac{1}{T_0} \int_0^{T_0} y(t) \, dt \tag{A.4}$$

which, upon examination, should be recognized as the time-average value of $y(t)$.

To find the coefficients, we require only that the integrals involved exist.

Trigonometric Fourier Series

When the signal $y(t)$ is a real function of time, then the negative and positive coefficients are related simply by complex conjugation:

$$c_{-n} = c_n^*$$

The exponential Fourier series can then be converted to a trigonometric form by regrouping pairwise all but the zeroth term in Eqn. (A.2a) so that the summation index is always positive:

$$y(t) = c_0 + \sum_{n=1}^{\infty} (c_n e^{jn\omega_0 t} + c_{-n} e^{-jn\omega_0 t})$$

Equation (A.3) can be manipulated to obtain

$$c_n = \left[\frac{1}{T_0} \int_0^{T_0} y(t) \cos n\omega_0 t \, dt \right] + j \left[-\frac{1}{T_0} \int_0^{T_0} y(t) \sin n\omega_0 t \, dt \right] \tag{A.5}$$

Manipulation of Eqns (A.3) and (A.5) gives the trigonometric Fourier series in the following form:

$$y(t) = \frac{a_0}{2} + \sum_{n=1}^{\infty} [a_n \cos n\omega_0 t + b_n \sin n\omega_0 t] \tag{A.6a}$$

$$= \frac{a_0}{2} + \sum_{n=1}^{\infty} Y_n \sin(n\omega_0 t + \phi_n) \tag{A.6b}$$

where

$$a_n = \frac{2}{T_0} \int_0^{T_0} y(t) \cos n\omega_0 t \, dt; \quad n = 0, 1, 2, \ldots \tag{A.6c}$$

$$b_n = \frac{2}{T_0} \int_0^{T_0} y(t) \sin n\omega_0 t \, dt; \quad n = 1, 2, \ldots \tag{A.6d}$$

$$Y_n = \sqrt{a_n^2 + b_n^2} \tag{A.6e}$$

$$\phi_n = \tan^{-1}\left(\frac{a_n}{b_n}\right) \tag{A.6f}$$

In Eqn. (A.6b), the term for $n = 1$ is called *fundamental* or *first-harmonic* and always has the same frequency as the repetition rate of the original periodic waveform, whereas $n = 2, 3, \ldots$, give second, third, and so forth harmonic frequencies as integer multiples of the fundamental frequency.

Introducing a change of variable, $\psi = \omega_0 t$, we obtain the following alternative equations for the coefficients of Fourier series:

$$a_n = \frac{1}{\pi}\int_0^{2\pi} y(t)\cos n\omega_0 t\, d(\omega_0 t); \quad n = 0, 1, 2, \ldots \tag{A.7a}$$

$$b_n = \frac{1}{\pi}\int_0^{2\pi} y(t)\sin n\omega_0 t\, d(\omega_0 t); \quad n = 1, 2, \ldots \tag{A.7b}$$

Certain simplifications are possible when $y(t)$ has a symmetry of one type or another.

(i) Even symmetry: $y(t) = y(-t)$ results in
$$b_n = 0; \quad n = 1, 2, \ldots \tag{A.7c}$$

(ii) Odd symmetry: $y(t) = -y(-t)$ results in
$$a_n = 0; \quad n = 0, 1, 2, \ldots \tag{A.7d}$$

(iii) Odd half-wave symmetry: $y(t \pm T_0/2) = -y(t)$ results in
$$a_n = b_n = 0; \quad n = 0, 2, 4, \ldots \tag{A.7e}$$

Fourier Transforms

We have seen that a periodic time function $y(t)$ could be expanded as the exponential Fourier series given by Eqns (A.2). A nonperiodic signal, not having an identifiable fundamental frequency, cannot be expressed in quite the same way. There is, however, a representation similar to (A.2) for the nonperiodic case, given by

$$y(t) = \frac{1}{2\pi}\int_{-\infty}^{\infty} Y(j\omega)e^{j\omega t}\, d\omega \tag{A.8a}$$

where
$$Y(j\omega) = \int_{-\infty}^{\infty} y(t)e^{-j\omega t}\, dt \tag{A.8b}$$

This representation is valid for any time function for which the integrals exist.

Integration over all time, given by Eqn. (A.8b), is referred to as the *Fourier transform* of $y(t)$, and is frequently denoted symbolically as $\mathscr{F}[y(t)]$.

Conversion back to the time domain by means of integration over all frequency, given by Eqn. (A.8a) is referred to as the *inverse Fourier transform* and is often denoted symbolically as $\mathscr{F}^{-1}[Y(j\omega)]$. Taken together, these two integrals are referred to as the *Fourier transform pair*:

$$\mathscr{F}[y(t)] = Y(j\omega) \triangleq \int_{-\infty}^{\infty} y(t)e^{-j\omega t}\, dt \qquad (A.9a)$$

$$\mathscr{F}^{-1}[Y(j\omega)] = y(t) \triangleq \frac{1}{2\pi}\int_{-\infty}^{\infty} Y(j\omega)e^{j\omega t}\, d\omega \qquad (A.9b)$$

Comparing Eqns (A.2a) and (A.9b) shows that they are of essentially the same form; one being a discrete summation over the index $n\omega_0$, the other an integration or 'continuous summation' over the variable ω. Previously, we interpreted the Fourier series as decomposing the periodic signal into a sum of harmonically related components at the discrete frequencies $n\omega_0$. This leads to frequency-domain representation of the periodic signal $y(t)$ by a line spectrum which can be found directly from the coefficients c_n given by Eqn. (A.2c). By parallel reasoning, the Fourier transform $Y(j\omega)$ given by Eqn. (A.9a) is the continuous spectrum of the nonperiodic signal $y(t)$.

A.3 LAPLACE TRANSFORMS

The Fourier transform is a technique of expressing a signal $y(t)$ as a continuous sum of exponential functions of the form $e^{j\omega t}$. The continuous summation is given by the integral in Eqn. (A.9a). This representation is valid for any signal for which the integral exists.

Unfortunately, there are many signals of interest that arise in linear system analysis problems for which Fourier transforms do not exist. To extend the transform technique to such signals, the signal is expressed as a continuous sum of exponentials of the type e^{st} with $s = \sigma + j\omega$. As is seen below, this introduces a convergence factor $e^{-\sigma t}$. This factor, in effect, makes the product $\{y(t)e^{-\sigma t}\}$ Fourier transformable even if the Fourier transform of $y(t)$ itself does not exist, thereby expanding the class of transformable functions.

The technique of expressing a signal as a continuous sum of exponentials e^{st} is known as the *Laplace transform* technique. The Laplace transform of $y(t)$, denoted symbolically as $\mathscr{L}[y(t)]$, is defined as follows:

$$\mathscr{L}[y(t)] = Y(s) \triangleq \int_{-\infty}^{\infty} y(t)e^{-st}\, dt \qquad (A.10)$$

By writing Eqn. (A.10) as

$$Y(s) = \int_{-\infty}^{\infty} [y(t)e^{-\sigma t}]\, e^{-j\omega t}\, dt$$

and comparing it with Eqn. (A.9a), we can see a strong similarity to the Fourier transform; the signal $y(t)$ possesses a Laplace transform if $\{y(t)e^{-\sigma t}\}$ possesses a Fourier transform.

Restricting our consideration to functions that are zero for $t < 0$, the transform integral in Eqn. (A.10) reduces to

$$Y(s) = \mathscr{L}[y(t)] = \int_0^\infty y(t)e^{-st}\,dt \qquad (A.11)$$

The integral in Eqn. (A.11) does not converge for all functions, and when it does, it does so for restricted values of s in the s-plane. Therefore, for each function for which Laplace transform exists, there is an associated *convergence region* in the complex s-plane.

Consider the exponential function

$$y(t) = 0 \quad \text{for } t < 0$$
$$= e^{-at} \quad \text{for } t \geq 0$$

where a is a constant.

The Laplace transform of this exponential can be obtained as follows:

$$Y(s) = \int_0^\infty e^{-at}e^{-st}\,dt = \frac{-1}{s+a}e^{-(s+a)t}\Big|_0^\infty = \frac{1}{s+a}\,; \text{Re}(s) > -a$$

The ability to recover a signal uniquely from its Laplace transform rests upon there being at least one value of s for which the infinite integral converges. For the exponential function, the integral converges for any complex value of s for which

$$\text{Re}(s) = \sigma > -a$$

There are similar regions of convergence on the s-plane for the Laplace transforms of other time functions.

The operation that changes $Y(s)$ back to $y(t)$ is referred to as the *inverse Laplace transformation* and is symbolized by $\mathscr{L}^{-1}[Y(s)]$. We observe from the inverse Fourier transform integral (A.9b), that

$$y(t)e^{-\sigma t} = \mathscr{F}^{-1}[Y(\sigma+j\omega)] = \frac{1}{2\pi}\int_{-\infty}^\infty Y(\sigma+j\omega)e^{j\omega t}\,d\omega$$

Multiplying both sides by $e^{\sigma t}$ and assuming σ constant, we obtain the inverse Laplace transform:

$$y(t) = \mathscr{L}^{-1}[Y(s)] = \frac{1}{2\pi j}\int_{\sigma-j\infty}^{\sigma+j\infty} Y(s)e^{st}\,ds \qquad (A.12)$$

where the change of variable: $s = \sigma + j\omega$ and $ds = j\,d\omega$, has been used.

The inverse Laplace transform operation involves closed contour integration on the complex plane within the region of convergence of $Y(s)$. We may make use of the residue theorem of the theory of complex variables to evaluate the integral.

Since comprehensive tables of *Laplace transform pairs* have been published, it is rarely necessary for a transform user to actually work out integrals in Eqn. (A.11) and Eqn. (A.12). A short table given in Section A.5 is adequate for the purposes of this text.

It may be noted that in the transform table of Section A.5, regions of convergence have not been specified. In our applications of system analysis which involve tranformation from time-domain to s-domain and inverse transformation, the variable s acts as a dummy operator; if transform pairs for functions of interest to us are available, we are not concerned with the regions of convergence.[1]

Transform Inversion by Partial Fractions

The inversion of Laplace transforms plays an important role in the study of control systems, primarily because of the emphasis on transient response. We frequently encounter transforms that are rational functions, i.e., the ratio of two polynomials in s. For simple rational functions, inverse Laplace transform is directly given by transform tables. For functions not listed in tranform tables, an approach based on partial fraction expansion is widely used for inverse transformation. This approach is based on the *linearity property* of the Laplace transforms:

$$\mathscr{L}^{-1}[a_1 Y_1(s) + a_2 Y_2(s)] = a_1 \mathscr{L}^{-1}[Y_1(s)] + a_2 \mathscr{L}^{-1}[Y_2(s)]$$

where a_1 and a_2 are constants.

The given rational function $Y(s)$ is expanded into partial fractions such that inverse Laplace transform of each fraction is available directly from the transform tables. The algebraic sum of the inverse Laplace transforms of the partial fractions yields the inverse Laplace transform of the given rational function.

Partial fraction expansion: In the analysis of linear time-invariant systems, we very often encounter rational fractions of the form:

$$F(s) = \frac{b_0 s^m + b_1 s^{m-1} + \cdots + b_{m-1} s + b_m}{s^n + a_1 s^{n-1} + \cdots + a_{n-1} s + a_n} = \frac{N(s)}{\Delta(s)} \quad (A.13)$$

where the coefficients a_i and b_j are real constants, and m and n are integers with $m < n$.

A fraction of the form (A.13) can be expanded into partial fractions. To do this, first of all we factorize the polynomial $\Delta(s)$ into n first-order factors. The roots of $\Delta(s)$, which are the poles of $F(s)$, can be real, complex, distinct or repeated. Various cases are discussed below.

---

1. Unlike *one-sided* Laplace transform (defined by Eqn. (A.11)) the *two-sided* Laplace transform (defined by Eqn. (A.10)) is not generally unique. That is, two different time functions can have the same two-sided Laplace transform, each with a different region of convergence. It is then necessary to additionally specify the region of convergence in order to specify the time function. We will not use two-sided transforms in this book.

Case I: $\Delta(s)$ *has distinct roots*

For this case, Eqn. (A.13) can be written as

$$F(s) = \frac{N(s)}{\Delta(s)} = \frac{N(s)}{(s+p_1)(s+p_2)\cdots(s+p_k)\cdots(s+p_n)} \qquad (A.14)$$

which when expanded, gives

$$F(s) = \frac{A_1}{s+p_1} + \frac{A_2}{s+p_2} + \cdots + \frac{A_k}{s+p_k} + \cdots + \frac{A_n}{s+p_n} \qquad (A.15)$$

The coefficient A_k is called the *residue* of $F(s)$ at the pole $s = -p_k$.

Multiplying both sides of Eqn. (A.15) by $(s+p_k)$ and letting $s = -p_k$, gives

$$A_k = [(s+p_k)F(s)]\bigg|_{s=-p_k} \qquad (A.16)$$

Case II: $\Delta(s)$ *has complex conjugate roots*

Suppose that there is a pair of complex conjugate roots in $\Delta(s)$ given by

$$s = -a - j\omega, \text{ and } s = -a + j\omega$$

Then $F(s)$ may be written as

$$F(s) = \frac{N(s)}{\Delta(s)} = \frac{N(s)}{(s+a+j\omega)(s+a-j\omega)(s+p_3)\cdots(s+p_n)} \qquad (A.17)$$

which, when expanded, gives

$$F(s) = \frac{A_1}{s+a+j\omega} + \frac{A_1^*}{s+a-j\omega} + \frac{A_3}{s+p_3} + \cdots + \frac{A_n}{s+p_n} \qquad (A.18)$$

where A_1 an A_1^* are residues of $F(s)$ at the poles $s = -a - j\omega$ and $s = -a + j\omega$ respectively. These residues form a complex conjugate pair.

As per Eqn. (A.16), the residue A_1 is given by

$$A_1 = [(s+a+j\omega)F(s)]\bigg|_{s=-(a+j\omega)} \qquad (A.19)$$

A_1^* is obtained by taking complex conjugate of A_1.

The residues $A_3, \ldots, A_n$ of $F(s)$ at the simple real poles $s = -p_3, \ldots, -p_n$, respectively, can be evaluated using the relation (A.16).

Case III: $\Delta(s)$ *has repeated roots*

Assume that the root p_1 of $\Delta(s)$ is of multiplicity r and all other roots are distinct. The function $F(s)$ can then be written as

$$F(s) = \frac{N(s)}{\Delta(s)} = \frac{N(s)}{(s+p_1)^r(s+p_{r+1})\cdots(s+p_n)} \qquad (A.20)$$

which, when expanded, gives

$$F(s) = \frac{A_{1(r)}}{(s+p_1)^r} + \frac{A_{1(r-1)}}{(s+p_1)^{r-1}} + \cdots + \frac{A_{12}}{(s+p_1)^2}$$

$$+ \frac{A_{11}}{s+p_1} + \frac{A_{r+1}}{s+p_{r+1}} + \cdots + \frac{A_n}{s+p_n} \qquad (A.21)$$

$A_{r+1}, \ldots, A_n$ are the residues of $F(s)$ at the simple poles $s = -p_{r+1}, \ldots, -p_n$, respectively, and can be evaluated using the relation (A.16).

Multiplying both sides of Eqn. (A.21) by $(s+p_1)^r$, we obtain

$$(s+p_1)^r F(s) = A_{1(r)} + (s+p_1)A_{1(r-1)} + \cdots + (s+p_1)^{r-2}A_{12}$$

$$+ (s+p_1)^{r-1}A_{11} + \cdots \qquad (A.22)$$

Letting $s = -p_1$ in Eqn. (A.22) gives

$$A_{1(r)} = [(s+p_1)^r F(s)]\Big|_{s=-p_1} \qquad (A.23)$$

Taking the derivative of both the sides in Eqn. (A.22) and letting $s = -p_1$, we obtain

$$A_{1(r-1)} = \frac{d}{ds}[(s+p_1)^r F(s)]\Big|_{s=-p_1}$$

Following relations easily follow from Eqn. (A.22).

$$A_{1(r-2)} = \frac{1}{2!}\frac{d^2}{ds^2}[(s+p_1)^r F(s)]\Big|_{s=-p_1} \qquad (A.24)$$

$$\vdots$$

$$A_{11} = \frac{1}{(r-1)!}\frac{d^{r-1}}{ds^{r-1}}[(s+p_1)^r F(s)]\Big|_{s=-p_1}$$

The general term is

$$A_{1(r-i)} = \frac{1}{i!}\frac{d^i}{ds^i}[(s+p_1)^r F(s)]\Big|_{s=-p_1} \qquad (A.25)$$

$$i = 0, 1, 2, \ldots, r-1$$

A_{11} is the residue of $F(s)$ at the multiple-order pole $s = -p_1$.

Some Laplace Transform Properties

In addition to Laplace transform pairs, some Laplace transform properties are also used in control system analysis. Properties of interest to us are given below:

Multiplication by t

From the basic definition (A.11),

$$\frac{dY(s)}{ds} = \int_0^\infty \frac{d}{ds}(y(t)e^{-st})\,dt = \int_0^\infty (-t)y(t)e^{-st}\,dt$$

Consequently,

$$\mathscr{L}[ty(t)] = -\frac{dY(s)}{ds} \qquad (A.26)$$

Transforms of derivatives

From the basic definition (A.11),

$$\mathscr{L}\left[\frac{dy(t)}{dt}\right] = \int_0^\infty \frac{dy(t)}{dt} e^{-st}\,dt$$

Integrating by parts, we get

$$\mathscr{L}\left[\frac{dy(t)}{dt}\right] = y(t)e^{-st}\bigg|_0^\infty + s\int_0^\infty y(t)e^{-st}\,dt$$

We assume that the function $y(t)$ is transformable; the existence of $Y(s)$ guarantees that $y(t)e^{-st}\bigg|_{t\to\infty} = 0$. Hence

$$\mathscr{L}\left[\frac{dy(t)}{dt}\right] = sY(s) - y(0)$$

Use of this result in a recursive manner results in the family of transform formulas:

$$\mathscr{L}\left[\frac{d^2 y(t)}{dt^2}\right] = s^2 Y(s) - sy(0) - \frac{dy}{dt}(0)$$

$$\mathscr{L}\left[\frac{d^n y(t)}{dt^n}\right] = s^n Y(s) - s^{n-1} y(0) - s^{n-2}\frac{dy}{dt}(0) - \cdots$$

$$- s\frac{d^{n-2} y}{dt^{n-2}}(0) - \frac{d^{n-1} y}{dt^{n-1}}(0) \qquad (A.27)$$

Transform of an integral

By definition,

$$\mathscr{L}\left[\int_{-\infty}^t y(\tau)\,d\tau\right] = \int_0^\infty \left[\int_{-\infty}^t y(\tau)\,d\tau\right] e^{-st}\,dt$$

Integrating by parts, we get

$$\mathscr{L}\left[\int_{-\infty}^{t} y(\tau)d\tau\right] = \frac{e^{-st}}{-s}\left[\int_{-\infty}^{t} y(\tau)d\tau\right]\bigg|_{0}^{\infty} + \frac{1}{s}\int_{0}^{\infty} y(t)e^{-st}\, dt$$

$$= \frac{Y(s)}{s} + \frac{\int_{-\infty}^{0} y(\tau)d\tau}{s} \tag{A.28}$$

Complex frequency shift
The Laplace transform of

$$y(t) = x(t)e^{-\alpha t} \tag{A.29a}$$

is

$$Y(s) = X(s + \alpha) \tag{A.29b}$$

where $X(s) = \mathscr{L}[x(t)]$, and α is a real constant.

The proof easily follows by substituting Eqn. (A.29a) into the definition of the Laplace transform and noting that the integral is the Laplace transform of $x(t)$ with the variable $s + \alpha$:

$$\mathscr{L}[x(t)e^{-\alpha t}] = \int_{0}^{\infty} e^{-\alpha t} x(t)e^{-st}\, dt = \int_{0}^{\infty} x(t)e^{-(s+\alpha)t}\, dt$$

$$= X(s + \alpha)$$

Time-delay
If the Laplace transform of $y(t)\mu(t)$ is $Y(s)$, then

$$\mathscr{L}[y(t-t_0)\mu(t-t_0)] = e^{-st_0}Y(s); \; t_0 > 0 \tag{A.30}$$

where $\mu(t)$ is a unit-step function.

The proof follows easily by using the definition of Laplace transform. Since $\mu(t - t_0) = 0$ for $t < t_0$, we obtain

$$\mathscr{L}[y(t-t_0)\mu(t-t_0)] = \int_{t_0}^{\infty} y(t-t_0)e^{-st}\, dt$$

Letting $\tau = t - t_0$ in the integrand, we get

$$\mathscr{L}[y(t-t_0)\mu(t-t_0)] = \int_{0}^{\infty} y(\tau)e^{-s(\tau+t_0)}\, d\tau = e^{-st_0}Y(s)$$

Final value theorem
In the analysis of dynamic systems, one often desires to know the value that a variable approaches as $t \to \infty$, assuming, of course, that it does approach a limit. Using partial fraction expansion, it is a simple matter to show that $y(t)$ approaches a limit as $t \to \infty$ if all the poles of $Y(s)$ lie in the left half s-plane with

the possible exception of a simple pole at the origin. A more compact way of phrasing this condition is to say that $sY(s)$ must be *analytic* on the imaginary axis and in the right-half s-plane. The final-value theorem states that when this condition on $sY(s)$ is satisfied, then

$$\lim_{t \to \infty} y(t) = \lim_{s \to 0} [sY(s)] \qquad (A.31)$$

The proof is as follows:

$$\mathscr{L}\left[\frac{dy}{dt}\right] = \int_0^\infty \frac{dy}{dt} e^{-st} \, dt \text{ or } sY(s) - y(0) = \int_0^\infty \frac{dy}{dt} e^{-st} \, dt$$

Letting $s \to 0$ on both sides, we have

$$\lim_{s \to 0} [sY(s)] = y(0) + \int_0^\infty \frac{dy}{dt} dt = y(0) + y(t)\Big|_0^\infty = y(\infty)$$

It is noted that when $y(t)$ is sinusoidal function $\sin \omega t$, $sY(s)$ has poles on $\pm j\omega$ and $\lim_{t \to \infty} y(t)$ does not exist. Therefore, the final value theorem is not valid for such a function. We must make sure that all the conditions for the final value theorem are satisfied before applying it to a given problem.

A.4 z-TRANSFORMS

The Fourier transform of a continuous-time signal $y(t)$ is a technique of expressing the signal as a continuous sum of exponential functions $e^{-j\omega t}$ (refer Eqn. (A.8b)). On the similar lines, the Fourier transform of a sequence $y(k)$ is a technique of expressing the signal as a discrete sum of exponential functions $e^{-j\Omega k}$:

$$\mathscr{F}[y(k)] = Y(e^{j\Omega}) = \sum_{k=-\infty}^{\infty} y(k) e^{-j\Omega k} \qquad (A.32)$$

$Y(e^{j\Omega})$ is a continuous function[2] of Ω—the frequency of the discrete-time exponential. Since k is a dimensionless integer, the dimension of Ω is radians.

The fact that k is always an integer leads to periodic nature of $Y(e^{j\Omega})$: it is periodic function of the continuous-valued variable Ω with period 2π. This peoperty follows directly from Eqn. (A.32) since

$$e^{j(\Omega + 2\pi)k} = e^{j\Omega k}$$

Equations (A.2) give the Fourier series expansion of $y(t)$—a periodic function of continuous-valued variable t with period T_0. These equations can be rearranged in the following form:

$$y(t) = \sum_{n=-\infty}^{\infty} c_n e^{-j\omega_n t}$$

---

2. The advantage of using the notation $Y(e^{j\Omega})$, and not $Y(j\Omega)$, will become obvious in the sequel.

where $\omega_n = \dfrac{2\pi n}{T_0} = n\omega_0$ is the nth harmonic frequency, and the coefficients c_n are given by

$$c_n = \dfrac{1}{T_0} \int_{-T_0/2}^{T_0/2} y(t) e^{j\omega_n t} \, dt$$

In exactly the same way, $Y(e^{j\Omega})$ can be expressed as the following Fourier series expansion:

$$Y(e^{j\Omega}) = \sum_{k=-\infty}^{\infty} c_k e^{-j\theta_k \Omega} \qquad (A.33a)$$

where $\theta_k = \dfrac{2\pi k}{2\pi} = k$ can be thought of as harmonic frequency, and

$$c_k = \dfrac{1}{2\pi} \int_{-\pi}^{\pi} Y(e^{i\Omega}) e^{j\Omega k} \, d\Omega \qquad (A.33b)$$

Note the similarity of the definition of Fourier transform in Eqn. (A.32) and the Fourier series expansion of $Y(e^{j\Omega})$ given by Eqn. (A.33a). This similarity allows us to define inverse Fourier transform

$$y(k) = \dfrac{1}{2\pi} \int_{-\pi}^{\pi} Y(e^{j\Omega}) e^{j\Omega k} \, d\Omega$$

For discrete-time signals, the Fourier transform pair is therefore given by the following equations:

$$\mathscr{F}[y(k)] = Y(e^{j\Omega}) = \sum_{k=-\infty}^{\infty} y(k) e^{-j\Omega k} \qquad (A.34a)$$

$$\mathscr{F}^{-1}[Y(e^{j\Omega})] = y(k) = \dfrac{1}{2\pi} \int_{-\pi}^{\pi} Y(e^{j\Omega}) e^{j\Omega k} \, d\Omega \qquad (A.34b)$$

The representation in frequency-domain given by Eqn. (A.34a) is valid for any signal $y(k)$ for which the summation converges.

Unfortunately, there are many signals of interest that arise in the study of discrete-time systems for which Fourier transforms do not exist. To extend the transform technique to such signals, the signal is expressed as a discrete sum of phasors of the type z^{-k}, where z is a complex variable. Let us express the complex variable z in polar form as

$$z = r e^{j\Omega} \qquad (A.35a)$$

with r as the magnitude of z and Ω as the angle of z. As is seen below, this introduces a convergence factor r^{-k}; this factor, in effect, makes the product $[y(k) r^{-k}]$ Fourier transformable even if the Fourier transform of $y(k)$ does not exist, thereby expanding the class of transformable functions.

The technique of expressing a signal as a discrete sum of phasors z^{-k} is known as the *z-transform technique*. The z-transform of $y(k)$, denoted symbolically as $\mathscr{Z}[y(k)]$, is defined as follows:

$$\mathscr{Z}[y(k)] = Y(z) \triangleq \sum_{k=-\infty}^{\infty} y(k) z^{-k} \qquad (A.35b)$$

By writing Eqn. (A.35b) as

$$Y(z) = \sum_{k=-\infty}^{\infty} (y(k) r^{-k}) e^{-j\Omega k}$$

and comparing it with Eqn. (A.34a), we can see a strong similarity to the Fourier transform; the signal $y(k)$ possesses a z-transform if $[y(k) r^{-k}]$ possesses a Fouries transform.

Restricting our consideration to functions that are zero for $k < 0$, the transform summation in Eqn. (A.35b) reduces to

$$Y(z) = \mathscr{Z}[y(k)] = \sum_{k=0}^{\infty} y(k) z^{-k} \qquad (A.36)$$

The summation in Eqn. (A.36) does not converge for all functions; and when it does, it does so for restricted values of z in the z-plane. Therefore, for each sequence for which z-transform exists, there is an associated *convergence region* in the complex z-plane. Consider the unit-sample sequence

$$\delta(k) = 1 \text{ for } k = 0$$
$$= 0 \text{ otherwise}$$

The z-transform of this elementary signal is

$$\mathscr{Z}[\delta(k)] = \sum_{k=0}^{\infty} \delta(k) z^{-k} = z^0 = 1; \ |z| > 0$$

The unit-step sequence

$$\mu(k) = 1 \text{ for } k \geq 0$$
$$= 0 \text{ otherwise}$$

has the z-transform

$$\mathscr{Z}[\mu(k)] = \sum_{k=0}^{\infty} \mu(k) z^{-k} = \sum_{k=0}^{\infty} z^{-k}$$

Using the geometric series formula

$$\sum_{k=0}^{\infty} x^k = \frac{1}{1-x}; \ |x| < 1,$$

this becomes

$$\mathscr{Z}[\mu(k)] = \frac{1}{1 - z^{-1}} = \frac{z}{z-1}; \ |z^{-1}| < 1$$

The ability to recover a sequence uniquely from its z-transform rests upon there being at least one value of z for which the infinite series converges. For unit-step sequence, the infinite series converges for any complex value of z with $|z| > 1$.

A geometric sequence
$$a^k; k \geq 0$$
has the z-transform
$$\mathscr{Z}[a^k] = \sum_{k=0}^{\infty} a^k z^{-k} = \sum_{k=0}^{\infty} (az^{-1})^k = \frac{1}{1-az^{-1}}; |az^{-1}| < 1$$

The transform converges for any complex value of z for which $|z| > |a|$.
A sampled exponential function
$$y(k) = e^{-\alpha kT} = (e^{-\alpha T})^k$$
is of the form a^k where the constant a is chosen to be $a = e^{\alpha T}$.

The transform of sampled sinusoids can be found by expanding the sequence into complex exponential components. For the sampled cosine,

$$\mathscr{Z}[\cos \Omega k] = \frac{1}{2}\mathscr{Z}[e^{j\Omega k}] + \frac{1}{2}\mathscr{Z}[e^{-j\Omega k}] = \frac{1}{2}\frac{z}{z-e^{j\Omega}} + \frac{1}{2}\frac{z}{z-e^{-j\Omega}}$$

$$= \frac{\frac{1}{2}z(z-e^{-j\Omega}) + \frac{1}{2}z(z-e^{j\Omega})}{(z-e^{j\Omega})(z-e^{-j\Omega})} = \frac{z^2 - \frac{1}{2}(e^{j\Omega}+e^{-j\Omega})z}{z^2 - (e^{j\Omega}+e^{-j\Omega})z + 1}$$

$$= \frac{z^2 - z\cos\Omega}{z^2 - 2z\cos\Omega + 1}$$

The operation that changes $Y(z)$ back to $y(k)$ is referred to as the *inverse z-transformation* and is symbolized by $\mathscr{Z}^{-1}[Y(z)]$. We observe from the inverse Fourier transform integral (A.34b) that

$$y(k)r^{-k} = \mathscr{F}^{-1}[Y(re^{j\Omega})] = \frac{1}{2\pi}\int_{-\pi}^{\pi} Y(re^{j\Omega})e^{j\Omega k}\, d\Omega$$

Multiplying both sides by r^k and assuming r constant, we obtain the inverse z-transform:

$$y(k) = \mathscr{Z}^{-1}[Y(z)] = \frac{1}{2\pi j}\oint Y(z)z^{k-1}\, dz \qquad (A.37)$$

where the change of variable: $z = re^{j\Omega}$ and $dz = jzd\Omega$, has been used.

The inverse z-transform operation involves closed contour integration on the complex plane within the region of convergence of $Y(z)$. We may make use of the residue theorem of the theory of complex variables to evaluate the integral.

Since comprehensive tables of *z-transform pairs* have been published, it is rarely necessary for a transform user to actually work out Eqns (A.36)–(A.37). A short table given in Section A.5 is adequate for the purposes of this text.

It may be noted that in the transform table of Section A.5, regions of convergence have not been specified. In our applications of system analysis which involve transformation from time-domain to z-domain and inverse transformation, the variable z acts as a dummy operator. If transform pairs for sequences of interest to us are available, we are not concerned with the region of convergence.[3]

Some z-Transform Properties

In addition to z-transform pairs, some z-transform properties are also used in the analysis of discrete-time systems. Properties of interest to us are given below.

Advance (forward shift)

z-transformation of difference equations written in terms of advanced versions of the input, output, and state variables requires the following results.

$$\mathscr{Z}[y(k+1)] = \sum_{k=0}^{\infty} y(k+1)z^{-k} = z \sum_{k=0}^{\infty} y(k+1)z^{-(k+1)}$$

Letting $k + 1 = m$, yields

$$\mathscr{Z}[y(k+1)] = z \sum_{m=1}^{\infty} y(m)z^{-m} = z \left[\sum_{m=0}^{\infty} y(m) z^{-m} - y(0) \right]$$

$$= zY(z) - zy(0)$$

Use of this result in a recursive manner results in the family of transform formulas:

$$\mathscr{Z}[y(k+2)] = z^2 Y(z) - z^2 y(0) - zy(1)$$

$$\vdots$$

$$\mathscr{Z}[y(k+n)] = z^n Y(z) - z^n y(0) - z^{n-1} y(1) - \ldots$$
$$- z^2 y(n-2) - zy(n-1) \quad (A.38)$$

Delay (backward shift)

The delay operation is of fundamental importance in the study of digital control systems. Here, we show that if the time sequence $y(k)$ is delayed by n sample periods, the effect in the z-domain is to multiply $Y(z)$ by z^{-n}.

$$\mathscr{Z}[y(k-n)] = \sum_{k=0}^{\infty} y(k-n) z^{-k}$$

Letting $m = k - n$, yields

---

[3]. Unlike *one-sided* z-transform (defined by Eqn. (A.36)), the *two-sided* z-transform (defined by Eqn. (A.35b)) is not generally unique. That is, two different sequences can have the same two-sided z-transform, each with a different region of convergence. It is then necessary to additionally specify the region of convergence in order to specify the sequence. We will not use two-sided transforms in this book.

$$\mathscr{Z}[y(k-n)] = \sum_{m=-n}^{\infty} y(m) z^{-(m+n)}$$

Since $y(m)$ is assumed zero for $m < 0$, we have

$$\mathscr{Z}[y(k-n)] = \sum_{m=0}^{\infty} y(m) z^{-(m+n)} = z^{-n} \left[\sum_{m=0}^{\infty} y(m) z^{-m} \right]$$

$$= z^{-n} Y(z); \; n \geq 0 \quad (A.39)$$

Summation

Analogous to the operation of integration, we can define the summation operation.

$$x(k) = \sum_{i=0}^{k} y(i) \quad (A.40)$$

In the course of deriving an expression for $X(z)$ in terms of $Y(z)$, we shall need the well-known infinite series sum:

$$\sum_{k=0}^{\infty} (az^{-1})^k = \frac{1}{1-az^{-1}} = \frac{z}{z-a} \quad (A.41)$$

which converges provided that $|az^{-1}| < 1$, or $|z| > a$.

Successive values of $x(k)$ are:

$$x(0) = y(0)$$

$$x(1) = y(0) + y(1)$$

$$\vdots$$

$$x(k) = y(0) + y(1) + \cdots + y(k)$$

Thus, $X(z)$ is the infinite sum given below.

$$X(z) = \sum_{k=0}^{\infty} x(k) z^{-k} = x(0) + z^{-1} x(1) + \cdots + z^{-k} x(k) + \cdots$$

$$= y(0) + z^{-1}[y(0) + y(1)] + \cdots + z^{-k}[y(0) + \cdots + y(k)] + \cdots$$

$$= y(0)[1 + z^{-1} + z^{-2} + \cdots] + y(1)[z^{-1} + z^{-2} + \cdots] + \cdots$$

$$+ y(k)[z^{-k} + z^{-k-1} + \cdots] + \cdots$$

$$= \left(\frac{z}{z-1} \right) [y(0) + z^{-1} y(1) + z^{-2} y(2) + \cdots]$$

$$= \left(\frac{z}{z-1} \right) \left[\sum_{k=0}^{\infty} y(k) z^{-k} \right]$$

Therefore,

$$X(z) = \frac{z}{z-1} Y(z) \qquad (A.42)$$

Complex frequency shift

The z-transform of

$$y(k) = a^k x(k) \qquad (A.43a)$$

is

$$Y(z) = X\left(\frac{z}{a}\right) \qquad (A.43b)$$

where $X(z) = \mathscr{Z}[x(k)]$ and a is a real constant.

The proof easily follows by substituting Eqn. (A.43a) into the definition of the z-transform and noting that the summation is the z-transform of $x(k)$ with the variable z/a:

$$\mathscr{Z}[a^k x(k)] + \sum_{k=0}^{\infty} a^k x(k) z^{-k} = \sum_{k=0}^{\infty} x(k)(z/a)^{-k} = X(z/a)$$

Multiplication by k

From the basic definition (A.36),

$$\frac{dY(z)}{dz} = \sum_{k=0}^{\infty} \frac{d}{dz}[y(k)z^{-k}] = \sum_{k=0}^{\infty} (-k) y(k) z^{-k-1}$$

Consequently

$$\mathscr{Z}[ky(k)] = -z \frac{dY(z)}{dz} \qquad (A.44)$$

z-transform inversion

The inversion of z-transform plays an important role in the study of digital control systems, primarily because of the emphasis on transient response. We frequently encounter transforms that are rational funtions, i.e., the ratio of two polynomials in z. For simple rational functions, inverse z-transform is directly given by transform tables.

For functions not listed in transform tables, an approach based on partial fraction expansion is widely used for inverse transformation. This approach is based on the *linearity property* of the z-transforms:

$$\mathscr{Z}^{-1}[a_1 Y_1(z) + a_2 Y_2(z)] = a_1 \mathscr{Z}^{-1}[Y_1(z)] + a_2 \mathscr{Z}^{-1}[Y_2(z)]$$

where a_1 and a_2 are constants.

The given rational function $Y(z)$ is expanded into partial fractions (refer Eqns (A.14)–(A.25)) such that inverse z-transform of each fraction is available directly from the transform tables. The algebraic sum of the inverse z-transforms of the partial fractions yields the inverse z-transform of the given rational function.

We observe that the transforms of the elementary functions (see the table of z-transforms in Section A.5) contain a factor of z in the numerator, e.g.,

$$\mathscr{Z}[\mu(k)] = \frac{z}{z-1}$$

where $\mu(k)$ is a unit-step sequence.

To ensure that the partial fraction expansion will yield terms corresponding to those tabulated, it is customary to first expand the function $Y(z)/z$ if $Y(z)$ has one or more zeros at the origin, and then multiply the resulting expansion by z.

For instance, if $Y(z)$ is given

$$Y(z) = \frac{2z^2 - 1.5z}{z^2 - 1.5z + 0.5} = \frac{2z^2 - 1.5z}{(z-0.5)(z-1)},$$

we are justified in writing

$$\frac{Y(z)}{z} = \frac{2z - 1.5}{(z-0.5)(z-1)} = \frac{A_1}{z-0.5} + \frac{A_2}{z-1}$$

Constants A_1 and A_2 can be evaluated by applying the conventional partial-fraction expansion rules. Applying rule (A.16), we obtain

$$\frac{Y(z)}{z} = \frac{1}{z-0.5} + \frac{1}{z-1} \text{ or } Y(z) = \frac{z}{z-0.5} + \frac{z}{z-1}$$

Using the transform pairs from Section A.5,

$$y(k) = (0.5)^k + 1; \; k \geq 0$$

▲▲

When $Y(z)$ does not have one or more zeros at the origin, we expand $Y(z)$, instead of $Y(z)/z$, into partial fractions and utilize shifting theorem given by Eqn. (A.39) to obtain inverse z-transform.

In applying the shifting theorem, notice that if

$$\mathscr{Z}^{-1}[Y(z)] = y(k)$$

then

$$\mathscr{Z}^{-1}[z^{-1}Y(z)] = y(k-1)$$

Let us consider an example:

$$Y(z) = \frac{10}{(z-1)(z-0.2)}$$

We first expand $Y(z)$ into partial fractions as follows:

$$Y(z) = \frac{12.5}{z-1} - \frac{12.5}{z-0.2}$$

Note that the inverse z-transform of $1/(z-1)$ is not available in the transform table of Section A.5. However, using the shifting theorem, we find that

$$\mathscr{Z}^{-1}\left[\frac{1}{z-1}\right] = \mathscr{Z}^{-1}\left[z^{-1}\left(\frac{z}{z-1}\right)\right] = \begin{cases} 1; & k = 1, 2, \ldots \\ 0; & k \leq 0 \end{cases}$$

Also

$$\mathcal{Z}^{-1}\left[\frac{1}{z-0.2}\right] = \mathcal{Z}^{-1}\left[z^{-1}\left(\frac{z}{z-0.2}\right)\right] = \begin{cases} (0.2)^{k-1}; & k=1,2,\ldots \\ 0 & ; k \leq 0 \end{cases}$$

Therefore

$$y(k) = \mathcal{Z}^{-1}[Y(z)] = \begin{cases} 12.5[1-(0.2)^{k-1}]; & k=1,2,\ldots \\ 0 & ; k \leq 0 \end{cases}$$

which can be rewritten as

$$y(k) = 12.5[1-(0.2)^{k-1}]\mu(k-1)$$

▲▲

When only the first few members in the sequence $\{y(0), y(1), \ldots\}$ are desired, rather than the function $y(k)$, it is often convenient to expand $Y(z)$ in the power series:

$$Y(z) = B_0 + B_1 z^{-1} + B_2 z^{-2} + \cdots$$

Referring to the basic definition of the z-transform given by Eqn. (A.36), we note that the coefficients of such a series must be the desired sequence values, i.e., $y(i) = B_i$. To obtain the first few members of the sequence, we can perform a long division of the numerator of $Y(z)$ by its denominator, where the divisor and dividend polynomials are arranged in descending power of z.

For

$$Y(z) = \frac{2z^2 - 1.5z}{z^2 - 1.5z + 0.5}$$

the long-division process is as follows:

$$\begin{array}{r}
2 + 1.5z^{-1} + 1.25z^{-2} + 1.125z^{-3} + \cdots \\
z^2 - 1.5z + 0.5 \overline{)\, 2z^2 - 1.5z } \\
\underline{2z^2 - 3z + 1} \\
1.5z - 1 \\
\underline{1.5z - 2.25 + 0.75z^{-1}} \\
1.25 - 0.75z^{-1} \\
\underline{1.25 - 1.875z^{-1} + 0.625z^{-2}} \\
1.125z^{-1} - 0.625z^{-2}
\end{array}$$

Thus, the first four values in the sequence are $\{2.0, 1.50, 1.25, 1.125\}$. Note that it is essential that the components of both the divisor and the dividend be placed in the order shown if the quotient is to have the required form. Other forms yield valid expansion of $Y(z)$ but do not have the numbers $\{y(k)\}$ as their coefficients and hence are of no interest to us.

Final value theorem

The final value theorem is concerned with the evaluation of $y(k)$ as $k \to \infty$ assuming, of course, that $y(k)$ does approach a limit. Using partial fraction expansion for inverting z-transforms, it is a simple matter to show that $y(k)$ approaches a limit as $k \to \infty$ if all the poles of $Y(z)$ lie inside the *unit circle* ($|z| < 1$) in the complex z-plane, with the possible exception of a simple pole on the boundary of the unit circle. A more compact way of phrasing this condition is to say that $(z-1)Y(z)$ must be *analytic* on the boundary and outside the unit circle in the complex z-plane. The final value theorem states that when this condition on $(z-1)Y(z)$ is satisfied, then

$$\lim_{k \to \infty} y(k) = \lim_{z \to 1} (z-1)Y(z) \qquad (A.45)$$

The proof is as follows:

$$\mathscr{Z}[y(k+1) - y(k)] = \lim_{m \to \infty} \sum_{k=0}^{m} [y(k+1) - y(k)]z^{-k}$$

or

$$zY(z) - zy(0) - Y(z) = \lim_{m \to \infty} \sum_{k=0}^{m} [y(k+1) - y(k)]z^{-k}$$

Letting $z \to 1$ on both sides,

$$\lim_{z \to 1} [(z-1)Y(z)] = y(0) + \lim_{z \to 1} \lim_{m \to \infty} \sum_{k=0}^{m} [y(k+1) - y(k)]z^{-k}$$

Interchanging the order of limits on the right-hand side, we have

$$\lim_{z \to 1} [(z-1)Y(z)] = y(0) + \lim_{m \to \infty} \sum_{k=0}^{m} [y(k+1) - y(k)] = y(\infty)$$

A.5   TABLE OF TRANSFORMS

For convenient reference, this section gives an abbreviated table of transforms. The table lists Laplace transforms of commonly encountered functions $f(t)$; $t \geq 0$. Also listed in the table are z-transforms of sampled version of the functions $f(t)$, given by $f(kT)$ where k is an integer and T is the sampling period.

If the indexing of the discrete-time signal is referred solely by k, its z-transform can easily be obtained from the table by letting $T = 1$. For example, $f(k) = d^k$ represents a signal without reference to an absolute sampling time. Its transform is $z/(z-d)$, which is obtained from the entry in the table corresponding to e^{-akT} with $T = 1$ and $d = e^{-a}$. For convenience, this alternative form is also given in the table for some discrete-time signals.

Appendix A: Mathematical Background

$F(s)$	$f(t); t \geq 0$	$f(kT)/f(k)$ $k = 0, 1, 2, \ldots$	$F(z)$
1	Unit impulse $\delta(t)$	—	1
$\dfrac{1}{s}$	Unit step $\mu(t)$	—	$\dfrac{z}{z-1}$
$\dfrac{1}{s^2}$	t	kT	$\dfrac{Tz}{(z-1)^2}$
$\dfrac{2}{s^3}$	t^2	$(kT)^2$	$\dfrac{T^2 z(z+1)}{(z-1)^3}$
$\dfrac{1}{s+a}$	e^{-at}	e^{-akT}	$\dfrac{z}{z-e^{-aT}}$
$\dfrac{1}{(s+a)^2}$	te^{-at}	$kT e^{-akT}$	$\dfrac{Te^{-aT} z}{(z-e^{-aT})^2}$
$\dfrac{a}{s(s+a)}$	$1-e^{-at}$	$1-e^{-akT}$	$\dfrac{(1-e^{-aT})z}{(z-1)(z-e^{-aT})}$
$\dfrac{b-a}{(s+a)(s+b)}$	$e^{-at}-e^{-bt}$	$e^{-akT}-e^{-bkT}$	$\dfrac{z}{z-e^{-aT}} - \dfrac{z}{z-e^{-bT}}$
$\dfrac{a^2}{s^2(s+a)}$	$at-1+e^{-at}$	$akT-1+e^{-akT}$	$\dfrac{aTz}{(z-1)^2} - \dfrac{(1-e^{-aT})z}{(z-1)(z-e^{-aT})}$
$\dfrac{a(s+b)}{s^2(s+a)}$	$\dfrac{a-b}{a}+bt -\left(\dfrac{a-b}{a}\right)e^{-at}$	$\dfrac{a-b}{a}+bkT -\left(\dfrac{a-b}{a}\right)e^{-akT}$	$\dfrac{bTz}{(z-1)^2} + \dfrac{(a-b)(1-e^{-aT})z}{a(z-1)(z-e^{-aT})}$
$\dfrac{ab}{s(s+a)(s+b)}$	$1+\dfrac{b}{a-b}e^{-at} -\dfrac{a}{a-b}e^{-bt}$	$1+\dfrac{b}{a-b}e^{-akT} -\dfrac{a}{a-b}e^{-bkT}$	$\dfrac{z}{z-1} + \dfrac{bz}{(a-b)(z-e^{-aT})} - \dfrac{az}{(a-b)(z-e^{-bT})}$
$\dfrac{a}{s^2+a^2}$	$\sin at$	$\sin akT$	$\dfrac{(\sin aT)z}{z^2 - (2\cos aT)z + 1}$
$\dfrac{s}{s^2+a^2}$	$\cos at$	$\cos akT$	$\dfrac{z^2 - (\cos aT)z}{z^2 - (2\cos aT)z + 1}$
$\dfrac{b}{(s+a)^2+b^2}$	$e^{-at}\sin bt$	$e^{-akT}\sin bkT$	$\dfrac{(e^{-aT}\sin bT)z}{z^2 - (2e^{-aT}\cos bT)z + e^{-2aT}}$
$\dfrac{s+a}{(s+a)^2+b^2}$	$e^{-at}\cos bt$	$e^{-akT}\cos bkT$	$\dfrac{z^2 - (e^{-aT}\cos bT)z}{z^2 - (2e^{-aT}\cos bT)z + e^{-2aT}}$
—	—	Unit sample $\delta(k)$	1

(contd)

$F(s)$	$f(t); t \geq 0$	$f(kT)/f(k)$ $k = 0, 1, 2, \ldots$	$F(z)$
	—	Unit step $\mu(k)$	$\dfrac{z}{z-1}$
—	—	a^k	$\dfrac{z}{z-a}$
—	—	k	$\dfrac{z}{(z-1)^2}$
—	—	k^2	$\dfrac{z(z+1)}{(z-1)^3}$
—	—	$k a^k$	$\dfrac{az}{(z-a)^2}$
—	—	$a^k \sin \Omega k$	$\dfrac{(a \sin \Omega)z}{z^2 - (2a \cos \Omega)z + a^2}$
—	—	$a^k \cos \Omega k$	$\dfrac{z^2 - (a \cos \Omega)z}{z^2 - (2a \cos \Omega)z + a^2}$

Appendix B

MATLAB Aided Control System Design: Conventional

Control theory, at the introductory level, deals primarily with linear time-invariant systems. Few real systems are exactly linear over their whole operating range, and few systems have parameter values that are precisely constant forever. But many systems approximately satisfy these conditions over a sufficiently narrow operating range. Our focus here is on a design procedure which is applicable to linear time-invariant systems.

Control system design begins with a proposed plant or process whose satisfactory dynamic performance depends on feedback for stability, disturbance regulation, tracking accuracy or reduction of the effects of parameter variations. We give here an outline of the design process that is general enough to be useful whether the plant is an electronic amplifier or a large structure to be placed in earth orbit. Obviously, to be widely applicable, our outline has to be vague with respect to physical details and specific only with respect to feedback-control problems.

To present our outline, we divide the control design problem into a sequence of characteristic steps. This sequence of steps is an approximation of good design practice; there will certainly be variations in the sequence depending upon the problem in hand. In some cases, we may carry out the steps in a different order; in others, we may omit a step or add one. Also, the design methods covered in our sequence of steps are not the only methods available in control engineer's toolbox. There are other powerful methods available in the literature. Our selection in the outline is based on the design methods covered in the present text.

This appendix also provides guidance to the reader to integrate the learning of control system analysis and design material with the learning of how one computes the answers with MATLAB software package. The appendix is not meant to be a substitute for the MATLAB software manuals; rather, it will help guide the reader to the appropriate place in the manuals for specific calculations. The manuals are the best source of learning the package; our attempt here is only to expedite the process of making full use of the power of the package for control system design problems.

The MATLAB statements/responses given in this appendix are from MATLAB version 6 and Control System Toolbox version 5, on the Windows platform. Refer [180] for an introduction to the MATLAB environment.

A sequence of characteristic steps of MATLAB-aided control system design follows.

1. *Make a system model*

When a control engineer is given a control problem, often one of the first tasks that he undertakes is the development of the mathematical model of the process to be controlled, in order to gain a clear understanding of the problem. Basically, there are only a few ways to actually generate the model. We can use the first principles of physics to write down a model. Another way is to perform "system identification" via the use of real plant data to produce a model of the system. Sometimes, a combined approach is used where we use physics to write down a general differential equation we believe represents plant behaviour, and then we perform experiments on the plant to determine certain model parameters or functions.

For time-invariant systems, mathematical model building based on physical laws normally results in a set of differential equations. These equations, when rearranged as a set of first-order differential equations, result in a state-space model of the following form:

$$\dot{\mathbf{x}} = \mathbf{f}(\mathbf{x}, \mathbf{u})$$
$$\mathbf{y} = \mathbf{h}(\mathbf{x}, \mathbf{u})$$
(B.1)

where $\mathbf{f}(\cdot)$ and $\mathbf{h}(\cdot)$ are nonlinear functions of their arguments; $\mathbf{x}(t)$ is the $n \times 1$ internal state vector, $\mathbf{u}(t)$ is the $p \times 1$ control input vector, and $\mathbf{y}(t)$ is the $q \times 1$ measured output vector. Overdot represents differentiation with respect to time t. The first equation, called the state equation, captures the dynamical portion of the system and has memory inherent in the n integrators. The second equation, called the output or measurement equation, represents how we chose to measure the system variables; it depends on the type and availability of sensors.

Model-building and its validation will require *simulation* of the model. The nonlinear state-space equations of the form (B.1) are very easy to simulate on a digital computer. MATLAB provides many Runge-Kutta numerical integration routines for solving ordinary differential equations; the function **ode23** usually sufffices for our applications. Use of **ode23** function for solving ordinary differential equations of the type (B.1) will be demonstrated later.

2. Determine the required performance specifications

This step requires an understanding of the process: what it is intended to do, how much error is permissible, how to describe the class of command and disturbance signals to be expected, what the physical capabilities and limitations are, etc. A control system normally operates in one of the two fashions:
  (i) The system is designed to regulate the output at a fixed level prescribed by a constant input (called the set-point) in the face of uncertainties in the plant behaviour and environmental disturbances to the plant.
  (ii) The system is designed to drive the output to track a changing input in the face of uncertainties in the plant behaviour and environmental disturbances to the plant. A system will be required to follow in practice a family of tracking functions.

As the nature of the transient response of a control system is dependent upon the system poles only and not on the type of the input, it is sufficient to analyse the transient response to one of the standard test signals; a step is generally used for this purpose.

In specifying the steady-state response characteristics, it is common to specify the steady-state error of the system to one or more of the standard test signals—step, ramp and parabola. Theoretically, it is desirable for a control system to have the capability of responding with zero error to all the polynomial inputs of degree $k : r(t) = (t^k/k!); t \geq 0, k = 0, 1, 2, \ldots$. The higher the value of k, the more stringent is the steady-state requirement, making it difficult to satisfy other specifications on the performance of the system. It is therefore necessary that tracking commands which a control system is expected to be subjected to, be carefully examined and steady-state specifications be formulated for inputs with minimum possible value of k.

Typical results of the second step in the design process are specifications that the system have (when subjected to command/disturbance inputs)
  (i) a step response inside some constraint boundaries—specified by rise time, settling time, peak overshoot, etc.; and
  (ii) steady-state error to step/ramp/parabolic input within prescribed limits under the constraints imposed by physical limitations of the selected plant, actuator, and sensor.

Another standard test signal of great importance is the sinusoidal signal: a sinusoidal function with variable frequency is used as input function; input frequency is swept from zero to beyond the significant range of system characteristics. Curves in terms of amplitude ratio and phase between input and output as a function of frequency, display the frequency response of the system.

Requirements that a system have a step response inside some constraint boundaries—specified by rise time, settling time, peak overshoot, etc., can equivalently be represented as requirements that the system have a frequency response satisfying certain constraints—specified by gain margin, phase margin, bandwidth, etc.

3. *Make a linear model*

From the specifications, identify the equilibrium point of interest and construct a small-signal dynamic model. Validate the model with experimental data where possible. Regrettably, we have a tendency to view the process as a linear time-invariant model, capable of responding to inputs of arbitrary size, and we tend to overlook the fact that the linear model is a very limited representation of the system, valid only for small signals, short times, and particular environmental conditions. We should not confuse the approximation with reality. We must be able to use the simplified model for the intended purpose and to return to an accurate model or the actual physical system to really verify the design performance.

Linearization of equations of the form (B.1) leads to a model of the form given below:

$$\dot{\mathbf{x}} = \mathbf{Ax} + \mathbf{Bu}$$
$$\mathbf{y} = \mathbf{Cx} + \mathbf{Du} \tag{B.2}$$

The $n \times 1$ vector $\mathbf{x}$ is the state of the system, $\mathbf{A}$ is the constant $n \times n$ system matrix, $\mathbf{B}$ is the constant $n \times p$ input matrix, $\mathbf{C}$ is the constant $q \times n$ output matrix, and $\mathbf{D}$ is the constant $q \times p$ matrix.

While considering only single-input, single-output (SISO) problems, we take number of inputs, p, and the number of outputs, q, to be one. For SISO problems, the state variable model may be expressed as

$$\dot{\mathbf{x}} = \mathbf{Ax} + \mathbf{b}u$$
$$y = \mathbf{cx} + du \tag{B.3}$$

Note that matrices $\mathbf{B}$, $\mathbf{C}$, and $\mathbf{D}$ of the MIMO (multi-input, multi-output) representation now become vectors $\mathbf{b}$ and $\mathbf{c}$, and scalar d respectively; and the input vector $\mathbf{u}$ and output vector $\mathbf{y}$ of the MIMO representation now become scalar variables u and y respectively.

Validation of the linear model with experimental data or simulation data obtained from (B.1) will require simulation of Eqns (B.2)/(B.3). MATLAB provides many functions for simulation of linear models. We consider here some functions important from the control-engineering perspective.

For the state-space representation (B.2)/(B.3), the data for the model consists of four matrices. For convenience, the MATLAB provides customized data structure (LTI object). This is called the **SS** object. This object encapsulates the model data and enables you to manipulate the LTI system as a single entity, rather than as a collection of data vectors and matrices.

An LTI object of the type **SS** is created whenever you invoke the construction function **ss**.

$$\text{sys} = \text{ss}(\mathbf{A}, \mathbf{B}, \mathbf{C}, \mathbf{D}) \tag{B.4}$$

creates the state-space model (B.2).

When the four matrices of the model (B.3) are entered,

$$\text{sys} = \text{ss}(\mathbf{A}, \mathbf{b}, \mathbf{c}, d) \tag{B.5}$$

creates a state-space model for SISO systems.

For model **sys** in (B.5), **step(sys)** will generate a plot of unit-step response $y(t)$ (with zero initial conditions). The time vector is automatically selected when **t** is not explicitly included in the step command.

If you wish to supply the time vector **t** at which the response will be computed, the following command is used.

$$\textbf{step(sys,t)}$$

You can specify either a final time **t = Tfinal** or a vector of evenly spaced time samples of the form

$$\textbf{t = 0: dt : Tfinal}$$

When invoked with left-hand arguments such as

$$\textbf{[y,t] = step(sys)}$$

$$\textbf{[y,t,X] = step(sys)}$$

$$\textbf{y = step(sys,t)}$$

no plot is generated on the screen. Hence it is necessary to use a **plot** command to see the response curves. The vector **y** and matrix **X** contain the output and state response of the system respectively, evaluated at the computation points returned in the time vector **t** (**X** has as many columns as states and one row for each element in vector **t**).

Other time-response functions of interest to us are

        impulse(sys)   % impulse response

        initial(sys,x0) % free response to initial state vector x0

        Isim(sys,u,t) % response to input time history in vector u

        Isim(sys,u,t,x0) % having length (t) rows

For MIMO models (B.4), these functions produce an array of plots.

4. *Make a design model*

Complex processes and machines often have several variables (outputs) that we wish to control, and several manipulated inputs are available to provide this control. In many situations, one input affects primarily one output and has only a weak effect on other outputs; it becomes possible to ignore weak interactions (coupling) and design controllers under the assumption that one input affects only one output. Input-output pairing, to minimize the effect of interactions and application of SISO control schemes to obtain separate controllers for each input-output pair, results in an acceptable performance. This, in fact, amounts to considering the multivariable system as consisting of an appropriate number of separate SISO systems. Coupling effects are considered as disturbances to the separate control systems. We will limit our discussion to SISO design models.

Depending on the type of model you use, the data for your model may consist of a simple numerator/denominator pair for transfer functions or four matrices for state-space models. MATLAB provides LTI objects **TF** and **SS** for transfer functions and state-space models respectively. We have already discussed the creation of LTI object of the type **SS**.

An LTI object of the type **TF** is created whenever you invoke the construction function **tf**.

$$\text{sys} = \text{tf(num,den)} \tag{B.6}$$

num and **den** vectors specify $n(s)$ and $d(s)$ respectively, of the transfer function $G(s) = n(s)/d(s)$.

MATLAB has the means to perform model conversions. Given the **SS** model **sys_ss**, the syntax for conversion to **TF** model is

$$\text{sys\_tf} = \text{tf(sys\_ss)}$$

Common pole-zero factors of $G(s)$ must be cancelled before we can claim that we have the transfer function representation of the system. To assist us in pole-zero cancellation, MATLAB provides **minreal** function.

$$\text{sysr} = \text{minreal(sys\_tf)}$$

Given the **TF** model **sys_tf**, the syntax for conversion to **SS** model is

$$\text{sys\_ss} = \text{ss(sys\_tf)}$$

MATLAB provides many functions for simulation of transfer functions. For the transfer function model **sys = tf(num,den)**, **step(sys)** will generate a plot of unit-step response $y(t)$ from the transfer function

$$\frac{Y(s)}{U(s)} = G(s) = \frac{\text{num}}{\text{den}}$$

The time vector is automatically selected when **t** is not explicitly included in the step command. If you wish to supply the time vector **t** at which the response will be computed, the following command is used.

$$\text{step(sys,t)}$$

When invoked with left-hand arguments such as

$$\text{[y,t]} = \text{step(sys)}$$

$$y = \text{step(sys,t)}$$

no plot is generated on the screen. Hence, it is necessary to use a **plot** command to see the response curve. The vector **y** has one column and one row for each element in time vector **t**.

Other time-response functions of interest to us are

impulse(sys) % imulse response

lsim(sys,u,t) % response to input time history in

% vector u having length (t) rows.

Select sensor and actuator and construct a design model for the feedback system. Process transfer function models frequently have deadtime (input-output delay). **TF** object for transfer functions with deadtime can be created using the syntax

$$\text{sys} = \text{tf(num,den,'InputDelay', value)}$$

5. *Discretize the design model*

Various forms of devices have been used for mechanization of industrial automatic controllers. Electronic controllers (based on op amp circuits) and computer-based controllers are commonly used in industrual applications.

Opting for computer-based controllers leads to two design alternatives.

(i) *Design using emulation:* The design is done in the continuous-time domain totally ignoring the fact that a sampler and a digital computer will eventually be used. Having the continuous-time controller, we then convert the design to a digital control.

This method produces a good controller for the case when the sampling rate is about 30 times faster than the bandwidth, but produces a controller needing further refinement for the case when the sampling rate is about 6 times the bandwidth. For a sampling period of T sec, sample-and-hold introduce a time delay of approximately $T/2$ sec. A method that has been frequently used by practising engineers is based on approximation of sampled-data system by a continuous-time system that includes the transfer function $e^{-sT/2}$ in the forward path in cascade with the plant transfer function $G(s)$. An analog controller $D(s)$ is then designed for the plant $G(s)e^{-sT/2}$; this design is then converted to a digital control.

Carrying out the initial design using continuous-time methods is a good idea, independent of whether it will be used in a subsequent emulation step or merely as a guide for direct digital design. Knowing how the system could perform if implemented with analog hardware provides a target for how well the digital system should perform, and helps in selecting the sampling rate.

(ii) *Direct digital design:* The other alternative is to design the controller directly in discrete-time domain. At the outset, the plant model is transformed into a discrete-time system and the design iterations to achieve the desired system performance are carried out in discrete-time domain. By carrying out the digitization on the plant model, instead of on continuous-time controller as was done for the emulation design method, the approximate nature of the process can be eliminated. This is so because the actual plant must be preceded by a hold (usually a zero-order hold) and, therefore, has an exact discrete equivalent that includes the lagging effect of the hold. The impact of using an exact discrete equivalent of the plant is that a digital controller found using direct digital design methods may yield performance close to the desired specifications for slow sample rates.

There are various standard methods for discretization of continuous-time models. None of these methods is exact for all types of input because no sampled system has access to the input time history between samples. In essence, each approximation makes a different assumption about what the continuous input is doing between samples. MATLAB software has functions that allow use of various approximations. We consider here couple of functions important from control-engineering perspective.

State-space model of a discrete-time SISO system is of the form

$$\mathbf{x}(k+1) = \mathbf{Fx}(k) + \mathbf{g}u(k)$$

$$y(k) = \mathbf{cx}(k) + du(k)$$

Construction of the **SS** object for this discrete-time model requires four matrices **F, g, c** and d, and the sampling interval T.

$$\mathbf{sysd = ss(F,g,c,d,'T')}$$

Transfer function model of a discrete-time SISO system is of the form

$$G(z) = \frac{n(z)}{d(z)} = \frac{num}{den}$$

Construction of the **TF** object for this discrete-time model requires *num* and *den* polynomials in z, and the sampling interval T.

$$\mathbf{sysd = tf(num, den,'T')}$$

For a continuous-time model **sysc**, the command

$$\mathbf{sysd = c2d(sysc,T)} \ \% \ \mathbf{T \ is \ sampling \ period \ in \ seconds}$$

performs ZOH conversion by default.

sysc is continuous-time state-space system (B.3) and **sysd** is the discrete-time system

$$\mathbf{x}(k+1) = \mathbf{Fx}(k) + \mathbf{g}u(k)$$

$$y(k) = \mathbf{cx}(k) + du(k)$$

assuming a zero-order hold on the input—the control input is assumed piecewise constant over the sample time T.

$$\mathbf{[F,g,c,d] = ssdata(sysd)}$$

c2d can also be used with transfer function models. Continuous-time system **sysc** representing

$$G(s) = \frac{num}{den}$$

gets converted to the discrete-time system **sysd**, representing

$$G_{h0}G(z) = \frac{numz}{denz}$$

$$\mathbf{[numd,dend] = tfdata(sysd,'v')}$$

returns numerator and denominator of transfer function as row vectors.

To use alternative conversion schemes, specify the method

$$\text{sysd} = \text{c2d(sysc,T,'tustin')} \quad \% \text{ Use Tustin approximation}$$

The function **c2d(sysc,T,'tustin')** converts a continuous-time system to discrete-time system using trapezoidal rule for integration, also called the bilinear transformation. (The function **d2c** is an inverse operation—it converts discrete-time models to continuous-time form).

MATLAB provides many functions for simulation of discrete-time systems. **step(sys)** will generate a plot of unit-step response $y(k)$ of the discrete-time system (transfer function model or state-space model) **sys**. Zero initial state is assumed if **sys** represents a state-space model. The number of sample points is automatically determined when time is not explicitly included in the command.

If you wish to supply the sample points vector **k** at which the response will be computed, the following command is used.

$$\text{step(sys,t)}$$

You can specify **t** as a vector of sample points:

$$t = 0\text{:T:tfinal}$$

When invoked with left-hand arguments, no plot is generated on the screen; it is necessary to use a **plot** command to see the response curves.

Other time-response functions of interest to us are

impulse(sys)	% impulse response
initial(sys,x0)	% free response to initial state vector x0
lsim(sys,u,t)	% response to input time history in
lsim(sys,u,t,x0)	% vector u having length (t) rows

You can analyze the time response using the **LTI Viewer**, which is a graphical user interface for viewing and manipulating response plots of LTI models. For example

$$\text{ltiview('step',sys)}$$

will open a window displaying the step response of the LTI model **sys**. Once initialized, the **LTI Viewer** assists you with the analysis of the response by facilitating such functions as zooming into regions of the response plots, calculating response characteristics such as peak response, settling time, rise time, steady-state, toggling the grid on or off the plot, and many other useful features.

6. *Try a simple lag-lead design*

To form an initial estimate of the complexity of the design problem, sketch frequency response plot (Bode plot) and root-locus plot with respect to plant gain. Try to meet the specifications with a simple controller of lag-lead variety. Do not overlook feedforward of the disturbances if the necessary sensor information is available. Consider the effect of sensor noise, and compare a lead network in the forward path to minor-loop feedback structure having direct feedback from velocity sensor, to see which gives a better design.

For design by root-locus method, the design specifications are translated into desired dominant closed-loop poles. Other closed-loop poles are required to be located at a large distance from the $j\omega$-axis. It may be noted that pole-placement methods do not allow the designer to judge how close the system performance is to the best possible. Also, there is lack of visibility into low-frequency disturbance rejection. This can cause many problems: the disturbance rejection may not be optimized, and the plant-parameter variations may cause large closed-loop response variations. Stability margins on Nyquist/Bode plots give a better robustness measure. For this reason, though the specifications on the closed-loop performance are often formulated in time domain, it is worthwhile to convert them into frequency-domain specifications and then design the compensator with frequency-domain methods. Root-locus plots are very impressive analysis tools for systems that have been designed by frequency-domain methods. For example, these plots can be valuable for the analysis of the effects of certain parameter variations on stability.

Note that the above comments reflect the opinion of practising engineers. Teaching in universities relies heavily on both the methods; our textbook also follows the standard pattern.

The function **bode(sys)** generates the Bode frequency-response plot for LTI model **sys**. This function automatically selects the frequency values by placing more points in regions where the frequency response is changing quickly. This range is user-selectable utilizing the **logspace** function. When invoked with left-hand arguments,

$$[mag,phase,w] = \textbf{bode(sys)}$$

$$[mag,phase] = \textbf{bode(sys,w)}$$

return the magnitude and phase of the frequency response at the frequencies **w**.

The function **margin** determines gain margin, phase margin, gain crossover frequency and phase crossover frequency.

The function **bode** handles both continuous-time and discrete-time models. For continuous-time models, it computes the frequency response by evaluating the transfer function $G(s)$ on the imaginary axis $s = j\omega$. For discrete-time models, the frequency response is obtained by evaluating the transfer function $G_{h0}G(z)$ on the unit circle $z = e^{j\omega T}$; where T is the sample time.

$$\text{mag}(\omega) = |G_{h0}G(e^{j\omega T})|$$

$$\text{phase}(\omega) = \angle G_{h0}G(e^{j\omega T})$$

These magnitude and phase relationship make useless the hand-plotting procedures developed by Bode. Also the ease with which a designer can predict the effect of pole and zero changes on the frequency response is lost.

The w-transform approach was developed so that Bode-plots plotting and interpretation are almost as easy for discrete-time systems as they are for continuous-time ones. Therefore, in practice, frequency-response design of systems using a discrete-time model is often carried out using the w-transformation; however, the need to replace the z-plane with the w-plane is less obvious in today's environment where good software tools are universally available for the designer to perform the plotting. The essential idea of this method is to retain many design features from the continuous-time systems.

The discrete-time model of the plant $G_{h0}G(z) = num/den$ is transformed with bilinear mapping using the MATLAB function **d2c**. Construction of discrete-time model requires numerator polynomial *num* in z, denominator polynomial *den* in z and value of the sampling interval T.

$$\text{sysd} = \text{tf(num,den,'T')}$$

Conversion to w-domain is given by

$$\text{sysw} = \text{d2c(sysd,'tustin')}$$

The Nichols frequency-response plot can be generated using **nichols** function; the frequency range is user-selectable. A Nichols chart grid is drawn on the existing plot with the **ngird** function. These functions are applicable to both continuous-time and discrete-time systems.

You can analyze the frequency response using the **LTI Viewer**, which is a graphical user interface for viewing and manipulating response plots of LTI models. For example,

$$\text{ltiview('nichols',sys)}$$

will open a window displaying the Nichols plot of the LTI system **sys**. Once initialized, the **LTI Viewer** assists you with the analysis of the response by facilitating such functions as zooming into regions of response plot; calculating response characteristics such as resonance peak, resonance frequency, bandwidth, stability margins, and many other useful features.

MATLAB function **rlocus(sys)** calculates and plots the root locus of the open-loop SISO model **sys**. If **sys** has transfer function

$$G(s) = \frac{n(s)}{d(s)}$$

rlocus adaptively selects a set of positive gains **K** and produces a smooth plot of the roots of

$$d(s) + Kn(s) = 0$$

Alternatively, **rlocus(sys,K)** uses the user-specified vector **K** of gains to plot the root locus.

The function **rlocfind** returns the feedback gain associated with a particular set of poles on the root locus. **[K,poles] = rlocfind(sys)** is used for interactive gain selection. The function **rlocfind** puts up a crosshair cursor on the root locus plot that you use to select a particular pole location. The root locus gain associated with this point is returned in **K** and the column vector **poles** contains the closed-loop poles for this gain. To use this command, the root locus must be present in the current figure window.

The functions **rlocus** and **rlocfind** work with both the continuous-time and discrete-time SISO systems. The functions **sgrid/spchart** and **zgrid/zpchart** are used for ω_n and ζ grid on continuous-time and discrete-time root locus, respectively.

You can design a compensator using the SISO Design tool, which is a Graphical User Interface (GUI). To initialize this GUI, type

sisotool(sys)

7. Try state-space design

Choices in the design methods are a design variable in the whole process. Many a times, the choice of a design method depends upon our selection of the sensed quantities to be used for feedback. If we decide to sense and feedback only the output, then the trial-and-error lead-lag type design methods is a good choice. However, if these compensators do not give entirely satisfactory performance, we may attempt to install sensors for all the state variables and consider design methods requiring full state feedback.

The approach we may take at this point is *pole placement*; that is, having picked a state-feedback control law with enough parameters to influence all the closed-loop poles, we will arbitrarily select the desired pole locations of the closed-loop system and see if the approach gives satisfactory performance as specified in Step 2.

Of course, used without thought, the method of pole placement can also result in a design that requires unreasonable levels of control effort, or is very sensitive to changes in plant model. In fact, there is no trial-and-error involved in calculating the parameters of state-feedback control law for a given set of desired closed-loop poles. The trial-and-error in the pole-placement design process is in selecting the closed-loop pole locations that meet the performance specifications of Step 2 under the constraints imposed on plant, actuator and sensors.

The optimal design method can also be used to design a state-feedback control law. It is based on minimizing a cost function that consists of the weighted sum of squares of the state errors and control. The relative weightings between the state variables and control are varied by the designer in order to meet all the system specifications of Step 2.

We use a state estimator to implement state-feedback control law when all the states cannot be sensed using hardware sensors. The state estimator is a software sensor. We select the closed-loop estimator error poles to be about eight times faster than the control poles. The reason for this is to keep the error-poles from reducing the robustness of the design; a fast estimator will have almost the same effect on the response as no estimator at all.

If we opt for optimal design method for the estimator, the relative weightings in the cost function are properly selected to meet the requirements of fast estimation.

Controllability and observability of a system in state variable form can be checked using the MATLAB functions **ctrb** and **obsv**, respectively. The inputs to be **ctrb** function are the system matrix **A** and the input matrix **b**; the output of **ctrb** function is the controllability matrix **U**. Similarly, the inputs to the **obsv** function are the system matrix **A** and the output matrix **c**; the output is observability matrix **V**. The function **rank** gives the controllability and observability properties.

Pole-placement design may be carried out using the MATLAB function **acker**. However, the computation of the controllability matrix has very poor numerical accuracy and this carries over to Ackermann's formula. The function **acker** can be used for the design of SISO systems with a small (≤ 5) number of state variables. For more complex cases a more reliable formula is available, implemented in MATLAB with the function **place**. A modest limitation on **place** is that none of the desired closed-loop poles may be repeated, that is, the poles must be distinct; a requirement that does not apply to **acker**.

The MATLAB function **lyap** solves Lyapunov matrix equation. The solution to discrete matrix Lyapunov equation is found with **dlyap**. The function **lqr** solves the linear quadratic regulator problem and associated Riccati equation. Discrete linear quadratic regulator design is carried out using the function **dlqr**.

8. *Compute a digital equivalent of the analog controller*

After reaching the best compromise on controller design choice, the next step is to build a computer model, and compute (simulate) the performance of the design. At this stage, we are required to compute a digital equivalent of the analog controller. This allows the final design to be implemented using digital processor logic.

Given the analog controller
$$D(s) = num/den$$
the following commands give the discrete equivalent
$$D(z) = numz/denz$$
of the controller, with T as the sampling period.

        sysc = tf(num,den)

        sysd = c2d(sysc,T,'tustin')

9. *Simulate the performance of the design*

After reaching the best compromise among process modification, actuator and sensor selection, and controller design choice, run a computer model of the system. This model should include important nonlinearities—such as actuator saturation, and the parameter variations you expect to find during operation of the system. The simulation will confirm stability and robustness and allow you to predict the true performance you can expect from the system.

To use the MATLAB numerical integration routines, the system dynamics must be written into an **M-file**. The state-space description makes this very direct; in fact, ordinary differential equation solver function **ode23** requires the dynamics in state-space form (B.1).

As an example, consider van der Pol oscillator which has dynamics

$$\ddot{y} + \alpha(y^2 - 1)\dot{y} + y = u$$

Defining the states as x_1 = position, x_2 = velocity, we get

$$\dot{x}_1 = x_2$$
$$\dot{x}_2 = \alpha(1 - x_1^2)x_2 - x_1 + u$$

For the van der Pol oscillator, the required M-file is

> **function xdot = vdpol (t, x)**
>
> **alpha = 0.8; u = 0;**
>
> **xdot = [x(2); alpha*(1 − x(1)^2)*x(2) − x(1) + u];**

where it is assumed that $u(t) = 0$. Now the sequence of commands required to invoke **ode23** and obtain time history plots, for instance, over a time horizon of 50 sec is

> **t0 = 0; tf = 50;**
>
> **x0 = [0.1;0.1];**
>
> **[t,x] = ode23('vdpol', [t0 tf], x0);**
>
> **plot(t, x)**

The phase-plane plot of $x2$ versus $x1$ is obtained using the command:

> **plot(x(:,1), x(:,2))**

10. Build a prototype

As a final step before production, it is common to build and test a prototype. At this point, we verify the quality of the model, discover unsuspected vibration and other modes, and consider ways to improve the design. After these tests, we may reconsider the sensor, actuator and process unless time, money or ideas have run out.

MATLAB/SIMULINK

In the simulation process, the computer is provided with appropriate input data and other information about system structure, operates on this input data and generates output data, which it subsequently displays. Several software packages that have been produced over the last two decades include computer programs that allow these simulation operations. Over the years, these simulation packages have become quite sophisticated, powerful and very "user-friendly". The usefulness and importance of these software packages is undeniable, because they greatly facilitate the analysis and design of control systems. They provide a tremendous tool in the hands of control engineers. However, a word of caution must be sounded. The availability of such packages and the ease with which one can use them should in no way detract one from learning the underlying concepts. Mastry of the theoretical foundation is a prerequisite for its correct implementation.

MATLAB/SIMULINK is one of the most successful software packages currently available, and is particularly suited for work in control. It is a powerful, comprehensive and user-friendly software package for simulation studies. Our objective here is to help the reader gain a basic understanding of this software package by showing how to set up and solve a simulation problem. Interested readers are encouraged to further explore this very complete and versatile mathematical computation package.

A very nice feature of SIMULINK is that it visually represents the simulation process by using *simulation block diagrams*. Especially, functions are represented by "subsystem blocks" that are then interconnected to form a SIMULINK block diagram that defines the system structure. Once the structure is defined, parameters are entered in the individual subsystem blocks that correspond to the given system data. Some additional simulation parameters must also be set to govern how the numerical computation will be carried out and how the output data will be displayed. As a matter of fact, the SIMULINK block diagrams are essentially the same we have used in the text to describe control system structures and signal flow.

Because Simulink is graphical and interactive, we encourage you to jump right in and try it. For a technical introduction to Simulink, read the document "Using Simulink". To help you start using Simulink quickly, we describe here the simulation process though a demonstration example on Microsoft Windows platform with MATLAB version 6, Control Toolbox version 5 and SIMULINK version 4.

To start SIMULINK, enter **simulink** command at the MATLAB prompt. SIMULINK Library Browser appears which displays tree-structured view of the SIMULINK block libraries. It contains several nodes; each of these nodes represents a library of subsystem blocks that is used to construct simulation block diagrams. You can expand/collapse the tree by clicking on the $\boxed{+}/\boxed{-}$ boxes beside each node.

Expand the node labeled "Simulink". Subnodes of this node (Continuous, Discrete, Functions & Tables, Math, Nonlinear, Signals & Systems, Sinks, Sources) are displayed. Expanding the "Sources" subnode displays a long list of Sources library blocks; contents are displayed in the diagram view. The purpose of the block "Step" is to generate a step function. The block "Constant" generates a specified real or complex value, independent of time.

You may now collapse the Sources subnode, and expand the "Sinks" subnode. A list of Sinks library blocks appears. The purpose of block labeled "XY Graph" is to display an X-Y plot of signals using a MATLAB figure window. The block has two scalar inputs; it plots data in the first input (the x

direction) against data in the second input (the *y* direction). This block is useful for phase-plane analysis. The block "Scope" displays its inputs (signals generated during a simulation) with respect to simulation time. The block "To Workspace" transfers the data to MATLAB workspace.

You may now collapse the Sinks subnode and expand the "Continuous" subnode. A list of library blocks corresponding to this subnode appears. The purpose of the "Derivative" block is to output the time derivative of the input. We will use this block for phase-plane analysis. The "State-Space" block implements a linear system whose behaviour is described by a state variable model. The "Transfer Fcn" block implements a transfer function.

The "Nonlinear" subnode has blockset of various nonlinearities: Backlash, Coulomb and Viscous Friction, Deadzone, Relay, Saturation, etc.

The "Math" subnode has several blocks. The block "Sum" generates the sum of inputs. It is useful as an error detector for control system simulations. The "Sign" block indicates the sign of the input (The output is 1 when the input is greater than zero: the output is 0 when the input is equal to zero; and the output is −1 when the input is less then zero). We can use this block to represent ideal relay nonlinearity.

Expand now the node "Control System Toolbox". The block "LTI system" accepts the continuous and discrete objects as defined in the Control System Toolbox. Transfer functions and state-space formats are supported in this block.

We have described some of the subsystem libraries available that contain the basic building blocks of simulation diagrams. The reader is encouraged to explore the other libraries as well. You can also customize and create your own blocks. For information on creating your own blocks, see the MATLAB documentation on "Writing S-Functions".

We are now ready to proceed to the next step, which is the construction of a simulation diagram. To do this, we need to open a new window. Click the **New** button on the Library Browser's toolbar. A new window opens up that will be used to build up an interconnection of SIMULINK blocks from the subsystem libraries. This is an "untitled" window; we call it the Simulation Window. We consider here phase-plane analysis of nonlinear system described in Review Example 10.5.

With the "Nonlinear" subnode of Simulink node expanded, move the pointer and click the block labeled "Saturation", and while keeping the mouse button pressed down, drag the block and place it inside the Simulation Window, and release the mouse button.

With the "Control System Toolbox" node expanded, click the block labeled "LTI system", drag to the Simulation Window and place it on one side of the Saturation block. Duplicate "LTI system" on the other side of Saturation block.

Drag the block labeled "Sum" from the "Math" subnode of Simulink node, the block "Constant" from the "Sources" subnode of Simulink node, the blocks "XY Graph" and "To Workspace" from "Sinks" subnode of Simulink node, and the block "Derivative" from the "Continuous" subnode of Simulink node.

We have now completed the process of dragging subsystem blocks from the appropriate libraries and placing them in the Simulation Window. The next step is to interconnect these subsystem blocks and obtain the structure of simulation block diagram. To do this, we just need to work in the Simulation Window.

The first step is to rearrange the blocks in the Simulation Window in a specified structure. This will require moving a block from one place to another within the Simulation Window. This can be done by clicking inside the block, keeping the mouse button pressed, dragging the block to the new desired location and releasing mouse button.

Lines are drawn to interconnect these blocks as per the desired structure. A line can connect the output port of one block with the input port of another block. A line can also connect the output port of one block with input ports of many blocks by using branch lines.

To connect the output port of one block to the input port of another block, position the pointer on the first block's output port; the pointer shape changes to a crosshair. Press and hold down the mouse button. Drag the pointer to the second block's input port. You can position the pointer on or near the port; the pointer shape changes to a double crosshair. Release the mouse button. SIMULINK replaces the port symbols by a connecting line with an arrow showing the direction of signal flow.

A *branch line* is a line that starts from an existing line and carries its signal to the input port of a block. Both the existing line and the branch line carry the same signal. To add a branch line, position the pointer on the line where you want the branch line to start. While holding down the **Ctrl** key, press and hold down the mouse button. Drag the pointer to the input port of the target block, then release the mouse button and the **Ctrl** key.

SIMULINK draws connecting lines using horizontal and vertical line segments (To draw a diagonal line, hold down the **Shift** key while drawing the line). The branch lines are usually an interconnection of line segments. With the **Ctrl** key pressed, identify the branch point and drag the mouse (horizontally/vertically) to an unoccupied area of the diagram and release the mouse button. An arrow appears on the unconnected end of the line. To add another line segment, position the pointer over the end of the segment and draw another segment.

To move a line segment, position the pointer on the segment you want to move. Press and hold down the left mouse button. Drag the pointer to the desired location and release.

To disconnect a block from its connecting lines, hold down the **Shift** key, then drag the block to a new location. You can insert a block in a line by dropping the block on the line.

You can duplicate blocks in a model as follows.

While holding down the **Ctrl** key, select the block with the mouse button, then drag it to the new location. The duplicated block has the same parameter values as the original block.

You can cancel the effects of an operation by choosing **Undo** from the **Edit** menu of the Simulation Window. You can thus undo the operations of adding/deleting a line/block. Effects of **Undo** command may be reversed by choosing **Redo** from the **Edit** menu.

To delete a block/line, select a block/line to be deleted and choose **Clear** or **Cut** from the **Edit** menu. The **Cut** command writes the block/line into the clipboard, which enables you to **Paste** it into a model. **Clear** command does not enable you to paste the block/line later.

This gives us a generic diagram because we have not yet specified the LTI system nor set parameter values of saturation, reference input, and error detector. Our next priority is to go into each of these blocks and set the parameters that correspond to our specific nonlinear system. In addition, we need to set some simulation parameters.

We begin with the reference input to the feedback system by double-clicking on the block labeled "Constant" in the Simulation Window. A dialog box pops up. Only one parameter need to be set: constant value. Set the value to 0 since our reference input is zero. When we are done, we click OK.

Next we set the "Sum" block. In the dialog box for this block, we enter Icon shape: round, and list of signs, + −. This gives us an error detector for negative feedback system.

Next we set the LTI system blocks. There are two blocks on the two sides of the saturation nonlinearity. The first block has the transfer function $1/(s + 1)$, and the second block has the transfer function $1/s$. Since our simulation study is with respect to initial condition on output, we convert the transfer functions to state-space form. Enter the initial condition -1.6 in the dialog box of the second block.

Saturation block dialog box requires upper limit and lower limit of saturation. 0.4 and -0.4 are the values as per our problem.

Next we need to set the parameters for the "XY Graph" block. Dialog box requires x-min, x-max, y-min and y-max. The values $[-1\ 2\ -2\ 1]$ may be entered.

"To Workspace" block requires variable name and the format. We use **matrix** format for our data and enter variable names x_1 and x_2 in the dialog boxes.

Finally, we need to set the parameters for the simulation run. We move the pointer to the menu labeled "Simulation", and enter parameters: start time, stop time, in the dialog box of submenu "Parameters". Also we tick against "Normal Simulation" submenu of "Simulation" menu.

All block names in a model must be unique and must contain at least one character. By default, block names appear below blocks. To edit a block name, click on the block name and insert/delete/write text. After you are done, click the pointer somewhere else in the model, the name is accepted or rejected. If you try to change the name of a block to a name that already exists, SIMULINK displays an error message.

At this point in the simulation process, we have generated the appropriate SIMULINK block diagram (shown in Fig. B.1) and entered the specific parameters for our system and simulation. We are now ready to execute the program, and have the computer perform the simulation. We move the pointer to the "Simulation" menu and choose "Start". A new window that shows the phase trajectory pops up.

You may now execute the following program in MATLAB workspace.

figure (1);
plot (x1,x2); grid;
hold on

Resimulate for an initial condition of -0.74 and plot the phase trajectory.

We have used an example to show how to enter data and carry out a simulation in the SIMULINK environment. The reader will agree that this is a very simple process.

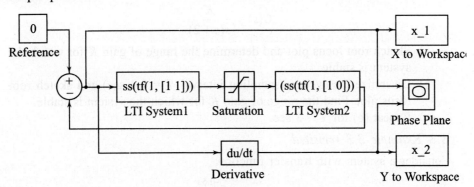

Fig. B.1

PROBLEMS

Each problem covers an important area of control system analysis or design. Important MATLAB commands are included as help to MATLAB problems, in the form of script files. Simulation Diagrams are included as help for the problems requiring SIMULINK.

Following each problem, one or more what-if's may be posed to examine the effect of variations of the key parameters. Comments to alert the reader to the special features of MATLAB commands are included in the script files to enhance the learning experience. Partial answers to the problems are given in the text.

The description of the MATLAB functions used in the script files can easily be accessed from the **help file** using **help** command.

B.1 *Example 3.1 revisited*

Consider a unity-feedback system with open-loop transfer function

$$G(s) = \frac{1}{s(s+1)}$$

(a) Plot the step response of the feedback system and determine error constants K_p, K_v and K_a.

(b) Discretize the system (sampling interval $T = 1$ sec) and plot the step response of the resulting feedback system. Also determine the error constants.

(c) Approximate the sampled system by an equivalent analog system with input delay of $T/2$. Plot the step response of the resulting analog feedback system.

(d) Using LTI Viewer, determine peak overshoot and settling time of analog and sampled systems.

B.2 Example 3.2 revisited

Consider a unity-feedback system with open-loop transfer function

$$G(s) = \frac{K}{s(s+2)}$$

(a) Sketch root locus plot and determine the range of gain K for which the system is stable.

(b) Discretize the system (sampling interval $T = 0.4$ sec) and sketch root locus plot. Find the range of gain K for which the system is stable.

(c) Repeat (b) for $T = 3$ sec.

B.3 Example 3.3 revisited

Consider a system with transfer function

$$G(s) = \frac{e^{-1.5s}}{s(s+1)}$$

Discretize this system (sampling time $T = 1$ sec) and report the result in zero-pole-gain form.

B.4 Example 2.10 revisited

A unity-feedback sampled-data system (sampling interval $T = 0.04$ sec) has plant transfer function

$$G(s) = \frac{10}{(1+0.5s)(1+0.1s)(1+0.05s)}$$

An approximating analog system is a unity-feedback system with plant transfer function $G(s)e^{-Ts/2}$. Show that the analog controller

$$D(s) = \frac{0.67s+1}{(2s+1)}$$

meets the specification: phase margin $\geq 40°$. Determine the bandwidth of the compensated system.

Discretize the design, and analyze the step response of the digital system using LTI Viewer (*Ans*: Peak overshoot 13%; Settling time 1.16 sec)

B.5 Example 4.3 revisited

A unity-feedback system has open-loop transfer function

$$G(s) = \frac{K}{s(s+5)}$$

It is desired to have the velocity error constant $K_v = 10$. Furthermore, we desire that the phase margin of the system be about 40° and bandwidth about 5.5 rad/sec. Design a digital control scheme ($T = 0.1$ sec) to meet these specifications.

Using LTI Viewer, determine peak overshoot and settling time from the step response of the feedback system (*Ans*: Peak overshoot 34%; Settling time 3 sec)

Are your results different from the ones given in the text? Why?

B.6 *Example 4.4 revisited*

Repeat problem B.5 under the constraint that we use phase-lead compensation to achieve the following performance specifications:
  (i) $K_v = 10$
  (ii) Phase margin = 40°
  (iii) Bandwidth = 12 rad/sec
      (*Ans*: Peak overshoot 35%; Settling time 1 sec)

B.7 *Example 4.7 revisited*

A unity-feedback system has open-loop transfer function

$$G(s) = \frac{K}{s(s+2)}$$

It is desired that dominant closed-loop poles provide damping ratio $\zeta = 0.5$ and have undamped natural frequency $\omega_n = 4$ rad/sec. Velocity error constant K_v is required to be about 2.5.

Design a digital control scheme ($T = 0.2$ sec) for the system to meet these specifications.

Perform simulation study on the compensated system using LTI Viewer (*Ans*: Peak overshoot 15%; Settling time 2.2 sec)

B.8 In the following, we point the reader to important matrix functions in MATLAB. Access the description of these functions from the **help file**, and execute each function, taking suitable data from the text.

Identity matrix	: **eye(n)**
Dimensions	: **size(A)**
Utility matrices	: **ones(n), ones(m,n), ones(size(A)), zeros(n), zeros(m,n), zeros(size(A))**
Complex conjugate transpose	: **ctranspose(A); A'**
Non-conjugate transpose	: **transpose(A); A.'**
Determinant	: **det(A)**
Inverse	: **inv(A)**
Rank	: **rank(A)**
Trace	: **trace(A)**

Spectral norm	: **norm(A)**
(Largest singular value)	
Euclidean norm of a vector	: **norm(x)**
Condition number with respect to inversion	: **cond(A)**
Eigenvalues	: **eig(A)**
Eigenvectors	: **[P,A1] = eig(A)**
Characteristic equation	: **poly(A)**
Matrix exponential	: **expm(A)**

B.9 Given the transfer function

$$G(s) = \frac{s}{s^3 + 2s^2 + 2.5s + 0.5}$$

(a) Obtain a state-space model **sys**, equivalent to the given $G(s)$.
(b) Discretize the model **sys** (sampling interval $T = 0.1$ sec) to obtain **sysd**.
(c) Simulate and plot the response of the models **sys** and **sysd**, when the input is

$$u(t) = \begin{cases} 2 \,; 0 \leq t \leq 2 \\ 0.5; t \geq 2 \end{cases}$$

and the initial condition is $x(0) = [1 \ 0 \ 2]^T$.

B.10 *Example 7.2 revisited*

Linearized equations governing the inverted pendulum system of Fig. 5.16 are

$$\dot{x} = Ax + bu$$

$$x = \begin{bmatrix} \theta & \dot{\theta} & z & \dot{z} \end{bmatrix}^T$$

$$A = \begin{bmatrix} 0 & 1 & 0 & 0 \\ 4.4537 & 0 & 0 & 0 \\ 0 & 0 & 0 & 1 \\ -0.5809 & 0 & 0 & 0 \end{bmatrix}; b = \begin{bmatrix} 0 \\ -0.3947 \\ 0 \\ 0.9211 \end{bmatrix}$$

(a) Show that the open-loop system is unstable.
(b) Design state feedback $u = -kx$ that results in closed-loop poles at $-1, -1, -1, -1$.
(c) Simulate the feedback system. Given initial state

$$x(0) = [0.1 \ 0 \ 0 \ 0]^T$$

B.11 *Example 7.7 revisited*

The plant model of a satellite attitude control system (Refer Figs 7.3–7.4) is

$$\dot{x} = Ax + bu$$

$$y = cx$$

with
$$A = \begin{bmatrix} 0 & 1 \\ 0 & 0 \end{bmatrix}; \; b = \begin{bmatrix} 0 \\ 1 \end{bmatrix}; \; c = [1 \; 0]$$

(a) Design state feedback $u = -kx$ that results in closed-loop poles at $-4 \pm j4$.
(b) Assuming that the state vector $x(t)$ is measurable, simulate the feedback system for $x(0) = [1 \; 0]^T$.
(c) Consider now that state measurements are not practical. Design a state observer that yields estimated states $\tilde{x}(t)$. Place the observer poles at $-10, -10$.
(d) Obtain state variable model of the compensator by cascading the state feedback control law and the state observer. Find the transfer function of the compensator.
(e) Set up state model of the form (7.57) for the observer-based regulator system and simulate the model (Note that $x(0) = [1 \; 0]^T$ leads to $\tilde{x}(0) = [1 \; 0]^T$ when $\hat{x}(0) = 0$).

B.12 Example 7.13 revisited

The plant model of a satellite attitude control system (Refer Figs 7.3–7.4) is
$$x(k+1) = Fx(k) + gu(k)$$
$$y(k) = cx(k)$$
with
$$F = \begin{bmatrix} 1 & T \\ 0 & 1 \end{bmatrix}; \; g = \begin{bmatrix} T^2/2 \\ T \end{bmatrix}; \; T = 0.1 \text{ sec}$$
$$c = [1 \; 0]$$

The reference input θ_r is a step function.
Design state feedback $u = -k_1(x_1(k) - \theta_r) - k_2 x_2(k)$ that results is deadbeat response.

Simulate the feedback system for a unit-step input θ_r.

B.13 Reconsider the inverted pendulum regulator problem raised in Problem B.10, wherein you designed state feedback control based on pole-placement.

Now design a state-feedback control law that minimizes the performance index
$$J = \frac{1}{2} \int_0^\infty (x^T Q x + u^T R u) dt$$
with
$$Q = \begin{bmatrix} 100 & 0 & 0 & 0 \\ 0 & 1 & 0 & 0 \\ 0 & 0 & 1 & 0 \\ 0 & 0 & 0 & 1 \end{bmatrix}; \; R = 0.01$$

Simulate the feedback system for initial state $\mathbf{x}(0) = [0.1 \ \ 0 \ \ 0 \ \ 0]^T$.

B.14 Reconsider the inverted pendulum system of problem B.10.
(a) Discretize the plant model (sampling interval $T = 0.1$ sec).
(b) Introduce integral state in the plant equations (Refer Eqn. (7.109)) and show that the augmented system is controllable.
(c) Design state feedback with integral control that minimizes the performance index

$$J = \frac{1}{2} \sum_{k=0}^{\infty} [\mathbf{x}^T(k)\mathbf{Q}\mathbf{x}(k) + u^T(k)Ru(k)]$$

with

$$\mathbf{Q} = \begin{bmatrix} 10 & 0 & 0 & 0 & 0 \\ 0 & 1 & 0 & 0 & 0 \\ 0 & 0 & 100 & 0 & 0 \\ 0 & 0 & 0 & 1 & 0 \\ 0 & 0 & 0 & 0 & 1 \end{bmatrix}; R = 1$$

(d) Simulate the digital servo (Refer Fig. 7.17) for a step input.

B.15 *Review Example 10.3 revisited*

Figure 10.40a shows the block diagram of a nonlinear system with saturation nonlinearity.
(a) Sketch the Nyquist plot for the linear transfer function

$$G(s) = \frac{1}{s(1+2s)(1+s)}$$

(b) Superimpose on this plot, the plot of describing function of saturation nonlinearity.
(c) Show the existence of a stable limit cycle and determine its amplitude and frequency.
(d) Drag the following blocks from SIMULINK block libraries:
  (i) "Sum" from Math subnode of Simulink node;
  (ii) "Saturation" from Nonlinear subnode of Simulink node;
  (iii) "LTI system" from Control System Toolbox node: and
  (iv) "Scope" from Sinks subnode of Simulink node.

Setup a simulation block diagram as per the feedback structure given in Fig. 10.40a. Simulate the system for an initial condition of $x0 = [5 \ \ 0 \ \ 0]^T$. The SIMULINK response shows a limit cycle. Determine the amplitude and frequency of the limit cycle and compare these parameters with the ones obtained in part (c).

Script PB.1

```
clear all
close all
disp('Refer Example 3.1')
%Effects of sampling on response
%Analog system
G=tf(1,[1 1 0]);
M=feedback(G,1);
%Transient response
figure(1)
step(M);
%Steady-state response
Kp=dcgain(G)
Gv=tf([1 0],1)*G;
Kv=dcgain(Gv)
Ga=tf([1 0],1)*Gv;
Ka=dcgain(Ga)
hold on
%Sampled system
Ts=input('Enter sampling time :')
Gz=c2d(G,Ts);
Mz=feedback(Gz,1);
%Transient response
Step(Mz);
%Steady-state response
Kpz=dcgain(Gz)
Gvz=tf([1 -1],1,Ts)*Gz/Ts;
Kvz=dcgain(Gvz)
Gaz=tf([1 -1],1,Ts)*Gvz/Ts;
Kaz=dcgain(Gaz)
hold off
%Delay approximation of ZOH.
%Refer Example 2.10.
figure(2)
step(M);
hold on
Gd=tf(1,[1 1 0],'InputDelay',Ts/2);
Gd1=pade(Gd,2);
M1=feedback(Gd1,1);
step(M1);
hold off
%Using LTI Viewer,determine peak overshoot
```

%& settling time of analog & sampled systems.
ltiview('step',M,M1)

Script PB.2

```
clear all;close all
disp('Refer Example 3.2')
%Effect of sampling on stability
G=tf(1,[1 2 0]);
figure(1)
rlocus(G);sgrid
axis equal
pause
%Waits for user to strike any key.
Ts=input('Enter sampling time:')
Gz=c2d(G,Ts);
figure(2)
rlocus(Gz);zgrid;
axis equal
[K1,p1]=rlocfind(Gz)
```

Script PB.3

```
clear all
close all
disp('Refer Example 3.3')
%Discretization of system with dead-time
T=1;N=1;delta=0.5;
G=tf(1,[1 1],'InputDelay',delta*T);
Gz1=c2d(G,1);
Gz2=tf(1,[1 0],1);
Gz=Gz1*Gz2;
zpk(Gz)
```

Script PB.4

```
clear all
close all
disp('Refer Example 2.10')
%Digital Implementation of Analog Design
%Approximation of sampled-data system
Ts=0.04;
```

Appendix B: MATLAB Aided Control System Design: Conventional   901

```
den=conv(conv([0.5 1],[0.1 1]),[0.05 1]);
G=tf(10,den,'InputDelay',Ts/2);
G1=pade(G,2);
%Analog Design
figure(1)
D=tf([0.67 1],[2 1]);
margin(D*G1)
figure(2)
%LTI Viewer functions are available
%in figure window.Zoom in region of -3dB
%contour & read bandwidth.
Gc=D*G1;
ngrid;
nichols(Gc);
%Discretization of Design
Dz=c2d(D,Ts,'tustin')
G0=tf(10,den);
Gz=c2d(G0,Ts);
Mz=feedback(Dz*Gz,1);
ltiview('step',Mz)
```

Script PB.5

```
clear all;
close all;
disp('Refer Example 4.3')
%Direct Digital Design(lag compensation)
%Uncompensated system
G=tf(50,[1 5 0]);
%Discretization
Gz=c2d(G,0.1);Gz1=zpk(Gz);
%Design in w-domain
Gw=d2c(Gz,'tustin');
figure(1);
margin(Gw);
%Read uncompensated phase margin from figure 1.
w=logspace(-1,1,100);
[mag,ph]=bode(Gw,w);
% mag & ph are arrays;MATLAB function interp1
%does not accept these arrays.Reshape the
%arrays into column vectors.
mag=reshape(mag,100,1);
ph=reshape(ph,100,1);
```

```
%Phi=-180+specified phase margin+epsilon=-130
Phi=input('Enter phase angle Phi:')
%Select frequency range properly;function interp1
%requires monotonic data.
wc2=interp1(ph,w,Phi)
%wc2 is the desired gain crossover .
beta=interp1(ph,mag,Phi);
%Upper corner frequency at wc2/4.
wcu=input('Enter upper corner frequency:')
tau=1/wcu;
Dw=tf([tau 1],[beta*tau 1])
%Digital controller
D=c2d(Dw,0.1,'tustin');
Dz=zpk(D)
figure(2)
%Read phase margin of compensated system
%from figure 2.
margin(Dw*Gw);
%Bandwidth(uncompensated &compensated)
w=[5:0.1:15];
[mag,ph,w]=bode(feedback(Gw,1));
magdB=20*log10(mag);
wb=interp1q(-magdB,w,3)
[mag1,ph1,w]=bode(feedback(Dw*Gw,1));
mag1dB=20*log10(mag1);
wb1=interp1q(-mag1dB,w,3)
%Analysis on Nichols chart
figure(3);
Gwc=Dw*Gw;
ngrid;
nichols(Gw,Gwc);
%Step-response analysis using LTI Viewer
Mz=feedback(D*Gz,1);
ltiview('step',Mz)
%Carry out design on GUI
```

Script PB.6

```
clear all;
close all;
disp('Refer Example 4.4')
%Direct Digital Design(lead Compensation)
%Uncompensated system
```

Appendix B: MATLAB Aided Control System Design: Conventional

```
G=tf(50,[1 5 0]);
%Discretization
Gz=c2d(G,0.1);
%Design in w-domain
Gw=d2c(Gz,'tustin');
figure(1)
margin(Gw);
%Read uncompensated phase margin from figure 1.
%PhiM=specified phase margin-uncompensated
%phase margin+epsilon=35
PhiM=input('Enter required phase lead :')
alpha=(1-sin(PhiM*pi/180))/(1+sin(PhiM*pi/180));
w=logspace(0,2,100);
[mag,ph]=bode(Gw,w);
magdB=20*log10(mag);
magdB=reshape(magdB,100,1);
wm=interp1(magdB,w,-20*log10(1/sqrt(alpha)));
tau=1/(wm*sqrt(alpha));
Dw=tf([tau,1],[alpha*tau,1])
%Digital controller
D=c2d(Dw,0.1,'tustin');
Dz=zpk(D)
figure(2);
%Read compensated phase margin from figure 2.
margin(Dw*Gw);
%Bandwidth(uncompensated & compensated)
w=[5:0.1:25];
[mag,ph,w]=bode(feedback(Gw,1));
magdB=20*log10(mag);
wb=interp1q(-magdB,w,3)
[mag1,ph1,w]=bode(feedback(Dw*Gw,1));
mag1dB=20*log10(mag1);
wb1=interp1q(-mag1dB,w,3)
%Analysis on Nichols chart
figure(3)
Gwc=Dw*Gw;
ngrid;
nichols(Gw,Gwc)
%Step-response analysis using LTI Viewer
Mz=feedback(D*Gz,1);
ltiview('step',Mz)
%Carry out the design on GUI
```

Script PB.7

```
close all
clear all
disp('Refer Example 4.7')
%Digital Controller Design on Rootlocus plots
%Plant model
G=tf(1,[1 2 0]);
T=0.2;
Gz=c2d(G,T);Gz1=zpk(Gz);
zeta=0.5;wn=4;
theta=wn*T*sqrt(1-zeta^2);
figure(1)
%Uncompensated root loci
rlocus(Gz);
%Take near-zero value for wn in zpchart funtion
%if you are not using wn locus in your design procedure.
zpchart(gca,0.5,.01);
%Replace zpchart command by zgrid(zeta,wn) if working with
%MATLAB 5.3 version.
axis equal;
hold on
%Locate desired dominant roots.
 polar(theta*[1 1],[0 1],'k');
%Gain adjustment does not satisfy specs.
%Lead compensator zero to cancel open-loop
%pole at z=0.67.Take pole at z=alpha.
alpha=input('Enter trial value of alpha:')
Dz=tf([1 -0.67],[1 -alpha],T);
figure(2)
rlocus(Dz*Gz);
zpchart(gca,zeta,0.01);
axis equal;
hold on
polar(theta*[1 1],[0 1],'k');
[KK,polesCL]=rlocfind(Dz*Gz)
%Place the crosshair cursor at desired root & click.
Gvz=tf([1 -1],1,T)*Dz*KK*Gz/T;
Kvz=dcgain(Gvz)
%Perform simulation study on compensated system
%using LTI Viewer.
Mz=feedback(KK*Dz*Gz,1);
ltiview('step',Mz)
%Carry out the design on GUI.
```

Script PB.9

```
clear all
close all
disp('State Space Analysis')
%Given transfer function
num=[1 0];
den=[1 2 2.5 0.5];
G=tf(num,den)
%Construct state variable model
sys=ss(G);
[A,b,c,d]=ssdata(sys)
%Discretization
T=0.1;
sysd=c2d(sys,T);
[F,g,c,d]=ssdata(sysd)
%Simulation
t=[0:0.1:20]';
x0=[1,0,2]';
%Input for 0<t<2.
u(1:20)=2*ones(20,1);
%Input for t>2.
u(21:201)=0.5*ones(181,1);
[y,t,X]=lsim(sys,u,t,x0);
[y1,t,X1]=lsim(sysd,u,t,x0);
figure(1)
plot(t,X(:,1),t,X1(:,1),'o');grid;
legend('x_1');
figure(2)
plot(t,X(:,2),t,X1(:,2),'o');grid;
legend('x_2');
figure(3)
plot(t,X(:,3),t,X1(:,3),'o');grid;
legend('x_3');
```

Script PB.10

```
clear all
close all
disp('Refer Example 7.2')
%Regulator Design(Pole-placement) for Inverted Pendulum
%Plant Model
A=[0,1,0,0;4.4537,0,0,0;0,0,0,1;-0.5809,0,0,0];
b=[0;-0.3947;0;0.9211];
```

```
c=[0 0 1 0];d=0;
sys=ss(A,b,c,d);
U=ctrb(A,b);
ControllablePoles=rank(U)
%Regulator Design
polesOL=pole(sys)
polesCL=[-1,-1,-1,-1];
k=acker(A,b,polesCL)
RegulatorPoles=eig(A-b*k)
%Simulation
x0=[0.1,0,0,0]';
t=[0:0.1:20];
sysCL=ss(A-b*k,[0;0;0;0],c,0);
[y,t,X]=lsim(sysCL,0*t,t,x0);
figure(1)
subplot(2,2,1);plot(t,X(:,1));grid;legend('x_1');
subplot(2,2,2);plot(t,X(:,2));grid;legend('x_2');
subplot(2,2,3);plot(t,X(:,3));grid;legend('x_3');
subplot(2,2,4);plot(t,X(:,4));grid;legend('x_4')
```

Script PB.11

```
clear all
close all
disp('Refer Example 7.7')
%Observer-based Regulator for Satellite Attitude Control
%Plant Model
A=[0 1;0 0];b=[0;1];c=[1 0];
%Regulator Design
polesCL=[-4+4i,-4-4i];
k=place(A,b,polesCL)
Regulatorpoles=eig(A-b*k)
sys=ss(A-b*k,[0;0],c,0);
%Simulation
t=[0:0.01:1];
x0=[1 0]';
[y,t,X]=initial(sys,x0,t);
figure(1)
plot(t,y);
hold on
%Observer Design
polesOB=[-10 -10];
kk=acker(A',c',polesOB);
```

```
m=kk'
Observerpoles=eig(A-m*c)
%Refer Eqns.(7.58)&(7.59)
sys_comp=ss(A-b*k-m*c,m,k,0);
D=tf(sys_comp)
%Simulation
%Refer Eqn.(7.57)
A1=A-b*k;A2=b*k;A3=zeros(size(A));A4=A-m*c;
A_cl=[A1(1,:),A2(1,:);A1(2,:),A2(2,:);A3(1,:),A4(1,:);A3(2,:),A4(2,:)];
sysCL=ss(A_cl,[0;0;0;0],[1 0 0 0],0);
x0=[1 0 1 0]';
[y1 t X1]=initial(sysCL,x0,t);
plot(t,y1,'r');grid;
legend('Output with no observer','Output with observer')
hold off
figure(2)
plot(t,X1(:,3),t,X1(:,4));grid;
legend('xtilde_1','xtilde_2')
```

Script PB.12

```
clear all;
close all;
disp('Refer Example 7.13')
%Deadbeat controller for Satellite Attitude Control
%Plant Model
T=0.1;
F=[1,T;0,1];g=[T^2/2;T];c=[1,0];
sysd=ss(F,g,c,0,T);
%Pole placement
polesCL=[0,0];
k=acker(F,g,polesCL)
%Step response
sysdCL=ss(F-g*k,100*g,c,0,'Ts',T);
step(sysdCL)
```

Script PB.13

```
clear all
close all
disp('Refer Example 7.2')
%Linear Quadratic Regulator for Inverted Pendulum
%Plant Model
```

```
A=[0,1,0,0;4.4537,0,0,0;0,0,0,1;-0.5809,0,0,0];
b=[0;-0.3947;0;0.9211];
c=[1,0,0,0];
sys=ss(A,b,c,0);
U=ctrb(A,b);
ControllablePoles=rank(U)
%Performance Index
Q=[100,0,0,0;0,1,0,0;0,0,1,0;0,0,0,1];R=0.01;
%State Feedback
[k P E]=lqr(A,b,Q,R);
PolesCL=E
GainMatrix=k
sysCL=ss(A-b*k,[0;0;0;0],c,0);
%Simulation
x0=[0.1,0,0,0]';
t=[0:0.1:20];
[y,t,X]=initial(sysCL,x0,t);
subplot(2,2,1);plot(t,X(:,1));grid;legend('x_1');
subplot(2,2,2);plot(t,X(:,2));grid;legend('x_2');
subplot(2,2,3);plot(t,X(:,3));grid;legend('x_3');
subplot(2,2,4);plot(t,X(:,4));grid;legend('x_4');
```

Script PB.14

```
clear all
close all
disp('Refer Example 7.2')
%Digital Servo for Inverted Pendulum
%Plant Model
A=[0,1,0,0;4.4537,0,0,0;;0,0,0,1;-0.5809,0,0,0];
b=[0;-0.3947;0;0.9211];
c=[0,0,1,0];
sys=ss(A,b,c,0);
T=0.1;
sysd=c2d(sys,T);
[F,g,c,d]=ssdata(sysd);
%Plant Model with integral state
%Refer Eqn.(7.109)
FI=[F(1,:),0;F(2,:),0;F(3,:),0;F(4,:),0;c*F,1];
gI=[g(1);g(2);g(3);g(4);c*g];
cI=[c,0];
U=ctrb(FI,gI);
```

```
rankU=rank(U)
%Performance Index
Q=[10,0,0,0,0;0,1,0,0,0;0,0,100,0,0;0,0,0,1,0;0,0,0,0,1];
R=1;
%State Feedback with Integral Control
K=dlqr(FI,gI,Q,R);
Kp=[K(1),K(2),K(3),K(4)]
Ki=K(5)
%Simulation(step response)
%Refer Eqns.(7.110)&(7.108)
FI_cl=FI-gI*K;
sysCL=ss(FI_cl,[0;0;0;0;-1],cI,0,'Ts',T);
t=[0:0.1:20];
[y,t,X]=step(sysCL,t);
plot(t,y);grid
```

Script PB.15

```
clear all;
close all;
disp('Refer Review Example 10.3')
%Describing Function Analysis
%Nyquist plot for G(s)
G=tf(3,conv(conv([2 1],[1 1]),[1 0]));
w=[1:0.1:100];
[re im w]=nyquist(G);
re=reshape(re,length(w),1);
im=reshape(im,length(w),1);
plot(re,im);
axis([-5 2 -5 2]);
hold on
%Describing Function plot for saturation
%Refer Eqns.(10.51)-(10.53)
S=1;
E=[1:0.5:10]';
N=[1.0,0.781,0.609,0.495,0.416,0.359,0.315,0.281,0.253,...
0.230,0.211,0.195,0.181,0.169,0.159,0.149,0.141,0.134,0.127]';
N1=-1./N;
re1=N1;
im1=0*length(N1);
plot(re1,im1,'ro');grid;
%Move the pointer to the intersection point & click.
[X,Y]=ginput(1)
```

```
E0=interp1(N1,E,X)
[mag ph w]=bode(G);
ph=reshape(ph,length(w),1);
w0=interp1(ph,w,-180)
```

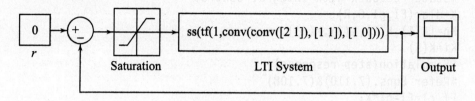

Fig. PB.15

Appendix C

MATLAB Aided Control System Design: Neural–Fuzzy

It is well-known that human control mechanisms defy exact mathematical modeling to perform a particular task. This is in sharp contrast with conventional approaches in the design of an automatic control system that often involve the construction of a mathematical model describing the dynamic behaviour of the plant to be controlled, and the application of analytical techniques to the model to derive an appropriate control law. Usually, such a mathematical model consists of a set of linear or nonlinear differential or difference equations, most of which are derived using some form of approximation and simplification. The traditional model-based control techniques may breakdown when a representative model is difficult to obtain due to uncertainty or sheer complexity, or when the model produced violates the underlying assumptions of the control law synthesis techniques. On the other hand, a human control mechanism uses imprecise and qualitative understanding of the processes to be controlled. *Knowledge-based control* or *intelligent control* is the name introduced to describe control systems in which the control strategies use behavioural (and not mathematical) description of the process, based on experience gathered by operators and process engineers. Actions are performed either as a result of evaluating rules (reasoning) or as unconscious actions based on presented process behaviour after a learning phase. Intelligence comes in as the capability to reason about facts and rules and to learn about presented behaviour.

Given a control problem, researchers working in the field of intelligent control often heuristically construct what turns out to be a nonlinear, perhaps adaptive, controller. While simulation results are typically used to "verify" the approach and some successful implementations have been achieved, it is often the case that no nonlinear analysis is performed to verify the behaviour of the closed-loop system. Although it is recognised that new ideas and techniques for control are being introduced by the intelligent control community, after careful examination of the results, control theorists and practitioners can often convincingly argue and/or demonstrate that they can obtain the same or better results with conventional techniques. We cannot throw out what has been done by control community in the excitement over the intelligent control. Overall, conventional control has a much better and well-developed track record than techniques from intelligent control, and this is important, especially to the practitioner seeking a reliable implementation for a control system. The notion that since a control system is "intelligent", it must automatically be better than other conventional approaches, is hype.

At the same time, it is bad for control engineers to simply ignore the field of intelligent control as being "sloppy". Perhaps it is not as "tidy" as conventional control, but this is due to the fact that the field of intelligent control is relatively new and unexplored. Intelligent control has certain techniques and concepts to offer; the challenge is to find out what it is good for, and perhaps more importantly, what it is not good for. From a control engineer's perspective, the best way to assess the contributions of intelligent control is to perform careful theoretical and experimental engineering analysis as has been done in the past for conventional control systems. Such an assessment will most likely tone down the often implied idea that "intelligent control can solve all of your problems". There is a need to build a bridge between conventional and intelligent control. Artificial intelligence field will drive the development of control theory and control technology by providing alternative strategies for the functionality and implementation of controllers for dynamic systems. On the other hand, certain results from control theory will allow the intelligent control technology to expand its role due to the fact that they will provide methods to "guarantee" that the technology would work in critical environments (e.g., the use of stability theory for ensuring safe operation of controllers for nuclear reactors and aircrafts).

From the control theory point of view, the ability of intelligent control techniques to deal with uncertain and nonlinear systems is perhaps most significant. The great diversity of nonlinear systems is the primary reason why no systematic and generally applicable theory for nonlinear control design has yet been evolved. A range of "traditional" methods for analysis and synthesis of nonlinear controllers for specific classes of nonlinear systems exist: phase-plane methods, linearization techniques, and describing function analysis are three examples. The ability of neutral networks/fuzzy systems to represent nonlinear mappings is the feature to be exploited in the synthesis of nonlinear controllers.

Model-based techniques may fail to yield acceptable design even for linear systems. The model for systems with uncertain dynamics and imprecisely known parameters may lead to a poor control solution obtained via traditional control design methods. During the past couple of decades, much progress has been made in the area of robust control of linear, time-invariant, dynamic systems. A succession of key concepts such as H_∞ optimisation developed during this period has made robust control an area rich in theory and potential applications.

Appendix C: MATLAB Aided Control System Design: Neural–Fuzzy

There are situations which are beyond the range of these robust control techniques. For example control of uncertain nonlinear dynamic systems or control of uncertain linear systems wherein uncertainty model is not available. Intelligent control techniques do not rely on a model and therefore have a great potential of solving these problems.

A further established area in control science with great relevance here is adaptive systems theory. This area has produced many fruitful theoretical results in the past fifteen years. Although the established adaptive theory is founded upon the assumption of linear time-invariant systems, the concepts and theoretical challenges addressed are likely to find strong parallels in the intelligent control field. However, the nonlinear nature of the subsystems involved means that the problems encountered are likely to be more complex and many fundamental theoretical questions will need to be addressed.

A tutorial introduction to knowledge-based tools—neural networks, fuzzy logic, genetic algorithms, for control system design was given in Chapters 11 and 12. This appendix is written to familiarize the reader with the use of Neural Network Toolbox and Fuzzy Logic Toolbox associated with MATLAB.

The Neural Network Toolbox (Version 4.0) is contained in a directory called **nnet**. Type **help nnet** for a listing of help topics.

The Neural Network Toolbox is a collection of functions built on the MATLAB numeric computing environment. It provides tools to create and edit neural network models, within the framework of MATLAB; or if we prefer, we can integrate our network models into simulations with SIMULINK. This toolbox relies heavily on graphical user interface (GUI) tools to help us accomplish our work, although we can work entirely from the command line if we prefer.

The toolbox provides three categories of tools:
- Command line functions
- Graphical interactive tools
- Simulink blocks

The first category of tools is made up of functions (M-files) that we can call from the command line. These functions are used to create, initialize, train and simulate neural networks. We can also extend the toolbox by adding our own M-files.

Creating a Network

The function **newff** creates a feedforward backpropagation network. It requires four inputs and returns the network object **net**. The first input is an $n \times 2$ matrix of maximum and minimum values for each of the n elements of the input vector. The second input is an array containing the sizes of each layer. The third input is a cell array containing the names of the transfer functions to be used in each layer, e.g., **logsig, tansig, purelin**, ... The final input contains the name of the training function to be used, e.g.,

Traingd : Gradient descent backpropagation

traingda : Gradient descent with adaptive learning rate backpropagation

traingdm : Gradient descent with momentum backpropagation

traingdx : Gradient descent with momentum and adaptive learning rate backpropagation

⋮

For example, the following command creates a two-layer network. There is one input vector with two elements. The values for the first element of the input vector range between –1 and 2, and the values of the second element range between 0 and 5. There are three neurons in the first layer and one neuron in the second (output) layer. The transfer function in the first layer is tan-sigmoid, and the output layer transfer function is linear. The training function is gradient descent backpropagation.

net = newff ([–1 2; 0 5], [3, 1], {'tansig', 'purelin'}, 'traingd');

Initializing Weights

The **newff** command will automatically initialize the weights and biases. Re-initialization may be done using the command **init**.

The network's weights and biases are referenced as follows to see how they have been initialized.

net.IW{i, j}: The weight matrix for the weights going to the ith layer (destination) from the jth input vector (source). The $i = j = 1$ situation will be frequently used in our applications.

net.LW{i, j}: The weight matrix for the weights going to the ith layer (destination) from the jth layer (source).

net.b{i}: The bias vector for the ith layer.

Training

The batch steepest descent training function is **traingd**. There are several training parameters associated with **traingd**. If we want to use the default values of these parameters, no additional commands are necessary. We might want to modify some of the default training parameters. This is done as is shown below by an example.

net.trainParam.show = 50;

net.trainParam.lr = 0.005;

net.trainParam.epochs = 300;

net.trainParam.goal = 1e−5;

The training status is displayed for every **show** iteration of the algorithm. **lr** is the specified learning rate. The other parameters determine when the training stops. The training stops if the number of iterations exceeds **epochs** or if the performance function drops below **goal**. The default performance function for the feedforward networks is mean square error **mse**—the average squared error between the network outputs and the target outputs.

Suppose that the network training set data consists of 3 concurrent 2-dimensional input vectors:

$$\mathbf{x}^{(1)} = \begin{pmatrix} x_1^{(1)} \\ x_2^{(1)} \end{pmatrix}; \mathbf{x}^{(2)} = \begin{pmatrix} x_1^{(2)} \\ x_2^{(2)} \end{pmatrix}; \mathbf{x}^{(3)} = \begin{pmatrix} x_1^{(3)} \\ x_2^{(3)} \end{pmatrix}$$

and the corresponding targets are

$$y^{(1)}, y^{(2)}, y^{(3)}$$

The concurrent input vectors are presented to the network as a single 2×3 matrix; ith row of this matrix is made up of ith element of each vector.

$$\mathbf{X}: [x_1^{(1)} \quad x_1^{(2)} \quad x_1^{(3)} ; x_2^{(1)} \quad x_2^{(2)} \quad x_2^{(3)}]$$

The concurrent targets are presented to the network as a single 1×3 vector.

$$\mathbf{y}: [y^{(1)} \quad y^{(2)} \quad y^{(3)}]$$

To perform batch training using the function **train**, the following command may be used:

[net, tr] = train(net, X, y)

The training record **tr** contains information about the progress of training.

The function **sim** simulates a network. It takes the network input **X** and the network object **net** and returns the network outputs $\hat{\mathbf{y}}$.

$$\hat{\mathbf{y}} = \mathbf{sim(net, X)}$$

A single matrix of concurrent vectors is presented to the network and the network produces a single matrix of concurrent vectors as output.

In the *batch mode*, the weights and biases of the network are updated only after the entire training set has been applied to the network. In the *incremental mode*, weights and biases are updated after each input is applied to the network. In this case, we use the function **adapt** and we present the inputs and targets as sequences.

The concurrent inputs $\mathbf{x}^{(1)}, \mathbf{x}^{(2)}, \mathbf{x}^{(3)}$ and targets $y^{(1)}, y^{(2)}, y^{(3)}$ are presented as a cell array of sequential vectors.

$$\mathbf{x}_s: \{[x_1^{(1)}; x_2^{(2)}] \, [x_1^{(2)}; x_2^{(2)}] \, [x_1^{(3)}; x_2^{(3)}]\}$$

$$\mathbf{y}_s: \{y^{(1)} \, y^{(2)} \, y^{(3)}\}$$

[net, netOutput, netError] = adapt(net, x$_s$, y$_s$)

The second category of tools is made up of a number of interactive tools that let us access many of the M-files through a GUI. This interface allows us to

- Create networks
- Enter data into the GUI
- Initialize, train and simulate networks
- Export the training results from the GUI to the command line workspace
- Import data from the command line workspace to the GUI

To open the **Network/Data Manager** window, type **nntool**. Click on **Help** to get started on a new problem and to see descriptions of the buttons and lists.

The third category of tools is a set of blocks for use with the Simulink simulation software. Bring up the Neural Network Toolbox blockset with the command **neural**. The result is a window that contains four blocks: Transfer Functions, Net Input Functions, Weight Functions, and Control Systems.

Each of the blocks of the **Transfer Functions** takes a net input vector and generates a corresponding output vector whose dimensions are the same as the input vector (Double-click on the **Transfer Functions** block). Each of the blocks of **Net Input Functions** takes any number of weighted input vectors, weighted layer output vectors, and bias vectors and returns a net-input vector. Each of the blocks of **Weight Functions** takes a neuron's weight vector and applies it to an input vector (or a layer output vector) to get a weighted input value for a neuron.

We can extend the blockset by adding our own S-functions. The blockset is used to build neural networks in Simulink. The function **gensim** generates the Simulink version of any network we have created in MATLAB. The call

$$\text{gensim(net, T)}$$

results in a screen that contains a Simulink system consisting of the network object **net** connected to an input block (a standard 'Constant' block from the Simulink blockset: 'Sources') and a scope. T is the discrete sample time. $T = -1$ tells **gensim** to generate a network with continuous sampling. To build a feedback control system with **net** as one of the components, we can replace the 'Constant' input block with a signal generator from the Simulink 'Sources' blockset, and use other blocks from the Simulink blockset/Neural Network Toolbox blockset/user-created S-function blocks.

Double-click on the **Control Systems** block in the **neural** window. This brings up the following three blocks:

- Model Reference Controller
- NARMA-L2 Controller
- NN Predictive Controller

The neutral model reference control architecture uses two neural networks: a controller network and a plant model network. The plant model is identified first, and then the **Model Reference Controller** is trained so that the plant output follows the reference model output.

The neurocontroller called NARMA-L2 Controller is also described by the name **Feedback Linearization Controller**. The central idea of this type of control is to transform nonlinear system dynamics into linear dynamics by cancelling the nonlinearities.

The **NN Predictive Controller** uses a neural network model of a nonlinear plant to predict future plant performance. The controller then calculates the control input that will optimize plant performance over a specified future time horizon. Refer [158] for description of the criterion and its minimization.

These three controllers are implemented as Simulink blocks. A controller block from the Neural Network Toolbox blockset is copied to Simulink model window. Double-clicking on the controller block brings up a window for designing the control. Performance of a closed-loop control system is evaluated by running its Simulink model.

▲▲

The Fuzzy Logic Toolbox (Version 2.1) is contained in a directory called **fuzzy**. Type **help fuzzy** for a listing of help topics.

The Fuzzy Logic Toolbox provides tools to create and edit fuzzy inference system (FIS) within the framework of MATLAB. We can integrate our fuzzy systems into simulations with SIMULINK.

Similar to the Neural Network Toolbox, the Fuzzy Logic Toolbox also provides three categories of tools:
- Command line functions
- Graphical interactive tools
- Simulink blocks

To build a system entirely from the command line, the commands **newfis**, **addvar**, **addmf** and **addrule** would be used.

The function **newfis** creates new FIS structures. It has upto seven input arguments, and the output argument is a FIS structure. The seven input arguments are as follows:
- **fisName** is the string name of FIS structure; **fisName.fis** you create.
- **fisType** is the type of FIS; Mamdani type is default.
- **andMethod, orMethod, impMethod, aggMethod,** and **defuzzMethod**, respectively, provide the methods for AND, OR, implication, aggregation, and defuzzification. The defaults are, respectively, min, max, min, max and centroid (centre of area).

The function **addvar** adds a variable to a FIS. It has four arguments in this order:
- The name of the FIS structure in the MATLAB workspace.
- The string representing the type of the variable we want to add ('input' or 'output').
- The string representing the name of the variable we want to add.
- The vector describing the limiting range values (universe of discourse) for the variable we want to add.

$$a = \mathbf{newfis('servo')}$$

$$a = \mathbf{addvar(a, 'varType', 'varName', 'varBounds')}$$

Indices are applied to variables in the order in which they are added; so the first input variable added to a system will always be known as input variable number one for that system. Input and output variables are numbered independently.

The Fuzzy Logic Toolbox includes many membership function types. The simplest membership functions are formed using straight lines. Of these, the simplest is the *triangular* membership function and it has the function name **trimf**. The triangular curve is a function of three scalar parameters a, b, and c; the parameters a and c locate the "feet" of the triangle and the parameter b locates the peak. The *trapezoidal* membership function, **trapmf**, has a flat top. The trapezoidal curve depends on four scalar parameters a, b, c, and d; the parameters a and d locate the "feet" of the trapezoid and the parameters b and c locate the "shoulders".

Each input/output variable existing in MATLAB workspace FIS (variables added by the function **addvar**) is resolved into a number of different fuzzy linguistic sets. The input/output variables must be fuzzified according to each of these linguistic sets.

A membership function can only be added to a variable in an existing MATLAB workspace FIS. Indices are assigned to membership functions in the order in which they are added; so the first membership function added to a variable will always be known as membership function number one for that variable. The function **addmf** adds a membership function to FIS. The function requires six input arguments in this order:

- A **MATLAB** variable name of a FIS structure in the workspace.
- A string representing the type of variable we want to add the membership function to ('input' or 'output').
- The index of the variable you want to add the membership function to (We cannot add a membership function to input variable number two of a system if only one input has been defined).
- A string representing the name of the new membership function.
- A string representing the type of the new membership function.
- The vector of parameters that specify the membership function.

a = newfis('servo')

a = addvar(a, 'varType', 'varName', 'varBounds')

a = addmf(a, 'varType', 'varIndex', 'mfName', 'mfType', 'mfParams')

Probably the trickiest part of the process of building a Fuzzy System is learning the short hand that the fuzzy inference systems use for building rules. This is accomplished using the command line function **addrule**.

Each variable, input, or, output, has an index number, and each membership function has an index number. The rules are built from statements like this:

IF input 1 is MF1 **and** input 2 is MF3 THEN Output is MF2

This rule is turned into a structure according to the following logic. If there are p inputs to a system and q outputs, then the first p vector entries of the rule structure correspond to inputs 1 through p. The entry in column 1 is the index number for the membership function associated with input 1. The entry in column 2 is the index number for the membership function associated with input 2, and so on. The next q columns work the same way for the outputs.

Column $p + q + 1$ is the weight associated with the rule (typically 1; all the rules have equal weightage) and column $p + q + 2$ specifies the connective used (where **and** = 1 and **or** = 2). The structure associated with the rule shown above is

$$1 \quad 3 \quad 2 \quad 1 \quad 1$$

The function **addrule** has two arguments. The first argument is the MATLAB workspace variable FIS name. The second argument is a matrix of one or more rows, each of which represents a given rule. The rule-list matrix takes the very specific format defined above.

ruleList = [
$$\begin{array}{ccccc} 1 & 1 & 1 & 1 & 1 \\ 1 & 2 & 2 & 1 & 1 \end{array}];$$

a = addrule(a, ruleList);

If the above system has two inputs and one output, the first rule can be interpreted as

"IF input 1 is MF1 **and** input 2 is MF1 THEN output 1 is MF1".

To evaluate the output of a fuzzy system for a given input, we use the function **evalfis**. It have the following arguments:

- A number or a matrix specifying the input values. If input is a $P \times p$ matrix, where p is the number of input variables, then **evalfis** takes each of the P rows of input as an input vector and returns the $P \times q$ matrix to the variable, output, where each row is an output vector and q is the number of output variables.
- The name of the FIS structure to be evaluated.

It is possible to use the Fuzzy Logic Toolbox by working strictly from the command line. However, in general, it is much easier to build a system graphically. These are five primary GUI tools for building, editing and observing fuzzy inference systems in the Fuzzy Logic Toolbox: the Fuzzy Inference System or FIS Editor, the Membership Function Editor, the Rule Editor, the Rule Viewer and the Surface Viewer.

The **FIS Editor** handles the high level issues for the system: How many input and output variables? What are their names?...

The **Membership Function Editor** is used to define the shapes of all the membership functions associated with each variable.

The **Rule Editor** is for editing the list of rules that defines the behaviour of the system.

The **Rule Viewer** is a display of the fuzzy inference diagram.

The **Surface Viewer** is used to display the dependency of one of the outputs on any one or two of the inputs.

The five primary GUIs can all interact and exchange information. For any fuzzy inference system, any or all of these five GUIs may be open. If more than one of these editors is open for a single system, the various GUI windows are aware of the existence of others and will, if necessary, update related windows. Thus if the names of the membership functions are changed using the

Membership Function Editor, these changes are reflected in the rules shown in the Rule Editor.

To start building a fuzzy inference system, type **fuzzy** at the MATLAB prompt. The generic untitled FIS Editor opens. At the top is a diagram of the system with input and output clearly labelled. By double-clicking on the input or output boxes, you can bring up the Membership Function Editor. Double-clicking on the fuzzy rule box in the centre of the diagram will bring up the Rule Editor.

Just below the diagram is a text field that displays the name of the current FIS. Lower left of window has a series of pop up menus, and the lower right has fields that provide information about the current variable.

GUI Editors:

fuzzy – Basic FIS Editor

mfedit – Membership Function Editor

ruleedit – Rule Editor

ruleview – Rule Viewer

surfview – Output Surface Viewer

When you save your fuzzy system to the MATLAB workspace, you are creating a variable (whose name you choose) that will act as a MATLAB structure for the FIS system.

Once you have created your fuzzy system entirely from the command line or using the GUI tools, you are ready to embed your system directly into Simulink and test it out in a simulation environment. The **Fuzzy Logic Toolbox** in the Simulink library contains the Fuzzy Logic Controller, and the Fuzzy Logic Controller with Rule Viewer blocks. It also includes a Membership Functions sub-library that contains Simulink blocks for the built-in membership functions. The Fuzzy Logic Controller with Rule Viewer block is an extension of the Fuzzy Logic Controller block. It allows you to visualize how rules are fired during simulation.

To start building a Simulink Fuzzy Model, drag the Fuzzy Logic Controller block (with or without the Rule Viewer) from the Simulink library to the simulation window (This can also be done by typing **fuzblock** at the MATLAB prompt). To initiate the Fuzzy Logic Controller block, double-click on the block and enter the name of the structure variable describing your FIS. This variable must be located in the MATLAB workspace. In most cases, the Fuzzy Logic Controller block automatically generates a hierarchical block diagram representation of your FIS. The block diagram representation only uses built-in Simulink blocks. This automatic, model-generation ability in called the **Fuzzy Wizard**. In cases where Fuzzy Wizard cannot handle FIS, the Fuzzy Logic Controller block uses the S-function **sffis** to simulate the FIS.

PROBLEMS

Each problem covers an important aspect of neural network/fuzzy logic/ genetic algorithm applications to function approximation, system identification, and control (motion control, and process control). Important MATLAB commands are included as help to these problems, in the form of script files. The description of the MATLAB functions used in the script files can easily be accessed from the **help file** using **help** command. Comments to the special features of commands are included in the script files to enhance the learning experience. To accelerate the learning process, brief descriptions are included with some problems.

Following each problem, one or more what-if's may be posed to examine the effect of variations of the key parameters.

C.1 *FunctionApproximation*

It is desired to design a 2-layer feedforward NN to approximate the function $y = f(x)$:

x	$-1 : 0.1 : 1$
y	$-0.960, -0.577, -0.073, 0.377, 0.641, 0.660, 0.461, 0.134,$ $-0.201, -0.434, -0.500, -0.393, -0.165, 0.099, 0.307, 0.396,$ $0.345, 0.182, -0.031, -0.219, -0.320$

The hidden layer with five neurons has hyperbolic activation functions, and the output layer is linear.

(a) Using batch gradient descent with momentum, train the network so that mean square error (mse) < 0.005. The learning rate $\eta = 0.01$. Repeat training three times with different initial weights. What is the conclusion of the experiment?

(b) Train the network using (i) batch gradient descent (ii) batch gradient descent with momentum, (iii) batch gradient descent with adaptive learning, and (iv) batch gradient descent with momentum and adaptive learning, and compare the convergence rates.

(c) Choose the fastest of the tested training algorithms. Train a network with 20 hidden units and compare its performance with one having 5 hidden units.

C.2 *SystemIdentification*

The nonlinear system to be identified is expressed by

$$y(k+1) = \frac{y(k)[y(k-1)+2][y(k)+2.5]}{8.5+[y(k)]^2+[y(k-1)]^2} + u(k)$$

where $y(k)$ is the output of the system at the kth time step and $u(k)$ is the input, which is a uniformly bounded function of time. The system is stable at $u(k) \in [-2, 2]$. Let the identification model be of the form

$$\hat{y}(k+1) = f(y(k), y(k-1)) + u(k)$$

where $f(y(k), y(k-1))$ represents the backpropagation network. The goal here is to train the network.

(a) Use the randomly created input data from the input space $[-2, 2]$ to obtain training patterns.
(b) Build the initial backpropagation network with 10 nodes (tansig) in the hidden layer and one (linear) output node.
(c) With batch training procedure, train the network to achieve mean square error (mse) < 0.005. The learning rate $\eta = 0.1$.
(d) After the backpropagation network is trained, test its prediction power for the input
$$u(k) = 2\cos(2\pi k/100), \; k \le 200$$
and
$$u(k) = 1.2\sin(2\pi k/20), \; 200 < k \le 500.$$
(e) Iterate on number of hidden layers, number of neurons in each layer, activation functions, mse goal, learning rate and training algorithm, if the learned network could not predict the nonlinear outputs quite well.

C.3 AdaptiveIdentifcation

Given the input signal
$$u(k) = 0.6\sin(2\pi k/10) + 1.2\cos(2\pi k/10); \; k \in \{1, 2, \ldots\}$$
This input is given to the linear system
$$y(k) = u(k) + 0.5u(k-1) - 1.5u(k-2)$$
Design a neural network that predicts the system output y, given the current and the previous two input signals u.

Parameters of the linear system are not robust. To take care of this problem, we go for adaptive identification. Train the network incrementally, assuming that the measurements are y_m:

randn('state', 0)

y_m = y + 0.1*randn(size (y))

('**randn**' is a MATLAB command for random number generation).

C.4 NNservocontrol

Consider the speed control system of Fig. 11.26 (Review Example 11.1).

(a) Using randomly generated inputs and the corresponding targets, train the NN to learn inverse dynamics of the plant. Evaluate the performance of the trained NN identifier by executing the dc motor model with the voltage
$$v_a(k) = 50\sin(2\pi kT/7) + 45\sin(2\pi kT/3); \; \forall \; kT \in [0, 20]$$
and comparing it with the estimated values given by NN identifier. What is the maximum prediction error?
(b) The tracking capability of the NN-based controller of Fig. 11.26 can be investigated for different arbitrarily specified trajectories. Simulate the performance of the controller in the configuration of Fig. 11.26 for an

arbitrarily selected speed track
$$\omega_r(k) = 10\sin(2\pi kT/4) + 16\sin(2\pi kT/7); \quad \forall \ kT \in [0, 20]$$

C.5 NNprocesscontrol

Consider the temperature control system of Fig. 11.29 (Review Example 11.2).

(a) Using a train of pulses as the input and the corresponding targets, train the NN to learn inverse dynamics of the plant. Evaluate the performance of the trained NN identifier on the training data set.

(b) From the initial condition $y(0) = Y_0$, the target is to follow a control reference, set to 35°C for $0 \leq t \leq 30$ samples, 55°C for $30 < t \leq 60$ samples, and 75°C for $60 < t \leq 90$ samples. Simulate the performance of the controller in the configuration of Fig. 11.29 for this temperature track.

C.6 *GeneticAlgorithm*

Write a set of functions built on the MATLAB numerical computing environment to implement the simple Genetic Algorithm described in Chapter 12. Using these MATLAB functions, solve Problem 12.16.

Help: The MATLAB Product Family does not include the Genetic Algorithm Toolbox. However, many research groups have developed such a toolbox as a collection of functions built on the MATLAB numerical computing environment; some of these toolboxes can be down loaded from the Internet.

The code in Script PC. 6 implements a simple Genetic Algorithm, constructed with the sole intent of clarifying the issues involved; generality, thus, is not the objective.

The MATLAB function **genetic** extremizes variables (within the limits imposed by minmax) as per fitness measure provided by **objfunc.m.**

The following notes will be helpful in appreciating the development of the program in Script PC. 6.

Note 1: Each population string encodes N parameters of the objective function. Number of bits for coding jth parameter is $nbit(j)$. The decimal value of this parameter varies between 0 and $2^{\wedge}nbit(j) - 1$. $inivar(i, j)$ is a random decimal integer value between these limits. This way we create a *popsize* × N initial-variable matrix whose elements are random decimal integers.

Note 2: The following example illustrates the conversion process from decimal to binary. Suppose the decimal value is 13.

2	13	
2	6	1
2	3	0
2	1	1
	0	1

Binary bits: 1101

Note 3: Actual parameter values are calculated as follows.

$$\text{Parameter value} = \text{min} + \frac{\text{max} - \text{min}}{2^{nbit} - 1} *(\text{decimal value})$$

Note 4: *randperm (popsize)* will return the shuffling of numbers from 1 to *popsize* and store them in *mplocation*. For example, if *popsize* = 4, we may get
mplocation = [1 4 3 2]
rand is a random number between 0 and 1. It is similar to a value we will get after a spun of wheel. Say
csum = [0.125 0.25 0.6 0.9 1], and *rand* = 0.5.
rand <= *csum*(3); the selected *j*th string (3rd in the example) is placed at random location in *mpool*.

Note 5: Since the mating pool is generated randomly, we pick the pairs of strings from the top of the list.

Note 6: Consider the *j*th parameter. Number of bits = *nbit*(*j*). Consider string *i*. There are *lchrom* bits in this string. Suppose bits *b*1 to *bn* on the string correspond to this parameter.

A bit at *n*th location has the decimal weight 2^{n-1}.

C.7 *FuzzySystem*

Using Fuzzy Logic Toolbox, realize the Fuzzy System specified in Problem 12.13. Determine the inferred value by the COA defuzzification method for the inputs $x_0 = 4$ and $y_0 = 8$.

C.8 *FLCservo*

Using Fuzzy Logic Toolbox, realize the Fuzzy System specified in Review Example 12.1.

From the initial condition $y(0) = 0$, the target is to follow a control reference, set to 20° for $0 \leq t \leq 250$ sec, and 23° for $250 < t \leq 500$ sec. Simulate the performance of the Fuzzy Controller in configuration of Fig. 12.32 for this position track. Choose the scaling constants GE, GV and GU' by trial-and-error.

C.9 *FLCprocess*

Using Fuzzy Logic Toolbox, realize the Fuzzy System specified in Review Example 12.2.

From the initial condition $y(0) = Y_0$, the target is to follow a control reference, set to 35°C for $0 \leq t \leq 360$ samples. Simulate the performance of the Fuzzy Controller for this temperature track. Choose the scaling constants GE, GV and GU' using genetic algorithm.

Script PC.1

```
%  FunctionApproximation
%  The given function  y=f(x)
clear all; close all;
% Training data:exemplar input pattern and target output vector
x=-1:0.1:1;
y=[-0.960,-0.577,-0.073,0.377,0.641,0.660,0.461,0.134,...
   -0.201,-0.434,-0.500,-0.393,-0.165,0.099,0.307,0.396,...
   0.345,0.182,-0.031,-0.219,-0.320];
% Create a NN and initialize weights
net=newff(minmax(x),[5 1],{'tansig','purelin'},'traingdx');
% Output of NN with initial weights
ycap1=sim(net,x);
figure(1);
plot(x,ycap1,'-',x,y,'o');
% Train the NN
figure(2)
net.trainParam.goal=0.005;
net.trainParam.epochs=500;
net.trainParam.show=50;
[net,tr]=train(net,x,y);
% Output of NN
figure(3);
% Generalization:input vector is different
%from the one used for training
x2=-1:0.01:1;
ycap2=sim(net,x2);
plot(x2,ycap2,'-',x,y,'o');
% Weights and Biases of the NN
w=net.IW{1,1}
bw=net.b{1}
v=net.LW{2,1}
bv=net.b{2}
```

Script PC.2

```
% SystemIdentification
clear all; close all;
% Define initial values
y(1)=0; y(2)=0;
% Create 500 random value data from input space  [-2,2]
%  and obtain training patterns
N=500;
```

```
for k=2:N
    f(k)=y(k)*(y(k-1)+2)*(y(k)+2.5)/(8.5+...
     y(k)*y(k)+y(k-1)*y(k-1));
    u(k)=(rand-0.5)*4;
    y(k+1)=f(k)+u(k);
end
for k=1:(N-1)
    Input(1,k)=y(k+1);
    Input(2,k)=y(k);
    Target(k)=f(k+1);
end
% Build the initial NN and train
HN=10;
net=newff(minmax(Input),[HN 1],{'tansig','purelin'});
net.trainParam.goal=0.005;
net=train(net,Input,Target);
% Test the NN
for k=1:200, u(k)=2*cos(2*pi*k/100); end
for k=201:500 u(k)=1.2*sin(2*pi*k/20); end
yp(1)=0; yp(2)=0; ycap(1)=0; ycap(2)=0;
% yp - from direct calculation
% ycap - NN simulation
for k=2:500
    yp(k+1)=yp(k)*(yp(k-1)+2)*(yp(k)+2.5)/(8.5+...
        yp(k)*yp(k)+yp(k-1)*yp(k-1))+u(k);
    ycap(k+1)= sim(net,[ycap(k); ycap(k-1)]) + u(k);
end
figure(2)
plot(1:501,yp,'r',1:501,ycap,'g');
```

Script PC.3

```
% AdaptiveIdentification
clear all;close all;
% Input signal
N=100;T=10;
index=1:N;
u=0.6*sin(2*pi*index/T)+1.2*cos(2*pi*index/T);
% Measured values of output
y(1)=u(1);y(2)=u(2)+0.5*u(1);
for k=3:N
    y(k)=u(k)+0.5*u(k-1)-1.5*u(k-2);
end;
randn('state',0);
```

```matlab
ym=y+0.1*randn(size(y));
% Training data
for k1=3:N
   k=k1-2;
   P(1,k)=u(k1);
   P(2,k)=u(k1-1);
   P(3,k)=u(k1-2);
   t(k)=ym(k1);
end;
% Initial NN
a=min(u);b=max(u);
net=newff([a b;a b;a b],[20 1],{'tansig','purelin'},'traingdm');
% Incremental training
net.trainParam.epochs=200;
net.trainParam.goal=1.0e-4;
net.trainParam.lr=0.5;
% Convert the input and output from numeric array
% to cell array for incremental training
P1=num2cell(P,1);
t1=num2cell(t,1);
disp('Starting adaptation.Please be patient...');
for i=1:10
    [net,netOut,netError]=adapt(net,P1,t1);
    disp(strcat('Pass',num2str(i),'Complete'));
end;
% Display mean squared adaptation error
figure(1)
% Combine a cell array into one matrix
plot(cell2mat(netError));
disp(strcat('Mean squared adaptation error=',num2str(mse(netError))));
% Plot measured output and NN estimate
figure(2)
plot(1:N-2,t,'b',1:N-2,sim(net,P),'r');
```

Script PC.4

```matlab
% NNservocontrol
clear all;close all;
% Plant model(Eqns.(11.95)-(11.96))
K1=0.34366;
K2=-0.1534068;
K3=-2.286928e-3;
K4=3.5193358e-4;
K5=0.2280595;
K6=-0.10518;
```

```
% Sampling Time
T=0.1;
% Generation of training data
w(1)=0;
w(2)=0;
for k=1:98
   va(k+1)=(rand-0.5)*200;
   w(k+2)=K1*w(k+1)+K2*w(k)+sign(w(k+1))*(K3*w(k+1)*w(k+1)+...
      K4*w(k)*w(k))+K5*va(k+1)+K6;
end;
y=[w(3:100);w(2:99);w(1:98)];
u=[va(2:99)];
% Preprocessing of training data
y=y/30;
u=u/100;
% Creation & training of NN
 net=newff(minmax(y),[4 1],{'tansig','purelin'},'traingdx');
net.trainParam.epochs=500;
net.trainParam.goal=0.005;
[net,tr]=train(net,y,u);
% Weights and biases of trained NN
w1=net.IW{1,1}
b1=net.b{1}
w2=net.LW{2,1}
b2=net.b{2}
% Testing the trained NN
kmax=floor(20/T);
for k=1:kmax,  va(k)=50*sin(2*pi*k*T/7)+45*sin(2*pi*k*T/3); end
wt(1)=0; wt(2)=0;
for  k=1:(kmax-1)
   wt(k+2)=K1*wt(k+1)+K2*wt(k)+sign(wt(k+1))*(K3*wt(k+1)*wt(k+1)+...
      K4*wt(k)*wt(k))+K5*va(k+1)+K6;
end
va_cap(1)=0;
for k=1:(kmax-1)
    Input=[wt(k+2); wt(k+1); wt(k)]/30;
    va_cap(k+1)=sim(net,Input)*100;
end
figure(2);
plot(1:kmax,va_cap,'r',1:kmax,va,'b');

%REFERENCE TRAJECTORY
for k=1:kmax
    wr(k)=10*sin(2*pi*k*T/4)+16*sin(2*pi*k*T/7);
end;
```

Appendix C: MATLAB Aided Control System Design: Neural–Fuzzy

```
%SIMULATION
% The command gensim(net,T), results in a Simulink screen
% containing
% the NN object net. We may build the feedback control system on
% this
% screen to simulate its performance.
%COMMAND LINE SIMULATION
w_sim(1)=0;
w_sim(2)=0;
for    k=1:kmax-2
    inp=[wr(k+2);w_sim(k+1);w_sim(k)]/30;
    va_sim(k+1)=sim(net,inp)*100;
    w_sim(k+2)=K1*w_sim(k+1)+K2*w_sim(k)+sign(w_sim(k+1))*...
    (K3*w_sim(k+1)*w_sim(k+1)+K4*w_sim(k)*w_sim(k))+K5*va_sim(k+1)+K6;
end
figure(3)
plot(1:kmax,wr,'b',1:kmax,w_sim,'r')
```

Script PC.5

```
% NNprocesscontrol
close all;clear all;
% Process Model(Equations  (11.101)-(11.103))
T=30;
Y0=25;
alpha=1.00151e-4;
beta=8.67973e-3;
F=exp(-alpha*T);
g=beta*(1-F)/alpha;
y(1)=0;
%  Generation  of  training  data
% Input is a train of pulses of random height
% between 0 and 5 & random width between 0 and 10.
s=0;
for j=1:10
    width(j)=rand*10;
    height(j)=rand*5;
    for k=1:10
      if(k<= width(j)) u(k+s)=height(j);
      else u(k+s)=0;
        end
y(k+s+1)=F*y(k+s)+g*u(k+s)/(1+exp(0.5*y(k+s)-40))+(1-F)*Y0;
    end;
    s=s+10;
```

```
end
yy=[y(2:101);y(1:100)];
u=[u(1:100)];
% Preprocessing of data
yy=yy/80;
uu=u/5;
% Creation and training of NN
net=newff(minmax(yy),[5 1],{'logsig' 'purelin'});
[net,tr]=train(net,yy,uu);
% Weights & biases of trained NN
w1=net.IW{1,1}
b1=net.b{1}
w2=net.LW{2,1}
b2=net.b{2}
% Testing the NN
ucap=sim(net,yy)*5;
figure(2)
k=1:100;
plot(k,ucap,'r',k,u,'b');

%REFERENCE TRAJECTORY
for i=1:90
   if(i<=30),yr(i)=35;
   elseif(i<=60),yr(i)=55;
     else,yr(i)=75;
     end;
end;
%COMMAND LINE SIMULATION
y_sim(1)=Y0;
for k=1:89
    inp=[yr(k+1);y_sim(k)]/80;
    u_sim(k)=sim(net,inp)*5;
y_sim(k+1)=F*y_sim(k)+g*u_sim(k)/(1+...
          exp(0.5*y_sim(k)-40))+(1-F)*Y0;
end
k=1:90;
figure(3)
plot(k,yr,'b',k,y_sim,'r')
```

Script PC.6

```
% GeneticAlgorithm
% Function 'genetic' is created in this program to find the
% optimum parameters for the system whose fitness function
% is stored in 'objfunc'. For a parameter optimization problem,
```

```
% functions 'genetic' and 'objfunc' must first be created as
% M-Files.

%   INPUT   VARIABLES:
%   popsize:Population   size(even   integer)
% objfunc:Function to be maximized;a valid existing M-file
% N :Number of variables in Objfunc.
% minmax :Nx2 matrix;jth row contains minimum and maximum values
% of the jth variable
% nbit :1xN row vector ;jth element contains number of bits
% for coding jth variable
% pc :Probabaility of crossover(pc<1)
% pm :Probability of mutation(pm<1)
% maxg :Maximum number of generations
% OUTPUT VARIABLE:
% opt_param:Optimal parameters
function[opt_param]=genetic(popsize,objfunc,N,minmax,nbit,pc,pm,maxg)
%  Initialize generation counter,'gc',and stopping flag 'stop'.
gc=1;
stop=0;
%  Select random initial decimal values of parameters in the
%  popsizexN matrix(refer Note 1)
for  i=1:popsize
    for j=1:N
        inivar(i,j)=round(rand*(2^nbit(j)-1));
    end;
end;
% Convert decimal values to binary bits and generate
% popsizexlchrom
% bitmatrix where lchrom is the length of a chromosome(refer Note
% 2)
lchrom=sum(nbit);
bitmat=zeros(popsize,lchrom);
for  i=1:popsize
    k=lchrom;
    for j=N:-1:1
        x=inivar(i,j);
        for m=1:nbit(j)
            bitmat(i,k)=rem(x,2);
            x=fix(x/2);
            k=k-1;
        end;
    end;
end;
```

```
% MAIN LOOP STARTS HERE
while (stop==0)
    gc
    % Calculate actual parameter values from bitmatrix(refer
    % Note 3)
    for j=1:N
      for i=1:popsize
          param(i,j)=minmax(j,1)...
              +(minmax(j,2)-minmax(j,1))/...
              (2^nbit(j)-1)*inivar(i,j);
      end;
    end;
    %   Calculate fitness values
    sumfit=0;
    for i=1:popsize
        argu=param(i,1:N);
         fitness(i)=feval(objfunc,argu);
         sumfit=sumfit+fitness(i);
    end
    avgfit=sumfit/popsize;
    % Parameters giving max fitness are given out as
        % optimum parameters
       [maxfit,index]=max(fitness);
    opt_param=param(index,:);
    % fit_ratio to be used in termination criteria
    fit_ratio=avgfit/maxfit;
    % Build an array of avg & max fitness values and
    % record them generation-wise
    af(gc)=avgfit;
     mf(gc)=maxfit;
    %  Generate mating pool through Roulette Wheel selection
    sum=0;
    for i=1:popsize
       % Compute probability of selection of the ith chromosome
          ps=fitness(i)/sumfit;
       % Obtain cumulative sum of the selection probabilities
          sum=sum+ps;
          csum(i)=sum;
    end;
    % randperm returns random sequencing of numbers from 1 to
    % popsize.
    mplocation=randperm(popsize);
    % The randomly spun RW selects jth chromosome & places it at
    % a
```

```matlab
% random location in mating pool.(refer Note 4)
for i=1:popsize
    rwspin=rand;
    for j=1:popsize
        if(rwspin<=csum(j))
            for k=1:lchrom
                mpool(mplocation(i),k)=bitmat(j,k);
            end;
            break;
        else;
        end;
    end;
end;
% Crossover operator
% Number of crossovers nc
nc=round(pc*popsize/2);
% Increment by 2 since crossover is pairwise(refer Note 5)
for i=1:2:nc
    xsite=round(rand*lchrom);
    for j=1:xsite
        temp=mpool(i,j);
        mpool((i+1),j)=temp;
    end;
end;
% Mutation operator
% Number of mutations nm
nm=round(pm*lchrom*popsize);
while(nm>0)
    for i=1:nm
        msite=round(rand*lchrom*popsize);
            % msite is being worked out not just within a
            % particular
            % chromosome,but over the entire bitmatrix
        if(msite==0)
      msite=1;
                % Since nm>0,make the 0 msite as 1 so that
                % atleast
                % one mutation does take place.
        end;
        % Location of the bit to be mutated
        colm=rem(msite,lchrom);
        if(colm==0)
            colm=lchrom;
            row=fix(msite/lchrom);
```

```
                else
                    row=fix(msite/lchrom)+1;
                end;
                if (mpool(row,colm)==0)
                    mpool(row,colm)=1;
            else
                    mpool(row,colm)=0;
            end;
        end;
        nm=0;
    end;
    % Decoding
    % Identify the bits in a string corresponding to a
    % parameter(refer Note 6)
    for i=1:popsize
        b1=0;
        bn=0;
      for j=1:N
         sum=0;
         n=nbit(j);
            if(j==1)
                b1=1;
            else
                b1=b1+nbit(j-1);
            end;
            bn=bn+nbit(j);
             % Calculate the decimal value
          for k=b1:bn
                n=n-1;
                sum=sum+(mpool(i,k))*(2^n);
            end;
         var(i,j)=sum;
      end;
    end;
          % Increment the generation counter
    gc=gc+1;
    inivar=var;
    bitmat=mpool;
    % Stopping criteria
    if((gc > maxg)|(fit_ratio > 0.9999)),stop=1;
    else stop=0;
    end;
end;
% MAIN LOOP ENDS
```

```
if (gc > maxg)
   disp('GOAL NOT REACHED');
end;
kf=1:gc-1;
plot(kf,mf,'b',kf,af,'r')
legend('maxfit','avgfit')

% Solution to Problem 12.16
% Create 'objfunc.m' file
function    [J]=objfunc(x)
J=(x(1)^2+x(2)-11)^2+(x(1)+x(2)^2-7)^2;
J=1/(1+J);

% To solve Problem 12.16, run the following commands after
% creating 'genetic' and 'objfunc' M-files.
clear all; close all;
[param]=genetic(20,'objfunc',2,[0 6;0 6],[10 10],0.8,0.05,100)
```

Script PC.7

```
% Fuzzy System
% Refer Problem 12.13
% Create new FIS with file name "FLCdemo.fis".
clear all; close all;
sys=newfis('FLCdemo');
% Add input variable x to FIS
sys=addvar(sys,'input','x',[2 9]);
% Add membership functions
sys=addmf(sys,'input',1,'A1','trimf',[2 5 8]);
sys=addmf(sys,'input',1,'A2','trimf',[3 6 9]);
% Add input variable y
sys=addvar(sys,'input','y',[4 11]);
% Add membership functions
sys=addmf(sys,'input',2,'B1','trimf',[5 8 11]);
sys=addmf(sys,'input',2,'B2','trimf',[4 7 10]);
% Add output variable to FIS
sys=addvar(sys,'output','z',[1 9]);
% Add membership functions
sys=addmf(sys,'output',1,'C1','trimf',[1 4 7]);
sys=addmf(sys,'output',1,'C2','trimf',[3 6 9]);
% Add rules to FIS
rule=[1 1 1 1 1;2 2 2 1 1];
sys=addrule(sys ,rule);
% Plot FIS I/O diagram
```

```
figure(1);
plotfis(sys);
% Perform fuzzy inference calculation(x=4;y=8)
z=evalfis([4 8],sys)
```

Script PC.8

```
% FLCservo
% Refer Review Example 12.1
clear all;close all;
% Create a new FIS with filename "FLCservo.fis"
sys=newfis('FLCservo');
% Define membership functions for the
% input variable "Rate of change of error"
sys=addvar(sys,'input','VELOCITY',[-1 1]);
sys=addmf(sys,'input',1,'N','trimf',[-1 -1 1]);
sys=addmf(sys,'input',1,'P','trimf',[-1 1 1]);
% Define membership functions for the input variable "Error"
sys=addvar(sys,'input','ERROR',[-1 1]);
sys=addmf(sys,'input',2,'N','trimf',[-1 -1 1]);
sys=addmf(sys,'input',2,'P','trimf',[-1 1 1]);
% Define membership functions for the
% output variable"Incremental change in input to the motor"
sys=addvar(sys,'output','CONTROL',[-1 1]);
sys=addmf(sys,'output',1,'N','trimf',[-1 -1 0]);
sys=addmf(sys,'output',1,'Z','trimf',[-1 0 1]);
sys=addmf(sys,'output',1,'P','trimf',[0 1 1]);
% Define fuzzy rules
rulelist=[1 2 2 1 1;1 1 1 1 1;2 2 3 1 1;2 1 2 1 1];
sys=addrule(sys,rulelist);
% Define gain constants
GE=0.03;
GV=0.05;
GU=60;
%Define reference signal
for k=1:500
    if(k<=250)yref(k)=20;
  else yref(k)=23;
    end;
end;
%COMMAND LINE SIMULATION
y(1)=0;y(2)=0;u(1)=0;
for k=1:499
    e(k)=yref(k)-y(k);
```

```
      if(k==1)
         v(1)=0;
      else
         v(k)=(e(k)-e(k-1))/0.25;
      end;
      einp(k)=e(k)*GE;
      vinp(k)=v(k)*GV;
      delu(k)=evalfis([vinp(k) einp(k)],sys);
      u(k+1)=u(k)+delu(k)*GU;
      if(k==1)
         y(2)=0;
      else
         y(k+1)=0.0237*u(k)+0.0175*u(k-1)+1.407*y(k)-0.407*y(k-1);
      end;
end;
k=1:500;
plot(k,yref,'b',k,y,'r')
% The fuzzy system created entirely from command line may be
% embeded
% directly into Simulink to test it out in a simulaion environment.
```

Script PC.9

```
% FLCprocess
% Refer Review Example 12.2
clear all; close all;
global alpha beta Y0 T F g sys yref
% Parameters of Plant Model(Equations (11.101)-
% (11.103))
alpha=1.00151e-4;
beta=8.67973e-3;
Y0=25;y(1)=Y0;u(1)=0;
T=30;
F=exp(-alpha*T);
g=(beta/alpha)*(1-F);
% Create a new FIS with filename "FLCprocess.fis"
sys=newfis('FLCprocess');
% Build the I/O membership functions and fuzzy rules
% Define membership functions for the
% input variable" Rate of change of error"
sys=addvar(sys,'input','VELOCITY',[-1 1]);
sys=addmf(sys,'input',1,'N','trimf',[-1 -1 0]);
```

```
sys=addmf(sys,'input',1,'NZ','trapmf',[-1 -0.1 0.1 1]);
sys=addmf(sys,'input',1,'P','trimf',[0 1 1]);
% Define membership functions for the input variable "Error"
sys=addvar(sys,'input','ERROR',[-1 1]);
sys=addmf(sys,'input',2,'N','trimf',[-1 -1 0]);
sys=addmf(sys,'input',2,'NZ','trapmf',[-1 -0.1 0.1 1]);
sys=addmf(sys,'input',2,'P','trimf',[0 1 1]);
% Define membership functions for the
% output variable "Incremental change in input to the thyristor"
sys=addvar(sys,'output','CONTROL',[-1 1]);
sys=addmf(sys,'output',1,'N','trimf',[-1 -1 0]);
sys=addmf(sys,'output',1,'Z','trimf',[-1 0 1]);
sys=addmf(sys,'output',1,'P','trimf',[0 1 1]);
% Define fuzzy rules
rulelist=[1 1 1 1 1;2 1 1 1 1;3 1 2 1 1;1 2 1 1 1;...
    2 2 2 1 1;3 2 3 1 1;1 3 2 1 1;2 3 3 1 1;3 3 3 1 1];
sys=addrule(sys,rulelist);
% Define reference signal
yref=ones(1,360)*35;
% Using Genetic Algorithm, obtain optimum values of Gain con
% stants.
% Creat MATLAB function 'objprocess' before executing the func
% tion genetic.
[param]=genetic(50,'objprocess',3,[0 1;0 1;0 500],[20 20 ...
40],0.8,0.05,10);
% Define gain constants
GE=param(1);GV=param(2);GU=param(3);
% Simulation results
for k=1:359
   e(k)=yref(k)-y(k);
    if(k==1)
        v(k)=0;
    else
        v(k)=(e(k)-e(k-1))/T;
    end;
   ee(k)=e(k)*GE;
    vv(k)=v(k)*GV;
    if(ee(k)>=-1 & ee(k)<=1 & vv(k)>=-1 & vv(k)<=1)
        delu(k)=evalfis([vv(k) ee(k)],sys)*GU;
    else delu(k)=0;
    end;
    u(k+1)=delu(k)+u(k);
    if(u(k+1)>5),u(k+1)=5;
    elseif(u(k+1)<0),u(k+1)=0;
```

```
        end
    y(k+1)=F*y(k)+(g*u(k+1)/(1+exp(0.5*y(k)-40)))+(1-F)*Y0;
end
k=1:360;
plot(k,yref,'r',k,y,'b')

% Create 'objprocess.m' file
function    [J]=objprocess(x)
global alpha beta Y0 T F g sys yref
% Define gain constants
GE=x(1);GV=x(2);GU=x(3);
y(1)=Y0;
u(1)=0;
for k=1:359
    e(k)=yref(k)-y(k);
    if(k==1)
        v(k)=0;
    else
      v(k)=(e(k)-e(k-1))/T;
    end;
    ee(k)=e(k)*GE;
    vv(k)=v(k)*GV;
    % If any input to the FLC is not in the range defined for it,
    % then the average of the output range is taken as FLC
    % output.
    if(ee(k)>=-1 & ee(k)<=1 & vv(k)>=-1 & vv(k)<=1)
        delu(k)=evalfis([vv(k) ee(k)],sys)*GU;
    else delu(k)=0;
    end
    u(k+1)=delu(k)+u(k);
        if(u(k+1)>5),u(k+1)=5;
    elseif(u(k+1)<0),u(k+1)=0;
    end
y(k+1)=F*y(k)+(g*u(k+1)/(1+exp(0.5*y(k)-40)))+(1-F)*Y0;
end;
J=0;
for i=1:359
    J=J+abs(e(i));
end;
J=1000/(1+J);
```

References

APPLICATIONS

1. Bogler, P.L., *Radar Principles with Applications to Tracking Systems*, New York: John Wiley and Sons, 1990.
2. Bose, B.K., *Power Electronics and AC Drives*, Englewood Cliffs, New Jersey: Prentice-Hall, 1986.
3. Sen, P.C., *Thyristor DC Drives*, New York: Wiley Interscience Publication, 1981.
4. Geiger, D.F., *Phaselock Loops for DC Motor Speed Control*, New York: Wiley Interscience Publication, 1981.
5. Rao, M., and H. Qiu, *Process Control Engineering*, Amsterdam; Gordon and Breach Science Publishers, 1993.
6. Coughanowr, D.R., *Process Systems Analysis and Control*, 2nd Edition, Singapore: McGraw-Hill Book Company, 1991.
7. Corripio, A.B., *Tuning of Industrial Control Systems*, Research Triangle Park, North Carolina: Instrument Society of America, 1990.
8. Deshpande, P.B., and R.H. Ash, *Computer Process Control*, 2nd Edition, Research Triangle Park, North Carolina: Instrument Society of America, 1989.
9. Seborg, D.E., T.F. Edgar, and D.A. Mellichamp, *Process Dynamics and Control*, New York: John Wiley and Sons, 1989.
10. Shinskey, F.G., *Process Control Systems*, 3rd Edition, New York: McGraw-Hill Book Company, 1988.
11. Astrom, K.J., and T. Hagglund, *Automatic Tuning of PID Regulators*, Research Triangle Park, North Carolina: Instrument Society of America, 1988.
12. Smith C.A., and A.B. Corripio, *Principles and Practice of Automatic Process Control*, New York: John Wiley and Sons, 1985.
13. Stephanopoulos, G., *Chemical Process Control—An Introduction to Theory and Practice*, Englewood Cliffs, New Jersey: Prentice-Hall, 1984.
14. Mclean, D., Automatic, *Flight Control Systems*, Hemel Hempstead: Prentice Hall International, 1990.
15. Ashley, H., *Engineering Analysis of Flight Vehicles*, Reading, Massachusetts: Addison-Wesley Publishing Company, 1974.

16. McRuer, D.T., I. Askenas, and D. Graham, *Aircraft Dynamics and Automatic Control*, Princeton, New Jersey: Princeton University Press, 1973.
17. Blakelock, J.H., *Automatic Control of Aircraft and Missiles*, New York: John Wiley and Sons, 1965.
18. Etkin, B., *Dynamics of Flight: Stability and Control*, New York: John Wiley and Sons, 1959.
19. Koivo, A.J., *Fundamentals for Control of Robotics Manipulators*, New York: John Wiley and Sons, 1989.
20. Spong, W., and M. Vidyasagar, *Robot Dynamics and Control*, New York: John Wiley and Sons, 1989.
21. Fu, K.S., R.C. Gonzalez, and C.S.G. Lee, *Robotics: Control, Sensing, Vision and Intelligence*, New York: McGraw-Hill Book Company, 1987.
22. Asada, H., and K. Youcef-Toumi, *Direct Drive Robots: Theory and Practice*, Cambridge, Massachusetts: The MIT Press, 1987.
23. Craig, J.J., *Introduction to Robotics: Mechanics and Control*, Reading, Massachusetts: Addition-Wesley Publishing Company, 1986.
24. Dorf, R.C., *Robotics and Automated Manufacturing*, Reston, Virginia: Reston Publishing Company, 1983.
25. Pessen, D.W., *Industrial Automation*, Singapore: John Wiley and Sons (SEA), 1990.
26. Hughes, T.A., *Programmable Controllers*, Research Triangle Park, North Carolina: Instrument Society of America, 1989.
27. Bryan, L.A., and E.A. Bryan, *Programmable Controllers: Theory and Implementation*, Chicago: Industrial Text Co., 1988.
28. Johnson, D.G., *Programmable Controllers for Factory Automation*, New York: Marcel Dekker, 1987.
29. Koren, Y., *Computer Control of Manufacturing Systems*, Singapore: McGraw-Hill Book Company, 1983.
30. Patton, W.J., *Numerical Control: Practice and Applications*, Reston, Virginia; Reston Publishing Company, 1970.
31. Beards, C.F., *Vibrations and Control Systems*, Chichester: Ellis Horwood Publishers, 1988.

MATHEMATICAL BACKGROUND

32. Noble, B., and J.W. Daniel, *Applied Linear Algebra*, 3rd Edition, Englewood Cliffs, New Jersey: Prentice-Hall, 1988.
33. Lancaster, P., and M. Tismenetsky, *The Theory of Matrices*, 2nd Edition, Orlando, Florida: Academic Press, 1985.
34. Oppenheim, A.V., A.S. Willsky, and I.T. Young, *Signals and Systems*. Englewood Cliffs, New Jersey: Prentice-Hall, 1983.
35. Churchill, R.V., J.W. Brown, and R.F. Verhev, *Complex Variables and Applications*, 3rd Edition, New York: McGraw-Hill Book Company, 1976.
36. Lefschetz, S., *Differential Equations: Geometric Theory*, New York: Wiley Interscience Publication, 1957.

MODELLING OF PHYSICAL SYSTEMS

37. Clark, R.N., *Control System Dynamics*, New York: Cambridge University Press, 1996.
38. Palm, W.J., III, *Modelling, Analysis and Control of Dynamics Systems*, New York: John Wiley and Sons, 1983.
39. Nagrath, I.J., and M. Gopal, *Systems: Modelling and Analysis*, New Delhi: Tata McGraw-Hill, 1982.
40. Doebelin, E.O., *System Modelling and Response*, New York: John Wiley and Sons, 1980.
41. Ogata, K., *System Dynamics*, Englewood Cliffs, New Jersey: Prentice-Hall, 1978.
42. Moschytz, G.S., *Linear Integrated Networks: Fundamentals*, New York: Van Nostrand Reinhold Company, 1974.
43. Schwarz, R.J., and B. Friedland, *Linear Systems*, New York: McGraw-Hill Book Company, 1965.
44. Mason, S.J., "Feedback Theory: Further Properties of Signal Flow Graphs", *Proc. IRE*, Vol. 44, No. 7, pp. 920-926, July 1956.
45. Mason, S.J., "Feedback Theory: Some Properties of Signal Flow Graphs", *Proc. IRE*, Vol. 41, No. 9, pp. 1144–1156, Sept. 1953.

INDUSTRIAL CONTROL DEVICES

46. De Silva, C.W., *Control Sensors and Actuators*, Englewood Cliffs, New Jersey: Prentice-Hall, 1989.
47. Anderson, W.R., *Controlling Electrohydraulic Systems*, New York: Marcel Dekker, 1988.
48. Parr, E.A., *Industrial Control Handbook*, Vol.1—Vol. 3, Oxford: BSP Professional Books, 1987.
49. Schuler., C.A., and W.L. McNamee, *Industrial Electronics and Robotics*, New York: McGraw-Hill Book Company, 1986.
50. Kenjo, T., and S. Nagamori, *Permanent-magnet and Brushless DC Motors*, Oxford: Clarendon Press, 1985.
51. Kenjo, T. *Stepping Motors and their Microprocessor Controls*, Oxford: Clarendon Press, 1984.
52. Morris, N.M., *Control Engineering*, 3rd Edition, London: McGraw-Hill Book Company, 1983.
53. Doebelin, E.O., *Measurement Systems*, 3rd Edition, New York: McGraw-Hill Book Company, 1983.
54. Singh, M.G., J-P Elloy, R. Mezencev, and N. Munro, *Applied Industrial Control*, Oxford: Pergamon Press, 1980.
55. Kuo, B.C. (ed.), *Incremental Motion Control*, Vol. 2: *Step Motors and Control Systems*, Champaign, Illinois: SRL Publishing Co., 1980.
56. Kuo, B.C. *Theory and Applications of Step Motors*, St. Paul, Minnesota: West Publishing Co., 1974.
57. Ahrendt, W.R., and C.J. Savant, Jr., *Servomechanism Practice*, 2nd Edition, New York: McGraw-Hill Book Company, 1960.

FEEDBACK CONTROL THEORY

58. Raven, F.H., *Automatic Control Engineering*, 5th Edition, New York: McGraw-Hill Book Company, 1995.
59. Wolovich, W.A., *Automatic Control Systems: Basic Analysis and Design*, Orlando, Florida: Saunders College Publishing, 1994.
60. Stafani, R.T., C.J. Savant, Jr., B. Shahian, and G.H. Hostetter, *Design of Feedback Control Systems*, 3rd Edition, Orlando, Florida: Saunders College Publishing, 1994.
61. Van De Vegte, J., *Feedback Control Systems*, 3rd Edition, Englewood Cliffs. New Jersey: Prentice-Hall, 1993.
62. Chen, C-T, *Control System Design*, Orlando, Florida: Saunders College Publishing, 1993.
63. Shinners, S.M., *Modern Control System Theory and Design*, New York: John Wiley and Sons, 1992.
64. Kuo, B.C., *Automatic Control Systems*, 6th Edition, Englewood Cliffs, New Jesey: Prentice-Hall, 1991.
65. Ogata, K., *Modern Control Engineering*, 2nd edition, Englewood Clifs, New Jersey: Prentice-Hall, 1990.
66. Dorf, R.C., *Modern Control Systems*, 5th Edition, Reading, Massachusetts: Addison-Wesley Publishing Company, 1989.
67. Thaler, G.J., *Automatic Control Systems*, St.Paul, Minnesota: West Publishing Co., 1989.
68. D'Azzo, J.J., and C.H. Houpis, *Linear Control System Analysis and Design*, 3rd Edition, New York: McGraw-Hill Book Company, 1988.
69. D'Souza, A.F., *Design of Control Systems*, Englewood Cliffs, New Jersey: Prentice-Hall, 1988.
70. Phillips, C.L., and R.D. Harbor, *Feedback Control Systems*, Englewood Cliffs, New Jersey: Prentice-Hall, 1988.
71. Franklin, G.F., J.D. Powell, and Abbas Emami-Naeini, *Feedback Control of Dynamic Systems*, Reading, Massachusetts: Addison-Wesley Publishing Company, 1986.
72. Nagrath, I.J., and M. Gopal, *Control Systems Engineering*, 2nd Edition, Singapore: John Wiley and Sons (SEA), 1986.
73. Palm, W.J., III, *Control Systems Engineering*, New York: John Wiley and Sons, 1986.
74. Doebelin, E.O., *Control System Principles and Design*, New York: John Wiley and Sons, 1985.
75. Kurman, K.J., *Feedback Control: Theory and Design*, Amsterdam: Elsevier Science Publishers, 1984.
76. Phelan, R.M., *Automatic Control Systems*, London: Cornell University Press, 1977.
77. Truxal, J.G., *Automatic Feedback Control System Synthesis*, New York: McGraw-Hill Book Company, 1955.

DIGITAL CONTROL

78. Santina, M.S., A.R. Stubberud, and G.H. Hostetter, *Digital Control System Design*, 2nd Edition, Orlando, Florida: Saunders College Publishing, 1994.
79. Jacquot, R.G., *Modern Digital Control Systems*, 2nd Edition, New York: Marcel Dekker, 1994.

80. Kuo, B.C., *Digital Control Systems*, 2nd Edition, Orlando, Florida: Saunders College Publishing, 1992.
81. Dote, Y., *Servo Motor and Motion Control Using Digital Signal Processors*, Englewood Cliffs, New Jersey: Prentice-Hall, 1990.
82. Astrom, K.J., and B. Wittenmark, *Computer Controlled Systems: Theory and Design*, 2nd Edition, Englewood Cliffs, New Jersey: Prentice-Hall, 1990.
83. Phillips, C.L., and H.T. Nagle, Jr., *Digital Control System Analysis and Design*, 2nd Edition, Englewood Cliffs, New Jersey: Prentice-Hall, 1990.
84. Franklin, G.F., J.G. Powell, and M.L. Workman, *Digital Control of Dynamic Systems*, 2nd Edition, Reading, Massachusetts: Addison-Wesley, 1990.
85. Isermann, R., *Digital Control Systems*, Vol. I, 2nd Edition, Berlin: Springer-Verlag, 1989.
86. Zikic, A.M., *Practical Digital Control*, Chichester: Ellis Horwood Publishers, 1989.
87. Gayakwad, R., and L. Sokoff, *Analog and Digital Control Systems*, Englewood Cliffs, New Jersey: Prentice-Hall, 1988.
88. Gopal, M., *Digital Control Engineering*, New Delhi: Wiley Eastern, 1988.
89. Warwick, K., and D. Rees (eds.), *Industrial Digital Control Systems*, Revised Edition, London: Peter Peregrinus, 1988.
90. Kuc, Roman, *Introduction to Digital Signal Processing*, New York: McGraw-Hill Book Company, 1988.
91. Ogata, K., *Discrete-Time Control Systems*, Englewood Clifs, New Jersey: Prentice-Hall, 1987.
92. Sinha, N.K. (ed.), *Microprocessor-Based Control Systems*, Dordrecht: D. Reidel Publishing Company, 1986.
93. Van Landingham, H.F., *Introduction to Digital Control Systems*, New York: MacMillan Publishing Company, 1985.
94. Ackermann, J., *Sampled-Data Control Systems*, Berlin: Springer-Verlag, 1985.
95. Leigh, J.R., *Applied Digital Control*, Englewood Cliffs, New Jersey: Prentice-Hall, 1985.
96. Houpis, C.H., and G.B. Lamont, *Digital Control Systems: Theory, Hardware, Software*, New York: McGraw-Hill Book Company, 1985.
97. Moroney, P., *Issues in the Implementation of Digital Feedback Compensators*, Cambridge, Massachusetts: The MIT Press, 1983.
98. Katz, P., *Digital Control Using Microprocessors*, Englewood Cliffs, New Jersey: Prentice-Hall, 1981.
99. Rao, G.V., *Complex Degitial Control Systems*. New York: Van Nostrand Reinhold Company, 1979.
100. Rao, M.V.C. and A.K. Subramanium "Elimination of singular cases in Jury's test". *IEEE Trans. Automatic Control.* Vol. AC-21, pp. 114–115. February 1976.
101. Jury, E.I., and J. Blanchard. "A Stability Test for Linear Discrete-time Systems in Table Form". *Proc. IRE.* Vol. 49, pp. 1947–1948, 1961.

STATE VARIABLE METHODS

102. Gopal M., *Modern Control System Theory*, 2nd Edition, New Delhi: Wiley Eastern, 1993.
103. Callier, F.M., and C.A. Desoer, *Linear System Theory*, New York: Springer-Verlag, 1991.

104. DeCarla, R.A., Linear Systems: *A State Variable Approach with Numerical Implementation*, Englewood Cliffs. New Jersey: Prentice-Hall, 1989.
105. Furuta, K., A Sano, and D. Atherton, *State Variable Methods in Automatic Control*, Chichester: John Wiley and Sons, 1988.
106. Friedland, B., *Control System Design: An Introduction to State-Space Methods*, New York: McGraw-Hill Book Company, 1986.
107. Brogan, W.L., *Modern Control Theory*, 2nd Edition, Englewood Cliffs, New Jersey: Prentice-Hall, 1985.
108. Chen, C-T., *Linear System Theory and Design*, New York: Holt, Rinehart and Winston, 1984.
109. Fortman, T.E., and K.L. Hitz. *An Introduction to Linear Control Systems*, New York: Marcel Dekker, 1977.
110. Wiberg, D.M., *State Space and Linear Systems* (Schaum's Outline Series), New York: McGraw-Hill Book Company, 1971.
111. Ogata, K., *State Space Analysis of Control Systems*, Englewood Cliffs, New Jersey: Prentice-Hall, 1967.
112. Schultz, D.G., and J.L. Melsa, *State Functions and Linear Control Systems*, New York: McGraw-Hill Book Company, 1967.
113. DeRusso, P.M., R.J. Roy, and C.M. Close, *State Variables for Engineers*, New York: John Wiley and Sons, 1965.
114. Dorf, R.C., *Time Domain Analysis and Design of Control Systems*, Reading, Massachusetts: Addison-Wesley Publishing Company, 1965.
115. Zadeh, L.A., and C.A. Desoer, *Linear System Theory: A State Space Approach*, New York: McGraw-Hill Book Company, 1963.

NONLINEAR CONTROL

116. Ching-Fang Lin, *Advanced Control Systems Design*, Englewood Cliffs, New Jersey: PTR Prentice-Hall, 1994.
117. Slotine, J-J.E., and W.Li., *Applied Nonlinear Control*, Englewood Cliffs, New Jersey: Prentice-Hall, 1991.
118. Mohler, R.R., *Nonlinear Systems*, Vol. I: *Dynamics and Control*, Englewood Cliffs, New Jersey: Prentice-Hall, 1991.
119. Atherton, D.P., *Stability of Nonlinear Systems*, New York: John Wiley and Sons, 1981.
120. Itkis, U., *Control Systems of Variable Structure*, New York: John Wiley and Sons, 1976.
121. Atherton, D.P., *Nonlinear Control Engineering*, London: Van Nostrand Reinhold Company, 1975.
122. Minorsky, N., *Theory of Nonlinear Control Systems*, New York: McGraw-Hill Book Company, 1969.

MULTIVARIABLE AND OPTIMAL CONTROL

123. Anderson, B.D.O., and J.B. Moore, *Optimal Control: Linear Quadratic Methods*, Englewood Cliffs, New Jersey: Prentice-Hall, 1990.
124. Grimble, M.J., and M.A. Johnson, *Optimal Control and Stochastic Estimation: Theory and Applications*, Vol. I, Chichester: John Wiley and Sons, 1988.

125. Kwakernaak, H., and R. Sivan, *Linear Optimal Control Systems*, New York: Wiley Interscience Publication, 1972.
126. Kirk, D.E., *Optimal Control Theory: An Introduction*, Englewood Cliffs, New Jersey: Prentice-Hall, 1970.
127. Sage, A.P., *Optimum Systems Control*, Englewood Cliffs, New Jersey: Prentice-Hall, 1968.
128. Kautsky, J., N.K. Nichols, and P. Dooren, "Robust pole assignment by linear state feedback", *Int., J. Control*, Vol. 41, No, 5, pp. 1129–1155, 1985.
129. Titli, A., and J. Bernussou, *Interconnected Dynamical Systems: Stability Decomposition and Decentralization*, System and Control Series, Vol. 5. Amsterdam: North Holland Publishing Company, 1982.
130. Patel, R.V., and N. Munro, *Multivariable System Theory and Design*, Oxford: Pergamon Press, 1982.
131. MacFarlane, A.J.G. (ed.), *Frequency Response Methods in Control Systems*, New York: IEEE Press, 1979.
132. Fisher, D.G., and D.E. Seborg (eds), *Mutlivariable Computer Control: A Case Study*, Amsterdam: North Holland Publishing Company, 1976.
133. Layton, J.M., *Multivariable Control Theory*, London: Peter Peregrinus, 1976.
134. Wolovich, W.A., *Linear Multivariable Systems*, New York: Springer-Verlag, 1974.

ROBUST CONTROL

135. Zhou, K., J.C. Doyle, and K. Glover, *Robust and Optimal Control*, New Jersey: Prentice-Hall, 1996.
136. Belanger, P.R., *Control Engineering: A Modern Approach*, Florida: Saunders College Publishing, 1995.
137. Saberi, A., B.M. Chen, and P. Sannuti, *Loop Transfer Recovery-Analysis and Design*, London: Springer-Verlag, 1993.
138. Doyle, J.C., B.A. Francis, and A.R. Tannenbaum, *Feedback Control Theory*. New York: MacMillan Publishing Company, 1992.
139. Morari. M., and E. Zafiriou, *Robust Process Control*, Englewood Cliffs. New Jersey: Prentice-Hall, 1989.
140. Bhattacharyya, S.P., *Robust Stabilization Against Structured Perturbations*. New York: Springer-Verlag, 1987.
141. Francis, B.A., *A Course in H_∞ Control Theory, Lecture Notes in Control and Information Sciences*, No. 88, Berlin: Springer-Verlag, 1987.
142. Doyle, J.C., and G. Stein, "Robustness with Observers". *IEEE Trans. Automatic Control*. Vol. AC-24, No. 4, pp. 607–611, Aug. 1979.
143. Dorato, P., L. Fortuna, and G. Muscato, *Robust Control for Unstructured Perturbations—An Introduction*, Lecture notes in control and information sciences, No. 168, Berlin: Springer-Verlag, 1992.
144. Frank, P.M., *Introduction to System Sensitivity Theory*, New York: Academic Press, 1978.

IDENTIFICATION AND ADAPTIVE CONTROL

145. Isermann, R., *Digital Control Systems*, Vol. II, 2nd Edition, Berlin: Springer-Verlag, 1991.

146. Landau, I.D., *System Identification and Control Design*, Englewood Cliffs. New Jersey: Prentice-Hall, 1990.
147. Sastry, S., and M. Bodson, *Adaptive Control: Stability, Convergence and Robustness*, Englewood Cliffs, New Jersey: Prentice-Hall, 1989.
148. Middleton. R.H., and G.C. Goodwin, *Digital Control and Estimation: A Unified Approach*. Englewood Cliffs, New Jersey: Prentice-Hall. 1989.
149. Narendra, K.S., and A.M. Annaswamy, *Stable Adaptive Systems*, Englewood Cliffs, New Jersey: Prentice-Hall, 1989.
150. Astrom., K.J., and B. Wittenmark, *Adaptive Control*, Reading, Massachusetts: Addison-Wesley Publishing Company, 1989.
151. Soderstorm, T., and P. Stoica, *System Identification*, Hemel Hempstead: Prentice-Hall International, 1989.
152. Grimble, M.J., and M.A. Johnson, *Optimal Control and Stochastic Estimation: Theory and Applications*, Vol. II, Chichester: John Wiley and Sons. 1988.
153. Ljung, L., *System Identification: Theory for the User*, Englewood Cliffs, New Jersey: Prentice-Hall, 1987.
154. Goodwin, G.C., and K.S. Sin, *Adaptive Filtering Prediction and Control*, Englewood Cliffs, New Jersey: Prentice-Hall, 1984.
155. Harris, C.J., and S.A. Billings (eds), *Self-Tuning and Adaptive Control: Theory and Applications*, Lodon: Peter Peregrinus, 1981.
156. Narendra K.S., and R.V. Monopoli (eds), *Applications of Adaptive Control*, New York: Academic Press, 1980.
157. Landau, Y.D., *Adaptive Control: The Model Reference Approach*, New York: Marcel Dekker, 1979.

INTELLIGENT CONTROL

158. Norgaard, M., O. Ravn, N.K. Poulsen, and L.K. Hansen, *Neural Networks for Modelling and Control of Dynamic Systems*, London: Springer-Verlag, 2000.
159. Lewis, F.L., S. Jagannathan, and A. Yesildirek, *Neural Network Control of Robot Manipulators and Nonlinear Systems*, London: Taylor & Francis, 1999.
160. Haykin, S., *Neural Networks: A Comprehensive Foundation*, 2nd Edition, Prentice-Hall, 1998.
161. Lin C-T, and C.S. George Lee, *Neural Fuzzy Systems*, Upper Saddle River, NJ: Prentice Hall, 1996.
162. Zurada, J.M., *Introduction to Artificial Neural Systems*, St Paul MN: West Publishing Company, 1992.
163. Ying, Hao., *Fuzzy Control and Modeling: Analytical Foundations and Applications*, New York: IEEE Press, 2000.
164. Passion, K.M., and S. Yurkovich, *Fuzzy Control,* California: Addison-Wesley, 1998.
165. Wang Li-Xin, *A Course in Fuzzy Systems and Control*, New Jersey: Prentice-Hall International, 1997.
166. Ross, T.J., *Fuzzy Logic with Engineering Applications*, New York, McGraw-Hill, 1995.
167. Driankov, D., H. Hellendoorn, and M. Reinfrank, *An Introduction to Fuzzy Control*, Berlin: Springer-Verlag, 1993.
168. Goldberg, D.E., *Genetic Algorithms in Search, Optimization, and Machine Learning*, Reading, Massachusetts, Addison-Wesley, 1989.

COMPUTER AIDED DESIGN

169. Linkens, D.A. (ed), *CAD for Control Systems*, New York: Marcel Dekker, 1993.
170. Chipperfield, A.J., and P.J. Fleming (eds), *MATLAB Toolboxes and Applications for Control*, London: Peter Peregrinus, 1993.
171. Jamshidi. M., and C.J. Herget (eds). *Advances in Computer-Aided Control Systems Engineering*, Amsterdam: North-Holland, 1985.
172. Korn, G.A., and J.V. Wait, *Digital Continuous System Simulation*. Englewood Cliffs, New Jersey: Prentice-Hall, 1978.
173. Speckhart, F.H., and W.L. Green, *A Guide to Using CSMP – The Continuous System Modelling Program (IBM Corp.)*, Englewood Cliffs, New Jersey: Prentice-Hall, 1976.
174. Program ACSL
    - Features: Advanced Continuous Simulation Language. The software has a section for simulation of sampled-data systems.
    - Source: Mitchell and Gauthier Associates, 73, Junction Square Drive, Concord, MA 01742-9990.
175. Program SIMNON
    - Features: Simulation of linear and nonlinear systems operating in continuous-time or discrete-time.
    - Source: Engineering Software Concepts, 436 Palo Alto Avenue, P.O. Box 66, Palo Alto, CA 94301.
176. Program CTRL-C
    - Features: Analysis and design of control systems; both the classical and the modern approaches. Permits direct interface to users pre-existing FORTRAN and C programs and to program ACSL.
    - Source: Systems Control Technology, Inc., 2300 Geng Road, P.O. Box 10180, Palo Alto, CA 94303-0888.
177. Program PC-MATLAB (with Control System Tool Box and SIMULAB)
    - Features: Classical and modern control tools; dynamic system simulation.
    - Source: The Math Works, Inc., Cochituate Place, 24 Prime Park Way. Natick, MA 01760.
    The Student Edition of MATLAB, Englewood Cliffs, New Jersey: Prentice-Hall, 1992.
178. Program MATRIX/PC
    - Features: Modelling, simulation and control design.
    - Source: Integrated Systems, Inc. Jay Street, Santa Clara, California 95054.
179. Program CC
    - Features: Analysis and design of linear control systems, analysis of stochastic systems; no facility for simulation of nonlinear systems.
    - Source: Systems Technology, Inc., 13766 South Hawthorne Blvd., Howthorne, CA 90250-7083.

COMPANION BOOK

180. Gopal, M., *Control Systems: Principles and Design*, 2nd Edition, New Delhi: Tata McGraw-Hill, 2002.

Companion Book

Gopal, M., *Control Systems: Principles and Design*, 2nd Edition, Tata McGraw-Hill, New Delhi, 2002.

International Edition, McGraw-Hill, Singapore, 2003.

1. **Introduction to the Control Problem**

    1.1 Control Systems: Terminology and Basic Structure; 1.2 The Genesis and Essence of the Feedback Control Theory; 1.3 Feedforward-Feedback Control Structure; 1.4 Multivariable Control Systems; 1.5 Scope and Organization of the Book

2. **Dynamic Models and Dynamic Response**

    2.1 Introduction; 2.2 State Variable Models; 2.3 Impulse Response Models; 2.4 Transfer Function Models; 2.5 Models of Disturbances and Standard Test Signals; 2.6 Dynamic Response; 2.7 Characteristic Parameters of First and Second-order Models; 2.8 Models of Mechanical Systems; 2.9 Models of Electrical Circuits; 2.10 Models of Thermal Systems; 2.11 Models of Hydraulic Systems; 2.12 Obtaining Models from Experimental Data; 2.13 Systems with Dead-time Elements; 2.14 Loading Effects in Interconnected Systems

3. **Models of Industrial Control Devices and Systems**

    3.1 Introduction; 3.2 Generalized Block Diagram of a Feedback System; 3.3 Block Diagram Manipulations; 3.4 Signal Flow Graphs and the Mason's Gain Rule; 3.5 DC and AC Motors in Control Systems; 3.6 Motion Control Systems; 3.7 Hydraulic Devices for Motion Control; 3.8 Pneumatic Devices for Process Control

4. **Basic Principles of Feedback Control**

    4.1 Introduction; 4.2 The Control Objectives; 4.3 Feedback Control System Characteristics; 4.4 Proportional Mode of Feedback Control; 4.5 Integral Mode of Feedback Control; 4.6 Derivative Mode of Feedback Control; 4.7 Alternative Control Configurations; 4.8 Multivariable Control Systems

5. **Concepts of Stability and the Routh Stability Criterion**

    5.1 Introduction; 5.2 Bounded-input Bounded-output Stability; 5.3 Zero-input Stability; 5.4 The Routh Stability Criterion; 5.5 Stability Range for a Parameter

6. The Performance of Feedback Systems

6.1 Introduction; 6.2 The Performance Specifications; 6.3 Response of a Standard Second-order System; 6.4 Effects of an Additional Zero and an Additional Pole; 6.5 Desired Closed-loop Pole Locations and the Dominance Condition; 6.6 Steady-state Error Constants and System-type Number; 6.7 Introduction to Design and Compensation

7. Compensator Design Using Root Locus Plots

7.1 Introduction; 7.2 The Root Locus Concept; 7.3 Guidelines for Sketching Root Loci; 7.4 Selected Illustrative Root Loci; 7.5 Reshaping the Root Locus; 7.6 Cascade Lead Compensation; 7.7 Cascade Lag Compensation; 7.8 Cascade Lag-Lead Compensation; 7.9 Minor-loop Feedback Compensation; 7.10 Compensation for Plants with Dominant Complex Poles; 7.11 The Root Locus of Systems with Dead-Time; 7.12 Sensitivity and the Root Locus

8. The Nyquist Stability Criterion and Stability Margins

8.1 Introduction; 8.2 Development of the Nyquist Criterion; 8.3 Selected Illustrative Nyquist Plots; 8.4 Stability Margins; 8.5 The Bode Plots; 8.6 Stability Margins on the Bode Plots; 8.7 Stability Analysis of Systems with Dead-Time; 8.8 Frequency Response Measurements

9. Feedback System Performance Based on the Frequency Response

9.1 Introduction; 9.2 Performance Specifications in Frequency Domain; 9.3 Correlation between Frequency-domain and Time-domain Specifications; 9.4 Constant-M Circles; 9.5 The Nichols Chart; 9.6 Sensitivity Analysis in Frequency Domain

10. Compensator Design Using Bode Plots

10.1 Introduction; 10.2 Reshaping the Bode Plot; 10.3 Cascade Lead Compensation; 10.4 Cascade Lag Compensation; 10.5 Cascade Lag-Lead Compensation; 10.6 Robust Control Systems

11. Hardware and Software Implementation of Common Compensators

11.1 Introduction; 11.2 Passive Electric Networks; 11.3 Operational Amplifier Usage; 11.4 Use of Digital Computer as a Compensator Device; 11.5 Configuration of the Basic Computer-control Scheme; 11.6 Principles of Signal Conversion; 11.7 Digital Implementation of Analog Compensators; 11.8 Tunable PID Controllers; 11.9 Ziegler-Nichols Methods for Controller Tuning

12. Control System Analysis Using State Variable Methods

12.1 Introduction; 12.2 Matrices; 12.3 State Variable Representation; 12.4 Conversion of State Variable Models to Transfer Functions; 12.5 Conversion of Transfer Functions to Canonical State Variable Models; 12.6 Solution of State Equations; 12.7 Concepts of Controllability and Observability; 12.8 Equivalence between Transfer Function and State Variable Representations

Appendix A: Mathematical Background

A.1 Introduction; A.2 Functions of a Complex Variable; A.3 Laplace Transforms; A.4 Table of Transforms

Appendix B: MATLAB Environment

Appendix C: Control Theory Quiz

Answers to Problems

Caution

For some problems (especially the design problems) of the book, many alternative solutions are possible. The answers given in the present section, correspond to only one possible solution for such problems.

▲▲

2.1 (a) x_1, x_2: Outputs of unit delayers, starting at the right and proceeding to the left.

$$\mathbf{F} = \begin{bmatrix} 0 & 1 \\ -0.368 & 1.368 \end{bmatrix}; \mathbf{g} = \begin{bmatrix} 0 \\ 1 \end{bmatrix}$$
$$\mathbf{c} = [0.264 \quad 0.368]$$

(b)

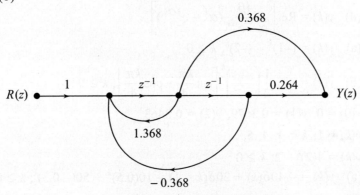

$$\frac{Y(z)}{R(z)} = \frac{0.368z + 0.264}{z^2 - 1.368z + 0.368}$$

2.2 (a) $y(k+2) + 5y(k+1) + 3y(k) = r(k+1) + 2r(k)$

(b) x_1, x_2: Outputs of unit delayers, starting at the right and proceeding to the left.

$$\mathbf{F} = \begin{bmatrix} 0 & 1 \\ -3 & -5 \end{bmatrix}; \mathbf{g} = \begin{bmatrix} 1 \\ -3 \end{bmatrix}; \mathbf{c} = \begin{bmatrix} 1 & 0 \end{bmatrix}$$

(c) $\dfrac{Y(z)}{R(z)} = \dfrac{z+2}{z^2+5z+3}$

2.3 (a) $y(k+1) + \dfrac{1}{2} y(k) = -r(k+1) + 2r(k)$

$$\dfrac{Y(z)}{R(z)} = \dfrac{-z+2}{z+\dfrac{1}{2}}$$

(b) $y(k) = \begin{cases} -1 & ; k = 0 \\ \dfrac{5}{2}\left(-\dfrac{1}{2}\right)^{k-1} & ; k \geq 1 \end{cases}$

(c) $y(k) = \dfrac{2}{3} - \dfrac{5}{3}\left(-\dfrac{1}{2}\right)^k ; k \geq 0$

2.4 (a) $y(k) = A\beta(\alpha)^{k-1}; k \geq 1$

(b) $y(k) = \dfrac{A\beta}{1-\alpha}[1-(\alpha)^k]; k \geq 0$

(c) $y(k) = \dfrac{A\beta}{(1-\alpha)^2}[\alpha^k + (1-\alpha)k - 1]; k \geq 0$

(d) $y(k) = \mathrm{Re}\left[\dfrac{A\beta}{\alpha - e^{j\Omega}}(\alpha^k - e^{j\Omega k})\right]$

2.5 (a) $y(k) = (-1)^k - (-2)^k; k \geq 0$

(b) $y(k) = 1 + \dfrac{1}{2}\left(\dfrac{1}{\sqrt{2}}\right)^k \left[\sin\dfrac{k\pi}{4} - \cos\dfrac{k\pi}{4}\right]; k \geq 0$

2.6 $y(0) = 0, y(1) = 0.3679, y(2) = 0.8463$
$y(k) = 1; k = 3, 4, 5, \ldots$

2.7 $y(k) = 3(2)^k - 2; k \geq 0$

2.8 (a) $y(k) = -40\delta(k) + 20\delta(k-1) - 10(0.5)^k + 50(-0.3)^k; k \geq 0$

(b) $y(k) = -16 + (0.56)^k[7.94 \sin(0.468k) + 16 \cos(0.468k)]; k \geq 0$

2.9 (a) $y(k) = -0.833(0.5)^k - 0.41(-0.3)^k + 0.476(-1)^k + 0.769; k \geq 0$

(b) $y(k) = -10k(0.5)^k + 2.5(0.5)^k - 6.94(0.1)^k + 4.44: k \geq 0$

2.10 $y(\infty) = \dfrac{K[1-a-a^2+a^3]}{(1-a)(1-a^2)}$

2.13 (a) No (b) Yes

Answers to Problems 953

2.14 Three

2.16 (b) $T = \pi/\sqrt{2}$

2.19

(i)

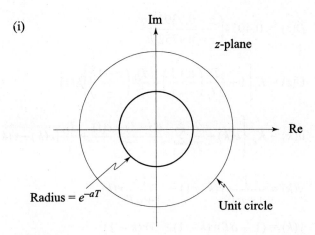

(ii)

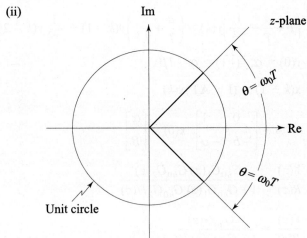

2.20 $G(z) = \dfrac{1 - e^{-T}}{z - e^{-T}}$

$y(k) = 1 - e^{-kT}; \; k \geq 0$

2.21 $\dfrac{Y(z)}{R(z)} = \dfrac{10}{16}\left[\dfrac{z + 0.76}{(z-1)(z-0.45)}\right]$

2.25 $u(k) - u(k-1) = K_c\left\{\left(1 + \dfrac{T}{T_I} + \dfrac{T_D}{T}\right)e(k)\right.$

$\left. - \left(1 + \dfrac{2T_D}{T}\right)e(k-1) + \dfrac{T_D}{T}e(k-2)\right\}$

$$U(z) = K_c\left[1 + \frac{T}{T_I}\left(\frac{1}{1-z^{-1}}\right) + \frac{T_D}{T}(1-z^{-1})\right]E(z)$$

2.26 (ii) $D(z) = 0.4074\left(\dfrac{z-0.9391}{z-0.9752}\right)$

2.27 (a) $U(z) = K_c\left[1 + \dfrac{T}{2T_I}\left(\dfrac{z+1}{z-1}\right) + \dfrac{T_D}{T}\left(\dfrac{z-1}{z}\right)\right]E(z)$

(b) $u(k) = K_c\left\{e(k) + \dfrac{T}{T_I}\sum_{i=1}^{k}\dfrac{e(i-1)+e(i)}{2} + \dfrac{T_D}{T}[e(k)-e(k-1)]\right\}$

2.28 (a) $y(k) = \dfrac{1}{1+aT}y(k-1) + \dfrac{T}{1+aT}r(k)$

(b) $y(k) = (1-aT)y(k-1) + Tr(k-1)$

2.29 (a) $\left(b + \dfrac{a}{T} + \dfrac{1}{T^2}\right)y(k) - \left(\dfrac{a}{T} + \dfrac{2}{T^2}\right)y(k-1) + \dfrac{1}{T^2}y(k-2) = 0;$

$y(0) = \alpha,\ y(-1) = \alpha - T\beta$

(b) $\mathbf{x}(k+1) = (\mathbf{I} + \mathbf{A}T)\mathbf{x}(k)$

$\mathbf{A} = \begin{bmatrix} 0 & 1 \\ -b & -a \end{bmatrix};\ \mathbf{x}(0) = \begin{bmatrix} \alpha \\ \beta \end{bmatrix}$

3.1 $\dfrac{Y(z)}{R(z)} = \dfrac{G_{h0}G_1(z)\,G_{h0}G_2(z)}{1 + G_{h0}G_1(z)\,G_{h0}G_2H(z)}$

3.2 $\dfrac{Y(z)}{R(z)} = \dfrac{G_{h0}G(z)}{1 + G_{h0}G(z)H(z)}$

3.3 $Y(z) = \dfrac{G_{h0}G_2(z)G_1R(z)}{1 + G_{h0}G_2HG_1(z)}$

3.4 $Y(z) = G_pH_2R(z) + \dfrac{D(z)G_{h0}G_p(z)}{1 + D(z)G_{h0}G_p(z)}[H_1R(z) - G_pH_2R(z)]$

3.5 $\dfrac{Y(z)}{R(z)} = \dfrac{D(z)G_{h0}G_1G_2(z)}{1 + D(z)[G_{h0}G_1(z) + G_{h0}G_1G_2(z)]}$

$\dfrac{X(z)}{R(z)} = \dfrac{D(z)G_{h0}G_1(z)}{1 + D(z)[G_{h0}G_1(z) + G_{h0}G_1G_2(z)]}$

3.6 $\quad Y(z) = \dfrac{GW(z)}{1 + D(z)G_{h0}G(z)}$

3.7 $\quad G_{h0}G(z) = 0.0288\left[\dfrac{z + 0.92}{(z-1)(z-0.7788)}\right]$

$\dfrac{\theta_L(z)}{\theta_R(z)} = \dfrac{G_{h0}G(z)}{1 + G_{h0}G(z)}$

3.8 $\quad \dfrac{\omega(z)}{\omega_r(z)} = 159.97\left(\dfrac{K_F + K_P}{z + 159.97K_P - 0.1353}\right)$

3.9 $\quad \dfrac{Y(z)}{R(z)} = 0.0095\left[\dfrac{z(z + 0.92)}{z^3 - 2.81z^2 + 2.65z - 0.819}\right]$

3.10 (i) $\quad \dfrac{Y(z)}{R(z)} = \dfrac{0.45z + 0.182}{z^2 + 0.082z + 0.182}$

(ii) $\quad \dfrac{Y(z)}{R(z)} = \dfrac{0.45z + 0.182}{z^2 - 0.368z^2 + 0.45z + 0.182}$

3.14 $0 < K < 4.293$

3.15 (i) The system is stable for $0 < K < 4.723$.
(ii) The system is stable for $0 < K < 3.315$.

3.16

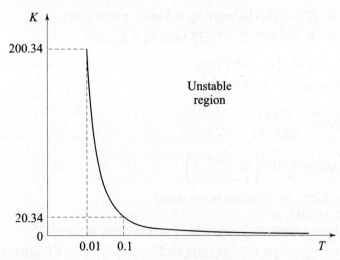

3.17 $0 < K < 0.785$.

3.18 For $T = 0.001$ sec, the response $y(k)$ is very close to $y(t)$.

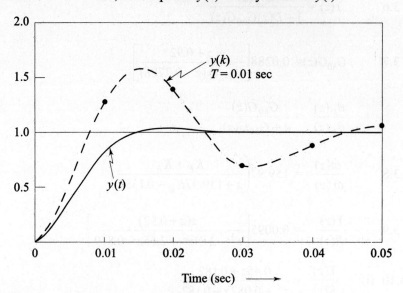

3.19 $y(k) = 1.02(0.795)^k \sin(0.89k); \ k \geq 0$

3.20 $y(0) = 0; \quad y(0.5T) = 0.393$

$y(T) = 0.632; \ y(1.5T) = 0.528$

$y(2T) = 0.465; \ y(2.5T) = 0.493$

$y(3T) = 0.509; \ y(3.5T) = 0.502$

$\vdots \qquad \vdots$

3.23 (a) $K = 30°C/(kg/min); \ \tau_D = 5$ min; $\tau = 60$ min

(b) $K_c = 0.545; \ T_I = 13.75$ min; $\tau_D = 2.2$ min

3.24 $K_c = 2.13; \ T_I = 666.66$ sec

4.1 $K_p = \infty; \ K_v = K_1/K_2; \ K_a = 0.$

4.2 $D_1(s) = \dfrac{25s+1}{62.5s+1}$

$D_1(z) = 0.4074\left(\dfrac{z-0.9391}{z-0.9752}\right)$

Velocity error constants are equal

4.3 $0, \ 1/3.041, \ \infty$

4.4 Underdamped response with $\zeta = 0.199$ and $\omega_n = 8.93$.

4.5 (a) $Y(z) = \dfrac{[D_2(z) + D_1(z)D_3(z)]G_{h0}G(z)}{1 + D_1(z)G_{h0}G(z)} R(z) + \dfrac{GW(z)}{1 + D_1(z)G_{h0}G(z)}$

(b) $Y(z) = D_3(z)R(z) + \dfrac{GW(z)}{1 + D_1(z)G_{h0}G(z)}$

(c) $D_1(z)$ can be made large to reject the disturbances

4.7 $S(z) = \dfrac{z - 0.607}{z - 0.214}$; $\omega_b = 2$ rad/sec

4.8 (b) $S_K^\lambda = -0.393$

(c) $S_\tau^\lambda = 0.6065$

4.9 $GM = 8$ dB; $\Phi M = 28°$

$v_b = 1.6$ rad/sec; $\omega_b = 1.35$ rad/sec

4.10 (a) Increase plant gain by a factor of 10; $\Phi M = 30°$;

(b) $D(z) = 4.2423\left(\dfrac{z - 0.8187}{z - 0.2308}\right)$

(c) $D(z) = 0.141\left(\dfrac{z - 0.98}{z - 0.998}\right)$

(d) $v_{b1} = 4.8$; $v_{b2} = 9.8$; $v_{b3} = 1.04$

(e) Yes

4.11 $D(z) = 37.333\left(\dfrac{z - 0.9048}{z - 0.1111}\right)$

$K_v = \infty$

In the low-frequency range, $\angle G_{h0}G(jv)$ is about $-180°$; therefore a lag compensator cannot fulfil the requirement of $50°$ phase margin.

4.12 (a) $K = 50$; $\Phi M = 20°$; $\omega_b = 9.27$ rad/sec.

(c) With lag compensator $D_1(z) = 0.342\dfrac{z - 0.923}{z - 0.973}$,

$\Phi M = 54°$, $\omega_b = 4.23$ rad/sec

With lag-lead compensator $D_1(z)D_2(z)$;

$D_2(z) = 2.49\dfrac{z - 0.793}{z - 0.484}$

$\Phi M = 60°$, $\omega_b = 7.61$ rad/sec

4.13 (a) $K = 61.25$

(b) Unstable

(c) $D(z) = 0.122K\left(\dfrac{z - 0.6}{z - 0.951}\right)$

4.14 (b) $0 < K < 2.1$; $K = 0.38$

4.15 (a) $K = 2.3925$

    (b) (i) $K = 1.4557$; (ii) $K = 0.9653$

4.16 (a) $K = 0.88$; 1.33 rad/sec

    (b) $K = 0.072$; $\tau = 2.3$ sec

    (c) $K = 0.18$; $\omega_n = 0.644$ rad/sec

4.17 $0 < A < 3.33$

4.18 (a) $\tau = 0.4854$

    (b) $K = 5.1223$

4.19 (a) $D_1(z) = 13.934\left(\dfrac{z - 0.8187}{z - 0.1595}\right)$

    (b) $K_v = 3$

    (c) $D_2(z) = \dfrac{z - 0.94}{z - 0.98}$

4.20 $D(z) = 1.91\left(\dfrac{z - 0.84}{z - 0.98}\right)$

4.21 (a) $z_{1,2} = 0.8 \pm j0.2$

    (b) Pole of $D(z)$ at $z = 0$; $(K/2) = 0.18$

    (c) $K_a = 0.072$

    (d) $z_3 = 0.2$; the third pole causes the response to slow down.

4.22 $D(z) = 150\left(\dfrac{z - 0.72}{z + 0.4}\right)$

4.23 (b) $D(z) = 208.33\left(\dfrac{z - 0.9048}{z + 0.9833}\right)$

4.24 (a) $D(z) = 135.22\left(\dfrac{(z - 0.9048)(z - 0.6135)}{(z + 0.9833)(z - 0.7491)}\right)$

    (b) $D(z) = 104.17\left(\dfrac{(z - 0.9048)(z + 1)}{(z + 0.9833)(z + 0.5)}\right)$; 0.15

4.25 $D(z) = \dfrac{4.8 - 3.9z^{-1}}{1 - z^{-1}}$

4.26 $D(z) = \dfrac{10 - 6z^{-1}}{1 + 0.75z^{-1}}$

4.27 $D(z) = \dfrac{1.8275 - 0.919z^{-1} + 0.09z^{-2}}{1 - 0.731z^{-1} - 0.269z^{-2}}$

4.28 $D(z) = \dfrac{1.582 - 0.582z^{-1}}{1 + 0.418z^{-1}}$

4.29 $D(z) = 1.582\left(\dfrac{1 - 0.3678z^{-2}}{1 - 0.6225z^{-2} - 0.3775z^{-3}}\right)$

No intersampling ripples in the output after the settling time is reached.

4.30 $D(z) = 2.5413\left[\dfrac{1 - 0.6065z^{-1}}{(1 - z^{-1})(1 + z^{-1})}\right]$

5.1 $x_1 = \theta_M$, $x_2 = \dot{\theta}_M$, $x_3 =$ motor armature current i_a, $x_4 =$ generator field current i_f; $y = \theta_L$

$$\mathbf{A} = \begin{bmatrix} 0 & 1 & 0 & 0 \\ 0 & -0.025 & 3 & 0 \\ 0 & -12 & -190 & 1000 \\ 0 & 0 & 0 & -4.2 \end{bmatrix}; \mathbf{b} = \begin{bmatrix} 0 \\ 0 \\ 0 \\ 0.2 \end{bmatrix}$$

$\mathbf{c} = [0.5 \ 0 \ 0 \ 0]$

5.2 $\mathbf{A} = \begin{bmatrix} 0 & 1 & 0 \\ 0 & -1 & 20 \\ 0 & 0 & -5 \end{bmatrix}; \mathbf{b} = \begin{bmatrix} 0 \\ 0 \\ 2.5 \end{bmatrix}$

$\mathbf{c} = [1 \ 0 \ 0]$

5.3 $x_1 = \theta_M$, $x_2 = \dot{\theta}_M$, $x_3 = i_a$

$$\mathbf{A} = \begin{bmatrix} 0 & 1 & 0 \\ 0 & -0.5 & 19 \\ -\dfrac{k_1}{40} & \dfrac{-(k_2 + 0.5)}{2} & \dfrac{-21}{2} \end{bmatrix}; \mathbf{b} = \begin{bmatrix} 0 \\ 0 \\ \dfrac{k_1}{2} \end{bmatrix}$$

$$\mathbf{c} = \begin{bmatrix} \dfrac{1}{20} & 0 & 0 \end{bmatrix}$$

5.4 $\mathbf{A} = \begin{bmatrix} \dfrac{-B}{J} & \dfrac{K_T}{J} \\ \dfrac{-(k_1 K_t K_c + K_b)}{L_a} & \dfrac{-(R_a + k_2 K_c)}{L_a} \end{bmatrix}$; $\mathbf{b} = \begin{bmatrix} 0 \\ \dfrac{k_1 K_c}{L_a} \end{bmatrix}$

$\mathbf{c} = \begin{bmatrix} 1 & 0 \end{bmatrix}$

5.5 (a) $\bar{\mathbf{A}} = \begin{bmatrix} -11 & 6 \\ -15 & 8 \end{bmatrix}$; $\bar{\mathbf{b}} = \begin{bmatrix} 1 \\ 2 \end{bmatrix}$; $\bar{\mathbf{c}} = \begin{bmatrix} 2 & -1 \end{bmatrix}$

(b)

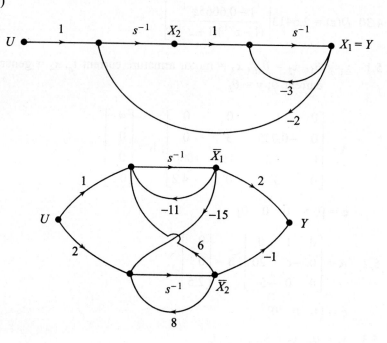

(c) $\dfrac{Y(s)}{U(s)} = \dfrac{1}{s^2 + 3s + 2}$

5.6 (a) $\mathbf{A} = \begin{bmatrix} 0 & 1 \\ 0 & 0 \end{bmatrix}$; $\mathbf{b} = \begin{bmatrix} 0 \\ 1 \end{bmatrix}$

(b) $\bar{\mathbf{A}} = \begin{bmatrix} 1 & 1 \\ -1 & -1 \end{bmatrix}$; $\bar{\mathbf{b}} = \begin{bmatrix} 0 \\ 1 \end{bmatrix}$

(c) $|\lambda\mathbf{I} - \mathbf{A}| = |\lambda\mathbf{I} - \overline{\mathbf{A}}| = \lambda^2$

5.7 $\mathbf{G}(s) = \dfrac{1}{\Delta}\begin{bmatrix} s(s+3) & s+3 & 1 \\ -1 & s(s+3) & s \\ -s & -1 & s^2 \end{bmatrix}$

$\mathbf{H}(s) = \dfrac{1}{\Delta}\begin{bmatrix} 1 \\ s \\ s^2 \end{bmatrix}$; $\Delta = s^3 + 3s^2 + 1$

5.9 x_1, x_2, x_3: outputs of integrators, starting at the right and proceeding to the left.

$\mathbf{A} = \begin{bmatrix} 0 & 1 & 0 \\ 0 & -2 & 1 \\ -2 & 1 & -2 \end{bmatrix}$; $\mathbf{b} = \begin{bmatrix} 0 \\ 0 \\ 1 \end{bmatrix}$; $\mathbf{c} = \begin{bmatrix} 2 & -2 & 1 \end{bmatrix}$

5.10 x_1, x_2, x_3, x_4: outputs of integrators
Top row: $x_1, x_2 = y_1$
Bottom row: $x_3, x_4 = y_2$

$\mathbf{A} = \begin{bmatrix} 0 & 0 & 0 & -4 \\ 1 & -3 & 0 & 0 \\ 0 & -1 & 0 & 0 \\ 0 & 0 & 1 & -4 \end{bmatrix}$; $\mathbf{B} = \begin{bmatrix} 3 & 0 \\ 1 & 2 \\ 0 & 3 \\ 0 & 0 \end{bmatrix}$; $\mathbf{x}(0) = \begin{bmatrix} 0 \\ y_1(0) \\ 0 \\ y_2(0) \end{bmatrix}$

$\mathbf{C} = \begin{bmatrix} 0 & 1 & 0 & 0 \\ 0 & 0 & 0 & 1 \end{bmatrix}$

5.11 (a) $G(s) = \dfrac{s+3}{(s+1)(s+2)}$

(b) $G(s) = \dfrac{1}{(s+1)(s+2)}$

5.12 $\mathbf{G}(s) = \dfrac{1}{\Delta}\begin{bmatrix} -3s+5 & 4(s-3) \\ -s^2+2s & 2(s^2-3s+1) \end{bmatrix}$

$\Delta = s^3 - 4s^2 + 6s - 5$

5.13 (a) $\mathbf{A} = \begin{bmatrix} -5 & 0.5 & -3.5 \\ 4 & -5 & 0 \\ 0 & 1 & 0 \end{bmatrix}; \mathbf{b} = \begin{bmatrix} 0 \\ 0 \\ -1 \end{bmatrix}$

$\mathbf{c} = \begin{bmatrix} 0 & 1 & 0 \end{bmatrix}$

(b) $G(s) = \dfrac{14}{(s+1)(s+2)(s+7)}$

5.14 (a) x_1 = output of lag $1/(s+2)$; x_2 = output of lag $1/(s+1)$

$\mathbf{A} = \begin{bmatrix} -2 & 1 \\ -1 & -1 \end{bmatrix}; \mathbf{b} = \begin{bmatrix} 0 \\ 1 \end{bmatrix}; \mathbf{c} = \begin{bmatrix} -1 & 1 \end{bmatrix}; d = 1$

(b) x_1 = output of lag $1/(s+2)$; x_2 = output of lag $1/s$; x_3 = output of lag $1/(s+1)$.

$\mathbf{A} = \begin{bmatrix} -2 & 1 & 1 \\ -1 & 0 & 0 \\ -1 & 0 & -1 \end{bmatrix}; \mathbf{b} = \begin{bmatrix} 0 \\ 1 \\ 1 \end{bmatrix}; \mathbf{c} = \begin{bmatrix} 0 & 1 & 1 \end{bmatrix}$

5.15 x_1 = output of lag $1/(s+1)$; x_2 = output of lag $5/(s+5)$; x_3 = output of lag $0.4/(s+0.5)$; x_4 = output of lag $4/(s+2)$.

$\mathbf{A} = \begin{bmatrix} -1-K_1 & -K_1 & 0 & 0 \\ 0 & -5 & -5K_2 & -5K_2 \\ -0.4K_1 & -0.4K_1 & -0.5 & 0 \\ 0 & 0 & -4K_1 & -2-4K_2 \end{bmatrix}$

$\mathbf{B} = \begin{bmatrix} K_1 & 0 \\ 0 & 5K_2 \\ 0.4K_1 & 0 \\ 0 & 4K_2 \end{bmatrix}; \mathbf{C} = \begin{bmatrix} 1 & 1 & 0 & 0 \\ 0 & 0 & 1 & 1 \end{bmatrix}$

5.16 (i) $\mathbf{A} = \begin{bmatrix} -1 & 0 \\ 0 & -2 \end{bmatrix}; \mathbf{b} = \begin{bmatrix} 1 \\ 1 \end{bmatrix}; \mathbf{c} = \begin{bmatrix} 2 & -1 \end{bmatrix}$

(ii) $\mathbf{A} = \begin{bmatrix} 0 & 0 & -2 \\ 1 & 0 & -5 \\ 0 & 1 & -4 \end{bmatrix}$; $\mathbf{b} = \begin{bmatrix} 5 \\ 0 \\ 0 \end{bmatrix}$; $\mathbf{c} = [0 \ \ 0 \ \ 1]$

(iii) $\mathbf{A} = \begin{bmatrix} 0 & 1 & 0 \\ 0 & 0 & 1 \\ -6 & -11 & -6 \end{bmatrix}$; $\mathbf{b} = \begin{bmatrix} 0 \\ 0 \\ 1 \end{bmatrix}$; $\mathbf{c} = [2 \ \ 6 \ \ 2]$; $d = 1$

5.17 (i) $\mathbf{A} = \begin{bmatrix} 0 & 0 & 0 \\ 1 & 0 & -2 \\ 0 & 1 & -3 \end{bmatrix}$; $\mathbf{b} = \begin{bmatrix} 1 \\ 1 \\ 0 \end{bmatrix}$; $\mathbf{c} = [0 \ \ 0 \ \ 1]$

(ii) $\mathbf{A} = \begin{bmatrix} 0 & 1 & 0 \\ 0 & 0 & 1 \\ -6 & -11 & -6 \end{bmatrix}$; $\mathbf{b} = \begin{bmatrix} 0 \\ 0 \\ 1 \end{bmatrix}$; $\mathbf{c} = [1 \ \ 0 \ \ 0]$

(iii) $\mathbf{\Lambda} = \begin{bmatrix} -1 & 0 & 0 \\ 0 & -2 & 0 \\ 0 & 0 & -3 \end{bmatrix}$; $\mathbf{b} = \begin{bmatrix} 1 \\ 1 \\ 1 \end{bmatrix}$; $\mathbf{c} = [-1 \ \ 2 \ \ 1]$; $d = 1$

5.18 (a) $\mathbf{A} = \begin{bmatrix} 0 & 1 & 0 \\ 0 & 0 & 1 \\ 0 & -100 & -52 \end{bmatrix}$; $\mathbf{b} = \begin{bmatrix} 0 \\ 0 \\ 1 \end{bmatrix}$; $\mathbf{c} = [5000 \ \ 1000 \ \ 0]$

(b) $\mathbf{\Lambda} = \begin{bmatrix} 0 & 0 & 0 \\ 0 & -2 & 0 \\ 0 & 0 & -50 \end{bmatrix}$; $\mathbf{b} = \begin{bmatrix} 50 \\ -31.25 \\ -18.75 \end{bmatrix}$; $\mathbf{c} = [1 \ \ 1 \ \ 1]$

5.19 (a) $\mathbf{\Lambda} = \begin{bmatrix} -1 & 1 & 0 \\ 0 & -1 & 0 \\ 0 & 0 & -2 \end{bmatrix}$; $\mathbf{b} = \begin{bmatrix} 0 \\ 1 \\ 1 \end{bmatrix}$; $\mathbf{c} = [1 \ \ 1 \ \ 1]$

(b) $y(t) = 2.5 - 2e^{-t} - te^{-t} - 0.5e^{-2t}$

5.20 (i) $\lambda_1 = 1, \lambda_2 = 2$

$$\mathbf{v}_1 = \begin{bmatrix} 1 \\ 0 \end{bmatrix}; \mathbf{v}_2 = \begin{bmatrix} 1 \\ 1 \end{bmatrix}$$

(ii) $\lambda_1 = -1, \lambda_2 = -2$

$$\mathbf{v}_1 = \begin{bmatrix} 1 \\ 1 \end{bmatrix}; \mathbf{v}_2 = \begin{bmatrix} 2 \\ 1 \end{bmatrix}$$

(iii) $\lambda_1 = -1, \lambda_2 = -2, \lambda_3 = -3$

$$\mathbf{v}_1 = \begin{bmatrix} 1 \\ -1 \\ -1 \end{bmatrix}; \mathbf{v}_2 = \begin{bmatrix} 1 \\ -2 \\ \frac{1}{2} \end{bmatrix}; \mathbf{v}_3 = \begin{bmatrix} 1 \\ -3 \\ 3 \end{bmatrix}$$

5.21 (b) $\lambda_1 = -2, \lambda_2 = -3, \lambda_3 = -4$

$$\mathbf{v}_1 = \begin{bmatrix} 1 \\ -2 \\ 4 \end{bmatrix}; \mathbf{v}_2 = \begin{bmatrix} 1 \\ -3 \\ 9 \end{bmatrix}; \mathbf{v}_3 = \begin{bmatrix} 1 \\ -4 \\ 16 \end{bmatrix}$$

5.22 (a) $\mathbf{P} = \begin{bmatrix} 1 & 1 & 1 \\ -1+j1 & -1-j1 & -1 \\ -j2 & j2 & 1 \end{bmatrix}$

(b) $\mathbf{Q} = \begin{bmatrix} \frac{1}{2} & -j\frac{1}{2} & 0 \\ \frac{1}{2} & j\frac{1}{2} & 0 \\ 0 & 0 & 1 \end{bmatrix}$

5.23 $\lambda_1 = -1, \lambda_2 = 2$

$$\mathbf{v}_1 = \begin{bmatrix} 1 \\ 1 \end{bmatrix}; \mathbf{v}_2 = \begin{bmatrix} 1 \\ 2 \end{bmatrix}$$

$$e^{\mathbf{A}t} = \begin{bmatrix} 2e^{-t} - e^{2t} & -e^{-t} + e^{2t} \\ 2e^{-t} - 2e^{2t} & -e^{-t} + 2e^{2t} \end{bmatrix}$$

5.24 (a) $e^{At} = \begin{bmatrix} \frac{3}{2}e^{-t} - \frac{1}{2}e^{-3t} & -\frac{3}{2}e^{-t} + \frac{3}{2}e^{-3t} \\ \frac{1}{2}e^{-t} - \frac{1}{2}e^{-3t} & -\frac{1}{2}e^{-t} + \frac{3}{2}e^{-3t} \end{bmatrix}$

(b) $e^{At} = \begin{bmatrix} \frac{3}{2}e^{-t} - \frac{1}{2}e^{-3t} & \frac{1}{2}e^{-t} - \frac{1}{2}e^{-3t} \\ -\frac{3}{2}e^{-t} + \frac{3}{2}e^{-3t} & -\frac{1}{2}e^{-t} + \frac{3}{2}e^{-3t} \end{bmatrix}$

5.25 (a) $e^{At} = \begin{bmatrix} 3e^{-2t} - 2e^{-3t} & e^{-2t} - e^{-3t} \\ -6e^{-2t} + 6e^{-3t} & -2e^{-2t} + 3e^{-3t} \end{bmatrix}$

(b) $e^{At} = \begin{bmatrix} (1+2t)e^{-2t} & 2te^{-2t} \\ -2te^{-2t} & (1-2t)e^{-2t} \end{bmatrix}$

5.26 (a)

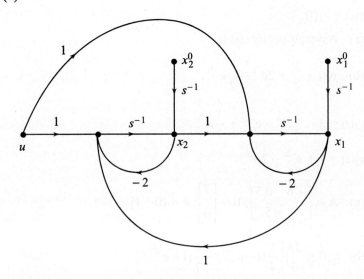

(b) $\dfrac{X_1(s)}{x_1^0} = G_{11}(s) = \dfrac{1/2}{s+1} + \dfrac{1/2}{s+3}$

$\dfrac{X_1(s)}{x_2^0} = G_{12}(s) = \dfrac{1/2}{s+1} + \dfrac{-1/2}{s+3}$

$\dfrac{X_2(s)}{x_1^0} = G_{21}(s) = \dfrac{1/2}{s+1} + \dfrac{-1/2}{s+3}$

$\dfrac{X_2(s)}{x_2^0} = G_{22}(s) = \dfrac{1/2}{s+1} + \dfrac{1/2}{s+3}$

$$\frac{X_1(s)}{U(s)} = H_1(s) = \frac{1}{s+1}; \quad \frac{X_2(s)}{U(s)} = H_2(s) = \frac{1}{s+1}$$

(c) (i) $\mathbf{x}(t) = \dfrac{1}{2}\begin{bmatrix} e^{-t}(x_1^0 + x_2^0) + e^{-3t}(x_1^0 - x_2^0) \\ e^{-t}(x_1^0 + x_2^0) + e^{-3t}(-x_1^0 - x_2^0) \end{bmatrix}$

(ii) $\mathbf{x}(t) = \begin{bmatrix} 1 - e^{-t} \\ 1 - e^{-t} \end{bmatrix}$

5.27 $x_1(t) = \dfrac{1}{3} - e^{-2t} + \dfrac{2}{3}e^{-3t}$

$x_2(t) = 2(e^{-2t} - e^{-3t})$

$x_3(t) = -2(2e^{-2t} - 3e^{-3t})$

$y(t) = x_1(t)$

5.28 (a) Asymptotically stable

(b) $y(t) = \dfrac{1}{2} + 2e^{-t} - \dfrac{3}{2}e^{-2t}$

5.29 $y_1(t) = 3 - \dfrac{5}{2}e^{-t} - e^{-2t} + \dfrac{1}{2}e^{-3t}$

$y_2(t) = 1 + e^{-2t} - 2e^{-3t}$

5.30 (a) $\mathbf{A} = \begin{bmatrix} -6 & 0.5 \\ 4 & -5 \end{bmatrix}; \mathbf{b} = \begin{bmatrix} 7 \\ 0 \end{bmatrix}; \mathbf{c} = [0 \quad 1]$

(b) $y(t) = \dfrac{28}{3}\left[\dfrac{1}{4}(1 - e^{-4t}) - \dfrac{1}{7}(1 - e^{-7t})\right]$

5.31 $\begin{bmatrix} x_1(1) \\ x_2(1) \end{bmatrix} = \begin{bmatrix} 2.7183 - k \\ 2k \end{bmatrix}$ for any $k \neq 0$

5.32 (a) Modes: $e^{\lambda_1 t}, e^{\lambda_2 t}, e^{\lambda_3 t}$

$\lambda_1 = -1, \lambda_2 = -2, \lambda_3 = -3$

(b) $\mathbf{x}(0) = \begin{bmatrix} k \\ -3k \\ 3k \end{bmatrix}; k \neq 0$

5.33 (b) $\mathbf{x}(0) = \begin{bmatrix} k \\ -k \end{bmatrix}; k \neq 0$

5.34 $e^{\mathbf{A}t} = \begin{bmatrix} 2e^{-t} - e^{-2t} & e^{-t} - e^{-2t} \\ 2e^{-2t} - 2e^{-t} & 2e^{-2t} - e^{-t} \end{bmatrix}$

$\mathbf{A} = \begin{bmatrix} 0 & 1 \\ -2 & -3 \end{bmatrix}$

5.37 Controllable but not observable.

5.38 (i) Controllable but not observable
 (ii) Controllable but not observable
 (iii) Both controllable and observable
 (iv) Both controllable and observable
 (v) Both controllable and observable

5.39 (i) Observable but not controllable
 (ii) Controllable but not observable
 (iii) Neither controllable nor observable

5.40 (i) $G(s) = \dfrac{1}{s+2}$. State model is not controllable

 (ii) $G(s) = \dfrac{s+4}{(s+2)(s+3)}$. State model is not observable

5.41 (a) $\lambda_1 = 1, \lambda_2 = -2, \lambda_3 = -1$; unstable

 (b) $G(s) = \dfrac{1}{(s+1)(s+2)}$; stable

5.42 (a) $\mathbf{A} = \begin{bmatrix} 0 & 1 \\ 0 & -1 \end{bmatrix}; \mathbf{b} = \begin{bmatrix} 0 \\ 1 \end{bmatrix}; \mathbf{c} = [10 \quad 0]$

 (b) $\mathbf{A} = \begin{bmatrix} 0 & 1 & 0 \\ 0 & 0 & 1 \\ 0 & -2 & -3 \end{bmatrix}; \mathbf{b} = \begin{bmatrix} 0 \\ 0 \\ 1 \end{bmatrix}; \mathbf{c} = [20 \quad 10 \quad 0]$

 (c) $\mathbf{A} = \begin{bmatrix} 0 & 0 & 0 \\ 1 & 0 & -2 \\ 0 & 1 & -3 \end{bmatrix}; \mathbf{b} = \begin{bmatrix} 20 \\ 10 \\ 0 \end{bmatrix}; \mathbf{c} = [0 \quad 0 \quad 1]$

6.1 $G(z) = \dfrac{1}{\Delta}\begin{bmatrix} z^2 & z & 1 \\ -(4z+1) & z(z+3) & z+3 \\ -z & -1 & z(z+3)+4 \end{bmatrix} z$

$\Delta = z^3 + 3z^2 + 4z + 1$

$H(z) = \dfrac{1}{\Delta}\begin{bmatrix} -3z^2 - 7z \\ -7z^2 - 9z + 3 \\ 3z + 7 \end{bmatrix}$

6.3 $G(z) = \dfrac{2z+2}{z^2 - z + \tfrac{1}{2}}$

6.4 $G(z) = \begin{bmatrix} \dfrac{2z+2}{\Delta} & 4 + \dfrac{-4z-14}{\Delta} & \dfrac{-30}{\Delta} \\ \dfrac{z+\tfrac{1}{2}}{\Delta} & \dfrac{-3z-4}{\Delta} & -2 + \dfrac{-3z-9}{\Delta} \end{bmatrix}$

$\Delta = z^2 - z + \tfrac{1}{2}$

6.5 x_1, x_2, x_3: Outputs of unit delayers starting at the top of the column of unit delayers and proceeding to the bottom.

$\mathbf{F} = \begin{bmatrix} \tfrac{1}{2} & \tfrac{1}{4} & 2 \\ 0 & -\tfrac{1}{2} & -1 \\ 0 & 3 & \tfrac{1}{3} \end{bmatrix}; \mathbf{g} = \begin{bmatrix} 1 \\ -1 \\ 2 \end{bmatrix}$

$\mathbf{c} = [5 \quad 6 \quad -7]; d = 8$

6.6 x_1, x_2, x_3: Outputs of unit delayers. x_1 and x_2 in first row, starting at the left and proceeding to the right.

$\mathbf{F} = \begin{bmatrix} 0 & 1 & 0 \\ 3 & 0 & 2 \\ -12 & -7 & -6 \end{bmatrix}; \mathbf{G} = \begin{bmatrix} 1 & 0 \\ 0 & 0 \\ 0 & 1 \end{bmatrix}$

$\mathbf{C} = \begin{bmatrix} 0 & 2 & 0 \\ 0 & 0 & 1 \end{bmatrix}; \mathbf{D} = \begin{bmatrix} 2 & 0 \\ 0 & 1 \end{bmatrix}$

6.7 (i) $\mathbf{F} = \begin{bmatrix} 0 & 1 \\ \frac{2}{3} & -\frac{1}{3} \end{bmatrix}$; $\mathbf{g} = \begin{bmatrix} 0 \\ 1 \end{bmatrix}$; $\mathbf{c} = [-1 \quad -2]$; $d = 3$

(ii) $\mathbf{F} = \begin{bmatrix} 0 & 0 & \frac{3}{4} \\ 1 & 0 & 1 \\ 0 & 1 & -1 \end{bmatrix}$; $\mathbf{g} = \begin{bmatrix} 0.5 \\ -3 \\ 4 \end{bmatrix}$

$\mathbf{c} = [0 \quad 0 \quad 1]$; $d = -2$

6.8 (i) $\mathbf{F} = \begin{bmatrix} -1 & 0 & 0 \\ 0 & -2 & 0 \\ 0 & 0 & -3 \end{bmatrix}$; $\mathbf{g} = \begin{bmatrix} 1 \\ 1 \\ 1 \end{bmatrix}$

$\mathbf{c} = [-1 \quad 2 \quad 1]$; $d = 1$

(ii) $\mathbf{F} = \begin{bmatrix} \frac{1}{3} & 1 & 0 \\ 0 & \frac{1}{3} & 1 \\ 0 & 0 & \frac{1}{3} \end{bmatrix}$; $\mathbf{g} = \begin{bmatrix} 0 \\ 0 \\ 1 \end{bmatrix}$

$\mathbf{C} = [5 \quad -2 \quad 3]$; $d = 0$

6.9 (i) $\mathbf{F} = \begin{bmatrix} 0 & 0 & -3 \\ 1 & 0 & -7 \\ 0 & 1 & -5 \end{bmatrix}$; $\mathbf{g} = \begin{bmatrix} 0 \\ 0 \\ 0 \end{bmatrix}$

$\mathbf{c} = [0 \quad 0 \quad 1]$; $d = 0$

(ii) $\mathbf{F} = \begin{bmatrix} -1 & 0 \\ 0 & -2 \end{bmatrix}$; $\mathbf{g} = \begin{bmatrix} 1 \\ 1 \end{bmatrix}$; $\mathbf{c} = [-2 \quad 7]$; $d = 0$

(iii) $\mathbf{F} = \begin{bmatrix} 0 & 1 & 0 \\ 0 & 0 & 1 \\ -3 & -7 & -5 \end{bmatrix}$; $\mathbf{g} = \begin{bmatrix} 0 \\ 0 \\ 1 \end{bmatrix}$

$\mathbf{c} = [2 \quad 1 \quad 0]$; $d = 0$

6.10 $\mathbf{F}^k = \begin{bmatrix} 1.5 - 0.5(3)^k & 0.5[(3)^k - 1] \\ -1.5[(3)^k - 1] & -0.5 + 1.5(3)^k \end{bmatrix}$

6.11 $y(k) = -\dfrac{17}{6}(-0.2)^k + \dfrac{22}{9}(-0.8)^k + \dfrac{25}{18}; k \geq 0$

6.12 $y_1(k) = 7\left(\dfrac{1}{2}\right)^k + 10\left(-\dfrac{1}{2}\right)^k + 2; k \geq 0$

$y_2(k) = 3\left(\dfrac{1}{2}\right)^k + 2\left(-\dfrac{1}{2}\right)^k + 1; k \geq 0$

6.13 (a) $\lambda_1 = -1, \lambda_2 = -1, \lambda_3 = -2$

Modes: $(-1)^k, k(-1)^{k-1}, (-2)^k$

(b) $\mathbf{x}(k) = \begin{bmatrix} k(-1)^{k-1} \\ (-1)^k \\ (-2)^k \end{bmatrix}$

6.14 (a) $G(z) = \mathscr{Z}[G_{h0}(s)G_a(s)] = \dfrac{0.2838z + 0.1485}{z^2 - 1.1353z + 0.1353}$

$\mathbf{F} = \begin{bmatrix} 0 & 1 \\ -0.1353 & 1.1353 \end{bmatrix}; \mathbf{g} = \begin{bmatrix} 0 \\ 1 \end{bmatrix}$

$\mathbf{c} = [0.1485 \quad 0.2838]$

(b) From controllable companion form continuous-time model:

$\mathbf{F} = \begin{bmatrix} 1 & 0.4323 \\ 0 & 0.1353 \end{bmatrix}; \mathbf{g} = \begin{bmatrix} 0.2838 \\ 0.4323 \end{bmatrix}$

$\mathbf{c} = [1 \quad 0]$

6.15 $\mathbf{F} = \begin{bmatrix} 0.696 & 0.246 \\ 0.123 & 0.572 \end{bmatrix}; \mathbf{g} = \begin{bmatrix} -0.021 \\ 0.747 \end{bmatrix}$

$\mathbf{c} = [2 \quad -4]; d = 6$

6.16 $\mathbf{x}(k+1) = \mathbf{F}\mathbf{x}(k) + \mathbf{g}_1 u(k) + \mathbf{g}_2 w(k)$

$\mathbf{F} = \begin{bmatrix} 1 & 0.1 \\ 0 & 0.99 \end{bmatrix}; \mathbf{g}_1 = \begin{bmatrix} 0.005 \\ 0.1 \end{bmatrix}; \mathbf{g}_2 = \begin{bmatrix} 0 \\ 0.01 \end{bmatrix}$

6.17 $\mathbf{F} = \begin{bmatrix} 0.741 & 0 \\ 0.222 & 0.741 \end{bmatrix}; \mathbf{G} = \begin{bmatrix} 259.182 & 0 \\ 36.936 & 259.182 \end{bmatrix}$

$\mathbf{C} = \begin{bmatrix} 1 & 0 \\ 0 & 1 \end{bmatrix}$

6.18 (a) $0.3679 \left[\dfrac{z + 0.7181}{(z-1)(z-0.3679)} \right]$

(b) $\dfrac{Y(z)}{R(z)} = \dfrac{0.3679z + 0.2642}{z^2 - z + 0.6321}$

$\mathbf{F} = \begin{bmatrix} 0 & 1 \\ -0.6321 & 1 \end{bmatrix}; \mathbf{g} = \begin{bmatrix} 0 \\ 1 \end{bmatrix}$

$\mathbf{c} = [0.2642 \quad 0.3679]$

6.19 (a) $G(z) = \mathscr{Z}[G_{h0}(s)G_a(s)] = \dfrac{0.4512z + 0.1809}{z^2 - 0.3679z}$

$\mathbf{F} = \begin{bmatrix} 0 & 1 \\ 0 & 0.3679 \end{bmatrix}; \mathbf{g} = \begin{bmatrix} 0 \\ 1 \end{bmatrix}$

$\mathbf{c} = [0.1809 \quad 0.4512]$

(b) $\dot{y}(t) = -y(t) + u(t - 0.4)$

$x_1(k) = y(k); x_2(k) = u(k-1)$

$\mathbf{F} = \begin{bmatrix} 0.3679 & 0.1809 \\ 0 & 0 \end{bmatrix}; \mathbf{g} = \begin{bmatrix} 0.4512 \\ 1 \end{bmatrix}$

$\mathbf{c} = [1 \quad 0]$

6.20 $x_1(k) = x(k); x_2(k) = u(k-3); x_3(k) = u(k-2); x_4(k) = u(k-1)$

$\mathbf{F} = \begin{bmatrix} 0.3679 & 0.2387 & 0.3935 & 0 \\ 0 & 0 & 1 & 0 \\ 0 & 0 & 0 & 1 \\ 0 & 0 & 0 & 0 \end{bmatrix}; \mathbf{g} = \begin{bmatrix} 0 \\ 0 \\ 0 \\ 1 \end{bmatrix}$

6.21 $x_1 = y$; $x_2 = \dot{y}$; $x_3(k) = u(k-1)$

$$\mathbf{F} = \begin{bmatrix} 1 & T & \tau_D(T-\tau_D/2) \\ 0 & 1 & \tau_D \\ 0 & 0 & 0 \end{bmatrix}; \mathbf{g} = \begin{bmatrix} (T-\tau_D)^2/2 \\ T-\tau_D \\ 1 \end{bmatrix}$$

$\mathbf{c} = \begin{bmatrix} 1 & 0 & 0 \end{bmatrix}$

6.23 (a) x_1 = output of lag $1/s$; x_2 = output of lag $1/(s+1)$; x_3 = output of lag $1/(s+2)$.

$$\mathbf{x}(k+1) = \begin{bmatrix} \frac{7}{4} - \frac{1}{2}T - e^{-T} + \frac{1}{4}e^{-2T} & 1 - e^{-T} & \frac{1}{2}(1 + e^{-2T} - 2e^{-T}) \\ -\frac{1}{2} + e^{-T} - \frac{1}{2}e^{-2T} & e^{-T} & e^{-T} - e^{-2T} \\ -\frac{1}{2} + \frac{1}{2}e^{-2T} & 0 & e^{-2T} \end{bmatrix} \mathbf{x}(k)$$

$$+ \begin{bmatrix} -\frac{3}{4} + \frac{1}{2}T + e^{-T} - \frac{1}{4}e^{-2T} \\ \frac{1}{2} - e^{-T} + \frac{1}{2}e^{-2T} \\ \frac{1}{2} - \frac{1}{2}e^{-2T} \end{bmatrix} r(k)$$

6.24 $\mathbf{x}(k+1) = \begin{bmatrix} 0.4 & 0.233 \\ -0.698 & -0.0972 \end{bmatrix} \mathbf{x}(k) + \begin{bmatrix} 0.2 \\ 0.233 \end{bmatrix} r(k)$

$y(k) = \begin{bmatrix} 1 & 0 \end{bmatrix} \mathbf{x}(k)$

6.25 $\mathbf{x}(k+1) = \begin{bmatrix} 0.6 & 0.233 & 0.2 \\ -0.465 & -0.0972 & 0.233 \\ -1 & 0 & -2 \end{bmatrix} \mathbf{x}(k) + \begin{bmatrix} 0 \\ 0 \\ 1 \end{bmatrix} r(k)$

6.26 $u(k) = 50e(k) - 41e(k-1)$

$x_1 = y$; $x_2 = \dot{y}$; $x_3(k) = e(k-1)$

$$\mathbf{x}(k+1) = \begin{bmatrix} 0.75 & 0.1 & -0.205 \\ -5 & 0.99 & -4.1 \\ -1 & 0 & 0 \end{bmatrix} \mathbf{x}(k) + \begin{bmatrix} 0.25 \\ 5 \\ 1 \end{bmatrix} r(k)$$

6.27 (a) Both controllable and observable
(b) Both controllable and observable

6.28 $T = n\pi; \; n = 1, 2, \ldots$

6.29 $T \neq n; \; n = 1, 2, 3, \ldots$

6.30 (a) $\lambda_1 = \dfrac{1}{4}, \; \lambda_2 = \dfrac{1}{2}$

(b) $G(z) = \dfrac{1}{z - \dfrac{1}{4}}$

(c) Controllable but not observable

7.2 (a) $k_1 = 74, \; k_2 = 25, \; k_3 = 3$

(b) $\dot{\hat{x}} = (A - mc)\hat{x} + bu + my$
$m^T = [3 \quad 7 \quad -1]$

(c) With reference to Fig. 7.7:

$$\hat{x}_e = \begin{bmatrix} \hat{x}_2 \\ \hat{x}_3 \end{bmatrix}; \quad \begin{bmatrix} a_{11} & a_{1e} \\ a_{e1} & A_{ee} \end{bmatrix} = \begin{bmatrix} 0 & 1 & 0 \\ 0 & 0 & 1 \\ -6 & -11 & -6 \end{bmatrix}$$

$$\begin{bmatrix} b_1 \\ b_e \end{bmatrix} = \begin{bmatrix} 0 \\ 0 \\ 1 \end{bmatrix}; \quad m = \begin{bmatrix} -2 \\ 17 \end{bmatrix}$$

7.3 (a) $k = [3 \quad 7 \quad -1]$
$\dot{x} = (A - bk)x$

(b) $m^T = [74 \quad 25 \quad 3]$

(c) $\begin{bmatrix} \dot{x}_3 \\ \dot{x}_1 \\ \dot{x}_2 \end{bmatrix} = \begin{bmatrix} -6 & 0 & 1 \\ -6 & 0 & 0 \\ -11 & 1 & 0 \end{bmatrix} \begin{bmatrix} x_3 \\ x_1 \\ x_2 \end{bmatrix} + \begin{bmatrix} 0 \\ 1 \\ 0 \end{bmatrix} u$

$= \begin{bmatrix} a_{11} & a_{1e} \\ a_{e1} & A_{ee} \end{bmatrix} \begin{bmatrix} x_3 \\ x_1 \\ x_2 \end{bmatrix} + \begin{bmatrix} b_1 \\ b_e \end{bmatrix} u$

With reference to Fig. 7.7:

$$\hat{\mathbf{x}}_e = \begin{bmatrix} \hat{x}_1 \\ \hat{x}_2 \end{bmatrix}; \; \mathbf{m} = \begin{bmatrix} 16 \\ 4 \end{bmatrix}$$

7.4 $\dot{\hat{\mathbf{x}}} = (\mathbf{A} - \mathbf{mc})\,\hat{\mathbf{x}} + \mathbf{B}u + \mathbf{m}(y - du)$

$\mathbf{m}^T = [-1 \quad 3]$

7.5 With reference to Fig. 7.7:

$$\hat{\mathbf{x}}_e = \hat{z}_2; \; \begin{bmatrix} a_{11} & a_{1e} \\ a_{e1} & A_{ee} \end{bmatrix} = \begin{bmatrix} 2 & 2 \\ \hdashline -1 & -1 \end{bmatrix}$$

$$\begin{bmatrix} b_1 \\ b_e \end{bmatrix} = \begin{bmatrix} 1 \\ 0 \end{bmatrix}; \; m = 4.5; \; \hat{\mathbf{x}} = \begin{bmatrix} y + \hat{z}_2 \\ y + 2\hat{z}_2 \end{bmatrix}$$

7.6 (a) $\mathbf{A} = \begin{bmatrix} 0 & 9 \\ 1 & 0 \end{bmatrix}; \; \mathbf{b} = \begin{bmatrix} 9 \\ 0 \end{bmatrix}; \; \mathbf{c} = [0 \quad 1]$

(b) $\mathbf{k} = \begin{bmatrix} \dfrac{2}{3} & 3 \end{bmatrix}$

(c) $\dot{\hat{\mathbf{x}}} = (\mathbf{A} - \mathbf{mc})\hat{\mathbf{x}} + \mathbf{b}u + \mathbf{m}y$

$\mathbf{m}^T = [81 \quad 12]$

(d) $\mathbf{k} = \begin{bmatrix} \dfrac{1}{9} & \dfrac{2}{9} \end{bmatrix}$

7.7 (a) $\mathbf{A} = \begin{bmatrix} 0 & 1 \\ -\omega_0^2 & 0 \end{bmatrix}; \; \mathbf{b} = \begin{bmatrix} 0 \\ 1 \end{bmatrix}; \; \mathbf{c} = [1 \quad 0]$

(b) $k_1 = 3\omega_0^2; \; k_2 = 4\omega_0$

(c) $\dot{\hat{\mathbf{x}}} = (\mathbf{A} - \mathbf{mc})\,\hat{\mathbf{x}} + \mathbf{b}u + \mathbf{m}y$

$\mathbf{m}^T = [20\omega_0 \quad 99\omega_0^2]$

(d) With reference to Fig. 7.7:

$$\hat{\mathbf{x}}_e = \hat{x}_2; \; \begin{bmatrix} a_{11} & a_{1e} \\ a_{e1} & A_{ee} \end{bmatrix} = \begin{bmatrix} 0 & 1 \\ \hdashline -\omega_0^2 & 0 \end{bmatrix}$$

$$\begin{bmatrix} b_1 \\ b_e \end{bmatrix} = \begin{bmatrix} 0 \\ \cdots \\ 1 \end{bmatrix}; \; m = 10\omega_0$$

7.8 (a) $\mathbf{k} = [29.6 \quad 3.6]$

(b) $\dot{\hat{\mathbf{x}}} = (\mathbf{A} - \mathbf{mc})\hat{\mathbf{x}} + \mathbf{b}u + \mathbf{m}y$
$\mathbf{m}^T = [16 \quad 84.6]$

(c) With reference to Fig. 7.9:

$$\frac{U(s)}{-Y(s)} = D(s) = \frac{778.16s + 3690.72}{s^2 + 19.6s + 151.2}$$

(d) $$\begin{bmatrix} \dot{x}_1 \\ \dot{x}_2 \\ \dot{\hat{x}}_1 \\ \dot{\hat{x}}_2 \end{bmatrix} = \begin{bmatrix} 0 & 1 & 0 & 0 \\ 20.6 & 0 & -29.6 & -3.6 \\ 16 & 0 & -16 & 1 \\ 84.6 & 0 & -93.6 & -3.6 \end{bmatrix} \begin{bmatrix} x_1 \\ x_2 \\ \hat{x}_1 \\ \hat{x}_2 \end{bmatrix}$$

7.9 (a) $\mathbf{A} = \begin{bmatrix} 0 & 1 \\ 0 & 0 \end{bmatrix}; \; \mathbf{b} = \begin{bmatrix} 0 \\ 1 \end{bmatrix}; \; \mathbf{c} = [1 \quad 0]$

(b) $\mathbf{k} = [1 \quad \sqrt{2}]$

(c) $\dot{\hat{\mathbf{x}}} = (\mathbf{A} - \mathbf{mc})\hat{\mathbf{x}} + \mathbf{b}u + \mathbf{m}y$
$\mathbf{m}^T = [5 \quad 25]$

(d) With reference to Fig. 7.9:

$$\frac{U(s)}{-Y(s)} = D(s) = \frac{40.4(s + 0.619)}{s^2 + 6.414s + 33.07}$$

(e) With reference to Fig. 7.7:

$$\hat{x}_e = \hat{x}_2; \; \begin{bmatrix} a_{11} & a_{1e} \\ a_{e1} & A_{ee} \end{bmatrix} = \begin{bmatrix} 0 & 1 \\ 0 & 0 \end{bmatrix}$$

$$\begin{bmatrix} b_1 \\ b_e \end{bmatrix} = \begin{bmatrix} 0 \\ \cdots \\ 1 \end{bmatrix}; \; m = 5$$

(f) $\dfrac{U(s)}{-Y(s)} = D(s) = \dfrac{8.07(s + 0.62)}{s + 6.41}$

7.10 (a) $k_1 = 4; \; k_2 = 3; \; k_3 = 1; \; N = k_1$

(b) $\mathbf{m}^T = [5 \quad 7 \quad 8]$

7.11 $\mathbf{k} = [-1.4 \quad 2.4]$; $k_i = 1.6$

7.12 $k_1 = 4$; $k_2 = 1.2$; $k_3 = 0.1$

7.13 $k_1 = 1.2$; $k_2 = 0.1$; $k_3 = 4$

7.14 $\mathbf{m}^T = [5 \quad 6 \quad 5]$

7.15 $K_A = 3.5$; $k_2 = 0.11$; $k_3 = 0.33$

7.16 $K_A = 40$; $k_2 = 0.325$; $k_3 = 3$

7.17 $k_1 = a_2/\beta$; $k_2 = (a_1 - \alpha)/\beta$; $N = k_1$

7.18 $k_1 = -0.38$; $k_2 = 0.6$; $k_3 = 6$

7.19 (a) $k_1 = 3$; $k_2 = 1.5$

  (b) For a unit-step disturbance, the steady-state error in the output is 1/7.

  (c) $k_1 = 2$; $k_2 = 1.5$; $k_3 = 3.5$
  Steady-state value of the output $= 0$

7.20 (a) $\mathbf{k} = [3 \quad 1.5]$

  (b) $N = 7$

  (c) For a unit-step disturbance, the steady-state error in the output is 1/7.

  (d) $k_1 = 2$; $k_2 = 1.5$; $k_3 = 3.5$

7.21 (a) $K = 0.095$; $N = 0.1$

  (b) For $A + \delta A = -0.6$, $\omega(\infty) = \dfrac{10}{10.1} r$

  (c) $K_1 = 0.105$; $K_2 = 0.5$

7.22 $k_1 = -4$; $k_2 = -3/2$; $k_3 = 0$

7.23 $\hat{\mathbf{x}}(k+1) = (\mathbf{F} - \mathbf{mc})\hat{\mathbf{x}}(k) + \mathbf{G}u(k) + \mathbf{m}[y(k) - du(k)]$

$$\mathbf{m}^T = \left[\frac{3}{2} \quad -\frac{11}{16} \quad 0\right]$$

7.24 $\mathbf{k} = [-0.5 \quad -0.2 \quad 1.1]$

  $\mathbf{x}(k+1) = (\mathbf{F} - \mathbf{gk})\mathbf{x}(k)$

7.25 $\bar{\mathbf{x}}(k+1) = \mathbf{F}\hat{\mathbf{x}}(k) + \mathbf{g}u(k)$

  $\hat{\mathbf{x}}(k+1) = \bar{\mathbf{x}}(k+1) + \mathbf{m}[y(k+1) - \mathbf{c}\bar{\mathbf{x}}(k+1)]$

  $\mathbf{m}^T = [6.25 \quad -5.25]$

7.26 $$\begin{bmatrix} x_2(k+1) \\ x_1(k+1) \\ x_3(k+1) \end{bmatrix} = \begin{bmatrix} 0 & 0 & 1 \\ 1 & 0 & 0 \\ -0.2 & -0.5 & 1.1 \end{bmatrix} \begin{bmatrix} x_2(k) \\ x_1(k) \\ x_3(k) \end{bmatrix} + \begin{bmatrix} 0 \\ 0 \\ 1 \end{bmatrix} u(k)$$

$$= \begin{bmatrix} f_{11} & \mathbf{f}_{1e} \\ \mathbf{f}_{e1} & \mathbf{F}_{ee} \end{bmatrix} \begin{bmatrix} x_2(k) \\ x_1(k) \\ x_3(k) \end{bmatrix} + \begin{bmatrix} g_1 \\ \mathbf{g}_e \end{bmatrix} u(k)$$

$$\hat{\mathbf{x}}_e(k) = [\hat{x}_1(k) \quad \hat{x}_3(k)]^T$$

$$\hat{\mathbf{x}}_e(k+1) = (\mathbf{F}_{ee} - \mathbf{m}\mathbf{f}_{1e})\hat{\mathbf{x}}_e(k) + (\mathbf{g}_e - \mathbf{m}g_1)u(k)$$
$$+ (\mathbf{f}_{e1} - \mathbf{m}f_{11})y(k) + \mathbf{m}y(k+1)$$

$$\mathbf{m}^T = [0 \quad 1.1]$$

7.27 (a) $\mathbf{k} = \begin{bmatrix} \dfrac{111}{76} & -\dfrac{18}{19} \end{bmatrix}$

(b) $\hat{\mathbf{x}}(k+1) = (\mathbf{F} - \mathbf{mc})\hat{\mathbf{x}}(k) + \mathbf{b}u(k) + \mathbf{m}[y(k) - du(k)]$

$\mathbf{m}^T = [8 \quad -5]$

$$\begin{bmatrix} x_1(k+1) \\ x_2(k+1) \\ \hat{x}_1(k+1) \\ \hat{x}_2(k+1) \end{bmatrix} = \begin{bmatrix} 2 & -1 & -5.84 & 3.79 \\ -1 & 1 & -4.38 & 2.84 \\ 8 & 8 & -11.84 & -5.21 \\ -5 & -5 & -0.38 & 8.84 \end{bmatrix} \begin{bmatrix} x_1(k) \\ x_2(k) \\ \hat{x}_1(k) \\ \hat{x}_2(k) \end{bmatrix}$$

7.28 (a) $\mathbf{F} = \begin{bmatrix} 0 & 1 \\ -0.16 & -1 \end{bmatrix}; \mathbf{g} = \begin{bmatrix} 0 \\ 1 \end{bmatrix}; \mathbf{c} = [1 \quad 0]$

(b) $k_1 = 0.36; k_2 = -2.2$

(c) $\hat{x}_2(k+1) = (-1 - m)\hat{x}_2(k) + u(k) - 0.16y(k) + my(k+1)$

$m = -1$

(d)

```
     U(z)  ┌─────────────────────────┐  Y(z)
   ──○────►│         z^-2            ├────●──►
     -↑    │ (1+0.8z^-1)(1+0.2z^-1)  │    │
      │    └─────────────────────────┘    │
      │    ┌─────────────────────────┐    │
      │    │   2.56(1 + 0.1375z^-1)  │    │
      └────┤   ─────────────────────  │◄───┘
           │        1 - 2.2z^-1       │
           └─────────────────────────┘
```

7.29 (a) $\mathbf{x}(k) = \mathbf{Q}\mathbf{z}(k);\ \mathbf{Q} = \begin{bmatrix} 1 & -1 \\ 0 & 1 \end{bmatrix}$

$$u = -0.36z_1(k) + 2.2z_2(k)$$

(b) $\hat{z}_2(k+1) = (-1-m)\hat{z}_2(k) + u(k) - 0.16y(k) + my(k+1)$

$$m = -1$$

(c) $\dfrac{U(z)}{-Y(z)} = D(z) = \dfrac{2.56(1 + 0.1375z^{-1})}{1 - 2.2z^{-1}}$

7.30 (a) $\mathbf{F} = \begin{bmatrix} 1 & 0.1 \\ 0 & 1 \end{bmatrix};\ \mathbf{g} = \begin{bmatrix} 0.005 \\ 0.1 \end{bmatrix}$

(b) $k_1 = 13;\ k_2 = 3.95$

(c) $\hat{\mathbf{x}}(k+1) = (\mathbf{F} - \mathbf{mc})\hat{\mathbf{x}}(k) + \mathbf{g}u(k) + \mathbf{m}y(k)$

$$\mathbf{m}^T = \begin{bmatrix} 2 & 10 \end{bmatrix}$$

(d) $\dfrac{U(z)}{-Y(z)} = D(z) = \dfrac{65.5(z - 0.802)}{z^2 + 0.46z + 0.26}$

7.31 (a) $\mathbf{F} = \begin{bmatrix} 1 & 0.0952 \\ 0 & 0.905 \end{bmatrix};\ \mathbf{g} = \begin{bmatrix} 0.00484 \\ 0.0952 \end{bmatrix};\ \mathbf{c} = \begin{bmatrix} 1 & 0 \end{bmatrix}$

(b) $k_1 = 105.1;\ k_2 = 14.625$

(c) $\hat{\mathbf{x}}(k+1) = (\mathbf{F} - \mathbf{mc})\hat{\mathbf{x}}(k) + \mathbf{g}u(k) + \mathbf{m}y(k)$

$$\mathbf{m}^T = \begin{bmatrix} 1.9 & 8.6 \end{bmatrix}$$

(d) $\hat{x}_2(k+1) = (0.905 - 0.0952m)\hat{x}_2(k)$
$\qquad + (0.0952 - 0.00484m)u(k) - my(k) + my(k+1)$
$\qquad m = 9.51$

7.32 (a) $y(k+1) = 0.368y(k) + 0.632u(k) + 0.632w(k)$
  (b) $K = 0.3687; N = 1.37$
  (c) Steady-state error for a unit-step disturbance is 0.73.
  (d) $K_1 = 0.553; K_2 = 2.013$

8.1 Asymptotically stable in-the-large.

8.2 Unstable.

8.3 Equilibrium state $\mathbf{x}^e = [2 \quad 0]^T$ is asymptotically stable.

8.4 $K > 0$

8.5 $0 < K < 8$

8.6 Asymptotically stable.

8.7 Asymptotically stable.

8.8 $|x_1| < 1$; origin is the equilibrium state.

8.9 Asymptotically stable in-the-large; origin is the equilibrium state.

8.10 Asymptotically stable; origin is the equilibrium state.

9.1 $\zeta = 0.5; J_{min} = 1$

9.2 (i) $\alpha = \sqrt{K}$
  (ii) $K \to \infty$

9.3 $k = 2; J_{min} = 3/2; S_k^{opt} = 0.107$

9.4 $K = 0.1; S_K^{opt} = 0.0556$

9.5 $k_1 = 4; k_2 = \sqrt{20}; J_{min} = \sqrt{5}/4$

9.6 $k \to \infty$

9.7 $k = 1; S_k^{opt} = 0.0275$

9.8 $\mathbf{K} = [1 \quad \sqrt{3}]$

9.9 $\mathbf{K} = [1 \quad \sqrt{2}]$

9.10 Sufficient conditions not satisfied.
$\qquad u = -[-1 \quad 4]\mathbf{x}$; optimal closed-loop system is asymptotically stable.

9.11 Asymptotically stable optimal solution does not exist.

9.12 $u = -x_1 - 0.23x_2 + r$

r = desired output $y_d = 1$

9.13 $u = -x_1 - \sqrt{6}x_2$

9.14 $u = -x_1 - x_2$

$\dot{\hat{\mathbf{x}}} = (\mathbf{A} - \mathbf{MC})\hat{\mathbf{x}} + \mathbf{B}u + \mathbf{M}y$

$\mathbf{M}^T = [5 \quad 4]$

9.15 $k_1 = \sqrt{2}$; $k_2 = 0.275$; $J_0 = 93.2375$

9.16 $\rho = 0.1$; $\mathbf{K} = [3.1623 \quad 0.5968]$

Poles: $-0.6377, -4.4592$

$\rho = 0.01$; $\mathbf{K} = [10 \quad 1.7082]$

Poles: $-2.2361, -4.4721$

$\rho = 0.001$; $\mathbf{K} = [31.6228 \quad 4.3939]$

Poles: $-4.6970 \pm j3.0921$

9.17 $k_1 = 1.345$; $k_2 = 0$; $J_{min} = 93.26$

9.18 (a) $K = \sqrt{2} - 1$

(b) $N = \sqrt{2}$

(c) Steady-state error to unit-step disturbance is $1/\sqrt{2}$

(d) $K = K_1 = 1$

(e)

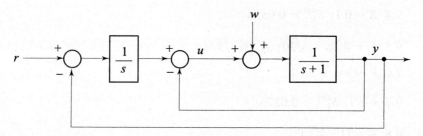

9.19 (a) $K = 0.095$; $N = 0.1$

(b) For $A + \delta A = -0.6$, $\omega(\infty) = \dfrac{10}{10.1}r$

(c) $K = 0.105$; $K_1 = 0.1$

9.20 $K = 2/3$; $S_K^{opt} = 0.1207$

9.21 $u(k) = -0.178x(k)$

9.22 (a) $K = 0.277$

(b) $y(\infty) = 0.217r$

(c) $N = 1.277$

9.23 (a) $K = 0.2$

(b) $N = 0.45$

(c) For $F + \delta F = 0.3$, $x(\infty) = \dfrac{0.9}{1.1} r$

10.2 $\dfrac{8M}{\pi}$; 1 rad/sec; $y(t) = \dfrac{-8M}{\pi} \sin t$

10.3 0.3; 10 rad/sec

10.4 $\Delta < 0.131$

10.5 0.42; 5.9 rad/sec

10.7 4.25; $\sqrt{2}$ rad/sec, stable limit cycle

10.8 3.75; 1 rad/sec

10.9 (a) Stable node; (0, 0) point in $(y, \dot{y})$-plane

(b) Stable node; (1,0) point in $(y, \dot{y})$-plane

(c) Unstable focus; (2,0) point in $(y, \dot{y})$-plane

10.10 Singularity at (0, 0) is either a stable node or a stable focus, depending upon the magnitudes of a and K. Singular point $(\pi, 0)$ is a saddle point.

10.11 (i) Singularity (1,0) in $(y, \dot{y})$-plane is a centre

(ii) Singularity (1,0) in $(y, \dot{y})$-plane is a stable focus.

10.12 Steady-state error to unit-step input $= -0.2$ rad

Maximum steady-state error $= \pm 0.3$ rad.

10.14 Saturation has a slowing down effect on the transient.

10.15 *Without Hysteresis:* Oscillatory behaviour with ever-decreasing amplitude and ever-increasing frequency.

With hysteresis: The system enters into a limit cycle.

10.17 The system has good damping and no oscillations but exhibits chattering behaviour. Steady-state error is zero.

10.18 (b) (i) Deadzone provides damping; oscillations get reduced.

(ii) Deadzone introduces steady-state error; maximum error $= \pm 0.2$.

(c) By derivative-control action, (i) settling time is reduced, but (ii) chattering effect appears.

10.20 $x_1 = e, x_2 = \dot{e}$
Switching curve:
$$x_1 = -x_2 + \frac{x_2}{|x_2|} \ln\left(1 + \frac{\dot{x}_2^2}{|x_2|}\right)$$

11.1 $\hat{y} = x_1 + x_2 - 1.5$

11.2 $\hat{y} = -0.5x_1 + 0.25x_2 + 0.5$

11.4 $w = 0.36$; $w_0 = 0.666$

11.6 (b) $\mathbf{w}(1) = [0.974 \quad -0.948 \quad 0 \quad 0.526]^T$
$\mathbf{w}(2) = [0.974 \quad -0.956 \quad 0.002 \quad 0.531]^T$
$\mathbf{w}(3) = [0.947 \quad -0.929 \quad 0.016 \quad 0.505]^T$

(c) $\mathbf{w} = [1.0518 \quad -0.943 \quad 0.0155 \quad 0.5005]^T$

11.7 (b) $w_{10} = 1.00043$; $w_{11} = 3.00043$;
$w_{12} = 4$; $w_{20} = -5.9878$; $w_{21} = 6.0123$;
$w_{22} = 5$; $v_0 = -3.9078$; $v_1 = 2.012$;
$v_2 = 4.0061$

(c) With initial weights, $\hat{y} = 0.51$
With updated weights, $\hat{y} = 0.5239$

12.1 (a) Supporting Interval: [1 3]
α-cut interval: $\left[2 - \sqrt{0.5} \quad 2 + \sqrt{0.5}\right]$

(b) Support: [1 3]
α-cut: [1.5 2.5]

12.2 (a) $\sigma = \dfrac{1}{\sqrt{2\pi}}$; $\mu = 2$
Support; unbounded $(-\infty \quad \infty)$
α-cut: [1.53 2.47]

(b) Support: $(-\infty \quad \infty)$
α-cut: [1 3]

12.3 $a = 3$ ft; $b = 6$ ft; $c = 9$ ft;
Support: [3 9]
Cross point: 4.5

12.4 (a) $\mu_A = \begin{cases} \dfrac{x+1}{5} & ; -1 \le x \le 4 \\ 1 & ; 4 \le x < 5 \\ \dfrac{x-9}{-4} & ; 5 \le x \le 9 \end{cases}$

(b) It is normal and convex.

12.5 (a) (i) Yes (ii) Yes (iii) No
(b) Support: [64 78]
Cross points: 68, 75
$\alpha\text{-cut}|_{\alpha=0.2}$: [65.6 76.8]
$\alpha\text{-cut}|_{\alpha=0.4}$: [67.2 75.6]

12.8 (i) $\mu_{\underset{\sim}{A}}(4) = 0.75$; (ii) $\mu_{\underset{\sim}{B}}(3) = 0.5$;

(iii) $\mu_{\underset{\sim}{A} \times \underset{\sim}{B}}(x, y) = 0.5$; (iv) $\mu_{\underset{\sim}{A} \to \underset{\sim}{B}}(x, y) = 0.5$;

12.11 $z^* = 6.76$

12.12 1857.28

12.13 $\mu_{\underset{\sim}{C'}}(z) = \max\left\{\min\left(\frac{2}{3}, \mu_{\underset{\sim}{C_1}}(z)\right), \min\left(\frac{1}{3}, \mu_{\underset{\sim}{C_2}}(z)\right)\right\}$

$z^*_{COA} = 4.7$

12.14 (i) 0.25, 0.62

(ii) $\mu_{agg}(z) = \max\left\{\min\left(0.25, \mu_{\underset{\sim}{PL}}(z)\right), \min\left(0.62, \mu_{\underset{\sim}{PM}}(z)\right)\right\}$

(iii) $z^* = 53.18$

12.15 34.40

Index

Acceleration error constant   236
Ackermann's formula   491, 499, 520, 522
Activation function;   728
   bipolar   730
   Gaussian   734
   hard limiting   732
   hyperbolic tangent   733
   linear   734
   log-sigmoid   734–735
   sigmoidal   732
   soft limiting   732
   tan-sigmoid   734–735
   unipolar   730
Adaptive control system;   540–546
   model-reference   545–546, 589–592, 768–771
   self-tuning   541–546, 589
Adaptive learning rate (NN)   756
A/D converter;   27, 32
   circuits   37–38
   model   129–130
Adjoint of a matrix   319
Aliasing   79–81
Alpha-cut; fuzzy set   797
Analog control design:
   discretization   108–110, 128
   optimal control   625–628
   parameter optimization   616–620
   pole-placement method   486–517
Analog prefilters   127
Analytic function   863, 872
Angle criterion; root locus   267
Antecedent;IF-THEN rule   785

Anti-aliasing filter   85, 127
Approximate reasoning   791, 807–811
Artificial neural network   735–744
   (also see Neural network)
Artificial neuron   726–730
   (also see Neuron model)
Atomic fuzzy proposition   799, 807
Asymptotic stability   67, 68, 403, 572
Autonomous systems   569

Backlash nonlinearity;   665–666
   describing function   671–673, 683–684
Backpropagation training;   753–757
   batch-mode   752
   gradient descent method   749–757
   hidden-layer size   744
   incremental-mode   752
   learning rate   750, 756
   Levenberg-Marquardt method   756
   momentum term   755–756
   weight initialization   755
Backward difference approximation of derivatives   98–102
Backward rectangular rule for integration   102–103
Bandwidth;   247, 248
   on Nichols chart   259
Bang-Bang control:
   optimal switching   700–703
   suboptimal switching   709–711
Batch-mode training   752
Bell-shaped fuzzy set   797
Bias (NN)   728

Index

BIBO stability 68–71, 403
Bilinear transformation; 105–108
  with frequency prewarping 253
Biological neuron 724–726
Block diagram manipulations 16–21, 133–135
Bode plots:
  lag compensation 257–258
  lag-lead compensation 258
  lead compensation 255–257

Cancellation compensation 275, 283, 285–286
Canonical state models; 338
  controllability form 400
  controllable companion form 426, 485
  first companion form 349–351, 432–434
  Jordan form 353–358, 435–438
  observability form 402
  observable companion form 426, 497
  second companion form 351–353, 434–435
Cascade programming
  of controllers 148–149
Cartesian product 803–805
Cayley-Hamilton theorem 346–347
Center of area defuzzification 824–825
Centre point; phase portrait 694
Characteristic
  equation 54
  function 791
  polynomial 54, 342
  roots 342
Chattering 711
Chromosome (GA) 826
Classical control design:
  Bode plot method 251–267
  discretization of analog
    design 108–110, 128
  root-locus method 267–283
  z-plane synthesis 284–296
Classical logic 791
Coding 32
Companion form of state model:
  controllable 426, 485
  first form 349–351, 432–434
  observable 426, 497
  second form 351–353, 434–435
Companion matrices 351
Compensation:
  by separation principle 505–508, 524–525
  cancellation 275, 283, 285–286
  lag on Bode plots 257–258
  lag on root-locus plots 277–281
  lead on Bode plots 255–257
  lead on root-locus plots 273–277
Complement; fuzzy set 802
Complimentary strips in s-plane 91
Compositional rule of inference 810
Composition control systems 405–407
Composition; max-min 810
Compound fuzzy proposition 808
Computational time delay 30
Computer control systems (see Digital control systems)
Computer Integrated Manufacturing System 12
Computer Integrated Process System 11
Computer simulation 42–43, 120, 315, 340, 438–442, 445–447
Condition number of a matrix 323
Conjugate of a matrix 318
Conjugate transpose 318
Conclusion; IF-THEN rule 785
Conjunction; fuzzy set 805
Consequent; IF-THEN rule 785
Constant-ω_n loci 94–95
Constant-ζ loci 94–95
Controllability:
  definition 389
  tests 391–393, 397, 407, 408, 455–456, 457, 461–462
Controllability canonical form of state model 400
Controllability loss due
  to sampling 458–460
Controllability matrix 391
Controllable companion form
  of state model 426, 485
Controllable eigenvalues (poles) 401

Controller (see also Compensation):
  bang-bang 700–703, 709–711
  fuzzy logic 812–825
  neuronal 765–772
  PID 154–157, 166–169
Controller programming (see Digital controller implementation)
Controller tuning:
  based on GA 837–839
  based on process reaction curve 160–165
  based on ultimate gain & period 158–160
  digital PID 169–170
  quarter-decay ratio response 159–160, 164–165
Control system:
  adaptive 540–546, 589–592
  deadbeat 293–295, 530–531
  multivariable 403–409, 461–464, 612–620, 625–628, 633–636
  optimal 616–620, 625–628, 633–636
  robust 250–251, 640–642
Control system examples:
  composition control 405–407
  liquid-level control 7–8, 608–610
  position control 3–6, 176–183, 336–337, 385–389, 483–486
  speed control 6–7, 334–336, 547–550, 772–776, 922, 924
  temperature control 141–143, 171–176, 777–779, 813–825, 841–843, 923, 924
Control system terminology:
  regulator 3
  servo 2
  set-point control 3
  tracking 3
Convex fuzzy set 796–797
Convolution sum 51
Coulomb friction 662, 698–700
Crisp set 791
Crossover (GA) 834–835
Cross point; fuzzy set 798
Cross product (see Cartesian product)
Cross site (GA) 834
Current state observer 522–523

Cylindrical extension; fuzzy relation 809

D/A converter; 27, 32
  circuits 34–37
  model 130
Damping ratio; correlation with
  peak overshoot 240
  phase margin 248
  resonance peak 248
Deadbeat control systems 293–295, 530–531
Deadbeat state observer 531
Deadtime in
  state equations 445–447
  transfer functions 139–141
Deadzone nonlinearity; 662
  describing function 669–671
  phase portrait 697
Decoding 33–34
Defuzzification 824–825
Derivative time; PID controller 155
Describing function method; 666–669
  sampled-data systems 685
  stability analysis 675–684
  table 674
Detectability 619
Determinant of a matrix 318
Diagonal matrix 316
Difference equations 49–50
Digital control design:
  Bode plot method 251–267
  discretization of analog design 108–110, 128
  fuzzy logic based 812–825
  neural network based 765–772
  optimal control 636
  parameter optimization 633–636
  pole-placement method 518–533
  root-locus method 267–283
  z-plane synthesis 284–296
Digital controller implementation; 143–147
  cascade realization 148–149
  direct realization 147–148
  nonrecursive 146, 151–152
  parallel realization 150–151
  recursive 146, 147–151

Digital control system:
  advantages of   26–27
  configuration   29–30
  implementation problems   27–28
Digital PID controllers:
  interacting position algorithm   169
  interacting velocity algorithm   169
  noninteracting position
    algorithm   166–167
  noninteracting velocity
    algorithm   168
  tuning   169–170
Digital signals (see Discrete-time
  signals)
Digital simulation (see Computer
  simulation)
Direct digital control   9
Direct digital design   128
Direct method of Lyapunov   576, 577,
  578–579, 593
Direct programming
  of controllers   147–148
Discrete-equivalent of continuous-time
  systems through:
    backward difference approximation
      of derivatives   98–102
    backward rectangular rule for
      integration   102–103
    bilinear transformation   105–108
    forward difference approximation of
      derivatives   100–102
    forward rectangular rule for
      integration   102–103
    impulse invariance   89–93
    step invariance   95–97
    trapezoidal rule for
      integration   106–108
    zero-order hold equivalence   97
Discrete-time impulse   39
Discrete-time signals:
  sinusoidal sequence   40
  unit-sample sequence   38–39
  unit-step sequence   40
Discretization of analog
  design   108–110, 128
Distributed computer control
  system   10
Disturbance rejection   248–250

Dominant poles   242–244
Duality   498

Early stopping; NN training   764
Eigenstructure assignment   641
Eigenvalue assignment
  (see Pole-placement by
  state feedback)
Eigenvalues:
  controllable   401
  definition   362
  observable   402
  uncontrollable   401
  unobservable   402
Eigenvectors:
  computation   366–372, 423
  definition   365
  generalized   370
Encoder; shaft   177–182
Encoding   32
Epoch; NN training   752
Equilibrium state   568
Equivalence transformation   338
Error constants:
  acceleration   236
  position   235
  velocity   235
Estimation of parameters (see Parameter
  estimation)
Estimators (see State observers)
Euclidean matrix norm   323
Euclidean vector norm   322
Euler method   120
Extended linearization   660–661

Feedback control systems:
  block diagram manipulations   16–21,
    133–135
  nonunity feedback   233
  unity feedback   19, 233
Feedback linearization   660, 719–721
Feedback network (see Recurrent
  network)
Feedfarward action:
  state-feedback servo   512–515,
    526–527, 615
Feedforward network:
  dynamic map   744

Index

hidden layer 735
input-output map 738
input terminals 735
multilayer 739
output layer 735
single layer 736
static map 744
Filter:
anti-aliasing 85, 127
finite impulse response 146
infinite impulse response 146
low pass 83–84
nonrecursive 146, 151–152
recursive 146, 147–151
zero-order hold 84–85
Final value theorem:
Laplace transform 862–863
z-transform 872
Finite impulse response system 146
Firing strength; IF-THEN rule 821, 822
First companion form of
state model 349–351, 432–434
First-harmonic approximation 667
First method of Lyapunov 571
First-order hold 74
First-order plus dead-time model 161
Fitness function 830–831
Focus; phase portrait 694
Frame of linguistic variables 800
Folding frequency 81
Forward difference approximation of
derivatives 100–102
Forward rectangular rule for
integration 102–103
Fourier series 852–855
Fourier transform 855–856
Frequency folding (see Aliasing)
Frequency prewarping 253
Frequency response
specifications 192–195
Frequency warping 100–101, 108, 253
Friction; nonlinear 662, 698–700
Full-order state observer 495–499, 521–523
current observer 522–523
prediction observer 521–522
Function approximation 757–758
Fuzzification 820

Fuzzy cartesian product 803–805
Fuzzy complement 802
Fuzzy conjunction 805
Fuzzy implication 806–807
Fuzzy inference; 807–811
compositional rule 810
Fuzzy intersection 802
Fuzzy logic 791
Fuzzy logic control 812–825
GA-based tuning 837–839
Fuzzy propositions 799, 807, 808
Fuzzy relation; 803–807
composition 810
cylindrical extension 809
projection 810
Fuzzy rules (IF-THEN):
Look-up table 788
Mamdani rules 807–811
singleton rules 810, 811
Takagi-Sugeno rules 811–812
Fuzzy singleton 798
Fuzzy sets; 795–807
a-cut 797
bell-shaped 797
convex 796–797
cross point 798
Gaussian 797
height 796
normal 796
singleton 798
support 798
trapezoidal 797
triangular 797
Fuzzy union 802

Gain margin 247
Gaussian activation (NN) 734
Gaussian fuzzy set 797
Generalization (NN) 763–764
Generalized eigenvectors 370
Genetic algorithm (GA):
chromosome 826
coding 828–830
controller tuning 837–839
crossover 834–835
cross site 834
fitness function 830–831
initialization 831–832

mating pool 832
mutation 835–836
reproduction 834–836
Roulette wheel parent selection 833
selection operator 832
stopping condition 836
Global stability 573
Gradient descent method 749–757
Grammian matrix 325, 326

Hard-limiting activation (NN) 732
H_∞ control 642
Height; fuzzy set 796
Hermitian matrix 318
Hessian matrix 601
Hidden layer (NN) 739, 744
Hidden oscillations 458–459
Hierarchical control systems 11
High loop-gain effects 251
Hold operation:
 first-order 74
 zero-order 84–85
Homogeneous state equations:
 solution 374
Hyperbolic tangent activation (NN) 733
Hysteresis nonlinearity; 664–665
 describing function 703–704
 phase portrait 709

IAE performance index 602
Ideal lowpass filter 83–84
Identification of models:
 least squres method 533–538
 NN-based 758–765
 recursive 538–540
Identity matrix 316
IF-THEN rule:
 antecedent 785
 conclusion 785
 consequent 785
 firing 821, 822
 implication 807–811
 premise 785
Implication; fuzzy set 806–807
Impulse; discrete-time 39

Impulse-invariance method for discretization 89–93
Impulse modulator model of sampling 71–73
Impulse response model 50–51
Incremental-mode training 752
Indefinite scalar function 576
Inference; fuzzy system 807–811
Infinite impulse response system 146
Initialization (GA) 831–832
Initialization (NN) 755
Inner product of vectors 321
Input-output map (NN) 738
Input terminals (NN) 735
Insensitivity 250–251
Instability theorem; Lyapunov 581
Integral action:
 state-feedback servo 515–517, 527–528, 615–616
Integral time; PID controller 154
Intelligent control (see Knowledge-based systems)
Interacting digital PID controller:
 position algorithm 169
 velocity algorithm 169
Interacting PID controller 155, 169
Intersample behaviour 214–216, 438–440
Intersample ripples 290, 292, 458–459
Intersection; fuzzy set 802
Inverse model (NN) 765–767
Inverse of a matrix 319
Inverted pendulum 385–389
ISE performance index 603
ITAE performance index 603
ITSE performance index 603
Jacobian matrix 584
Jordan canonical form of state model 353–358, 435–438
Jury stability criterion 69–71

Krasovskii method 583–585
Knowledge-based systems; 15, 718–850
 MATLAB aided design 911–939

Ladder diagram 203–212

Lag compensation on:
   Bode plots 257–258
   root-locus plots 277–281
Lag-lead compensation 258
Lag space (NN) 761
Laplace transforms:
   definition 856
   final value theorem 862–863
   inverse 857
   pairs 873–874
   properties 860–863
Layers of neurons:
   hidden 735
   multiple 739
   output 735
   single 736
Lead compensation on:
   Bode plots 255–257
   root-locus plots 273–277
Learning rate; 750
   adaptive 756
Learning in NN; 745–749
   (see also Neural Network training)
      reinforcement 748
      supervised 748
      unsupervised 748–749
Least squares estimation; 533–538
   recursive 538–540
Levenberg-Marquardt algorithm 756
Lie Derivatives 719
Limit cycles 570, 571
Linear activation (NN) 734
Linearization:
   extended linearization 660–661
   feedback linearization 660, 719–721
   first-harmonic approximation 667
   NN based 721–722
   Taylor's series 333, 659–660
Linear dependence of
   vectors 324–326
Linear independence of
   vectors 324–326
Linear system stability tests:
   Jury 69–71
   Lyapunov 586–587, 593–594
   Routh 111
Linguistic variables:
   frame 800

   terms 800
Liquid-level control system 7–8,
   608–610
Local stability 573
Logic:
   classical 791
   fuzzy 791
Log-sigmoid activation (NN) 734–735
Lower triangular matrix 317
Lowpass filter 83–84
Luenberger state observer 496
Lyapunov equations 587, 594, 635
Lyapunov functions; 576
   for linear systems 586–587,
      593–594
   for nonlinear systems 577, 578–579,
      586–587, 593–594
Lyapunov instability theorem 581
Lyapunov stability analysis:
   direct method 577, 578–579, 593
   first method 571
   second method 577, 578–579, 593

Magnitude criterion; root locus 267
Mapping of s-plane to z-plane; 91
   constant-ω_n loci 94–95
   constant-ζ loci 94–95
Mapping of z-plane to r-plane 111
Mapping of z-plane to w-plane 111, 252
Mapping of w-plane to z-plane 252
Marginal stability 68
Mating pool (GA) 832
MATLAB aided control design:
   knowledge-based 911–939
   model-based 875–910
Matrix:
   adjoint 319
   condition number 323
   conjugate 318
   conjugate transpose 318
   determinant 318
   diagonal 316
   eigenvalues 362
   Grammian 325, 326
   Hermitian 318
   Hessian 601
   identity 316
   inverse 319

Jacobian 584
lower triangular 317
negative definite 327
negative semidefinite 327
nonsingular 319
norm; Euclidean 323
norm; spectral 323
null 317
nullity 310
orthogonal 324
partitioned 320
positive definite 327
positive semidefinite 327
rank 320, 325
singular 319
singular values 323
skew-Hermitian 318
skew-symmetric 318
symmetric 317
trace 320
transpose 317
triangular 317
unit 316
upper triangular 317
zero 317
Matrix exponential:
   definition 372
   properties 373–374
Matrix exponential evaluation by:
   Cayley-Hamilton
      technique 379–380, 413–414
   inverse Laplace transform 375–376
   numerical algorithm 441–442
   similarity transformation 377–378
Matrix Riccati equation 627, 636
Max-min composition 810
Measurement noise 249–250
Membership functions (see Fuzzy sets)
MIMO systems; definition 14
Minor-loop feedback 6
Model reference adaptive
   control 545–546, 589–592, 768–771
Model-based control design; 15,
   231–296, 480–546, 601–642
      MATLAB aided design 875–910
Modern control design:
   optimal control 616–620, 625–628,
      636

parameter optimization 616–620,
   633–636
pole-placement method 486–517,
   518–533
Modes 398
Momentum gradient algorithm 755–756
Multilayer NN 739–741
Multilevel control systems 10
Multiloop control systems 6
Multiple-rate sampling 30
Multivariable control
   systems 403–409, 461–464,
      612–620, 625–628, 633–636
Mutation (GA) 835–836

Negative definite matrix 327
Negative definite scalar
   function 576
Negative semidefinite matrix 327
Negative semidefinite scaler
   function 576
Neural Network:
   feedforward 735–741
for control 765–772
for function approximation 757–758
   for model identification 758–765
   hidden layer 739, 744
   input output map 738
   lag space 761
   multilayer 739–741
   RBF 734
   recurrent 741–744
   single layer 735–738
Neural Network Training; 762
   (see also Learning in NN)
      early stopping 764
      epoch 752
      generalization 763–764
      initialization 755
      regularization 764
      validation 764–765
Neuron:
   artificial 726–730
   biological 724–726
   model 730–735
Neuro-control:
   feedback linearization 771–772
   feedforward-feedback 767–768

inverse model 765–767
model-reference adaptive 768–771
Nichols chart; bandwidth determination 259
Nodal point; phase portrait 694
Nonhomogeneous state equations:
   solution 382–383, 404, 449, 461
Noninteracting digital PID controller:
   position algorithm 166–167
   velocity algorithm 168
Noninteracting PID controller 155, 166–168
Nonlinearities:
   backlash 665–666
   bang-bang 663–665
   deadzone 662
   describing function table 674–675
   friction 662
   hysteresis 664–666
   on-off (see bang-bang)
   saturation 662
Nonlinear systems:
   extended linearization 660–661
   feedback linearization 660, 719–721
   first-harmonic approximation 667
   first-order approximation 333, 659–660
   sampled-data 685
Nonlinear system stability:
   describing function 675–685
   Lyapunov 581–585
Nonminimum-phase transfer function 255
Nonrecursive controller 146, 151–152
Nonrecursive filter 146, 151–152
Nonsingular matrix 319
Nonsingleton fuzzy system 811
Nonunity feedback system 233
Norm:
   Euclidean; matrix 323
   Euclidean; vector 322
   H_∞ 642
   spectral; matrix 323
Normal fuzzy set 796
Normalized dead-time 165
Normalized union of discourse 842
Nullity 364

Null matrix 317

Observability:
   definition 390
   tests 395, 397, 407, 408, 456–457, 461–462
Observability canonical form of state model 402
Observability loss due to sampling 458–460
Observability matrix 395
Observable companion form of state model 426, 497
Observable eigenvalues (poles) 402
Observer (see State observer)
On-off controllers; 663–665
   describing functions 668–671, 679–683, 703–704
   phase portraits 695–697, 709
Optimal control systems 616–620, 625–628, 633–636
Optimal servo system; with integral control 615–616
Optimal state estimators (see Optimal state observers)
Optimal state observers 613–615
Optimal state regulator 625–628, 636
Optimization of parameters; 601–605, 616–621, 633–636
   GA-based 825–836
Order of a system 45, 54
Orthogonal matrix 324
Orthogonal vectors 323
Orthonormal vectors 324
Output feedback (see Partial state feedback)
Output layer (NN) 735
Output regulator 611–612
Overfitting (NN) 763

Parallel programming of controllers 150–151
Parameter estimation:
   least squares method 533–538
   recursive 538–540
Parameter optimization; 601–605, 616–621, 633–636
   GA-based 825–836

Parameter sensitivity 217–219, 250–251
Parameter uncertainty 217–219, 250–251, 640–642
Partial fraction expansion 858–860
Partial state feedback 613
Partitioned matrix 320
PD controller 784
Peak overshoot; 239
   correlation with damping ratio 240
Peak resonance; 246
   correlation with damping ratio 248
Peak time 239, 240
Perceptron 722, 731
Performance index (also see Quadratic performance index):
   IAE 602
   ISE 603
   ITAE 603
   ITSE 603
Performance specifications:
   frequency-response 244–248
   time-response 239–242
Permanent-magnet stepping motors 184–186
Phase-lag compensation (see Lag compensation)
Phase-lead compensation (see Lead compensation)
Phase margin; 247
   correlation with damping ratio 248
Phase-plane analysis 687, 695–700
Phase portraits 686, 688–695
Phase trajectory 686
PI controller 154–155, 839–841
PID controller:
   derivative time 155
   integral time 154
   interacting 155, 169
   noninteracting 155, 166–168
   rate time 155
   reset time 154
PID controller, digital (see Digital PID controller)
Pole-placement by state feedback; 486–490, 518–520
   Ackermann's formula 491, 520
   multi-input systems 494, 641
Poles (also see Eigenvalues) 54

Poles and zeros 54
Pole-zero cancellation 275, 283, 285–286, 393–403, 458
Population initialization (GA) 831–832
Position control systems 3–6, 176–183, 336–337, 385–389, 483–486
Position error constant 235
Position form of digital PID algorithm 166–167, 169
Positive definite matrix 327
Positive definite scalar function 576
Positive semidefinite matrix 327
Positive semidefinite scalar function 576
Prediction state observer 521–522
Prefiltering 127
Premise; IF-THEN rule 785
Prewarping 253
Primary strip in s-plane 91
Principle of duality; 498
Process reaction curve 160–165
Programmable logic controller; 192
   applications 213–214
   building blocks 196–202
   ladder diagram 203–212
   programming 212–213
Projection; fuzzy relation 810
Proper transfer function 342
Proportional band 154
Proportional controller 154
PWM control 172–173

Quadratic forms of scalar functions:
   negative definite 327
   negative semidefinite 327
   positive definite 327
   positive semidefinite 327
Quadratic performance index:
   ISE 603
   output regulator 611–612
   state regulator 610–613
   sum square error 634
Quantization errors 27–28, 30–34
Quarter-decay ratio response 159–160, 164–165

Radially unbounded function 575
Rank of a matrix 320, 325

Rate time; PID controller 154
RBF network 734
Realization of a transfer function:
    cascade programming 148–149
    direct programming 147–148
    first companion form 349–351, 432–434
    Jordan form 353–358, 435–438
    parallel programming 150–151
    second companion form 351–353, 434–435
Recurrent networks 741–744
Recursive controller 146, 147–151
Recursive filter 146, 147–151
Recursive least squares estimation 538–540
Reduced-order state observer 502–505, 523–524
Regularization (NN) 764
Regulator; definition 3
Reinforcement learning 748
Relaxed system 50
Relay nonlinearity:
    bang-bang 663
    with deadzone 664
    with deadzone and hysteresis 665
Reproduction (GA) 834–836
Reset time; PID controller 154
Resolvent algorithm 345–346
Resolvent matrix 345
Resonance frequency 247, 248
Resonance peak; 246
    correlation with damping ratio 248
Riccati equation 627, 636
Ringing poles 288
Rise time 239, 240
Robust control systems 15, 250–251, 640–642
Robust observers 511, 640–642
Root locus method:
    angle criterion 267
    construction rules table 268–269
    magnitude criterion 267
Root locus plots:
    lag compensation 277–281
    lead compensation 273–277
Root sensitivity 217–219

Roulette-wheel parent selection (GA) 833
Routh stability criterion 111
    r-plane:
        z-plane mapping 111
Rule antecedent 785
Rule base; 785–789
    look-up table 788
    Mamdani 807–811
    PD 784
    PI 839–841
    singleton 810, 811
    Takagi-Sugeno 811–812
Rule conclusion 785
Rule consequent 785
Rule premise 785

Saddle point; phase portrait 695
Saturation nonlinearity; 662
    describing function 705–706
Sampled-data control systems; 26–30
    state model 438–442, 445–447
    transfer function 129–135, 139–141
Sampled data nonlinear system:
    describing function analysis 685
Sampling:
    impulse modulator model 71–73
    multiple rate 30
    uniform 32
Sampling effects; 28
    on controllability and observability 458–460
    on stability 139
    on steady-state error 238–239
Sampling frequency 71
Sampling period; 71
    selection 86–89
Sampling rate 71
Sampling theorem 82
Satellite altitude control system 483–486
Scalar product of vectors 321
Scaling gain (FLC) 818–820
SCR controller 7, 172–173
Second companion form 351–353, 434–435
Second method of Lyapunov 577, 578–579, 593

Selection operator (GA) 832
Self-tuning control 541–546, 589
Sensitivity analysis 217–219, 250–251
Sensitivity function 251
Separation principle 505–508, 524–525
Servo design with state feedback:
   with feedforward control 512–515,
     526–527, 615
   with integral control 515–517,
     527–528, 615–616
Servo system; definition 2
Set-point control system:
   definition 3
Settling time 239, 240
Shaft encoder 177–182
S/H device:
   circuit 76–77
   model 74–76
Sigmoid activation (NN) 732
Similarity transformation 338
Simulation (see Computer simulation)
SIMULINK 888–893
Single layer NN 735–738
Singleton:
   fuzzy set 798
   rules 810, 811
Singular matrix 319
Singular points:
   center 694
   focus 694
   node 694
   saddle 695
   vortex 694
Singular values of a matrix 323
Sinusoidal sequence 40
SISO systems; definition 14
Skew-Hermitian matrix 318
Skew-symmetric matrix 318
s-norm 802
Soft-limiting activation (NN) 732
Solution of:
   homogeneous state equations 374
   nonhomogeneous state
     equations 382–383, 404, 449, 461
Specifications (see Performance
   specifications)
Spectral norm of a matrix 323

Speed control systems 6–7, 334–336,
   547–550, 772–776, 923, 924
s-plane to z-plane mapping 91
Stability:
   asymptotic 67, 68, 403, 572
   BIBO 68–71, 403
   describing function
     method 675–685
   global 573
   in-the-large 573
   in the sense of Lyapunov 571–573
   in-the-small 573
   Jury test 69–71
   local 573
   Lyapunov test 577, 578–579, 593
   marginal 68
   Routh test 111
   sampling effects 139
   zero-input 67
Stability tests for linear systems:
   Jury 69–71
   Lyapunov 586–587, 593–594
   Routh 111
Stability tests for nonlinear systems:
   describing function 675–685
   Lyapunov 581–585
Stability theorems; Lyapunov 577,
   578–579, 586–587, 593–594
Stabilizability 626
State diagram 340
State model:
   conversion to transfer
     function 341–343, 404, 431–432,
     461
   equivalence with transfer
     function 398–403, 458
   sampled plant 438–442, 445–447
   system with dead-time 445–447
State models; canonical 338
   controllability form 400
   controllable companion form 426,
     485
   first companion form 349–351,
     432–434
   Jordan form 353–358, 435–438
   observability form 402
   observable companion form 426,
     497

second companion form 351–353, 434–435
State models; companion form 351
   controllable 426, 485
   first 349–351, 432–434
   observable 426, 497
   second 351–353, 434–435
State observers; 481
   current 522–523
   deadbeat 531
   full-order 495–499, 521–523
   prediction 521–522
   reduced-order 502–505, 523–524
   robust 511, 640–642
State observer design through:
   Ackermann's formula 499, 522
   matrix Riccati equation 613–615
State regulator 482, 610–613
State regulator design through:
   Ackermann's formula 491, 520
   Lyapunov equation 616–619, 633–635
   matrix Riccati equation 625–628, 636
State transition equation 383, 453
State transition matrix:
   definition 374, 450
   properties 374–375
State transition matrix evaluation by:
   Cayley-Hamilton technique 379–380, 413–414
   inverse Laplace transform 375–376
   inverse z-transform 450
   numerical algorithm 441–442
   similarity transformation 377–378, 451, 452–453
Steady-state error; 234–237
   sampling effects 238–239
Steady-state error constants (see Error constants)
Steepest descent optimization 749–757
Step-invariance method for discretization 95–97
Step motors (see Stepping motors)
Stepper motors (see Stepping motors)
Stepping motors:
   in feedback loop 8
   interfacing to microprocessors 189–190
   permanent magnet 184–186
   torque-speed curves 188
   variable-reluctance 187–189
Strictly proper transfer function 342
Suboptimal state regulator (also see Parameter (optimization) 613
Sum-square error performance index 634
Supervised learning (NN) 748
Support; fuzzy set 798
Sylvester's test 328
Symmetric matrix 317
System identification:
   least squares method 533–538
   NN-based 758–765
   recursive 538–540
System terminology:
   autonomous 569
   MIMO 14
   relaxed 50
   sampled-data 26–30
   SISO 14
   type-1 236–237
   type-2 237
   type-0 236

Takagi-Sugeno fuzzy rules 811–812
Tan-sigmoid activation (NN) 734–735
Taylor series 333
t-conorm; fuzzy sets 802
Temperature control systems 141–143, 171–176, 777–779, 813–825, 841–843, 923, 924
Terms of linguistic variables 800
Thyristor controller (see SCR controller)
Time-delay approximation of zero-order hold 85, 170
Time-response specifications 239–242
t-norm; fuzzy sets 802
Trace of a matrix 320
Tracking control systems:
   definition 3
Training NN:
   adaptive learning 756
   backpropagation training 753–757

batch updating 752
gradient descent algorithm 749–757
incremental updating 752
initialization 755
learning constant 750
Levenberg-Marquardt
  algorithm 756
momentum gradient
  algorithm 755–756
multilayer 739–741
single layer 735–738
Transfer function:
  definition 52
  derived from state model 341–343, 404, 431–432, 461
  equivalence with state
    model 398–403, 458
  nonminimum-phase 255
  poles and zeros 54
  proper 342
  sampled-data systems 129–135, 139–141
  strictly proper 342
  systems with dead-time 139–141
  zero-order hold 74–76
Transfer function matrix 404, 461
Transportation lag (see Dead-time)
Transpose of a matrix 317
Trapezoidal fuzzy set 797
Trapezoidal rule for
  integration 106–108
Triangular fuzzy set 797
Triangular matrix 317
Tuning of process controller (see Controller tuning)
Type number of a system 235
Type-1 system 236–237
Type-2 system 237
Type-0 system 236

Ultimate gain 159
Ultimate period 159
Uncertainties in control 217–219, 250–251, 640–642
Uncontrollable eigenvalues (poles) 401
Uncontrollable system 389, 455
Underfitting (NN) 763

Uniform sampling 32
Union; fuzzy set 802
Unit circle in z-plane 54–55
Unit delayer 42, 55
Unit matrix 316
Unit-sample sequence 38–39
Unit-step sequence 40
Unit vector 323
Unity feedback systems 19, 233
Universal approximation
  property (NN) 757
Universe of discourse 791, 795, 842
Unobservable eigenvalues (poles) 402
Unstable system 67
Unsupervised learning (NN) 748–749
Upper triangular matrix 317

Validation (NN) 764–765
Variable reluctance stepping
  motors 187–189
Variable structure control 661–662
Vectors:
  inner product 321
  linearly dependent 324–326
  linearly independent 324–326
  norm; Euclidean 322
  orthogonal 323
  orthonormal 324
  scalar product 321
  unit 323
Velocity error constant 235
Velocity form of digital PID
  algorithm 168, 169
Vortex point; phase portrait 694

Warping 100–102, 108, 253
Weights (NN) 728
w-plane; 111
  z-plane mapping 252
w-transform 111, 252–255

Zero-input stability 67
Zero matrix 317
Zero-order hold:
  circuit 76–77
  equivalence for discretization 97
  filtering characteristic 84–85
  time-delay approximation 85, 170

transfer function model 74–76
Zeros and poles 54
Ziegler-Nichols tuning:
  based on process reaction
    curve 160–165
  based on ultimate gain and
    period 158–160
  for quarter decay ratio
    response 159–160, 164–165
z-plane:
  r-plane mapping 111
  s-plane mapping 91
  unit circle 54–55
  w-plane mapping 111
z-plane synthesis 284–296
z-transfer function:
  definition 52
  derived from state model 341–343,
    404, 431–432, 461
  equivalence with state
    model 398–403, 458
  poles and zeros 54
  sampled-data systems 129–135,
    139–141
  systems with dead-time 139–141
z-transform:
  definition 863
  final value theorem 872
  inverse 866
  pairs 873–874
  pairs for systems with
    dead-time 141
  properties 867–869